DNA Replication

A subject collection from *Cold Spring Harbor Perspectives in Biology*

OTHER SUBJECT COLLECTIONS FROM *COLD SPRING HARBOR PERSPECTIVES IN BIOLOGY*

Wnt Signaling
Protein Synthesis and Translational Control
The Synapse
Extracellular Matrix Biology
Protein Homeostasis
Calcium Signaling
The Golgi
Germ Cells
The Mammary Gland as an Experimental Model
The Biology of Lipids: Trafficking, Regulation, and Function
Auxin Signaling: From Synthesis to Systems Biology
The Nucleus
Neuronal Guidance: The Biology of Brain Wiring
Cell Biology of Bacteria
Cell–Cell Junctions
Generation and Interpretation of Morphogen Gradients
Immunoreceptor Signaling
NF-κB: A Network Hub Controlling Immunity, Inflammation, and Cancer
Symmetry Breaking in Biology
The Origins of Life
The p53 Family

SUBJECT COLLECTIONS FROM *COLD SPRING HARBOR PERSPECTIVES IN MEDICINE*

Addiction
Parkinson's Disease
Type 1 Diabetes
Angiogenesis: Biology and Pathology
HIV: From Biology to Prevention and Treatment
The Biology of Alzheimer Disease

DNA Replication

A subject collection from *Cold Spring Harbor Perspectives in Biology*

EDITED BY

Stephen D. Bell
Indiana University, Bloomington

Marcel Méchali
Institute of Human Genetics, CNRS, Montpellier

Melvin L. DePamphilis
National Institutes of Health, Bethesda

COLD SPRING HARBOR LABORATORY PRESS
Cold Spring Harbor, New York • www.cshlpress.org

DNA Replication

A Subject Collection from *Cold Spring Harbor Perspectives in Biology*
Articles online at www.cshperspectives.org

Printed in the United States of America

Executive Editor	Richard Sever
Managing Editor	Maria Smit
Project Manager	Barbara Acosta
Permissions Administrator	Carol Brown
Production Editor	Diane Schubach
Production Manager/Cover Designer	Denise Weiss
Publisher	John Inglis

Front cover artwork: DNA combing showing the progression of the replication forks from origins of DNA replication. DNA was combed on silanized glasses after two consecutive pulses of two different fluorescent nucleotides. Each line is a microscopy photograph of a single DNA molecule in which replication proceeds bidirectionally from its center (the replication origin). Different molecules are aligned and superposed at their replication origins, showing the progression of the two replication forks from the origin. Image courtesy of Slavica Stanojcik and Marcel Méchali, Institute of Human Genetics, Centre National de la Recherche Scientifique (CNRS).

Library of Congress Cataloging-in-Publication Data

DNA replication / edited by Stephen D. Bell, Marcel Méchali, and Melvin L. DePamphilis.
p. ; cm
"A subject collection from Cold Spring Harbor perspectives in biology."
Includes bibliographical references and index.
ISBN 978-1-936113-48-4 (hardcover : alk. paper)
I. Bell, Stephen D. II. Méchali, Marcel, 1949- III. DePamphilis, Melvin L. IV. Cold Spring Harbor perspectives in biology.
[DNLM: 1. DNA Replication--Collected Works. 2. DNA-Directed DNA Polymerase--physiology--Collected Works. QU 475]

572.8′645--dc23

2012025936

10 9 8 7 6 5 4 3 2 1

All World Wide Web addresses are accurate to the best of our knowledge at the time of printing.

For a complete catalog of all Cold Spring Harbor Laboratory Press publications, visit our website at www.cshlpress.org.

Contents

Preface

THE FUNDAMENTAL PRINCIPLES THAT GOVERN DNA replication are elegant and simple. Take a DNA double helix, unzip it, and, following the chemical rules of base complementarity, use the single strands as templates to generate new daughter molecules. Yet to accomplish this task in appropriate time and space, and with sufficient fidelity, requires the coordinated interplay and regulation of a multitude of complex protein assemblies. In the ensuing pages, 77 authors describe the exquisite complexity of the macromolecular machines that drive this conceptually simple process. This is a truly exciting field in which to work—the rate of progress of the development of techniques and concepts is remarkable. This is reflected by the fact that this book comes just 6 years after the last Cold Spring Harbor Laboratory Press volume on the subject. During that time we have learned much about the core mechanisms of replication-associated processes and have gained a much fuller appreciation of the interplay between the regulatory circuits that drive cell cycle progression and the replication apparatus itself. As will be apparent from the contents of the book, the full complement of state-of-the-art techniques, from structural biology through biophysical analyses, single-molecule studies, biochemistry, genetics, genomics, imaging, and cell biology, have been exploited with remarkable effect to tease apart these intricate processes.

DNA replication is, of course, fundamental to the propagation of all life on the planet. It is a process that when it goes awry can have profound consequences for the organism. In the case of humans, as detailed by Abbas et al. and A.P. Jackson et al., errors in replication can lead to cancer, yet, conversely, the very presence of elevated levels of replication-associated proteins can be a powerful indicator of cancerous or precancerous conditions. Human pathogens, whether bacterial or viral, need to replicate their genomes within their host. Authors Cotmore and Tattersall, Chow and Broker, Hoeben and Uil, Weller and Coen, Hammerschmidt and Sugden, and Moss all deal with the mechanisms of viral DNA replication. In many cases virus-specific proteins facilitate initiation of replication but then co-opt components of the cellular machinery for elongation. By characterizing the virus-specific components, potential candidates for drug development can be identified. At least part of the reason viruses encode their own initiator proteins is to circumvent the cellular circuitry that controls DNA replication and, in most cases, limits it to occurring once per cell division cycle. The chapter by Zielke et al. deals with some important exceptions to this once per cell division cycle rule. The interface between control circuitry and core machinery is complex and tightly interwoven in eukaryotes, and Leonard and Mechali, McIntosh and Blow, Bell and Kaguni, Tanaka and Araki, D. Jackson et al., Rhind and Gilbert, Siddiqui et al., and de la Paz Sanchez et al. all deal with the various aspects of this interplay. Indeed, although a number of aspects linked to the mechanisms of DNA synthesis appear to be conserved in all living organisms, new regulatory events have been introduced in the process of the initiation of DNA replication in metazoans. Origins of DNA replication appear to be more complex structures, involving different sequence constraints and epigenetic controls, and these are probably tightly linked to cell cycle controls and adaptations to cell identity. New factors involved in the assembly and control of replication origin complexes thus appeared during evolution. These aspects are treated in chapters by Leonard and Mechali, Tanaka and Araki, Siddiqui et al., and de la Paz Sanchez et al.

Given the essential and mechanistically conserved nature of DNA replication, it is perhaps surprising that bacteria utilize a set of proteins that are not orthologous to their counterparts in archaea and eukaryotes. As detailed in chapters by O'Donnell et al., Duderstadt et al., and Skarstad and

Katayama, although the basic tenets of replication are similar between the three domains of life, the machineries have some important differences. Makarova and Koonin explore the evolution of these distinct apparatuses and the relationship between archaeal and eukaryal replication-associated proteins. However, it must not be forgotten that eukaryotic cells harbor remnant bacteria in the form of mitochondria with their own replication proteins and processes, and these are described in the chapter by Holt and Reyes.

Despite the variation in the precise nature of the proteins that mediate DNA replication, all cellular organisms replicate their DNA via a common structure—the replication fork. The structure is propagated by the action of the replicative helicase, built around the minichromosome maintenance (MCM) complex in archaea and eukaryotes. The nature and activation of this assembly is discussed in chapters by Tanaka and Araki and Bell and Botchan. The helicase provides single-stranded DNA that acts as a template for synthesis of new DNA by the replicative DNA polymerases on the exposed single-stranded templates on both leading and lagging strand (chapters by Johansson and Dixon and Peters and Nishiyama). Because of the low inherent processivity of the polymerases, they require an interaction with a sliding clamp that, in turn, must be actively loaded onto DNA, a conserved process that is discussed by Hedglin et al. Because of the discontinuous nature of lagging-strand replication, this is a highly dynamic assembly. As detailed in the chapter by Duderstadt et al., recent single-molecule studies both in vivo and in vitro have yielded significant insight into the coordination of events during replication-fork progression in a variety of model systems. Balakrishnan and Bambara discuss the interplay of a variety of pathways that lead to maturation of the lagging-strand DNA from RNA-primed Okazaki fragments to covalently intact DNA molecules.

During the life of a replication fork, DNA lesions or other impediments to its progress may be encountered, potentially resulting in fork stalling or even collapse. Cells have evolved complex checkpoint pathways to deal with such events and a variety of mechanisms can be brought into play to rescue stalled or damaged replication forks (Yeeles et al.). These can include the co-option of specialized lesion bypass polymerases that have the capacity to synthesize over even quite bulky lesions in DNA; however, this has the potential to introduce mutation into DNA and so must be tightly controlled (Goodman and Woodgate). Another DNA repair pathway that utilizes the replication apparatus is break-induced replication (Anand et al.).

The extraordinary degree of compaction of eukaryotic genomes into chromatin, combined with the importance of epigenetic regulation of gene expression, has led to a tight association of chromatin assembly proteins with the replication fork. The coordination between these pathways is described in MacAlpine and Almouzni. Another eukaryotic-specific issue lies in the replication of telomeres. Since the last volume, tremendous progress has been made in understanding the protein complexes that carry out this specialized task (Pfeiffer and Lingner). Another process that in eukaryotes is inextricably interwoven with replication is the establishment of sister chromatid cohesion (Peters and Nishiyama).

It has been a true pleasure for the editors to work with leading members of the field to put this book together and we would like to express our gratitude to the authors for their contributions. We are also profoundly grateful to Barbara Acosta, Inez Sialiano, and Diane Schubach at Cold Spring Harbor Laboratory Press for their skillful and dedicated assistance.

Stephen D. Bell
Marcel Méchali
Melvin L. DePamphilis

Dedication to Arthur Kornberg

Arthur Kornberg (1918–2007)

THIS BOOK IS DEDICATED TO Arthur Kornberg, who was more than a pioneer in the field of DNA replication—he was a legend in his own lifetime. Arthur was a great source of inspiration to scientists interested in DNA replication for an unusually long period of time. In his laboratory, outstanding discoveries were made that paved the way in this field for decades that followed. Among them, three milestones must be remembered. First and foremost was the discovery that DNA synthesis is an enzymatic process carried out by a protein that Arthur purified with the assistance of two postdoctoral fellows, Maurice Bessman and Robert Lehman, and named DNA polymerase (1956–1958) (Kornberg et al. 1956; Lehman et al. 1958). In 1959, one year after publication of this discovery, Arthur Kornberg was awarded the Nobel Prize in Physiology or Medicine. The second milestone was the complete synthesis of a biologically active viral DNA (Goulian and Kornberg 1967) that was carried out in collaboration with Mehran Goulian and Robert Sinsheimer, and hailed in the popular press as "creation of life in a test tube." The third milestone was the in vitro reproduction of the initiation of DNA synthesis from the *Escherichia coli* origin of DNA replication (Fuller et al. 1981). During this period, scientists in Arthur's laboratory identified most of the proteins involved in bacterial DNA synthesis, a feat that alone would also have deserved a Nobel Prize. The work generated by his laboratory was prodigious. Everyone in this field can probably remember following the DNA replication enzymology series published nearly every month in *The Journal of Biological Chemistry* (173 papers in all), sometimes with up to 10 episodes in a single issue! This avalanche of pioneering results often left people with the impression that every important mechanism had been solved in DNA replication. However, as this book reveals, DNA replication in archaea and multicellular eukaryotes is more than just an "interesting variation" of what has been observed in bacteria (Kornberg 1979).

Research performed in the Kornberg laboratory was also a magnificent example of the power of biochemistry and enzymology, and Arthur was always very keen to promote this field. Arthur Kornberg is also well known for his quotations, in particular his famous version of "The Ten Commandments" (Kornberg 2003), of which number III, "Thou shalt not believe something just because you can explain it," and number IV, "Thou shalt not waste clean thinking on dirty enzymes," are often quoted. Arguably Arthur's greatest contribution to science was the host of students and postdoctoral fellows he mentored, many of whom became outstanding scientists in their own right. He infected all of us with his love of science. In an editorial Arthur wrote in 1995 (Kornberg 1995), he said, "[R]ich or poor, science is great! To frame a question and arrive at an answer that opens a window to yet another question, and to do this in the company of like-minded people with whom one can share the thrill of unanticipated and extended vistas, is what science is all about. That is what will sustain us in the days and years ahead."

MARCEL MÉCHALI
STEPHEN D. BELL
MELVIN L. DEPAMPHILIS

References

Fuller RS, Kaguni JM, Kornberg A. 1981. Enzymatic replication of the origin of the *Escherichia coli* chromosome. *Proc Natl Acad Sci* **78:** 7370–7374.

Goulian M, Kornberg A. 1967. Enzymatic synthesis of DNA, XXIV. Synthesis of infectious phage phi-X174 DNA. *Proc Natl Acad Sci* **58:** 1723–1730.

Kornberg A. 1979. Aspects of DNA replication. *Cold Spring Harb Symp Quant Biol* **43:** 1–9.

Kornberg A. 1995. Science in the stationary phase. *Science* **269:** 1799.

Kornberg A. 2003. Ten Commandments of enzymology, amended. *Trends Biochem Sci* **10:** 515–517.

Kornberg A, Lehman IR, Bessman M, Simms E. 1956. Enzymic synthesis of deoxyribonucleic acid. *Biochem Biophys Acta* **21:** 197–198.

Lehman IR, Bessman MJ, Simms ES, Kornberg A. 1958. Enzymatic synthesis of deoxyribonucleic acid. I. Preparation of substrates and partial purification of an enzyme from *Escherichia coli. J Biol Chem* **233:** 163–170.

In Memoriam

WE FONDLY REMEMBER Arturo Falaschi as a colleague of extraordinary energy, creativity, and dedication to the field of DNA replication and to the development of European science. Bolstered by the 3 years (1962–1965) he spent as a postdoctoral fellow with Arthur Kornberg, Arturo turned his attention to the complexities of DNA replication in eukaryotic cells. His laboratory developed pioneering methods to map DNA replication origins along chromosomes and characterized in detail the lamin B2 origin of DNA replication. He was also deeply involved in the biochemistry of DNA replication, with the characterization of several proteins, including DNA helicases, and topoisomerases. Arturo Falaschi was also responsible for the International Centre for Genetic Engineering and Biotechnology (ICGEB), an international research organization conceived within the United Nations, creating two laboratories in Trieste and New Dehli, promoting research and training young scientists from developing countries.

Arturo Falaschi (1933–2010)

Principles and Concepts of DNA Replication in Bacteria, Archaea, and Eukarya

Michael O'Donnell[1], Lance Langston[1], and Bruce Stillman[2]

[1]The Rockefeller University and Howard Hughes Medical Institute, New York, New York 10065

[2]Cold Spring Harbor Laboratory, Cold Spring Harbor, New York 11724

Correspondence: odonnell@mail.rockefeller.edu

The accurate copying of genetic information in the double helix of DNA is essential for inheritance of traits that define the phenotype of cells and the organism. The core machineries that copy DNA are conserved in all three domains of life: bacteria, archaea, and eukaryotes. This article outlines the general nature of the DNA replication machinery, but also points out important and key differences. The most complex organisms, eukaryotes, have to coordinate the initiation of DNA replication from many origins in each genome and impose regulation that maintains genomic integrity, not only for the sake of each cell, but for the organism as a whole. In addition, DNA replication in eukaryotes needs to be coordinated with inheritance of chromatin, developmental patterning of tissues, and cell division to ensure that the genome replicates once per cell division cycle.

The genetic information within the cells of our body is stored in the double helix of DNA, a long cylinderlike structure with a radius that is only 10 Å or one billionth of a meter but can be of considerable length. A single DNA molecule within a bacterium that grows in our gut flora is approximately 5 million base pairs in length and when stretched out, is about 1.6 mm in length, roughly the diameter of a pinhead. In contrast, the single DNA molecule in the largest human chromosome is 245,203,898 base pairs or about 8.33 cm long. The entire human genome, consisting of its 24 different chromosomes in a male is about 3 billion base pairs or 1 m long. Each cell in our body, with rare exceptions, contains two copies of the genome and thus 2 m of total DNA. Thus the scale and complexity of duplicating genomes is remarkable. For example, ~2200 human cells can sit on the top of a 1.5 mm pinhead and when extracted and laid out in a line, the DNA from these cells would be ~4.5 km (2.8 miles) long. In our body, about 500–700 million new blood cells are born every minute in the bone marrow (Doulatov et al. 2012), containing a total of about 1 million km of DNA, or enough DNA to wrap around the equator of the earth 25 times. Thus DNA replication is a serious business in our body, occurring from the time that a fertilized egg first begins duplicating DNA to yield the many trillions of cells that make up an adult body and continuing in all tissues of the adult body throughout our life. The amount of DNA duplicated in an entire human body represents

Cite this article as *Cold Spring Harb Perspect Biol* doi: 10.1101/cshperspect.a010108

an unimaginable amount of information transfer. Moreover, each round of duplication needs to be highly accurate, making one mistake in less than 100 million bases copied per cell division. How copying of the double helix occurs and how it is so highly accurate is the topic of this collection. Inevitably the processes of accurate copying of the genome can go awry, yielding mutations that affect our lives, and thus the collection outlines the disorders that accelerate human disease.

However, the problem of copying DNA is much more complicated than indicated above. The 2 m of DNA in each human cell is wrapped up with histone proteins within the cell's nucleus that is only about 5 μm wide, presenting a compaction in DNA length of about 2 million-fold. How can the copying process deal with the fact that the DNA is wrapped around proteins and scrunched into a volume that creates a spatial organization problem of enormous magnitude? Not only is the DNA copied, but the proteins associated with the DNA need to be duplicated, along with all the chemical modifications attached to DNA and histones that greatly influence developmental patterning of gene expression. The protein machineries that replicate DNA and duplicate proteins within the chromosomes are some of the most complex and intriguing machineries known. Furthermore, the regulations of the processes are some of the most complex because they need to ensure that each DNA molecule in each chromosome is copied once, and only once each time before a cell divides. Errors in the regulation of DNA replication lead to accelerated mutation rates, often associated with increased rates of cancer and other diseases.

The process of accurately copying a genome can be broken down into various subprocesses that combine to provide efficient genome duplication. Central to the entire process is the machinery that actually copies the DNA with high fidelity, including proteins that start the entire process and the proteins that actually copy one helix to produce two. Superimposed on this fundamental process are mechanisms that detect and repair errors and damage to the DNA. Also associated with the DNA replication apparatus are the proteins that ensure that the histone proteins and their modifications in chromatin are inherited along with the DNA. Finally, other machineries cooperate with the DNA replication apparatus to ensure that the resulting two DNA molecules, the sister chromatids, are tethered together until the cell completes duplicating all of its DNA and segregates the sister chromatids evenly to the two daughter cells. Only by combining all of these processes can genetic inheritance ensure that each cell has a faithful copy of its parent's genome.

WHERE TO START

Replication begins at particular positions in chromosomes called "origins" where designated initiator proteins bind to DNA to start the process of replication. There are important differences among bacteria, archaea, and eukaryotes in this process, but there are also many striking similarities that suggest the process dates back to the last universal cellular ancestor (Stillman 2005; Kaguni 2011). Bacteria often contain only one chromosome with one origin at which two replication forks assemble and move in opposite directions (Fig. 1A). Although not all bacteria follow this paradigm, this is the case for the *Escherichia coli* circular 4.4 Mb genome, forming a single replicon or unit of replication from a single origin. At a rate of 1 Kb/s for each fork, this genome is replicated within 30 min. In contrast, eukaryotes typically have multiple linear chromosomes, each with many origins. Multiple origins are a necessity for eukaryotes as they have much larger genomes than bacteria and eukaryotic replication forks move about 20 times more slowly than bacterial replication forks. As an example, the largest human chromosome (chromosome #1) is 250 Mb and if it had only one origin, it would require more than 50 days to replicate compared to the typical 24 h division time of a eukaryotic cell and approximately 8 h for copying DNA in S phase. Initiation at each origin produces two divergent DNA replication forks along the chromosome to create a replicon that is duplicated only once per cell division. The duplication of many replicons eventually yields two daughter chromosomes

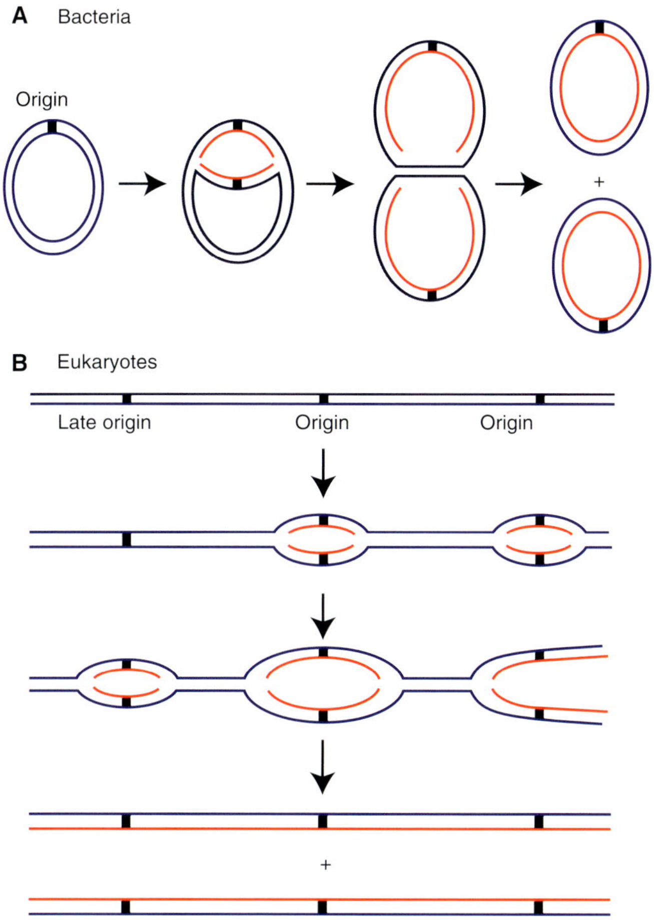

Figure 1. Replication initiation in bacteria and eukaryotes. (*A*) Most bacteria have a circular chromosome with one origin, although there are exceptions to this. Illustrated here is the *E. coli* chromosome that has one origin from which two replication forks proceed in opposite directions. (*B*) Eukaryotes have long linear chromosomes. Bidirectional replication is initiated at multiple origins along each chromosome.

called sister chromatids that are tethered together until they separate during mitosis (Fig. 1B). Although few archaea species have been characterized, they appear to be evolutionary hybrids between bacteria and eukaryotes, because some species have a single chromosome with a single origin, whereas other species have multiple origins per chromosome (Samson and Bell 2011). Moreover, the ploidy of the genome in archaea varies considerably, with some species having a 1C–2C distribution throughout their cell cycle, whereas others have up to 25 copies of their genome in proliferating cells. The rate of DNA replication fork progression also appears to be in between that in bacteria and eukaryotes, at about 20 kB/min, ~10 times faster than that in eukaryotes. Although some bacteria like *E. coli* replicate their genomes much faster, others such as *Caulobacter crescentus* replicate at roughly the same rate as some archaea, such as *Pyrococcus abyssi* (Dingwall and Shapiro 1989; Myllykallio et al. 2000).

Bacterial origins are well-defined sequences to which the replication initiator proteins bind. In contrast, eukaryotic origins are not typically defined at the level of DNA sequence (with the important exception of the budding yeast *Saccharomyces cerevisiae*). Significant recent progress has indicated that eukaryotic origins are defined less by DNA sequence than by chromatin organization, with many origins corresponding to regions of DNA with transcriptional activity or other features that allow access to origin-binding proteins (Masai et al. 2010). Many human cell origins occur in sequences that are evolutionarily conserved among mammals, suggesting that they are far from arbitrary (Cadoret et al. 2008). In most eukaryotes, a small subset of potential origins is used in a typical cell cycle in individual cells, but origin utilization can be greatly increased to facilitate astonishingly rapid cell division, as seen in the fertilized eggs of many animals (Rhind 2006). Whether an origin is used or not is a stochastic process that depends on the chromatin context and in some cases the developmental state of cells in multicellular organisms. Furthermore the multiple origins in eukaryotic chromosomes are organized into clusters that are activated at specific times during S phase of the cell cycle, and the temporal patterning varies again with developmental patterning of cells (Gilbert et al. 2010). Origins of replication are discussed in Leonard and Méchali (2013).

To begin the process of activating an origin for replication, bacterial, archaeal, and eukaryotic cells use origin-binding proteins composed of AAA^+ family subunit(s) (Erzberger and Berger 2006). AAA^+ proteins generally function as multimeric machines. For example in bacteria, multiple copies of the DnaA origin-binding protein form a helical filament that binds the origin (Kaguni 2011). The DnaA filament binds ATP to unwind an A/T-rich region of the origin, resulting in a single-strand DNA (ssDNA) "bubble" onto which the replicative helicase loads (described in the next section).

Eukaryotes contain a six subunit origin-binding protein referred to as ORC (origin recognition complex) (Stillman 2005). Five of the ORC subunits are related to AAA^+ proteins and together with another AAA^+ protein called Cdc6 that is highly related in sequence to the largest ORC subunit, Orc1, they form a ring-shaped hexamer that binds DNA (Sun et al. 2012). However, unlike bacterial DnaA, ORC does not unwind DNA at regions to which it binds. Archaeal cells also use AAA^+ proteins that are related to the largest subunit of ORC, Orc1 and to Cdc6, but the number of these subunits varies depending on the particular type of archaeal cell (Barry and Bell 2006). Both DnaA and ORC are used in other processes besides replication (see Bell and Kaguni 2013).

LOADING THE HELICASE

The objective of origin-binding proteins in bacteria, archaea, and eukaryotes is the loading of two helicases onto DNA, which eventually give rise to two DNA replication forks that move in opposite directions from each origin. In all three domains of life, the helicase is a six subunit complex that unwinds DNA by encircling one strand of the parental duplex (Gai et al. 2010). Each helicase uses ATP hydrolysis to translocate along the single strand, acting as a moving wedge to force the parental duplex apart. The cellular DNA helicases are similar to hexameric enzymes present in several eukaryotic cell viruses such as the simian virus 40 T antigen and the human papillomavirus E1 helicase. Beyond these important similarities lie many differences among the replicative helicases of bacteria, archaea, and eukaryotic cells and their viruses. The bacterial helicase is a homohexamer that is placed around ssDNA generated by the DnaA protein at the origin; it travels $5'-3'$ along the strand onto which it is bound. This directionality places the bacterial helicase around the lagging-strand template at a replication fork. In contrast, the eukaryotic helicase is a heterohexamer known as the MCM2-7 complex. Each of the six MCM subunits is encoded by a separate gene but they are related in sequence and are AAA^+ protein ATPases, whereas the bacterial helicase ATPase architecture is based on a RecA-like fold (Enemark and Joshua-Tor 2008). Eukaryotic MCM encircles the leading strand template at a replication fork and tracks along ssDNA $3'-5'$, the opposite polarity of the bacterial helicase. Another distinctive feature of the eukaryotic MCM2-7 helicase is that it is initially loaded onto the origin as a head-to-head double hexamer with the double-strand DNA (dsDNA) passing through the hexamer channel and therefore must transition to ssDNA to function as a helicase (Masai et al. 2010). This transition, although not well understood, is an important feature regulating replication initiation and involves addition of the Cdc45 and GINS proteins to form an active helicase called the CMG (Cdc45-Mcm2-7-GINS) (Ilves et al. 2010). Without these accessory proteins, the MCM2-7 is inactive as a helicase. Interestingly, the archaeal helicase is also a double hexamer, but in this case made up from a single protein called MCM that is related to the eukaryotic cell helicase. It also travels on ssDNA in the $3'-5'$ direction and hence on the leading strand template and does not require accessory proteins for its helicase activity (Barry and Bell 2006).

All cells require other factors in addition to the origin-binding protein to load the helicase onto DNA. Before loading, bacterial DnaB and eukaryotic MCM are bound by DnaC and Cdt1, respectively, which facilitate delivery of the helicase complexes to the origin. DnaC is an AAA^+ protein that uses ATP to bind the DnaB helicase in an inactive form and it cooperates with DnaA to load DnaB onto the ssDNA bubble formed at the origin by DnaA. ATP hydrolysis ejects DnaC after the loading step, enabling the helicase to become active in DNA unwinding (Kaguni 2011). In eukaryotes, Cdt1 brings the MCM2-7 helicase to the ORC-Cdc6 complex that is bound to the origin DNA (Masai et al. 2010). MCM loading triggers ATP hydrolysis by Cdc6, ejecting it from the DNA and promoting release of Cdt1. Archaea have the AAA^+ Orc1/Cdc6 origin-binding protein, but to date no archaeal Cdt1 homolog has been identified, so the MCM hexamer may bind directly to the initiator protein (Barry and Bell 2006). The precise mechanism by which these proteins load the helicase is unknown in any system. MCM2-7 loading by ORC, Cdc6, and Cdt1 forms a prereplicative complex (the Pre-RC) in which MCM2-7 surrounds duplex DNA, but it remains inactive for DNA unwinding until cells commit to enter S phase of the cell cycle (Masai et al. 2010).

Loading the helicase and activating it to unwind DNA are central replication control points in all cell types, but the way bacteria and eukaryotes regulate this process is fundamentally different. *E. coli* DnaA binding at the origin is regulated by SeqA, which sequesters the origin and prevents access to DnaA (Dame et al. 2011). Sequestration is dependent on the methylation state of the origin DNA (SeqA can only bind newly replicated, hemimethylated DNA). Under optimal growth conditions, *E. coli* reinitiates DNA synthesis at the origin before completing the previous round of replication, yielding multiple chromosomes in one cell; the chromosomes eventually segregate into individual cells. Eukaryotes cannot afford the luxury of rereplication because of their requirement for multiple origins on each chromosome. Reinitiation at some origins and not others would lead to copy number differences within regions of a chromosome and problems with chromosome segregation during mitosis. Hence, under most circumstances, eukaryotic origin initiation is tightly regulated so that origins initiate once, and only once, per cell division. Eukaryotes achieve this exquisite level of control by separating initiation events into different phases of the cell cycle and imposing multiple regulatory processes on the mechanism of initiation of DNA replication (Diffley 2011), whereas bacteria lack a well-defined cell cycle (see Fig. 2) (Morgan 2007). Progression from one eukaryotic cell phase to the next is driven by many regulated events including the synthesis of new proteins, the destruction of others, and protein modification such as phosphorylation by kinases, a modification that is largely absent among bacterial replication proteins. An additional level of control, also distinct from bacteria, is that eukaryotic replication occurs in the nucleus and this compartmentalization allows for tight regulation by excluding key proteins from the nucleus when their activity is not required or when it might be detrimental.

Formation of the pre-RC occurs during mitotic exit in rapidly proliferating eukaryotic cells or during the G_1 phase of the cell cycle, but the MCM2-7 helicase remains inactive after it is loaded onto the dsDNA. The establishment of active replication forks is regulated by kinases that drive the cell into S phase by directing the chromatin association of many factors, most importantly Cdc45 and GINS, which are now

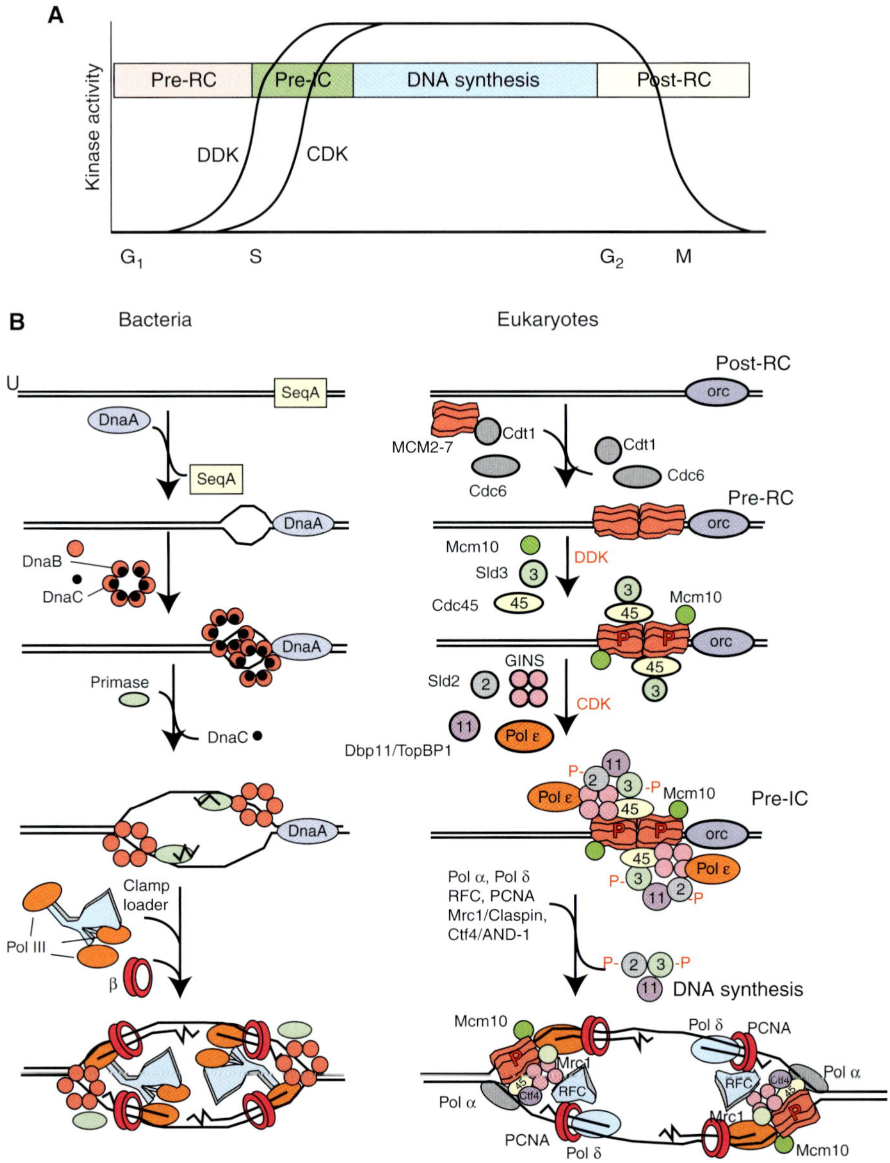

Figure 2. Origin activation and replisome assembly in bacteria and eukaryotes. (*A*) Origin activation in eukaryotes is regulated by DDK (Dbf4-Cdc7 kinase) and CDK (cyclin-dependent protein kinase) kinases that have low activity in G_1 phase, and high activity in S phase. (*B*) Steps in origin activation and replisome assembly in bacteria and eukaryotes. The relatively more complex process of DNA replication in eukaryotes is reflected in the larger number of proteins required to initiate and elongate DNA synthesis from each origin. See text for details.

known to activate the MCM2-7 helicase activity (Ilves et al. 2010). During S phase, the Cdc6 and Cdt1 proteins are eliminated by either selective protein degradation or nuclear exclusion, thereby preventing further MCM2-7 loading and reinitiation in S phase. These events underlie the phenomenon known as "licensing" of origins (Masai et al. 2010). Origins are licensed by MCM2-7 loading during mitotic exit or during G_1, and then activated in S phase. By separating the helicase loading and DNA synthesis steps into two different phases of the cell cycle, eukaryotes prevent origins from initiating more than once (Diffley 2011). After replication, cells enter G_2 phase and then M (mitosis) phase, in which the duplicated chromosomes are segregated into new daughter cells (see Fig. 2). Cell-cycle phases and their regulation are explained in McIntosh and Blow (2012), Siddiqui et al. (2013), Skarstad and Katayama (2013), and Zielke et al. (2013).

REACTIONS LEADING TO PRIMING

In all cell types studied so far, DNA polymerases cannot initiate new chains of nucleic acids and thus the synthesis of a primer by a DNA-dependent RNA polymerase is needed to begin cellular DNA replication (Frick and Richardson 2001). Priming occurs only on ssDNA, which requires prior helicase loading and unwinding activity. In *E. coli*, the primase is a single-subunit enzyme, DnaG, which transiently binds DnaB helicase to synthesize an RNA primer of ~12 nucleotides (nt). Binding of DnaG primase to DnaB also stimulates release of the regulatory protein DnaC from DnaB, indicating that initial priming and unwinding are tightly coordinated (Kaguni 2011). Priming in eukaryotes is performed by the four-subunit DNA polymerase α-primase complex (Pol α/primase) that synthesizes an RNA of ~12 nt and extends this primer RNA with ~25 nt of DNA to form an RNA/DNA hybrid primer. Priming in eukaryotes occurs only after G_1 phase transits to the S phase (see Fig. 2). How the Pol α/primase is recruited is not known, but in the case of SV40 DNA replication it binds directly to the helicase. In eukaryotic cells, recruitment of Pol α/primase may be mediated by the MCM10 protein that is required for initiation of DNA replication, but MCM10 then dissociates so that it cannot keep Pol α/primase on the lagging strand (Waga and Stillman 1998; van Deursen et al. 2012). The G_1/S transition in eukaryotes, and eventual priming of DNA, involves activation of the S-phase CDK (cyclin-dependent protein kinase) and DDK (Dbf4-Cdc7 kinase) kinases (Diffley 2010). In all cells, elongation of the first primer becomes the leading strand. Subsequent priming events occur on the lagging strand to form Okazaki fragments. Okazaki fragments are 1–2 kb in bacteria, and only 100–200 bp in eukaryotes (Balakrishnan and Bambara 2013).

The reactions leading to priming in eukaryotes are still rather mysterious, and several proteins are involved that have no clear homologs in bacteria including Sld2, Sld3, Sld7, Cdc45, the four-subunit GINS complex, and Dpb11/TopBP1 (Masai et al. 2010). An S-phase-specific CDK phosphorylates Sld2 and Sld3 that allows them to bind to Dbp11, whereas the DDK phosphorylation of the MCM4 subunit is necessary for initiation of DNA replication (Diffley 2010; Labib 2010; Sheu and Stillman 2010). Additional details on the complex transactions that activate a licensed origin in eukaryotes are in Tanaka and Araki (2013). Current studies indicate that DNA polymerase ε (Pol ε), a replicative DNA polymerase, associates with the origin before Pol α/primase in a complex with GINS, Dpb11, and CDK-phosphorylated Sld2 (Araki 2010). Thus the initial priming event may be performed by some protein other than Pol α/primase. Alternatively, Pol ε may act as a structural component that helps recruit Pol α/primase. At an undefined point in the process, the MCM2-7 complex transitions from encircling dsDNA to encircling ssDNA. Only when this transition is complete can DNA be unwound by the CMG and provide an ssDNA template for primase activity to initiate replication.

THE REPLICATION MACHINE

Replication is performed by a multiprotein replication machine that synthesizes both daughter duplexes simultaneously. Replication machines have the same core components in all cells: DNA polymerases, circular sliding clamps, a pentameric clamp loader, helicase, primase, and SSB (single-strand binding protein) (Waga and Stillman 1998; Garg and Burgers 2005; Johnson and O'Donnell 2005; Barry and Bell 2006). The way in which these proteins are arranged and connect to one another differs among cell types. The replication machine is often referred to as a

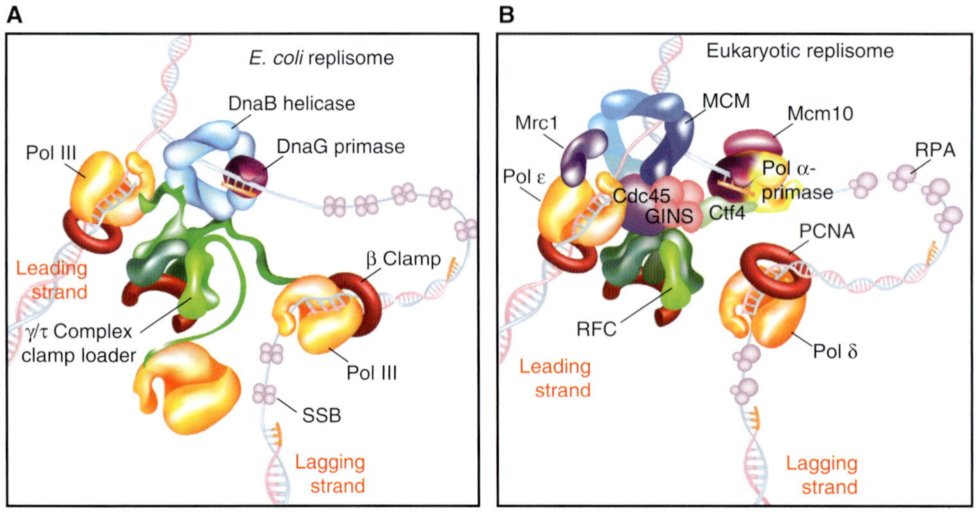

Figure 3. Organization of bacterial and eukaryotic replisome machines. (*A*) Replisome architecture in *E. coli*. The helicase (DnaB) encircles the lagging strand. Three molecules of Pol III are attached to one clamp loader. The clamp loader binds the helicase and repeatedly assembles β clamps onto primed sites as they are formed by primase. (*B*) Proposed architecture of a eukaryotic replisome. The MCM2-7 helicase encircles the leading strand; unwinding is aided by association of GINS and Cdc45 with MCM2-7 to form the CMG complex. The RFC clamp loader repeatedly loads PCNA (proliferating cell nuclear antigen) clamps onto lagging-strand primers formed by Pol α-primase. Unlike *E. coli*, the clamp loader may not form stabile attachments to the replisome. The leading-strand polymerase (Pol ε) is stabilized on DNA by Mrc1. Pol δ replicates the lagging strand. Contacts between Pol δ and other components of the replisome are not yet defined. Mcm10 and Ctf4 contact Pol α-primase.

"replisome." The bacterial replisome, illustrated in Figure 3A, is organized by the clamp loader, which contains three identical τ subunits that bind three C-family DNA polymerase III (Pol III) polymerases (see Fig. 3). The τ subunit also binds the homohexameric helicase (i.e., *E. coli* DnaB). As primase forms RNA primers, the clamp loader repeatedly loads new circular β clamps onto the primer/template for use by the lagging-strand Pol III(s). This primase/polymerase switch is facilitated by SSB, which binds ssDNA and enables the clamp loader to dislodge primase from the primed site. SSB protects the ssDNA from nucleases and facilitates elongation by Pol III-β on ssDNA. The bacterial replisome is highly processive, meaning that in *E. coli* it can synthesize ~86 kb on the leading strand without dissociating from the template (Georgescu et al. 2010). However, high processivity is a disadvantage on the lagging strand, which is synthesized as numerous short ~1000 nt Okazaki fragments and requires the polymerase to dissociate from the template DNA and reassociate at a new primed site for each Okazaki fragment. To overcome this "processivity barrier," specific mechanisms have evolved that pry the polymerase from the clamp to release an Okazaki fragment, on which the polymerase reassociates with a new clamp at the next RNA primed site. The three Pol IIIs in the bacterial replisome aid the production of multiple Okazaki fragments. These mechanisms will be further discussed in Hedglin et al. (2013).

Although eukaryotic and archaeal replisomes have similar components, the connections among the components are quite different from bacterial replisomes (Fig. 3B). The hexameric MCM2-7 encircles the leading strand, not the lagging strand and needs the Cdc45 and GINS proteins for it to function in unwinding DNA. Also, the eukaryotic replisome contains two different B-family polymerases that function separately for the leading and lagging strands, Pol ε and Pol δ, respectively, in addition to Pol α/primase, which starts each Okazaki fragment. The lagging strand is coated by replication protein A (RPA), a heterotrimer SSB that is structurally and functionally analogous to the bacterial SSB tetramer. As in bacteria, a primase/polymerase switch is mediated by RPA. Both Pol ε and Pol δ function with a ring-shaped proliferating cell nuclear antigen (PCNA) clamp of similar structure to *E. coli* β. PCNA is assembled on DNA by the pentameric replication factor C (RFC) clamp loader, composed of subunits with sequence homology and similar structure to the bacterial clamp loader (Garg and Burgers 2005). Unlike bacteria, RFC does not appear to contact the polymerases or the helicase, and the connections among replisome components remain unclear. There are numerous candidates among the many proteins known to be required for efficient eukaryotic replication that have no homolog in bacteria. For example, the GINS heterotetramer and the Cdc45 protein that form a complex with the MCM2-7 helicase to yield the CMG complex may bind Pol α/primase. GINS and Cdc45 are required for efficient DNA unwinding activity and may bind other factors in addition to the MCM2-7 complex to organize the replisome. In budding yeast, a replication progression complex has been identified containing potential links between the CMG and Pol α/primase via Ctf4/AND-1, and between CMG and Pol ε via Mrc1/Claspin (Masai et al. 2010). Detailed evidence for these and other possible connections, the functions of analogous proteins in other eukaryotes, and their importance for fork progression can be found in Bell and Botchan (2013).

The oligomeric structures of sliding clamps enable multiple proteins to bind one clamp simultaneously, referred to as the "toolbelt" hypothesis (Pages and Fuchs 2002). An extreme case is an archaeal PCNA heterotrimer in which each different subunit binds a different partner protein (i.e., polymerase, ligase, and Fen1) that is involved in the synthesis and maturation of Okazaki fragments (Barry and Bell 2006). These features of sliding clamps are discussed in Hedglin et al. (2013), but one aspect will be mentioned here as it has important implications for replisome structure and function. In particular, all cells contain a variety of low-fidelity DNA polymerases that can bypass lesions in the DNA, referred to as "TLS" Pols (translesion synthesis polymerases). Different Pols can bind the clamp simultaneously and trade places with one another, making the replisome a much more dynamic machine than originally thought (Langston et al. 2009). *E. coli* contains three "translesion" DNA polymerases (TLS Pols), which are induced on DNA damage. Studies have shown that at high concentrations such as those induced on DNA damage, the TLS Pols bind the β clamp and trade places with Pol III at a moving fork, yet retain the helicase to form a "TLS replisome." TLS replisomes move much more slowly than Pol III and dictate the speed of the helicase. An obvious advantage of forming TLS replisomes on DNA damage is to slow the replication fork, giving time for DNA repair before a replication fork encounters a lesion. In the event a lesion is encountered, it can be bypassed by the TLS polymerase.

DEALING WITH DNA DAMAGE DURING REPLICATION

In bacteria and eukaryotes there are mechanisms to activate a stalled DNA replication fork that may be caused by the replisome running into DNA damage, a double-strand break, or a protein block. There are many repair mechanisms, such as error-free and error-prone DNA synthesis at the DNA replication fork, or postreplicative repair by nucleotide excision repair or base excision repair (Lazzaro et al. 2009; Hubscher and Maga 2011; Lehmann 2011). The multiple DNA polymerases involved in DNA repair are discussed in Goodman and Woodgate (2013) and Yeeles et al. (2013). Alternatively, branch migration followed by DNA synthesis or intersister chromatid recombination can be used to allow replicative bypass of DNA lesions. If these repair events occur while DNA replication is still occurring, stalled DNA replication forks must either be reactivated or bypassed, events that are discussed in McIntosh and Blow (2012) and Yeeles et al. (2013). Unlike bacteria, eukaryotes have multiple origins and thus failure to replicate a complete replicon can be compensated by activation of adjacent origins of DNA replication, another way of bypassing a stalled DNA replication fork (Blow et al. 2011).

If significant DNA damage occurs either before or during DNA replication in eukaryotes, so-called checkpoint mechanisms signal to the cell-cycle regulatory machinery, principally the CDK and DDK protein kinases, that subsequent events in the cell cycle should wait until DNA damage is repaired. The biochemistry of these varied signaling events is still being worked out, but one common signal is the stable presence of RPA-coated ssDNA. Normally ssDNA should not exist in a cell, but its presence signals that a stalled replication fork exists, that resection of damaged DNA has occurred, or that recombination is taking place. In any event, the cell does not want to progress until the damage has been repaired. In both bacteria and eukaryotes, restarting a stalled DNA replication fork is a key process that involves DNA helicases and priming events that are unique to stalled fork recovery. For example, in eukaryotes cancer-prone disorders called Bloom's syndrome and Warner's syndrome have defects in specialized DNA helicases that are involved in dealing with DNA damage at stalled replication forks. Another cancer-prone syndrome called Fanconi anemia has revealed defects in a family of proteins that handle cross-links in DNA. These pathways, as well as the recovery of stalled forks are discussed in Abbas et al. (2013), Jackson et al. (2013), and Yeeles et al. (2013).

SPATIAL AND TEMPORAL ORGANIZATION OF DNA REPLICATION

In bacteria, the DNA replication machinery is assembled at the single origin of DNA replication in a characteristic location. In *Caulobacter*, the replisome is located at one end of the rod-shaped bacterium but in *E. coli*, it is located in the middle of the cell (Toro and Shapiro 2010). After initiation of DNA replication, the new replicated origins spool out of the replisome, which stays in place, and the origins move to predetermined locations. In *E. coli*, the origins move to the quarters of the cell but in *Caulobacter*, the origins move to the other end of the cell. Thus the origins of DNA replication are associated with DNA elements that can move the DNA and prepare the cell for separating the two daughter chromosomes during cell division, thereby remotely resembling centromeres in eukaryotic cells. Interestingly, pre-RC proteins in eukaryotes play a role at centromeres and centrosomes, perhaps reflecting an evolutionary link between chromosome replication and segregation (Saffery et al. 2000; Prasanth et al. 2004; Hemerly et al. 2009; Hossain and Stillman 2012; Varma et al. 2012).

As the two DNA bacterial forks move in opposite directions away from the origin, they eventually meet at a termination site that locates to the center of the cell before cell division. In rapidly growing bacteria, it is possible to reinitiate DNA replication from the origin before cell division actually takes place and this is made possible because of the spatial separation of the origin and termination regions of the genome and their placement relative to the plane of cell division. Such reinitiation is lethal in eukaryotic cells because it causes aneuploidy and genome instability.

In contrast to bacteria, eukaryotes have to deal with multiple chromosomes and numerous origins. The fastest way to replicate the multiple chromosomes in eukaryotes is for all origins to fire at the same time, but this is rare and occurs in the early and rapid cell divisions of *Xenopus* embryos and during the replication of the syncytial nuclei in the *Drosophila* embryo. In somatic cells origins of DNA replication fire at specific times during S phase, with some firing early and others late in a specific temporal pattern that is characteristic for each cell type (Fig. 1B) (Toro and Shapiro 2010). Within a chromosome, clusters of origins of DNA replication that are located adjacent to each other initiate DNA replication at the same time, creating megabase-pair-sized domains of chromosomes that are copied synchronously. Remarkably, whole genome analysis of these domains of chromosomes that are replicated at the same time has revealed that they correspond to regions of the genome that are spatially associated with each other in the nucleus (Ryba et al. 2010). Thus, the genome within the nucleus is organized into domains that are physically associated with each other and these domains replicate at characteristic times during S phase. Such an arrangement explains why replication proteins such as the DNA polymerase clamp PCNA form temporally regulated patterns in the nucleus during S phase that correspond to sites of DNA synthesis of large regions of the genome (Kill et al. 1991; O'Keefe et al. 1992; Leonhardt et al. 2000).

Pre-RCs are assembled at all origins of DNA replication before the start of S phase and the temporal patterning of actual DNA synthesis from each origin is governed by the chromatin context, including the nature of histone modifications, and by rate-limiting proteins that establish the preinitiation complex (Douglas and Diffley 2012). The activation of pre-RCs is regulated by at least two independent protein kinase signaling systems, the CDK and the DDK (Labib 2010; Tanaka and Araki 2013). Once the pre-RC is used or once passive DNA replication passes through a pre-RC, the pre-RC is destroyed and cannot be re-formed during the time in the cell cycle that CDKs are active. Thus pre-RCs cannot be established until the cells pass through the metaphase to anaphase transition when the cyclin moieties of CDKs are destroyed. Such a mechanism limits DNA replication from all origins to once per cell division cycle (Diffley 2011). So, unlike bacterial chromosome replication, it is clear that the complex genomes of eukaryotes have demanded more complex regulatory systems to maintain genome integrity. These issues are discussed in Siddiqui et al. (2013) and Zielke et al. (2013).

WHAT NEEDS TO BE DONE?

Although the broad outline and many important details of DNA replication have been identified, many important aspects of this central process remain to be discovered. In large part, we still do not know how origins function. How do origin-binding proteins organize the DNA? Exactly how do helicase loaders function? How and when does the MCM helicase transition from encircling dsDNA to encircling ssDNA? The functions of several proteins required for origin activation and priming in eukaryotes are still shrouded in mystery. Priming and replisome assembly require numerous proteins that lack homologs in bacteria. What are the functions of Sld2, Sld3, Dbp11, Mcm10, GINS, and Cdc45 and how is their function influenced by phosphorylation? We lack an understanding of how the multiple origins in eukaryotes are coordinated and how the domain structure is established and maintained through multiple cell divisions. For example, just what are nuclear foci and how are replication foci organized within them? Are origins within one replicon clustered into one focus? Once replication forks are established, we know little about how they are regulated. If one replication fork in a focus were to stop, would it halt the other forks within that focus? How do replisomes move through nucleosomes, especially in highly condensed DNA and how are the parental nucleosomes inherited to the sister chromatids? How do replisomes deal with cohesin rings and how are these loaded? We have barely scratched the surface on questions surrounding the interface of replication with repair and recombination. For example, how can replication forks form during break-induced replication in S and G_2 phase when the MCMs are thought to be loaded only in G_1? How do checkpoint mechanisms act on moving replication forks? The newly revealed coordination of DNA metabolism with chromatin establishment, gene silencing, and epigenetic control is only beginning to be explored. Most of what we know about DNA replication has been learned in organisms with stable karyotypes and ploidy. However some organisms, particularly microbial eukaryotes, have extreme variations in ploidy and variable numbers of chromosomes. What mechanisms exist to facilitate this yet maintain order in this apparent chaos? Finally, and perhaps most important, some types of human disease, including certain cancers, have their basis in replication. Clearly many important questions remain, despite the enormous progress of recent years. We hold hope that understanding the mechanistic details of these processes may lead to cures, or at least treatment of human disease in the future.

ACKNOWLEDGMENTS

The authors' research is supported by the Howard Hughes Medical Institute (M.O.) and the National Institutes of Health (GM38839 to M.O.; CA13016 and GM45436 to B.S.).

REFERENCES

*Reference is also in this collection.

* Abbas T, Keaton MA, Dutta A. 2013. Genomic instability in cancer. *Cold Spring Harb Perspect Biol* doi: 10.1101/cshperspect.a012914.

Araki H. 2010. Cyclin-dependent kinase-dependent initiation of chromosomal DNA replication. *Curr Opin Cell Biol* **22:** 766–771.

* Balakrishnan L, Bambara RA. 2013. Okazaki fragment metabolism. *Cold Spring Harb Perspect Biol* **5:** a010173.

Barry ER, Bell SD. 2006. DNA replication in the archaea. *Microbiol Mol Biol Rev* **70:** 876–887.

* Bell SD, Botchan MR. 2013. The minichromosome maintenance replicative helicase. *Cold Spring Harb Perspect Biol* doi: 10.1101/cshperspect.a012807.

* Bell SP, Kaguni JM. 2013. Helicase loading at chromosomal origins of replication. *Cold Spring Harb Perspect Biol* doi: 10.1101/cshperspect.a010124.

Blow JJ, Ge XQ, Jackson DA. 2011. How dormant origins promote complete genome replication. *Trends Biochem Sci* **36:** 405–414.

Cadoret JC., Meisch F, Hassan-Zadeh V, Luyten I, Guillet C, Duret L, Quesneville H, Prioleau MN. 2008. Genome-wide studies highlight indirect links between human replication origins and gene regulation. *Proc Natl Acad Sci* **105:** 15837–15842.

Dame RT, Kalmykowa OJ, Grainger DC. 2011. Chromosomal macrodomains and associated proteins: Implications for DNA organization and replication in gram negative bacteria. *PLoS Genet* **7:** e1002123.

Diffley JF. 2010. The many faces of redundancy in DNA replication control. *Cold Spring Harb Symp Quant Biol* **75:** 135–142.

Diffley JF. 2011. Quality control in the initiation of eukaryotic DNA replication. *Philos Trans R Soc Lond B Biol Sci* **366:** 3545–3553.

Dingwall A, Shapiro L. 1989. Rate, origin, and bidirectionality of *Caulobacter* chromosome replication as determined by pulsed-field gel electrophoresis. *Proc Natl Acad Sci* **86:** 119–123.

Douglas ME, Diffley JF. 2012. Replication timing: The early bird catches the worm. *Curr Biol* **22:** R81–R82.

Doulatov S, Notta F, Laurenti E, Dick JE. 2012. Hematopoiesis: A human perspective. *Cell Stem Cell* **10:** 120–136.

Enemark EJ, Joshua-Tor L. 2008. On helicases and other motor proteins. *Curr Opin Struct Biol* **18:** 243–257.

Erzberger JP, Berger JM. 2006. Evolutionary relationships and structural mechanisms of AAA^+ proteins. *Annu Rev Biophys Biomol Struct* **35:** 93–114.

Frick DN, Richardson CC. 2001. DNA primases. *Annu Rev Biochem* **70:** 39–80.

Gai D, Chang YP, Chen XS. 2010. Origin DNA melting and unwinding in DNA replication. *Curr Opin Struct Biol* **20:** 756–762.

Garg P, Burgers PM. 2005. DNA polymerases that propagate the eukaryotic DNA replication fork. *Crit Rev Biochem Mol Biol* **40:** 115–128.

Georgescu RE, Yao NY, O'Donnell M. 2010. Single-molecule analysis of the *Escherichia coli* replisome and use of clamps to bypass replication barriers. *FEBS Lett* **584:** 2596–2605.

Gilbert DM, Takebayashi SI, Ryba T, Lu J, Pope BD, Wilson KA, Hiratani I. 2010. Space and time in the nucleus: Developmental control of replication timing and chromosome architecture. *Cold Spring Harb Symp Quant Biol* **75:** 143–153.

* Goodman MF, Woodgate R. 2013. Translesion DNA polymerases. *Cold Spring Harb Perspect Biol* doi: 10.1101/cshperspect.a010363.

* Hedglin M, Kumar R, Benkovic SJ. 2013. Replication clamps and clamp loaders. *Cold Spring Harb Perspect Biol* doi: 10.1101/cshperspect.a010165.

Hemerly AS, Prasanth SG, Siddiqui K, Stillman B. 2009. Orc1 controls centriole and centrosome copy number in human cells. *Science* **323:** 789–793.

Hossain M, Stillman B. 2012. Meier-Gorlin syndrome mutations disrupt an Orc1 CDK inhibitory domain and cause centrosome re-duplication. *Genes Dev* **26:** 1797–1810.

Hubscher U, Maga G. 2011. DNA replication and repair bypass machines. *Curr Opin Chem Biol* **15:** 627–635.

Ilves I, Petojevic T, Pesavento JJ, Botchan MR. 2010. Activation of the MCM2-7 helicase by association with Cdc45 and GINS proteins. *Mol Cell* **37:** 247–258.

* Jackson AP, Laskey RA, Coleman N. 2013. Replication proteins and human disease. *Cold Spring Harb Perspect Biol* doi: 10.1101/cshperspect.a013060.

Johnson A, O'Donnell M. 2005. Cellular DNA replicases: Components and dynamics at the replication fork. *Annu Rev Biochem* **74:** 283–315.

Kaguni JM. 2011. Replication initiation at the *Escherichia coli* chromosomal origin. *Curr Opin Chem Biol* **15:** 606–613.

Kill IR, Bridger JM, Campbell KH, Maldonado-Codina G, Hutchison CJ. 1991. The timing of the formation and usage of replicase clusters in S-phase nuclei of human diploid fibroblasts. *J Cell Sci* **100:** 869–876.

Labib K. 2010. How do Cdc7 and cyclin-dependent kinases trigger the initiation of chromosome replication in eukaryotic cells? *Genes Dev* **24:** 1208–1219.

Langston LD, Indiani C, O'Donnell M. 2009. Whither the replisome: Emerging perspectives on the dynamic nature of the DNA replication machinery. *Cell Cycle* **8:** 2686–2691.

Lazzaro F, Giannattasio M, Puddu F, Granata M, Pellicioli A, Plevani P, Muzi-Falconi M. 2009. Checkpoint mechanisms at the intersection between DNA damage and repair. *DNA Repair (Amst)* **8:** 1055–1067.

Lehmann AR. 2011. DNA polymerases and repair synthesis in NER in human cells. *DNA Repair (Amst)* **10:** 730–733.

* Leonard AC, Méchali M. 2013. DNA replication origins. *Cold Spring Harb Perspect Biol* doi: 10.1101/cshperspect.a010116.

Leonhardt H, Rahn HP, Weinzierl P, Sporbert A, Cremer T, Zink D, Cardoso MC. 2000. Dynamics of DNA replication factories in living cells. *J Cell Biol* **149:** 271–280.

Masai H, Matsumoto S, You Z, Yoshizawa-Sugata N, Oda M. 2010. Eukaryotic chromosome DNA replication: Where, when, and how? *Annu Rev Biochem* **79:** 89–130.

* McIntosh D, Blow JJ. 2012. Dormant origins, the licensing checkpoint, and the response to replicative stresses. *Cold Spring Harb Perspect Biol* **4:** a012955.

Morgan DO. 2007. *The cell cycle: Principles of control.* Oxford University Press, London.

Myllykallio H, Lopez P, Lopez-Garcia P, Heilig R, Saurin W, Zivanovic Y, Philippe H, Forterre P. 2000. Bacterial mode of replication with eukaryotic-like machinery in a hyperthermophilic archaeon. *Science* **288:** 2212–2215.

O'Keefe RT, Henderson SC, Spector DL. 1992. Dynamic organization of DNA replication in mammalian cell nuclei: Spatially and temporally defined replication of chromosome-specific α-satellite DNA sequences. *J Cell Biol* **116:** 1095–1110.

Pages V, Fuchs RP. 2002. How DNA lesions are turned into mutations within cells? *Oncogene* **21:** 8957–8966.

Prasanth SG, Prasanth KV, Siddiqui K, Spector DL, Stillman B. 2004. Human Orc2 localizes to centrosomes, centromeres and heterochromatin during chromosome inheritance. *Embo J* **23:** 2651–2663.

Rhind N. 2006. DNA replication timing: Random thoughts about origin firing. *Nat Cell Biol* **8:** 1313–1316.

Ryba T, Hiratani I, Lu J, Itoh M, Kulik M, Zhang J, Schulz TC, Robins AJ, Dalton S, Gilbert DM. 2010. Evolutionarily conserved replication timing profiles predict long-range chromatin interactions and distinguish closely related cell types. *Genome Res* **20:** 761–770.

Saffery R, Irvine DV, Griffiths B, Kalitsis P, Wordeman L, Choo KH. 2000. Human centromeres and neocentromeres show identical distribution patterns of >20 functionally important kinetochore-associated proteins. *Hum Mol Genet* **9:** 175–185.

Samson RY, Bell SD. 2011. Cell cycles and cell division in the archaea. *Curr Opin Microbiol* **14:** 350–356.

Sheu YJ, Stillman B. 2010. The Dbf4-Cdc7 kinase promotes S phase by alleviating an inhibitory activity in Mcm4. *Nature* **463:** 113–117.

* Siddiqui K, On KF, Diffley JFX. 2013. Regulating DNA replication in eukarya. *Cold Spring Harb Perspect Biol* doi: 10.1101/cshperspect.a012930.

* Skarstad K, Katayama T. 2013. Regulating DNA replication in bacteria. *Cold Spring Harb Perspect Biol* doi: 10.1101/cshperspect.a012922.

Stillman B. 2005. Origin recognition and the chromosome cycle. *FEBS Lett* **579:** 877–884.

Sun J, Kawakami H, Zech J, Speck C, Stillman B, Li H. 2012. Cdc6-induced conformational changes in ORC bound to origin DNA revealed by cryo-electron microscopy. *Structure* **20:** 534–544.

Tanaka S, Araki H. 2010. Regulation of the initiation step of DNA replication by cyclin-dependent kinases. *Chromosoma* **119:** 565–574.

* Tanaka S, Araki H. 2013. Helicase activation and establishment of replication forks at chromosomal origins of replication. *Cold Spring Harb Perspect Biol* doi: 10.1101/cshperspect.a010371.

Toro E, Shapiro L. 2010. Bacterial chromosome organization and segregation. *Cold Spring Harb Perspect Biol* **2:** a000349.

van Deursen F, Sengupta S, De Piccoli G, Sanchez-Diaz A, Labib K. 2012. Mcm10 associates with the loaded DNA helicase at replication origins and defines a novel step in its activation. *Embo J* **31:** 2195–2206.

Varma D, Chandrasekaran S, Sundin LJ, Reidy KT, Wan X, Chasse DA, Nevis KR, Deluca JG, Salmon ED, Cook JG. 2012. Recruitment of the human Cdt1 replication licensing protein by the loop domain of Hec1 is required for stable kinetochore-microtubule attachment. *Nat Cell Biol* **14:** 593–603.

Waga S, Stillman B. 1998. The DNA replication fork in eukaryotic cells. *Annu Rev Biochem* **67:** 721–751.

* Yeeles JTP, Poli J, Marians KJ, Pasero P. 2013. Rescuing stalled or damaged replication forks. *Cold Spring Harb Perspect Biol* doi: 10.1101/cshperspect.a012815.

* Zielke N, Edgar BA, DePamphilis ML. 2013. Endoreplication. *Cold Spring Harb Perspect Biol* **5:** a012948.

DNA Replication Origins

Alan C. Leonard[1] and Marcel Méchali[2]

[1]Department of Biological Sciences, Florida Institute of Technology, Melbourne, Florida 32901

[2]Institute of Human Genetics, CNRS, Montpellier, France

Correspondence: aleonard@fit.edu; mechali@igh.cnrs.fr

The onset of genomic DNA synthesis requires precise interactions of specialized initiator proteins with DNA at sites where the replication machinery can be loaded. These sites, defined as replication origins, are found at a few unique locations in all of the prokaryotic chromosomes examined so far. However, replication origins are dispersed among tens of thousands of loci in metazoan chromosomes, thereby raising questions regarding the role of specific nucleotide sequences and chromatin environment in origin selection and the mechanisms used by initiators to recognize replication origins. Close examination of bacterial and archaeal replication origins reveals an array of DNA sequence motifs that position individual initiator protein molecules and promote initiator oligomerization on origin DNA. Conversely, the need for specific recognition sequences in eukaryotic replication origins is relaxed. In fact, the primary rule for origin selection appears to be flexibility, a feature that is modulated either by structural elements or by epigenetic mechanisms at least partly linked to the organization of the genome for gene expression.

Timely duplication of the genome is an essential step in the reproduction of any cell, and it is not surprising that chromosomal DNA synthesis is tightly regulated by mechanisms that determine precisely where and when new replication forks are assembled. The first model for a DNA synthesis regulatory circuit was described about 50 years ago (Jacob et al. 1963), based on the idea that an early, key step in building new replication forks was the binding of a chromosomally encoded initiator protein to specialized DNA regions, termed replication origins (Fig. 1). The number of replication origins in a genome is, for the most part, dependent on chromosome size. Bacterial and archaeal genomes, which usually consist of a small circular chromosome, frequently have a single replication origin (Barry and Bell 2006; Gao and Zhang 2007). In contrast, eukaryotic genomes contain significantly more origins, ranging from 400 in yeast to 30,000–50,000 in humans (Cvetic and Walter 2005; Méchali 2010), because timely duplication of their larger linear chromosomes requires establishment of replication forks at multiple locations. The interaction of origin DNA and initiator proteins (Fig. 1) ultimately results in the assembly of prereplicative complexes (pre-RCs), whose role is to load and activate the DNA helicases necessary to unwind DNA before replication (Remus and Diffley 2009; Kawakami and Katayama 2010). Following helicase-catalyzed DNA unwinding, replisomal proteins become associated with the single-stranded DNA, and new replication forks

Cite this article as *Cold Spring Harb Perspect Biol* doi: 10.1101/cshperspect.a010116

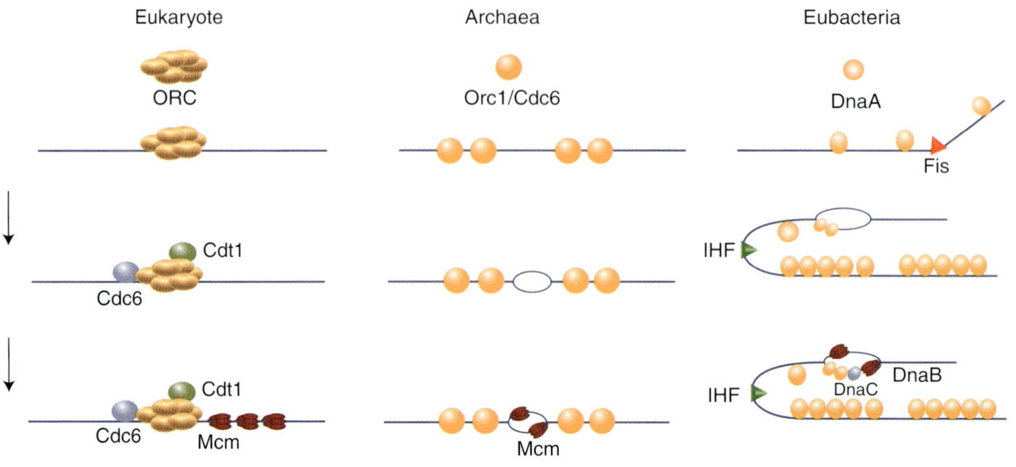

Figure 1. Revised versions of the replicon model for all domains of life. For cells of each domain type, *trans*-acting initiators recognize replication origins to assemble prereplicative complexes required to unwind the DNA and load DNA helicase. Eukaryotic initiators are preassembled into hexameric origin recognition complexes (ORCs) before interacting with DNA. In prokaryotes, single initiators (archaeal Orc1/Cdc6 or bacterial DnaA) bind to recognition sites and assemble into complexes on DNA. In all cases, the DNA helicases (MCMs or DnaB) are recruited to the origin and loaded onto single DNA strands. In bacteria, DNA-bending proteins, such as Fis or IHF, may modulate the assembly of pre-RC by bending the origin DNA. Two activities of DnaA are described in the figure. The larger version binds to recognition sites, and the smaller version represents DnaA required to assist DnaC in loading DnaB helicase on single-stranded DNA.

proceed bidirectionally along the genome until every region is duplicated (for review, see O'Donnell 2006; Masai et al. 2010).

Initiator proteins from all forms of life share structural similarities, including membership in the AAA$^+$ family of proteins (ATPases associated with various cellular activities) (Duderstadt and Berger 2008; Wigley 2009) that are activated by ATP binding and inactivated by ATP hydrolysis (Duderstadt and Berger 2008; Duncker et al. 2009; Kawakami and Katayama 2010). Despite these similarities, initiators assemble into prereplicative complexes in two fundamentally different ways (Fig. 2). In prokaryotes, initiator monomers interact with the origin at multiple repeated DNA sequence motifs, and the arrangement of these motifs (see below) can direct assembly of oligomers that mediate strand separation (Erzberger et al. 2006; Rozgaja et al. 2011). In eukaryotes, a hexameric origin recognition complex (ORC) binds to replication origins and then recruit additional factors (as Cdc6 and Cdt1) that will themselves recruit the hexameric MCM2-7 DNA helicase to form a prereplicative complex (for review, see Diffley 2011). This process occurs during mitosis and along G_1 and is called "DNA replication licensing," a crucial regulation of eukaryotic DNA replication (for review, see Blow and Gillespie 2008). Importantly, this complex is still inactive, and only a subset of these preassembled origins will be activated in S phase. This process is, therefore, fundamentally different from initiation of replication in bacteria. Moreover, because sequence specificity appears more relaxed in large eukaryotic genomes, prokaryotic mechanisms that regulate initiator–DNA site occupation must be replaced by alternative mechanisms, such as structural elements or the use of epigenetic factors.

Here, we describe replication origins on prokaryotic and eukaryotic genomes below, with a particular focus on the attributes responsible for orderly initiator interactions and origin selection specificity, as well as on the shift from origin sequence-dependent regulation to epigenetic regulation. You are also referred to other related articles in this collection and several

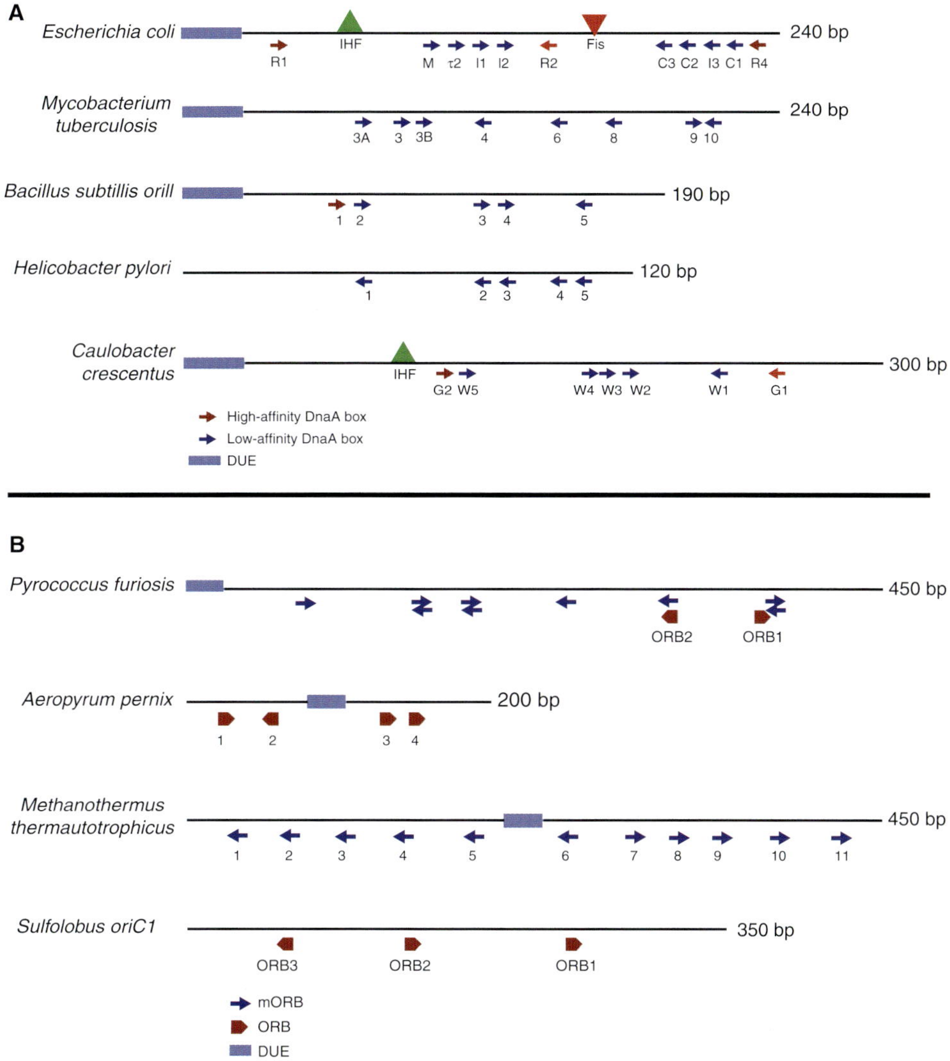

Figure 2. Functional elements in some well-studied prokaryotic replication origins. (*A*) Bacterial *oriC*s. The DNA elements described in the text are (arrows) DnaA recognition boxes or (boxes) DNA unwinding elements (DUEs). When recognition site affinities are known, colored arrows designate high- ($K_d > 100$ nM) and low- ($K_d < 100$ nM) affinity sites. (*B*) Archaeal *oriC*s. Arrows and boxes designate DNA elements as in *A*, but the initiator protein is Orc1/Cdc6 rather than DnaA. (Thick arrows) Long origin recognition boxes (ORBs); (thin arrows) shorter versions (miniORBs). Both ORBs and miniORBs are identified in *Pyrococcus*. DUEs are not yet well defined for *Helicobacter* or *Sulfolobus* genera and are not labeled in this figure.

recent reviews covering the topics of DNA replication initiation in more detail (Méchali 2010; Beattie and Bell 2011; Blow et al. 2011; Bryant and Aves 2011; Ding and MacAlpine 2011; Dorn and Cook 2011; Kaguni 2011; Leonard and Grimwade 2011; Sequeira-Mendes and Gomez 2012).

REPLICATION ORIGINS IN BACTERIA

Bacterial Origins Share Similar Nucleotide Sequence Motifs

The locations of replication origins are now predicted for more than 1300 bacterial genomes (Worning et al. 2006; Gao and Zhang 2007,

2008; see also http://tubic.tju.edu.cn/doric/). Putative *oriC*s are generally located at a unique site on the circular chromosome, often adjacent to the gene encoding the conserved bacterial initiator protein, DnaA (Mackiewicz et al. 2004; Gao and Zhang 2007). Among the small fraction of *oriC*s that are functionally characterized, there is considerable variability in both origin length (ranging from 100 to 1000 bp) and nucleotide sequence. Common features include an extremely A-T-rich DNA unwinding element (DUE) that facilitates DNA strand separation (Kowalski and Eddy 1989), and 9-bp motifs that are recognized by DnaA (DnaA boxes) (see Fig. 2A; Table 1) (for review, see Zakrzewska-Czerwinska et al. 2007). The number and arrangement of DnaA boxes vary among bacterial types, but DnaA boxes are often clustered within *oriC*s (e.g., see Fig. 2A) (Mackiewicz et al. 2004; Zawilak-Pawlik et al. 2005). The clustering of DnaA boxes shown among the examples in Figure 2A suggests that all bacterial origins play a role in directing the assembly of DnaA oligomers (see below), although the length and orientation of these oligomers vary.

In *Escherichia coli*, the sequence 5′-TTATC/ACACA-3′ (boxes R1, R2, and R4) is occupied by the active initiator form, DnaA-ATP with high affinity ($K_d = 4-100$ nM) (Schaper and Messer 1995; Nievera et al. 2006). However, boxes deviating from this sequence at two or more positions coexist with high-affinity sites, and these nonconsensus boxes have a dramatically lower DnaA affinity ($K_d > 200$ nM) (Rozgaja et al. 2011). Many low-affinity DnaA boxes in *E. coli* also show a strong preference (fourfold) for DnaA-ATP compared with inactive DnaA-ADP (McGarry et al. 2004), suggesting a mechanism to couple pre-RC assembly to new DnaA-ATP synthesis. The DnaA recognition boxes presented in Table 1 suggest that the nucleotide sequence variations observed for *E. coli* are common features among bacterial *oriC*, and the affinity for DnaA at each site is likely to be important for directed pre-RC assembly and its coupling to cell growth (see below).

Replication origins of *E. coli* and other bacteria also carry recognition sites for proteins that modulate DnaA interactions within *oriC* (e.g., see Fig. 2A) (Hwang et al. 1992; Slater et al. 1995; Polaczek et al. 1997; Quon et al. 1998; Brassinga et al. 2002). These modulators ensure properly timed activation and inactivation of origin activity during the cell cycle via several modes of action, including bending origin DNA into configurations that promote or inhibit DnaA subassemblies, and by directly blocking DnaA access to low-affinity recognition sites (Marczynski and Shapiro 2002; Ryan et al. 2004; Nievera et al. 2006; Leonard and Grimwade 2011). In *E. coli*, the DNA-bending proteins Fis and IHF act as repressors and stimulators, respectively, of the initiation step, producing a switch in *oriC* DNA conformation as the origin loads with DnaA (Ryan et al. 2004). An interesting repercussion of these conformational changes is that high- and low-affinity DnaA recognition sites are moved into proximal positions to allow nucleation of new DnaA oligomers (Rozgaja et al. 2011). Although the DNA-bending proteins are not required for cell viability, their loss disturbs the timing of pre-RC assembly (Ryan et al. 2004).

Bacterial Origins Encode the Directions for Ordered pre-RC Assembly

In *E. coli*, high-affinity DnaA boxes in *oriC* are occupied throughout the majority of the cell cycle, whereas all low-affinity recognition sites become occupied only at the time of initiation of DNA synthesis (Nievera et al. 2006). Once all recognition sites become fully occupied, the resulting DnaA complex is sufficient both to unwind the DUE and assist in loading of DNA helicase (DnaB) onto the exposed single strands, completing bacterial pre-RC assembly (for review, see Kaguni 2011). The persistent occupation of DnaA at high-affinity boxes during the cell cycle serves as the bacterial equivalent of the eukaryotic ORC (origin recognition complex) (Miller et al. 2009), by both marking the origin throughout the cell cycle and by serving as a scaffold to recruit additional initiator molecules (Fig. 1).

The continuous occupation of high-affinity boxes raises questions regarding the mechanism responsible for ordered filling of low-affinity

Table 1. Initiator recognition sequences in bacterial and archaeal *oriC*s

	Site	Affinity
Bacteria representative DnaA boxes		
Escherichia coli		
TGTGGATAA	R1/R4	High
TGTGTATAA	R2	High
TGTCAGGAA	C1	Low
TGTGAATGA	R5	Low
TGGGATCAG	I2	Low with DnaA-ATP preference
TTGGATCAA	I3	Low with DnaA-ATP preference
GGTGATCCT	τ2	Low with DnaA-ATP preference
AAAGAACAA	C3	Low with DnaA-ATP preference
ACCGTATAA	I1	Low with DnaA-ATP preference
GGTAGTTAT	C2	Low with DnaA-ATP preference
Caulobacter crescentus		
TGTGGATCA	G1	Medium
TGTTAATCA	G2	Medium
TCATCCCCA	W1	Low
TCATCCCCA	W2	Low
GCATCCCCC	W3	Low
GAATCCCGA	W4	Low
TCATCCCCA	W5	Low
Mycobacterium tuberculosis		
TGTGAGTGT	3A	
TGTGCACAC	3	
TGTGCGCGG	3B	
TGTGAACGG	4	
TGTGGGGTG	6	
TGTGGATGA	8	
TGTCCCCAA	9	
TGTGCAGTT	10	
Bacillus subtilis		
oriII region		
TGTGGATAA		
TGTGAAAAA		
TGTGGACAG		
TGTGTATGA		
TGTGAATAA		
Helicobacter pylori		
TGTCTATAA	H1	
TGTGAATGA	H2/H3	
TCATTCACC	H4	
CGTGAATAA	H5	
Representative archaeal miniORB (as 12-mer) and ORB sites (as 23-mer)		
Pyrococcus furiosus		
TTCCAATGGAGA		
TTCCTCTGGAAA		
TTCCAGTGGAAA		
CTCCAGTGGAAATGAAACTCTGG (ORB2)		

Continued

Table 1. *Continued*

Site	Affinity
Aeropyrum pernix	
CTCCACAGGAAA	
CTCCAGAGGAAC	
ATCCAGCGGAAA	
CTCCACAGGAAACGGAGGGGTCG (ORB4)	
Sulfolobus solfataricus	
CTCCAGTGGAAA	
TTCCAAACGAAA	
TTCCAGAGGAAA	
TTCCAGAGGAAATAGATGGGTCC (ORB3 in *oriC1*)	
Methanothermobacter thermoautotrophicus	
TTACAGTTGAAA	
TTACACTTGAAA	
TTACAGTTGAAC	

boxes during the cell cycle. A key component of this mechanism is encoded by *oriC* as a specific arrangement of low-affinity DnaA boxes (Fig. 2A) (Rozgaja et al. 2011; Taylor et al. 2011). DnaA occupying each high-affinity DnaA box is sufficient to promote DnaA interactions at proximal low-affinity boxes (Rozgaja et al. 2011). However, these initial steps in pre-RC assembly are limited by distance, and complete complexes of 10–20 DnaA molecules require additional DnaA–DNA interactions to span larger stretches of origin DNA (for review, see Leonard and Grimwade 2011). In *E. coli*, these DnaA interactions are guided along *oriC* DNA by cryptic low-affinity DnaA boxes separated by 2 nt and clustered into sets of two helically phased arrays facing in opposite directions (see Fig. 2A) (Rozgaja et al. 2011). Using this configuration, two DnaA oligomers (Duderstadt et al. 2010) are extended along the DNA gap regions from both R1 and R4 and ultimately anchored at R2 (Rozgaja et al. 2011).

This arrangement explains how protein modulators are able to regulate DnaA oligomer formation by altering the proximity of high- and low-affinity sites on bent DNA or by blocking the addition of monomers onto a growing oligomeric chain. Placing high-affinity DnaA boxes in or out of helical phase with low-affinity arrays, as is the case with *E. coli*, may also play a role in generating the torsional stress needed for initial unwinding of the DUE, because DNA twisting during oligomer assembly might be required to link a continuous DnaA oligomer to its high-affinity anchors at both ends of the origin. Further comparative analysis of low-affinity DnaA interactions in bacterial origins will be important to determine how different bacterial types use origin nucleotide sequences to direct pre-RC assembly.

REPLICATION ORIGINS IN ARCHAEA

Studies of the archaeal DNA replication machinery reveal a mixture of the attributes found in bacteria and eukarya (for review, see Grabowski and Kelman 2003; Barry and Bell 2006). Although most small circular archaeal genomes carry one copy of *oriC*, several genera carry multiple *oriC* copies (Kelman and Kelman 2004; Robinson et al. 2004; Robinson and Bell 2007), which may respond to distinct initiator complexes (SD Bell, pers. comm.). The archaeal initiator Cdc6/Orc1 (Matsunaga et al. 2010) shares extensive sequence homology with eukaryotic ORC proteins (for review, see Wigley 2009), but also shares attributes of DnaA, including regulation by bound ATP (SD Bell, pers. comm.) and interacting with *oriC* as a monomer rather than preassembling into a hexameric complex (for review, see Lundgren and Bernander 2005).

Cdc6/Orc1 contacts nucleotide sequence motifs termed "origin recognition boxes" (Taagepera et al. 1993) that are clustered within archaeal *oriC*s (Fig. 2B). Although number, orientation, and spacing of ORBs vary among archaeal genera, two predominant ORB motifs exist (Table 1). A long ORB motif (22–35 bp) with dyad symmetry is mapped at two or more locations in origins from *Sulfolobus*, *Pyrococcus*, and *Aeropyrum* (Grainge et al. 2006; Dueber et al. 2007; Gaudier et al. 2007). The orientation of each bound Cdc6/Orc1 is determined by the ORB sequence (Dueber et al. 2007; Gaudier et al. 2007). A shorter (12–13 bp) version, termed miniORB, coexists with ORBs in *Pyrococcus* (Matsunaga et al. 2010) and *Sulfolobus* (Dueber et al. 2007), or in the absence ORBs in *Methanothermobacter oriC*s (Capaldi and Berger 2004). MiniORBs are often found as multiple (seven to 15) direct repeats (Capaldi and Berger 2004; Matsunaga et al. 2010), analogous to some bacterial origins. Cdc6/Orc1 binding to proximal miniORBs is also cooperative (Capaldi and Berger 2004; Grainge et al. 2006), suggesting the assembly of oligomeric structures analogous to those formed by DnaA.

Although Cdc6/Orc1 higher-order complexes are reported to alter DNA topology and lead to origin unwinding (Matsunaga et al. 2010), Cdc6/Orc1 contact with ORBs differs from DnaA-*oriC* interactions in several ways. Although DnaA and Cdc6/Orc1 both contact double-stranded DNA through the carboxy-terminal domain, Cdc6/Orc1 make additional contacts with ORBS through the AAA^{+} domain of the protein. These contacts appear to rely less on specific base sequence contacts and more on DNA deformability than is the case for DnaA (Gaudier et al. 2007; Dueber et al. 2011). Each Cdc6/Orc1 interaction also distorts the DNA helix to a greater degree than is seen during bacterial pre-RC assembly (Gaudier et al. 2007; Dueber et al. 2011). It remains unclear whether the distinction between ORBs and miniORBS is due to the inclusion of sequences recognized or distorted by the AAA^{+} domain.

DNA REPLICATION ORIGINS IN EUKARYOTES

In *Saccharomyces cerevisiae*, DNA replication origins were first defined by their ability to confer autonomous replication to plasmids (Stinchcomb et al. 1979; Bell and Stillman 1992; Theis and Newlon 1994). These elements were called autonomous replication sequences (ARSs) and contain the specific consensus element autonomous consensus sequences (ACSs) of 11 bp further extended to 17 bp (for review, see Stinchcomb et al. 1979; Theis and Newlon 1997; Dhar et al. 2012). ACS is the binding site for ORC, the main factor that subsequently serves as a landing platform for the assembly of the other pre-RC proteins. In metazoans, despite the identification of several proteins involved in initiation of DNA replication (most, but not all, are conserved from yeast to humans), it is still not known how they target DNA replication origins. Studies are hampered by the absence of a single, really efficient genetic assay to isolate replication origins because DNA transferred into metazoan cells replicates very poorly. Thus, different biochemical approaches have been developed over the years, including two-dimensional (2D) gel electrophoresis of replication intermediates (Brewer and Fangman 1987) and quantitative PCR analysis of nascent strands (Giacca et al. 1994), a method greatly improved by the exploitation of the RNA-primed nature of nascent strands at origins (Gerbi and Bielinsky 1997; Bielinsky and Gerbi 1998). An interesting, but less used, bubble-trap method is based on the structure of replication origins (Mesner et al. 2011). Genome-wide sequencing of RNA-primed nascent DNAs presently appears as the more precise method.

DNA REPLICATION ORIGINS IN YEAST

Sequence Specificity

Specific consensus sequences for origins are identified only in *S. cerevisiae*. Although this 11-bp sequence [5/(A/T)TTTA(T/C)(A/G)TTT (A/T)-3] defines all the 12,000 ACSs present in this genome, only a few of them (~400) are used

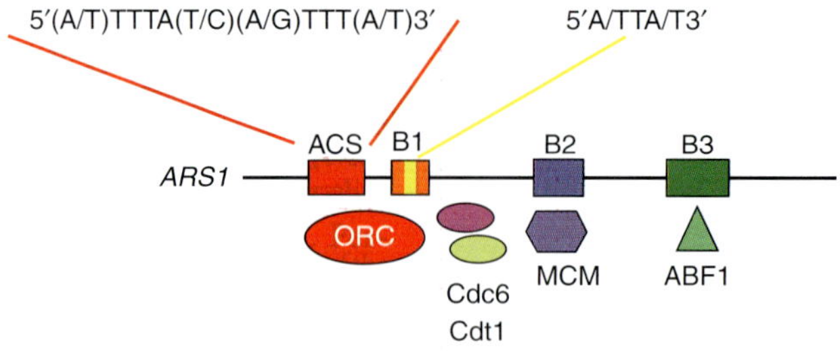

Figure 3. *S. cerevisiae* replication origins. Elements found at *S. cerevisiae* origins are represented, including the ACS and B elements. The main proteins that bind to these elements and that constitute the pre-RC are indicated.

in normal conditions (Linskens and Huberman 1988). Other elements close to the ACS motif contribute to its activity and provide a modular structure to origins (Fig. 3) (Marahrens and Stillman 1992). One example, the B domain, might serve as a DNA unwinding element (Huang and Kowalski 1993) and can be subdivided into three subdomains. B1 is adjacent to the ACS and is important for ORC binding (Rao and Stillman 1995; Rowley et al. 1995). It also contains an A/TTA/T motif that is found in 228 ARSs conserved in six *Saccharomyces* species and could be involved in ORC binding activity (Chang et al. 2008). The B2 element is suggested to be the binding site for the MCM2-7 helicase (Zou and Stillman 2000; Wilmes and Bell 2002). It may also contain a degenerate ANA/TA/TAAAT common sequence that contributes to the efficiency of origin recognition (Chang et al. 2011). B3 binds to Abf1 (ARS binding factor 1), a DNA-binding protein that plays a role in transcription and can influence chromatin structure and nucleosome assembly (Miyake et al. 2002; Ganapathi et al. 2011).

In yeast strains other than *S. cerevisiae*, a clear consensus sequence is not found, and origin locations are not conserved in budding yeast species (Liachko et al. 2010; Xu et al. 2012). A 9-bp putative motif, similar to the ACS in *S. cerevisiae*, has been identified in *Lachancea kluyveri*, but it is not sufficient to confer ARS activity and must be embedded in a much larger DNA sequence (Liachko et al. 2011). The preference for poly(A) tracks was confirmed in *Schizosaccharomyces pombe* and *Schizosaccharomyces octosporus*, and AT-rich sequences were also found in *Lachancea kluyveri* (Liachko et al. 2011). In *S. pombe*, large DNA replication origins were found composed of several short AT-rich modules capable of initiating replication from ectopic chromosome positions (Cotobal et al. 2010). However, different motifs can be used in different unicellular organisms. Surprisingly, the A-T content is a strong negative predictor of origin function in *Schizosaccharomyces japonicus* (Xu et al. 2012) in which origins have a high G-C content. Because poly(G) motifs, like AT-rich elements, exclude nucleosomes (Tsankov et al. 2011; Fenouil et al. 2012), they are expected to have a similar role in the chromatin environment (see below).

Altogether, these results suggest more plasticity for origin recognition in unicellular eukaryotes than previously thought based only on the analysis of the *S. cerevisiae* genome. However, plasticity does not exclude site specificity. In *S. pombe*, there is a clear association of DNA replication origins with ORC-binding sites and nucleosome-free regions (Hayashi et al. 2007; Xu et al. 2012).

Chromatin Environment

In addition to sequence specificity, a second important feature of *S. cerevisiae* origins is nucleosome exclusion. The patterns of nucleosome occupancy around the ACS are similar to open transcription promoters, and genome-wide analyses of nucleosome positioning support the conclusions from studies on single origins (Thoma et al. 1984; Berbenetz et al. 2010; Eaton et al. 2010). The ACS is necessary to confer a nucleosome-free region. ORC binding is then required to induce a periodic positioning of nucleosomes adjacent to the ACS and might act as a barrier element to position and phase nucleosomes at precise intervals on both sides of the origin. Interestingly, nucleosome positioning around the ACS characterizes origins that will be active. The nucleosome-free feature of this region might be due at least in part to the A-T richness of this domain (Kaplan et al. 2009; Zhang et al. 2009). The size of the nucleosome-free region (~130 bp) provides room for the assembly of the other pre-RC proteins and does not seem to vary during the cell cycle. However, if the ORC is depleted, the size of the nucleosome-free region decreases, and the positioning of adjacent nucleosomes is affected.

Although nucleosomes are depleted at the ACS, interaction of the ORC complex with adjacent chromatin regions and histone modifications are likely to contribute to the general mechanism of origin recognition in yeast. In agreement, if nucleosomes are moved away from the ORC-binding site, origin function is reduced (Lipford and Bell 2001), emphasizing the importance of precise nucleosome positioning for replication initiation. The exact set of histone modifications involved in DNA replication origin recognition or firing remains largely unknown in yeast as well as in multicellular organisms. This is partly because not all potential origins are active in each individual cell, making it difficult to identify the modifications specifically associated with origin recognition or firing. Nevertheless, using a mini-chromosome that contains an efficient yeast origin, increased histone H3 and H4 acetylation was correlated with origin activity (Unnikrishnan et al. 2010).

DNA REPLICATION ORIGINS IN MULTICELLULAR ORGANISMS

Organization of Metazoan Origins

In metazoans, because of the larger size of their genomes, thousands of replication origins are activated at each cell cycle (30,000–50,000 in human or mouse cells). Until very recently, only a few origins were clearly identified, but improved genome-wide analyses permit a better understanding of their nature (Cadoret et al. 2008; Sequeira-Mendes et al. 2009; Cayrou et al. 2011; Costas et al. 2011; Mesner et al. 2011). Several features are now emerging. First, there are many more origins than needed at each cell cycle. Only one out of five potential replication origins within a replicon is used in any cell during a given cell cycle, and, therefore, DNA replication origin activation appears to be a relatively inefficient process (Cayrou et al. 2011). This feature is not unique to metazoans because it is shared by yeast origins, including *S. cerevisiae*, where origins are defined by specific motifs (Friedman et al. 1997; Heichinger et al. 2006). In theory, this abundance of potential replication origins in each replication unit or replicon might favor the activation of at least one origin per replicon (Karschau et al. 2012). This low usage of replication origins results from a flexible choice of replication origins in each cell (Fig. 4). Thus, within a given cell population, each cell has the potential to activate a large number of origins (100,000 in mouse fibroblasts) and might use a different cohort of origins in an apparent stochastic manner. This situation results in initiation zones with an average of five potential origins that constitute the first order of organization of a replication unit: the flexible replicon (Cayrou et al. 2011). The second hierarchical level of organization is the cluster of replicons. DNA replication origins are activated in clusters where several replication origins are often activated simultaneously, as first observed by autoradiography and more recently by DNA combing methods (Lebofsky et al. 2006; Cayrou et al. 2011). In origin clusters, early- and late-firing origins are often separated by long, origin-poor regions, also called "temporal transition regions" (TTRs). TTRs were first detected in the mouse Ig heavy-chain locus (Norio et al. 2005) and then in genome-wide analysis of DNA replication origins (Cadoret et al. 2008; Cayrou et al. 2011). However, TTR replication might differ in different cell types because replication origins in TRR might be silenced in some cells, but gradually activated

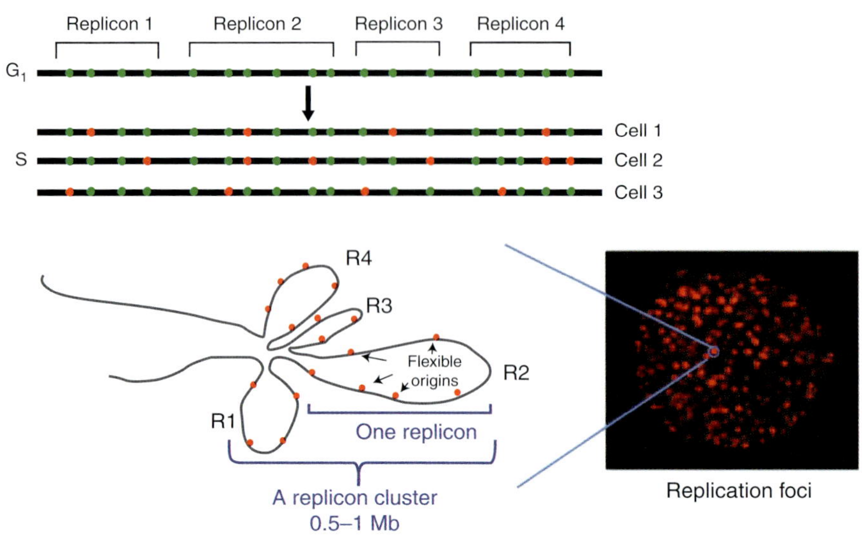

Figure 4. From replication origins to replicon clusters. Replicons are organized as functional domains that contain several potential DNA replication origins (on average five), already licensed. A single origin is activated per replicon during the S phase. The origin choice within each replicon can occur stochastically or can be influenced by specific cell fates or transcriptional activity. Replicon clusters include several consecutive replicons that are activated simultaneously (Berezney et al. 2000). Replicons could be organized in chromatin loops where activation of one origin silences the other origins within the same replicon.

in others during the S-phase progression (Guilbaud et al. 2011).

The large number and the flexible activation of DNA replication origins have several advantages. Replication origins in excess (spare origins) can be used in case of poor growing conditions or when the replication fork experiences difficulties in passing through higher-order DNA or chromatin structures. Spare origins can also be activated in specific physiological conditions, such as during early embryonic development. Indeed, during early *Drosophila* or *Xenopus* development, the S phase lasts $<$15 min, and replication origins are activated every 10-15 kb (Blumenthal et al. 1974; Hyrien et al. 1995). It would be interesting to reinvestigate the site specificity of these origins in early embryos to determine whether all potential origins are used in every cell, as opposed to completely random initiation. The large number of flexible origins also include dormant origins that normally are not activated unless the cell is under replication stress (such as DNA damage and in response to checkpoint signals) (Branzei and Foiani 2005; Woodward et al. 2006; Blow et al. 2011). Finally, the large number of potential origins also includes those that allow the replication program to adapt to the transcription program. The number of identified DNA replication origins that are cell or developmental stage specific and that are normally inactive in other cell types is relatively limited at present (Hyrien et al. 1995; Norio et al. 2005), but they are likely to increase in the next years.

Active transcription may have two potentially opposite consequences on the choice of DNA replication origins. Indeed, it can provide a nucleosome-free area at the promoter region to which the replication complex might bind opportunely. On the other hand, active transcription may also not be favorable to the assembly of a replication complex (for review, see Méchali 2010; Sequeira-Mendes and Gomez 2012). This cross talk between genome organization for DNA replication and gene expression is likely to be crucial for the maintenance of cell identity during development and differentiation.

Sequence Specificity in Metazoans

The question of sequence specificity remains unresolved in metazoan cells. The ORC is the first replication factor known to bind to replication origins, but it does not recognize a specific sequence, although in vitro it shows a preference for supercoiled DNA and poly(dA)–poly(dT) tracts (Vashee et al. 2003; Remus et al. 2004). A recent genome-wide distribution of ORCs in *Drosophila* could identify motifs that have a predictive value for origins (MacAlpine et al. 2010). However, ORC also binds to sites other than replication origins, such as centrosomes, centromeres, or heterochromatin (Prasanth et al. 2004; Hemerly et al. 2009). Sequence specificity for replication origins is more directly addressed by recent genome-wide identification of active sites of initiation of DNA replication. They confirmed a preference for CpG islands (Delgado et al. 1998; Cadoret et al. 2008; Sequeira-Mendes et al. 2009; Cayrou et al. 2011). Association with GC-rich regions was also observed in plant replication origins (Costas et al. 2011). In addition, in mouse and human cells, an origin G-rich repeat element (OGRE) is present in 80%–90% of DNA replication origins and can form G quadruplexes (Cayrou et al. 2011, 2012) in which guanines can form unusual four-stranded DNA structures through hydrogen bonding. This finding suggests that metazoan origins may be rather different from yeast origins, which are mainly A-T rich. At metazoan origins, a DNA structure formed by a specific positioning of bases, similar to the case for G quadruplexes, might be more important than strict sequence specificity for initiation of DNA replication. It now remains to show experimentally that these conserved elements are components of the metazoan replication origins.

Finally, a completely different approach to map DNA replication origins, based on in silico analysis of the nucleotide composition skew, showed an abrupt change in the skew profile at known replication origins, possibly because of different mutation rates at the leading or lagging strand of DNA synthesis (Chen et al. 2011). Based on this feature, very large interacting chromatin domains with efficient origins at the borders could be defined.

Transcription, Chromatin, and Nuclear Organization in the Selection of DNA Replication Origins in Metazoans

A permissive chromatin environment is a major characteristic of active transcription promoters and has been considered an important feature of replication origins for many years (for review, see Sequeira-Mendes and Gomez 2011). This may explain the general correlation between transcription activity and enrichment in DNA replication origins found by several of the recent genome-wide analyses of DNA replication. Nucleosome-free or unstable chromatin regions of active promoter regions are frequently associated with origins because AT-rich or G-rich regions are not favorable to nucleosome formation. These regions are also marked by specific histone variants, such as H2AZ or H3.3 (Jin et al. 2009; MacAlpine et al. 2010; Stroud et al. 2012), and by the association of chromatin remodelers that contribute to the recruitment of the pre-RC complex, such as the histone acetylase HBO1 (Miotto and Struhl 2010), the histone methyl transferase PR-Set7 (Tardat et al. 2010), or the chromatin-remodeling complex SNF2H (Sugimoto et al. 2011).

However, a nucleosome-free region does not exclude the possibility that interactions of pre-RC proteins with adjacent nucleosomes might be important, at least in specific chromatin environments. For instance, ORC interacts with heterochromatin and HP1 in *Drosophila* (Pak et al. 1997) in agreement with the finding that on *Drosophila* chromosome *4*, which is organized mainly into heterochromatin, origins are associated with HP1 sites (Cayrou et al. 2011). Similarly, ORC interactions (regulated by the BAH domain of the ORC1 subunit) with chromatin and with adjacent nucleosomes were also described in *S. cerevisiae* and in human cells (Noguchi et al. 2006; Muller et al. 2010). The BAH domain in human Orc1 facilitates the ability of Orc1 to activate replication origins in vivo (Noguchi et al. 2006) and links H4K20me2 to DNA replication licensing and to Meier–Gorlin syndrome, a genetic problem leading to primordial dwarfism (Kuo et al. 2012). However, the role of this domain in ORC association with chromatin varies at different origins, emphasizing the non-unique nature of replication origins.

Histone modifications are essential for modulating gene expression and might also play an important role in the choice of DNA replication origins and in the regulation of their activation. Histone lysine acetylation has often been associated with the regulation of replication timing, but links with DNA replication origins activation have also been reported (Aggarwal and Calvi 2004; Liu et al. 2012). Histone lysine methylations have been recently reviewed (Dorn and Cook 2011), and H4K20 methylation appears to be involved in the control of DNA replication origins. H4K20me1 is high during pre-RC formation, from mitosis to G_1, but low in S phase (Rice et al. 2002), and is found at some human DNA replication origins (Tardat et al. 2010). In addition, tethering PR-Set7, the methylase responsible for H4K20 methylation, to a specific genomic locus promotes pre-RC assembly. Another methylation mark found at replication origins is H3K4 di- and trimethylation (Costas et al. 2011; Eaton et al. 2011).

At a higher level of chromosomal organization, the timing of replication domains strongly correlates with long-range chromatin interactions, as measured by using the Hi-C technique (Ryba et al. 2010), supporting the hypothesis that replication domains localized near each other replicate at similar times. These domains may correspond to clusters of replication origins forming the DNA replication foci or factories that are observed in all nuclei (Fig. 4) (for review, see Jackson and Pombo 1998; Berezney et al. 2000; Maya-Mendoza et al. 2010). How DNA replication origins are organized in such structures remains hypothetical. Each replicon, which is constituted by several potential origins, might form a chromatin loop (Cayrou et al. 2011), and several clustered chromatin loops could constitute replication domains or replication foci. The chromatin loop hypothesis is supported by the correlation between the size of the loop and the size of replicons in different species (Buongiorno-Nardelli et al. 1982) and under different growing conditions (Lemaitre et al. 2005; Courbet et al. 2008). Cohesin, a protein involved in sister-chromatid cohesion, might be involved in the stabilization of these loops and is enriched at some replication origins (Guillou et al. 2010), although it appears to be loaded via pre-RC assembly-independent mechanisms. A concordance between ORC-binding sites and cohesin sites confirmed this relationship between replication origins, cohesin, and chromatin loops (MacAlpine et al. 2010).

CONCLUDING REMARKS

The differences in the amount of information encoded by prokaryotic and eukaryotic replication origins and the contribution of chromatin structure have obvious impacts on both the nucleotide stringency and length of required chromosomal DNA. In the case of bacteria, much of the encoded information has emerged only recently, and as the nucleotide sequences of origins are examined in different species, additional instructions for assembling ordered DnaA prereplicative complexes and even components of the cell cycle timing mechanism will be revealed. It is likely that metazoan replication origins are characterized by a combination of different elements, including some DNA motifs or structures, and that different classes of origins coexist in the metazoan genomes. This modular nature of replication origins might provide the flexibility required to organize chromosomes for their replication, depending on the cell fate, as well as to provide the ability to complete DNA replication of the whole genome in a wide range of physiological conditions. Further analysis of archaeal origins is needed to understand the important bridge from origins with high information content to more streamlined versions, ultimately leading to the minimal sequence specificity found among the higher eukaryotes. The study of eukaryotic genomes must continue in order to determine the requirements of "origin-ness," and the bulk of the effort will surely be focused on unraveling the epigenetic regulators that allow ORC to gain DNA access. It is clear that a great deal of the story remains to be told.

ACKNOWLEDGMENTS

We are grateful to many colleagues for their helpful discussions and insights. Research in our laboratories is supported by National Institutes of Health grant GM054042 (to A.C.L.) and the European Research Council (FP7/2007-2013 grant agreement no. 233339). This work is also supported by the ANR, the ARC, and the Ligue Nationale Contre le Cancer (LNCC).

REFERENCES

Aggarwal BD, Calvi BR. 2004. Chromatin regulates origin activity in *Drosophila* follicle cells. *Nature* **430:** 372–376.

Barry ER, Bell SD. 2006. DNA replication in the archaea. *Microbiol Mol Biol Rev* **70:** 876–887.

Beattie TR, Bell SD. 2011. Molecular machines in archaeal DNA replication. *Curr Opin Chem Biol* **15:** 614–619.

Bell SP, Stillman B. 1992. ATP-dependent recognition of eukaryotic origins of DNA replication by a multiprotein complex. *Nature* **357:** 128–134.

Berbenetz NM, Nislow C, Brown GW. 2010. Diversity of eukaryotic DNA replication origins revealed by genome-wide analysis of chromatin structure. *PLoS Genet* **6:** e1001092.

Berezney R, Dubey DD, Huberman JA. 2000. Heterogeneity of eukaryotic replicons, replicon clusters, and replication foci. *Chromosoma* **108:** 471–484.

Bielinsky AK, Gerbi SA. 1998. Discrete start sites for DNA synthesis in the yeast ARS1 origin. *Science* **279:** 95–98.

Blow JJ, Gillespie PJ. 2008. Replication licensing and cancer—A fatal entanglement? *Nat Rev Cancer* **8:** 799–806.

Blow JJ, Ge XQ, Jackson DA. 2011. How dormant origins promote complete genome replication. *Trends Biochem Sci* **36:** 405–414.

Blumenthal AB, Kriegstein HJ, Hogness DS. 1974. The units of DNA replication in *Drosophila melanogaster* chromosomes. *Cold Spring Harb Symp Quant Biol* **38:** 205–223.

Branzei D, Foiani M. 2005. The DNA damage response during DNA replication. *Curr Opin Cell Biol* **17:** 568–575.

Brassinga AK, Siam R, McSween W, Winkler H, Wood D, Marczynski GT. 2002. Conserved response regulator CtrA and IHF binding sites in the α-proteobacteria *Caulobacter crescentus* and *Rickettsia prowazekii* chromosomal replication origins. *J Bacteriol* **184:** 5789–5799.

Brewer BJ, Fangman WL. 1987. The localization of replication origins on ARS plasmids in *S. cerevisiae*. *Cell* **51:** 463–471.

Bryant JA, Aves SJ. 2011. Initiation of DNA replication: Functional and evolutionary aspects. *Ann Bot* **107:** 1119–1126.

Buongiorno-Nardelli M, Micheli G, Carri MT, Marilley M. 1982. A relationship between replicon size and supercoiled loop domains in the eukaryotic genome. *Nature* **298:** 100–102.

Cadoret JC, Meisch F, Hassan-Zadeh V, Luyten I, Guillet C, Duret L, Quesneville H, Prioleau MN. 2008. Genome-wide studies highlight indirect links between human replication origins and gene regulation. *Proc Natl Acad Sci* **105:** 15837–15842.

Capaldi SA, Berger JM. 2004. Biochemical characterization of Cdc6/Orc1 binding to the replication origin of the euryarchaeon *Methanothermobacter thermoautotrophicus*. *Nucleic Acids Res* **32:** 4821–4832.

Cayrou C, Coulombe P, Vigneron A, Stanojcic S, Ganier O, Peiffer I, Rivals E, Puy A, Laurent-Chabalier S, Desprat R, et al. 2011. Genome-scale analysis of metazoan replication origins reveals their organization in specific but flexible sites defined by conserved features. *Genome Res* **21:** 1438–1449.

Cayrou C, Coulombe P, Puy A, Rialle S, Kaplan N, Segal E, Mechali M. 2012. New insights into replication origin characteristics in metazoans. *Cell Cycle* **11:** 658–667.

Chang F, Theis JF, Miller J, Nieduszynski CA, Newlon CS, Weinreich M. 2008. Analysis of chromosome III replicators reveals an unusual structure for the ARS318 silencer origin and a conserved WTW sequence within the origin recognition complex binding site. *Mol Cell Biol* **28:** 5071–5081.

Chang F, May CD, Hoggard T, Miller J, Fox CA, Weinreich M. 2011. High-resolution analysis of four efficient yeast replication origins reveals new insights into the ORC and putative MCM binding elements. *Nucleic Acids Res* **39:** 6523–6535.

Chen CL, Duquenne L, Audit B, Guilbaud G, Rappailles A, Baker A, Huvet M, d'Aubenton-Carafa Y, Hyrien O, Arneodo A, et al. 2011. Replication-associated mutational asymmetry in the human genome. *Mol Biol Evol* **28:** 2327–2337.

Costas C, de la Paz Sanchez M, Stroud H, Yu Y, Oliveros JC, Feng S, Benguria A, Lopez-Vidriero I, Zhang X, Solano R, et al. 2011. Genome-wide mapping of *Arabidopsis thaliana* origins of DNA replication and their associated epigenetic marks. *Nat Struct Mol Biol* **18:** 395–400.

Cotobal C, Segurado M, Antequera F. 2010. Structural diversity and dynamics of genomic replication origins in *Schizosaccharomyces pombe*. *EMBO J* **29:** 934–942.

Courbet S, Gay S, Arnoult N, Wronka G, Anglana M, Brison O, Debatisse M. 2008. Replication fork movement sets chromatin loop size and origin choice in mammalian cells. *Nature* **455:** 557–560.

Cvetic C, Walter JC. 2005. Eukaryotic origins of DNA replication: Could you please be more specific? *Semin Cell Dev Biol* **16:** 343–353.

Dhar MK, Sehgal S, Kaul S. 2012. Structure, replication efficiency and fragility of yeast ARS elements. *Res Microbiol* **163:** 243–253.

Delgado S, Gomez M, Bird A, Antequera F. 1998. Initiation of DNA replication at CpG islands in mammalian chromosomes. *EMBO J* **17:** 2426–2435.

Diffley JF. 2011. Quality control in the initiation of eukaryotic DNA replication. *Philos Trans R Soc Lond B Biol Sci* **366:** 3545–3553.

Ding Q, MacAlpine DM. 2011. Defining the replication program through the chromatin landscape. *Crit Rev Biochem Mol Biol* **46:** 165–179.

Dorn ES, Cook JG. 2011. Nucleosomes in the neighborhood: New roles for chromatin modifications in replication origin control. *Epigenetics* **6:** 552–559.

Duderstadt KE, Berger JM. 2008. AAA$^+$ ATPases in the initiation of DNA replication. *Crit Rev Biochem Mol Biol* **43:** 163–187.

Duderstadt KE, Mott ML, Crisona NJ, Chuang K, Yang H, Berger JM. 2010. Origin remodeling and opening in bacteria rely on distinct assembly states of the DnaA initiator. *J Biol Chem* **285:** 28229–28239.

Dueber EL, Corn JE, Bell SD, Berger JM. 2007. Replication origin recognition and deformation by a heterodimeric archaeal Orc1 complex. *Science* **317:** 1210–1213.

Dueber EC, Costa A, Corn JE, Bell SD, Berger JM. 2011. Molecular determinants of origin discrimination by Orc1 initiators in archaea. *Nucleic Acids Res* **39:** 3621–3631.

Duncker BP, Chesnokov IN, McConkey BJ. 2009. The origin recognition complex protein family. *Genome Biol* **10:** 214.

Eaton ML, Galani K, Kang S, Bell SP, MacAlpine DM. 2010. Conserved nucleosome positioning defines replication origins. *Genes Dev* **24:** 748–753.

Eaton ML, Prinz JA, MacAlpine HK, Tretyakov G, Kharchenko PV, MacAlpine DM. 2011. Chromatin signatures of the *Drosophila* replication program. *Genome Res* **21:** 164–174.

Erzberger JP, Mott ML, Berger JM. 2006. Structural basis for ATP-dependent DnaA assembly and replication-origin remodeling. *Nat Struct Mol Biol* **13:** 676–683.

Fenouil R, Cauchy P, Koch F, Descostes N, Cabeza JC, Innocenti C, Ferrier P, Spicuglia S, Gut M, Andrau JC. 2012. CpG islands and GC contents dictate nucleosome depletion in a transcription independent manner at mammalian promoters. *Genome Res* doi: 10.1101/gr.138776.112.

Friedman KL, Brewer BJ, Fangman WL. 1997. Replication profile of *Saccharomyces cerevisiae* chromosome VI. *Genes Cells* **2:** 667–678.

Ganapathi M, Palumbo MJ, Ansari SA, He Q, Tsui K, Nislow C, Morse RH. 2011. Extensive role of the general regulatory factors, Abf1 and Rap1, in determining genome-wide chromatin structure in budding yeast. *Nucleic Acids Res* **39:** 2032–2044.

Gao F, Zhang CT. 2007. DoriC: A database of oriC regions in bacterial genomes. *Bioinformatics* **23:** 1866–1867.

Gao F, Zhang CT. 2008. Ori-Finder: A web-based system for finding oriCs in unannotated bacterial genomes. *BMC Bioinformatics* **9:** 79.

Gaudier M, Schuwirth BS, Westcott SL, Wigley DB. 2007. Structural basis of DNA replication origin recognition by an ORC protein. *Science* **317:** 1213–1216.

Gerbi SA, Bielinsky AK. 1997. Replication initiation point mapping. *Methods* **13:** 271–280.

Giacca M, Zentilin L, Norio P, Diviacco S, Dimitrova D, Contreas G, Biamonti G, Perini G, Weighardt F, Riva S, et al. 1994. Fine mapping of a replication origin of human DNA. *Proc Natl Acad Sci* **91:** 7119–7123.

Grabowski B, Kelman Z. 2003. Archeal DNA replication: Eukaryal proteins in a bacterial context. *Annu Rev Microbiol* **57:** 487–516.

Grainge I, Gaudier M, Schuwirth BS, Westcott SL, Sandall J, Atanassova N, Wigley DB. 2006. Biochemical analysis of a DNA replication origin in the archaeon *Aeropyrum pernix*. *J Mol Biol* **363:** 355–369.

Guilbaud G, Rappailles A, Baker A, Chen CL, Arneodo A, Goldar A, d'Aubenton-Carafa Y, Thermes C, Audit B, Hyrien O. 2011. Evidence for sequential and increasing activation of replication origins along replication timing gradients in the human genome. *PLoS Comput Biol* **7:** e1002322.

Guillou E, Ibarra A, Coulon V, Casado-Vela J, Rico D, Casal I, Schwob E, Losada A, Mendez J. 2010. Cohesin organizes chromatin loops at DNA replication factories. *Genes Dev* **24:** 2812–2822.

Hayashi M, Katou Y, Itoh T, Tazumi A, Yamada Y, Takahashi T, Nakagawa T, Shirahige K, Masukata H. 2007. Genome-wide localization of pre-RC sites and identification of replication origins in fission yeast. *EMBO J* **26:** 1327–1339.

Heichinger C, Penkett CJ, Bahler J, Nurse P. 2006. Genome-wide characterization of fission yeast DNA replication origins. *EMBO J* **25:** 5171–5179.

Hemerly AS, Prasanth SG, Siddiqui K, Stillman B. 2009. Orc1 controls centriole and centrosome copy number in human cells. *Science* **323:** 789–793.

Huang RY, Kowalski D. 1993. A DNA unwinding element and an ARS consensus comprise a replication origin within a yeast chromosome. *EMBO J* **12:** 4521–4531.

Hwang DS, Thony B, Kornberg A. 1992. IciA protein, a specific inhibitor of initiation of *Escherichia coli* chromosomal replication. *J Biol Chem* **267:** 2209–2213.

Hyrien O, Maric C, Mechali M. 1995. Transition in specification of embryonic metazoan DNA replication origins. *Science* **270:** 994–997.

Jackson DA, Pombo A. 1998. Replicon clusters are stable units of chromosome structure: Evidence that nuclear organization contributes to the efficient activation and propagation of S phase in human cells. *J Cell Biol* **140:** 1285–1295.

Jacob F, Brenner J, Cuzin F. 1963. On the regulation of DNA replication in bacteria. *Cold Spring Harbor Symp Quant Biol* **28:** 329–348.

Jin C, Zang C, Wei G, Cui K, Peng W, Zhao K, Felsenfeld G. 2009. H3.3/H2A.Z double variant-containing nucleosomes mark "nucleosome-free regions" of active promoters and other regulatory regions. *Nat Genet* **41:** 941–945.

Kaguni JM. 2011. Replication initiation at the *Escherichia coli* chromosomal origin. *Curr Opin Chem Biol* **15:** 606–613.

Kaplan N, Moore IK, Fondufe-Mittendorf Y, Gossett AJ, Tillo D, Field Y, LeProust EM, Hughes TR, Lieb JD, Widom J, et al. 2009. The DNA-encoded nucleosome organization of a eukaryotic genome. *Nature* **458:** 362–366.

Karschau J, Blow JJ, de Moura AP. 2012. Optimal placement of origins for DNA replication. *Phys Rev Lett* **108:** 058101.

Kawakami H, Katayama T. 2010. DnaA, ORC, and Cdc6: Similarity beyond the domains of life and diversity. *Biochem Cell Biol* **88:** 49–62.

Kelman LM, Kelman Z. 2004. Multiple origins of replication in archaea. *Trends Microbiol* **12:** 399–401.

Kowalski D, Eddy MJ. 1989. The DNA unwinding element: A novel, *cis*-acting component that facilitates opening of the *Escherichia coli* replication origin. *EMBO J* **8:** 4335–4344.

Kuo AJ, Song J, Cheung P, Ishibe-Murakami S, Yamazoe S, Chen JK, Patel DJ, Gozani O. 2012. The BAH domain of ORC1 links H4K20me2 to DNA replication licensing and Meier–Gorlin syndrome. *Nature* **484:** 115–119.

Lebofsky R, Heilig R, Sonnleitner M, Weissenbach J, Bensimon A. 2006. DNA replication origin interference increases the spacing between initiation events in human cells. *Mol Biol Cell* **17:** 5337–5345.

Lemaitre JM, Danis E, Pasero P, Vassetzky Y, Mechali M. 2005. Mitotic remodeling of the replicon and chromosome structure. *Cell* **123:** 1–15.

Leonard AC, Grimwade JE. 2011. Regulation of DnaA assembly and activity: Taking directions from the genome. *Annu Rev Microbiol* **65:** 19–35.

Liachko I, Bhaskar A, Lee C, Chung SC, Tye BK, Keich U. 2010. A comprehensive genome-wide map of autonomously replicating sequences in a naive genome. *PLoS Genet* **6:** e1000946.

Liachko I, Tanaka E, Cox K, Chung SC, Yang L, Seher A, Hallas L, Cha E, Kang G, Pace H, et al. 2011. Novel features of ARS selection in budding yeast *Lachancea kluyveri*. *BMC Genomics* **12:** 633.

Linskens M, Huberman JA. 1988. Organization of replication in ribosomal DNA in *Saccharomyces cerevisiae*. *Mol Cell Biol* **8:** 4927–4935.

Lipford JR, Bell SP. 2001. Nucleosomes positioned by ORC facilitate the initiation of DNA replication. *Mol Cell* **7:** 21–30.

Liu J, McConnell K, Dixon M, Calvi BR. 2012. Analysis of model replication origins in *Drosophila* reveals new aspects of the chromatin landscape and its relationship to origin activity and the prereplicative complex. *Mol Biol Cell* **23:** 200–212.

Lundgren M, Bernander R. 2005. Archaeal cell cycle progress. *Curr Opin Microbiol* **8:** 662–668.

MacAlpine HK, Gordan R, Powell SK, Hartemink AJ, MacAlpine DM. 2010. *Drosophila* ORC localizes to open chromatin and marks sites of cohesin complex loading. *Genome Res* **20:** 201–211.

Mackiewicz P, Zakrzewska-Czerwinska J, Zawilak A, Dudek MR, Cebrat S. 2004. Where does bacterial replication start? Rules for predicting the oriC region. *Nucleic Acids Res* **32:** 3781–3791.

Marahrens Y, Stillman B. 1992. A yeast chromosomal origin of DNA replication defined by multiple functional elements. *Science* **255:** 817–823.

Marczynski GT, Shapiro L. 2002. Control of chromosome replication in *Caulobacter crescentus*. *Annu Rev Microbiol* **56:** 625–656.

Masai H, Matsumoto S, You Z, Yoshizawa-Sugata N, Oda M. 2010. Eukaryotic chromosome DNA replication: Where, when, and how? *Annu Rev Biochem* **79:** 89–130.

Matsunaga F, Takemura K, Akita M, Adachi A, Yamagami T, Ishino Y. 2010. Localized melting of duplex DNA by Cdc6/Orc1 at the DNA replication origin in the hyperthermophilic archaeon *Pyrococcus furiosus*. *Extremophiles* **14:** 21–31.

Maya-Mendoza A, Olivares-Chauvet P, Shaw A, Jackson DA. 2010. S phase progression in human cells is dictated by the genetic continuity of DNA foci. *PLoS Genet* **6:** e1000900.

McGarry KC, Ryan VT, Grimwade JE, Leonard AC. 2004. Two discriminatory binding sites in the *Escherichia coli* replication origin are required for DNA strand opening by initiator DnaA-ATP. *Proc Natl Acad Sci* **101:** 2811–2816.

Méchali M. 2010. Eukaryotic DNA replication origins: Many choices for appropriate answers. *Nature Rev Mol Cell Biol* **11:** 728–738.

Mesner LD, Valsakumar V, Karnani N, Dutta A, Hamlin JL, Bekiranov S. 2011. Bubble-chip analysis of human origin distributions demonstrates on a genomic scale significant clustering into zones and significant association with transcription. *Genome Res* **21:** 377–389.

Miller DT, Grimwade JE, Betteridge T, Rozgaja T, Torgue JJ, Leonard AC. 2009. Bacterial origin recognition complexes direct assembly of higher-order DnaA oligomeric structures. *Proc Natl Acad Sci* **106:** 18479–18484.

Miotto B, Struhl K. 2010. HBO1 histone acetylase activity is essential for DNA replication licensing and inhibited by Geminin. *Mol Cell* **37:** 57–66.

Miyake T, Loch CM, Li R. 2002. Identification of a multifunctional domain in autonomously replicating sequence-binding factor 1 required for transcriptional activation, DNA replication, and gene silencing. *Mol Cell Biol* **22:** 505–516.

Muller P, Park S, Shor E, Huebert DJ, Warren CL, Ansari AZ, Weinreich M, Eaton ML, MacAlpine DM, Fox CA. 2010. The conserved bromo-adjacent homology domain of yeast Orc1 functions in the selection of DNA replication origins within chromatin. *Genes Dev* **24:** 1418–1433.

Nievera C, Torgue JJ, Grimwade JE, Leonard AC. 2006. SeqA blocking of DnaA–oriC interactions ensures staged assembly of the *E. coli* pre-RC. *Mol Cell* **24:** 581–592.

Noguchi K, Vassilev A, Ghosh S, Yates JL, DePamphilis ML. 2006. The BAH domain facilitates the ability of human Orc1 protein to activate replication origins in vivo. *EMBO J* **25:** 5372–5382.

Norio P, Kosiyatrakul S, Yang Q, Guan Z, Brown NM, Thomas S, Riblet R, Schildkraut CL. 2005. Progressive activation of DNA replication initiation in large domains of the immunoglobulin heavy chain locus during B cell development. *Mol Cell* **20:** 575–587.

O'Donnell M. 2006. Replisome architecture and dynamics in *Escherichia coli*. *J Biol Chem* **281:** 10653–10656.

Pak DT, Pflumm M, Chesnokov I, Huang DW, Kellum R, Marr J, Romanowski P, Botchan MR. 1997. Association of the origin recognition complex with heterochromatin and HP1 in higher eukaryotes. *Cell* **91:** 311–323.

Polaczek P, Kwan K, Liberies DA, Campbell JL. 1997. Role of architectural elements in combinatorial regulation of initiation of DNA replication in *Escherichia coli*. *Mol Microbiol* **26:** 261–275.

Prasanth SG, Prasanth KV, Siddiqui K, Spector DL, Stillman B. 2004. Human Orc2 localizes to centrosomes, centromeres and heterochromatin during chromosome inheritance. *EMBO J* **23:** 2651–2663.

Quon KC, Yang B, Domian IJ, Shapiro L, Marczynski GT. 1998. Negative control of bacterial DNA replication by a cell cycle regulatory protein that binds at the chromosome origin. *Proc Natl Acad Sci* **95:** 120–125.

Rao H, Stillman B. 1995. The origin recognition complex interacts with a bipartite DNA binding site within yeast replicators. *Proc Natl Acad Sci* **92:** 2224–2228.

Remus D, Diffley JF. 2009. Eukaryotic DNA replication control: Lock and load, then fire. *Curr Opin Cell Biol* **21:** 771–777.

Remus D, Beall EL, Botchan MR. 2004. DNA topology, not DNA sequence, is a critical determinant for *Drosophila* ORC–DNA binding. *EMBO J* **23:** 897–907.

Rice JC, Nishioka K, Sarma K, Steward R, Reinberg D, Allis CD. 2002. Mitotic-specific methylation of histone H4 Lys 20 follows increased PR-Set7 expression and its localization to mitotic chromosomes. *Genes Dev* **16:** 2225–2230.

Robinson NP, Bell SD. 2007. Extrachromosomal element capture and the evolution of multiple replication origins in archaeal chromosomes. *Proc Natl Acad Sci* **104:** 5806–5811.

Robinson NP, Dionne I, Lundgren M, Marsh VL, Bernander R, Bell SD. 2004. Identification of two origins of replication in the single chromosome of the archaeon *Sulfolobus solfataricus*. *Cell* **116:** 25–38.

Rowley A, Cocker JH, Harwood J, Diffley JF. 1995. Initiation complex assembly at budding yeast replication origins begins with the recognition of a bipartite sequence by limiting amounts of the initiator, ORC. *EMBO J* **14:** 2631–2641.

Rozgaja TA, Grimwade JE, Iqbal M, Czerwonka C, Vora M, Leonard AC. 2011. Two oppositely oriented arrays of low-affinity recognition sites in oriC guide progressive binding of DnaA during *Escherichia coli* pre-RC assembly. *Mol Microbiol* **82:** 475–488.

Ryan VT, Grimwade JE, Camara JE, Crooke E, Leonard AC. 2004. *Escherichia coli* prereplication complex assembly is regulated by dynamic interplay among Fis, IHF and DnaA. *Mol Microbiol* **51:** 1347–1359.

Ryba T, Hiratani I, Lu J, Itoh M, Kulik M, Zhang J, Schulz TC, Robins AJ, Dalton S, Gilbert DM. 2010. Evolutionarily conserved replication timing profiles predict long-range chromatin interactions and distinguish closely related cell types. *Genome Res* **20:** 761–770.

Schaper S, Messer W. 1995. Interaction of the initiator protein DnaA of *Escherichia coli* with its DNA target. *J Biol Chem* **270:** 17622–17626.

Sequeira-Mendes J, Gomez M. 2011. On the opportunistic nature of transcription and replication initiation in the metazoan genome. *Bioessays* **34:** 119–125.

Sequeira-Mendes J, Gomez M. 2012. On the opportunistic nature of transcription and replication initiation in the metazoan genome. *Bioessays* **34:** 119–125.

Sequeira-Mendes J, Diaz-Uriarte R, Apedaile A, Huntley D, Brockdorff N, Gomez M. 2009. Transcription initiation activity sets replication origin efficiency in mammalian cells. *PLoS Genet* **5:** e1000446.

Slater S, Wold S, Lu M, Boye E, Skarstad K, Kleckner N. 1995. *E. coli* SeqA protein binds oriC in two different methyl-modulated reactions appropriate to its roles in DNA replication initiation and origin sequestration. *Cell* **82:** 927–936.

Stinchcomb DT, Struhl K, Davis RW. 1979. Isolation and characterisation of a yeast chromosomal replicator. *Nature* **282:** 39–43.

Stroud H, Otero S, Desvoyes B, Ramirez-Parra E, Jacobsen SE, Gutierrez C. 2012. Genome-wide analysis of histone H3.1 and H3.3 variants in *Arabidopsis thaliana*. *Proc Natl Acad Sci* **109:** 5370–5375.

Sugimoto N, Yugawa T, Iizuka M, Kiyono T, Fujita M. 2011. Chromatin remodeler sucrose nonfermenting 2 homolog (SNF2H) is recruited onto DNA replication origins through interaction with Cdc10 protein-dependent transcript 1 (Cdt1) and promotes pre-replication complex formation. *J Biol Chem* **286:** 39200–39210.

Taagepera S, Rao PN, Drake FH, Gorbsky GJ. 1993. DNA topoisomerase IIa is the major chromosomal protein recognized by the mitotic phosphoprotein antibody MPM-2. *Proc Natl Acad Sci* **90:** 8407–8411.

Tardat M, Brustel J, Kirsh O, Lefevbre C, Callanan M, Sardet C, Julien E. 2010. The histone H4 Lys 20 methyltransferase PR-Set7 regulates replication origins in mammalian cells. *Nat Cell Biol* **12:** 1086–1093.

Taylor JA, Ouimet MC, Wargachuk R, Marczynski GT. 2011. The *Caulobacter crescentus* chromosome replication origin evolved two classes of weak DnaA binding sites. *Mol Microbiol* **82:** 312–326.

Theis JF, Newlon CS. 1994. Domain B of ARS307 contains two functional elements and contributes to chromosomal replication origin function. *Mol Cell Biol* **14:** 7652–7659.

Theis JF, Newlon CS. 1997. The ARS309 chromosomal replicator of *Saccharomyces cerevisiae* depends on an exceptional ARS consensus sequence. *Proc Natl Acad Sci* **94:** 10786–10791.

Thoma F, Bergman LW, Simpson RT. 1984. Nuclease digestion of circular TRP1ARS1 chromatin reveals positioned nucleosomes separated by nuclease-sensitive regions. *J Mol Biol* **177:** 715–733.

Tsankov A, Yanagisawa Y, Rhind N, Regev A, Rando OJ. 2011. Evolutionary divergence of intrinsic and *trans*-regulated nucleosome positioning sequences reveals plastic rules for chromatin organization. *Genome Res* **21:** 1851–1862.

Unnikrishnan A, Gafken PR, Tsukiyama T. 2010. Dynamic changes in histone acetylation regulate origins of DNA replication. *Nat Struct Mol Biol* **17:** 430–437.

Vashee S, Cvetic C, Lu W, Simancek P, Kelly TJ, Walter JC. 2003. Sequence-independent DNA binding and replication initiation by the human origin recognition complex. *Genes Dev* **17:** 1894–1908.

Wigley DB. 2009. ORC proteins: Marking the start. *Curr Opin Struct Biol* **19:** 72–78.

Wilmes GM, Bell SP. 2002. The B2 element of the *Saccharomyces cerevisiae* ARS1 origin of replication requires specific sequences to facilitate pre-RC formation. *Proc Natl Acad Sci* **99:** 101–106.

Woodward AM, Gohler T, Luciani MG, Oehlmann M, Ge X, Gartner A, Jackson DA, Blow JJ. 2006. Excess Mcm2-7 license dormant origins of replication that can be used under conditions of replicative stress. *J Cell Biol* **173:** 673–683.

Worning P, Jensen LJ, Hallin PF, Staerfeldt HH, Ussery DW. 2006. Origin of replication in circular prokaryotic chromosomes. *Environ Microbiol* **8:** 353–361.

Xu J, Yanagisawa Y, Tsankov AM, Hart C, Aoki K, Kommajosyula N, Steinmann KE, Bochicchio J, Russ C, Regev A, et al. 2012. Genome-wide identification and characterization of replication origins by deep sequencing. *Genome Biol* **13:** R27.

Zawilak-Pawlik A, Kois A, Majka J, Jakimowicz D, Smulczyk-Krawczyszyn A, Messer W, Zakrzewska-Czerwinska J. 2005. Architecture of bacterial replication initiation complexes: Orisomes from four unrelated bacteria. *Biochem J* **389:** 471–481.

Zakrzewka-Czerwinska J, Jalimowicz D, Zawilak-Pawlik A, Messer W. 2007. Regulation of the initiation of chromosomal replication in bacteria. *FEMS Microbiol Rev* **31:** 378–387.

Zhang Y, Moqtaderi Z, Rattner BP, Euskirchen G, Snyder M, Kadonaga JT, Liu XS, Struhl K. 2009. Intrinsic histone–DNA interactions are not the major determinant of nucleosome positions in vivo. *Nat Struct Mol Biol* **16:** 847–852.

Zou L, Stillman B. 2000. Assembly of a complex containing Cdc45p, replication protein A, and Mcm2p at replication origins controlled by S-phase cyclin-dependent kinases and Cdc7p–Dbf4p kinase. *Mol Cell Biol* **20:** 3086–3096.

Dormant Replication Origins

Debbie McIntosh and J. Julian Blow

Centre for Gene Regulation & Expression, University of Dundee, Dundee DD1 5EH, United Kingdom

Correspondence: j.j.blow@dundee.ac.uk

Only ~10% of replication origins that are licensed by loading minichromosome maintenance 2-7 (MCM2-7) complexes are normally used, with the majority remaining dormant. If replication fork progression is inhibited, nearby dormant origins initiate to ensure that all of the chromosomal DNA is replicated. At the same time, DNA damage-response kinases are activated, which preferentially suppress the assembly of new replication factories. This diverts initiation events away from completely new areas of the genome toward regions experiencing replicative stress. Mice hypomorphic for MCM2-7, which activate fewer dormant origins in response to replication inhibition, are cancer-prone and are genetically unstable. The licensing checkpoint delays entry into S phase if an insufficient number of origins have been licensed. In contrast, humans with Meier-Gorlin syndrome have mutations in pre-RC proteins and show defects in cell proliferation that may be a consequence of chronic activation of the licensing checkpoint.

Replicating the large amount of DNA in eukaryotic cells is a complex task, requiring the activation of hundreds or thousands of origins spread throughout the genome. To maintain genetic stability, it is essential that during S phase genomic DNA is precisely duplicated, with no sections of DNA left unreplicated and no section of DNA replicated more than once. To prevent re-replication, cells divide the process of DNA replication into two non-overlapping phases. Prior to S phase, origins are licensed by the binding of minichromosome maintenance 2-7 (MCM2-7) double hexamers (Gillespie et al. 2001; Blow and Dutta 2005; Arias and Walter 2007). During S phase, these are activated as the core of the CMG (Cdc45-MCM-GINS) replicative helicase (Moyer et al. 2006; Ilves et al. 2010). Prior to the onset of S phase, licensing proteins are down-regulated or inhibited, so that no more origins can be licensed (Wohlschlegel et al. 2000; Tada et al. 2001; Li et al. 2003; Li and Blow 2005). One consequence of using this mechanism for preventing re-replication of DNA is that it is imperative that enough origins are licensed prior to S-phase entry, so that no regions of the genome remain unreplicated, even if some replication forks stall or some origins fail to initiate (Blow et al. 2011). Metazoan cells employ a licensing checkpoint to monitor that sufficient origins are licensed, inhibiting S-phase entry until this is established (Shreeram et al. 2002; Blow and Gillespie 2008).

Here we review recent research showing how cells ensure complete genome duplication by licensing more replication origins in G_1 than are normally used during S phase. The otherwise dormant replication origins become important for ensuring the completion of DNA

Cite this article as *Cold Spring Harb Perspect Biol* doi: 10.1101/cshperspect.a012955

replication if replication forks stall or are inhibited during S phase. We also review research showing how the licensing checkpoint ensures that a large enough number of origins are licensed before cells embark on S phase.

DORMANT REPLICATION ORIGINS

MCM2-7 complexes are loaded onto DNA in excess over the number normally used, with only ~10% being used in an unperturbed S phase (Burkhart et al. 1995; Donovan et al. 1997; Mahbubani et al. 1997; Blow et al. 2011; Wong et al. 2011). The rest normally remain dormant and are passively replicated by oncoming forks (Woodward et al. 2006; Ge et al. 2007). Dormant origins are fully capable of initiating, but are not essential for normal S-phase progression (Woodward et al. 2006). The existence of these dormant origins means that origin usage can change depending on the circumstance. Levels of chromatin-bound MCM2-7 can be reduced several-fold in *Xenopus* egg extracts, *C. elegans*, and in human cell lines without causing a significant effect on overall replication rates (Mahbubani et al. 1997; Edwards et al. 2002; Oehlmann et al. 2004; Ge et al. 2007; Ibarra et al. 2008). However, these "minimally licensed" cells fail to complete replication when challenged with exogenous replicative stress, suggesting a role for dormant origins in overcoming such stresses (Woodward et al. 2006; Ge et al. 2007; Ibarra et al. 2008).

Replication forks can be inhibited either by exposure to exogenous inhibitors or when they encounter impediments such as proteins tightly bound to DNA, chemically modified bases or DNA strand breaks. Often the problems can be overcome, and the inhibited forks can resume DNA synthesis. However, there are occasions where replication forks irreversibly collapse, which may be associated with the disassembly of replisome proteins from the DNA (Sirbu et al. 2011). Since new origins cannot be licensed after entry into S phase, cells must rely on previously licensed origins to overcome this issue. Replication forks travel bidirectionally, meaning that if a single fork stalls, a fork coming from the opposite direction can rescue this (Fig. 1A,B). However, when two converging forks stall, the genome between them is left in a compromised state, and if there is no dormant origin between the stalled forks, completing DNA replication will be difficult (Fig. 1C), possibly depending on recombination-dependent fork restart (Blow et al. 2011). However, if a dormant origin is activated between the stalled forks, this provides a simple mechanism for replication of all the intervening DNA (Fig. 1D) (Woodward et al. 2006; Ge et al. 2007).

The exact mechanism for regulating dormant origins is unclear. It has been known for many years that in response to fork slowing, there is an increased density of origins within clusters (Ockey and Saffhill 1976; Taylor 1977; Francis et al. 1985; Griffiths and Ling 1985; Anglana et al. 2003; Ge et al. 2007; Gilbert 2007; Blow et al. 2011; Borowiec and Schildkraut 2011; Karnani and Dutta 2011), indicating dormant origins have fired within these clusters. It is likely that firing of particular origins within a single cluster is a stochastic process, which can explain the observed cell-to-cell variation in origin use (Anglana et al. 2003; Blow and Ge 2009; Borowiec and Schildkraut 2011). Since licensed origins are inactivated when they are passively replicated by oncoming forks, they are normally inactivated before they have time to initiate. Therefore, a reduction in fork elongation rate increases the length of time before inactivation of dormant origins, increasing the probability that they will initiate. Computer simulation of a 250-kb origin cluster showed that stochastic firing of origins can provide a good approximation to observed patterns of dormant origin firing if there are ~3 dormant origins for each origin that fires under unstressed conditions (Blow and Ge 2009). The model also shows how dormant origins can allow replication to complete even if some forks stall irreversibly. This provides a simple explanation for how replication origin firing is regulated, which simultaneously provides protection against replicative stress while minimizing the cost of using large numbers of replication forks (Blow and Ge 2009; Blow et al. 2011). Consistent with this interpretation, there is a correlation between the degree of fork slowing and the overall increase in

Cite this article as *Cold Spring Harb Perspect Biol* doi: 10.1101/cshperspect.a012955

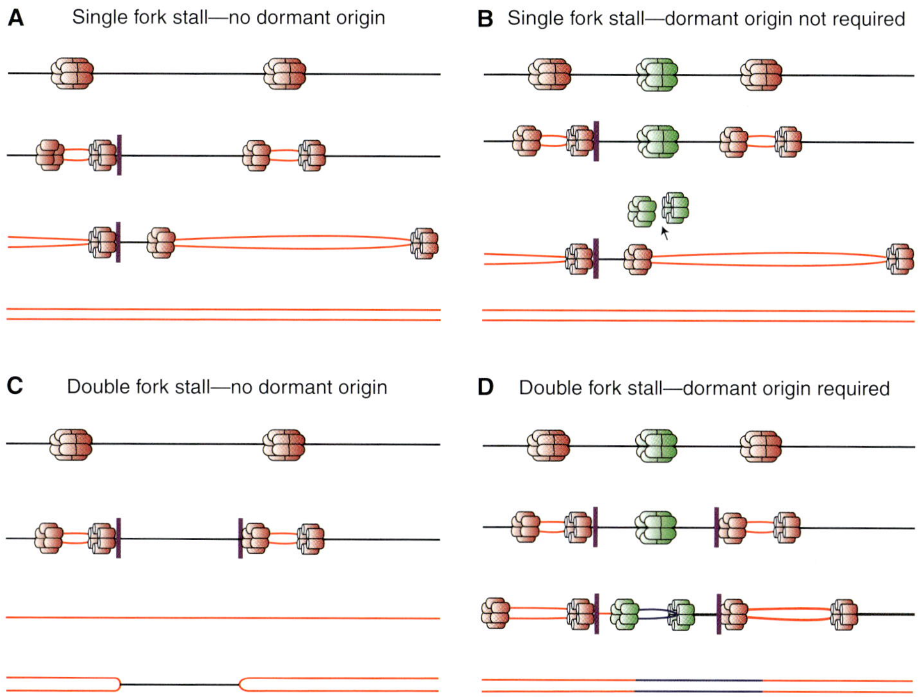

Figure 1. The effect of fork stalling on completion of replication. A small segment of chromosomal DNA is shown with two efficient licensed origins (red hexamers) plus an optional dormant origin (green hexamers). Fork stalling is indicated by a vertical purple bar. (*A*) One fork stalls, but all the intervening DNA is replicated by the fork originating at an adjacent origin. (*B*) A dormant origin is inactivated by a fork coming from the right. (*C*) Each of the two converging forks stall without a dormant origin between them. Replication cannot be completed because no new MCM2-7 complexes can be loaded onto DNA once S phase has begun. (*D*) Two converging forks stall, but a dormant origin between them allows replication to be completed.

origin density (Ge et al. 2007; Courbet et al. 2008).

Although dormant origin firing can occur as a consequence of purely stochastic processes, it is likely that it is also modulated by other regulatory mechanisms. In response to fork stalling, the DNA damage kinases ATR and Chk1 are activated (Branzei and Foiani 2005; Lambert and Carr 2005). Inhibition of ATR or Chk1 results in an increase in origin density, even in the absence of stress (Miao et al. 2003; Marheineke and Hyrien 2004; Syljuasen et al. 2005; Woodward et al. 2006; Ge et al. 2007; Karnani and Dutta 2011). One explanation for this behavior is that dormant origins are activated as a consequence of increased fork stalling that occurs in the absence of normal ATR and Chk1 activity (Zachos et al. 2003). However, it is also possible that the low levels ATR/Chk1 active even in the absence of exogenous replicative stresses, play a role in suppressing dormant origins. In addition, low levels of Chk1 preferentially suppresses new origin firing in clusters yet to engage in replication, allowing dormant origins the chance to initiate in the clusters where fork stalling has occurred (see next section).

Once cells have progressed through S phase in the presence of replicative stresses, there is evidence that the usage of origins in the next S phase will be changed (Courbet et al. 2008). After growing cells under conditions in which forks could only move slowly, which promoted a high rate of initiation at dormant origins, cells were synchronized in mitosis and replated into fresh medium that allowed fast fork progression. In the first S phase after replating, the overall initiation density dropped, as expected by the stochastic mechanism described above. When a particular locus around the AMPD gene was examined, initiation did not occur predominantly at what is normally the most efficient origin, oriGNAI3, but was distributed among the dormant origins as well as oriGNAI3. Only in the second S phase after the increase in fork speed was the dominance of oriGNAI3 regained and the relative efficiency of the dormant origins decreased. Therefore, it appears that cells acquire a memory of dormant origin use and can adapt to replicative stress by raising the efficiency of origins that are normally dormant (less efficient) (Blow and Ge 2008; Courbet et al. 2008).

The use of dormant origins probably provides the first line of defense against slowed fork elongation and fork stalling. The importance of dormant origins in ensuring genome stability has been shown by their role in suppressing tumorigenesis. Mice hypomorphic for either MCM2 or MCM4 have reduced levels of chromatin-bound MCM2-7, resulting in a reduction in the number of dormant origins (Pruitt et al. 2007; Shima et al. 2007; Kunnev et al. 2010; Kawabata et al. 2011; Klotz-Noack and Blow 2011). Cells from MCM hypomorphic mice suffer from increased fork stalling and increased Rad51 and FancD2 foci, indicative of DNA damage, even in the absence of exogenous stress (Kunnev et al. 2010; Kawabata et al. 2011). These mice fail to deal normally with spontaneous damage caused by endogenous factors and this is likely to explain their dramatic increase in tumor formation. This suggests that dormant origins play an important role as tumor suppressors and that failure to activate dormant origins in response to endogenous replicative stresses can result in DNA damage and tumor formation.

REPLICATION FACTORIES

In metazoan cells, clusters of 2–5 adjacent origins initiate simultaneously during S phase (Jackson and Pombo 1998; Gillespie and Blow 2010). One or more clusters are arranged into DNA domains comprising ~1 Mb of DNA and roughly 6–12 replicons, each replicon typically replicating 50–150 kb. Different clusters of origins initiate at varying times throughout S phase, giving rise to a timing program (Dimitrova and Gilbert 1999; Zink 2006). The DNA within each domain can be visualized as individual foci, thousands of which can be visualized during each S phase. As each focus becomes activated for replication, it becomes associated with replication proteins to form a replication factory (Hozak et al. 1993; Leonhardt et al. 2000; Kitamura et al. 2006). Early S-phase factories are all similar in size with a diameter of roughly 150 nm, and similar numbers are detected in all early S-phase nuclei (Blow et al. 2011). It is currently unknown if factories are physical structures where the fork proteins are organized into a superstructure or whether the foci as formed as a result of the organization of the DNA into higher-order chromatin structure.

It has recently been shown that these factories are regulated in response to replicative stresses. In response to low levels of replicative stress, there is an increase in the number of replication forks per factory that corresponds to the activation of dormant origins (Ge and Blow 2010). This increase in initiation leads to the increased fork density seen by DNA fiber analysis. If the overall rate of replication of DNA domains were reduced by replication fork inhibition, and if the rate at which new factories are activated were to remain constant, the result should be an increase in the number of active factories. However, the opposite is actually observed: inhibition of replication forks causes a reduction in the number of active factories (Cseresnyes et al. 2009; Ge and Blow 2010). Experiments examining factory dynamics after replication forks are inhibited showed that the reduction in factory number is a result of a reduction in the rate at which new replication factories are activated and depends on Chk1 (Fig. 2) (Ge and Blow 2010). ATR is activated in response to slowed fork progression or stalling and in turn activates Chk1 kinase activity (Branzei and Foiani 2005; Lambert and Carr 2005). By preferentially inhibiting the activation of new replication factories, Chk1 suppresses activation of origins in clusters yet to begin replicating but allows dormant origins to fire in currently replicating clusters that may

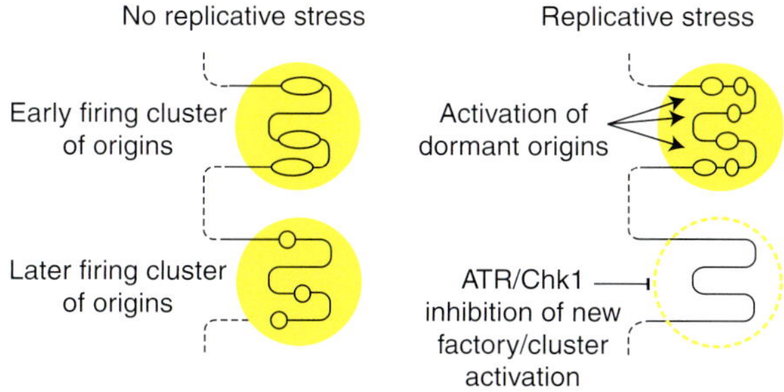

Figure 2. Model for how cells respond to low levels of replicative stress. Two adjacent clusters of origins (factories bounded by yellow circles) are shown on a single piece of DNA (black lines). Under normal circumstances (*left*), the upper factory is activated slightly earlier than the factory below, and each initiates three origins. Under low levels of replicative stress (*right*), replication forks are inhibited in the earlier replicating cluster, which promotes the firing of dormant origins as a direct consequence of stochastic origin firing. Replicative stress activates DNA damage checkpoint kinases, which preferentially inhibit the activation of the unfired later clusters/new factories.

have encountered replication stress (Ge and Blow 2010; Blow et al. 2011).

Very little is known about how replication factories are regulated and it is currently not known how Chk1 is able to distinguish between active clusters and those already replicating. The activation of new factories is highly sensitive to varying cyclin-dependent kinase (CDK) activity. When CDK activity was reduced using the drug roscovitine, the number of factories was also reduced (Thomson et al. 2010). Two different possibilities have been proposed to explain this effect (Ge and Blow 2010; Gillespie and Blow 2010; Thomson et al. 2010; Blow et al. 2011). CDKs are directly required for individual replication origins to fire, and it is possible that the firing of the first origin within a cluster might propagate a change throughout the cluster to facilitate initiation at other origins within the cluster. A reduction in CDK level would lower the overall initiation rate and thereby reduce the activation of new factories. Downstream targets of Chk1, such as p53 and Cdc25 could reduce CDK activity and thereby reduce the rate of activation of new factories. An alternative possibility is that CDK activity plays a direct role in factory regulation, by phosphorylating factory-specific substrates (such as, for example, proteins involved in chromatin remodeling) distinct from substrates that are required for the firing of individual origins. These factory-specific substrates might then be directly inhibited by Chk1 in response to replicative stresses (Ge and Blow 2010; Gillespie and Blow 2010; Thomson et al. 2010; Blow et al. 2011).

The regulation of replication at the factory level is important for ensuring that the genome is completely replicated. By preferentially inhibiting the activation of new replication factories, Chk1 can still allow dormant origins to fire stochastically in factories that are experiencing replicative stress. This effectively directs replication activity to regions of the genome encountering stress, by activating dormant origins in factories where forks are encountering problems. Correct regulation of this process requires the proper operation of DNA damage and cell cycle checkpoint responses. As these systems are often compromised in cancer cells, they may be unable to correctly regulate dormant origins and replication factories when replication fork progress is inhibited. It is possible that this may explain the sensitivity of some cancer cells to chemotherapeutic drugs that target the process of DNA replication (Blow and Gillespie 2008; Blow et al. 2011). It is also possible that chromosome fragile sites, which are prone to breakage as a result of replication fork failure, represent regions where dormant origins are either inefficiently licensed or inefficiently activated (Letessier et al. 2011; Ozeri-Galai et al. 2011; Debatisse et al. 2012).

THE LICENSING CHECKPOINT

To prevent re-replication of regions of chromosomal DNA, it is essential that once cells pass the G_1-S-phase boundary, no further origins are licensed (Blow and Dutta 2005; Arias and Walter 2007). Entering S phase with too few origins would result in a reduced number of dormant origins and would require forks to replicate longer sections of DNA without stalling (Blow and Ge 2009). It would therefore be better for cells to delay entry into S phase if they do not have a sufficiently large number of licensed origins. It has been shown that several different cell types employ a "licensing checkpoint" to achieve this (Shreeram et al. 2002; Gillespie et al. 2007; Blow and Gillespie 2008). The S phase delay characteristic of the licensing checkpoint has been observed when the loading of MCM2-7 onto DNA is reduced in a range of different ways: by knockdown of ORC, Cdc6 Ctd1 or MCM2-7 by RNAi, or inhibition of Cdt1 by geminin (Shreeram et al. 2002; Feng et al. 2003; Machida et al. 2005; Gillespie et al. 2007; Liu et al. 2009; Nevis et al. 2009). Many cancer cells are defective in this checkpoint and can enter S phase despite a severe reduction in the number of origins that have been licensed (Shreeram et al. 2002; Feng et al. 2003; Liu et al. 2009; Nevis et al. 2009), a feature that may contribute to the genetic instability typically detected in cancer cells (Fig. 3) (Blow and Gillespie 2008).

Under normal circumstances, progression through the restriction point into S phase depends on the CDK-dependent phosphorylation of Rb, which abolishes its inhibition of E2F and promotes transcription of S-phase genes. The G_1 arrest observed in cells that have engaged the licensing checkpoint is associated with low CDK4/6-Cyclin D activity, low CDK2-Cyclin E activity and Rb hypophosphorylation, suggesting that the licensing checkpoint prevents progression through the restriction point (Fig. 4) (Shreeram et al. 2002; Machida et al. 2005; Teer et al. 2006; Liu et al. 2009; Nevis et al. 2009). The way that the degree of licensing is sensed by the checkpoint is currently unknown, but it appears to converge on CDK down-regulation by several different routes. One common route appears to be the p53-dependent activation of CDK

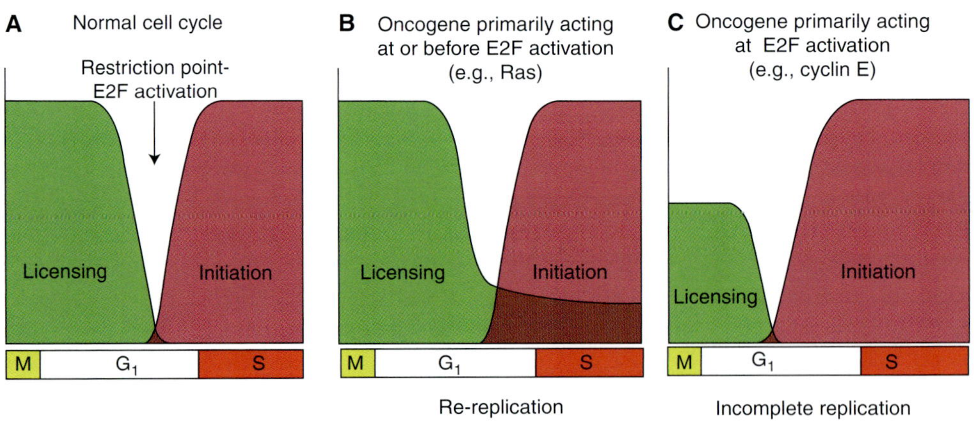

Figure 3. How oncogene activation might normally disrupt licensing control. Progression through the cell cycle (as shown in the lower bar) depends on sequential activation of the licensing system (green) and initiation activities (fuchsia). (*A*) In normal cells, there is no significant overlap between licensing and initiation activities. The transition from the licensing state to the initiation state occurs around the restriction point. (*B*) If an oncogene is constitutively expressed that acts upstream of the restriction point, cells may be unable to fully repress licensing activity during S phase, potentially leading to re-replication of DNA. (*C*) If an oncogene is constitutively expressed that acts downstream of the restriction point, cells may not be able to license a sufficiently large number of origins before progressing into S phase, potentially leading to incomplete replication of DNA. See Blow and Gillespie (2008) for more discussion of these potential effects.

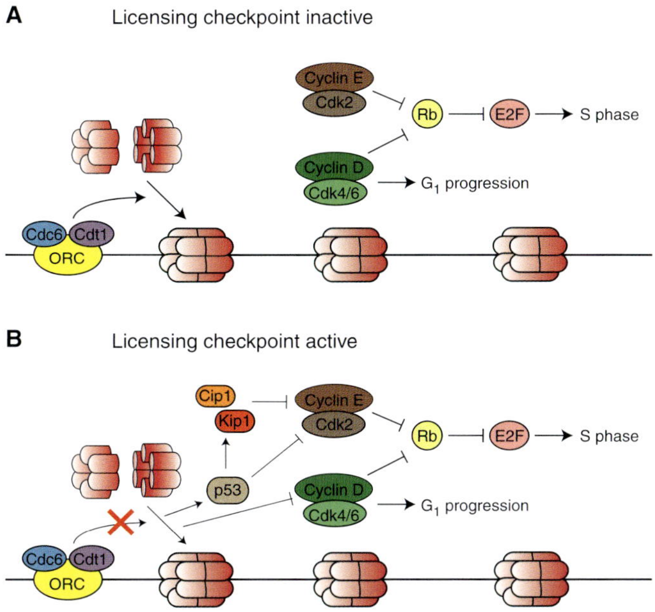

Figure 4. The licensing checkpoint. A piece of DNA containing several origins is denoted by the black line. (*A*) When licensing occurs, ORC, Cdc6, and Cdt1 promote the loading of MCM2-7 (red hexamers) onto DNA. This allows normal activation of Cdk4/6 and Cdk2 kinases, and subsequent activation of E2F by Rb phosphorylation. (*B*) When the licensing is inhibited, the licensing checkpoint is activated, causing suppression of cyclin D transcription and activation of p53, resulting in induction and Cip1 and Kip1 and inhibition of activating Cdk2 phosphorylation.

inhibitors $p21^{Cip1}$ and $p27^{Kip1}$ (Shreeram et al. 2002; Machida et al. 2005; Teer et al. 2006; Liu et al. 2009). A second p53-dependent pathway involves loss of phosphorylation at Thr160 in the activation T loop of CDK2, accompanied by delayed nuclear accumulation of CDK2 protein in late G_1 (Nevis et al. 2009). Depletion of p53 in primary cells abolished the licensing checkpoint, promoting Rb hyperphosphorylation and E2F transcriptional activity. There also appears to be other p53-independent mechanisms by which CDK activity is reduced when the licensing checkpoint is engaged, as RNAi knockdown of components of the licensing system directly suppressed cyclin D1 transcription and lowered CDK4/6 kinase activity (Liu et al. 2009). Another way by which defective origin licensing delays S-phase CDK activation has been demonstrated in *Xenopus* egg extract. The nucleoporin ELYS associates with MCM2-7 on chromatin and if licensing is inhibited (for example, by geminin addition or MCM2-7 depletion), the association of ELYS with chromatin is delayed (Gillespie et al. 2007; Khoudoli et al. 2008). Since ELYS is required on chromatin for nuclear envelope formation, it is ultimately required for CDK activation, the initiation of DNA replication, and the activation of geminin, thereby creating a feed-forward loop to promote S-phase progression only once adequate origin licensing has occurred.

Mutations in genes encoding pre-RC proteins required for replication licensing (ORC1, ORC4, ORC6, CDT1, and CDC6) have been identified in patients with the genetic disorder Meier-Gorlin syndrome (Bicknell et al. 2011a,b; Klingseisen and Jackson 2011; de Munnik et al. 2012; Kuo et al. 2012). Cells from these patients have a delayed G_1 progression, most likely due to the activation of the licensing checkpoint, and a prolonged S phase. These delays in cell cycle progression result in reduced cell numbers, potentially explaining the primordial dwarfism phenotype. This provides evidence that the licensing checkpoint is operational in humans.

It is currently unclear why humans with Meier-Gorlin syndrome show cell proliferation defects while mice hypomorphic for MCM2-7 are cancer-prone. One possibility is that this represents a difference between mice and humans. Alternatively, the licensing checkpoint might respond less strongly to defects in MCM2-7 than to defects in other pre-RC proteins (ORC, Cdc6, or Cdt1). A third possibility is that the licensing defect in the MCM2-7 hypomorphic mice might be weaker than in Meier-Gorlin syndrome and may not be severe enough to reproducibly activate the licensing checkpoint (Klotz-Noack and Blow 2011). Indeed, MCM2-7 hypomorphic mice do show some defects in cell proliferation associated with a reduction in the number of stem cells (Pruitt et al. 2007).

The reliance of the licensing checkpoint on p53 and Rb control systems that are often lost in cancer cells can potentially explain why the licensing checkpoint is not operational in cancer-derived cell lines. Since E2F cannot be repressed by the licensing checkpoint, cancer cells lose the ability to inhibit progression into S phase. Defects in the licensing checkpoint could contribute to the genomic instability often detected in cancer cells, as cells entering S phase with too few origins have increased chances of fork stalling and DNA damage. The lack of licensing checkpoint in cancer cells also makes it a good anti-cancer target (Shreeram et al. 2002; Blow and Gillespie 2008). Small-molecule inhibitors of the replication licensing system would kill cancer cells by allowing them to progress into S phase with an insufficient number of licensed origins. Normal cells would instead activate the licensing checkpoint and arrest temporarily in G_1-like state, from where they could resume cell cycle progression once the drug is removed or metabolized.

ACKNOWLEDGMENT

J.J.B. is supported by CRUK grant C303/A7399.

REFERENCES

Anglana M, Apiou F, Bensimon A, Debatisse M. 2003. Dynamics of DNA replication in mammalian somatic cells: Nucleotide pool modulates origin choice and interorigin spacing. *Cell* **114:** 385–394.

Arias EE, Walter JC. 2007. Strength in numbers: Preventing rereplication via multiple mechanisms in eukaryotic cells. *Genes Dev* **21:** 497–518.

Bicknell LS, Bongers EM, Leitch A, Brown S, Schoots J, Harley ME, Aftimos S, Al-Aama JY, Bober M, Brown PA, van Bokhoven H, Dean J, et al. 2011a. Mutations in the pre-replication complex cause Meier-Gorlin syndrome. *Nat Genet* **43:** 356–359.

Bicknell LS, Walker S, Klingseisen A, Stiff T, Leitch A, Kerzendorfer C, Martin CA, Yeyati P, Al Sanna N, Bober M, Johnson D, Wise C, et al. 2011b. Mutations in ORC1, encoding the largest subunit of the origin recognition complex, cause microcephalic primordial dwarfism resembling Meier-Gorlin syndrome. *Nat Genet* **43:** 350–355.

Blow JJ, Dutta A. 2005. Preventing re-replication of chromosomal DNA. *Nat Rev Mol Cell Biol* **6:** 476–486.

Blow JJ, Ge XQ. 2008. Replication forks, chromatin loops and dormant replication origins. *Genome Biol* **9:** 244.

Blow JJ, Ge XQ. 2009. A model for DNA replication showing how dormant origins safeguard against replication fork failure. *EMBO Rep* **10:** 406–412.

Blow JJ, Gillespie PJ. 2008. Replication licensing and cancer—a fatal entanglement? *Nat Rev Cancer* **8:** 799–806.

Blow JJ, Ge XQ, Jackson DA. 2011. How dormant origins promote complete genome replication. *Trends Biochem Sci* **36:** 405–414.

Borowiec JA, Schildkraut CL. 2011. Open sesame: activating dormant replication origins in the mouse immunoglobulin heavy chain (Igh) locus. *Curr Op Cell Biol* **23:** 284–292.

Branzei D, Foiani M. 2005. The DNA damage response during DNA replication. *Curr Op Cell Biol* **17:** 568–575.

Burkhart R, Schulte D, Hu D, Musahl C, Gohring F, Knippers R. 1995. Interactions of human nuclear proteins P1Mcm3 and P1Cdc46. *Eur J Biochem* **228:** 431–438.

Courbet S, Gay S, Arnoult N, Wronka G, Anglana M, Brison O, Debatisse M. 2008. Replication fork movement sets chromatin loop size and origin choice in mammalian cells. *Nature* **455:** 557–560.

Cseresnyes Z, Schwarz U, Green CM. 2009. Analysis of replication factories in human cells by super-resolution light microscopy. *BMC Cell Biol* **10:** 88.

Debatisse M, Le Tallec B, Letessier A, Dutrillaux B, Brison O. 2012. Common fragile sites: Mechanisms of instability revisited. *Trends Genet* **28:** 22–32.

de Munnik SA, Bicknell LS, Aftimos S, Al-Aama JY, van Bever Y, Bober MB, Clayton-Smith J, Edrees AY, Feingold M, Fryer A, van Hagen JM, Hennekam RC, et al. 2012. Meier-Gorlin syndrome genotype-phenotype studies: 35 individuals with pre-replication complex gene mutations and 10 without molecular diagnosis. *Eur J Hum Genet* **6:** 598–606.

Dimitrova DS, Gilbert DM. 1999. The spatial position and replication timing of chromosomal domains are both established in early G1 phase. *Mol Cell* **4:** 983–993.

Donovan S, Harwood J, Drury LS, Diffley JF. 1997. Cdc6p-dependent loading of Mcm proteins onto pre-replicative chromatin in budding yeast. *Proc Natl Acad Sci* **94:** 5611–5616.

Edwards MC, Tutter AV, Cvetic C, Gilbert CH, Prokhorova TA, Walter JC. 2002. MCM2–7 complexes bind chromatin in a distributed pattern surrounding the origin recognition complex in Xenopus egg extracts. *J Biol Chem* **277:** 33049–33057.

Feng D, Tu Z, Wu W, Liang C. 2003. Inhibiting the expression of DNA replication-initiation proteins induces apoptosis in human cancer cells. *Cancer Res* **63:** 7356–7364.

Francis D, Davies ND, Bryant JA, Hughes SG, Sibson DR, Fitchett PN. 1985. Effects of psoralen on replicon size and mean rate of DNA synthesis in partially synchronized cells of Pisum sativum L. *Exp Cell Res* **158:** 500–508.

Ge XQ, Blow JJ. 2010. Chk1 inhibits replication factory activation but allows dormant origin firing in existing factories. *J Cell Biol* **191:** 1285–1297.

Ge XQ, Jackson DA, Blow JJ. 2007. Dormant origins licensed by excess Mcm2 7 are required for human cells to survive replicative stress. *Genes Dev* **21:** 3331–3341.

Gilbert DM. 2007. Replication origin plasticity, Taylor-made: Inhibition vs recruitment of origins under conditions of replication stress. *Chromosoma* **116:** 341–347.

Gillespie PJ, Blow JJ. 2010. Clusters, factories and domains—the complex structure of S phase comes into focus. *Cell Cycle* **9:** 3218–3226.

Gillespie PJ, Li A, Blow JJ. 2001. Reconstitution of licensed replication origins on Xenopus sperm nuclei using purified proteins. *BMC Biochem* **2:** 15.

Gillespie PJ, Khoudoli GA, Stewart G, Swedlow JR, Blow JJ. 2007. ELYS/MEL-28 chromatin association coordinates nuclear pore complex assembly and replication licensing. *Curr Biol* **17:** 1657–1662.

Griffiths TD, Ling SY. 1985. Effect of ultraviolet light on thymidine incorporation, DNA chain elongation and replicon initiation in wild-type and excision-deficient Chinese hamster ovary cells. *Biochim Biophys Acta* **826:** 121–128.

Hozak P, Hassan AB, Jackson DA, Cook PR. 1993. Visualization of replication factories attached to nucleoskeleton. *Cell* **73:** 361–373.

Ibarra A, Schwob E, Mendez J. 2008. Excess MCM proteins protect human cells from replicative stress by licensing backup origins of replication. *Proc Natl Acad Sci* **105:** 8956–8961.

Ilves I, Petojevic T, Pesavento JJ, Botchan MR. 2010. Activation of the MCM2–7 helicase by association with Cdc45 and GINS proteins. *Mol Cell* **37:** 247–258.

Jackson DA, Pombo A. 1998. Replicon clusters are stable units of chromosome structure: Evidence that nuclear organization contributes to the efficient activation and propagation of S phase in human cells. *J Cell Biol* **140:** 1285–1295.

Karnani N, Dutta A. 2011. The effect of the intra-S-phase checkpoint on origins of replication in human cells. *Genes Dev* **25:** 621–633.

Kawabata T, Luebben S, Yamaguchi S, Ilves I, Matise I, Buske T, Botchan M, Shima N. 2011. Stalled fork rescue via dormant replication origins in unchallenged S phase promotes proper chromosome segregation and tumor suppression. *Mol Cell* **41:** 543–553.

Khoudoli GA, Gillespie PJ, Stewart G, Andersen JS, Swedlow JR, Blow JJ. 2008. Temporal profiling of the chromatin proteome reveals system-wide responses to replication inhibition. *Curr Biol* **18:** 838–843.

Kitamura E, Blow JJ, Tanaka TU. 2006. Live-cell imaging reveals replication of individual replicons in eukaryotic replication factories. *Cell* **125:** 1297–1308.

Klingseisen A, Jackson AP. 2011. Mechanisms and pathways of growth failure in primordial dwarfism. *Genes Dev* **25:** 2011–2024.

Klotz-Noack K, Blow JJ. 2011. A role for dormant origins in tumor suppression. *Mol Cell* **41:** 495–496.

Kunnev D, Rusiniak ME, Kudla A, Freeland A, Cady GK, Pruitt SC. 2010. DNA damage response and tumorigenesis in Mcm2-deficient mice. *Oncogene* **29:** 3630–3638.

Kuo AJ, Song J, Cheung P, Ishibe-Murakami S, Yamazoe S, Chen JK, Patel DJ, Gozani O. 2012. The BAH domain of ORC1 links H4K20me2 to DNA replication licensing and Meier-Gorlin syndrome. *Nature* **484:** 115–119.

Lambert S, Carr AM. 2005. Checkpoint responses to replication fork barriers. *Biochimie* **87:** 591–602.

Leonhardt H, Rahn HP, Weinzierl P, Sporbert A, Cremer T, Zink D, Cardoso MC. 2000. Dynamics of DNA replication factories in living cells. *J Cell Biol* **149:** 271–280.

Letessier A, Millot GA, Koundrioukoff S, Lachages AM, Vogt N, Hansen RS, Malfoy B, Brison O, Debatisse M. 2011. Cell-type-specific replication initiation programs set fragility of the FRA3B fragile site. *Nature* **470:** 120–123.

Li A, Blow JJ. 2005. Cdt1 downregulation by proteolysis and geminin inhibition prevents DNA re-replication in Xenopus. *EMBO J* **24:** 395–404.

Li X, Zhao Q, Liao R, Sun P, Wu X. 2003. The SCF(Skp2) ubiquitin ligase complex interacts with the human replication licensing factor Cdt1 and regulates Cdt1 degradation. *J Biol Chem* **278:** 30854–30858.

Liu P, Slater DM, Lenburg M, Nevis K, Cook JG, Vaziri C. 2009. Replication licensing promotes cyclin D1 expression and G1 progression in untransformed human cells. *Cell Cycle* **8:** 125–136.

Machida YJ, Teer JK, Dutta A. 2005. Acute reduction of an origin recognition complex (ORC) subunit in human cells reveals a requirement of ORC for Cdk2 activation. *J Biol Chem* **280:** 27624–27630.

Mahbubani HM, Chong JP, Chevalier S, Thömmes P, Blow JJ. 1997. Cell cycle regulation of the replication licensing system: involvement of a Cdk-dependent inhibitor. *J Cell Biol* **136:** 125–135.

Marheineke K, Hyrien O. 2004. Control of replication origin density and firing time in Xenopus egg extracts: Role of a caffeine-sensitive, ATR-dependent checkpoint. *J Biol Chem* **279:** 28071–28081.

Miao H, Seiler JA, Burhans WC. 2003. Regulation of cellular and SV40 virus origins of replication by Chk1-dependent intrinsic and UVC radiation-induced checkpoints. *J Biol Chem* **278:** 4295–4304.

Moyer SE, Lewis PW, Botchan MR. 2006. Isolation of the Cdc45/Mcm2–7/GINS (CMG) complex, a candidate for the eukaryotic DNA replication fork helicase. *Proc Natl Acad Sci* **103:** 10236–10241.

Nevis KR, Cordeiro-Stone M, Cook JG. 2009. Origin licensing and p53 status regulate Cdk2 activity during G1. *Cell Cycle* **8:** 1952–1963.

Ockey CH, Saffhill R. 1976. The comparative effects of short-term DNA Inhibition on replicon synthesis in mammalian cells. *Exp Cell Res* **103:** 361–373.

Oehlmann M, Score AJ, Blow JJ. 2004. The role of Cdc6 in ensuring complete genome licensing and S phase checkpoint activation. *J Cell Biol* **165:** 181–190.

Ozeri-Galai E, Lebofsky R, Rahat A, Bester AC, Bensimon A, Kerem B. 2011. Failure of origin activation in response to fork stalling leads to chromosomal instability at fragile sites. *Mol Cell* **43:** 122–131.

Pruitt SC, Bailey KJ, Freeland A. 2007. Reduced Mcm2 expression results in severe stem/progenitor cell deficiency and cancer. *Stem Cells* **25:** 3121–3132.

Shima N, Alcaraz A, Liachko I, Buske TR, Andrews CA, Munroe RJ, Hartford SA, Tye BK, Schimenti JC. 2007. A viable allele of Mcm4 causes chromosome instability and mammary adenocarcinomas in mice. *Nat Genet* **39:** 93–98.

Shreeram S, Sparks A, Lane DP, Blow JJ. 2002. Cell type-specific responses of human cells to inhibition of replication licensing. *Oncogene* **21:** 6624–6632.

Sirbu BM, Couch FB, Feigerle JT, Bhaskara S, Hiebert SW, Cortez D. 2011. Analysis of protein dynamics at active, stalled, and collapsed replication forks. *Genes Dev* **25:** 1320–1327.

Syljuasen RG, Sorensen CS, Hansen LT, Fugger K, Lundin C, Johansson F, Helleday T, Sehested M, Lukas J, Bartek J. 2005. Inhibition of human Chk1 causes increased initiation of DNA replication, phosphorylation of ATR targets, and DNA breakage. *Mol Cell Biol* **25:** 3553–3562.

Tada S, Li A, Maiorano D, Mechali M, Blow JJ. 2001. Repression of origin assembly in metaphase depends on inhibition of RLF-B/Cdt1 by geminin. *Nat Cell Biol* **3:** 107–113.

Taylor JH. 1977. Increase in DNA replication sites in cells held at the beginning of S phase. *Chromosoma* **62:** 291–300.

Teer JK, Machida YJ, Labit H, Novac O, Hyrien O, Marheineke K, Zannis-Hadjopoulos M, Dutta A. 2006. Proliferating human cells hypomorphic for origin recognition complex 2 and pre-replicative complex formation have a defect in p53 activation and Cdk2 kinase activation. *J Biol Chem* **281:** 6253–6260.

Thomson AM, Gillespie PJ, Blow JJ. 2010. Replication factory activation can be decoupled from the replication timing program by modulating Cdk levels. *J Cell Biol* **188:** 209–221.

Wohlschlegel JA, Dwyer BT, Dhar SK, Cvetic C, Walter JC, Dutta A. 2000. Inhibition of eukaryotic replication by geminin binding to Cdt1. *Science* **290:** 2309–2312.

Wong PG, Winter SL, Zaika E, Cao TV, Oguz U, Koomen JM, Hamlin JL, Alexandrow MG. 2011. Cdc45 limits replicon usage from a low density of preRCs in mammalian cells. *PLoS ONE* **6:** e17533.

Woodward AM, Gohler T, Luciani MG, Oehlmann M, Ge X, Gartner A, Jackson DA, Blow JJ. 2006. Excess Mcm2–7 license dormant origins of replication that can be used under conditions of replicative stress. *J Cell Biol* **173:** 673–683.

Zachos G, Rainey MD, Gillespie DA. 2003. Chk1-deficient tumour cells are viable but exhibit multiple checkpoint and survival defects. *EMBO J* **22:** 713–723.

Zink D. 2006. The temporal program of DNA replication: New insights into old questions. *Chromosoma* **115:** 273–287.

Break-Induced DNA Replication

Ranjith P. Anand, Susan T. Lovett, and James E. Haber

Rosenstiel Basic Medical Sciences Research Center and Department of Biology, Brandeis University, Waltham, Massachusetts 02254-9110

Correspondence: lovett@brandeis.edu; haber@brandeis.edu

Recombination-dependent DNA replication, often called break-induced replication (BIR), was initially invoked to explain recombination events in bacteriophage but it has recently been recognized as a fundamentally important mechanism to repair double-strand chromosome breaks in eukaryotes. This mechanism appears to be critically important in the restarting of stalled and broken replication forks and in maintaining the integrity of eroded telomeres. Although BIR helps preserve genome integrity during replication, it also promotes genome instability by the production of loss of heterozygosity and the formation of nonreciprocal translocations, as well as in the generation of complex chromosomal rearrangements.

The break-copy mode of recombination (as opposed to break-join), was initially proposed by Meselson and Weigle (1961). Break-copy recombination, now more commonly known as recombination-dependent DNA replication or break-induced replication (BIR), is believed to account for restarting replication at broken replication forks and may also play a central role in the maintenance of telomeres in the absence of telomerase. BIR has been studied in various model systems and has been invoked to explain chromosome rearrangements in humans. This review focuses primarily on mechanistic studies in *Escherichia coli* and its bacteriophages, T4 and λ, in the budding yeasts *Saccharomyces cerevisiae* and *Kluyveromyces lactis* and on apparently similar, but less well-documented, mechanisms in mammalian cells.

Homology-dependent repair of DNA double-strand breaks (DSBs) occur by three major repair pathways (Pâques and Haber 1999) (Fig. 1). When both ends of the DNA share substantial homology with a donor template (a sister chromatid, a homologous chromosome, or an ectopically located segment), repair occurs almost exclusively by gene conversion (GC). If the DSB is flanked by direct repeats, then a second repair process, single-strand annealing (SSA), can occur as 5′ to 3′ resection of the DSB ends exposes complementary sequences that can anneal to each other and repair the break by the formation of a deletion. However, when only one DSB end shares homology with a donor sequence, repair occurs by BIR. There are two BIR pathways, one dependent on Rad51 recombinase and the other independent of Rad51.

Cite this article as *Cold Spring Harb Perspect Biol* doi: 10.1101/cshperspect.a010397

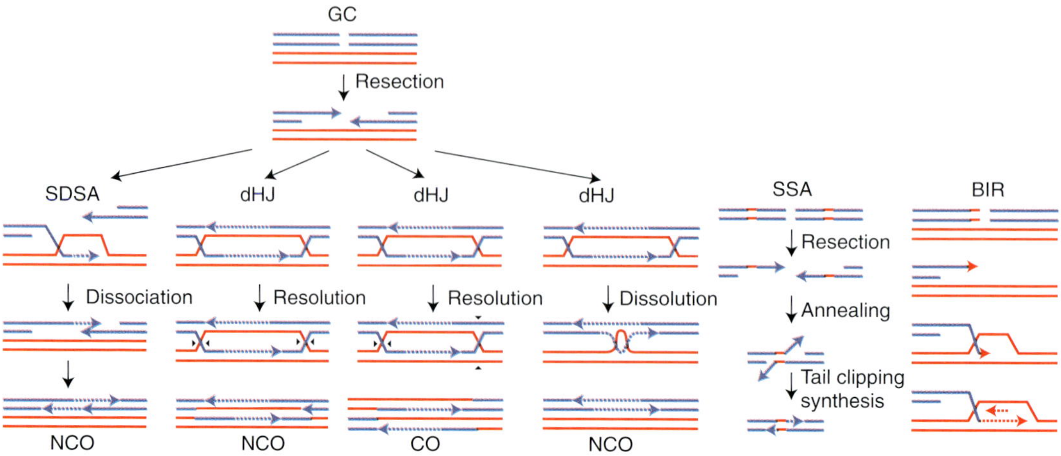

Figure 1. Three major repair pathways of homology-dependent recombination. Noncrossover (NCO) and crossover (CO) events are indicated. Black triangles represent resolution of Holliday junctions (HJs). Dashed lines represent new DNA synthesis. GC, gene conversion; SSA, single-strand annealing; BIR, break-induced replication.

BIR IN BACTERIA AND BACTERIOPHAGES

Recombination-Dependent Replication in Phage T4

Recombination-dependent replication (RDR) is an intrinsic part of the T4 life cycle (reviewed in Mosig 1987; Kreuzer 2000). Early studies of this process, both in vivo and in vitro, provided a molecular framework and paved the way for the study of similar processes in bacterial and eukaryotic cells. T4 is a linear dsDNA bacteriophage; its packaged nucleic acid is slightly more than one genome's worth, causing its ends to be direct repeats. Because of the 5′ to 3′ directionality of replication and a requirement for RNA primers to initiate DNA synthesis, the T4 genome, like all linear chromosomes, has a problem with replication of its termini. RDR provides a solution to complete terminal replication.

The initial phase of T4 replication begins at one of several discrete origins of replication. This initiation requires host RNA polymerase, involves the formation of RNA:DNA hybrid molecules (R-loops), and is blocked by the RNAP inhibitor rifampicin (reviewed in Kreuzer and Brister 2010). Later, T4 proteins modify the specificity of RNAP, and the later phase is a rifampicin-resistant mode of recombination-dependent replication (RDR) (Luder and Mosig 1982). Mutations in T4's recombination functions not only lead to poor recovery of genetic recombinants, but also block the secondary, late pathway of replication (reviewed in Mosig 1987; Kreuzer 2000). One of these recombination functions, UvsX, operates much like the *E. coli* RecA protein in the catalysis of strand-pairing, invasion, and strand-transfer reactions (reviewed in Liu and Morrical 2010) (Table 1). Another recombination function, UvsY, acts as a mediator protein in promoting UvsX binding and replacement of the single-strand DNA binding protein gp32 on single-strand DNA substrates, similar to *E. coli*'s RecFOR and eukaryotic Rad52. A third recombination function, UvsW, is a DNA helicase that promotes stability of D-loops by branch migration of the junction between the two homologous molecules (Carles-Kinch et al. 1997), analogous to the *E. coli* RecG helicase. In vivo, recombination is initiated at T4 ends, at which 3′ ssDNA should be persistent because of the end-replication problem. These ends appear to be bound or further processed by gp46/47 nuclease, proteins related to the SbcCD complex of bacteria and the Mre11 Rad50 complexes of eukaryotes (Sharples and Leach 1995). Mutants in gp46/47 exhibit a profound arrest of late DNA replication (RDR) (Mosig 1998). Replication is

Table 1. Recombination-dependent replication functions in T4 bacteriophage, *E. coli,* and *S. cerevisiae*

Function	T4	*E. coli*	Budding yeast
Strand invasion and exchange	UvsX	RecA	Rad51
Mediator	UvsY	RecFOR RecBCD	Rad52 Rad55/57
Single-strand DNA binding protein	gp32	SSB	RPA
End processing nuclease	gp46/47	SbcCD RecBCD	Mre11/Rad50/Xrs2 Sae2 Dna2/Sgs1 Exo1
Replicative helicase	gp41	DnaB	Mcm2-7 complex
Helicase reloader	gp59	PriA, PriB PriC DnaC DnaT	?
Branch migration helicase	gp41, UvsW	RuvAB RecG	Sgs1 (RecQ helicase family) Srs2
Branched molecule cleavage	gp49 (Endo VII)	RuvC	Mus81/Mms4 Yen1

initiated at these recombination intermediates, leading to a network of branched molecules (Kemper and Brown 1976) that are resolved to linear packageable DNA molecules by the gp49 resolvase function (Mizuuchi et al. 1982).

By the 1980s, Mosig had codified a mechanism for late replication, initiated by homologous recombination (Mosig 1987). The key intermediate in this mechanism is the displacement loop (D-loop) (Fig. 2), a heteroduplex joint molecule formed by the strand invasion of a 3′ tailed linear molecule, catalyzed by UvsX, UvsY, and gp32. Such D-loops would be recognized as substrates for the recruitment of the replicative helicase/primase complex, gp61/41, and loading of the DNA polymerase/clamp complex. This model has been further validated by in vitro reconstitution reactions and by in vivo experiments monitoring the processing of T4 DSBs.

Formosa and Alberts (1986) reconstituted an in vitro system for replication initiated by recombination between a single-strand linear molecule and a homologous covalently closed circle. These reactions required UvsX, gp32, gp43 (DNA polymerase), and gp45 and gp44/62 (clamp and clamp loader complex). The products of such reactions were long, single-stranded DNA concatemers. UvsX/gp32 could not be replaced with their cognate functions from *E. coli*, RecA/SSB, indicating that there must be a species-specific interaction between the replisome and the recombination machinery. In these reactions, the nature of the products and the absence of any requirement for topoisomerase activity led investigators to propose the bubble migration model, a conservative mechanism of DNA synthesis, by which DNA is polymerized and then displaced as a single strand from its template (much like the mechanism of RNA transcription from DNA) (Fig. 3). The bubble migration mechanism corresponds to the priming of leading- but not lagging-strand synthesis, which likely does not reflect the natural mechanism (see below).

Full establishment of bona fide replication forks from recombination intermediates requires helicase loading via gp59. Because gene 59 mutants display a late replication arrest phenotype (Wu and Yeh 1974), gp59 must play an essential role in RDR, presumably in the loading of gp41 onto recombination-generated

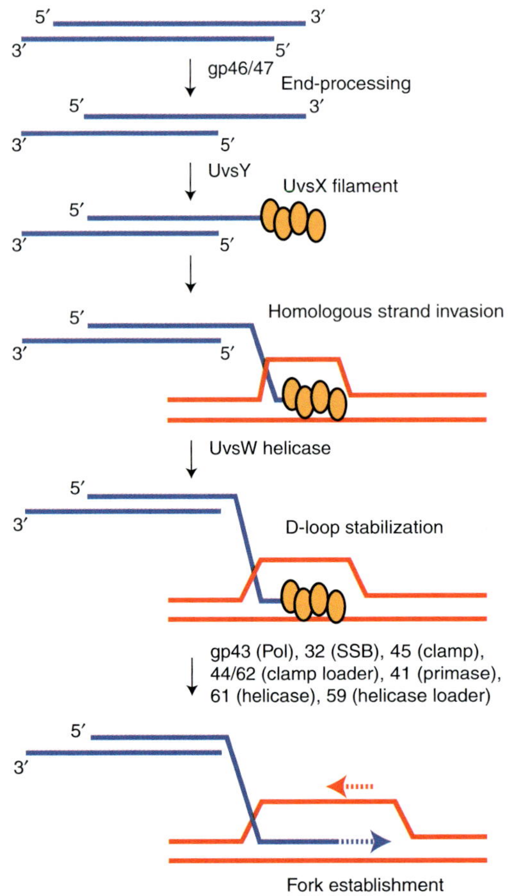

Figure 2. Scheme for recombination-dependent replication of T4. The end-replication problem leaves 3′ ssDNA ends on dsDNA (blue), further processed by gp46/47 nuclease. Mediator UvsY assists UvsX to form a presynaptic filament, competent for strand invasion with a homolgous duplex (red) to form D-loop intermediates. UvsW helicase stabilizes this structure. Full assembly of the replisome, consisting of gp43 (DNA polymerase), gp32 (single-strand DNA binding protein), gp45 (clamp) and gp44/62 (clamp loader), primase (gp41), and (gp61) fork helicase and helicase loader (gp59), establishing bidirectional replication.

D-loops, in a manner reminiscent of the primosome PriA protein of *E. coli* (see below). Inclusion of gp59/41/61 to the reconstituted system promotes establishment of full leading- and lagging-strand semiconservative DNA synthesis from UvsX-generated D-loops (Barry and Alberts 1994) in contrast to the conservative replication mode of bubble migration seen in the absence of gp59/41/61 (Formosa and Alberts 1986). Gp59 promotes the coupling of lagging-strand DNA synthesis with that of the leading strand (Dudas and Kreuzer 2005) by acting as a gatekeeper to block engagement of the 3′ invading strand by DNA polymerase. Indicative of this gatekeeper property of gp59, specific interactions between gp59 and DNA polymerase gp43 (Xi et al. 2005) inhibit 3′ end extension (Nossal et al. 2001) in vitro. This gatekeeper function presumably prevents premature leading-strand synthesis prior to assembly of the helicase/primase complex that will provoke coupled lagging/leading-strand replication.

Insights into the RDR mechanism have been provided by systems in which site-specific DSBs induce coupled recombination and replication of a homologous partner molecule in vivo (Kreuzer et al. 1995; George and Kreuzer 1996; George et al. 2001). Such reactions require the suite of T4 recombination proteins (UvsX, UvsY, UvsW, gp46/47) and T4 replication functions (gp43, gp45, gp44/62), gp59, primase (gp41), and helicase (gp61). Consistent with the extensive chromosome replication mechanism of RDR, the two broken ends of a plasmid molecule can engage two separate homologous partners, independently provoking their replication (Stohr and Kreuzer 2002).

Break-Induced Replication in *E. coli*

Even before elucidation of T4 RDR, connections between replication and recombination in bacteria had been suspected. Lederberg (1955) proposed a break-copy model of recombination, whereby a broken chromosome would stimulate replication by recombination. Skalka (1974), through her studies of bacteriophage lambda, suggested that convergence of replication forks onto preexisting nicks in the template strand leads to DSBs, which might be repaired by recombination with a homologous chromosome through a D-loop intermediate.

By the mid-1980s, the homologous recombination pathways in *E. coli* had been defined genetically: the RecBCD helicase/nuclease, in conjunction with the RecA strand transfer

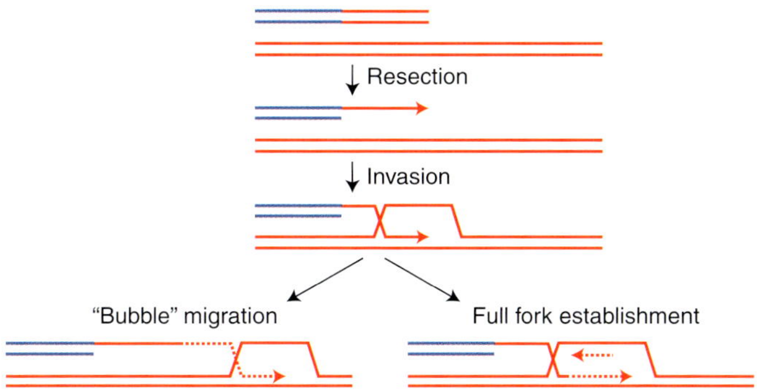

Figure 3. Key intermediates of replication initiated by homologous recombination. The double-stranded DNA end is resected, exposing the single-stranded 3′ end. Strand invasion results in a D-loop intermediate. Bubble migration involves priming of leading- but not lagging-strand DNA synthesis (*left*), whereas full fork establishment involves coupling of leading- and lagging-strand synthesis (*right*). The solid line with arrowhead represents the invading 3′ end. Dashed lines represent new DNA synthesis.

protein, constituted a mechanism for DSB repair (Wang and Smith 1983; Clark and Sandler 1994). The sequence of the *E. coli* chromosome revealed that Chi sites (GCTGGTGG) that stimulate RecBCD-mediated recombination and attenuate its nuclease activity (reviewed in Persky and Lovett 2008) are found overrepresented in the *E. coli* genome, in the orientation that protects broken forks from extensive degradation (Burland et al. 1993) and promotes their processing into D-loops that could restore a replication fork. The preferential orientation of Chi sites suggests that repair of collapsed replication forks, rather than simple repair of DSBs, is the primary role of the RecBCD repair system (Kuzminov 1995). Smith (1991) proposed that even the classical recombination processes in *E. coli* (involving DNA transfer via conjugation or dsDNA introduced by phage transduction) set up obligate replication forks to mediate inheritance of genetic markers.

Despite earlier suggestions (Asai et al. 1994), it took several years to establish conclusively that DNA DBSs could prime replication in *E. coli* (Kuzminov and Stahl 1999). When recombination between bacteriophage lambda is catalyzed by the *E. coli* recombination systems, the break-copy mechanism accounts for approximately one-half of the detected recombinants, and the nonreplicative break-join mechanism accounts for the remainder (Motamedi et al. 1999). These two mechanisms are distinguished by the sedimentation properties of recombinant phage grown in light (^{14}N) medium from heavy (^{15}N)-labeled parents. Break-join phage remain heavy (HH), depend on HJ resolvase activity, and can appear in the absence of DNA polymerase III activity, whereas break-copy outcomes show extensive replication with an HL density, are independent of resolvase, and are dependent on DNA polymerase III.

Ascertainment of recombination-associated replication in *E. coli* in vivo has been aided by its requirement for a special set of replication restart or primosome assembly factors that are not required for normal replication initiation at the origin of replication, *oriC*. During normal replication, the DnaA protein, a AAA$^+$ ATPase, is required, along with DnaC protein, to load the replicative helicase for initiation at *oriC*. In contrast, for replication restart, an alternative set of factors (PriA, PriB, PriC, and DnaT), along with DnaC, loads DnaB. These factors were initially characterized for the role in the initiation of replication of phiX174 bacteriophage (Shlomai and Kornberg 1980a,b) at an origin consisting of a region of DNA secondary structure. Phage phiX174 has evidently commandeered this special restart system to initiate its replication.

The PriA protein, an AAA$^+$ ATPase and 3′ to 5′ helicase, initiates the assembly of replication forks from recombinational D-loop structures (Liu et al. 1999; Xu and Marians 2000). PriA's binding to DNA substrates appears to be enhanced by PriB, potentially through an interaction with SSB; it then recruits DnaT, which aids the loading of DnaB helicase from a DnaBC complex and concomitantly the DnaG primase (reviewed in Gabbai and Marians 2010). In addition to orchestrating the primosome assembly process through protein recruitment, PriA 3′ helicase activity may process intermediates to provide a single-strand assembly site on the lagging-strand template for DnaB loading. Interestingly, like its counterpart gp59 of bacteriophage T4, PriA binds and occludes the 3′ invading strand from DNA polymerase extension, potentially ensuring that a full fork supporting both leading and lagging strand is established (Xu and Marians 2003). PriA can also support restart from some forklike structures, suggesting a role for restart both with and without associated replication. For forks with a leading-strand gap, an alternative pathway involving PriC and Rep appears to be more efficient (Heller and Marians 2006).

In vivo, mutants in *priA* have reduced viability, especially under fast growth conditions that support multiple ongoing replication forks. In *priA* mutants, a subpopulation of cells is filamentous and is induced for the SOS response to DNA damage, indicating an abnormal response to spontaneous source of damage (Lee and Kornberg 1991; Nurse et al. 1991). Mutants in *priA* are sensitive to DNA damaging agents, such as UV and gamma irradiation and mitomycin C (Nurse et al. 1991; Masai et al. 1994; Kogoma et al. 1996). These phenotypes can be fully complemented by the Walker A K230R mutant, which is defective for helicase activity but fully competent for primosome assembly (Zavitz and Marians 1992). In *priA* mutants, there is a high level of RecFOR-mediated recombination, indicative of persistent DNA gaps (Grompone et al. 2004). PriA mutants also show reduced recombinational inheritance during transduction and conjugation (Kogoma et al. 1996), implicating establishment of replication forks are invoked during these processes as proposed earlier by Smith (1991). In addition to one-ended recombination events involving repair of a broken fork with its sister chromosome, PriA is required for two-ended recombination events that repair a DSB caused by SbcCD nuclease cleavage of secondary structure elements (Eykelenboom et al. 2008).

A phenomenon known as induced stable DNA replication (iSDR) has been studied in some detail and reflects a mode of BIR in *E. coli*. DNA replication, as detected by radioactive thymine precursor incorporation, normally ceases after inhibition of protein synthesis such as caused by the antibiotic chloramphenicol. However, induction of the SOS response via treatments such as UV irradiation, nalidixic acid treatment, or thymine starvation permits a mode of DnaA-independent, chloramphenicol-resistant replication (reviewed in Kogoma 1997), which depends on the primosomal assembly factor, PriA (Masai et al. 1994), and DNA recombination proteins, RecA and RecBCD (Magee and Kogoma 1990). The requirement for RecBCD, an enzyme specific for recombination initiated at DSBs (see Persky and Lovett 2008), implicates chromosomal breaks in the mechanism although the RecF pathway is required for a minor component of iSDR is (Kogoma 1997). The role of the helicase function activity of PriA in iSDR is unclear since the Walker A K230R mutant has been reported both to restore (Kogoma et al. 1996) and to lack (Tanaka et al. 2003) iSDR function. The above observations support a model by which replication is initiated at RecABCD-dependent D-loops (Kogoma 1997); thus, defects in the proteins that process D-loops, RecG, and RuvABC elevate iSDR in vivo (Asai and Kogoma 1994). iSDR is apparently mutagenic (Lark and Lark 1979) but the molecular nature of these mutations has not been fully investigated.

BIR in Budding Yeast

The best-studied BIR system in eukaryotes is budding yeast *S. cerevisiae*. Three different assays have been used to explore BIR: (1) transformation of ends-out double-stranded DNA fragments, (2) induction of site-specific HO or I-SceI endonucleases to create chromosomal

DSBs that share only one-ended homology, and (3) creation of strains that lack functional telomerase but maintain their telomeres by recombination-dependent replication (Fig. 4). None of these systems directly address the problem of repair of a stalled and broken replication fork where there is a broken chromosome end and an intact sister chromatid. However, ways to address this central question have been developed by Aguilera (Cortes-Ledesma et al. 2007) and by Bjergbaek (Nielsen et al. 2009), using site-specific enzymes to create single-strand nicks in G_1 cells, which are converted to DSBs as cells enter S phase. Although all of these events require the Rad52 protein, there are both Rad51-dependent and Rad51-independent BIR processes.

Rad51-Dependent BIR during Transformation

Dunn et al. (1984) first showed that a transformed linearized plasmid sharing homology at one end with subtelomeric sequences could acquire new telomeres by Rad52-dependent recombination. Evidence of much more extensive recombination-dependent replication was provided by Morrow et al. (1997) using chromosome fragmentation vectors (CFV), carrying an origin of replication (ARS) and a centromere (Fig. 4A). These studies also found that centromeres pose major barriers to BIR, so that most transformants actually contained half-crossover outcomes most likely representing strand invasion events that branch-migrate into an intermediate that can be resolved as a crossover but cause loss of the template chromosome. Why centromeres are not barriers to bulk replication but block BIR is an unsolved question. Surprisingly, BIR can progress through the entire array of more than 100 9-kb rDNA repeats (B Llorente, pers. comm.).

Another BIR-related phenomenon occurs during gene targeting. In a significant fraction of transformations, one finds hit-and-run events in which one or both ends invade and set up a BIR event that can copy substantial lengths of DNA adjacent to the integration site (Adair et al. 1989). If the copying includes an ARS and the two ends then rejoin, the apparent gene knockout will prove to be an autonomously replicating circle carrying the selectable marker (Kraus et al. 2001).

DSB-Induced Rad51-Dependent BIR

HO endonuclease cleaves a 24-bp site to catalyze mating-type (*MAT*) switching, an efficient GC event in which both sides of the DSB share homology with a donor (Haber 2012). However, the normal donors can be deleted and the cleavage site inserted at other locations to establish situations in which only the centromere-proximal end of the DSB shares homology with an ectopic sequence (in a haploid) or allelic sequence (in a diploid). HO-induced BIR can occur with as little as 70 bp of homology to create a nonreciprocal translocation (Bosco and Haber 1998; Ira and Haber 2002). Although Rad51-dependent BIR is quite efficient, it represents only a tiny fraction of outcomes in the

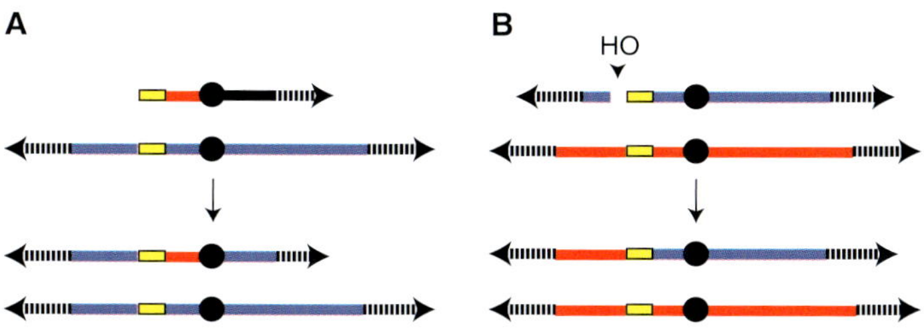

Figure 4. Three different assays used to explore BIR. BIR assays involving (*A*) a typical chromosome fragmentation vector (CFV), (*B*) HO-endonuclease-induced DSB where only one end shares homology with the donor (where homology is indicated in yellow), and (*C*) telomere maintenance in the absence of telomerase by recombination-dependent replication.

situation where both ends of the break share homology and GC dominates.

Because HO induction is rapid and highly efficient, it is possible to follow repair kinetics on Southern blots (Malkova et al. 2005) (Fig. 5). Whereas interchromosomal GC is completed in 2–3 h, BIR events are delayed until 4–6 h. This delay is enforced after Rad51 promotes strand invasion (measured by a chromatin immunoprecipitation assay) but before the initiation of new DNA synthesis (assayed by PCR). Once BIR is initiated, its rate of DNA synthesis over the 100 kb to the end of the chromosome, 3–4 kb/min, is comparable to normal replication.

The delay in initiating BIR after strand invasion reflects the action of a recombination execution checkpoint that can sense whether the two ends of a DSB can synapse with the same donor sequence within a critical distance and in the proper orientation to complete repair (Jain et al. 2009). Thus, if the two halves of the donor sequence are separated by about 1 kb, repair is rapid and efficient; but if the two homologous targets are 5 kb or more apart, the initiation of repair DNA synthesis is—like BIR—delayed by several hours. As further evidence that there is a transition from GC to BIR, the large gap repair events become dependent on Pol32 (Jain et al. 2009).

DSB-Induced Rad51-Independent BIR

In a diploid where MAT cleavage on one chromosome can be repaired by allelic recombination, nearly all repair occurs by GC. Without Rad51, there is still surprisingly efficient repair by Rad51-independent BIR, retaining a marker on the opposite arm of the chromosome but losing heterozygosity to the right of the DSB. These events are distinctly different from Rad51-dependent BIR. First, many of the repair events occur after the cells have escaped from arrest imposed by the DNA damage checkpoint so that the colonies are sectored, containing both BIR outcomes and 2n-1 cells that have lost the broken chromosome. Second, repair depends primarily on a pair of inverted Ty retrotransposons elements lying 30 kb proximal to the break. Most repair events prove to be nonreciprocal translocations involving Ty sequences on other chromosomes. This process requires the MRX complex, Rad59, and Rdh54/Tid1, as well as Rad52 (Signon et al. 2001).

A second instance in which Rad51-independent BIR predominates is when both ends of an HO-induced DSB share very limited homology with a donor, for example, in a centromeric plasmid containing sequences in inverted orientation, with only 33 bp homology shared between the donor and recipient (Ira and Haber 2002). Repair significantly increases without rad51. Presumably, Rad51 binds to the ssDNA ends containing the 33 bp homology, but it is incapable of carrying out the next steps in recombination. The repair events that are recovered in the absence of Rad51 are also distinctive in 50% having an apparent reciprocal crossover. As suggested by Symington (Bartsch et al. 2000),

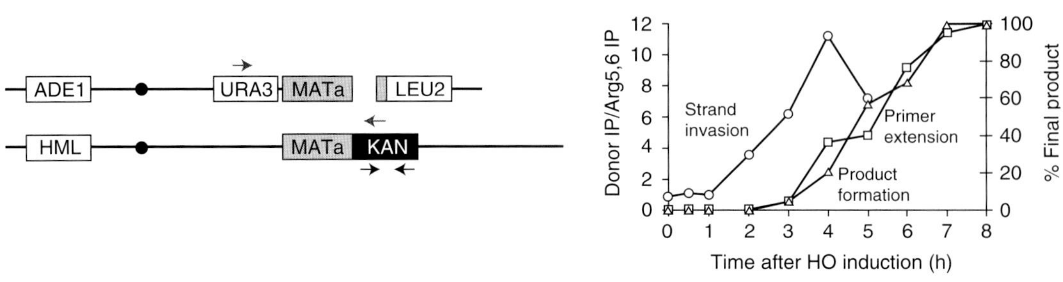

Figure 5. Kinetics of BIR. Schematic representation of a diploid BIR strain used to measure the kinetics of BIR, where there is only homology to the left of the DSB (*left*). Black arrows represent the position of primers used for ChIP analysis. Gray arrows represent the position of primers used for primer extension assay. The graph (*right*) shows the kinetics of strand invasion (circles), initiation of new DNA synthesis (squares), and product formation (triangles). (Figure modified from Jain et al. 2009.)

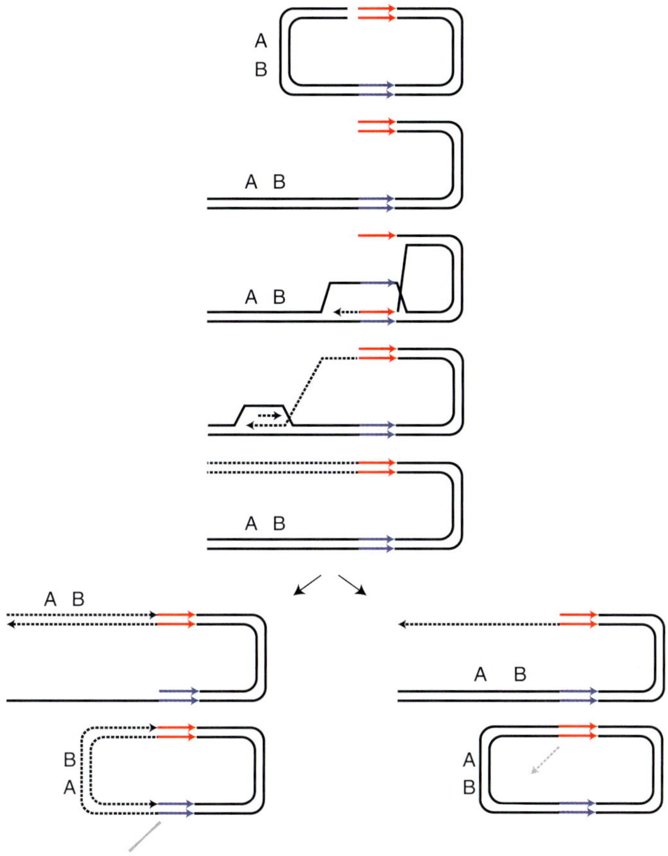

Figure 6. Schematic of events that involve a combination of BIR and SSA. Inverted repeats are shown as blue and red arrows, respectively, which are separated by marker AB. DSB at one of the inverted repeats is followed by resection and invasion into the other (homologous) inverted repeat. DNA synthesis (dashed line) proceeds until the end of the break. Depending on which strand gets resected, SSA results in different orientation of the marker (AB). (Figure adapted from Bartsch et al. 2000.)

these repair events most likely arise by BIR followed by single-strand annealing (Fig. 6).

Telomere Maintenance without Telomerase

BIR also allows yeast telomeres to be maintained in the absence of telomerase (Lundblad and Blackburn 1993; McEachern and Blackburn 1996; Teng and Zakian 1999; McEachern and Haber 2006). In mammalian cells, such alternative lengthening of telomeres (ALT) has been shown to involve recombination, but how it occurs is not well defined. In budding yeast, there are two recombination-dependent mechanisms, both requiring Rad52, but one which is Rad51-independent. In cells lacking either telomerase RNA or protein components, telomeres erode progressively until the vast majority of cells undergo senescence and cell death. However, a tiny proportion of cells become survivors and proliferate. There are two distinct pathways for telomere maintenance. One pathway (Type I) involves recombination with and amplification of Y′ subtelomeric sequences (Lundblad and Blackburn 1993; Louis et al. 1994). Type I survivors depend on the canonical recombination proteins including Rad52, Rad51, Rad55/57, and Rad54 (Le et al. 1999; Teng and Zakian 1999). A second mechanism (Type II) leads to long extensions of the telomere repeats themselves; these survivors arise independent of

Rad51 but require Rad52, the MRX complex, and Rad59 (Lundblad and Blackburn 1993; McEachern and Blackburn 1996; Teng and Zakian 1999; Teng et al. 2000). Type II survivors also require Sgs1 helicase (not required for HO-induced Rad51-independent BIR) and the checkpoint proteins Mec1 and Tel1 (Cohen and Sinclair 2001; Huang et al. 2001; Johnson et al. 2001; Signon et al. 2001; Tsai et al. 2002). Both pathways require the nonessential DNA polymerase Pol δ, Pol32 (Lydeard et al. 2007), and thus are replicative processes rather than half-crossovers that would elongate one telomere at the expense of a sister chromatid.

In Vivo Biochemical Analysis of Rad51-Dependent BIR

A detailed investigation of the factors needed for Rad51-dependent BIR compared to those for GC conducted by Lydeard et al. (2010b) revealed that BIR requires nearly all the factors for normal replication, except some of those required to establish the prereplication complex at origins.

Recombination-Mediated Establishment of Replication

The initial stages of Rad51-dependent BIR are shared with gene conversion, namely, 5′ to 3′ resection of the DSB, homology searching, and strand invasion. Resection is initiated by the MRX-Sae2 proteins but is then taken over by two different activities, the exonuclease Exo1 or the helicase-endonuclease complex of Sgs1-Rmi1-Top3 and Dna2 (Neale et al. 2005; Gravel et al. 2008; Mimitou and Symington 2008; Zhu et al. 2008; Symington and Gautier 2011). Surprisingly, deletion of these different proteins had opposing effects on the efficiencies of BIR; *mre11Δ* or *rad50Δ* caused a 50% reduction in BIR, whereas *sae2Δ* resulted in a slight increase (Lydeard et al. 2010a). This difference could reflect other roles of MRX and Sae2 in the establishment and maintenance of the DNA damage checkpoint that prevents mitosis before BIR is complete or in later steps of recombination. However, a transformation-based assay found no role for MRX in BIR (Davis and Symington 2004).

Deleting Exo1 or Sgs1 results in a nearly 50% increase in BIR; conversely, their overexpression severely inhibits BIR (Lydeard et al. 2010a). The increase in BIR in the absence of Exo1 or Sgs1 is also seen in transformation-based assay (Marrero and Symington 2010). Reducing the extent of ssDNA may ensure that Rad51 is bound to the homologous sequences at the end of the DSB rather than being attracted to more distant unrelated ssDNA. Supporting this idea, overexpressing Rad51 nearly doubles the efficiency of BIR and partially suppresses overexpressing Sgs1 or Exo1 (Lydeard et al. 2010a). Alternatively, Sgs1 may discourage strand invasion and synapsis with short regions of homology, similar to its documented role in dismantling heteroduplex DNA containing mismatches, but how Exo1 would affect this is unclear. Similar to Sgs1, deleting another 3′ to 5′ helicase, Mph1, causes significant increases in BIR efficiency (JE Haber and S Jain, unpubl.). More strikingly, an *sgs1 mph1* double mutant additively increases BIR efficiency and unexpectedly also seems to eliminate the delay in initiating new DNA synthesis. How this occurs remains under investigation.

As noted above, the initial steps in recombination appear to be the same for BIR as for GC: 5′ to 3′ resection of DSB ends, loading of RPA, and then its displacement by Rad51, aided by Rad52, Rad55, and Rad57 (Krogh and Symington 2004). However, another set of Rad51-aiding proteins, Psy3-Csm2-Shu1-Shu2 (PCSS), are much more important in BIR than in GC (Lydeard et al. 2010b). The idea that the PCSS proteins might be an alternative Rad51-facilitating pathway is supported by the finding that overexpressing Rad51 suppresses this defect in both *shu2Δ rad55Δ* and *shu2Δ*.

Initiation of New DNA Synthesis

One of the least-understood steps in BIR is the mechanism by which the strand-invasion D-loop is converted into a replication fork with leading- and lagging-strand synthesis. There is no homolog of bacterial PriA. A striking finding is that nearly all BIR events require Pol32, a nonessential subunit of DNA Pol δ, whereas GC is not affected by *pol32Δ* (Lydeard et al. 2007). It is possible that Pol32, similar to PriA in prokaryotes, acts in some way to recruit a replicative helicase and assemble an origin-independent replication fork.

A second novel requirement for BIR that has little effect on normal replication or GC was found as a suppressor of *pol32Δ*'s cold-sensitivity for growth. Surprisingly, two *pol30* mutations that suppress *pol32Δ*'s cold sensitivity do not suppress *pol32Δ*'s BIR defect; in fact, these *pol30* mutations, by themselves, are defective in BIR (Lydeard et al. 2010b). Moreover, these mutations are dominant, suggesting that even one such mutant subunit of the PCNA trimer may be sufficient to disrupt BIR. These PCNA mutations could be defective in the steps of establishing a full replication fork. The defect of *pol30-FF248, 249AA* is much more severe than the 50% reduction seen in a PCNA mutation that cannot be ubiquitylated or SUMOylated (*pol30-K164R*) or when the ubiquitylation and SUMOylation pathways are disrupted (*rad18Δ siz1Δ*)(Lydeard et al. 2010b).

BIR requires virtually all of the components of normal DNA replication, including the replicative helicase complex consisting of the Mcm2-7 heterohexamer, Cdc45, and the GINS complex (Lydeard et al. 2010b). BIR also requires the DDK kinase complex, Cdc7-Dbf4. These components are not needed for GC. Although Cdc7 can be bypassed by mutations in Mcm proteins, these suppressors do not relieve the requirement for Cdc7 in BIR. BIR involves both leading- and lagging-strand synthesis; consequently, BIR requires Pol α-primase and two factors that help Pol α act processively: Mcm10 and Ctf4. BIR also requires components that were previously identified as prereplication origin-loading factors: Dpb11-Sld2-Sld3 and Cdt1. Interestingly, without these factors, even the initial presumably leading-strand extension from the 3′ end of the invading strand does not occur. That no leading-strand synthesis occurs without Pol α or its cofactors again suggests that leading- and lagging-strand synthesis are tightly coordinated and that there is not an initial step of primer extension (as in GC) that is later converted into a full replication fork.

One remarkable difference between normal replication and BIR is that the initiation of BIR requires Pol δ, but proceeds without Pol ε (Lydeard et al. 2007). However Pol ε is required for replication of the template to continue after the first several kilobases. This separation of functions between Pol δ and Pol ε is strikingly different from what has been seen in replication, where Pol ε appears to direct leading-strand synthesis, and Pol δ is concerned with lagging-strand copying. However, it is known that S-phase replication can be completed in the absence of catalytic activity of Pol ε so long as the carboxy-terminal portion of the Pol 2 (Pol ε) protein is present (Kesti et al. 1999). The nonessential subunits of Pol ε (Dpb3 and Dpb4) are not required for BIR (Lydeard et al. 2010b).

Replication during BIR Is Far More Mutagenic than Normal Replication

The idea that there is a transition between Pol ε-independent and Pol ε-dependent replication complexes after the first several kilobases of new DNA synthesis resonates with the finding from the Symington lab that there are frequent template switches when a linearized fragment can initiate BIR with either of two polymorphic homologous chromosomes of a diploid (Smith et al. 2007). These template switches are confined to the first several kilobases, after which copying is apparently processive. This observation could be explained if initial DNA synthesis involved only the primer-extended leading strand, which then matured into a full and more processive replication fork. However, as noted above, physical monitoring of BIR does not support this hypothesis, as there was no primer extension when Pol α primase was inactivated. It is possible that there are mechanistic differences in CFV events compared to a chromosomal DSB.

Despite the fact that BIR uses the complete replication machinery that is employed in normal S phase, the rate of mutation accompanying BIR is as much as 2800 times higher than spontaneous events. By placing a *lys2* gene containing an out-of-frame run of As at several locations along the template chromosome arm, Deem et al. (2011) demonstrated that frameshift mutations increased both close to the point of strand invasion and much further down the template chromosome. Another set of reversion events involved a template jump between quasi-palindromic sequences. Mutations were independent of the error-prone DNA polymerase Pol ς or another translesion polymerase, Pol η. However, 50% or more of possible frameshift mutations were corrected by the mismatch repair machinery because the rate was two- to 11-fold higher in an *msh2Δ* mutant. Similarly, in a Pol δ proofreading mutant (*pol3-5DV*), there is a significantly higher level of BIR-associated frameshifts, suggesting that Pol δ is a major player in creating, but then correcting, errors. One contribution to the high mutation rate is a DNA damage checkpoint regulated, Dun1-dependent increase in dNTP pools, causing DNA polymerases to be error prone.

These results are quite different from the high level of mutations seen in GC, where the high rate of frameshift mutagenesis and base pair substitutions were independent of mismatch repair (Hicks et al. 2010). Moreover, a different Pol δ mutation, *pol3-01*, had a dramatically opposite effect on mutations, namely eliminating essentially all the template jumps, be they −1 frameshifts, quasipalindrome events, or interchromosomal template switches.

Resolution of BIR

A major unsolved question concerns the resolution of Holliday junctions (HJ) and the final disposition of newly synthesized strands during BIR. If BIR proceeds by formation of a unidirectional replication fork inside a D-loop, there should be an HJ near the point of strand invasion. If it is cleaved by an HJ resolvase, the newly synthesized DNA should remain base paired to its template. In this case, one should find a new DNA strand on both the donor and the new translocated region. If this branch migrates toward the terminus, the two strands of the donor will be re-paired, as will the two newly made strands, so that the replication will appear to be conservative, as was found for short patch DNA synthesis in GC (Ira et al. 2006).

Whether BIR leads to semiconservative or conservative DNA synthesis can be answered by DNA combing or by methods in which base analogs such as BrdU are incorporated into DNA and detected using immunofluorescence. By analyzing the location of the Lys^+ revertants in the BIR assay described above, Malkova (A Malkova, pers. comm.) has preliminary evidence that they are all in the recipient (and hence conservatively replicated) chromosome.

Budding yeast has at least four HJ resolvases. Two of these—Mus81-Mms4 and Yen1—appear to play the major role in generating crossovers in interchromosomal mitotic recombination (Ho et al. 2010); however, even the *mus81Δ yen1Δ* double mutant did not affect the success of BIR.

BIR in Other Organisms

Kluyveromyces lactis

Studies in *K. lactis* have provided useful mechanistic details of telomere maintenance in the absence of telomerase (McEachern and Blackburn 1995; McEachern and Iyer 2001; Natarajan and McEachern 2002; Groff-Vindman et al. 2005; Cesare et al. 2008) and also helped in establishing the generality of telomere maintenance by means of BIR (McEachern and Haber 2006). Unlike in *S. cerevisiae,* all *K. lactis* survivors are of Type II, except when telomeric sequences are placed internally to the chromosome ends, when Type I survivors are also observed (Natarajan et al. 2003). However, the unit telomere repeat in *K. lactis* is 25 nt, compared to the degenerate $TG_{1\text{-}3}$ repeats in *S. cerevisiae.* A key intermediate that enables telomere maintenance in the absence of telomerase is a telomeric circle, which can be excised by recombination from its original location. By a mechanism termed roll and spread, the telomeric circle is copied onto the telomere at the chromosome end by rolling circle replication, involving BIR. Two main observations support this hypothesis: visualization of telomeric circles in strains having a dysfunctional telomerase and the spreading of a marker from a transformed plasmid containing telomeric sequences into a majority of the telomeric ends (Natarajan and McEachern 2002). However, an alternative mechanism—the reinsertion of amplified circles back into the telomeric ends by recombination—has not been ruled out.

Schizosaccharomyces pombe

BIR has been documented in *S. pombe* by Humphrey's lab, using an HO-induced DSB on a minichromosome that was derived from Chr III (Cullen et al. 2007). In addition to gene conversion events, it was possible to recover repair events in which the distal portion of the minichromosome was lost, resulting in loss of heterozygosity (LOH). Some of these events must have arisen by BIR. These events required the fission yeast homologs of Rad51, Rad52, Rad55-Rad57, Rad54, and the MRX proteins. Unlike budding yeast, BIR in fission yeast also needs the branched structure resolvase, Mus81.

Drosophila

Drosophila deals with the end-replication problem by means of a pathway utilizing transpositions of retrotransposons Het-A and TART to the chromosome ends. However, it has been proposed that a terminal break induced by P-element excision can result in copying of sequences from the homologous chromosomes and proceeding to the end of the chromosome. BIR has not yet been studied at sites of P-element excision or when site-specific endonucleases create DSBs, although this should be possible. Data from the Golic lab suggest that a few gene-targeting outcomes could involve BIR (Engels 2000; Rong and Golic 2000).

BIR in Mammals

BIR has not been formally demonstrated in mammals. One possible example is the nonreciprocal translocation that is often associated with facioscapularhumeral muscular dystrophy (FSHD) in humans. This involves recombination between sets of diverged 3.3-kb repeated sequences that are located within 100 kb of the ends of chromosomes 4 and 10 (Tawil and Van Der Maarel 2006). Whether these occur by BIR is presently unknown, but the nonreciprocal nature of the events and their asymmetry (more often chromosome 10 repeats transferred to chromosome 4 than vice versa) are compatible with such a mechanism.

Recently, Halazonetis (TD Halazonetis, pers. comm.) has found that when mammalian cells are placed under replication stress by overexpression of an oncogene, cell growth is severely impaired if the Pol32 homolog, Pol3D, is ablated. There is no effect in untransformed cells.

The main focus of attention in thinking about BIR in humans involves the maintenance of telomeres by ALT (Scheel and Poremba 2002; Reddel 2003). A majority of human tumor cells escape senescence by reactivating telomerase, but many transformed cell lines and some cancers avoid cell death by telomere shortening by a recombination-dependent mechanism (Dunham et al. 2000). There are multiple pieces of evidence, direct and indirect, supporting recombination in telomere maintenance and ALT. Cells deficient in Rad51D or Rad54 exhibit telomere shortening (Jaco et al. 2003; Tarsounas et al. 2004). Telomeres in ALT cells are associated with ALT-associated promyoleucytic leukemia (PML) bodies (APBs) that contain—in addition to telomere DNA sequences—Rad51, the MRN proteins, and the BLM and WRN helicases (Henson et al. 2002). Overexpressing Sp100, a constituent of PML bodies but not APBs, cause the relocalization of MRN proteins away from APBs and result in the inactivation of ALT and the progressive shortening of telomeres (Jiang et al. 2005). Because Mre11, Rad50, and Xrs2 are essential for type II telomere maintenance in budding yeast, it is of course tempting to imagine these proteins are carrying out a similar role in mammalian ALT. In addition, depletion of the replication-associated flap endonuclease FEN1 causes the senescence of ALT cells, except those expressing telomerase (Saharia and Stewart 2009). Similar to results in *K. lactis*, a marker integrated into a telomeric region can be found at multiple chromosomal sites in ALT cells (Dunham et al. 2000). There is also a significantly elevated rate of telomere sister-chromatid exchange (T-SCE) relative to sister-chromatid exchange (SCE) rates elsewhere in the genome (Londono-Vallejo et al. 2004). This high rate is not seen in telomerase-positive cells or in BLM syndrome cells that exhibit ~10-fold elevation in SCE rates. SCEs could arise in BIR by resolution of an HJ. Mus81 HJ resolvase appears to be required for ALT (Zeng et al. 2009), analogous to BIR dependence on Mus81 in *S. pombe*. Thus, although it is clear that recombination is involved in ALT, the implication of BIR awaits further experimentation.

VARIATIONS ON A THEME: MM-BIR

Recent human-genome-sequencing studies attribute 12% of the variation among genomes to copy number variation (CNV) (Redon et al. 2006; Zhang et al. 2009; Conrad et al. 2010). These studies have revealed many examples of duplications and triplications, often in quite very complex rearrangements, the most dramatic of which are the apparent shattering and rejoining within a single chromosome, termed chromothripsis (Stephens et al. 2011). These rearrangements are partially explained by forkstalling and template switching (FoSTeS), now renamed microhomology-mediated BIR (MM-BIR) (Hastings et al. 2009). Although some nonrecurrent CNV events may represent multiple breakages and rejoinings by nonhomologous end joining (NHEJ), those where the copy number is increased to three or more copies must have involved some type of replication, presumably MM-BIR. The initial steps that trigger MM-BIR remain speculative, but recent evidence suggests that replicative stress leading to broken forks might be an important factor. Indeed, agents that perturb replication, such as HU and aphidicolin, are potent inducers of CNVs (Durkin et al. 2008; Arlt et al. 2009, 2011).

The most direct study of nonrecurrent CNV has been performed in budding yeast, where the slow growth caused by deleting one of two copies of an essential ribosomal protein gene, *RPL20A*, provides a strong selection for duplications of *RPL20B* (Koszul et al. 2004; Payen et al. 2008). These rearrangements can be tandem duplications of the chromosomal region containing *RPL20B* or segmental duplications (SDs), in which part of the chromosome containing *RPL20B* is inserted into another chromosome. SDs arise spontaneously at a rate of 1 $\times$ 10^{-7} cells; but defects in DNA replication increases the incidence of SD 700 fold. Sequences that naturally stall replication fork progression, such as the tRNA genes, are also hotspots of chromosome breakage leading to SDs. Most tandem duplications in wild-type cells are recurrent, arising by what was first assumed to be unequal crossing over between two divergent 300-bp-long terminal repeat (LTR) sequences flanking *RPL20B*. Surprisingly, even these tandem duplication events are Pol32 dependent, as if they have not occurred by a simple unequal crossover but involve some kind of BIR (Payen et al. 2008). Most nonrecurrent intrachromosomal and interchromosomal SDs do not involve LTR sequences, have junctions containing microhomologies and are Pol32 dependent. Most likely they involve MM-BIR.

Analysis of SDs in various mutant backgrounds has revealed other distinct mechanisms of SD formation. Deletion of Rad51 leads to an increase in interchromosomal BIR events that appear to occur between the LTR centromere proximal to RPL20B and divergent LTRs elsewhere in the genome. This finding fits a pattern of observations reviewed above that the great majority of Rad51-independent BIR events use these diverged sequences.

An even more striking discovery is that there is a remarkably efficient Rad52-independent pathway of SD formation because both Rad51-dependent and Rad51-independent BIR require Rad52. Deleting Rad52 completely eliminates intrachromosomal and interchromosomal rearrangements involving the LTR repeats, but in their place are found very short regions of microhomology located at both intrachromosomal tandem duplications and interchromosomal insertions (Payen et al. 2008). Moreover, these events are Pol32 dependent. These outcomes seem most compatible with some sort of perturbation of DNA replication that may involve MM-BIR.

The key feature of MM-BIR is that recombination-induced replication is not processive and may dissociate a partially replicated strand and invade and anneal at another location where replication proceeds. These exits and entries may happen many times. The exact causes of the jumps and whether these dissociations and invasions involve still-attached DNA polymerases or would involve a homologous recombination step (recognizing unusually tiny lengths of homology) have yet to be answered.

ACKNOWLEDGMENTS

Research in the Haber and Lovett Laboratories are supported by National Institutes of Health grants GM20056, GM61766, GM76020, 2T32GM007122, GM51753, and earlier grants from the National Science Foundation and Department of Energy. R.P.A. is grateful to the Ruth L. Kirschstein National Research Service Award 1F32GM096690-01.

REFERENCES

Adair GM, Nairn RS, Wilson JH, Seidman MM, Brotherman KA, MacKinnon C, Scheerer JB. 1989. Targeted homologous recombination at the endogenous adenine phosphoribosyltransferase locus in Chinese hamster cells. *Proc Natl Acad Sci* **86:** 4574–4578.

Arlt MF, Mulle JG, Schaibley VM, Ragland RL, Durkin SG, Warren ST, Glover TW. 2009. Replication stress induces genome-wide copy number changes in human cells that resemble polymorphic and pathogenic variants. *Am J Hum Genet* **84:** 339–350.

Arlt MF, Ozdemir AC, Birkeland SR, Wilson TE, Glover TW. 2011. Hydroxyurea induces de novo copy number variants in human cells. *Proc Natl Acad Sci* **108:** 17360–17365.

Asai T, Kogoma T. 1994. Roles of ruvA, ruvC and recG gene functions in normal and DNA damage-inducible replication of the Escherichia coli chromosome. *Genetics* **137:** 895–902.

Asai T, Bates DB, Kogoma T. 1994. DNA replication triggered by double-stranded breaks in *E. coli*: Dependence on homologous recombination functions. *Cell* **78:** 1051–1061.

Barry J, Alberts B. 1994. Purification and characterization of bacteriophage T4 gene *59* protein: A DNA helicase assembly protein involved in DNA replication. *J Biol Chem* **269:** 33049–33062.

Bartsch S, Kang LE, Symington LS. 2000. RAD51 is required for the repair of plasmid double-stranded DNA gaps from either plasmid or chromosomal templates. *Mol Cell Biol* **20:** 1194–1205.

Bosco G, Haber JE. 1998. Chromosome break-induced DNA replication leads to nonreciprocal translocations and telomere capture. *Genetics* **150:** 1037–1047.

Burland V, Plunkett G III, Daniels DL, Blattner FR. 1993. DNA sequence and analysis of 136 kilobases of the *Escherichia coli* genome: Organizational symmetry around the origin of replication. *Genomics* **16:** 551–561.

Carles-Kinch K, George JW, Kreuzer KN. 1997. Bacteriophage T4 UvsW protein is a helicase involved in recombination, repair and the regulation of DNA replication origins. *EMBO J* **16:** 4142–4151.

Cesare AJ, Groff-Vindman C, Compton SA, McEachern MJ, Griffith JD. 2008. Telomere loops and homologous recombination-dependent telomeric circles in a *Kluyveromyces lactis* telomere mutant strain. *Mol Cell Biol* **28:** 20–29.

Clark AJ, Sandler SJ. 1994. Homologous genetic recombination: The pieces begin to fall into place. *Crit Rev Microbiol* **20:** 125–142.

Cohen H, Sinclair DA. 2001. Recombination-mediated lengthening of terminal telomeric repeats requires the Sgs1 DNA helicase. *Proc Natl Acad Sci* **98:** 3174–3179.

Conrad DF, Pinto D, Redon R, Feuk L, Gokcumen O, Zhang Y, Aerts J, Andrews TD, Barnes C, Campbell P, et al. 2010. Origins and functional impact of copy number variation in the human genome. *Nature* **464:** 704–712.

Cortes-Ledesma F, Tous C, Aguilera A. 2007. Different genetic requirements for repair of replication-born double-strand breaks by sister-chromatid recombination and break-induced replication. *Nucleic Acids Res* **35:** 6560–6570.

Cullen JK, Hussey SP, Walker C, Prudden J, Wee BY, Dave A, Findlay JS, Savory AP, Humphrey TC. 2007. Break-induced loss of heterozygosity in fission yeast: Dual roles for homologous recombination in promoting translocations and preventing de novo telomere addition. *Mol Cell Biol* **27:** 7745–7757.

Davis AP, Symington LS. 2004. RAD51-dependent break-induced replication in yeast. *Mol Cell Biol* **24:** 2344–2351.

Deem A, Keszthelyi A, Blackgrove T, Vayl A, Coffey B, Mathur R, Chabes A, Malkova A. 2011. Break-induced replication is highly inaccurate. *PLoS Biol* **9:** e1000594.

Dudas KC, Kreuzer KN. 2005. Bacteriophage T4 helicase loader protein gp59 functions as gatekeeper in origin-dependent replication in vivo. *J Biol Chem* **280:** 21561–21569.

Dunham MA, Neumann AA, Fasching CL, Reddel RR. 2000. Telomere maintenance by recombination in human cells. *Nat Genet* **26:** 447–450.

Dunn B, Szauter P, Pardue ML, Szostak JW. 1984. Transfer of yeast telomeres to linear plasmids by recombination. *Cell* **39:** 191–201.

Durkin SG, Ragland RL, Arlt MF, Mulle JG, Warren ST, Glover TW. 2008. Replication stress induces tumor-like microdeletions in FHIT/FRA3B. *Proc Natl Acad Sci* **105:** 246–251.

Engels WR. 2000. Reversal of fortune for *Drosophila* geneticists? *Science* **288:** 1973–1975.

Eykelenboom JK, Blackwood JK, Okely E, Leach DR. 2008. SbcCD causes a double-strand break at a DNA palindrome in the *Escherichia coli* chromosome. *Mol Cell* **29:** 644–651.

Formosa T, Alberts BM. 1986. DNA synthesis dependent on genetic recombination: Characterization of a reaction catalyzed by purified bacteriophage T4 proteins. *Cell* **47:** 793–806.

Gabbai CB, Marians KJ. 2010. Recruitment to stalled replication forks of the PriA DNA helicase and replisome-loading activities is essential for survival. *DNA Repair (Amst)* **9:** 202–209.

George JW, Kreuzer KN. 1996. Repair of double-strand breaks in bacteriophage T4 by a mechanism that involves extensive DNA replication. *Genetics* **143:** 1507–1520.

George JW, Stohr BA, Tomso DJ, Kreuzer KN. 2001. The tight linkage between DNA replication and double-strand break repair in bacteriophage T4. *Proc Natl Acad Sci* **98:** 8290–8297.

Gravel S, Chapman JR, Magill C, Jackson SP. 2008. DNA helicases Sgs1 and BLM promote DNA double-strand break resection. *Genes Dev* **22:** 2767–2772.

Groff-Vindman C, Cesare AJ, Natarajan S, Griffith JD, McEachern MJ. 2005. Recombination at long mutant telomeres produces tiny single- and double-stranded telomeric circles. *Mol Cell Biol* **25:** 4406–4412.

Grompone G, Sanchez N, Dusko Ehrlich S, Michel B. 2004. Requirement for RecFOR-mediated recombination in priA mutant. *Mol Microbiol* **52:** 551–562.

Haber JE. 2012. Mating-type genes and *MAT* switching in *Saccharomyces cerevisiae*. *Genetics* **191:** 33–64.

Hastings PJ, Ira G, Lupski JR. 2009. A microhomology-mediated break-induced replication model for the origin of human copy number variation. *PLoS Genet* **5:** e1000327.

Heller RC, Marians KJ. 2006. Replication fork reactivation downstream of a blocked nascent leading strand. *Nature* **439:** 557–562.

Henson JD, Neumann AA, Yeager TR, Reddel RR. 2002. Alternative lengthening of telomeres in mammalian cells. *Oncogene* **21:** 598–610.

Hicks WM, Kim M, Haber JE. 2010. Increased mutagenesis and unique mutation signature associated with mitotic gene conversion. *Science* **329:** 82–85.

Ho CK, Mazon G, Lam AF, Symington LS. 2010. Mus81 and Yen1 promote reciprocal exchange during mitotic recombination to maintain genome integrity in budding yeast. *Mol Cell* **40:** 988–1000.

Huang P, Pryde FE, Lester D, Maddison RL, Borts RH, Hickson ID, Louis EJ. 2001. SGS1 is required for telomere elongation in the absence of telomerase. *Curr Biol* **11:** 125–129.

Ira G, Haber JE. 2002. Characterization of RAD51-independent break-induced replication that acts preferentially with short homologous sequences. *Mol Cell Biol* **22:** 6384–6392.

Ira G, Satory D, Haber JE. 2006. Conservative inheritance of newly synthesized DNA in double-strand break-induced gene conversion. *Mol Cell Biol* **26:** 9424–9429.

Jaco I, Munoz P, Goytisolo F, Wesoly J, Bailey S, Taccioli G, Blasco MA. 2003. Role of mammalian Rad54 in telomere length maintenance. *Mol Cell Biol* **23:** 5572–5580.

Jain S, Sugawara N, Lydeard J, Vaze M, Tanguy Le Gac N, Haber JE. 2009. A recombination execution checkpoint regulates the choice of homologous recombination pathway during DNA double-strand break repair. *Genes Dev* **23:** 291–303.

Jiang WQ, Zhong ZH, Henson JD, Neumann AA, Chang AC, Reddel RR. 2005. Suppression of alternative lengthening of telomeres by Sp100-mediated sequestration of the MRE11/RAD50/NBS1 complex. *Mol Cell Biol* **25:** 2708–2721.

Johnson FB, Marciniak RA, McVey M, Stewart SA, Hahn WC, Guarente L. 2001. The *Saccharomyces cerevisiae* WRN homolog Sgs1p participates in telomere maintenance in cells lacking telomerase. *EMBO J* **20:** 905–913.

Kemper B, Brown DT. 1976. Function of gene *49* of bacteriophage T4: II. Analysis of intracellular development and the structure of very fast-sedimenting DNA. *J Virol* **18:** 1000–1015.

Kesti T, Flick K, Keranen S, Syvaoja JE, Wittenberg C. 1999. DNA polymerase epsilon catalytic domains are dispensable for DNA replication, DNA repair, and cell viability. *Mol Cell* **3:** 679–685.

Kogoma T. 1997. Stable DNA replication: Interplay between DNA replication, homologous recombination, and transcription. *Microbiol Mol Biol Rev* **61:** 212–238.

Kogoma T, Cadwell GW, Barnard KG, Asai T. 1996. The DNA replication priming protein, PriA, is required for homologous recombination and double-strand break repair. *J Bacteriol* **178:** 1258–1264.

Koszul R, Caburet S, Dujon B, Fischer G. 2004. Eucaryotic genome evolution through the spontaneous duplication of large chromosomal segments. *EMBO J* **23:** 234–243.

Kraus E, Leung WY, Haber JE. 2001. Break-induced replication: A review and an example in budding yeast. *Proc Natl Acad Sci* **98:** 8255–8262.

Kreuzer KN. 2000. Recombination-dependent DNA replication in phage T4. *Trends Biochem Sci* **25:** 165–173.

Kreuzer KN, Brister JR. 2010. Initiation of bacteriophage T4 DNA replication and replication fork dynamics: A review in the *Virology Journal* series on bacteriophage T4 and its relatives. *Virol J* **7:** 358.

Kreuzer KN, Saunders M, Weislo LJ, Kreuzer HW. 1995. Recombination-dependent DNA replication stimulated by double-strand breaks in bacteriophage T4. *J Bacteriol* **177:** 6844–6853.

Krogh BO, Symington LS. 2004. Recombination proteins in yeast. *Annu Rev Genet* **38:** 233–271.

Kuzminov A. 1995. Collapse and repair of replication forks in *Escherichia coli*. *Mol Microbiol* **16:** 373–384.

Kuzminov A, Stahl FW. 1999. Double-strand end repair via the RecBC pathway in *Escherichia coli* primes DNA replication. *Genes Dev* **13:** 345–356.

Lark KG, Lark CA. 1979. recA-dependent DNA replication in the absence of protein synthesis: Characteristics of a dominant lethal replication mutation, dnaT, and requirement for recA+ function. *Cold Spring Harb Symp Quant Biol* **43:** 537–549.

Le S, Moore JK, Haber JE, Greider CW. 1999. RAD50 and RAD51 define two pathways that collaborate to maintain telomeres in the absence of telomerase. *Genetics* **152:** 143–152.

Lederberg J. 1955. Genetic recombination in bacteria. *Science* **122:** 920.

Lee EH, Kornberg A. 1991. Replication deficiencies in *priA* mutants of *Escherichia coli* lacking the primosomal replication n′ protein. *Proc Natl Acad Sci* **88:** 3029–3032.

Liu J, Morrical SW. 2010. Assembly and dynamics of the bacteriophage T4 homologous recombination machinery. *Virol J* **7:** 357.

Liu J, Xu L, Sandler SJ, Marians KJ. 1999. Replication fork assembly at recombination intermediates is required for bacterial growth. *Proc Natl Acad Sci* **96:** 3552–3555.

Londono-Vallejo JA, Der-Sarkissian H, Cazes L, Bacchetti S, Reddel RR. 2004. Alternative lengthening of telomeres is characterized by high rates of telomeric exchange. *Cancer Res* **64:** 2324–2327.

Louis EJ, Naumova ES, Lee A, Naumov G, Haber JE. 1994. The chromosome end in yeast: Its mosaic nature and influence on recombinational dynamics. *Genetics* **136:** 789–802.

Luder A, Mosig G. 1982. Two alternative mechanisms for initiation of DNA replication forks in bacteriophage T4: Priming by RNA polymerase and by recombination. *Proc Natl Acad Sci* **79:** 1101–1105.

Lundblad V, Blackburn EH. 1993. An alternative pathway for yeast telomere maintenance rescues est1 senescence. *Cell* **73:** 347–360.

Lydeard JR, Jain S, Yamaguchi M, Haber JE. 2007. Break-induced replication and telomerase-independent telomere maintenance require Pol32. *Nature* **448:** 820–823.

Lydeard JR, Lipkin-Moore Z, Jain S, Eapen VV, Haber JE. 2010a. Sgs1 and Exo1 redundantly inhibit break-induced replication and de novo telomere addition at broken chromosome ends. *PLoS Genet* **6:** e1000973.

Lydeard JR, Lipkin-Moore Z, Sheu YJ, Stillman B, Burgers PM, Haber JE. 2010b. Break-induced replication requires all essential DNA replication factors except those specific for pre-RC assembly. *Genes Dev* **24:** 1133–1144.

Magee TR, Kogoma T. 1990. Requirement of RecBC enzyme and an elevated level of activated RecA for induced stable DNA replication in *Escherichia coli*. *J Bacteriol* **172:** 1834–1839.

Malkova A, Naylor ML, Yamaguchi M, Ira G, Haber JE. 2005. RAD51-dependent break-induced replication differs in kinetics and checkpoint responses from RAD51-mediated gene conversion. *Mol Cell Biol* **25:** 933–944.

Marrero VA, Symington LS. 2010. Extensive DNA end processing by Exo1 and Sgs1 inhibits break-induced replication. *PLoS Genet* **6:** e1001007.

Masai H, Asai T, Kubota Y, Arai K, Kogoma T. 1994. *Escherichia coli* PriA protein is essential for inducible and constitutive stable DNA replication. *EMBO J* **13:** 5338–5345.

McEachern MJ, Blackburn EH. 1995. Runaway telomere elongation caused by telomerase RNA gene mutations. *Nature* **376:** 403–409.

McEachern MJ, Blackburn EH. 1996. Cap-prevented recombination between terminal telomeric repeat arrays (telomere CPR) maintains telomeres in *Kluyveromyces lactis* lacking telomerase. *Genes Dev* **10:** 1822–1834.

McEachern MJ, Haber JE. 2006. Break-induced replication and recombinational telomere elongation in yeast. *Annu Rev Biochem* **75:** 111–135.

McEachern MJ, Iyer S. 2001. Short telomeres in yeast are highly recombinogenic. *Mol Cell* **7:** 695–704.

Meselson M, Weigle JJ. 1961. Chromosome breakage accompanying genetic recombination in bacteriophage. *Proc Natl Acad Sci* **47:** 857–868.

Mimitou EP, Symington LS. 2008. Sae2, Exo1 and Sgs1 collaborate in DNA double-strand break processing. *Nature* **455:** 770–774.

Mizuuchi K, Kemper B, Hays J, Weisberg RA. 1982. T4 endonuclease VII cleaves Holliday structures. *Cell* **29:** 357–365.

Morrow DM, Connelly C, Hieter P. 1997. "Break copy" duplication: A model for chromosome fragment formation in *Saccharomyces cerevisiae*. *Genetics* **147:** 371–382.

Mosig G. 1987. The essential role of recombination in phage T4 growth. *Annu Rev Genet* **21:** 347–371.

Mosig G. 1998. Recombination and recombination-dependent DNA replication in bacteriophage T4. *Annu Rev Genet* **32:** 379–413.

Motamedi MR, Szigety SK, Rosenberg SM. 1999. Double-strand-break repair recombination in *Escherichia coli*: Physical evidence for a DNA replication mechanism in vivo. *Genes Dev* **13:** 2889–2903.

Natarajan S, McEachern MJ. 2002. Recombinational telomere elongation promoted by DNA circles. *Mol Cell Biol* **22:** 4512–4521.

Natarajan S, Groff-Vindman C, McEachern MJ. 2003. Factors influencing the recombinational expansion and spread of telomeric tandem arrays in *Kluyveromyces lactis*. *Eukaryot Cell* **2:** 1115–1127.

Neale MJ, Pan J, Keeney S. 2005. Endonucleolytic processing of covalent protein-linked DNA double-strand breaks. *Nature* **436:** 1053–1057.

Nielsen I, Bentsen IB, Lisby M, Hansen S, Mundbjerg K, Andersen AH, Bjergbaek L. 2009. A Flp-nick system to study repair of a single protein-bound nick in vivo. *Nat Methods* **6:** 753–757.

Nossal NG, Dudas KC, Kreuzer KN. 2001. Bacteriophage T4 proteins replicate plasmids with a preformed R loop at the T4 ori(uvsY) replication origin in vitro. *Mol Cell* **7:** 31–41.

Nurse P, Zavitz KH, Marians KJ. 1991. Inactivation of the *Escherichia coli* priA DNA replication protein induces the SOS response. *J Bacteriol* **173:** 6686–6693.

Pâques F, Haber JE. 1999. Multiple pathways of recombination induced by double-strand breaks in *Saccharomyces cerevisiae*. *Microbiol Mol Biol Rev* **63:** 349–404.

Payen C, Koszul R, Dujon B, Fischer G. 2008. Segmental duplications arise from Pol32-dependent repair of broken forks through two alternative replication-based mechanisms. *PLoS Genet* **4:** e1000175.

Persky NS, Lovett ST. 2008. Mechanisms of recombination: Lessons from *E. coli*. *Crit Rev Biochem Mol Biol* **43:** 347–370.

Reddel RR. 2003. Alternative lengthening of telomeres, telomerase, and cancer. *Cancer Lett* **194:** 155–162.

Redon R, Ishikawa S, Fitch KR, Feuk L, Perry GH, Andrews TD, Fiegler H, Shapero MH, Carson AR, Chen W, et al. 2006. Global variation in copy number in the human genome. *Nature* **444:** 444–454.

Rong YS, Golic KG. 2000. Gene targeting by homologous recombination in *Drosophila*. *Science* **288:** 2013–2018.

Saharia A, Stewart SA. 2009. FEN1 contributes to telomere stability in ALT-positive tumor cells. *Oncogene* **28:** 1162–1167.

Scheel C, Poremba C. 2002. Telomere lengthening in telomerase-negative cells: The ends are coming together. *Virchows Arch* **440:** 573–582.

Sharples GJ, Leach DR. 1995. Structural and functional similarities between the SbcCD proteins of *Escherichia coli* and the RAD50 and MRE11 (RAD32) recombination and repair proteins of yeast. *Mol Microbiol* **17:** 1215–1217.

Shlomai J, Kornberg A. 1980a. A prepriming DNA replication enzyme of *Escherichia coli*. I. Purification of protein n′: A sequence-specific, DNA-dependent ATPase. *J Biol Chem* **255:** 6789–6793.

Shlomai J, Kornberg A. 1980b. A prepriming DNA replication enzyme of *Escherichia coli*. II. Actions of protein n′: A sequence-specific, DNA-dependent ATPase. *J Biol Chem* **255:** 6794–6798.

Signon L, Malkova A, Naylor ML, Klein H, Haber JE. 2001. Genetic requirements for RAD51- and RAD54-independent break-induced replication repair of a chromosomal double-strand break. *Mol Cell Biol* **21:** 2048–2056.

Skalka AM. 1974. A replicator's view of recombination (and repair). In *Mechanism and recombination* (ed. Grell RF), pp. 421–432. Plenum, New York.

Smith GR. 1991. Conjugational recombination in *E. coli*: Myths and mechanisms. *Cell* **64:** 19–27.

Smith CE, Llorente B, Symington LS. 2007. Template switching during break-induced replication. *Nature* **447:** 102–105.

Stephens PJ, Greenman CD, Fu B, Yang F, Bignell GR, Mudie LJ, Pleasance ED, Lau KW, Beare D, Stebbings LA, et al. 2011. Massive genomic rearrangement acquired in a single catastrophic event during cancer development. *Cell* **144:** 27–40.

Stohr BA, Kreuzer KN. 2002. Coordination of DNA ends during double-strand-break repair in bacteriophage T4. *Genetics* **162:** 1019–1030.

Symington LS, Gautier J. 2011. Double-strand break end resection and repair pathway choice. *Annu Rev Genet* **45:** 247–271.

Tanaka T, Taniyama C, Arai K, Masai H. 2003. ATPase/helicase motif mutants of *Escherichia coli* PriA protein essential for recombination-dependent DNA replication. *Genes Cells* **8:** 251–261.

Tarsounas M, Munoz P, Claas A, Smiraldo PG, Pittman DL, Blasco MA, West SC. 2004. Telomere maintenance requires the RAD51D recombination/repair protein. *Cell* **117:** 337–347.

Tawil R, Van Der Maarel SM. 2006. Facioscapulohumeral muscular dystrophy. *Muscle Nerve* **34:** 1–15.

Teng SC, Zakian VA. 1999. Telomere-telomere recombination is an efficient bypass pathway for telomere maintenance in *Saccharomyces cerevisiae*. *Mol Cell Biol* **19:** 8083–8093.

Teng SC, Chang J, McCowan B, Zakian VA. 2000. Telomerase-independent lengthening of yeast telomeres occurs by an abrupt Rad50p-dependent, Rif-inhibited recombinational process. *Mol Cell* **6:** 947–952.

Tsai YL, Tseng SF, Chang SH, Lin CC, Teng SC. 2002. Involvement of replicative polymerases, Tel1p, Mec1p, Cdc13p, and the Ku complex in telomere-telomere recombination. *Mol Cell Biol* **22:** 5679–5687.

Wang TC, Smith KC. 1983. Mechanisms for recF-dependent and recB-dependent pathways of postreplication repair in UV-irradiated *Escherichia coli* uvrB. *J Bacteriol* **156:** 1093–1098.

Wu R, Yeh YC. 1974. DNA arrested mutants of gene *59* of bacteriophage T4. II. Replicative intermediates. *Virology* **59:** 108–122.

Xi J, Zhang Z, Zhuang Z, Yang J, Spiering MM, Hammes GG, Benkovic SJ. 2005. Interaction between the T4 helicase loading protein (gp59) and the DNA polymerase (gp43): Unlocking of the gp59-gp43-DNA complex to initiate assembly of a fully functional replisome. *Biochemistry* **44:** 7747–7756.

Xu L, Marians KJ. 2000. Purification and characterization of DnaC810, a primosomal protein capable of bypassing PriA function. *J Biol Chem* **275:** 8196–8205.

Xu L, Marians KJ. 2003. PriA mediates DNA replication pathway choice at recombination intermediates. *Mol Cell* **11:** 817–826.

Zavitz KH, Marians KJ. 1992. ATPase-deficient mutants of the *Escherichia coli* DNA replication protein PriA are capable of catalyzing the assembly of active primosomes. *J Biol Chem* **267:** 6933–6940.

Zeng S, Xiang T, Pandita TK, Gonzalez-Suarez I, Gonzalo S, Harris CC, Yang Q. 2009. Telomere recombination requires the MUS81 endonuclease. *Nat Cell Biol* **11:** 616–623.

Zhang F, Gu W, Hurles ME, Lupski JR. 2009. Copy number variation in human health, disease, and evolution. *Annu Rev Genomics Hum Genet* **10:** 451–481.

Zhu Z, Chung WH, Shim EY, Lee SE, Ira G. 2008. Sgs1 helicase and two nucleases Dna2 and Exo1 resect DNA double-strand break ends. *Cell* **134:** 981–994.

Helicase Loading at Chromosomal Origins of Replication

Stephen P. Bell[1] and Jon M. Kaguni[2]

[1]Howard Hughes Medical Institute, Massachusetts Institute of Technology, Cambridge, Massachusetts 02139

[2]Department of Biochemistry and Molecular Biology, Michigan State University, East Lansing, Michigan 48824

Correspondence: kaguni@msu.edu

Loading of the replicative DNA helicase at origins of replication is of central importance in DNA replication. As the first of the replication fork proteins assemble at chromosomal origins of replication, the loaded helicase is required for the recruitment of the rest of the replication machinery. In this work, we review the current knowledge of helicase loading at *Escherichia coli* and eukaryotic origins of replication. In each case, this process requires both an origin recognition protein as well as one or more additional proteins. Comparison of these events shows intriguing similarities that suggest a similar underlying mechanism, as well as critical differences that likely reflect the distinct processes that regulate helicase loading in bacterial and eukaryotic cells.

Replicative DNA helicases are ring-shaped molecules with a central cavity through which DNA passes as they unwind DNA. Their loading at replication origins is a critical and highly regulated event in chromosomal replication. The DNA helicase is the first of the replication fork proteins recruited to and loaded onto origins of replication, and the loaded helicase is required for the recruitment of the rest of the replication machinery (Remus and Diffley 2009; Kaguni 2011). Indeed, the replicative DNA helicase links the replication machinery to the parental DNA (O'Donnell 2006). In *Escherichia coli* cells, the DnaB replicative helicase binds to primase (DnaG) and the sliding clamp loader, which in turn binds the DNA polymerases. Although the polymerases are also linked to the template DNA by sliding clamps, when these interactions are broken, the polymerases' association with the sliding clamp loader and the helicase keeps them at the site of replication. The interactions that tether the DNA polymerases to the eukaryotic replication fork are less clear but very likely involve direct and indirect interactions with the Mcm2–7 replicative helicase (Calzada et al. 2005).

Helicase loading is carefully regulated to control the location and frequency of replication initiation. In eukaryotic cells, helicase loading is tightly restricted to the G_1 phase of the cell cycle. This constraint is a key part of the mechanisms that ensure that no origin can initiate more than once per cell cycle (Siddiqui et al. 2013; Zielke et al. 2013). In addition, the sites

Cite this article as *Cold Spring Harb Perspect Biol* doi: 10.1101/cshperspect.a010124

of eukaryotic replicative helicase loading define the potential sites of replication initiation in the cell (but not all loaded helicases are used during a given S phase [Rhind and Gilbert 2012]). Although the central regulated event in bacterial chromosome duplication is the recruitment of the ATP-bound initiator protein DnaA (Skarstad and Katayama 2013), the loading of the replicative helicase represents a key committed step during initiation.

Here we will discuss the mechanism of helicase loading in bacteria and eukaryotic cells. Much of the discussion will focus on studies in the bacterium *E. coli*, the yeast *Saccharomyces cerevisiae*, and the frog *Xenopus laevis*, in which the events of helicase loading are best understood. Comparison of these mechanisms shows important similarities and differences between the domains of life. In both bacteria and eukaryotic cells, multiple AAA^+ proteins use ATP binding and hydrolysis to direct helicase loading and both helicases are initially loaded in an inactive form. On the other hand, the eukaryotic helicase is loaded around double-stranded DNA (dsDNA) and as a double hexamer, whereas the bacterial helicase is loaded around single-stranded DNA (ssDNA) as a single hexamer. These distinctions are very likely due to the very different regulatory mechanisms of DNA replication in bacteria and eukaryotic cells.

Before describing the process of helicase loading, it is relevant to know that the essential function of replicative helicases is to unwind the parental duplex DNA using the energy provided by the hydrolysis of a nucleoside triphosphate (Table 1) (reviewed in Patel et al. 2011). Replicative DNA polymerases then copy each parental DNA strand to duplicate the genome. That replicative DNA helicases are hexameric (and sometimes heptameric) structures that have a positively charged, central channel provides a framework for how these molecular machines work. Several models that describe the mechanism of unwinding have been considered. One is the steric or strand-exclusion model, in which the helicase excludes one strand of DNA while the other passes through the central cavity as the enzyme moves. Because these enzymes can bind and translocate on duplex DNA without unwinding (Kaplan et al. 2003), two additional models have been proposed. In the ploughshare model, duplex DNA enters the central cavity and exits as unwound DNA by virtue of a domain/protein that acts as a ploughshare or pin, which disrupts the hydrogen bonds of DNA as it is pumped through the enzyme (reviewed in Takahashi et al. 2005). A second model, the DNA-pumping model, was proposed on the basis of the double-hexameric forms of SV40, Mcm2–7, and other replicative helicases (Mastrangelo et al. 1989; Remus et al. 2009). This model proposes that the two helicases pump duplex DNA toward one another, resulting in torsional strain that forces the two strands apart, at which point they exit the central channel as two ssDNA loops. Current evidence supports the steric exclusion model (Jezewska et al. 1998b; Kaplan 2000; Galletto et al. 2004; Fu et al. 2011). Of interest, the *E. coli* enzyme moves in the 5′-to-3′ direction relative to the engaged ssDNA that passes through the central cavity, whereas archaeal and eukaryotic enzymes move in the 3′-to-5′ direction.

HELICASE LOADING IN *E. coli*

The *E. coli* replication origin (*oriC*) has two essential functions. One is to serve as a site where

Table 1. Replicative DNA helicases of free-living organisms are hexameric

Domain	Model organism(s)	Helicase[a]	Direction of movement
Bacteria	*Escherichia coli*	DnaB	5′ → 3′
Archaea	*Sulfolobus solfataricus*	MCM	3′ → 5′
Eukarya	*Saccharomyces cerevisiae* *Drosophila melanogaster* *Xenopus laevis*	Mcm2–7	3′ → 5′

[a]DnaB and *S. solfataricus* MCM are homohexamers, whereas eukaryotic Mcm2–7 is composed of six nonidentical subunits.

the replication fork machinery assembles. DNA sequence motifs in this chromosomal locus and their roles are described in detail elsewhere (Leonard and Méchali 2013). The second is to act as a site where the process of DNA replication is controlled. As described in Skarstad and Katayama (2013), several separate mechanisms modulate the frequency of initiation of DNA replication. At this stage, a critical event is the loading of the replicative helicase named DnaB (Bell and Botchan 2013), which must be in complex with its partner, DnaC. A summary of the functions of DnaA, DnaB, and DnaC serves as the foundation for a description of the helicase loading process.

DnaA IS THE REPLICATION INITIATOR IN BACTERIA

E. coli DnaA performs a central role in the initiation of DNA replication at *oriC*, recognizing specific DNA sequences named the DnaA box, I-site, τ-site, and C-site within *oriC* (Leonard and Méchali 2013). Studies reveal that this protein, which is highly conserved among bacteria, can be divided into four domains designated 1 to 4 from amino to carboxyl terminus (http://www.molgen.mpg.de/~messer/) (reviewed in Mott and Berger 2007; Ozaki and Katayama 2009; Leonard and Grimwade 2011). The interactions of these domains with specific proteins, DNA sequences, and nucleotides are summarized below.

The amino-terminal domain 1 (amino acids 1–90 of *E. coli* DnaA) interacts with a variety of proteins, including DnaB (Sutton et al. 1998), and DnaA itself in the assembly of the DnaA oligomer at *oriC* (Simmons et al. 2003; Felczak and Kaguni 2004; Abe et al. 2007). These interactions are critical for helicase loading and are described in more detail below. Other proteins that interact with domain 1 are HU (Chodavarapu et al. 2008a), DiaA (Keyamura et al. 2007, 2009), Dps (Chodavarapu et al. 2008b), ribosomal protein L2 (Chodavarapu et al. 2011), and RNA polymerase (Flåtten et al. 2009). The interaction with RNA polymerase correlates with the role of DnaA as a transcriptional activator in stabilizing the binding of RNA polymerase at the *gidA* and λP_R promoters (Szalewska-Pałasz et al. 1998; Flåtten et al. 2009), and possibly at the *glpD* and *fliC* promoters (Mizushima et al. 1994; Messer and Weigel 1997). With HU or DiaA, their respective binding stabilizes DnaA oligomerized at *oriC* to stimulate initiation in vitro (Keyamura et al. 2007, 2009; Chodavarapu et al. 2008a). By comparison, L2 inhibits initiation by impairing the formation of a DnaA oligomer at *oriC* (Chodavarapu et al. 2011). Like L2, Dps bound to domain 1 interferes with strand opening of *oriC* in vitro, which correlates with less frequent initiation in vivo when the Dps level is elevated (Chodavarapu et al. 2008b). These observations suggest that domain 1 acts as a sensor, responding to proteins that modulate the initiation process by affecting the assembly of DnaA at *oriC*.

Among DnaAs of bacteria (http://www.molgen.mpg.de/~messer/), domain 2 (residues 90–130 of *E. coli* DnaA) is not conserved in amino acid sequence or length. For *E. coli* DnaA, the removal of residues 96–120 in domain 2 leads to reduced activity in DNA replication, but deletion of other consecutive sequences from this region has no apparent effect on activity (Nozaki and Ogawa 2008; Molt et al. 2009). These observations suggest that domain 2 acts as a flexible link between domains 1 and 3.

The third domain of DnaA (residues 130–347 of *E. coli* DnaA) is responsible for ATP binding and hydrolysis (Duderstadt and Berger 2008). Like other members of the AAA^+ family of ATPases, this region of DnaA carries the Walker A (P-loop) and B boxes that are involved in ATP binding. The sensor I, II (box VII), and box VIII motifs are believed to coordinate ATP hydrolysis with a change in conformation (reviewed in Erzberger and Berger 2006). The adenine nucleotide bound to a shortened variant of *Aquifex aeolicus* DnaA lacking domains 1 and 2 has a striking effect on its X-ray crystal structure (Erzberger et al. 2002, 2006). With ADP, six monomers of this DnaA assemble as a closed ring. With ATP instead, an open ring with a right-handed pitch forms by virtue of a distended conformation of domain 3.

Acidic phospholipids in a fluid bilayer bind to a segment spanning the junction between domains 3 and 4 (reviewed in Boeneman and Crooke 2005). Intriguingly, this interaction promotes nucleotide release from DnaA (Sekimizu and Kornberg 1988). As the concentration of ATP is higher than that of ADP in vivo, this effect may favor the ATP-bound state of DnaA and is the basis for speculation that this interaction may regulate the activity of DnaA. Although fluorescence microscopy indicates that DnaA associates with the inner membrane in vivo (Boeneman et al. 2009) as well as with the *oriC* region (Nozaki et al. 2009), it is not known if interactions with the inner membrane promote nucleotide exchange in living cells.

Domain 4 (residues 347–467) binds to the DnaA box. Nuclear magnetic resonance and X-ray crystallographic analysis of this domain reveal a helix-turn-helix amino acid motif and a basic loop (Erzberger et al. 2002; Fujikawa et al. 2003). Combined with the biochemical characterization of mutant DnaAs bearing specific amino acid substitutions that impair the recognition of the DnaA box, we have an atomic understanding of how DnaA binds to the DnaA box (Sutton and Kaguni 1997; Blaesing et al. 2000).

DnaB IS THE REPLICATIVE HELICASE, AND INTERACTS WITH DnaC TO FORM THE DnaB–DnaC COMPLEX

In *E. coli*, DnaB is the replicative helicase whose native structure is a toroid of six identical subunits oriented in the same direction (Reha-Krantz and Hurwitz 1978; Arai et al. 1981; Donate et al. 2000; Bailey et al. 2007). Whereas Mg^{2+} ion stabilizes this native structure, removal of the metal ion by dialysis or chelation causes the DnaB hexamer to dissociate into trimers and monomers (Bujalowski et al. 1994). X-ray crystallography of the homologous proteins from *Geobacillus kaustophilus* and *Geobacillus stearothermophilus* provides a detailed view of this helicase and the relative arrangement of its smaller amino-terminal and larger carboxy-terminal domains (Bailey et al. 2007; Lo et al. 2009).

Studies of *E. coli* DnaB and others in the DnaB helicase family indicate that the enzyme unwinds DNA by translocating in the 5′-to-3′ direction on the ssDNA to which it is bound (Venkatesan et al. 1982; Matson et al. 1983; LeBowitz and McMacken 1986; Lee et al. 1989; Richardson and Nossal 1989). Fluorescent energy transfer experiments suggest that 20 ± 3 nucleotides of ssDNA pass through the central cavity of DnaB during translocation (Jezewska et al. 1996; Kaplan 2000). DnaB is uniquely oriented on a forked DNA molecule with a 5′ ssDNA tail (Jezewska et al. 1998a,b). On the basis of this orientation, the smaller amino-terminal domain is upstream of the carboxy-terminal domain, which is nearer the apex of the replication fork (Fig. 1, step 3).

To participate in helicase loading at *oriC*, DnaB must be in a complex with its partner, DnaC. Early studies suggested that ATP bound to DnaC is necessary for DnaC to form an isolable complex with DnaB and to protect DnaC from inactivation by *N*-ethylmaleimide (Wickner and Hurwitz 1975; Kobori and Kornberg 1982). More recent studies indicate that ATP is not needed for DnaC to interact with DnaB (Davey et al. 2002; Galletto et al. 2003; Mott et al. 2008). Both cryoelectron microscopy of the DnaB–DnaC complex in comparison with DnaB and fluorescence energy transfer experiments indicate that DnaC is bound to the larger (carboxy-terminal) end of the DnaB toroid (Bárcena et al. 2001; Galletto et al. 2003). Relative to the ensuing direction of helicase movement, DnaC in the DnaB–DnaC complex is positioned proximal to the replication fork junction (Fig. 1).

Because amino acid substitutions within residues 10–44 near the amino terminus of DnaC impair the interaction between DnaC and DnaB, this region of DnaC apparently binds directly to DnaB to form the DnaB–DnaC complex (Ludlam et al. 2001). Initial studies showed that six DnaC monomers bind per DnaB hexamer (Wickner and Hurwitz 1975; Kobori and Kornberg 1982; Lanka and Schuster 1983). Subsequent sedimentation velocity analysis to determine the affinity of DnaC for DnaB in combination with estimates of the in vivo levels of DnaC and DnaB suggest that the cellular DnaB–DnaC complex is heterogeneous in composition, with a fraction containing less than the

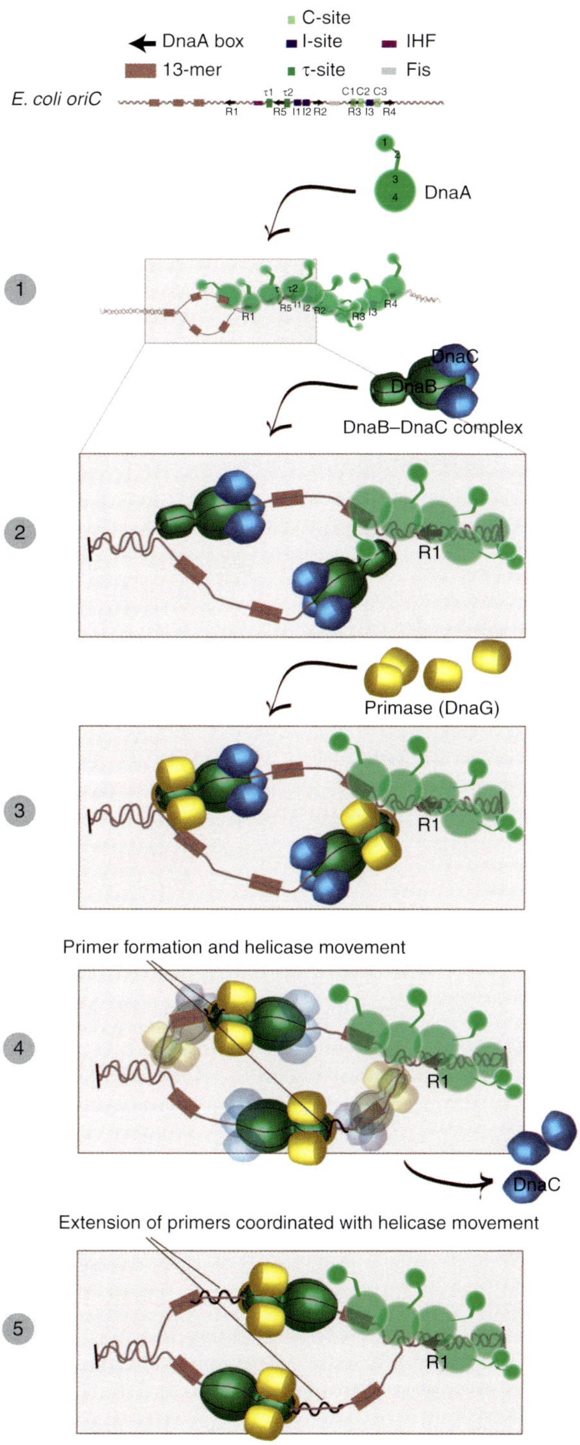

Figure 1. Helicase loading at the *E. coli* replication origin (*oriC*) is a stepwise process. Near the *top* of the figure, DNA sequence elements in *E. coli oriC* are shown (see Leonard and Méchali 2013 for details). Binding sites for Fis and IHF are shown, which are believed to alter the architecture of *oriC*, and their roles are described in detail in Leonard and Méchali (2013). The four domains of DnaA shown in green are represented by the overlaid numbers. In step 1, DnaA complexed to ATP binds to the DnaA boxes, I-, τ-, and C-sites to form a DnaA oligomer. Following the unwinding of the region of *oriC* containing the 13-mers, DnaA then loads a DnaB–DnaC complex on each of the separated strands (step 2). In step 3, primase (DnaG) interacts with the amino-terminal region of DnaB. Primer formation by primase (steps 4 and 5) leads to the dissociation of DnaC from the carboxy-terminal domain of DnaB, which is necessary to activate DnaB as a DNA helicase (see Bell and Botchan 2013). Although not shown, the primers displayed in black are then extended by the cellular replicase, DNA polymerase III holoenzyme, for the continuous synthesis of each leading strand at each replication fork (see Johansson and Dixon 2013). A dimer of this DNA polymerase is believed to be at each replication fork. DnaB is proposed to interact with one unit of a dimer of DNA polymerase III holoenzyme, coordinating the unwinding of the parental DNA with DNA synthesis. As this DNA helicase translocates in the 5′-to-3′ direction to support replication fork movement, primase occasionally interacts with DnaB to synthesize additional primers. These primers are used by the unit of the DNA polymerase dimer that synthesizes Okazaki fragments.

maximum of six DnaC monomers per DnaB hexamer (Galletto et al. 2003).

DnaC CONTROLS THE ACTIVITY OF DnaB

DnaC, like DnaA, is a member of the AAA$^+$ family of ATPases (Makarova and Koonin 2013). Despite the presence in DnaC of motifs shared by AAA$^+$ family members and the high affinity of other AAA$^+$ proteins for ATP, DnaC binds weakly to this nucleotide; the K_d is about 8 μM (Davey et al. 2002; Galletto et al. 2003; Biswas et al. 2004). By itself, DnaC is also a feeble ATPase. The influence of ATP on DnaC function has been an enigma, but new evidence suggests that its ATPase activity is activated at a specific step of the initiation process (see below). Stimulated by ATP, DnaC also interacts weakly with ssDNA (Learn et al. 1997; Davey et al. 2002; Mott et al. 2008), supporting the idea that this activity may be involved in helicase loading. Because DnaC complexed to DnaB inhibits its ATPase and helicase activities (Wahle et al. 1989a,b; Allen and Kornberg 1991; Davey et al. 2002; Mott et al. 2008), DnaC must dissociate from DnaB for helicase activation to occur.

DnaA UNWINDS A REGION NEAR THE LEFT BORDER OF *oriC*

With a negatively supercoiled plasmid carrying *oriC*, DnaA bound to ATP unwinds an AT-rich region containing the M and R 13-mers as judged by potassium permanganate sensitivity (Fig. 1) (Sekimizu et al. 1987; Bramhill and Kornberg 1988; Gille and Messer 1991). Considering that isolated plasmids have a negative superhelix density ranging from −0.05 to −0.08, as much as 330 to 530 nucleotides can be single-stranded if these DNAs are supercoiled solely as a result of unwinding of the duplex DNA instead of the writhing of the DNA upon itself (Cozzarelli 1980; Bliska and Cozzarelli 1987). These observations suggest that DnaA confines some of the single-stranded character of the plasmid to this AT-rich region of *oriC*. To provide a biochemical mechanism of unwinding, recent studies on the assembly of a DnaA oligomer as a right-handed filament suggest that positively charged and hydrophobic amino acids that line the interior of the DnaA filament interact with the negatively charged ssDNA (Erzberger et al. 2006; Ozaki et al. 2008; Duderstadt et al. 2010, 2011; Ozaki and Katayama 2011). Consistent with the ATP requirement for DnaA-dependent origin unwinding, the formation of this DnaA filament requires the ATP-bound form of the enzyme.

HELICASE LOADING AT *oriC* BY DnaA

After unwinding of *oriC*, DnaA loads the DnaB–DnaC complex onto this unwound DNA, which apparently becomes enlarged to include the left 13-mer and additional sequences beyond it (Fig. 1, step 2) (Davey et al. 2002). Footprinting studies suggest that the DnaB–DnaC complex protects a region on the top strand near the left border of *oriC*. On the bottom strand the area protected is to the left of DnaA box R1 that overlaps the right 13-mer. Quantitative analysis of this protein complex assembled at *oriC* supports the conclusion from footprinting studies that two DnaB–DnaC complexes are bound (Fang et al. 1999; Carr and Kaguni 2001).

Helicase loading appears to involve two regions of DnaA. One region is within domain 3, based on studies of a monoclonal antibody that interferes with the interaction between DnaA and DnaB measured in solid-phase binding assays (Marszalek and Kaguni 1994; Sutton et al. 1998). This antibody recognizes an epitope within residues 111–148 of DnaA. Deletion analysis localized the interacting region to amino acids 135–148 (Seitz et al. 2000). A region within domain 1 is also needed because an alanine substitution for phenylalanine 46 abrogates this interaction (Keyamura et al. 2009). On the basis that DnaC from *A. aeolicus* can interact with DnaA, and like DnaA forms a helical filament as determined by X-ray crystallography, a specific proposal is that DnaA oligomerized at *oriC* loads one DnaB–DnaC complex on the top strand via the interaction between DnaA molecules at the left end of the DnaA filament and DnaC in the DnaB–DnaC complex (Mott et al. 2008). The second DnaB–DnaC complex loads on the bottom strand through the interaction between the left end of the DnaA oligomer and DnaB complexed to DnaC. During loading, it is interesting to consider that one of the interfaces between DnaB protomers must open in this ring-shaped protein in order for the helicase to encircle the ssDNA. The role of DnaA or DnaC in this ring opening remains unclear.

After DnaA loads one DnaB–DnaC complex on each of the separated strands of *oriC*, DnaC must dissociate for DnaB to be active as a helicase. Of interest, recent studies suggest that three DnaC monomers are bound to each DnaB hexamer (Makowska-Grzyska and Kaguni 2010), whereas earlier experiments suggested that the DnaB–DnaC complex contains six DnaC monomers per DnaB (Wickner and Hurwitz 1975; Kobori and Kornberg 1982; Lanka and Schuster 1983; Wahle et al. 1989a). Perhaps the different forms of the DnaB–DnaC complex function at the initiation stage of DNA replication and in restarting stalled or collapsed replication forks.

ATP or ATPγS bound to DnaC supports the DnaA-dependent loading of the DnaB–DnaC complex at *oriC* (Davey et al. 2002; Makowska-Grzyska and Kaguni 2010). That DnaC remains bound to DnaB in the presence of either ATP or the analog suggests that the act of helicase loading does not stimulate the hydrolysis of ATP bound to DnaC or its release from DnaB (Makowska-Grzyska and Kaguni 2010). In the presence of ATP and other ribonucleotides, the inclusion of primase (DnaG), which interacts with the amino-terminal domain of DnaB as it forms primers for DNA replication, induces the release of DnaC from the carboxy-terminal (large) domain of DnaB (Fig. 1, steps 3–5). Apparently the interaction of primase with DnaB while it synthesizes a primer alters the conformation of DnaB that causes DnaC to dissociate. Other results with mutant forms of DnaC that are speculated to be defective in ATP hydrolysis suggest that this interaction of primase with DnaB stimulates the hydrolysis of ATP bound by DnaC. This step of helicase activation is comparable in function to the activation of the Mcm2–7 complex at a eukaryotic replication origin (Labib 2010).

HELICASE LOADING IN EUKARYOTIC CELLS

Origin Licensing, Pre-RC Formation, and Helicase Loading

Early work on eukaryotic DNA replication revealed that the events of replication initiation were separated into separate cell-cycle stages. Mammalian cell fusion studies suggested the existence of positive factors in S-phase cells that cause G_1 nuclei to initiate DNA replication (Johnson and Rao 1970). Analyses of cycling *Xenopus* egg extracts suggested the existence of a "licensing factor" that gained access to the DNA only after nuclear envelope breakdown (during mitosis) and was consumed during S phase (Blow and Laskey 1988). Finally, in vivo footprinting studies at *S. cerevisiae* origins showed a distinct and more extensive protection pattern during G_1 (called the prereplicative complex or pre-RC) than in S, G_2, and M phase (Diffley et al. 1994).

We now know that replication origin licensing and pre-RC formation are both related to the helicase loading event that only occurs in G_1 cells (Bell and Dutta 2002; Sclafani and Holzen 2007). The licensing factor that is excluded from the nucleus is probably not identical among different cell types (Arias and Walter 2007), but origin licensing is clearly equivalent to helicase loading. The pre-RC footprint that is associated with *S. cerevisiae* origins during G_1 is due to binding of helicase loading factors to the origin DNA (Perkins and Diffley 1998; Speck et al. 2005). Much of this footprint is attributed to origin recognition complex (ORC) and Cdc6, but it additionally reflects helicase loading. For this reason the term "pre-RC formation" has become synonymous with helicase loading. We will refer to the G_1 events at origins as helicase loading and the events that occur in S phase as helicase activation and replisome assembly because these terms most accurately reflect the events that are occurring at the origin.

Mcm2–7 HELICASE

The eukaryotic replicative DNA helicase is the Mcm2–7 complex. Each Mcm2–7 complex contains one copy of the six essential and related

Mcm2, Mcm3, Mcm4, Mcm5, Mcm6, and Mcm7 proteins (reviewed in Bochman and Schwacha 2009). Although high-resolution structural data are not available, electron microscopy (EM) and subunit interaction studies indicate that the Mcm2–7 proteins interact in a defined order (Davey et al. 2003; Yu et al. 2004; Bochman et al. 2008) to form either a ring (Remus et al. 2009) or a gapped ring (Costa et al. 2011) with a positively charged central channel (Fig. 2). The carboxy-terminal half of each Mcm2–7 subunit contains a conserved AAA^+ domain that includes insertions specific to the Mcm2–7 proteins (Iyer et al. 2004) that are predicted to form β-hairpins that may interact with the ssDNA during unwinding (Bochman and Schwacha 2009). As with other AAA^+ multimers, the interface between subunits in the Mcm2–7 complex forms an ATPase active site. The amino-terminal half of each subunit contains an OB-fold-related motif that is found in many ssDNA-binding proteins, and several of the subunits also contain a Zn-finger motif. Although the AAA^+ domain is highly conserved among Mcm2–7 subunits, the characteristic amino- and carboxy-terminal extensions allow homologs for each of the six Mcm2–7 proteins to be readily identified in all eukaryotes studied. As described above, as a 3′ → 5′ helicase the Mcm2–7 helicase moves on the leading-strand DNA template at a replication fork.

EUKARYOTIC HELICASE LOADING PROTEINS

Eukaryotic helicase loading is directed by three proteins: ORC, Cdc6, and Cdt1. ORC is a heterohexamer with five of the six subunits (Orc1, -2, -3, -4, and -5) showing homology with AAA^+ proteins, although only two of these subunits (Orc1 and Orc5) are known to bind ATP (Klemm et al. 1997). Four ORC subunits (Orc1, -2, -4, and -5) also contain potential winged-helix domains at their carboxyl termini. The smallest ORC subunit, Orc6, is related to the RNA Pol II auxilliary factor TFIIB in metazoans (Liu et al. 2011), although fungal Orc6 lacks this similarity. Like Mcm2–7 and ORC, Cdc6 is an AAA^+ protein in the same initiator clade as the AAA^+ ORC subunits, DnaA and DnaC (Iyer et al. 2004). In addition, the carboxy-terminal portion of Cdc6 folds into a winged-helix domain (Liu et al. 2000). Both Orc6 and Cdt1 are poorly conserved across eukaryotes, but all Cdt1

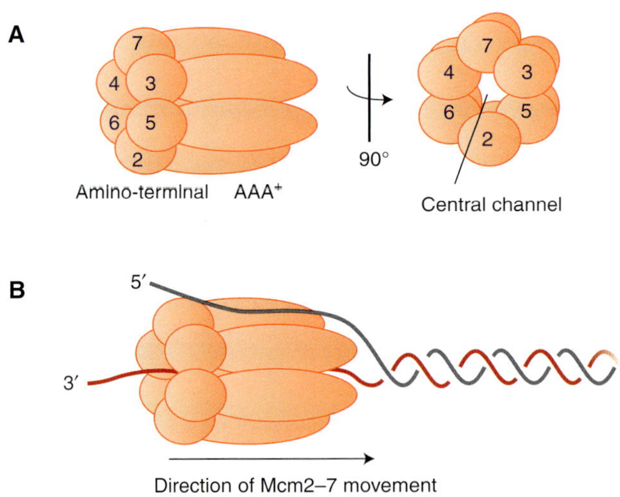

Figure 2. Structure of the Mcm2–7 complex. (*A*) The Mcm2–7 proteins assemble into a hexameric toroid. Each Mcm2–7 complex includes one copy of the Mcm2–7 proteins that are arranged in the indicated order around a central channel. (*B*) Directionality of Mcm2–7 movement. The Mcm2–7 proteins move in a 3′ → 5′ direction along ssDNA. By analogy with the archaea homologs, the carboxy-terminal AAA^+ motif is proximal to the replication fork.

proteins carry a pair of winged-helix domains in their carboxy-terminal half (Lee et al. 2004; Khayrutdinov et al. 2009; Jee et al. 2010) and the most carboxy-terminal winged helix binds to Mcm2–7 (Yanagi et al. 2002; Teer and Dutta 2008; You and Masai 2008; Jee et al. 2010). Interestingly, structural studies suggest that the winged-helix domains associated with ORC, Cdc6, and Cdt1 are related (Khayrutdinov et al. 2009; Jee et al. 2010). Mcm8 and Mcm9 are Mcm2–7-related proteins found in metazoan cells, but are not found to associate with any of the Mcm2–7 proteins either in or out of the Mcm2–7 complex (Maiorano et al. 2005; Lutzmann and Méchali 2008). Although these proteins have been implicated in helicase loading (Volkening and Hoffmann 2005; Lutzmann and Méchali 2008), this view remains controversial (Blanton et al. 2005; Hartford et al. 2011) and a direct mechanistic involvement in helicase loading has not been shown.

INITIAL RECRUITMENT OF THE HELICASE TO THE ORIGIN DNA

Before helicase loading, ORC, Cdc6, Cdt1, and Mcm2–7 are recruited to the origin DNA. This complex is short-lived and is only detected in vitro when Cdc6 ATP hydrolysis is inhibited (Randell et al. 2006). In vivo, chromatin immunoprecipitation assays only detect ORC and Mcm2–7 at specific origins (Aparicio et al. 1997; Tanaka et al. 1997), presumably because of the rapid release of Cdc6 and Cdt1 after helicase loading.

ORIGIN RECOGNITION

The first step of helicase loading is the binding of ORC to origin DNA. In the yeast *S. cerevisiae*, ORC recognizes a conserved sequence within the origin of replication and this event is ATP-dependent (Bell and Stillman 1992). This sequence is not sufficient to direct ORC binding in vivo. An important additional determinant is an asymmetric nucleosome-free region surrounding the consensus sequence (Eaton et al. 2010). *S. pombe* ORC shows limited sequence specificity that is mediated by the AT-rich DNA-binding specificity of an AT-hook domain within the Orc4 subunit (Chuang and Kelly 1999; Kong and DePamphilis 2001; Lee et al. 2001).

The determinants for ORC binding in metazoan cells are less clear. ORC from metazoan cells does not exhibit sequence-specific DNA binding in vitro (Vashee et al. 2003; Remus et al. 2004), but ORC is localized to specific sites along metazoan chromosomes, indicating the existence of other determinants for ORC localization (Karnani et al. 2010; MacAlpine et al. 2010). Genome-wide studies of metazoan ORC binding support a number of determinants including nucleosome-depleted regions, promoter proximity, and dynamic nucleosomes (reviewed in Ding and MacAlpine 2011). In addition, DNA-binding studies show that ORC prefers to bind negatively supercoiled DNA (Remus et al. 2004; Houchens et al. 2008), which may be related to the wrapping of DNA around ORC (Clarey et al. 2006, 2008). Perhaps the most intriguing potential mechanism of ORC localization is the recognition of specifically modified nucleosomes. Human ORC binds a protein called ORCA or LRWD1 that interacts with methylated nucleosomes that are associated with heterochromatin, suggesting an alternative method to localize ORC to compacted chromatin (Bartke et al. 2010; Shen et al. 2010; Vermeulen et al. 2010). In addition, the amino terminus of the Orc1 protein contains a conserved bromo-adjacent homology (BAH) domain. In metazoan cells this domain interacts with histone H4 tails methylated at lysine 20, suggesting another mechanism for ORC to be localized along the chromatin (Kuo et al. 2012).

INTERACTION OF Cdc6, Cdt1, AND Mcm2–7

As cells enter G_1 phase, ORC recruits Cdc6, Cdt1, and Mcm2–7 to the origin DNA. Biochemical studies support a model in which ORC first interacts with Cdc6 and this complex then recruits Cdt1 and Mcm2–7. Loss of Cdt1 does not prevent Cdc6 chromatin association in vivo or origin recruitment in vitro (Maiorano et al. 2000; Nishitani et al. 2000; Tsuyama et al. 2005; Randell et al. 2006; Remus et al. 2009). In contrast, elimination of Cdc6 interferes with Cdt1 origin binding in vitro (Randell et al. 2006; Remus et al. 2009). Studies in *Xenopus* extracts have shown that Cdt1 associates with chromatin in the absence of Cdc6 (Gillespie et al. 2001; Tsuyama et al. 2005); however, only Cdt1 associated in the presence of Cdc6 is able to contribute to Mcm2–7 loading (Tsuyama et al. 2005). Consistent with a robust interaction between Cdc6 and ORC, a complex between the proteins has been structurally characterized (Speck et al. 2005; Sun et al. 2012) and origin-bound ORC stimulates Cdc6 ATP hydrolysis (Randell et al. 2006).

Mcm2–7 and Cdt1 are recruited to the origin as a complex. The carboxy-terminal winged-helix domain of Cdt1 binds Mcm2–7 (Yanagi et al. 2002; Ferenbach et al. 2005; Teer and Dutta 2008; You and Masai 2008; Zhang et al. 2010; Takara and Bell 2011). Cdt1 mutants lacking this region prevent or dominantly inhibit Mcm2–7 loading (Ferenbach et al. 2005; Zhang et al. 2010; Takara and Bell 2011). In *S. cerevisiae* cells, Cdt1 and Mcm2–7 are imported to the nucleus as a complex (Tanaka and Diffley 2002), and formation of this complex is necessary for the recruitment of either protein to the origin (Takara and Bell 2011). The primary Cdt1-binding site on Mcm2–7 is the carboxy-terminal region of Mcm6, and structural studies suggest that this region also folds into a winged-helix domain (Yanagi et al. 2002; Wei et al. 2010).

Characterization of the helicase loading intermediate composed of ORC, Cdc6, Cdt1, and Mcm2–7 suggests how a single asymmetric ORC protein can direct the assembly of a bidirectional replication fork. Binding of Cdt1/Mcm2–7 to the origin requires the smallest ORC subunit, Orc6, which includes two binding sites for Cdt1 (Chen et al. 2007). Analysis of the protein complex formed at the origin before helicase loading (when Cdc6 ATP hydrolysis is inhibited) indicates that multiple Cdt1 proteins and their associated Mcm2–7 complexes are initially recruited to each ORC–Cdc6 complex (Takara and Bell 2011). It is most likely that this complex contains two Cdt1 molecules, one bound to each of the two Cdt1-binding sites on Orc6. Importantly, interfering with the formation of this multi-Cdt1 intermediate prevents subsequent helicase loading, and it is likely that this intermediate facilitates the simultaneous loading of an Mcm2–7 double hexamer (see below).

How Cdc6 contributes to Cdt1 and Mcm2–7 recruitment is less well understood. It is possible that there are direct interactions between Cdc6 and Cdt1 and/or Mcm2–7. Alternatively, Cdc6 could alter the conformation of ORC to allow Cdt1 and Mcm2–7 to bind. The latter hypothesis would explain why ORC does not bind either Cdt1 or Mcm2–7 in the absence of Cdc6 (Takara and Bell 2011) but isolated Orc6 can bind Cdt1 (Chen et al. 2007). Consistent with this model, recent EM studies of ORC bound to DNA in the presence and absence of Cdc6 suggested that Orc6 becomes exposed upon Cdc6 binding (Sun et al. 2012).

LOADING OF RECRUITED HELICASES

After recruitment of the helicase loading factors and Mcm2–7 to the origin, loading of the recruited Mcm2–7 requires ATP hydrolysis and involves significant changes of both the proteins associated with the origin and the structure of the Mcm2–7 ring. Loaded Mcm2–7 complexes form double-hexameric pairs that encircle dsDNA. Thus, helicase loading necessarily requires the establishment of strong interactions between the amino termini of two Mcm2–7 complexes and opening and closing of the Mcm2–7 ring to allow DNA access to the Mcm2–7 central channel. Finally, to ensure the proper regulation of replication initiation, as initially loaded, the Mcm2–7 complexes are inactive for unwinding the parental duplex.

STRUCTURE OF LOADED Mcm2–7

The reconstitution of Mcm2–7 loading using four purified proteins has led to important advances in the understanding of the architecture of loaded Mcm2–7. EM of loaded Mcm2–7 shows head-to-head double hexamers, with the amino termini of the Mcm2–7 subunits mediating hexamer interactions (Evrin et al. 2009; Remus et al. 2009). This structure is similar to the double hexamers observed for the homohexameric archaea Mcm complexes (Brewster and Chen 2010). Unlike the archaea homologs, Mcm2–7 double hexamers are only detected after loading. Mcm2–7 double hexamers survive treatment with DNase and gel filtration, indicating that once formed, these complexes are very stable (Evrin et al. 2009; Gambus et al. 2011). Based on structural studies of the archaea Mcm complex, these interactions likely involve the Zn-finger domains in the amino termini of Mcm2–7 subunits (Fletcher et al. 2003). The inter-Mcm2–7 contacts within the double hexamer are unknown but must be largely heterotypic owing to their head-to-head interaction. A lack of loaded single hexamers and the presence of multiple Cdt1/Mcm2–7 complexes before Mcm2–7 loading supports a model in which both hexamers are loaded in a concerted fashion (Remus et al. 2009; Takara and Bell 2011). Importantly, both EM and topological linkage studies indicate that loaded Mcm2–7 complexes encircle and slide nondirectionally on dsDNA (Evrin et al. 2009; Remus et al. 2009). Thus, loaded Mcm2–7 is topologically linked to the DNA but is neither active as a helicase nor tightly engaged with the DNA.

ROLE OF ATP DURING HELICASE LOADING

ATP binding and hydrolysis play critical roles during helicase loading. As described above, ATP binding but not hydrolysis is required for the initial recruitment of the helicase to the origin (Gillespie et al. 2001; Randell et al. 2006). ATP hydrolysis by Cdc6 and ORC plays distinct and ordered roles in helicase loading. Cdc6 is activated to bind and hydrolyze ATP when it associates with origin-bound ORC, and this hydrolysis event is required to observe Mcm2–7 loading (Perkins and Diffley 1998; Randell et al. 2006). ORC ATP hydrolysis is also essential but functions after Cdc6 hydrolysis and appears to regulate repeated Mcm2–7 loading (Bowers et al. 2004; Randell et al. 2006). One likely role for ORC ATP hydrolysis is to cause the release of the other helicase loading factors from the origin (Tsakraklides and Bell 2010). Evidence from studies of *S. cerevisiae* suggests that a second consequence of ATP hydrolysis is to disable ORC DNA binding until it is able to bind another molecule of ATP. An important unanswered question is how many Mcm2–7 hexamers are loaded per round of ATP hydrolysis by Cdc6 and ORC. Although current evidence suggests two hexamers are loaded coordinately, it remains possible that Cdc6 or ORC could require multiple rounds of ATP hydrolysis to accomplish the loading of an Mcm2–7 double hexamer. Overall, the coordinated and ordered set of ATP hydrolysis reactions ensures that Cdc6 only acts on Cdt1/Mcm2–7 when bound to ORC at the origin and a new round of loading cannot occur until the previous round is complete.

Mcm2–7 GATE FUNCTION AND IMPLICATIONS FOR LOADING

The Mcm2–7 ring structure must be opened during loading to provide access to the central DNA-binding channel. Binding of purified Mcm2–7 to ssDNA circles suggests that the interface between Mcm2 and Mcm5 acts as a "gate" to the central channel (Bochman and Schwacha 2008). EM studies of purified *Drosophila* Mcm2–7 support this conclusion (Costa et al. 2011). Although ATP binding is believed to close the Mcm2/5 gate, it is likely that some combination of Cdt1 binding and Cdc6 ATP hydrolysis modulates Mcm2–7 ring opening during helicase loading. Because Cdt1 is released from Mcm2–7 after loading (Randell et al. 2006), it is tempting to speculate that Cdt1 binding opens the Mcm2/5 gate and Cdc6 ATP hydrolysis separates Cdt1 from Mcm2–7, leaving behind a closed form of the Mcm2–7 ring. The role of ATP binding and hydrolysis by Mcm2–7 during helicase loading remains uncertain despite studies to address this issue. Mutations in the Mcm6 and Mcm7 ATP-binding sites show normal chromatin association but are defective for replication initiation (Ying and Gautier 2005). It is unclear, however, whether the observed chromatin association reflects loading or just Mcm2–7 recruitment. In addition, the distinct functions of six Mcm2–7 ATPase active sites (reviewed in Bochman and

Schwacha 2009) suggest that ATP bound to other sites could be required for helicase loading.

REGULATION AND DYNAMICS OF HELICASE LOADING

Helicase loading is tightly restricted to G_1 phase to ensure that no origin of replication can initiate DNA replication more than once per cell cycle (reviewed in Arias and Walter 2007; Siddiqui et al. 2013; Zielke et al. 2013). This regulation is mediated by the inhibition of helicase loading outside of G_1 phase. The mechanisms of inhibition are diverse and vary among organisms. In budding yeast, inhibition is primarily mediated by cyclin-dependent kinase (CDK) phosphorylation of Cdc6, which leads to its degradation (Drury et al. 2000), Mcm2–7 nuclear export (Labib et al. 1999; Nguyen et al. 2000), and the inhibition of Cdt1/Mcm2–7 binding to ORC (Chen and Bell 2011). In metazoan organisms, Cdt1 is a primary target for inhibition. Cdt1 is inhibited by geminin binding outside of G_1 (McGarry and Kirschner 1998; Wohlschlegel et al. 2000) and degraded in a proliferating cell nuclear antigen (PCNA)/DNA-dependent (Arias and Walter 2005) or CDK-dependent (Li et al. 2003; Liu et al. 2004) fashion in S phase. In addition, in many metazoans, CDK activity inhibits ORC DNA binding during G_2/M phase (reviewed in DePamphilis 2005). Although it is tempting to think that the multiple mechanisms that inhibit helicase loading are redundant, analysis of mutants that are defective for a subset of mechanisms shows partial rereplication (Green et al. 2006). Importantly, even limited rereplication is lethal to most dividing cells and can induce gene amplification associated with cancer (Green et al. 2010).

The role of the helicase loading proteins after loading is unclear. In vitro studies show that after loading, Mcm2–7 no longer requires ORC, Cdc6, or Cdt1 to associate with origin DNA (Donovan et al. 1997; Randell et al. 2006). Consistent with this observation, in vitro studies show that the helicase loading proteins are released from the origin upon Mcm2–7 loading (Randell et al. 2006; Tsakraklides and Bell 2010). In contrast to these findings, in vivo inactivation of *S. cerevisiae* ORC or Cdc6 in late G_1 (after Mcm2–7 loading) results in loss of Mcm2–7 origin association (Aparicio et al. 1997; Semple et al. 2006; Chen et al. 2007). This suggests either that these proteins are required to maintain Mcm2–7 association with the origin DNA or that there is an activity that removes loaded Mcm2–7 from the DNA in G_1 cells and the helicase loading proteins are required to restore Mcm2–7 to the origin. In either case, this reliance on helicase loading proteins changes upon entry into S phase, as Mcm2–7 is tightly associated with origins at this stage (unless initiation has occurred) even though Cdc6 is eliminated from S phase cells by degradation. It is noteworthy that studies in mammalian G_1 cells suggest that there is little exchange between chromatin-associated and free Mcm2–7 (Kuipers et al. 2011), arguing for a model in which helicase loading proteins stabilize loaded Mcm2–7 rather than a model in which Mcm2–7 release is balanced with new loading.

A MODEL FOR HELICASE LOADING

Based on current knowledge, we propose the following model for the events of eukaryotic helicase loading (Fig. 3). ORC recruitment to origin DNA is directed by a combination of DNA affinity (sequence-specific in budding yeast), local chromatin structure, and interaction with other proteins (e.g., ORCA). Upon entry into G_1 phase, ORC recruits Cdc6. In *S. cerevisiae* cells, we propose that this interaction reveals Cdt1-binding sites on Orc6 that are bound by two Cdt1/Mcm2–7 complexes. The distinct structure of Orc6 in metazoans raises the possibility that the initial recruitment is mediated by different interactions in these organisms. The connection of these events to Mcm2–7 ring opening is unclear; however, it seems most likely that the Mcm2/5 gate is opened at some point during these events. Alternatively, the ring could be opened before recruitment (e.g., after Cdt1 binds to Mcm2–7), although this would risk Mcm2–7 interacting with nonorigin DNA. Assuming two Mcm2–7 complexes are loaded coordinately, we suggest that the Mcm2/5 gates of the two helicases are aligned to form a single continuous

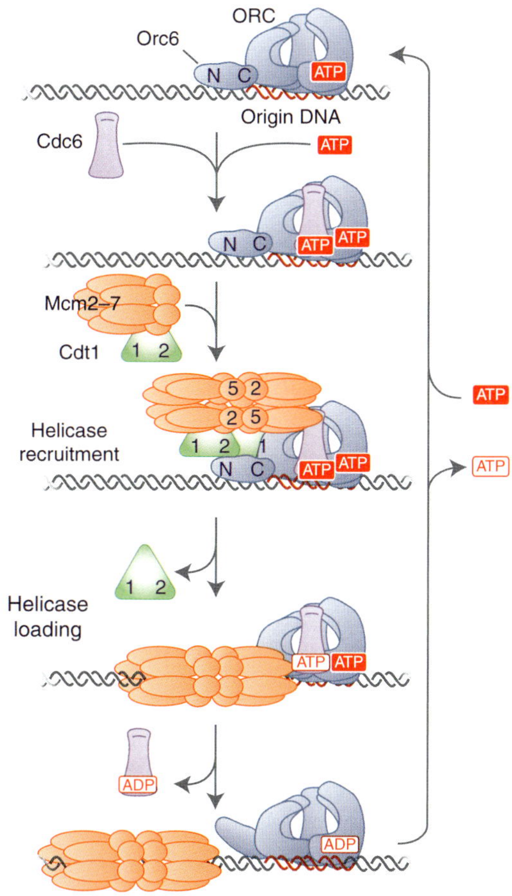

Figure 3. Model for eukaryotic replicative DNA helicase loading. After localization to the origin DNA, ATP-bound ORC recruits Cdc6 bound to ATP. The resulting ORC–Cdc6 complex then recruits two Cdt1/Mcm2–7 complexes via interactions between Cdt1 and Orc6. Although this illustration suggests that the Mcm2–7 complexes have initiated interactions at their amino termini at this stage, it is also possible that these interactions only occur during or after helicase loading. The interactions between ORC, Cdc6, Cdt1, and Mcm2–7 are proposed to result in the opening of the Mcm2–7 ring at the Mcm2/5 gate. Cdc6 ATP hydrolysis results in the loading of an Mcm2–7 double hexamer around double-stranded origin DNA and the release of Cdt1. Whether the DNA enters the Mcm2–7 central channel before (upon initial ring opening) or after (as illustrated) Cdc6 ATP hydrolysis is unknown. ORC ATP hydrolysis is proposed to lead to the release of Cdc6–ADP and loaded Mcm2–7 from ORC. ORC ADP/ATP exchange leads to resetting of the loading machinery, allowing a new round of helicase loading to initiate.

opening/gate. By analogy to the function of the sliding clamp loader (Kelch et al. 2011), we speculate that the Mcm2–7 complexes can encircle the DNA at this stage but remain open. The next stage of loading is triggered by Cdc6 ATP hydrolysis. This event leads to the release of Cdc6 and Cdt1 from the helicase, which is coupled with entrapment of the DNA within the Mcm2–7 central channel and closure of the Mcm2/5 gate. Finally, we propose that ATP hydrolysis by ORC leads to the release of the loaded helicases from ORC to prepare for a new round of loading after ADP → ATP exchange.

Once loaded, the Mcm2–7 complex awaits activation. Biochemical studies show that activation requires the association of two helicase activating proteins (Cdc45 and GINS) with Mcm2–7 (Ilves et al. 2010). Association of these proteins is triggered by the action of two kinases: S-phase CDK (S-CDK) and the Dbf4-dependent Cdc7 kinase (DDK). DDK phosphorylates Mcm2–7 (Sheu and Stillman 2006, 2010; Randell et al. 2010). As a result of Mcm2–7 phosphorylation, Cdc45 associates with Mcm2–7 (Heller et al. 2011; Tanaka et al. 2011). Interestingly, this event occurs at a subset of origins before entry into S phase (Aparicio et al. 1999; Kanemaki and Labib 2006; Heller et al. 2011; Tanaka et al. 2011). S-CDK targets two additional proteins, Sld2 and Sld3, causing them to associate with Dpb11 and eventually resulting in the recruitment of the GINS proteins as well as DNA Pol ε (Tanaka et al. 2007; Zegerman and Diffley 2007; Muramatsu et al. 2010). The subsequent activation of DNA unwinding is required for the recruitment of the lagging-strand DNA polymerases (DNA Pol α-primase and DNA Pol δ (Mimura et al. 2000; Heller et al. 2011).

COMPARISON OF HELICASES AND HELICASE LOADING IN *E. coli* AND EUKARYOTIC CELLS

There are many similarities between helicase loading at *E. coli* and eukaryotic chromosomal origins, suggesting that the fundamental mechanism of these events is related. Both *E. coli* DnaB and eukaryotic Mcm2–7 are composed of six subunits that assemble into ring-shaped molecules. Second, both events rely on origin recognition proteins (DnaA and ORC) that contact the origin DNA over an extended region. Third, loading of each DNA helicase requires at least one additional helicase loading factor that binds to the helicase. In *E. coli*, DnaC is required for loading and also restrains DnaB helicase activity. In eukaryotes, both Cdc6 and Cdt1 share properties with DnaC. Like DnaC, Cdc6 is an AAA^+ protein and is a weak ATPase on its own, whereas Cdt1 shares DnaC's association with its cognate helicase. Finally, both helicases must be activated after loading, either by release of an inhibitor (DnaC in *E. coli*) or by recruiting helicase activators (Cdc45 and GINS in eukaryotic cells).

There are also a number of important differences. First, although similar in overall structure, DnaB and the Mcm2–7 complexes have opposite polarities: DnaB moves on the lagging-strand template, whereas Mcm2–7 moves on the leading-strand template. DnaB is a homohexamer that has a RecA-like ATPase domain (Bailey et al. 2007), whereas Mcm2–7 is a heterohexamer that has AAA^+-like ATPase domains (Iyer et al. 2004). Second, although DnaA origin binding leads to local DNA melting, ORC DNA binding does not. Third, two separate single hexamers of DnaB are loaded around ssDNA after loading, whereas a head-to-head double hexamer of Mcm2–7 is loaded around dsDNA.

These differences are most likely related to the rate of helicase activation after loading. Loading of DnaB around ssDNA at *oriC* is consistent with the rapid activation of DnaB after loading. In contrast, the need to restrain helicase activation until S phase in eukaryotic cells suggests the need for a loaded helicase that is less easily activated. Recent findings indicate that Mcm2–7 surrounds ssDNA and acts as a single hexamer at the replication fork (Yardimci et al. 2010; Fu et al. 2011). This indicates that Mcm2–7 activation requires two major changes in the loaded double hexamer. First, each Mcm2–7 complex must dissociate from its double-hexamer partner. Second, the Mcm2–7 ring originally bound to duplex DNA must open to expel the lagging-strand ssDNA template before DNA unwinding. Although we have an initial understanding of how DnaC expulsion leads to activation of DnaB (Makowska-Grzyska and Kaguni 2010), how the helicase activating proteins Cdc45 and GINS and the proteins that direct them to the origin (Sld2, Sld3, and Dpb11) catalyze these events remains unknown.

Although 25 years of research has provided an increasingly detailed understanding of the events of helicase loading, fundamental questions remain. How is the opening and closing of the DNA helicase ring controlled? How is the origin DNA properly positioned to ensure that the helicase encircles the appropriate form of the DNA at the correct position? How are these events regulated by the AAA^+ ATPases that are conserved among the different realms of life? These and other questions propel studies of this central event in replication initiation.

ACKNOWLEDGMENTS

We thank members of our labs for helpful discussions. This work is supported by the Michigan Agricultural Station and by grant RO1 GM090063 from the National Institutes of Health to J.M.K. and by the Howard Hughes Medical Institute and grant RO1 GM052339 from the National Institutes of Health to S.P.B.

REFERENCES

*Reference is also in this collection.

Abe Y, Jo T, Matsuda Y, Matsunaga C, Katayama T, Ueda T. 2007. Structure and function of DnaA N-terminal domains: Specific sites and mechanisms in inter-DnaA interaction and in DnaB helicase loading on *oriC*. *J Biol Chem* **282:** 17816–17827.

Allen GC, Kornberg A. 1991. Fine balance in the regulation of DnaB helicase by DnaC protein in replication in *Escherichia coli*. *J Biol Chem* **266:** 22096–22101.

Aparicio OM, Weinstein DM, Bell SP. 1997. Components and dynamics of DNA replication complexes in *S. cerevisiae*: Redistribution of MCM proteins and Cdc45p during S phase. *Cell* **91:** 59–69.

Aparicio OM, Stout AM, Bell SP. 1999. Differential assembly of Cdc45p and DNA polymerases at early and late origins of DNA replication. *Proc Natl Acad Sci* **96:** 9130–9135.

Arai K, Yasuda S, Kornberg A. 1981. Mechanism of *dnaB* protein action. I. Crystallization and properties of *dnaB* protein, an essential replication protein in *Escherichia coli*. *J Biol Chem* **256:** 5247–5252.

Arias EE, Walter JC. 2005. PCNA functions as a molecular platform to trigger Cdt1 destruction and prevent re-replication. *Nat Cell Biol* **8:** 84–90.

Arias EE, Walter JC. 2007. Strength in numbers: Preventing rereplication via multiple mechanisms in eukaryotic cells. *Genes Dev* **21:** 497–518.

Bailey S, Eliason WK, Steitz TA. 2007. Structure of hexameric DnaB helicase and its complex with a domain of DnaG primase. *Science* **318:** 459–463.

Bárcena M, Ruiz T, Donate LE, Brown SE, Dixon NE, Radermacher M, Carazo JM. 2001. The DnaB·DnaC complex: A structure based on dimers assembled around an occluded channel. *EMBO J* **20:** 1462–1468.

Bartke T, Vermeulen M, Xhemalce B, Robson SC, Mann M, Kouzarides T. 2010. Nucleosome-interacting proteins regulated by DNA and histone methylation. *Cell* **143:** 470–484.

* Bell SD, Botchan MR. 2013. The minichromosome maintenance replicative helicase. *Cold Spring Harb Perspect Biol* doi: 10.1101/cshperspect.a012807.

Bell SP, Dutta A. 2002. DNA replication in eukaryotic cells. *Annu Rev Biochem* **71:** 333–374.

Bell SP, Stillman B. 1992. ATP-dependent recognition of eukaryotic origins of DNA replication by a multiprotein complex. *Nature* **357:** 128–134.

Biswas SB, Flowers S, Biswas-Fiss EE. 2004. Quantitative analysis of nucleotide modulation of DNA binding by DnaC protein of *Escherichia coli*. *Biochem J* **379:** 553–562.

Blaesing F, Weigel C, Welzeck M, Messer W. 2000. Analysis of the DNA-binding domain of *Escherichia coli* DnaA protein. *Mol Microbiol* **36:** 557–569.

Blanton HL, Radford SJ, McMahan S, Kearney HM, Ibrahim JG, Sekelsky J. 2005. REC, *Drosophila* MCM8, drives formation of meiotic crossovers. *PLoS Genet* **1:** e40.

Bliska JB, Cozzarelli NR. 1987. Use of site-specific recombination as a probe of DNA structure and metabolism in vivo. *J Mol Biol* **194:**205–218.

Blow JJ, Laskey RA. 1988. A role for the nuclear envelope in controlling DNA replication within the cell cycle. *Nature* **332:** 546–548.

Bochman ML, Schwacha A. 2008. The Mcm2-7 complex has in vitro helicase activity. *Mol Cell* **31:** 287–293.

Bochman ML, Schwacha A. 2009. The Mcm complex: Unwinding the mechanism of a replicative helicase. *Microbiol Mol Biol Rev* 73: 652–683.

Bochman ML, Bell SP, Schwacha A. 2008. Subunit organization of Mcm2-7 and the unequal role of active sites in ATP hydrolysis and viability. *Mol Cell Biol* **28:** 5865–5873.

Boeneman K, Crooke E. 2005. Chromosomal replication and the cell membrane. *Curr Opin Microbiol* **8:** 143–148.

Boeneman K, Fossum S, Yang Y, Fingland N, Skarstad K, Crooke E. 2009. *Escherichia coli* DnaA forms helical structures along the longitudinal cell axis distinct from MreB filaments. *Mol Microbiol* **72:** 645–657.

Bowers JL, Randell JCW, Chen S, Bell SP. 2004. ATP hydrolysis by ORC catalyzes reiterative Mcm2-7 assembly at a defined origin of replication. *Mol Cell* **16:** 967–978.

Bramhill D, Kornberg A. 1988. Duplex opening by dnaA protein at novel sequences in initiation of replication at the origin of the *E. coli* chromosome. *Cell* **52:** 743–755.

Brewster AS, Chen XS. 2010. Insights into the MCM functional mechanism: Lessons learned from the archaeal MCM complex. *Crit Rev Biochem Mol Biol* **45:** 243–256.

Bujalowski W, Klonowska MM, Jezewska MJ. 1994. Oligomeric structure of *Escherichia coli* primary replicative helicase DnaB protein. *J Biol Chem* **269:** 31350–31358.

Calzada A, Hodgson B, Kanemaki M, Bueno A, Labib K. 2005. Molecular anatomy and regulation of a stable replisome at a paused eukaryotic DNA replication fork. *Genes Dev* **19:** 1905–1919.

Carr KM, Kaguni JM. 2001. Stoichiometry of DnaA and DnaB protein in initiation at the *Escherichia coli* chromosomal origin. *J Biol Chem* **276:** 44919–44925.

Chen S, Bell SP. 2011. CDK prevents Mcm2–7 helicase loading by inhibiting Cdt1 interaction with Orc6. *Genes Dev* **25:** 363–372.

Chen S, de Vries MA, Bell SP. 2007. Orc6 is required for dynamic recruitment of Cdt1 during repeated Mcm2–7 loading. *Genes Dev* **21:** 2897–2907.

Chodavarapu S, Felczak MM, Yaniv JR, Kaguni JM. 2008a. *Escherichia coli* DnaA interacts with HU in initiation at the *E. coli* replication origin. *Mol Microbiol* **67:** 781–792.

Chodavarapu S, Gomez R, Vicente M, Kaguni JM. 2008b. *Escherichia coli* Dps interacts with DnaA protein to impede initiation: A model of adaptive mutation. *Mol Microbiol* **67:** 1331–1346.

Chodavarapu S, Felczak MM, Kaguni JM. 2011. Two forms of ribosomal protein L2 of *Escherichia coli* that inhibit DnaA in DNA replication. *Nucleic Acids Res* **39:** 4180–4191.

Chuang RY, Kelly TJ. 1999. The fission yeast homologue of Orc4p binds to replication origin DNA via multiple AT-hooks. *Proc Natl Acad Sci* **96:** 2656–2661.

Clarey MG, Erzberger JP, Grob P, Leschziner AE, Berger JM, Nogales E, Botchan M. 2006. Nucleotide-dependent conformational changes in the DnaA-like core of the origin recognition complex. *Nat Struct Mol Biol* **13:** 684–690.

Clarey MG, Botchan M, Nogales E. 2008. Single particle EM studies of the *Drosophila melanogaster* origin recognition complex and evidence for DNA wrapping. *J Struct Biol* **164:** 241–249.

Costa A, Ilves I, Tamberg N, Petojevic T, Nogales E, Botchan MR, Berger JM. 2011. The structural basis for MCM2–7 helicase activation by GINS and Cdc45. *Nat Struct Mol Biol* **18:** 471–477.

Cozzarelli NR. 1980. DNA gyrase and the supercoiling of DNA. *Science* **207:** 953–960.

Davey MJ, Fang L, McInerney P, Georgescu RE, O'Donnell M. 2002. The DnaC helicase loader is a dual ATP/ADP switch protein. *EMBO J* **21:** 3148–3159.

Davey MJ, Indiani C, O'Donnell M. 2003. Reconstitution of the Mcm2-7p heterohexamer, subunit arrangement, and ATP site architecture. *J Biol Chem* **278:** 4491–4499.

DePamphilis ML. 2005. Cell cycle dependent regulation of the origin recognition complex. *Cell Cycle* **4:** 70–79.

Diffley JF, Cocker JH, Dowell SJ, Rowley A. 1994. Two steps in the assembly of complexes at yeast replication origins in vivo. *Cell* **78:** 303–316.

Ding Q, MacAlpine DM. 2011. Defining the replication program through the chromatin landscape. *Crit Rev Biochem Mol Biol* **46:** 165–179.

Donate LE, Llorca O, Bárcena M, Brown SE, Dixon NE, Carazo JM. 2000. pH-controlled quaternary states of hexameric DnaB helicase. *J Mol Biol* **303:** 383–393.

Donovan S, Harwood J, Drury LS, Diffley JF. 1997. Cdc6p-dependent loading of Mcm proteins onto pre-replicative chromatin in budding yeast. *Proc Natl Acad Sci* **94:** 5611–5616.

Drury LS, Perkins G, Diffley JF. 2000. The cyclin-dependent kinase Cdc28p regulates distinct modes of Cdc6p proteolysis during the budding yeast cell cycle. *Curr Biol* **10:** 231–240.

Duderstadt KE, Berger JM. 2008. AAA$^+$ ATPases in the initiation of DNA replication. *Crit Rev Biochem Mol Biol* **43:** 163–187.

Duderstadt KE, Mott ML, Crisona NJ, Chuang K, Yang H, Berger JM. 2010. Origin remodeling and opening in bacteria rely on distinct assembly states of the DnaA initiator. *J Biol Chem* **285:** 28229–28239.

Duderstadt KE, Chuang K, Berger JM. 2011. DNA stretching by bacterial initiators promotes replication origin opening. *Nature* **478:** 209–213.

Eaton ML, Galani K, Kang S, Bell SP, MacAlpine DM. 2010. Conserved nucleosome positioning defines replication origins. *Genes Dev* **24:** 748–753.

Erzberger JP, Berger JM. 2006. Evolutionary relationships and structural mechanisms of AAA$^+$ proteins. *Annu Rev Biophys Biomol Struct* **35:** 93–114.

Erzberger JP, Pirruccello MM, Berger JM. 2002. The structure of bacterial DnaA: Implications for general mechanisms underlying DNA replication initiation. *EMBO J* **21:** 4763–4773.

Erzberger JP, Mott ML, Berger JM. 2006. Structural basis for ATP-dependent DnaA assembly and replication-origin remodeling. *Nat Struct Mol Biol* **13:** 676–683.

Evrin C, Clarke P, Zech J, Lurz R, Sun J, Uhle S, Li H, Stillman B, Speck C. 2009. A double-hexameric MCM2-7 complex is loaded onto origin DNA during licensing of eukaryotic DNA replication. *Proc Natl Acad Sci* **106:** 20240–20245.

Fang L, Davey MJ, O'Donnell M. 1999. Replisome assembly at *oriC*, the replication origin of *E. coli*, reveals an explanation for initiation sites outside an origin. *Mol Cell* **4:** 541–553.

Felczak MM, Kaguni JM. 2004. The box VII motif of *Escherichia coli* DnaA protein is required for DnaA oligomerization at the *E. coli* replication origin. *J Biol Chem* **279:** 51156–51162.

Ferenbach A, Li A, Brito-Martins M, Blow JJ. 2005. Functional domains of the *Xenopus* replication licensing factor Cdt1. *Nucleic Acids Res* **33:** 316–324.

Flåtten I, Morigen, Skarstad K. 2009. DnaA protein interacts with RNA polymerase and partially protects it from the effect of rifampicin. *Mol Microbiol* **71:** 1018–1030.

Fletcher RJ, Bishop BE, Leon RP, Sclafani RA, Ogata CM, Chen XS. 2003. The structure and function of MCM from archaeal *M. thermoautotrophicum*. *Nat Struct Biol* **10:** 160–167.

Fu YV, Yardimci H, Long DT, Guainazzi A, Bermudez VP, Hurwitz J, van Oijen A, Schärer OD, Walter JC. 2011. Selective bypass of a lagging strand roadblock by the eukaryotic replicative DNA helicase. *Cell* **146:** 931–941.

Fujikawa N, Kurumizaka H, Nureki O, Terada T, Shirouzu M, Katayama T, Yokoyama S. 2003. Structural basis of replication origin recognition by the DnaA protein. *Nucleic Acids Res* **31:** 2077–2086.

Galletto R, Jezewska MJ, Bujalowski W. 2003. Interactions of the *Escherichia coli* DnaB helicase hexamer with the replication factor the DnaC protein. Effect of nucleotide cofactors and the ssDNA on protein–protein interactions and the topology of the complex. *J Mol Biol* **329:** 441–465.

Galletto R, Jezewska MJ, Bujalowski W. 2004. Unzipping mechanism of the double-stranded DNA unwinding by a hexameric helicase: The effect of the 3′ arm and the stability of the dsDNA on the unwinding activity of the *Escherichia coli* DnaB helicase. *J Mol Biol* **343:** 101–114.

Gambus A, Khoudoli GA, Jones RC, Blow JJ. 2011. MCM2-7 form double hexamers at licensed origins in *Xenopus* egg extract. *J Biol Chem* **286:** 11855–11864.

Gille H, Messer W. 1991. Localized DNA melting and structural pertubations in the origin of replication, *oriC*, of *Escherichia coli* in vitro and in vivo. *EMBO J* **10:** 1579–1584.

Gillespie PJ, Li A, Blow JJ. 2001. Reconstitution of licensed replication origins on *Xenopus* sperm nuclei using purified proteins. *BMC Biochem* **2:** 15.

Green BM, Morreale RJ, Ozaydin B, Derisi JL, Li JJ. 2006. Genome-wide mapping of DNA synthesis in *Saccharomyces cerevisiae* reveals that mechanisms preventing reinitiation of DNA replication are not redundant. *Mol Biol Cell* **17:** 2401–2414.

Green BM, Finn KJ, Li JJ. 2010. Loss of DNA replication control is a potent inducer of gene amplification. *Science* **329:** 943–946.

Hartford SA, Luo Y, Southard TL, Min IM, Lis JT, Schimenti JC. 2011. Minichromosome maintenance helicase paralog MCM9 is dispensable for DNA replication but functions in germ-line stem cells and tumor suppression. *Proc Natl Acad Sci* **108:** 17702–17707.

Heller RC, Kang S, Lam WM, Chen S, Chan CS, Bell SP. 2011. Eukaryotic origin-dependent DNA replication in vitro reveals sequential action of DDK and S-CDK kinases. *Cell* **146:** 80–91.

Houchens CR, Lu W, Chuang R-Y, Frattini MG, Fuller A, Simancek P, Kelly TJ. 2008. Multiple mechanisms contribute to *Schizosaccharomyces pombe* origin recognition complex-DNA interactions. *J Biol Chem* **283:** 30216–30224.

Ilves I, Petojevic T, Pesavento JJ, Botchan MR. 2010. Activation of the MCM2-7 helicase by association with Cdc45 and GINS proteins. *Mol Cell* **37:** 247–258.

Iyer LM, Leipe DD, Koonin EV, Aravind L. 2004. Evolutionary history and higher order classification of AAA$^+$ ATPases. *J Struct Biol* **146:** 11–31.

Jee J, Mizuno T, Kamada K, Tochio H, Chiba Y, Yanagi K-I, Yasuda G, Hiroaki H, Hanaoka F, Shirakawa M. 2010.

Structure and mutagenesis studies of the C-terminal region of licensing factor Cdt1 enable the identification of key residues for binding to replicative helicase Mcm proteins. *J Biol Chem* **285:** 15931–15940.

Jezewska MJ, Kim US, Bujalowski W. 1996. Binding of *Escherichia coli* primary replicative helicase DnaB protein to single-stranded DNA. Long-range allosteric conformational changes within the protein hexamer. *Biochemistry* **35:** 2129–2145.

Jezewska MJ, Rajendran S, Bujalowska D, Bujalowski W. 1998a. Does single-stranded DNA pass through the inner channel of the protein hexamer in the complex with the *Escherichia coli* DnaB helicase? Fluorescence energy transfer studies. *J Biol Chem* **273:** 10515–10529.

Jezewska MJ, Rajendran S, Bujalowski W. 1998b. Complex of *Escherichia coli* primary replicative helicase DnaB protein with a replication fork: Recognition and structure. *Biochemistry* **37:** 3116–3136.

* Johansson E, Dixon N. 2013. Replicative DNA polymerases. *Cold Spring Harb Perspect Biol* doi: 10.1101/cshperspect.a012799.

Johnson RT, Rao PN. 1970. Mammalian cell fusion: Induction of premature chromosome condensation in interphase nuclei. *Nature* **226:** 717–722.

Kaguni JM. 2011. Replication initiation at the *Escherichia coli* chromosomal origin. *Curr Opin Chem Biol* **15:** 606–613.

Kanemaki M, Labib K. 2006. Distinct roles for Sld3 and GINS during establishment and progression of eukaryotic DNA replication forks. *EMBO J* **25:** 1753–1763.

Kaplan DL. 2000. The 3′-tail of a forked-duplex sterically determines whether one or two DNA strands pass through the central channel of a replication-fork helicase. *J Mol Biol* **301:** 285–299.

Kaplan DL, Davey MJ, O'Donnell M. 2003. Mcm4, 6, 7 uses a "pump in ring" mechanism to unwind DNA by steric exclusion and actively translocate along a duplex. *J Biol Chem* **278:** 49171–49182.

Karnani N, Taylor CM, Malhotra A, Dutta A. 2010. Genomic study of replication initiation in human chromosomes reveals the influence of transcription regulation and chromatin structure on origin selection. *Mol Biol Cell* **21:** 393–404.

Kelch BA, Makino DL, O'Donnell M, Kuriyan J. 2011. How a DNA polymerase clamp loader opens a sliding clamp. *Science* **334:** 1675–1680.

Keyamura K, Fujikawa N, Ishida T, Ozaki S, Su'etsugu M, Fujimitsu K, Kagawa W, Yokoyama S, Kurumizaka H, Katayama T. 2007. The interaction of DiaA and DnaA regulates the replication cycle in *E. coli* by directly promoting ATP DnaA-specific initiation complexes. *Genes Dev* **21:** 2083–2099.

Keyamura K, Abe Y, Higashi M, Ueda T, Katayama T. 2009. DiaA dynamics are coupled with changes in initial origin complexes leading to helicase loading. *J Biol Chem* **284:** 25038–25050.

Khayrutdinov BI, Bae WJ, Yun YM, Lee JH, Tsuyama T, Kim JJ, Hwang E, Ryu K-S, Cheong H-K, Cheong C, et al. 2009. Structure of the Cdt1 C-terminal domain: Conservation of the winged helix fold in replication licensing factors. *Protein Sci* **18:** 2252–2264.

Klemm RD, Austin RJ, Bell SP. 1997. Coordinate binding of ATP and origin DNA regulates the ATPase activity of the origin recognition complex. *Cell* **88:** 493–502.

Kobori JA, Kornberg A. 1982. The *Escherichia coli dnaC* gene product. III. Properties of the dnaB–dnaC protein complex. *J Biol Chem* **257:** 13770–13775.

Kong D, DePamphilis ML. 2001. Site-specific DNA binding of the *Schizosaccharomyces pombe* origin recognition complex is determined by the Orc4 subunit. *Mol Cell Biol* **21:** 8095–8103.

Kuipers MA, Stasevich TJ, Sasaki T, Wilson KA, Hazelwood KL, McNally JG, Davidson MW, Gilbert DM. 2011. Highly stable loading of Mcm proteins onto chromatin in living cells requires replication to unload. *J Cell Biol* **192:** 29–41.

Kuo AJ, Song J, Cheung P, Ishibe-Murakami S, Yamazoe S, Chen JK, Patel DJ, Gozani O. 2012. The BAH domain of ORC1 links H4K20me2 to DNA replication licensing and Meier–Gorlin syndrome. *Nature* **484:** 115–119.

Labib K. 2010. How do Cdc7 and cyclin-dependent kinases trigger the initiation of chromosome replication in eukaryotic cells? *Genes Dev* **24:** 1208–1219.

Labib K, Diffley JF, Kearsey SE. 1999. G1-phase and B-type cyclins exclude the DNA-replication factor Mcm4 from the nucleus. *Nat Cell Biol* **1:** 415–422.

Lanka E, Schuster H. 1983. The *dnaC* protein of *Escherichia coli*. Purification, physical properties and interaction with *dnaB* protein. *Nucleic Acids Res* **11:** 987–997.

Learn BA, Um SJ, Huang L, McMacken R. 1997. Cryptic single-stranded-DNA binding activities of the phage λ P and *Escherichia coli* DnaC replication initiation proteins facilitate the transfer of *E. coli* DnaB helicase onto DNA. *Proc Natl Acad Sci* **94:** 1154–1159.

LeBowitz JH, McMacken R. 1986. The *Escherichia coli* dnaB replication protein is a DNA helicase. *J Biol Chem* **261:** 4738–4748.

Lee EH, Kornberg A, Hidaka M, Kobayashi T, Horiuchi T. 1989. *Escherichia coli* replication termination protein impedes the action of helicases. *Proc Natl Acad Sci* **86:** 9104–9108.

Lee JK, Moon KY, Jiang Y, Hurwitz J. 2001. The *Schizosaccharomyces pombe* origin recognition complex interacts with multiple AT-rich regions of the replication origin DNA by means of the AT-hook domains of the spOrc4 protein. *Proc Natl Acad Sci* **98:** 13589–13594.

Lee C, Hong B, Choi JM, Kim Y, Watanabe S, Ishimi Y, Enomoto T, Tada S, Kim Y, Cho Y. 2004. Structural basis for inhibition of the replication licensing factor Cdt1 by geminin. *Nature* **430:** 913–917.

Leonard AC, Grimwade JE. 2011. Regulation of DnaA assembly and activity: Taking directions from the genome. *Annu Rev Microbiol* **65:** 19–35.

* Leonard AC, Méchali M. 2013. DNA replication origins. *Cold Spring Harb Perspect Biol* doi: 10.1101/cshperspect.a010116.

Li X, Zhao Q, Liao R, Sun P, Wu X. 2003. The SCF^{Skp2} ubiquitin ligase complex interacts with the human replication licensing factor Cdt1 and regulates Cdt1 degradation. *J Biol Chem* **278:** 30854–30858.

Liu J, Smith CL, DeRyckere D, DeAngelis K, Martin GS, Berger JM. 2000. Structure and function of Cdc6/

Cdc18: Implications for origin recognition and checkpoint control. *Mol Cell* **6:** 637–648.

Liu E, Li X, Yan F, Zhao Q, Wu X. 2004. Cyclin-dependent kinases phosphorylate human Cdt1 and induce its degradation. *J Biol Chem* **279:** 17283–17288.

Liu S, Balasov M, Wang H, Wu L, Chesnokov IN, Liu Y. 2011. Structural analysis of human Orc6 protein reveals a homology with transcription factor TFIIB. *Proc Natl Acad Sci* **108:** 7373–7378.

Lo Y-H, Tsai K-L, Sun Y-J, Chen W-T, Huang C-Y, Hsiao C-D. 2009. The crystal structure of a replicative hexameric helicase DnaC and its complex with single-stranded DNA. *Nucleic Acids Res* **37:** 804–814.

Ludlam AV, McNatt MW, Carr KM, Kaguni JM. 2001. Essential amino acids of *Escherichia coli* DnaC protein in an N-terminal domain interact with DnaB helicase. *J Biol Chem* **276:** 27345–27353.

Lutzmann M, Méchali M. 2008. MCM9 binds Cdt1 and is required for the assembly of prereplication complexes. *Mol Cell* **31:** 190–200.

MacAlpine HK, Gordân R, Powell SK, Hartemink AJ, MacAlpine DM. 2010. *Drosophila* ORC localizes to open chromatin and marks sites of cohesin complex loading. *Genome Res* **20:** 201–211.

Maiorano D, Moreau J, Méchali M. 2000. XCDT1 is required for the assembly of pre-replicative complexes in *Xenopus laevis*. *Nature* **404:** 622–625.

Maiorano D, Cuvier O, Danis E, Méchali M. 2005. MCM8 is an MCM2-7-related protein that functions as a DNA helicase during replication elongation and not initiation. *Cell* **120:** 315–328.

* Makarova KS, Koonin EV. 2013. Archaeology of eukaryotic DNA replication. *Cold Spring Harb Perspect Biol* doi: 10.1101/cshperspect.a012963.

Makowska-Grzyska M, Kaguni JM. 2010. Primase directs the release of DnaC from DnaB. *Mol Cell* **37:** 90–101.

Marszalek J, Kaguni JM. 1994. DnaA protein directs the binding of DnaB protein in initiation of DNA replication in *Escherichia coli*. *J Biol Chem* **269:** 4883–4890.

Mastrangelo IA, Hough PV, Wall JS, Dodson M, Dean FB, Hurwitz J. 1989. ATP-dependent assembly of double hexamers of SV40 T antigen at the viral origin of DNA replication. *Nature* **338:** 658–662.

Matson SW, Tabor S, Richardson CC. 1983. The gene 4 protein of bacteriophage T7. Characterization of helicase activity. *J Biol Chem* **258:** 14017–14024.

McGarry TJ, Kirschner MW. 1998. Geminin, an inhibitor of DNA replication, is degraded during mitosis. *Cell* **93:** 1043–1053.

Messer W, Weigel C. 1997. DnaA initiator—also a transcription factor. *Mol Microbiol* **24:** 1–6.

Mimura S, Masuda T, Matsui T, Takisawa H. 2000. Central role for Cdc45 in establishing an initiation complex of DNA replication in *Xenopus* egg extracts. *Genes Cells* **5:** 439–452.

Mizushima T, Tomura A, Shinpuku T, Miki T, Sekimizu K. 1994. Loss of flagellation in *dnaA* mutants of *Escherichia coli*. *J. Bacteriol.* **176:** 5544–5546.

Molt KL, Sutera VA, Moore KK, Lovett ST. 2009. A role for nonessential domain II of initiator protein, DnaA, in replication control. *Genetics* **183:** 39–49.

Mott ML, Berger JM. 2007. DNA replication initiation: Mechanisms and regulation in bacteria. *Nat Rev Microbiol* **5:** 343–354.

Mott ML, Erzberger JP, Coons MM, Berger JM. 2008. Structural synergy and molecular crosstalk between bacterial helicase loaders and replication initiators. *Cell* **135:** 623–634.

Muramatsu S, Hirai K, Tak Y-S, Kamimura Y, Araki H. 2010. CDK-dependent complex formation between replication proteins Dpb11, Sld2, Pol ε, and GINS in budding yeast. *Genes Dev* **24:** 602–612.

Nguyen VQ, Co C, Irie K, Li JJ. 2000. Clb/Cdc28 kinases promote nuclear export of the replication initiator proteins Mcm2–7. *Curr Biol* **10:** 195–205.

Nishitani H, Lygerou Z, Nishimoto T, Nurse P. 2000. The Cdt1 protein is required to license DNA for replication in fission yeast. *Nature* **404:** 625–628.

Nozaki S, Ogawa T. 2008. Determination of the minimum domain II size of *Escherichia coli* DnaA protein essential for cell viability. *Microbiology* **154:** 3379–3384.

Nozaki S, Niki H, Ogawa T. 2009. Replication initiator DnaA of *Escherichia coli* changes its assembly form on the replication origin during the cell cycle. *J. Bacteriol.* **191:** 4807–4814.

O'Donnell M. 2006. Replisome architecture and dynamics in *Escherichia coli*. *J Biol Chem* **281:** 10653–10656.

Ozaki S, Katayama T. 2009. DnaA structure, function, and dynamics in the initiation at the chromosomal origin. *Plasmid* **62:** 71–82.

Ozaki S, Katayama T. 2011. Highly organized DnaA–*oriC* complexes recruit the single-stranded DNA for replication initiation. *Nucleic Acids Res* **40:** 1648–1665.

Ozaki S, Kawakami H, Nakamura K, Fujikawa N, Kagawa W, Park S-Y, Yokoyama S, Kurumizaka H, Katayama T. 2008. A common mechanism for the ATP-DnaA-dependent formation of open complexes at the replication origin. *J Biol Chem* **283:** 8351–8362.

Patel SS, Pandey M, Nandakumar D. 2011. Dynamic coupling between the motors of DNA replication: Hexameric helicase, DNA polymerase, and primase. *Curr Opin Chem Biol* **15:** 595–605.

Perkins G, Diffley JF. 1998. Nucleotide-dependent prereplicative complex assembly by Cdc6p, a homolog of eukaryotic and prokaryotic clamp-loaders. *Mol Cell* **2:** 23–32.

Randell JCW, Bowers JL, Rodríguez HK, Bell SP. 2006. Sequential ATP hydrolysis by Cdc6 and ORC directs loading of the Mcm2-7 helicase. *Mol Cell* **21:** 29–39.

Randell JCW, Fan A, Chan C, Francis LI, Heller RC, Galani K, Bell SP. 2010. Mec1 is one of multiple kinases that prime the Mcm2-7 helicase for phosphorylation by Cdc7. *Mol Cell* **40:** 353–363.

Reha-Krantz LJ, Hurwitz J. 1978. The *dnaB* gene product of *Escherichia coli*. II. Single stranded DNA-dependent ribonucleoside triphosphatase activity. *J Biol Chem* **253:** 4051–4057.

Remus D, Diffley JFX. 2009. Eukaryotic DNA replication control: Lock and load, then fire. *Curr Opin Cell Biol* **21:** 771–777.

Remus D, Beall EL, Botchan MR. 2004. DNA topology, not DNA sequence, is a critical determinant for *Drosophila* ORC–DNA binding. *EMBO J* **23:** 897–907.

Remus D, Beuron F, Tolun G, Griffith JD, Morris EP, Diffley JFX. 2009. Concerted loading of Mcm2–7 double hexamers around DNA during DNA replication origin licensing. *Cell* **139:** 719–730.

* Rhind N, Gilbert DM. 2013. DNA replication timing. *Cold Spring Harb Perspect Biol* doi: 10.1101/cshperspect.a010132.

Richardson RW, Nossal NG. 1989. Characterization of the bacteriophage T4 gene 41 DNA helicase. *J Biol Chem* **264:** 4725–4731.

Sclafani RA, Holzen TM. 2007. Cell cycle regulation of DNA replication. *Annu Rev Genet* **41:** 237–280.

Seitz H, Weigel C, Messer W. 2000. The interaction domains of the DnaA and DnaB replication proteins of *Escherichia coli*. *Mol Microbiol* **37:** 1270–1279.

Sekimizu K, Kornberg A. 1988. Cardiolipin activation of dnaA protein, the initiation protein of replication in *Escherichia coli*. *J Biol Chem* **263:** 7131–7135.

Sekimizu K, Bramhill D, Kornberg A. 1987. ATP activates dnaA protein in initiating replication of plasmids bearing the origin of the *E. coli* chromosome. *Cell* **50:** 259–265.

Semple JW, Da-Silva LF, Jervis EJ, Ah-Kee J, Al-Attar H, Kummer L, Heikkila JJ, Pasero P, Duncker BP. 2006. An essential role for Orc6 in DNA replication through maintenance of pre-replicative complexes. *EMBO J* **25:** 5150–5158.

Shen Z, Sathyan KM, Geng Y, Zheng R, Chakraborty A, Freeman B, Wang F, Prasanth KV, Prasanth SG. 2010. A WD-repeat protein stabilizes ORC binding to chromatin. *Mol Cell* **40:** 99–111.

Sheu Y-J, Stillman B. 2006. Cdc7-Dbf4 phosphorylates MCM proteins via a docking site-mediated mechanism to promote S phase progression. *Mol Cell* **24:** 101–113.

Sheu Y-J, Stillman B. 2010. The Dbf4-Cdc7 kinase promotes S phase by alleviating an inhibitory activity in Mcm4. *Nature* **463:** 113–117.

* Siddiqui K, On KF, Diffley JFX. 2013. Regulating DNA replication in eukarya. *Cold Spring Harb Perspect Biol* doi: 10.1101/cshperspect.a012930.

Simmons LA, Felczak M, Kaguni JM. 2003. DnaA protein of *Escherichia coli*: Oligomerization at the *E. coli* chromosomal origin is required for initiation and involves specific N-terminal amino acids. *Mol Microbiol* **49:** 849–858.

* Skarstad K, Katayama T. 2013. Regulating DNA replication in bacteria. *Cold Spring Harb Perspect Biol* doi: 10.1101/cshperspect.a012922.

Speck C, Chen Z, Li H, Stillman B. 2005. ATPase-dependent cooperative binding of ORC and Cdc6 to origin DNA. *Nat Struct Mol Biol* **12:** 965–971.

Sun J, Kawakami H, Zech J, Speck C, Stillman B, Li H. 2012. Cdc6-induced conformational changes in ORC bound to origin DNA revealed by cryo-electron microscopy. *Structure* **20:** 534–544.

Sutton MD, Kaguni JM. 1997. Threonine 435 of *Escherichia coli* DnaA protein confers sequence-specific DNA binding activity. *J Biol Chem* **272:** 23017–23024.

Sutton MD, Carr KM, Vicente M, Kaguni JM. 1998. *Escherichia coli* DnaA protein. The N-terminal domain and loading of DnaB helicase at the *E. coli* chromosomal origin. *J Biol Chem* **273:** 34255–34262.

Szalewska-Pałasz A, Wegrzyn A, Błaszczak A, Taylor K, Wegrzyn G. 1998. DnaA-stimulated transcriptional activation of *ori*λ: *Escherichia coli* RNA polymerase β subunit as a transcriptional activator contact site. *Proc Natl Acad Sci* **95:** 4241–4246.

Takahashi TS, Wigley DB, Walter JC. 2005. Pumps, paradoxes and ploughshares: Mechanism of the MCM2–7 DNA helicase. *Trends Biochem Sci* **30:** 437–444.

Takara TJ, Bell SP. 2011. Multiple Cdt1 molecules act at each origin to load replication-competent Mcm2–7 helicases. *EMBO J* **30:** 4885–4896.

Tanaka S, Diffley JFX. 2002. Interdependent nuclear accumulation of budding yeast Cdt1 and Mcm2–7 during G1 phase. *Nat Cell Biol* **4:** 198–207.

Tanaka T, Knapp D, Nasmyth K. 1997. Loading of an Mcm protein onto DNA replication origins is regulated by Cdc6p and CDKs. *Cell* **90:** 649–660.

Tanaka S, Umemori T, Hirai K, Muramatsu S, Kamimura Y, Araki H. 2007. CDK-dependent phosphorylation of Sld2 and Sld3 initiates DNA replication in budding yeast. *Nature* **445:** 328–332.

Tanaka S, Nakato R, Katou Y, Shirahige K, Araki H. 2011. Origin association of Sld3, Sld7, and Cdc45 proteins is a key step for determination of origin-firing timing. *Curr Biol* **21:** 2055–2063.

Teer JK, Dutta A. 2008. Human Cdt1 lacking the evolutionarily conserved region that interacts with MCM2–7 is capable of inducing re-replication. *J Biol Chem* **283:** 6817–6825.

Tsakraklides V, Bell SP. 2010. Dynamics of pre-replicative complex assembly. *J Biol Chem* **285:** 9437–9443.

Tsuyama T, Tada S, Watanabe S, Seki M, Enomoto T. 2005. Licensing for DNA replication requires a strict sequential assembly of Cdc6 and Cdt1 onto chromatin in *Xenopus* egg extracts. *Nucleic Acids Res* **33:** 765–775.

Vashee S, Cvetic C, Lu W, Simancek P, Kelly TJ, Walter JC. 2003. Sequence-independent DNA binding and replication initiation by the human origin recognition complex. *Genes Dev* **17:** 1894–1908.

Venkatesan M, Silver LL, Nossal NG. 1982. Bacteriophage T4 gene 41 protein, required for the synthesis of RNA primers, is also a DNA helicase. *J Biol Chem* **257:** 12426–12434.

Vermeulen M, Eberl HC, Matarese F, Marks H, Denissov S, Butter F, Lee KK, Olsen JV, Hyman AA, Stunnenberg HG, et al. 2010. Quantitative interaction proteomics and genome-wide profiling of epigenetic histone marks and their readers. *Cell* **142:** 967–980.

Volkening M, Hoffmann I. 2005. Involvement of human MCM8 in prereplication complex assembly by recruiting hcdc6 to chromatin. *Mol Cell Biol* **25:** 1560–1568.

Wahle E, Lasken RS, Kornberg A. 1989a. The dnaB-dnaC replication protein complex of *Escherichia coli*. I. Formation and properties. *J Biol Chem* **264:** 2463–2468.

Wahle E, Lasken RS, Kornberg A. 1989b. The dnaB-dnaC replication protein complex of *Escherichia coli*. II. Role of the complex in mobilizing dnaB functions. *J Biol Chem* **264:** 2469–2475.

Wei Z, Liu C, Wu X, Xu N, Zhou B, Liang C, Zhu G. 2010. Characterization and structure determination of the

Cdt1 binding domain of human minichromosome maintenance (Mcm) 6. *J Biol Chem* **285:** 12469–12473.

Wickner S, Hurwitz J. 1975. Interaction of *Escherichia coli dnaB* and *dnaC(D)* gene products *in vitro*. *Proc Natl Acad Sci* **72:** 921–925.

Wohlschlegel JA, Dwyer BT, Dhar SK, Cvetic C, Walter JC, Dutta A. 2000. Inhibition of eukaryotic DNA replication by geminin binding to Cdt1. *Science* **290:** 2309–2312.

Yanagi K-I, Mizuno T, You Z, Hanaoka F. 2002. Mouse geminin inhibits not only Cdt1-MCM6 interactions but also a novel intrinsic Cdt1 DNA binding activity. *J Biol Chem* **277:** 40871–40880.

Yardimci H, Loveland AB, Habuchi S, van Oijen AM, Walter JC. 2010. Uncoupling of sister replisomes during eukaryotic DNA replication. *Mol Cell* **40:** 834–840.

Ying CY, Gautier J. 2005. The ATPase activity of MCM2–7 is dispensable for pre-RC assembly but is required for DNA unwinding. *EMBO J* **24:** 4334–4344.

You Z, Masai H. 2008. Cdt1 forms a complex with the minichromosome maintenance protein (MCM) and activates its helicase activity. *J Biol Chem* **283:** 24469–24477.

Yu Z, Feng D, Liang C. 2004. Pairwise interactions of the six human MCM protein subunits. *J Mol Biol* **340:** 1197–1206.

Zegerman P, Diffley JFX. 2007. Phosphorylation of Sld2 and Sld3 by cyclin-dependent kinases promotes DNA replication in budding yeast. *Nature* **445:** 281–285.

Zhang J, Yu L, Wu X, Zou L, Sou KKL, Wei Z, Cheng X, Zhu G, Liang C. 2010. The interacting domains of hCdt1 and hMcm6 involved in the chromatin loading of the MCM complex in human cells. *Cell Cycle* **9:** 4848–4857.

* Zielke N, Edgar BA, DePamphilis ML. 2013. Endoreplication. *Cold Spring Harb Perspect Biol* **5:** a012948.

Helicase Activation and Establishment of Replication Forks at Chromosomal Origins of Replication

Seiji Tanaka and Hiroyuki Araki

Division of Microbial Genetics, National Institute of Genetics, and Department of Genetics, SOKENDAI, Mishima, Shizuoka 411-8540, Japan

Correspondence: hiaraki@lab.nig.ac.jp

Many replication proteins assemble on the pre-RC-formed replication origins and constitute the pre-initiation complex (pre-IC). This complex formation facilitates the conversion of Mcm2–7 in the pre-RC to an active DNA helicase, the Cdc45–Mcm–GINS (CMG) complex. Two protein kinases, cyclin-dependent kinase (CDK) and Dbf4-dependent kinase (DDK), work to complete the formation of the pre-IC. Each kinase is responsible for a distinct step of the process in yeast; Cdc45 associates with origins in a DDK-dependent manner, whereas the association of GINS with origins depends on CDK. These associations with origins also require specific initiation proteins: Sld3 for Cdc45; and Dpb11, Sld2, and Sld3 for GINS. Functional homologs of these proteins exist in metazoa, although pre-IC formation cannot be separated by requirement of DDK and CDK because of experimental limitations. Once the replicative helicase is activated, the origin DNA is unwound, and bidirectional replication forks are established.

The main events at the initiation step of DNA replication are the unwinding of double-stranded DNA and subsequent recruitment of DNA polymerases, to start DNA synthesis. Eukaryotic cells require an active DNA helicase to unwind the origin DNA. The core components of the replicative helicase, Mcm2–7, are loaded as a head-to-head double hexamer connected via their amino-terminal rings (Evrin et al. 2009; Remus et al. 2009; Gambus et al. 2011) onto Orc-associated origins, to form the pre-RC in late M and G_1 phases (see Bell and Kaguni 2013). However, Mcm2–7 alone does not show DNA helicase activity at replication origins. After the formation of the pre-RC, other replication factors assemble on origins, and the pre-initiation complex (pre-IC) is formed. The pre-IC is defined as a complex formed just before the initiation of DNA replication (Zou and Stillman 1998); in yeast, it contains at least seven additional factors: Cdc45, GINS, Dpb11, Sld2, Sld3, Cdc45, and DNA polymerase ε (Pol ε) (Muramatsu et al. 2010). The formation of the pre-IC is a prerequisite for the activation of the Mcm2–7 helicase; two additional factors, Cdc45 and GINS, associate with Mcm2–7 and form a tight complex, the Cdc45–Mcm–GINS (CMG) complex (Gambus et al. 2006; Moyer

Cite this article as *Cold Spring Harb Perspect Biol* doi: 10.1101/cshperspect.a010371

et al. 2006). This reaction requires components of the pre-IC and two protein kinases, cyclin-dependent kinase (CDK) and Dbf4-dependent kinase (DDK) (for reviews, see Labib 2010; Masai et al. 2010; Tanaka and Araki 2010). In this article, we summarize and discuss the manner via which the pre-IC is formed in yeasts and metazoa. Although there are some discrepancies, the process of formation of the pre-IC is conserved fairly well in these organisms.

OVERVIEW OF THE FORMATION OF THE PRE-IC IN THE BUDDING YEAST *Saccharomyces cerevisiae*

The initiation reaction is best understood in a unicellular model eukaryote, the budding yeast *Saccharomyces cerevisiae*. The outline of the reaction is shown schematically in Figure 1. The reactions involved in the formation of the pre-IC can be further separated into two steps in yeast cells according to the requirements of the two protein kinases, CDK and DDK, which are essential for the initiation of chromosomal DNA replication. In the first step, Sld3–Sld7–Cdc45 associates with the pre-RC-formed origins in a DDK-dependent manner (Heller et al. 2011; Tanaka et al. 2011a). In the second step, Sld2, Dpb11, Pol ε, and GINS associate with the origins that bind to Sld3 and Cdc45, in a CDK-dependent manner (Masumoto et al. 2002; Tak et al. 2006; Tanaka et al. 2007b; Zegerman and Diffley 2007; Muramatsu et al. 2010; Heller et al. 2011). After assembly, the CMG complex, which is a holoreplicative helicase, is formed (Gambus et al. 2006; Moyer et al. 2006). Because the Mcm2–7 complex embraces the double-stranded origin DNA in the pre-RC (Evrin et al. 2009; Remus et al. 2009) and the single-stranded template of the leading strand at the replication forks (Fu et al. 2011), the CMG complex somehow encircles the single-stranded DNA to unwind the double-stranded DNA. Then, DNA polymerase α (Pol α), which associates tightly

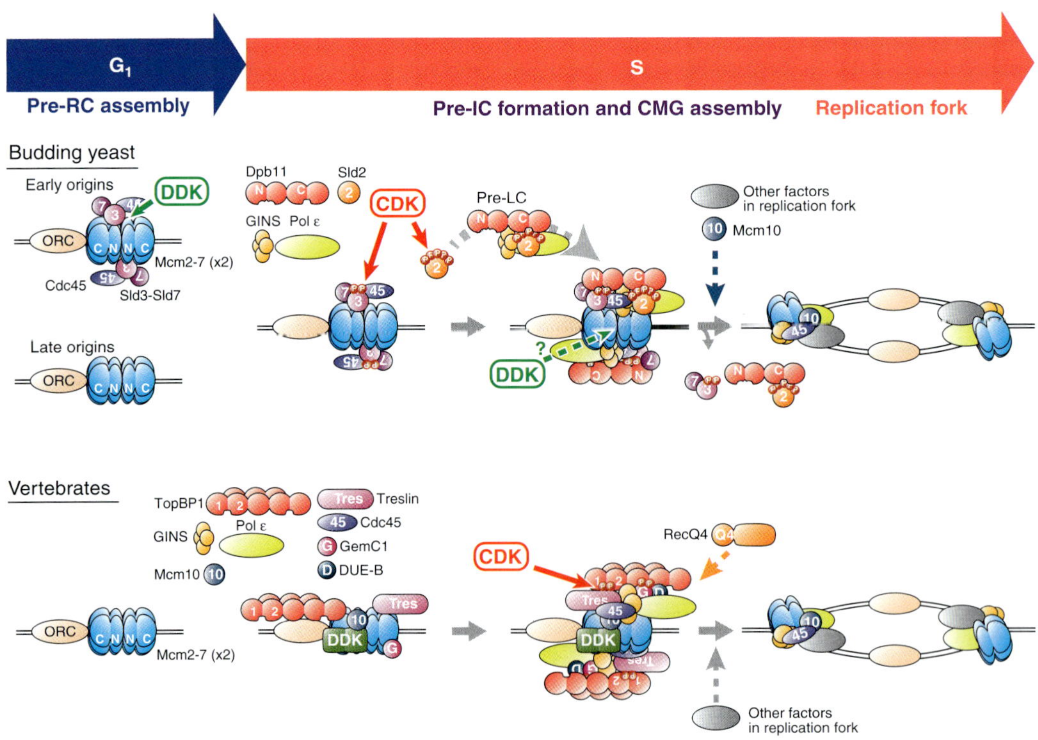

Figure 1. Schematic drawings of the formation of the pre-IC and the initiation of DNA replication. See text for details.

with primase, is recruited to the unwound origins, to start DNA synthesis.

DDK-Dependent Association of Cdc45 with Origins

DDK consists of Cdc7 catalytic and Dbf4 regulatory subunits (Sclafani 2000). Its activity is regulated by the level of the Dbf4 protein, which increases at G_1/S boundary; is kept high in S phase; and decreases from late M phase (Cheng et al. 1999; Oshiro et al. 1999; Ferreira et al. 2000). DDK facilitates the association of Sld3, Sld7, and Cdc45 with the pre-RC-formed origins, in vivo as well as in vitro (Heller et al. 2011; Tanaka et al. 2011a). Sld3 and Sld7 are the subunits of the tight complex (the Sld3–Sld7 complex), which loosely associates with Cdc45 (Tanaka et al. 2011b). Sld3 and Cdc45 associate with origins in a mutually dependent manner (Kamimura et al. 2001), whereas Sld7 is dispensable but is required for efficient initiation of DNA replication (Tanaka et al. 2011b). The DDK-dependent association of Sld3–Sld7–Cdc45 with origins seems to be weak, because this association is detected only via chromatin-immunoprecipitation assay using cross-linking reagents (Kamimura et al. 2001; Kanemaki and Labib 2006). When CDK is activated at the G_1/S boundary, association of Cdc45 with chromatin, which is usually examined in metazoa, is detected in the absence of cross-linking reagents (Zou and Stillman 1998); however, this association requires Dpb11, Sld2, and GINS.

Accumulating evidence strongly suggests that DDK phosphorylates the Mcm2–7 complex and that this phosphorylation step enhances the association between Sld3–Sld7–Cdc45 and origins. Eukaryotic Mcm2, Mcm4, and Mcm6 have extended unstructured serine/threonine-rich domains (NSD) at their amino termini (Fig. 2), which are extensively phosphorylated by DDK in vitro (Randell et al. 2010). Consistent with the in vitro phosphorylation, the combination of mutant Mcm4 and Mcm6, both of which have multiple alanine substitutions at DDK phosphorylation sites, does not support cell growth (Fig. 2) (Randell et al. 2010). Moreover, mutations of Mcm4 and Mcm5 are reported to bypass the requirement of DDK (Hardy et al. 1997; Sheu and Stillman 2010). Furthermore, the results of a recently reported in vitro replication assay suggest that DDK-catalyzed phosphorylation of the pre-RC is required for the association of the pre-RC with Sld3–Sld7–Cdc45 (Heller et al. 2011). This assay consists of three consecutive steps: incubation of immobilized template DNA with a G_1-phase extract (formation of the pre-RC), treatment of the pre-RC with DDK after separation from the extract, and incubation with an S-phase extract (DNA synthesis). The S-phase extract is prepared from cells that bear the *cdc7–4* temperature-sensitive mutation and are arrested at a

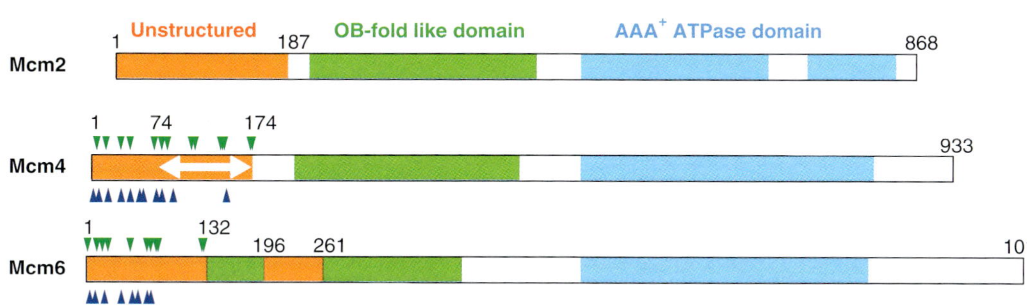

Figure 2. Schematic drawings of the structure of budding yeast Mcm2, Mcm4, and Mcm6. (Orange) Unstructured amino-terminal portion, (light green) conserved OB-fold like domain, (light blue) conserved AAA^+-type ATPase domain. The numbers on *top* indicate the positions of amino acid residues. (White arrow) The amino-terminal portion of Mcm4 corresponding to the Δ74–174 mutation. The positions of alanine substitutions in the phosphorylation site mutants (Randell et al. 2010) are shown by arrowheads; (green) AD/E + ASP/Q, (blue) AP + AQ.

nonpermissive temperature. The step of treatment of the pre-RC with DDK is indispensable for the initiation of DNA replication and for association of Sld3 and Cdc45 with the template DNA in this assay (Heller et al. 2011).

The manner via which phosphorylation of the Mcm2–7 complex promotes the interaction between this complex and Sld3–Sld7–Cdc45 remains obscure. However, it is likely that the phosphorylation prevents unknown inhibitory effects during the interaction. In budding yeast, this inhibitory effect is intrinsic, at least in the case of Mcm4 (Sheu and Stillman 2010). The NSD (1–174) of Mcm4 contains redundant phosphorylation sites for DDK, followed by a "docking region" for DDK in the structured amino-terminal domain (Fig. 2) (Sheu and Stillman 2006). Furthermore, the region located between residues 74 and 174 has inhibitory activity for the initiation (Fig. 2), because Mcm4 lacking this region (*mcm4*$^{\Delta 74-174}$) bypasses the requirement of DDK (Sheu and Stillman 2010). Thus, phosphorylation of the Mcm4 NSD by DDK promotes the initiation of DNA replication by alleviating inhibitory activity in Mcm4.

Currently, the mechanisms underlying this inhibitory effect are not understood. The study of DDK in fission yeast provides clues regarding this inhibition. In fission yeast, DDK activity is not required for the initiation of DNA replication at high temperature or in the absence of Mrc1 or Rif1 (Matsumoto et al. 2011; Hayano et al. 2012) (see below). Although the same is not true in budding yeast, it is conceivable that the obstacles to the interaction between the pre-RC and Sld3–Sld7–Cdc45, such as chromatin structure, are removed by DDK-dependent phosphorylation.

Interestingly, the association of DDK with Mcm2–7 and the phosphorylation of Mcm2–7 in the pre-RC require prior (priming) phosphorylation of the pre-RC (Francis et al. 2009). DDK is an acidophilic kinase; thus, priming phosphorylation at the +1 position of the DDK phosphorylation site will be favored (Masai et al. 2006; Montagnoli et al. 2006; Sasanuma et al. 2008; Wan et al. 2008). Recently, the DDK phosphorylation sites on Mcm2, Mcm4, and Mcm6 in the pre-RC were determined in vitro, as were the "priming" phosphorylation sites required for DDK phosphorylation in the same proteins (Randell et al. 2010). Two types of priming sites were identified: S/T-P and S/T-Q (Randell et al. 2010). A member of the phosphoinositide 3-kinase family, Mec1 (ATR kinase in budding yeast), is responsible for SQ phosphorylation in S phase. However, the identity of the SP kinase remains unclear (Randell et al. 2010). CDK may be a candidate priming kinase for S/T-P sites because this motif matches the CDK recognition site (S/T-P), and a role for CDKs in modifying Mcm2–7 has been proposed in eukaryotes (Masai et al. 2000; Montagnoli et al. 2006; Devault et al. 2008; Sheu and Stillman 2010).

CDK-Dependent Association of GINS with Origins

GINS is recruited in a CDK-dependent manner to the Sld3–Sld7–Cdc45-associated pre-RC to form the CMG complex. CDK phosphorylates two essential replication proteins, Sld2 and Sld3, and promotes the formation of the Sld2–Dpb11–Sld3 complex (Fig. 1).

At the onset of S phase, S-phase CDK (S-CDK; Clb5–Cdc28 and Clb6–Cdc28 in the budding yeast) is activated. Phosphorylation of Sld2 and Sld3 by S-CDK promotes the interaction with Dpb11 (Masumoto et al. 2002; Tak et al. 2006; Tanaka et al. 2007b; Zegerman and Diffley 2007). Dpb11 has four BRCA1 carboxy-terminal (BRCT) domains. A tandem pair of BRCT domains constitutes a phosphopeptide-binding domain (Glover et al. 2004), and the amino-terminal BRCT pair of Dpb11 interacts with phosphorylated Sld3 and the carboxy-terminal BRCT pair interacts with phosphorylated Sld2 (Tak et al. 2006; Tanaka et al. 2007b; Zegerman and Diffley 2007). The phosphorylation-dependent Dpb11–Sld2 interaction leads to the formation of the pre-loading complex (pre-LC) (Muramatsu et al. 2010). The pre-LC contains phosphorylated Sld2, Dpb11, GINS, and Pol ε. Importantly, formation of the pre-LC requires CDK, whereas it does not require the pre-RC. Once S-CDK is activated, association of pre-LC components with replication origins is observed. Therefore, the pre-LC seems to assemble first via Sld2 phosphorylation, followed by association with replication origins.

On the other hand, Sld3, together with Sld7 and Cdc45, associates with the pre-RC-formed origins in a DDK-dependent manner, as described above (Heller et al. 2011; Tanaka et al. 2011a). This association requires neither CDK activity nor components of the pre-LC; rather, the association of Sld3 and Cdc45 with origins is a prerequisite for that of the pre-LC (Muramatsu et al. 2010). The copy numbers of Sld3, Sld7, and Cdc45 (especially Sld3) are limited in comparison with the number of origins (Mantiero et al. 2011; Tanaka et al. 2011a); thus, most of these proteins associate with origins in G_1 phase. The origins associating with these proteins in G_1 fire early in S phase (Tanaka et al. 2011a). Therefore, it is conceivable that Sld3 on origins is phosphorylated by CDK, associates with the amino-terminal BRCT domains of Dpb11 in the pre-LC, and eventually recruits GINS to origins. After pre-LC association, the origin-DNA–protein complex might be remodeled, and the replicative helicase, the CMG complex, is formed and activated, to establish bidirectional replication forks. Sld3–Sld7, Dpb11, and Sld2 do not move with the replication forks (Fig. 1).

Pol ε in the pre-LC is important for the retention of GINS in this complex (Muramatsu et al. 2010). Pol ε consists of four subunits: Pol2, Dpb2, Dpb3, and Dpb4. Pol2 is a catalytic subunit, Dpb2 is an essential subunit of unknown function, and Dpb3 and Dpb4 are nonessential subunits that contain a histone fold (Hamatake et al. 1990; Morrison et al. 1990; Araki et al. 1991; Ohya et al. 2000). The amino-terminal polymerase domain of Pol2 is dispensable for DNA replication (Dua et al. 1999; Kesti et al. 1999), probably because DNA polymerase δ (Pol δ) compensates for the lack of catalytic activity of Pol ε. However, the carboxy-terminal half of Pol2, which interacts with other subunits, is essential for DNA replication and cell growth (Dua et al. 1999; Kesti et al. 1999). Thus, it is likely that the formation of the pre-LC at the initiation step of DNA replication is an essential function of Pol ε.

Subsequent Events

After origin unwinding, the ssDNA-binding protein RPA associates with origin DNA and the Pol α/primase complex is recruited. Then primase synthesizes primer RNA, and Pol α synthesizes short DNA strands using the primers. Pol ε and Pol δ elongate the leading and lagging strands, respectively (Burgers 2009). The Mcm10 protein is evolutionarily conserved in the eukaryotic domain of life and is required for DNA replication (Tye 1999). Noticeably, it was identified as a component of replication forks (Gambus et al. 2006) and is required for the stability of the large subunit of Pol α in budding yeast (Ricke and Bielinsky 2004). Studies in yeasts and vertebrates show that Mcm10 is required for the chromatin loading of Cdc45, which suggests a role in the formation of CMG (Wohlschlegel et al. 2002; Gregan et al. 2003; Ricke and Bielinsky 2004; Sawyer et al. 2004). However, a recent in vitro study showed that Mcm10 is loaded to origins after GINS loading, which might indicate that the loading of Mcm10 occurs after CMG formation and is required for the further loading of Pol α and Pol δ (Heller et al. 2011). Mcm10 is not required for the loading of Pol ε in vitro, because Pol ε is loaded to origins as a component of the pre-LC (Muramatsu et al. 2010; Heller et al. 2011). Recent in vivo studies indicate that Mcm10 is required for the unwinding of origin DNA and further loading of Pol α to origins after Cdc45, Mcm2–7, and GINS form a complex (Kanke et al. 2012; van Deursen et al. 2012; Watase et al. 2012). Because CDK and DDK are required for the activation of the replicative helicase, the question of whether CMG formation is sufficient for the loading of Mcm10 is intriguing.

FORMATION OF THE PRE-IC IN THE FISSION YEAST *Schizosaccharomyces pombe*

The orthologs of Dpb11, Sld2, and Sld3 in the fission yeast *Schizosaccharomyces pombe* are Cut5 (also called Rad4), Drc1, and Sld3, respectively (Saka et al. 1994; Nakajima and Masukata 2002; Noguchi et al. 2002). These proteins are well conserved and, thus, easily identified by sequence similarity. The CDK-catalyzed phosphorylation of Drc1 and Sld3 facilitates the interaction between Cut5 and Drc1 and Sld3, as observed in budding yeast (Saka et al. 1994; Nakajima and Masukata 2002; Noguchi et al. 2002; Fukuura et al. 2011). However, phosphorylation of Sld3 is not essential, although it is important, for the association of Cut5 with origins; the disruption of phosphorylation sites reduces the interaction between Cut5 and Sld3 and renders cells sensitive to low temperature (Nakajima and Masukata 2002; Fukuura et al. 2011). Interestingly, the interaction between Cut5 and Drc1 further enhances the interaction with Sld3 (Fukuura et al. 2011). Sld3 associates with replication origins before initiation in a DDK-dependent manner (Yabuuchi et al. 2006), and its association with origins is a prerequisite for the subsequent recruitment of Cut5, GINS, Drc1, and Cdc45 (Yabuuchi et al. 2006). Pol ε plays an essential role in the assembly of the CMG complex (Handa et al. 2012). Moreover, the DNA polymerase domain of the catalytic subunit of Pol ε is dispensable for replication, whereas the carboxy-terminal half of the catalytic subunit, which functions in the recruitment of GINS as a component of the pre-LC in budding yeast, is essential, as observed in budding yeast (Feng and D'Urso 2001). Therefore, the mechanism of formation of the pre-IC is conserved well in fission and budding yeasts, although the manner via which Cdc45 and GINS are recruited onto origins in fission yeast remains obscure.

In fission yeast, DDK is also required for the initiation of replication. As in budding yeast, Mcm4 and Mcm6 are phosphorylated in a DDK-dependent manner, and alanine-substitution mutations in their DDK phosphorylation sites cannot be combined (Masai et al. 2006). Although several mutations in Mcm4 and Mcm5 can bypass the requirement of DDK in budding yeast, a similar rescue has not been reported in fission yeast. Noticeably, high temperature or deletion of *mrc1* or *rif1* can bypass the requirement of DDK in fission yeast (Matsumoto et al. 2011; Hayano et al. 2012). High temperature may ease the unwinding of double-stranded DNA. Mrc1 functions at checkpoints and stabilization of paused replication forks (Alcasabas et al. 2001; Tanaka and Russell 2001; Katou et al. 2003; Osborn and Elledge 2003). The bypass of the DDK requirement in *mrc1Δ* is independent of its checkpoint function (Matsumoto et al. 2011). Mrc1 associates with early-firing origins and is likely to mark these origins, although *mrc1Δ* early-firing origins fire earlier than do those of the wild type (Hayano et al. 2011). In the budding yeast, Mrc1 facilitates the phosphorylation of the priming sites of chromatin-bound Mcm2–7 by Mec1 in S phase, although this might be part of the checkpoint function of Mrc1 (Randell et al. 2010). Rif1 binds to many chromosome-arm loci as well as telomeres and affects the timing of origin firing (Hayano et al. 2012). In summary, although DDK is required for the association of Sld3 with origins, the requirement of DDK in fission yeast may be less than that observed in budding yeast. Further analyses are needed to understand the DDK-dependent process in this organism.

FORMATION OF THE PRE-IC IN METAZOA

The replication machinery, such as DNA polymerases and DNA helicase components (Mcm2–7, Cdc45, and GINS), is well conserved in eukaryotic organisms. In contrast, the regulatory components have diverged during evolution. Dpb11, Sld2, and Sld3 are examples of this divergence. These factors sense CDK activity and promote the initiation of chromosomal DNA replication, as described above. Because the cellular environment differs among organisms, the sensing mechanism of CDK activity may have diverged. Nonetheless, TopBP1 (also denoted Cut5 and Mus101), RecQ4 (also denoted RecQL4), and Treslin/Ticrr are reported as functional homologs of Dpb11, Sld2, and Sld3, respectively. Each of them shows limited sequence similarity to their yeast counterparts, respectively, and all of them are much larger in size than their yeast counterparts (Fig. 3) (Makiniemi et al. 2001; Van Hatten et al. 2002; Hashimoto and Takisawa 2003; Garcia et al. 2005; Sangrithi et al. 2005; Matsuno et al. 2006;

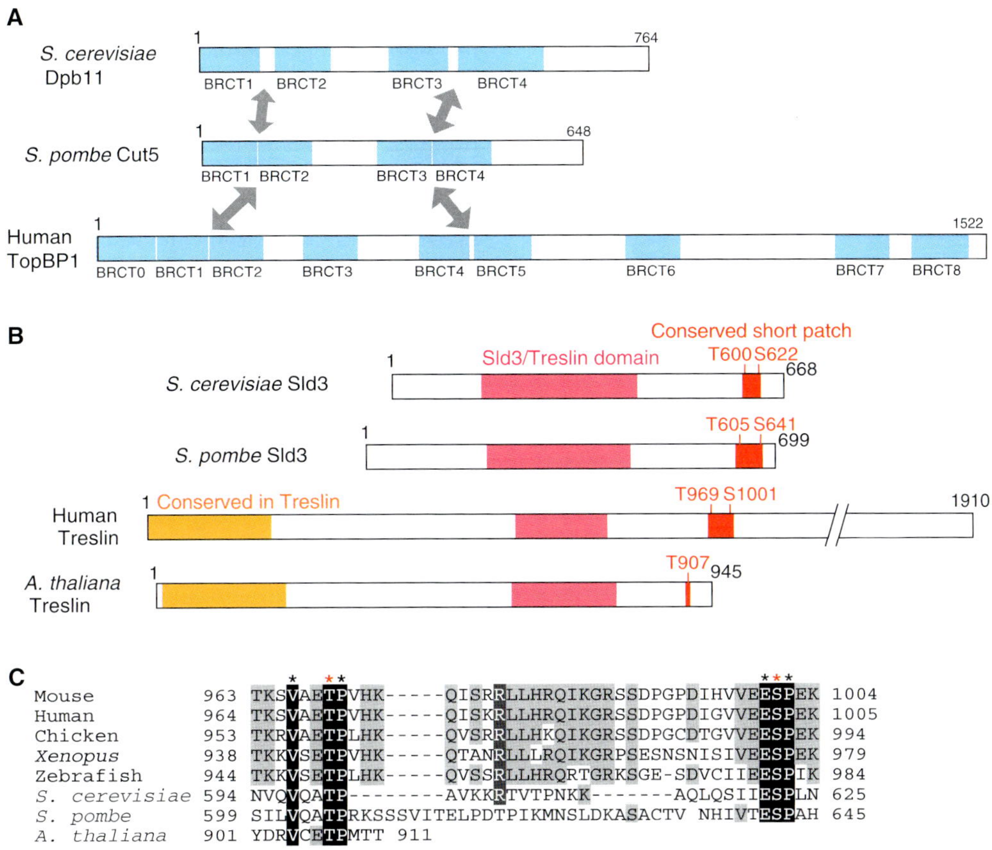

Figure 3. Schematic drawings of Dpb11/TopBP1 and Sld3/Treslin/Ticrr proteins. (*A*) Dpb11/TopBP1 proteins. (Light blue) BRCT domains. (*B*) Sld3/Treslin/Ticrr proteins. (Magenta) The Sld3/Treslin domain; (red) the conserved short patch that contains conserved CDK phosphorylation sites; (orange) the amino-terminal region that is conserved in animal and plant Treslin. (*C*) Multiple alignments of amino acid sequences that contain conserved CDK phosphorylation sites. (Red asterisks in the *top* row) Conserved serine and threonine.

Kumagai et al. 2010; Sansam et al. 2010). Although these proteins are required for the initiation of DNA replication, the requirement for, and order of, association with chromatin vary slightly among them.

TopBP1

TopBP1 is involved in DNA replication and checkpoints, as observed for its yeast counterparts (Garcia et al. 2005; Walter and Araki 2006; Tanaka et al. 2007a; Labib 2010; Tanaka and Araki 2010). Although yeast Dpb11/Cut5 has four BRCT domains, metazoan TopBP1s have multiple BRCTs (more than six). Moreover, vertebrate TopBP1 has eight BRCTs (BRCT1–8), and its amino-terminal portion, which contains five BRCTs (BRCT1–5), shows sequence similarity to the yeast counterparts, with the exception of BRCT3 (Fig. 3A) (Garcia et al. 2005). A recent structural analysis of human TopBP1 showed that vertebrate TopBP1 has one additional BRCT domain at the very end of its amino terminus; this BRCT domain is called BRCT0, to maintain consistency with the existing nomenclature (Rappas et al. 2011).

TopBP1 is required for the loading of Cdc45, RecQ4, and Pol ε onto chromatin, but not for pre-RC assembly, chromatin loading of Treslin/Ticrr, or the elongation stage of DNA replication (Van Hatten et al. 2002; Hashimoto and Takisawa 2003; Kumagai et al. 2010). Intriguingly, TopBP1 associates with chromatin in two distinctive modes: CDK-independent low-level and CDK-dependent high-level associations. The CDK-independent association of TopBP1 with chromatin depends on ORC but not on the pre-RC, whereas CDK-dependent association depends on the pre-RC. The CDK-independent low-level association of TopBP1 with chromatin fully supports DNA replication (Van Hatten et al. 2002; Hashimoto and Takisawa 2003). A recent study showed that the amino-terminal portion of TopBP1, which contains BRCT0–3 (originally denoted as BRCT I–III), is sufficient to fully support DNA replication in *Xenopus* egg extracts, and that BRCT4–5, which shows similarity with the carboxy-terminal BRCT pair of the yeast counterparts, are dispensable for DNA replication (Kumagai et al. 2010). Such dispensability of BRCT4–5 of TopBP1 might reflect the fact that vertebrate RecQ4 does not require CDK for its association with TopBP1 and chromatin (see below). Moreover, Cdc45 binds to BRCT0–2 and BRCT6 of TopBP1 (Schmidt et al. 2008), whereas such bindings have not been reported in yeast.

Treslin/Ticrr

Treslin/Ticrr was identified via screening for TopBP1-binding proteins (Treslin) in *Xenopus* (Kumagai et al. 2010) and for G_2/M checkpoint regulators in zebrafish (Ticrr) (Sansam et al. 2010). Subsequently, the sequence homology between Sld3 and Treslin/Ticrr was shown using a sophisticated method, and the Sld3–Treslin/Ticrr domain (STD), which is conserved across eukaryotic species, including animals, fungi, and plants, was identified (Fig. 3B) (Sanchez-Pulido et al. 2010). Interestingly, Treslin/Ticrr family members in animals and plants have an additional conserved region in their amino terminus (Sanchez-Pulido et al. 2010), although its function is unknown (see below).

Treslin/Ticrr shows many functional similarities to Sld3. A central region of Treslin, located proximal to the carboxy-terminal side of the STD domain, binds to BRCT0–2 of TopBP1, which is similar to BRCT1–2 of Dpb11, a binding domain for phosphorylated Sld3 in budding yeast (Boos et al. 2011; Kumagai et al. 2011). This binding requires CDK-catalyzed phosphorylation of specific serine and threonine. The short stretches of amino acid sequences around these phosphorylation sites are well conserved in animals and fungi (Fig. 3C). Moreover, CDK phosphorylation of these sites is required for DNA replication in vivo (Boos et al. 2011; Kumagai et al. 2011). Treslin also interacts with Cdc45 (Kumagai et al. 2011) and is required for the association of Cdc45 with chromatin. Moreover, the Treslin–TopBP1 interaction is reduced in cells treated with hydroxyurea in a checkpoint kinase 1 (Chk1)-dependent manner (Boos et al. 2011). This is reminiscent of the situation in which the Rad53 checkpoint kinase phosphorylates Sld3 to inhibit initiation by disrupting the Sld3–Dpb11 interaction in budding yeast (Lopez-Mosqueda et al. 2010; Zegerman and Diffley 2010).

The association of Treslin/Ticrr with chromatin, however, is slightly different from that observed for Sld3, because it depends mostly on the pre-RC and not on TopBP1 or CDK (Kumagai et al. 2010). TopBP1 associates with chromatin in a Treslin-independent manner. Thus, it is proposed that the CDK phosphorylation-dependent interaction between TopBP1 and Treslin/Ticrr occurs on chromatin, which renders the complex competent for recruiting Cdc45 (Kumagai et al. 2010). Moreover, the amino terminus of Treslin, which is specific in animal and plant orthologs (Fig. 3B) (Sanchez-Pulido et al. 2010), shows a weak interaction with TopBP1 (Kumagai et al. 2011).

RecQ4

Whether RecQ4 is a functional homolog of Sld2 remains controversial because it seems to work at a subsequent stage of the initiation step in a CDK-independent manner, unlike yeast Sld2/Drc1 (Fig. 1). RecQ4 belongs to the RecQ helicase family of vertebrates, which plays various roles in DNA metabolism (Bachrati and Hickson 2008). Mutations in the carboxy-terminal RecQ helicase domain are responsible for the subset of Rothmund–Thomson, Baller–Gerold, and Rapadilino syndromes, which show increased genetic instability and share key symptoms, such as growth deficiency (Capp et al. 2010). However, the RecQ helicase domain is dispensable for DNA replication, whereas the amino-terminal portion, which has a weak similarity to Sld2, is essential for cell growth and DNA replication (Sangrithi et al. 2005; Matsuno et al. 2006). RecQ4 associates with chromatin in a TopBP1- and pre-RC-dependent manner. Pol α and RPA require RecQ4 for association with chromatin, whereas TopBP1, Cdc45, GINS, and Pol ε are recruited to chromatin in the absence of RecQ4, unlike what is observed in yeast. Moreover, the association of RecQ4 with chromatin and the association between TopBP1 and the RecQ4 amino-terminal domain do not require CDK (Sangrithi et al. 2005; Matsuno et al. 2006). Furthermore, the formation of the CMG complex in HeLa cells requires RecQ4 and Mcm10 (Im et al. 2009), and RecQ4 expressed in human 293 cells binds to Mcm10 and associates with the CMG complex in an Mcm10-dependent manner (Xu et al. 2009). These results suggest that RecQ4 functions later in the initiation step compared with Sld2.

Other Factors

Two additional proteins, GEMC1 (Balestrini et al. 2010) and DUE-B (Chowdhury et al. 2010), that bind to both TopBP1 and Cdc45 are required for the recruitment of Cdc45 to chromatin. However, their actual functions have not been elucidated. The function of metazoan Mcm10 is also controversial; it may function in the formation of the pre-IC or collaborate with AND-1 (the counterpart of yeast Ctf4/Mcl1) in recruiting Pol α to chromatin (Wohlschlegel et al. 2002; Zhu et al. 2007). The function of metazoan Pol ε in the initiation step is obscure. Depletion of Pol ε in *Xenopus* egg extracts reduces DNA replication, but not the association of Cdc45 with chromatin (Waga et al. 2001; Fukui et al. 2004). Knockdown of the catalytic subunit of Pol ε in *Drosophila* inhibits DNA replication in mitotic and endocycling cells (Suyari et al. 2012). Moreover, deletion of the DNA polymerase domain in the catalytic subunit of Pol ε abolishes endoreplication in the salivary gland but does not affect DNA replication in the mitotic cycle (Suyari et al. 2012). Thus, the DNA polymerase activity of Pol ε is dispensable for replication in the *Drosophila* mitotic cycle, similar to what is observed in yeast. This suggests an essential function for the carboxy-terminal half of the catalytic subunit of Pol ε, similar to that of yeast.

In budding yeast, Sld7 was recently identified as an Sld3-interacting protein that is not essential for DNA replication; however, *sld7* deletion slows DNA replication (Tanaka et al. 2011b). Although Sld7 does not show a clear sequence similarity to metazoan proteins, it will be intriguing to determine whether Treslin/Ticrr associates with the Sld7-like protein in vertebrate cells, as do Sld3 and Sld7 in budding yeast.

In addition to CDK, DDK is required for the initiation of DNA replication in metazoa. Studies using *Xenopus* egg extracts showed that DDK could complete its function in the absence of CDK (Jares and Blow 2000; Walter 2000). This is consistent with the recent finding in an in vitro replication assay of budding yeast: The requirement of DDK activity precedes that of CDK activity. As in budding yeast, MCM2–7 is a target of DDK, and their phosphorylation facilitates the loading of Cdc45 onto chromatin in mammalian cells (Masai et al. 2006). An in vitro study showed that DDK phosphorylates mainly the amino-terminal tail domain of MCM2, MCM4, and MCM6 and that this phosphorylation is stimulated by prior phosphorylation by CDK (Masai et al. 2000). As observed in budding yeast, *Xenopus* Mcm2–7 assemble as a head-to-head double hexamer in the pre-RC (Gambus et al. 2011). These results suggest that DDK phosphorylates the NTD of Mcm2/4/6 commonly in eukaryotes to activate the replicative helicase, although the exact role of phosphorylation requires further investigation.

REGULATORY ASPECTS OF THE FORMATION OF THE PRE-IC

DNA replication in eukaryotes initiates from multiple origins of replication (see Leonard and Méchali 2013). It is widely known that the timing of the activation of individual origins in the S phase differs among origins, with the exception of some embryonic cells: Some origins fire early in the S phase (early-firing origins), whereas others fire late in S phase (late-firing origins). The pre-RCs are assembled at all potential origins in G_1 phase, and formation of the pre-ICs determines the timing of origin firing in budding yeast because the replication proteins that assemble at this step are limited. Origins with a high affinity for such factors fire earlier in S phase (Patel et al. 2008; Wu and Nurse 2009; Mantiero et al. 2011; Tanaka et al. 2011a). For details regarding the temporal regulation of replication, see Rhind and Gilbert (2013).

The formation of the pre-IC is also the target of checkpoint kinases in yeast and humans. When the replication fork is stalled by DNA damage or dNTP depletion after hydroxyurea treatment, the DNA damage- or intra-S-phase-checkpoint pathways are activated. Under such conditions, further activation of unfired origins is inhibited (Painter and Young 1980; Santocanale and Diffley 1998; Shirahige et al. 1998; Larner et al. 1999). Studies using budding yeast showed that the checkpoint kinase Rad53 inhibits both the CDK- and DDK-dependent pathways, which are essential for the initiation of replication, by phosphorylating Sld3 and Dbf4, respectively (Lopez-Mosqueda et al. 2010; Zegerman and Diffley 2010). Human Treslin/Ticrr, a functional homolog of Sld3, is also inhibited by the phosphorylation checkpoint kinase Chk1 when cells are exposed to DNA-replication stress (Boos et al. 2011).

Because DNA replication in eukaryotes occurs as a two-step reaction (origin licensing and initiation), these reactions are temporally separated in the cell cycle, to limit the activation of replication origins to once per cell cycle. Failures in this regulation cause rereplication, which is genotoxic (Nguyen et al. 2001; Mimura et al. 2004; Tanaka and Araki 2011). It is well established that multiple regulatory mechanisms prevent relicensing at activated origins after S phase (Arias and Walter 2007) (also see Siddiqui et al. 2013). Because budding yeast G_1 cells have DDK activity (although weak), the bypass of CDK activity can induce DNA replication and cause a loss in cell viability (Tanaka and Araki 2011). To prevent the premature initiation of licensed origins in G_1 phase, budding yeast cells have multiple regulatory mechanisms: The targets of S-CDK, Sld2, and Sld3 are not phosphorylated, and the protein levels of pre-LC components Dpb11 and Sld2 are low. These mechanisms contribute to the prevention of the reaction promoted by S-CDK (Tanaka and Araki 2011). As in the case of prevention of relicensing, the presence of multiple inhibitory mechanisms is important for the prevention of the untimely initiation of chromosomal replication, to preserve the stability of the genome over generations. Although much less is known regarding how premature initiation is prevented during the G_1 phase in other eukaryotes, similar multiple mechanisms might contribute to the prevention of premature initiation because the molecular mechanisms involved in replication initiation seem to be largely conserved.

GENERAL VIEW OF THE FORMATION OF THE PRE-IC

Several replication proteins assemble on replication origins during the formation of the pre-IC to form the active helicase (the CMG complex). This reaction requires two protein kinases: DDK and CDK. DDK facilitates the association of Cdc45 with origins, and CDK promotes the association of GINS with origins. However, the DDK-dependent and CDK-independent association of Cdc45 with origins is detected only via a chromatin-immunoprecipitation assay using cross-linking reagents. Thus, this association of Cdc45 with origins has not been reported in metazoa. Rather, the CDK-dependent association of GINS with origins is well conserved in eukaryotic cells. The interaction between BRCT-containing proteins and CDK-phosphorylated initiation proteins seems to recruit GINS to origins, although the CDK-dependent association of GINS with origins has not been well elucidated in metazoa.

The association of Cdc45 and GINS with origins requires initiation-specific proteins, such as Sld3 (Treslin/Ticrr), Sld2, and BRCT-containing proteins (Dpb11, Cut5, and TopBP1). These proteins are important for connecting DNA replication to the cell cycle via DDK- and CDK-dependent phosphorylation. Moreover, it is conceivable that these proteins are also required for the conversion of the assembly of Mcm2–7, Cdc45, and GINS to the tight CMG complex. Furthermore, the manner via which the CMG complex encircles single-stranded DNA, rather than the interaction with double-stranded DNA by Mcm2–7, in the pre-RC remains unknown.

The order of association of the initiation proteins varies slightly among different eukaryotic cells. However, we must be aware of the execution points of these proteins. The apparent association may vary according to the assay method used. For example, TopBP1 associates with chromatin before CDK activation; however, the execution point seems to occur after the time of association. Future studies will clarify these points.

ACKNOWLEDGMENTS

We apologize to the authors whose work could not be cited here because of space limitations. S.T. and H.A. are supported by the Ministry of Education, Culture, Sports, Science, and Technology of Japan.

REFERENCES

Reference is also in this collection.

Alcasabas AA, Osborn AJ, Bachant J, Hu F, Werler PJ, Bousset K, Furuya K, Diffley JF, Carr AM, Elledge SJ. 2001. Mrc1 transduces signals of DNA replication stress to activate Rad53. *Nat Cell Biol* **3:** 958–965.

Araki H, Hamatake RK, Morrison A, Johnson AL, Johnston LH, Sugino A. 1991. Cloning *DPB3*, the gene encoding the third subunit of DNA polymerase II of *Saccharomyces cerevisiae*. *Nucleic Acids Res* **19:** 4867–4872.

Arias EE, Walter JC. 2007. Strength in numbers: Preventing rereplication via multiple mechanisms in eukaryotic cells. *Genes Dev* **21:** 497–518.

Bachrati CZ, Hickson ID. 2008. RecQ helicases: Guardian angels of the DNA replication fork. *Chromosoma* **117:** 219–233.

Balestrini A, Cosentino C, Errico A, Garner E, Costanzo V. 2010. GEMC1 is a TopBP1-interacting protein required for chromosomal DNA replication. *Nat Cell Biol* **12:** 484–491.

* Bell SD, Botchan MR. 2013. The minichromosome maintenance replicative helicase. *Cold Spring Harb Perspect Biol* doi: 10.1101/cshperspect.a012807.

Boos D, Sanchez-Pulido L, Rappas M, Pearl LH, Oliver AW, Ponting CP, Diffley JF. 2011. Regulation of DNA replication through Sld3–Dpb11 interaction is conserved from yeast to humans. *Curr Biol* **21:** 1152–1157.

Burgers PM. 2009. Polymerase dynamics at the eukaryotic DNA replication fork. *J Biol Chem* **284:** 4041–4045.

Capp C, Wu J, Hsieh TS. 2010. RecQ4: The second replicative helicase? *Crit Rev Biochem Mol Biol* **45:** 233–242.

Cheng L, Collyer T, Hardy CF. 1999. Cell cycle regulation of DNA replication initiator factor Dbf4p. *Mol Cell Biol* **19:** 4270–4278.

Chowdhury A, Liu G, Kemp M, Chen X, Katrangi N, Myers S, Ghosh M, Yao J, Gao Y, Bubulya P, et al. 2010. The DNA unwinding element binding protein DUE-B interacts with Cdc45 in preinitiation complex formation. *Mol Cell Biol* **30:** 1495–1507.

Devault A, Gueydon E, Schwob E. 2008. Interplay between S-cyclin-dependent kinase and Dbf4-dependent kinase in controlling DNA replication through phosphorylation of yeast Mcm4 N-terminal domain. *Mol Biol Cell* **19:** 2267–2277.

Dua R, Levy DL, Campbell JL. 1999. Analysis of the essential functions of the C-terminal protein/protein interaction domain of *Saccharomyces cerevisiae* Pol ε and its unexpected ability to support growth in the absence of the DNA polymerase domain. *J. Biol Chem* **274:** 22283–22288.

Evrin C, Clarke P, Zech J, Lurz R, Sun J, Uhle S, Li H, Stillman B, Speck C. 2009. A double-hexameric MCM2–7 complex is loaded onto origin DNA during licensing of eukaryotic DNA replication. *Proc Natl Acad Sci* **106:** 20240–20245.

Feng W, D'Urso G. 2001. *Schizosaccharomyces pombe* cells lacking the amino-terminal catalytic domains of DNA polymerase epsilon are viable but require the DNA damage checkpoint control. *Mol Cell Biol* **21:** 4495–4504.

Ferreira MF, Santocanale C, Drury LS, Diffley JF. 2000. Dbf4p, an essential S phase-promoting factor, is targeted for degradation by the anaphase-promoting complex. *Mol Cell Biol* **20:** 242–248.

Francis LI, Randell JC, Takara TJ, Uchima L, Bell SP. 2009. Incorporation into the prereplicative complex activates the Mcm2–7 helicase for Cdc7–Dbf4 phosphorylation. *Genes Dev* **23:** 643–654.

Fu YV, Yardimci H, Long DT, Ho TV, Guainazzi A, Bermudez VP, Hurwitz J, van Oijen A, Scharer OD, Walter JC. 2011. Selective bypass of a lagging strand roadblock by the eukaryotic replicative DNA helicase. *Cell* **146:** 931–941.

Fukui T, Yamauchi K, Muroya T, Akiyama M, Maki H, Sugino A, Waga S. 2004. Distinct roles of DNA polymerases δ and ε at the replication fork in *Xenopus* egg extracts. *Genes Cells* **9:** 179–191.

Fukuura M, Nagao K, Obuse C, Takahashi TS, Nakagawa T, Masukata H. 2011. CDK promotes interactions of Sld3 and Drc1 with Cut5 for initiation of DNA replication in fission yeast. *Mol Biol Cell* **22:** 2620–2633.

Gambus A, Jones RC, Sanchez-Diaz A, Kanemaki M, van Deursen F, Edmondson RD, Labib K. 2006. GINS maintains association of Cdc45 with MCM in replisome progression complexes at eukaryotic DNA replication forks. *Nat Cell Biol* **8:** 358–366.

Gambus A, Khoudoli GA, Jones RC, Blow JJ. 2011. MCM2–7 form double hexamers at licensed origins in *Xenopus* egg extract. *J Biol Chem* **286:** 11855–11864.

Garcia V, Furuya K, Carr AM. 2005. Identification and functional analysis of TopBP1 and its homologs. *DNA Repair (Amst)* **4:** 1227–1239.

Glover JN, Williams RS, Lee MS. 2004. Interactions between BRCT repeats and phosphoproteins: Tangled up in two. *Trends Biochem Sci* **29:** 579–585.

Gregan J, Lindner K, Brimage L, Franklin R, Namdar M, Hart EA, Aves SJ, Kearsey SE. 2003. Fission yeast Cdc23/Mcm10 functions after pre-replicative complex formation to promote Cdc45 chromatin binding. *Mol Biol Cell* **14:** 3876–3887.

Hamatake RK, Hasegawa H, Clark AB, Bebenek K, Kunkel TA, Sugino A. 1990. Purification and characterization of DNA polymerase II from the yeast *Saccharomyces cerevisiae*. Identification of the catalytic core and a possible holoenzyme form of the enzyme. *J Biol Chem* **265:** 4072–4083.

Handa T, Kanke M, Takahashi TS, Nakagawa T, Masukata H. 2012. DNA polymerization-independent functions of DNA polymerase ε in assembly and progression of the replisome in fission yeast. *Mol Biol Cell* **23:** 3240–3253.

Hardy CF, Dryga O, Seematter S, Pahl PM, Sclafani RA. 1997. mcm5/cdc46-bob1 bypasses the requirement for the S phase activator Cdc7p. *Proc Natl Acad Sci* **94:** 3151–3155.

Hashimoto Y, Takisawa H. 2003. *Xenopus* Cut5 is essential for a CDK-dependent process in the initiation of DNA replication. *EMBO J* **22:** 2526–2535.

Hayano M, Kanoh Y, Matsumoto S, Masai H. 2011. Mrc1 marks early-firing origins and coordinates timing and efficiency of initiation in fission yeast. *Mol Cell Biol* **31:** 2380–2391.

Hayano M, Kanoh Y, Matsumoto S, Renard-Guillet C, Shirahige K, Masai H. 2012. Rif1 is a global regulator of timing of replication origin firing in fission yeast. *Genes Dev* **26:** 137–150.

Heller RC, Kang S, Lam WM, Chen S, Chan CS, Bell SP. 2011. Eukaryotic origin-dependent DNA replication in vitro reveals sequential action of DDK and S-CDK kinases. *Cell* **146:** 80–91.

Im JS, Ki SH, Farina A, Jung DS, Hurwitz J, Lee JK. 2009. Assembly of the Cdc45–Mcm2-7–GINS complex in human cells requires the Ctf4/And-1, RecQL4, and Mcm10 proteins. *Proc Natl Acad Sci* **106:** 15628–15632.

Jares P, Blow JJ. 2000. *Xenopus* cdc7 function is dependent on licensing but not on XORC, XCdc6, or CDK activity and is required for XCdc45 loading. *Genes Dev* **14:** 1528–1540.

Kamimura Y, Tak YS, Sugino A, Araki H. 2001. Sld3, which interacts with Cdc45 (Sld4), functions for chromosomal DNA replication in *Saccharomyces cerevisiae*. *EMBO J* **20:** 2097–2107.

Kanemaki M, Labib K. 2006. Distinct roles for Sld3 and GINS during establishment and progression of eukaryotic DNA replication forks. *EMBO J* **25:** 1753–1763.

Kanke M, Kodama Y, Takahashi ST, Nakagawa T, Masukata H. 2012. Mcm10 plays an essential role in origin DNA unwinding after loading of the CMG components. *EMBO J* **31:** 2182–2194.

Katou Y, Kanoh Y, Bando M, Noguchi H, Tanaka H, Ashikari T, Sugimoto K, Shirahige K. 2003. S-phase checkpoint proteins Tof1 and Mrc1 form a stable replication-pausing complex. *Nature* **424:** 1078–1083.

Kesti T, Flick K, Keranen S, Syvaoja JE, Wittenberg C. 1999. DNA polymerase ε catalytic domains are dispensable for DNA replication, DNA repair, and cell viability. *Mol Cell* **3:** 679–685.

Kumagai A, Shevchenko A, Dunphy WG. 2010. Treslin collaborates with TopBP1 in triggering the initiation of DNA replication. *Cell* **140:** 349–359.

Kumagai A, Shevchenko A, Dunphy WG. 2011. Direct regulation of Treslin by cyclin-dependent kinase is essential for the onset of DNA replication. *J Cell Biol* **193:** 995–1007.

Labib K. 2010. How do Cdc7 and cyclin-dependent kinases trigger the initiation of chromosome replication in eukaryotic cells? *Genes Dev* **24:** 1208–1219.

Larner JM, Lee H, Little RD, Dijkwel PA, Schildkraut CL, Hamlin JL. 1999. Radiation down-regulates replication origin activity throughout the S phase in mammalian cells. *Nucleic Acids Res* **27:** 803–809.

* Leonard AC, Méchali M. 2013. DNA replication origins. *Cold Spring Harb Perspect Biol* doi: 10.1101/cshperspect.a010116.

Lopez-Mosqueda J, Maas NL, Jonsson ZO, Defazio-Eli LG, Wohlschlegel J, Toczyski DP. 2010. Damage-induced phosphorylation of Sld3 is important to block late origin firing. *Nature* **467:** 479–483.

Makiniemi M, Hillukkala T, Tuusa J, Reini K, Vaara M, Huang D, Pospiech H, Majuri I, Westerling T, Makela TP, et al. 2001. BRCT domain-containing protein TopBP1 functions in DNA replication and damage response. *J Biol Chem* **276:** 30399–30406.

Mantiero D, Mackenzie A, Donaldson A, Zegerman P. 2011. Limiting replication initiation factors execute the temporal programme of origin firing in budding yeast. *EMBO J* **30:** 4805–4814.

Masai H, Matsui E, You Z, Ishimi Y, Tamai K, Arai K. 2000. Human Cdc7-related kinase complex. In vitro phosphorylation of MCM by concerted actions of Cdks and Cdc7 and that of a criticial threonine residue of Cdc7 by Cdks. *J Biol Chem* **275:** 29042–29052.

Masai H, Taniyama C, Ogino K, Matsui E, Kakusho N, Matsumoto S, Kim JM, Ishii A, Tanaka T, Kobayashi T, et al. 2006. Phosphorylation of MCM4 by Cdc7 kinase facilitates its interaction with Cdc45 on the chromatin. *J Biol Chem* **281:** 39249–39261.

Masai H, Matsumoto S, You Z, Yoshizawa-Sugata N, Oda M. 2010. Eukaryotic chromosome DNA replication: Where, when, and how? *Annu Rev Biochem* **79:** 89–130.

Masumoto H, Muramatsu S, Kamimura Y, Araki H. 2002. S-Cdk-dependent phosphorylation of Sld2 essential for chromosomal DNA replication in budding yeast. *Nature* **415:** 651–655.

Matsumoto S, Hayano M, Kanoh Y, Masai H. 2011. Multiple pathways can bypass the essential role of fission yeast Hsk1 kinase in DNA replication initiation. *J Cell Biol* **195:** 387–401.

Matsuno K, Kumano M, Kubota Y, Hashimoto Y, Takisawa H. 2006. The N-terminal noncatalytic region of *Xenopus* RecQ4 is required for chromatin binding of DNA polymerase α in the initiation of DNA replication. *Mol Cell Biol* **26:** 4843–4852.

Mimura S, Seki T, Tanaka S, Diffley JF. 2004. Phosphorylation-dependent binding of mitotic cyclins to Cdc6 contributes to DNA replication control. *Nature* **431:** 1118–1123.

Montagnoli A, Valsasina B, Brotherton D, Troiani S, Rainoldi S, Tenca P, Molinari A, Santocanale C. 2006. Identification of Mcm2 phosphorylation sites by S-phase-regulating kinases. *J Biol Chem* **281:** 10281–10290.

Morrison A, Araki H, Clark AB, Hamatake RK, Sugino A. 1990. A third essential DNA polymerase in *S. cerevisiae*. *Cell* **62:** 1143–1151.

Moyer SE, Lewis PW, Botchan MR. 2006. Isolation of the Cdc45/Mcm2-7/GINS (CMG) complex, a candidate for the eukaryotic DNA replication fork helicase. *Proc Natl Acad Sci* **103:** 10236–10241.

Muramatsu S, Hirai K, Tak YS, Kamimura Y, Araki H. 2010. CDK-dependent complex formation between replication proteins Dpb11, Sld2, Pol ε, and GINS in budding yeast. *Genes Dev* **24:** 602–612.

Nakajima R, Masukata H. 2002. SpSld3 is required for loading and maintenance of SpCdc45 on chromatin in DNA replication in fission yeast. *Mol Biol Cell* **13:** 1462–1472.

Nguyen VQ, Co C, Li JJ. 2001. Cyclin-dependent kinases prevent DNA re-replication through multiple mechanisms. *Nature* **411:** 1068–1073.

Noguchi E, Shanahan P, Noguchi C, Russell P. 2002. CDK phosphorylation of Drc1 regulates DNA replication in fission yeast. *Curr Biol* **12:** 599–605.

Ohya T, Maki S, Kawasaki Y, Sugino A. 2000. Structure and function of the fourth subunit (Dpb4p) of DNA polymerase ε in *Saccharomyces cerevisiae*. *Nucleic Acids Res* **28:** 3846–3852.

Osborn AJ, Elledge SJ. 2003. Mrc1 is a replication fork component whose phosphorylation in response to DNA replication stress activates Rad53. *Genes Dev* **17:** 1755–1767.

Oshiro G, Owens JC, Shellman Y, Sclafani RA, Li JJ. 1999. Cell cycle control of Cdc7p kinase activity through regulation of Dbf4p stability. *Mol Cell Biol* **19:** 4888–4896.

Painter RB, Young BR. 1980. Radiosensitivity in ataxia-telangiectasia: A new explanation. *Proc Natl Acad Sci* **77:** 7315–7317.

Patel PK, Kommajosyula N, Rosebrock A, Bensimon A, Leatherwood J, Bechhoefer J, Rhind N. 2008. The Hsk1(Cdc7) replication kinase regulates origin efficiency. *Mol Biol Cell* **19:** 5550–5558.

Randell JC, Fan A, Chan C, Francis LI, Heller RC, Galani K, Bell SP. 2010. Mec1 is one of multiple kinases that prime the Mcm2–7 helicase for phosphorylation by Cdc7. *Mol Cell* **40:** 353–363.

Rappas M, Oliver AW, Pearl LH. 2011. Structure and function of the Rad9-binding region of the DNA-damage checkpoint adaptor TopBP1. *Nucleic Acids Res* **39:** 313–324.

Remus D, Beuron F, Tolun G, Griffith JD, Morris EP, Diffley JF. 2009. Concerted loading of Mcm2–7 double hexamers around DNA during DNA replication origin licensing. *Cell* **139:** 719–730.

* Rhind N, Gilbert DM. 2013. DNA replication timing. *Cold Spring Harb Perspect Biol* doi: 10.1101/cshperspect.a010132.

Ricke RM, Bielinsky AK. 2004. Mcm10 regulates the stability and chromatin association of DNA polymerase-α. *Mol Cell* **16:** 173–185.

Saka Y, Fantes P, Sutani T, McInerny C, Creanor J, Yanagida M. 1994. Fission yeast cut5 links nuclear chromatin and M phase regulator in the replication checkpoint control. *EMBO J* **13:** 5319–5329.

Sanchez-Pulido L, Diffley JF, Ponting CP. 2010. Homology explains the functional similarities of Treslin/Ticrr and Sld3. *Curr Biol* **20:** R509–R510.

Sangrithi MN, Bernal JA, Madine M, Philpott A, Lee J, Dunphy WG, Venkitaraman AR. 2005. Initiation of DNA replication requires the RECQL4 protein mutated in Rothmund–Thomson syndrome. *Cell* **121:** 887–898.

Sansam CL, Cruz NM, Danielian PS, Amsterdam A, Lau ML, Hopkins N, Lees JA. 2010. A vertebrate gene, *ticrr*, is an essential checkpoint and replication regulator. *Genes Dev* **24:** 183–194.

Santocanale C, Diffley JF. 1998. A Mec1- and Rad53-dependent checkpoint controls late-firing origins of DNA replication. *Nature* **395:** 615–618.

Sasanuma H, Hirota K, Fukuda T, Kakusho N, Kugou K, Kawasaki Y, Shibata T, Masai H, Ohta K. 2008. Cdc7-dependent phosphorylation of Mer2 facilitates initiation of yeast meiotic recombination. *Genes Dev* **22:** 398–410.

Sawyer SL, Cheng IH, Chai W, Tye BK. 2004. Mcm10 and Cdc45 cooperate in origin activation in *Saccharomyces cerevisiae*. *J Mol Biol* **340:** 195–202.

Schmidt U, Wollmann Y, Franke C, Grosse F, Saluz HP, Hanel F. 2008. Characterization of the interaction between the human DNA topoisomerase IIβ-binding protein 1 (TopBP1) and the cell division cycle 45 (Cdc45) protein. *Biochem J* **409:** 169–177.

Sclafani RA. 2000. Cdc7p–Dbf4p becomes famous in the cell cycle. *J Cell Sci* **113:** 2111–2117.

Sheu YJ, Stillman B. 2006. Cdc7–Dbf4 phosphorylates MCM proteins via a docking site-mediated mechanism to promote S phase progression. *Mol Cell* **24:** 101–113.

Sheu YJ, Stillman B. 2010. The Dbf4–Cdc7 kinase promotes S phase by alleviating an inhibitory activity in Mcm4. *Nature* **463:** 113–117.

Shirahige K, Hori Y, Shiraishi K, Yamashita M, Takahashi K, Obuse C, Tsurimoto T, Yoshikawa H. 1998. Regulation of DNA-replication origins during cell-cycle progression. *Nature* **395:** 618–621.

* Siddiqui K, On KF, Diffley JFX. 2013. Regulating DNA replication in eukarya. *Cold Spring Harb Perspect Biol* doi: 10.1101/cshperspect.a012930.

Suyari O, Kawai M, Ida H, Yoshida H, Sakaguchi K, Yamaguchi M. 2012. Differential requirement for the N-terminal catalytic domain of the DNA polymerase ε p255 subunit in the mitotic cell cycle and the endocycle. *Gene* **495:** 104–114.

Tak YS, Tanaka Y, Endo S, Kamimura Y, Araki H. 2006. A CDK-catalysed regulatory phosphorylation for formation of the DNA replication complex Sld2–Dpb11. *EMBO J* **25:** 1987–1996.

Tanaka S, Araki H. 2010. Regulation of the initiation step of DNA replication by cyclin-dependent kinases. *Chromosoma* **119:** 565–574.

Tanaka S, Araki H. 2011. Multiple regulatory mechanisms to inhibit untimely initiation of DNA replication are important for stable genome maintenance. *PLoS Genet* **7:** e1002136.

Tanaka K, Russell P. 2001. Mrc1 channels the DNA replication arrest signal to checkpoint kinase Cds1. *Nat Cell Biol* **3:** 966–972.

Tanaka S, Tak YS, Araki H. 2007a. The role of CDK in the initiation step of DNA replication in eukaryotes. *Cell Div* **2:** 16.

Tanaka S, Umemori T, Hirai K, Muramatsu S, Kamimura Y, Araki H. 2007b. CDK-dependent phosphorylation of Sld2 and Sld3 initiates DNA replication in budding yeast. *Nature* **445:** 328–332.

Tanaka S, Nakato R, Katou Y, Shirahige K, Araki H. 2011a. Origin association of sld3, sld7, and cdc45 proteins is a key step for determination of origin-firing timing. *Curr Biol* **21:** 2055–2063.

Tanaka T, Umemori T, Endo S, Muramatsu S, Kanemaki M, Kamimura Y, Obuse C, Araki H. 2011b. Sld7, an Sld3-associated protein required for efficient chromosomal DNA replication in budding yeast. *EMBO J* **30:** 2019–2030.

Tye BK. 1999. MCM proteins in DNA replication. *Annu Rev Biochem* **68:** 649–686.

van Dursen F, Sengupta S, De Piccoli G, Sanchez-Diaz A, Labib K. 2012. Mcm10 associates with the loaded DNA helicase at replication origins and defines a novel step in its activation. *EMBO J* **31:** 2195–2206.

Van Hatten RA, Tutter AV, Holway AH, Khederian AM, Walter JC, Michael WM. 2002. The *Xenopus* Xmus101 protein is required for the recruitment of Cdc45 to origins of DNA replication. *J Cell Biol* **159:** 541–547.

Waga S, Masuda T, Takisawa H, Sugino A. 2001. DNA polymerase ε is required for coordinated and efficient chromosomal DNA replication in *Xenopus* egg extracts. *Proc Natl Acad Sci* **98:** 4978–4983.

Walter JC. 2000. Evidence for sequential action of cdc7 and cdk2 protein kinases during initiation of DNA replication in *Xenopus* egg extracts. *J Biol Chem* **275:** 39773–39778.

Walter JC, Araki H. 2006. Activation of pre-replication complexes. In *DNA replication and human disease* (ed. DePamphilis ML), pp. 89–104. Cold Spring Harbor Laboratory Press, Cold Spring Harbor, NY.

Wan L, Niu H, Futcher B, Zhang C, Shokat KM, Boulton SJ, Hollingsworth NM. 2008. Cdc28–Clb5 (CDK-S) and Cdc7–Dbf4 (DDK) collaborate to initiate meiotic recombination in yeast. *Genes Dev* **22:** 386–397.

Watase G, Takisawa H, Kanemaki MT. 2012. Mcm10 plays a role in functioning of the eukaryotic replicative DNA helicase, Cdc45–Mcm–GINS. *Curr Biol* **22:** 343–349.

Wohlschlegel JA, Dhar SK, Prokhorova TA, Dutta A, Walter JC. 2002. *Xenopus* Mcm10 binds to origins of DNA replication after Mcm2–7 and stimulates origin binding of Cdc45. *Mol Cell* **9:** 233–240.

Wu PY, Nurse P. 2009. Establishing the program of origin firing during S phase in fission yeast. *Cell* **136:** 852–864.

Xu X, Rochette PJ, Feyissa EA, Su TV, Liu Y. 2009. MCM10 mediates RECQ4 association with MCM2–7 helicase complex during DNA replication. *EMBO J* **28:** 3005–3014.

Yabuuchi H, Yamada Y, Uchida T, Sunathvanichkul T, Nakagawa T, Masukata H. 2006. Ordered assembly of Sld3, GINS and Cdc45 is distinctly regulated by DDK and CDK for activation of replication origins. *EMBO J* **25:** 4663–4674.

Zegerman P, Diffley JF. 2007. Phosphorylation of Sld2 and Sld3 by cyclin-dependent kinases promotes DNA replication in budding yeast. *Nature* **445:** 281–285.

Zegerman P, Diffley JF. 2010. Checkpoint-dependent inhibition of DNA replication initiation by Sld3 and Dbf4 phosphorylation. *Nature* **467:** 474–478.

Zhu W, Ukomadu C, Jha S, Senga T, Dhar SK, Wohlschlegel JA, Nutt LK, Kornbluth S, Dutta A. 2007. Mcm10 and And-1/CTF4 recruit DNA polymerase α to chromatin for initiation of DNA replication. *Genes Dev* **21:** 2288–2299.

Zou L, Stillman B. 1998. Formation of a preinitiation complex by S-phase cyclin CDK-dependent loading of Cdc45p onto chromatin. *Science* **280:** 593–596.

The Minichromosome Maintenance Replicative Helicase

Stephen D. Bell[1] and Michael R. Botchan[2]

[1]Molecular and Cellular Biochemistry Department, Biology Department, Indiana University, Bloomington, Indiana 47405

[2]Department of Molecular and Cell Biology, Division of Biochemistry and Molecular Biology, University of California, Berkeley, Berkeley, California 94720

Correspondence: stedbell@indiana.edu; mbotchan@uclink.berkeley.edu

The eukaryotic replicative helicase, the minichromosome maintenance (MCM) complex, is composed of six distinct, but related, subunits MCM(2–7). The relationship between the sequences of the subunits indicates that they are derived from a common ancestor and indeed, present-day archaea possess a homohexameric MCM. Recent progress in the biochemical and structural studies of both eukaryal and archaeal MCM complexes are beginning to shed light on the mechanisms of action of this key component of the replisome.

The minichromosome maintenance (MCM) complex subunits are members of the AAA$^+$ superfamily of ATPases and thus use energy derived from cycles of ATP binding and hydrolysis to move or reorganize bound substrates. In the case of the MCM complex, the energy is harnessed to effect DNA unwinding. The AAA$^+$ proteins can be classified into seven distinct clades, based on the topography of their active sites. MCMs are members of clade 7, being characterized by the presence of an additional α-helix when compared with the classical AAA$^+$ fold (Iyer et al. 2004; Erzberger and Berger 2006). In addition to the AAA$^+$ domain, MCMs also have an amino-terminal domain (NTD) that plays a role in higher-order structure assembly. Finally, following the AAA$^+$ domain is a degenerate winged helix (wH) structure (Fig. 1). Although the archaeal MCMs possess this simple NTD–AAA$^+$–wH domain architecture, many of the eukaryal MCM(2–7) subunits are embellished with amino- or carboxy-terminal extensions that play roles in the regulation or recruitment of MCM(2–7). The eukaryal MCM complex is an important target for regulatory posttranslational modifications; the nature and consequences of these modifications are dealt with in Bell and Kaguni (2013), Tanaka and Araki (2013), and Siddiqui et al. (2013). Much of what we know regarding the inner workings of the MCM helicase has been learned from structural and mechanistic studies of the simple archaeal model (Sakakibara et al. 2009). Additionally, the crystal structures of distantly related superfamily three helicases, SV40 LTAg, and the E1 helicase of bovine papilloma virus, have provided important structural frameworks for understanding the mode of action of hexameric helicases (Gai et al. 2004; Enemark and Joshua-Tor 2006, 2008).

Cite this article as *Cold Spring Harb Perspect Biol* doi: 10.1101/cshperspect.a012807

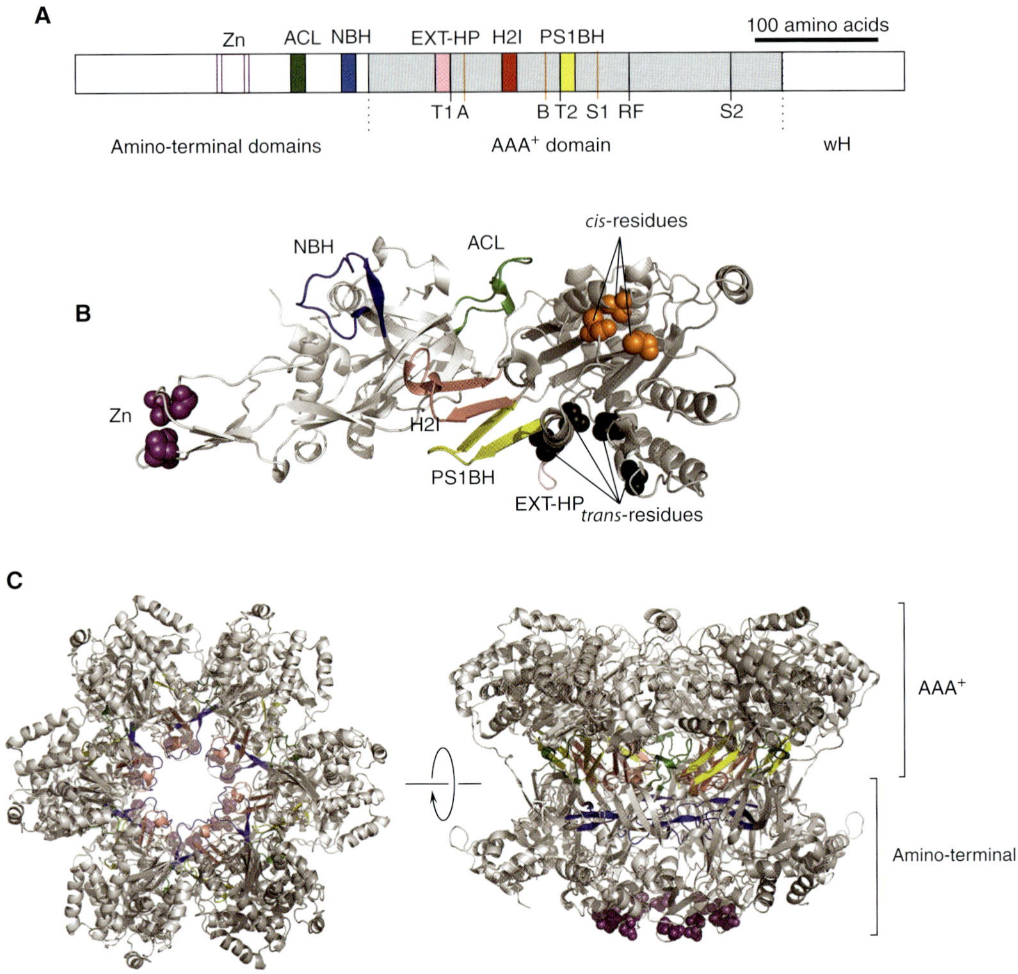

Figure 1. (*A*) Linear representation of a monomer of the archaeal MCM. (Gray) The central AAA$^+$ domain; (white) the flanking amino-terminal domains and winged helix (wH). The position of key secondary structural elements—(Zn) zinc-binding; (ACL) allosteric communication loop; (NBH) amino-terminal β-hairpin; (EXT-HP) external β-hairpin; (H2I) helix 2 insert; and (PS1BH) pre-sensor 1 β-hairpin—are indicated above the figure and shown by colored blocks, the colors corresponding to those used in panels *B* and *C*. Key residues involved in the ATPase active site are indicated below the figure. (Orange lines) A, B, and S1, shown as orange lines are the Walker A lysine (K346), Walker B glutamate (E404), and Sensor 1 asparagine (N448), respectively, and constitute "*cis*"-acting residues. (Black lines) "*Trans*"-acting residues T1 (R331), T2 (Q423), arginine finger (R473), and Sensor 2 (R560). Numbering is from SsoMCM.(*B*) Structure of a monomer of SsoMCM (lacking detail of the wH domain). Secondary structure elements are labeled and colored in cartoon format, using the color scheme in panel *A*. (Purple spheres) The atoms of the zinc-coordinating residues; (orange spheres) the *cis*-acting residues; (black spheres) the *trans*-residues.(*C*) Model of a symmetric hexamer of SsoMCM. (*Left*) View down the central cavity of SsoMCM, looking from the carboxy-terminal face. (*Right*) The same hexamer rotated 90° to show a side view. The two tiers corresponding to the amino-terminal and AAA$^+$ domains are indicated. The color scheme is as in panels *A* and *B*. Panels *B* and *C* were generated from PDB entry 3F9V using PyMOL (http://www.pymol.org).

In the following, we shall describe structural and mechanistic insights derived from studies of the simpler archaeal MCMs before extending our discussion to the eukaryotic assembly.

INSIGHTS FROM STRUCTURAL BIOLOGY

Initial electron microscopy studies of *Methanothermobacter thermautotrophicus* (Mth) MCM revealed a double hexamer structure with the component hexamers joined in a head-to-head manner (Chong et al. 2000). Subsequent studies revealed a range of different forms of the same protein, including single hexamer and single heptamer rings and filaments (Yu et al. 2002; Pape et al. 2003; Chen et al. 2005). The ring structures all have a wide central pore, large enough to potentially accommodate single- or double-stranded DNA. An important step forward came with X-ray crystallographic studies that revealed the structure of residues 2–286 comprising the amino-terminal domains of the MthMCM (Fletcher et al. 2003). Strikingly, this formed a head-to-head double hexamer ring. The central channel, of minimal diameter of 23 Å, was ringed by β-hairpin elements, one contributed by each protomer. The channel is highly positively charged, and conserved basic residues are located at the tips of the β-hairpins. DNA-binding studies revealed that MthMCM (2–286) was able to bind to both dsDNA and ssDNA in a manner dependent on the basic residues at the tips of the hairpins (Fletcher et al. 2003).

In addition to the β-hairpin structures, the amino-terminal domains of MCM possess a zinc-binding motif. Biochemical studies with MthMCM have implicated this motif in stabilizing the amino-terminal domains and thus facilitating interhexamer interactions (Poplawski et al. 2001; Fletcher et al. 2003, 2005; Kasiviswanathan et al. 2004). However, the physiological relevance of the double-hexamer form of MthMCM has been drawn into question by the observation that other all other archaeal MCMs characterized to date exist as single hexamers. A study by Kelman and colleagues has helped resolve this issue (Shin et al. 2009). Using gel filtration approaches, it was revealed that low temperature helped stabilize the double hexamer, although higher temperatures favored a single hexameric form. The experiments were somewhat constrained by the inability of conventional chromatography equipment to operate at temperatures above 50°C, and it was therefore not possible to perform the experiment at 60°C, the normal growth temperature of Mth! Nevertheless, these experiments, in conjunction with studies of multimer formation under physiological salt concentrations (~800 mM potassium) and helicase activity assays, provided strong support for the active form of MthMCM, as found for other archaeal MCMs, to be a single hexamer (Shin et al. 2009).

To date, there are no high-resolution structures available for MCM hexamers from any archaea; however, EM reconstructions of hexameric rings of MthMCM have been generated, revealing a two-tiered structure (Costa and Onesti 2009). Modeling of the amino-terminal domains into the EM envelope has indicated that one tier is composed of the amino-terminal domains, and the other tier by the AAA$^+$ domain. More recently, crystal structures of a monomer of an inactive MCM from *Methanopyrus kandleri* and of a monomer of *Sulfolobus solfataricus* MCM residues 7–601 (lacking detail of the 85 wH domain) have been determined to 1.9 Å and 4.35 Å, respectively (Brewster et al. 2008; Bae et al. 2009), allowing the generation of a series of first-generation models for the hexameric assembly (Fig. 1B,C). These models use a sixfold symmetric rotation of the monomer around an axis, and from what is understood from other AAA$^+$ assemblies, it is likely that important asymmetry with regard to mechanism will be uncovered with high-resolution structures of the hexamer (Martin et al. 2005).

HAIRPINS IN MCM

In common with the distantly related superfamily three (SF3) helicases encoded by several viruses (Enemark and Joshua-Tor 2008), the AAA$^+$ fold of MCM is embellished by DNA-interacting β-hairpin motifs. However, in contrast to the single β-hairpin in the SF3 helicases, three such elements have been described in the MCM AAA$^+$ core. Starting from the carboxy-terminal end and proceeding toward the amino terminus of the AAA$^+$ domain of the Mcms, the first hairpin encountered is the presensor 1 β-hairpin (PS1BH). This hairpin, with homology to the hairpin of the SF3 family members, extends into the central pore of the SF3 enzymes and plays a pivotal role in the helicase activity (McGeoch et al. 2005). Substitution by alanine of a conserved basic residue found near the tip of the β-hairpin only modestly impaired DNA binding by SsoMCM. Notably, however, the helicase activity of the MCM was abrogated by the mutation. The analogous β-hairpin in SF3 helicases has been shown to undergo significant movement (up to 17 Å in the case of SV40 LTAg) during the ATP utilization cycle of the helicase (Gai et al. 2004). Furthermore, the PS1BH elements in the crystal structure of the BPV E1 helicase bound to DNA stack in a staircase, the position of a given β-hairpin being dictated by the nucleotide status of the relevant subunit (Enemark and Joshua-Tor 2006). As a subunit goes through its ATP binding, hydrolysis, and release cycle, the β-hairpin repositions, thereby driving the helicase along DNA. The conservation of the hairpin in MCM and its essential nature for the helicase activity of the enzyme strongly hint at a similar mode of action for this element. Indeed, evidence has been found for a relative repositioning of the PS1BH in a nucleotide-dependent manner (Barry et al. 2009). Interestingly, the second β-hairpin element in MCM, the H2I (helix 2 insert) has also been shown to undergo nucleotide-dependent repositioning (Jenkinson and Chong 2006). Studies in MthMCM revealed changes in fluorescence of the single tryptophan residue in this protein, fortuitously located in the H2I, in the presence of nucleotide. Modulation of fluorescence was also observed in the presence of DNA, leading to the proposal that the H2I could project into the central cavity of the MCM hexamer (Jenkinson and Chong 2006). This proposal has been substantiated and extended by subsequent structural work (Brewster et al. 2008). Indeed, the crystal structure of SsoMCM reveals very close apposition of H2I and PS1BH, suggesting that these two elements may function as a single unit (Beattie and Bell 2011).

The third β-hairpin element, the EXT-hairpin, in MCM's AAA$^+$ domain lies on the outer surface of the hexamer. It is relatively poorly conserved but may play a further role in DNA binding, as discussed below.

MCM is a 3′-to-5′ helicase; thus, it translocates along DNA in a 3′-to-5′ direction, and this leads to DNA melting. Several models have been proposed for how the helicase activity is effected. Several fluorescence resonance energy transfer (FRET) studies have provided support for a simple steric exclusion model in which a hexamer of MCM translocates along single-stranded DNA, displacing a second strand of DNA ahead of it (Fig. 2) (McGeoch et al. 2005; Rothenberg et al. 2007; Graham et al. 2011). Initial ensemble FRET studies indicated that MCM could slide onto the 3′-protruding strand of

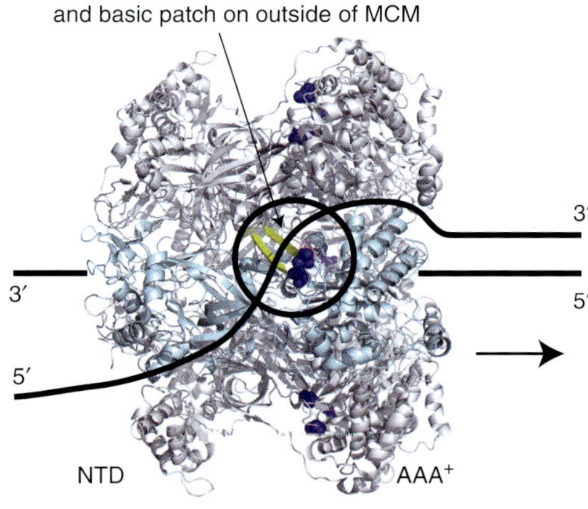

Figure 2. Model for MCM's helicase activity. MCM is shown to 3′-to-5′ translocate along a single strand of DNA that passes through the central cavity of the hexamer, one subunit of which is shown in pale blue. The displaced strand makes transient contacts with the outer surface of the hexamer (Rothenberg et al. 2007; Graham et al. 2011). Residues K323 and R440 of SsoMCM have been implicated in this contact (dark blue spheres) (Graham et al. 2012). (Yellow ribbon) The adjacent PS1BH. Aside from the potential role of these charged residues in contacting the displaced strand, nothing is known regarding the details of the trajectory of the displaced strand on the outer surface of the helicase.

a Y-shaped DNA substrate with the AAA$^+$ domains of the protein facing the duplex junction and the amino-terminal domains trailing behind (McGeoch et al. 2005). More recent single-molecule FRET experiments have revealed that MCM interacts more stably with a fork substrate than with a simple 3′-extended molecule—suggesting that the displaced strand might somehow stabilize the assembly and perhaps bind this strand on surfaces topologically external to the central channel (Rothenberg et al. 2007). Significantly, single-molecule FRET studies examining interactions between the 3′- and 5′-extended arms of the substrate revealed that the displaced strand did, indeed, undergo a dynamic interaction with MCM (Rothenberg et al. 2007). Subsequent work has confirmed the initial observations and provided evidence that charged residues on the outer surface of MCM (K323, located at the base of the EXT-hairpin, and R440, near the base of the PS1BH, of *Sso*MCM) (Fig. 2), were important for this interaction with the displaced strand (Graham et al. 2011). These transient contacts between the displaced strand and the outer surface of MCM could have several consequences —first, they might stabilize the hexameric assembly on DNA; second, they could facilitate processive DNA unwinding by the helicase; and third, they may help channel the displaced strand, the lagging strand template, to other components of the replisome (see below).

INTERSUBUNIT COMMUNICATION

Although the AAA$^+$ domains of MCM are the motors that drive the helicase along its substrate, the amino-terminal domains also play pivotal roles in governing helicase action. Deletion studies have revealed that the amino-terminal domains are not required for helicase activity in vitro, but their absence alters the range of DNA substrates that the MCM can melt (Barry et al. 2007). The presence of the amino-terminal domains also enhances the processivity of the helicase in a manner dependent on the ability of the amino-terminal domains to interact with DNA via the amino-terminal β-hairpin (Barry et al. 2007). The contacts between this fourth MCM β-hairpin and DNA are distal from the AAA$^+$ core where ATP binding and hydrolysis cycles drive the conformational changes in the motor that, in turn, exert the forces required for the power stroke; nevertheless, such contacts are critical for substrate affinity. It thus seems likely that the amino-terminal domains are not acting simply as a passive sliding clamp trailing behind the AAA$^+$ domains but rather play a more active role coordinating interprotomer interactions within the MCM hexamer.

Like many AAA$^+$ proteins, the ATP-binding sites in MCM are located at the interface between protomers (Iyer et al. 2004; Erzberger and Berger 2006). Residues from both flanking subunits coordinate ATP, the classical examples being the residues constituting the Walker A, Walker B, and Sensor 1 motif of a given active site (termed *cis*-residues in the following) and elements from the flanking protomer that contribute *trans*-acting residues, the classic example being the arginine finger. The intersubunit nature of the active sites in conjunction with the ring-shaped assembly opens the door to cooperative interactions between active sites around the MCM ring. Indeed, several other ring-shaped helicases and translocases show high cooperativity between active sites, such that incorporation of a single mutant site into an otherwise wild-type ring results in complete abrogation of enzyme activity (Crampton et al. 2006). However, not all ring-shaped AAA$^+$ machines show such high cooperativity. For example, studies of the protein unfoldase ClpX revealed that ATP hydrolysis by a given active site was essentially probabilistic in nature and that there was no strict requirement for cooperative or sequential hydrolysis events around the ring (Martin et al. 2005). Mutant doping experiments with SsoMCM revealed an intriguing intermediate behavior, with localized cooperativity between adjacent pairs of active sites in SsoMCM (Moreau et al. 2007). More specifically, mutation of an individual active site also impaired function at the neighboring site, but the toxic effect spread no further around the ring. These results reveal that despite the homohexameric nature of the archaeal MCM, asymmetries exist at the functional level.

Intriguingly, deletion of the amino-terminal domains abolished the cooperativity between AAA$^+$ domains (Barry et al. 2009). Mutational studies have revealed a role for a conserved loop, termed the allosteric communication loop (ACL), at the AAA$^+$-proximal face of the amino-terminal domains in facilitating communication between protomers (Sakakibara et al. 2008; Barry et al. 2009). This loop was found to lie in close proximity to the PS1BH of the neighboring protomer. Furthermore, nucleotide-modulated contacts between the PS1BH and the ACL may, in turn, reposition the amino-terminal β-hairpin of the ACL-containing subunit (Barry et al. 2009). The precise details of the mechanochemistry of the archaeal MCM awaits further structural elucidation; however, mutagenesis studies have implicated a series of conserved charged and polar residues in transducing information between the active site of the helicase and the PS1BH/H2I assembly (Liew and Bell 2011). It is eagerly anticipated that higher-resolution structures of active forms of archaeal MCMs will be forthcoming in the near future.

Of course, the archaeal MCM does not function in isolation in unwinding DNA at the replication fork; rather, it will be integrated into a higher-order replisome assembly. Details of the organization of the archaeal replisome remain sketchy thus far, but several laboratories have characterized archaeal counterparts of the eukaryotic GINS complex as interacting with MCM (Marinsek et al. 2006; Labib and Gambus 2007; Yoshimochi et al. 2008; Ishino et al. 2011). Initially, archaeal GINS homologs were identified by bioinformatics approaches (Makarova et al. 2005). A subsequent yeast two-hybrid screen identified Gins23 (homologous to eukaryotic GINS components Psf2 and Psf3) as an interaction partner for the amino-terminal domains of SsoMCM (Marinsek et al. 2006). Purification of the endogenous Gins23 revealed association with a second GINS component homolog, Gins15 (homologous to eukaryotic Psf1 and Sld5). Gins23:Gins15 formed a 2:2 tetrameric complex. The *Sulfolobus* GINS complex copurified with a further protein, related to the ssDNA binding domain of RecJ (Marinsek et al. 2006). Importantly, recent bioinformatics studies have revealed a clear relationship between RecJ and the eukaryotic replication factor Cdc45 (Sanchez-Pulido and Ponting 2011). This association of a Cdc45 homolog with archaeal GINS is conserved across diverse archaea, because recent work has identified an RecJ/Cdc45 homolog, GAN (GINS-associated nuclease), as a partner of the *Thermococcus* GINS complex (Li et al. 2010, 2011). Despite the association of the GINS/Cdc45 with SsoMCM, no effect on the helicase or ATPase activities of the MCM could be observed in vitro on mixing of the two complexes (Marinsek et al. 2006). In contrast, studies in *Pyrococcus* and *Thermococcus* have indicated a stimulation of the basal helicase activity of these species' MCM by their corresponding GINS complex (Yoshimochi et al. 2008; Ishino et al. 2011).

THE EUKARYOTIC MCM COMPLEX

Yeast plasmids that harbor only single origins of replication are sensitized for loss in cells with hypomorphic mutations in DNA replication genes. Tye first used this insight to identify genes required for "minichromosome maintenance." Therefore, genes encoding MCMs of widely different functions were initially identified through screens for plasmid maintenance, chromosome segregation, and cell cycle progression. The six genes called MCM 2–7 encode subunits of the eukaryal MCM complex and are now known to be the core of the eukaryotic replicative helicase. For excellent reviews of the identification of MCM(2–7) and evidence for their role as the replicative helicase, see Kearsey and Labib (1998) and Bochman and Schwacha (2009).

The six subunits of MCM assemble in a ring with a defined geometry of MCM5-3-7-4-6-2 (Fig. 3) (Crevel et al. 2001; Schwacha and Bell 2001; Davey et al. 2003). Although it has been known for a number of years that the subassembly of MCM(4,6,7) possesses robust helicase activity in vitro (Ishimi 1997) and despite the convincing evidence that MCM(2–7) was required for initiation and elongation of DNA replication in vivo (Labib et al. 2000),

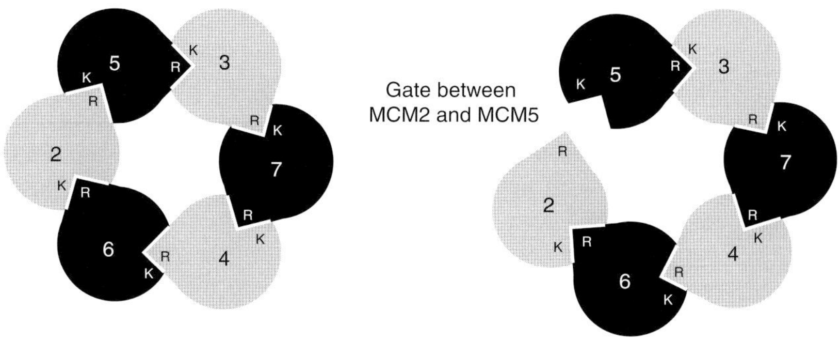

Figure 3. Organization of eukaryotic MCM(2–7). Interactions between protomers of MCM result in a heterohexamer of defined geometry (Ishimi 1997; Schwacha and Bell 2001; Davey et al. 2003). In the cartoon, the Ks indicate the *cis*-acting Walker A lysine residue of a given subunit, and the Rs show the *trans*-acting arginine fingers. Thus, the active site between protomers is contributed to by the *cis* surface of one protomer and the *trans* surface of its neighbor. As detailed in the text, considerable biochemical and structural data support the existence of a gate between subunits Mcm2 and Mcm5 (*right* panel).

in vitro evidence directly showing helicase activity of MCM(2–7) was lacking. However, two seminal advances have unambiguously shown that MCM(2–7) is a functional helicase. First, MCM(2–7) was shown to be the core of a larger complex with GINS and Cdc45, termed CMG. CMG possessed robust helicase activity in vitro (Moyer et al. 2006; Ilves et al. 2010). Second, studies of budding yeast MCM(2–7) revealed that the assembly had helicase activity but that this was dependent on specific buffer conditions, not previously tested (Bochman and Schwacha 2008). More specifically, the helicase activity is highly dependent on the choice of anion in the buffer, with activity being stimulated by glutamate and to a lesser extent acetate. There appears to be a degree of species specificity to these results, because studies with *Drosophila* MCM(2–7), purified as recombinant proteins from Sf9 cells, revealed this complex to have minimal helicase activity in glutamate buffer, contrasting with the robust activity of the analogous CMG complex (Ilves et al. 2010). The elevated helicase activity of CMG is probably in part due to the ~300-fold elevation of the V_{max} for ATP hydrolysis conferred on *Drosophila* MCM(2–7) by the association with GINS and Cdc45. Perhaps the association of the budding yeast GINS and Cdc45 proteins with the core MCM(2–7) proteins will also increase the specific activity of the yeast complex. Hence, the apparent differences may simply reflect different basal activities.

A GATE IN MCM(2–7)

Studies with purified *Saccharomyces cerevisiae* MCM(2–7) revealed that pre-incubation of the complex with ATP or a poorly hydrolyzable analog resulted in an impairment of the helicase to bind to covalently closed circular single-stranded DNA but had negligible effects on the ability of the complex to bind linear DNA (Bochman and Schwacha 2008). This result suggested that a gate may exist in MCM that is closed in the presence of ATP, thereby topologically excluding DNA from the central pore of the toroid. As described above, the ATPase sites of MCM lie between subunits, and thus a series of mutagenesis studies was performed to target individual interfaces systematically, rendering them incapable of binding ATP. Strikingly, mutation of either the Walker A lysine residue of MCM5 or the arginine finger of MCM2 rendered the MCM(2–7) insensitive to ATP when binding circular single-stranded DNA. Thus, it appears that the interface between MCM2 and MCM 5 serves as an ATP-modulated gate through which ssDNA can pass (Bochman and Schwacha 2008). These conclusions, drawn from biochemical experiments, have received structural confirmation with the elucidation of EM reconstructions of purified recombinant *Drosophila* MCM(2–7). Like the archaeal MCM, a two-tier structure was observed, corresponding to amino-terminal domains and AAA^+ domains (Costa et al. 2011). However, two distinct conformations of MCM(2–7) were observed. In one form, a near symmetric ring structure was seen that nevertheless possessed a notch in the AAA^+ tiers. The second, and more abundant, form had an asymmetric gapped or a lock washer configuration with an opening extending the full length of the MCM. In these conformations, the individual subunits were nonplanar and assumed a slight helical twist. Experiments using MCM(2–7) with individually tagged subunits revealed that the gap lay between MCM2 and MCM5 (Costa et al. 2011). Further work illustrated the structure of the *Drosophila* CMG complex. This revealed a more symmetrical ring-shaped configuration of the MCMs where closer interaction between surfaces for each of the subunits was detected, embellished by a handle of additional density on one side of the ring (Costa et al. 2011). More detailed analyses indicated that the MCM adopted the notched ring conformation seen with the isolated MCM(2–7). Incubation with ADP·BeF_3 resulted in narrowing of the notch. Furthermore, the handle-like protuberance, corresponding to GINS/CDC45, wrapped around several subunits of MCM, including those, MCM2 and 5, contributing to the gap (Fig. 4). These observations led to the conclusion that the enhanced ATPase activity and helicase function of the CMG relative to the MCM(2–7) complex was likely affected allosterically by the closing of the MCM2 and 5 gate. This induced conformational change would enable the appropriate juxtaposition of the Walker A and Arg-finger surfaces of the different subunits to form an active site. It is tempting to speculate that the yeast MCM complex may also assume open and closed complex configurations (Remus et al. 2009). Furthermore, it is possible that, in the absence of yeast GINS and CDC45, an equilibrium with more of the complex in the closed form may exist than that observed with the *Drosophila* complex.

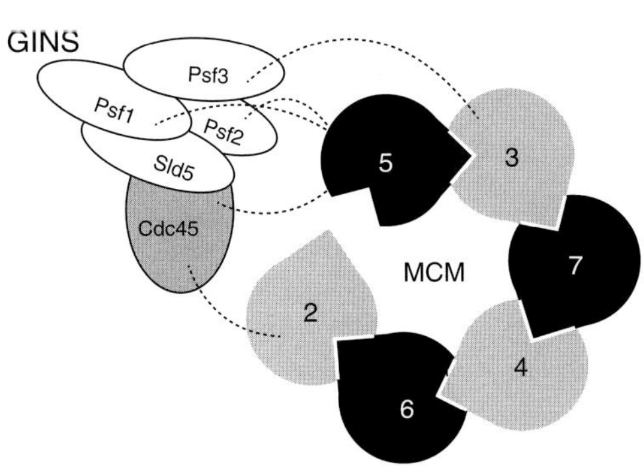

Figure 4. Interaction of the GINS components and Cdc45 with MCM. The cartoon indicates the open form of MCM(2–7) with dotted lines revealing the interactions between MCM subunits and Psf1, 2, 3, and Cdc45.

Taken together, these EM snapshots suggest a tantalizing model for how MCM might become activated at replication origins (Costa et al. 2011). At eukaryotic origins of replication, the MCM complex is loaded before its association with GINS and CDC45. To date, all of the available evidence indicates that MCM is loaded onto double-stranded DNA at origins (Remus et al. 2009). Thus, the open lock washer configuration observed for MCM(2–7) may correspond to a loading-proficient form of the helicase. Once loaded, MCM may isomerize to a planar form, possibly concomitant with localized DNA melting. The leading-strand template will be of the correct polarity to be stably bound by the β-hairpins in the heart of the MCM's pore, whereas the lagging-strand template could be displaced out via the gate. The extruded single-stranded DNA would then be trapped by the flanking GINS/CDC45 handle, recruited to the outer surface of the core helicase (Fig. 5). In this regard, it is enticing to note that the archaeal MCM has been shown to have DNA-binding surfaces on the outside of the helicase (see Fig. 2). Perhaps these could play a role in channeling the extruded DNA to the GINS/Cdc45 addendum on the core helicase (Rothenberg et al. 2007; Graham et al. 2011). The actual path of the "lagging" 5′ strand in or around the external surface of the helicase remains an important area for investigation.

Thus, the CMG complex may have two pores, one passing through the center of the toroid of MCM by which the helicase motor

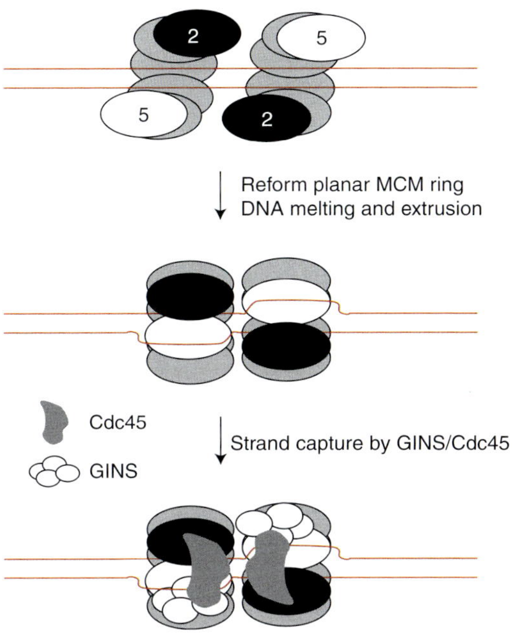

Figure 5. A speculative model for MCM loading and DNA melting at replication origins in eukarya and, by inference, archaea. Two open-gate hexamers of MCM are loaded at an origin. Ring closure occurs, concomitant with localized melting of DNA. One strand, the leading-strand template, is retained in the heart of the helicase; the other, the extruded lagging-strand template, is captured by the GINS/Cdc45 on the outer surface of the helicase assembly.

engages with the leading-strand template; and the other between the outer surface of MCM and GINS/CDC45 entrapping the lagging-strand template. It is possible that this additional binding may contribute significantly to the processivity of the helicase. We also note that such a model could explain why CMG has been observed to bind fork DNA substrates preferentially over 3′-extended molecules and with a 10-fold higher affinity than the isolated MCM(2–7). Unpublished studies with the *Drosophila* CMG complex substantiate the anticipated parallels with the archaeal MCM complex. Mutations in the presensor-1 β-hairpins affect the affinity for substrate binding and collectively are required for helicase function. Furthermore, DNA/protein cross-linking experiments reveal a distinct pattern of MCM:DNA interactions with the leading 3′ strand when DNA probes are engineered with UV-sensitive T residues within the single-strand regions of a fork. However, no cross-linking is detected with protein when such residues are placed within the duplex stretch proximal to the nexus of the fork (J Pesavento, T Petojevic, and MR Botchan, unpubl.). These results indicate that the CMG translocates on the 3′ strand through tight contacts within the interior of the toroid, without duplex in intimate association with the protein or entering the channel.

Recent work using a *Xenopus* cell-free system has provided strong evidence that the eukaryotic MCM helicase preferentially translocates along a single strand of DNA (Fu et al. 2011). It was observed that the CMG complex could be stalled very effectively by a roadblock in the form of an intrastrand cross-link, a strand-specific biotin–streptavidin moiety, or a 20-nm quantum dot attached to the leading-strand template. In contrast, such roadblocks had much less effect when placed on the lagging-strand template. However, the lagging-strand blocks did still cause transient stalling, albeit to a lower level than seen with the leading strand. Furthermore, because the experiments were performed with cell extracts, it is conceivable that additional factors present in the extract facilitate bypass of a lagging-strand block. The lack of a strong effect of the lagging-strand template blocks indicates that the contacts that may occur with the secondary channel or the "outside surfaces" of the MCM ring are as not sterically constrained and are of lower affinity than those contacts made interior to the main channel. Understanding how other components such as the DNA polymerases, RPA, RFC, and the clamp loader are organized in contact with the CMG and perhaps affect the rates of unwinding will be challenges for the future.

ACKNOWLEDGMENTS

Work in S.D.B.'s laboratory is funded by the Wellcome Trust and the Biotechnology and Biological Sciences Research Council (UK). Work in the Botchan laboratory is funded by NIH CA R37-30490. We apologize to those of our colleagues whose work we have not had space to cite.

REFERENCES

*Reference is also in this collection.

Bae B, Chen Y-H, Costa A, Onesti S, Brunzelle JS, Lin Y, Cann IKO, Nair SK. 2009. Insights into the architecture of the replicative helicase from the structure of an Archaeal MCM homolog. *Structure* **17:** 211–222.

Barry ER, McGeoch AT, Kelman Z, Bell SD. 2007. Archaeal MCM has separable processivity, substrate choice and helicase domains. *Nucleic Acids Res* **35:** 988–998.

Barry ER, Lovett JE, Costa A, Lea SM, Bell SD. 2009. Intersubunit allosteric communication mediated by a conserved loop in the MCM helicase. *Proc Natl Acad Sci* **106:** 1051–1056.

Beattie TR, Bell SD. 2011. Molecular machines in Archaeal DNA replication. *Curr Opin Chem Biol* **15:** 614–619.

* Bell SP, Kaguni JM. 2013. Helicase loading at chromosomal origins of replication. *Cold Spring Harb Perspect Biol* doi: 10.1101/cshperspect.a010124.

Bochman ML, Schwacha A. 2008. The Mcm2–7 complex has in vitro helicase activity. *Mol Cell* **31:** 287–293.

Bochman ML, Schwacha A. 2009. The Mcm complex: Unwinding the mechanism of a replicative helicase. *Microbiol Mol Biol Rev* **73:** 652–683.

Brewster AS, Wang G, Yu X, Greenleaf WB, Maria Carazo J, Tjajadia M, Klein MG, Chen XS. 2008. Crystal structure of a near-full-length archaeal MCM: Functional insights for an AAA plus hexameric helicase. *Proc Natl Acad Sci* **105:** 20191–20196.

Chen YJ, Yu XO, Kasiviswanathan R, Shin JH, Kelman Z, Egelman EH. 2005. Structural polymorphism of *Methanothermobacter thermautotrophicus* MCM. *J Mol Biol* **346:** 389–394.

Chong JPJ, Hayashi MK, Simon MN, Xu RM, Stillman B. 2000. A double-hexamer archaeal minichromosome maintenance protein is an ATP-dependent DNA helicase. *Proc Natl Acad Sci* **97:** 1530–1535.

Costa A, Onesti S. 2009. Structural biology of MCM helicases. *Crit Rev Biochem Mol Biol* **44:** 326–342.

Costa A, Ilves I, Tamberg N, Petojevic T, Nogales E, Botchan MR, Berger JM. 2011. The structural basis for MCM2–7 helicase activation by GINS and Cdc45. *Nat Struct Mol Biol* **18:** 471–477.

Crampton DJ, Mukherjee S, Richardson CC. 2006. DNA-induced switch from independent to sequential dTTP hydrolysis in the bacteriophage T7 DNA helicase. *Mol Cell* **21:** 165–174.

Crevel G, Ivetic A, Ohno K, Yamaguchi M, Cotterill S. 2001. Nearest neighbour analysis of MCM protein complexes in *Drosophila melanogaster. Nucleic Acids Res* **29:** 4834–4842.

Davey MJ, Indiani C, O'Donnell M. 2003. Reconstitution of the Mcm2–7p heterohexamer, subunit arrangement, and ATP site architecture. *J Biol Chem* **278:** 4491–4499.

Enemark EJ, Joshua-Tor L. 2006. Mechanism of DNA translocation in a replicative hexameric helicase. *Nature* **442:** 270–275.

Enemark EJ, Joshua-Tor L. 2008. On helicases and other motor proteins. *Curr Opin Struct Biol* **18:** 243–257.

Erzberger JP, Berger JM. 2006. Evolutionary relationships and structural mechanisms of AAA plus proteins. *Annu Rev Biophys Biomol Struct* **35:** 93–114.

Fletcher RJ, Bishop BE, Leon RP, Sclafani RA, Ogata CM, Chen XJS. 2003. The structure and function of MCM from archaeal *M. thermoautotrophicum. Nat Struct Biol* **10:** 160–167.

Fletcher RJ, Shen JP, Gomez-Llorente Y, San Martin C, Carazo JM, Chen XJS. 2005. Double hexamer disruption and biochemical activities of *Methanobacterium thermoautotrophicum* MCM. *J Biol Chem* **280:** 42405–42410.

Fu YV, Yardimci H, Long DT, Ho TV, Guainazzi A, Bermudez VP, Hurwitz J, van Oijen A, Schaerer OD, Walter JC. 2011. Selective bypass of a lagging strand roadblock by the eukaryotic replicative DNA helicase. *Cell* **146:** 930–940.

Gai DH, Zhao R, Li DW, Finkielstein CV, Chen XS. 2004. Mechanisms of conformational change for a replicative hexameric helicase of SV40 large tumor antigen. *Cell* **119:** 47–60.

Graham BW, Schauer GD, Leuba SH, Trakselis MA. 2011. Steric exclusion and wrapping of the excluded DNA strand occurs along discrete external binding paths during MCM helicase unwinding. *Nucleic Acids Res* **39:** 6585–6595.

Ilves I, Petojevic T, Pesavento JJ, Botchan MR. 2010. Activation of the MCM2–7 helicase by association with Cdc45 and GINS proteins. *Mol Cell* **37:** 247–258.

Ishimi Y. 1997. A DNA helicase activity is associated with an MCM4, -6, and -7 protein complex. *J Biol Chem* **272:** 24508–24513.

Ishino S, Fujino S, Tomita H, Ogino H, Takao K, Daiyasu H, Kanai T, Atomi H, Ishino Y. 2011. Biochemical and genetical analyses of the three *mcm* genes from the hyperthermophilic archaeon, *Thermococcus kodakarensis. Genes Cells* **16:** 1176–1189.

Iyer LM, Leipe DD, Koonin EV, Aravind L. 2004. Evolutionary history and higher order classification of AAA plus ATPases. *J Struct Biol* **146:** 11–31.

Jenkinson ER, Chong JPJ. 2006. Minichromosome maintenance helicase activity is controlled by N- and C-terminal motifs and requires the ATPase domain helix-2 insert. *Proc Natl Acad Sci* **103:** 7613–7618.

Kasiviswanathan R, Shin JH, Melamud E, Kelman Z. 2004. Biochemical characterization of the *Methanothermobacter thermautotrophicus* minichromosome maintenance (MCM) helicase N-terminal domains. *J Biol Chem* **279:** 28358–28366.

Kearsey SE, Labib K. 1998. MCM proteins: Evolution, properties, and role in DNA replication. *Biochim Biophys Acta* **1398:** 113–136.

Labib K, Gambus A. 2007. A key role for the GINS complex at DNA replication forks. *Trends Cell Biol* **17:** 271–278.

Labib K, Tercero JA, Diffley JFX. 2000. Uninterrupted MCM2–7 function required for DNA replication fork progression. *Science* **288:** 1643–1647.

Li Z, Santangelo TJ, Cubonova Lu, Reeve JN, Kelman Z. 2010. Affinity purification of an Archaeal DNA replication protein network. *MBio* **1:** e00221-10.

Li Z, Pan M, Santangelo TJ, Chemnitz W, Yuan W, Edwards JL, Hurwitz J, Reeve JN, Kelman Z. 2011. A novel

DNA nuclease is stimulated by association with the GINS complex. *Nucleic Acids Res* **39:** 6114–6123.

Liew LP, Bell SD. 2011. The interplay of DNA binding, ATP hydrolysis and helicase activities of the archaeal MCM helicase. *Biochem J* **436:** 409–414.

Makarova KS, Wolf YI, Mekhedov SL, Mirkin BG, Koonin EV. 2005. Ancestral paralogs and pseudoparalogs and their role in the emergence of the eukaryotic cell. *Nucleic Acids Res* **33:** 4626–4638.

Marinsek N, Barry ER, Makarova KS, Dionne I, Koonin EV, Bell SD. 2006. GINS, a central nexus in the archaeal DNA replication fork. *EMBO Rep* **7:** 539–545.

Martin A, Baker TA, Sauer RT. 2005. Rebuilt AAA plus motors reveal operating principles for ATP-fuelled machines. *Nature* **437:** 1115–1120.

McGeoch AT, Trakselis MA, Laskey RA, Bell SD. 2005. Organization of the Archaeal MCM complex on DNA and implications for the helicase mechanism. *Nat Struct Mol Biol* **12:** 756–762.

Moreau MJ, McGeoch AT, Lowe AR, Itzhaki LS, Bell SD. 2007. ATPase site architecture and helicase mechanism of an Archaeal MCM. *Mol Cell* **28:** 304–314.

Moyer SE, Lewis PW, Botchan MR. 2006. Isolation of the Cdc45/Mcm2-7/GINS (CMG) complex, a candidate for the eukaryotic DNA replication fork helicase. *Proc Natl Acad Sci* **103:** 10236–10241.

Pape T, Meka H, Chen SX, Vicentini G, van Heel M, Onesti S. 2003. Hexameric ring structure of the full-length archaeal MCM protein complex. *EMBO Rep* **4:** 1079–1083.

Poplawski A, Grabowski B, Long SF, Kelman Z. 2001. The zinc finger domain of the archaeal minichromosome maintenance protein is required for helicase activity. *J Biol Chem* **276:** 49371–49377.

Remus D, Beuron F, Tolun G, Griffith JD, Morris EP, Diffley JFX. 2009. Concerted loading of Mcm2–7 double hexamers around DNA during DNA replication origin licensing. *Cell* **139:** 719–730.

Rothenberg E, Trakselis MA, Bell SD, Ha T. 2007. MCM forked substrate specificity involves dynamic interaction with the 5′-tail. *J Biol Chem* **282:** 34229–34234.

Sakakibara N, Kasiviswanathan R, Melamud E, Han M, Schwarz FP, Kelman Z. 2008. Coupling of DNA binding and helicase activity is mediated by a conserved loop in the MCM protein. *Nucleic Acids Res* **36:** 1309–1320.

Sakakibara N, Kelman LM, Kelman Z. 2009. Unwinding the structure and function of the archaeal MCM helicase. *Mol Microbiol* **72:** 286–296.

Sanchez-Pulido L, Ponting CP. 2011. Cdc45: The missing RecJ ortholog in eukaryotes? *Bioinformatics* **27:** 1885–1888.

Schwacha A, Bell SP. 2001. Interactions between two catalytically distinct MCM subgroups are essential for coordinated ATP hydrolysis and DNA replication. *Mol Cell* **8:** 1093–1104.

Shin J-H, Heo G-Y, Kelman Z. 2009. The *Methanothermobacter thermautotrophicus* MCM helicase is active as a hexameric ring. *J Biol Chem* **284:** 540–546.

* Siddiqui K, On KF, Diffley JFX. 2013. Regulating DNA replication in eukarya. *Cold Spring Harb Perspect Biol* doi: 10.1101/cshperspect.a012930.

* Tanaka S, Araki H. 2013. Helicase activation and establishment of replication forks at chromosomal origins of replication. *Cold Spring Harb Perspect Biol* doi: 10.1101/cshperspect.a010371.

Yoshimochi T, Fujikane R, Kawanami M, Matsunaga F, Ishino Y. 2008. The GINS complex from *Pyrococcus furiosus* stimulates the MCM helicase activity. *J Biol Chem* **283:** 1601–1609.

Yu X, VanLoock MS, Poplawski A, Kelman Z, Xiang T, Tye BK, Egelman EH. 2002. The *Methanobacterium thermoautotrophicum* MCM protein can form heptameric rings. *EMBO Rep* **3:** 792–797.

Spatial and Temporal Organization of DNA Replication in Bacteria and Eukarya

Dean Jackson[1], Xindan Wang[2], and David Z. Rudner[2]

[1]University of Manchester, Faculty of Life Sciences, Michael Smith Building, Oxford Road, Manchester M13 9PT, United Kingdom

[2]Harvard Medical School, Department of Microbiology and Immunobiology, Boston, Massachusetts 02115

Correspondence: dean.jackson@manchester.ac.uk; rudner@hms.harvard.edu

Here we discuss the spatial and temporal organization of replication in eubacteria and eukaryotes. Although there are significant differences in how replication is organized in cells that contain nuclei from those that do not, you will see that organization of replication in all organisms is principally dictated by the structured arrangement of the chromosome. We will begin with how replication is organized in eubacteria with particular emphasis on three well studied model organisms. We will then discuss spatial and temporal organization of replication in eukaryotes highlighting the similarities and differences between these two domains of life.

During the bacterial cell cycle, the chromosome exists as a condensed and highly ordered structure that is known as the nucleoid. Most bacteria that have been actively studied have a single circular chromosome that ranges in size from 2 Mb to 8 Mb. DNA replication initiates at a unique origin (*oriC*) and proceeds bidirectionally around the chromosome to a site opposite *oriC* called the replication terminus (*ter*). Unlike eukaryotes, segregation of the newly replicated DNA occurs in lockstep with replisome progression. After a period of cohesion, the replicated DNA is resolved into spatially separate lobes that become fully segregated on the completion of DNA synthesis.

Temporal control of replication initiation is dictated by growth rate and is principally mediated by nutrient dependent changes in the synthesis and activity of the replication initiator protein DnaA. Accumulation of ATP-bound DnaA above a critical threshold triggers initiation (Katayama et al. 2010; Leonard and Grimwade 2011). On origin firing, a diverse collection of control mechanisms ensure that cells do not prematurely reinitiate a new round of synthesis and serve to "reset" DnaA to appropriately monitor nutrient status (Boye et al. 2000; Wang and Levin 2009). In cases in which new rounds of replication initiate prior to the completion of an earlier round (so called, multiforked replication), origins fire synchronously (Skarstad et al. 1986). Thus, unlike eukaryotes in which there is temporal organization of initiation among the myriad of chromosomal origins (see below), in bacteria, spatial and temporal organization of replication is reduced to spatial organization. The exception to this rule is *Vibrio cholerae*, which contains two unique chromosomes. In

Cite this article as *Cold Spring Harb Perspect Biol* doi: 10.1101/cshperspect.a010389

this bacterium, it appears that replication of the two chromosomes is initiated at different but stereotypical times in the cell cycle (Rasmussen et al. 2007). The molecular mechanisms underlying the difference in origin firing is only beginning to be unraveled (Duigou et al. 2008), and will not be discussed in this article. Here we focus on the spatial organization of bacterial replication.

Early cytological experiments suggested that replication in bacteria occurs at a specific mid-cell position akin to the replication factories observed in eukaryotes (see below) (Lemon and Grossman 1998, 2000). However, accumulating evidence over the last decade suggests that after replication initiation, the replisomes are not stationary nor do they cluster (Jensen et al. 2001; Bates and Kleckner 2005; Berkmen and Grossman 2006; Reyes-Lamothe et al. 2008). Instead, it appears that the nucleoid serves as both the template and track for the replication machinery. Accordingly, our current view is that the nucleoid itself is the major organizer of DNA replication in bacteria. Our premise is that understanding the spatial organization of bacterial replication requires an understanding of the structure and organization of the bacterial chromosome. We will, therefore, begin by describing how the chromosome is organized within the bacterial nucleoid and how the nucleoid is organized within the cell with particular emphasis on the three model organisms for which we have the most information: *Escherichia coli*, *Bacillus subtilis*, and *Caulobacter crescentus*. We will then discuss spatial organization of replication in the context of this highly organized structure.

ORGANIZATION OF THE BACTERIAL CHROMOSOME

The bacterial chromosome is compacted ~1000-fold within the bacterial cell. The principal mechanism of chromosome condensation is through DNA supercoiling. In eukaryotes, nearly all supercoils are constrained by the wrapping of DNA around histone octamers called nucleosomes. In bacteria, approximately half of the DNA supercoils are constrained by abundant nucleoid associated proteins (NAPs) that bend or twist the DNA. The rest of the supercoils are unconstrained arising from under-winding the DNA helix (Vologodskii and Cozzarelli 1994; Holmes and Cozzarelli 2000; Dillon and Dorman 2010). The resulting interwound loops or superhelical branches play a key role in compacting and organizing the genome. Evidence suggests that the smallest unit of organization of the bacterial chromosome is an independent supercoiled loop. In *E. coli*, these unconstrained supercoiled domains have an average size of ~10 kb (Postow et al. 2004; Stein et al. 2005). Accordingly, its 4.6 Mb genome is thought to contain ~450 of these topologically isolated domains. Our current view is that these plectonemic loops emanate from a central chromosome core like the bristles of a bottlebrush (Fig. 1A). In addition to their role in the condensation of the chromosome, these topological domains protect the chromosome from DNA relaxation, assist in decatenation of chromosomal links, and have been proposed to aid in the repair of double strand breaks by maintaining broken ends in close proximity.

There do not appear to be specific sequence elements that define the domain boundaries for these supercoiled loops making it a challenge to characterize them molecularly. It is thought that abundant nucleoid associated proteins with relaxed sequence specificity are responsible for constraining the DNA loops at the chromosome core (Hardy and Cozzarelli 2005). However, definitive evidence for this has been difficult to obtain. RNA polymerase and the SMC/Muk condensin complexes have also been suggested to serve as boundary proteins (Deng et al. 2005; Cui et al. 2008). The current view is that these topologically isolated loops and their boundaries are highly dynamic, changing in response to DNA transactions that occur within and between them. Thus, these small fluid domains provide structure to the chromosome without imposing rigidity.

The next level of organization within the nucleoid involves chromosomal regions that are physically insulated from each other. These regions, called macrodomains, are 800 kb–1 Mb in size and have, thus far, only been described in *E. coli*. We suspect that this higher-order organization exists in many bacteria. Organization of the chromosome in macrodomains was first described by Niki and colleagues

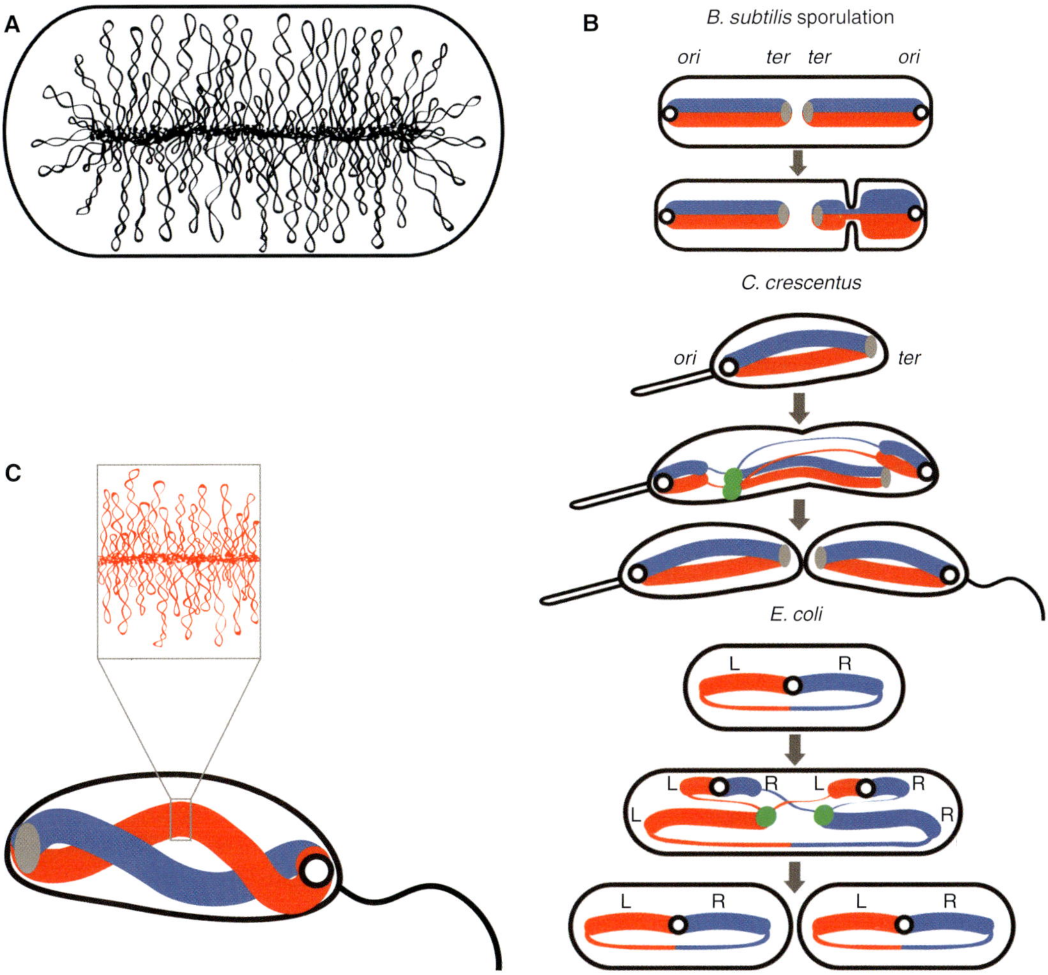

Figure 1. The bacterial nucleoid. (*A*) Diagram of the bottlebrush model of the nucleoid. The image shows a schematic representation of the interwoven super-helical loops emanating from a chromosome core. These flexible, topologically isolated domains are thought to be on average 10 kb in size. (*B*) Chromosome organization in *B. subtilis* during spore formation, and *C. crescentus* and slow growing *E. coli* during the vegetative cell cycle (gray arrows). Replication origins are illustrated as open circles, termini as solid gray ovals, the left (L) and right (R) replication arms as thick blue and red lines, the *E. coli* terminus region and newly replicated DNA as thinner lines, and the replisomes as green dots. (*C*) A model for the gently twisted arms of the *C. crescentus* chromosome based on recent chromosome conformation capture analysis. The *inset* shows the plectonemic loops within one of the arms.

based on cytological observations (Niki et al. 2000) using fluorescence in situ hybridization (FISH). Loci in large regions flanking the origin and terminus had specific localizations distinct from the rest of the chromosome suggesting that loci in these regions cluster and that these large domains are spatially separate. A genetic chromosome conformation assay involving recombination to monitor the frequency of random collisions between different sites on the chromosome independently identified four large regions of the chromosome that do not interact with each other (Valens et al. 2004). Strikingly, this genetic assay identified the same macrodomains surrounding the origin and terminus. It also defined two additional physically insulated domains. These domains flank the Ter macrodomain and are called the left and right macrodomains. Finally, two flexible or unstructured regions were identified that lie to the left and right of the Ori macrodomain. Loci in these unstructured regions can interact with loci in adjacent domains but not with other regions of the chromosome (Valens et al. 2004).

Further support for the presence of structured macrodomains and unstructured flexible chromosomal regions came from time-lapse imaging of fluorescently labeled chromosomal loci (Espeli et al. 2008). These studies revealed that macrodomains and the unstructured regions have strikingly different dynamic behavior. Loci in the flexible regions were significantly more mobile than those found in the macrodomains. The larger territory occupied by loci in the unstructured regions is consistent with the ability of these regions to support long distance interactions with loci present in flanking macrodomains.

Until recently, macrodomains, like the smaller topologically isolated loops, were inferred to exist based on indirect assays. The molecular mechanism underlying this higher-order organization was lacking. In a triumph of bioinformatics analysis and the careful delineation of the Ter macrodomain, Merceir et al. (2008), identified a sequence motif (*matS*) that was present 23 times in this region of the chromosome but was virtually absent in the rest of the genome. This motif led to discovery of the site-specific DNA binding protein MatP that interacts with it. MatP binds to the *matS* sites in the Ter macrodomain in vivo. Moreover, cytological studies indicate that MatP localizes as a focus that overlaps loci in the Ter macrodomain, suggesting that it gathers or organizes *matS* sites together. In support of the idea that MatP is indeed the Ter macrodomain organizer, loci in this domain become more mobile in cells lacking MatP. Moreover, these loci recombine with neighboring domains more frequently when MatP is absent.

Although no experiment has connected the smaller flexible supercoiled domains with the larger constrained macrodomains, we and others imagine that the plectonemic loops (the bottlebrush bristles that emanate from the nucleoid core) are organized-gathered together via site-specific DNA binding proteins like MatP bridging these topologically isolated loops. Based on their size, individual macrodomains are expected to contain ~100 plectonemic loops. *matS* sites are well-spaced throughout the Ter macrodomain: there is approximately one every 40 kb. Thus, MatP is well-positioned to gather many of these dynamic bottlebrush bristles into a spatially distinct chromosome lobe. We suspect that analogous DNA binding proteins organize the other macrodomains in *E. coli* and that similar, yet to be discovered, architectural proteins function in higher-order organization in other bacteria.

The emergence of fluorescence microscopy in bacteria has had a transformative effect on the study of bacterial chromosome organization and the spatial organization of replication. Prior to the cytological studies, it was appreciated that the bacterial chromosome was highly condensed and likely structured but how this organization related to the body plan of the cell was unclear. The ability to visualize individual chromosomal loci in live cells using fluorescent fusions to DNA binding proteins and their target sequences revealed a degree of spatial organization of the chromosome that had not been previously appreciated (Robinett et al. 1996; Straight et al. 1996; Michaelis et al. 1997; Webb et al. 1997; Li and Austin 2002; Lau et al. 2003; Viollier et al. 2004).

The earliest cytological studies aimed at defining the localization of chromosomal loci within the bacterial cell were performed in *B. subtilis*. These studies revealed that replication initiates at or near mid-cell and that the newly replicated origins move apart to the outer edges of the nucleoid (Lin et al. 1997; Webb et al. 1997; Lemon and Grossman 1998). Analysis of four chromosomal loci suggests that the chromosome adopts an organization in which the replicated origins reside at the cell quarters, the termini at mid-cell, and left and right chromosome arms lie between them (Teleman et al. 1998). The proposed "ori-ter ter-ori" organization of the sister chromatids requires further investigation; however, a similar layout has been established during spore-formation in *B.*

subtilis using an elegant genetic strategy (Wu and Errington 1998). During sporulation, the replicated chromosomes adopt an elongated structure that extends from one cell pole to the other (Fig. 1B) (Lin et al. 1997; Webb et al. 1997; Ben-Yehuda et al. 2003). Within this structure, the replicated origins are anchored at opposite cell poles and the termini reside at mid-cell.

More recent analysis of chromosome organization in *C. crescentus* and *E. coli* benefited from new tools, improved technology, and system-wide approaches. In *C. crescentus*, analysis of 112 individual loci revealed that the physical position of a locus inside the cell recapitulates the genetic map (Viollier et al. 2004). Prior to replication initiation, the origin and terminus reside near opposite cell poles while other loci lie along the long axis of the cell in an order that correlates with their position in the genome (Fig. 1B). On replication initiation, one of the origins is segregated to the opposite cell pole. When replication is complete, the replicated chromosomes reside in an ori-ter ter-ori organization.

In *E. coli*, similar systematic analyses were performed using slow-growing cells in which newborn cells have a single copy of the chromosome and a G_1-like period prior to the initiation of replication. Strikingly, these studies revealed that the *E. coli* chromosome has a markedly different organization from *C. crescentus*. In cells with a single chromosome, the origin resides at mid-cell with the left and right chromosome arms in opposite cell halves (Fig. 1B) (Nielsen et al. 2006; Wang et al. 2006). The terminus region appears to connect the two arms to complete the circle. After chromosome replication and prior to cytokinesis, the replicated chromosomes reside in a left-ori-right left-ori-right organization. The distinct spatial localization of the left and right chromosome arms observed in these studies is consistent with low frequency of recombination between loci in the right and left macrodomains (Valens et al. 2004). However, the cytological analysis that indicates that the *ter* region connects the left and right arms appears to be at odds with a structured Ter macrodomain. How the Ter macrodomain fits in the context of the cellular organization of the chromosome remains to be discovered.

The picture that emerges of this highly condensed and structured nucleoid is of a bottlebrush with plectonemic loops emanating along its length in an order that corresponds with the linear order of loci on the chromosome. In *E. coli* during slow growth, in which the left and right arms are located in separate cell halves, we envision two bottlebrush arms flanking the origin. In the case of *C. crescentus* and *B. subtilis*, the localization studies are not of high enough resolution to say whether there is single core for the left and right chromosome arms or whether the left and right arms represent spatially distinct bottlebrushes.

Recent experiments aimed at defining the three-dimensional structure of the *C. crescentus* chromosome has shed light on this issue. Umbarger and colleagues used a high throughput chromosome conformation capture assay (called, 5C) to quantify thousands of interloci interaction frequencies (Umbarger et al. 2011). These frequencies were then used to infer the average three-dimensional distances between each locus. Computational modeling of the entire dataset generated the first three-dimensional model of the bacterial chromosome. In agreement with earlier cytological studies, the resulting model suggests that the *C. crescentus* chromosome is ellipsoidal, with left and right arms roughly symmetric about the ori-ter axis. The two highly condensed arms are spatially distinct but are wound around each other approximately one and half times (Fig. 1C). Thus, it appears that this bacterium has two bottlebrushes, one for each chromosome arm, that are gently wound around each other along their lengths. This rendering of the *C. crescentus* chromosome is reminiscent of the nucleoid structures observed by fluorescence microscopy in *B. subtilis* in which the nucleoid appears to gently twist around itself (Berlatzky et al. 2008).

SPATIAL ORGANIZATION OF REPLICATION

Now that we have presented a picture of how the bacterial chromosome is organized within the nucleoid (and how the nucleoid is arranged within the bacterial cell), we return to the central issue of how replication is organized. As mentioned at the beginning of this article, when bacterial cell biology emerged as a field and the technology and tools were in their infancy, replisomes were observed to localize in a mid-cell focus and it was proposed that DNA synthesis occurred in eukaryotic-like replication factories. Experiments over the last decade have challenged this view and it is now generally accepted that after replication initiation, the polymerases that replicate the left and right chromosomal arms function independently of each other and use the chromosome as both a template and a track.

In accordance with the organization of the chromosome within the bacterial cell, replication in *E. coli* and *B. subtilis* initiates at or near mid-cell (Lemon and Grossman 1998; Bates and Kleckner 2005; Reyes-Lamothe et al. 2008), whereas in *C. crescentus* replication initiates at the cell pole (Jensen et al. 2001). The cellular "marks" that direct the replisomes to these positions appear to be the origin itself rather than extrinsic localization determinants. Experiments in *B. subtilis* in which the origin region was tethered to the cell pole resulted in replication initiation near the cell poles (Berkmen and Grossman 2006). Similarly, in *E. coli* MukB mutants, the origin localizes close to the cell pole and replication initiates at this position (Danilova et al. 2007). Although assembly of a prereplicative complex is dictated by the localization of *oriC*, there is strong evidence, at least in *B. subtilis*, that origin firing occurs at the cytoplasmic membrane. Membrane fractions are enriched in origin DNA compared to other regions of the chromosome and an essential component of the helicase loader is a peripherally associated membrane protein (Beeson and Sueoka 1979; Hoshino et al. 1987; Rokop et al. 2004). We envision that the subcellular localization of *oriC* directs the initiator protein DnaA to this position. However, once DnaA assembles at this site, transient association of the complex with the membrane triggers origin unwinding and replication initiation. Initiation of replication at the membrane ensures that newly replicated origins are generated at the nucleoid periphery. We hypothesize that this facilitates origin segregation and ensures compartmentalization of replicated and unreplicated DNA.

Once DNA synthesis has begun, the replisomes independently replicate the two arms moving out from mid-cell toward the cell poles and back again in *E. coli* and from one cell pole toward the other in *C. crescentus* (Jensen et al. 2001; Reyes-Lamothe et al. 2008). In *B. subtilis*, the movement of the replisome during a replication cycle has not been described, but we suspect that replication proceeds along the arms in a pattern that is dictated by the organization of the chromosome. In all cases, topologically isolated domains and the larger macrodomains must be disrupted and then reformed during passage of the replisomes, but the path of the polymerases is thought to follow the path of the DNA.

In line with this theme, termination occurs at the location of the terminus DNA. In *E. coli*, *B. subtilis*, and *C. crescentus*, this is at mid-cell. In *B. subtilis* and *E. coli*, the terminus DNA appears to remain at mid-cell throughout the replication cycle and only on its duplication do the two replicated chromosomes become spatially separated from each other (Teleman et al. 1998; Lau et al. 2003). In *C. crescentus*, the terminus starts out close to the cell pole opposite the origin, but at the time when replication is near complete, the replicated origins are present at opposite cell poles with the terminus DNA between them (Jensen et al. 2001).

In summary, the spatial organization of replication in bacteria mirrors the organization of the chromosome within the nucleoid. With the replication machinery following the track of the nucleoid, it stands to reason that the replication process itself could serve as a key player in setting up the organization of the resulting sister chromatids. Indeed, recent work in *E. coli* suggests that replication along the nucleoid and the coupled segregation and compaction of the newly replicated DNA helps to reproduce the same nucleoid pattern in the daughter cells (Nielsen et al. 2006; Wang et al. 2006).

SPATIAL AND TEMPORAL ORGANIZATION OF EUKARYOTIC REPLICATION

Unlike bacteria, eukaryotes temporally separate the replication of its genome (S phase) and its segregation (M phase). Accordingly, the discussion of spatial and temporal control of DNA replication in eukaryotes will be restricted to the S phase of the cell cycle. Eukaryotes face three major challenges in replicating their genomes that are not encountered by bacteria. First, the much larger genomes of eukaryotes are fragmented into genetically independent chromosomes, which must be replicated together during S phase. Second, eukaryotic genomes are folded into complex higher-order chromatin structures in which numerous DNA–protein interactions regulate function. This higher-ordered structure is replicated with an average replication fork rate of ~1.5 kb/min, which is much slower than the bacterial rate of ~40 kb/min. Thus, eukaryotic cells initiate replication at hundreds (e.g., yeasts) to many thousands (e.g., mammalian cells) of sites to complete synthesis within an appropriate time of their cell cycle. Third, replication must occur with a degree of precision that ensures not only genetic but also epigenetic information is preserved and transmitted to daughter cells. The requirement to activate replication at so many locations on the genome and also preserve the epigenetic code implies that DNA replication must be highly coordinated both temporally and spatially.

STRUCTURING THE REPLICATION PROGRAM

Seminal experiments performed in the 1960s showed two key features of the mammalian replication process: the existence of replicon clusters and that these clusters fire at stereotypical times in the cell cycle. DNA fiber autoradiography showed that replicons in eukaryotes were often activated in groups in which clusters of adjacent replicons (typically 4–10) initiated replication at very similar times (Huberman and Riggs 1966; Edenberg and Huberman 1975). This implied that although individual replicons are discrete synthetic units, the higher-order targets for replication initiation are replicon clusters. Replicon clusters contain ~1 Mb of contiguous DNA and are typically 5- to 10-fold smaller than cytogenetically defined subchromosomal bands, which were also shown around this time to replicate at predictable times during the S phase.

THE NUCLEAR MATRIX AND NUCLEOIDS

How might spatial information contribute to the replication process? During the 1970s, two groups developed biochemical approaches to address this question. In seminal work, Berezney and Coffey prepared a subnuclear structural framework, which they called the nuclear matrix (Berezney and Coffey 1974); nuclei were extracted with 2M NaCl after first fragmenting DNA with DNase. This nuclear remnant was highly enriched in pulse-labeled nascent DNA (Berezney and Coffey 1975) and pulse-chase experiments inferred that synthesis was taking place within complexes that were associated with the matrix during synthesis (Pardoll et al. 1980). The idea that sites of active DNA synthesis were associated with the nuclear substructure was supported by work from Cook and colleagues (McCready et al. 1980), who prepared "nucleoids" from HeLa cells by direct lysis in a hypertonic (2M NaCl) lysis mix. Nucleoids contain all nuclear DNA, which can be revealed as a "halo" of supercoiled (hence intact) DNA when nucleoids are spread on a water meniscus and visualized by electron microscopy. Using nucleoids, pulse-labeled nascent DNA was shown to be bound to the nuclear substructure—called the nuclear cage—and then move away from the structure as synthesis proceeds (McCready et al. 1980).

Nucleoids provide an analytical tool, which because of the extraction used may or may not faithfully represent the structures present in vivo. To address this point, physiological extraction protocols were developed using cells grown in agarose microbeads (Jackson and Cook 1986). Critically, with this approach, the conditions used preserve the natural chromatin architecture and all synthetic activity that was present in the cell at the time of lysis. As with studies using hypertonic extraction, the nuclear framework—here called the "nucleoskeleton" to reflect the physiological entity—was shown to provide an organizing center on which the active sites of nascent replication could be revealed as replication factories (Hozak et al. 1993). Within these active sites, many replicons were labeled together and this correlated with the coreplication of adjacent replicons within clusters (see above). Importantly, this concept imposes structural constraints on the replication process in which the chromatin must move during synthesis while the synthetic machinery remains fixed. Though this model is clearly different from that seen in bacteria, the basic idea has been supported by live cell imaging of tagged genomic loci in budding yeast (Kitamura et al. 2006) in which factories are also observed (Meister et al. 2006). Collectively, these early studies provided the first evidence for spatial organization of replication within the nucleus and the idea that active sites of DNA synthesis associate with the nuclear substructure.

VISUALIZING REPLICATION IN THE NUCLEUS

The structural organization of replication described above likely reflects a topological imperative to maintain genetic integrity when many thousands of initiation events are required for efficient DNA synthesis. However, the synthetic process in higher eukaryotes must also replicate the epigenetic features of chromatin that maintain cell differentiation. This appears to be achieved by establishing a strict temporal order of DNA replication.

The ability to analyze replication of specific regions of the mammalian genome at defined times during the S phase was facilitated by our ability to label sites of DNA replication and visualize these sites in fixed or living cells both in nuclei and mitotic chromosomes. In early studies, the analysis of chromosomes prepared from cells labeled with the thymidine analog bromo-deoxyuridine (BrdU)—in the previous S phase—and visualized by indirect immunolabeling showed that euchromatin was always replicated before the heterochromatin (Holmquist and Ashley 2006). Using this approach, indirect immunofluorescence was used to visualize replication sites within nuclei of rat fibroblasts (Nakamura et al. 1986). Early in S phase, these cells were found to contain ~125 nascent replication sites, each with an estimated ~1 Mb of DNA. These sites were proposed to represent domains of replicon clusters that engaged synthesis over periods of ~1 h before synthesis terminated and switched to adjacent sites. In subsequent studies, this labeling approach was used to define a strict temporal program of S phase synthesis (Fig. 2) (Zink 2006).

Many studies have now confirmed the importance of higher-order chromatin domains as synthetic units during DNA synthesis and recent work has begun to reveal how different features of chromosome structure contribute to the S phase timing program. Protocols for labeling nascent sites of DNA synthesis with different halogenated replication precursors (e.g., CldU and IdU) provided early clues that the nascent sites within chromatin foci might be activated according to a temporal program (Manders et al. 1992, 1996). These observations were used to develop "next-in-line" models of S-phase progression and studies on mid/late replicating chromatin domains were used to develop this idea into a "domino hypothesis" (Sporbert et al. 2002). A direct test of this hypothesis, using replication domains in individual chromosome territories and DNA fibers, confirmed that the sequential activation of adjacent chromatin domains along chromosomes is a major mechanism of S-phase progression (Maya-Mendoza et al. 2010).

Over many years, the analysis of replication in fixed or live mammalian cells showed that clusters of adjacent replicons were replicated together throughout the S phase, and that these were contained in higher-order chromatin domains and most probably replicated by machinery within dedicated replication factories. If true, factory level control might explain how synchronous initiation of active replicons and suppression of potential origins within the same chromatin domains could be achieved. However, recent experiments using super-resolution light microscopy indicate that replication factories rather than being distinct structures might better be considered as nuclear domains with specific functional roles. Hence, the ~1200 factories typically seen in diploid mammalian cells using confocal microscopy (Ma

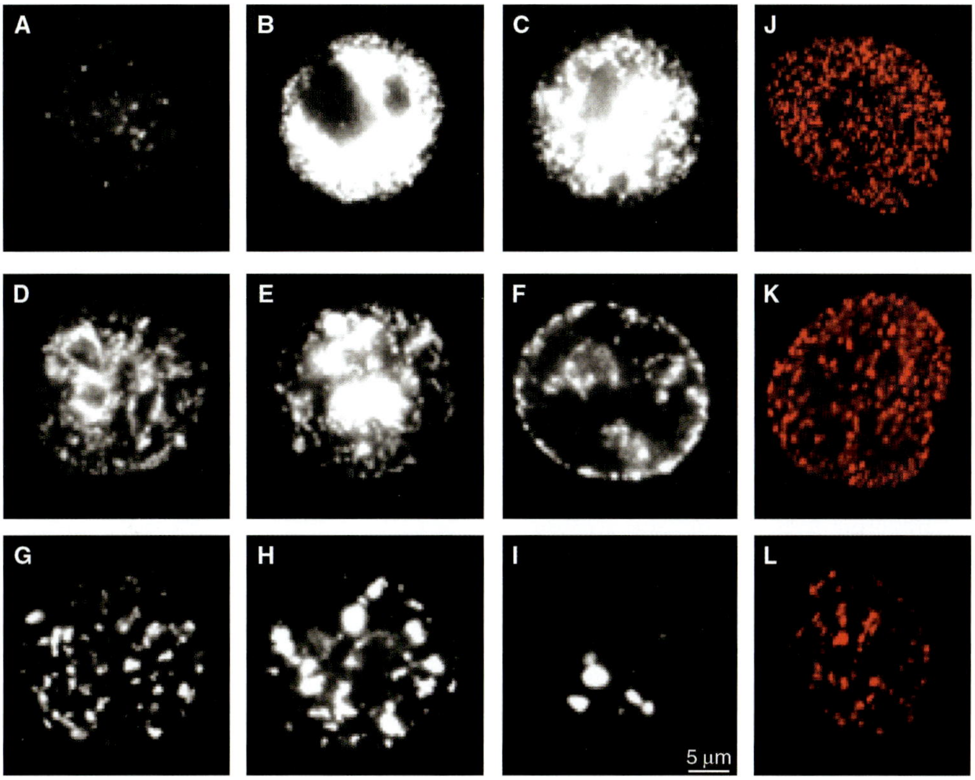

Figure 2. Patterns of DNA synthesis change throughout the S phase. DNA replication precursor analogs can be used to monitor DNA synthesis. HeLa cells (*A–I*) were synchronized in mitosis using nocodazol and at the onset of the S phase using aphidicolin, a DNA synthesis inhibitor. After removing the inhibitor, cells were grown in medium and nascent sites of DNA replication visualized at 1-h intervals (*A–I*) during the ~9-h S phase. At each time point, incorporation of biotin-dUTP into nascent sites was monitored in permeabilized cells by indirect immuno-labeling. Typical patterns seen at different times change as S phase proceeds, with classical early- (*A–C*), mid- (*D–F*), and late-S phase (*G–I*) sites correlating broadly with the synthesis of euchromatin, facultative heterochromatin, and constitutive heterochromatin within different nuclear compartments. Diploid human fibroblasts (*J–L*) show the same general patterns when nascent sites are visualized in unsynchronized cells after labeling with Cy3-dUTP in vivo.

et al. 1998; Cseresnyes et al. 2009) could be resolved into 2–3 times more active sites using super-resolution systems (Cseresnyes et al. 2009; Baddeley et al. 2010).

CHROMOSOME STRUCTURE AND REPLICATION TIMING

In bacteria, the time of replication of any specific chromosomal locus will be defined by its proximity to *oriC*. In higher eukaryotes where replication initiates at ~50,000 sites, initiation cannot be the only control point. In fact, in mammalian cells, while all potential origins are established well before the onset of the S phase (Blow and Dutta 2005), only a subset of these is selected for activation when the S phase begins. The analysis of replication foci in nuclei and replicon clusters on DNA fibers shows that the same loci are targeted for synthesis at the onset of the S phase with high efficiency (Jackson and Pombo 1998; Ma et al. 1998). However, it is important to note that at any time during the S phase, only about 15% of the genome is engaged in synthesis, so that initiation must occur throughout the S phase. As noted above, chromosome structure plays a key role in regulating origin activation in the downstream replication-timing program.

What are the basic features of this structure–function relationship? Individual chromosomes form chromosome territories (CT) that occupy discrete nuclear domains (Cremer and Cremer 2010). Within these structures, DNA binds around histone octamers as nucleosomes and the resulting nucleosome fiber folds in space through a series of protein–protein interaction, which then defines DNA loops. Locally, the chromatin loops are constrained within higher-order structures that typically contain 250–1000 kb of DNA. As in bacteria, this level of chromatin organization provides structure without imposing a rigid chromatin environment. These chromatin domains or DNA foci provide the fundamental subunits of higher-order chromatin folding within CTs and have been shown to reflect stable units of higher-order chromatin folding in fixed (Jackson and Pombo 1998) and live (Bornfleth et al. 1999) mammalian cells. Based on patterns of labeling with replication precursor analogs, these ~1 Mbp chromatin domains clearly correlate with the independent replicon clusters described above.

The ~1 Mbp replicon clusters were recognized as major determinants of the S phase structure, even though the molecular mechanisms underlying these domains were unknown. Importantly, recent developments in chromosome conformation capture (3C) techniques have confirmed the basic features of chromosome architecture by demonstrating the existence of steady-state, higher-order chromatin entities in cell populations. Two recent studies used unbiased, genome-wide 3C—called Hi-C—to elaborate chromosome architecture (Lieberman-Aiden et al. 2009; Kalhor et al. 2012). This analysis makes three key points. First, steady-state structures seen at the Mbp scale have a "fractal" architecture allowing these fractal globules to fold and unfold without knotting. Second, globules within the active and inactive chromatin compartments have distinct structural and predicted dynamic properties with a preference to interact with structures in the same chromatin class. And third, DNA within the active chromatin class in particular interacts promiscuously with many other part of the genome in the same class. These conclusions emphasize how chromatin compartments can represent pseudo-stable higher-order domains that are sufficiently dynamic to engage in a variety of interactions across cell populations. The underlying mechanisms that define these structures are likely to be very complex, so that the structure of chromatin domains correlates with numerous features of chromatin function. This is especially evident during cell differentiation, when subtle changes in higher-order chromatin architecture during differentiation provide an excellent correlation with changes in replication timing (Ryba et al. 2010).

S-PHASE PROGRESSION

Replication must be performed with 100% precision to preserve genome integrity. As we have seen, many studies infer that chromosome structure and nuclear architecture contribute to this process. However, to fully understand the replication process, it is necessary to recognize the very complex nature of the regulatory mechanisms and explore how the random nature of the initiation process can be directed by structural features. In doing this, it is also important to realize that in eukaryotic cells as in bacteria the process of DNA replication must occur in such a way that the replication machinery displaces proteins that define the structure of local DNA loops while allowing these to reform once synthesis is complete. This imperative to reform chromatin structure following replication explains why local chromosome structure must be a complex composite of many thousands of dynamic interactions.

In the unicellular model eukaryote budding yeast, origins are genetically defined and are activated according to a temporal program (Raghuraman et al. 2001), which in part at least is controlled by origin activation by different cyclins (McCune et al. 2008). But even in this system, analysis of DNA from single cells shows that initiation can be intrinsically disordered in time (Czajkowsky et al. 2008), implying that origin selection has a significant stochastic component (Patel et al. 2006). In mammalian cells, origin selection is also known to be highly stochastic (see Maya-Mendoza et al. 2009 for review). However, this innate feature of the replication process arises as potential origins outnumber origins that are used in each cell cycle by ~5:1. Despite this flexibility, there is excellent evidence from many sources that replication in mammalian cells is programmed so that different parts of the genome are replicated at preferred times (Zink 2006; Aladjem 2007; Sclafani and Holzen 2007).

As in yeast, in higher eukaryotes euchromatin is replicated early in the S phase and late in heterochromatin. But how ordered is the timing program and what are the underlying mechanisms? As noted above, studies of nascent sites within nuclei show that a specific subset of potential initiation targets are select with good efficiency at the onset of the S phase (Jackson and Pombo 1998; Ma et al. 1998). Numerous genome-wide studies have shown that active genes are preferentially replicated at the beginning of the S phase and the process of selection correlates with patterns of gene expression (Cadoret et al. 2008; Farkash-Amar et al. 2008), active chromatin marks (Hiratani et al. 2008), and the mostly highly interactive gene regions defined by Hi-C (Ryba et al. 2010). These observations imply that the local chromatin context is a key determinant of initiation at the onset of the S phase, with arguably the most mobile chromatin having the highest probability of engaging synthesis.

Exactly how these concepts feed into the selection of potential origins is unknown. However, the key step in the initiation of synthesis must involve enzyme-mediate activation of potential origins to generate preinitiation complexes (pre-ICs), which are then able to bind the synthetic machinery and initiate replication. The early steps in this process must occur within chromatin domains and subsequent reorganization of these domains is thought to be required for pre-ICs to interact within replication factories. The synchronous activation of contiguous replicons within clusters emphasizes the importance of coordination during synthesis, and although the regulatory mechanisms are unknown, these observations are at least consistent with a role for some form of a factory level control process. As the S phase proceeds, changes in expression of different cyclin–CDK complexes serve to increase the potency of potential origins in heterochromatin at late times of the S phase (Katsuno et al. 2009). Hence, the combination of accessibility of potential origins in open chromatin during early S phase, subsequent increase in activity of remaining potential origins in late S phase and robust checkpoints for DNA completion ensures that synthesis is performed in a way that defines the preservation of both genetic and epigenetic information.

Genomic approaches to the analysis of replication timing support the view that synthesis spreads along chromosomes from dominant initiations sites that operate at the onset of the S phase (reviewed in Cadoret and Prioleau 2010; Farkash-Amar and Simon 2010), with subtle differences in timing occurring during cell differentiation (Hiratani et al. 2008; Ryba et al. 2010). Importantly, the demonstration of replication timing zones with sizes typically in the range ~0.5–1.0 Mb of DNA across cell populations shows that the replication domains correlate with structural chromatin foci, which also correspond with replicon clusters (Maya-Mendoza et al. 2010). Together, these observations show that cells with a specific differentiation status will follow unique temporal programs of S phase progression, which then become an innate feature of DNA synthesis in the corresponding different cell types. In conclusion, in both eubacteria and eukaryotes the process of DNA replication is principally controlled by chromosome organization. In the different bacterial models discussed above, a key feature of the replication process is to ensure that the sister nucleoids are properly positioned to ensure efficient segregation. Eukaryotes have evolved precise mechanisms to ensure the preservation of genetic integrity during cell division. However, eukaryotes face additional challenges in maintaining their genetic and epigenetic integrity during DNA synthesis, in large part because of the huge size of their genomes. As in eubacteria, fundamental aspects of the synthetic process are defined by chromosome structure and function, with many features of organization contributing to the temporal program through which genetic and epigenetic integrity is maintained. Accordingly, one of the key challenges for the future is to define the molecular basis of higher-order chromosome organization that dictates the spatial and temporal organization of DNA replication.

REFERENCES

Aladjem MI. 2007. Replication in context: Dynamic regulation of DNA replication patterns in metazoans. *Nat Rev Genet* **8:** 588–600.

Baddeley D, Chagin VO, Schermelleh L, Martin S, Pombo A, Carlton PM, Gahl A, Domaing P, Birk U, Leonhardt H, et al. 2010. Measurement of replication structures at the nanometer scale using super-resolution light microscopy. *Nucleic Acids Res* **38:** e8.

Bates D, Kleckner N. 2005. Chromosome and replisome dynamics in *E. coli*: Loss of sister cohesion triggers global chromosome movement and mediates chromosome segregation. *Cell* **121:** 899–911.

Beeson J, Sueoka N. 1979. Membrane enrichment of genetic markers close to the origin and terminus during the deoxyribonucleic acid replication cycle in *Bacillus subtilis*. *J Bacteriol* **139:** 911–916.

Ben-Yehuda S, Rudner DZ, Losick R. 2003. RacA, a bacterial protein that anchors chromosomes to the cell poles. *Science* **299:** 532–536.

Berezney R, Coffey DS. 1974. Identification of a nuclear protein matrix. *Biochem Biophys Res Commun* **60:** 1410–1417.

Berezney R, Coffey DS. 1975. Nuclear protein matrix: Association with newly synthesized DNA. *Science* **189:** 291–293.

Berkmen MB, Grossman AD. 2006. Spatial and temporal organization of the *Bacillus subtilis* replication cycle. *Mol Microbiol* **62:** 57–71.

Berlatzky IA, Rouvinski A, Ben-Yehuda S. 2008. Spatial organization of a replicating bacterial chromosome. *Proc Natl Acad Sci* **105:** 14136–14140.

Blow JJ, Dutta A. 2005. Preventing re-replication of chromosomal DNA. *Nat Rev Mol Cell Biol* **6:** 476–486.

Bornfleth H, Edelmann P, Zink D, Cremer T, Cremer C. 1999. Quantitative motion analysis of subchromosomal foci in living cells using four-dimensional microscopy. *Biophys J* **77:** 2871–2886.

Boye E, Lobner-Olesen A, Skarstad K. 2000. Limiting DNA replication to once and only once. *EMBO Rep* **1:** 479–483.

Cadoret JC, Prioleau MN. 2010. Genome-wide approaches to determining origin distribution. *Chromosome Res* **18:** 79–89.

Cadoret JC, Meisch F, Hassan-Zadeh V, Luyten I, Guillet C, Duret L, Quesneville H, Prioleau MN. 2008. Genome-wide studies highlight indirect links between human replication origins and gene regulation. *Proc Natl Acad Sci* **105:** 15837–15842.

Cremer T, Cremer M. 2010. Chromosome territories. *Cold Spring Harb Perspect Biol* **2:** a003889.

Cseresnyes Z, Schwarz U, Green CM. 2009. Analysis of replication factories in human cells by super-resolution light microscopy. *BMC Cell Biol* **10:** 88.

Cui Y, Petrushenko ZM, Rybenkov VV. 2008. MukB acts as a macromolecular clamp in DNA condensation. *Nat Struct Mol Biol* **15:** 411–418.

Czajkowsky DM, Liu J, Hamlin JL, Shao Z. 2008. DNA combing reveals intrinsic temporal disorder in the replication of yeast chromosome VI. *J Mol Biol* **375:** 12–19.

Danilova O, Reyes-Lamothe R, Pinskaya M, Sherratt D, Possoz C. 2007. MukB colocalizes with the oriC region and is required for organization of the two *Escherichia coli* chromosome arms into separate cell halves. *Mol Microbiol* **65:** 1485–1492.

Deng S, Stein RA, Higgins NP. 2005. Organization of supercoil domains and their reorganization by transcription. *Mol Microbiol* **57:** 1511–1521.

Dillon SC, Dorman CJ. 2010. Bacterial nucleoid-associated proteins, nucleoid structure and gene expression. *Nat Rev Microbiol* **8:** 185–195.

Duigou S, Yamaichi Y, Waldor MK. 2008. ATP negatively regulates the initiator protein of *Vibrio cholerae* chromosome II replication. *Proc Natl Acad Sci* **105:** 10577–10582.

Edenberg HJ, Huberman JA. 1975. Eukaryotic chromosome replication. *Annu Rev Genet* **9:** 245–284.

Espeli O, Mercier R, Boccard F. 2008. DNA dynamics vary according to macrodomain topography in the *E. coli* chromosome. *Mol Microbiol* **68:** 1418–1427.

Farkash-Amar S, Simon I. 2010. Genome-wide analysis of the replication program in mammals. *Chromosome Res* **18:** 115–125.

Farkash-Amar S, Lipson D, Polten A, Goren A, Helmstetter C, Yakhini Z, Simon I. 2008. Global organization of replication time zones of the mouse genome. *Genome Res* **18:** 1562–1570.

Hardy CD, Cozzarelli NR. 2005. A genetic selection for supercoiling mutants of *Escherichia coli* reveals proteins implicated in chromosome structure. *Mol Microbiol* **57:** 1636–1652.

Hiratani I, Ryba T, Itoh M, Yokochi T, Schwaiger M, Chang CW, Lyou Y, Townes TM, Schubeler D, Gilbert DM. 2008. Global reorganization of replication domains during embryonic stem cell differentiation. *PLoS Biol* **6:** e245.

Holmes VF, Cozzarelli NR. 2000. Closing the ring: Links between SMC proteins and chromosome partitioning, condensation, and supercoiling. *Proc Natl Acad Sci* **97:** 1322–1324.

Holmquist GP, Ashley T. 2006. Chromosome organization and chromatin modification: Influence on genome function and evolution. *Cytogenet Genome Res* **114:** 96–125.

Hoshino T, McKenzie T, Schmidt S, Tanaka T, Sueoka N. 1987. Nucleotide sequence of *Bacillus subtilis* dnaB: A gene essential for DNA replication initiation and membrane attachment. *Proc Natl Acad Sci* **84:** 653–657.

Hozak P, Hassan AB, Jackson DA, Cook PR. 1993. Visualization of replication factories attached to nucleoskeleton. *Cell* **73:** 361–373.

Huberman JA, Riggs AD. 1966. Autoradiography of chromosomal DNA fibers from Chinese hamster cells. *Proc Natl Acad Sci* **55:** 599–606.

Jackson DA, Cook PR. 1986. Replication occurs at a nucleoskeleton. *EMBO J* **5:** 1403–1410.

Jackson DA, Pombo A. 1998. Replicon clusters are stable units of chromosome structure: Evidence that nuclear organization contributes to the efficient activation and propagation of S phase in human cells. *J Cell Biol* **140:** 1285–1295.

Jensen RB, Wang SC, Shapiro L. 2001. A moving DNA replication factory in *Caulobacter crescentus*. *EMBO J* **20:** 4952–4963.

Kalhor R, Tjong H, Jayathilaka N, Alber F, Chen L. 2012. Genome architectures revealed by tethered chromosome conformation capture and population-based modeling. *Nat Biotechnol* **30:** 90–98.

Katayama T, Ozaki S, Keyamura K, Fujimitsu K. 2010. Regulation of the replication cycle: Conserved and diverse regulatory systems for DnaA and oriC. *Nat Rev Microbiol* **8:** 163–170.

Katsuno Y, Suzuki A, Sugimura K, Okumura K, Zineldeen DH, Shimada M, Niida H, Mizuno T, Hanaoka F, Nakanishi M. 2009. Cyclin A-Cdk1 regulates the origin firing program in mammalian cells. *Proc Natl Acad Sci* **106:** 3184–3189.

Kitamura E, Blow JJ, Tanaka TU. 2006. Live-cell imaging reveals replication of individual replicons in eukaryotic replication factories. *Cell* **125:** 1297–1308.

Lau IF, Filipe SR, Soballe B, Okstad OA, Barre FX, Sherratt DJ. 2003. Spatial and temporal organization of replicating *Escherichia coli* chromosomes. *Mol Microbiol* **49:** 731–743.

Lemon KP, Grossman AD. 1998. Localization of bacterial DNA polymerase: Evidence for a factory model of replication. *Science* **282:** 1516–1519.

Lemon KP, Grossman AD. 2000. Movement of replicating DNA through a stationary replisome. *Mol Cell* **6:** 1321–1330.

Leonard AC, Grimwade JE. 2011. Regulation of DnaA assembly and activity: Taking directions from the genome. *Annu Rev Microbiol* **65:** 19–35.

Li Y, Austin S. 2002. The P1 plasmid is segregated to daughter cells by a "capture and ejection" mechanism coordinated with *Escherichia coli* cell division. *Mol Microbiol* **46:** 63–74.

Lieberman-Aiden E, van Berkum NL, Williams L, Imakaev M, Ragoczy T, Telling A, Amit I, Lajoie BR, Sabo PJ, Dorschner MO, et al. 2009. Comprehensive mapping of long-range interactions reveals folding principles of the human genome. *Science* **326:** 289–293.

Lin DC, Levin PA, Grossman AD. 1997. Bipolar localization of a chromosome partition protein in *Bacillus subtilis*. *Proc Natl Acad Sci* **94:** 4721–4726.

Ma H, Samarabandu J, Devdhar RS, Acharya R, Cheng PC, Meng C, Berezney R. 1998. Spatial and temporal dynamics of DNA replication sites in mammalian cells. *J Cell Biol* **143:** 1415–1425.

Manders EM, Stap J, Brakenhoff GJ, van Driel R, Aten JA. 1992. Dynamics of three-dimensional replication patterns during the S-phase, analysed by double labelling of DNA and confocal microscopy. *J Cell Sci* **103:** 857–862.

Manders EM, Stap J, Strackee J, van Driel R, Aten JA. 1996. Dynamic behavior of DNA replication domains. *Exp Cell Res* **226:** 328–335.

Maya-Mendoza A, Tang CW, Pombo A, Jackson DA. 2009. Mechanisms regulating S phase progression in mammalian cells. *Front Biosci* **14:** 4199–4213.

Maya-Mendoza A, Olivares-Chauvet P, Shaw A, Jackson DA. 2010. S phase progression in human cells is dictated by the genetic continuity of DNA foci. *PLoS Genet* **6:** e1000900.

McCready SJ, Godwin J, Mason DW, Brazell IA, Cook PR. 1980. DNA is replicated at the nuclear cage. *J Cell Sci* **46:** 365–386.

McCune HJ, Danielson LS, Alvino GM, Collingwood D, Delrow JJ, Fangman WL, Brewer BJ, Raghuraman MK. 2008. The temporal program of chromosome replication: Genomewide replication in clb5Δ *Saccharomyces cerevisiae*. *Genetics* **180:** 1833–1847.

Meister P, Taddei A, Gasser SM. 2006. In and out of the replication factory. *Cell* **125:** 1233–1235.

Mercier R, Petit MA, Schbath S, Robin S, El Karoui M, Boccard F, Espeli O. 2008. The MatP/matS site-specific system organizes the terminus region of the *E. coli* chromosome into a macrodomain. *Cell* **135:** 475–485.

Michaelis C, Ciosk R, Nasmyth K. 1997. Cohesins: Chromosomal proteins that prevent premature separation of sister chromatids. *Cell* **91:** 35–45.

Nakamura H, Morita T, Sato C. 1986. Structural organizations of replicon domains during DNA synthetic phase in the mammalian nucleus. *Exp Cell Res* **165:** 291–297.

Nielsen HJ, Ottesen JR, Youngren B, Austin SJ, Hansen FG. 2006. The *Escherichia coli* chromosome is organized with the left and right chromosome arms in separate cell halves. *Mol Microbiol* **62:** 331–338.

Niki H, Yamaichi Y, Hiraga S. 2000. Dynamic organization of chromosomal DNA in *Escherichia coli*. *Genes Dev* **14:** 212–223.

Pardoll DM, Vogelstein B, Coffey DS. 1980. A fixed site of DNA replication in eucaryotic cells. *Cell* **19:** 527–536.

Patel PK, Arcangioli B, Baker SP, Bensimon A, Rhind N. 2006. DNA replication origins fire stochastically in fission yeast. *Mol Biol Cell* **17:** 308–316.

Postow L, Hardy CD, Arsuaga J, Cozzarelli NR. 2004. Topological domain structure of the *Escherichia coli* chromosome. *Genes Dev* **18:** 1766–1779.

Raghuraman MK, Winzeler EA, Collingwood D, Hunt S, Wodicka L, Conway A, Lockhart DJ, Davis RW, Brewer BJ, Fangman WL. 2001. Replication dynamics of the yeast genome. *Science* **294:** 115–121.

Rasmussen T, Jensen RB, Skovgaard O. 2007. The two chromosomes of *Vibrio cholerae* are initiated at different time points in the cell cycle. *EMBO J* **26:** 3124–3131.

Reyes-Lamothe R, Possoz C, Danilova O, Sherratt DJ. 2008. Independent positioning and action of *Escherichia coli* replisomes in live cells. *Cell* **133:** 90–102.

Robinett CC, Straight A, Li G, Willhelm C, Sudlow G, Murray A, Belmont AS. 1996. In vivo localization of DNA sequences and visualization of large-scale chromatin organization using lac operator/repressor recognition. *J Cell Biol* **135:** 1685–1700.

Rokop ME, Auchtung JM, Grossman AD. 2004. Control of DNA replication initiation by recruitment of an essential initiation protein to the membrane of *Bacillus subtilis*. *Mol Microbiol* **52:** 1757–1767.

Ryba T, Hiratani I, Lu J, Itoh M, Kulik M, Zhang J, Schulz TC, Robins AJ, Dalton S, Gilbert DM. 2010. Evolutionarily conserved replication timing profiles predict long-range chromatin interactions and distinguish closely related cell types. *Genome Res* **20:** 761–770.

Sclafani RA, Holzen TM. 2007. Cell cycle regulation of DNA replication. *Annu Rev Genet* **41:** 237–280.

Skarstad K, Boye E, Steen HB. 1986. Timing of initiation of chromosome replication in individual *Escherichia coli* cells. *EMBO J* **5:** 1711–1717.

Sporbert A, Gahl A, Ankerhold R, Leonhardt H, Cardoso MC. 2002. DNA polymerase clamp shows little turnover at established replication sites but sequential de novo assembly at adjacent origin clusters. *Mol Cell* **10:** 1355–1365.

Stein RA, Deng S, Higgins NP. 2005. Measuring chromosome dynamics on different time scales using resolvases with varying half-lives. *Mol Microbiol* **56:** 1049–1061.

Straight AF, Belmont AS, Robinett CC, Murray AW. 1996. GFP tagging of budding yeast chromosomes reveals that protein–protein interactions can mediate sister chromatid cohesion. *Curr Biol* **6:** 1599–1608.

Teleman AA, Graumann PL, Lin DC, Grossman AD, Losick R. 1998. Chromosome arrangement within a bacterium. *Curr Biol* **8:** 1102–1109.

Umbarger MA, Toro E, Wright MA, Porreca GJ, Bau D, Hong SH, Fero MJ, Zhu LJ, Marti-Renom MA, McAdams HH, et al. 2011. The three-dimensional architecture of a bacterial genome and its alteration by genetic perturbation. *Mol Cell* **44:** 252–264.

Valens M, Penaud S, Rossignol M, Cornet F, Boccard F. 2004. Macrodomain organization of the *Escherichia coli* chromosome. *EMBO J* **23:** 4330–4341.

Viollier PH, Thanbichler M, McGrath PT, West L, Meewan M, McAdams HH, Shapiro L. 2004. Rapid and sequential movement of individual chromosomal loci to specific subcellular locations during bacterial DNA replication. *Proc Natl Acad Sci* **101:** 9257–9262.

Vologodskii AV, Cozzarelli NR. 1994. Conformational and thermodynamic properties of supercoiled DNA. *Annu Rev Biophys Biomol Struct* **23:** 609–643.

Wang JD, Levin PA. 2009. Metabolism, cell growth and the bacterial cell cycle. *Nat Rev Microbiol* **7:** 822–827.

Wang X, Liu X, Possoz C, Sherratt DJ. 2006. The two *Escherichia coli* chromosome arms locate to separate cell halves. *Genes Dev* **20:** 1727–1731.

Webb CD, Teleman A, Gordon S, Straight A, Belmont A, Lin DC, Grossman AD, Wright A, Losick R. 1997. Bipolar localization of the replication origin regions of chromosomes in vegetative and sporulating cells of *B. subtilis*. *Cell* **88:** 667–674.

Wu LJ, Errington J. 1998. Use of asymmetric cell division and spoIIIE mutants to probe chromosome orientation and organization in *Bacillus subtilis*. *Mol Microbiol* **27:** 777–786.

Zink D. 2006. The temporal program of DNA replication: New insights into old questions. *Chromosoma* **115:** 273–287.

DNA Replication Timing

Nicholas Rhind[1] and David M. Gilbert[2]

[1]Department of Biochemistry and Molecular Pharmacology, University of Massachusetts Medical School, Worcester, Massachusetts 01605

[2]Department of Biological Science, Florida State University, Tallahassee, Florida 32306

Correspondence: nick.rhind@umassmed.edu

Patterns of replication within eukaryotic genomes correlate with gene expression, chromatin structure, and genome evolution. Recent advances in genome-scale mapping of replication kinetics have allowed these correlations to be explored in many species, cell types, and growth conditions, and these large data sets have allowed quantitative and computational analyses. One striking new correlation to emerge from these analyses is between replication timing and the three-dimensional structure of chromosomes. This correlation, which is significantly stronger than with any single histone modification or chromosome-binding protein, suggests that replication timing is controlled at the level of chromosomal domains. This conclusion dovetails with parallel work on the heterogeneity of origin firing and the competition between origins for limiting activators to suggest a model in which the stochastic probability of individual origin firing is modulated by chromosomal domain structure to produce patterns of replication. Whether these patterns have inherent biological functions or simply reflect higher-order genome structure is an open question.

Replication of eukaryotic chromosomes takes place in segments that generally replicate in a predictable temporal order. Because the rate of elongation of replication forks varies little throughout S phase, this "replication timing program" is largely mediated by the time of initiation of replication within the corresponding segments. However, it is the temporal order of replication, not the sites of initiation, that is conserved among species (Aladjem et al. 2002; Farkash-Amar et al. 2008; Liachko et al. 2010; Ryba et al. 2010; Yaffe et al. 2010; Di Rienzi et al. 2012; Muller and Nieduszynski 2012; Xu et al. 2012), suggesting that replication timing is regulated independently of mechanisms specifying origins. Despite this evolutionary conservation, the biological significance of replication timing has remained elusive. In multicellular but not unicellular organisms, early replication is correlated with transcriptional activity and is developmentally regulated (Hiratani et al. 2009), but causal links have not been established. Recent findings establish the importance of large-scale chromatin folding in the regulation of replication timing in both yeasts and mammals and provide new promise for our understanding of both the significance and mechanism of the replication program. These results suggest a unifying model of how structural compartmentalization of the genome in the

Cite this article as *Cold Spring Harb Perspect Biol* doi: 10.1101/cshperspect.a010132

eukaryotic nucleus can influence overall functional output.

In this chapter, we will begin by summarizing the important contributions of genome-scale methods to the field. Next we will examine the relationships between replication timing and the three-dimensional (3D) organization of chromosomes in the nucleus. We will follow this with a discussion of mechanistic insights and conclude with speculation on the potential biological significance of a replication timing program. Along the way, we will refer the reader to many outstanding recent reviews for more detailed discussion of various aspects of this complex topic.

CONTRIBUTIONS FROM GENOME-SCALE METHODS

Although pioneering genome-scale studies of replication timing were performed more than 10 years ago (Donaldson and Schildkraut 2006), they have more recently been applied to many different cell types and experimental conditions, providing a comprehensive view of the temporal program and a robust tool for experimentation. In fact, replication timing profiles are such a stable characteristic of particular cell types that they can be used for cell type identification (Pope et al. 2011; Ryba et al. 2011). Details of these methods have been described in several recent reviews (Farkash-Amar and Simon 2010; Raghuraman and Brewer 2010; Ryba et al. 2011). Here we will focus on the salient findings from these studies. Table 1 compiles a list of published genome-wide replication timing data sets for various species at the time of this writing.

Interpreting Genome-Wide Replication Timing Profiles

Figure 1 shows exemplary profiles of replication timing in human, *Drosophila*, and yeast cells, and Figure 2 shows a schematic to help with the interpretation of these profiles. In the large genomes of multicellular organisms, the profiles consist of broad, computationally defined constant timing regions (CTRs), or "replication domains," punctuated by slopes of progressive change in timing, commonly referred to as timing transition regions (TTRs). The appearance of constant timing across large replication domains may in part be due to the resolution of existing methodology, but the prevailing working model (Fig. 2) is that the similar timing on the population level is due to the averaging of heterogeneous firing on the individual cell level from clusters of multiple origins or initiation zones (regions within which different cells initiate at different sites) that fire with similar timing. The evidence for this model is indirect, inferred from a combination of genome-scale data that show similar average replication times at 100-kb resolution and individual DNA fiber origin-mapping data that reveal evidence for clusters of initiation at sites that are heterogeneous between cells (Lebofsky et al. 2006; Cayrou et al. 2011; Letessier et al. 2011; Besnard et al. 2012).

TTRs are the regions between CTRs that have a gradient of replication timing from their early-replicating sides to their late-replicating sides (Fig. 2). Their replication timing is consistent with either unidirectional forks emanating from the early CTRs toward the late CTRs (Farkash-Amar et al. 2008; Hiratani et al. 2008; Desprat et al. 2009; Ryba et al. 2010) or sequentially firing origins (Guilbaud et al. 2011). At least part of one TTR has been shown to replicate predominantly as a single long unidirectional fork in different human and mouse cell lines (Norio et al. 2005; Schultz et al. 2010), and sequences that can function as origins at their normal locations rarely fire when inserted into this TTR (Guan et al. 2009). Even strong transcription or tethering of a potent histone acetyltransferase near an origin within a TTR did not alter its firing rate, whereas tethering similar molecules near origins that do not reside within a TTR strongly affects origin firing (Goren et al. 2008; Hassan-Zadeh et al. 2012). On the other hand, the rate of replication through TTRs genome-wide has been estimated to vary by two- to threefold more than global rate of fork elongation (Takebayashi et al. 2005; Frum et al. 2009; Ryba et al. 2010; Guilbaud et al. 2011), suggesting that sequential firing of origins occurs

Table 1. Genome-wide replication profiles

Species	Cell type	Cell line	References
Homo sapiens	hESC	H1	Desprat et al. 2009; Ryba et al. 2011
		H7	Ryba et al. 2010
		H9	Ryba et al. 2010
		BG01	Ryba et al. 2010
		BG02	Hansen et al. 2010; Ryba et al. 2010
	hiPSC	iPSC4	Ryba et al. 2010
		iPSC5	Ryba et al. 2010
	Definitive endoderm	BG02-derived	Ryba et al. 2011
	Mesendoderm	BG02-derived	Ryba et al. 2011
	Mesoderm	BG02-derived	Ryba et al. 2011
	Smooth muscle	BG02-derived	Ryba et al. 2011
	Neural precursors	BG01-derived	Ryba et al. 2010
	Myoblast	7 individuals	Pope et al. 2011
	Lymphoblastoid	CO2O2 (male)	Ryba et al. 2010
		GM06990 (female)	Hansen et al. 2010; Ryba et al. 2011
		TL010	Hansen et al. 2010
		HO287 (male)	Hansen et al. 2010
		MOLT-4	Yaffe et al. 2010
	Fibroblast	IMR90	Pope et al. 2011
		BJ	Hansen et al. 2010
		FFT	Yaffe et al. 2010
		HeLa	Chen et al. 2010
	Erythroid	K562	Hansen et al. 2010
		$CD4^+$-derived basophilic erythroblasts	Desprat et al. 2009
Mus musculus	mESC	D3	Hiratani et al. 2008
		46C	Hiratani et al. 2008
		TT2	Hiratani et al. 2008
		TT2 flox G9a ESC: mock and OHT	Yokochi et al. 2009
	miPSC	iPS	Hiratani et al. 2008
		iPSC1D4	Hiratani et al. 2010
		iPSC2D4	Hiratani et al. 2010
	pmiPSC	piPSC1A2	Hiratani et al. 2010
		piPSC1B3	Hiratani et al. 2010
		piPSC1V3	Hiratani et al. 2010
	Early primitive ectoderm	D3-derived as embryoid bodies	Hiratani et al. 2010
	Early primitive ectoderm	D3-derived as monolayer	Hiratani et al. 2010
	Definitive ectoderm	D3-derived as embryoid bodies	Hiratani et al. 2010
	Neural precursors	D3-derived as embryoid bodies	Hiratani et al. 2008
		46C-derived as monolayer	Hiratani et al. 2008
		TT2-derived as monolayer	Hiratani et al. 2008

Continued

Table 1. *Continued*

Species	Cell type	Cell line	References
		TT2 flox G9a-derived: mock and OHT	Yokochi et al. 2009
	Endoderm	GscSox17-EB5-derived	Hiratani et al. 2010
	Mesoderm	GscSox17-EB5-derived	Hiratani et al. 2010
	mEpiSC	EpiSC5	Hiratani et al. 2010
		EpiSC7 (female)	Hiratani et al. 2010
	Lymphoblastoid	L1210	Farkash-Amar et al. 2008; Hiratani et al. 2010
	Fibroblast	C127 mammary	Lu et al. 2010
		MEF male and female	Hiratani et al. 2010; Yaffe et al. 2010
	Myoblast	J185a	Hiratani et al. 2010
Drosophila melanogaster	Wing imaginal disc	Cl8	Schwaiger et al. 2009
	Embryonic	Kc	Schwaiger et al. 2009; Eaton et al. 2011
		Kc HP1 knockdown	Schwaiger et al. 2010
		S2	Eaton et al. 2011
	Neuronal	Bg3	Eaton et al. 2011
Chicken	Lymphoid	DT40	Hassan-Zadeh et al. 2012
Arabidopsis thalania		Col-0 (chromosome 4 only)	Lee et al. 2010
Saccharomyces cerevisiae	wt		Raghuraman et al. 2001
	wt		Yabuki et al. 2002
	wt		Feng et al. 2006
	*rad35*Δ		Feng et al. 2006
	wt		Alvino et al. 2007
	wt + HU		Alvino et al. 2007
	wt		McCune et al. 2008
	*clb5*Δ		McCune et al. 2008
	*clb5*Δ *clb6*Δ		McCune et al. 2008
	wt		Knott et al. 2009
	*rpd3*Δ		Knott et al. 2009
	*eaf3*Δ (Rpd3S)		Knott et al. 2009
	*set2*Δ (Rpd3S)		Knott et al. 2009
	*rco1*Δ (Rpd3S)		Knott et al. 2009
	*dep1*Δ (Rpd3L)		Knott et al. 2009
	*cti6*Δ (Rpd3L)		Knott et al. 2009
	*rpd3*Δ *set2*Δ		Knott et al. 2009
	*rpd3*Δ *cti6*Δ		Knott et al. 2009
	*rpd3*Δ *eaf3*Δ		Knott et al. 2009
	*cti6*Δ *eaf3*Δ		Knott et al. 2009
	*dep1*Δ *eaf3*Δ		Knott et al. 2009
	*rco1*Δ *eaf3*Δ		Knott et al. 2009
	*eaf3*Δ *set2*Δ		Knott et al. 2009
	wt		Crabbe et al. 2010
	rad9		Crabbe et al. 2010
	rev3 rad30		Crabbe et al. 2010
	eco1		Crabbe et al. 2010
	ctf4		Crabbe et al. 2010
	ddc1		Crabbe et al. 2010
	rad24		Crabbe et al. 2010

Continued

Table 1. *Continued*

Species	Cell type	Cell line	References
	pol2-12		Crabbe et al. 2010
	elg1		Crabbe et al. 2010
	tof1		Crabbe et al. 2010
	mrc1^{AQ}		Crabbe et al. 2010
	mrc1^{AQ} *rad9*		Crabbe et al. 2010
	ctf8		Crabbe et al. 2010
	ctf18		Crabbe et al. 2010
	mrc1		Crabbe et al. 2010
	ddc1		Crabbe et al. 2010
	ctf18 rad9		Crabbe et al. 2010
	mec1-100		Crabbe et al. 2010
	mec1-100		Crabbe et al. 2010
	mec1-100		Crabbe et al. 2010
	mec1-100		Crabbe et al. 2010
	rad53		Crabbe et al. 2010
	wt		Lian et al. 2011
	*yku70*Δ		Lian et al. 2011
	wt		Koren et al. 2010
	*sic1*Δ		Koren et al. 2010
	*dia2*Δ		Koren et al. 2010
	*clb5*Δ		Koren et al. 2010
	*mrc1*Δ		Koren et al. 2010
	*rrm3*Δ		Koren et al. 2010
	*dpb3*Δ		Koren et al. 2010
	*dpb4*Δ		Koren et al. 2010
	*rnr1*Δ		Koren et al. 2010
	*rad27*Δ		Koren et al. 2010
	*ura7*Δ		Koren et al. 2010
	*tda3*Δ		Koren et al. 2010
	*met7*Δ		Koren et al. 2010
	*gln3*Δ		Koren et al. 2010
	*arg2*Δ		Koren et al. 2010
	wt		Tanaka et al. 2011
	Overexpression of Sld3, Sld7, and Cdc45		Tanaka et al. 2011
	Overexpression of Dbf4		Tanaka et al. 2011
	Overexpression of Cdc7 and Dbf4		Tanaka et al. 2011
	wt		Knott et al. 2012
	*fkh1*Δ		Knott et al. 2012
	*fkh2*Δ		Knott et al. 2012
	*fkh1*Δ *fkh2*Δ		Knott et al. 2012
Schizosaccharomyces pombe	wt		Heichinger et al. 2006
	wt		Feng et al. 2006
	wt		Hayashi et al. 2007
	wt		Eshaghi et al. 2007
	wt		Mickle et al. 2007b
	wt		Mickle et al. 2007a

Continued

Table 1. *Continued*

Species	Cell type	Cell line	References
	wt		Kumar and Huberman 2009
	wt		Wu and Nurse 2009
	wt		Hayano et al. 2011
	wt		Hayano et al. 2011
	wt		Xu et al. 2012

hESC, human embryonic stem cells; hiPSC, human-induced pluripotent stem cells; mESC, mouse embryonic stem cells; OHT, 4-hydoxytamoxifen; miPSC, mouse-induced pluripotent stem cells; pmiPSC, partially reprogrammed mouse-induced pluripotent stem cells; mEpiSC, mouse epiblast-derived stem cells; wt, wild type; HU, hydroxyurea.

within at least some TTRs (Guilbaud et al. 2011). Furthermore, origin activity has been detected within TTRs (Cadoret et al. 2008; Cayrou et al. 2011). The frequency of initiation within TTRs may be influenced by factors that slow replication fork movement, which may induce the firing of dormant origins within the TTRs. TTRs terminate by merging with forks coming from an opposing TTR or a later-replicating CTR (Fig. 2). Merging TTRs create a U-shaped valley in the replication profiles resulting from the population average of forks terminating at different sites, without the need to invoke late-firing origins. Some of these late-replicating regions coincide with common fragile sites and fail to activate origins even under conditions of replicational stress (Le Tallec et al. 2011; Letessier et al. 2011; Debatisse et al. 2012). Large

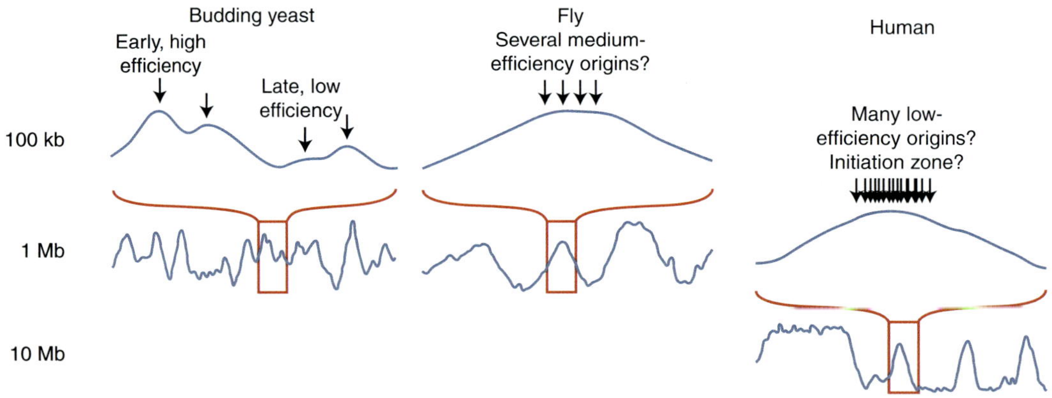

Figure 1. Scales of replication timing in species with different-sized genomes. Smoothed replication profiles of segments of the human (Ryba et al. 2010), fly (Schwaiger et al. 2009), and budding yeast genomes (Alvino et al. 2007; Yang et al. 2010). Although the profiles look qualitatively similar, they show features on very different scales. In yeasts, the peaks in the replication profiles represent individual origins and the average replication times of origins are determined by a combination of their average firing times and the frequency with which they are passively replicated by forks originating at neighboring origins. In mammals, what appear to be sharp peaks of early replication flatten at higher resolution to broad, near-megabase-sized domains, which contain many unresolved individual replicons. This lack of resolution can be accounted for by spatial or temporal heterogeneity in origin firing within each domain. Fly genomes are an order of magnitude smaller than human, and their domains of coordinate replication are similarly smaller but still probably contain multiple unresolved replicons. The slope of the curves moving away from early-replicating regions is often interpreted as being proportional to the rate of replication in that region. However, even in budding yeasts, where origins can be very efficient, this correlation is not strong; the slope is determined more by the ratio of fork directions than by the rate of those forks (Sekedat et al. 2010; Retkute et al. 2011).

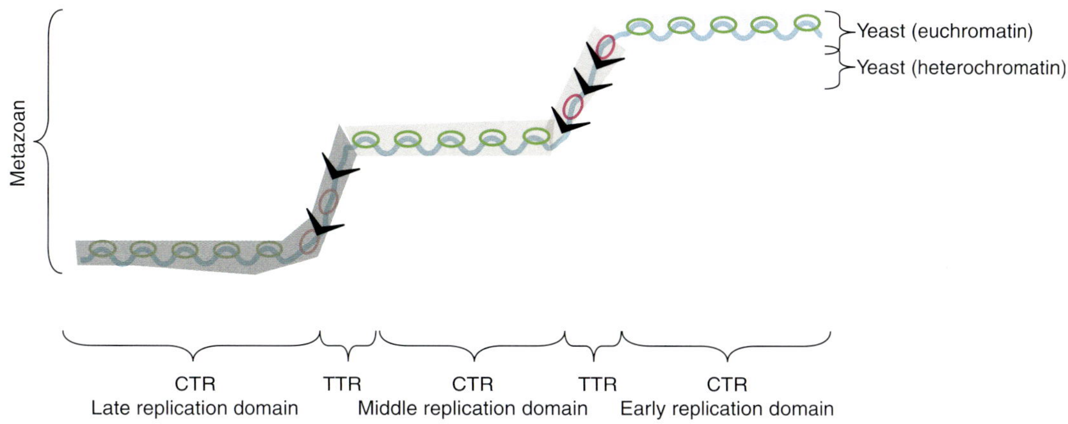

Figure 2. Interpreting replication profiles in different species. A hypothetical segment of a replication timing profile contains regions of constant timing (CTRs), or "replication domains," that replicate at different times during S phase, and regions of temporal transition (TTRs). However, these regions can be interpreted differently depending on genome size and computational parameters. In metazoan genomes, replication domains can be operationally defined regions where replication timing differs by 10%–20% of the length of S phase. The similarity of replication timing within such domains is proposed to be due in part to the heterogeneous, population-averaged firing of adjacent origins (green bubbles) with similar firing times, but in the case of large CTRs can be due to the nearly simultaneous firing of adjacent but independently regulated replication domains. The actual number of initiation sites within each domain that can potentially be used in a population of cells is believed to be in the dozens (origin clusters) to hundreds (initiation zones). TTRs are regions of suppressed origin activity (indicated as a gray "slime"), which may be replicated either by a single fork (black arrows) moving unidirectionally through time (*y*-axis) or—if slow-moving forks stimulate firing of inefficient or "dormant" origins—by sequentially activated origins (red bubbles). In contrast, the entire genome in less complex organisms such as yeasts can be thought of as a single replication domain, with the majority of regulation controlled by more origin-proximal mechanisms. The exception is the late-replicating heterochromatin, such as telomeres, which form the equivalents of TTRs, being passively replicated by forks originating in neighboring euchromatic regions.

late replication domains, however, likely contain late-firing origins that fire heterogeneously but with similar average timing, much as early CTRs do (Frum et al. 2009; Guilbaud et al. 2011; Le Tallec et al. 2011; Letessier et al. 2011; Debatisse et al. 2012). Ultimately, accurate interpretation of these temporal profiles will require more extensive studies of origin site selection that combine ensemble and DNA fiber origin-mapping methods (Cayrou et al. 2011).

In multicellular organisms with smaller genomes, such as *Drosophila* and *Arabidopsis* (Lee et al. 2010), replication timing profiles are qualitatively similar but at a different scale. Whereas replication domains in mammals range from several hundred kilobases to many megabases (Hiratani et al. 2008; Ryba et al. 2010), domains in these organisms range from 75 to 250 kb (MacAlpine et al. 2004; Schwaiger et al. 2009), which is small enough to be initiated from one or a few closely spaced origins (Fig. 1). Unfortunately, there are no studies of fork rates in *Drosophila* or *Arabidopsis* that can inform the interpretation of replication timing profiles in these species.

In the budding and fission yeasts, in which the genomes are 100-fold smaller than in mammals, the replication profiles look qualitatively similar to those of metazoans (Fig. 1). However, because origins are well-defined loci in yeasts, it is possible to resolve individual origins. Therefore the peaks in the yeast profiles correspond to actual origins, not just early-replicating regions. This resolution allows information about the efficiency and timing of individual origin firing to be extracted from yeast replication timing profiles. Although at the population level origins show distinct average firing times, at the single-cell level origin firing is heterogeneous, in that a different subset of origins is activated in a different timing pattern in each cell (Patel et al. 2006; Czajkowsky et al. 2008). Therefore, forks from neighboring origins can passively replicate even the earliest origins at some frequency, and caution must be used in interpreting the profiles. For instance, variations in the slope of replication profiles is generally due to the ratio of fork directions, not due to variations in individual fork speed (Sekedat et al. 2010; Retkute et al. 2011). Likewise, the probability of an origin firing is not directly proportional to its efficiency (i.e., the fraction of cells in which it fires); efficiency is determined by a combination of an origin's firing probability and its proximity to other origins with higher or lower probabilities of firing earlier and generating a replication fork that passively replicates the origin (de Moura et al. 2010). As in metazoans, there are early- and late-replicating segments of the yeast genome (Heichinger et al. 2006; McCune et al. 2008); however, these segments are not due to the clustering of early or late origins per se (Yang et al. 2010). Instead, the early-replicating segments have both early and late origins, but the late origins are usually passively replicated by forks from their early neighbors. In another similarity to metazoans, the latest-replicating sequences in the budding and fission yeast genomes are heterochromatin. However, because heterochromatin constitutes such a small portion of yeast genomes, these regions do not form late-replicating CTRs, as in metazoans, but rather form TTRs, which are often passively replicated by forks originating from neighboring euchromatic segments.

The availability of high-resolution, genome-wide replication timing profiles has permitted the development of mathematical models of replication kinetics. These models have been used to interpret population-averaged replication timing profiles, taking into account single-molecule DNA fiber data, and have generated testable hypotheses (Lygeros et al. 2008; de Moura et al. 2010; Hyrien and Goldar 2010; Luo et al. 2010; Yang et al. 2010). Many of these models have concluded that replication timing profiles can be accounted for by a combination of the probability with which origins in a region fire and their proximity to other origins with higher or lower probabilities, which determines the frequency of inactivation of neighboring origins by passive replication. In such models, factors that establish the probability of origin firing determine the replication timing program.

Replication Timing and DNA Sequence Composition

In multicellular organisms there are statistically significant links between replication timing and features of primary genomic sequence. Gene density correlates with early replication, while the density of repetitive elements associated with heterochromatin correlates with late replication. In vertebrates a general correlation exists between early replication and gene-rich, LINE-poor, GC-rich regions of the genome (Fig. 3, top), often segmented into regions called isochores (Schmegner et al. 2007; Costantini et al. 2009). However, these DNA sequence features are clearly not sufficient to dictate replication timing. Changes in replication timing during development (see below) can strongly influence the genome-wide alignment of replication timing to isochores (Hiratani et al. 2010; Ryba et al. 2010). Moreover, homologous loci can replicate at different times in the same cells (Karnani et al. 2007; Farkash-Amar et al. 2008; Hansen et al. 2010). Isochores that exhibit the most extreme of the above-mentioned sequence compositions tend to replicate at the same time in all tissues, whereas regions with intermediate or mixed sequence features are more prone to change replication timing during development (Fig. 3, top), suggesting that sequence composition has some indirect influence on replication timing. There is also evidence that replication timing can influence sequence composition. At about 1000 sites in the human genome, there are significant transitions in nucleotide composition that coincide with peaks of earliest regional replication timing (Touchon et al. 2005; Huvet et al. 2007); a similar phenomenon is seen in yeast genomes (Agier and Fischer 2012; Marsolier-Kergoat and

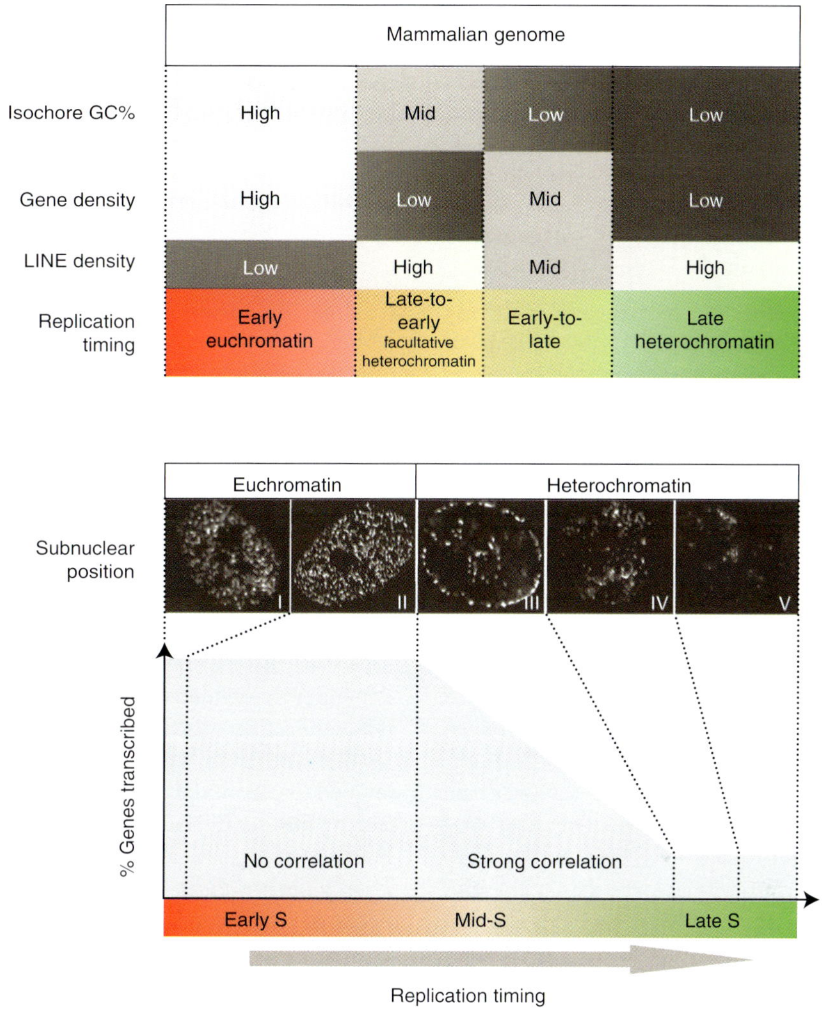

Figure 3. Relationship between isochore properties and replication timing regulation, subnuclear positioning, and transcription. (*Top*) Isochores with unusual sequence properties are subject to replication timing regulation. The mammalian genome is partitioned into isochores with different GC content, LINE composition, and gene density, which are generally correlated. Isochores that are high in GC and gene density but low in LINE density are replicated early in S phase, whereas the alternate extremes are replicated late. Isochores with intermediate or mixed sequence features are frequently subject to replication timing regulation during differentiation (speculatively labeled "Facultative heterochromatin") (Hiratani et al. 2008; Ryba et al. 2010). (*Bottom*) Changes in replication timing that traverse the middle of S phase accompany changes in subnuclear position and transcriptional potential. Replication early in S phase (patterns I and II) takes place within the interior euchromatic compartment, whereas replication later in S phase (patterns III, IV, and V) takes place at the nuclear periphery (pattern III), at the nucleolar periphery (pattern III), and at internal blocks of heterochromatin (patterns IV and V) (Berezney et al. 2000; McNairn and Gilbert 2003). Note the dramatic transition from euchromatic to heterochromatic replication pattern during mid-S phase (from pattern II to III). In addition, a strong relationship between earlier replication timing and transcription is observed for genes that replicate during mid- to late stages of S phase ("strong correlation"; corresponds to pattern III) (Hiratani et al. 2008). Very few genes are replicated at the end of S phase. In contrast, genes that are replicated in the first third of S phase have equally high probability of being expressed ("no correlation"), and thus even large changes within this period may not accompany changes in subnuclear repositioning and/or transcriptional competence may be inconsequential for transcription. (*Legend continues on following page.*)

Goldar 2012). These transitions are believed to arise from different mutation frequencies experienced by leading versus lagging strands, leaving an evolutionary imprint of replication fork polarity.

Relationships to Transcription: Divergence of Multicellular and Single-Celled Organisms

Genome-wide studies have revealed a strong positive correlation between early replication and transcriptional activity in multicellular organisms, including plants, particularly when transcriptional output is integrated across replication domains (MacAlpine et al. 2004; Hiratani et al. 2008; Lee et al. 2010). In contrast, no such correlation can be found in budding or fission yeasts, except for the fact that the few silent heterochromatic regions of yeasts, such as telomeres, are late-replicating (Raghuraman et al. 2001; Heichinger et al. 2006). One way of reconciling this discrepancy is to consider that in higher eukaryotes there is no correlation between transcription and replication timing throughout the entire first half of S phase; the observed correlation is accounted for by genes that replicate in the second half of S (Fig. 3, bottom). Moreover, in multicellular organisms, any given cell type replicates 75% of genes in the first half of S phase. In this sense the bulk of the yeast genome can be likened to early-replicating domains in multicellular organisms, with only the heterochromatic regions, which comprise a small fraction of the yeast genome, acting as later-replicating domains. Perhaps the "exploitation" of heterochromatin by multicellular organisms led to the expansion of the heterochromatic chromatin compartment and more extensive late-replicating domains (Gilbert 2002).

Developmental Control and Stable Epigenetic States

Another property of replication timing in multicellular organisms is developmental control. At least half the mammalian genome is subject to developmentally regulated changes in replication timing that occur in units of 400–800 kb and are generally coordinated with changes in transcriptional regulation of a certain class of promoters (Hiratani et al. 2008; Ryba et al. 2010). The smaller and more uniform size of these developmentally regulated replication domains suggests that larger domains of constant replication timing observed in any given cell type consist of multiple units that can be independently regulated during development. In fact, replication domains in pluripotent cells are smaller and their replication timing becomes aligned to consolidate them into larger domains during differentiation. Some differences in replication timing have also been observed between *Drosophila melanogaster* cell lines, and these also occur in smaller units than the replication domains found within a cell type (Schwaiger et al. 2009). As expected from the correlation to transcription, changes in replication timing generally coincide with changes in transcription. However, this coordination is promoter-specific in mammals; changes in transcription of genes with low- but not high-CpG-content promoters

Figure 3. (*Continued*) The figure illustrates how the durations of different spatial replication patterns are likely to relate to the probability of transcription based on published results (Dimitrova and Gilbert 1999; Hiratani et al. 2008). These relationships imply that genome-wide replication timing analyses provide a means to infer changes in subnuclear position and transcriptional potential (Hiratani et al. 2008). Together these observations strongly suggest that changes in replication timing that traverse the mid-S phase (i.e., pattern II to III or vice versa) accompany subnuclear repositioning and altered transcriptional potential, with the latter likely confined to certain classes of genes. We submit that although genome-wide analysis of subnuclear position changes by fluorescence in situ hybridization is impractical, these spatiotemporal patterns of replication provide a proxy of their 3D distribution. Genome-wide replication timing analyses provide a means to infer changes in subnuclear position and even transcriptional potential using these relationships (Hiratani et al. 2008). Photos are deconvolved images from Chinese hamster cells pulse-labeled with bromodeoxyuridine at different times during S phase (J Lu and DM Gilbert, unpubl.). (From Hiratani et al. 2009; adapted, with permission, from Elsevier.)

correlate with replication timing changes (Hiratani et al. 2008, 2010). Given that replication domains are 400–800 kb, many genes can change replication timing coordinately, with some showing coordinated transcriptional changes and others not, which has made dissection of this relationship difficult. However, when the analysis is focused on low-CpG-content promoters during the course of stem cell differentiation, transcription and replication switches are so closely coordinated that it is still difficult to answer the long-standing question of which change occurs first (Hiratani et al. 2010).

What is clear, however, is that switches in replication timing are correlated with epigenetically stable transitions in cell state and gene responsiveness. Dosage compensation in mammals occurs by the inactivation of one of the two female X chromosomes, accompanied by a switch to late replication. Interestingly, monotremes, marsupials, and eutherian mammals have widely different molecular mechanisms to silence transcription on the inactive X, but a switch to late replication is common to all of these clades (Hiratani and Gilbert 2010). Moreover, genome-wide studies uncovered a specific set of autosomal replication domains that switch from early to late replication coincident with X inactivation (Hiratani et al. 2010). These regions remain late-replicating in all committed tissues queried, such that their early replication constitutes a "pluripotency fingerprint" (Ryba et al. 2011). These timing changes coincide with a loss in the ability of cells to colonize blastocysts, make chimeras, or revert to the embryonic state. Importantly, they reflect a discrete epigenetic barrier to cellular reprogramming; both early replication and transcription of genes within these regions are blocked in somatic cells that fail to reprogram to induced pluripotent stem cells (iPSCs), and the expressed transcription factors Oct4, Sox2, Nanog, and Klf4 fail to bind to their promoters (Hiratani et al. 2010).

Relationships to Chromatin Proteins

Early studies reported a close correlation of replication timing with R and G chromosome banding patterns, which presumably arise from heterogeneous protein composition along the lengths of chromosomes, but genomic analyses revealed that this alignment breaks down at molecular resolution (Pope et al. 2010). However, such studies prompted the search for other known properties of chromatin that correlate with replication timing. As expected given the correlation to transcriptional activity, early replication correlates well with general nuclease accessibility (Gilbert et al. 2004; Bell et al. 2010), histone modification, and chromatin proteins associated with active transcription (Hiratani et al. 2008; Ryba et al. 2010). However, the correlation with repressive or heterochromatin histone marks was found to be highly cell-type-specific and quite poor in some cell types (Hiratani et al. 2008; Ryba et al. 2010), consistent with earlier cytogenetic findings (Wu et al. 2005). To date, none of these chromatin marks were confined to the boundaries of replication timing, although some were found enriched at the borders of early replication domains (Ryba et al. 2010). Surprisingly, despite a strong correlation of early replication to chromatin accessibility (Prendergast et al. 2007; Bell et al. 2010), accessibility does not change coordinately with replication timing; regions that change replication timing remain relatively inaccessible whether early- or late-replicating (Takebayashi et al. 2012). It is important to point out, however, that computational methods that integrate the density of proteins or protein modifications could miss important localized features of chromatin.

A more stringent test of the role of a protein or histone modification is to determine whether it is necessary to maintain replication timing. In the past such studies were limited to testing effects of gene knockouts on the replication timing of a few specific regions (Stevenson and Gottschling 1999; Vogelauer et al. 2002; Aparicio et al. 2004; Jorgensen et al. 2007). Genomic methods are now robust enough to routinely perform an unbiased search for regions of the genome affected by mutations. In mammals the timing program is surprisingly resilient to such mutations (Jorgensen et al. 2007; Yokochi et al. 2009). For example, a strong statistical and spatial correlation between late replication and H3K9me2 in mouse embryonic stem cells (ESCs) was tested using a conditional knockout of the histone methyltransferase G9a. This mutation caused a near complete ablation of this mark on late-replicating chromatin but had no effect on replication timing or developmentally regulated changes in replication timing during differentiation (Yokochi et al. 2009). It is possible that multiple redundant modifications maintain a robust replication timing program, and studies combining mutations with chemical inhibitors of chromatin-modifying enzymes are consistent with this conclusion (Casas-Delucchi et al. 2012). Moreover, replication timing is not a binary early or late phenomenon but rather a continuum throughout S phase; hence it is not surprising that many mutations in both mammals and yeasts usually have only minor effects on replication timing. However, some chromatin mutations do exhibit profound localized effects on replication timing (Li et al. 2005; Wu et al. 2006; Bergstrom et al. 2007; Hayashi et al. 2009; Schwaiger et al. 2010), and a few examples of gene disruptions causing broad misregulation of mammalian replication timing are emerging (Cornacchia et al. 2012; Yamazaki et al. 2012).

Results measuring replication timing in yeasts that have been disrupted for chromatin modification enzymes have revealed similar complex effects. Disruption of the histone acetyltransferase Rpd3L advances the firing time of approximately 100 origins (about 30% of detected origins) in *Saccharomyces cerevisiae* (Knott et al. 2009), and deletion of budding yeast Rif1 causes earlier replication of telomeres (Lian et al. 2011). Moreover, recent results in fission yeasts suggest that deletion of either of the telomeric proteins Taz1 (Tazumi et al. 2012) or Rif1 (Hayano et al. 2012) results in a dramatic shift in the time of initiation of a set of non-telomeric origins. A similar effect occurs when Rif1 is knocked down in mammalian cells (Cornacchia et al. 2012; Yamazaki et al. 2012), although the mammalian Rif1 protein has a different function than its yeast homolog (de Lange 2004). A profound effect on replication timing was recently reported after knockout of the forkhead transcription factors in budding yeasts (Knott et al. 2012).

3D CHROMATIN ORGANIZATION AND REPLICATION TIMING

A far-reaching conclusion from genome-wide studies is the uncanny alignment of replication timing to maps of 3D chromatin interactions. This observation has shifted paradigms in our understanding of a long-standing enigmatic relationship between replication timing and large-scale chromosome structure. Moreover, very recent results have shed light on the similarities and differences in this level of regulation between yeasts and metazoans. In this section, we briefly summarize the classic observations and propose a unified working model in which the structural organization of chromosomes can regulate replication timing in unicellular and multicellular eukaryotes.

Spatial Compartmentalization of Early and Late Replication

X-chromosome inactivation in mammals is accompanied by a chromosome-wide switch to late replication, dramatic compaction, and movement of the entire chromosome to the nuclear periphery (Hiratani and Gilbert 2010). Similarly, autosomal chromatin that replicates at different times during S phase is globally compartmentalized, with early replication taking place in the interior and chromatin at the periphery being replicated later in S phase (Figs. 3, bottom, 4B). Compartmentalization appears to result from the anchorage of chromatin to immobile structures such as the nuclear periphery (Gilbert and Gasser 2006; Steglich et al. 2012) or nucleolus (Nemeth and Langst 2011). In organisms with large nuclei, chromatin movement is constrained throughout the nucleus. In yeasts, the entire nucleus is similar in scale to the range of chromatin motion in mammals, but some regions, particularly those containing late-replicating origins, are more constrained and molecules that anchor chromatin to specific subnuclear locations have been elucidated (Gilbert and Gasser 2006). The significance of compartmentalization to the replication timing program remains elusive, and under some experimental conditions, tethering budding yeast replication origins to the nuclear periphery is neither necessary (Hiraga et al. 2006) nor sufficient (Lian et al. 2011) for late replication.

Determinants of Replication Timing Established during G_1 Are Lost before G_2 Phase

A compelling link between replication timing and the spatial compartmentalization of chromatin is the finding that replication timing is established coincident with the global repositioning and anchorage of chromatin as the nucleus is reassembled during early G_1 phase (Dimitrova and Gilbert 1999). The molecular events occurring at this timing decision point (TDP) remain elusive. Recently it was shown that when a single complete round of replication is initiated during G_2 phase, replication proceeds in a random spatial and temporal order despite maintenance of chromatin spatial organization (Lu et al. 2010), implying that determinants of replication timing are lost during S phase. Possible mechanisms include the dilution of preexisting chromatin components during replication or the disruption of normal chromatin interactions when two identical sister chromatids share the same nuclear space. Together these results suggest that the global spatial organization of chromatin is not sufficient to dictate timing during G_2 phase, although it may contribute to the establishment of replication timing. This observation is consistent with a similar finding in budding yeasts that chromosomal context is important during early G_1 to establish late replication but is not necessary thereafter (Raghuraman et al. 1997; Heun et al. 2001).

Changes in Subnuclear Position Linked to Replication Timing and Cell-Fate Transitions

The spatial patterns of DNA synthesis change dramatically and abruptly during the middle of S phase in all multicellular species that have been examined (Fig. 3, bottom). Likewise, there seems to be a similar change in subnuclear position when replication domains change replication timing during development. Fluorescence in situ hybridization detects the movement of domains toward or away from the periphery during stem cell differentiation when replication timing becomes later or earlier, respectively (Williams et al. 2006; Hiratani et al. 2008). As predicted from global spatial patterns, these changes only occur when replication timing changes through mid- to late S phase (Hiratani et al. 2010), which is also the same period of S phase during which a correlation exists between replication timing and transcription (Fig. 3, bottom). Hence, replication timing can change substantially during the first half of S phase without a detectable change in subnuclear position, but a switch through mid- to late S phase is highly predictive of a subnuclear position change, suggesting that replication timing may provide a convenient genome-wide readout of subnuclear position changes (Fig. 3, bottom). What is not clear, however, is which occurs first during the course of differentiation, and whether the subnuclear position change is confined to the region of replication timing change. Intuitively it seems that without major chromatin unfolding, subnuclear position changes would need to involve chromosomal regions larger than the size of a 400- to 800-kb replication domain, but this remains to be tested.

Replication Foci: Cytogenetic Units of Replication Timing Regulation?

A common feature of the spatial patterns of DNA synthesis is a punctate appearance of the labeled sites of DNA synthesis, or "replication foci." In mammals each of these foci takes 45–60 min to complete replication (Berezney et al. 2000). Based on the number of foci observed over the course of S phase (hundreds at any one time), it is estimated that each contains 0.5–1 Mb of DNA (Berezney et al. 2000), close to the size of developmentally regulated replication domains. When pulse-labeled cells are chased through multiple cell cycles, labeled foci do not change size, shape, or intensity of label, indicating that the DNA that is synthesized together remains together as a structural unit (Sadoni et al. 2004). In many cases, adjacent foci are genetically contiguous and replicate sequentially (Sporbert et al. 2002; Maya-Mendoza et al. 2010).

Models for the structure of chromatin within replication foci have been rapidly evolving. A long-standing model is that replication foci represent "factories" of DNA organized in such a way that multiple replicons throughout hundreds of kilobases are simultaneously spooled through a single common replication complex (Fig. 4) (Gilbert and Gasser 2006), although evidence for this model has been inconclusive. Recently evidence has been presented for the presence of replication foci containing multiple replication forks in budding (Kitamura et al. 2006) and fission (Meister et al. 2007) yeasts, suggesting that they may share a similar structural organization. However, a study in cell-free *Xenopus* egg extracts has shown that individual molecules tethered at both ends can be efficiently replicated, showing that neither large immobilized complexes nor complexes simultaneously engaged with both bidirectional replication forks are necessary for in vitro DNA replication (Yardimci et al. 2010). Moreover, high-resolution light microscopy methods (Cseresnyes et al. 2009; Baddeley et al. 2010) applied to the study of replication foci reveal that each "focus" appears to resolve into many separate but clustered subfoci that are consistent with the amount of DNA in a single bidirectional replicon (Fig. 4D). In short, although replication foci likely represent the cytological manifestation of replication domains, their structure remains unresolved.

Importantly for our discussion, although the ultrastructure of replication foci is not resolved, evidence suggests that replication timing is regulated at the level of replication foci. As discussed above, coordinated changes in replication timing during differentiation are regulated at the level of 400- to 800-kb replication domains, which are in line with the sizes of foci defined cytogenetically. Moreover, studies in *Xenopus* egg extracts have shown that the timing program can be compressed or elongated, without modulating the temporal order of replication, by increasing or decreasing the number of simultaneously activated replication foci through modulation of cyclin-dependent kinase (CDK) activity, independent of the numbers of origins or rate of elongation (Thomson et al. 2010). Together these results have strengthened the view of the replication focus as a unit of replication regulation.

An Uncanny Link between Replication Timing and Chromatin Interaction Maps

Replication timing correlates better with genome-wide maps of 3D chromatin interactions than any other chromosomal property analyzed to date (Ryba et al. 2010); in fact, it is one of the strongest correlations in genomics (Fig. 4A). Chromatin interaction maps are made by cross-linking cellular chromatin, digesting it with a restriction enzyme, and then religating to covalently link all pieces of DNA that are close together in the nucleus. In Hi-C (Lieberman-Aiden et al. 2009), a genome-wide chromosome-interaction mapping technique, the junctions of all of these novel ligation events are sequenced to reveal all the interactions (near the restriction sites) in a population of cells. These studies have shown that chromatin is organized into two spatially distinct compartments such that chromatin in one compartment rarely interacts with chromatin in the alternate compartment, with distinct boundaries separating interactions between alternate compartments. Hi-C data sets not only correlate strongly with replication timing but align precisely at both large and small replication timing transition boundaries, suggesting that the chromatin between segments of self-interacting chromatin may be the TTRs. The fact that chromosome interaction correlates better with replication timing than either of them do with any single histone modification or chromatin-binding protein suggests that replication timing is connected to chromatin structure through a complex combination of histone modifications and chromatin-binding proteins, no one of which alone captures the overall large-scale structure. This observation also suggests that replication timing may be a convenient proxy for chromatin interaction maps.

These results provide molecular evidence supporting conclusions from cytogenetic data that early- and late-replicating DNAs reside in

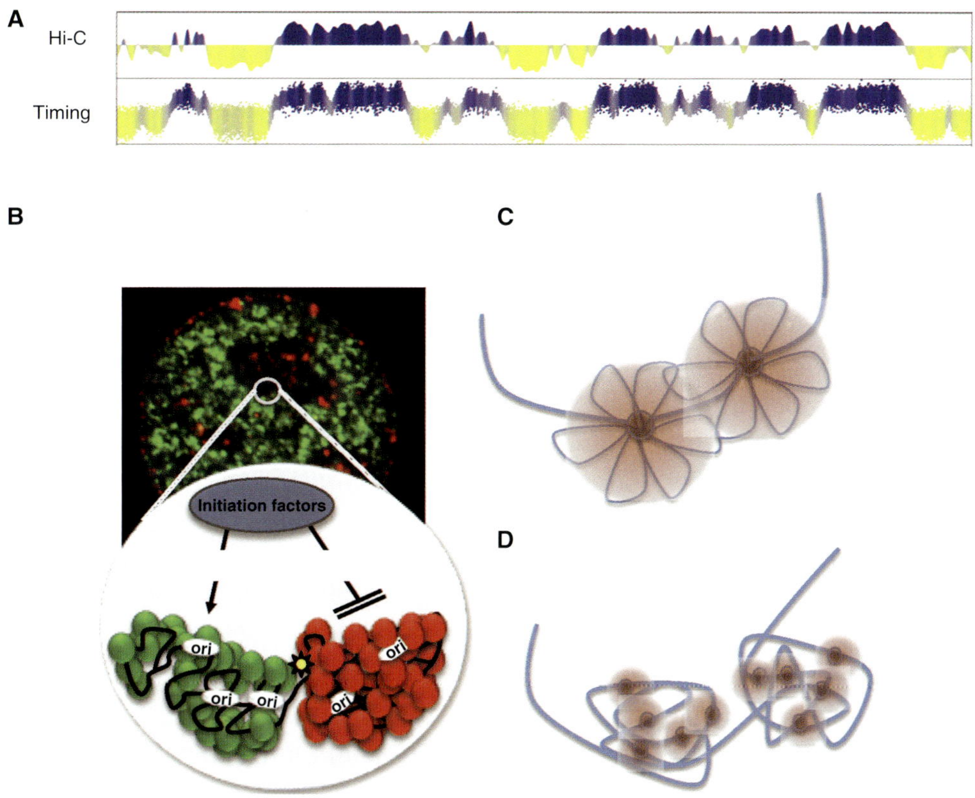

Figure 4. Units of replication timing regulation correspond to units of large-scale chromatin organization. (*A*) Replication timing profiles align better to eigenvector displays of chromatin conformation capture profiles than to any other structural or functional property of chromosomes examined to date (Ryba et al. 2010). (*B*) Regions that replicate at different times are spatially segregated and may form self-interacting domains that may set thresholds for the accessibility of S-phase promoting factors. (*C,D*) Two interpretations of replication foci. (*C*) Concept of a replication factory where several replicons in a spatially contiguous chromosome region are replicated by a common, fixed replication protein complex. (*D*) Current super-resolution microscopy methods suggest that replication foci consist of clusters of smaller foci that were not resolved by prior light microscopy. Together the data suggest that foci are the result of several separate replication complexes, possibly replicating both bidirectional leading and lagging strands together (Heun et al. 2001; Kitamura et al. 2006; Meister et al. 2007), but rather than being a single fixed complex, they consist of a spatially clustered group of independent replication complexes that initiate replication nearly synchronously, possibly as a result of their common presence within a single self-interacting unit of large-scale chromatin organization. (*A* and *B* from Ryba et al. 2010; reprinted, with permission, © Cold Spring Harbor Laboratory Press.)

spatially separate compartments of the nucleus and that the DNA that is in close spatial proximity is replicated together as a structural and functional unit. Data from Hi-C can be fit to models in which chromosomes are organized into self-interacting domains that are remarkably similar in size to replication domains and are predicted to fold into structures resembling replication foci (Fig. 4B). It remains to be seen whether this correlation will also be found in yeasts, but chromatin interaction maps in budding yeasts show that early replication origins frequently interact (Duan et al. 2010). Moreover, a recent study shows that forkhead transcription factors bind near a subset of early replication origins in budding yeasts, interact to organize early origins in nuclear space, and are required for their early replication (Knott et al. 2012).

MECHANISMS REGULATING REPLICATION TIMING

The strong correlation between replication timing and chromatin structure suggests a mechanistic connection. However, timing appears to be controlled at the level of large replication domains, whereas the regulation of replication initiation is at the level of individual origins. How the chromatin structure of large domains is related to the firing times of individual origins within those domains remains one of the central questions in the field.

Establishment versus Execution

The regulation of replication timing can be thought of as having two mechanistically distinct processes: establishment and execution. Establishment, which occurs during late mitosis and G_1, sets the timing program by committing origins to their characteristic firing times. Execution, which occurs during S, realizes the timing program by activating origins at those times. Manipulating either process can affect the timing of replication, but in mechanistically different ways. Manipulations that affect establishment tend to affect large-scale attributes, such as chromatin structure and subnuclear localization, whereas manipulations that affect execution affect the firing of individual origins without affecting chromatin structure and subnuclear localization.

Execution Reflects Competition for Rate-Limiting Factors

Recent work has shed light on the execution of replication timing: how and when individual origins are selected to fire. Underlying this work is the heterogeneous nature of origin firing. In budding and fission yeasts, in which origins are well-defined genetic loci and the replication timing pattern of each origin is known, the execution of the replication program during S phase involves the stochastic firing of individual origins, which nonetheless results in reproducible patterns of average firing times (Patel et al. 2006; Rhind 2006; Czajkowsky et al. 2008; Lygeros et al. 2008; de Moura et al. 2010; Yang et al. 2010). Likewise in metazoans, origin firing is heterogeneous, with individual origins often firing with very low efficiency (Anglana et al. 2003; Lebofsky et al. 2006; Labit et al. 2008; Cayrou et al. 2011). It is important to emphasize that stochastic firing of origins does not imply that all origins fire with equal probability; indeed, different origins can have very different probabilities of firing. Rather, the term "stochastic" simply implies that the firing of origins is governed by probability, such that an origin may fire in one cell but not another. Such behavior can be explained if each origin has a characteristic probability of firing (Rhind 2006; Rhind et al. 2010). Origins with high firing probabilities are more likely to fire early in S and therefore, on average, have earlier firing times. Likewise, origins with lower firing probabilities will, on average, have later replication times. Mathematical models based on stochastic origin firing account well for observed replication kinetics in yeasts, frog embryos, and mammalian cells (Goldar et al. 2008; Lygeros et al. 2008; de Moura et al. 2010; Yang et al. 2010; Gauthier et al. 2012), and recent high-resolution DNA fiber analysis of replication kinetics at the mouse *Igh* locus is consistent with stochastic origin firing (Demczuk et al. 2012).

The probability of origin firing is affected by competition for rate-liming factors. In budding and fission yeasts, initiation factors—such as Cdc45, Sld2, and Sld3—and the Dbf4-dependent kinase Cdc7(DDK) initiation kinase are limiting for origin firing; overexpression of these factors increases the probability of origin firing, and tethering them to specific loci can increase the firing of local origins (Patel et al. 2008; Wu and Nurse 2009; Mantiero et al. 2011; Tanaka et al. 2011). In addition, limiting the expression of S-phase cyclins can limit the firing of late origins (Donaldson et al. 1998; McCune et al. 2008; Katsuno et al. 2009). The idea that varying the level of limiting activators affects the execution but not the establishment of replication timing is supported by the observation that slowing replication, either by slowing fork rates or deleting replication factors, elongates the replication timing program but preserves the relative order of average replication times (Alvino et al. 2007; Koren et al. 2010). Stochastic firing of origins in mammalian cells suggests that they are also regulated by rate-limiting factors (Lebofsky et al. 2006; Cayrou et al. 2011; Letessier et al. 2011). Consistent with this possibility, varying the level of CDK affects replication timing in vertebrates (Krasinska et al. 2008; Katsuno et al. 2009; Thomson et al. 2010).

The activation of origins by limiting factors predicts that as S phase progresses and fewer unfired origins remain, the remaining origins will be able to compete more efficiently for the limiting activators and so their firing probability will increase (Herrick et al. 2000; Hyrien et al. 2003; Rhind 2006). Such an increase in firing probabilities as S phase progresses has been observed in yeasts and metazoans (Herrick et al. 2000; Patel et al. 2006; Eshaghi et al. 2007) and has been proposed to be a universal feature of eukaryotic replication (Goldar et al. 2009). The probability of origin firing is predicted to increase until it becomes limited by the diffusion rates of the activators, a prediction consistent with quantitative analysis (Gauthier and Bechhoefer 2009). The increasing probability of origin firing has been also been invoked as a guarantee that late-replicating regions of the genome will replicate efficiently and that replication will finish in a timely manner, despite stochastic origin firing (Yang and Bechhoefer 2008; Rhind et al. 2010).

Establishment Is Set at the Level of Chromosome Domains

If the execution of origin timing reflects the probability of origins competing for limiting initiation factors, the establishment of origin timing sets those probabilities. The difference in timing between different replication domains correlates with the chromatin structure of those domains and is not intrinsic to the origins themselves. As described above, targeting chromatin regulators near an origin or moving an origin to new chromatin contexts can affect origin timing (Ferguson and Fangman 1992; Friedman et al. 1996; Vogelauer et al. 2002; Zappulla et al. 2002; Goren et al. 2008). Furthermore, the mammalian TDP occurs before the point at which specific origins are selected (origin decision point; ODP), suggesting that timing is regulated independently of individual origins. This separation is consistent with the observation that timing patterns are more strongly conserved across species than origin locations (Ryba et al. 2010; Di Rienzi et al. 2012; Muller and Nieduszynski 2012; Xu et al. 2012). For instance, mice and humans have a different distribution of origins at the β-globin locus but similar developmentally regulated timing programs (Aladjem et al. 2002). Likewise, different growth conditions can change origin usage without affecting timing patterns (Courbet et al. 2008).

These observations have led to a model in which the establishment of replication timing involves regulation of the chromatin structure affecting large chromosomal regions. Chromatin context established during G_1 could affect the ability of origins to compete for limiting factors during S. In fission yeasts, centromeric heterochromatin replicates early, in contrast to the late replication of most heterochromatin (Kim et al. 2003). However, early replication is dependent on the recruitment of the DDK replication kinase to heterochromatin by interaction with the fission yeast HP1 homolog (Hayashi et al. 2009). In the absence of this interaction, centromeres replicate late, in a heterochromatin-dependent manner, suggesting that heterochromatin generally enforces late replication by reducing access of origins to initiation factors such as DDK.

In addition, chromatin context during G_1 could affect the licensing of origins in ways that affect the timing of origin firing in S. For instance, the time at which the origin-recognition complex (ORC) binds to origins in fission yeasts correlates with origin timing, with origins binding ORC earlier in G_1 also firing earlier in S (Wu and Nurse 2009). Origin function correlates well with nucleosome-free regions (Eaton et al. 2010; MacAlpine et al. 2010; Lubelsky et al. 2011), so nucleosome density may play an important role in regulating ORC binding. How the regulation of ORC binding could affect the timing of origin firing is unclear. However, it has been proposed that firing time is regulated by the number of minichromosome maintenance (MCM) complexes loaded at an origin, with early origins having more MCMs loaded and thus being more likely to fire early (Yang et al. 2010). In such a scenario, the amount of time ORC spends bound to an origin in G_1 could affect the number of MCMs loaded at an origin, which could in turn affect the average firing time of the origin. In mammals, MCMs are bound to both early- and late-replicating chromatin before the timing program is set (Dimitrova and Gilbert 1999), but MCMs continue to load during the course of G_1 phase and it is not clear whether that continued loading is homogeneous (Kuipers et al. 2011).

There can also be significant differences in timing between origins within one replication domain. The lower resolution of mammalian timing profiles and origin locations has not allowed for systematic analysis of intradomain timing. However, in budding yeasts, substantial differences in origin timing are detected between euchromatic origins (Fig. 1) (Friedman et al. 1997; Raghuraman et al. 2001). Furthermore, deviations from the correlation between timing and chromatin contexts, and experimental manipulations that can modify specific origin firing (Lin et al. 2003; Goren et al. 2008; Hassan-Zadeh et al. 2012), make it clear that although regional chromatin context strongly biases replication timing, it does not impose absolute control. It seems likely that the timing of firing of individual origins is a combination of intrinsic probability of the origin firing modified by the origin's chromosomal context.

Checkpoints Affect Replication Timing

In addition to the execution of the replication timing program established during G_1, replication timing can also be affected by feedback regulation during S phase. The best-understood of these mechanisms involve the S-phase DNA damage checkpoint, which is activated by fork stalling and inhibits subsequent origin firing (Shirahige et al. 1998; Costanzo et al. 2003; Merrick et al. 2004). Such a mechanism could act to balance fork termination with origin firing to maintain a uniform fork density (Diaz-Martinez and Clarke 2003). The damage checkpoint acts to restrain origin firing by inhibiting the function of limiting activators, such as Cdc45, Sld2, Sld3, and DDK (Lopez-Mosqueda et al. 2010; Zegerman and Diffley 2010). In budding yeasts, a similar checkpoint-dependent mechanism acts to maintain the temporal order of origin firing when S phase is elongated by limiting nucleotides or mutant replication factors (Alvino et al. 2007; Koren et al. 2010). However, these intra-S-phase timing mechanisms function primarily to maintain the appropriate fork density by enforcing the relative order of origin firing, but are not involved in determining replication timing patterns per se.

WHAT IS THE BIOLOGICAL FUNCTION OF A TIMING PROGRAM?

Replication timing is universally observed in eukaryotes, suggesting an important and conserved function. Nonetheless, that function is largely mysterious. Moreover, it is possible that replication timing per se does not contribute to fitness but is an inevitable consequence of some other factor, such as the domain architecture of the genome. To understand the biological function of replication timing, we have to consider two questions. First, why is S phase longer than its minimal possible length? Second, what is the reason for the particular conserved order of replication? Although there is little direct evidence to address either question, several plausible hypotheses are consistent with existing data.

Possible Reasons for an Extended S Phase

S phase in metazoan somatic cells can be more than 10 times longer than embryonic S phase. In fly and frog embryos, replication takes less than 20 min (Graham 1966; Blumenthal et al. 1974). Yet in somatic cells of the same species, replication takes as long as 8 h (Blumenthal et al. 1974). A slower S phase may be more economical; by extending the length of S phase, cells reduce the number of concurrent replication forks and thus the number of fork proteins that need to be synthesized. Even in frog embryos, which are transcriptionally silent and in which cellular energy is focused on genome duplication, replication kinetics appear to be optimized to use the fewest number of forks required to complete replication in the allotted time (Yang and Bechhoefer 2008). However, the theoretical energy savings may not be biologically significant compared with the total energy budget of the cell.

Another possible advantage to pacing the rate of replication is to allow for late-firing origins to serve as backups. Licensing more origins than are normally required for S phase provides a robust solution to replication fork stalling (Blow et al. 2011). If two approaching forks both stall, the only way to replicate the intervening sequence is to activate new forks from an origin between them. And because origins can only be established during G_1, such backup origins must be positioned throughout the genome and not fired early in S phase. Thus, dramatically reducing the number of licensed origins does not affect normal replication, but does make cells much more sensitive to replication stress (Woodward et al. 2006; Ge et al. 2007). In a system of redundant, inefficient origins, such backup origins need not be qualitatively different "dormant" origins, they can simply be whichever origins do not happen to fire in a particular cell.

Reducing the number of concurrent forks may also enhance genome stability by avoiding overtaxing DNA repair and fork stability mechanisms. In bacteria, increasing the number of replication forks can saturate the mismatch repair capacity and lead to increased mutation rates (Schaaper and Radman 1989). Also, concurrent fork collapses may increase the chance of illegitimate repair events involving nonhomologous replication structures, which could lead to translocations and other genome rearrangements. In addition, higher numbers of active forks increases the flux in nucleotide pool levels, which, because polymerase fidelity is sensitive to nucleotide concentration (Kumar et al. 2010), may lead increased misincorporation. Although there is little direct evidence to support the idea the extending S phase reduces genome instability, increasing the rate of origin firing moderately increases genome instability (Patel et al. 2008).

Finally, extending S phase may facilitate specific regulatory mechanisms that control gene expression or other aspects of chromatin metabolism (Gilbert 2002; Hiratani et al. 2009). If extending S phase does have a regulatory role, the order in which various loci are replicated becomes an important parameter. Therefore, the fact that patterns of replication timing are conserved has led to speculation that these patterns have regulatory significance.

Possible Reasons for Temporal Order of Replication

The most compelling model for a regulatory role for replication timing is that the time during S phase that a locus is replicated can influence its chromatin structure. Because chromatin is assembled at the replication fork, if different chromatin conformations are favored during different parts of S phase, the replication timing of a locus could influence its chromatin structure. This model is supported by the observation that the time at which a plasmid is injected into a cell during S phase influences its chromatin structure and its expression level (Zhang et al. 2002; Lande-Diner et al. 2009). An attractive feature of this model is its self-reinforcing nature. Chromatin context influences replication time, while replication at that time may reestablish chromatin context. This model also provides a mechanism by which replication timing and chromatin context could spread. If a localized alteration causes early replication of a small region in an otherwise late-replicating domain, the passive early replication of surrounding loci could alter chromatin context and reprogram the replication timing of the entire replication domain. Thus, this model suggests mechanisms for both the epigenetic inheritance of replication timing and its developmental reprogramming.

It is also possible that replication timing regulates other aspects of genome metabolism. For instance, mutation rates are lower in early-replicating regions (Stamatoyannopoulos et al. 2009; Chen et al. 2010; Lang and Murray 2011). If replication timing directly influences mutation rates, perhaps by concentrating repair capacity early in S phase or altering the balance of nucleotide pools during S, there might be evolutionary pressure to replicate gene-rich regions earlier in S phase. Likewise, recombination rates are higher in earlier-replicating regions, allowing the possibility that the time of replication during premeiotic S phase could regulate crossover frequency.

The inevitable effect of replication on gene dosage could also have regulatory consequences. Genes that are duplicated early will be present at twice the copy number of late-replicating genes for most of the duration of S phase, increasing transcriptional output. Early-replicating sequences are also the first to have sister-chromatid templates for homologous recombination. The ability to perform DNA repair by homologous recombination for a greater fraction of the cell cycle may contribute to the fact the early-replicating DNA has lower rates of mutation. These dosage effects would be most pronounced in cells in which S phase is a substantial fraction of the cell cycle, such as mammalian ESCs.

It is important to keep in mind that the strong correlation between replication timing and important aspects of genome metabolism may simply be an indirect consequence of how closely replication timing reflects the structure of chromatin. For example, mutation rates are also known to be lowest in open chromatin (Prendergast et al. 2007). It is possible then that a particular temporal sequence to replication may serve no specific biological function, but rather just be a reflection of the large-scale organization of the genome. This model explains the bulk of available data with no extraneous assumptions or mechanisms and thus should be considered the null hypothesis against which any more complicated model for functional significance should be weighed.

CONCLUDING REMARKS

In comparison with our sophisticated knowledge of most fundamental cellular processes, our understanding of replication timing—both its biological significance and mechanism—has remained primitive and experimental progress has mainly underscored its complexity. The timing program is established at a minimum of two levels, both of which act heterogeneously in different individual cells to establish a probability for any given region to replicate at a particular time. At one level local sequence and chromatin composition influence whether a prereplicative complex is formed, the number of MCM complexes loaded, or the local affinity for initiation proteins such as Cdc45, Sld2, Sld3, and DDK. Mutations in several gene products affecting this local level of regulation have been identified, which exhibit partial or localized effects on the timing program. At a second level the global timing program of large replication domains appears to be established by the large-scale folding of chromosomes, a more challenging problem to dissect. The execution of the timing program results from an integration of these two levels of influence, which may play more or less important roles in different species. For example, local regulation likely plays a more important role in the small genomes of single-celled organisms. In fact, we propose that the majority of the yeast genome can be thought of as the equivalent of early replication domains in mammals (Fig. 1), with the large-scale structure of early-replicating regions spatially clustered by a few central organizers such as the Fkh proteins and centromere-proximal regions (Knott et al. 2012), whereas late-replicating regions such as telomeres and mating-type loci are organized into spatially separate clusters. In progressively larger genomes, large-scale folding is a major organizing principle, diminishing the importance of localized effects to the overall timing program. Elucidating these complex influences will remain a challenge, but perhaps even more challenging and important will be assessing the degree to which cell-to-cell heterogeneity plays a role in the overall timing program. This challenge will require advancements in our ability to evaluate replication of individual molecules.

As new experimental systems to probe mechanism emerge, the elephant in the room remains our complete lack of appreciation of the biological significance of this program. We have summarized many of the hypotheses that have been proposed; however, any proposed biological function for replication timing must be tested against the simple possibility that the program results from a combination of limiting resources for replication, executed in a temporal sequence that reflects the assembly of chromatin for other cellular functions. Even if the simple possibility should be the case, replication timing has emerged as a robust means to assess the large-scale epigenetic state of specific cell types and identify regions of large-scale structural reorganization during differentiation or disease (Ryba et al. 2012).

ACKNOWLEDGMENTS

We thank Oscar Aparicio, John Bechhoefer, David MacAlpine, Conrad Nieduszynski, and Paolo Norio for valuable comments on the manuscript. Research in the Gilbert lab is supported by National Institutes of Health (NIH) grants GM083337 and GM085354. Research in the Rhind lab is supported by NIH grants GM069957 and GM098815 and an American Cancer Society Research Scholar Grant.

REFERENCES

Agier N, Fischer G. 2012. The mutational profile of the yeast genome is shaped by replication. *Mol Biol Evol* **29:** 905–913.

Aladjem MI, Rodewald LW, Lin CM, Bowman S, Cimbora DM, Brody LL, Epner EM, Groudine M, Wahl GM. 2002. Replication initiation patterns in the β-globin loci of totipotent and differentiated murine cells: Evidence for multiple initiation regions. *Mol Cell Biol* **22:** 442–452.

Alvino GM, Collingwood D, Murphy JM, Delrow J, Brewer BJ, Raghuraman MK. 2007. Replication in hydroxyurea: It's a matter of time. *Mol Cell Biol* **27:** 6396–6406.

Anglana M, Apiou F, Bensimon A, Debatisse M. 2003. Dynamics of DNA replication in mammalian somatic cells: Nucleotide pool modulates origin choice and interorigin spacing. *Cell* **114:** 385–394.

Aparicio JG, Viggiani CJ, Gibson DG, Aparicio OM. 2004. The Rpd3-Sin3 histone deacetylase regulates replication timing and enables intra-S origin control in *Saccharomyces cerevisiae. Mol Cell Biol* **24:** 4769–4780.

Baddeley D, Chagin VO, Schermelleh L, Martin S, Pombo A, Carlton PM, Gahl A, Domaing P, Birk U, Leonhardt H, et al. 2010. Measurement of replication structures at the nanometer scale using super-resolution light microscopy. *Nucleic Acids Res* **38:** e8.

Bell O, Schwaiger M, Oakeley EJ, Lienert F, Beisel C, Stadler MB, Schubeler D. 2010. Accessibility of the *Drosophila* genome discriminates PcG repression, H4K16 acetylation and replication timing. *Nat Struct Mol Biol* **17:** 894–900.

Berezney R, Dubey DD, Huberman JA. 2000. Heterogeneity of eukaryotic replicons, replicon clusters, and replication foci. *Chromosoma* **108:** 471–484.

Bergstrom R, Whitehead J, Kurukuti S, Ohlsson R. 2007. CTCF regulates asynchronous replication of the imprinted H19/Igf2 domain. *Cell Cycle* **6:** 450–454.

Besnard E, Babled A, Lapasset L, Milhavet O, Parrinello H, Dantec C, Marin JM, Lemaitre JM. 2012. Unraveling cell type–specific and reprogrammable human replication origin signatures associated with G-quadruplex consensus motifs. *Nat Struct Mol Biol* **19:** 837–844.

Blow JJ, Ge XQ, Jackson DA. 2011. How dormant origins promote complete genome replication. *Trends Biochem Sci* **36:** 405–414.

Blumenthal AB, Kriegstein HJ, Hogness DS. 1974. The units of DNA replication in *Drosophila melanogaster* chromosomes. *Cold Spring Harb Symp Quant Biol* **38:** 205–223.

Cadoret JC, Meisch F, Hassan-Zadeh V, Luyten I, Guillet C, Duret L, Quesneville H, Prioleau MN. 2008. Genome-wide studies highlight indirect links between human replication origins and gene regulation. *Proc Natl Acad Sci* **105:** 15837–15842.

Casas-Delucchi CS, van Bemmel JG, Haase S, Herce HD, Nowak D, Meilinger D, Stear JH, Leonhardt H, Cardoso MC. 2012. Histone hypoacetylation is required to maintain late replication timing of constitutive heterochromatin. *Nucleic Acids Res* **40:** 159 169.

Cayrou C, Coulombe P, Vigneron A, Stanojcic S, Ganier O, Peiffer I, Rivals E, Puy A, Laurent-Chabalier S, Desprat R, et al. 2011. Genome-scale analysis of metazoan replication origins reveals their organization in specific but flexible sites defined by conserved features. *Genome Res* **21:** 1438–1449.

Chen CL, Rappailles A, Duquenne L, Huvet M, Guilbaud G, Farinelli L, Audit B, d'Aubenton-Carafa Y, Arneodo A, Hyrien O, et al. 2010. Impact of replication timing on non-CpG and CpG substitution rates in mammalian genomes. *Genome Res* **20:** 447–457.

Cornacchia D, Dileep V, Quivy JP, Foti R, Tili F, Santarella-Mellwig R, Anthony C, Almouzni G, Gilbert DM, Buonomo SB. 2012. Mouse Rif1 is a key regulator of the replication-timing programme in mammalian cells. *EMBO J* **31:** 3678–3690.

Costantini M, Cammarano R, Bernardi G. 2009. The evolution of isochore patterns in vertebrate genomes. *BMC Genomics* **10:** 146.

Costanzo V, Shechter D, Lupardus PJ, Cimprich KA, Gottesman M, Gautier J. 2003. An ATR- and Cdc7-dependent DNA damage checkpoint that inhibits initiation of DNA replication. *Mol Cell* **11:** 203–213.

Courbet S, Gay S, Arnoult N, Wronka G, Anglana M, Brison O, Debatisse M. 2008. Replication fork movement sets chromatin loop size and origin choice in mammalian cells. *Nature* **455:** 557–560.

Crabbe L, Thomas A, Pantesco V, De Vos J, Pasero P, Lengronne A. 2010. Analysis of replication profiles reveals key role of RFC-Ctf18 in yeast replication stress response. *Nat Struct Mol Biol* **17:** 1391–1397.

Cseresnyes Z, Schwarz U, Green CM. 2009. Analysis of replication factories in human cells by super-resolution light microscopy. *BMC Cell Biol* **10:** 88.

Czajkowsky DM, Liu J, Hamlin JL, Shao Z. 2008. DNA combing reveals intrinsic temporal disorder in the replication of yeast chromosome VI. *J Mol Biol* **375:** 12–19.

Debatisse M, Le Tallec B, Letessier A, Dutrillaux B, Brison O. 2012. Common fragile sites: Mechanisms of instability revisited. *Trends Genet* **28:** 22–32.

Demczuk A, Gauthier MG, Veras I, Kosiyatrakul S, Schildkraut CL, Busslinger M, Bechhoefer J, Norio P. 2012. Regulation of DNA replication within the immunoglobulin heavy-chain locus during B cell commitment. *PLoS Biol* **10:** e1001360.

de Moura AP, Retkute R, Hawkins M, Nieduszynski CA. 2010. Mathematical modelling of whole chromosome replication. *Nucleic Acids Res* **38:** 5623–5633.

Desprat R, Thierry-Mieg D, Lailler N, Lajugie J, Schildkraut C, Thierry-Mieg J, Bouhassira EE. 2009. Predictable dynamic program of timing of DNA replication in human cells. *Genome Res* **19:** 2288–2299.

Diaz-Martinez L, Clarke DJ. 2003. Self-regulating model for control of replication origin firing in budding yeast. *Cell Cycle* **2:** 576–578.

Dimitrova DS, Gilbert DM. 1999. The spatial position and replication timing of chromosomal domains are both established in early G1 phase. *Mol Cell* **4:** 983–993.

Di Rienzi SC, Lindstrom KC, Mann T, Noble WS, Raghuraman MK, Brewer BJ. 2012. Maintaining replication origins in the face of genomic change. *Genome Res* **22:** 1940–1952.

Donaldson AD, Schildkraut CL. 2006. Temporal order of DNA replication. In *DNA replication and human disease* (ed. DePamphilis ML), pp. 197–216. Cold Spring Harbor Laboratory Press, Cold Spring Harbor, NY.

Donaldson AD, Raghuraman MK, Friedman KL, Cross FR, Brewer BJ, Fangman WL. 1998. CLB5-dependent activation of late replication origins in *S. cerevisiae*. *Mol Cell* **2:** 173–182.

Duan Z, Andronescu M, Schutz K, McIlwain S, Kim YJ, Lee C, Shendure J, Fields S, Blau CA, Noble WS. 2010. A three-dimensional model of the yeast genome. *Nature* **465:** 363–367.

Eaton ML, Galani K, Kang S, Bell SP, Macalpine DM. 2010. Conserved nucleosome positioning defines replication origins. *Genes Dev* **24:** 748–753.

Eaton ML, Prinz JA, MacAlpine HK, Tretyakov G, Kharchenko PV, MacAlpine DM. 2011. Chromatin signatures of the *Drosophila* replication program. *Genome Res* **21:** 164–174.

Eshaghi M, Karuturi RK, Li J, Chu Z, Liu ET, Liu J. 2007. Global profiling of DNA replication timing and efficiency reveals that efficient replication/firing occurs late during S-phase in *S. pombe*. *PLoS ONE* **2:** e722.

Farkash-Amar S, Simon I. 2010. Genome-wide analysis of the replication program in mammals. *Chromosome Res* **18:** 115–125.

Farkash-Amar S, Lipson D, Polten A, Goren A, Helmstetter C, Yakhini Z, Simon I. 2008. Global organization of replication time zones of the mouse genome. *Genome Res* **18:** 1562–1570.

Feng W, Collingwood D, Boeck ME, Fox LA, Alvino GM, Fangman WL, Raghuraman MK, Brewer BJ. 2006. Genomic mapping of single-stranded DNA in hydroxyurea-challenged yeasts identifies origins of replication. *Nat Cell Biol* **8:** 148–155.

Ferguson BM, Fangman WL. 1992. A position effect on the time of replication origin activation in yeast. *Cell* **68:** 333–339.

Friedman KL, Diller JD, Ferguson BM, Nyland SV, Brewer BJ, Fangman WL. 1996. Multiple determinants controlling activation of yeast replication origins late in S phase. *Genes Dev* **10:** 1595–1607.

Friedman KL, Brewer BJ, Fangman WL. 1997. Replication profile of *Saccharomyces cerevisiae* chromosome VI. *Genes Cells* **2:** 667–678.

Frum RA, Khondker ZS, Kaufman DG. 2009. Temporal differences in DNA replication during the S phase using single fiber analysis of normal human fibroblasts and glioblastoma T98G cells. *Cell Cycle* **8:** 3133–3148.

Gauthier MG, Bechhoefer J. 2009. Control of DNA replication by anomalous reaction-diffusion kinetics. *Phys Rev Lett* **102:** 158104.

Gauthier MG, Norio P, Bechhoefer J. 2012. Modeling inhomogeneous DNA replication kinetics. *PLoS ONE* **7:** e32053.

Gilbert DM. 2002. Replication timing and transcriptional control: Beyond cause and effect. *Curr Opin Cell Biol* **14:** 377–383.

Gilbert DM, Gasser SM. 2006. Nuclear structure and DNA replication. In *DNA replication and human disease* (ed. DePamphilis ML), pp. 175–196. Cold Spring Harbor Laboratory Press, Cold Spring Harbor, NY.

Gilbert N, Boyle S, Fiegler H, Woodfine K, Carter NP, Bickmore WA. 2004. Chromatin architecture of the human genome: Gene-rich domains are enriched in open chromatin fibers. *Cell* **118:** 555–566.

Goldar A, Labit H, Marheineke K, Hyrien O. 2008. A dynamic stochastic model for DNA replication initiation in early embryos. *PLoS ONE* **3:** e2919.

Goldar A, Marsolier-Kergoat MC, Hyrien O. 2009. Universal temporal profile of replication origin activation in eukaryotes. *PLoS ONE* **4:** e5899.

Goren A, Tabib A, Hecht M, Cedar H. 2008. DNA replication timing of the human β-globin domain is controlled by histone modification at the origin. *Genes Dev* **22:** 1319–1324.

Graham CF. 1966. The regulation of DNA synthesis and mitosis in multinucleate frog eggs. *J Cell Sci* **1:** 363–374.

Guan Z, Hughes CM, Kosiyatrakul S, Norio P, Sen R, Fiering S, Allis CD, Bouhassira EE, Schildkraut CL. 2009. Decreased replication origin activity in temporal transition regions. *J Cell Biol* **187:** 623–635.

Guilbaud G, Rappailles A, Baker A, Chen CL, Arneodo A, Goldar A, d'Aubenton-Carafa Y, Thermes C, Audit B, Hyrien O. 2011. Evidence for sequential and increasing activation of replication origins along replication timing gradients in the human genome. *PLoS Comput Biol* **7:** e1002322.

Hansen RS, Thomas S, Sandstrom R, Canfield TK, Thurman RE, Weaver M, Dorschner MO, Gartler SM, Stamatoyannopoulos JA. 2010. Sequencing newly replicated DNA reveals widespread plasticity in human replication timing. *Proc Natl Acad Sci* **107:** 139–144.

Hassan-Zadeh V, Chilaka S, Cadoret JC, Ma MK, Boggetto N, West AG, Prioleau MN. 2012. USF binding sequences from the HS4 insulator element impose early replication timing on a vertebrate replicator. *PLoS Biol* **10:** e1001277.

Hayano M, Kanoh Y, Matsumoto S, Masai H. 2011. Mrc1 marks early-firing origins and coordinates timing and efficiency of initiation in fission yeast. *Mol Cell Biol* **31:** 2380–2391.

Hayano M, Kanoh Y, Matsumoto S, Renard-Guillet C, Shirahige K, Masai H. 2012. Rif1 is a global regulator of timing of replication origin firing in fission yeast. *Genes Dev* **26:** 137–150.

Hayashi M, Katou Y, Itoh T, Tazumi A, Yamada Y, Takahashi T, Nakagawa T, Shirahige K, Masukata H. 2007. Genome-wide localization of pre-RC sites and identification of replication origins in fission yeast. *EMBO J* **26:** 1327–1339.

Hayashi MT, Takahashi TS, Nakagawa T, Nakayama J, Masukata H. 2009. The heterochromatin protein Swi6/HP1 activates replication origins at the pericentromeric region and silent mating-type locus. *Nat Cell Biol* **11:** 357–362.

Heichinger C, Penkett CJ, Bahler J, Nurse P. 2006. Genome-wide characterization of fission yeast DNA replication origins. *EMBO J* **25:** 5171–5179.

Herrick J, Stanislawski P, Hyrien O, Bensimon A. 2000. Replication fork density increases during DNA synthesis in *X. laevis* egg extracts. *J Mol Biol* **300:** 1133–1142.

Heun P, Laroche T, Raghuraman MK, Gasser SM. 2001. The positioning and dynamics of origins of replication in the budding yeast nucleus. *J Cell Biol* **152:** 385–400.

Hiraga S, Robertson ED, Donaldson AD. 2006. The Ctf18 RFC-like complex positions yeast telomeres but does not specify their replication time. *EMBO J* **25:** 1505–1514.

Hiratani I, Gilbert DM. 2010. Autosomal lyonization of replication domains during early mammalian development. *Adv Exp Med Biol* **695:** 41–58.

Hiratani I, Ryba T, Itoh M, Yokochi T, Schwaiger M, Chang CW, Lyou Y, Townes TM, Schubeler D, Gilbert DM. 2008. Global reorganization of replication domains during embryonic stem cell differentiation. *PLoS Biol* **6:** e245.

Hiratani I, Takebayashi SI, Lu J, Gilbert DM. 2009. Replication timing and transcriptional control: Beyond cause and effect—part II. *Curr Opin Genet Dev* **19:** 142–149.

Hiratani I, Ryba T, Itoh M, Rathjen J, Kulik M, Papp B, Fussner E, Bazett-Jones DP, Plath K, Dalton S, et al. 2010. Genome-wide dynamics of replication timing revealed by in vitro models of mouse embryogenesis. *Genome Res* **20:** 155–169.

Huvet M, Nicolay S, Touchon M, Audit B, d'Aubenton-Carafa Y, Arneodo A, Thermes C. 2007. Human gene organization driven by the coordination of replication and transcription. *Genome Res* **17:** 1278–1285.

Hyrien O, Goldar A. 2010. Mathematical modelling of eukaryotic DNA replication. *Chromosome Res* **18:** 147–161.

Hyrien O, Marheineke K, Goldar A. 2003. Paradoxes of eukaryotic DNA replication: MCM proteins and the random completion problem. *Bioessays* **25:** 116–125.

Jorgensen HF, Azuara V, Amoils S, Spivakov M, Terry A, Nesterova T, Cobb BS, Ramsahoye B, Merkenschlager M, Fisher AG. 2007. The impact of chromatin modifiers on the timing of locus replication in mouse embryonic stem cells. *Genome Biol* **8:** R169.

Karnani N, Taylor C, Malhotra A, Dutta A. 2007. Pan-S replication patterns and chromosomal domains defined by genome-tiling arrays of ENCODE genomic areas. *Genome Res* **17:** 865–876.

Katsuno Y, Suzuki A, Sugimura K, Okumura K, Zineldeen DH, Shimada M, Niida H, Mizuno T, Hanaoka F, Nakanishi M. 2009. Cyclin A–Cdk1 regulates the origin firing program in mammalian cells. *Proc Natl Acad Sci* **106:** 3184–3189.

Kim SM, Dubey DD, Huberman JA. 2003. Early-replicating heterochromatin. *Genes Dev* **17:** 330–335.

Kitamura E, Blow JJ, Tanaka TU. 2006. Live-cell imaging reveals replication of individual replicons in eukaryotic replication factories. *Cell* **125:** 1297–1308.

Knott SR, Viggiani CJ, Tavare S, Aparicio OM. 2009. Genome-wide replication profiles indicate an expansive role for Rpd3L in regulating replication initiation timing or efficiency, and reveal genomic loci of Rpd3 function in *Saccharomyces cerevisiae*. *Genes Dev* **23:** 1077–1090.

Knott SR, Peace JM, Ostrow AZ, Gan Y, Rex AE, Viggiani CJ, Tavare S, Aparicio OM. 2012. Forkhead transcription factors establish origin timing and long-range clustering in *S. cerevisiae*. *Cell* **148:** 99–111.

Koren A, Soifer I, Barkai N. 2010. *MRC1*-dependent scaling of the budding yeast DNA replication timing program. *Genome Res* **20:** 781–790.

Krasinska L, Besnard E, Cot E, Dohet C, Mechali M, Lemaitre JM, Fisher D. 2008. Cdk1 and Cdk2 activity levels determine the efficiency of replication origin firing in *Xenopus*. *EMBO J* **27:** 758–769.

Kuipers MA, Stasevich TJ, Sasaki T, Wilson KA, Hazelwood KL, McNally JG, Davidson MW, Gilbert DM. 2011. Highly stable loading of Mcm proteins onto chromatin in living cells requires replication to unload. *J Cell Biol* **192:** 29–41.

Kuma D, Viberg J, Nilsson AK, Chabes A. 2010. Highly mutagenic and severely imbalanced dNTP pools can escape detection by the S-phase checkpoint. *Nucleic Acids Res* **38:** 3975–3983.

Kumar S, Huberman JA. 2009. Checkpoint-dependent regulation of origin firing and replication fork movement in response to DNA damage in fission yeast. *Mol Cell Biol* **29:** 602–611.

Kumar D, Viberg J, Nilsson AK, Chabes A. 2010. Highly mutagenic and severely imbalanced dNTP pools can escape detection by the S-phase checkpoint. *Nucleic Acids Res* **38:** 3975–3983.

Labit H, Perewoska I, Germe T, Hyrien O, Marheineke K. 2008. DNA replication timing is deterministic at the level of chromosomal domains but stochastic at the level of replicons in *Xenopus* egg extracts. *Nucleic Acids Res* **36:** 5623–5634.

Lande-Diner L, Zhang J, Cedar H. 2009. Shifts in replication timing actively affect histone acetylation during nucleosome reassembly. *Mol Cell* **34:** 767–774.

Lang GI, Murray AW. 2011. Mutation rates across budding yeast chromosome VI are correlated with replication timing. *Genome Biol Evol* **3:** 799–811.

Lebofsky R, Heilig R, Sonnleitner M, Weissenbach J, Bensimon A. 2006. DNA replication origin interference increases the spacing between initiation events in human cells. *Mol Biol Cell* **17:** 5337–5345.

Lee TJ, Pascuzzi PE, Settlage SB, Shultz RW, Tanurdzic M, Rabinowicz PD, Menges M, Zheng P, Main D, Murray JA, et al. 2010. *Arabidopsis thaliana* chromosome 4 replicates in two phases that correlate with chromatin state. *PLoS Genet* **6:** e1000982.

Le Tallec B, Dutrillaux B, Lachages AM, Millot GA, Brison O, Debatisse M. 2011. Molecular profiling of common fragile sites in human fibroblasts. *Nat Struct Mol Biol* **18:** 1421–1423.

Letessier A, Millot GA, Koundrioukoff S, Lachages AM, Vogt N, Hansen RS, Malfoy B, Brison O, Debatisse M. 2011. Cell-type-specific replication initiation programs set fragility of the *FRA3B* fragile site. *Nature* **470:** 120–123.

Li J, Santoro R, Koberna K, Grummt I. 2005. The chromatin remodeling complex NoRC controls replication timing of rRNA genes. *EMBO J* **24:** 120–127.

Liachko I, Bhaskar A, Lee C, Chung SC, Tye BK, Keich U. 2010. A comprehensive genome-wide map of autonomously replicating sequences in a naive genome. *PLoS Genet* **6:** e1000946.

Lian HY, Robertson ED, Hiraga S, Alvino GM, Collingwood D, McCune HJ, Sridhar A, Brewer BJ, Raghuraman MK, Donaldson AD. 2011. The effect of Ku on telomere replication time is mediated by telomere length but is independent of histone tail acetylation. *Mol Biol Cell* **22:** 1753–1765.

Lieberman-Aiden E, van Berkum NL, Williams L, Imakaev M, Ragoczy T, Telling A, Amit I, Lajoie BR, Sabo PJ, Dorschner MO, et al. 2009. Comprehensive mapping of long-range interactions reveals folding principles of the human genome. *Science* **326:** 289–293.

Lin CM, Fu H, Martinovsky M, Bouhassira E, Aladjem MI. 2003. Dynamic alterations of replication timing in mammalian cells. *Curr Biol* **13:** 1019–1028.

Lopez-Mosqueda J, Maas NL, Jonsson ZO, Defazio-Eli LG, Wohlschlegel J, Toczyski DP. 2010. Damage-induced phosphorylation of Sld3 is important to block late origin firing. *Nature* **467:** 479–483.

Lu J, Li F, Murphy CS, Davidson MW, Gilbert DM. 2010. G2 phase chromatin lacks determinants of replication timing. *J Cell Biol* **189:** 967–980.

Lubelsky Y, Sasaki T, Kuipers MA, Lucas I, Le Beau MM, Carignon S, Debatisse M, Prinz JA, Dennis JH, Gilbert DM. 2011. Pre-replication complex proteins assemble at regions of low nucleosome occupancy within the Chinese hamster dihydrofolate reductase initiation zone. *Nucleic Acids Res* **39:** 3141–3155.

Luo H, Li J, Eshaghi M, Liu J, Karuturi RK. 2010. Genome-wide estimation of firing efficiencies of origins of DNA replication from time-course copy number variation data. *BMC Bioinformatics* **11:** 247.

Lygeros J, Koutroumpas K, Dimopoulos S, Legouras I, Kouretas P, Heichinger C, Nurse P, Lygerou Z. 2008. Stochastic hybrid modeling of DNA replication across a complete genome. *Proc Natl Acad Sci* **105:** 12295–12300.

MacAlpine DM, Rodriguez HK, Bell SP. 2004. Coordination of replication and transcription along a *Drosophila* chromosome. *Genes Dev* **18:** 3094–3105.

MacAlpine HK, Gordan R, Powell SK, Hartemink AJ, MacAlpine DM. 2010. *Drosophila* ORC localizes to open chromatin and marks sites of cohesin complex loading. *Genome Res* **20:** 201–211.

Mantiero D, Mackenzie A, Donaldson A, Zegerman P. 2011. Limiting replication initiation factors execute the temporal programme of origin firing in budding yeast. *EMBO J* **30:** 4805–4814.

Marsolier-Kergoat MC, Goldar A. 2012. DNA replication induces compositional biases in yeast. *Mol Biol Evol* **29:** 893–904.

Maya-Mendoza A, Olivares-Chauvet P, Shaw A, Jackson DA. 2010. S phase progression in human cells is dictated by the genetic continuity of DNA foci. *PLoS Genet* **6:** e1000900.

McCune HJ, Danielson LS, Alvino GM, Collingwood D, Delrow JJ, Fangman WL, Brewer BJ, Raghuraman MK. 2008. The temporal program of chromosome replication: Genomewide replication in *clb5Δ Saccharomyces cerevisiae*. *Genetics* **180:** 1833–1847.

McNairn AJ, Gilbert DM. 2003. Epigenomic replication: Linking epigenetics to DNA replication. *Bioessays* **25:** 647–656.

Meister P, Taddei A, Ponti A, Baldacci G, Gasser SM. 2007. Replication foci dynamics: Replication patterns are modulated by S-phase checkpoint kinases in fission yeast. *EMBO J* **26:** 1315–1326.

Merrick CJ, Jackson D, Diffley JF. 2004. Visualization of altered replication dynamics after DNA damage in human cells. *J Biol Chem* **279:** 20067–20075.

Mickle KL, Oliva A, Huberman JA, Leatherwood J. 2007a. Checkpoint effects and telomere amplification during DNA re-replication in fission yeast. *BMC Mol Biol* **8:** 119.

Mickle KL, Ramanathan S, Rosebrock A, Oliva A, Chaudari A, Yompakdee C, Scott D, Leatherwood J, Huberman JA. 2007b. Checkpoint independence of most DNA replication origins in fission yeast. *BMC Mol Biol* **8:** 112.

Muller CA, Nieduszynski CA. 2012. Conservation of replication timing reveals global and local regulation of replication origin activity. *Genome Res* **22:** 1953–1962.

Nemeth A, Langst G. 2011. Genome organization in and around the nucleolus. *Trends Genet* **27:** 149–156.

Norio P, Kosiyatrakul S, Yang Q, Guan Z, Brown NM, Thomas S, Riblet R, Schildkraut CL. 2005. Progressive activation of DNA replication initiation in large domains of the immunoglobulin heavy chain locus during B cell development. *Mol Cell* **20:** 575–587.

Patel PK, Arcangioli B, Baker SP, Bensimon A, Rhind N. 2006. DNA replication origins fire stochastically in fission yeast. *Mol Biol Cell* **17:** 308–316.

Patel PK, Kommajosyula N, Rosebrock A, Bensimon A, Leatherwood J, Bechhoefer J, Rhind N. 2008. The Hsk1(Cdc7) replication kinase regulates origin efficiency. *Mol Biol Cell* **19:** 5550–5558.

Pope BD, Hiratani I, Gilbert DM. 2010. Domain-wide regulation of DNA replication timing during mammalian development. *Chromosome Res* **18:** 127–136.

Pope BD, Tsumagari K, Battaglia D, Ryba T, Hiratani I, Ehrlich M, Gilbert DM. 2011. DNA replication timing is maintained genome-wide in primary human myoblasts independent of D4Z4 contraction in FSH muscular dystrophy. *PLoS ONE* **6:** e27413.

Prendergast JG, Campbell H, Gilbert N, Dunlop MG, Bickmore WA, Semple CA. 2007. Chromatin structure and evolution in the human genome. *BMC Evol Biol* **7:** 72.

Raghuraman MK, Brewer BJ. 2010. Molecular analysis of the replication program in unicellular model organisms. *Chromosome Res* **18:** 19–34.

Raghuraman MK, Winzeler EA, Collingwood D, Hunt S, Wodicka L, Conway A, Lockhart DJ, Davis RW, Brewer BJ, Fangman WL. 2001. Replication dynamics of the yeast genome. *Science* **294:** 115–121.

Retkute R, Nieduszynski CA, de Moura A. 2011. Dynamics of DNA replication in yeast. *Phys Rev Lett* **107:** 068103.

Rhind N. 2006. DNA replication timing: Random thoughts about origin firing. *Nat Cell Biol* **8:** 1313–1316.

Rhind N, Yang SC, Bechhoefer J. 2010. Reconciling stochastic origin firing with defined replication timing. *Chromosome Res* **18:** 35–43.

Ryba T, Hiratani I, Lu J, Itoh M, Kulik M, Zhang J, Schulz TC, Robins AJ, Dalton S, Gilbert DM. 2010. Evolutionarily conserved replication timing profiles predict long-range chromatin interactions and distinguish closely related cell types. *Genome Res* **20:** 761–770.

Ryba T, Battaglia D, Pope BD, Hiratani I, Gilbert DM. 2011. Genome-scale analysis of replication timing: From bench to bioinformatics. *Nat Protoc* **6:** 870–895.

Ryba T, Battaglia D, Chang BH, Shirley JW, Buckley Q, Pope BD, Devidas M, Druker BJ, Gilbert DM. 2012. Abnormal developmental control of replication-timing domains in pediatric acute lymphoblastic leukemia. *Genome Res* **22:** 1833–1844.

Sadoni N, Cardoso MC, Stelzer EH, Leonhardt H, Zink D. 2004. Stable chromosomal units determine the spatial and temporal organization of DNA replication. *J Cell Sci* **117:** 5353–5365.

Schaaper RM, Radman M. 1989. The extreme mutator effect of *Escherichia coli mutD5* results from saturation of mismatch repair by excessive DNA replication errors. *EMBO J* **8:** 3511–3516.

Schmegner C, Hameister H, Vogel W, Assum G. 2007. Isochores and replication time zones: A perfect match. *Cytogenet Genome Res* **116:** 167–172.

Schultz SS, Desbordes SC, Du Z, Kosiyatrakul S, Lipchina I, Studer L, Schildkraut CL. 2010. Single-molecule analysis reveals changes in the DNA replication program for the *POU5F1* locus upon human embryonic stem cell differentiation. *Mol Cell Biol* **30:** 4521–4534.

Schwaiger M, Stadler MB, Bell O, Kohler H, Oakeley EJ, Schubeler D. 2009. Chromatin state marks cell-type- and gender-specific replication of the *Drosophila* genome. *Genes Dev* **23:** 589–601.

Schwaiger M, Kohler H, Oakeley EJ, Stadler MB, Schubeler D. 2010. Heterochromatin protein 1 (HP1) modulates replication timing of the *Drosophila* genome. *Genome Res* **20:** 771–780.

Sekedat MD, Fenyo D, Rogers RS, Tackett AJ, Aitchison JD, Chait BT. 2010. GINS motion reveals replication fork progression is remarkably uniform throughout the yeast genome. *Mol Syst Biol* **6:** 353.

Shirahige K, Hori Y, Shiraishi K, Yamashita M, Takahashi K, Obuse C, Tsurimoto T, Yoshikawa H. 1998. Regulation of DNA-replication origins during cell-cycle progression. *Nature* **395:** 618–621.

Silverman J, Takai H, Buonomo SB, Eisenhaber F, de Lange T. 2004. Human Rif1, ortholog of a yeast telomeric protein, is regulated by ATM and 53BP1 and functions in the S-phase checkpoint. *Genes Dev* **18:** 2108–2119.

Sporbert A, Gahl A, Ankerhold R, Leonhardt H, Cardoso MC. 2002. DNA polymerase clamp shows little turnover at established replication sites but sequential de novo assembly at adjacent origin clusters. *Mol Cell* **10:** 1355–1365.

Stamatoyannopoulos JA, Adzhubei I, Thurman RE, Kryukov GV, Mirkin SM, Sunyaev SR. 2009. Human mutation rate associated with DNA replication timing. *Nat Genet* **41:** 393–395.

Steglich B, Filion G, van Steensel B, Ekwall K. 2012. The inner nuclear membrane proteins Man1 and Ima1 link to two different types of chromatin at the nuclear periphery in *S. pombe*. *Nucleus* **3:** 77–87.

Stevenson JB, Gottschling DE. 1999. Telomeric chromatin modulates replication timing near chromosome ends. *Genes Dev* **13:** 146–151.

Takebayashi S, Sugimura K, Saito T, Sato C, Fukushima Y, Taguchi H, Okumura K. 2005. Regulation of replication at the R/G chromosomal band boundary and pericentromeric heterochromatin of mammalian cells. *Exp Cell Res* **304:** 162–174.

Takebayashi S, Dileep V, Ryba T, Dennis JH, Gilbert DM. 2012. Chromatin-interaction compartment switch at developmentally regulated chromosomal domains reveals an unusual principle of chromatin folding. *Proc Natl Acad Sci* **109:** 12574–12579.

Tanaka S, Nakato R, Katou Y, Shirahige K, Araki H. 2011. Origin association of Sld3, Sld7, and Cdc45 proteins is a key step for determination of origin-firing timing. *Curr Biol* **21:** 2055–2063.

Tazumi A, Fukuura M, Nakato R, Kishimoto A, Takenaka T, Ogawa S, Song JH, Takahashi TS, Nakagawa T, Shirahige K, et al. 2012. Telomere-binding protein Taz1 controls global replication timing through its localization near late replication origins in fission yeast. *Genes Dev* **26:** 2050–2062.

Thomson AM, Gillespie PJ, Blow JJ. 2010. Replication factory activation can be decoupled from the replication timing program by modulating Cdk levels. *J Cell Biol* **188:** 209–221.

Touchon M, Nicolay S, Audit B, Brodie of Brodie EB, d'Aubenton-Carafa Y, Arneodo A, Thermes C. 2005. Replication-associated strand asymmetries in mammalian genomes: Toward detection of replication origins. *Proc Natl Acad Sci* **102:** 9836–9841.

Vogelauer M, Rubbi L, Lucas I, Brewer BJ, Grunstein M. 2002. Histone acetylation regulates the time of replication origin firing. *Mol Cell* **10:** 1223–1233.

Williams RR, Azuara V, Perry P, Sauer S, Dvorkina M, Jorgensen H, Roix J, McQueen P, Misteli T, Merkenschlager M, et al. 2006. Neural induction promotes large-scale chromatin reorganisation of the *Mash1* locus. *J Cell Sci* **119:** 132–140.

Woodward AM, Gohler T, Luciani MG, Oehlmann M, Ge X, Gartner A, Jackson DA, Blow JJ. 2006. Excess Mcm2–7 license dormant origins of replication that can be used under conditions of replicative stress. *J Cell Biol* **173:** 673–683.

Wu PY, Nurse P. 2009. Establishing the program of origin firing during S phase in fission yeast. *Cell* **136:** 852–864.

Wu R, Terry AV, Singh PB, Gilbert DM. 2005. Differential subnuclear localization and replication timing of histone H3 lysine 9 methylation states. *Mol Biol Cell* **16:** 2872–2881.

Wu R, Singh PB, Gilbert DM. 2006. Uncoupling global and fine-tuning replication timing determinants for mouse pericentric heterochromatin. *J Cell Biol* **174:** 185–194.

Xu J, Yanagisawa Y, Tsankov AM, Hart C, Aoki K, Kommajosyula N, Steinmann KE, Bochicchio J, Russ C, Regev A, et al. 2012. Genome-wide identification and characterization of replication origins by deep sequencing. *Genome Biol* **13:** R27.

Yabuki N, Terashima H, Kitada K. 2002. Mapping of early firing origins on a replication profile of budding yeast. *Genes Cells* **7:** 781–789.

Yaffe E, Farkash-Amar S, Polten A, Yakhini Z, Tanay A, Simon I. 2010. Comparative analysis of DNA replication timing reveals conserved large-scale chromosomal architecture. *PLoS Genet* **6:** e1001011.

Yamazaki S, Ishii A, Kanoh Y, Oda M, Nishito Y, Masai H. 2012. Rif1 regulates the replication timing domains on the human genome. *EMBO J* **31:** 3667–3677.

Yang SC, Bechhoefer J. 2008. How *Xenopus laevis* embryos replicate reliably: Investigating the random-completion problem. *Phys Rev E Stat Nonlin Soft Matter Phys* **78:** 041917.

Yang SC, Rhind N, Bechhoefer J. 2010. Modeling genome-wide replication kinetics reveals a mechanism for regulation of replication timing. *Mol Syst Biol* **6:** 404.

Yardimci H, Loveland AB, Habuchi S, van Oijen AM, Walter JC. 2010. Uncoupling of sister replisomes during eukaryotic DNA replication. *Mol Cell* **40:** 834–840.

Yokochi T, Poduch K, Ryba T, Lu J, Hiratani I, Tachibana M, Shinkai Y, Gilbert DM. 2009. G9a selectively represses a class of late-replicating genes at the nuclear periphery. *Proc Natl Acad Sci* **106:** 19363–19368.

Zappulla DC, Sternglanz R, Leatherwood J. 2002. Control of replication timing by a transcriptional silencer. *Curr Biol* **12:** 869–875.

Zegerman P, Diffley JF. 2010. Checkpoint-dependent inhibition of DNA replication initiation by Sld3 and Dbf4 phosphorylation. *Nature* **467:** 474–478.

Zhang J, Xu F, Hashimshony T, Keshet I, Cedar H. 2002. Establishment of transcriptional competence in early and late S phase. *Nature* **420:** 198–202.

Replication-Fork Dynamics

Karl E. Duderstadt[1,4], Rodrigo Reyes-Lamothe[2,4], Antoine M. van Oijen[1], and David J. Sherratt[3]

[1]Zernike Institute for Advanced Materials, University of Groningen, 9747 AG, Groningen, Netherlands

[2]Department of Biology, McGill University, Montreal H3G 0B1, Canada

[3]Department of Biochemistry, University of Oxford, Oxford OX1 3QU, United Kingdom

Correspondence: a.m.van.oijen@rug.nl; david.sherratt@bioch.ox.ac.uk

The proliferation of all organisms depends on the coordination of enzymatic events within large multiprotein replisomes that duplicate chromosomes. Whereas the structure and function of many core replisome components have been clarified, the timing and order of molecular events during replication remains obscure. To better understand the replication mechanism, new methods must be developed that allow for the observation and characterization of short-lived states and dynamic events at single replication forks. Over the last decade, great progress has been made toward this goal with the development of novel DNA nanomanipulation and fluorescence imaging techniques allowing for the direct observation of replication-fork dynamics both reconstituted in vitro and in live cells. This article reviews these new single-molecule approaches and the revised understanding of replisome operation that has emerged.

The duplication of chromosomes in cells is conducted with high speed and accuracy by the concerted efforts of many cellular factors operating in concert within large replication complexes. A typical example is the replication machinery of *Escherichia coli*: It is composed of more than a dozen components forming an assembly with a molecular weight over 1 MDa that unwinds and copies DNA at a rate approaching 1000 base pairs per second (Chandler et al. 1975) while making less than one mistake per 10^9 nucleotide incorporations (Drake et al. 1969). What makes this achievement even more remarkable is that the process depends on the coordination of molecular events over a broad range of time and length scales, from the breaking and reformation of chemical bonds within DNA polymerases to the large-scale structural rearrangements of chromosomes. This high degree of efficiency is maintained even in eukaryotic cells where the process is more complex, involving additional regulation and a larger number of cellular components (Bell and Dutta 2002). Although decades of intense research has revealed the function and architecture of many core replisome subassemblies, the coordination of enzymatic events at the replication fork remains poorly understood because of

[4]Karl E. Duderstadt and Rodrigo Reyes-Lamothe contributed equally to this manuscript.

Cite this article as *Cold Spring Harb Perspect Biol* doi: 10.1101/cshperspect.a010157

the challenge of directly observing replication complexes in action.

Recent work has further complicated our view of the processes at the replication fork by revealing that replisome composition is highly dynamic, with components rapidly exchanging during synthesis. A partial explanation for these observations emerges when considering that robust replication in the cellular environment depends not only on the efficiency and accuracy of synthesis, but also on the ability of replisomes to overcome obstacles encountered on parental chromosomes, such as transcription complexes and DNA lesions (Cox et al. 2000). The recent discoveries of novel mechanisms for replication restart have provided insight into how replisomes cope with these obstacles (Heller and Marians 2006; Lopes et al. 2006; Pomerantz and O'Donnell 2008; Yeeles and Marians 2011) and revealed that replication-fork collapse and reassembly may be a common occurrence even within a single cell cycle (Cox et al. 2000; Indiani et al. 2009; Langston et al. 2009). These studies support the notion that replisomes do not operate as static assemblies but instead have a dynamic composition that is responsive to the cellular environment.

To construct a complete understanding of the replication cycle, we must develop and employ new methods that allow for the observation and characterization of short-lived states and dynamic events at the replication fork. The last decade has seen great progress toward these goals with the advent of new techniques for observing and characterizing single replisomes both reconstituted in vitro and in live cells. These efforts have started to provide a dynamic view of the events that occur during chromosome duplication. Flow-stretching nanomanipulation and fluorescence-based assays have revealed the formation and release of loops during Okazaki fragment synthesis and the dynamic exchange of polymerases during replication (Hamdan et al. 2009; Loparo et al. 2011; Tanner et al. 2011; Lia et al. 2012). The development of new fluorescence imaging techniques has allowed for the real-time observation of individual replisomes in vivo, which has led to a revised understanding of replisome composition in bacteria (Reyes-Lamothe et al. 2010). In this article, we review our current understanding of replisome function while highlighting a broad range of new methods and technologies that allow for the direct observation of events at single replication forks.

EXPERIMENTAL TOOLBOXES

Mechanical Stretching

The mechanical manipulation of individual DNA molecules by means of optical trapping, magnetic tweezing, and flow stretching has provided powerful tools for probing the function of nucleic-acid processing enzymes involved in DNA replication. Optical trapping and magnetic tweezing allow for the direct application of force on individual DNA molecules and the read out of enzymatic function using feedback control of force (Fig. 1A,B). These techniques have been used to study the mechanism used by helicases to unwind DNA (Johnson et al. 2007; Lionnet et al. 2007; Sun et al. 2011) and the forces at work within polymerases during nucleotide incorporation (Maier et al. 2000; Wuite et al. 2000). Although these methods provide the highest resolution of enzymatic function on DNA, in most cases (Ribeck and Saleh 2008), they are not compatible with multiplexing, which makes collection of statistically significant samples sizes difficult and time consuming. An alternative approach that can be multiplexed is a technique where individual DNA molecules are stretched by laminar flow, and their lengths are monitored by tracking the position of small beads attached to their ends (Fig. 1C) (van Oijen et al. 2003; Tanner and van Oijen 2009). Both leading-strand and coordinated synthesis by replisomes have been observed using this method. In the former case, bead movement reports on the conversion from an extended double-stranded state to a contracted single-stranded state (Lee et al. 2006; Tanner et al. 2008). In the latter case, bead movement reports on the formation and release of replication loops on the lagging strand (Hamdan et al. 2009).

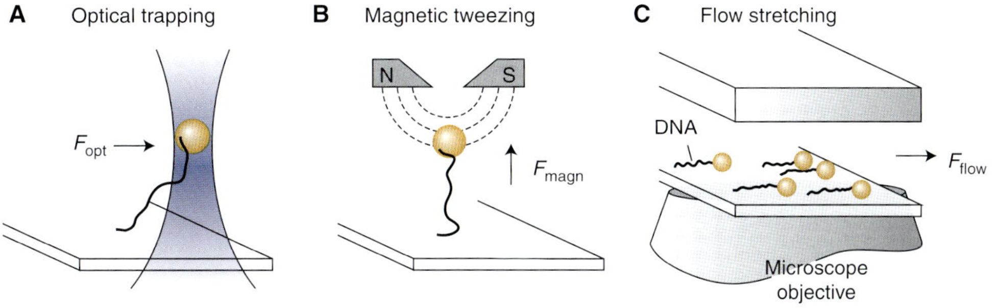

Figure 1. DNA stretching techniques. Three methods of DNA nanomanipulation are shown. (*A*) Optical trapping: A focused beam of light traps a bead that is tethered to DNA. Feedback control allows the position and force on the bead to be precisely monitored and controlled. (*B*) Magnetic tweezing: Permanent magnets located above the sample generate a vertical extending force on DNA through a magnetic bead. (*C*) Flow stretching: DNA-tethered beads are extended using laminar flow and imaged using wide-field optical microscopy. In all cases, the end of the DNA opposite the beads is attached to a functionalized glass surface.

Fluoresence Imaging

Although mechanical manipulation techniques provide detailed kinetic information about enzymatic function, they do not provide direct information about the composition of protein assemblies or conformational states adopted by single enzymes during operation. A complete understanding of the events at the replication fork will only be possible by simultaneously observing the composition and function of single replisomes in operation. Fortunately, many single-molecule fluorescence techniques have emerged that allow for the characterization of these properties. Single-molecule fluorescence resonance energy transfer (FRET) has been used to distinguish between different replisome assembly pathways (Xi et al. 2005a,b; Zhang et al. 2005; Smiley et al. 2006) and to observe the conformational states adopted by polymerases during substrate engagement (Luo et al. 2007). Single-molecule fluorescence techniques have also been used to visualize replication products and short-lived changes in DNA structure during synthesis (Hamdan et al. 2009; Pandey et al. 2009). Using these methods, the activity of fully assembled *E. coli* replisomes has been observed in real time on rolling-circle substrates by flow stretching the products and staining with a DNA-intercalating dye (Fig. 2A,B) (Georgescu et al. 2009; Tanner et al. 2009, 2011; Georgescu et al. 2012). Finally, both the composition and structure of the phage T7 replisome have been observed simultaneously by combining nanomanipulation techniques with the labeling of individual polymerases and replication substrates (Fig. 2C) (Loparo et al. 2011).

Observing Replication-Fork Dynamics In Vivo

Understanding the action of molecular machines, like replisomes, in their natural cellular environment is the ultimate goal of biochemical research. The use of single-molecule fluorescence microscopy in live cells promises to reveal details of composition, architecture, and dynamics of active replisomes. The main limitation of this approach is the difficulty in detecting single fluorophores over a high background, but use of strategies to illuminate small sections of the cells can provide the desired sensitivity, as is the case for TIRF, HILO, and SPIM (Fig. 3A) (Tokunaga et al. 2008; Li and Xie 2011). Bacterial cells (with widths <1 μm) have much lower background fluorescence derived from autofluorescence and out-of-focus fluorophores, thereby permitting the detection of single molecules using standard epifluorescence (Li and Xie 2011). The small number of replication forks and range of genetic tools make bacteria perfect models for this approach. The use of single-molecule techniques has permitted the

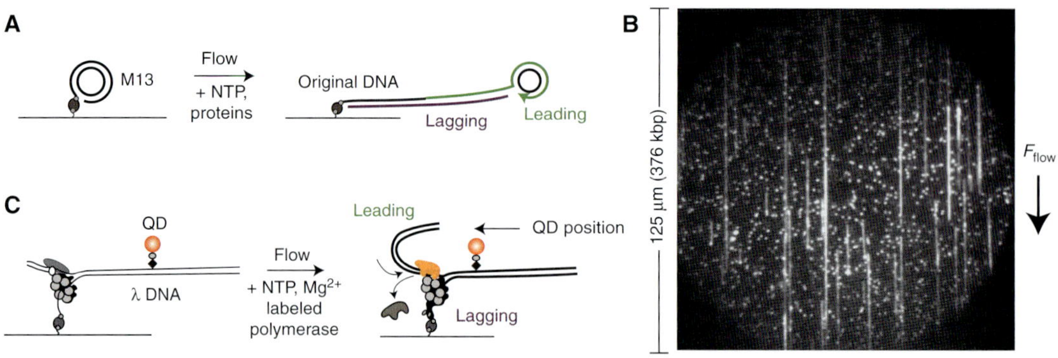

Figure 2. Study of reconstituted replisomes. (*A*) Schematic of single-molecule rolling-circle replication assay. Leading-strand synthesis displaces DNA from the circle as the replisome rolls around the template. The displaced tail is then used for lagging-strand synthesis. The DNA that couples the M13 circle to the surface increases in length as replication proceeds. Synthesis is monitored by fluorescently staining and flow stretching the newly synthesized products. (*B*) Example image of rolling-circle replication products collected using total internal reflection fluorescence (TIRF) microscopy. (*C*) Schematic of single-molecule assay for simultaneously measuring replisome activity and composition. Leading-strand synthesis is performed using a replication fork introduced at one end of λ DNA. During synthesis, the template is unwound and a compact single-stranded product is generated on the lagging strand. The conversion from double-stranded to single-stranded DNA is monitored using a quantum dot (QD) attached to the template. Fluorescently labeled polymerases are monitored as they enter and exit the replisome during synthesis. All replication substrates are attached to functionalized glass surfaces for TIRF imaging.

determination of stoichiometry and the study of dynamics of individual molecules in bacteria. Two strategies have been adopted for these studies. In the first technique, the sample is illuminated by short flashes of light at intervals limited by the camera reading time (strobing illumination) (Xie et al. 2008). Strobing illumination has the advantage of allowing large fields of view. The second technique used consists in concentrating the laser beam in the sample so it only illuminates a very small region of the slide, originally ~6 μm diameter (slimfield microscopy) (Reyes-Lamothe et al. 2010). Slimfield microscopy can only be used for small cells but has the advantage of permitting capture rates of 3 msec and up (Fig. 3B).

Fluorescence recovery after photobleaching (FRAP) is widely used to probe molecular dynamics in cells (Lippincott-Schwartz et al. 2001). The technique consists of irreversibly bleaching a region of the sample and monitoring the diffusion and turnover of fluorescent molecules expressed as recovery of fluorescence (Fig. 3C). This technique has been used to study both the diffusion of free replisome components and turnover of active replisome components in bacteria and eukaryotes (Sporbert et al. 2002; Solovjeva et al. 2005; Su'etsugu and Errington 2011). Use of photoswitchable proteins (Patterson et al. 2010) promises to offer an alternative to FRAP in studies of component turnover, and as it helps to provide single-molecule data of densely packed replisomes like those of eukaryotic cells.

LIFE OF A REPLISOME

Given the importance of faithful replication, the assembly and activity of the replisome must be stringently orchestrated. The unit of replication, the replicon, is controlled through initiation by action of a specific initiator acting at the replicon's origin of replication (Jacob et al. 1963). The key step during replication initiation is the helicase loader-mediated loading of the ring-shaped replicative helicase to an active replication origin bound by the initiator. The function of the helicase loader is to open the otherwise

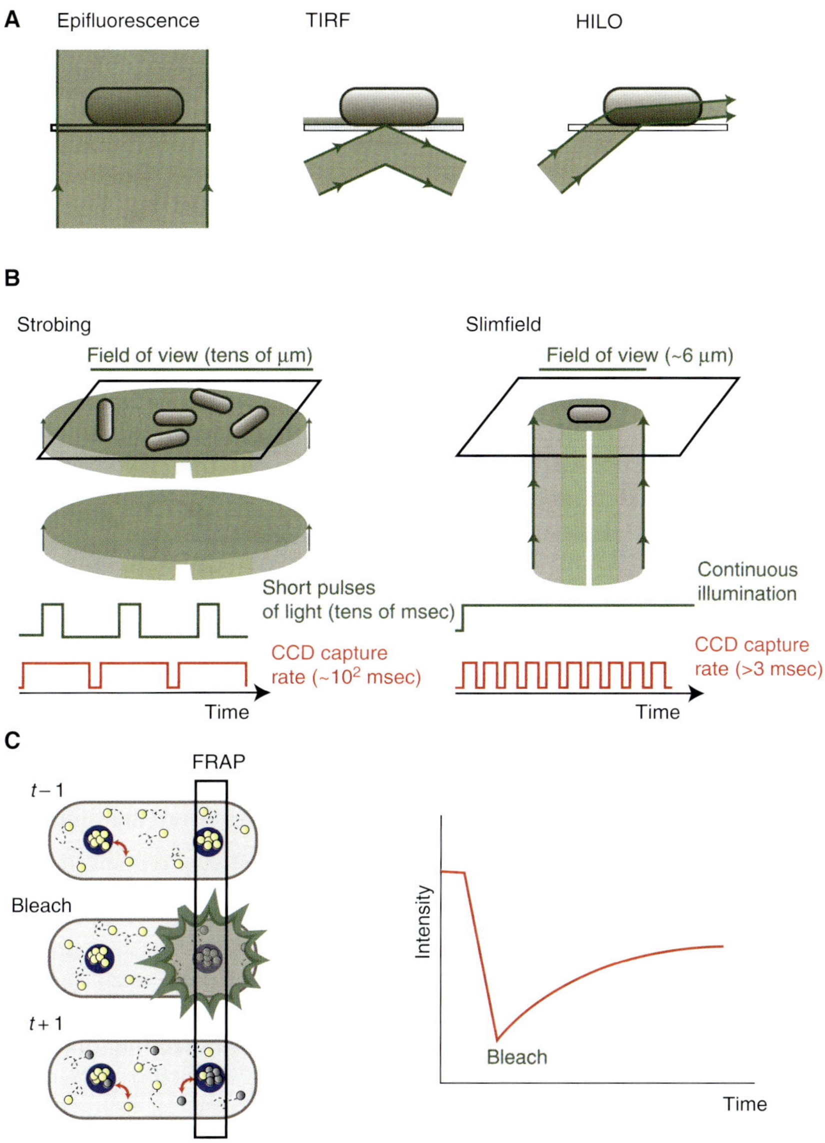

Figure 3. Study of replisomes in live cells. (*A*) Two examples of techniques used to enhance the contrast of fluorescence microscopy: TIRF and high inclined and laminated optical sheet (HILO) microscopy. Both techniques use a tilted light beam, as opposed to epifluorescence, where the beam is perpendicular to the slide. Depending on the angle used, the beam either results in the illumination of only the ~100 nm closer to the slide (TIRF) or the creation of a sheet of light that illuminates a slice of the sample (HILO). By reducing the volume illuminated, the contribution of out-of-focus fluorescent or autofluorescent molecules is also reduced. (*B*) Two strategies to detect single molecules in cells: illuminating the sample using short pulses of light followed by intervals with a length no shorter than the camera's processing time (strobing) and, alternatively, illuminating a small region of the sample with the signal captured continuously (slimfield). (*C*) Use of fluorescence recovery after photobleaching (FRAP) to study dynamics of molecular machines (enclosed in blue circles). Fluorescent molecules (yellow balls) are either relatively immobile, when part of a machine and generating fluorescent spots, or rapidly diffusing in the cytoplasm. Bleaching fluorescence of a region of the cell causes a fraction of molecules to lose their fluorescence (grey balls). The exchange between fluorescent and bleached molecules can be followed by changes in the intensity of spots.

stable toroidal structure of the helicase, so it can entrap DNA at its center. Both initiator and helicase loader proteins are members of the ATPase associated with various activities (AAA^+) superfamily (Duderstadt and Berger 2008), and their action is regulated by nucleotide binding. The initiator in bacteria, DnaA, destabilizes the duplex by binding and oligomerizing at *oriC*, permitting the recruitment of the DnaB-DnaC helicase-loader complex onto ssDNA (Mott and Berger 2007). In contrast, eukaryotes, and possibly archaea, load the MCM helicase onto dsDNA (Evrin et al. 2009; Remus et al. 2009).

Movement of the replicative helicase away from the origin occurs concomitantly with dissociation of the helicase loader and recruitment of primase, the protein responsible for the synthesis of RNA primers on the lagging-strand template. Recruitment of primase permits the loading of the remaining replisome components, including the replicative polymerase, processivity clamp, and clamp loader (Fig. 4). Replication initiation results in the assembly of two sister replisomes, each facing outward from the origin. Synchronous loading and firing of sister replisomes protects chromosomes from exposing recombinogenic replisome-free forked structures. Eukaryotes achieve coordination of these events through the activity of cell-cycle-regulated phosphorylation of replisome components, with an inactive MCM helicase loaded on the origin long before it gets activated (Masai et al. 2010). In bacteria, coordination is achieved by coupling cell-cycle-regulated origin opening by DnaA-ATP to immediate loading of two DnaB helicases, which in turn stimulate the assembly of two sister replisomes (Katayama et al. 2010).

An idea that has lingered for decades is that, after replisome assembly at origins, sister replisomes remain attached to each other as replication progresses, with their action potentially coordinated. This idea was born as part of a segregation model proposed before the concept of bidirectional elongation had been established (Jacob et al. 1963), and it was later formulated explicitly by Dingman (1974). This concept of a replication factory containing sister replisomes was expanded to accommodate the observation that each eukaryote replication fluorescent focus needed to contain up to 40 replisomes to accommodate the total number of replication forks per cell predicted by other means (Cook 1999). One could imagine that in such putative mega-factories, coordination of the activities of all origins could be coupled, or that any coordination would be solely between sister replisomes. Support for at least the sister-replisome factory model came from structural and biochemical data of SV40 helicase, T-ag (it is loaded onto the origin as double hexamer, and its action is dependent on the maintenance of this link between hexamers) (Weisshart et al. 1999; Valle et al. 2000; Alexandrov et al. 2002). Early fluorescence microscopy studies of replisomes and chromosomal loci within *Bacillus subtillis* and *Saccharomyces cerevisiae* also supported the existence of a replication factory that pulled in the unreplicated DNA and extruded the replicated DNA (Lemon and Grossman 1998, 2000; Kitamura et al. 2006).

Nevertheless, with better resolution assays, evidence for independent action and cellular positioning of replisomes has accumulated. Fluorescence microscopy of *E. coli* cells showed that single replisomes track independently on the chromosome during replication, although closely spaced sister replisomes are not always spatially separable in epifluorescence (Reyes-Lamothe et al. 2008). The situation may be the same in eukaryotes (e.g., in *Xenous* egg extracts, where sister replisomes separated from each other as replication proceeded on a stretched template without any visible effect on their rate) (Yardimci et al. 2010). Consistent with the independent progression of sister forks, their functional independence has been demonstrated in *E. coli*, *B. subtillis*, and *S. cerevisiae* (Breier et al. 2005; Wang et al. 2007; Doksani et al. 2009), where blocking or slowing down one of them does not prevent the normal progression of its sister. There is no a priori reason for sister replisomes to stay together, and if they were complexed into a fixed structure, it would seem difficult to generate the precatenane interlinks between newly replicated sisters that result from individual fork rotation as

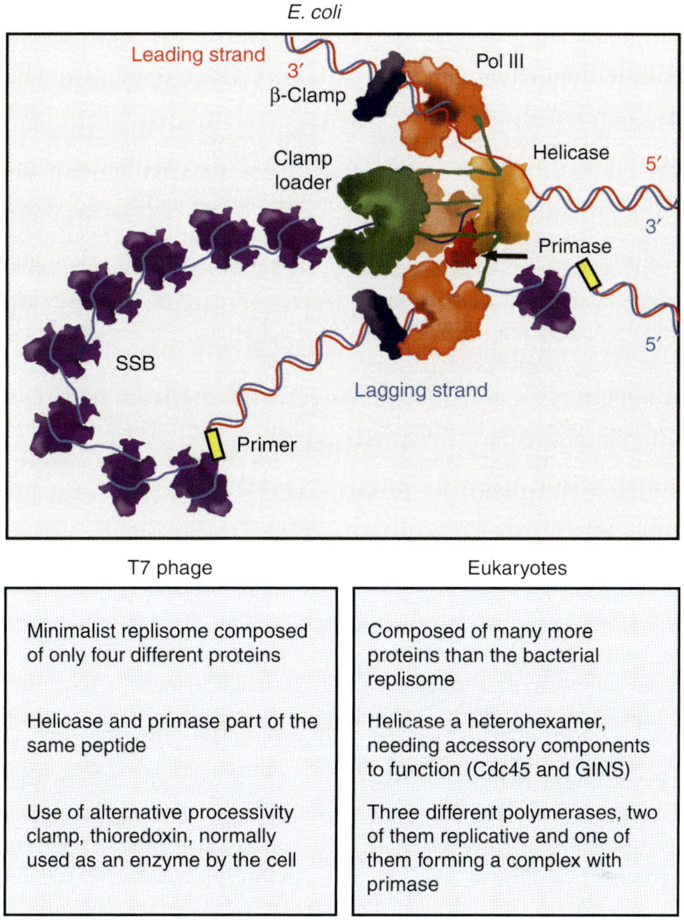

Figure 4. Replisome architecture. The arrangement of the different components of the *E. coli* replisome is shown. At the front of the replication fork, a hexameric helicase melts the dsDNA. Helicase binds primase, and it is connected to the three replicative polymerases (Pol III) via the τ subunit of the clamp loader. A leading-strand polymerase travels in the same direction of the fork, while the two remaining polymerases can potentially act on the lagging strand. Each active polymerase is bound to the homodimeric β-clamp, which acts as a processivity factor. The leading strand is synthesized continuously, whereas the lagging strand requires cycles of chain elongation. Progression of helicase causes the accumulation of ssDNA, which is covered by the homotetrameric SSB. The clamp loader (heteropentamer) loads β-clamp, mediates primase handover to polymerase, and connects the polymerase with helicase. The architecture of the replisome in other systems is thought to be similar to that of *E. coli* (Yao and O'Donnell 2010). Some differences with phage and eukaryote replisomes are listed.

replication proceeds (Hearst et al. 1998; Wang et al. 2008). It seems likely that, in many cases, the proposed factories result from diffraction-limited light microscopy of densely replisome-populated chromosomal regions. Indeed, super-resolution microscopy of human cells showed the number of replisomes per fluorescent spot to be between two and three (Cseresnyes et al. 2009).

Elongation of replicating DNA strands occurs with a 5′ to 3′ directionality (Kornberg and Baker 1992) (Fig. 4). The anti-parallel nature of DNA permits continuous synthesis only in the strand elongated in the same direction as the migration of the replication fork, the leading strand. Meanwhile, the copying of the complementary strand, lagging strand, occurs discontinuously and requires cycles of priming, clamp loading, polymerase loading, and polymerase release at every Okazaki fragment (Fig. 5). Single-stranded DNA binding proteins have to cover and protect the stretches of ssDNA until they are displaced by polymerase. Other auxiliary activities, like removal of RNA primers and ligation of Okazaki fragments, mediated by proteins not generally thought of as part of the replisome, have to be accomplished to obtain an uninterrupted copy of DNA. The cytoplasmic pool of proteins involved in lagging-strand synthesis is limited, and this imposes their need to be recycled (Fig. 5).

Once replication has initiated, the two diverging forks normally continue until they each meet a converging fork and replication terminates. Precisely what occurs mechanistically when replication forks converge, how the torsional tension is handled, and how overreplication is avoided still needs to be determined. Replicon size varies from a few kb to many Mb (Table 1), and is determined by the spacing of potentially active origins within a chromosome; in eukaryotes, replicon size can change dramatically during development (Méchali 2010). Replisomes are not always loaded in pairs. Replication of some plasmids, some of them ~100 kb long (Nordstrom 2006), is unidirectional, presumably because only a single replisome is loaded by a single replicative helicase, showing that there is no physical impediment to a single replisome forming and acting at a given initiation.

COORDINATION OF LEADING- AND LAGGING-STRAND SYNTHESIS

Processive replication depends on the spatial and temporal coordination of enzymatic events at the replication fork. An essential characteristic of this coordination is the reorientation of the lagging-strand DNA through the formation of a loop during Okazaki-fragment synthesis, which allows the leading- and lagging-strand polymerases to move coordinately in parallel. Electron-microscopy studies have visualized the formation of these lagging-strand loops and characterized their length distributions (Park et al. 1998; Lee et al. 2002; Chastain et al. 2003), but the timing and mechanisms underlying the transitions that occur during Okazaki-fragment synthesis have remained less clear. There are at least three enzymatic events that must occur during each cycle: the synthesis of a primer, its handoff to the lagging-strand polymerase, and loop release (Fig. 5). Several recent studies have begun to elucidate the kinetics of this process, providing experimental observations that discriminate between different mechanisms of regulation for each stage of the cycle.

Two mechanisms have been proposed for ensuring efficient loop release and a fluid transition from the synthesis of one Okazaki fragment to the next. In the signaling mechanism, primer synthesis triggers loop release even if nascent Okazaki-fragment synthesis is unfinished. Supporting this model is the observation that Okazaki-fragment size is dependent on primase activity in the *E. coli*, T7, and T4 systems (Li and Marians 2000; Lee et al. 2002; Yang et al. 2006) and sensitive to helicase-primase interactions in *E. coli* (Tougu and Marians 1996). Additional support comes from the observation of ssDNA gaps on the lagging strands in *E. coli* and T4 (Yang et al. 2006; Georgescu et al. 2012). An alternative model, the collision mechanism, proposes that the lagging-strand polymerase dissociates once it has encountered the 5′ terminus of the previously synthesized Okazaki fragment. This model is supported by the observation that the T4 DNA polymerase and *E. coli* DNA polymerase III holoenzyme dissociate rapidly when they encounter a 5′ terminus (Hacker and Alberts 1994; Carver et al. 1997; Li and Marians 2000; Lopez de Saro et al. 2003; Yao et al. 2009). However, a recent *E. coli* study suggests the collision mechanism is too slow to explain the observed rate of replication (Dohrmann et al. 2011).

Direct observation of coordinated synthesis using single-molecule methods has provided compelling evidence that reconciles the seemingly contradictory findings of past work. The formation and release of single replication loops has been observed by fully reconstituted T7 replisomes by tracking length changes in flow-stretched DNA substrates (Hamdan et al. 2009).

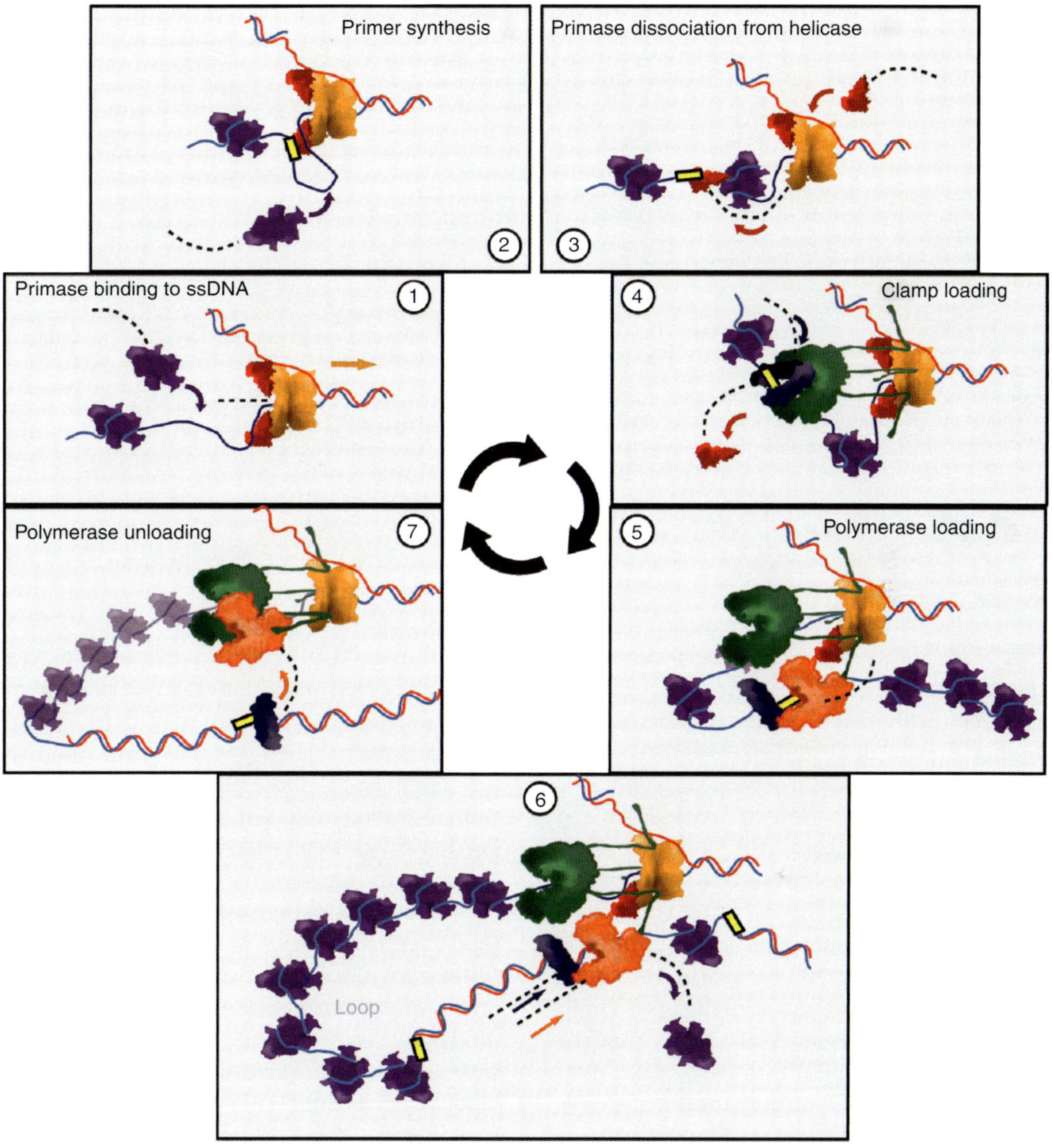

Figure 5. Replisome dynamics. The major source of molecular exchange during replication is produced during the synthesis of the lagging strand. Primase binds to ssDNA (*1*) and produces a primer forming a short loop in the process (*2*). The primase then dissociates from helicase, possibly remaining bound to the primer (*3*). Loading of β-clamp displaces the primase (*4*) and recruits one of the idle polymerase in the replisome (*5*). Polymerase elongates the chain causing the growth of a DNA loop (*6*). Finally, the lagging-strand polymerase detaches from the β-clamp and causes the release of the DNA loop, while the clamp remains bound to DNA (*7*). For simplicity, only the collision mechanism for loop release is depicted. SSB, β-clamp, primase, and likely polymerase (during loop release, not shown here) are exchanged with the diffusing pool of proteins in the cytoplasm at different stages of the cycle. For clarity, only a part of the replisome is represented in the different steps.

Table 1. Facts on DNA replication dynamics

	T4	T7	*E. coli*	Archaea	*S. cerevisiae*	Human
Length of Okazaki fragment (kb)[a]	1.5–2.2	2.3	1.5	0.09–0.16	0.16[b]	0.1–0.2[c]
Primer length (nt)[d]	4–5	4–5	10–12	~10	8–10	11–14
SSB/RPA binding (nt)[e,f]	10	14	32 or 64	4.5–5	19 or 27	8 or 30
Rate of fork progression (bp sec^{-1})[g]	600	370	600–1000	88–330	26–48	25
Replicon size (kb)[h]	40	166	4600	1000, 700, 500	36	~100

[a]Edenberg and Huberman 1975; Kornberg and Baker 1992; Debyser et al. 1994; Chastain et al. 2000; Matsunaga et al. 2003; Smith and Whitehouse 2012.

[b]Okazaki fragment length matches that of nucleosome binding in yeast. Multiples of 165 nt, 330 nt, and 495 nt were reported.

[c]Extrapolation from the study of SV40 system.

[d]Frick and Richardson 2001; Matsunaga et al. 2003.

[e]Kim et al. 1992; Kornberg and Baker 1992; Blackwell and Borowiec 1994; Kumaran et al. 2006.

[f]Extent of binding by a functional unit; T4 gp32 is a monomer, T7 gp2.5 is a homodimer, bacterial SSB is a homotetramer, archaeal SSB is a monomer, and the eukaryotic RPAs are heterotrimers.

[g]Edenberg and Huberman 1975; Rabkin and Richardson 1990; Kornberg and Baker 1992; Myllykallio et al. 2000; Raghuraman et al. 2001; Kerr et al. 2003; Duggin et al. 2008.

[h]Edenberg and Huberman 1975; Kornberg and Baker 1992; Duggin et al. 2008; Liachko et al. 2010.

This work shows that loop formation occurs only half the time during replication, a finding consistent with previous electron microscopy studies (Park et al. 1998; Lee et al. 2002; Chastain et al. 2003). Analysis of loop sizes and delay times between loops, a step unobservable in bulk-phase assays, shows that both the initiation of primer synthesis and the completion of an Okazaki fragment serve as triggers for loop release. Single-molecule studies conducted with *E. coli* replisomes have also revealed a mixture of loop-release mechanisms (Tanner et al. 2011; Georgescu et al. 2012). Furthermore, *E. coli* replisomes composed of two polymerases prematurely terminate Okazaki synthesis, whereas replisomes composed of three polymerases almost always complete synthesis (Georgescu et al. 2012). These studies show that replisomes use both signaling and collision mechanisms to regulate lagging-strand synthesis. The combination of these two mechanisms is likely critical for highly processive and coordinated synthesis. While the signaling mechanism allows for rapid transitions from Okazaki-fragment synthesis to the site of a new primer, the collision mechanism ensures that polymerases do not stall if a new primer has not been synthesized before they encounter the previous Okazaki fragment.

Although lagging-strand synthesis involves many more enzymatic steps than leading-strand synthesis does, the duplication of DNA on both strands is coordinated by mechanisms that are not completely understood. Single-molecule techniques have allowed for detailed investigations of this issue, which have provided new evidence in support of several explanations. Flow-stretching experiments, conducted using the T7 replisome, have shown that activation of primase, by the addition of required ribonucleotides, causes transient pauses in leading-strand synthesis (Lee et al. 2006; Hamdan et al. 2009) with an average duration consistent with the kinetics of primer synthesis (Kusakabe and Richardson 1997; Frick et al. 1999). These observations reveal a possible mechanism for ensuring that the slow enzymatic events on the lagging strand do not cause leading-strand synthesis to proceed faster than lagging-strand synthesis. An alternative explanation is supported by recent single-molecule experiments investigating the T7 leading-strand synthesis reaction (Pandey et al. 2009) and T4 helicase-primase complex (Manosas et al. 2009). These studies demonstrate the formation of a small ssDNA loop between the helicase and primase that may allow the replisome to continue during primer synthesis. In this case, a faster rate of lagging-strand synthesis has been suggested as a mechanism to ensure efficient coupling of daughter-strand synthesis.

A complete understanding of how primer synthesis influences coordinated replication depends on careful consideration of the steps involved in RNA polymerization. Several studies have demonstrated that primer synthesis proceeds in two distinct stages (Kusakabe and Richardson 1997; Frick et al. 1999; Qimron et al. 2006): the formation of a dinucleotide and the slow extension of that dinucleotide to form a full-length primer. During replication, this is a stochastic process in which only a fraction of priming sequences are recognized and utilized by primase (Lee et al. 2006). In between these actual primer-synthesis events, primase scans the lagging strand for initiation sites. Whereas it has been demonstrated that dinucleotide formation and primer synthesis can trigger replication loop release and cause pausing (Lee et al. 2006; Hamdan et al. 2009), the molecular events that cause the primase to utilize one priming site and bypass others are not understood. The formation of a small ssDNA priming loop may allow the primase to efficiently utilize the primase-recognition sequences for dinucleotide formation while allowing the helicase to continue unwinding (Fig. 5). Then, once a ribodinucleotide has formed, slow extension into a full-length primer may occur, a process that could cause the replisome to pause. This dynamic balance between priming-loop formation and replisome pausing could provide an optimal compromise between speed and coordination and would provide an explanation for the current experimental evidence. Work conducted with *E. coli* and T7 systems has shown that primase activity depends on in trans interactions between adjacent primases bound to the helicase (Lee and Richardson 2002; Corn et al. 2005), providing a structural framework that could explain the observed pausing behavior.

Primase action not only regulates the length of Okazaki fragments, but also the transfer of the replicative polymerase to the primer. A single Toprim-family protein comprises bacterial and phage primases, whereas two subunits (which share no homology to bacteria primases) comprise archaeal and eukaryotic primases (Kuchta and Stengel 2010). Nevertheless, in all cases, the result of their function is a primer that is just a few bp long (Table 1). What determines the distribution of Okazaki-fragment sizes (Table 1) is not known, but it is likely that the interaction between primase and helicase plays a role. Binding of bacterial DnaG primase to DnaB helicase stimulates primase ssDNA binding and priming (Corn and Berger 2006). In turn, association of up to three primase molecules increases the processivity of helicase, probably by stabilizing its structure (Bailey et al. 2007). *E. coli* primase, present at 50–100 copies per cell (Kornberg and Baker 1992), remains bound to DNA after primer synthesis every cycle (Yuzhakov et al. 1999), implying the need for fast turnover of the primase. Finally, two primase proteins may need to interact to function (Corn and Berger 2006). Considering all these constraints, the cellular primase pool only accounts for the elongation of less than 75 kb of DNA. It is therefore possible that the activity of the otherwise idle primase is fine-tuned by the frequency of its association to helicase (Corn and Berger 2006). Such a scenario is supported by the observation that priming frequency increases with primase concentration, this holding true for T4 and even T7 phage systems, the latter having primase and helicase as parts of a single peptide. The situation is less clear for archaea and eukaryotes. However, primase interacts with the MCM helicase through GINS, mediated by Ctf4 in *S. cerevisiae*, suggesting a role of this interaction for the coordination of their respective activities (Marinsek et al. 2006; Gambus et al. 2009; Tanaka et al. 2009).

Primase remains bound to the primer-template to stabilize, protect, and selectively transfer it to the replicative polymerase (Yuzhakov et al. 1999). Handing off the primer to the polymerase seems to occur in a number of ways according to the different replisome architectures. In *E. coli*, as with other bacteria, the handing off of primer seems to occur simultaneously with the loading of the clamp onto the primer. After synthesizing the primer, primase forms a ternary complex with DNA, and binding of primase to SSB stabilizes this interaction

(Yuzhakov et al. 1999). The nonessential component of the clamp loader, χ, was suggested to compete for the interaction with SSB, releasing primase from the template (Yuzhakov et al. 1999). It is nevertheless likely this model is incomplete given that χ is not essential for viability (Kelman and O'Donnell 1995) and because genetic and structural evidence suggests rather a structural role for χ maintaining processivity of leading-strand synthesis (Marceau et al. 2011). The question of how the switch from primase to polymerase is regulated in bacteria remains open. In eukaryotes, primase is part of a heterotetramer that also includes DNA polymerase Pol α. In this system, the length of primer and the handover to the DNA polymerase subunit is regulated by the noncatalytic subunit of primase (p58) (Kuchta and Stengel 2010). A typical RNA primer would be ~10 nt in length, onto which Pol α adds ~20 nt more before it is replaced by Pol ε via RFC clamp-loader-mediated handoff (Waga and Stillman 1998). For binding to RPA, RFC competes with Pol α; for binding to the primer, PCNA gets loaded, and then Pol ε displaces RFC again through a competition for RPA (Yuzhakov et al. 1999; Maga et al. 2000). The theme of a stable interaction between primase and polymerase facilitating primer transfer has also been proposed in the replisome of phage T7 (Kato et al. 2004).

PERSPECTIVES

Although great progress has been made in understanding the coordination of enzymatic events at the replication fork, many mechanistic questions remain unanswered. In particular, how conformational and structural changes within replisomes are coupled to replication-fork progression remains largely unknown. The studies discussed in this article demonstrate that a complex network of interactions between different replisome components regulates each stage of lagging-strand synthesis. However, detailed kinetic studies are needed to clarify the timing and importance of each interaction within active replication complexes. For example, although primer synthesis has been shown to cause polymerase pausing events and the formation of small priming loops, the importance of these events and the conformational states involved still need to be determined. Likewise, the mechanism by which newly synthesized primers are handed off to polymerases on the lagging strand remains obscure.

Until recently, the replisome was thought to operate as a stable multiprotein machine with a defined structure throughout replication. Now it is clear that replisome architecture is actually highly dynamic with individual components rapidly exchanging with their counterparts in solution. In particular, during lagging-strand synthesis, primase, SSB/RPA, the lagging-strand replicative polymerase, and the enzymes that remove the RNA primers and ligate the new Okazaki fragments are known to dissociate and turnover locally or to be recycled into the cellular pool. The turnover frequency of clamp-loader components is not known, but recent work has shown that leading-strand polymerases frequently exchange in contrast to previously proposed models (Loparo et al. 2011). Sliding clamps are recruited continuously during lagging-strand synthesis, with single molecule in vitro analysis showing that they can be recycled locally (Tanner et al. 2011), but the situation in vivo remains unclear. The influence of these turnover events on replication-fork progression is not fully understood. However, subunit exchange is emerging as a general feature of the action of many large multiprotein complexes, as evidenced by the demonstration that flagellar motor components are continually turned over as the bacterial flagella rotates (Delalez et al. 2010; Li and Xie 2011).

Although the importance of continual subunit exchange for efficient replisome operation is not known in all cases, these events appear to play a critical role when the replisome encounters obstacles. During replication, DNA lesions and other barriers to fork progression lead to replication stalling or breaking of the template. In response to these events, a range of different machineries can be recruited specifically to the replisome through association with the sliding clamp, SSB/RPA, or single-stranded DNA or though recognition of other structures at the stalled or broken fork. Recently, many novel mechanisms of replication restart have been discovered, revealing a web of possible reaction pathways to overcome these obstacles (Heller and Marians 2006; Lopes et al. 2006; Pomerantz and O'Donnell 2008; Yeeles and Marians 2011). In particular, specialized translesion polymerases, which are capable of synthesizing across damaged DNA bases (Fujii and Fuchs 2004), enter the replisome through exchange events and allow for lesion bypass. Even though these specialized polymerases are capable of bypassing DNA damage, they synthesize at slower rates and introduce more errors making them poorly suited for normal replication. To alleviate this problem, these specialized polymerases exchange after bypass, returning the replisome to normal operation.

The single-molecule in vivo and in vitro techniques highlighted in this article provide a powerful set of tools to address complex mechanistic questions about replisome function. In vivo imaging allows for the direct visualization of single replication complexes during operation under the most physiologically relevant conditions. These techniques have been used to determine replisome composition in *E. coli* cells, revealing for the first time that the replisome operates with three polymerases, one more than is required to synthesize the two daughter strands. In vivo imaging methods have also been used to visualize lagging-strand polymerase exchange events within active replisome complexes, demonstrating that highly dynamic processes can be probed (Lia et al. 2012). However, many remaining questions about replisome operation cannot be addressed using in vivo techniques alone because of the difficulty of directly monitoring synthesis rate, the introduction of components with severe mutations or truncations, and the lack of control over reaction conditions. In vitro single-molecule approaches are ideal for these types of situations because both the replication rate and replisome composition can be probed simultaneously and a wide range of reaction conditions can be tested (Loparo et al. 2011; Tanner et al. 2011). The combination of both in vivo and in vitro techniques discussed in this article promise to address a wide range of complex mechanistic questions about DNA replication previously inaccessible using traditional approaches.

ACKNOWLEDGMENTS

Work in the Sherratt laboratory is funded by a Wellcome Trust Programme grant to D.J.S. and by a Todd-Bird Junior Research Fellowship of New College (Oxford) to R.R.-L. The research on DNA replication in the van Oijen group is funded by an ERC Starting grant (281098; SINGLEREPLISOME), an NWO Vici grant (680-47-607) to A.M.vO. and a Human Frontier Science Program Fellowship to K.E.D.

REFERENCES

Alexandrov AI, Botchan MR, Cozzarelli NR. 2002. Characterization of simian virus 40 T-antigen double hexamers bound to a replication fork: The active form of the helicase. *J Biol Chem* **277:** 44886–44897.

Bailey S, Eliason WK, Steitz TA. 2007. Structure of hexameric DnaB helicase and its complex with a domain of DnaG primase. *Science* **318:** 459–463.

Bell SP, Dutta A. 2002. DNA replication in eukaryotic cells. *Annu Rev Biochem* **71:** 333–374.

Blackwell LJ, Borowiec JA. 1994. Human replication protein A binds single-stranded DNA in two distinct complexes. *Mol Cell Biol* **14:** 3993–4001.

Breier AM, Weier H-UG, Cozzarelli NR. 2005. Independence of replisomes in *Escherichia coli* chromosomal replication. *Proc Natl Acad Sci* **102:** 3942–3947.

Carver TE Jr, Sexton DJ, Benkovic SJ. 1997. Dissociation of bacteriophage T4 DNA polymerase and its processivity clamp after completion of Okazaki fragment synthesis. *Biochemistry* **36:** 14409–14417.

Chandler M, Bird RE, Caro L. 1975. The replication time of the *Escherichia coli* K12 chromosome as a function of cell doubling time. *J Mol Biol* **94:** 127–132.

Chastain PD II, Makhov AM, Nossal NG, Griffith J. 2000. Analysis of the Okazaki fragment distributions along single long DNAs replicated by the bacteriophage T4 proteins. *Mol Cell* **6:** 803–814.

Chastain PD II, Makhov AM, Nossal NG, Griffith J. 2003. Architecture of the replication complex and DNA loops at the fork generated by the bacteriophage t4 proteins. *J Biol Chem* **278:** 21276–21285.

Cook PR. 1999. The organization of replication and transcription. *Science* **284:** 1790–1795.

Corn JE, Berger JM. 2006. Regulation of bacterial priming and daughter strand synthesis through helicase-primase interactions. *Nucleic Acids Res* **34:** 4082–4088.

Corn JE, Pease PJ, Hura GL, Berger JM. 2005. Crosstalk between primase subunits can act to regulate primer synthesis in trans. *Mol Cell* **20:** 391–401.

Cox MM, Goodman MF, Kreuzer KN, Sherratt DJ, Sandler SJ, Marians KJ. 2000. The importance of repairing stalled replication forks. *Nature* **404:** 37–41.

Cseresnyes Z, Schwarz U, Green CM. 2009. Analysis of replication factories in human cells by super-resolution light microscopy. *BMC Cell Biol* **10:** 88.

Debyser Z, Tabor S, Richardson CC. 1994. Coordination of leading and lagging strand DNA synthesis at the replication fork of bacteriophage T7. *Cell* **77:** 157–166.

Delalez NJ, Wadhams GH, Rosser G, Xue Q, Brown MT, Dobbie IM, Berry RM, Leake MC, Armitage JP. 2010. Signal-dependent turnover of the bacterial flagellar switch protein FliM. *Proc Natl Acad Sci* **107:** 11347–11351.

Dingman CW. 1974. Bidirectional chromosome replication: Some topological considerations. *J Theor Biol* **43:** 187–195.

Dohrmann PR, Manhart CM, Downey CD, McHenry CS. 2011. The rate of polymerase release upon filling the gap between Okazaki fragments is inadequate to support cycling during lagging strand synthesis. *J Mol Biol* **414:** 15–27.

Doksani Y, Bermejo R, Fiorani S, Haber JE, Foiani M. 2009. Replicon dynamics, dormant origin firing, and terminal fork integrity after double-strand break formation. *Cell* **137:** 247–258.

Drake JW, Allen EF, Forsberg SA, Preparata RM, Greening EO. 1969. Genetic control of mutation rates in bacteriophageT4. *Nature* **221:** 1128–1132.

Duderstadt KE, Berger JM. 2008. AAA^+ ATPases in the initiation of DNA replication. *Crit Rev Biochem Mol Biol* **43:** 163–187.

Duggin IG, McCallum SA, Bell SD. 2008. Chromosome replication dynamics in the archaeon *Sulfolobus acidocaldarius*. *Proc Natl Acad Sci* **105:** 16737–16742.

Edenberg HJ, Huberman JA. 1975. Eukaryotic chromosome replication. *Annu Rev Genet* **9:** 245–284.

Evrin C, Clarke P, Zech J, Lurz R, Sun J, Uhle S, Li H, Stillman B, Speck C. 2009. A double-hexameric MCM2–7 complex is loaded onto origin DNA during licensing of eukaryotic DNA replication. *Proc Natl Acad Sci* **106:** 20240–20245.

Frick DN, Richardson CC. 2001. DNA primases. *Annu Rev Biochem* **70:** 39–80.

Frick DN, Kumar S, Richardson CC. 1999. Interaction of ribonucleoside triphosphates with the gene 4 primase of bacteriophage T7. *J Biol Chem* **274:** 35899–35907.

Fujii S, Fuchs RP. 2004. Defining the position of the switches between replicative and bypass DNA polymerases. *EMBO J* **23:** 4342–4352.

Gambus A, van Deursen F, Polychronopoulos D, Foltman M, Jones RC, Edmondson RD, Calzada A, Labib K. 2009. A key role for Ctf4 in coupling the MCM2–7 helicase to DNA polymerase alpha within the eukaryotic replisome. *EMBO J* **28:** 2992–3004.

Georgescu RE, Kurth I, Yao NY, Stewart J, Yurieva O, O'Donnell M. 2009. Mechanism of polymerase collision release from sliding clamps on the lagging strand. *EMBO J* **28:** 2981–2991.

Georgescu RE, Kurth I, O'Donnell ME. 2012. Single-molecule studies reveal the function of a third polymerase in the replisome. *Nat Struct Mol Biol* **19:** 113–116.

Hacker KJ, Alberts BM. 1994. The rapid dissociation of the T4 DNA polymerase holoenzyme when stopped by a DNA hairpin helix: A model for polymerase release following the termination of each Okazaki fragment. *J Biol Chem* **269:** 24221–24228.

Hamdan SM, Loparo JJ, Takahashi M, Richardson CC, van Oijen AM. 2009. Dynamics of DNA replication loops reveal temporal control of lagging-strand synthesis. *Nature* **457:** 336–339.

Hearst JE, Kauffman L, McClain WM. 1998. A simple mechanism for the avoidance of entanglement during chromosome replication. *Trends Genet* **14:** 244–247.

Heller RC, Marians KJ. 2006. Replication fork reactivation downstream of a blocked nascent leading strand. *Nature* **439:** 557–562.

Indiani C, Langston LD, Yurieva O, Goodman MF, O'Donnell M. 2009. Translesion DNA polymerases remodel the replisome and alter the speed of the replicative helicase. *Proc Natl Acad Sci U S A* **106:** 6031–6038.

Jacob F, Brenner S, Cuzin F. 1963. On the regulation of DNA replication in bacteria. *Cold Spring Harbor Symp Quant Biol* **28:** 329–347.

Johnson DS, Bai L, Smith BY, Patel SS, Wang MD. 2007. Single-molecule studies reveal dynamics of DNA unwinding by the ring-shaped T7 helicase. *Cell* **129:** 1299–1309.

Katayama T, Ozaki S, Keyamura K, Fujimitsu K. 2010. Regulation of the replication cycle: Conserved and diverse regulatory systems for DnaA and oriC. *Nat Rev Microbiol* **8:** 163–170.

Kato M, Ito T, Wagner G, Ellenberger T. 2004. A molecular handoff between bacteriophage T7 DNA primase and T7 DNA polymerase initiates DNA synthesis. *J Biol Chem* **279:** 30554–30562.

Kelman Z, O'Donnell M. 1995. DNA polymerase III holoenzyme: Structure and function of a chromosomal replicating machine. *Annu Rev Biochem* **64:** 171–200.

Kerr ID, Wadsworth RI, Cubeddu L, Blankenfeldt W, Naismith JH, White MF. 2003. Insights into ssDNA recognition by the OB fold from a structural and thermodynamic study of Sulfolobus SSB protein. *EMBO J* **22:** 2561–2570.

Kim YT, Tabor S, Bortner C, Griffith JD, Richardson CC. 1992. Purification and characterization of the bacteriophage T7 gene 2.5 protein: A single-stranded DNA-binding protein. *J Biol Chem* **267:** 15022–15031.

Kitamura E, Blow JJ, Tanaka TU. 2006. Live-cell imaging reveals replication of individual replicons in eukaryotic replication factories. *Cell* **125:** 1297–1308.

Kornberg A, Baker T. 1992. *DNA replication.* Freeman, New York.

Kuchta RD, Stengel G. 2010. Mechanism and evolution of DNA primases. *Biochim Biophys Acta* **1804:** 1180–1189.

Kumaran S, Kozlov AG, Lohman TM. 2006. *Saccharomyces cerevisiae* replication protein A binds to single-stranded DNA in multiple salt-dependent modes. *Biochemistry* **45:** 11958–11973.

Kusakabe T, Richardson CC. 1997. Template recognition and ribonucleotide specificity of the DNA primase of bacteriophage T7. *J Biol Chem* **272:** 5943–5951.

Langston LD, Indiani C, O'Donnell M. 2009. Whither the replisome: Emerging perspectives on the dynamic nature of the DNA replication machinery. *Cell Cycle* **8:** 2686–2691.

Lee SJ, Richardson CC. 2002. Interaction of adjacent primase domains within the hexameric gene 4 helicase-primase of bacteriophage T7. *Proc Natl Acad Sci* **99:** 12703–12708.

Lee J, Chastain PD II, Griffith JD, Richardson CC. 2002. Lagging strand synthesis in coordinated DNA synthesis by bacteriophage T7 replication proteins. *J Mol Biol* **316:** 19–34.

Lee JB, Hite RK, Hamdan SM, Xie XS, Richardson CC, van Oijen AM. 2006. DNA primase acts as a molecular brake in DNA replication. *Nature* **439:** 621–624.

Lemon KP, Grossman AD. 1998. Localization of bacterial DNA polymerase: Evidence for a factory model of replication. *Science* **282:** 1516–1519.

Lemon KP, Grossman AD. 2000. Movement of replicating DNA through a stationary replisome. *Mol Cell* **6:** 1321–1330.

Li X, Marians KJ. 2000. Two distinct triggers for cycling of the lagging strand polymerase at the replication fork. *J Biol Chem* **275:** 34757–34765.

Li GW, Xie XS. 2011. Central dogma at the single-molecule level in living cells. *Nature* **475:** 308–315.

Lia G, Michel B, Allemand JF. 2012. Polymerase exchange during Okazaki fragment synthesis observed in living cells. *Science* **335:** 328–331.

Liachko I, Bhaskar A, Lee C, Chung SC, Tye BK, Keich U. 2010. A comprehensive genome-wide map of autonomously replicating sequences in a naive genome. *PLoS Genet* **6:** e1000946.

Lionnet T, Spiering MM, Benkovic SJ, Bensimon D, Croquette V. 2007. Real-time observation of bacteriophage T4 gp41 helicase reveals an unwinding mechanism. *Proc Natl Acad Sci* **104:** 19790–19795.

Lippincott-Schwartz J, Snapp E, Kenworthy A. 2001. Studying protein dynamics in living cells. *Nat Rev Mol Cell Biol* **2:** 444–456.

Loparo JJ, Kulczyk AW, Richardson CC, van Oijen AM. 2011. Simultaneous single-molecule measurements of phage T7 replisome composition and function reveal the mechanism of polymerase exchange. *Proc Natl Acad Sci* **108:** 3584–3589.

Lopes M, Foiani M, Sogo JM. 2006. Multiple mechanisms control chromosome integrity after replication fork uncoupling and restart at irreparable UV lesions. *Mol Cell* **21:** 15–27.

Lopez de Saro FJ, Georgescu RE, O'Donnell M. 2003. A peptide switch regulates DNA polymerase processivity. *Proc Natl Acad Sci* **100:** 14689–14694.

Luo G, Wang M, Konigsberg WH, Xie XS. 2007. Single-molecule and ensemble fluorescence assays for a functionally important conformational change in T7 DNA polymerase. *Proc Natl Acad Sci* **104:** 12610–12615.

Maga G, Stucki M, Spadari S, Hubscher U. 2000. DNA polymerase switching: I. Replication factor C displaces DNA polymerase alpha prior to PCNA loading. *J Mol Biol* **295:** 791–801.

Maier B, Bensimon D, Croquette V. 2000. Replication by a single DNA polymerase of a stretched single-stranded DNA. *Proc Natl Acad Sci* **97:** 12002–12007.

Manosas M, Spiering MM, Zhuang Z, Benkovic SJ, Croquette V. 2009. Coupling DNA unwinding activity with primer synthesis in the bacteriophage T4 primosome. *Nat Chem Biol* **5:** 904–912.

Marceau AH, Bahng S, Massoni SC, George NP, Sandler SJ, Marians KJ, Keck JL. 2011. Structure of the SSB-DNA polymerase III interface and its role in DNA replication. *EMBO J* **30:** 4236–4247.

Marinsek N, Barry ER, Makarova KS, Dionne I, Koonin EV, Bell SD. 2006. GINS, a central nexus in the archaeal DNA replication fork. *EMBO Rep* **7:** 539–545.

Masai H, Matsumoto S, You Z, Yoshizawa-Sugata N, Oda M. 2010. Eukaryotic chromosome DNA replication: Where, when, and how? *Annu Rev Biochem* **79:** 89–130.

Matsunaga F, Norais C, Forterre P, Myllykallio H. 2003. Identification of short "eukaryotic" Okazaki fragments synthesized from a prokaryotic replication origin. *EMBO Rep* **4:** 154–158.

Méchali M. 2010. Eukaryotic DNA replication origins: Many choices for appropriate answers. *Nat Rev Mol Cell Biol* **11:** 728–738.

Mott ML, Berger JM. 2007. DNA replication initiation: Mechanisms and regulation in bacteria. *Nat Rev Microbiol* **5:** 343–354.

Myllykallio H, Lopez P, Lopez-Garcia P, Heilig R, Saurin W, Zivanovic Y, Philippe H, Forterre P. 2000. Bacterial mode of replication with eukaryotic-like machinery in a hyperthermophilic archaeon. *Science* **288:** 2212–2215.

Nordstrom K. 2006. Plasmid R1—replication and its control. *Plasmid* **55:** 1–26.

Pandey M, Syed S, Donmez I, Patel G, Ha T, Patel SS. 2009. Coordinating DNA replication by means of priming loop and differential synthesis rate. *Nature* **462:** 940–943.

Park K, Debyser Z, Tabor S, Richardson CC, Griffith JD. 1998. Formation of a DNA loop at the replication fork generated by bacteriophage T7 replication proteins. *J Biol Chem* **273:** 5260–5270.

Patterson G, Davidson M, Manley S, Lippincott-Schwartz J. 2010. Superresolution imaging using single-molecule localization. *Annu Rev Phys Chem* **61:** 345–367.

Pomerantz RT, O'Donnell M. 2008. The replisome uses mRNA as a primer after colliding with RNA polymerase. *Nature* **456:** 762–766.

Qimron U, Lee SJ, Hamdan SM, Richardson CC. 2006. Primer initiation and extension by T7 DNA primase. *EMBO J* **25:** 2199–2208.

Rabkin SD, Richardson CC. 1990. In vivo analysis of the initiation of bacteriophage T7 DNA replication. *Virology* **174:** 585–592.

Raghuraman MK, Winzeler EA, Collingwood D, Hunt S, Wodicka L, Conway A, Lockhart DJ, Davis RW, Brewer BJ, Fangman WL. 2001. Replication dynamics of the yeast genome. *Science* **294:** 115–121.

Remus D, Beuron F, Tolun G, Griffith JD, Morris EP, Diffley JF. 2009. Concerted loading of Mcm2–7 double hexamers around DNA during DNA replication origin licensing. *Cell* **139:** 719–730.

Reyes-Lamothe R, Possoz C, Danilova O, Sherratt DJ. 2008. Independent positioning and action of *Escherichia coli* replisomes in live cells. *Cell* **133:** 90–102.

Reyes-Lamothe R, Sherratt DJ, Leake MC. 2010. Stoichiometry and architecture of active DNA replication machinery in *Escherichia coli*. *Science* **328:** 498–501.

Ribeck N, Saleh OA. 2008. Multiplexed single-molecule measurements with magnetic tweezers. *Rev Sci Instrum* **79:** 094301.

Smiley RD, Zhuang Z, Benkovic SJ, Hammes GG. 2006. Single-molecule investigation of the T4 bacteriophage DNA polymerase holoenzyme: Multiple pathways of holoenzyme formation. *Biochemistry* **45:** 7990–7997.

Smith DJ, Whitehouse I. 2012. Intrinsic coupling of lagging-strand synthesis to chromatin assembly. *Nature* **483:** 434–438.

Solovjeva L, Svetlova M, Sasina L, Tanaka K, Saijo M, Nazarov I, Bradbury M., 2005. High mobility of flap endonuclease 1 and DNA polymerase eta associated with replication foci in mammalian S-phase nucleus. *Mol Biol Cell* **16:** 2518–2528.

Sporbert A, Gahl A, Ankerhold R, Leonhardt H, Cardoso MC. 2002. DNA polymerase clamp shows little turnover at established replication sites but sequential de novo assembly at adjacent origin clusters. *Mol Cell* **10:** 1355–1365.

Su'etsugu M, Errington J. 2011. The replicase sliding clamp dynamically accumulates behind progressing replication forks in *Bacillus subtilis* cells. *Mol Cell* **41:** 720–732.

Sun B, Johnson DS, Patel G, Smith BY, Pandey M, Patel SS, Wang MD. 2011. ATP-induced helicase slippage reveals highly coordinated subunits. *Nature* **478:** 132–135.

Tanaka H, Katou Y, Yagura M, Saitoh K, Itoh T, Araki H, Bando M, Shirahige K. 2009. Ctf4 coordinates the progression of helicase and DNA polymerase alpha. *Genes Cells* **14:** 807–820.

Tanner NA, van Oijen AM. 2009. Single-molecule observation of prokaryotic DNA replication. *Methods Mol Biol* **521:** 397–410.

Tanner NA, Hamdan SM, Jergic S, Loscha KV, Schaeffer PM, Dixon NE, van Oijen AM. 2008. Single-molecule studies of fork dynamics in *Escherichia coli* DNA replication. *Nat Struct Mol Biol* **15:** 998.

Tanner NA, Loparo JJ, Hamdan SM, Jergic S, Dixon NE, van Oijen AM. 2009. Real-time single-molecule observation of rolling-circle DNA replication. *Nucleic Acids Res* **37:** e27.

Tanner NA, Tolun G, Loparo JJ, Jergic S, Griffith JD, Dixon NE, van Oijen AM. 2011. *E. coli* DNA replication in the absence of free beta clamps. *EMBO J* **30:** 1830–1840.

Tokunaga M, Imamoto N, Sakata-Sogawa K. 2008. Highly inclined thin illumination enables clear single-molecule imaging in cells. *Nat Methods* **5:** 159–161.

Tougu K, Marians KJ. 1996. The interaction between helicase and primase sets the replication fork clock. *J Biol Chem* **271:** 21398–21405.

Valle M, Gruss C, Halmer L, Carazo JM, Donate LE. 2000. Large T-antigen double hexamers imaged at the simian virus 40 origin of replication. *Mol Cell Biol* **20:** 34–41.

van Oijen AM, Blainey PC, Crampton DJ, Richardson CC, Ellenberger T, Xie XS. 2003. Single-molecule kinetics of λ exonuclease reveal base dependence and dynamic disorder. *Science* **301:** 1235–1238.

Waga S, Stillman B. 1998. The DNA replication fork in eukaryotic cells. *Annu Rev Biochem* **67:** 721–751.

Wang JD, Berkmen MB, Grossman AD. 2007. Genome-wide coorientation of replication and transcription reduces adverse effects on replication in *Bacillus subtilis*. *Proc Natl Acad Sci* **104:** 5608–5613.

Wang X, Reyes-Lamothe R, Sherratt DJ. 2008. Modulation of *Escherichia coli* sister chromosome cohesion by topoisomerase IV. *Genes Dev* **22:** 2426–2433.

Weisshart K, Taneja P, Jenne A, Herbig U, Simmons DT, Fanning E. 1999. Two regions of simian virus 40 T antigen determine cooperativity of double-hexamer assembly on the viral origin of DNA replication and promote hexamer interactions during bidirectional origin DNA unwinding. *J Virol* **73:** 2201–2211.

Wuite GJ, Smith SB, Young M, Keller D, Bustamante C. 2000. Single-molecule studies of the effect of template tension on T7 DNA polymerase activity. *Nature* **404:** 103–106.

Xi J, Zhang Z, Zhuang Z, Yang J, Spiering MM, Hammes GG, Benkovic SJ. 2005a. Interaction between the T4 helicase loading protein (gp59) and the DNA polymerase (gp43): Unlocking of the gp59-gp43-DNA complex to initiate assembly of a fully functional replisome. *Biochemistry* **44:** 7747–7756.

Xi J, Zhuang Z, Zhang Z, Selzer T, Spiering MM, Hammes GG, Benkovic SJ. 2005b. Interaction between the T4 helicase-loading protein (gp59) and the DNA polymerase (gp43): A locking mechanism to delay replication during replisome assembly (erratum *Biochemistry* [2005] **44:** 12264; **44:** 4600). *Biochemistry* **44:** 2305–2318.

Xie XS, Choi PJ, Li GW, Lee NK, Lia G. 2008. Single-molecule approach to molecular biology in living bacterial cells. *Annu Rev Biophys* **37:** 417–444.

Yang J, Nelson SW, Benkovic SJ. 2006. The control mechanism for lagging strand polymerase recycling during bacteriophage T4 DNA replication. *Mol Cell* **21:** 153–164.

Yao NY, O'Donnell M. 2010. SnapShot: The replisome. *Cell* **141:** 1088.

Yao NY, Georgescu RE, Finkelstein J, O'Donnell ME. 2009. Single-molecule analysis reveals that the lagging strand increases replisome processivity but slows replication fork progression. *Proc Natl Acad Sci* **106:** 13236–13241.

Yardimci H, Loveland AB, Habuchi S, van Oijen AM, Walter JC. 2010. Uncoupling of sister replisomes during eukaryotic DNA replication. *Mol Cell* **40:** 834–840.

Yeeles JT, Marians KJ. 2011. The *Escherichia coli* replisome is inherently DNA damage tolerant. *Science* **334:** 235–238.

Yuzhakov A, Kelman Z, Hurwitz J, O'Donnell M. 1999a. Multiple competition reactions for RPA order the assembly of the DNA polymerase delta holoenzyme. *EMBO J* **18:** 6189–6199.

Yuzhakov A, Kelman Z, O'Donnell M. 1999b. Trading places on DNA—a three-point switch underlies primer handoff from primase to the replicative DNA polymerase. *Cell* **96:** 153–163.

Zhang Z, Spiering MM, Trakselis MA, Ishmael FT, Xi J, Benkovic SJ, Hammes GG. 2005. Assembly of the bacteriophage T4 primosome: Single-molecule and ensemble studies. *Proc Natl Acad Sci* **102:** 3254–3259.

Replication Clamps and Clamp Loaders

Mark Hedglin, Ravindra Kumar, and Stephen J. Benkovic

Department of Chemistry, The Pennsylvania State University, University Park, Pennsylvania 16802

Correspondence: sjb1@psu.edu

To achieve the high degree of processivity required for DNA replication, DNA polymerases associate with ring-shaped sliding clamps that encircle the template DNA and slide freely along it. The closed circular structure of sliding clamps necessitates an enzyme-catalyzed mechanism, which not only opens them for assembly and closes them around DNA, but specifically targets them to sites where DNA synthesis is initiated and orients them correctly for replication. Such a feat is performed by multisubunit complexes known as clamp loaders, which use ATP to open sliding clamp rings and place them around the 3′ end of primer–template (PT) junctions. Here we discuss the structure and composition of sliding clamps and clamp loaders from the three domains of life as well as T4 bacteriophage, and provide our current understanding of the clamp-loading process.

During each round of DNA replication, thousands to billions of nucleotides must be faithfully copied in a short period of time. However, by themselves, replicative DNA polymerases are distributive, synthesizing only ten or so nucleotides of complementary DNA before dissociating. To achieve the high degree of processivity required for efficient DNA replication, replicative DNA polymerases associate with ring-shaped sliding clamps that encircle the template DNA and slide freely along it. Such an association effectively tethers the polymerase to DNA, substantially increasing the amount of continuous replication. The closed circular structure of sliding clamps necessitates an enzyme-catalyzed mechanism, which not only opens them for assembly and closes them around DNA, but specifically targets them to sites where DNA synthesis is initiated and orients them correctly for interaction with DNA polymerases. Such a feat is performed by multisubunit complexes known as clamp loaders, which use ATP to open sliding clamp rings and place them around the 3′ end of primer–template (PT) junctions. Here we discuss the structure and composition of sliding clamps and clamp loaders from the three domains of life as well as T4 bacteriophage, and provide our current understanding of the clamp-loading process.

SLIDING CLAMPS

Crystal Structures

Based on early biochemical assays, it had been predicted that sliding clamps encircle DNA in a sequence-independent manner that allows them to slide freely along DNA (Stukenberg et al. 1991). In 1992, this hypothesis was

Cite this article as *Cold Spring Harb Perspect Biol* doi: 10.1101/cshperspect.a010165

confirmed when the first crystal structure of a sliding clamp, *E. coli* β clamp, was solved (Fig. 1A, left). In the structure, two semicircular β subunits are aligned head-to-tail, forming a ring with an inner diameter large enough to accommodate duplex DNA. As a result of the head-to-tail arrangement, the ring has two structurally distinct faces, one of which, the "carboxy-terminal face," contains both carboxyl termini and serves as a platform for interaction with other proteins, including polymerases and the clamp loader, as discussed below. Each β subunit consists of three independent domains connected by a long loop, referred to as the interdomain connecting loop (IDCL), which protrudes from the carboxy-terminal face. Although the amino acid sequence of each domain is quite different, all fold into similar three-dimensional structures with identical chain topologies; two adjacent α-helices flanked by two, four-stranded antiparallel β sheets (Fig. 1A, middle). Thus, the β clamp has pseudo sixfold symmetry (Kong et al. 1992). The 12 α helices line the inner surface of the ring while the β sheets form a continuous outer surface. Although the β clamp has a net negative charge, a strong positive electrostatic potential lines the inner surface of the ring where water molecules mediate protein–DNA interactions, allowing the β clamp to move freely along DNA in a sequence-independent manner (Fig. 1A, right).

The functional equivalents of the β subunit in eukaryotes (proliferating cell nuclear antigen, PCNA) and T4 bacteriophage (gene 45 protein, gp45) do not share any significant sequence similarity (<10%) with the β subunit and each is approximately two-thirds its size (Kong et al. 1992; Krishna et al. 1994; Gulbis et al. 1996; Moarefi et al. 2000). Early biochemical analyses suggested that these proteins trimerize to form a six-domain ring like the β clamp (Bauer and Burgers 1988; Jarvis et al. 1989b; Yao et al. 1996). Subsequent crystallographic studies confirmed that each sliding clamp is a homotrimeric ring with each monomer contributing two nearly superimposable domains with chain topologies identical to that in the β subunit (Fig. 1B,D,E). The three monomers are aligned head-to-tail, forming a ring with two structurally

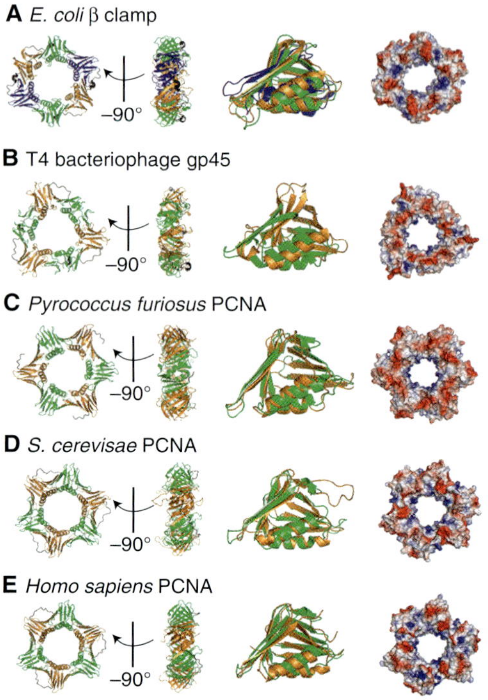

Figure 1. Sliding clamps from different organisms. Crystal structures of (*A*) the *Escherichia coli* β clamp (Protein Data Bank [PDB] code 2POL), (*B*) gene 45 protein (gp45) from T4 bacteriophage (PDB code 1CZD), (*C*) *Pyrococcus furiosus* proliferating cell nuclear antigen (PCNA, PDB code 1GE8), (*D*) *Saccharomyces cerevisiae* PCNA (PDB code 1PLQ), and (*E*) *Homo sapiens* PCNA (PDB code 1AXC). For each organism, the top and side views of the three-dimensional crystal structure are shown on the *left* and alignment of the globular domains in the *middle*. The electrostatic surface potential for each is shown on the *right* with red and blue representing negative and positive electrostatic potentials, respectively. All images were constructed using PyMOL. The sliding clamps from all domains of life have a similar architecture comprised of six domains arranged in a circle. The chain-folding topologies of each domain are the same. The *E. coli* β clamp is homodimeric (three domains per monomer), whereas T4 gp45 and PCNA from eukaryotes and archaea are each homotrimeric (two domains per monomer). The monomers are arranged head-to-tail in all clamps, resulting in structurally distinct faces. The face from which the carboxyl termini protrude ("carboxyl-terminus face") interacts with clamp loaders and polymerases. A strong positive electrostatic potential lines the inner surface of each ring where double-stranded DNA resides, favoring nonspecific interactions with the negatively charged phosphate backbone.

distinct faces, one of which contains the carboxyl termini and the IDCLs. The negatively charged outer surface of each ring is composed of continuous β sheets, whereas α helices comprise the positively charged inner surface. Each ring has a diameter large enough to completely encircle duplex DNA (Krishna et al. 1994; Gulbis et al. 1996; Moarefi et al. 2000).

Most archaeal genomes, such as *Pyrococcus furiosus* (*Pfu*), encode a homotrimeric sliding clamp (*Pfu*PCNA in Fig. 1C) similar to their evolutionary counterparts (Cann et al. 1999; Iwai et al. 2000; Matsumiya et al. 2001). However, in archaea of the crenarchaeal subdomain, three PCNA homologs have been identified (Cann and Ishino 1999; Iwai et al. 2000; Daimon et al. 2002; Dionne et al. 2003). In *Sulfolobus solfataricus*, the three PCNA homologs (ssoPCNAs 1–3) share only 8%–22% sequence identity and do not form homotrimers due to electrostatic and steric clashes (Pascal et al. 2006). Rather, they exist as monomers in solution and self-assemble into a functional heterotrimer (ssoPCNA123) in a defined order (Dionne et al. 2003; Williams et al. 2006; Hlinkova et al. 2008). Unlike its evolutionary counterparts, each interface within the ssPCNA123 heterotimer is unique, which has implications on the clamp-loading mechanism, as discussed below. However, the ring-shaped heterotrimer has an overall shape and dimensions similar to the homotrimeric sliding clamps in Figure 1 (Pascal et al. 2006; Williams et al. 2006; Hlinkova et al. 2008). Thus, despite the various subunit compositions and lack of any significant sequence homology, the three-dimensional structure of the sliding clamp has been highly conserved throughout evolution.

Solution Structures

Crystal structures are static representations of a particular conformation and may not accurately reflect the solution structure of a given protein. For sliding clamps, one or more of the subunit interfaces may dissociate because of thermal fluctuations and flexibility of the protein, causing the closed ring to transiently or permanently open in solution. Such behavior has profound implications in both the clamp-loading and unloading pathways, as discussed in future sections.

The stability of a closed sliding clamp is largely governed by the intermolecular interactions at subunit interfaces. Although the subunits of the β clamp appear to be under "spring tension" caused by strenuous bending within the β monomers, continuation of the β-sheet structures across subunit interfaces contributes six intermolecular ion pairs and four strong hydrogen bonds to each interface, suggesting that the β-clamp ring remains closed in solution (Kong et al. 1992; Jeruzalmi et al. 2001b). Indeed, biochemical analyses have shown that the closed β-clamp ring remains stable in solution, with a K_d for subunit dissociation less than 60 pM, and only opens in the presence of the clamp loader (Yao et al. 1996; Paschall et al. 2011). Similar behavior is observed for *Pfu*PCNA, where four hydrogen bonds and 10 intermolecular ion pairs are present at each subunit interface. Such enhancement of electrostatic interactions is thought to be responsible for the increased thermostability of the sliding clamp from *P. furiosus*, a hyperthermophilic archaeon that grows optimally at 100°C (Matsumiya et al. 2001, 2003). However, it is interesting to note that a recent report suggests that PCNA from *Methanosarcina acetivorans* (*Mac*), a mesophilic archaeon that grows best in moderate temperatures (25°C–40°C), may exist as a monomer in solution (Liu et al. 2011).

In contrast to β clamp and *Pfu*PCNA, various experimental approaches have collectively shown that gp45 exists exclusively in an open state in solution, with one interface opened within the plane of the ring (Soumillion et al. 1998; Alley et al. 1999b; Millar et al. 2004). Such behavior may be ascribed to differences in the architecture of the gp45 sliding clamp that weakens the binding affinity at subunit interfaces. Indeed, contacts at the subunit interfaces are far less extensive; the hydrogen bonding networks are less regular, no intermolecular ion pairs are present, and the buried surface area is roughly half that for β clamp and *Pfu*PCNA (Moarefi et al. 2000). Consequently, the K_d for subunit dissociation in gp45 (~210 nM) is approximately 3500-fold greater than that for β clamp (Yao et al. 1996).

Eukaryotic PCNA seems to exist as a closed ring in solution (Schurtenberger et al. 1998; Zhuang et al. 2006b). However, recent molecular dynamics (MD) simulations suggest that the PCNA ring spontaneously relaxes in-plane to adopt more open conformations and may transiently sample out-of-plane, spiral conformations (Kazmirski et al. 2005; Adelman et al. 2010; Tainer et al. 2010). Importantly, the barrier to reclose is essentially nonexistent and occurs on a nanosecond timescale in the absence of clamp loader, preventing DNA from entering the PCNA ring. Although the total surface area buried at each interface of eukaryotic PCNA is comparable that for β clamp and *Pfu*PCNA, only one ion pair is present at each interface compared with six in β clamp (Kong et al. 1992; Krishna et al. 1994; Gulbis et al. 1996). In agreement, the K_d for subunit dissociation in human PCNA (21 nM) is ~350-fold greater than that for β clamp (Yao et al. 1996). Thus, the closed ring structure may be favored in eukaryotic PCNA but weakened intermolecular contacts may permit spontaneous and transient opening.

CLAMP LOADER COMPOSITION AND ARCHITECTURE

Bacterial Clamp Loaders

The *E. coli* clamp loader is comprised of three different subunits, δ, δ′, and γ/τ, which assemble into a heteropentamer with stoichiometry $\delta\delta'(\gamma/\tau)_3$ (Onrust et al. 1995; Pritchard et al. 2000; Jeruzalmi et al. 2001a). Two other subunits, χ and ψ, are associated with the clamp loader but these subunits are not required for its assembly nor its clamp-loading activity and will not be discussed further (Onrust and O'Donnell 1993; Xiao et al. 1993; Naktinis et al. 1995; Olson et al. 1995; Glover and McHenry 1998; Kelman et al. 1998; Turner et al. 1999). The γ and τ subunits are encoded by the same gene. The τ subunit is the full-length product and, when present within the clamp loader, binds the DNA polymerase III core and helicase via its carboxyl terminus, effectively trimerizing the polymerase at the replication fork. However, the carboxyl terminus of the τ subunit is not required for clamp loading and its truncation by a translational frameshift produces the γ subunit (Leu et al. 2003; Reyes-Lamothe et al. 2010). Thus, clamp loader comprised of $\delta\delta'\gamma_3$ has been predominantly used in in vitro studies on the bacterial clamp loader and is referred to as the minimal clamp loader or simply the γ complex. Herein, our description of the bacterial clamp loader refers only to the γ complex. Each subunit of the γ complex has the same general architecture consisting of three distinct domains (Fig. 2A). The amino-terminal domain (Domain I) is a RecA-Type ATPase domain containing Walker A (P-Loop), Walker B, sensor 1, and Ser-Arg-Cys (SRC) motifs (Guenther et al. 1997; Jeruzalmi et al. 2001a; Kazmirski et al. 2004). Walker A and B motifs are common among NTPases and are important for nucleotide binding and hydrolysis (Walker et al. 1982). The sensor 1 motif functions to monitor and relay information regarding the state of bound ATP through structural rearrangements of the γ complex. The highly conserved arginine residue of the SRC motif functions as an "arginine finger" within an interfacial ATPase site, discussed further below (Davey et al. 2002). Domain II contains a sensor 2 motif that is analogous to sensor 1. Together, Domains I and II form an ATP-binding module, referred to as an AAA^+ module, which is structurally conserved among members of the AAA^+ superfamily of ATPases. The carboxy-terminal domains (Domain III) are unique to clamp loaders among AAA^+ ATPases and pack together to form a ring-shaped "collar" that holds the complex together (Fig. 2B). The five subunits are arranged such that the γ subunits are adjacent to one another, in positions B–D, with the δ and δ′ subunits flanking each side, in the A and E positions, respectively. The amino-terminal domains are arranged in an open "horseshoe" below the carboxy-terminal collar with a gap present between the AAA^+ modules of the δ(A) and δ′(E) subunits to allow entry of PT DNA into the complex, as discussed further below. This arrangement positions the SRC

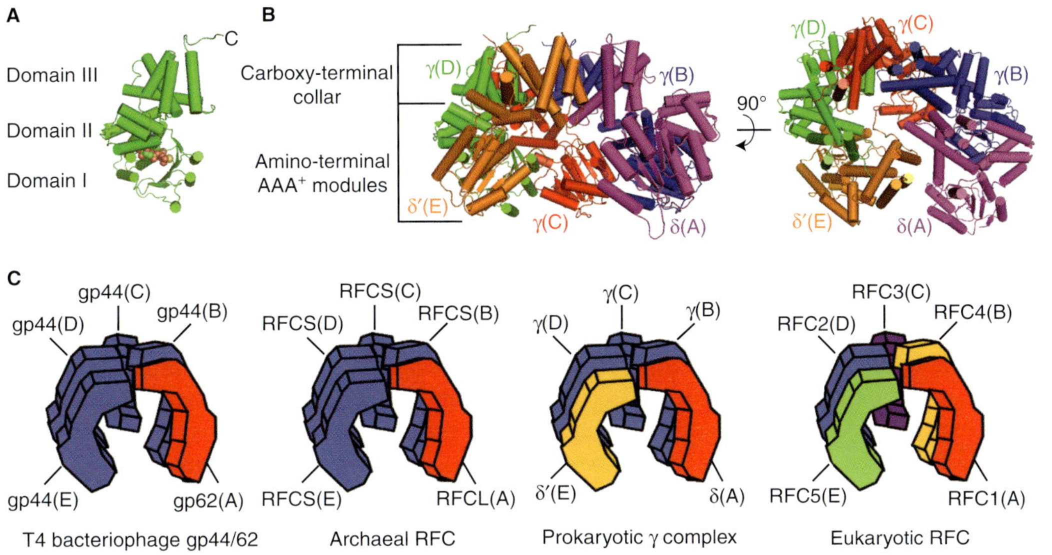

Figure 2. Clamp loader structure. (*A*) Close-up of the *E. coli* γ_D subunit (from PDB code 1XXH) bound to ATPγS (shown in space-filling format). The carboxyl terminus and the three domains (I, II, and III) are indicated. (*B*) Front (*left*) and carboxy-terminal/top (*right*) views of the $\gamma_3\delta\delta'$ minimal clamp loader structure (PDB code 1XXH). The "carboxy-terminal collar" and amino-terminal AAA$^+$ ATPase modules are indicated. All images were generated using PyMOL. (*C*) Stylized images of the clamp loaders from bacteria (γ complex), eukaryotes (RFC), bacteriophage T4 (gp44/62), and archaea (RFC). The positions of the five subunits of each clamp loader complex are denoted by the letters A–E, with analogous subunits indicated by matching colors.

motif of a given subunit near the ATP-binding module of the next such that each ATPase site is located at a subunit interface and faces the inner surface of the complex (Fig. 2A). However, only the γ subunits bind ATP as the Walker A motifs of the δ and δ′ subunits are degenerate. The δ subunit also lacks an SRC motif (Xiao et al. 1995; Guenther et al. 1997; Jeruzalmi et al. 2001a; Kazmirski et al. 2004). Thus, only three functional ATPase sites are present: between subunits δ′(E) and γ(D), γ(D) and γ(C), and γ(C) and γ(B) (Tsuchihashi and Kornberg 1989; Hingorani et al. 1999; Johnson and O'Donnell 2003). The SRC motif of γ(B) does not abut a nucleotide in the δ(A) subunit.

T4 Bacteriophage Clamp Loader

In T4, four gp44 subunits and one gp62 subunit associate to form the gp44/62 clamp loader complex (Spicer et al. 1984; Jarvis et al. 1989b). The gp44 subunits contain AAA$^+$ modules and are homologous to the *E. coli* γ subunits (Spicer et al. 1984; Jarvis et al. 1989b; Zhuang et al. 2006a). However, gp62 lacks both SRC and Walker A motifs, analogous to the *E. coli* δ subunit (Spicer et al. 1984; O'Donnell et al. 1993). Thus, based on homology between the subunits of the γ complex and gp44/62, it was presumed that the subunit arrangement within gp44/62 is analogous to that in the γ complex, with the gp62 subunit occupying position A while the four gp44 subunits occupy positions B–E (Fig. 2C). This has recently been verified in a crystal structure of gp44/62 complexed to both gp45 and PT DNA. Such an arrangement provides for only three interfacial ATPase sites as the gp62 subunit in position A does not present an "arginine finger" to the gp44 subunit in position E (Kelch et al. 2011). However, it should be noted that several studies have each independently showed that gp44/62 hydrolyzes four ATP molecules during the clamp-loading process, suggesting that each gp44 subunit contains an equally active ATPase site (Jarvis et al. 1989a; Berdis and Benkovic 1996; Sexton et al. 1996;

Young et al. 1996; Zhuang et al. 2006a; Pietroni and von Hippel 2008). In the aforementioned crystal structure, ADP bound in the E subunit packs closely against a carboxy-terminal extension of the gp62 (A) collar domain, referred to as the Á domain. This domain has the same fold topology as a corresponding domain in the A subunit of eukaryotic RFC (discussed below), tempting one to speculate that a novel interfacial ATPase site may be present between the A and E subunits of both T4 and eukaryotic clamp loaders (Bowman et al. 2004; Kelch et al. 2011).

Eukaryal Clamp Loaders

The eukaryotic clamp loader, referred to as replication factor C (RFC) or activator 1, is highly conserved from yeast to humans and is composed of five distinct subunits, RFCs 1–5, all of which share a similar three-domain core architecture reminiscent of their ancestral counterparts (Burbelo et al. 1993; Cullmann et al. 1995; Uhlmann et al. 1997b; Bowman et al. 2004). RFCs 2–4 are homologous to each other and to the ATP-hydrolyzing subunits of the T4 (gp44 subunits) and *E. coli* (γ subunits) clamp loaders. Thus, each contains an AAA$^+$ module and an SRC motif (Chen et al. 1992a,b; O'Donnell et al. 1993; Li and Burgers 1994a,b; Noskov et al. 1994; Yao et al. 2003). The RFC5 subunit is analogous to the δ' subunit of the *E. coli* γ complex in that it contains an SRC motif but lacks several key residues that are conserved at the active sites of functional ATPases (Gary and Burgers 1995; Cai et al. 1998; Podust et al. 1998a; Schmidt et al. 2001; Yao et al. 2003; Bowman et al. 2004). However, it is still capable of binding ATP (Bowman et al. 2004). Collectively, the RFC2–5 subunits are often referred to as the "small subunits" and occupy positions analogous to their bacterial counterparts in the five-subunit complex (Fig. 2C); RFC5 in position E followed by RFCs 2, 3, and 4 in positions D, C, and B, respectively (Pan et al. 1993; Uhlmann et al. 1996; Cai et al. 1997; Ellison and Stillman 1998; Yao et al. 2003; Bowman et al. 2004). The remaining subunit, RFC1, occupies position A and is similar to the gp62 subunit of T4 gp44/62 in that it lacks an SRC motif but contains an A′ domain, which abuts the nucleotide bound in position E (RFC5 [Li and Burgers 1994a; Cullmann et al. 1995; Bowman et al. 2004; Kelch et al. 2011]). However, unlike gp62, RFC1 contains a consensus Walker A motif (Bunz et al. 1993; Cullmann et al. 1995; Cai et al. 1998; Podust et al. 1998a; Schmidt et al. 2001; Yao et al. 2003). Thus, four interfacial ATPase sites are present within RFC; between RFC5 (E) and RFC2 (D), RFC2 (D) and RFC3 (C), RFC3 (C) and RFC4 (B), and RFC4 (B) and RFC1 (A) (Gomes et al. 2001; Yao et al. 2003). Furthermore, the RFC1 (A) subunit also contains unique amino- and carboxy-terminal extensions beyond the three-domain core. Hence, RFC1 is often referred to as the "large subunit." The carboxy-terminal extensions are required for complex formation with the small RFC subunits whereas the amino-terminal extensions are dispensable for PCNA loading. In fact, their deletion actually increases the PCNA-loading activity of RFC and subsequent DNA synthesis by polymerase δ (Burbelo et al. 1993; Uhlmann et al. 1996, 1997a; Allen et al. 1998; Podust et al. 1998b; Gomes et al. 2000; Gomes and Burgers 2001). Apart from a minor role in DNA damage repair, the function of the amino-terminal extension remains largely unknown (Gomes et al. 2000). Interestingly, it shares homology with all known bacterial DNA ligases and several eukaryotic poly-ADP ribose polymerases (PARPs) although there is no evidence of either activity. Rather, this region of RFC1, designated the ligase homology domain, is most likely responsible for the nonspecific DNA-binding affinity of RFC, discussed further below (Bunz et al. 1993; Burbelo et al. 1993; Cullmann et al. 1995; Fotedar et al. 1996; Allen et al. 1998).

Archaeal Clamp Loaders

Similar to their eukaryotic counterparts, archaeal clamp loaders are heteropentameric complexes comprised of a single large subunit in position A and four smaller subunits in positions B–E, as shown in Figure 2C (Pisani et al. 2000; Cann et al. 2001; Seybert et al. 2002; Miyata et al. 2005; Chen et al. 2009b). Each subunit adopts the conserved three-domain core architecture (Mayanagi et al. 2001; Oyama et al. 2001; Miyata et al. 2004, 2005; Seybert et al. 2006). The large subunit, RFCL, is homologous to the carboxy-terminal portion of the large subunit of eukaryotic RFC (RFC1) and contains a conserved Walker A motif but lacks an SRC motif (Cann and Ishino 1999; Pisani et al. 2000; Cann et al. 2001; Oyama et al. 2001; Henneke et al. 2002; Chen et al. 2005). Based on low-resolution structures of RFC from *Pyrococcus furiosus*, RFCL may also contain an A′ domain, similar to yeast RFC1 and T4 gp62 (Miyata et al. 2005). In most archaea, a single small subunit (RFCS) is present, which contains both Walker A and SRC motifs, homologous to RFCs 2–4 of eukaryotic RFC (Cann and Ishino 1999; Kelman and Hurwitz 2000; Cann et al. 2001; Oyama et al. 2001; Henneke et al. 2002). However, in several archaea, such as *Methanosarcina acetivorans* (*Mac*), an additional small subunit is present, which contains an SRC motif and a degenerate Walker A motif. This subunit, referred to as RFCS2, is homologous to RFC5 from eukaryotic RFC and occupies an analogous position (E) in the five-subunit complex (Chen et al. 2005, 2009b). Thus, the five-subunit archaeal clamp loader contains four interfacial ATPase sites, similar to eukaryotic RFC (Seybert and Wigley 2004; Seybert et al. 2006). However, it should be noted that a unique heterohexameric clamp loader in which two large subunits (RFCL) are complexed to four small subunits (RFCS) has been reported for *Methanobacterium thermoautotrophicum* (Kelman and Hurwitz 2000).

CLAMP LOADERS ARE DNA-DEPENDENT ATPases

Amid the numerous nucleic acid structures within cells, clamp loaders must target their respective clamps to sites where DNA synthesis initiates. Thus, clamp loaders must have a higher affinity for PT junctions than any other nucleic acid structure. However, if the high affinity was static, clamp loaders would compete with replicative polymerases for binding to the same PT junctions. On the contrary, if their binding affinity were dynamic, switching reversibly between high- and low-affinity states, clamp loaders could associate with PT junctions to load their respective clamps and then dissociate to allow polymerase binding (Ason et al. 2000). Detailed analyses over the years have revealed that the affinity of clamp loaders for PT junctions is indeed dynamic and modulated by ATP. Recent crystal structures of the *E. coli* γ complex have provided the structural basis for such behavior.

In the absence of DNA, the γ complex binds tightly to ATP but the "arginine fingers" of the SRC motifs are held out of reach (Fig. 3A) (Hingorani et al. 1999; Johnson and O'Donnell 2003; Kazmirski et al. 2004; Thompson et al. 2009). These motifs increase the rate of ATP hydrolysis

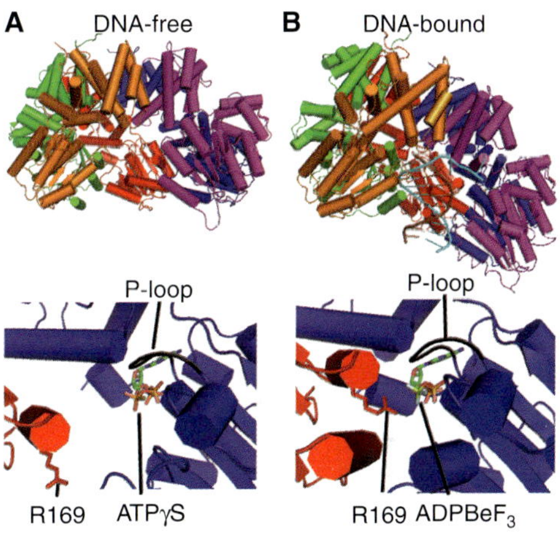

Figure 3. DNA promotes ATP hydrolysis by triggering a structural rearrangement of the interfacial ATPase sites of the γ complex. The DNA-free (*A*, PDB code 1XXH) and DNA-bound (*B*, PDB code 3GLF) forms of the γ complex are shown. For each, the structure of the γ complex is shown on *top* with a close-up of the γ_C-γ_B interfacial ATPase site shown at the *bottom*. All images were generated using PyMOL. Nucleotide is bound in the AAA$^+$ module of the γ_B subunit. The "P-loop" is indicated and shown in black. In the absence of DNA, the "arginine finger" (Arg169) from the γ_C subunit is too far away to interact with the γ phosphate of ATP in the γ_B subunit. On binding DNA, the γ complex adopts a right-handed spiral configuration with a uniform rise and rotation that closely mimics that of the bound DNA. Such a structural rearrangement presents the "arginine finger" (Arg169) from one subunit (γ_C) to the γ phosphate of ATP bound in the next subunit (γ_B), promoting ATP hydrolysis.

by stabilizing the transition state (Ahmadian et al. 1997; Kazmirski et al. 2004). Consequently, ATPase activity of the γ complex alone is negligible (Onrust et al. 1991; Hingorani et al. 1999; Ason et al. 2000). This ATP-bound form of the γ complex has high affinity for PT junctions and binds rapidly (Reems et al. 1995; Bloom et al. 1996; Hingorani and O'Donnell 1998; Ason et al. 2000). The single-stranded portion of the template-strand is presumed to enter through the gap between the AAA$^+$ modules of the δ (A) and δ′ (E) subunits, inducing a right-handed spiral conformation that closely mimics the helical geometry of the DNA duplex and "screws" onto the double-stranded portion. In this "notched screw cap" arrangement, the γ complex interacts intimately with only the template strand; conserved positively charged/polar residues within the AAA$^+$ modules contact the duplex region and track along the minor groove while the 5′ single-stranded overhang exits the central chamber and binds to the exterior surface of the δ (A) subunit. In contrast, the primer strand is mostly surface-exposed. Apart from its 3′ terminal nucleotide, which runs into the interior wall of the carboxy-terminal collar, the primer strand is disengaged from the γ complex. This lack of intimate contact is vital to the clamp loader function as it allows recognition of both RNA and DNA primers during DNA replication and repair (Simonetta et al. 2009). The "notched screw cap" arrangement of the γ complex subunits realigns the interfacial ATPase sites such that the SRC motif from one subunit is presented to the ATP binding site of the next, promoting ATP hydrolysis (Fig. 3B) (Hingorani et al. 1999). The sensor 1 and sensor 2 motifs, "sensing" the absence of the γ phosphate of ATP, subsequently trigger a conformational change converting the γ complex to a low-affinity state and promoting rapid dissociation (Ason et al. 2003; Simonetta et al. 2009; Thompson et al. 2009). On release from DNA, the γ complex can exchange ADP for ATP, allowing the dynamic DNA binding cycle to continue (Ason et al. 2000, 2003; Thompson et al. 2009).

All clamp loaders must bind and hydrolyze multiple ATP molecules to power site-specific loading of their respective clamps to PT junctions. The DNA-dependent ATPase activity discussed above for the *E. coli* γ complex is conserved among clamp loaders from T4 (Piperno et al. 1978; Rush et al. 1989; Capson et al. 1991; Munn and Alberts 1991b; Zhuang et al. 2006a), eukaryotes (Lee and Hurwitz 1990; Lee et al. 1991; Tsurimoto and Stillman 1991; Yoder and Burgers 1991; Fien and Stillman 1992; Cai et al. 1996; Ellison and Stillman 1998; Hingorani and Coman 2002; Sakato et al. 2011a), and archaea (Kelman and Hurwitz 2000; Pisani et al. 2000; Cann et al. 2001; Seybert et al. 2002; Seybert and Wigley 2004; Chen et al. 2005). Historically, it had been assumed that clamp loaders use the energy from ATP hydrolysis solely to apply mechanical force to open sliding clamps. Although ATP hydrolysis may contribute to ring opening within T4 (discussed below), detailed studies over the last 20 years or so have revealed that ATP hydrolysis primarily serves to modulate the interaction between clamp loaders and PT DNA by allowing interconversion between high- and low-affinity DNA-binding states. Such a dynamic DNA-binding mechanism specifically targets clamp loaders to PT junctions yet triggers their departure to prevent competition with replicative polymerases, as discussed further below.

Clamp-Loading Mechanism

Detailed mechanistic studies on the clamp-loading process from all domains of life have revealed a similar sequential mechanism that is outlined in Figure 4. Given the conservation of the clamp and clamp loader structures depicted in Figures 1 and 2, this is no surprise. Nonetheless, subtle differences have emerged that distinguish each system. These are discussed below.

Opening of the Sliding Clamp Ring

In the absence of ATP, clamp loaders interact with their respective clamps very weakly. On binding ATP, clamp loaders undergo a conformational change that permits optimal interaction with the carboxy-terminal face of their respective clamp and subsequent opening of the clamp ring (Fig. 4A). In *E. coli*, all subunits of

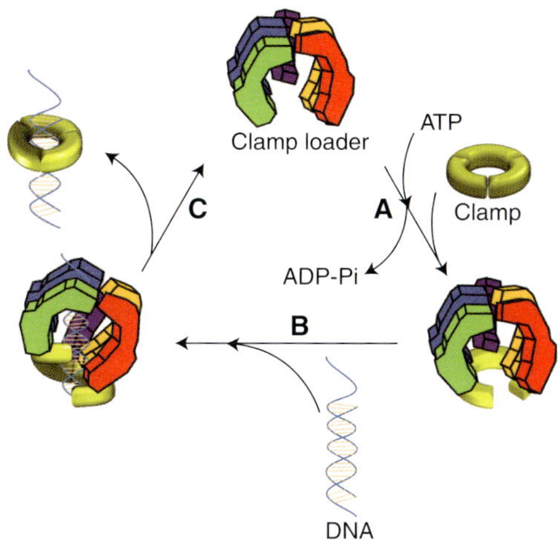

Figure 4. The clamp-loading mechanism. The catalytic reaction cycle for the loading of sliding clamps onto DNA by clamp loaders is shown as a schematic diagram. (*A*) Opening of the sliding clamp ring. In the absence of ATP, clamp loaders bind their respective clamps very weakly. On binding ATP, clamp loaders undergo a conformational change, which permits optimal interaction with the carboxy-terminal face of their respective clamp and subsequent opening of the clamp ring. (*B*) PT junction binding. The open clamp–clamp loader complex together specifically recognizes and binds a PT junction, adopting a "notched screw cap arrangement," which matches the helical geometry of the DNA duplex and properly aligns the interfacial ATPase sites for hydrolysis. (*C*) Closure of the clamp ring. On hydrolysis of ATP, clamp loaders revert back to a low-affinity DNA-binding state and eject, leaving the PT DNA positioned within an opened sliding clamp ring. Concurrent or subsequent to ejection, electrostatic interactions between the positively charged inner surface of the sliding clamp ring and the negatively charged DNA drives closure of the open sliding clamp around DNA.

the γ complex bind the β clamp but the δ subunit is the major contact point and is mainly responsible for ring opening (Naktinis et al. 1995; Turner et al. 1999; Leu et al. 2000; Jeruzalmi et al. 2001b; Leu and O'Donnell 2001; Johnson and O'Donnell 2003; Paschall et al. 2011). In the absence of ATP, the δ subunit is sequestered within the γ complex, largely because of the interaction with the nearby δ′ subunit, and the γ complex interacts with the β clamp very weakly (Onrust and O'Donnell 1993; Naktinis et al. 1995; Onrust et al. 1995; Leu and O'Donnell 2001; Indiani and O'Donnell 2003). On binding ATP, the interaction between δ and δ′ is disrupted, presenting δ for interaction with the carboxy-terminal face of a closed β clamp (Naktinis et al. 1995; Bloom et al. 1996; Goedken et al. 2004; Snyder et al. 2004; Thompson et al. 2009; Paschall et al. 2011). A "hydrophobic plug" of the δ subunit subsequently wedges into a hydrophobic pocket on either β monomer, destabilizing the adjacent dimer interface and prying open the β-clamp ring. This process is favored by the release of "spring tension" between domains within the β monomers and occurs in the absence of ATP hydrolysis (Onrust et al. 1991; Naktinis et al. 1995; Hingorani and O'Donnell 1998; Hingorani et al. 1999; Turner et al. 1999; Jeruzalmi et al. 2001b; Stewart et al. 2001; Goedken et al. 2004; Williams et al. 2004; Anderson et al. 2009; Paschall et al. 2011). Thus, ATP binding but not hydrolysis is required for γ-complex binding to and opening the β clamp. The same is true for RFC-catalyzed opening of PCNA rings in both eukaryotes (Lee and Hurwitz 1990; Shiomi et al. 2000; Gomes and Burgers 2001; Schmidt et al. 2001; Johnson et al. 2006; Yao et al. 2006; Zhuang et al. 2006b; Chen et al. 2009a; Sakato et al. 2011a,b; Thompson et al. 2011) and archaea (Henneke et al. 2002; Matsumiya et al. 2002; Seybert et al. 2002, 2006; Seybert and Wigley 2004), suggesting a similar mechanism applies to these domains of life. However, it remains to be seen whether ATP-bound RFC "actively" pries open the PCNA ring as in *E. coli* or "passively" waits for it open (Kazmirski et al. 2005; Thompson et al. 2009). Also, subtle variations in some archaea should be noted. As alluded to above, three unique interfaces are present within the *sso*PCNA123 heterotrimer of *Sulfolobus solfataricus* and only the PCNA3–PCNA1 interface is opened by *sso*RFC (Dionne et al. 2003, 2008). On the contrary, each subunit interface within a homopolymeric sliding clamp is identical and, thus, the one selected by the clamp loader for opening is random. Furthermore, a recent report suggests that, in *Methanosarcina acetivorans* (*Mac*), PCNA exists as monomers in solution and RFC functions to assemble an open

PCNA homotrimer by binding individual monomers in an ATP-dependent manner (Liu et al. 2011). Regardless of these slight differences, it is tempting to speculate that sliding clamps within bacteria, eukaryotes, and archaea are opened predominantly in-plane in the clamp–clamp loader complexes as tandem out-of-plane movements are presumed to induce the right-handed spiral conformation of clamp loaders that is competent for ATP hydrolysis.

Like RFC and the γ complex, T4 gp44/62 also undergoes an ATP-dependent conformational change that permits optimal interaction with the carboxy-terminal face of an open gp45 sliding clamp (Latham et al. 1997; Pietroni et al. 1997, 2001; Zhuang et al. 2006a; Pietroni and von Hippel 2008). However, on interaction with ATP-bound gp44/62, the open gp45 sliding clamp opens further, stimulating ATP hydrolysis at two of the four ATPase sites within gp44/62 (Mace and Alberts 1984; Rush et al. 1989; Berdis and Benkovic 1996; Sexton et al. 1996, 1998; Young et al. 1996; Alley et al. 2000; Trakselis et al. 2003). Such partial ATP hydrolysis is required to open and stabilize gp45 in an "activated" conformation that is competent for optimal loading onto DNA, in stark contrast to that described above (Sexton et al. 1996; Pietroni et al. 1997; Alley et al. 2000; Trakselis et al. 2003). This implies that the gp45 ring opens out-of plane on interacting with ATP-bound gp44/62, permitting the gp45–gp44/62 complex to adopt an intermediate, right-handed spiral conformation in which only two of the four ATP sites of gp44/62 are aligned properly for hydrolysis. Because of the unique open structure of gp45 in solution, perhaps gp44/62 uses ATP hydrolysis to passively "catch" and stabilize gp45 rings that have spontaneously adopted this productive conformation. However, it is also possible that gp44/62 uses ATP hydrolysis to "actively" sculpt gp45 rings into this intermediate conformation.

PT Junction Binding

Once formed, the open clamp–clamp loader complex together specifically recognizes and binds a PT junction (Fig. 4B) undergoing another conformational change, which is illustrated in the recent crystal structure of the gp45–gp44/62–DNA complex (Jarvis et al. 1989a; Munn and Alberts 1991b; Hockensmith et al. 1993; Pietroni et al. 2001). The PT DNA is positioned such that the duplex region is threaded through the central pore of gp45 and is sequestered within the inner chamber of gp44/62 (Fig. 5). The gp45 ring is held open by extensive interactions between its carboxy-terminal face and the entire undersurface of gp44/62 and the gp45–gp44/62 complex together adopts a "notched screw cap arrangement" that matches the helical geometry of the DNA duplex. In this orientation, all four interfacial ATPase sites are aligned properly for optimal hydrolysis (Piperno et al. 1978; Mace and Alberts 1984; Young et al. 1996; Kelch et al. 2011). Furthermore, the gap between domains in the open clamp is not wide enough to allow passage of duplex DNA. Hence, it is presumed that the single-stranded portion of the template strand enters through the gaps in the gp45 clamp and between the A′ and AAA$^+$ domains of the gp62 (A) subunit of gp44/62 and the complex "screws" onto the double-stranded portion of the DNA. Similar to

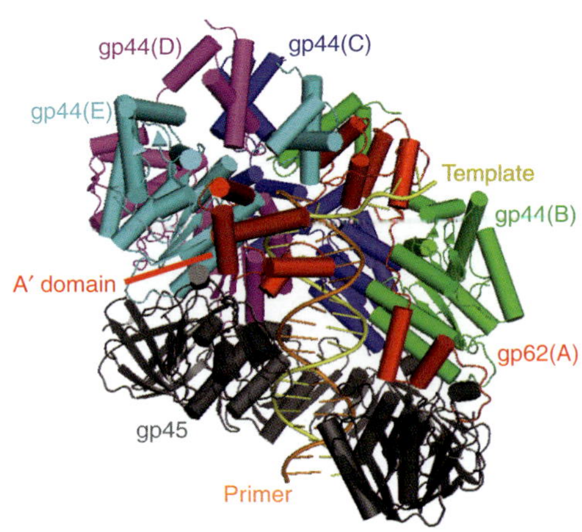

Figure 5. The clamp–clamp loader–DNA complex from T4 bacteriophage. Structure of the T4 clamp loader (gp44/62) bound to an open clamp (gp45) and PT DNA (from PDB 3U60) generated using PyMOL. PT DNA, gp44/62 (multicolored), and gp45 (gray) are shown in cartoon form. The A′ domain (red) as well as the primer (orange) and template (yellow) strands of the duplex DNA are indicated.

that observed in the crystal structure of the *E. coli* clamp loader–DNA complex described above, the closest contacts between gp44/62 and the DNA involve the template strand (Jarvis et al. 1989a; Hockensmith et al. 1993; Kelch et al. 2011). Positively-charged/polar residues from the inner surface of the clamp and the AAA$^+$ modules within the inner chamber of gp44/62 track the minor groove of the duplex region. The single-stranded template overhang protrudes through the gap between the A and A′ domains of the gp62 (A) subunit and binds to its collar domain. The 3′ end of the primer is blocked by the collar domain of the gp44$_B$ (B) subunit. This structure depicted in Figure 5 closely resembles a lower resolution, electron microscopy (EM) structure from the archaeon *Pyrococcus furiosus* in which *Pfu*RFC is bound to DNA and an open *Pfu*PCNA ring. Indeed, most of the T4 crystal structure fits well into the archaeal EM structure, showing that the open clamp–clamp loader–DNA complex is a common intermediate in the clamp-loading process for both T4 and archaea (Miyata et al. 2004, 2005; Kelch et al. 2011).

For bacteria and eukaryotes, it is suggested that the ATP-bound clamp loader holds the clamp in an open, right-handed spiral around DNA as depicted in Figure 5 but such complexes have yet to be captured. However, substantial evidence supports their existence. The DNA-dependent ATPase activity of the *E. coli* γ complex (Onrust et al. 1991; Xiao et al. 1995; Williams et al. 2004) and eukaryotic RFC (Tsurimoto and Stillman 1990; Lee et al. 1991; Podust et al. 1992; Cai et al. 1996; Ellison and Stillman 1998; Zhang et al. 1999; McNally et al. 2010) is stimulated by their respective sliding clamp, suggesting that an open clamp stabilizes the "notched screw cap" arrangement of the clamp loader on DNA. Furthermore, the experimentally-measured distance between PCNA subunits in the PCNA–RFC–DNA complex from eukaryotes (34 Å) agrees fairly well with that determined from a MD-based model (39 Å) of the open PCNA–RFC–DNA complex (Kazmirski et al. 2005; Kelch et al. 2011; Sakato et al. 2011b). Finally, the asymmetric DNA footprint of the eukaryotic PCNA–RFC complex agrees with that predicted by the MD-based model (Tsurimoto and Stillman 1991). Thus, the clamp-loading process within all domains of life likely proceeds through a common open clamp–clamp loader–DNA intermediate.

Closure of the Clamp Ring

In the "notched screw cap" arrangement of the clamp–clamp loader complex on DNA, the interfacial ATPase sites are aligned properly for optimal hydrolysis. As described above, the clamp loaders revert back to a low-affinity DNA-binding state on hydrolysis of ATP and eject (Fig. 4C), leaving the PT DNA positioned within the opened sliding clamp ring (Tsurimoto and Stillman 1991; Kaboord and Benkovic 1995, 1996; Berdis and Benkovic 1996; Zhang et al. 1999; Gomes and Burgers 2001; Stewart et al. 2001; Trakselis et al. 2003; Seybert and Wigley 2004; Johnson et al. 2006; Seybert et al. 2006; Zhuang et al. 2006b; Chen et al. 2009a; Park and O'Donnell 2009). Concurrent or subsequent to ejection of the clamp loader, electrostatic interactions between the positively charged inner surface of the sliding clamp ring and the negatively charged DNA drive contraction of the open sliding clamp around DNA. For bacteria, eukaryotes, and presumably archaea, such a process results in complete closure of the sliding clamp ring (Tinker et al. 1994; Fukuda et al. 1995; Johnson et al. 2006; Zhuang et al. 2006b; Georgescu et al. 2008; Anderson et al. 2009; McNally et al. 2010). However, the T4 gp45 ring remains partially open and T4 gp43 DNA polymerase subsequently fills the void by inserting its carboxyl terminus into the open subunit interface and docks onto the carboxy-terminal face of gp45 (Pietroni et al. 1997; Alley et al. 1999a, 2000; Latham et al. 1999; Trakselis et al. 2001). Regardless, in all domains of life, ATP hydrolysis is required for ejection of the clamp loader from DNA and closure of the sliding clamp ring. Indeed, ATP hydrolysis is necessary for association of replicative polymerases with loaded sliding clamps, a process presumed to require closure of the clamp ring as well as displacement of the clamp loader as clamp loaders and polymerases share common binding sites on sliding clamps (Piperno and Alberts

1978; Tsurimoto and Stillman 1990; Burgers 1991; Lee et al. 1991; Munn and Alberts 1991a; Fotedar et al. 1996; Mossi et al. 1997; Oku et al. 1998; Zhang et al. 1999; Gomes et al. 2001; Seybert and Wigley 2004; Liu et al. 2011). However, it should be noted that numerous studies have reported that eukaryotic RFC remains engaged with DNA throughout DNA replication (Waga and Stillman 1998; Masuda et al. 2007; Kumar et al. 2010). Although such behavior may be ascribed to nonspecific binding of RFC either to DNA through the ligase homology domain of RFC1 (Podust et al. 1998b) or to replication protein A (RPA) through RFCs 1, 4, and 5 (Waga and Stillman 1998; Yuzhakov et al. 1999), this raises the possibility that RFC may remain in the vicinity of the PT junction after ATP hydrolysis to either "hold" or reload PCNA at the primer terminus on dissociation of polymerase δ (Masuda et al. 2007; Waga and Stillman 1998). Regardless, once clamp loaders exchange ADP for ATP, the catalytic cycle may continue (Bertram et al. 1998, 2000; Hingorani and O'Donnell 1998; Pietroni et al. 2001).

The aforementioned mechanism for assembly of sliding clamps onto DNA is most likely the predominant pathway in all domains of life (Gomes et al. 2001; Thompson et al. 2009; Sakato et al. 2011b). However, the unique open structure of free gp45 clamp may also provide alternative pathways. ATP-bound gp44/62 may first bind to a PT junction and then load gp45 (Smiley et al. 2006; Zhuang et al. 2006a). For other domains of life, the geometry of the clamp loader–DNA complex likely prevents the clamp loader from productively binding a closed clamp (Thompson et al. 2009). Although this pathway is productive in T4 bacteriophage, it is most likely insignificant as binding of clamp loaders to PT DNA triggers rapid ATP hydrolysis and their subsequent dissociation such that clamp loader–DNA complexes are very short-lived (Chen et al. 2009a; Thompson et al. 2009). Alternatively, the open gp45 ring may spontaneously load onto DNA in a nonproductive manner and subsequently be "corrected" by gp44/62 before entry of gp43 polymerase. However, this pathway is highly inefficient and requires high concentrations of gp45 and/or molecular crowding (Reddy et al. 1993; Sanders et al. 1994; Smiley et al. 2006).

CLAMP UNLOADING

In a replicating cell, Okazaki fragments are in excess of sliding clamps, necessitating an efficient unloading mechanism to recycle sliding clamps during DNA replication (Yao et al. 1996; Leu et al. 2000). For example, in an *E. coli* cell, there are only about 350 β clamps for 2000–4000 Okazaki fragments. With each round of replication taking only 40 or so min, this requires each β clamp to recycle every 3.5 to 7 min (Leu et al. 2000). Similar arguments apply to other domains of life (Yao et al. 1996). In T4, the open gp45 ring is not stable on DNA and rapidly dissociates on departure of gp43 polymerase. Such a process occurs with a half-life of approximately 1.4 min, well within the allotted time window for clamp recycling. Thus, enzyme-catalyzed clamp removal is not necessary (Kaboord and Benkovic 1996; Yao et al. 1996; Soumillion et al. 1998). However, as mentioned above, the closed circular structure of sliding clamps from bacteria, eukaryotes, and archaea is very stable. Consequently, spontaneous dissociation from DNA occurs much slower than that for gp45 with half-lives of 24 and 72 min reported for eukaryotic PCNA and *E. coli* β clamp, respectively (Podust et al. 1995; Yao et al. 1996; Leu et al. 2000). Thus, sliding clamps must be catalytically recycled within these domains of life to keep up with ongoing DNA replication (Lee and Hurwitz 1990; Yao et al. 2006). Surprisingly, clamp loaders seem to perform this feat. In addition to loading their respective clamps onto DNA, clamp loaders from bacteria (γ complex), archaea (*Pfu*RFC), and eukaryotes (RFC) have been shown to remove loaded clamps from DNA in an ATP-dependent process (Naktinis et al. 1995; Cai et al. 1996; Yao et al. 1996; Shibahara and Stillman 1999; Gomes et al. 2000; Leu et al. 2000; Cann et al. 2001). Although the mechanistic details have yet to be revealed, this suggests that replicative polymerases prevent unwanted clamp unloading during ongoing replication by blocking the clamp loaders access to loaded sliding clamps.

On completion of an Okazaki fragment, the polymerase dissociates, providing an opportunity for the clamp loader to recycle the sliding clamp (Naktinis et al. 1995; Oku et al. 1998).

CONCLUDING REMARKS

The twenty-first century has been an exciting time for the study of clamps and clamp loaders. Three-dimensional structures of sliding clamps and clamp loaders from all domains of life as well as T4 bacteriophage have now been reported, showing that their basic architecture and composition has been conserved throughout evolution. Also, thorough biochemical analyses along with ground-breaking crystal structures have revealed the primary role of ATP as a modulator of the interaction between clamp loaders and PT DNA and outlined similar sequential mechanisms for the clamp-loading process within the different life forms that are sprinkled with subtle differences. However, many important questions still remain unanswered. For instance, what is the nature of the ATP-dependent conformational change within archaeal, eukaryotic, and T4 clamps loaders that permit optimal interaction with their respective clamps? Do these clamp loaders "actively" pry open sliding clamp rings like the bacterial clamp loader or "passively" wait for them to open once bound? What is the timing of clamp loader ejection and ring closure? How is the unloading activity of clamp loaders controlled during DNA replication? It will be interesting to see what answers future studies will provide for these lingering questions.

ACKNOWLEDGMENTS

We thank Dr. Jarrod French of the Benkovic Laboratory for critically reading this chapter. This work is supported by National Institutes of Health (NIH) grant GM13306 (S.J.B.). M.H. is supported by the National Cancer Institute of the NIH under Award Number F32CA165471. The content is solely the responsibility of the authors and does not necessarily represent the official views of the National Institutes of Health.

REFERENCES

Adelman JL, Chodera JD, Kuo IF, Miller TF III, Barsky D. 2010. The mechanical properties of PCNA: Implications for the loading and function of a DNA sliding clamp. *Biophys J* **98:** 3062–3069.

Ahmadian MR, Stege P, Scheffzek K, Wittinghofer A. 1997. Confirmation of the arginine-finger hypothesis for the GAP-stimulated GTP-hydrolysis reaction of Ras. *Nat Struct Biol* **4:** 686–689.

Allen BL, Uhlmann F, Gaur LK, Mulder BA, Posey KL, Jones LB, Hardin SH. 1998. DNA recognition properties of the N-terminal DNA binding domain within the large subunit of replication factor C. *Nucleic Acids Res* **26:** 3877–3882.

Alley SC, Jones AD, Soumillion P, Benkovic SJ. 1999a. The carboxyl terminus of the bacteriophage T4 DNA polymerase contacts its sliding clamp at the subunit interface. *J Biol Chem* **274:** 24485–24489.

Alley SC, Shier VK, Abel-Santos E, Sexton DJ, Soumillion P, Benkovic SJ. 1999b. Sliding clamp of the bacteriophage T4 polymerase has open and closed subunit interfaces in solution. *Biochemistry* **38:** 7696–7709.

Alley SC, Abel-Santos E, Benkovic SJ. 2000. Tracking sliding clamp opening and closing during bacteriophage T4 DNA polymerase holoenzyme assembly. *Biochemistry* **39:** 3076–3090.

Anderson SG, Thompson JA, Paschall CO, O'Donnell M, Bloom LB. 2009. Temporal correlation of DNA binding, ATP hydrolysis, and clamp release in the clamp loading reaction catalyzed by the Escherichia coli γ complex. *Biochemistry* **48:** 8516–8527.

Ason B, Bertram JG, Hingorani MM, Beechem JM, O'Donnell M, Goodman MF, Bloom LB. 2000. A model for *Escherichia coli* DNA polymerase III holoenzyme assembly at primer/template ends. DNA triggers a change in binding specificity of the γ complex clamp loader. *J Biol Chem* **275:** 3006–3015.

Ason B, Handayani R, Williams CR, Bertram JG, Hingorani MM, O'Donnell M, Goodman MF, Bloom LB. 2003. Mechanism of loading the *Escherichia coli* DNA polymerase III β sliding clamp on DNA. Bona fide primer/templates preferentially trigger the γ complex to hydrolyze ATP and load the clamp. *J Biol Chem* **278:** 10033–10040.

Bauer GA, Burgers PM. 1988. The yeast analog of mammalian cyclin/proliferating-cell nuclear antigen interacts with mammalian DNA polymerase δ. *Proc Natl Acad Sci* **85:** 7506–7510.

Berdis AJ, Benkovic SJ. 1996. Role of adenosine 5′-triphosphate hydrolysis in the assembly of the bacteriophage T4 DNA replication holoenzyme complex. *Biochemistry* **35:** 9253–9265.

Bertram JG, Bloom LB, Turner J, O'Donnell M, Beechem JM, Goodman MF. 1998. Pre-steady state analysis of the assembly of wild type and mutant circular clamps of *Escherichia coli* DNA polymerase III onto DNA. *J Biol Chem* **273:** 24564–24574.

Bertram JG, Bloom LB, Hingorani MM, Beechem JM, O'Donnell M, Goodman MF. 2000. Molecular mechanism and energetics of clamp assembly in *Escherichia coli*. The role of ATP hydrolysis when γ complex loads β on DNA. *J Biol Chem* **275:** 28413–28420.

Bloom LB, Turner J, Kelman Z, Beechem JM, O'Donnell M, Goodman MF. 1996. Dynamics of loading the β sliding clamp of DNA polymerase III onto DNA. *J Biol Chem* **271:** 30699–30708.

Bowman GD, O'Donnell M, Kuriyan J. 2004. Structural analysis of a eukaryotic sliding DNA clamp-clamp loader complex. *Nature* **429:** 724–730.

Bunz F, Kobayashi R, Stillman B. 1993. cDNAs encoding the large subunit of human replication factor C. *Proc Natl Acad Sci* **90:** 11014–11018.

Burbelo PD, Utani A, Pan ZQ, Yamada Y. 1993. Cloning of the large subunit of activator 1 (replication factor C) reveals homology with bacterial DNA ligases. *Proc Natl Acad Sci* **90:** 11543–11547.

Burgers PM. 1991. *Saccharomyces cerevisiae* replication factor C II formation and activity of complexes with the proliferating cell nuclear antigen and with DNA polymerases δ and ε. *J Biol Chem* **266:** 22698–22706.

Cai J, Uhlmann F, Gibbs E, Flores-Rozas H, Lee CG, Phillips B, Finkelstein J, Yao N, O'Donnell M, Hurwitz J. 1996. Reconstitution of human replication factor C from its five subunits in baculovirus-infected insect cells. *Proc Natl Acad Sci* **93:** 12896–12901.

Cai J, Gibbs E, Uhlmann F, Phillips B, Yao N, O'Donnell M, Hurwitz J. 1997. A complex consisting of human replication factor C p40, p37, and p36 subunits is a DNA-dependent ATPase and an intermediate in the assembly of the holoenzyme. *J Biol Chem* **272:** 18974–18981.

Cai J, Yao N, Gibbs E, Finkelstein J, Phillips B, O'Donnell M, Hurwitz J. 1998. ATP hydrolysis catalyzed by human replication factor C requires participation of multiple subunits. *Proc Natl Acad Sci* **95:** 11607–11612.

Cann IK, Ishino Y. 1999. Archaeal DNA replication: Identifying the pieces to solve a puzzle. *Genetics* **152:** 1249–1267.

Cann IK, Ishino S, Hayashi I, Komori K, Toh H, Morihawa K, Ishino Y. 1999. Functional interactions of a homolog of proliferating cell nuclear antigen with DNA polymerases in *Archaea*. *J Bacteriol* **181:** 6591–6599.

Cann IK, Ishino S, Yuasa M, Daiyasu H, Toh H, Ishino Y. 2001. Biochemical analysis of replication factor C from the hyperthermophilic archaeon *Pyrococcus furiosus*. *J Bacteriol* **183:** 2614–2623.

Capson TL, Benkovic SJ, Nossal NG. 1991. Protein-DNA cross-linking demonstrates stepwise ATP-dependent assembly of T4 DNA polymerase and its accessory proteins on the primer-template. *Cell* **65:** 249–258.

Chen M, Pan ZQ, Hurwitz J. 1992a. Sequence and expression in *Escherichia coli* of the 40-kDa subunit of activator 1 (replication factor C) of HeLa cells. *Proc Natl Acad Sci* **89:** 2516–2520.

Chen M, Pan ZQ, Hurwitz J. 1992b. Studies of the cloned 37-kDa subunit of activator 1 (replication factor C) of HeLa cells. *Proc Natl Acad Sci* **89:** 5211–5215.

Chen YH, Kocherginskaya SA, Lin Y, Sriratana B, Lagunas AM, Robbins JB, Mackie RI, Cann IK. 2005. Biochemical and mutational analyses of a unique clamp loader complex in the archaeon *Methanosarcina acetivorans*. *J Biol Chem* **280:** 41852–41863.

Chen S, Levin MK, Sakato M, Zhou Y, Hingorani MM. 2009a. Mechanism of ATP-driven PCNA clamp loading by *S. cerevisiae* RFC. *J Mol Biol* **388:** 431–442.

Chen YH, Lin Y, Yoshinaga A, Chhotani B, Lorenzini JL, Crofts AA, Mei S, Mackie RI, Ishinio Y, Cann IK. 2009b. Molecular analyses of a three-subunit euryarchaeal clamp loader complex from *Methanosarcina acetivorans*. *J Bacteriol* **191:** 6539–6549.

Cullmann G, Fien K, Kobayashi R, Stillman B. 1995. Characterization of the five replication factor C genes of *Saccharomyces cerevisiae*. *Mol Cell Biol* **15:** 4661–4671.

Daimon K, Kawarabayasi Y, Kikuchi H, Sako Y, Ishino Y. 2002. Three proliferating cell nuclear antigen-like proteins found in the hyperthermophilic archaeon *Aeropyrum pernix*: Interactions with the two DNA polymerases. *J Bacteriol* **184:** 687–694.

Davey MJ, Jeruzalmi D, Kuriyan J, O'Donnell M. 2002. Motors and switches: AAA$^+$ machines within the replisome. *Nat Rev Mol Cell Biol* **3:** 826–835.

Dionne I, Nookala RK, Jackson SP, Doherty AJ, Bell SD. 2003. A heterotrimeric PCNA in the hyperthermophilic archaeon *Sulfolobus solfataricus*. *Mol Cell* **11:** 275–282.

Dionne I, Brown NJ, Woodgate R, Bell SD. 2008. On the mechanism of loading the PCNA sliding clamp by RFC. *Mol Microbiol* **68:** 216–222.

Ellison V, Stillman B. 1998. Reconstitution of recombinant human replication factor C (RFC) and identification of an RFC subcomplex possessing DNA-dependent ATPase activity. *J Biol Chem* **273:** 5979–5987.

Fien K, Stillman B. 1992. Identification of replication factor C from *Saccharomyces cerevisiae*: A component of the leading-strand DNA replication complex. *Mol Cell Biol* **12:** 155–163.

Fotedar R, Mossi R, Fitzgerald P, Rousselle T, Maga G, Brickner H, Messier H, Kasibhatia S, Hübscher U, Fotedar A. 1996. A conserved domain of the large subunit of replication factor C binds PCNA and acts like a dominant negative inhibitor of DNA replication in mammalian cells. *EMBO J* **15:** 4423–4433.

Fukuda K, Morioka H, Imajou S, Ikeda S, Ohtsuka E, Tsurimoto T. 1995. Structure-function relationship of the eukaryotic DNA replication factor, proliferating cell nuclear antigen. *J Biol Chem* **270:** 22527–22534.

Gary SL, Burgers MJ. 1995. Identification of the fifth subunit of *Saccharomyces cerevisiae* replication factor C. *Nucleic Acids Res* **23:** 4986–4991.

Georgescu RE, Kim SS, Yurieva O, Kuriyan J, Kong XP, O'Donnell M. 2008. Structure of a sliding clamp on DNA. *Cell* **132:** 43–54.

Glover BP, McHenry CS. 1998. The χ ψ subunits of DNA polymerase III holoenzyme bind to single-stranded DNA-binding protein (SSB) and facilitate replication of an SSB-coated template. *J Biol Chem* **273:** 23476–23484.

Goedken ER, Levitus M, Johnson A, Bustamante C, O'Donnell M, Kuriyan J. 2004. Fluorescence measurements on the *E.coli* DNA polymerase clamp loader: Implications for conformational changes during ATP and clamp binding. *J Mol Biol* **336:** 1047–1059.

Gomes XV, Burgers PM. 2001. ATP utilization by yeast replication factor C. I. ATP-mediated interaction with DNA and with proliferating cell nuclear antigen. *J Biol Chem* **276:** 34768–34775.

Gomes XV, Gary SL, Burgers PM. 2000. Overproduction in *Escherichia coli* and characterization of yeast replication factor C lacking the ligase homology domain. *J Biol Chem* **275:** 14541–14549.

Gomes XV, Schmidt SL, Burgers PM. 2001. ATP utilization by yeast replication factor C. II. Multiple stepwise ATP binding events are required to load proliferating cell nuclear antigen onto primed DNA. *J Biol Chem* **276:** 34776–34783.

Guenther B, Onrust R, Sali A, O'Donnell M, Kuriyan J. 1997. Crystal structure of the δ′ subunit of the clamp-loader complex of *E. coli* DNA polymerase III. *Cell* **91:** 335–345.

Gulbis JM, Kelman Z, Hurwitz J, O'Donnell M, Kuriyan J. 1996. Structure of the C-terminal region of p21(WAF1/CIP1) complexed with human PCNA. *Cell* **87:** 297–306.

Henneke G, Gueguen Y, Flament D, Azam P, Querellou J, Dietrich H, Hübscher U, Raffin JP. 2002. Replication factor C from the hyperthermophilic archaeon *Pyrococcus abyssi* does not need ATP hydrolysis for clamp-loading and contains a functionally conserved RFC PCNA-binding domain. *J Mol Biol* **323:** 795–810.

Hingorani MM, Coman MM. 2002. On the specificity of interaction between the *Saccharomyces cerevisiae* clamp loader replication factor C and primed DNA templates during DNA replication. *J Biol Chem* **277:** 47213–47224.

Hingorani MM, O'Donnell M. 1998. ATP binding to the *Escherichia coli* clamp loader powers opening of the ring-shaped clamp of DNA polymerase III holoenzyme. *J Biol Chem* **273:** 24550–24563.

Hingorani MM, Bloom LB, Goodman MF, O'Donnell M. 1999. Division of labor—sequential ATP hydrolysis drives assembly of a DNA polymerase sliding clamp around DNA. *EMBO J* **18:** 5131–5144.

Hlinkova V, Xing G, Bauer J, Shin YJ, Dionne I, Rajashankar KR, Bell SD, Ling H. 2008. Structures of monomeric, dimeric and trimeric PCNA: PCNA-ring assembly and opening. *Acta Crystallogr D Biol Crystallogr* **64:** 941–949.

Hockensmith JW, Kubasek WL, Evertsz EM, Mesner LD, von Hippel PH. 1993. Laser cross-linking of proteins to nucleic acids. II. Interactions of the bacteriophage T4 DNA replication polymerase accessory proteins complex with DNA. *J Biol Chem* **268:** 15721–15730.

Indiani C, O'Donnell M. 2003. Mechanism of the δ wrench in opening the β sliding clamp. *J Biol Chem* **278:** 40272–40281.

Iwai T, Kurosawa N, Itoh YH, Horiuchi T. 2000. Phylogenetic analysis of archaeal PCNA homologues. *Extremophiles* **4:** 357–364.

Jarvis TC, Paul LS, Hockensmith JW, von Hippel PH. 1989a. Structural and enzymatic studies of the T4 DNA replication system. II. ATPase properties of the polymerase accessory protein complex. *J Biol Chem* **264:** 12717–12729.

Jarvis TC, Paul LS, von Hippel PH. 1989b. Structural and enzymatic studies of the T4 DNA replication system. I. Physical characterization of the polymerase accessory protein complex. *J Biol Chem* **264:** 12709–12716.

Jeruzalmi D, O'Donnell M, Kuriyan J. 2001a. Crystal structure of the processivity clamp loader gamma (γ) complex of *E. coli* DNA polymerase III. *Cell* **106:** 429–441.

Jeruzalmi D, Yurieva O, Zhao Y, Young M, Stewart J, Hingorani M, O'Donnell M, Kuriyan J. 2001b. Mechanism of processivity clamp opening by the δ subunit wrench of the clamp loader complex of *E. coli* DNA polymerase III. *Cell* **106:** 417–428.

Johnson A, O'Donnell M. 2003. Ordered ATP hydrolysis in the γ complex clamp loader AAA$^+$ machine. *J Biol Chem* **278:** 14406–14413.

Johnson A, Yao NY, Bowman GD, Kuriyan J, O'Donnell M. 2006. The replication factor C clamp loader requires arginine finger sensors to drive DNA binding and proliferating cell nuclear antigen loading. *J Biol Chem* **281:** 35531–35543.

Kaboord BF, Benkovic SJ. 1995. Accessory proteins function as matchmakers in the assembly of the T4 DNA polymerase holoenzyme. *Curr Biol* **5:** 149–157.

Kaboord BF, Benkovic SJ. 1996. Dual role of the 44/62 protein as a matchmaker protein and DNA polymerase chaperone during assembly of the bacteriophage T4 holoenzyme complex. *Biochemistry* **35:** 1084–1092.

Kazmirski SL, Podobnik M, Weitze TF, O'Donnell M, Kuriyan J. 2004. Structural analysis of the inactive state of the *Escherichia coli* DNA polymerase clamp-loader complex. *Proc Natl Acad Sci* **101:** 16750–16755.

Kazmirski SL, Zhao Y, Bowman GD, O'Donnell M, Kuriyan J. 2005. Out-of-plane motions in open sliding clamps: Molecular dynamics simulations of eukaryotic and archaeal proliferating cell nuclear antigen. *Proc Natl Acad Sci* **102:** 13801–13806.

Kelch BA, Makino DL, O'Donnell M, Kuriyan J. 2011. How a DNA polymerase clamp loader opens a sliding clamp. *Science* **334:** 1675–1680.

Kelman Z, Hurwitz J. 2000. A unique organization of the protein subunits of the DNA polymerase clamp loader in the archaeon *Methanobacterium thermoautotrophicum* δH. *J Biol Chem* **275:** 7327–7336.

Kelman Z, Yuzhakov A, Andjelkovic J, O'Donnell M. 1998. Devoted to the lagging strand—the subunit of DNA polymerase III holoenzyme contacts SSB to promote processive elongation and sliding clamp assembly. *EMBO J* **17:** 2436–2449.

Kong XP, Onrust R, O'Donnell M, Kuriyan J. 1992. Three-dimensional structure of the β subunit of *E. coli* DNA polymerase III holoenzyme: A sliding DNA clamp. *Cell* **69:** 425–437.

Krishna TS, Kong XP, Gary S, Burgers PM, Kuriyan J. 1994. Crystal structure of the eukaryotic DNA polymerase processivity factor PCNA. *Cell* **79:** 1233–1243.

Kumar R, Nashine VC, Mishra PP, Benkovic SJ, Lee TH. 2010. Stepwise loading of yeast clamp revealed by ensemble and single-molecule studies. *Proc Natl Acad Sci* **107:** 19736–19741.

Latham GJ, Bacheller DJ, Pietroni P, von Hippel PH. 1997. Structural analyses of gp45 sliding clamp interactions during assembly of the bacteriophage T4 DNA polymerase holoenzyme. II. The Gp44/62 clamp loader interacts with a single defined face of the sliding clamp ring. *J Biol Chem* **272:** 31677–31684.

Latham GJ, Dong F, Pietroni P, Dozono JM, Bacheller DJ, von Hippel PH. 1999. Opening of a monomer-monomer interface of the trimeric bacteriophage T4-coded GP45 sliding clamp is required for clamp loading onto DNA. *Proc Natl Acad Sci* **96:** 12448–12453.

Lee SH, Hurwitz J. 1990. Mechanism of elongation of primed DNA by DNA polymerase δ, proliferating cell nuclear antigen, and activator 1. *Proc Natl Acad Sci* **87:** 5672–5676.

Lee SH, Kwong AD, Pan ZQ, Hurwitz J. 1991. Studies on the activator 1 protein complex, an accessory factor for proliferating cell nuclear antigen-dependent DNA polymerase δ. *J Biol Chem* **266:** 594–602.

Leu FP, O'Donnell M. 2001. Interplay of clamp loader subunits in opening the β sliding clamp of *Escherichia coli* DNA polymerase III holoenzyme. *J Biol Chem* **276:** 47185–47194.

Leu FP, Hingorani MM, Turner J, O'Donnell M. 2000. The δ subunit of DNA polymerase III holoenzyme serves as a sliding clamp unloader in *Escherichia coli. J Biol Chem* **275:** 34609–34618.

Leu FP, Georgescu R, O'Donnell M. 2003. Mechanism of the *E. coli* τ processivity switch during lagging-strand synthesis. *Mol Cell* **11:** 315–327.

Li X, Burgers PM. 1994a. Cloning and characterization of the essential *Saccharomyces cerevisiae* RFC4 gene encoding the 37-kDa subunit of replication factor C. *J Biol Chem* **269:** 21880–21884.

Li X, Burgers PM. 1994b. Molecular cloning and expression of the *Saccharomyces cerevisiae* RFC3 gene, an essential component of replication factor C. *Proc Natl Acad Sci* **91:** 868–872.

Liu C, McKinney MC, Chen YH, Earnest TM, Shi X, Lin LJ, Ishino Y, Dahmen K, Cann IK, Ha T. 2011. Reverse-chaperoning activity of an AAA$^+$ protein. *Biophys J* **100:** 1344–1352.

Mace DC, Alberts BM. 1984. The complex of T4 bacteriophage gene 44 and 62 replication proteins forms an ATPase that is stimulated by DNA and by T4 gene 45 protein. *J Mol Biol* **177:** 279–293.

Masuda Y, Suzuki M, Piao J, Gu Y, Tsurimoto T, Kamiya K. 2007. Dynamics of human replication factors in the elongation phase of DNA replication. *Nucleic Acids Res* **35:** 6904–6916.

Matsumiya S, Ishino Y, Morikawa K. 2001. Crystal structure of an archaeal DNA sliding clamp: Proliferating cell nuclear antigen from *Pyrococcus furiosus. Protein Sci* **10:** 17–23.

Matsumiya S, Ishino S, Ishino Y, Morikawa K. 2002. Physical interaction between proliferating cell nuclear antigen and replication factor C from *Pyrococcus furiosus. Genes Cells* **7:** 911–922.

Matsumiya S, Ishino S, Ishino Y, Morikawa K. 2003. Intermolecular ion pairs maintain the toroidal structure of *Pyrococcus furiosus* PCNA. *Protein Sci* **12:** 823–831.

Mayanagi K, Miyata T, Oyama T, Ishino Y, Morikawa K. 2001. Three-dimensional electron microscopy of the clamp loader small subunit from *Pyrococcus furiosus. J Struct Biol* **134:** 35–45.

McNally R, Bowman GD, Goedken ER, O'Donnell M, Kuriyan J. 2010. Analysis of the role of PCNA-DNA contacts during clamp loading. *BMC Struct Biol* **10:** 3.

Millar D, Trakselis MA, Benkovic SJ. 2004. On the solution structure of the T4 sliding clamp (gp45). *Biochemistry* **43:** 12723–12727.

Miyata T, Oyama T, Mayanagi K, Ishino S, Ishino Y, Morikawa K. 2004. The clamp-loading complex for processive DNA replication. *Nat Struct Mol Biol* **11:** 632–636.

Miyata T, Suzuki H, Oyama T, Mayanagi K, Ishino Y, Morikawa K. 2005. Open clamp structure in the clamp-loading complex visualized by electron microscopic image analysis. *Proc Natl Acad Sci* **102:** 13795–13800.

Moarefi I, Jeruzalmi D, Turner J, O'Donnell M, Kuriyan J. 2000. Crystal structure of the DNA polymerase processivity factor of T4 bacteriophage. *J Mol Biol* **296:** 1215–1223.

Mossi R, Jonsson ZO, Allen BL, Hardin SH, Hubscher U. 1997. Replication factor C interacts with the C-terminal side of proliferating cell nuclear antigen. *J Biol Chem* **272:** 1769–1776.

Munn MM, Alberts BM. 1991a. DNA footprinting studies of the complex formed by the T4 DNA polymerase holoenzyme at a primer-template junction. *J Biol Chem* **266:** 20034–20044.

Munn MM, Alberts BM. 1991b. The T4 DNA polymerase accessory proteins form an ATP-dependent complex on a primer-template junction. *J Biol Chem* **266:** 20024–20033.

Naktinis V, Onrust R, Fang L, O'Donnell M. 1995. Assembly of a chromosomal replication machine: Two DNA polymerases, a clamp loader, and sliding clamps in one holoenzyme particle. II. Intermediate complex between the clamp loader and its clamp. *J Biol Chem* **270:** 13358–13365.

Noskov V, Maki S, Kawasaki Y, Leem SH, Ono B, Araki H, Pavlov Y, Sugino A. 1994. The RFC2 gene encoding a subunit of replication factor C of *Saccharomyces cerevisiae. Nucleic Acids Res* **22:** 1527–1535.

O'Donnell M, Onrust R, Dean FB, Chen M, Hurwitz J. 1993. Homology in accessory proteins of replicative polymerases—*E. coli* to humans. *Nucleic Acids Res* **21:** 1–3.

Oku T, Ikeda S, Sasaki H, Fukuda K, Morioka H, Ohtsuka E, Yoshikawa H, Tsurimoto T. 1998. Functional sites of human PCNA which interact with p21 (Cip1/Waf1), DNA polymerase δ and replication factor C. *Genes Cells* **3:** 357–369.

Olson MW, Dallmann HG, McHenry CS. 1995. DnaX complex of *Escherichia coli* DNA polymerase III holoenzyme. The χ ψ complex functions by increasing the affinity of tau and γ for δ.δ′ to a physiologically relevant range. *J Biol Chem* **270:** 29570–29577.

Onrust R, O'Donnell M. 1993. DNA polymerase III accessory proteins. II. Characterization of δ and δ′. *J Biol Chem* **268:** 11766–11772.

Onrust R, Stukenberg PT, O'Donnell M. 1991. Analysis of the ATPase subassembly which initiates processive DNA synthesis by DNA polymerase III holoenzyme. *J Biol Chem* **266:** 21681–21686.

Onrust R, Finkelstein J, Naktinis V, Turner J, Fang L, O'Donnell M. 1995. Assembly of a chromosomal replication machine: Two DNA polymerases, a clamp loader, and sliding clamps in one holoenzyme particle. I. Organization of the clamp loader. *J Biol Chem* **270:** 13348–13357.

Oyama T, Ishino Y, Cann IK, Ishino S, Morikawa K. 2001. Atomic structure of the clamp loader small subunit from *Pyrococcus furiosus*. *Mol Cell* **8:** 455–463.

Pan ZQ, Chen M, Hurwitz J. 1993. The subunits of activator 1 (replication factor C) carry out multiple functions essential for proliferating-cell nuclear antigen-dependent DNA synthesis. *Proc Natl Acad Sci* **90:** 6–10.

Park MS, O'Donnell M. 2009. The clamp loader assembles the β clamp onto either a 3′ or 5′ primer terminus: The underlying basis favoring 3′ loading. *J Biol Chem* **284:** 31473–31483.

Pascal JM, Tsodikov OV, Hura GL, Song W, Cotner EA, Classen S, Tomkinson AE, Tainer JA, Ellenberger T. 2006. A flexible interface between DNA ligase and PCNA supports conformational switching and efficient ligation of DNA. *Mol Cell* **24:** 279–291.

Paschall CO, Thompson JA, Marzahn MR, Chiraniya A, Hayner JN, O'Donnell M, Robbins AH, McKenna R, Bloom LB. 2011. The *E. coli* clamp loader can actively pry open the β-sliding clamp. *J Biol Chem* **286:** 42704–42714.

Pietroni P, von Hippel PH. 2008. Multiple ATP binding is required to stabilize the "activated" (clamp open) clamp loader of the T4 DNA replication complex. *J Biol Chem* **283:** 28338–28353.

Pietroni P, Young MC, Latham GJ, von Hippel PH. 1997. Structural analyses of gp45 sliding clamp interactions during assembly of the bacteriophage T4 DNA polymerase holoenzyme. I. Conformational changes within the gp44/62-gp45-ATP complex during clamp loading. *J Biol Chem* **272:** 31666–31676.

Pietroni P, Young MC, Latham GJ, von Hippel PH. 2001. Dissection of the ATP-driven reaction cycle of the bacteriophage T4 DNA replication processivity clamp loading system. *J Mol Biol* **309:** 869–891.

Piperno JR, Alberts BM. 1978. An ATP stimulation of T4 DNA polymerase mediated via T4 gene 44/62 and 45 proteins. The requirement for ATP hydrolysis. *J Biol Chem* **253:** 5174–5179.

Piperno JR, Kallen RG, Alberta BM. 1978. Analysis of a T4 DNA replication protein complex. Studies of the DNA recognition site for T4 gene 44/62 and 45 protein-catalyzed ATP hydrolysis. *J Biol Chem* **253:** 5180–5185.

Pisani FM, De Felice M, Carpentieri F, Rossi M. 2000. Biochemical characterization of a clamp-loader complex homologous to eukaryotic replication factor C from the hyperthermophilic archaeon *Sulfolobus solfataricus*. *J Mol Biol* **301:** 61–73.

Podust VN, Georgaki A, Strack B, Hubscher U. 1992. Calf thymus RF-C as an essential component for DNA polymerase δ and ε holoenzymes function. *Nucleic Acids Res* **20:** 4159–4165.

Podust LM, Podust VN, Sogo JM, Hubscher U. 1995. Mammalian DNA polymerase auxiliary proteins: Analysis of replication factor C-catalyzed proliferating cell nuclear antigen loading onto circular double-stranded DNA. *Mol Cell Biol* **15:** 3072–3081.

Podust VN, Tiwari N, Ott R, Fanning E. 1998a. Functional interactions among the subunits of replication factor C potentiate and modulate its ATPase activity. *J Biol Chem* **273:** 12935–12942.

Podust VN, Tiwari N, Stephan S, Fanning E. 1998b. Replication factor C disengages from proliferating cell nuclear antigen (PCNA) upon sliding clamp formation, and PCNA itself tethers DNA polymerase δ to DNA. *J Biol Chem* **273:** 31992–31999.

Pritchard AE, Dallmann HG, Glover BP, McHenry CS. 2000. A novel assembly mechanism for the DNA polymerase III holoenzyme DnaX complex: Association of δδ′ with DnaX(4) forms DnaX(3)δδ′. *EMBO J* **19:** 6536–6545.

Reddy MK, Weitzel SE, von Hippel PH. 1993. Assembly of a functional replication complex without ATP hydrolysis: A direct interaction of bacteriophage T4 gp45 with T4 DNA polymerase. *Proc Natl Acad Sci* **90:** 3211–3215.

Reems JA, Wood S, McHenry CS. 1995. *Escherichia coli* DNA polymerase III holoenzyme subunits α, β, and γ directly contact the primer-template. *J Biol Chem* **270:** 5606–5613.

Reyes-Lamothe R, Sherratt DJ, Leake MC. 2010. Stoichiometry and architecture of active DNA replication machinery in *Escherichia coli*. *Science* **328:** 498–501.

Rush J, Lin TC, Quinones M, Spicer EK, Douglas I, Williams KR, Konigsberg WH. 1989. The 44P subunit of the T4 DNA polymerase accessory protein complex catalyzes ATP hydrolysis. *J Biol Chem* **264:** 10943–10953.

Sakato M, O'Donnell M, Hingorani MM. 2011a. A central swivel point in the RFC clamp loader controls PCNA opening and loading on DNA. *J Mol Biol* **416:** 163–175.

Sakato M, Zhou Y, Hingorani MM. 2011b. ATP-binding- and hydrolysis-driven rate-determining events in the RFC-catalyzed PCNA clamp loading reaction. *J Mol Biol* **416:** 176–191.

Sanders GM, Kassavetis GA, Geiduschek EP. 1994. Use of a macromolecular crowding agent to dissect interactions and define functions in transcriptional activation by a DNA-tracking protein: Bacteriophage T4 gene 45 protein and late transcription. *Proc Natl Acad Sci* **91:** 7703–7707.

Schmidt SL, Gomes XV, Burgers PM. 2001. ATP utilization by yeast replication factor C. III. The ATP-binding domains of Rfc2, Rfc3, and Rfc4 are essential for DNA recognition and clamp loading. *J Biol Chem* **276:** 34784–34791.

Schurtenberger P, Egelhaaf SU, Hindges R, Maga G, Jónsson ZO, May RP, Glatter O, Hübscher U. 1998. The solution structure of functionally active human proliferating cell nuclear antigen determined by small-angle neutron scattering. *J Mol Biol* **275:** 123–132.

Sexton DJ, Carver TE, Berdis AJ, Benkovic SJ. 1996. Protein–protein and protein-DNA interactions at the bacteriophage T4 DNA replication fork. Characterization of a fluorescently labeled DNA polymerase sliding clamp. *J Biol Chem* **271:** 28045–28051.

Sexton DJ, Kaboord BF, Berdis AJ, Carver TE, Benkovic SJ. 1998. Dissecting the order of bacteriophage T4 DNA polymerase holoenzyme assembly. *Biochemistry* **37:** 7749–7756.

Seybert A, Wigley DB. 2004. Distinct roles for ATP binding and hydrolysis at individual subunits of an archaeal clamp loader. *EMBO J* **23:** 1360–1371.

Seybert A, Scott DJ, Scaife S, Singleton MR, Wigley DB. 2002. Biochemical characterisation of the clamp/clamp loader proteins from the euryarchaeon *Archaeoglobus fulgidus*. *Nucleic Acids Res* **30:** 4329–4338.

Seybert A, Singleton MR, Cook N, Hall DR, Wigley DB. 2006. Communication between subunits within an archaeal clamp-loader complex. *EMBO J* **25:** 2209–2218.

Shibahara K, Stillman B. 1999. Replication-dependent marking of DNA by PCNA facilitates CAF-1-coupled inheritance of chromatin. *Cell* **96:** 575–585.

Shiomi Y, Usukura J, Masamura Y, Takeyasu K, Nakayama Y, Obuse C, Yoshikawa H, Tsurimoto T. 2000. ATP-dependent structural change of the eukaryotic clamp-loader protein, replication factor C. *Proc Natl Acad Sci* **97:** 14127–14132.

Simonetta KR, Kazmirski SL, Goedken ER, Cantor AJ, Kelch BA, McNally R, Seyedin SN, Makino DL, O'Donnell M, Kuriyan J. 2009. The mechanism of ATP-dependent primer-template recognition by a clamp loader complex. *Cell* **137:** 659–671.

Smiley RD, Zhuang Z, Benkovic SJ, Hammes GG. 2006. Single-molecule investigation of the T4 bacteriophage DNA polymerase holoenzyme: Multiple pathways of holoenzyme formation. *Biochemistry* **45:** 7990–7997.

Snyder AK, Williams CR, Johnson A, O'Donnell M, Bloom LB. 2004. Mechanism of loading the *Escherichia coli* DNA polymerase III sliding clamp: II. Uncoupling the β and DNA binding activities of the γ complex. *J Biol Chem* **279:** 4386–4393.

Soumillion P, Sexton DJ, Benkovic SJ. 1998. Clamp subunit dissociation dictates bacteriophage T4 DNA polymerase holoenzyme disassembly. *Biochemistry* **37:** 1819–1827.

Spicer EK, Nossal NG, Williams KR. 1984. Bacteriophage T4 gene 44 DNA polymerase accessory protein. Sequences of gene 44 and its protein product. *J Biol Chem* **259:** 15425–15432.

Stewart J, Hingorani MM, Kelman Z, O'Donnell M. 2001. Mechanism of β clamp opening by the δ subunit of *Escherichia coli* DNA polymerase III holoenzyme. *J Biol Chem* **276:** 19182–19189.

Stukenberg PT, Studwell-Vaughan PS, O'Donnell M. 1991. Mechanism of the sliding β-clamp of DNA polymerase III holoenzyme. *J Biol Chem* **266:** 11328–11334.

Tainer JA, McCammon JA, Ivanov I. 2010. Recognition of the ring-opened state of proliferating cell nuclear antigen by replication factor C promotes eukaryotic clamp-loading. *J Am Chem Soc* **132:** 7372–7378.

Thompson JA, Paschall CO, O'Donnell M, Bloom LB. 2009. A slow ATP-induced conformational change limits the rate of DNA binding but not the rate of β clamp binding by the *Escherichia coli* γ complex clamp loader. *J Biol Chem* **284:** 32147–32157.

Thompson JA, Marzahn MR, O'Donnell M, Bloom LB. 2011. Replication factor C is a more effective PCNA opener than the checkpoint clamp loader, RAD24-RFC. *J Biol Chem* **287:** 2203–2209.

Tinker RL, Kassavetis GA, Geiduschek EP. 1994. Detecting the ability of viral, bacterial and eukaryotic replication proteins to track along DNA. *EMBO J* **13:** 5330–5337.

Trakselis MA, Alley SC, Abel-Santos E, Benkovic SJ. 2001. Creating a dynamic picture of the sliding clamp during T4 DNA polymerase holoenzyme assembly by using fluorescence resonance energy transfer. *Proc Natl Acad Sci* **98:** 8368–8375.

Trakselis MA, Berdis AJ, Benkovic SJ. 2003. Examination of the role of the clamp-loader and ATP hydrolysis in the formation of the bacteriophage T4 polymerase holoenzyme. *J Mol Biol* **326:** 435–451.

Tsuchihashi Z, Kornberg A. 1989. ATP interactions of the tau and γ subunits of DNA polymerase III holoenzyme of *Escherichia coli*. *J Biol Chem* **264:** 17790–17795.

Tsurimoto T, Stillman B. 1990. Functions of replication factor C and proliferating-cell nuclear antigen: Functional similarity of DNA polymerase accessory proteins from human cells and bacteriophage T4. *Proc Natl Acad Sci* **87:** 1023–1027.

Tsurimoto T, Stillman B. 1991. Replication factors required for SV40 DNA replication in vitro. I. DNA structure-specific recognition of a primer-template junction by eukaryotic DNA polymerases and their accessory proteins. *J Biol Chem* **266:** 1950–1960.

Turner J, Hingorani MM, Kelman Z, O'Donnell M. 1999. The internal workings of a DNA polymerase clamp-loading machine. *EMBO J* **18:** 771–783.

Uhlmann F, Cai J, Flores-Rozas H, Dean FB, Finkelstein J, O'Donnell M, Hurwitz J. 1996. In vitro reconstitution of human replication factor C from its five subunits. *Proc Natl Acad Sci* **93:** 6521–6526.

Uhlmann F, Cai J, Gibbs E, O'Donnell M, Hurwitz J. 1997a. Deletion analysis of the large subunit p140 in human replication factor C reveals regions required for complex formation and replication activities. *J Biol Chem* **272:** 10058–10064.

Uhlmann F, Gibbs E, Cai J, O'Donnell M, Hurwitz J. 1997b. Identification of regions within the four small subunits of human replication factor C required for complex formation and DNA replication. *J Biol Chem* **272:** 10065–10071.

Waga S, Stillman B. 1998. Cyclin-dependent kinase inhibitor p21 modulates the DNA primer-template recognition complex. *Mol Cell Biol* **18:** 4177–4187.

Walker JE, Saraste M, Runswick MJ, Gay NJ. 1982. Distantly related sequences in the α- and β-subunits of ATP synthase, myosin, kinases and other ATP-requiring enzymes and a common nucleotide binding fold. *EMBO J* **1:** 945–951.

Williams CR, Snyder AK, Kuzmic P, O'Donnell M, Bloom LB. 2004. Mechanism of loading the *Escherichia coli* DNA polymerase III sliding clamp: I. Two distinct activities for individual ATP sites in the γ complex. *J Biol Chem* **279:** 4376–4385.

Williams GJ, Johnson K, Rudolf J, McMahon SA, Carter L, Oke M, Liu H, Taylor GL, White MF, Naismith JH. 2006. Structure of the heterotrimeric PCNA from *Sulfolobus solfataricus*. *Acta Crystallogr Sect F Struct Biol Cryst Commun* **62:** 944–948.

Xiao H, Dong Z, O'Donnell M. 1993. DNA polymerase III accessory proteins. IV. Characterization of χ and ψ. *J Biol Chem* **268:** 11779–11784.

Xiao H, Naktinis V, O'Donnell M. 1995. Assembly of a chromosomal replication machine: Two DNA polymerases, a clamp loader, and sliding clamps in one holoenzyme particle. IV. ATP-binding site mutants identify the clamp loader. *J Biol Chem* **270:** 13378–13383.

Yao N, Turner J, Kelman Z, Stukenberg PT, Dean F, Shechter D, Pan ZQ, Hurwitz J, O'Donnell M. 1996. Clamp loading, unloading and intrinsic stability of the PCNA, β and gp45 sliding clamps of human. *E coli* and T4 replicases. *Genes Cells* **1:** 101–113.

Yao N, Coryell L, Zhang D, Georgescu RE, Finkelstein J, Coman MM, Hingorani MM, O'Donnell M. 2003. Replication factor C clamp loader subunit arrangement within the circular pentamer and its attachment points to proliferating cell nuclear antigen. *J Biol Chem* **278:** 50744–50753.

Yao NY, Johnson A, Bowman GD, Kuriyan J, O'Donnell M. 2006. Mechanism of proliferating cell nuclear antigen clamp opening by replication factor C. *J Biol Chem* **281:** 17528–17539.

Yoder BL, Burgers PM. 1991. *Saccharomyces cerevisiae* replication factor C. I. Purification and characterization of its ATPase activity. *J Biol Chem* **266:** 22689–22697.

Young MC, Weitzel SE, von Hippel PH. 1996. The kinetic mechanism of formation of the bacteriophage T4 DNA polymerase sliding clamp. *J Mol Biol* **264:** 440–452.

Yuzhakov A, Kelman Z, Hurwitz J, O'Donnell M. 1999. Multiple competition reactions for RPA order the assembly of the DNA polymerase δ holoenzyme. *EMBO J* **18:** 6189–6199.

Zhang G, Gibbs E, Kelman Z, O'Donnell M, Hurwitz J. 1999. Studies on the interactions between human replication factor C and human proliferating cell nuclear antigen. *Proc Natl Acad Sci* **96:** 1869–1874.

Zhuang Z, Berdis AJ, Benkovic SJ. 2006a. An alternative clamp loading pathway via the T4 clamp loader gp44/62-DNA complex. *Biochemistry* **45:** 7976–7989.

Zhuang Z, Yoder BL, Burgers PM, Benkovic SJ. 2006b. The structure of a ring-opened proliferating cell nuclear antigen-replication factor C complex revealed by fluorescence energy transfer. *Proc Natl Acad Sci* **103:** 2546–2551.

Okazaki Fragment Metabolism

Lata Balakrishnan[1] and Robert A. Bambara[2]

[1]Department of Biochemistry and Biophysics, University of Rochester Medical Center, Rochester, New York 14642

[2]Department of Microbiology and Immunology, University of Rochester Medical Center, Rochester, New York 14642

Correspondence: robert_bambara@urmc.rochester.edu

Cellular DNA replication requires efficient copying of the double-stranded chromosomal DNA. The leading strand is elongated continuously in the direction of fork opening, whereas the lagging strand is made discontinuously in the opposite direction. The lagging strand needs to be processed to form a functional DNA segment. Genetic analyses and reconstitution experiments identified proteins and multiple pathways responsible for maturation of the lagging strand. In both prokaryotes and eukaryotes the lagging-strand fragments are initiated by RNA primers, which are removed by a joining mechanism involving strand displacement of the primer into a flap, flap removal, and then ligation. Although the prokaryotic fragments are ~1200 nucleotides long, the eukaryotic fragments are much shorter, with lengths determined by nucleosome periodicity. The prokaryotic joining mechanism is simple and efficient. The eukaryotic maturation mechanism involves many enzymes, possibly three pathways, and regulation that can shift from high efficiency to high fidelity.

Replication of cellular chromosomal DNA is initiated by the multienzyme replisome machinery, which unwinds the DNA helix to create a replication fork. The antiparallel structure of double-helical DNA and the 3′ end extension specificity of all DNA polymerases confine the mechanisms that can be used by the cell for DNA duplication. One copied strand, called leading, can conveniently be extended in a continuous manner in the same direction that the helix must open to allow exposure of templates for polymerization. The other, or lagging strand, must be periodically extended away from the opening helix. This can only be accomplished if the strand is made discontinuously (Kornberg and Baker 1992). The strand is synthesized in short segments, named Okazaki fragments, after their discoverer (Sakabe and Okazaki 1966; Okazaki et al. 1968) and the segments are then joined. This requirement has two fundamental consequences: (1) The lagging strand must have evolved priming and fragment joining mechanisms involving many additional steps and reactions than needed for leading-strand extension. (2) Mechanisms of lagging-strand replication must have developed means of avoiding mutagenesis while handling the necessary strand manipulations.

Cite this article as *Cold Spring Harb Perspect Biol* doi: 10.1101/cshperspect.a010173

PARALLELS BETWEEN PROKARYOTIC AND EUKARYOTIC REPLICATION

Although replication of eukaryotic DNA on a chromatinized DNA template is a relatively more complex process than replication in bacteria, involving more types of proteins and reactions, the fundamental processes of DNA duplication have striking parallels in all cells. The replisome machineries of both organisms are minimally composed of helicases, which unwind the duplex strands, primase, which initiates synthesis and DNA polymerases, which duplicate the parental strands of the DNA. In bacteria, DNA replication proceeds within a fork, wherein the lagging strand loops into a "trombonelike" structure allowing for the replication enzymes to be continually recycled on the DNA for repeated synthesis and joining (Alberts et al. 1982). Presumably, use of a similar mechanism in eukaryotes allows coordination of synthesis between the leading and lagging strands (Fig. 1) (Pandey et al. 2009). Because DNA polymerases cannot incorporate dNTPs without a primer terminated by a 3′ hydroxyl, the leading strand and each Okazaki fragment are primed by RNA to initiate synthesis (Hubscher et al. 2002). On the lagging strand the primer is extended by the addition of dNMPs to form short segments of DNA. These segments need to be further processed to form a fully functional strand of DNA. Specifically, the RNA primers have to be excised from the fragments. Removal of RNA primers is performed partly by a ribonuclease H (RNase H). There are two distinct classes of RNase H enzymes in bacterial and eukaryotic systems (type 1: RNase HI [*Escherichia coli*] and RNase H1 [eukaryotes] and type 2: RNase HII [*E. coli*] and RNase H2 [eukaryotes]) (Cerritelli and Crouch 2009). Although

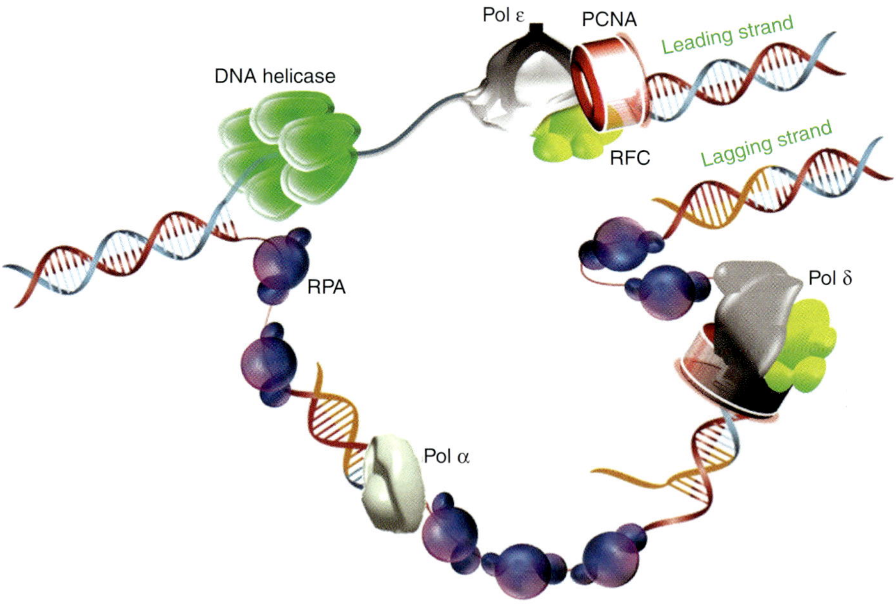

Figure 1. Eukaryotic replisome. A DNA helicase initially unwinds the duplex DNA (red and blue strands) to separate the DNA and form a replication fork. The single-stranded DNA (ssDNA) is coated by the single-strand binding protein, replication protein A (RPA). On the leading strand, replication factor C (RFC) loads proliferating cell nuclear antigen (PCNA) and DNA polymerase ε to continuously synthesize the leading strand. The lagging strand is initially primed by DNA polymerase α (Pol α), which synthesizes a short RNA/DNA initiator primer (orange strand). RFC then displaces Pol α from the lagging strand to initiate the switch from the priming mode to the elongation mode. The initiator primer is extended by PCNA/DNA polymerase δ complex to form short segments of DNA known as Okazaki fragments.

 Cite this article as *Cold Spring Harb Perspect Biol* doi: 10.1101/cshperspect.a010173

type 1 RNases H require a minimum of four ribonucleotides for hydrolysis, type 2 RNases H can recognize a single ribonucleotide (Cerritelli and Crouch 2009). In eukaryotes, the initiator RNA primers are removed, apparently by two partly redundant processes (Kao and Bambara 2003). The first is the action of cellular RNase H2, which can begin degrading the primer as soon as it is made, but presumably not so soon that it interferes with initiation of DNA synthesis (Turchi et al. 1994; Murante et al. 1998). RNase H2 degrades between ribonucleotides of an RNA strand annealed to DNA. It cannot cleave between the 3′-most ribonucleotide and the initial deoxynucleotide. A second nuclease is needed for this. The second nuclease operates after strand displacement synthesis resulting from extension of the 3′ end of the adjacent fragment. This synthesis raises what remains of the RNA primer into a single-stranded flap structure that is removed by endonuclease action. The resulting nick is then ligated to make a continuous strand. The absence of RNase HI in *E. coli* leads to initiation of replication at sites other than the replication initiation site oriC (Hong and Kogoma 1993). Genetic deletions of the RNase H enzymes in *Saccharomyces cerevisiae* did not yield a distinct phenotype, leading to the suggestion that RNase H is not the primary pathway for RNA removal in those cells (Frank et al. 1998; Qiu et al. 1999). Fundamental differences among organisms include structural variations in the proteins involved, and length variations in the fragments. There are also as many as three pathways in eukaryotes, which involve different but overlapping sets of proteins (Balakrishnan and Bambara 2011b). Most recently, there is evidence that flow through these pathways is regulated to optimize fidelity and rate of synthesis (Balakrishnan and Bambara 2011b).

KORNBERG SIXTH COMMANDMENT: "DEPEND ON VIRUSES TO OPEN WINDOWS"

Most of our initial understanding of the process of DNA replication was obtained from studies in vitro using cellular extracts to replicate Simian Virus 40 (SV40) DNA, a short double-stranded circle (Waga et al. 1994; Bambara et al. 1997; Waga and Stillman 1998). Because SV40 encodes only the minimal proteins required for a viral assembly, it utilizes the cellular machinery from the host cell for replication (Tooze 1981). Studies by Stillman and coworkers (Waga et al. 1994; Waga and Stillman 1998) using SV40 as a model system showed that T antigen (encoded by the early viral genes), along with host single-stranded DNA-binding protein, called replication protein A (RPA), and topoisomerases, initiates DNA replication by binding to the origin (ori) and unwinding the duplex DNA (Tsurimoto et al. 1990). Reconstitution of SV40 DNA replication using purified proteins helped in the identification of specific enzymatic mechanisms used by the eukaryotic replication fork (Dean et al. 1992; Eki et al. 1992; Ishimi et al. 1992). Individual steps of this process and the enzymes involved are described in the sections below. Although the SV40 replication system in vitro provided valuable insights into eukaryotic replication, the absence of Pol ε as the leading-strand polymerase and the use of viral T antigen as the helicase for SV40 replication made it difficult to consider it as an absolutely analogous system to the actual more complex mammalian eukaryotic system.

PRIMING THE LAGGING STRAND—THE INITIAL STEP

Because DNA polymerases cannot initiate synthesis de novo, the replicative polymerase has evolved a partnership with primase to enable synthesis on a DNA template. DnaG acts as the primase in bacteria and in eukaryotes the primase is part of DNA polymerase α (Pol α) (Hubscher et al. 2002; Kuchta and Stengel 2010). Pol α is a four-subunit complex, comprised of the polymerase subunit (Pol I, p140), a regulatory subunit (Pol 12, p79), and two primase subunits. The Pri1 subunit (p48) acts as the fundamental initiator of replication by catalyzing the formation of the short RNA primers, which are subsequently elongated by Pol I (Hubscher et al. 2002). The second primase subunit, Pri2 (p58) acts as a scaffold, which holds together the primase and the polymerase subunits.

Primases frequently associate with helicases, greatly improving their affinities for ssDNA and increasing the number of primer initiation sites (Kuchta and Stengel 2010). Priming does not occur randomly and, with the exception of archael and *Aquifex aeolicus* primase, priming is generally initiated at pyrimidine residues because the eukaryotic primase binds more tightly to pyrimidine-rich than to purine-rich DNA (Holmes et al. 1985). The RNA/DNA primer synthesized by the primase/polymerase is known as the "initiator primer." Typically, during eukaryotic replication, primase synthesizes an RNA segment known as "initiator RNA (iRNA)," which is ~8–10 nt in length. A shorter primer may increase the possibility of slippage in the polymerase active site causing mutagenesis (Zhang and Grosse 1990; Kornberg and Baker 1992). Concentration of nucleotide triphosphates (NTPs) and deoxynucleotide triphosphates (dNTPs) in the cell can also account for varying primer lengths (Hauschka 1973). Priming of the DNA is the rate-limiting step in lagging-strand replication, with the rate of NTP polymerization by primase being at least two orders of magnitude slower than the rate of dNTP polymerization by Pol α (Sheaff and Kuchta 1993).

TRANSFER FROM PRIMASE TO DNA POLYMERASE

The ability of the primase to count the number of NTPs incorporated allows for the switch from the primase subunit to elongation by the DNA polymerizing part of Pol α (Qimron et al. 2006). Although the T7 primase synthesizes only a 4-nt-long primer, eukaryotic primases typically synthesize primers longer than 7 nt to enable efficient hand off of the primer-template from the primase to DNA Pol α (Sheaff et al. 1994). This is performed by the p58 subunit, which acts as the point of contact between the primase and the polymerase (Copeland and Wang 1993). The *E. coli* primase also switches the primer-template to Pol III directly but requires the assembly of the β clamp and the ssDNA-binding protein (SSB) (Yuzhakov et al. 1999). Because it lacks 3′–5′ exonuclease proofreading activity, Pol α is considered to be an error-prone polymerase. Pol α elongates the initiator RNA primer by the addition of 20–22 nt of "initiator DNA" (iDNA).

THE LEADING-STRAND REPLICATIVE POLYMERASE

Although studies using SV40 identified DNA polymerase δ (Pol δ) as the polymerase responsible for replicating both the leading and lagging strands, much recent evidence from the Kunkel laboratory has definitively shown DNA polymerase ε (Pol ε) to be the polymerase involved in leading-strand replication (Pursell et al. 2007). On account of its high processivity and association with PCNA, Pol ε can continually synthesize the leading strand (Waga and Stillman 1998). The lagging-strand polymerase, Pol δ, is made up of three subunits in *S. cerevisiae* (Pol 3, Pol 31, and Pol 32), and with the addition of a fourth subunit (Cdm1) in *Schizosaccharomyces pombe* and humans (Garg and Burgers 2005a). This fourth subunit acts to stabilize the polymerase holoenzyme (Podust et al. 2002). In addition to its polymerase function, Pol δ also possesses 3′–5′ exonuclease activity, allowing it to be a higher fidelity polymerase than Pol α (Pavlov et al. 2006).

THE MINIMAL ENZYMES INVOLVED IN BASIC OKAZAKI FRAGMENT PROCESSING

Helicases initially unwind the double-stranded DNA at specific sequences on the genome known as origins. In eukaryotes the replication protein A (RPA) coats the ssDNA to prevent it from reannealing, degradation by nucleases, recombination with other cellular ssDNA, or formation of hairpin structures that would normally obstruct replication fork progression (Wold 1997). RPA is a heterotrimeric protein composed of three subunits (70, 32, and 14 kDa). RPA also coordinates the assembly and disassembly of replication-associated proteins. The ATP-dependent replication factor C (RFC) binds to Pol α and triggers the switch from the priming mode to the elongation mode, the making of iDNA (Tsurimoto and Stillman 1990). RFC loads the proliferating cell nuclear antigen (PCNA) along with Pol δ to initiate the elongation on the lagging-strand DNA template (Tsurimoto and Stillman 1990). PCNA, a functional homolog of the prokaryotic β clamp, having a similar ring-shaped structure, is opened up with the help of RFC to be loaded onto the double-stranded iDNA (Burgers 2009). PCNA binds to the back of the Pol δ and acts as a sliding clamp, increasing the processivity of the polymerase. Pol δ adds ~100 nt of DNA in humans and ~250 nt of DNA in *S. cerevisiae* to form short Okazaki fragments, which need to be further matured to form a functional strand of DNA. During a single round of nuclear DNA replication in *S. cerevisiae* ~100,000 Okazaki fragments are made and matured (Garg and Burgers 2005b).

SIGNIFICANCE OF FRAGMENT SIZE

Despite the much larger DNA content of eukaryotic compared with prokaryotic cells, Okazaki fragments are ~1200 nt long in bacteria but only about 200 nt long in eukaryotes (Ogawa and Okazaki 1980). This means that to prepare for every human cell division, >10 million fragments must be made and joined. The bacterial fragments might be larger because the ligation process is much slower than the time needed to synthesize each fragment. Pol III holoenzyme elongates primers at 1200 nt/sec. This is consistent with the ability of bacteria to double in the range of a half-hour. Possibly the joining of the lagging strand could not keep up this pace if the fragments were shorter. Another possibility is that the nucleosomal structure of DNA influences the frequency of fragment priming. Is it coincidence that the average fragment size is similar to the length of DNA associated with a mononucleosome? Recent evidence using high-resolution analysis has shown that Okazaki fragments are sized according to chromatin repeats (Smith and Whitehouse 2012).

PATHWAYS OF EUKARYOTIC OKAZAKI FRAGMENT PROCESSING

Short Flap Pathway

Because eukaryotic lagging-strand DNA is primed at short intervals, Pol δ frequently encounters the downstream primed Okazaki fragment and displaces the RNA/DNA initiator primer into a 5′ flap structure. This strand displacing activity is very similar to that reported for bacterial Pol I. However, unlike Pol I, Pol δ does not possess a nuclease activity to cleave the displaced flap. Flap endonuclease 1 (FEN1; or scRad27), a structure-specific 5′-3′ endonuclease, recognizes the displaced 5′ flap and cleaves at the base creating a nicked substrate for ligation (Bambara et al. 1997; Lieber 1997). This is the predominant method of removing the Pol α-synthesized initiator primer during the maturation process. Initial studies characterizing the mechanism of FEN1 suggested a "tracking" model for FEN1 in which the nuclease moves from the 5′ end of the flap to its base where it performs a specific cleavage (Bambara et al. 1997). This mechanism was proposed because studies in vitro showed that endonucleolytic cleavage by FEN1 is inhibited when the 5′ end of the flap is blocked either with a complementary primer or a biotin conjugated streptavidin moiety (Murante et al. 1995). However, recent work has shown that FEN1 initially binds to the base of the flap causing a change in the substrate conformation that orients FEN1 in a manner that allows for precise cleavage (Gloor et al. 2010; Tsutakawa et al. 2011). The free 5′ end of the flap is then threaded past or through the helical arch and active site of FEN1 permitting a single cleavage event (Gloor et al. 2010).

DNA ligase I (cdc9 in *S. cerevisiae*) seals the nick generated by FEN1 to create a fully functional continuous double-stranded DNA (Bambara et al. 1997). PCNA interacts with both FEN1 and DNA ligase I and stimulates the enzymatic functions of both these proteins (Rossi et al. 2006). Interactions with PCNA are critical for the creation and proper ligation of the lagging-strand DNA. It has been previously suggested that PCNA serves to recruit the core enzymes to the replication fork and functions to sequentially hand off the proteins to perform their enzymatic tasks during the maturation process (Kao and Bambara 2003). Sequential strand displacement and cleavage by Pol δ and FEN1, respectively, helps to remove the entire initiator RNA before nick ligation. The wild-type Pol δ shows reduced strand displacement activity compared with its exonuclease mutants (Pol δ-5DV and Pol δ-01) (Garg and Burgers 2005b). Apparently the exonuclease regulates displacement (Garg and Burgers 2005b). It is still not very clear how the wild-type enzyme senses how many nucleotides of the RNA/DNA primer it has displaced. However, analysis of lagging-strand replication using purified *S. cerevisiae* proteins has shown that the majority of the substrates Pol δ exonuclease activity and FEN1 cleavage functions cooperate to create and cleave flaps shorter than 10 nt long (Ayyagari et al. 2003; Rossi and Bambara 2006). Several displacement and cleavage reactions are required to remove the initiator primer. A flap created and processed via this mechanism has matured by the "short flap pathway" (Fig. 2).

Long Flap Pathway

In some instances FEN1 transiently disengages from the replication complex. This can cause enough of a delay in cleavage that flaps displaced by Pol δ become long. When they reach lengths >22 nt, RPA can bind stably (Rossi and Bambara 2006). RPA-bound flaps are refractory to FEN1 cleavage, requiring the action of another nuclease for proper processing (Bae et al. 2001). Budd and Campbell identified this alternate nuclease, Dna2, in a genetic screen in *S. cerevisiae* (Budd and Campbell 1995). Dna2 nuclease/helicase is a multifunctional protein containing 5′–3′ endonuclease and minor 3′–5′ exonuclease activities, plus 5′–3′ helicase and ATPase functions (Kang et al. 2010). Campbell and colleagues showed that overexpression of Dna2 compensated for defects in FEN1 and overexpression of FEN1 did the same for Dna2 (Budd and Campbell 1997). More recent work showed that the double mutant of 5′–3′ nuclease-defective Dna2 (*dna2-1*) and 3′ nuclease-deficient Pol δ (*pol3-01*), which has augmented strand displacement activity, is lethal (Budd et al. 2005). These results imply that Dna2 works with FEN1 specifically to process long flaps. In addition, Dna2 was recently found to be complexed with FEN1 in human cell extracts, suggesting it as a physical and functional partner of FEN1 (Balakrishnan et al. 2010).

S. cerevisiae Dna2 can dissociate the RPA from a long flap (Bae et al. 2001; Stewart et al. 2008). Using a mechanism very similar to that of FEN1, Dna2 binds the flap base, and then threads the free 5′ end of the flap (Stewart et al. 2010). However, this nuclease cleaves periodically up to a terminal product flap ~5–6 nt in length. This is too short to bind RPA, so the short flap is readily available for cleavage by FEN1 and subsequent ligation. This method of flap processing is known as the "long flap pathway" (Fig. 2). Reconstitution experiments have shown that although the majority of the displaced flaps are processed by the short flap pathway, a minority require the long flap pathway. Seo and colleagues originally proposed the Dna2 pathway as the primary means of Okazaki fragment processing (Bae et al. 2001). Their proposal was likely influenced by genetic evidence in *S. cerevisiae* that Dna2 inactivation in cells was lethal, whereas FEN1 mutants only showed a slow-growing phenotype. However, these observations can be misleading. Exo1, a 5′–3′ exonuclease interacts with both FEN1 and Dna2 and can specifically act as a backup for FEN1 nuclease activity. Although this is an inefficient process, and probably not biologically relevant, it explains why *rad27Δ* (FEN1) mutants are viable. Also, recent work from the Campbell laboratory showed that Dna2 interacts with Rad9, the damage checkpoint activator, participating in the double-strand break repair response. The double deletion *dna2Δ* and *rad9Δ* rescued *dna2Δ* lethality suggesting that Rad9-dependent activation of the checkpoint contributed to the lethality in *dna2Δ* cells (Budd et al. 2011). In the absence of Dna2 and Rad9, the damage response utilizes the Exo1 pathway for repairing the damaged DNA (Balakrishnan and Bambara 2011a).

EVOLVED INTERACTIONS OF FEN1, Dna2, AND Pif1

Why did both FEN1 and Dna2 develop a mechanism in which the nucleases bind to the base of the flap and thread the free 5′ end through their

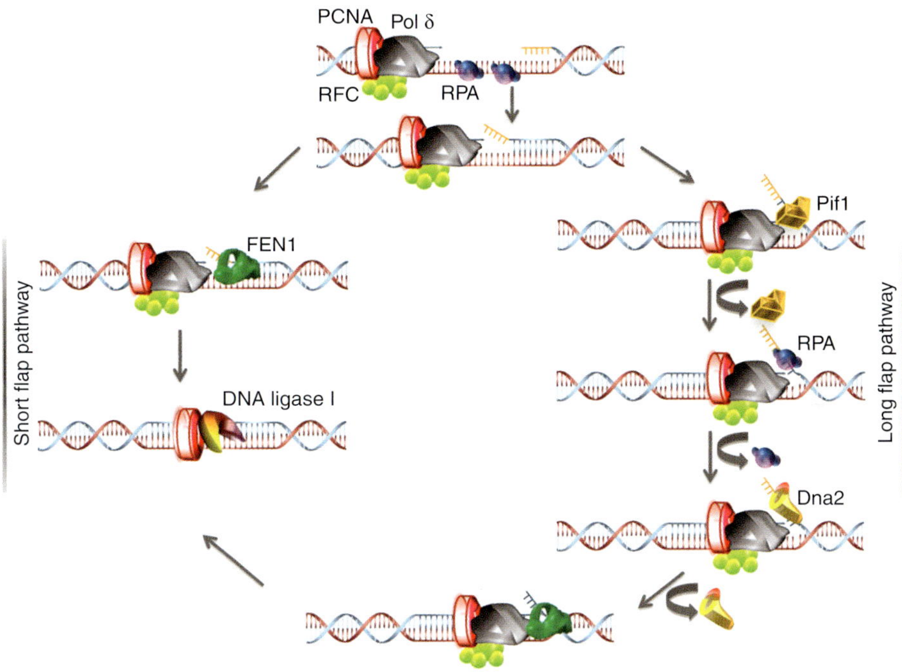

Figure 2. Okazaki fragment maturation. During Okazaki fragment maturation, (i) Pol δ displaces a short segment of the initiator primer into a 5′ flap; (ii) FEN1 recognizes the displaced flap, binds to the base of the flap and cleaves the flap; (iii) DNA ligase seals the nick; (iv) certain flaps are elongated by the action of the 5′–3′ helicase, Pif1; (v) the long flaps are stably coated by RPA; and (vi) Dna2 displaces RPA and cleaves the flap at multiple sites leaving a terminal product ~5–6 nt in length.

active site? A reasonable explanation is that the requirement to enter of a free 5′ end of a flap prevents these very active endonucleases from cleaving the single-stranded templates between Okazaki fragments, resulting in dangerous double-stranded breaks in the chromosome. Why should this minor long flap pathway have evolved? Reconstitution experiments showed that if flaps were being displaced in the presence of FEN1 and RPA, then FEN1 was able to overcome the inhibition by RPA (Rossi and Bambara 2006). This was likely owing to FEN1 binding the base of the flap and orienting itself in a position allowing for cleavage, before RPA binding could be inhibitory. This observation appeared to remove the need for Dna2 in the processing pathway. Hence, why did this protein evolve to interact with the replication proteins?

The answer to this question was provided by genetic studies in *S. pombe* wherein Seo and colleagues showed that Pfh1 (a homolog of *S. cerevisiae* Pif1) enhanced the strand displacement capabilities of Pol δ. Pif1 was capable of binding ahead of Pol δ to enhance flap creation in the downstream Okazaki fragment creating a longer 5′ flap substrate that would attract RPA binding. Importantly, in *S. cerevisiae*, deletion of the *PIF1* gene rescued the lethality of the *dna2Δ* strains (Budd et al. 2006). This result indicates that expression of Pif1 creates a need for Dna2. A reasonable interpretation is that Pif1 makes Dna2 necessary for long flap processing. In fact, inclusion of Pif1 in reconstitution assays augmented the amount of long flaps, and resulted in inhibition of FEN1 by RPA. Because Pif1 complicates fragment processing, why has it evolved to interact with the lagging-strand synthesis machinery? An answer was also suggested by Seo and colleagues (Ryu et al. 2004). Pif1 may fully displace some fragments. Additional reconstitution experiments suggest that fragments with sequences having the potential to form 5′ end region secondary structure are difficult to process. When the flap is created, it folds in a way that prevents cleavage by either FEN1 or Dna2. However, Pif1 can bind between the structure and the flap base and fully displace the fragment. The upstream fragment can then extend through the structured region. This is effectively a third pathway of fragment processing (Pike et al. 2010). Although the long flap pathway has evolved elegantly to process such flaps, evidence in vitro suggests that the short flap pathway is more commonly used, with each protein from the long flap pathway (Dna2, Pif1, and RPA) stimulating the function of FEN1, and so promoting the use of the short flap pathway (Henry et al. 2010).

MECHANISTIC SIMILARITIES OF OKAZAKI FRAGMENT PROCESSING AND LONG PATCH BASE EXCISION REPAIR

Cells are constantly being exposed to endogenous and exogenous stresses that cause oxidative damage to DNA bases. Base excision repair (BER) is the most commonly utilized means of dealing with these damaged bases. It proceeds via two pathways. In the short patch pathway (SP-BER) a damaged base is recognized by a DNA glycosylase and removed. The abasic site is cleaved by apurinic/apyrimidinic endonuclease 1 (APE1). The one-nucleotide gap is filled by DNA polymerase β (Pol β), which also removes the 5′ deoxyribose phosphate (dRP) using an intrinsic lyase activity. However, if the dRP is oxidized, reduced, or otherwise altered, the lyase function does not work. The Pol β then displaces the damaged site into a 2–12 nt 5′ flap (Balakrishnan et al. 2009). These flaps are directed down the long patch pathway (LP-BER) in which they are cleaved by FEN1 and sealed by DNA ligase I. It is not clear whether the longest flaps would bind RPA with sufficient avidity to require Dna2; however, the recent report of Dna2 involvement in mitochondrial LP-BER suggests that some do (Zheng et al. 2008). The overlap of proteins used for Okazaki fragment processing and LP-BER suggests that the two processes evolved from the same ancestral basic pathway. The overlap also suggests that regulatory mechanisms for one process will similarly influence the other.

POSTTRANSLATIONAL REGULATION OF REPLICATION PROTEINS

Most of the proteins involved in eukaryotic DNA replication experience various posttranslational modifications, regulating their enzymatic functions, subcellular localizations, or participation in a specific pathway. Pol α/primase has been reported to be phosphorylated both in humans and *S. cerevisiae*. Phosphorylation of Pol α occurs late in S phase, thereby possibly coordinating the S phase with the mitotic phase. Results obtained in vitro suggest that although the polymerase activity of Pol α is not altered by phosphorylation, the primase function is slightly stimulated without affecting the length of the initiator primer (Waga and Stillman 1998). Recent evidence from the Lee laboratory has shown that phosphorylation of Pol δ on the p68 subunit decreases the binding affinity of the polymerase to PCNA (Rahmeh et al. 2012). This in turn decreases the processivity of polymerization. Pol δ was also found to be acetylated on the catalytic subunit in a mass spectrometric analysis (Choudhary et al. 2009). PCNA is modified by a diverse range of modifications such as acetylation, phosphorylation, and ubiquitination. Although acetylation improves the binding affinity to PCNA to DNA polymerases (Naryzhny and Lee 2004), tyrosine phosphorylation of PCNA controls the protein stability (Wang et al. 2006). Ubiquitination and sumoylation alter the pathways in which PCNA functions (Papouli et al. 2005). FEN1 is posttranslationally modified by phosphorylation, methylation, and acetylation. Phosphorylation of FEN1 decreases its binding affinity to PCNA, whereas methylation prevents phosphorylation of FEN1 (Zheng and Shen 2011). Acetylation greatly diminishes the cleavage function of FEN1 (Hasan et al. 2001). Interestingly, a recent report shows that phosphorylation of FEN1 stimulates its sumoylation. This subsequently helps in the ubiquitination of the protein leading to degradation via the proteosome pathway, thereby regulating the levels of FEN1 in the cell

(Guo et al. 2012). Dna2 is also phosphorylated and acetylated. Phosphorylation of Dna2 improves recruitment to sites of double-strand breaks (DSBs) (Chen et al. 2011), whereas acetylation of Dna2 greatly alters its enzymatic activities (Balakrishnan et al. 2010). RPA is also modified by phosphorylation (on the 70 and 32 kDa subunits) in response to DSBs, which in turn allows the RPA to help in the recruitment of DSB response proteins (Wold 1997). Finally, RPA was found to be acetylated on the 70 kDa subunit (Choudhary et al. 2009).

REGULATION OF PATHWAYS BY ACETYLATION

Although most posttranslational modifications alter the function of individual proteins, recent evidence of the regulation of proteins by acetylation suggests a coherent hypothesis in which acetylation of replication proteins regulates the choice of a specific pathway for Okazaki fragment maturation. It has been known for some time that FEN1 can be acetylated by the histone acetyltransferase p300 (Hasan et al. 2001). Complete acetylation reduces the cleavage activity by about 90%. The reason why the cell would want to down-regulate FEN1 activity to that degree was initially unclear, because when expression of FEN1 was knocked out on one of the two chromosomes in diploid cells, the 50% reduction in cleavage activity resulted in genomic damage (Kucherlapati et al. 2002). We now know that this phenomenon makes more sense when viewed in the context of regulation of most lagging-strand replication proteins by acetylation. Suggestions of a more global regulation mechanism came from analyses of the effects of acetylation on the properties of other lagging-strand replication proteins.

The p300 acetylase also reacts with Dna2, with multifold stimulation of nuclease, helicase, and ATPase activities (Balakrishnan et al. 2010). Notably, alteration of the helicase activity allows Dna2 to drive the nuclease active site to the base of the original flap, and on some flaps even farther. This results in a shifted cleavage distribution farther downstream. Moreover, Pol δ is acetylated on the catalytic subunit. Preliminary results show that acetylation of the polymerase greatly augments its ability to perform strand displacement synthesis. The combined effect is that flaps are created faster by the modified Pol δ. This would normally be a genome stability problem because long flaps can form secondary structures that inhibit processing, and can recombine at ectopic sites. However, the increased efficiency of Dna2 must prevent the flaps from actually achieving great length. Instead, displacement will occur for a greater distance, and ultimate ligation will be delayed because the lowered activity of FEN1 will not be able to rapidly create nicks. The overall consequence of the modification is that, without actually making long flap intermediates, a longer patch of the downstream fragment would be removed and replaced.

Why would this be desirable? In higher eukaryotes, millions of 150- to 200-nt fragments are needed to make the lagging strand. If an average of 40 nucleotides are removed and replaced from each, then ~23% of the lagging strand has to be made twice. This would only be desirable if it protects DNA that provides the organism with a selective advantage. A reasonable interpretation is that lagging-strand replication is selectively regulated for fidelity. Pol α makes the primer for each fragment, but then adds an additional ~20 nt of DNA. Because Pol α lacks a proofreading exonuclease, those nucleotides are added at lower fidelity than the later nucleotides added by the much more accurate Pol δ. As a consequence, each fragment can be viewed as having a gradient of potential errors decreasing from the 5′ to the 3′ end. Because of this gradient, replacement of a long patch of the Pol α-synthesized nucleotides with extension from an upstream primer by Pol δ would correct a very high proportion of replication errors in the lagging strand.

Why isn't this corrective effect constitutive? The p300 acetylase activates selected areas of chromatin for gene expression. Its distribution to these areas suggests that it preferentially acetylates replication/repair proteins for synthesis of active genes. Possibly Pol α synthesis patch full replacement has evolved into a regulated process because the cell tries to protect actively transcribed DNA but replicates most other DNA with unacetylated proteins, in an efficient but less accurate manner.

SUMMARY

Although Okazaki fragment processing is one of the fundamental processes of life, it can be optimized in any particular organism for speed, fidelity, energy consumption, or some combination. Speed and energy consumption would appear to be most important in bacteria because they are competing with other rapidly growing cells. Moreover, occasional lethal mutations should not affect the success of the population. The result is the evolution of a long fragment mechanism. Higher eukaryotes appear to have developed processing that is optimized for fidelity in active genes. This would appear best for survival through development, delay of cancers, and a long average life span.

ACKNOWLEDGMENTS

This work was supported by National Institutes of Health grant GM098328 to L.B. and GM024441 to R.A.B. We especially thank Christopher Petrides and Athena Kantartzis for assistance with the figures.

REFERENCES

Alberts BM, Barry K, Bedinger P, Formosa T, Jongeneel CV, Kreuzer KN. 1982. Studies on DNA replication in the T4 bacteriophage in vitro system. *Cold Spring Harb Symp Quant Biol* **47:** 655–668.

Ayyagari R, Gomes XV, Gordenin DA, Burgers PM. 2003. Okazaki fragment maturation in yeast. I. Distribution of functions between FEN1 AND DNA2. *J Biol Chem* **278:** 1618–1625.

Bae SH, Bae KH, Kim JA, Seo YS. 2001. RPA governs endonuclease switching during processing of Okazaki fragments in eukaryotes. *Nature* **412:** 456–461.

Balakrishnan L, Bambara RA. 2011a. The changing view of Dna2. *Cell Cycle* **10:** 2620–2621.

Balakrishnan L, Bambara RA. 2011b. Eukaryotic lagging strand DNA replication employs a multi-pathway mechanism that protects genome integrity. *J Biol Chem* **286:** 6865–6870.

Balakrishnan L, Brandt PD, Lindsey-Boltz LA, Sancar A, Bambara RA. 2009. Long patch base excision repair proceeds via coordinated stimulation of the multienzyme DNA repair complex. *J Biol Chem* **284:** 15158–15172.

Balakrishnan L, Stewart J, Polaczek P, Campbell JL, Bambara RA. 2010. Acetylation of Dna2 endonuclease/helicase and flap endonuclease 1 by p300 promotes DNA stability by creating long flap intermediates. *J Biol Chem* **285:** 4398–4404.

Bambara RA, Murante RS, Henricksen LA. 1997. Enzymes and reactions at the eukaryotic DNA replication fork. *J Biol Chem* **272:** 4647–4650.

Budd ME, Campbell JL. 1995. A yeast gene required for DNA replication encodes a protein with homology to DNA helicases. *Proc Natl Acad Sci* **92:** 7642–7646.

Budd ME, Campbell JL. 1997. A yeast replicative helicase, Dna2 helicase, interacts with yeast FEN-1 nuclease in carrying out its essential function. *Mol Cell Biol* **17:** 2136–2142.

Budd ME, Tong AH, Polaczek P, Peng X, Boone C, Campbell JL. 2005. A network of multi-tasking proteins at the DNA replication fork preserves genome stability. *PLoS Genet* **1:** e61.

Budd ME, Reis CC, Smith S, Myung K, Campbell JL. 2006. Evidence suggesting that Pif1 helicase functions in DNA replication with the Dna2 helicase/nuclease and DNA polymerase delta. *Mol Cell Biol* **26:** 2490–2500.

Budd ME, Antoshechkin IA, Reis C, Wold BJ, Campbell JL. 2011. Inviability of a DNA2 deletion mutant is due to the DNA damage checkpoint. *Cell Cycle* **10:** 1690–1698.

Burgers PM. 2009. Polymerase dynamics at the eukaryotic DNA replication fork. *J Biol Chem* **284:** 4041–4045.

Cerritelli SM, Crouch RJ. 2009. Ribonuclease H: The enzymes in eukaryotes. *FEBS J* **276:** 1494–1505.

Chen X, Niu H, Chung WH, Zhu Z, Papusha A, Shim EY, Lee SE, Sung P, Ira G. 2011. Cell cycle regulation of DNA double-strand break end resection by Cdk1-dependent Dna2 phosphorylation. *Nat Struct Mol Biol* **18:** 1015–1019.

Choudhary C, Kumar C, Gnad F, Nielsen ML, Rehman M, Walther TC, Olsen JV, Mann M. 2009. Lysine acetylation targets protein complexes and co-regulates major cellular functions. *Science* **325:** 834–840.

Copeland WC, Wang TS. 1993. Enzymatic characterization of the individual mammalian primase subunits reveals a biphasic mechanism for initiation of DNA replication. *J Biol Chem* **268:** 26179–26189.

Dean FB, Borowiec JA, Eki T, Hurwitz J. 1992. The simian virus 40 T antigen double hexamer assembles around the DNA at the replication origin. *J Biol Chem* **267:** 14129–14137.

Eki T, Matsumoto T, Murakami Y, Hurwitz J. 1992. The replication of DNA containing the simian virus 40 origin by the monopolymerase and dipolymerase systems. *J Biol Chem* **267:** 7284–7294.

Frank G, Qiu J, Somsouk M, Weng Y, Somsouk L, Nolan JP, Shen B. 1998. Partial functional deficiency of E160D flap endonuclease-1 mutant in vitro and in vivo is due to defective cleavage of DNA substrates. *J Biol Chem* **273:** 33064–33072.

Garg P, Burgers PM. 2005a. DNA polymerases that propagate the eukaryotic DNA replication fork. *Crit Rev Biochem Mol Biol* **40:** 115–128.

Garg P, Burgers PM. 2005b. How the cell deals with DNA nicks. *Cell Cycle* **4:** 221–224.

Gloor JW, Balakrishnan L, Bambara RA. 2010. Flap endonuclease 1 mechanism analysis indicates flap base binding prior to threading. *J Biol Chem* **285:** 34922–34931.

Guo Z, Kanjanapangka J, Liu N, Liu S, Liu C, Wu Z, Wang Y, Loh T, Kowolik C, Jamsen J, et al. 2012. Sequential post-translational modifications program FEN1 degradation during cell-cycle progression. *Mol Cell* **47:** 444–456.

Hasan S, Stucki M, Hassa PO, Imhof R, Gehrig P, Hunziker P, Hubscher U, Hottiger MO. 2001. Regulation of human flap endonuclease-1 activity by acetylation through the transcriptional coactivator p300. *Mol Cell* **7:** 1221–1231.

Hauschka PV. 1973. Analysis of nucleotide pools in animal cells. *Methods Cell Biol* **7:** 361–462.

Henry RA, Balakrishnan L, Ying-Lin ST, Campbell JL, Bambara RA. 2010. Components of the secondary pathway stimulate the primary pathway of eukaryotic Okazaki fragment processing. *J Biol Chem* **285:** 28496–28505.

Holmes AM, Cheriathundam E, Bollum FJ, Chang LM. 1985. Initiation of DNA synthesis by the calf thymus DNA polymerase-primase complex. *J Biol Chem* **260:** 10840–10846.

Hong X, Kogoma T. 1993. Absence of a direct role for RNase HI in initiation of DNA replication at the oriC site on the *Escherichia coli* chromosome. *J Bacteriol* **175:** 6731–6734.

Hubscher U, Maga G, Spadari S. 2002. Eukaryotic DNA polymerases. *Annu Rev Biochem* **71:** 133–163.

Ishimi Y, Sugasawa K, Hanaoka F, Eki T, Hurwitz J. 1992. Topoisomerase II plays an essential role as a swivelase in the late stage of SV40 chromosome replication in vitro. *J Biol Chem* **267:** 462–466.

Kang YH, Lee CH, Seo YS. 2010. Dna2 on the road to Okazaki fragment processing and genome stability in eukaryotes. *Crit Rev Biochem Mol Biol* **45:** 71–96.

Kao HI, Bambara RA. 2003. The protein components and mechanism of eukaryotic Okazaki fragment maturation. *Crit Rev Biochem Mol Biol* **38:** 433–452.

Kornberg A, Baker TA. 1992. *DNA replication*. W.H. Freeman, New York.

Kucherlapati M, Yang K, Kuraguchi M, Zhao J, Lia M, Heyer J, Kane MF, Fan K, Russell R, Brown AM, et al. 2002. Haploinsufficiency of Flap endonuclease (Fen1) leads to rapid tumor progression. *Proc Natl Acad Sci* **99:** 9924–9929.

Kuchta RD, Stengel G. 2010. Mechanism and evolution of DNA primases. *Biochim Biophys Acta* **1804:** 1180–1189.

Lieber MR. 1997. The FEN-1 family of structure-specific nucleases in eukaryotic DNA replication, recombination and repair. *Bioessays* **19:** 233–240.

Murante RS, Rust L, Bambara RA. 1995. Calf 5′ to 3′ exo/endonuclease must slide from a 5′ end of the substrate to perform structure-specific cleavage. *J Biol Chem* **270:** 30377–30383.

Murante RS, Henricksen LA, Bambara RA. 1998. Junction ribonuclease: An activity in Okazaki fragment processing. *Proc Natl Acad Sci* **95:** 2244–2249.

Naryzhny SN, Lee H. 2004. The post-translational modifications of proliferating cell nuclear antigen: Acetylation, not phosphorylation, plays an important role in the regulation of its function. *J Biol Chem* **279:** 20194–20199.

Ogawa T, Okazaki T. 1980. Discontinuous DNA replication. *Annu Rev Biochem* **49:** 421–457.

Okazaki R, Okazaki T, Sakabe K, Sugimoto K, Sugino A. 1968. Mechanism of DNA chain growth. I. Possible discontinuity and unusual secondary structure of newly synthesized chains. *Proc Natl Acad Sci* **59:** 598–605.

Pandey M, Syed S, Donmez I, Patel G, Ha T, Patel SS. 2009. Coordinating DNA replication by means of priming loop and differential synthesis rate. *Nature* **462:** 940–943.

Papouli E, Chen S, Davies AA, Huttner D, Krejci L, Sung P, Ulrich HD. 2005. Crosstalk between SUMO and ubiquitin on PCNA is mediated by recruitment of the helicase Srs2p. *Mol Cell* **19:** 123–133.

Pavlov YI, Frahm C, Nick McElhinny SA, Niimi A, Suzuki M, Kunkel TA. 2006. Evidence that errors made by DNA polymerase α are corrected by DNA polymerase δ. *Curr Biol* **16:** 202–207.

Pike JE, Henry RA, Burgers PM, Campbell JL, Bambara RA. 2010. An alternative pathway for Okazaki fragment processing: Resolution of fold-back flaps by Pif1 helicase. *J Biol Chem* **285:** 41712–41723.

Podust VN, Chang LS, Ott R, Dianov GL, Fanning E. 2002. Reconstitution of human DNA polymerase δ using recombinant baculoviruses: The p12 subunit potentiates DNA polymerizing activity of the four-subunit enzyme. *J Biol Chem* **277:** 3894–3901.

Pursell ZF, Isoz I, Lundstrom EB, Johansson E, Kunkel TA. 2007. Yeast DNA polymerase epsilon participates in leading-strand DNA replication. *Science* **317:** 127–130.

Qimron U, Lee SJ, Hamdan SM, Richardson CC. 2006. Primer initiation and extension by T7 DNA primase. *EMBO J* **25:** 2199–2208.

Qiu J, Qian Y, Frank P, Wintersberger U, Shen B. 1999. *Saccharomyces cerevisiae* RNase H(35) functions in RNA primer removal during lagging-strand DNA synthesis, most efficiently in cooperation with Rad27 nuclease. *Mol Cell Biol* **19:** 8361–8371.

Rahmeh AA, Zhou Y, Xie B, Li H, Lee EY, Lee MY. 2012. Phosphorylation of the p68 subunit of Pol δ acts as a molecular switch to regulate its interaction with PCNA. *Biochemistry* **51:** 416–424.

Rossi ML, Bambara RA. 2006. Reconstituted Okazaki fragment processing indicates two pathways of primer removal. *J Biol Chem* **281:** 26051–26061.

Rossi ML, Purohit V, Brandt PD, Bambara RA. 2006. Lagging strand replication proteins in genome stability and DNA repair. *Chem Rev* **106:** 453–473.

Ryu GH, Tanaka H, Kim DH, Kim JH, Bae SH, Kwon YN, Rhee JS, MacNeill SA, Seo YS. 2004. Genetic and biochemical analyses of Pfh1 DNA helicase function in fission yeast. *Nucleic Acids Res* **32:** 4205–4216.

Sakabe K, Okazaki R. 1966. A unique property of the replicating region of chromosomal DNA. *Biochem Biophys Acta* **129:** 651–654.

Sheaff RJ, Kuchta RD. 1993. Mechanism of calf thymus DNA primase: Slow initiation, rapid polymerization, and intelligent termination. *Biochemistry* **32:** 3027–3037.

Sheaff RJ, Kuchta RD, Ilsley D. 1994. Calf thymus DNA polymerase α-primase: "Communication" and primer-template movement between the two active sites. *Biochemistry* **33:** 2247–2254.

Smith D, Whitehouse I. 2012. Intrinsic coupling of lagging strand synthesis to chromatin assembly. *Nature* **483:** 434–438.

Stewart JA, Miller AS, Campbell JL, Bambara RA. 2008. Dynamic removal of replication protein A by Dna2 facilitates primer cleavage during Okazaki fragment processing in *Saccharomyces cerevisiae*. *J Biol Chem* **283:** 31356–31365.

Stewart JA, Campbell JL, Bambara RA. 2010. Dna2 is a structure-specific nuclease, with affinity for 5′-flap intermediates. *Nucleic Acids Res* **38:** 920–930.

Tooze J. 1981. *DNA tumor viruses*, 2nd ed. Cold Spring Harbor Laboratory Press, Cold Spring Harbor, NY.

Tsurimoto T, Stillman B. 1990. Functions of replication factor C and proliferating-cell nuclear antigen: Functional similarity of DNA polymerase accessory proteins from human cells and bacteriophage T4. *Proc Natl Acad Sci* **87:** 1023–1027.

Tsurimoto T, Melendy T, Stillman B. 1990. Sequential initiation of lagging and leading strand synthesis by two different polymerase complexes at the SV40 DNA replication origin. *Nature* **346:** 534–539.

Tsutakawa SE, Classen S, Chapados BR, Arvai AS, Finger LD, Guenther G, Tomlinson CG, Thompson P, Sarker AH, Shen B, et al. 2011. Human flap endonuclease structures, DNA double-base flipping, and a unified understanding of the FEN1 superfamily. *Cell* **145:** 198–211.

Turchi JJ, Huang L, Murante RS, Kim Y, Bambara RA. 1994. Enzymatic completion of mammalian lagging-strand DNA replication. *Proc Natl Acad Sci* **91:** 9803–9807.

Waga S, Stillman B. 1998. The DNA replication fork in eukaryotic cells. *Annu Rev Biochem* **67:** 721–751.

Waga S, Bauer G, Stillman B. 1994. Reconstitution of complete SV40 DNA replication with purified replication factors. *J Biol Chem* **269:** 10923–10934.

Wang SC, Nakajima Y, Yu YL, Xia W, Chen CT, Yang CC, McIntush EW, Li LY, Hawke DH, Kobayashi R, et al. 2006. Tyrosine phosphorylation controls PCNA function through protein stability. *Nat Cell Biol* **8:** 1359–1368.

Wold MS. 1997. Replication protein A: A heterotrimeric, single-stranded DNA-binding protein required for eukaryotic DNA metabolism. *Annu Rev Biochem* **66:** 61–92.

Yuzhakov A, Kelman Z, O'Donnell M. 1999. Trading places on DNA—A three-point switch underlies primer handoff from primase to the replicative DNA polymerase. *Cell* **96:** 153–163.

Zhang SS, Grosse F. 1990. Accuracy of DNA primase. *J Molec Biol* **216:** 475–479.

Zheng L, Shen B. 2011. Okazaki fragment maturation: Nucleases take centre stage. *J Mol Cell Biol* **3:** 23–30.

Zheng L, Zhou M, Guo Z, Lu H, Qian L, Dai H, Qiu J, Yakubovskaya E, Bogenhagen DF, Demple B, et al. 2008. Human DNA2 is a mitochondrial nuclease/helicase for efficient processing of DNA replication and repair intermediates. *Mol Cell* **32:** 325–336.

Chromatin and DNA Replication

David M. MacAlpine[1] and Geneviève Almouzni[2]

[1]Department of Pharmacology and Cancer Biology, Duke University, Durham, North Carolina 27710
[2]Institut Curie, Section de Recherche, UMR218 du CNRS, 75231 Paris Cedex 05, France
Correspondence: david.macalpine@duke.edu

The size of a eukaryotic genome presents a unique challenge to the cell: package and organize the DNA to fit within the confines of the nucleus while at the same time ensuring sufficient dynamics to allow access to specific sequences and features such as genes and regulatory elements. This is achieved via the dynamic nucleoprotein organization of eukaryotic DNA into chromatin. The basic unit of chromatin, the nucleosome, comprises a core particle with 147 bp of DNA wrapped 1.7 times around an octamer of histones. The nucleosome is a highly versatile and modular structure, both in its composition, with the existence of various histone variants, and through the addition of a series of posttranslational modifications on the histones. This versatility allows for both short-term regulatory responses to external signaling, as well as the long-term and multigenerational definition of large functional chromosomal domains within the nucleus, such as the centromere. Chromatin organization and its dynamics participate in essentially all DNA-templated processes, including transcription, replication, recombination, and repair. Here we will focus mainly on nucleosomal organization and describe the pathways and mechanisms that contribute to assembly of this organization and the role of chromatin in regulating the DNA replication program.

HISTONE VARIANTS AND MODIFICATIONS

The histone octamer consists of two molecules of each of the core histones H3, H4, H2A, and H2B. The H3 and H4 histones form a tetramer $(H3-H4)_2$, that organizes the central 70 bp of the DNA for the further addition of the two flanking H2A–H2B dimers (Luger et al. 1997). The constituent core histones (H2A, H2B, H3, and H4) represent the bulk of nucleosome-associated histones. However, multiple histone variants contribute to the diversity of chromatin structure and function (Ahmad and Henikoff 2002a; Kamakaka and Biggins 2005; Probst et al. 2009). Sequence variants are found in metazoans for all of the histones, except for histone H4. Distinct assembly mechanisms exist for the different variants, allowing them to be deposited in a replication-dependent or -independent manner (see below), and may contribute to their role in defining epigenetic states.

H3 Variants

Centromere-specific H3 variants are found in all eukaryotes and are referred to as CenH3s. The mammalian centromere-associated protein (CENP-A) is an essential protein that defines the location and function of centromeres

Cite this article as *Cold Spring Harb Perspect Biol* doi: 10.1101/cshperspect.a010207

(Stimpson and Sullivan 2010). The various CenH3s have a conserved globular domain but feature unique amino-terminal tails that are important for kinetochore function. The rapid evolution of centromeric repeats coupled with the adaptive evolution of CenH3 (Henikoff et al. 2001) underscores the importance of variant histones in maintaining epigenetic identity and state.

All metazoans possess a replication-dependent histone variant corresponding to H3.2 in mammals. Additional replicative variants are also present in mammals, such as H3.1, which differs from H3.2 by a single amino acid. The expression of H3.1 and H3.2 is tightly coupled to the cell cycle, showing a peak of synthesis during S phase and providing the major source of histones for deposition behind the replication fork.

H3.3 is a replacement variant that differs from H3.2 by four amino acids and is enriched in specific regions of the genome. Constitutively expressed throughout the cell cycle and in quiescent cells, it is deposited onto DNA in a replication-independent manner (Ahmad and Henikoff 2002b; Tagami et al. 2004; Ray-Gallet et al. 2011). The eviction of nucleosomes by transcription and chromatin remodeling events provides opportunities for deposition of newly synthesized H3.3. This can explain how H3.3 marks active and dynamic chromatin and accumulates in transcribed regions, enhancers, promoters (Mito et al. 2007), and origins of DNA replication (Deal et al. 2010; MacAlpine et al. 2010). This deposition depends primarily on a separate assembly pathway (ASF1/HIRA) discussed below.

H2A/H2B Variants

H2A.Z is the most conserved H2A variant across different species, even more so than the canonical H2A. H2A.Z is enriched at the flanks of nucleosome-depleted regions surrounding active promoters and promotes gene activation (Raisner and Madhani 2006). In *Saccharomyces cerevisiae*, H2A.Z acts as a barrier at promoters, preventing the spread of silent heterochromatin (Meneghini et al. 2003).

H2A.X represents 2%–25% of the mammalian H2A histone pool and is structurally similar to the other H2A variants except for a serine (Ser-139) positioned four amino acids from the carboxyl terminus (Kinner et al. 2008). Ser-139 is phosphorylated in response to DNA damage, resulting in the accumulation of phospho-H2A.X (also referred to as γ-H2A.X) in chromatin surrounding the DNA damage lesion. Subsequent repair activities are recruited to the lesion via interactions with γ-H2A.X (Stucki et al. 2005). Of note, *S. cerevisiae* has a single H2A mostly related to H2A.X, whereas *Drosophila* has a single bifunctional histone variant, H2A.v, that combines the properties of H2A.Z and H2A.X.

MacroH2A is a histone variant with a large carboxy-terminal ligand-binding domain called the macro domain (Pehrson and Fried 1992). MacroH2A is found enriched on the inactive X chromosome (Costanzi and Pehrson 1998). Conversely, H2A.BbD (Barr body deficient) is a divergent variant specific to vertebrates that is excluded from the inactive X chromosome and colocalizes with acetylated histone H4, suggesting a role in maintaining euchromatic function (Chadwick and Willard 2001).

Temporal Expression of Histones and Variants

In each cell cycle, a sufficient amount of core histones must be synthesized to provide ~20 million new nucleosomes for packaging the newly replicated daughter strands. Not surprisingly, the synthesis of the canonical core histones is tightly coupled to the cell cycle (Marzluff and Duronio 2002; Gunjan et al. 2005). Both transcriptional and posttranscriptional regulatory controls exist to ensure sufficient histone levels and prevent the accumulation of excess histones. Interestingly, histone mRNAs lack a poly(A) tail and instead have a stem-loop structure at the 3′ end, which is important for stability and translation in mammals (Marzluff et al. 2008). Failure to regulate histone levels can have profound consequences on cell-cycle progression and genome stability. Insufficient histone levels can trigger a cell-cycle arrest in *S. cerevisiae* (Han et al. 1987; Kim et al. 1988) and impair S-phase progression in mammalian systems (Nelson et al. 2002). Similarly, excess histones are also harmful, resulting in DNA damage and genome instability (Gunjan and Verreault 2003).

Posttranslational Modifications

The abundance of lysine residues within the nonconserved histone tails and their ability to be posttranslationally modified provides for a massive combinatorial repertoire, denoted a "histone code," and has the potential to regulate many chromatin-templated functions (Jenuwein and Allis 2001). These regulatory functions potentially include transcription, replication, repair, recombination, and chromatin condensation and segregation. Posttranslational modification (PTM) of histones includes, but is not limited to, methylation, acetylation, SUMOylation, ubiquitination, and ribosylation of lysine residues and phosphorylation on serine and threonine residues (Kouzarides 2007). Notably, these modifications are reversible. Finally, several of these PTMs (and histone variants) can contribute to the specific marking of chromatin states, which in some cases can be stably propagated through multiple cell divisions; as such, they are believed to be purveyors of epigenetic information. Given the importance of not only the nucleosomal structure itself for genome functions but also the dynamic regulatory properties of chromatin, it is important to understand the mechanisms that govern the disassembly and reassembly of chromatin states following passage of the replication fork.

ASSEMBLY OF CHROMATIN

An immediate consequence of DNA replication is the disruption of the existing chromatin structure by passage of the replication fork. Two pathways contribute to the reassembly of chromatin on the nascent DNA. In the first, disruption of the DNA–histone octamer interaction ahead of the replication fork generates parental histones that can be recycled behind the fork, and the second pathway facilitates the deposition of newly synthesized histones onto nascent DNA (Fig. 1). The segregation of parental histones combined with the de novo deposition of newly synthesized histones are critical for genome stability and the inheritance of chromatin states.

Replication-Dependent Deposition of Histones

Our understanding of the mechanisms and factors involved in histone deposition began with pioneering cell-free systems enabling chromatin assembly in *Xenopus laevis* egg extracts (Laskey et al. 1977). Next, the in vitro replication system using SV40 origin-containing plasmids along with the SV40 large T antigen and human cytosolic extracts provided a complementation assay in which addition of nuclear extracts enabled their efficient assembly into minichromosomes (Stillman 1986). This assay led to the biochemical identification of human chromatin assembly factor 1 (CAF-1), a histone chaperone with the unique property of promoting deposition of histones H3–H4 onto replicating DNA (Smith and Stillman 1989), a function that is evolutionarily conserved. The CAF-1 complex comprises three subunits (p150, p60, and RbAp48 in mammals) (Kaufman et al. 1995; Verreault et al. 1996). CAF-1 is targeted to replication forks through an interaction with proliferating cell nuclear antigen (PCNA), a ring-shaped homotrimeric protein that serves as a processivity factor for the DNA polymerases (Shibahara and Stillman 1999; Moggs et al. 2000). This CAF-1–PCNA interaction depends on phosphorylation of the large subunit of CAF-1 (p150) by the replicative kinase Cdc7-Dbf4 in human cells (Gerard et al. 2006), which offers a potential means to ensure a tight coordination between histone deposition and ongoing DNA replication. The importance of CAF-1 in vivo is shown by loss-of-function studies leading to loss of viability during development in mouse (Houlard et al. 2006), *Xenopus* (Quivy et al. 2001), and *Drosophila* (Song et al. 2007; Klapholz et al. 2009) and impaired S-phase progression in human cells (Hoek and Stillman 2003).

Together with CAF-1, another H3–H4 histone chaperone, antisilencing function 1 (ASF1), identified initially in a yeast screen for silencing

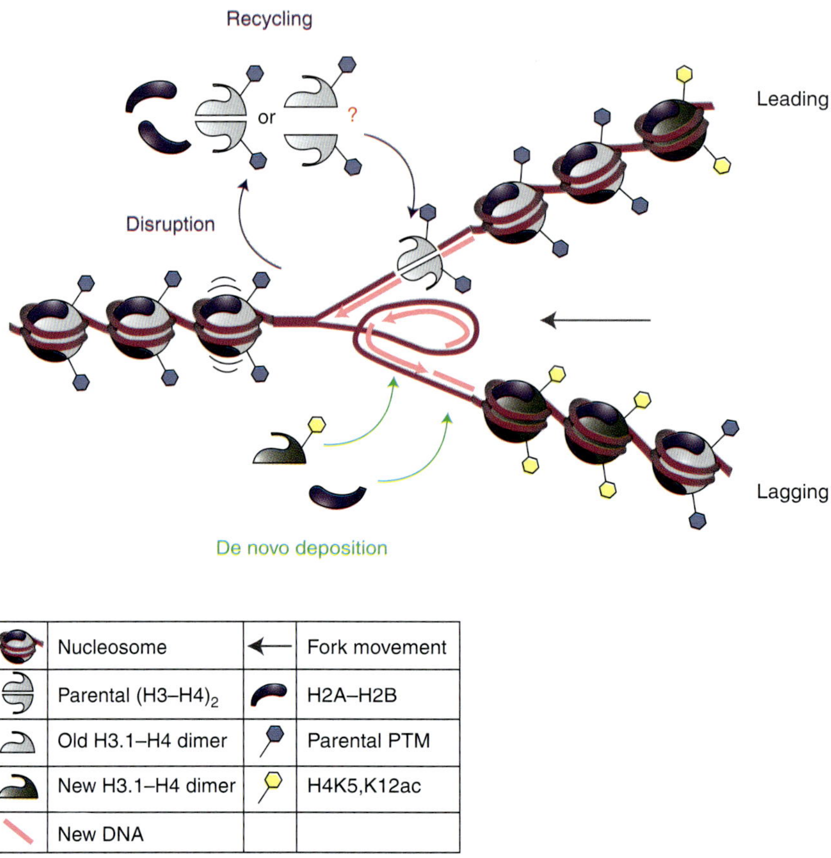

Figure 1. Histone dynamics at the replication fork. New and parental histones are incorporated into chromatin behind the replication fork. The disassembly of nucleosomes ahead of the replication fork provides a parental pool of H3–H4 tetramers or dimers for assembly by histone chaperones. Newly synthesized dimers of H3–H4 histones are also deposited at the fork. The recycling of parental histones provides a means to maintain and propagate distinct chromatin states. H2A–H2B dimers are assembled into chromatin following the deposition of the H3–H4 tetramer.

defects on overexpression (Le et al. 1997), facilitates chromatin assembly coupled to DNA synthesis in vitro (Tyler et al. 1999; Mello et al. 2002). However, the addition of purified ASF1 to human cell extracts or *X. laevis* egg extracts depleted for histone cell cycle regulation defective homolog A (HIRA), CAF-1, and ASF1 is not sufficient to promote histone deposition. This indicates that ASF1 is unlikely to play a direct role in either the replication-coupled or -independent chromatin assembly pathways (Mello et al. 2002; Ray-Gallet et al. 2007). Instead, ASF1 acts as a histone donor for the histone chaperone CAF-1 during DNA replication or repair, a collaboration conserved in various organisms. ASF1 interacts with the B-domain of the p60 subunit of CAF-1 (Tyler et al. 2001; Mello et al. 2002; Sanematsu et al. 2006; Tang et al. 2006; Malay et al. 2008) at a site opposite to that of its interaction with H3–H4 (English et al. 2006; Natsume et al. 2007). Formation of a ternary complex (CAF-1–ASF1–H3–H4) could thus function as an intermediate enabling histones to be handed over from one chaperone to the next. The transfer of histones from ASF1 to CAF-1 as part of an "assembly line" would ensure an efficient histone deposition coupled to DNA replication (Fig. 2).

 Cite this article as *Cold Spring Harb Perspect Biol* doi: 10.1101/cshperspect.a010207

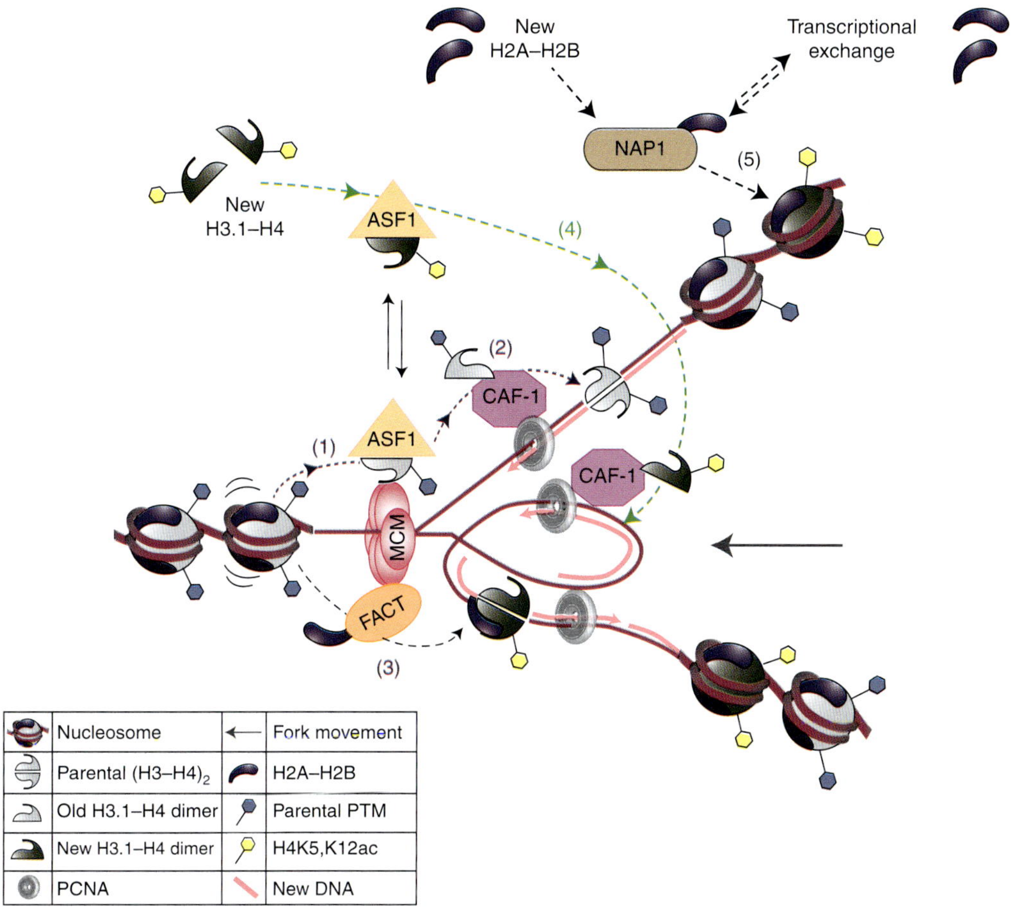

Figure 2. Chromatin assembly is mediated by a network of histone chaperones. Disruption of chromatin in front of the fork aided by ATP-dependent chromatin remodeling activities and the Mcm2–7 helicase results in release of parental histones. Interactions between the Mcm2–7 complex and the histone chaperones ASF1 and FACT may aid in the disassembly of the nucleosome and provide a means for sequestering parental histones at the fork. ASF1 may function in a histone chaperone "assembly line" to split $(H3–H4)_2$ tetramers into H3–H4 dimers (*1*) for deposition by CAF-1 on either the leading or lagging strand (*2*). CAF-1 is tethered to the leading and lagging strands via an interaction with PCNA, thereby providing a potential mechanism for semiconservative deposition of parental histones. Similarly, FACT would facilitate the retention and assembly of H2A–H2B dimers onto the nascent DNA (*3*). Newly synthesized histone H3–H4 dimers are delivered to the replication fork by ASF1 for deposition by CAF-1 (*4*) or, in the case of histone H2A–H2B dimers, by NAP1 (*5*).

To complete nucleosome formation following the delivery of two dimers of H3–H4 onto newly synthesized DNA, the subsequent addition of histones H2A–H2B involves the nucleosome assembly protein 1 (NAP1) chaperone (Zlatanova et al. 2007). Given that the FACT (facilitates chromatin transcription) complex also acts as an H2A–H2B chaperone in transcription, DNA replication, and DNA repair, it could perhaps help to provide a connection with NAP1 (Krogan et al. 2006). However, it should be noted that incorporation of new H2A–H2B does not necessarily have to be tightly linked to DNA replication, as significant H2A–H2B exchange also occurs outside replication (Kimura and Cook 2001).

Other histone chaperones should also be considered as players in chromatin assembly and histone dynamics; for example, nuclear autoantigenic sperm protein (NASP), initially reported as a histone H1 linker chaperone (Finn et al. 2008), was later found to function as an H3–H4 chaperone (Osakabe et al. 2010) as part of a multichaperone complex (Tagami et al. 2004; Groth et al. 2005). Recent work indicates that NASP may act to fine-tune soluble H3 levels by counteracting degradation involving the chaperone-mediated autophagy pathway (Cook et al. 2011).

Replication-Independent Deposition of Histone Variants H3.3 and CenH3

In contrast to the replicative histone variants H3.1 and H3.2, which are deposited throughout the genome during replication, the replacement variant H3.3 accumulates in actively transcribed chromatin regions, as first shown in *Drosophila* (Ahmad and Henikoff 2002b). H3.3 nucleosomes are enriched with "active" PTMs compared with the replicative variants (McKittrick et al. 2004; Hake et al. 2006; Loyola et al. 2006). The accumulation of histone H3.3 at promoters of active genes or at regulatory elements exploits a replication-independent mechanism involving the histone chaperone HIRA (Ray-Gallet et al. 2011). However, H3.3 is not only confined to sites of active transcription but is also enriched at other genomic regions depending on the developmental context. This is illustrated at the time of fertilization with a massive and global accumulation of H3.3 onto sperm DNA (Loppin et al. 2005), and in embryonic stem cells with the accumulation of H3.3 at telomeres (Goldberg et al. 2010; Wong et al. 2010). How these events are controlled and which factors are involved are beginning to be unraveled (for review, see Elsaesser et al. 2010). Recent data showing that HIRA can directly bind to DNA suggest a mechanism whereby the HIRA-dependent pathway could act as a gap-filling mechanism to restore nucleosomal organization wherever it may be compromised (Ray-Gallet et al. 2011), for example, as a result of faulty CAF-1 deposition. Thus, one should consider these pathways as interlinked to restore nucleosome density after fork passage. Other H3.3 chaperones have also been uncovered, and their exact role in H3.3 deposition may depend on the cell type and context (Elsaesser et al. 2010).

The mammalian CenH3, CENP-A, also called the deviant H3 (Wolffe and Pruss 1996), is highly divergent from H3 and stands out as the best example of a histone H3 variant that specifies a functional genomic locus. Specifically, CenH3 defines the centromere (Warburton et al. 1997), where it serves as an essential platform for kinetochore assembly (Allshire and Karpen 2008). During replication of centromeric chromatin, CenH3 nucleosomes become diluted to half the initial concentration on daughter chromatin (Shelby et al. 2000; Jansen et al. 2007). It is not until the next G_1 phase that new CenH3 gets incorporated again (Jansen et al. 2007; Schuh et al. 2007). This case illustrates a situation in which the disruption of chromatin during replication is clearly separated from the functional reassembly of the chromatin outside of S phase. With respect to the associated mechanism of the human CenH3 deposition, the recent identification of Holliday junction recognition protein (HJURP) is particularly enlightening. HJURP is a CenH3 chaperone, localized at centromeres precisely from late telophase to early G_1, which promotes the specific targeting/incorporation and maintenance of CenH3 at centromeres (Dunleavy et al. 2009; Foltz et al. 2009). So far, we have considered the situation from the point of view of how to restore the initial chromatin state after a disruptive event such as replication. However, the following alternative consideration is equally valid: Incorporation of CenH3 in G_1 may be programmed in anticipation of the disruptive event during replication, rather than being a restoration of half the pool of CENP-A. Whichever way one looks at this issue, it provides a general conceptual framework for the mechanism by which chromatin marks can be dealt with during the cell cycle.

DISRUPTION OF CHROMATIN AT THE REPLICATION FORK

Transcription and DNA replication require that the chromatin be disrupted ahead of the RNA or

DNA polymerase complex. Although the replicative helicase, Mcm2–7 complex, can in theory provide the means to disrupt nucleosomes ahead of the replication fork, it is important to consider how distinct ATP-dependent chromatin remodeling complexes facilitate the sliding or removal of nucleosomes from DNA (Flaus and Owen-Hughes 2011). A number of these chromatin remodeling complexes have been identified either biochemically or genetically as important for progression of the replication fork. ATP-utilizing chromatin assembly and remodeling factor (ACF) was initially discovered in *Drosophila* extracts and helps to reconstitute long arrays of regularly spaced nucleosomes in vitro (Ito et al. 1997; Varga-Weisz et al. 1997). One of the ACF subunits is ISWI, (imitation switch), an ATP-dependent chromatin remodeling protein found in a number of different complexes. Loss of another ACF subunit in *Drosophila*, ACF1, results in a more rapid progression through S phase and disrupts the polycomb-mediated silencing of facultative heterochromatin (Fyodorov et al. 2004). In mammalian systems ACF1 is required for the replication of pericentromeric heterochromatin (Collins et al. 2002). In vitro nucleosomal array reconstitution experiments indicate that ACF may facilitate the maturation of a nonnucleosomal histone octamer–DNA intermediate into a mature nucleosome (Torigoe et al. 2011). Another ISWI-containing complex, Williams syndrome transcription factor (WSTF), is targeted to the replication fork via an interaction with the DNA polymerase processivity factor PCNA to promote efficient DNA replication throughout S phase (Poot et al. 2004). The loss of WSTF results in an increased accumulation of the heterochromatin protein HP1 and increased chromatin compaction. In addition, WSTF is also important for the DNA damage response (Xiao et al. 2009). Together these data highlight how ATP-dependent chromatin remodelers can contribute to both the disruption of chromatin ahead of the forks as well as the reassembly of chromatin structure and maintenance of epigenetic states behind the fork.

Additional ATP-dependent chromatin remodeling complexes have been found to mediate histone dynamics at the replication fork in S phase. The INO80 complex and its catalytic component (SNF2) have been implicated in transcription, DNA damage repair, and fork progression during replicative stress (Bao and Shen 2007). INO80 also associates with stalled forks and early origins of replication, suggesting a possible role in initiation and replication fork restart (Shimada et al. 2008). In addition, INO80 and ISW2 can function in parallel to promote efficient fork progression through late-replicating regions of the *S. cerevisiae* genome (Vincent et al. 2008). It remains to be determined how these factors are targeted to the replication fork, as well as their relative contribution to chromatin disassembly ahead of the fork or reassembly behind the fork.

Histone chaperones are also likely involved in the disassembly of chromatin, where they may act as histone acceptors. A local concentration of parental histones may be maintained at the replication fork by the chaperone-mediated acceptance and sequestration of disassembled histones. Increasing evidence suggests that ASF1 and FACT can also act as histone acceptors at the replication fork (Groth et al. 2007; Jasencakova et al. 2010) for subsequent transfer to CAF-1 and NAP1 for incorporation into chromatin.

The H3–H4 chaperone, ASF1, is clearly linked to DNA replication, as is evident by the S-phase defects observed in a variety of organisms on ASF1 depletion (reviewed in Mousson et al. 2007), including replication fork defects and accumulation of cells in S phase (Groth et al. 2007). The function of ASF1 during DNA replication appears to be independent of CAF-1, as replication-coupled assembly is not impaired in chicken DT40 cells lacking ASF1 (Sanematsu et al. 2006). Recent experiments have shown that during replication stress ASF1 can associate with parental histones, suggesting a role as a histone acceptor (Groth et al. 2007; Jasencakova et al. 2010). In support of this idea, structural studies suggest that ASF1 binds the carboxyl terminus of H4 and that this "strand capture" can facilitate the splitting of the H3–H4 tetramer into dimers (English et al. 2006). Thus, a model begins to emerge of a histone chaperone "assembly line" at the replication fork, with ASF1 accepting parental H3–H4 tetramers, splitting them, and then passing them onto CAF-1 for deposition (Corpet et al. 2010). Importantly, defects in the recycling of parental histones could impact the transmission of parental PTMs, thereby compromising the stable maintenance of a given chromatin state.

An assembly line model for the disruption of nucleosomes and recycling of parental histones would require mechanisms to keep a population of histone chaperones at the replication fork. The FACT complex has been shown to interact with multiple fork components, including the Mcm2–7 helicase complex and replication protein A (RPA) (Gambus et al. 2006; Tan et al. 2006; VanDemark et al. 2006). Thus, FACT is well positioned to accept evicted H2A–H2B dimers and perhaps cooperate with NAP1 for their subsequent reassembly. Interestingly, organism-specific mechanisms appear to target ASF1 to the active replication fork. In *S. cerevisiae* ASF1 interacts with replication factor C (RFC), the PCNA clamp loader (Franco et al. 2005), whereas in mammalian cells ASF1 is connected to the Mcm2–7 helicase via an H3–H4 histone bridge (Groth et al. 2007). Together these results suggest a tight coupling between DNA unwinding at the replication fork and the disassembly and assembly of chromatin. Underscoring this connection, it has recently been shown in *S. cerevisiae* that the deposition of Okazaki fragments during lagging-strand synthesis occurs in nucleosomal-sized steps and is tightly coupled to assembly (Smith and Whitehouse 2012).

STOICHIOMETRY OF HISTONE DEPOSITION

The stability of the histone $(H3–H4)_2$ tetramer when it is free of DNA in solution (Baxevanis et al. 1991) led to the long-standing assumption that these histones were deposited directly as a tetramer. Recent data have now challenged this view. Specifically, histones H3 and H4 were found as dimers together with histone chaperones within predeposition complexes in human cells (Tagami et al. 2004). Moreover, the investigation of CenH3 (centromeric-specific H3) nucleosomes identified hemisomes containing one copy each of CenH3, H2A, H2B, and H4 (Dalal et al. 2007). Together these data suggest that histones H3–H4 or CenH3–H4 can be first provided as dimers. Two H3–H4 dimers then associate during deposition onto replicating DNA to form tetramers. In addition, the resolution of the crystal structure of ASF1 interacting with a dimer of histones H3–H4 revealed that ASF1 physically blocks the formation of an $(H3–H4)_2$ tetramer (English et al. 2006; Natsume et al. 2007). It is unclear whether both new histone H3–H4 dimers are provided by ASF1 and then deposited by CAF-1 onto DNA or whether additional chaperones are also involved. In vivo metabolic labeling studies led to the general view that ahead of the replication fork the core histone octamer is disrupted into two H2A–H2B dimers and a histone $(H3–H4)_2$ tetramer (Annunziato 2005). Recent studies using isotope labeling combined with mass spectrometry analysis of histone content have shed light on whether the histone $(H3–H4)_2$ tetramer remains intact during transfer (Xu et al. 2010). Although the vast majority of H3.1–H4 tetramers do not split, the investigators observed a significant number of splitting events for the H3.3-containing tetramers (Xu et al. 2010). Thus, although newly synthesized histones are provided as dimers (Tagami et al. 2004), three alternative modes for H3–H4 partitioning during nucleosome assembly can now be considered as real (Fig. 3). It will be important to determine whether mixing events are exclusively variant specific or if they reflect particular histone dynamics associated with specific chromatin regions.

PTMs AND THE ESTABLISHMENT/MAINTENANCE OF EPIGENETIC STATES

PTMs of histones are crucial for defining and maintaining the epigenetic state of chromatin (Kouzarides 2007). A diverse array of histone-modifying enzymes regulates the PTM of both newly synthesized nascent histones as well as parental nucleosomal histones. A consequence of the combinatorial nature of potential histone modifications is that the preexisting

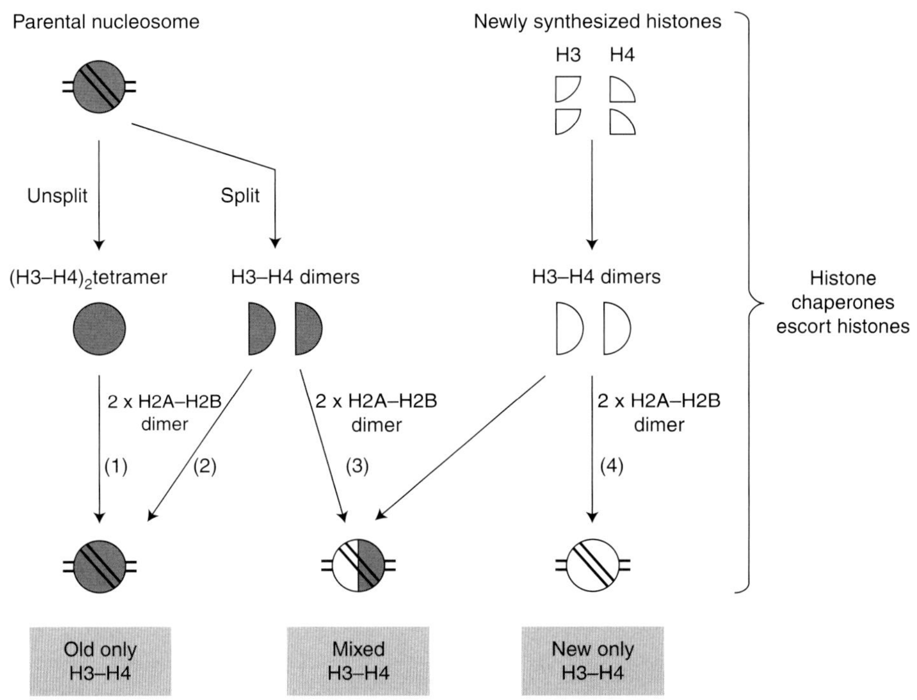

Figure 3. Partitioning of parental and newly synthesized histones. The deposition of newly synthesized histones or parental histones with existing PTMs can affect the inheritance and maintenance of specific chromatin states. Following disruption of nucleosomes at the replication fork, there are three possible outcomes for the deposition of parental and newly synthesized histones in the reassembled chromatin: deposition of parental H3–H4 only, deposition of mixed H3–H4 molecules composed of parental and nascent histones, or deposition of only newly synthesized H3–H4. On disassembly of the parental nucleosome, the $(H3–H4)_2$ tetramer can either remain intact (*1*) or split into two H3–H4 dimers (*2*). Deposition of the $(H3–H4)_2$ tetramer or deposition of two parental H3–H4 dimers followed by addition of two H2A–H2B dimers will result in the inheritance of a nucleosome with a parental H3–H4 tetramer core. Alternatively, the split H3–H4 dimers may associate with newly synthesized H3–H4 dimers (*3*), resulting in a nucleosome with a mixed H3–H4 tetramer core. Finally, the deposition of two newly synthesized H3–H4 dimers (*4*) will result in a nucleosome devoid of any parental histone PTMs.

modifications on a histone may dictate the subsequent action or inhibition of downstream histone-modifying enzymes. Thus, it is important to consider the initial PTMs of histones before deposition onto DNA to understand the specification and maintenance of specific chromatin states.

PTM of Newly Synthesized Histones

Almost all eukaryotic organisms show diacetylation of lysines 5 and 12 on newly synthesized histone H4 (Sobel et al. 1995; Loyola et al. 2006). The histone acetyltransferase (HAT) responsible for H4K5,K12ac is HAT1 (Parthun 2007), which forms a complex with the H3–H4 dimer, CAF-1, and ASF1 before deposition (Tagami et al. 2004; Loyola et al. 2006). The function of H4K5,K12ac remains an enigma, as the acetylation of H4K5,K12 is not required for histone deposition in vitro as unmodified H3–H4 histones are readily deposited on chromatin by CAF-1 in SV40 DNA replication assays (Shibahara et al. 2000) and in vivo (Ma et al. 1998). These results suggest that instead of facilitating chromatin deposition per se, the tight coupling of H4K5,K12ac with chromatin assembly may be important for marking a transient state (just having been replicated), which can impact the maintenance and propagation of heterochromatin if not removed (Taddei et al. 1999).

In *S. cerevisiae*, there is a peak of acetylation on lysine 56 on newly synthesized histone H3 that is observed during S phase (Kuo et al. 1996; Masumoto et al. 2005). H3K56 is located in the globular domain near the DNA entry/exit point of the nucleosome. Mutation of K56 to glutamine, commonly used as a constitutive acetylation mimic, results in an increased sensitivity to digestion by micrococcal nuclease (Masumoto et al. 2005), suggesting that H3K56ac weakens histone DNA interactions at the entry and exit points of the nucleosome. Consistent with the increased accessibility of nucleosomal DNA by H3K56ac, it promotes access to replication-coupled DNA damage to facilitate the repair of stalled replication forks and double-strand breaks (Driscoll et al. 2007; Han et al. 2007; Celic et al. 2008).

The acetylation of H3K56 in *S. cerevisiae* is mediated by the HAT Rtt109 in complex with either of two histone chaperones, ASF1 or Vps75 (a NAP1-related protein) (Han et al. 2007; Tsubota et al. 2007). The acetylation of H3K56 increases the affinity of CAF-1 for histone H3 and promotes the assembly of nucleosomes during S phase (Li et al. 2008). Thus, in the case of *S. cerevisiae*, there is a clear role for posttranslational histone modifications in promoting chromatin assembly. In contrast, the role of H3K56ac is less clear in mammals, as the abundance of the mark is very low (Garcia et al. 2007; Das et al. 2009; Xie et al. 2009) and not likely to be found on the bulk of newly synthesized histones. However, it remains formally possible that H3K56ac may be rapidly deacetylated following deposition. Further studies on the role of H3K56ac during replication-coupled chromatin assembly in mammals are clearly warranted.

The chromatin marks associated with nascent histones H3 and H4 are transient and rapidly removed during chromatin maturation as detected in heterochromatin regions (Taddei et al. 1999). The removal of H4K5 and K12 diacetylation is critical for the association of HP1 and the maintenance of silenced heterochromatin (Taddei et al. 1999), as well as proper centromere function (Ekwall et al. 1997; Taddei et al. 2001). Indeed, global defects in the removal of acetyl groups, mediated by the Cre recombinase-driven inactivation of histone deacetylase 3 (HDAC3), can lead to impaired S-phase progression and increased sensitivity to DNA damage (Bhaskara et al. 2008). Similarly, the deacetylation of H3K56 in *S. cerevisiae* during chromatin maturation is also important for genome stability. Loss of the two HDACs, Hst3 and Hst4, which target H3K56ac, leads to DNA damage (Celic et al. 2006) and sensitivity to replication stress (Celic et al. 2008). Thus, the proper maturation of chromatin is essential for genome stability.

In mammalian systems, newly synthesized histones are typically devoid of lysine methylation; however, it has recently been shown that there is a significant population of H3K9 monomethylation on predeposited H3.1 and H3.3 (Loyola et al. 2006). The histone methyltransferase SetDB1 targets H3K9 for monomethylation in complex with CAF-1 (Loyola et al. 2009). Deposition of nascent H3K9me may serve as a seed for the subsequent di- and trimethylation events mediated by the methyltransferase Suv39h, which recognizes the H3K9me1 specifically as its substrate (Loyola et al. 2009). In this manner, the mark incorporated before deposition may impact the final chromatin state. Of importance will be understanding the mechanisms of when and where a predeposition "seed" modification may be incorporated and the role of histone chaperones in specifying the deposition of the modified histone to influence chromatin state.

Epigenetics at the Replication Fork

DNA methylation and PTMs of histones are widely viewed to be conveyors of epigenetic information that is independent from the underlying genomic sequence. As the chromatin is disassembled on passage of the fork, there must be mechanisms to ensure the faithful reestablishment of epigenetic information following chromatin assembly. Finally, these mechanisms have to be plastic and able to dynamically respond to tissue-specific and developmental demands.

DNA methylation of cytosine at CpG sequences is important for the maintenance of a repressive chromatin structure (Weber and Schubeler 2007). The symmetrical nature of CpG methylation implies that following DNA replication, each daughter molecule of DNA will be hemimethylated. The DNA methyltransferase Dnmt1 has a preferred affinity for hemimethylated CpG motifs (Pradhan et al. 1999) and can be targeted to the active replication fork via an interaction with PCNA, which suggests a straightforward model for the semiconservative inheritance of DNA methylation. However, it should be noted that the interaction between PCNA and Dmnt1 is not strictly required for the maintenance of DNA methylation following DNA replication (Schermelleh et al. 2007; Spada et al. 2007).

As with DNA methylation, PCNA serves as the common link between the replication fork and factors that propagate histone PTMs. For example, PCNA tethers CAF-1 to the fork (Shibahara and Stillman 1999; Moggs et al. 2000), and in addition to functioning as a histone chaperone, the p150 subunit of CAF-1 binds HP1 (Murzina et al. 1999) and promotes the redistribution of HP1 on chromatin during replication (Quivy et al. 2004). As the p150 subunit of CAF-1 is required for the replication of pericentric heterochromatin (Quivy et al. 2008), it may also function in the disassembly of chromatin and removal of HP1 ahead of the fork, followed by the subsequent deposition behind the fork of recycled parental histones presenting H3K9me3. HP1 serves to recruit the H3K9 methyltransferase Suv39h, thus establishing a positive-feedback loop to reestablish the pericentric heterochromatin state following passage of the replication fork. In addition, a number of other histone-modifying enzymes interact with PCNA, including the methyltransferase PR-Set7/Set8 (Jorgensen et al. 2007; Huen et al. 2008), as well as HDAC enzymes (Milutinovic et al. 2002), to ensure the local activity of these enzymes at the replication fork.

The maintenance of H3–H4 PTMs following DNA replication is complicated by the deposition of both parental and nascent histones behind the fork. The random model posits that parental $(H3–H4)_2$ with mature PTMs are mixed on either daughter strand with the newly synthesized histones. A consequence of this mixing is the dilution of the parental histone PTMs. On long arrays of nucleosomes, the relative local density of nucleosomes with parental histones relative to nascent histones could serve as a template to recruit histone-modifying enzymes to propagate the mark to the newly synthesized histones (Probst et al. 2009; Margueron and Reinberg 2010). This model is attractive for broad regions of similar chromatin states, such as the repressive H3K9m2/me3 marks found at pericentromeric heterochromatin. However, when one considers specific PTMs confined to only one or two nucleosomes, it becomes increasingly unlikely that the parental tetramer can serve as a template. However, the recent evidence of split H3–H4 dimers (Xu et al. 2010) may represent a mechanism to ensure that each daughter strand receives an H3–H4 dimer with the parental PTM. Thus, the modification would be propagated via an intraparticle mechanism. Other models (Probst et al. 2009) should also be considered, and it is possible that unique mechanisms for the propagation of chromatin states are associated with specific loci and particular developmental stages.

NUCLEOSOME POSITIONING

In the wake of the replication fork, chromatin is assembled into nucleosomes that are often deposited in specific locations relative to functional regulatory elements. This nucleosome positioning can be maintained between cell divisions and throughout the population. The location and occupancy of nucleosomes throughout the genome partly governs the accessibility of the DNA for binding by *trans*-acting factors. A series of classic experiments at the *PHO5* locus in *S. cerevisiae* highlighted the importance of nucleosome occupancy in modulating the transcriptional response and the binding of *trans*-acting factors (Almer and Horz 1986; Almer et al. 1986; Raser and O'Shea 2004). The regulatory consequence of nucleosome positioning also impacts other nuclear processes including DNA recombination (Roth and Roth 2000) and replication (Simpson 1990; Lipford and Bell 2001; Eaton et al. 2010).

The advent of genomic technologies has made it possible to systematically survey nucleosome positioning throughout eukaryotic genomes (for review, see Radman-Livaja and Rando 2010). Briefly, chromatin is digested by micrococcal nuclease, mononucleosomes are isolated, and the 147-bp DNA fragments are recovered. Genomic tiling microarrays or next-generation sequencing is used to map the short DNA fragments back to the genome and determine nucleosome occupancy. The locations of nucleosomes relative to annotated chromosomal features, such as transcription start sites (TSSs), are remarkably similar across different eukaryotic organisms (Yuan et al. 2005; Whitehouse et al. 2007; Mavrich et al. 2008b; Schones et al. 2008; Valouev et al. 2008). Typically, an array of well-positioned nucleosomes starting from the TSS and progressing into the gene body is observed, with a region of low nucleosome occupancy immediately upstream of the TSS. This nucleosome-depleted region is most pronounced at actively transcribed genes and is often enriched for *cis*-acting regulatory elements and *trans*-acting factors.

Nucleosome positioning throughout the genome is mediated by both *cis*- and *trans*-acting factors. The histone octamer core does not bind specific sequences through a recognizable DNA-binding domain. Rather, flexibility constraints of the DNA, imposed by the necessity of wrapping the DNA almost two times around the histone core, specify preferential sequences. For example, dinucleotides of AA, TT, or TA at 10-bp intervals, by favoring a proper DNA curvature, increased affinity for the histone octamer (Anselmi et al. 1999; Thastrom et al. 1999). In contrast, poly(A)-rich tracts are relatively inflexible and do not favor nucleosome occupancy (Iyer and Struhl 1995). Analysis of the DNA sequence properties from the in vivo nucleosome positioning studies has led to the proposal of a nucleosome positioning code (Segal et al. 2006)—that the underlying sequence is predictive of nucleosome positioning throughout the genome. This nucleosome positioning code was tested in part by depositing reconstituted chicken histone octamers onto yeast DNA in vitro in the absence of any *trans*-acting factors such as transcription factors or chromatin remodeling activities (Kaplan et al. 2010). Indeed, the reconstituted nucleosomes were highly correlated (0.74) with the in vivo nucleosome positioning data, although they did not capture all of the in vivo nuances. Although specific sequences do contribute to the exclusion and positioning of nucleosomes, *trans*-acting factors also have an important role in specifying nucleosome position. *Trans*-acting factors may act as a fixed barrier, with the nucleosomes nearest the barrier showing the highest degree of positioning, followed by a gradual loss of positioning with increasing distance from the barrier (Mavrich et al. 2008a). Finally, it will be important to understand how initial histone deposition and chromatin assembly are coordinated with the reassociation of *trans*-acting factors behind the replication fork to reestablish nucleosome positioning for proper genome function.

CHROMATIN AND THE REGULATION OF REPLICATION ORIGINS

The selection and activation of DNA replication origins must occur within the context of the local chromatin environment. Early studies examining the patterns of radiolabeled thymidine incorporation found that specific chromosomal domains were replicated at discrete times in S phase (Goldman et al. 1984). Transcriptionally active euchromatin replicated early in S phase, whereas the condensed and gene-poor heterochromatin was copied at the end of S phase. These and similar experiments (Stambrook and Flickinger 1970) suggested that the local chromatin environment affected not only the transcription program but also the DNA replication program.

In *S. cerevisiae*, origins of DNA replication were first identified as short autonomously replicating sequence (ARS) elements that were required to propagate and maintain an episome. Each ARS is defined, in part, by a degenerate *cis*-acting sequence element termed the ARS consensus sequence (ACS). The ACS is necessary but not sufficient for origin function (Celniker et al. 1984), as there are >10,000 matches to the ACS motif in the yeast genome (Breier et al. 2004). However, <400 of those potential ACS matches function as bona fide binding sites for the origin-recognition complex (ORC) (Xu et al. 2006). Thus, other chromosomal features are likely involved in defining replication origins.

Despite the conservation of ORC and other *trans*-acting factors required for the assembly of the prereplicative complex (pre-RC) and loading of the Mcm2–7 helicase at origins, no conserved *cis*-acting elements directing DNA replication have been identified in higher eukaryotes (Gilbert 2004). Unlike in *S. cerevisiae*, ORC purified from higher eukaryotes shows little or no sequence specificity in vitro (Vashee et al. 2003; Remus et al. 2004). Together these results might suggest that origin selection and replication initiation are random or stochastic events in higher eukaryotes, but numerous studies have identified specific origins of replication as well as ORC binding at specific and reproducible chromosomal locations (Austin et al. 1999; Ladenburger et al. 2002; Karnani et al. 2010; MacAlpine et al. 2010). Thus, in contrast to the sequence cues that contribute to origin selection in yeast, the determinants of ORC binding in higher eukaryotes appear to be primarily dependent on the local chromatin environment.

Together these results suggest that in both lower and higher eukaryotes the local chromatin organization and structure are important features of origin selection. Indeed, recent studies in multiple experimental model systems have highlighted the role of histone modifications and chromatin organization in regulating the DNA replication program.

Histone Modifications and Origin Regulation

Experiments in *S. cerevisiae* showed the importance of the local chromatin environment in regulating origin function. In yeasts, the genes near the ends of the chromosome are often silenced by Sir2p, an HDAC, and are typically late-replicating (reviewed in Rusche et al. 2003). The delay in origin activation near the telomeres is not caused by sequence, as transposition of ARS501, a telomeric late-activating origin, to another region of the genome relieved the telomeric suppression of origin activation (Ferguson and Fangman 1992). Instead, the repression of origin activation near the telomere is because of the local chromatin structure, and disruption of that structure led to an advancement of origin firing earlier in S phase (Stevenson and Gottschling 1999). More recently, the genome-wide analysis of the time at which specific sequences are replicated in S phase has identified a clear correlation with transcriptional activity and replication timing in human (Birney et al. 2007; Ryba et al. 2010), mouse (Hiratani et al. 2010), *Drosophila* (Schübeler et al. 2002; MacAlpine et al. 2004), and chicken cells (Hassan-Zadeh et al. 2012). Gene-dense, transcriptionally active regions of the genome are replicated before regions of the genome with sparse gene activity. The correlation between replication timing and transcriptional activity is not at the level of individual genes but rather is defined by the broad transcriptional activity of large chromosomal domains (MacAlpine et al. 2004). Not surprisingly, there is also a strong correlation between chromatin modifications associated with active transcription and early replication. Experiments from the encyclopedia of DNA elements (ENCODE) project found that sequences replicated in early S phase from HeLa cells were enriched for activating chromatin marks (H3K4Me2 and H3K4Me2) as well as the hyperacetylation of histones H3 and H4 (Birney et al. 2007). Similar broad correlations between time of replication and activating histone modifications have been reported in numerous independent studies from a number of eukaryotic model systems (Hiratani et al. 2008; Schwaiger et al. 2009; Lee et al. 2010). The apparent coordination between the transcription and replication programs may, in part, be caused by the frequent localization of replication origins near the start sites of transcription (Cadoret et al. 2008; Cayrou et al. 2011).

Perturbation of histone acetylation impacts the replication program. In the absence of Rpd3, an HDAC, the length of S phase is shortened, presumably because of the earlier activation of a subset of replication origins (Vogelauer et al. 2002). Indeed, genome-wide analysis of origin activity in the absence of Rpd3 resulted in the earlier activation of almost 100 late-firing origins of replication (Knott et al. 2009). A local increase in histone acetylation mediated by tethering Gcn5, a histone acetyltransferase, to a late-activating origin also promoted origin initiation significantly earlier in S phase (Vogelauer et al. 2002). The role of histone acetylation in promoting replication is not specific to yeast, as the recruitment of Chameau (Hbo1), a *Drosophila* histone acetyltransferase (see below), to the chorion locus stimulated replication activity (Aggarwal and Calvi 2004). Similarly, the targeting of a HAT or HDAC activity to the B-globin locus can shift the time of replication from late to early and vice versa, respectively (Goren et al. 2008). These results clearly indicate that local changes in histone acetylation are able to fine-tune origin activity.

The recent finding that the Orc1 bromo-adjacent homology (BAH) domain specifically binds dimethylated lysine 20 of histone H4 (Kuo et al. 2012) suggests that H4K20me2 may be important for ORC recruitment to chromatin and defining replication origins. Mutations in the Orc1 BAH domain as well as other pre-RC components result in Meier–Gorlin syndrome, a rare primordial form of dwarfism (Bicknell et al. 2011). Furthermore, the loss of Suv4-20h1/h2, the methyltransferase responsible for H4K20me2, results in similar developmental defects to Meier–Gorlin syndrome in zebrafish, suggesting a critical role for H4K20 dimethylation and the DNA replication program during normal development (Kuo et al. 2012). However, it should be noted that H4K20me2 is the most abundant histone modification, accounting for >85% of histone H4 levels in the mouse (Schotta et al. 2008). On average, 97% of all nucleosomes will contain at least one histone H4K20me2, which makes it hard to argue that H4K20me2 is a specificity factor for ORC. Instead, H4K20me2 may simply help to stabilize ORC on chromatin, with other chromosomal features acting as origin specificity factors.

Instead of influencing ORC localization, new experiments suggest that the local chromatin environment may regulate the loading of the replicative helicase in G_1. Hbo1 (histone acetylase binding to ORC) is an abundant histone acetyltransferase responsible for the bulk of histone H4 acetylation in mammalian genomes and was initially identified based on its interaction with ORC, Mcm2, and Cdt1 (Iizuka and Stillman 1999; Burke et al. 2001; Iizuka et al. 2006). In *Xenopus* extracts Hbo1 is required for pre-RC assembly (Iizuka et al. 2006), and the artificial recruitment of catalytically nonfunctional Hbo1 to a mammalian origin of replication negatively impacts the loading of the Mcm2–7 helicase (Miotto and Struhl 2010). Interestingly, like many proteins with HAT activity, Hbo1 has the ability to acetylate nonhistone proteins including Orc2, Mcm2, and Cdc6 in vitro (Burke et al. 2001). Thus, it is not entirely clear if the role of Hbo1 in regulating origins is because of the specific acetylation of histones or, alternatively, the acetylation of pre-RC components.

Similar experiments also suggest that the levels of H4K20 monomethylation are important for helicase loading and pre-RC formation (Tardat et al. 2007). Set8, also known as PR-Set7, is a cell-cycle-regulated methyltransferase that monomethylates histone H4 at lysine 20 (Fang et al. 2002; Nishioka et al. 2002). Set8 is targeted to the proteasome during S phase by the E3 ubiquitin ligase Crl4 in a PCNA-dependent manner (Abbas et al. 2010; Centore et al. 2010; Oda et al. 2010). Set8 levels are critical for maintaining genomic stability, as the loss of Set8 function results in an S-phase delay, chromosome decondensation, increased DNA damage, G_2 arrest, and centrosome amplification (Karachentsev et al. 2005; Jorgensen et al. 2007). Similarly, the stabilization and overexpression of Set8 is also detrimental to the cell, resulting in premature chromatin compaction and rereplication (Tardat et al. 2007). Presumably, the rereplication induced by Set8 stabilization is due in part to promiscuous pre-RC formation, as a local increase in H4K20me and the recruitment of pre-RC components are observed when Set8 is tethered to a specific locus (Tardat et al. 2007). Understanding how the methylation state of lysine 20 (mono-, di-, or trimethylation) is regulated through the cell cycle and its impact on ORC localization, pre-RC assembly, and genome stability will undoubtedly be an active and important area of future research.

Nucleosome Positioning and Origin Selection

Early experiments on an episome containing ARS1, an *S. cerevisiae* origin of DNA replication, identified well-positioned nucleosomes flanking the origin sequence (Simpson 1990). The analysis of genome-wide nucleosome positioning data sets from *S. cerevisiae* revealed that almost all ARS elements are depleted of bulk nucleosomes (Mavrich et al. 2008a). Subsequent analysis of nucleosome positioning relative to the orientation of the 17-bp ACS, and not the broadly mapped ARS elements, identified precisely positioned nucleosomes flanking almost all yeast origins of replication (Berbenetz et al. 2010; Eaton et al. 2010). Thus, the positioned nucleosomes first observed at ARS1 on a plasmid appear to be a defining feature of *S. cerevisiae* origins of DNA replication. Furthermore, decreased nucleosome occupancy at origins of replication has also been observed in a variety of different eukaryotic organisms outside of *S. cerevisiae*, including *Drosophila* (MacAlpine et al. 2010), mammalian cells (Lubelsky et al. 2011), and *Schizosaccharomyces pombe* (Givens et al. 2012).

An emerging model is that in *S. cerevisiae*, the ACS and downstream sequence cues keep the origin region free from encroaching nucleosomes and that this large nucleosome-free region facilitates ORC localization (Fig. 4). Once bound, ORC and presumably an ATP-dependent chromatin remodeling activity are required to generate the array of well-positioned nucleosomes flanking the origin of replication (Eaton et al. 2010). Interestingly, if the positioning of the upstream nucleosome is altered, Mcm2–7 loading and initiation are impaired, suggesting that the chromatin architecture is critical for pre-RC assembly and origin

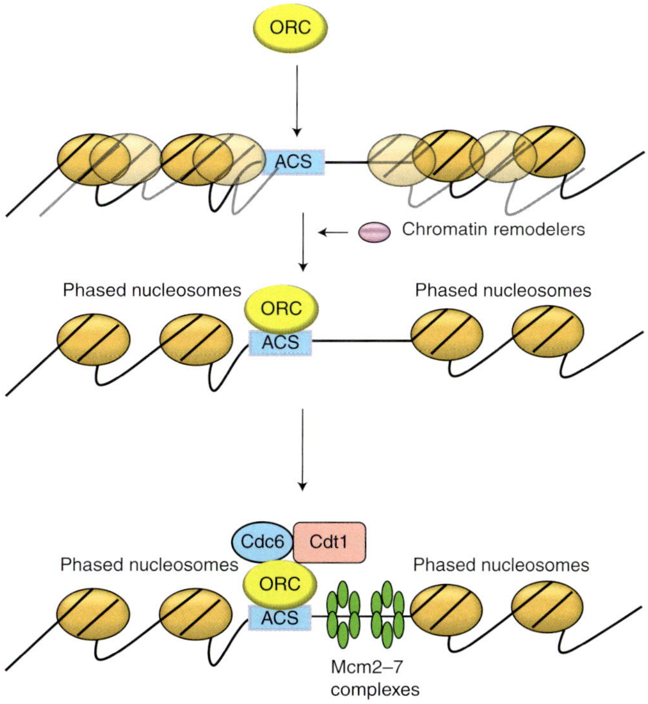

Figure 4. Nucleosome organization at origins of replication. The AT-rich nature of the ACS and flanking sequences at *S. cerevisiae* origins of replication prevent encroachment of nucleosomes into the origin. The nucleosome-free region at origins of replication is observed in yeast and higher eukaryotes and may function as a primary determinant for ORC binding. On ORC binding the flanking nucleosomes become precisely positioned, and this positioning is dependent on ORC and an ATP-dependent chromatin remodeling activity. The nucleosome-free region at the origin may facilitate the loading of multiple Mcm2–7 complexes and subsequent DNA unwinding events before initiation.

activation (Lipford and Bell 2001). However, several important questions remain. How do nucleosome positioning, turnover, and eviction facilitate Mcm2–7 loading and origin activation? Which ATP-dependent chromatin remodeling activities are required, and how do they contribute to origin activation, timing, and efficiency? Understanding how the local chromatin structure impacts and regulates the relatively well-defined *S. cerevisiae* DNA replication program will be critical for understanding how the local chromatin environment and dynamics contribute to the plasticity of origin selection in higher eukaryotes.

ACKNOWLEDGMENTS

This work is supported by an American Cancer Society Research Scholar grant (120222-RSG-11-048-01-DMC) to D.M.M. G.A.'s laboratory is supported by la Ligue Nationale contre le Cancer (Equipe labellisée Ligue 2010), the European Commission Network of Excellence EpiGeneSys (HEALTH-F4-2010-257082), ERC Advanced grant 2009-AdG_20090506 "Eccentric," and ANR. The authors wish to thank Heather MacAlpine and Zack Gurard-Levin for comments and suggestions.

REFERENCES

Abbas T, Shibata E, Park J, Jha S, Karnani N, Dutta A. 2010. CRL4[Cdt2] regulates cell proliferation and histone gene expression by targeting PR-Set7/Set8 for degradation. *Mol Cell* **40**: 9–21.

Aggarwal BD, Calvi BR. 2004. Chromatin regulates origin activity in *Drosophila* follicle cells. *Nature* **430**: 372–376.

Ahmad K, Henikoff S. 2002a. Histone H3 variants specify modes of chromatin assembly. *Proc Natl Acad Sci* **99** (Suppl 4): 16477–16484.

Ahmad K, Henikoff S. 2002b. The histone variant H3.3 marks active chromatin by replication-independent nucleosome assembly. *Mol Cell* **9**: 1191–1200.

Allshire RC, Karpen GH. 2008. Epigenetic regulation of centromeric chromatin: Old dogs, new tricks? *Nat Rev Genet* **9**: 923–937.

Almer A, Horz W. 1986. Nuclease hypersensitive regions with adjacent positioned nucleosomes mark the gene boundaries of the PHO5/PHO3 locus in yeast. *EMBO J* **5**: 2681–2687.

Almer A, Rudolph H, Hinnen A, Horz W. 1986. Removal of positioned nucleosomes from the yeast PHO5 promoter upon PHO5 induction releases additional upstream activating DNA elements. *EMBO J* **5**: 2689–2696.

Annunziato AT. 2005. Split decision: What happens to nucleosomes during DNA replication? *J Biol Chem* **280**: 12065–12068.

Anselmi C, Bocchinfuso G, De Santis P, Savino M, Scipioni A. 1999. Dual role of DNA intrinsic curvature and flexibility in determining nucleosome stability. *J Mol Biol* **286**: 1293–1301.

Austin RJ, Orr-Weaver TL, Bell SP. 1999. *Drosophila* ORC specifically binds to *ACE3*, an origin of DNA replication control element. *Genes Dev* **13**: 2639–2649.

Bao Y, Shen X. 2007. INO80 subfamily of chromatin remodeling complexes. *Mutat Res* **618**: 18–29.

Baxevanis AD, Godfrey JE, Moudrianakis EN. 1991. Associative behavior of the histone (H3-H4)2 tetramer: Dependence on ionic environment. *Biochemistry* **30**: 8817–8823.

Berbenetz NM, Nislow C, Brown GW. 2010. Diversity of eukaryotic DNA replication origins revealed by genome-wide analysis of chromatin structure. *PLoS Genet* **6**: e1001092.

Bhaskara S, Chyla BJ, Amann JM, Knutson SK, Cortez D, Sun ZW, Hiebert SW. 2008. Deletion of histone deacetylase 3 reveals critical roles in S phase progression and DNA damage control. *Mol Cell* **30**: 61–72.

Bicknell LS, Walker S, Klingseisen A, Stiff T, Leitch A, Kerzendorfer C, Martin CA, Yeyati P, Al Sanna N Bober M, et al. 2011. Mutations in *ORC1*, encoding the largest subunit of the origin recognition complex, cause microcephalic primordial dwarfism resembling Meier-Gorlin syndrome. *Nat Genet* **43**: 350–355.

Birney E, Stamatoyannopoulos JA, Dutta A, Guigo R, Gingeras TR, Margulies EH, Weng Z, Snyder M, Dermitzakis ET, Thurman RE, et al. 2007. Identification and analysis of functional elements in 1% of the human genome by the ENCODE pilot project. *Nature* **447**: 799–816.

Breier AM, Chatterji S, Cozzarelli NR. 2004. Prediction of *Saccharomyces cerevisiae* replication origins. *Genome Biol* **5**: R22.

Burke TW, Cook JG, Asano M, Nevins JR. 2001. Replication factors MCM2 and ORC1 interact with the histone acetyltransferase HBO1. *J Biol Chem* **276**: 15397–15408.

Cadoret JC, Meisch F, Hassan-Zadeh V, Luyten I, Guillet C, Duret L, Quesneville H, Prioleau MN. 2008. Genome-wide studies highlight indirect links between human replication origins and gene regulation. *Proc Natl Acad Sci* **105**: 15837–15842.

Cayrou C, Coulombe P, Vigneron A, Stanojcic S, Ganier O, Peiffer I, Rivals E, Puy A, Laurent-Chabalier S, Desprat R, et al. 2011. Genome-scale analysis of metazoan replication origins reveals their organization in specific but flexible sites defined by conserved features. *Genome Res* **21**: 1438–1449.

Celic I, Masumoto H, Griffith WP, Meluh P, Cotter RJ, Boeke JD, Verreault A. 2006. The sirtuins Hst3 and Hst4p preserve genome integrity by controlling histone H3 lysine 56 deacetylation. *Curr Biol* **16**: 1280–1289.

Celic I, Verreault A, Boeke JD. 2008. Histone H3 K56 hyperacetylation perturbs replisomes and causes DNA damage. *Genetics* **179**: 1769–1784.

Celniker SE, Sweder K, Srienc F, Bailey JE, Campbell JL. 1984. Deletion mutations affecting autonomously replicating sequence *ARS1* of *Saccharomyces cerevisiae*. *Mol Cell Biol* **4**: 2455–2466.

Centore RC, Havens CG, Manning AL, Li JM, Flynn RL, Tse A, Jin J, Dyson NJ, Walter JC, Zou L. 2010. $CRL4^{Cdt2}$-mediated destruction of the histone methyltransferase Set8 prevents premature chromatin compaction in S phase. *Mol Cell* **40**: 22–33.

Chadwick BP, Willard HF. 2001. A novel chromatin protein, distantly related to histone H2A, is largely excluded from the inactive X chromosome. *J Cell Biol* **152**: 375–384.

Collins N, Poot RA, Kukimoto I, Garcia-Jimenez C, Dellaire G, Varga-Weisz PD. 2002. An ACF1–ISWI chromatin-remodeling complex is required for DNA replication through heterochromatin. *Nat Genet* **32**: 627–632.

Cook AJ, Gurard-Levin ZA, Vassias I, Almouzni G. 2011. A specific function for the histone chaperone NASP to fine-tune a reservoir of soluble H3-H4 in the histone supply chain. *Mol Cell* **44:** 918–927.

Corpet A, Almouzni G. 2012. DNA replication and inheritance of epigenetic states. In *Genome organization and function in the mammalian cell nucleus*, pp. 365–394. Wiley-VCH, Weinheim.

Costanzi C, Pehrson JR. 1998. Histone macroH2A1 is concentrated in the inactive X chromosome of female mammals. *Nature* **393**: 599–601.

Dalal Y, Wang H, Lindsay S, Henikoff S. 2007. Tetrameric structure of centromeric nucleosomes in interphase *Drosophila* cells. *PLoS Biol* **5**: e218.

Das C, Lucia MS, Hansen KC, Tyler JK. 2009. CBP/p300-mediated acetylation of histone H3 on lysine 56. *Nature* **459**: 113–117.

Deal RB, Henikoff JG, Henikoff S. 2010. Genome-wide kinetics of nucleosome turnover determined by metabolic labeling of histones. *Science* **328**: 1161–1164.

Driscoll R, Hudson A, Jackson SP. 2007. Yeast Rtt109 promotes genome stability by acetylating histone H3 on lysine 56. *Science* **315**: 649–652.

Dunleavy EM, Roche D, Tagami H, Lacoste N, Ray-Gallet D, Nakamura Y, Daigo Y, Nakatani Y, Almouzni-Pettinotti G. 2009. HJURP is a cell-cycle-dependent maintenance and deposition factor of CENP-A at centromeres. *Cell* **137**: 485–497.

Eaton ML, Galani K, Kang S, Bell SP, MacAlpine DM. 2010. Conserved nucleosome positioning defines replication origins. *Genes Dev* **24**: 748–753.

Ekwall K, Olsson T, Turner BM, Cranston G, Allshire RC. 1997. Transient inhibition of histone deacetylation alters the structural and functional imprint at fission yeast centromeres. *Cell* **91**: 1021–1032.

Elsaesser SJ, Goldberg AD, Allis CD. 2010. New functions for an old variant: No substitute for histone H3.3. *Curr Opin Genet Dev* **20**: 110–117.

English CM, Adkins MW, Carson JJ, Churchill ME, Tyler JK. 2006. Structural basis for the histone chaperone activity of Asf1. *Cell* **127**: 495–508.

Fang J, Feng Q, Ketel CS, Wang H, Cao R, Xia L, Erdjument-Bromage H, Tempst P, Simon JA, Zhang Y. 2002. Purification and functional characterization of SET8, a nucleosomal histone H4-lysine 20-specific methyltransferase. *Curr Biol* **12**: 1086–1099.

Ferguson BM, Fangman WL. 1992. A position effect on the time of replication origin activation in yeast. *Cell* **68**: 333–339.

Finn RM, Browne K, Hodgson KC, Ausio J. 2008. sNASP, a histone H1-specific eukaryotic chaperone dimer that facilitates chromatin assembly. *Biophys J* **95**: 1314–1325.

Flaus A, Owen-Hughes T. 2011. Mechanisms for ATP-dependent chromatin remodelling: The means to the end. *FEBS J* **278**: 3579–3595.

Foltz DR, Jansen LE, Bailey AO, Yates JR III, Bassett EA, Wood S, Black BE, Cleveland DW. 2009. Centromere-specific assembly of CENP-a nucleosomes is mediated by HJURP. *Cell* **137**: 472–484.

Franco AA, Lam WM, Burgers PM, Kaufman PD. 2005. Histone deposition protein Asf1 maintains DNA replisome integrity and interacts with replication factor C. *Genes Dev* **19**: 1365–1375.

Fyodorov DV, Blower MD, Karpen GH, Kadonaga JT. 2004. Acf1 confers unique activities to ACF/CHRAC and promotes the formation rather than disruption of chromatin in vivo. *Genes Dev* **18**: 170–183.

Gambus A, Jones RC, Sanchez-Diaz A, Kanemaki M, van Deursen F, Edmondson RD, Labib K. 2006. GINS maintains association of Cdc45 with MCM in replisome progression complexes at eukaryotic DNA replication forks. *Nat Cell Biol* **8**: 358–366.

Garcia BA, Hake SB, Diaz RL, Kauer M, Morris SA, Recht J, Shabanowitz J, Mishra N, Strahl BD, Allis CD, et al. 2007. Organismal differences in post-translational modifications in histones H3 and H4. *J Biol Chem* **282**: 7641–7655.

Gerard A, Koundrioukoff S, Ramillon V, Sergere JC, Mailand N, Quivy JP, Almouzni G. 2006. The replication kinase Cdc7-Dbf4 promotes the interaction of the p150 subunit of chromatin assembly factor 1 with proliferating cell nuclear antigen. *EMBO Rep* **7**: 817–823.

Gilbert DM. 2004. In search of the holy replicator. *Nat Rev Mol Cell Biol* **5**: 848–855.

Givens RM, Lai WK, Rizzo JM, Bard JE, Mieczkowski PA, Leatherwood J, Huberman JA, Buck MJ. 2012. Chromatin architectures at fission yeast transcriptional promoters and replication origins. *Nucleic Acids Res* **40**: 7176–7189.

Goldberg AD, Banaszynski LA, Noh KM, Lewis PW, Elsaesser SJ, Stadler S, Dewell S, Law M, Guo X, Li X, et al. 2010. Distinct factors control histone variant H3.3 localization at specific genomic regions. *Cell* **140**: 678–691.

Goldman MA, Holmquist GP, Gray MC, Caston LA, Nag A. 1984. Replication timing of genes and middle repetitive sequences. *Science* **224**: 686–692.

Goren A, Tabib A, Hecht M, Cedar H. 2008. DNA replication timing of the human β-globin domain is controlled by histone modification at the origin. *Genes Dev* **22**: 1319–1324.

Groth A, Ray-Gallet D, Quivy JP, Lukas J, Bartek J, Almouzni G. 2005. Human Asf1 regulates the flow of S phase histones during replicational stress. *Mol Cell* **17**: 301–311.

Groth A, Corpet A, Cook AJ, Roche D, Bartek J, Lukas J, Almouzni G. 2007. Regulation of replication fork progression through histone supply and demand. *Science* **318**: 1928–1931.

Gunjan A, Verreault A. 2003. A Rad53 kinase-dependent surveillance mechanism that regulates histone protein levels in *S. cerevisiae*. *Cell* **115**: 537–549.

Gunjan A, Paik J, Verreault A. 2005. Regulation of histone synthesis and nucleosome assembly. *Biochimie* **87**: 625–635.

Hake SB, Garcia BA, Duncan EM, Kauer M, Dellaire G, Shabanowitz J, Bazett-Jones DP, Allis CD, Hunt DF. 2006. Expression patterns and post-translational modifications associated with mammalian histone H3 variants. *J Biol Chem* **281**: 559–568.

Han M, Chang M, Kim UJ, Grunstein M. 1987. Histone H2B repression causes cell-cycle-specific arrest in yeast: Effects on chromosomal segregation, replication, and transcription. *Cell* **48**: 589–597.

Han J, Zhou H, Horazdovsky B, Zhang K, Xu RM, Zhang Z. 2007. Rtt109 acetylates histone H3 lysine 56 and functions in DNA replication. *Science* **315**: 653–655.

Hassan-Zadeh V, Chilaka S, Cadoret JC, Ma MK, Boggetto N, West AG, Prioleau MN. 2012. USF binding sequences from the HS4 insulator element impose early replication timing on a vertebrate replicator. *PLoS Biol* **10**: e1001277.

Henikoff S, Ahmad K, Malik HS. 2001. The centromere paradox: Stable inheritance with rapidly evolving DNA. *Science* **293**: 1098–1102.

Hiratani I, Ryba T, Itoh M, Yokochi T, Schwaiger M, Chang CW, Lyou Y, Townes TM, Schubeler D, Gilbert DM. 2008. Global reorganization of replication domains during embryonic stem cell differentiation. *PLoS Biol* **6**: e245.

Hiratani I, Ryba T, Itoh M, Rathjen J, Kulik M, Papp B, Fussner E, Bazett-Jones DP, Plath K, Dalton S, et al. 2010. Genome-wide dynamics of replication timing revealed by in vitro models of mouse embryogenesis. *Genome Res* **20**: 155–169.

Hoek M, Stillman B. 2003. Chromatin assembly factor 1 is essential and couples chromatin assembly to DNA replication in vivo. *Proc Natl Acad Sci* **100**: 12183–12188.

Houlard M, Berlivet S, Probst AV, Quivy JP, Hery P, Almouzni G, Gerard M. 2006. CAF-1 is essential for heterochromatin organization in pluripotent embryonic cells. *PLoS Genet* **2**: e181.

Huen MS, Sy SM, van Deursen JM, Chen J. 2008. Direct interaction between SET8 and proliferating cell nuclear antigen couples H4-K20 methylation with DNA replication. *J Biol Chem* **283**: 11073–11077.

Iizuka M, Stillman B. 1999. Histone acetyltransferase HBO1 interacts with the ORC1 subunit of the human initiator protein. *J Biol Chem* **274**: 23027–23034.

Iizuka M, Matsui T, Takisawa H, Smith MM. 2006. Regulation of replication licensing by acetyltransferase Hbo1. *Mol Cell Biol* **26**: 1098–1108.

Ito T, Bulger M, Pazin MJ, Kobayashi R, Kadonaga JT. 1997. ACF, an ISWI-containing and ATP-utilizing chromatin assembly and remodeling factor. *Cell* **90**: 145–155.

Iyer V, Struhl K. 1995. Poly(dA:dT), a ubiquitous promoter element that stimulates transcription via its intrinsic DNA structure. *EMBO J* **14**: 2570–2579.

Jansen LE, Black BE, Foltz DR, Cleveland DW. 2007. Propagation of centromeric chromatin requires exit from mitosis. *J Cell Biol* **176**: 795–805.

Jasencakova Z, Scharf AN, Ask K, Corpet A, Imhof A, Almouzni G, Groth A. 2010. Replication stress interferes with histone recycling and predeposition marking of new histones. *Mol Cell* **37**: 736–743.

Jenuwein T, Allis CD. 2001. Translating the histone code. *Science* **293**: 1074–1080.

Jorgensen S, Elvers I, Trelle MB, Menzel T, Eskildsen M, Jensen ON, Helleday T, Helin K, Sorensen CS. 2007. The histone methyltransferase SET8 is required for S-phase progression. *J Cell Biol* **179**: 1337–1345.

Kamakaka RT, Biggins S. 2005. Histone variants: Deviants? *Genes Dev* **19**: 295–310.

Kaplan N, Moore I, Fondufe-Mittendorf Y, Gossett AJ, Tillo D, Field Y, Hughes TR, Lieb JD, Widom J, Segal E. 2010. Nucleosome sequence preferences influence in vivo nucleosome organization. *Nat Struct Mol Biol* **17**: 918–920; author reply 920–922.

Karachentsev D, Sarma K, Reinberg D, Steward R. 2005. PR-Set7-dependent methylation of histone H4 Lys 20 functions in repression of gene expression and is essential for mitosis. *Genes Dev* **19**: 431–435.

Karnani N, Taylor CM, Malhotra A, Dutta A. 2010. Genomic study of replication initiation in human chromosomes reveals the influence of transcription regulation and chromatin structure on origin selection. *Mol Biol Cell* **21**: 393–404.

Kaufman PD, Kobayashi R, Kessler N, Stillman B. 1995. The p150 and p60 subunits of chromatin assembly factor I: A molecular link between newly synthesized histones and DNA replication. *Cell* **81**: 1105–1114.

Kim UJ, Han M, Kayne P, Grunstein M. 1988. Effects of histone H4 depletion on the cell cycle and transcription of *Saccharomyces cerevisiae*. *EMBO J* **7**: 2211–2219.

Kimura H, Cook PR. 2001. Kinetics of core histones in living human cells: Little exchange of H3 and H4 and some rapid exchange of H2B. *J Cell Biol* **153**: 1341–1353.

Kinner A, Wu W, Staudt C, Iliakis G. 2008. γ-H2AX in recognition and signaling of DNA double-strand breaks in the context of chromatin. *Nucleic Acids Res* **36**: 5678–5694.

Klapholz B, Dietrich BH, Schaffner C, Heredia F, Quivy JP, Almouzni G, Dostatni N. 2009. CAF-1 is required for efficient replication of euchromatic DNA in *Drosophila* larval endocycling cells. *Chromosoma* **118**: 235–248.

Knott SRV, Viggiani CJ, Tavaré S, Aparicio OM. 2009. Genome-wide replication profiles indicate an expansive role for Rpd3L in regulating replication initiation timing or efficiency, and reveal genomic loci of Rpd3 function in *Saccharomyces cerevisiae*. *Genes Dev* **23**: 1077–1090.

Kouzarides T. 2007. Chromatin modifications and their function. *Cell* **128**: 693–705.

Krogan NJ, Cagney G, Yu H, Zhong G, Guo X, Ignatchenko A, Li J, Pu S, Datta N, Tikuisis AP, et al. 2006. Global landscape of protein complexes in the yeast *Saccharomyces cerevisiae. Nature* **440**: 637–643.

Kuo MH, Brownell JE, Sobel RE, Ranalli TA, Cook RG, Edmondson DG, Roth SY, Allis CD. 1996. Transcription-linked acetylation by Gcn5p of histones H3 and H4 at specific lysines. *Nature* **383**: 269–272.

Kuo AJ, Song J, Cheung P, Ishibe-Murakami S, Yamazoe S, Chen JK, Patel DJ, Gozani O. 2012. The BAH domain of ORC1 links H4K20me2 to DNA replication licensing and Meier–Gorlin syndrome. *Nature* **484**: 115–119.

Ladenburger EM, Keller C, Knippers R. 2002. Identification of a binding region for human origin recognition complex proteins 1 and 2 that coincides with an origin of DNA replication. *Mol Cell Biol* **22**: 1036–1048.

Laskey RA, Mills AD, Morris NR. 1977. Assembly of SV40 chromatin in a cell-free system from *Xenopus* eggs. *Cell* **10**: 237–243.

Le S, Davis C, Konopka JB, Sternglanz R. 1997. Two new S-phase-specific genes from *Saccharomyces cerevisiae. Yeast* **13**: 1029–1042.

Lee T-J, Pascuzzi PE, Settlage SB, Shultz RW, Tanurdzic M, Rabinowicz PD, Menges M, Zheng P, Main D, Murray JAH, et al. 2010. *Arabidopsis thaliana* chromosome 4 replicates in two phases that correlate with chromatin state. *PLoS Genet* **6**: e1000982.

Li Q, Zhou H, Wurtele H, Davies B, Horazdovsky B, Verreault A, Zhang Z. 2008. Acetylation of histone H3 lysine 56 regulates replication coupled nucleosome assembly. *Cell* **134**: 244–255.

Lipford JR, Bell SP. 2001. Nucleosomes positioned by ORC facilitate the initiation of DNA replication. *Mol Cell* **7**: 21–30.

Loppin B, Bonnefoy E, Anselme C, Laurencon A, Karr TL, Couble P. 2005. The histone H3.3 chaperone HIRA is essential for chromatin assembly in the male pronucleus. *Nature* **437**: 1386–1390.

Loyola A, Bonaldi T, Roche D, Imhof A, Almouzni G. 2006. PTMs on H3 variants before chromatin assembly potentiate their final epigenetic state. *Mol Cell* **24**: 309–316.

Loyola A, Tagami H, Bonaldi T, Roche D, Quivy JP, Imhof A, Nakatani Y, Dent SY, Almouzni G. 2009. The HP1α–CAF1–SetDB1-containing complex provides H3K9me1 for Suv39-mediated K9me3 in pericentric heterochromatin. *EMBO Rep* **10**: 769–775.

Lubelsky Y, Sasaki T, Kuipers MA, Lucas I, Le Beau MM, Carignon S, Debatisse M, Prinz JA, Dennis JH, Gilbert DM. 2011. Pre-replication complex proteins assemble at regions of low nucleosome occupancy within the Chinese hamster dihydrofolate reductase initiation zone. *Nucleic Acids Res* **39**: 3141–3155.

Luger K, Mader AW, Richmond RK, Sargent DF, Richmond TJ. 1997. Crystal structure of the nucleosome core particle at 2.8 Å resolution. *Nature* **389**: 251–260.

Ma XJ, Wu J, Altheim BA, Schultz MC, Grunstein M. 1998. Deposition-related sites K5/K12 in histone H4 are not required for nucleosome deposition in yeast. *Proc Natl Acad Sci* **95**: 6693–6698.

MacAlpine DM, Rodriguez HK, Bell SP. 2004. Coordination of replication and transcription along a *Drosophila* chromosome. *Genes Dev* **18**: 3094–3105.

MacAlpine HK, Gordan R, Powell SK, Hartemink AJ, MacAlpine DM. 2010. *Drosophila* ORC localizes to open chromatin and marks sites of cohesin complex loading. *Genome Res* **20**: 201–211.

Malay AD, Umehara T, Matsubara-Malay K, Padmanabhan B, Yokoyama S. 2008. Crystal structures of fission yeast histone chaperone Asf1 complexed with the Hip1 B-domain or the Cac2 C terminus. *J Biol Chem* **283**: 14022–14031.

Margueron R, Reinberg D. 2010. Chromatin structure and the inheritance of epigenetic information. *Nat Rev Genet* **11**: 285–296.

Marzluff WF, Duronio RJ. 2002. Histone mRNA expression: Multiple levels of cell cycle regulation and important developmental consequences. *Curr Opin Cell Biol* **14**: 692–699.

Marzluff WF, Wagner EJ, Duronio RJ. 2008. Metabolism and regulation of canonical histone mRNAs: Life without a poly(A) tail. *Nature Rev Genet* **9**: 843–854.

Masumoto H, Hawke D, Kobayashi R, Verreault A. 2005. A role for cell-cycle-regulated histone H3 lysine 56 acetylation in the DNA damage response. *Nature* **436**: 294–298.

Mavrich TN, Ioshikhes IP, Venters BJ, Jiang C, Tomsho LP, Qi J, Schuster SC, Albert I, Pugh BF. 2008a. A barrier nucleosome model for statistical positioning of nucleosomes throughout the yeast genome. *Genome Res* **18**: 1073–1083.

Mavrich TN, Jiang C, Ioshikhes IP, Li X, Venters BJ, Zanton SJ, Tomsho LP, Qi J, Glaser RL, Schuster SC, et al. 2008b. Nucleosome organization in the *Drosophila* genome. *Nature* **453**: 358–362.

McKittrick E, Gafken PR, Ahmad K, Henikoff S. 2004. Histone H3.3 is enriched in covalent modifications associated with active chromatin. *Proc Natl Acad Sci* **101**: 1525–1530.

Mello JA, Sillje HH, Roche DM, Kirschner DB, Nigg EA, Almouzni G. 2002. Human Asf1 and CAF-1 interact and synergize in a repair-coupled nucleosome assembly pathway. *EMBO Rep* **3**: 329–334.

Meneghini MD, Wu M, Madhani HD. 2003. Conserved histone variant H2A.Z protects euchromatin from the ectopic spread of silent heterochromatin. *Cell* **112**: 725–736.

Milutinovic S, Zhuang Q, Szyf M. 2002. Proliferating cell nuclear antigen associates with histone deacetylase activity, integrating DNA replication and chromatin modification. *J Biol Chem* **277**: 20974–20978.

Miotto B, Struhl K. 2010. HBO1 histone acetylase activity is essential for DNA replication licensing and inhibited by Geminin. *Mol Cell* **37**: 57–66.

Mito Y, Henikoff JG, Henikoff S. 2007. Histone replacement marks the boundaries of *cis*-regulatory domains. *Science* **315**: 1408–1411.

Moggs JG, Grandi P, Quivy JP, Jonsson ZO, Hubscher U, Becker PB, Almouzni G. 2000. A CAF-1–PCNA-mediated chromatin assembly pathway triggered by sensing DNA damage. *Mol Cell Biol* **20**: 1206–1218.

Mousson F, Ochsenbein F, Mann C. 2007. The histone chaperone Asf1 at the crossroads of chromatin and DNA checkpoint pathways. *Chromosoma* **116**: 79–93.

Murzina N, Verreault A, Laue E, Stillman B. 1999. Heterochromatin dynamics in mouse cells: Interaction between chromatin assembly factor 1 and HP1 proteins. *Mol Cell* **4**: 529–540.

Natsume R, Eitoku M, Akai Y, Sano N, Horikoshi M, Senda T. 2007. Structure and function of the histone chaperone CIA/ASF1 complexed with histones H3 and H4. *Nature* **446**: 338–341.

Nelson DM, Ye X, Hall C, Santos H, Ma T, Kao GD, Yen TJ, Harper JW, Adams PD. 2002. Coupling of DNA synthesis and histone synthesis in S phase independent of cyclin/cdk2 activity. *Mol Cell Biol* **22**: 7459–7472.

Nishioka K, Rice JC, Sarma K, Erdjument-Bromage H, Werner J, Wang Y, Chuikov S, Valenzuela P, Tempst P, Steward R, et al. 2002. PR-Set7 is a nucleosome-specific methyltransferase that modifies lysine 20 of histone H4 and is associated with silent chromatin. *Mol Cell* **9**: 1201–1213.

Oda H, Hubner MR, Beck DB, Vermeulen M, Hurwitz J, Spector DL, Reinberg D. 2010. Regulation of the histone H4 monomethylase PR-Set7 by $CRL4^{Cdt2}$-mediated PCNA-dependent degradation during DNA damage. *Mol Cell* **40**: 364–376.

Osakabe A, Tachiwana H, Matsunaga T, Shiga T, Nozawa RS, Obuse C, Kurumizaka H. 2010. Nucleosome formation activity of human somatic nuclear autoantigenic sperm protein (sNASP). *J Biol Chem* **285**: 11913–11921.

Parthun MR. 2007. Hat1: The emerging cellular roles of a type B histone acetyltransferase. *Oncogene* **26**: 5319–5328.

Pehrson JR, Fried VA. 1992. MacroH2A, a core histone containing a large nonhistone region. *Science* **257**: 1398–1400.

Poot RA, Bozhenok L, van den Berg DL, Steffensen S, Ferreira F, Grimaldi M, Gilbert N, Ferreira J, Varga-Weisz PD. 2004. The Williams syndrome transcription factor interacts with PCNA to target chromatin remodelling by ISWI to replication foci. *Nat Cell Biol* **6**: 1236–1244.

Pradhan S, Bacolla A, Wells RD, Roberts RJ. 1999. Recombinant human DNA (cytosine-5) methyltransferase. I. Expression, purification, and comparison of de novo and maintenance methylation. *J Biol Chem* **274**: 33002–33010.

Probst AV, Dunleavy E, Almouzni G. 2009. Epigenetic inheritance during the cell cycle. *Nat Rev Mol Cell Biol* **10**: 192–206.

Quivy JP, Grandi P, Almouzni G. 2001. Dimerization of the largest subunit of chromatin assembly factor 1: Importance in vitro and during *Xenopus* early development. *EMBO J* **20**: 2015–2027.

Quivy JP, Roche D, Kirschner D, Tagami H, Nakatani Y, Almouzni G. 2004. A CAF-1 dependent pool of HP1 during heterochromatin duplication. *EMBO J* **23**: 3516–3526.

Quivy JP, Gerard A, Cook AJ, Roche D, Almouzni G. 2008. The HP1-p150/CAF-1 interaction is required for pericentric heterochromatin replication and S-phase progression in mouse cells. *Nat Struct Mol Biol* **15**: 972–979.

Radman-Livaja M, Rando OJ. 2010. Nucleosome positioning: How is it established, and why does it matter? *Dev Biol* **339**: 258–266.

Raisner RM, Madhani HD. 2006. Patterning chromatin: Form and function for H2A.Z variant nucleosomes. *Curr Opin Genet Dev* **16**: 119–124.

Raser JM, O'Shea EK. 2004. Control of stochasticity in eukaryotic gene expression. *Science* **304**: 1811–1814.

Ray-Gallet D, Quivy JP, Sillje HW, Nigg EA, Almouzni G. 2007. The histone chaperone Asf1 is dispensable for direct de novo histone deposition in *Xenopus* egg extracts. *Chromosoma* **19**: 19.

Ray-Gallet D, Woolfe A, Vassias I, Pellentz C, Lacoste N, Puri A, Schultz DC, Pchelintsev NA, Adams PD, Jansen LE, et al. 2011. Dynamics of histone H3 deposition in vivo reveal a nucleosome gap-filling mechanism for H3.3 to maintain chromatin integrity. *Mol Cell* **44**: 928–941.

Remus D, Beall EL, Botchan MR. 2004. DNA topology, not DNA sequence, is a critical determinant for *Drosophila* ORC-DNA binding. *EMBO J* **23**: 897–907.

Roth DB, Roth SY. 2000. Unequal access: Regulating V(D)J recombination through chromatin remodeling. *Cell* **103**: 699–702.

Rusche LN, Kirchmaier AL, Rine J. 2003. The establishment, inheritance, and function of silenced chromatin in *Saccharomyces cerevisiae. Annu Rev Biochem* **72**: 481–516.

Ryba T, Hiratani I, Lu J, Itoh M, Kulik M, Zhang J, Schulz TC, Robins AJ, Dalton S, Gilbert DM. 2010. Evolutionarily conserved replication timing profiles predict long-range chromatin interactions and distinguish closely related cell types. *Genome Res* **20**: 761–770.

Sanematsu F, Takami Y, Barman HK, Fukagawa T, Ono T, Shibahara K, Nakayama T. 2006. Asf1 is required for viability and chromatin assembly during DNA replication in vertebrate cells. *J Biol Chem* **281**: 13817–13827.

Schermelleh L, Haemmer A, Spada F, Rosing N, Meilinger D, Rothbauer U, Cardoso MC, Leonhardt H. 2007. Dynamics of Dnmt1 interaction with the replication machinery and its role in postreplicative maintenance of DNA methylation. *Nucleic Acids Res* **35**: 4301–4312.

Schones DE, Cui K, Cuddapah S, Roh TY, Barski A, Wang Z, Wei G, Zhao K. 2008. Dynamic regulation of nucleosome positioning in the human genome. *Cell* **132**: 887–898.

Schotta G, Sengupta R, Kubicek S, Malin S, Kauer M, Callen E, Celeste A, Pagani M, Opravil S, De La Rosa-Velazquez IA, et al. 2008. A chromatin-wide transition to H4K20 monomethylation impairs genome integrity and programmed DNA rearrangements in the mouse. *Genes Dev* **22**: 2048–2061.

Schübeler D, Scalzo D, Kooperberg C, Steensel Bv, Delrow J, Groudine M. 2002. Genome-wide DNA replication profile for *Drosophila melanogaster*: A link between transcription and replication timing. *Nat Genet* **32**: 438–442.

Schuh M, Lehner CF, Heidmann S. 2007. Incorporation of *Drosophila* CID/CENP-A and CENP-C into centromeres during early embryonic anaphase. *Curr Biol* **17**: 237–243.

Schwaiger M, Stadler MB, Bell O, Kohler H, Oakeley EJ, Schubeler D. 2009. Chromatin state marks cell-type- and gender-specific replication of the *Drosophila* genome. *Genes Dev* **23**: 589–601.

Segal E, Fondufe-Mittendorf Y, Chen L, Thastrom A, Field Y, Moore IK, Wang JP, Widom J. 2006. A genomic code for nucleosome positioning. *Nature* **442**: 772–778.

Shelby RD, Monier K, Sullivan KF. 2000. Chromatin assembly at kinetochores is uncoupled from DNA replication. *J Cell Biol* **151**: 1113–1118.

Shibahara K, Stillman B. 1999. Replication-dependent marking of DNA by PCNA facilitates CAF-1-coupled inheritance of chromatin. *Cell* **96**: 575–585.

Shibahara K, Verreault A, Stillman B. 2000. The N-terminal domains of histones H3 and H4 are not necessary for chromatin assembly factor-1-mediated nucleosome assembly onto replicated DNA in vitro. *Proc Natl Acad Sci* **97**: 7766–7771.

Shimada K, Oma Y, Schleker T, Kugou K, Ohta K, Harata M, Gasser SM. 2008. Ino80 chromatin remodeling complex promotes recovery of stalled replication forks. *Curr Biol* **18**: 566–575.

Simpson RT. 1990. Nucleosome positioning can affect the function of a *cis*-acting DNA element in vivo. *Nature* **343**: 387–389.

Smith S, Stillman B. 1989. Purification and characterization of CAF-I, a human cell factor required for chromatin assembly during DNA replication in vitro. *Cell* **58**: 15–25.

Smith DJ, Whitehouse I. 2012. Intrinsic coupling of lagging-strand synthesis to chromatin assembly. *Nature* **483**: 434–438.

Sobel RE, Cook RG, Perry CA, Annunziato AT, Allis CD. 1995. Conservation of deposition-related acetylation sites in newly synthesized histones H3 and H4. *Proc Natl Acad Sci* **92**: 1237–1241.

Song Y, He F, Xie G, Guo X, Xu Y, Chen Y, Liang X, Stagljar I, Egli D, Ma J, et al. 2007. CAF-1 is essential for *Drosophila* development and involved in the maintenance of epigenetic memory. *Dev Biol* **311**: 213–222.

Spada F, Haemmer A, Kuch D, Rothbauer U, Schermelleh L, Kremmer E, Carell T, Langst G, Leonhardt H. 2007. DNMT1 but not its interaction with the replication machinery is required for maintenance of DNA methylation in human cells. *J Cell Biol* **176**: 565–571.

Stambrook PJ, Flickinger RA. 1970. Changes in chromosomal DNA replication patterns in developing frog embryos. *J Exp Zool* **174**: 101–113.

Stevenson JB, Gottschling DE. 1999. Telomeric chromatin modulates replication timing near chromosome ends. *Genes Dev* **13**: 146–151.

Stillman B. 1986. Chromatin assembly during SV40 DNA replication in vitro. *Cell* **45**: 555–565.

Stimpson KM, Sullivan BA. 2010. Epigenomics of centromere assembly and function. *Curr Opin Cell Biol* **22**: 772–780.

Stucki M, Clapperton JA, Mohammad D, Yaffe MB, Smerdon SJ, Jackson SP. 2005. MDC1 directly binds phosphorylated histone H2AX to regulate cellular responses to DNA double-strand breaks. *Cell* **123**: 1213–1226.

Taddei A, Roche D, Sibarita JB, Turner BM, Almouzni G. 1999. Duplication and maintenance of heterochromatin domains. *J Cell Biol* **147**: 1153–1166.

Taddei A, Maison C, Roche D, Almouzni G. 2001. Reversible disruption of pericentric heterochromatin and centromere function by inhibiting deacetylases. *Nat Cell Biol* **3**: 114–120.

Tagami H, Ray-Gallet D, Almouzni G, Nakatani Y. 2004. Histone H3.1 and H3.3 complexes mediate nucleosome assembly pathways dependent or independent of DNA synthesis. *Cell* **116**: 51–61.

Tan BC, Chien CT, Hirose S, Lee SC. 2006. Functional cooperation between FACT and MCM helicase facilitates initiation of chromatin DNA replication. *EMBO J* **25**: 3975–3985.

Tang Y, Poustovoitov MV, Zhao K, Garfinkel M, Canutescu A, Dunbrack R, Adams PD, Marmorstein R. 2006. Structure of a human ASF1a–HIRA complex and insights into specificity of histone chaperone complex assembly. *Nat Struct Mol Biol* **13**: 921–929.

Tardat M, Murr R, Herceg Z, Sardet C, Julien E. 2007. PR-Set7-dependent lysine methylation ensures genome replication and stability through S phase. *J Cell Biol* **179**: 1413–1426.

Thastrom A, Lowary PT, Widlund HR, Cao H, Kubista M, Widom J. 1999. Sequence motifs and free energies of selected natural and non-natural nucleosome positioning DNA sequences. *J Mol Biol* **288**: 213–229.

Torigoe SE, Urwin DL, Ishii H, Smith DE, Kadonaga JT. 2011. Identification of a rapidly formed nonnucleosomal histone-DNA intermediate that is converted into chromatin by ACF. *Mol Cell* **43**: 638–648.

Tsubota T, Berndsen CE, Erkmann JA, Smith CL, Yang L, Freitas MA, Denu JM, Kaufman PD. 2007. Histone H3-K56 acetylation is catalyzed by histone chaperone-dependent complexes. *Mol Cell* **25**: 703–712.

Tyler JK, Adams CR, Chen SR, Kobayashi R, Kamakaka RT, Kadonaga JT. 1999. The RCAF complex mediates chromatin assembly during DNA replication and repair. *Nature* **402**: 555–560.

Tyler JK, Collins KA, Prasad-Sinha J, Amiott E, Bulger M, Harte PJ, Kobayashi R, Kadonaga JT. 2001. Interaction between the *Drosophila* CAF-1 and ASF1 chromatin assembly factors. *Mol Cell Biol* **21**: 6574–6584.

Valouev A, Ichikawa J, Tonthat T, Stuart J, Ranade S, Peckham H, Zeng K, Malek JA, Costa G, McKernan K, et al. 2008. A high-resolution, nucleosome position map of *C. elegans* reveals a lack of universal sequence-dictated positioning. *Genome Res* **18**: 1051–1063.

VanDemark AP, Blanksma M, Ferris E, Heroux A, Hill CP, Formosa T. 2006. The structure of the yFACT Pob3-M domain, its interaction with the DNA replication factor RPA, and a potential role in nucleosome deposition. *Mol Cell* **22**: 363–374.

Varga-Weisz PD, Wilm M, Bonte E, Dumas K, Mann M, Becker PB. 1997. Chromatin-remodelling factor CHRAC contains the ATPases ISWI and topoisomerase II. *Nature* **388**: 598–602.

Vashee S, Cvetic C, Lu W, Simancek P, Kelly TJ, Walter JC. 2003. Sequence-independent DNA binding and replication initiation by the human origin recognition complex. *Genes Dev* **17**: 1894–1908.

Verreault A, Kaufman PD, Kobayashi R, Stillman B. 1996. Nucleosome assembly by a complex of CAF-1 and acetylated histones H3/H4. *Cell* **87**: 95–104.

Vincent JA, Kwong TJ, Tsukiyama T. 2008. ATP-dependent chromatin remodeling shapes the DNA replication landscape. *Nat Struct Mol Biol* **15**: 477–484.

Vogelauer M, Rubbi L, Lucas I, Brewer BJ, Grunstein M. 2002. Histone acetylation regulates the time of replication origin firing. *Mol Cell* **10**: 1223–1233.

Warburton PE, Cooke CA, Bourassa S, Vafa O, Sullivan BA, Stetten G, Gimelli G, Warburton D, Tyler-Smith C, Sullivan KF, et al. 1997. Immunolocalization of CENP-A suggests a distinct nucleosome structure at the inner kinetochore plate of active centromeres. *Curr Biol* **7**: 901–904.

Weber M, Schubeler D. 2007. Genomic patterns of DNA methylation: Targets and function of an epigenetic mark. *Curr Opin Cell Biol* **19**: 273–280.

Whitehouse I, Rando OJ, Delrow J, Tsukiyama T. 2007. Chromatin remodelling at promoters suppresses antisense transcription. *Nature* **450**: 1031–1035.

Wolffe AP, Pruss D. 1996. Deviant nucleosomes: The functional specialization of chromatin. *Trends Genet* **12**: 58–62.

Wong LH, McGhie JD, Sim M, Anderson MA, Ahn S, Hannan RD, George AJ, Morgan KA, Mann JR, Choo KH. 2010. ATRX interacts with H3.3 in maintaining telomere structural integrity in pluripotent embryonic stem cells. *Genome Res* **20**: 351–360.

Xiao A, Li H, Shechter D, Ahn SH, Fabrizio LA, Erdjument-Bromage H, Ishibe-Murakami S, Wang B, Tempst P, Hofmann K, et al. 2009. WSTF regulates the H2A.X DNA damage response via a novel tyrosine kinase activity. *Nature* **457**: 57–62.

Xie W, Song C, Young NL, Sperling AS, Xu F, Sridharan R, Conway AE, Garcia BA, Plath K, Clark AT, et al. 2009. Histone H3 lysine 56 acetylation is linked to the core transcriptional network in human embryonic stem cells. *Mol Cell* **33**: 417–427.

Xu W, Aparicio JG, Aparicio OM, Tavare S. 2006. Genome-wide mapping of ORC and Mcm2p binding sites on tiling arrays and identification of essential ARS consensus sequences in *S. cerevisiae. BMC Genomics* **7**: 276.

Xu M, Long C, Chen X, Huang C, Chen S, Zhu B. 2010. Partitioning of histone H3-H4 tetramers during DNA replication–dependent chromatin assembly. *Science* **328**: 94–98.

Yuan GC, Liu YJ, Dion MF, Slack MD, Wu LF, Altschuler SJ, Rando OJ. 2005. Genome-scale identification of nucleosome positions in *S. cerevisiae. Science* **309**: 626–630.

Zlatanova J, Seebart C, Tomschik M. 2007. Nap1: Taking a closer look at a juggler protein of extraordinary skills. *FASEB J* **21**: 1294–1310.

Sister Chromatid Cohesion

Jan-Michael Peters[1] and Tomoko Nishiyama[1,2]

[1]IMP, Dr. Bohr-Gasse 7, A-1030 Vienna, Austria

[2]Nagoya University, Furo-cho, Chikusa-ku, Nagoya 464-8602, Japan

Correspondence: peters@imp.ac.at

During S phase, not only does DNA have to be replicated, but also newly synthesized DNA molecules have to be connected with each other. This sister chromatid cohesion is essential for the biorientation of chromosomes on the mitotic or meiotic spindle, and is thus an essential prerequisite for chromosome segregation. Cohesion is mediated by cohesin complexes that are thought to embrace sister chromatids as large rings. Cohesin binds to DNA dynamically before DNA replication and is converted into a stably DNA-bound form during replication. This conversion requires acetylation of cohesin, which in vertebrates leads to recruitment of sororin. Sororin antagonizes Wapl, a protein that is able to release cohesin from DNA, presumably by opening the cohesin ring. Inhibition of Wapl by sororin therefore "locks" cohesin rings on DNA and allows them to maintain cohesion for long periods of time in mammalian oocytes, possibly for months or even years.

DNA replication during the synthesis (S) phase generates identical DNA molecules, which, in their chromatinized form, are called sister chromatids. The pairs of sister chromatids remain united as part of one chromosome during the subsequent gap (G_2) phase and during early mitosis, in prophase, prometaphase, and metaphase. During these stages of mitosis chromosomes condense, in most eukaryotes the nuclear envelope breaks down, and in all species chromosomes are ultimately attached to both poles of the mitotic spindle. Only once this biorientation has been achieved for all chromosomes, the sister chromatids are separated from each other in anaphase and transported toward opposite spindle poles of the mother cell, enabling its subsequent division into two genetically identical daughter cells.

This series of events critically depends on the fact that sister chromatids remain physically connected with each other from S phase until metaphase. This physical connection, called sister chromatid cohesion, opposes the pulling forces that are generated by microtubules that attach to kinetochores and thereby enables the biorientation of chromosomes on the mitotic spindle (Tanaka et al. 2000b). Without cohesion, sister chromatids could therefore not be segregated symmetrically between the forming daughter cells, resulting in aneuploidy. For the same reasons, cohesion is essential for chromosome segregation in meiosis I and meiosis II. Cohesion defects in human oocytes can lead to aneuploidy, which is thought to be the major cause of spontaneous abortion, because only a few types of aneuploidy are compatible with

Cite this article as *Cold Spring Harb Perspect Biol* doi: 10.1101/cshperspect.a011130

viability, such as trisomy 21 (Down syndrome), trisomy 18 (Edwards syndrome), and trisomy 13 (Patau syndrome) (Hunt and Hassold 2010). Studying the mechanisms of cohesion is therefore essential for understanding how the genome is passed properly from one cell generation to the next.

In addition, sister chromatid cohesion facilitates the repair of DNA double-strand breaks in cells that have replicated their DNA, where such breaks can be repaired by a homologous recombination mechanism that uses the undamaged sister chromatid as a template (for review, see Watrin et al. 2006). Furthermore, mutations in the proteins that are required for sister chromatid cohesion can cause defects in chromatin structure and gene regulation, and can in rare cases lead to congenital developmental disorders, called Cornelia de Lange syndrome, Roberts/SC Phocomelia syndrome, and Warsaw Breakage syndrome (for review, see Mannini et al. 2010).

SISTER CHROMATID COHESION IS MEDIATED BY COHESIN

Sister chromatid cohesion depends on a multisubunit protein complex called cohesin (Fig. 1). Three of cohesin's subunits—Smc1, Smc3, and Scc1/Rad21/Mcd1—form an unusual ring-shaped structure (Anderson et al. 2002; Haering et al. 2002). Smc1 and Smc3 are members of the "structural maintenance of chromosomes" (SMC) protein family. Members of this family are also found in other DNA-binding protein complexes, such as condensins, the Smc5/6 complex, and in bacterial SMC complexes. Like other SMC proteins, Smc1 and Smc3 are highly elongated molecules almost 50 nm in length that contain a central rod-shaped domain that is composed of anti-parallel coiled coils. Smc1 and Smc3 bind directly to each other at one of their ends via a "hinge" domain. At the other end of their coiled-coil domains, Smc1 and Smc3 form a nucleotide-binding domain (NBD). These NBDs form two ATP-binding sites, where one molecule of Mg^{2+}-ATP is simultaneously bound to Walker A and Walker B motifs in the NBD of one SMC subunit and to an ABC signature motif in the other SMC subunit (Haering et al. 2004). The NBDs of Smc1 and Smc3 are connected by the kleisin ("bridging") subunit Scc1, with Scc1's amino terminus being bound to Smc3 and its carboxy-terminal Winged helix domain being bound to Smc1 (Haering et al. 2002). These interactions between Smc1, Smc3, and Scc1 lead to the formation of a tripartite ring with an inner diameter of 30–40 nm.

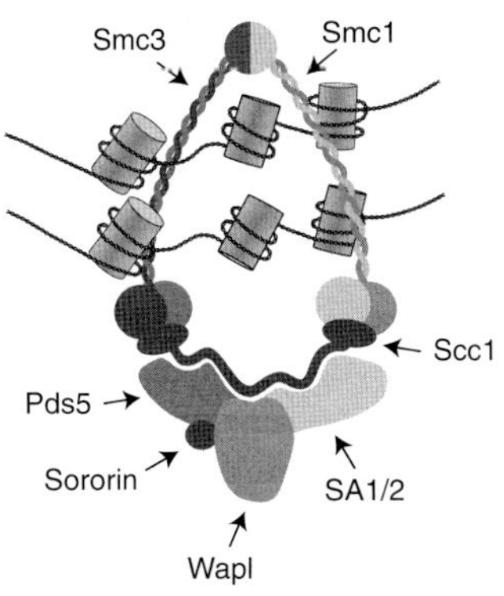

Figure 1. The cohesin complex. A schematic illustration of how cohesin subunits might be arranged in vertebrate cohesin complexes that mediate sister chromatid cohesion by embracing the two sister chromatid strands.

The integrity of this ring structure is essential for its ability to mediate sister chromatid cohesion (Gruber et al. 2003), and experimental observations using yeast mini-chromosomes indicate that cohesin mediates cohesion by topologically embracing the two sister chromatids (Ivanov and Nasmyth 2005; Haering et al. 2008). It has therefore been proposed that cohesin mediates cohesion by embracing both sister chromatids as a single ring (for review, see Nasmyth 2011). Other models propose that only a single DNA molecule can be embraced by a cohesin ring, and that two cohesin rings have to interact to connect two sister chromatids (Huang et al. 2005; Zhang et al. 2008b).

A fourth cohesin subunit, called Scc3 in yeast and stromal antigen (SA/STAG) in higher eukaryotes, binds directly to Scc1 (Haering et al. 2002). Scc3 is also essential for sister chromatid cohesion (Michaelis et al. 1997; Nishiyama et al. 2010). In vertebrate somatic cells, cohesin complexes contain either one of two different SA/STAG proteins, called SA1 and SA2 (Losada et al. 2000; Sumara et al. 2000). In most organisms, Scc1 and Scc3/SA interact with several other proteins called Wapl/Rad61, Pds5, and Sororin (Table 1). These proteins are believed to determine how stably the cohesin ring interacts with DNA, and if these interactions are able to mediate cohesion (see below).

In many organisms, different kleisin subunits and isoforms of Smc1, Scc3/SA, and Pds5 exist, some of which are differentially expressed between somatic and meiotic cells. For example, in mammalian oocytes and spermatocytes, many cohesin complexes contain Smc1β instead of Smc1 (the latter is therefore also called Smc1α), and the kleisins Rec8 or Rad21L instead of Scc1. Table 1 provides an overview regarding the different isoforms of these and other cohesin subunits and their regulators that have been identified in different species.

LOADING OF COHESIN ONTO DNA DEPENDS ON ADHERIN, COHESIN'S ATPASE ACTIVITY, AND POSSIBLY OPENING OF THE COHESIN RING

Cohesin is loaded onto DNA before DNA replication, in vertebrates already at the end of mitosis in telophase (Fig. 2). This process depends on cohesin's ATPase activity (Arumugam et al. 2003; Weitzer et al. 2003) and on a separate complex, called adherin (Furuya et al. 1998) or kollerin (Nasmyth 2011). Because cohesin is thought to embrace DNA as a ring, it has furthermore been proposed that cohesin rings have to be opened transiently to allow entry of the DNA into the ring (Gruber et al. 2003).

Adherin is a chromatin-bound protein complex composed of two subunits, a large protein (with a molecular mass of 316 kDa in humans), called Scc2 in budding yeast and NIPBL in mammals, and a smaller subunit, called Scc4 in yeast and MAU2 in mammals (Table 1). In *Xenopus* eggs, adherin is recruited to chromatin by binding to pre-replicative complexes (pre-RCs) (Gillespie and Hirano 2004; Takahashi et al. 2004), whereas in yeast, adherin is particularly enriched at centromeres but is also found on chromosome arms and near telomeres (Lengronne et al. 2004). Both pre-RC components in *Xenopus* and centromere-associated kinetochore proteins in yeast are also required for efficient cohesin loading (for review, see Peters et al. 2008). At least in the case of the yeast kinetochore, its role in cohesin loading might go beyond the recruitment of adherin to chromatin, because tethering Scc2 to a chromosomal locus where it is normally not present is not sufficient for recruiting cohesin to such a site (Hu et al. 2011).

Both subunits of adherin are composed of repetitive sequence elements, and no enzymatic functions are known for them. Because loading of cohesin onto DNA depends on both adherin and on cohesin's ATPase activity, and because the loading reaction is thought to require opening of the cohesin ring, adherin could be an allosteric regulator of cohesin's ATPase activity, and it is conceivable that this activity would facilitate ring opening. If this were true, how might ring opening be achieved, and how might ATP hydrolysis mediate this process? Because mutation of Smc1's and Smc3's hinge domains, or artificially tethering these domains to each other, blocks the loading of cohesin onto DNA, it has been proposed that Smc1's and Smc3's hinge domains have to be separated from each other to create an "entry gate" for DNA (Fig. 3) (Gruber et al. 2006; Mishra et al. 2010; Kurze et al. 2011; Nasmyth 2011). Studies of a cohesin-related bacterial SMC complex support this view because the hinge domain of this complex is also required for DNA binding (Hirano and Hirano 2006).

Ideas for how this entry gate might be opened have come from structural and biochemical analyses of proteins whose NBDs are related in structure to the NBDs of SMC proteins. Such NBDs are found in ATP-binding cassette (ABC) transporters and in Rad50 (Hopfner and Tainer 2003). ABC transporters are homodimeric or heterodimeric ATPases that are using cycles of ATP binding and hydrolysis to transport small molecules across membranes, whereas Rad50 is a homodimeric protein required for DNA repair that contains both NBDs and coiled-coil domains that are similar to the ones of SMC proteins. In these proteins, ATP binding leads to a transient physical interaction between

Table 1. Subunits and regulators of cohesin in yeasts, fruit flies, and humans

	Saccharomyces cerevisiae		*Schizosaccharomyces pombe*		*Drosophila melanogaster*		*Homo sapiens*	
	Mitosis	Meiosis[a]	Mitosis	Meiosis[a]	Mitosis	Meiosis[a]	Mitosis	Meiosis[a]
SMC	Smc1		Psm1		Smc1		Smc1	Smc1b
	Smc3		Psm3		Smc3		Smc3	
Kleisin	Mcd1/Scc1	Rec8	Rad21	Rec8	Rad21	C(2)M	Scc1	Rec8,[b] Rad21L
Kleisin-associated	Scc3		Psc3	Rec11	SA/Scc3	SNM	SA1/STAG1	SA3/STAG3
HEAT repeat protein							SA2/STAG2	
Releasin	Rad61/Wpl1		Wpl1		Wapl		Wapl	
	Pds5		Spo76/Pds5		Pds5		Pds5A, Pds5B	
Sororin					Dalmatian		Sororin	
Adherin/kollerin	Scc2		Mis4		Nipped-B		NIPBL/Scc2	
Cohesin loading	Scc4		Ssl3				MAU-2/Scc4	
Cohesin positioning					CTCF		CTCF	BORIS/CTCFL[c]
Cohesin cleavage	Esp1		Cut1		Separase		Separase	
Acetyltransferase	Eco1		Eso1		Deco		Esco1, Esco2	
Deacetylase	Hos1						HDAC8	

[a]Expressed specifically in meiosis and largely/partially replace mitotic subunit.
[b]Also expressed in B cells.
[c]Paralog of CTCF specifically expressed in testis.

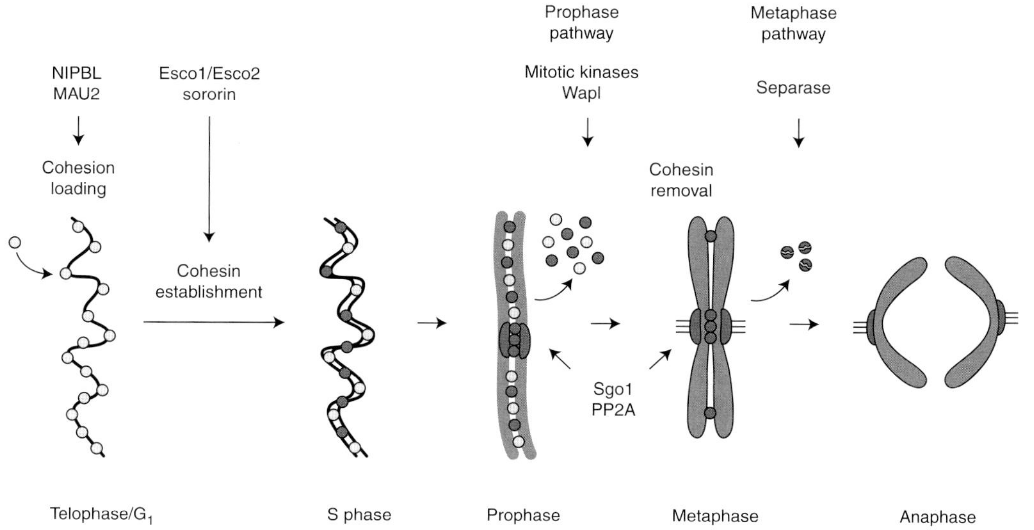

Figure 2. The cohesin cycle in vertebrate cells. A schematic illustration showing cohesin loading, cohesion establishment, and cohesin removal from chromosomes in telophase/G_1 phase, S phase, and mitosis, respectively. (Small circles) Cohesin complexes; (light gray circles) cohesin complexes that are dynamically bound to DNA; (dark gray circles) "cohesive" complexes that are stably bound to DNA (Gerlich et al. 2006).

the two NBDs (Hopfner et al. 2000), and in the case of ABC transporters, this "engagement" step facilitates transport of substrate molecules through the transmembrane channel (for review, see Rees et al. 2009). ATP hydrolysis reverts these conformational changes by leading to disengagement of the NBDs. Similarly to how one cycle of ATP binding and hydrolysis leads to

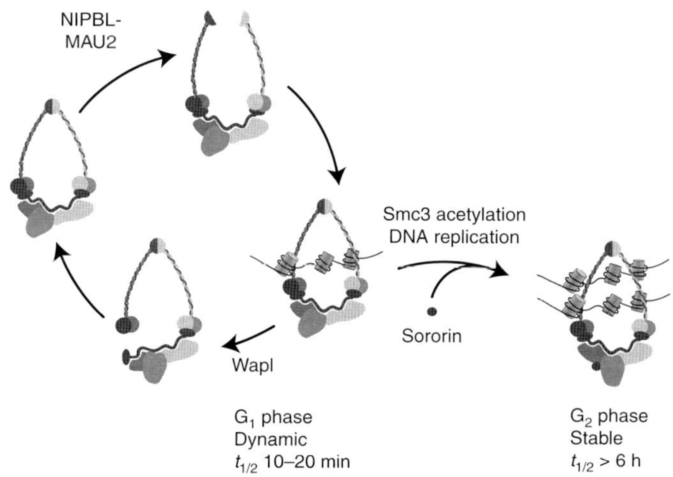

Figure 3. Dynamic and stable binding modes of cohesin to DNA. A schematic drawing illustrating how adherin-mediated separation of the hinge domains of Smc1 and Smc3 might create an entry gate for DNA into the cohesin ring, whereas a transient Wapl-mediated dissociation of Scc1 from the SMC subunits may release cohesin from DNA by generating an exit gate (Nasmyth 2011; Chan et al. 2012). Wapl's ability to open this exit gate might be inhibited by sororin, which binds to cohesin in vertebrates following Smc3 acetylation and possibly other unidentified events during DNA replication (Nishiyama et al. 2010).

transient opening of the transmembrane channel in ABC transporters, it is therefore possible that ATP binding and hydrolysis would lead to transient opening of an entry gate for DNA in the cohesin ring.

But how could binding of ATP to the NBDs of Smc1 and Smc3 lead to separation of the hinge domains of Smc1 and Smc3, which are almost 50 nm away from the NBDs? Atomic force microscopy and crystallographic studies have revealed that ATP binding to the signature motif of Rad50 is not only accompanied by NBD engagement but also induces a dramatic conformational change that alters the angle at which the coiled-coil domain of Rad50 emanates from the NBD (Moreno-Herrero et al. 2005; Lammens and Hopfner 2010; Lim et al. 2011; Williams et al. 2011). This conformational change is mediated by helices in Rad50, called the signature-coupling helices, which move relative to each other and thereby function as a lever that exerts force on the coiled-coil domain. It is therefore conceivable that ATP binding and hydrolysis also induce changes in the coiled-coil domains of cohesin, and thereby could affect interactions between Smc1's and Smc3's dimerization domains over relatively long distances.

Further support for the hypothesis that cohesin loading onto DNA is controlled by ATP hydrolysis comes from the recent identification of a cohesin mutant in the Walker B motif of Smc1 and Smc3 that might be locked in a transition state of the loading reaction (Hu et al. 2011). This mutant binds to DNA in an Scc2-dependent manner, in particular at centromeres, but is unable to associate with DNA stably and to mediate cohesion. Based on analogy with ABC transporters, it has been speculated that this mutant might be locked in a state where the NBDs are engaged but the hinge domains are physically separated, which would explain why these mutants can interact with DNA only transiently.

COHESIN IS POSITIONED AT SPECIFIC SITES IN THE GENOME

Chromatin immunoprecipitation (ChIP) experiments performed in several different organisms have revealed that cohesin can predominantly be detected at specific sites in the genome, indicating that cohesin is not distributed randomly along the length of chromosomes (for review, see Peters et al. 2008). For example, in budding yeast and fission yeast, large amounts of cohesin are found at centromeres and pericentric regions, and smaller amounts are found in discrete cohesin attachment regions (CARs) along chromosome arms, which on average are 10 kb apart from each other. Because adherin is essential for the ability of cohesin to associate with DNA, it is generally assumed that cohesin loading occurs at adherin-binding sites. However, ChIP experiments in budding yeast and fission yeast indicate that most adherin complexes are bound to different regions of the genome from cohesin, with the exception of centromeres where both complexes are abundant (Lengronne et al. 2004; Schmidt et al. 2009; Hu et al. 2011; for a conflicting view, see Kogut et al. 2009). It has therefore been proposed that cohesin is first loaded onto DNA at adherin sites and that it is subsequently translocated from these sites to peri-centric regions or CARs (Lengronne et al. 2004). Consistent with this notion, the Walker B mutant of cohesin, which is thought to be locked in a transition state of the loading reaction, accumulates at yeast centromeres but to a much lesser degree than wild-type cohesin at peri-centric regions and CARs, as if cohesin were normally loaded predominantly at centromeres and subsequently moved from there to chromosomal arm regions (Hu et al. 2011).

How cohesin is moved from one site to the other remains unknown, but many CARs in budding and fission yeast are found in regions of convergent transcription, and it has been shown that the induction of regulatable genes leads to removal of cohesin from an intragenic position to the end of the transcription unit (Lengronne et al. 2004). This finding and biochemical experiments that suggest that cohesin rings are able to "slide off" short linearized yeast mini-chromosomes in vitro (Ivanov and Nasmyth 2005) have led to the proposal that cohesin is able to move laterally along DNA while encircling it. According to this idea, RNA polymerases or the native transcripts and RNPs produced by them would simply push cohesin out of actively transcribed regions. If this is true, the pattern of cohesin distribution that is seen along yeast chromosome arms could largely be an indirect consequence of gene activity, rather than a pattern that is specified because cohesin could only fulfill its function in particular chromosomal locations. An exception to this rule could be the centromere, where a high number of cohesin complexes are thought to resist the pulling forces that are generated by the mitotic spindle.

Because cohesin mediates sister chromatid cohesion from yeast to men, the principle mechanisms of how cohesin associates with DNA might have evolved early during the evolution of eukaryotes, or possibly even before, because DNA-binding SMC complexes related to cohesin exist in all kingdoms of life. Although this may well be the case for the initial loading process in which the cohesin rings entrap DNA, it is less clear if the subsequent positioning of cohesin in the genome occurs according to the same principles in different organisms. In contrast to the situation in yeasts, where cohesin and adherin can be found in different chromosomal locations, the two complexes largely colocalize in the *Drosophila* genome, where in many cases they occupy the entire length of actively transcribed genes (Misulovin et al. 2008). Yet a different situation is found in mammalian cells, where NIPBL is frequently located at the promoters of active genes, often together with mediator, a large transcriptional coactivator complex, and cohesin (Kagey et al. 2010). However, although some cohesin can be detected at the majority of NIPBL sites (Kagey et al. 2010), the opposite is not true. There are many more cohesin-binding sites detectable in ChIP experiments (estimates range up to 55,000 sites for the nonrepetitive part of mammalian genomes) (Schmidt et al. 2010) than NIPBL sites, implying that also in mammals, cohesin loading may occur at sites that are distinct from the sites where cohesin finally accumulates. But unlike in yeasts, there is presently no evidence that cohesin is moved to these positions by the transcriptional machinery.

Instead, there are numerous observations that indicate that the pattern of cohesin distribution in mammalian genomes is specified by DNA sequence elements that are recognized by CCCTC binding factor (CTCF), a protein that binds to DNA via one or several of its 11 zinc finger motifs. CTCF is found at the majority, but possibly not at all cohesin-binding sites (for review, see Merkenschlager 2010). Interestingly, partial depletion of CTCF or mutation of CTCF-binding sites reduces cohesin binding at these sites, without detectably reducing the total amount of cohesin on DNA (Parelho et al. 2008; Stedman et al. 2008; Wendt et al. 2008). The implication is that CTCF is dispensable for loading cohesin onto DNA but required for a subsequent positioning step. This situation is particularly evident at the H19-Igf2 locus. In this chromosomal location, the maternal and paternal alleles are differentially methylated, leading to different gene expression patterns on the two alleles within one and the same cell. This "imprinted" gene expression pattern depends on CTCF, which can bind to an imprinting control region (ICR) on the maternal allele but not on the paternal allele, where cytosine methylation prevents CTCF binding to the ICR. Interestingly, cohesin binds only to the ICR on the maternal allele, where CTCF is present, but not on the paternal allele, where CTCF is not bound, further supporting the hypothesis that CTCF specifies where cohesin accumulates once it has been loaded onto DNA (Rubio et al. 2008; Stedman et al. 2008; Wendt et al. 2008; Nativio et al. 2009).

Recent experiments have shown that the carboxy-terminal region of CTCF can directly bind to the SA2 subunit of mammalian cohesin (Xiao et al. 2011), suggesting that cohesin may be recruited to CTCF-binding sites through direct protein–protein interactions. But how cohesin moves from its presumed loading sites where NIPBL is present to CTCF sites remains unknown. Either cohesin can somehow be translocated along DNA, or loading would normally have to occur directly at CTCF sites. In the latter case, one would have to postulate that NIPBL is present, at least transiently, at many more CTCF sites than could be detected in ChIP experiments so far, or NIPBL might be able to promote cohesin loading in *cis*. In this case, an NIPBL site would have to be able to interact with multiple CTCF sites, possibly in a dynamic fashion that depends on chromatin folding.

RECRUITMENT OF COHESIN TO CTCF SITES MAY DETERMINE CHROMATIN STRUCTURE

There is presently no evidence that cohesin has to accumulate at CTCF sites to be able to mediate sister chromatid cohesion, because no cohesion defects have been observed after depletion of CTCF, despite the fact that less cohesin is detected at CTCF sites under these conditions (Wendt et al. 2008). Furthermore, cohesin does not colocalize with CTCF on *Drosophila* chromosomes, and there is no evidence that a CTCF-related protein colocalizes with cohesin in the genome of budding yeast or fission yeast, further indicating that cohesin's role in cohesion is independent of CTCF. However, cohesin depletion does cause defects in gene expression (for review, see Dorsett and Ström 2012), and in mammalian cells, some of these defects resemble the defects that are seen after depletion of CTCF. For example, cohesin depletion changes the levels of H19 and Igf2 transcripts in a similar way as depletion of CTCF, implying that both proteins are required for imprinted gene expression at the H19-Igf2 locus (Wendt et al. 2008; Nativio et al. 2009). Because CTCF is required for recruitment of cohesin to the ICR, whereas the opposite is not true, CTCF appears not to be sufficient for imprinted gene expression. Instead, the main function of CTCF in this process might be to recruit cohesin, and it could therefore be cohesin that affects transcription more directly than CTCF. If this were true, then cohesin positioning in mammalian cells might not be an indirect consequence of gene expression, as has been proposed to be the case in yeast, but may on the contrary control patterns of gene expression.

It is believed that cohesin controls gene expression by mediating long-range chromatin interactions, for example, by forming physical contacts between different CTCF sites that could lead to the formation of chromatin loops. This hypothesis is based on the notion that activity of gene expression depends on promoter–enhancer interactions, and that the proximity between promoters and enhancers might be controlled topologically through the formation of chromatin loops or more complex higher-order chromatin structures. For example, CTCF is thought to function as a transcriptional insulator on the maternal allele of the H19-Igf2 locus by inducing the formation of a chromatin loop, which topologically separates the H19 enhancer from the promoters of the Igf2 gene. Given that cohesin is able to mediate sister chromatid cohesion, that is, it can physically connect two DNA molecules in *trans*, it is conceivable that cohesin also tethers different sequence elements (CTCF sites) on the same chromosome in *cis*, leading to the formation of intrachromatid loops. Support for this hypothesis has come from "chromatin conformation capture" (3C) experiments (for review, see Seitan et al. 2012), but how generally important cohesin and CTCF are for regulating chromatin structure remains to be determined.

SISTER CHROMATID COHESION IS ESTABLISHED DURING DNA REPLICATION AND COINCIDES WITH STABILIZATION OF COHESIN ON DNA

Cohesin can bind to DNA both before and after DNA replication, but it can establish sister chromatid cohesion normally only during S phase (Uhlmann and Nasmyth 1998). This and other observations indicate that cohesion establishment is coupled to DNA replication. It is attractive to think that cohesion establishment must occur at or in the direct vicinity of replication forks, because such a scenario could explain how cohesion is only established between sister chromatids, and not between non-sister DNA molecules. Consistent with this possibility, numerous replication fork components are required for cohesion (Table 2), and Eco1/Ctf7, an acetyltransferase required for establishment of sister chromatid cohesion (Skibbens et al. 1999; Toth et al. 1999), is thought to travel along DNA with replication forks (Lengronne et al. 2004). Even though cohesion may normally only be established at replication forks, recent experiments in budding yeast have revealed that cohesion can also be established de novo after

Table 2. DNA replication proteins required for sister chromatid cohesion

	S. cerevisiae	*S. pombe*	Vertebrates	Function	References
Rfc^{Ctf18}	Ctf18 Ctf8 Dcc1	Ctf18 Ctf8 Dcc1	Ctf18 Ctf8 Dcc1	PCNA clamp loader	Hanna et al. 2001; Mayer et al. 2001; Ansbach et al. 2008; Terret et al. 2009
Rfc^{Elg1}	Elg1	Elg1	Elg1	PCNA clamp loader	Maradeo et al. 2009
Chl1/ChlR1/Ddx11	Chl1	Chl1	ChlR1/Ddx11	DNA helicase	Mayer et al. 2004; Petronczki et al. 2004; Parish et al. 2006; Ansbach et al. 2008
Ctf4/And1	Ctf4	Mcl1	And1	Recruitment of Pol α to chromatin	Hanna et al. 2001; Williams et al. 2002; Mayer et al. 2004; Petronczki et al. 2004; Errico et al. 2009
Tof1/Timeless/Tim	Tof1	Swi1	Tim1	Tim1–Tipin complex stabilizes replication fork	Mercier et al. 2001; Mayer et al. 2004; Errico et al. 2009
Csm3/Tipin	Csm3	Swi3	Tipin		Mayer et al. 2004; Ansbach et al. 2008; Errico et al. 2009
Mrc1/Claspin	Mrc1	Mrc1	Claspin	S-phase checkpoint mediator	Xu et al. 2004
Rad27/Rad2/Fen1	Rad27	Rad2	Fen1	Maturation of Okazaki fragment	Farina et al. 2008

DNA replication if DNA has been damaged by double-strand breaks (see below). In addition, in this case, cohesion establishment depends on Eco1. However, in the absence of Eco1 and either DNA replication or DNA damage, the association of cohesin with DNA is not sufficient to establish sister chromatid cohesion. The implication is that it is not the binding of cohesin to DNA per se but an Eco1-dependent process and possibly other DNA replication or DNA damage-dependent events that make cohesin "cohesive." An interesting hint as to what may determine the "cohesiveness" of cohesin has come from fluorescence recovery after photobleaching (FRAP) experiments in mammalian cells. These experiments revealed that cohesin binds to DNA dynamically in G_1 phase with a residence time of several minutes, whereas many cohesin complexes bind to DNA much more stably after DNA replication, with a residence time of at least many hours (Gerlich et al. 2006). Thus, the ability of cohesin to mediate cohesion coincides with stabilization of cohesin on DNA, implying that stable binding of cohesin to DNA may be an inherent property of cohesive cohesin, and that this property may be required to maintain cohesion from S phase until the subsequent metaphase. Cohesin's ability to bind to DNA for long periods of time may be particularly important in mammalian oocytes. In these cells cohesion is established during pre-meiotic S phase, which occurs before birth, and then has to be maintained until completion of meiosis, which only occurs during ovulation many months or years later (Hunt and Hassold 2010).

COHESIN ACETYLATION IS ESSENTIAL FOR SISTER CHROMATID COHESION

Eco1's essential role in sister chromatid cohesion is the acetylation of cohesin's Smc3 subunit on two lysine residues, K112 and K113 in budding yeast, and K105 and K106 in human Smc3

(Ben-Shahar et al. 2008; Unal et al. 2008; Zhang et al. 2008a; Rowland et al. 2009). Both residues are well conserved in many eukaryotes, as are orthologs of Eco1, of which there are two different ones in vertebrates, called Esco1 and Esco2 (Table 1) (Hou and Zou 2005). Eco1 orthologs are also required for sister chromatid cohesion in fission yeast (Tanaka et al. 2000a), *Drosophila* (Williams et al. 2003), and human cells (Hou and Zou 2005), and in the latter case, both enzymes contribute to Smc3 acetylation (Zhang et al. 2008a; Nishiyama et al. 2010). The acetylation of Smc3 can be reverted by specific enzymes. In human cells, the major enzyme responsible for Smc3 deacetylation is HDAC8 (Deardorff et al. 2012). In budding yeast, the deacetylase Hos1 is required for this process (Beckouet et al. 2010; Borges et al. 2010; Xiong et al. 2010). Hos1 deletion does not cause cohesion defects; however, it becomes crucial for cohesion either when Eco1 is inactivated (Beckouet et al. 2010) or when de novo Smc3 synthesis is suppressed (Borges et al. 2010). These observations raise the interesting possibility that cohesin may have to be acetylated during the process of cohesion establishment, whereas it may be impossible to convert previously acetylated cohesin into a cohesive state.

Although the importance of Smc3 acetylation for sister chromatid cohesion is well established, it is less clear how this modification enables cohesin to become cohesive. The acetylated lysine residues in Smc3 are predicted to be located close to the ATP-binding pocket in the NBD domain. Interestingly, mutations that are thought to mimic the acetylated state of Smc3 abolish a dominant-negative effect that Walker B mutations in Smc3 have on wild-type cohesin (Heidinger-Pauli et al. 2010). Because Smc3 mutations that prevent ATP binding also suppress this dominant effect, it has been proposed that Smc3 acetylation regulates the ATP-binding and hydrolysis cycle of cohesin, either by reducing ATP binding to cohesin or by preventing ATP-dependent interactions between cohesin and its loading machinery. Furthermore, cohesin is normally only acetylated when bound to chromatin. Together with the genetic interactions between acetyl-mimicking and Walker B mutations, this observation raises the possibility that Smc3 acetylation on chromatin prevents rebinding of ATP to cohesin that has entrapped DNA. According to this hypothesis, Smc3 acetylation could stabilize cohesin on DNA by blocking cycles of ATP binding and hydrolysis.

An alternative explanation for how acetylation may contribute to cohesion establishment has come from the observation that the essential function of Eco1 becomes dispensable in fission yeast and budding yeast if the cohesin-associated proteins Wapl and Pds5 are inactivated, or if specific residues in the cohesin core subunits Smc3 and Scc3 are mutated (Tanaka et al. 2001; Ben-Shahar et al. 2008; Unal et al. 2008; Zhang et al. 2008a; Rowland et al. 2009; Sutani et al. 2009). Because Wapl promotes the release of cohesin from DNA in vertebrates (see below), it has been proposed that Smc3 acetylation by Eco1 might stabilize cohesin on DNA by inactivating Wapl (Ben-Shahar et al. 2008; Unal et al. 2008). Alternatively, it has been argued that Wapl, Pds5, Smc3, and Scc3 are required for an activity that prevents cohesion establishment, and that Smc3 acetylation does not prevent the Wapl-mediated release of cohesin from DNA but instead overcomes this "anti-establishment" activity (Rowland et al. 2009). In vertebrates, the existing evidence supports the idea that Smc3 acetylation mediates cohesion by preventing dissociation of cohesin from DNA because in these species, cohesin acetylation leads to the recruitment of sororin, a protein that prevents release of cohesin from DNA (see below).

SORORIN STABILIZES COHESIN–DNA INTERACTIONS AND MEDIATES SISTER CHROMATID COHESION BY ANTAGONIZING WAPL

Sororin is a protein that has only been identified in metazoans so far, where it is essential for sister chromatid cohesion (Rankin et al. 2005) and for stabilizing cohesin on chromatin during DNA replication (Schmitz et al. 2007). Sororin only binds to cohesin once Smc3 has been acetylated. However, this modification is not sufficient for sororin recruitment because DNA replication is also necessary for this event (Lafont et al. 2010; Nishiyama et al. 2010). Remarkably, sororin's essential function in sister chromatid cohesion becomes dispensable when Wapl is depleted from cells or *Xenopus* egg extracts, indicating that sororin might mediate cohesion by inhibiting Wapl (Nishiyama et al. 2010). This hypothesis could explain how sororin stabilizes cohesin on DNA (Schmitz et al. 2007), because the release of cohesin from DNA depends on Wapl (see below).

In vitro, sororin can displace Wapl from its binding partner Pds5, but there is no evidence that sororin can fully dissociate Wapl from cohesin (Nishiyama et al. 2010). It has therefore been proposed that sororin changes the topology of how Wapl and Pds5 interact with cohesin, and that these conformational changes inhibit Wapl's ability to release cohesin from DNA. As a result, cohesin may become stably "locked" on DNA and may thus be able to mediate sister chromatid cohesion for prolonged periods of time (Fig. 3).

It remains unknown if cohesion depends on sororin in all species because related proteins have been identified in some but not all eukaryotes. Sororin exists in vertebrates and *Drosophila*, where it is called Dalmatian, and related proteins containing a conserved carboxy-terminal "sororin domain" have been identified in many other metazoans, including cephalochordates, echinoderms, cnidaria, placozoa, and in plants, but not yet in yeasts and worms (Nishiyama et al. 2010). It therefore remains to be understood if Smc3 acetylation regulates cohesin also in yeast and worms by recruiting a sororin-related protein or by a different mechanism.

DNA DOUBLE-STRAND BREAKS ALSO INDUCE COHESION ESTABLISHMENT

Although cohesion is normally established during DNA replication, DNA double-strand breaks also trigger de novo cohesion establishment even in postreplicative budding yeast cells in an Eco1-dependent manner (Strom et al. 2007; Unal et al. 2007). In this case, it is thought that acetylation of Scc1 by Eco1, but not of Smc3, is required for cohesion establishment (Heidinger-Pauli et al. 2009). This cohesion can be established independently of DNA replication and is induced by double-strand breaks not only on the damaged but also on undamaged chromosomes that are present in the same cells. These observations imply that a diffusible cohesion establishment factor, which is inactivated after DNA replication, is somehow reactivated in response to DNA damage. This factor is presumably Eco1 itself because Eco1 is normally degraded after S phase, but stabilized following DNA damage, and because overexpression of Eco1 is sufficient to induce cohesion establishment after DNA replication (Lyons and Morgan 2011).

COHESIN IS REMOVED FROM MITOTIC CHROMOSOMES IN TWO STEPS

Because cohesin physically connects sister chromatids, sister chromatids can only be separated from each other and transported toward opposite spindle poles in anaphase if cohesin is removed from chromosomes and cohesion is dissolved. This can be achieved by either of two mechanisms. In vertebrates, and presumably in most other metazoans, most cohesin complexes are removed from chromosome arms already in prophase and prometaphase, whereas cohesin at centromeres is protected from this "prophase pathway" and maintains cohesion until anaphase onset (Waizenegger et al. 2000). It is this differential removal of cohesin from chromosome arms and centromeres, respectively, that causes the characteristic "X-shape" (or "V-shape" in species where centromeres are close to one of the telomeres) of chromosomes in prometaphase and metaphase (Gimenez-Abian et al. 2004).

WAPL RELEASES COHESIN FROM DNA

The dissociation of cohesin from chromosomes in prophase and prometaphase depends on Wapl (Gandhi et al. 2006; Kueng et al. 2006) and on phosphorylation of cohesin's SA2 subunit (Hauf et al. 2005) by Polo-like kinase

(Plk1) and possibly other mitotic kinases, such as cyclin-dependent kinase 1 (Cdk1) and Aurora B (Losada et al. 2002; Sumara et al. 2002; Hegemann et al. 2011). This pathway of removing cohesin from mitotic chromosomes might use the same mechanism by which cohesin is released from DNA in interphase because depletion of Wapl also increases the residence time and the steady-state levels of cohesin on DNA in interphase (Kueng et al. 2006). In mitosis, this release mechanism might become activated by mitotic kinases, either by phosphorylation of SA2, which is located next to Wapl in the cohesin complex (Gandhi et al. 2006; Kueng et al. 2006), or by phosphorylation of Wapl and Pds5, which are also phosphorylated in mitosis in a Plk1-dependent manner (Hegemann et al. 2011). Because sororin antagonizes Wapl, sororin also has to be inactivated in mitosis to enable Wapl to release cohesin from DNA. In addition, sororin inactivation is mediated by phosphorylation (Nishiyama et al. 2010), which has been reported to depend on Cdk1 (Dreier et al. 2011) or Plk1 (Zhang et al. 2011). At the end of mitosis, sororin is further degraded following ubiquitylation by the ubiquitin ligase APC/C^{Cdh1}, possibly to prevent precocious formation of cohesin complexes that become too stably bound to DNA (Rankin et al. 2005). Activation of the prophase pathway of cohesin dissociation may therefore depend on mitotic phosphorylation of several proteins: sororin, SA2 (and presumably also SA1), and possibly Wapl and Pds5. In addition, adherin is also removed from chromosomes in prophase (Gillespie and Hirano 2004; Watrin et al. 2006), implying that cohesin cannot be loaded onto DNA any longer in early mitosis. Together, these events may shift the equilibrium between soluble and DNA-bound cohesin complexes so that during prophase and prometaphase most cohesin is removed from chromosome arms.

Although the majority of cohesin complexes are removed from mitotic chromosomes in vertebrate cells by Wapl, some cohesin complexes can only be removed by a different mechanism that depends on the protease separase, which cleaves the Scc1 subunit and thereby opens the cohesin ring (see below). Unlike separase, Wapl releases cohesin from DNA presumably by enabling opening of the cohesin ring without cleaving one of its subunits (Waizenegger et al. 2000; Kueng et al. 2006; Peters et al. 2008). It has been proposed that the Wapl-mediated opening of the cohesin ring occurs via an "exit gate" for DNA that is different from the proposed "entry gate" (Fig. 3) (Nasmyth 2011; Chan et al. 2012). Whereas the latter is thought to be created by separation of the hinge domains of Smc1 and Smc3 (see above), the former has been proposed to require dissociation of Scc1 from the SMC subunits (Nasmyth 2011). There is some precedence that such a conformational change could occur from structural studies of MukB, an SMC-related protein required for chromosome segregation in bacteria (Woo et al. 2009). MukB forms a homodimer whose NBDs interact with two carboxy-terminal Winged helix domains of the MukF protein, similarly to how Scc1's carboxy-terminal Winged helix binds to Smc1 (Haering et al. 2002). Interestingly, ATP-mediated engagement of the MukB's NBDs results in detachment of one of MukF's Winged helix domains from MukB (Woo et al. 2009), implying that conformational changes could also lead to transient dissociation of Scc1 from cohesin's SMC subunits.

POSSIBLE COEVOLUTION OF COHESIN ACETYLATION AND WAPL-MEDIATED COHESIN RELEASE FROM DNA

In budding yeast, it has been reported that a Plk1-dependent mechanism removes cohesin from meiotic chromosomes (Yu and Koshland 2005), but it is unknown if this pathway depends on Wapl, and in mitotic yeast cells, most if not all cohesin is cleaved by separase (Uhlmann et al. 1999). It is therefore unclear if there is a prophase pathway of cohesin dissociation in budding yeast that is related to the one in vertebrates. However, because Wapl antagonizes the function of Eco1 (see above), Wapl may also be required for releasing cohesin from DNA in budding yeast. Direct support for this notion has come from studies in fission yeast, where, as in vertebrates, Wapl is required for dissociation of cohesin from DNA (Feytout et al. 2011).

Nasmyth and Schleiffer (2004) pointed out that the genome of the simple eukaryote *Encephalitozoon cuniculi* lacks a gene that encodes an Eco1-related enzyme. We found 19 other species in whose genome we could not identify an Eco1-related gene (Table 3). Interestingly, in seven of these 20 species without a detectable Eco1 gene, at least one of the two lysine residues that are acetylated by Eco1 in yeast is replaced by another amino acid residue or missing. In contrast, we found only a single species (*Trypanosoma brucei*) with an Eco1 gene in which the KK motif in Smc3 was not conserved. More interestingly, of the seven species in whose genomes we could not identify an Eco1 gene but could find a mutated or missing KK motif in Smc3, six species have genes encoding either Pds5 or Wapl orthologs. The presence of Pds5 and Wapl therefore coincides in many cases either with the presence of Eco1, or with mutations in the lysine residues that are acetylated by Eco1 in other species. These observations raise the interesting possibility that the ability of Pds5 and/or Wapl to release cohesin from DNA may have coevolved with Smc3 acetylation by Eco1.

SEPARASE CLEAVES COHESIN IN METAPHASE AND ENABLES SISTER CHROMATID SEPARATION IN ANAPHASE

The Wapl-dependent prophase pathway of cohesin dissociation removes most cohesin from

Table 3. Coevolution of Smc3 acetylation by Eco1 and cohesin release by Wapl

	Organism	Smc3	Pds5	Wapl	Eco1
Species with an Eco1 ortholog	*Homo sapiens*	KK	+	+	+
	Ciona intestinalis	KK	+	+	+
	Xenopus laevis	KK	+	+	+
	Arabidopsis thaliana	KK	+	+	+
	Saccharomyces cerevisiae	KK	+	+	+
	Schizosaccharomyces pombe	KK	+	+	+
	Drosophila melanogaster	KK	+	+	+
	Trypanosoma brucei TREU927	KQ	+		+
Species without an identifiable Eco1 ortholog	*Acromyrmex echinatior*	KK	+	+	
	Dictyostelium fasciculatum	NK	+	+	
	Gibberella zeae PH-1	KK	+	+	
	Glomerella graminicola M1.001	KK	+	+	
	Grosmannia clavigera kw1407	RK	+	+	
	Moniliophthora perniciosa FA553	KK	+	+	
	Monosiga brevicollis MX1	HK	+	+	
	Trichinella spiralis	KK	+	+	
	Tuber melanosporum	KK	+	+	
	Ustilago maydis 521	KK	+	+	
	Chlamydomonas reinhardtii	KK	+		
	Hydra magnipapillata	KK	+		
	Malassezia globosa CBS 7966	TR	+		
	Metarhizium acridum CQMa 102	KK	+		
	Trichomonas vaginalis	RK	+		
	Volvox carteri f. nagariensis	—	+		
	Cryptosporidium parvum	QK			
	Aureococcus anophagefferens	KK			
	Encephalitozoon cuniculi GB-M1	KK			
	Schistosoma japonicum	KK			

(KK) Lysine residues that correspond to the ones that have been shown to be acetylated by Eco1 in *S. cerevisiae* Smc3. In the sequence of *Volvox carteri f. nagariensis* Smc3, the region containing these lysine residues is missing.

Gene or protein sequences are available upon request.

chromosomes arms but leaves cohesin behind at centromeres, where cohesion is maintained until the onset of anaphase (Waizenegger et al. 2000). These centromeric cohesin complexes are protected from the prophase pathway by a protein called shugoshin, or Sgo1, which recruits the protein phosphatase 2A complex (PP2A) to centromeres (for review, see Gutiérrez-Caballero et al. 2012). PP2A has been proposed to protect cohesin from the prophase pathway by reverting the phosphorylation of SA2 at centromeres (Kitajima et al. 2006), but it is conceivable that PP2A also has other critical substrates whose dephosphorylation is required to maintain centromeric cohesion, such as sororin, Wapl, or Pds5. A second protein that has been shown to protect cohesin at centromeres is a protein kinase called Haspin. Haspin's only known substrate is the core histone H3, which can be phosphorylated on threonine 5, but how Haspin contributes to protection of cohesin at centromeres remains to be understood (for review, see Higgins 2010).

Those cohesin complexes that are protected from the prophase pathway by shugoshin-PP2A can only be removed from chromosomes by separase, a protease that cleaves the Scc1 subunit of cohesin at two distinct sites and thereby opens the cohesin ring (Uhlmann et al. 1999, 2000; Waizenegger et al. 2000). This process is essential for sister chromatid separation because either expression of non-cleavable Scc1 or inactivation of separase prevents proper sister chromatid separation (for review, see Peters et al. 2008).

In budding yeast, Scc1 cleavage by separase is also a prerequisite for Smc3 deacetylation by Hos1 (Beckouet et al. 2010; Borges et al. 2010). If cohesin contains non-cleavable Scc1, Smc3 is not deacetylated at the time of anaphase onset even though cyclin B is degraded. Vice versa, Scc1 cleavage triggers Smc3 deacetylation in metaphase arrested cells in the absence of cyclin B degradation. Although Scc1 cleavage is a prerequisite for cohesin dissociation from chromatin, cleavage per se is not necessary for deacetylation in vitro because release of cohesin from DNA by nuclease digestion is sufficient for Smc3 deacetylation even if Scc1 remains intact (Borges et al. 2010). Separase may therefore enable cohesin deacetylation by simply releasing cohesin from chromatin, with which Hos1 is not associated (Borges et al. 2010).

CONCLUDING REMARKS

It is one of the most fundamental aspects of life that new cells are generated by division of older cells and that the newly formed daughter cells share most if not all properties of their mothers because the daughters inherit an identical copy of the mother cell's genome. The two key processes that constitute this chromosome cycle are the precise duplication of the genome by DNA replication and the subsequent segregation of the two copies of the genome by chromosome segregation. Although the basic logic of this cycle was understood already during the early days of cell biology in the 19th century, it has only become obvious in the more recent past that understanding the mechanisms of sister chromatid cohesion will also be essential for understanding the eukaryotic cell division cycle. It is now well established that sister chromatid cohesion is mediated by cohesin, but exactly how cohesin performs this and possibly other functions and how these are controlled during the cell cycle and during cell differentiation remain incompletely understood. Numerous important questions will have to be answered to build a comprehensive molecular model of how cohesin mediates these functions: How and where does adherin load cohesin onto DNA? Is cohesin translocated from loading sites to its final binding sites? If so, how, and by which mechanism does CTCF recruit cohesin to these sites in vertebrates? How does Wapl release cohesin from DNA, and how is this process prevented by sororin or possibly by other mechanisms that depend on cohesin acetylation? Is cohesin converted into a cohesive form at replication forks? If so, how are cohesin acetyltransferases and sororin recruited to these forks, and exactly how do the two sister DNA strands become connected by cohesin at or in the vicinity of replication forks? What happens to cohesin when DNA or RNA polymerases or other enzymes move along DNA? Can these enzymes move through cohesin rings, or bypass them, or push them along DNA, or do they have to evict cohesin transiently from DNA? Is cohesin really able to mediate cohesion for months or years, as has been speculated to be the case in mammalian oocytes? How does cohesin's ability to mediate cohesion relate to its role in gene regulation? Does cohesin mediate the latter function through the formation of intrachromatid loops? If yes, how is the ability of cohesin to connect DNA strands targeted to sister chromatids when cohesion is established, but to different DNA sequences in *cis* during the process of gene regulation? Answering these and further questions will be an exciting task for the future, and an essential prerequisite for understanding how chromosomes are organized and passed from one cell generation to the next.

ACKNOWLEDGMENTS

We thank T. Hirano and K. Shintomi for inspiring the ideas regarding the coevolution of Eco1 and Wapl, and Hannes Tkadletz for help with preparing the figures. Research in the laboratory of J.-M.P. is supported by Boehringer Ingelheim, the Vienna Science and Technology Fund (WWTF LS09-13), the Austrian Science Fund (FWF Special Research Program SFB F34, "Chromosome Dynamics," and Wittgenstein award Z196-B20), and the European Community's Seventh Framework Programme (FP7/2007-2013) under grant agreement number 241548 (MitoSys).

REFERENCES

Anderson DE, Losada A, Erickson HP, Hirano T. 2002. Condensin and cohesin display different arm conformations with characteristic hinge angles. *J Cell Biol* **156:** 419–424.

Ansbach AB, Noguchi C, Klansek IW, Heidlebaugh M, Nakamura TM, Noguchi E. 2008. RFCCtf18 and the Swi1–Swi3 complex function in separate and redundant pathways required for the stabilization of replication forks to facilitate sister chromatid cohesion in *Schizosaccharomyces pombe*. *Mol Biol Cell* **19:** 595–607.

Arumugam P, Gruber S, Tanaka K, Haering CH, Mechtler K, Nasmyth K. 2003. ATP hydrolysis is required for cohesin's association with chromosomes. *Curr Biol* **13:** 1941–1953.

Beckouet F, Hu B, Roig MB, Sutani T, Komata M, Uluocak P, Katis VL, Shirahige K, Nasmyth K. 2010. An Smc3 acetylation cycle is essential for establishment of sister chromatid cohesion. *Mol Cell* **39:** 689–699.

Ben-Shahar T, Heeger S, Lehane C, East P, Flynn H, Skehel M, Uhlmann F. 2008. Eco1-dependent cohesin acetylation during establishment of sister chromatid cohesion. *Science* **321:** 563–566.

Borges V, Lehane C, Lopez-Serra L, Flynn H, Skehel M, Rolef Ben-Shahar T, Uhlmann F. 2010. Hos1 deacetylates Smc3 to close the cohesin acetylation cycle. *Mol Cell* **39:** 677–688.

Chan KL, Roig MB, Hu B, Beckouet F, Metson J, Nasmyth K. 2012. Cohesin's DNA exit gate is distinct from its entrance gate and is regulated by acetylation. *Cell* **150:** 961–974.

Deardorff MA, Bando M, Nakato R, Watrin E, Itoh T, Minamino M, Saitoh K, Komata M, Katou Y, Clark D, et al. 2012. HDAC8 mutations in Cornelia de Lange syndrome affect the cohesin acetylation cycle. *Nature* doi: 10.1038/nature11316.

Dorsett D, Strom L. 2012. The ancient and evolving roles of cohesin in gene expression and DNA repair. *Curr Biol* **22:** R240–R250.

Dreier MR, Bekier ME II, Taylor WR. 2011. Regulation of sororin by Cdk1-mediated phosphorylation. *J Cell Sci* **124:** 2976–2987.

Errico A, Cosentino C, Rivera T, Losada A, Schwob E, Hunt T, Costanzo V. 2009. Tipin/Tim1/And1 protein complex promotes Pol α chromatin binding and sister chromatid cohesion. *EMBO J* **28:** 3681–3692.

Farina A, Shin JH, Kim DH, Bermudez VP, Kelman Z, Seo YS, Hurwitz J. 2008. Studies with the human cohesin establishment factor, ChlR1. Association of ChlR1 with Ctf18-RFC and Fen1. *J Biol Chem* **283:** 20925–20936.

Feytout A, Vaur S, Genier S, Vazquez S, Javerzat JP. 2011. Psm3 acetylation on conserved lysine residues is dispensable for viability in fission yeast but contributes to Eso1-mediated sister chromatid cohesion by antagonizing Wpl1. *Mol Cell Biol* **31:** 1771–1786.

Furuya K, Takahashi K, Yanagida M. 1998. Faithful anaphase is ensured by Mis4, a sister chromatid cohesion molecule required in S phase and not destroyed in G_1 phase. *Genes Dev* **12:** 3408–3418.

Gandhi R, Gillespie PJ, Hirano T. 2006. Human Wapl is a cohesin-binding protein that promotes sister-chromatid resolution in mitotic prophase. *Curr Biol* **16:** 2406–2417.

Gerlich D, Koch B, Dupeux F, Peters JM, Ellenberg J. 2006. Live-cell imaging reveals a stable cohesin–chromatin interaction after but not before DNA replication. *Curr Biol* **16:** 1571–1578.

Gillespie PJ, Hirano T. 2004. Scc2 couples replication licensing to sister chromatid cohesion in *Xenopus* egg extracts. *Curr Biol* **14:** 1598–1603.

Gimenez-Abian JF, Sumara I, Hirota T, Hauf S, Gerlich D, de la Torre C, Ellenberg J, Peters JM. 2004. Regulation of sister chromatid cohesion between chromosome arms. *Curr Biol* **14:** 1187–1193.

Gruber S, Haering CH, Nasmyth K. 2003. Chromosomal cohesin forms a ring. *Cell* **112:** 765–777.

Gruber S, Arumugam P, Katou Y, Kuglitsch D, Helmhart W, Shirahige K, Nasmyth K. 2006. Evidence that loading of cohesin onto chromosomes involves opening of its SMC hinge. *Cell* **127:** 523–537.

Gutierrez-Caballero C, Cebollero LR, Pendas AM. 2012. Shugoshins: From protectors of cohesion to versatile adaptors at the centromere. *Trends Genet* **28:** 351–360.

Haering CH, Lowe J, Hochwagen A, Nasmyth K. 2002. Molecular architecture of SMC proteins and the yeast cohesin complex. *Mol Cell* **9:** 773–788.

Haering CH, Schoffnegger D, Nishino T, Helmhart W, Nasmyth K, Lowe J. 2004. Structure and stability of cohesin's Smc1–kleisin interaction. *Mol Cell* **15:** 951–964.

Haering CH, Farcas AM, Arumugam P, Metson J, Nasmyth K. 2008. The cohesin ring concatenates sister DNA molecules. *Nature* **454:** 297–301.

Hanna JS, Kroll ES, Lundblad V, Spencer FA. 2001. *Saccharomyces cerevisiae* CTF18 and CTF4 are required for sister chromatid cohesion. *Mol Cell Biol* **21:** 3144–3158.

Hauf S, Roitinger E, Koch B, Dittrich CM, Mechtler K, Peters JM. 2005. Dissociation of cohesin from chromosome arms and loss of arm cohesion during early mitosis depends on phosphorylation of SA2. *PLoS Biol* **3:** e69.

Hegemann B, Hutchins JR, Hudecz O, Novatchkova M, Rameseder J, Sykora MM, Liu S, Mazanek M, Lenart P, Heriche JK, et al. 2011. Systematic phosphorylation analysis of human mitotic protein complexes. *Sci Signal* **4:** rs12.

Heidinger-Pauli JM, Unal E, Koshland D. 2009. Distinct targets of the Eco1 acetyltransferase modulate cohesion in S phase and in response to DNA damage. *Mol Cell* **34:** 311–321.

Heidinger-Pauli JM, Onn I, Koshland D. 2010. Genetic evidence that the acetylation of the Smc3p subunit of cohesin modulates its ATP-bound state to promote cohesion establishment in *Saccharomyces cerevisiae. Genetics* **185:** 1249–1256.

Higgins JM. 2010. Haspin: A newly discovered regulator of mitotic chromosome behavior. *Chromosoma* **119:** 137–147.

Hirano M, Hirano T. 2006. Opening closed arms: Long-distance activation of SMC ATPase by hinge–DNA interactions. *Mol Cell* **21:** 175–186.

Hopfner KP, Tainer JA. 2003. Rad50/SMC proteins and ABC transporters: Unifying concepts from high-resolution structures. *Curr Opin Struct Biol* **13:** 249–255.

Hopfner KP, Karcher A, Shin DS, Craig L, Arthur LM, Carney JP, Tainer JA. 2000. Structural biology of Rad50 ATPase: ATP-driven conformational control in DNA double-strand break repair and the ABC-ATPase superfamily. *Cell* **101:** 789–800.

Hou F, Zou H. 2005. Two human orthologues of Eco1/Ctf7 acetyltransferases are both required for proper sister-chromatid cohesion. *Mol Biol Cell* **16:** 3908–3918.

Hu B, Itoh T, Mishra A, Katoh Y, Chan KL, Upcher W, Godlee C, Roig MB, Shirahige K, Nasmyth K. 2011. ATP hydrolysis is required for relocating cohesin from sites occupied by its Scc2/4 loading complex. *Curr Biol* **21:** 12–24.

Huang CE, Milutinovich M, Koshland D. 2005. Rings, bracelet or snaps: Fashionable alternatives for Smc complexes. *Philos Trans R Soc Lond B Biol Sci* **360:** 537–542.

Hunt P, Hassold T. 2010. Female meiosis: Coming unglued with age. *Curr Biol* **20:** R699–R702.

Ivanov D, Nasmyth K. 2005. A topological interaction between cohesin rings and a circular minichromosome. *Cell* **122:** 849–860.

Kagey MH, Newman JJ, Bilodeau S, Zhan Y, Orlando DA, van Berkum NL, Ebmeier CC, Goossens J, Rahl PB, Levine SS, et al. 2010. Mediator and cohesin connect gene expression and chromatin architecture. *Nature* **467:** 430–435.

Kitajima TS, Sakuno T, Ishiguro K, Iemura S, Natsume T, Kawashima SA, Watanabe Y. 2006. Shugoshin collaborates with protein phosphatase 2A to protect cohesin. *Nature* **441:** 46–52.

Kogut I, Wang J, Guacci V, Mistry RK, Megee PC. 2009. The Scc2/Scc4 cohesin loader determines the distribution of cohesin on budding yeast chromosomes. *Genes Dev* **23:** 2345–2357.

Kueng S, Hegemann B, Peters BH, Lipp JJ, Schleiffer A, Mechtler K, Peters JM. 2006. Wapl controls the dynamic association of cohesin with chromatin. *Cell* **127:** 955–967.

Kurze A, Michie KA, Dixon SE, Mishra A, Itoh T, Khalid S, Strmecki L, Shirahige K, Haering CH, Lowe J, et al. 2011. A positively charged channel within the Smc1/Smc3 hinge required for sister chromatid cohesion. *EMBO J* **30:** 364–378.

Lafont AL, Song J, Rankin S. 2010. Sororin cooperates with the acetyltransferase Eco2 to ensure DNA replication-dependent sister chromatid cohesion. *Proc Natl Acad Sci* **107:** 20364–20369.

Lammens A, Hopfner KP. 2010. Structural basis for adenylate kinase activity in ABC ATPases. *J Mol Biol* **401:** 265–273.

Lengronne A, Katou Y, Mori S, Yokobayashi S, Kelly GP, Itoh T, Watanabe Y, Shirahige K, Uhlmann F. 2004. Cohesin relocation from sites of chromosomal loading to places of convergent transcription. *Nature* **430:** 573–578.

Lim HS, Kim JS, Park YB, Gwon GH, Cho Y. 2011. Crystal structure of the Mre11–Rad50–ATPγS complex: Understanding the interplay between Mre11 and Rad50. *Genes Dev* **25:** 1091–1104.

Losada A, Yokochi T, Kobayashi R, Hirano T. 2000. Identification and characterization of SA/Scc3p subunits in the *Xenopus* and human cohesin complexes. *J Cell Biol* **150:** 405–416.

Losada A, Hirano M, Hirano T. 2002. Cohesin release is required for sister chromatid resolution, but not for condensin-mediated compaction, at the onset of mitosis. *Genes Dev* **16:** 3004–3016.

Lyons NA, Morgan DO. 2011. Cdk1-dependent destruction of Eco1 prevents cohesion establishment after S phase. *Mol Cell* **42:** 378–389.

Mannini L, Menga S, Musio A. 2010. The expanding universe of cohesin functions: A new genome stability caretaker involved in human disease and cancer. *Hum Mutat* **31:** 623–630.

Maradeo ME, Skibbens RV. 2009. The Elg1–RFC clamp-loading complex performs a role in sister chromatid cohesion. *PLoS ONE* **4:** e4707.

Mayer ML, Gygi SP, Aebersold R, Hieter P. 2001. Identification of RFC(Ctf18p, Ctf8p, Dcc1p): An alternative RFC complex required for sister chromatid cohesion in *S. cerevisiae*. *Mol Cell* **7:** 959–970.

Mayer ML, Pot I, Chang M, Xu H, Aneliunas V, Kwok T, Newitt R, Aebersold R, Boone C, Brown GW, et al. 2004. Identification of protein complexes required for efficient sister chromatid cohesion. *Mol Biol Cell* **15:** 1736–1745.

Mercier R, Vezon D, Bullier E, Motamayor JC, Sellier A, Lefevre F, Pelletier G, Horlow C. 2001. SWITCH1 (SWI1): A novel protein required for the establishment of sister chromatid cohesion and for bivalent formation at meiosis. *Genes Dev* **15:** 1859–1871.

Merkenschlager M. 2010. Cohesin: A global player in chromosome biology with local ties to gene regulation. *Curr Opin Genet Dev* **20:** 555–561.

Michaelis C, Ciosk R, Nasmyth K. 1997. Cohesins: Chromosomal proteins that prevent premature separation of sister chromatids. *Cell* **91:** 35–45.

Mishra A, Hu B, Kurze A, Beckouet F, Farcas AM, Dixon SE, Katou Y, Khalid S, Shirahige K, Nasmyth K. 2010. Both interaction surfaces within cohesin's hinge domain are essential for its stable chromosomal association. *Curr Biol* **20:** 279–289.

Misulovin Z, Schwartz YB, Li XY, Kahn TG, Gause M, MacArthur S, Fay JC, Eisen MB, Pirrotta V, Biggin MD, et al. 2008. Association of cohesin and Nipped-B with transcriptionally active regions of the *Drosophila melanogaster* genome. *Chromosoma* **117:** 89–102.

Moreno-Herrero F, de Jager M, Dekker NH, Kanaar R, Wyman C, Dekker C. 2005. Mesoscale conformational changes in the DNA-repair complex Rad50/Mre11/Nbs1 upon binding DNA. *Nature* **437:** 440–443.

Nasmyth K. 2011. Cohesin: A catenase with separate entry and exit gates? *Nat Cell Biol* **13:** 1170–1177.

Nasmyth K, Schleiffer A. 2004. From a single double helix to paired double helices and back. *Philos Trans R Soc Lond B Biol Sci* **359:** 99–108.

Nativio R, Wendt KS, Ito Y, Huddleston JE, Uribe-Lewis S, Woodfine K, Krueger C, Reik W, Peters JM, Murrell A. 2009. Cohesin is required for higher-order chromatin conformation at the imprinted IGF2-H19 locus. *PLoS Genet* **5:** e1000739.

Nishiyama T, Ladurner R, Schmitz J, Kreidl E, Schleiffer A, Bhaskara V, Bando M, Shirahige K, Hyman AA, Mechtler K, et al. 2010. Sororin mediates sister chromatid cohesion by antagonizing Wapl. *Cell* **143:** 737–749.

Parelho V, Hadjur S, Spivakov M, Leleu M, Sauer S, Gregson HC, Jarmuz A, Canzonetta C, Webster Z, Nesterova T, et al. 2008. Cohesins functionally associate with CTCF on mammalian chromosome arms. *Cell* **132:** 422–433.

Parish JL, Rosa J, Wang X, Lahti JM, Doxsey SJ, Androphy EJ. 2006. The DNA helicase ChlR1 is required for sister chromatid cohesion in mammalian cells. *J Cell Sci* **119:** 4857–4865.

Peters JM, Tedeschi A, Schmitz J. 2008. The cohesin complex and its roles in chromosome biology. *Genes Dev* **22:** 3089–3114.

Petronczki M, Chwalla B, Siomos MF, Yokobayashi S, Helmhart W, Deutschbauer AM, Davis RW, Watanabe Y, Nasmyth K. 2004. Sister-chromatid cohesion mediated by the alternative RF-CCtf18/Dcc1/Ctf8, the helicase Chl1 and the polymerase-α-associated protein Ctf4 is essential for chromatid disjunction during meiosis II. *J Cell Sci* **117:** 3547–3559.

Rankin S, Ayad NG, Kirschner MW. 2005. Sororin, a substrate of the anaphase-promoting complex, is required for sister chromatid cohesion in vertebrates. *Mol Cell* **18:** 185–200.

Rees DC, Johnson E, Lewinson O. 2009. ABC transporters: The power to change. *Nat Rev Mol Cell Biol* **10:** 218–227.

Rowland BD, Roig MB, Nishino T, Kurze A, Uluocak P, Mishra A, Beckouet F, Underwood P, Metson J, Imre R, et al. 2009. Building sister chromatid cohesion: smc3 acetylation counteracts an antiestablishment activity. *Mol Cell* **33:** 763–774.

Rubio ED, Reiss DJ, Welcsh PL, Disteche CM, Filippova GN, Baliga NS, Aebersold R, Ranish JA, Krumm A. 2008. CTCF physically links cohesin to chromatin. *Proc Natl Acad Sci* **105:** 8309–8314.

Schmidt CK, Brookes N, Uhlmann F. 2009. Conserved features of cohesin binding along fission yeast chromosomes. *Genome Biol* **10:** R52.

Schmidt D, Schwalie PC, Ross-Innes CS, Hurtado A, Brown GD, Carroll JS, Flicek P, Odom DT. 2010. A CTCF-independent role for cohesin in tissue-specific transcription. *Genome Res* **20:** 578–588.

Schmitz J, Watrin E, Lenart P, Mechtler K, Peters JM. 2007. Sororin is required for stable binding of cohesin to chromatin and for sister chromatid cohesion in interphase. *Curr Biol* **17:** 630–636.

Seitan VC, Krangel MS, Merkenschlager M. 2012. Cohesin, CTCF and lymphocyte antigen receptor locus rearrangement. *Trends Immunol* **33:** 153–159.

Skibbens RV, Corson LB, Koshland D, Hieter P. 1999. Ctf7p is essential for sister chromatid cohesion and links mitotic chromosome structure to the DNA replication machinery. *Genes Dev* **13:** 307–319.

Stedman W, Kang H, Lin S, Kissil JL, Bartolomei MS, Lieberman PM. 2008. Cohesins localize with CTCF at the KSHV latency control region and at cellular c-myc and H19/Igf2 insulators. *EMBO J* **27:** 654–666.

Strom L, Karlsson C, Lindroos HB, Wedahl S, Katou Y, Shirahige K, Sjogren C. 2007. Postreplicative formation of cohesion is required for repair and induced by a single DNA break. *Science* **317:** 242–245.

Sumara I, Vorlaufer E, Gieffers C, Peters BH, Peters JM. 2000. Characterization of vertebrate cohesin complexes and their regulation in prophase. *J Cell Biol* **151:** 749–762.

Sumara I, Vorlaufer E, Stukenberg PT, Kelm O, Redemann N, Nigg EA, Peters JM. 2002. The dissociation of cohesin from chromosomes in prophase is regulated by Polo-like kinase. *Mol Cell* **9:** 515–525.

Sutani T, Kawaguchi T, Kanno R, Itoh T, Shirahige K. 2009. Budding yeast Wpl1(Rad61)–Pds5 complex counteracts sister chromatid cohesion-establishing reaction. *Curr Biol* **19:** 492–497.

Takahashi TS, Yiu P, Chou MF, Gygi S, Walter JC. 2004. Recruitment of Xenopus Scc2 and cohesin to chromatin requires the pre-replication complex. *Nat Cell Biol* **6:** 991–996.

Tanaka K, Yonekawa T, Kawasaki Y, Kai M, Furuya K, Iwasaki M, Murakami H, Yanagida M, Okayama H. 2000a. Fission yeast Eso1p is required for establishing sister chromatid cohesion during S phase. *Mol Cell Biol* **20:** 3459–3469.

Tanaka T, Fuchs J, Loidl J, Nasmyth K. 2000b. Cohesin ensures bipolar attachment of microtubules to sister centromeres and resists their precocious separation. *Nat Cell Biol* **2:** 492–499.

Tanaka K, Hao Z, Kai M, Okayama H. 2001. Establishment and maintenance of sister chromatid cohesion in fission yeast by a unique mechanism. *EMBO J* **20:** 5779–5790.

Terret ME, Sherwood R, Rahman S, Qin J, Jallepalli PV. 2009. Cohesin acetylation speeds the replication fork. *Nature* **462:** 231–234.

Toth A, Ciosk R, Uhlmann F, Galova M, Schleiffer A, Nasmyth K. 1999. Yeast cohesin complex requires a conserved protein, Eco1p(Ctf7), to establish cohesion between sister chromatids during DNA replication. *Genes Dev* **13:** 320–333.

Uhlmann F, Nasmyth K. 1998. Cohesion between sister chromatids must be established during DNA replication. *Curr Biol* **8:** 1095–1101.

Uhlmann F, Lottspeich F, Nasmyth K. 1999. Sister-chromatid separation at anaphase onset is promoted by cleavage of the cohesin subunit Scc1. *Nature* **400:** 37–42.

Uhlmann F, Wernic D, Poupart MA, Koonin EV, Nasmyth K. 2000. Cleavage of cohesin by the CD clan protease separin triggers anaphase in yeast. *Cell* **103:** 375–386.

Unal E, Heidinger-Pauli JM, Koshland D. 2007. DNA double-strand breaks trigger genome-wide sister-chromatid cohesion through Eco1 (Ctf7). *Science* **317:** 245–248.

Unal E, Heidinger-Pauli JM, Kim W, Guacci V, Onn I, Gygi SP, Koshland DE. 2008. A molecular determinant for the establishment of sister chromatid cohesion. *Science* **321:** 566–569.

Waizenegger IC, Hauf S, Meinke A, Peters JM. 2000. Two distinct pathways remove mammalian cohesin from chromosome arms in prophase and from centromeres in anaphase. *Cell* **103:** 399–410.

Watrin E, Schleiffer A, Tanaka K, Eisenhaber F, Nasmyth K, Peters JM. 2006. Human Scc4 is required for cohesin binding to chromatin, sister chromatid cohesion, and mitotic progression. *Curr Biol* **16:** 863–874.

Weitzer S, Lehane C, Uhlmann F. 2003. A model for ATP hydrolysis-dependent binding of cohesin to DNA. *Curr Biol* **13:** 1930–1940.

Wendt KS, Yoshida K, Itoh T, Bando M, Koch B, Schirghuber E, Tsutsumi S, Nagae G, Ishihara K, Mishiro T, et al. 2008. Cohesin mediates transcriptional insulation by CCCTC-binding factor. *Nature* **451:** 796–801.

Williams DR, McIntosh JR. 2002. mcl1$^+$, the *Schizosaccharomyces pombe* homologue of CTF4, is important for chromosome replication, cohesion, and segregation. *Eukaryot Cell* **1:** 758–773.

Williams BC, Garrett-Engele CM, Li Z, Williams EV, Rosenman ED, Goldberg ML. 2003. Two putative acetyltransferases, san and deco, are required for establishing sister chromatid cohesion in *Drosophila*. *Curr Biol* **13:** 2025–2036.

Williams GJ, Williams RS, Williams JS, Moncalian G, Arvai AS, Limbo O, Guenther G, SilDas S, Hammel M, Russell P, et al. 2011. ABC ATPase signature helices in Rad50 link nucleotide state to Mre11 interface for DNA repair. *Nat Struct Mol Biol* **18:** 423–431.

Woo JS, Lim JH, Shin HC, Suh MK, Ku B, Lee KH, Joo K, Robinson H, Lee J, Park SY, et al. 2009. Structural studies of a bacterial condensin complex reveal ATP-dependent disruption of intersubunit interactions. *Cell* **136:** 85–96.

Xiao T, Wallace J, Felsenfeld G. 2011. Specific sites in the C terminus of CTCF interact with the SA2 subunit of the cohesin complex, and are required for cohesin dependent insulation activity. *Mol Cell Biol* **31:** 2174–2183.

Xiong B, Lu S, Gerton JL. 2010. Hos1 is a lysine deacetylase for the Smc3 subunit of cohesin. *Curr Biol* **20:** 1660–1665.

Xu H, Boone C, Klein HL. 2004. Mrc1 is required for sister chromatid cohesion to aid in recombination repair of spontaneous damage. *Mol Cell Biol* **24:** 7082–7090.

Yu HG, Koshland D. 2005. Chromosome morphogenesis: Condensin-dependent cohesin removal during meiosis. *Cell* **123:** 397–407.

Zhang J, Shi X, Li Y, Kim BJ, Jia J, Huang Z, Yang T, Fu X, Jung SY, Wang Y, et al. 2008a. Acetylation of Smc3 by Eco1 is required for S phase sister chromatid cohesion in both human and yeast. *Mol Cell* **31:** 143–151.

Zhang N, Kuznetsov SG, Sharan SK, Li K, Rao PH, Pati D. 2008b. A handcuff model for the cohesin complex. *J Cell Biol* **183:** 1019–1031.

Zhang N, Panigrahi AK, Mao Q, Pati D. 2011. Interaction of Sororin protein with polo-like kinase 1 mediates resolution of chromosomal arm cohesion. *J Biol Chem* **286:** 41826–41837.

Replicative DNA Polymerases

Erik Johansson[1] and Nicholas Dixon[2]

[1]Department of Medical Biochemistry and Biophysics, Umeå University, SE-90187 Umeå, Sweden

[2]School of Chemistry, University of Wollongong, NSW 2522, Australia

Correspondence: erik.johnsson@medchem.umu.se; nickd@uow.edu.au

In 1959, Arthur Kornberg was awarded the Nobel Prize for his work on the principles by which DNA is duplicated by DNA polymerases. Since then, it has been confirmed in all branches of life that replicative DNA polymerases require a single-stranded template to build a complementary strand, but they cannot start a new DNA strand de novo. Thus, they also depend on a primase, which generally assembles a short RNA primer to provide a 3′-OH that can be extended by the replicative DNA polymerase. The general principles that (1) a helicase unwinds the double-stranded DNA, (2) single-stranded DNA-binding proteins stabilize the single-stranded DNA, (3) a primase builds a short RNA primer, and (4) a clamp loader loads a clamp to (5) facilitate the loading and processivity of the replicative polymerase, are well conserved among all species. Replication of the genome is remarkably robust and is performed with high fidelity even in extreme environments. Work over the last decade or so has confirmed (6) that a common two-metal ion-promoted mechanism exists for the nucleotidyltransferase reaction that builds DNA strands, and (7) that the replicative DNA polymerases always act as a key component of larger multiprotein assemblies, termed replisomes. Furthermore (8), the integrity of replisomes is maintained by multiple protein–protein and protein–DNA interactions, many of which are inherently weak. This enables large conformational changes to occur without dissociation of replisome components, and also means that in general replisomes cannot be isolated intact.

The genomes, from the smallest to the largest, provide an enormous challenge for the replicative DNA polymerases to faithfully copy to give the many generations that follow a comparable condition for life. In this article, we discuss the structural and functional bases by which replicative DNA polymerases are able to efficiently and faithfully build new copies of genomes in eubacteria, archaea, and eukaryotes.

BUILDING DNA

Considerable progress in the description of the chemical and structural basis for DNA synthesis has been made in recent years. A large variety of DNA polymerases have been isolated and studied since the initial discovery of *Escherichia coli* DNA Pol I. The DNA polymerases have been divided into families based on homology of their primary sequences. The replicative DNA

Cite this article as *Cold Spring Harb Perspect Biol* doi: 10.1101/cshperspect.a012799

polymerases from eukarya are found in family B, bacteria in families A and C, and archaea in families B and D. The structures of the catalytic subunits of these polymerases share a similar organization, and the nucleotidyl transferase reaction of adding nucleotides to the 3′-OH of the growing strand is conserved (Table 1). All known structures of DNA polymerases (examples in Fig. 1) appear to resemble a right hand, where the functional domains are depicted as fingers, palm, and thumb domains, and the replicative polymerases often contain additional domains for 3′–5′ exonucleolytic proofreading, interactions with other proteins, or other functions.

Both the growing DNA and template strands are integral components of their active sites, providing information about which nucleotide should be added, and the 3′-hydroxyl group acts as a nucleophile to attack the α phosphate of the correct incoming deoxyribonucleoside triphosphate (dNTP). This all occurs in the palm domain, which is the most highly conserved subdomain. Here, two magnesium

Table 1. A current view of subunit compositions and roles of subunits of replicative DNA polymerases

	Human	Yeast (*Sc*)	Yeast (*Sp*)	(Eury)archaea	Eubacteria (*Eco*)	Firmicutes (*Bsu*)
Chromosomal replicases	Pol δ (δ) and Pol ε (ε)	Pol δ (δ) and Pol ε (ε)	Pol δ (δ) and Pol ε (ε)	Pol B (B) Pol D (D)	Pol III HE (DnaE + nine other subunits)	Pol C
Replicase core(s)	(δ): PolD1/ PolD2/ PolD3/ PolD4 (ε): Pol E/ PolE2/ PolE3/ PolE4	(δ): Pol3/ Pol31/ Pol32 (ε): Pol2/ Dpb2/ Dpb3/ Dpb4	(δ): Pol3/ Cdc1/ Cdc27/ Cdm1 (ε): Pol2/ Dpb2/ Dpb3/ Dpb4	(B): Pol B (D): (DP1/ DP2)$_2$	αεθ	Pol C
Polymerase subunit	(δ): PolD1 (ε): Pol E	(δ): Pol3 (ε): Pol2	(δ): Pol3 (ε): Pol2	(B): Pol B (D): DP2	α (DnaE)	Pol C
Proofreading subunit	(δ): PolD1 (ε): Pol E	(δ) Pol3 (ε) Pol2	(δ): Pol3 (ε): Pol2	(B): Pol B (D): DP1	ε (DnaQ)	Pol C
Primase (subunits)	Pol α (Prim1/ Prim2A)	Pol α (Pri1/ Pri2)	Pol α (Pri1/ Spp2)	PriSL	DnaG	DnaG
Primer extension (subunits)	Pol α (PolA1/ Pol2A)	Pol α (Pol1/ Pol12)	Pol α (Pol1/ Pol12)			DnaE/ DnaQ
Leading-strand Pol	Pol ε	Pol ε	Pol ε	Pol B	αεθ	Pol C
Lagging-strand Pol	Pol δ	Pol δ	Pol δ	Pol B or Pol D (?)	αεθ	Pol C
Clamp	PCNA	PCNA	PCNA	PCNA	β	β
Clamp loader	RFC (pentamer of five different subunits)	RFC (pentamer of five different subunits)	RFC (pentamer of five different subunits)	RFC (pentamer of two different subunits)	$\tau_2\gamma\delta\delta'\chi\psi$	$\tau_3\delta\delta''$
Okazaki fragment processing	Pol δ/FEN1/ PCNA, Dna2	Pol δ/FEN1/ PCNA, Dna2	Pol δ/FEN1/ PCNA, Dna2	FEN1	Pol I	Pol I

Isolable complexes are color coded. PCNA, proliferating cell nuclear antigen; RFC, replication factor C.

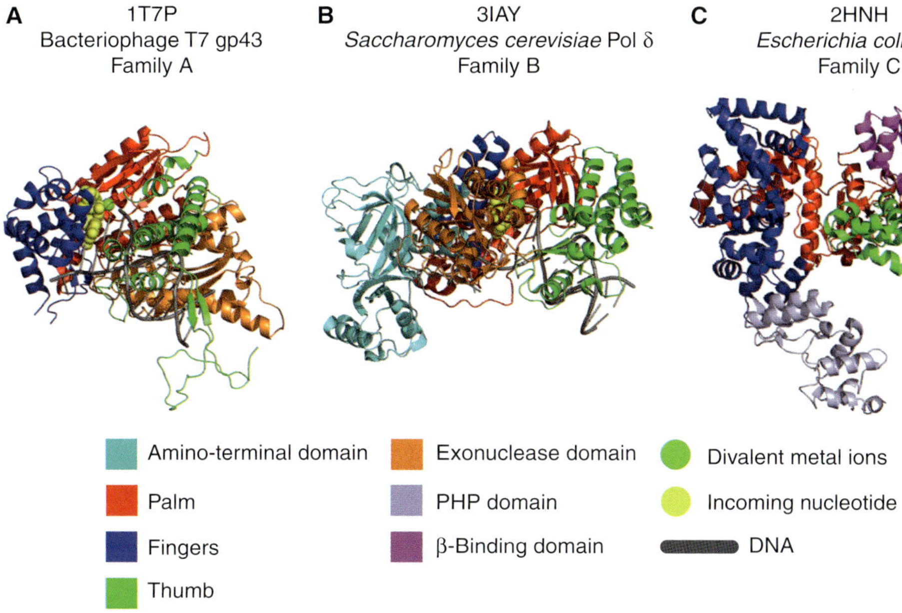

Figure 1. Representative structures and domain architecture of DNA polymerases from the A, B, and C families. (*A*) Bacteriophage T7 gene 5 protein with primer-template DNA (family A; pdb 1T7P). (Data from Doublié et al. 1998.) (*B*) *Saccharomyces cerevisiae* Pol δ (family B; pdb 3IAY). (Data from Swan et al. 2009.) (*C*) *E. coli* Pol III α subunit, residues 1–917 (family C; pdb 2HNH). (Data from Lamers et al. 2006.) Figure drawn using PyMOL.

ions are coordinated by two invariable aspartic acids. This is the basis for the two-metal catalytic mechanism (Steitz 1993), conserved among all replicative polymerases studied so far. The fingers domain is adjacent to the palm domain, contacts the incoming nucleotide, and goes through large conformational changes when the nucleotide is positioned near the metals, template, and the 3′ end of the growing strand in the active site (Yang et al. 1999; Franklin et al. 2001). The thumb domain, which interacts with the duplex DNA upstream of the polymerase active site, has been shown in A- and B-family polymerases to influence the partitioning between editing and polymerization modes (Doublié et al. 1998; Franklin et al. 2001). Mutations in the thumb domain can act both as antimutators and mutators, depending on how the balance has been shifted either toward the editing or the polymerization mode (Stocki et al. 1995; Wu et al. 1998). The thumb domain in C-family polymerases may have a similar function to that in A- and B-family polymerases, because it also contacts the duplex DNA (Bailey et al. 2006; Lamers et al. 2006; Wing et al. 2008).

Replicative DNA polymerases are highly accurate during the synthesis of DNA. In addition they have either a built-in 3′–5′ exonuclease site, located at a distance from the polymerase active site, or an associated subunit with 3′–5′ exonuclease activity (reviewed in Patel and Loeb 2001; Kunkel 2004; McHenry 2011). In either case, the distance from the polymerase site to the exonuclease site necessitates that at least three nucleotides of the double-stranded DNA are unwound to allow editing. This is a particularly important facet of the editing function because the exonuclease sites only accommodate single-stranded DNA.

At every replication fork, replicative polymerases cooperate in copying both DNA strands. Because the directions of the two template strands are opposite, and polymerases can only extend preexisting primers from their 3′-OH end, only one strand can be made continuously (the leading strand). The other (lagging) strand has to be synthesized in the direction opposite the fork movement as a series of short (Okazaki) fragments that are processed and subsequently joined together. In general, leading- and lagging-strand synthesis is coordinated, but how this is achieved varies somewhat among organisms.

Replicases from all domains of life function within larger molecular machines called replisomes that contain helicase(s), which separate the two DNA strands at the apex of the replication fork, a primase for repeated RNA priming in lagging-strand synthesis, sliding clamps that encircle the nascent double-stranded product to tether polymerases onto the DNA to ensure they dissociate rarely, clamp loaders that load the clamps, various single-stranded DNA-binding proteins, etc. The ring-shaped sliding clamps, called the β subunit (a homodimer) in eubacteria and PCNA (a homotrimer) in archaea and eukaryotes, have very similar structures (Kong et al. 1992; Krishna et al. 1994). They interact at the same site in each of their subunits with many other proteins including the replicative polymerases, via short peptide motifs called the clamp-binding motif in eubacteria (Dalrymple et al. 2001) and the PIP box in higher organisms (Warbrick 1998). The structures and functions of the sliding clamps and other accessory replisomal components are described in detail in the literature.

So far we have highlighted a few well-conserved features of the replicative polymerases. However, as we now proceed with a more detailed comparison, you will find that they have special features to function in their specific environments.

QUATERNARY STRUCTURE

The quaternary structures of replicative polymerases vary significantly (Table 1). In part, this variation rather arbitrarily depends on which subunits are isolated together to define the polymerase. From some species (e.g., Pol III in *Escherichia coli* [*Eco*]), a holoenzyme can be purified as an entity with 10 different subunits. In eukaryotes, the replicative polymerases appear to be more loosely associated within the replisome, and they are thus isolated in assemblies that contain fewer accessory subunits. We will discuss here the core catalytic subunit and accessory subunits that are considered to be parts of the DNA polymerase.

Eubacteria

Pol III holoenzyme is the *E. coli* chromosomal replicase that synthesizes both leading and lagging strands simultaneously. As isolated directly from cells, it has an average composition close to $(\alpha\varepsilon\theta)_2$–$(\tau_2\gamma\delta\delta'\psi\chi)$–$(\beta_2)_2$ (17 subunits), where $\alpha\varepsilon\theta$ is the polymerase core discussed in more detail below, β_2 is the sliding clamp, and $\tau_2\gamma\delta\delta'\psi\chi$ is the clamp loader complex that may contain two to three τ and one to zero γ subunits (McHenry 2011).

The $\alpha\varepsilon\theta$ core of Pol III is a tightly associated complex. The large α subunit is a family C polymerase, and ε is a separate $3'$–$5'$ editing exonuclease subunit from the DnaQ family. The small θ subunit has a role in stabilizing ε, but it only occurs in a limited range of bacterial species. The α subunit is made up of a series of domains (Fig. 1C) (Bailey et al. 2006; Lamers et al. 2006): the amino-terminal PHP domain seems to be a vestigial exonuclease domain that may still be functional as a proofreader in some species (Stano et al. 2006). In *E. coli*, it has evolved to be the site of interaction of the ε subunit (Wieczorek and McHenry 2006). This domain is followed by the usual polymerase palm, thumb, and fingers domains, and a β-binding domain that contains a conserved clamp-binding motif that interacts with the β_2 clamp to tether α to the product DNA to ensure its high processivity. This is followed by an OB-fold domain that is likely to interact with the single-stranded template DNA, and a carboxy-terminal domain that interacts tightly with the carboxy-terminal domain of the τ subunit of the clamp loader. Because the clamp loader contains two (or three) τ subunits, at least two $\alpha\varepsilon\theta$ cores are maintained in the replicase complex, one each for leading- and lagging-strand synthesis.

The ε subunit has a globular amino-terminal exonuclease domain (residues 7–180), the structure of which is reminiscent of many editing exonuclease domains from other polymerases (Hamdan et al. 2002). It contains a binuclear metal site for processing nucleotides misincorporated by α, followed by a poorly conserved 63-residue carboxy-terminal segment that by itself is unstructured. The last ~40 residues are now known to be involved directly in binding to the PHP domain of α (Ozawa et al. 2008); a short segment that immediately follows the exonuclease domain contains a second clamp-binding motif (S Jergic and N Dixon, unpubl.), and the unstructured segment in between remains flexible even in the αεθ complex (Ozawa et al. 2008). This may suggest an unusual mechanism in which ε can occupy quite different positions in the complex when it is or is not being used for proofreading.

Most (or all) eubacterial species seem to have replicase subunits closely related to α (DnaE family) and ε, but some Gram-positive bacteria like *Bacillus subtilis* (*Bsu*) also have a second more distantly related polymerase/exonuclease called Pol C (Table 1). These latter enzymes have the proofreading and polymerase activities in a single polypeptide, and a different domain organization, with putative τ-binding and OB-fold domains preceding a discontinuous PHP domain that incorporates the proofreader, and the β-binding motif being right at the carboxyl terminus (Evans et al. 2008). Although it was for some time believed that Pol C was dedicated to leading-strand synthesis and DnaE to the lagging strand (Dervyn et al. 2001), recent work suggests that most chromosomal DNA synthesis in these bacteria is performed by Pol C. However, because only the DnaE polymerase can extend RNA primers like those made by the DnaG primase, it has a critical role in lagging-strand synthesis. The current model is that DnaE extends RNA primers for some length before passing them to Pol C (Sanders et al. 2010). This has some parallel with the eukaryotic Pol α (below). Note that in eubacteria, RNA priming is performed by a dedicated primase (DnaG) that does not associate directly with the polymerase; it is associated, sometimes firmly and often more transiently, with the replicative helicase to enable primer synthesis on both the leading and lagging strands.

There is no crystal structure available of a complete Pol III core from any bacterium, but available structures of α subunits include *E. coli* α(1–917), which misses its internal clamp-binding motif and domains that follow it (Lamers et al. 2006), and full-length *Thermus aquaticus* (*Taq*) α (DnaE), both by itself and in complex with primer-template DNA (Bailey et al. 2006; Wing et al. 2008). The most recent structure is of a ternary complex containing primer-template DNA and incoming dNTP bound to Pol C from *Geobacillus kaustophilus* (*Gka*) (Evans et al. 2008). The *Taq* and *Eco* structures without DNA are very similar in the regions that can be compared (Fig. 1C), but DNA binding induces large conformational changes to a closed state, especially in the β-binding region of the *Taq* protein. The *Gka* Pol C ternary complex is also similar to the *Taq* α–DNA structure in this respect. Interestingly, the structure of a β_2 DNA complex (Georgescu et al. 2008a) can be docked neatly into both the α and Pol C–DNA structures to give plausible models of the α or Pol C–β_2–DNA replicases in the polymerization mode (Evans et al. 2008; Wing et al. 2008), and it has been suggested that the open structures might mimic the replicase structure in the editing mode (Fig. 2A). The position of the proofreading exonuclease has not been defined (it was deleted from the Pol C construct used for crystallization), but it is presumably between the PHP domains and β_2 in the modeled structures. A second interesting aspect of the α/Pol C structures is that the palm domain (active site) architecture is distinct from that found in the A- and B-family polymerases, being closer to that of X family members like the eukaryotic repair polymerase, Pol β. This suggests that the replicative polymerases of eubacteria evolved independently of those in eukaryotes and archaea.

The other polymerase that plays a significant role in bacterial DNA replication is Pol I, the founding member of the family A polymerases (representative Pol A structure is shown in Fig. 1A). Its primary function in replication is in Okazaki fragment processing on the lagging

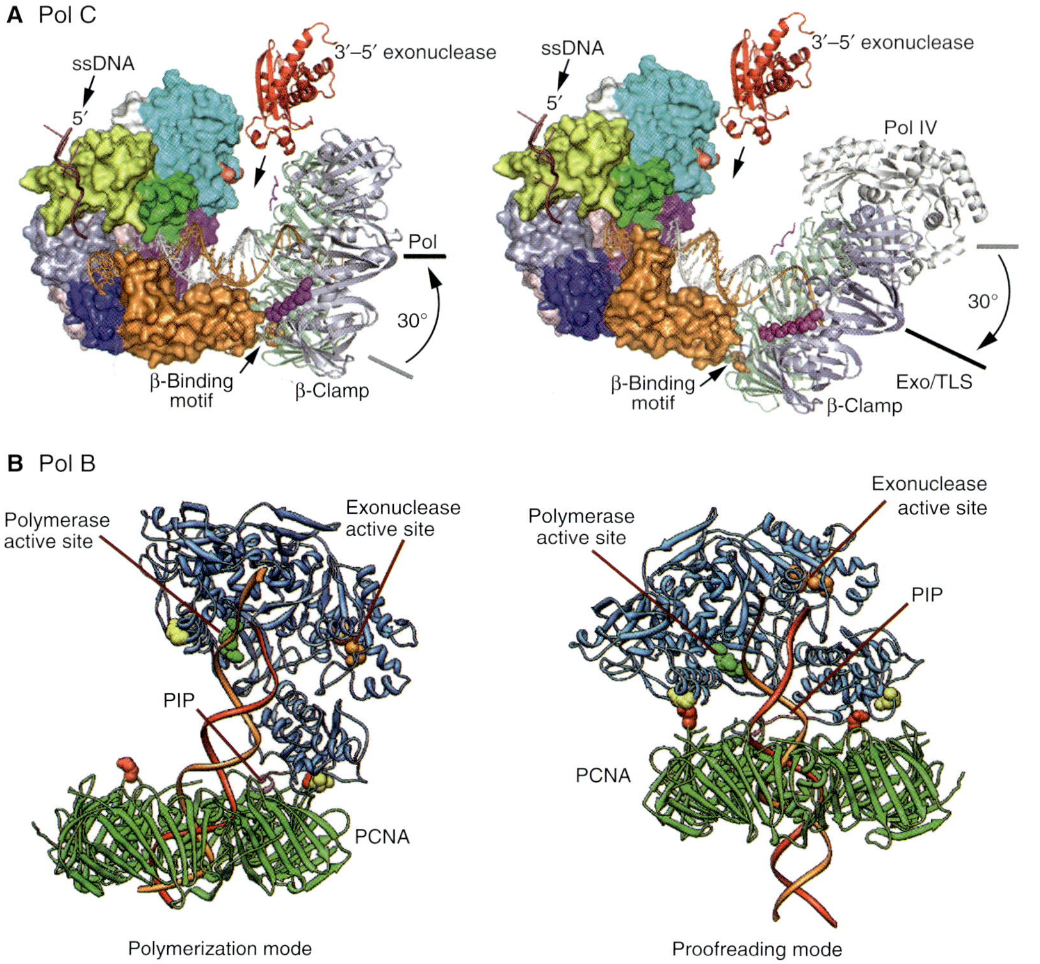

Figure 2. Modeled structures of ternary complexes of polymerases, clamps, and DNA in the polymerization and editing modes. (*A*) *Geobacillus kaustophilus* Pol C with DNA and β sliding clamp. (Panel *A* is from Evans et al. 2008; reprinted, with express permission, from the authors.) (*B*) *Pyrococcus furiosus* Pol B with DNA and PCNA (Mayanagi et al. 2011). (Panel *B* is from Mayanagi et al. 2011; reprinted, with permission, from The National Academy of Science © 2011.)

strand. Pol III is capable of synthesizing Okazaki fragments right up to the 5′ end of a preceding RNA primer, whereupon it is recycled to a new primer terminus, leaving behind a nick or short gap (discussed in Dohrmann et al. 2011). Pol I has three separate activities in a single polypeptide chain. The amino-terminal domain is a 5′–3′ exonuclease capable of excising the RNA primers at the same time as the carboxy-terminal polymerase domain (with thumb, palm, and fingers subdomains) extends the DNA primer behind it. The central domain is a DnaQ family proofreading exonuclease used to ensure high fidelity. Thus, Pol I uses a process of "nick translation" to replace RNA primers with DNA, leaving a nick with a 3′-OH and 5′-phosphate that is a substrate for DNA ligase, which seals the nick to make a contiguous lagging strand.

Archaea

The replisome in archaebacteria is less complex than in eubacteria, and in most aspects is a simplified version of the eukaryotic system.

All archaea contain single-subunit Pol B polymerases, with the proofreading activity included in the catalytic subunit as a domain near the amino terminus (Table 1). Some genera have multiple Pol Bs (family B), whose separate roles in chromosomal replication are yet to be clearly defined, whereas the euryarchaea uniquely also contain a more complex heterotetrameric Pol D polymerase, with subunit composition $(DP1)_2-(DP2)_2$. The smaller DP1 subunit contains a $3'-5'$ exonuclease domain, whereas the larger DP2 is the (family D) polymerase. Although Pol D is a processive and efficient enzyme that, in common with Pol B, also interacts with the PCNA sliding clamp, the role of Pol D as a chromosomal replicase in euryarchaeal species is still uncertain (Tori et al. 2007). It has been suggested that in *Pyrococcus abyssi*, Pol B synthesizes the leading and Pol D the lagging strand (Henneke et al. 2005). High-resolution structural information on Pol D is limited to the crystal structure of the amino-terminal domain of *Pyrococcus horikoshii* DP2, which has a role in Pol D subunit oligomerization (Matsui et al. 2011).

The crenarchaea do not have Pol Ds, but often have multiple Pol Bs. For example, *Sulfolobus solfataricus* has three of them, and Pol B1 (Dpo1) is a high-fidelity polymerase thought to carry out both leading- and lagging-strand replication (Choi et al. 2011). The archaeal Pol Bs have been studied extensively, in part because some of them from thermophilic species are used extensively for high-fidelity PCR applications. Although crystal structures of Pol Bs from several archaeal species have been determined (the earliest are in Hopfner et al. 1999, Zhao et al. 1999, Rodriguez et al. 2000, and Hashimoto et al. 2001), there is none yet that shows primer-template DNA in the polymerization or proofreading modes. These structures can, however, be modeled using the structures of the T4-related phage RB69 Pol B, bound to DNA in both of these modes (Franklin et al. 2001).

The archaeal Pol Bs seem generally and uniquely among family B polymerases (Wardle et al. 2008) to contain an additional amino-terminal domain that precedes the proofreading exonuclease domain, whose function is to scan the template DNA ahead of the polymerase for the uracil base in dUMP produced by cytosine deamidation (and hypoxanthine, from adenosine) that has escaped repair by uracil-*N*-glycosylase (Greagg et al. 1999, Fogg et al. 2002). On recognition of dUMP in the template, the polymerase stalls four nucleotides before the lesion is encountered. Although how this might lead to subsequent repair of the lesion is not known, crystal structures have been reported of stalled complexes containing primer templates with appropriately incorporated dUMP or dIMP (Firbank et al. 2008, Killelea et al. 2010). The bases are buried within a specific pocket in the amino-terminal domain, and the primer-template DNA is thought to be in a position in the structure close to that in the editing mode, by comparison to the RB69 Pol B structure (Franklin et al. 2001).

In a recent study (Mayanagi et al. 2011), a series of available crystal structures was combined with low-resolution reconstructions from electron microscopy and single-particle analysis to generate plausible models of the structures of archaeal Pol B–DNA–clamp ternary complexes in both the polymerization and proofreading modes (Fig. 2B). As for the bacterial Pol III complex (Fig. 2A), these models generate new hypotheses that will be tested in the future, leading to further progressive improvement of our understanding of replisome structure and function.

A significant difference between the archaeal replisomes and those in eukaryotes (see below) is that archaea, like the bacteria, use primases that are not physically associated with the polymerases. Nevertheless, the archaeal primases contain two subunits (p41 and p46) that are homologs of the PriS and PriL subunits of the eukaryotic Pol α–primase complex (Table 1). Recent in vitro studies with purified proteins have shown quite clearly that the *Thermococcus kodakarensis* primase can carry out (DNA) primer synthesis in the total absence of ribonucleoside triphosphates (rNTPs), and in a rolling-circle assay with a synthetically primed minicircle DNA template, it can prime efficient lagging-strand Okazaki-fragment synthesis dependent on the hexameric MCM replicative helicase and Pol B (Chemnitz Galal et al. 2012). Thus, the stage is now set to use in vitro studies with the archaeal replisome to probe mechanistic questions in the way that has already been performed with the model bacterial systems.

Eukarya

Eukaryotes have evolved a more complex replisome that depends on three different B-family DNA polymerases: Pol α, Pol δ, and Pol ε (Kunkel and Burgers 2008; Burgers 2009). Each of these multisubunit polymerases (Table 1) has one catalytic core subunit and two to four accessory subunits. Pol α is built up by four different subunits: PriS, PriL, Pol1, and Pol12 (Muzi-Falconi et al. 2003); PriL has primase activity and Pol1 has DNA polymerase activity. Negative-stain EM reconstructions combined with high-resolution structures of domains of the *Saccharomyces cerevisiae* (*Sc*) Pol1–Pol12 complex have revealed two globular domains connected via a flexible linker (Klinge et al. 2009). It was suggested that an intramolecular hand-off occurs from the primase domain to the polymerase domain when the initial RNA primer has reached a specific length (Klinge et al. 2009; Nunez-Ramirez et al. 2011). The final result is a 30–35 nucleotide primer with RNA at the 5′ end and DNA at the 3′ end. Thus, Pol α synthesizes a substantial amount of DNA because the eukaryotic Okazaki fragments are on average only about 165 nucleotides long (Smith and Whitehouse 2012). Thus, the absence of a proofreading activity in Pol α requires that errors made by it are corrected by other mechanisms. For instance Pol δ, but not Pol ε, proofreads errors made by Pol α, in addition to error correction by the mismatch repair system (Pavlov et al. 2006; Nick McElhinny et al. 2010a).

Pol ε and Pol δ are responsible for the bulk synthesis of DNA on the leading and lagging strands (Kunkel and Burgers 2008; Burgers 2009). They are both highly accurate DNA polymerases with built-in 3′–5′ exonuclease activities for proofreading (Morrison et al. 1991; Morrison and Sugino 1994). They are also highly processive enzymes when interacting with the clamp, PCNA (Burgers 1991; Lee et al. 1991; Chilkova et al. 2007). The exact roles of Pols ε and δ at the replication fork have been debated (Kunkel and Burgers 2008; Burgers 2009). Over the past few years, evidence has accumulated that in *S. cerevisiae* and *Schizosaccharomyces pombe* (*Sp*), Pol ε is primarily responsible for leading-strand and Pol δ for lagging-strand synthesis (Pursell et al. 2007; Nick McElhinny et al. 2008; Miyabe et al. 2011). This suggests that the roles of Pol ε and Pol δ are evolutionarily conserved among eukaryotes.

Pol δ is constituted of three subunits (Pol3, Pol31, and Pol32) in *S. cerevisiae*, and four in *S. pombe* and *H. sapiens* (Table 1) (Gerik et al. 1998; Reynolds et al. 1998; Liu et al. 2000). This may be species variations but could also reflect how tightly specific subunits are associated with the catalytic core subunit. The hydrodynamic properties of *Sc*Pol δ and *Sp*Pol δ suggested that Pol δ has an elongated structure, heavily influenced by the Pol32 subunit (Johansson et al. 2001; Bermudez et al. 2002). This was later confirmed by small-angle X-ray scattering (SAXS) analysis of *Sc*Pol3 in complex with Pol31 and the amino terminus of Pol32 and a high-resolution structure of the human accessory subunits Pold2 together with the amino terminus of Pold3 (Baranovskiy et al. 2008; Jain et al. 2009). The catalytic subunit *Sc*Pol3 is located at one end of the complex; Pol31 acts as a bridge to the very elongated Pol32 that carries an important interaction motif with PCNA at the far end of the elongated structure extending away from the catalytic core subunit Pol3 (Johansson et al. 2004; Baranovskiy et al. 2008; Jain et al. 2009). A high-resolution structure of *Sc*Pol3 (amino acid residues 68–985) was solved which included both the exonuclease and polymerase domains (Fig. 1B), but excluded the carboxy-terminal domain with the CysA and CysB motifs discussed below. Overall, the structure of *Sc*Pol3 is highly reminiscent of a typical family B polymerase, with a fold very similar to that of RB69 despite the limited sequence identity (Swan et al. 2009).

Pol ε consists of four subunits. In *S. cerevisiae*, these are called Pol2, Dpb2, Dpb3, and Dpb4 (Hamatake et al. 1990; Chilkova et al. 2003). Structural information on Pol ε is limited to a low-resolution cryo-electron microscopy (cryo-EM) structure. *Sc*Pol ε has two domains, a globular domain with the catalytic subunit and an extended tail domain that has been suggested to harbor the three accessory subunits Dpb2, Dpb3, and Dpb4 (Asturias et al. 2006). Dpb2 is essential in *S. cerevisiae* and *S. pombe* (Araki et al. 1991; Feng et al. 2003). However, the Dpb2 subunit does not appear to influence the polymerase activity in human Pol ε (Bermudez et al. 2011). It is possible that the essential function of Dpb2 is during the initiation of DNA replication, as described in Tanaka and Araki (2013) (also reviewed in Araki 2010). Dpb3 and Dpb4 form a heterodimer that has a high affinity for double-stranded DNA (Tsubota et al. 2006). In contrast to Dpb2, *Sc*Pol ε depends on Dpb3 and Dpb4 for full processivity in the absence of PCNA (Aksenova et al. 2010).

The eukaryotic replicative polymerases all have two conserved motifs with cysteines (CysA and CysB) located at the carboxyl terminus of the catalytic subunit (Netz et al. 2012). For a long time they have been considered to be Zn-finger motifs and Zn was found in both Pol ε and Pol α (Dua et al. 2002; Klinge et al. 2009). Their primary role was suggested to form an interaction surface with the B subunit of the DNA polymerase. Recently it was shown that the two motifs CysA and CysB have differential functions and coordinate different metals in *Sc*Pol δ (Netz et al. 2012). CysA coordinates Zn and is required to support processive synthesis by Pol δ via a direct interaction with PCNA, whereas CysB coordinates a Fe–S center and is important for the interaction with the Pol31 subunit. Loss of interaction with Pol31 led to loss of Pol32, and as a result an important interaction with the PCNA clamp was lost. Thus, Pol δ has at least two separate interactions with PCNA, CysA in Pol3 and the PIP box in Pol32, to support efficient loading and synthesis of DNA (Netz et al. 2012). Experiments with carboxy-terminal domains of Pol α and Pol ε implied that these replicative polymerases may also have an Fe–S cluster. However, conflicting results with expression of the carboxy-terminal domains of the catalytic subunits from human Pols α and ε in *E. coli* were recently published (Baranovskiy et al. 2012) and it remains to be shown whether the *Sc*Pols α and ε holoenzymes carry Fe–S clusters that mediate the interactions between the catalytic subunit and Pol12 and Dpb2, respectively. The study of the carboxy-terminal domain of the catalytic subunit in human Pol δ confirmed that there is an Fe–S center coordinated by CysB (Baranovskiy et al. 2012).

As discussed above, *Sc*Pol δ has at least two well-defined motifs that mediate the contact with PCNA. Two additional motifs, one in Pol3 and one in Pol31, were recently reported, emphasizing that *Sc*Pol δ has multiple surfaces to stabilize the interaction with PCNA when synthesizing DNA (Acharya et al. 2011). Less is known about how Pol ε interacts with PCNA. In *S. cerevisiae*, the catalytic subunit and Dpb2 both have a PIP box, but they are not located at the amino or carboxyl terminus as found typically in PCNA-interacting proteins (Dua et al. 2002). Biochemical characterizations and genetic experiments in yeast suggested that these are not functional motifs during DNA replication (Dua et al. 2002; Chilkova et al. 2007). Nevertheless, Pol ε is stimulated by PCNA in vitro, suggesting that either the exact motif(s) remains to be found or there are multiple weak interactions that together stabilize the interaction with PCNA (Maga et al. 1999). Pol ε also has an unusually high intrinsic processivity in part mediated by the Dpb3 and Dpb4 subunits (Aksenova et al. 2010). In addition, the catalytic subunit has by itself an unusually high processivity and it was speculated that a 66 amino acid insertion in the vicinity of the palm domain might give Pol ε this unusual property (Shcherbakova et al. 2003).

FIDELITY OF THE REPLICATIVE POLYMERASES

The high fidelity of replicative polymerases is determined by the tolerance for the incoming nucleotide when performing chemistry and the 3′–5′ exonucleolytic proofreading activity. In addition, the concentration of each dNTP in the cell and the quality of the template influence the rate at which errors are made (Kunkel 2011). Until recently, not much attention was paid to how the relatively large pools of rNTPs in eukaryotic cells influence the fidelity of the replicative DNA polymerases. The polymerases have a built-in steric gate that should select for dNTPs and against rNTPs (Joyce 1997). However, the concentration of dNTPs is very low in comparison to the rNTPs and it appears that Pols α, δ, and ε misincorporate ribonucleotides at a surprisingly high rate when synthesizing DNA (Nick McElhinny et al. 2010c). All three Pols also incorporated rNTPs in vivo in *S. cerevisiae* and *S. pombe* and gave a significantly increased elevation of 2–5 nucleotide deletions (Nick McElhinny et al. 2010b; Miyabe et al. 2011). To avoid mutagenic events, ribonucleotides must be removed by a repair mechanism that involves RNase H2 before the next S phase (Nick McElhinny et al. 2010b; Miyabe et al. 2011). The incorporation of ribonucleotides and the dependence on RNase H2 for the removal of ribonucleotides was also found in mice, suggesting that this repair mechanism is conserved among all eukaryotes (Reijns et al. 2012). The exact mechanism behind the error signature in *S. cerevisiae* is unknown, but Pol α, Pol δ, and Pol ε stall when encountering a ribonucleotide in the template and this may lead to deletions in repetitive sequences (Watt et al. 2011). In case ribonucleotides are not removed, postreplication repair will be activated during the next S phase (Lazzaro et al. 2012). Pol ζ is capable of replicating across 1–4 ribonucleotides in the template and also MMS2-dependent template switching contributes to the tolerance for unrepaired ribonucleotide lesions (Lazzaro et al. 2012).

It is yet unclear if archaeal DNA polymerases also frequently misinsert ribonucleotides in the genome. However, as described earlier, the archaeal replicative polymerases (in, e.g., *P. furiosus* and *S. solfataricus*) have a unique capacity to proofread the template for uracil and other deamidated bases (Greagg et al. 1999; Choi et al. 2011). A binding pocket for uracil is found in their unique amino-terminal domains where the single-stranded template is scanned before entering the polymerase catalytic site (Fogg et al. 2002; Shuttleworth et al. 2004; Firbank et al. 2008).

ANTIBIOTICS/INHIBITORS

DNA replication is an essential process in all organisms. Although there are sufficient variations in structures among replisomal proteins from bacteria and humans to make replisomes a very good target for discovery of new antibacterial therapeutics, it is a target that is surprisingly underexploited both by pharmaceutical companies and in nature by other organisms (reviewed in Robinson et al. 2010, 2012). Although there are no known inhibitors of DnaE-type Pol IIIs, there have been substantial efforts to target Pol C from Gram-positive bacteria, including *Staphylococcus aureus*, using a range of 6-anilinouracils (6-AUs) and quinazolin-2-ylamino-quinazolin-4-ols (BisQuinols). The 6-AUs act as Pol C inhibitors by competing with dGTP for base pairing with cytosines in the template DNA, whereas the BisQuinols appear to interfere with template DNA binding. Another promising target is the β clamp, which makes essential interactions with the clamp-binding motifs in a large number of other proteins (including all five DNA polymerases, the clamp loader, and DNA ligase in *E. coli*, for example) at a single highly conserved binding site that is structurally distinct from the PIP-binding sites in PCNA (Dalrymple et al. 2001). Development of resistance to antibacterials that inhibit interactions at this site by target mutagenesis would be expected to be slow because it would require simultaneous mutations to arise in genes encoding several different proteins. There have been recent reports of identification through screening campaigns of new inhibitors of interactions with β, including two crystal structures (Georgescu et al. 2008b; Wijffels et al. 2011), but there is much to be done to develop these compounds into useful leads.

CONCLUDING REMARKS

Almost 60 years have passed since the first isolation of a replicative DNA polymerase by Arthur Kornberg in the mid-1950s. In this long history of research, many modern tools in structural and molecular biology have provided fragments of information for progressive synthesis of coherent pictures of how these complex multifunctional enzymes work in organisms from all three domains of life. There are common aspects that are well established, like the common catalytic mechanism of the nucleotidyltransferase reaction, but our pictures are still far from complete. The main reason for this is that these enzymes never work alone. They are key parts of larger, very dynamic nucleoprotein machines (replisomes) that undergo large structural changes as they function, with proteins coming and going and often changing binding partners. Sometimes this occurs at common binding sites (as on the sliding clamps) to ensure ordered sequences of events, and often interactions involve intrinsically unstructured regions of proteins that only become ordered as they interact with their binding partners. This latter fact alone has made structural studies slow and difficult, and it also means that multiple structures need to be solved to really understand function. A second basic operating principle for replisomes is that their integrity is maintained by a very large number of protein–protein and protein–nucleic acid interactions, many of which are inherently weak and can easily be transiently broken. These interactions are often difficult to detect individually, but sum to be strong enough to hold the replisomes together as they function.

ACKNOWLEDGMENTS

We apologize to those colleagues whose work is not cited because of formatting and space restrictions. We thank Matthew Hogg for help with Figure 1, and Rick Lewis for comments on the manuscript. Research in our laboratories is funded by the Swedish Cancer Society, the Swedish Research Council, the Knut and Alice Wallenberg foundation (E.J.), the Australian Research Council, including an Australian Professorial Fellowship, and the Australian National Health and Medical Research Council (N.D.).

REFERENCES

**Reference is also in this collection.*

Acharya N, Klassen R, Johnson RE, Prakash L, Prakash S. 2011. PCNA binding domains in all three subunits of yeast DNA polymerase δ modulate its function in DNA replication. *Proc Natl Acad Sci* **108:** 17927–17932.

Aksenova A, Volkov K, Maceluch J, Pursell ZF, Rogozin IB, Kunkel TA, Pavlov YI, Johansson E. 2010. Mismatch repair—Independent increase in spontaneous mutagenesis in yeast lacking non-essential subunits of DNA polymerase ε. *PLoS Genet* **6:** e1001209.

Araki H. 2010. Cyclin-dependent kinase-dependent initiation of chromosomal DNA replication. *Curr Opin Cell Biol* **22:** 766–771.

Araki H, Hamatake RK, Johnston LH, Sugino A. 1991. DPB2, the gene encoding DNA polymerase II subunit B, is required for chromosome replication in *Saccharomyces cerevisiae. Proc Natl Acad Sci* **88:** 4601–4605.

Asturias FJ, Cheung IK, Sabouri N, Chilkova O, Wepplo D, Johansson E. 2006. Structure of *Saccharomyces cerevisiae* DNA polymerase epsilon by cryo-electron microscopy. *Nat Struct Mol Biol* **13:** 35–43.

Bailey S, Wing RA, Steitz TA. 2006. The structure of *T. aquaticus* DNA polymerase III is distinct from eukaryotic replicative DNA polymerases. *Cell* **126:** 893–904.

Baranovskiy AG, Babayeva ND, Liston VG, Rogozin IB, Koonin EV, Pavlov YI, Vassylyev DG, Tahirov TH. 2008. X-ray structure of the complex of regulatory subunits of human DNA polymerase δ. *Cell Cycle* **7:** 3026–3036.

Baranovskiy AG, Lada AG, Siebler HM, Zhang Y, Pavlov YI, Tahirov TH. 2012. DNA polymerase δ and ζ switch by sharing accessory subunits of DNA polymerase δ. *J Biol Chem* **287:** 17281–17287.

Bermudez VP, MacNeill SA, Tappin I, Hurwitz J. 2002. The influence of the Cdc27 subunit on the properties of the *Schizosaccharomyces pombe* DNA polymerase δ. *J Biol Chem* **277:** 36853–36862.

Bermudez VP, Farina A, Raghavan V, Tappin I, Hurwitz J. 2011. Studies on human DNA polymerase ε and GINS complex and their role in DNA replication. *J Biol Chem* **286:** 28963–28977.

Burgers PM. 1991. *Saccharomyces cerevisiae* replication factor C. II. Formation and activity of complexes with the proliferating cell nuclear antigen and with DNA polymerases δ and ε. *J Biol Chem* **266:** 22698–22706.

Burgers PM. 2009. Polymerase dynamics at the eukaryotic DNA replication fork. *J Biol Chem* **284:** 4041–4045.

Chemnitz Galal W, Pan M, Kelman Z, Hurwitz J. 2012. Characterization of DNA primase complex isolated from the archaeon, *Thermococcus kodakearaensis*, *J Biol Chem* **287:** 16209–16219.

Chilkova O, Jonsson BH, Johansson E. 2003. The quaternary structure of DNA polymerase ε from *Saccharomyces cerevisiae. J Biol Chem* **278:** 14082–14086.

Chilkova O, Stenlund P, Isoz I, Stith CM, Grabowski P, Lundstrom EB, Burgers PM, Johansson E. 2007. The eukaryotic leading and lagging strand DNA polymerases are loaded onto primer-ends via separate mechanisms but have comparable processivity in the presence of PCNA. *Nucleic Acids Res* **35:** 6588–6597.

Choi JY, Eoff RL, Pence MG, Wang J, Martin MV, Kim EJ, Folkmann LM, Guengerich FP. 2011. Roles of the four DNA polymerases of the crenarchaeon *Sulfolobus solfataricus* and accessory proteins in DNA replication. *J Biol Chem* **286:** 31180–31193.

Dalrymple BP, Kongsuwan K, Wijffels G, Dixon NE, Jennings PA. 2001. A universal protein-protein interaction motif in the eubacterial DNA replication and repair systems. *Proc Natl Acad Sci* **98:** 11627–11632.

Dervyn E, Suski C, Daniel R, Bruand C, Chapuis J, Errington J, Jannière L, Ehrlich SD. 2001. Two essential DNA polymerases at the bacterial replication fork. *Science* **294:** 1716–1719.

Dohrmann PR, Manhart CM, Downey CD, McHenry CS. 2011. The rate of polymerase release upon filling the gap between Okazaki fragments is inadequate to support cycling during lagging strand synthesis. *J Mol Biol* **414:** 15–27.

Doublié S, Tabor S, Long AM, Richardson CC, Ellenberger T. 1998. Crystal structure of a bacteriophage T7 DNA replication complex at 2.2 Å resolution. *Nature* **391:** 251–258.

Dua R, Levy DL, Li CM, Snow PM, Campbell JL. 2002. In vivo reconstitution of *Saccharomyces cerevisiae* DNA polymerase ε in insect cells. Purification and characterization. *J Biol Chem* **277:** 7889–7896.

Evans RJ, Davies DR, Bullard JM, Christensen J, Green LS, Guiles JW, Pata JD, Ribble WK, Janjic N, Jarvis TC. 2008. Structure of PolC reveals unique DNA binding and fidelity determinants. *Proc Natl Acad Sci* **105:** 20695–20700.

Feng W, Rodriguez-Menocal L, Tolun G, D'Urso G. 2003. *Schizosacchromyces pombe* Dpb2 binds to origin DNA early in S phase and is required for chromosomal DNA replication. *Mol Biol Cell* **14:** 3427–3436.

Firbank SJ, Wardle J, Heslop P, Lewis RJ, Connolly BA. 2008. Uracil recognition in archaeal DNA polymerases captured by X-ray crystallography. *J Mol Biol* **381:** 529–539.

Fogg MJ, Pearl LH, Connolly BA. 2002. Structural basis for uracil recognition by archaeal family B DNA polymerases. *Nat Struct Biol* **9:** 922–927.

Franklin MC, Wang J, Steitz TA. 2001. Structure of the replicating complex of a pol α family DNA polymerase. *Cell* **105:** 657–667.

Georgescu RE, Kim S-S, Yurieva O, Kuriyan J, Kong X-P, O'Donnell M. 2008a. Structure of a sliding clamp on DNA. *Cell* **132:** 43–54.

Georgescu RE, Yurieva O, Kim S-S, Kuriyan J, Kong X-P, O'Donnell M. 2008b. Structure of a small-molecule inhibitor of a DNA polymerase sliding clamp. *Proc Natl Acad Sci* **105:** 11116–11121.

Gerik KJ, Li X, Pautz A, Burgers PM. 1998. Characterization of the two small subunits of *Saccharomyces cerevisiae* DNA polymerase δ. *J Biol Chem* **273:** 19747–19755.

Greagg MA, Fogg MJ, Panayotou G, Evans SJ, Connolly BA, Pearl LH. 1999. A read-ahead function in archaeal DNA polymerases detects promutagenic template-strand uracil. *Proc Natl Acad Sci* **96:** 9045–9050.

Hamatake RK, Hasegawa H, Clark AB, Bebenek K, Kunkel TA, Sugino A. 1990. Purification and characterization of DNA polymerase II from the yeast *Saccharomyces cerevisiae*. Identification of the catalytic core and a possible holoenzyme form of the enzyme. *J Biol Chem* **265:** 4072–4083.

Hamdan S, Carr PD, Brown SE, Ollis DL, Dixon NE. 2002. Structural basis for proofreading during replication of the *Escherichia coli* chromosome. *Structure* **10:** 535–546.

Hashimoto H, Nishioka M, Fujiwara S, Takagi M, Imanaka T, Inoue T, Kai Y. 2001. Crystal structure of DNA polymerase from hyperthermophilic archaeon *Pyrococcus kodakaraensis* KOD1. *J Mol Biol* **306:** 469–477.

Henneke G, Flament D, Hübscher U, Querellou J, Raffin JP. 2005. The hyperthermophilic euryarchaeota *Pyrococcus abyssi* likely requires the two DNA polymerases D and B for DNA replication. *J Mol Biol* **350:** 53–64.

Hopfner K-P, Eichinger A, Engh RA, Laue F, Ankenbauer W, Huber R, Angerer B. 1999. Crystal structure of a thermostable type B DNA polymerase from *Thermococcus gorgonarius*. *Proc Natl Acad Sci* **96:** 3600–3605.

Jain R, Hammel M, Johnson RE, Prakash L, Prakash S, Aggarwal AK. 2009. Structural insights into yeast DNA polymerase δ by small angle X-ray scattering. *J Mol Biol* **394:** 377–382.

Johansson E, Majka J, Burgers PM. 2001. Structure of DNA polymerase δ from *Saccharomyces cerevisiae*. *J Biol Chem* **276:** 43824–43828.

Johansson E, Garg P, Burgers PM. 2004. The Pol32 subunit of DNA polymerase δ contains separable domains for processive replication and proliferating cell nuclear antigen (PCNA) binding. *J Biol Chem* **279:** 1907–1915.

Joyce CM. 1997. Choosing the right sugar: How polymerases select a nucleotide substrate. *Proc Natl Acad Sci* **94:** 1619–1622.

Killelea T, Ghosh S, Tan SS, Heslop P, Firbank SJ, Kool ET, Connolly BA. 2010. Probing the interaction of archaeal DNA polymerases with deaminated bases using X-ray crystallography and non-hydrogen bonding isosteric base analogues. *Biochemistry* **49:** 5772–5781.

Klinge S, Nunez-Ramirez R, Llorca O, Pellegrini L. 2009. 3D architecture of DNA Pol α reveals the functional core of multi-subunit replicative polymerases. *Embo J* **28:** 1978–1987.

Kong X-P, Onrust R, O'Donnell M, Kuriyan J. 1992. Three-dimensional structure of the β subunit of *E. coli* DNA polymerase III holoenzyme: A sliding DNA clamp. *Cell* **69:** 425–437.

Krishna TSR, Kong X-P, Gary S, Burgers PM, Kuriyan J. 1994. Crystal structure of the eukaryotic DNA polymerase processivity factor PCNA. *Cell* **79:** 1233–1243.

Kunkel TA. 2004. DNA replication fidelity. *J Biol Chem* **279:** 16895–16898.

Kunkel TA. 2011. Balancing eukaryotic replication asymmetry with replication fidelity. *Curr Opin Chem Biol* **15:** 620–626.

Kunkel TA, Burgers PM. 2008. Dividing the workload at a eukaryotic replication fork. *Trends Cell Biol* **18:** 521–527.

Lamers MH, Georgescu RE, Lee SG, O'Donnell M, Kuriyan J. 2006. Crystal structure of the catalytic α subunit of *E. coli* replicative DNA polymerase III. *Cell* **126:** 881–892.

Lazzaro F, Novarina D, Amara F, Watt DL, Stone JE, Costanzo V, Burgers PM, Kunkel TA, Plevani P, Muzi-Falconi M. 2012. RNase H and postreplication repair protect cells from ribonucleotides incorporated in DNA. *Mol Cell* **45:** 99–110.

Lee SH, Pan ZQ, Kwong AD, Burgers PM, Hurwitz J. 1991. Synthesis of DNA by DNA polymerase ε in vitro. *J Biol Chem* **266:** 22707–22717.

Liu L, Mo J, Rodriguez-Belmonte EM, Lee MY. 2000. Identification of a fourth subunit of mammalian DNA polymerase delta. *J Biol Chem* **275:** 18739–18744.

Maga G, Jonsson ZO, Stucki M, Spadari S, Hubscher U. 1999. Dual mode of interaction of DNA polymerase ε with proliferating cell nuclear antigen in primer binding and DNA synthesis. *J Mol Biol* **285:** 259–267.

Matsui I, Urushibata Y, Shen Y, Matsui E, Yokoyama H. 2011. Novel structure of an N-terminal domain that is crucial for the dimeric assembly and DNA-binding of an archaeal DNA polymerase D large subunit from *Pyrococcus horikoshii*. *FEBS Lett* **585:** 452–458.

Mayanagi K, Kiyonari S, Nishida H, Saito M, Kohda D, Ishino Y, Shirai T, Morikawa K. 2011. Architecture of the DNA polymerase B-proliferating cell nuclear antigen (PCNA)-DNA ternary complex. *Proc Natl Acad Sci* **108:** 1845–1849.

McHenry CS. 2011. DNA replicases from a bacterial perspective. *Annu Rev Biochem* **80:** 403–436.

Miyabe I, Kunkel TA, Carr AM. 2011. The major roles of DNA polymerases ε and δ at the eukaryotic replication fork are evolutionarily conserved. *PLoS Genet* **7:** e1002407.

Morrison A, Sugino A. 1994. The 3′ → 5′ exonucleases of both DNA polymerases δ and ε participate in correcting errors of DNA replication in *Saccharomyces cerevisiae*. *Mol Gen Genet* **242:** 289–296.

Morrison A, Bell JB, Kunkel TA, Sugino A. 1991. Eukaryotic DNA polymerase amino acid sequence required for 3′–5′ exonuclease activity. *Proc Natl Acad Sci* **88:** 9473–9477.

Muzi-Falconi M, Giannattasio M, Foiani M, Plevani P. 2003. The DNA polymerase α-primase complex: Multiple functions and interactions. *ScientificWorldJournal* **3:** 21–33.

Netz DJ, Stith CM, Stumpfig M, Kopf G, Vogel D, Genau HM, Stodola JL, Lill R, Burgers PM, Pierik AJ. 2012. Eukaryotic DNA polymerases require an iron-sulfur cluster for the formation of active complexes. *Nat Chem Biol* **8:** 125–132.

Nick McElhinny SA, Gordenin DA, Stith CM, Burgers PM, Kunkel TA. 2008. Division of labor at the eukaryotic replication fork. *Mol Cell* **30:** 137–144.

Nick McElhinny SA, Kissling GE, Kunkel TA. 2010a. Differential correction of lagging-strand replication errors made by DNA polymerases α and δ. *Proc Natl Acad Sci* **107:** 21070–21075.

Nick McElhinny SA, Kumar D, Clark AB, Watt DL, Watts BE, Lundstrom EB, Johansson E, Chabes A, Kunkel TA. 2010b. Genome instability due to ribonucleotide incorporation into DNA. *Nat Chem Biol* **6:** 774–781.

Nick McElhinny SA, Watts BE, Kumar D, Watt DL, Lundstrom EB, Burgers PM, Johansson E, Chabes A, Kunkel TA. 2010c. Abundant ribonucleotide incorporation into DNA by yeast replicative polymerases. *Proc Natl Acad Sci* **107:** 4949–4954.

Nunez-Ramirez R, Klinge S, Sauguet L, Melero R, Recuero-Checa MA, Kilkenny M, Perera RL, Garcia-Alvarez B, Hall RJ, Nogales E, et al. 2011. Flexible tethering of primase and DNA Pol α in the eukaryotic primosome. *Nucleic Acids Res* **39:** 8187–8199.

Ozawa K, Jergic S, Park AY, Dixon NE, Otting G. 2008. The proofreading exonuclease subunit ε of *Escherichia coli* DNA polymerase III is tethered to the polymerase subunit alpha via a flexible linker. *Nucleic Acids Res* **36:** 5074–5082.

Patel PH, Loeb LA. 2001. Getting a grip on how DNA polymerases function. *Nat Struct Biol* **8:** 656–659.

Pavlov YI, Frahm C, Nick McElhinny SA, Niimi A, Suzuki M, Kunkel TA. 2006. Evidence that errors made by DNA polymerase α are corrected by DNA polymerase δ. *Curr Biol* **16:** 202–207.

Pursell ZF, Isoz I, Lundstrom EB, Johansson E, Kunkel TA. 2007. Yeast DNA polymerase ε participates in leading-strand DNA replication. *Science* **317:** 127–130.

Reijns MA, Rabe B, Rigby RE, Mill P, Astell KR, Lettice LA, Boyle S, Leitch A, Keighren M, Kilanowski F, et al. 2012. Enzymatic removal of ribonucleotides from DNA is essential for mammalian genome integrity and development. *Cell* **149:** 1008–1022.

Reynolds N, Watt A, Fantes PA, MacNeill SA. 1998. Cdm1, the smallest subunit of DNA polymerase d in the fission yeast *Schizosaccharomyces pombe*, is non-essential for growth and division. *Curr Genet* **34:** 250–258.

Robinson A, Brzoska AJ, Turner KM, Withers R, Harry EJ, Lewis PJ, Dixon NE. 2010. Essential biological processes of an emerging pathogen: DNA replication, transcription and cell division in *Acinetobacter* spp. *Microbiol Mol Biol Rev* **74:** 273–297.

Robinson A, Causer RJ, Dixon NE. 2012. Architecture and conservation of the bacterial DNA replication machinery, an underexploited drug target. *Curr Drug Targets* **13:** 352–372.

Rodriguez AC, Park HW, Mao C, Beese LS. 2000. Crystal structure of a pol α family DNA polymerase from the hyperthermophilic archaeon *Thermococcus sp.* 9° N-7. *J Mol Biol* **299:** 447–462.

Sanders GM, Dallmann HG, McHenry CS. 2010. Reconstitution of the *B. subtilis* replisome with 13 proteins including two distinct replicases. *Mol Cell* **37:** 273–281.

Shcherbakova PV, Pavlov YI, Chilkova O, Rogozin IB, Johansson E, Kunkel TA. 2003. Unique error signature of the four-subunit yeast DNA polymerase ε. *J Biol Chem* **278:** 43770–43780.

Shuttleworth G, Fogg MJ, Kurpiewski MR, Jen-Jacobson L, Connolly BA. 2004. Recognition of the pro-mutagenic base uracil by family B DNA polymerases from archaea. *J Mol Biol* **337:** 621–634.

Smith DJ, Whitehouse I. 2012. Intrinsic coupling of lagging-strand synthesis to chromatin assembly. *Nature* **483:** 434–438.

Stano NM, Chen J, McHenry CS. 2006. A coproofreading Zn^{2+}-dependent exonuclease within a bacterial replicase. *Nat Struct Mol Biol* **13:** 458–459.

Steitz TA. 1993. DNA- and RNA-dependent DNA polymerases. *Curr Opin Struct Biol* **3:** 31–38.

Stocki SA, Nonay RL, Reha-Krantz LJ. 1995. Dynamics of bacteriophage T4 DNA polymerase function: Identification of amino acid residues that affect switching between polymerase and 3′ → 5′ exonuclease activities. *J Mol Biol* **254:** 15–28.

Swan MK, Johnson RE, Prakash L, Prakash S, Aggarwal AK. 2009. Structural basis of high-fidelity DNA synthesis by yeast DNA polymerase δ. *Nat Struct Mol Biol* **16:** 979–986.

* Tanaka S, Araki H. 2013. Helicase activation and establishment of replication forks at chromosomal origins of replication. *Cold Spring Harb Perspect Biol* doi: 10.1101/cshperspect.a010371.

Tori K, Kimizu M, Ishino S, Ishino Y. 2007. DNA polymerase B1 and D from the hyperthermophilic archaeon *Pyrococcus furiosus* both bind proliferating cell nuclear antigen with their C-terminal PIP-box motifs. *J Bacteriol* **189:** 5652–5657.

Tsubota T, Tajima R, Ode K, Kubota H, Fukuhara N, Kawabata T, Maki S, Maki H. 2006. Double-stranded DNA binding, an unusual property of DNA polymerase ε, promotes epigenetic silencing in *Saccharomyces cerevisiae*. *J Biol Chem* **281:** 32898–32908.

Warbrick E. 1998. PCNA binding through a conserved motif. *BioEssays* **20:** 195–199.

Wardle J, Burgers PM, Cann IK, Darley K, Heslop P, Johansson E, Lin LJ, McGlynn P, Sanvoisin J, Stith CM, et al. 2008. Uracil recognition by replicative DNA polymerases is limited to the archaea, not occurring with bacteria and eukarya. *Nucleic Acids Res* **36:** 705–711.

Watt DL, Johansson E, Burgers PM, Kunkel TA. 2011. Replication of ribonucleotide-containing DNA templates by yeast replicative polymerases. *DNA Repair (Amst)* **10:** 897–902.

Wieczorek A, McHenry CS. 2006. The NH2-terminal php domain of the α subunit of the *Escherichia coli* replicase binds the ε proofreading subunit. *J Biol Chem* **281:** 12561–12567.

Wijffels G, Johnson WM, Oakley AJ, Turner K, Epa VC, Briscoe SJ, Polley M, Liepa A, Hofmann A, Buchardt J, et al. 2011. Binding inhibitors of the bacterial sliding clamp by design. *J Med Chem* **54:** 4831–4838.

Wing RA, Bailey S, Steitz TA. 2008. Insights into the replisome from the structure of a ternary complex of the DNA polymerase III α-subunit. *J Mol Biol* **382:** 859–869.

Wu P, Nossal N, Benkovic SJ. 1998. Kinetic characterization of a bacteriophage T4 antimutator DNA polymerase. *Biochemistry* **37:** 14748–14755.

Yang G, Lin T, Karam J, Konigsberg WH. 1999. Steady-state kinetic characterization of RB69 DNA polymerase mutants that affect dNTP incorporation. *Biochemistry* **38:** 8094–8101.

Zhao Y, Jeruzalmi D, Moarefi I, Leighton L, Lasken R, Kuriyan J. 1999. Crystal structure of an archaebacterial DNA polymerase. *Structure* **7:** 1189–1199.

Translesion DNA Polymerases

Myron F. Goodman[1] and Roger Woodgate[2]

[1]Department of Biological Sciences and Department of Chemistry, University of Southern California, University Park, Los Angeles, California 90089-2910

[2]Laboratory of Genomic Integrity, National Institute of Child Health and Human Development, National Institutes of Health, Bethesda, Maryland 20892-3371

Correspondence: woodgate@nih.gov

Living cells are continually exposed to DNA-damaging agents that threaten their genomic integrity. Although DNA repair processes rapidly target the damaged DNA for repair, some lesions nevertheless persist and block genome duplication by the cell's replicase. To avoid the deleterious consequence of a stalled replication fork, cells use specialized polymerases to traverse the damage. This process, termed "translesion DNA synthesis" (TLS), affords the cell additional time to repair the damage before the replicase returns to complete genome duplication. In many cases, this damage-tolerance mechanism is error-prone, and cell survival is often associated with an increased risk of mutagenesis and carcinogenesis. Despite being tightly regulated by a variety of transcriptional and posttranslational controls, the low-fidelity TLS polymerases also gain access to undamaged DNA where their inaccurate synthesis may actually be beneficial for genetic diversity and evolutionary fitness.

The Watson–Crick (WC) DNA structure was published in 1953 (Watson and Crick 1953). Coincidentally, it was in 1953 when Weigle discovered that λ bacteriophage that had been killed by exposure to UV radiation could, in essence, be brought back to life by irradiation of its *Escherichia coli* host (Weigle 1953). The reactivation of λ, however, was accompanied by a sizable increase in phage mutagenesis (Weigle 1953). With the benefit of nearly five decades of hindsight, the increase in phage survival and mutagenesis observed by Weigle turns out to have been caused by the action of translesion DNA synthesis (TLS) polymerases. Currently, there are three TLS polymerases (Pols) in *E. coli*, and many of the 15 known polymerases in eukaryotes have the capacity to promote some degree of TLS (Table 1) (Goodman 2002; Bebenek and Kunkel 2004; Waters et al. 2009; Lange et al. 2011; Sale et al. 2012).

In this review, we revisit the concepts and experiments in *E. coli* centered on the induction and regulation of mutagenesis that began in the mid 1970s, continuing throughout the 1980s, and culminating with the identification of a new family (Y-family) of error-prone DNA polymerases (Ohmori et al. 2001). We review early TLS models based principally on genetic data from *E. coli* that were generated in the 1980s and discuss recent biochemical data leading to current models of TLS in *E. coli* and humans. The typically poor fidelity of TLS polymerases

Cite this article as *Cold Spring Harb Perspect Biol* doi: 10.1101/cshperspect.a010363

Table 1. Properties of key TLS polymerases across all domains of life

Organism	Gene name	Protein name	Characteristics[a]
E. coli	*polB*	Pol II	B-family polymerase Regulated by LexA; damage inducible Creates −2 frameshift mutations during TLS of N^2-dG-AAF adducts
E. coli	*dinB*	Pol IV	Regulated by LexA; damage inducible Prone to making −1 frameshifts when overexpressed in vivo Bypasses N^2-dG adducts efficiently and accurately Involved in TLS of alkylation damage in vivo Major polymerase involved in stress-induced mutagenesis
E. coli	*umuDC*	Pol V	Regulated by LexA; damage inducible Major TLS polymerase in *E. coli* Composed of a heterotrimer of $UmuD'_2$ (∼24 kDa) and UmuC (∼48 kDa) to form an ∼72-kDa $UmuD'_2C$ complex Interacts with RecA and ATP to form Pol V Mut
Sulfolobus solfataricus	*dpo4*	Dpo4	Archaeal ortholog of *E. coli* Pol IV Numerous crystal structures of Dpo4 in the process of TLS have been solved
Sulfolobus acidocaldarius	*dbh*	Dbh	First archaeal DinB homology identified Less processive than Dpo4
Saccharomyces cerevisiae	*REV3*	Rev3p	B-family polymerase Together with Rev7, Pol31 and Pol32 forms Pol ζ_4 Can bypass lesions unassisted Works with other TLS polymerases to facilitate the extension step of TLS Stimulated by an interaction with Rev1
S. cerevisiae	*REV1*	Rev1p	Specifically incorporates dCMP opposite abasic sites and undamaged template dG Interacts with Pol ζ to stimulate Pol ζ–dependent TLS in vivo
S. cerevisiae	*RAD30*	Pol η	Bypasses a thymine–thymine CPD relatively accurately and efficiently
Homo sapiens	*REV3L*	REV3	B-family polymerase Catalytic subunit of human Pol ζ Very large protein consisting of 3130 amino acids Murine homozygous knockout is embryonic lethal
H. sapiens	*REV1*	REV1	Similar to *S. cerevisiae* Rev1 protein, it specifically incorporates dCMP opposite dG and abasic sites Acts as scaffold protein that interacts with TLS polymerases ζ, η, ι, and κ Generates mutations at G-C base pairs during immunoglobulin gene somatic hypermutation
H. sapiens	*POLH/ XPV/ RAD30A*	Pol η	Bypasses a thymine–thymine CPD relatively efficiently and accurately Defects lead to the sunlight-sensitive and cancer-prone xeroderma pigmentosum variant (XP-V) phenotype Regulated by ubiquitination and phosphorylation Generates mutations at A-T base pairs during immunoglobulin gene somatic hypermutation

Continued

Table 1. *Continued*

Organism	Gene name	Protein name	Characteristics[a]
H. sapiens	*POLI/RAD30B*	Pol ι	Has unique replication fidelity; incorporates opposite template dA reasonably accurately, but opposite template dT in a highly error-prone manner In vivo functions remain unknown
H. sapiens	*POLK/DINB1*	Pol κ	Prone to making −1 frameshift mutations, but can accurately and efficiently bypass a number of N^2-dG lesions Plays additional roles in repair synthesis steps of nucleotide excision repair

[a]Unless specified, the TLS polymerase belongs to the Y-family of DNA polymerases (Ohmori et al. 2001).

can now be understood thanks to high-resolution X-ray structures that indicate a more spacious active cleft that can accommodate a variety of non-WC base pairs. Owing to their potential to cause excessive numbers of deleterious mutations compared with replication polymerases, we describe how TLS Pol expression is tightly regulated, at both the transcriptional and posttranslational levels. However, owing to the need for generating beneficial mutations, we also review recent data showing that TLS Pols are involved in providing evolutionary fitness in bacteria and in generating immunological diversity in higher vertebrates. Recognizing that the cellular function(s) of many of the human TLS Pols remain unknown, it may turn out that lesion bypass is but one important property of TLS Pols, and perhaps not even their most salient one.

DNA DAMAGE-INDUCED REGULATION OF GENE EXPRESSION IN *E. coli*

Weigle's 1953 phage reactivation experiment suggested the possibility of an induced cellular DNA repair process. The presence of a bacterial repressor system that could be inactivated by excessive DNA damage was proposed by Witkin in 1967 (Witkin 1967) as a molecular model to explain UV-induced phage reactivation. The pivotal connection to mutagenesis was made by Radman, who proposed the SOS model for an inducible error-prone mechanism in *E. coli* for the repair of DNA (Radman 1974). The conceptual framework provided by Witkin and Radman paved the way for several decades of genetic and biochemical studies aimed at elucidating the molecular basis of "SOS error-prone repair" (Witkin 1976; Walker 1984; Friedberg et al. 2006; Schlacher and Goodman 2007).

The damage-inducible SOS regulon requires interplay of the LexA and RecA proteins. The *lexA* gene encodes the LexA transcriptional repressor, which binds to sequences in the operator region of the >40 genes under its control (Fernández de Henestrosa et al. 2000; Courcelle et al. 2001). There is a minimal 6-bp operator consensus sequence, but it is the nonconsensus surrounding bases that determine repressor-operator binding constants, so that the earliest expressed genes are those whose operators are bound weakly by LexA (Lewis et al. 1994; Fernández de Henestrosa et al. 2000). It is estimated that there are roughly 8000 RecA molecules in an undamaged cell (Boudsocq et al. 1997). When the cellular DNA is damaged, RecA assembles as a nucleoprotein filament on single-stranded DNA (ssDNA), requiring the presence of ATP (Kuzminov 1999). This nucleoprotein filament, commonly referred to as RecA*, acts as a coprotease to mediate self-cleavage of LexA molecules in solution (Little et al. 1980). As the overall cellular concentration of intact LexA diminishes, only those genes that bind LexA tightly remain repressed.

The earliest genes to be induced after DNA damage are those involved in error-free repair, such as nucleotide excision repair (*uvrA*, *uvrB*, *uvrD*, and *cho*), as well as recombinational repair (*recA*, *recN*). *E. coli*'s three TLS polymerases are also regulated by LexA. The operators for Pol II (*polB*) and Pol IV (*dinB*) bind LexA weakly (Fernández de Henestrosa et al. 2000) and are induced early in the SOS response. In contrast, the *umuDC* operon (encoding Pol V) has one of the tightest LexA-binding sites and is induced late in the response, 30–40 min after DNA damage (Sommer et al. 1993). Given their early induction and relatively high basal expression levels in the absence of DNA damage, it is thought that Pol II and Pol IV are likely to participate in mostly error-free TLS of specific DNA lesions. In contrast, Pol V can traverse a wide range of DNA lesions and does so in a much more error-prone manner. Its induction late in the SOS response therefore suggests that *E. coli* only uses Pol V as a last resort, once all other error-free repair pathways have been exhausted (Sommer et al. 1993).

E. coli AS A MODEL SYSTEM FOR TLS AND MUTAGENESIS

By the mid-1970s, early 1980s, genetic experiments had led to the identification of the principal players involved in damage-induced mutagenesis, *lexA*, *recA*, along with *umuD* and *umuC*. The same mechanism used by RecA* to mediate LexA cleavage (Little et al. 1980) is also used to cleave UmuD to form UmuD′ (Burckhardt et al. 1988; Shinagawa et al. 1988) and is an absolute requirement for SOS mutagenesis (Nohmi et al. 1988). By isolating RecA mutants that were UV non-mutable, despite retaining the ability to cleave LexA and UmuD, Devoret and colleagues showed that RecA* also had a separate *direct* role in SOS mutagenesis (Dutreix et al. 1989). This mutagenic role for RecA* remained a mystery until 2009 (Jiang et al. 2009); its biochemical mechanism is discussed below.

Early TLS Models

The first model to address the function of the Umu proteins during UV-induced TLS was proposed by Bridges and Woodgate in 1985 (Bridges and Woodgate 1985a,b). According to this model, TLS occurred in a two-step process in which Pol III incorporates a nucleotide opposite the first (3′) T of a T-T CPD, requiring the presence of RecA protein bound to the template proximal to the lesion. Pol III would subsequently interact with the UmuDC proteins to incorporate another nucleotide at the second (5′) T of the CPD (Fig. 1A). At least one of the two incorporations would be non-WC, thereby causing a mutation targeted at the site of the CPD.

A model proposed by Echols and Goodman in 1990 (Echols and Goodman 1990) envisioned complete blockage of the Pol III holoenzyme (Pol III core, β-sliding clamp, γ-clamp-loading complex) when encountering a template lesion, followed by the assembly of a damage-localized nucleoprotein complex involving RecA, UmuC, UmuD′, SSB, and Pol III holoenzyme—a "mutasome"—to copy past a template lesion (Woodgate et al. 1989; Echols and Goodman 1991). This model took into account the finding that RecA* facilitates cleavage of UmuD to a mutagenically active form, UmuD′ (Burckhardt et al. 1988; Nohmi et al. 1988; Shinagawa et al. 1988). Subsequently, it was shown that it was actually a dimeric $UmuD_2$ that is cleaved to $UmuD'_2$ and that subsequently interacts with UmuC to form a stable complex of $UmuD'_2C$ (Woodgate et al. 1989; Bruck et al. 1996), identified in 1999 as Pol V (Tang et al. 1999). Whereas a Pol III replisome performs rapid and processive "error-free" genome replication on undamaged DNA (Johnson and O'Donnell 2005; McHenry 2011), a mutasome would perform slow, poorly processive, "error-prone" TLS.

Biochemical Characterization of the *E. coli* Umu Proteins and Evolution of TLS Models

At the time of the mutasome model, UmuD′ could be obtained in high purity and yield; not so for UmuC, which formed insoluble inclusion bodies in *E. coli* cell lysates. Woodgate, working with Echols, denatured and then renatured UmuC to obtain soluble UmuC that was included in a biochemical reconstitution assay along with UmuD′, SSB, β-sliding clamp, γ-clamp loader, and Pol III core. Using this combination of proteins to copy a template DNA containing an abasic moiety, a small

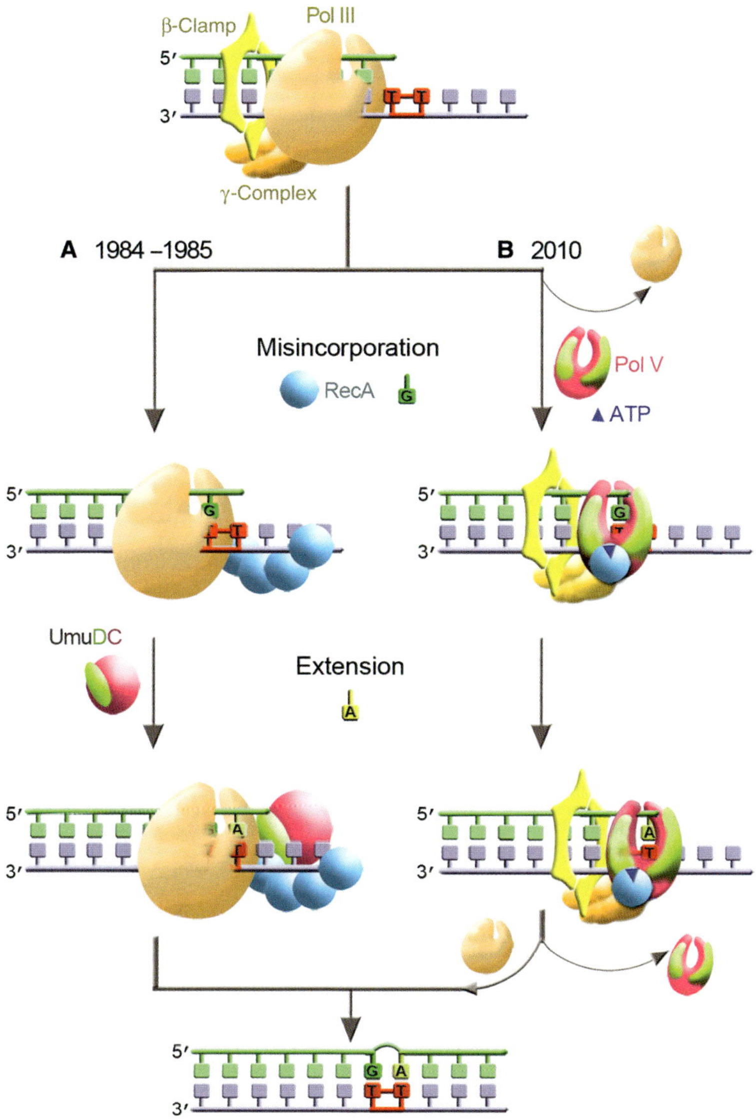

Figure 1. Two-step model for UV mutagenesis in *E. coli.* (*A*) The two-step model for UV mutagenesis proposed in 1984/1985 assumed that TLS in vivo is performed by the replicative Pol III. The first step, nucleotide misincorporation opposite a 3′ T of a CPD, was hypothesized to be mediated by the RecA protein (represented as blue sphere). The misincorporated base was subsequently fixed as a mutation in a second extension/bypass step that was dependent on UmuC (purple) and UmuD (lime) (Bridges and Woodgate 1984, 1985a,b). (*B*) Subsequent studies revealed that rather than being accessory factors of Pol III, the products of *umuDC* genes encode a bona fide DNA polymerase, Pol V (shown as purple UmuC and two lime UmuD′ subunits assembled in the shape of a right hand) (Reuven et al. 1999; Tang et al. 1999), which interacts with RecA* in *trans* (not shown) from which molecules of RecA (blue sphere) and ATP (dark blue triangle) are transferred from the 3′-filament tip to generate Pol V Mut (Jiang et al. 2009). Pol V Mut can perform both the (mis)insertion and extension steps of TLS. After traversing the damaged DNA, Pol V Mut is replaced by Pol III holenzyme, which resumes high-fidelity chromosomal duplication.

amount of primer was extended past the lesion (Rajagopalan et al. 1992). The difficulties with obtaining sufficient quantities of UmuC for detailed biochemical characterization were circumvented in 1996, by the isolation and purification of the $UmuD'_2C$ complex (from 200 L of *E. coli*!) (Bruck et al. 1996). When $UmuD'_2C$ was used in replication assays, robust TLS was observed. Remarkably, TLS also occurred in the *absence* of Pol III core (Tang et al. 1998). It appeared, therefore, that $UmuD'_2C$ by itself had the ability to copy undamaged and damaged DNA templates, and, about a year later, $UmuD'_2C$ was conclusively shown to be a DNA polymerase, Pol V (Reuven et al. 1999; Tang et al. 1999).

A key feature of the mutasome model is the assembly of RecA* on ssDNA proximal to the lesion (Fig. 1A). When a replication fork encounters a lesion, an uncoupling of leading- and lagging-strand synthesis may ensue. Then, one of the TLS Pols can replace Pol III on the β-clamp and copy the damaged DNA. If the lesion were present on the leading strand, then DNA unwinding by the DnaB helicase downstream from the lesion would leave a region of ssDNA on which a RecA* filament could form. In the case of a lagging-strand lesion, ssDNA is likely to be present as a consequence of Okazaki fragment synthesis and processing. Thus, for both leading- and lagging-strand lesions, one can easily envision how RecA* could assemble in *cis*, that is, on the template strand being copied, proximal to the lesion. However, the presence of RecA bound in *cis* to the template blocks DNA synthesis and would assuredly have to be removed to permit synthesis to occur beyond the lesion (Pham et al. 2001). Because both RecA* assembly and disassembly normally occur in a $5' \rightarrow 3'$ direction (Kuzminov 1999), removal of a blocking RecA* ahead of an advancing Pol V would entail filament disassembly in an "abnormal" $3' \rightarrow 5'$ direction (Pham et al. 2001).

The problematic issue of RecA* formation and dissolution on the template strand would be irrelevant if RecA* were absent from the DNA template, but rather acted in *trans* by forming on a non-template ssDNA strand (Schlacher and Goodman 2007; Patel et al. 2010). And that appears to be what is happening. Transactivation of Pol V by RecA* to perform TLS has been shown to take place in vitro (Schlacher et al. 2006; Jiang et al. 2009), giving rise to a Pol V mutasome model for TLS (Fig. 1B). The unanticipated role for RecA* in the transactivation of Pol V provided the key to establishing its biochemical function. By incubating Pol V with RecA* bound to resin and then spinning down the resin-bound RecA*, what remained in solution along with "unreacted" Pol V was a new form of Pol V, named Pol V Mut=$UmuD'_2C$-RecA-ATP (Jiang et al. 2009). Once formed, Pol V Mut copies undamaged and damaged DNA (i.e., performs TLS) in the *absence* of RecA*.

Thus, the direct role of RecA* in SOS-mediated TLS, described by Devoret and colleagues in 1989 (Dutreix et al. 1989), is to transfer a molecule of RecA from its 3′-filament tip along with a molecule of ATP to convert $UmuD'_2C$ into Pol V Mut, which can traverse a variety of DNA lesions unassisted (Fig. 1B). Pol V Mut has the unique property that it exists in two conformational states, either activated (able to copy DNA) or deactivated (unable to copy DNA), perhaps depending on where RecA-ATP is bound in relation to the polymerase subunits $UmuD'_2$ and UmuC (Jiang et al. 2009). A deactivated Pol V Mut is completely reactivated when re-exposed to RecA*, which is used to replace the old RecA-ATP with new RecA-ATP from the 3′-filament tip (Jiang et al. 2009). This type of repeating on–off switching has not been observed for any other DNA polymerase, and perhaps provides a way for *E. coli* to protect undamaged DNA from gratuitous mutation, enabling the cell to activate Pol V Mut whenever replication forks have stalled at DNA template damage sites.

POL V ORTHOLOGS

E. coli Pol V has been the subject of intense study, but many orthologs are now known to exist in a wide range of prokaryotes and self-transmissible plasmids that they harbor (Vaisman et al. 2012). Of particular interest are Pol V orthologs MucA′B and RumA′B found on the IncN R-plasmid R46/pKM101 (Perry and Walker 1982) and integrating conjugative element

(ICE) R391 (Kulaeva et al. 1995), respectively. The ability of MucA′B to increase mutagenesis in strains of *Salmonella typhimurium* exposed to a variety of DNA-damaging agents plays an important role in increasing the efficacy of the "Ames test" that was proposed as a possible way to identify human carcinogens (McCann et al. 1975). Although the MucA′B proteins appear to be significantly more mutagenic than their *E. coli* counterparts, the RumA′B proteins are the most potent at promoting cellular mutagenesis than any other Pol V ortholog identified so far (Mead et al. 2007). This is especially worrisome, because the *rum* genes are found on natural isolates of the STX/R391 ICE family of mobile elements that are involved in antibiotic resistance spread among γ-proteobacteria, including virulent strains of *Vibrio cholera* and *Proteus mirabilis*, and their ability to provide genetic diversification to the host bacterium may pose a significant risk to human health (Beaber et al. 2004).

E. coli TLS POL II AND POL IV

Special mention should be made regarding Pol II in relation to SOS and TLS. The discovery in 1970 of Pol II (Knippers 1970) in the De Lucia and Cairns *polA* mutant (De Lucia and Cairns 1969) that had a severely compromised Pol I protein was followed shortly thereafter by the discovery of Pol III (Gefter et al. 1971). The isolation of a conditionally lethal temperature-sensitive Pol III mutant showed that Pol III was absolutely required for *E. coli* survival (Gefter et al. 1971), acting as the principal replicative polymerase responsible for copying almost the entire bacterial genome (Johnson and O'Donnell 2005; McHenry 2011).

Mutations in Pol II, however, proved to be non-informative, at least not until relatively recently (Foster et al. 1995; Rangarajan et al. 1997; Frisch et al. 2010). Pol II was "rediscovered" in 1988, as a polymerase activity isolated from UV-irradiated cells that could replicate past abasic template lesions (Bonner et al. 1988). This activity was induced sevenfold in response to UV radiation that depended on expression of the *lexA* gene. It became evident upon purification that the induced lesion-copying protein was Pol II (Bonner et al. 1988) and that the DNA damage-inducible *dinA* gene identified in 1980 by Kenyon and Walker (Kenyon and Walker 1980) encodes Pol II (Bonner et al. 1990; Iwasaki et al. 1990) and is allelic with the *polB* gene that had been identified in 1970 (Qiu and Goodman 1997). In addition to bypassing abasic sites, Pol II appears adept at bypassing N^2-deoxyguanosine-acetyl aminofluorene (AAF) adducts in an error-prone manner that produces −2 frameshift mutations (Table 1) (Fuchs and Fujii 2007).

The *dinB* gene was identified by Kenyon and Walker in 1980 as being induced as part of the cellular SOS response to DNA damage (Kenyon and Walker 1980). However, its function remained largely unknown for many years aside from a paper in 1986 implicating a role in producing "untargeted" mutagenesis of bacteriophage λ (Brotcorne-Lannoye and Maenhaut-Michel 1986). Some 15 years after the *dinB* locus was first identified, Ohmori et al. (1995) discovered a damage-inducible gene, *dinP*, in *E. coli* that mapped to the same general region of the *E. coli* chromosome as *dinB*. A short time thereafter, the *dinB/P* genes were also shown to be allelic (Kim et al. 1997) and encode the Y-family DNA Pol IV (Wagner et al. 1999). DinB orthologs are the most ubiquitous of Y-family polymerases, because they are found in all domains of life (Ohmori et al. 2001). However, like many other TLS polymerases, they are not absolutely essential for life, and some organisms (such as *Saccharomyces cerevisiae*) lack any obvious *dinB* ortholog. DinB's role in TLS was largely overlooked because it is not highly mutagenic (unlike Pol V). Indeed, Pol IV appears to bypass certain N^2-dG adducts (such as N^2-furfuryl-dG) in an error-free manner (Table 1) (Jarosz et al. 2006) and can even copy past N^2-N^2-guanine interstrand cross-links with high fidelity (Kumari et al. 2008).

TLS POLYMERASES IN ARCHAEA AND EUKARYOTES

Ever since the discovery of "reversionless" (*REV*) strains of *S. cerevisiae* (Lemontt 1971) and non-mutable (*umu/uvm*) strains of *E. coli* (Kato and Shinoura 1977; Steinborn 1978) in the 1970s, it was known that cellular mutagenesis is not a passive process but instead required the active participation of certain so-called mutagenesis proteins, which we now know are bona fide DNA polymerases. Over the years, the number of mutagenesis proteins discovered increased steadily, and with the advent of whole-genome sequencing projects, orthologs were identified at an ever-increasing pace. By the mid 1990s, orthologs of the Rev1, Rev3, UmuC, and DinB proteins had been identified in bacteria, archaea, and eukaryotes.

The first archaeal ortholog was reported in the literature in 1996, as a DinB homolog (Dbh) from what was originally believed at the time to be *Sulfolobus solfataricus* (Table 1) (Kulaeva et al. 1996). However, analysis of the sequenced *S. solfataricus* genome (She et al. 2001) identified a different DinB ortholog (Dpo4), which is only 53% identical to the Dbh protein (Boudsocq et al. 2001). Subsequent studies suggest that the original *dbh* gene appears to be from the closely related organism *Sulfolobus acidocaldarius* instead of *S. solfataricus* (Chen et al. 2005; McDonald et al. 2006). Both archaeal Dbh and Dpo4 proteins have proven extremely useful for subsequent structure–function studies into TLS (Table 1) (Ling et al. 2001, 2003, 2004a,b; Silvian et al. 2001; Zhou et al. 2001; Boudsocq et al. 2004; Yang and Woodgate 2007; Wilson and Pata 2008; Pata 2010) (see below).

S. cerevisiae RAD30 was identified in 1997, as an ortholog of *E. coli* UmuC and DinB proteins that is involved in error-free repair of UV damage (McDonald et al. 1997; Roush et al. 1998). In 1999, the *RAD30* gene was subsequently shown to encode Pol η, which can bypass a T-T CPD efficiently and with the same accuracy as undamaged DNA (Johnson et al. 1999b). A human ortholog of *S. cerevisiae* Pol η was also identified in 1999 and shown to possess similar properties to the yeast enzyme (Masutani et al. 1999a; Johnson et al. 2000b; Matsuda et al. 2001), but perhaps more importantly, was found to be defective in humans with the sunlight-sensitive and cancer-prone xeroderma pigmentosum variant (XP-V) syndrome (Johnson et al. 1999a; Masutani et al. 1999a,b). A second Rad30 ortholog (McDonald et al. 1999) and *E. coli* DinB ortholog (Gerlach et al. 1999; Ogi et al. 1999) were identified in humans and mice and were subsequently shown to encode DNA polymerases ι and κ, respectively (Johnson et al. 2000a; Ohashi et al. 2000; Tissier et al. 2000; Gerlach et al. 2001; Zhang et al. 2001). Thus, within a period of a roughly 18 mo, a significant number of the phylogenetically related "mutagenesis proteins," which were once thought to be merely accessory factors to replicative polymerases, evolved into the "Y-family" of DNA polymerases (Ohmori et al. 2001) that can perform unassisted TLS.

To date, only defects in Pol η have been associated with a human disease (XP-V), and this phenotype has been recapitulated in mouse models (Dumstorf et al. 2006; Ohkumo et al. 2006). A loss of Pol η function is also responsible for the absence of mutations at A-T sites in immunoglobulin variable regions during somatic hypermutation (Rogozin et al. 2001; Zeng et al. 2001). Thus, Pol η is responsible for copying damaged DNA (i.e., pyrimidine dimers) accurately, in the avoidance of skin cancer, and for copying undamaged DNA (i.e., Ig V-gene repair gaps) inaccurately, to generate mutations that contribute to antibody diversity (Peled et al. 2008). Mice with defects in Pol η, Pol ι, or Rev1 are viable but do not show any gross phenotypic changes (Ogi et al. 2002; Schenten et al. 2002; McDonald et al. 2003; Jansen et al. 2006). In contrast, homozygous inactivation of murine Rev3 (catalytic subunit of Pol ζ) is embryonic lethal, suggesting that Pol ζ plays additional cellular roles besides TLS (Bemark et al. 2000; Esposito et al. 2000; Wittschieben et al. 2000, 2006). Distinct roles for Pol ζ is supported by the recent discovery of a four-subunit complex, Pol ζ_4 (Rev3-Rev7-Pol31-Pol32) (Baranovskiy et al. 2012; Johnson et al. 2012; Makarova et al. 2012), where the binding of the Pol31 and Pol32 accessory subunits of Pol δ are required for damaged-induced mutagenesis in yeast (Johnson et al. 2012; Makarova et al. 2012).

The molecular mechanism of TLS appears to be conserved from *E. coli* to humans (Fig. 2). Similar to the original TLS model proposed by

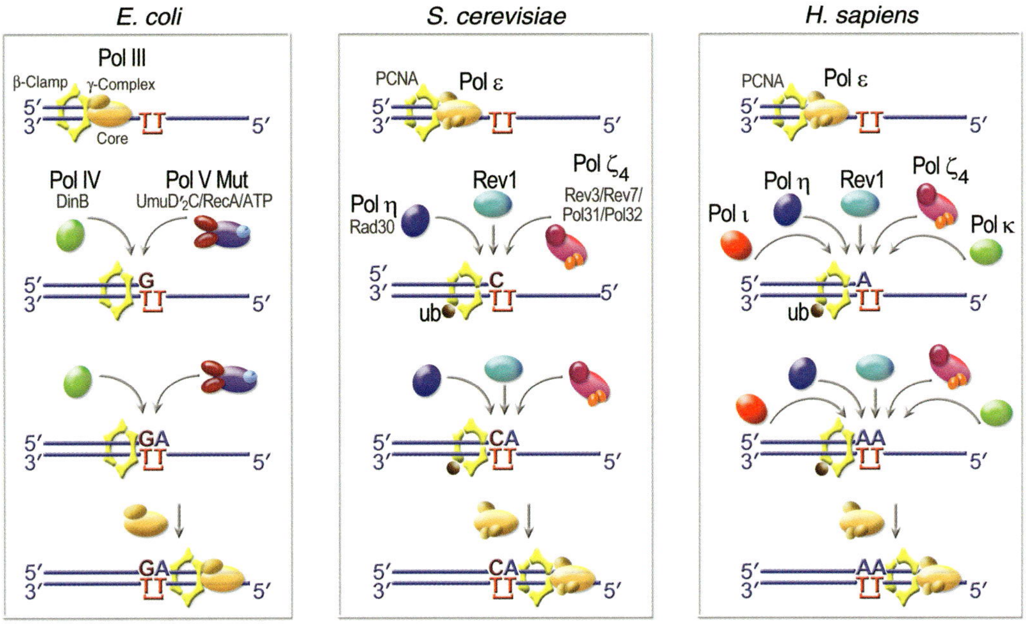

Figure 2. Evolutionarily conserved roles of TLS polymerases. Replicative polymerases such as *E. coli* Pol III or eukaryotic Pol ε stall at the site of a DNA lesion. TLS polymerases are recruited to the site via interactions with the replicative sliding processivity clamp (β-subunit in *E. coli* and PCNA in eukaryotes). *E. coli* has a choice of three TLS polymerases: Pol II, Pol IV, and Pol V Mut (composed of UmuD$'_2$C-RecA-ATP) (Jiang et al. 2009). *S. cerevisiae* also has a choice of three TLS polymerases: Pol ζ_4, Rev1, and Pol η. In humans, at least five TLS polymerases can be recruited to sites of arrested replication, including Pol η, Pol ι, Pol κ, Rev1, and Pol ζ_4. Additional human DNA polymerases, such as Pol β, Pol λ, Pol θ, and/or Pol ν (Maga et al. 2007; Seki and Wood 2008; Shtygasheva et al. 2008; Yamanaka et al. 2010; Hogg et al. 2011; Villani et al. 2011), may also facilitate TLS under certain conditions, but these have been omitted for clarity. The likelihood that this insertion step is error-prone, as shown for *E. coli* and *S. cerevisiae*, or error-free, as shown for humans, will depend on the DNA lesion encountered and the polymerase used for TLS. The extension step may be facilitated by the same enzyme that performed the (mis)insertion or by a different polymerase. Once the nascent DNA chain has been extended beyond the lesion, the TLS polymerase is replaced by the cell's replicative DNA polymerase so as to complete genome duplication.

Bridges and Woodgate in 1985, it is generally thought to occur in two kinetically separable steps: misincorporation followed by extension. In principle, a single TLS enzyme (such as *E. coli* Pol V or human Pol η) can facilitate both steps. However, it is also evident that some TLS polymerases are more proficient at extension (such as Pol η and Pol ζ) than incorporation; thus, TLS not only occurs in two steps but may also require two polymerases for complete lesion bypass to occur (Fig. 2). Support for the proposed polymerase switching model comes from biochemical studies that reveal a dynamic exchange process in which *E. coli* Pol IV (Indiani et al. 2005, 2009; Furukohri et al. 2008) and Pol II (Indiani et al. 2009) can replace Pol III on the β-sliding clamp at a stalled replication fork, with a subsequent resumption of rapid DNA synthesis following the displacement of Pol IV and Pol II by Pol III.

STRUCTURAL INSIGHTS INTO TLS AND MUTAGENIC SPECIFICITY

Probably the biggest surprise to researchers studying TLS was the fact that the mutagenesis proteins turned out to be bona fide (Y-family) DNA polymerases, because the primary amino acid sequence of the proteins show virtually no sequence homology to DNA polymerases from other families. However, quite remarkably, crystallographical studies of archaeal Dbh and Dpo4 revealed a topology similar to replicases, including domains likened to a right-hand with palm, thumb, and fingers subdomains (Ling et al. 2001; Silvian et al. 2001; Zhou et al. 2001). The TLS polymerases also have an additional domain that has been called the little finger (to coincide with the analogy to a hand) (Ling et al. 2001) or polymerase-associated domain (Fig. 3A,B) (Trincao et al. 2001). As with high-fidelity replicases, residues required for catalysis are located in the palm domain of the Y-family polymerases. In contrast to the replicases, the remaining domains are much smaller and are stubby.

As a consequence, the active site of the Y-family TLS polymerases is much more spacious

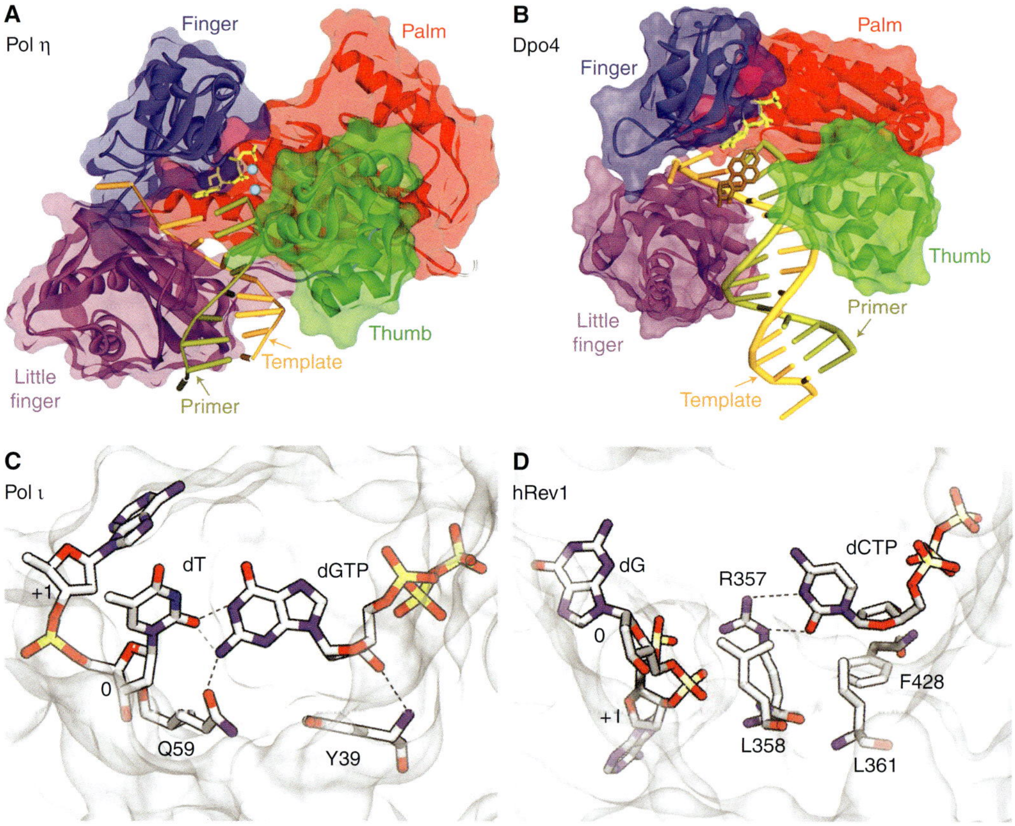

Figure 3. Structural insights into TLS polymerases and their mutagenic specificity. In panels *A* and *B*, the main domains are color-coded: (red) palm; (green) thumb; (blue) fingers; (purple) little finger. (*A*) Crystal structure of human Pol η in a ternary complex with a CPD. In this view, the 3′T of the CPD is in the active site and is correctly paired with incoming dATP (PDB: 3MR3) (Biertümpfel et al. 2010). (Rust) The template strand; (olive green) the primer; (yellow) the incoming dNTP. (Burgundy stick) The position of the CPD; (small blue spheres) the metal ions. The protein backbone is represented by the ribbon surrounded by the semitransparent solvent-accessible surface. (*B*) Crystal structure of the *S. solfataricus* Dpo4 in a ternary complex with DNA containing a benzo[*a*]pyrene lesion (brown) and incoming nucleotide (yellow) (PDB: 1S0M_BP-2) (Ling et al. 2004b). As can be seen, the benzo[*a*]pyrene lesion is flipped into the major groove, so as to accommodate base pairing. (*C*) Human Pol ι making a G:T mispair (PDB:3GV8); note that the template dT and incoming dG are both in an *anti* conformation (Kirouac and Ling 2009) and the mispair is stabilized through hydrogen bonds with Gln59. (*D*) Arg357-directed dC incorporation by human Rev1 (PDB: 3GQC) (Swan et al. 2009). (*C,D*) (White/gray) The protein surface; (dotted lines) hydrogen bonds.

and solvent exposed (Ling et al. 2001), and it can often accommodate large bulky DNA lesions that are unable to fit into the tighter constraints of the high-fidelity replicases. Based on a thermodynamic model, it has been proposed that high-fidelity Pols enhance WC base-pairing specificities by partially excluding bulk water from the Pol active cleft (Petruska et al. 1986; Petruska and Goodman 1995). By having a more open architecture that enables water to compete with nucleobases in forming H-bonds, the error-prone Pols generally show less ability to discriminate between forming WC and non-WC base pairs. Having said that, "not all Y-family TLS polymerases are created equal." Pol η appears to have evolved to bypass CPDs efficiently and accurately. It does so by acting as a "molecular splint" straightening the normally kinked damaged DNA backbone, so that the covalently linked pyrimidines of the CPD can be "read" and accurately copied (Fig. 3A) (Biertümpfel et al. 2010).

Structural studies have also revealed that key residues within the active sites of Pol ι and Rev1 contribute to their unique mutagenic "signature." For example, Pol ι is best characterized in vitro for its ability to misincorporate dG opposite template T threefold to 10-fold better than the normal base dA (Table 1) (Johnson et al. 2000a; Tissier et al. 2000; Zhang et al. 2000). This is in contrast to efficient and accurate (for a Y-family polymerase) incorporation of T opposite dA (Johnson et al. 2000a; Tissier et al. 2000; Zhang et al. 2000).

Such diverse properties can be explained by the presence of amino acids (most notably Gln59) of the finger domain that protrude into the active-site cavity and direct the orientation of the templating base. In the case of the G:T mispair, the template base, dT, is always held in an *anti* conformation irrespective of the incoming dNTP. The mispaired dG also adopts an *anti* conformation and is further stabilized by hydrogen bonds with Gln59 (Fig. 3C) (Kirouac and Ling 2009). In contrast to a dT, a template dA is driven into a *syn* conformation by the incoming dTTP, which remains in an *anti* conformation. As a consequence, base pairing is limited so that dA makes a Hoogsteen, rather than WC base pair with the incoming dTTP (Nair et al. 2004, 2006). Similarly, Rev1 is characterized as a dCMP transferase (Nelson et al. 1996a), but this specificity is not dependent on bonding of the incoming dCMP with a template base, but rather is determined by a particular arginine residue (K324 in *S. cerevisiae* and K357 in humans) of the little finger domain of the Rev1 protein (Fig. 3D) (Nair et al. 2005; Swan et al. 2009).

Structural studies with archaeal Dpo4 have been particularly enlightening because the protein readily crystallizes in a ternary complex with an incoming nucleotide and lesion-containing template DNA. There are more than 60 different structures in the PDB database with individual "snapshots" of Dpo4 in the process of facilitating TLS (Vaisman et al. 2012). What is evident is that the enzyme can perform a wide range of molecular gymnastics with either the template base or incoming nucleotide so as to allow TLS to occur. An example is shown in Figure 3B, in which a large bulky benzo[*a*]pyrene adduct linked to dA is flipped into the major groove of the template DNA, so as to allow base pairing of the adducted dA with the incoming T.

Last, we mention *E. coli* Pol II. The apo structure of Pol II was first reported in 1994 (Anderson et al. 1994), but only recently has the ternary complex with template DNA and incoming nucleotide been solved. It appears that Pol II contains many molecular "nooks and crannies" near the active site of the polymerase that allow for manipulation of the lesion-containing template DNA (Wang and Yang 2009). In the case of an N^2-dG-AAF adduct, the template loops out in such a manner that two bases are skipped, leading to the -2 frameshift mutagenesis that is a characteristic of Pol II bypass of AAF adducts (Fuchs and Fujii 2007).

REGULATION OF TLS POLYMERASES

The ability to traverse a wide array of DNA lesions often comes at the price of reduced fidelity when replicating undamaged DNA. With the exception of *E. coli* Pol II, which has intrinsic $3'-5'$ exonucleolytic proofreading, none of the other TLS polymerases has the capacity to excise errors made when duplicating an undamaged DNA template. It is therefore not surprising that cells use several mechanisms to keep the error-prone TLS polymerases in check, so that they are only used as, and when required. Perhaps the best-studied example is in *E. coli*, where the three TLS polymerases are first negatively regulated at the transcriptional level by the LexA repressor. Although no additional regulation of Pol II is required because of its high fidelity ensured by intrinsic proofreading, the Y-family polymerases, Pol IV and Pol V, are exonuclease deficient and are subject to a variety of posttranslational controls.

Most notably, the UmuD′ subunit of Pol V has to be activated via damage-inducible cleavage of UmuD (Nohmi et al. 1988), and the UmuD, UmuD′, and UmuC proteins are all subject to rapid proteolytic degradation by the Lon and ClpXP proteases (Frank et al. 1996; Gonzalez et al. 1998, 2000; Neher et al. 2003). Any $UmuD'_2C$ complex that escapes degradation is essentially catalytically inactive until activated through protein interactions with RecA* to form Pol V Mut, whose activity is enhanced by binding to the β-clamp (Schlacher et al. 2006; Karata et al. 2012). Pol IV activity is enhanced through protein interactions with UmuD, RecA, and the β-clamp (Wagner et al. 2001; Becherel et al. 2002; Godoy et al. 2007).

Although the *S. cerevisiae RAD30* (Pol η) transcript appears to be induced roughly threefold after DNA damage (McDonald et al. 1997), much of the regulation of TLS polymerases in eukaryotes appears to rely on their posttranslational modification and/or specific protein–protein interactions that target them to the site of DNA damage. A key player in this process is ubiquitin. In *S. cerevisiae*, ubiquitination of Rad30 (Pol η) leads to its rapid degradation by the proteasome (Skoneczna et al. 2007). Similarly, ubiquitination of human Pol η by the E3 ubiquitin ligase Pirh2 results in its degraded by the 20S proteasome (Jung et al. 2010). In addition to ubiquitination via conventional E3 ligases, the human Y-family polymerases can undergo self-ubiquitination that is mediated by their UBMs (found in Pol ι or Rev1) or UBZ motifs (found in Pol η and Pol κ) (Fig. 4) (Bienko et al. 2005, 2010; Guo et al. 2006; Plosky et al. 2006; Jung et al. 2011). These motifs bind ubiquitin in a noncovalent manner and induce a conformational change in the TLS polymerase that prevents any subsequent interaction with other ubiquitinated proteins, such as PCNA. As a consequence, the enzymes are no longer correctly targeted to sites of DNA damage, which precludes their ability to facilitate TLS (Bienko et al. 2005, 2010; Plosky et al. 2006).

The eukaryotic Y-family TLS polymerases are also regulated through key protein–protein interactions. One such interaction is with the replicative clamp, PCNA. The Y-family polymerases possess specific PCNA-binding motifs (PIP-box) that allow for a direct interaction with the clamp in a manner similar to β-clamp binding by the bacterial polymerases (Fig. 4) (Warbrick 1998, 2000; Dalrymple et al. 2001). Upon DNA damage, PCNA is monoubquitinated (Hoege et al. 2002), and this further strengthens the interaction with the TLS polymerases because they bind to the monoubiquitinated PCNA via the PIP box *and* the UBM/UBZ motifs (Bienko et al. 2005, 2010; Guo et al. 2006; Plosky et al. 2006). Another important protein interaction regulating the eukaryotic TLS polymerases is with the Rev1 protein. In addition to its catalytic dCMP transferase activity, Rev1 plays a crucial structural function by interacting with Pol ζ (Lawrence and Hinkle 1996; Nelson et al. 1996b) and with Pol η, Pol ι, and Pol κ (Fig. 4) (Guo et al. 2003; Ohashi et al. 2004; Tissier et al. 2004) and targeting them to sites of DNA damage.

CONCLUDING REMARKS: IS TLS PERHAPS THE TIP OF THE ICEBERG?

Although the TLS polymerases are best characterized by their ability to traverse lesions that would otherwise block replicases, it is becoming increasingly obvious that they also participate in other cellular pathways. In some cases, such as Pol IV–dependent stress-induced mutagenesis in *E. coli* (Bull et al. 2001; Tompkins et al. 2003) or Pol η/Rev1-dependent somatic hypermutation of immunoglobulin genes in humans

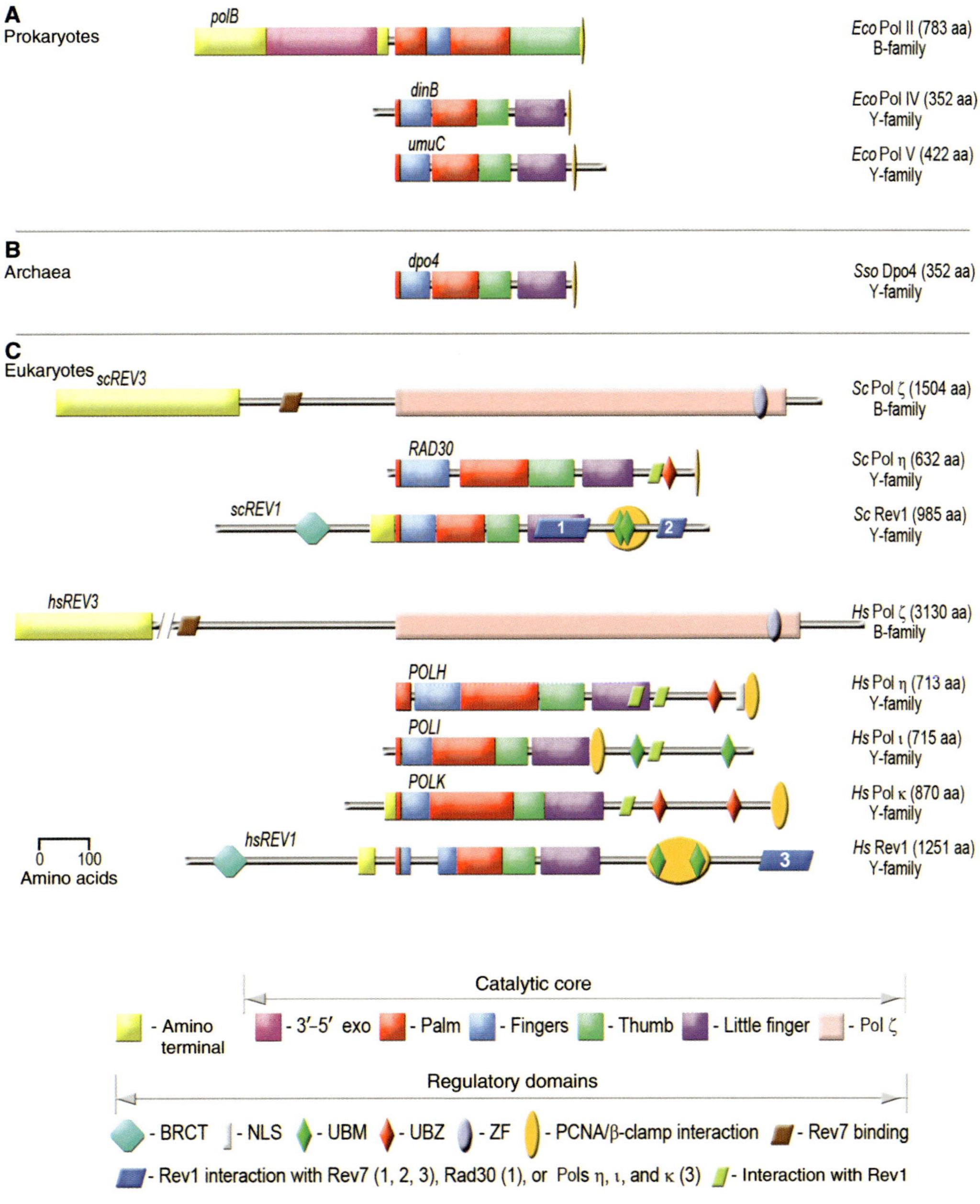

Figure 4. Domain organization of selected TLS polymerases. (*A*) Prokaryotes; (*B*) archaea; (*C*) eukaryotes. The name and phylogenetic family relationship of each of the TLS polymerases along with number of amino acid residues in each polymerase are indicated on the right-hand side of the figure. The structural catalytic domains present in all of the polymerases are color-coded as follows: (red) palm; (green) thumb; (blue) finger; (purple) little finger; (violet) 3′–5′ exonuclease of Pol II; (yellow) amino-terminal domain of Pol II and Rev3; (pink) uncharacterized catalytic domain of Rev3. Additional domains involved in localization and regulation of the TLS polymerases are as follows: (teal octagon) breast cancer-associated protein-1 carboxy-terminal domain (BRCT); (gray rectangle) nuclear localization signal (NLS); (green diamond) ubiquitin binding motif (UBM); (red diamond) ubiquitin binding zinc-finger motif (UBZ); (grayish oval) zinc finger (ZF); (gold oval) PCNA/β-clamp motif; (brown or blue rhomboid) Rev7 binding region; (olive rhomboid) Rev1-interacting region. Species: *Ec*, *E. coli*; *Ss*, *S. solfataricus*; *Sc*, *S. cerevisiae*; *Hs*, *H. sapiens*.

(Rogozin et al. 2001; Zeng et al. 2001), this occurs during error-prone replication of undamaged DNA and generates various mutations, the ultimate source of genetic diversity (Peled et al. 2008). However, there is also evidence that human Pol κ participates in nucleotide excision repair (Ogi et al. 2010) and human Pol η in recombinational repair (Kawamoto et al. 2005; McIlwraith et al. 2005), both of which are thought to be essentially error-free processes.

The heavy emphasis that has been accorded as principal roles for SOS *E. coli* Pols, and more generally for eukaryotic TLS Pols, namely, copying damaged DNA templates, has to a large extent ignored what could well turn out to be an even more important biological function in long-term survival and evolutionary fitness in the absence of exogenous DNA damage. In *E. coli*, eliminating any one of the *E. coli* SOS Pols (Pol II, IV, or V) causes a severe loss of cell fitness, so that when wild-type cells are grown in the presence of cells lacking either Pol II, Pol IV, or Pol V, the wild-type cells take over as 100% of the population of surviving cells, thus driving each of the strains lacking one of the SOS Pols to extinction within ∼6–7 days of continuous growth in rich medium (Yeiser et al. 2002). For this relative fitness determination, wild-type and mutant cells compete for nutrient resources when both are inoculated at low but equal population densities (∼10^2–10^3 cells/mL) and undergo co-outgrowth while growing exponentially. Thus, when competing for nutrient resources that are diminishing with time, the elimination of any one of the SOS Pols is sufficient to ensure the mutant's eventual demise when competing against wild-type *E. coli*.

Our understanding of the molecular mechanisms of TLS has increased dramatically in the past few years, but there is still much to be learned. Although we now know of several regulatory pathways that target the TLS polymerase sites of DNA damage, how the physical switch between the replicative and TLS polymerases occurs in a living cell is still largely unknown, and insights into this process should be forthcoming with the advent of single-molecule visualization of DNA polymerases at replication forks in vivo (Reyes-Lamothe et al. 2010; Lia et al. 2012). Crystal structures of the catalytic domains of the eukaryotic TLS polymerase have been reported along with isolated PIP-box and UBM domains (Bomar et al. 2007, 2010; Hishiki et al. 2009), but to date, there have been no reports of a full-length eukaryotic TLS polymerase. Indeed, such a structure would provide a major advance by allowing us to visualize how the various regulatory domains interact with the catalytic core of the TLS polymerase. It is evident that the TLS polymerases are regulated through a multitude of direct protein–protein interactions and structures of such multiprotein complexes as Pol V Mut ($UmuD'_2C$-RecA-ATP) (Jiang et al. 2009) and Pol ζ_4 (Rev3-Rev7-Pol31-Pol32) (Baranovskiy et al. 2012; Nelson et al. 1996b; Makarova et al. 2012) are critical to further our understanding of TLS.

ACKNOWLEDGMENTS

We thank Alexandra Vaisman (National Institutes of Health/National Institute of Child Health and Human Development) for the graphic artwork depicted in this review. R.W. is supported by funds from the NIH/NICHD Intramural Research Program and M.F.G. by NIH grants GM21422 and ES012259.

REFERENCES

Anderson WF, Prince DB, Yu H, McEntee K, Goodman MF. 1994. Crystallization of DNA polymerase II from *Escherichia coli*. *J Mol Biol* **238:** 120–122.

Baranovskiy AG, Lada AG, Siebler HM, Zhang Y, Pavlov YI, Tahirov TH. 2012. DNA polymerase δ and ζ switch by sharing accessory subunits of DNA polymerase δ. *J Biol Chem* **287:** 17281–17287.

Beaber JW, Hochhut B, Waldor MK. 2004. SOS response promotes horizontal dissemination of antibiotic resistance genes. *Nature* **427:** 72–74.

Bebenek K, Kunkel TA. 2004. Functions of DNA polymerases. *Adv Protein Chem* **69:** 137–165.

Becherel OJ, Fuchs RP, Wagner J. 2002. Pivotal role of the β-clamp in translesion DNA synthesis and mutagenesis in *E. coli* cells. *DNA Repair* **1:** 703–708.

Bemark M, Khamlichi AA, Davies SL, Neuberger MS. 2000. Disruption of mouse polymerase ζ (Rev3) leads to embryonic lethality and impairs blastocyst development in vitro. *Curr Biol* **10:** 1213–1216.

Bienko M, Green CM, Crosetto N, Rudolf F, Zapart G, Coull B, Kannouche P, Wider G, Peter M, Lehmann AR, et al. 2005. Ubiquitin-binding domains in Y-family polymerases regulate translesion synthesis. *Science* **310:** 1821–1824.

Bienko M, Green CM, Sabbioneda S, Crosetto N, Matic I, Hibbert RG, Begovic T, Niimi A, Mann M, Lehmann AR, et al. 2010. Regulation of translesion synthesis DNA polymerase η by monoubiquitination. *Mol Cell* **37:** 396–407.

Biertümpfel C, Zhao Y, Kondo Y, Ramon-Maiques S, Gregory M, Lee JY, Masutani C, Lehmann AR, Hanaoka F, Yang W. 2010. Structure and mechanism of human DNA polymerase η. *Nature* **465:** 1044–1048.

Bomar MG, Pai MT, Tzeng SR, Li SS, Zhou P. 2007. Structure of the ubiquitin-binding zinc finger domain of human DNA Y-polymerase η. *EMBO Rep* **8:** 247–251.

Bomar MG, D'Souza S, Bienko M, Dikic I, Walker GC, Zhou P. 2010. Unconventional ubiquitin recognition by the ubiquitin-binding motif within the Y family DNA polymerases ι and Rev1. *Mol Cell* **37:** 408–417.

Bonner CA, Randall SK, Rayssiguier C, Radman M, Eritja R, Kaplan BE, McEntee K, Goodman MF. 1988. Purification and characterization of an inducible *Escherichia coli* DNA polymerase capable of insertion and bypass at abasic lesions in DNA. *J Biol Chem* **263:** 18946–18952.

Bonner CA, Hays S, McEntee K, Goodman MF. 1990. DNA polymerase II is encoded by the DNA damage-inducible *dinA* gene of *Escherichia coli. Proc Natl Acad Sci* **87:** 7663–7667.

Boudsocq F, Campbell M, Devoret R, Bailone A. 1997. Quantitation of the inhibition of Hfr × F^- recombination by the mutagenesis complex UmuD′C. *J Mol Biol* **270:** 201–211.

Boudsocq F, Iwai S, Hanaoka F, Woodgate R. 2001. *Sulfolobus solfataricus* P2 DNA polymerase IV (Dpo4): An archaeal DNA polymerase with lesion-bypass properties akin to eukaryotic polη. *Nucleic Acids Res* **29:** 4607–4616.

Boudsocq F, Kokoska RJ, Plosky BS, Vaisman A, Ling H, Kunkel TA, Yang W, Woodgate R. 2004. Investigating the role of the little finger domain of Y-family DNA polymerases in low-fidelity synthesis and translesion replication. *J Biol Chem* **279:** 32932–32940.

Bridges BA, Woodgate R. 1984. Mutagenic repair in *Escherichia coli*, X. The *umuC* gene product may be required for replication past pyrimidine dimers but not for the coding error in UV mutagenesis. *Mol Gen Genet* **196:** 364–366.

Bridges BA, Woodgate R. 1985a. Mutagenic repair in *Escherichia coli*: Products of the *recA* gene and of the *umuD* and *umuC* genes act at different steps in UV-induced mutagenesis. *Proc Natl Acad Sci* **82:** 4193–4197.

Bridges BA, Woodgate R. 1985b. The two-step model of bacterial UV mutagenesis. *Mutat Res* **150:** 133–139.

Brotcorne-Lannoye A, Maenhaut-Michel G. 1986. Role of RecA protein in untargeted UV mutagenesis of bacteriophage λ: Evidence for the requirement for the *dinB* gene. *Proc Natl Acad Sci* **83:** 3904–3908.

Bruck I, Woodgate R, McEntee K, Goodman MF. 1996. Purification of a soluble UmuD′C complex from *Escherichia coli*: Cooperative binding of UmuD′C to single-stranded DNA. *J Biol Chem* **271:** 10767–10774.

Bull HJ, Lombardo MJ, Rosenberg SM. 2001. Stationary-phase mutation in the bacterial chromosome: Recombination protein and DNA polymerase IV dependence. *Proc Natl Acad Sci* **98:** 8334–8341.

Burckhardt SE, Woodgate R, Scheuermann RH, Echols H. 1988. UmuD mutagenesis protein of *Escherichia coli*: Overproduction, purification and cleavage by RecA. *Proc Natl Acad Sci* **85:** 1811–1815.

Chen L, Brugger K, Skovgaard M, Redder P, She Q, Torarinsson E, Greve B, Awayez M, Zibat A, Klenk HP, et al. 2005. The genome of *Sulfolobus acidocaldarius*, a model organism of the Crenarchaeota. *J Bacteriol* **187:** 4992–4999.

Courcelle J, Khodursky A, Peter B, Brown PO, Hanawalt PC. 2001. Comparative gene expression profiles following UV exposure in wild-type and SOS-deficient *Escherichia coli. Genetics* **158:** 41–64.

Dalrymple BP, Kongsuwan K, Wijffels G, Dixon NE, Jennings PA. 2001. A universal protein–protein interaction motif in the eubacterial DNA replication and repair systems. *Proc Natl Acad Sci* **98:** 11627–11632.

De Lucia P, Cairns J. 1969. Isolation of an *E. coli* strain with a mutation affecting DNA polymerase. *Nature* **224:** 1164–1166.

Dumstorf CA, Clark AB, Lin Q, Kissling GE, Yuan T, Kucherlapati R, McGregor WG, Kunkel TA. 2006. Participation of mouse DNA polymerase ι in strand-biased mutagenic bypass of UV photoproducts and suppression of skin cancer. *Proc Natl Acad Sci* **103:** 18083–18088.

Dutreix M, Moreau PL, Bailone A, Galibert F, Battista JR, Walker GC, Devoret R. 1989. New *recA* mutations that dissociate the various RecA protein activities in *Escherichia coli* provide evidence for an additional role for RecA protein in UV mutagenesis. *J Bacteriol* **171:** 2415–2423.

Echols H, Goodman MF. 1990. Mutation induced by DNA damage: A many protein affair. *Mutat Res* **236:** 301–311.

Echols H, Goodman MF. 1991. Fidelity mechanisms in DNA replication. *Annu Rev Biochem* **60:** 477–511.

Esposito G, Godindagger I, Klein U, Yaspo M, Cumano A, Rajewsky K. 2000. Disruption of the Rev3l-encoded catalytic subunit of polymerase ζ in mice results in early embryonic lethality. *Curr Biol* **10:** 1221–1224.

Fernández de Henestrosa AR, Ogi T, Aoyagi S, Chafin D, Hayes JJ, Ohmori H, Woodgate R. 2000. Identification of additional genes belonging to the LexA-regulon in *Escherichia coli. Mol Microbiol* **35:** 1560–1572.

Foster PL, Gudmundsson G, Trimarchi JM, Cai H, Goodman MF. 1995. Proofreading-defective DNA polymerase II increases adaptive mutation in *Escherichia coli. Proc Natl Acad Sci* **92:** 7951–7955.

Frank EG, Ennis DG, Gonzalez M, Levine AS, Woodgate R. 1996. Regulation of SOS mutagenesis by proteolysis. *Proc Natl Acad Sci* **93:** 10291–10296.

Friedberg EC, Walker GC, Siede W, Wood R, Schultz RA, Ellenberger T. 2006. *DNA repair and mutagenesis.* ASM Press, Washington, DC.

Frisch RL, Su Y, Thornton PC, Gibson JL, Rosenberg SM, Hastings PJ. 2010. Separate DNA Pol II– and Pol IV–dependent pathways of stress-induced mutation during double-strand-break repair in *Escherichia coli* are controlled by RpoS. *J Bacteriol* **192:** 4694–4700.

Fuchs RP, Fujii S. 2007. Translesion synthesis in *Escherichia coli*: Lessons from the *Nar*I mutation hot spot. *DNA Repair* **6:** 1032–1041.

Furukohri A, Goodman MF, Maki H. 2008. A dynamic polymerase exchange with *Escherichia coli* DNA polymerase IV replacing DNA polymerase III on the sliding clamp. *J Biol Chem* **283:** 11260–11269.

Gefter ML, Hirota Y, Kornberg T, Wechsler JA, Barnoux C. 1971. Analysis of DNA polymerases II and III in mutants of *Escherichia coli* thermosensitive for DNA synthesis. *Proc Natl Acad Sci* **68:** 3150–3153.

Gerlach VL, Aravind L, Gotway G, Schultz RA, Koonin EV, Friedberg EC. 1999. Human and mouse homologs of *Escherichia coli* DinB (DNA polymerase IV), members of the UmuC/DinB superfamily. *Proc Natl Acad Sci* **96:** 11922–11927.

Gerlach VL, Feaver WJ, Fischhaber PL, Friedberg EC. 2001. Purification and characterization of polκ, a DNA polymerase encoded by the human *DINB1* gene. *J Biol Chem* **276:** 92–98.

Godoy VG, Jarosz DF, Simon SM, Abyzov A, Ilyin V, Walker GC. 2007. UmuD and RecA directly modulate the mutagenic potential of the Y family DNA polymerase DinB. *Mol Cell* **28:** 1058–1070.

Gonzalez M, Frank EG, Levine AS, Woodgate R. 1998. Lon-mediated proteolysis of the *Escherichia coli* UmuD mutagenesis protein: In vitro degradation and identification of residues required for proteolysis. *Genes Dev* **12:** 3889–3899.

Gonzalez M, Rasulova F, Maurizi MR, Woodgate R. 2000. Subunit-specific degradation of the UmuD/D′ heterodimer by the ClpXP protease: The role of *trans* recognition in UmuD′ stability. *EMBO J* **19:** 5251–5258.

Goodman MF. 2002. Error-prone repair DNA polymerases in prokaryotes and eukaryotes. *Annu Rev Biochem* **71:** 17–50.

Guo C, Fischhaber PL, Luk-Paszyc MJ, Masuda Y, Zhou J, Kamiya K, Kisker C, Friedberg EC. 2003. Mouse Rev1 protein interacts with multiple DNA polymerases involved in translesion DNA synthesis. *EMBO J* **22:** 6621–6630.

Guo C, Tang TS, Bienko M, Parker JL, Bielen AB, Sonoda E, Takeda S, Ulrich HD, Dikic I, Friedberg EC. 2006. Ubiquitin-binding motifs in REV1 protein are required for its role in the tolerance of DNA damage. *Mol Cell Biol* **26:** 8892–8900.

Hishiki A, Hashimoto H, Hanafusa T, Kamei K, Ohashi E, Shimizu T, Ohmori H, Sato M. 2009. Structural basis for novel interactions between human translesion synthesis polymerases and proliferating cell nuclear antigen. *J Biol Chem* **284:** 10552–10560.

Hoege C, Pfander B, Moldovan GL, Pyrowolakis G, Jentsch S. 2002. RAD6-dependent DNA repair is linked to modification of PCNA by ubiquitin and SUMO. *Nature* **419:** 135–141.

Hogg M, Seki M, Wood RD, Doublie S, Wallace SS. 2011. Lesion bypass activity of DNA polymerase θ (POLQ) is an intrinsic property of the Pol domain and depends on unique sequence inserts. *J Mol Biol* **405:** 642–652.

Indiani C, McInerney P, Georgescu R, Goodman MF, O'Donnell M. 2005. A sliding-clamp toolbelt binds high- and low-fidelity DNA polymerases simultaneously. *Mol Cell* **19:** 805–815.

Indiani C, Langston LD, Yurieva O, Goodman MF, O'Donnell M. 2009. Translesion DNA polymerases remodel the replisome and alter the speed of the replicative helicase. *Proc Natl Acad Sci* **106:** 6031–6038.

Iwasaki H, Nakata A, Walker GC, Shinagawa H. 1990. The *Escherichia coli polB* gene, which encodes DNA polymerase II, is regulated by the SOS system. *J Bacteriol* **172:** 6268–6273.

Jansen JG, Langerak P, Tsaalbi-Shtylik A, van den Berk P, Jacobs H, de Wind N. 2006. Strand-biased defect in C/G transversions in hypermutating immunoglobulin genes in Rev1-deficient mice. *J Exp Med* **203:** 319–323.

Jarosz DF, Godoy VG, Delaney JC, Essigmann JM, Walker GC. 2006. A single amino acid governs enhanced activity of DinB DNA polymerases on damaged templates. *Nature* **439:** 225–228.

Jiang Q, Karata K, Woodgate R, Cox MM, Goodman MF. 2009. The active form of DNA polymerase V is UmuD′$_2$C-RecA-ATP. *Nature* **460:** 359–363.

Johnson A, O'Donnell M. 2005. Cellular DNA replicases: Components and dynamics at the replication fork. *Annu Rev Biochem* **74:** 283–315.

Johnson RE, Kondratick CM, Prakash S, Prakash L. 1999a. *hRAD30* mutations in the variant form of xeroderma pigmentosum. *Science* **285:** 263–265.

Johnson RE, Prakash S, Prakash L. 1999b. Efficient bypass of a thymine–thymine dimer by yeast DNA polymerase, polη. *Science* **283:** 1001–1004.

Johnson RE, Washington MT, Haracska L, Prakash S, Prakash L. 2000a. Eukaryotic polymerases ι and ζ act sequentially to bypass DNA lesions. *Nature* **406:** 1015–1019.

Johnson RE, Washington MT, Prakash S, Prakash L. 2000b. Fidelity of human DNA polymerase η. *J Biol Chem* **275:** 7447–7450.

Johnson RE, Prakash L, Prakash S. 2012. Pol 31 and Pol 32 subunits of yeast DNA polymerase δ are also essential subunits of DNA polymerase ζ. *Proc Natl Acad Sci* **109:** 12455–12460.

Jung YS, Liu G, Chen X. 2010. Pirh2 E3 ubiquitin ligase targets DNA polymerase η for 20S proteasomal degradation. *Mol Cell Biol* **30:** 1041–1048.

Jung YS, Hakem A, Hakem R, Chen X. 2011. Pirh2 E3 ubiquitin ligase monoubiquitinates DNA polymerase η to suppress translesion DNA synthesis. *Mol Cell Biol* **19:** 3997–4006.

Karata K, Vaisman A, Goodman MF, Woodgate R. 2012. Simple and efficient purification of *E. coli* DNA polymerase V: Cofactor requirements for optimal activity and processivity in vitro. *DNA Repair* **11:** 431–440.

Kato T, Shinoura Y. 1977. Isolation and characterization of mutants of *Escherichia coli* deficient in induction of mutations by ultraviolet light. *Mol Gen Genet* **156:** 121–131.

Kawamoto T, Araki K, Sonoda E, Yamashita YM, Harada K, Kikuchi K, Masutani C, Hanaoka F, Nozaki K, Hashimoto N, et al. 2005. Dual roles for DNA polymerase η in homologous DNA recombination and translesion DNA synthesis. *Mol Cell* **20:** 793–799.

Kenyon CJ, Walker GC. 1980. DNA-damaging agents stimulate gene expression at specific loci in *Escherichia coli*. *Proc Natl Acad Sci* **77:** 2819–2823.

Kim SR, Maenhaut-Michel G, Yamada M, Yamamoto Y, Matsui K, Sofuni T, Nohmi T, Ohmori H. 1997. Multiple pathways for SOS-induced mutagenesis in *Escherichia coli*: An overexpression of *dinB/dinP* results in strongly enhancing mutagenesis in the absence of any exogenous treatment to damage DNA. *Proc Natl Acad Sci* **94:** 13792–13797.

Kirouac KN, Ling H. 2009. Structural basis of error-prone replication and stalling at a thymine base by human DNA polymerase ι. *EMBO J* **28:** 1644–1654.

Knippers R. 1970. DNA polymerase II. *Nature* **228:** 1050–1053.

Kulaeva OI, Wootton JC, Levine AS, Woodgate R. 1995. Characterization of the *umu*-complementing operon from R391. *J Bacteriol* **177:** 2737–2743.

Kulaeva OI, Koonin EV, McDonald JP, Randall SK, Rabinovich N, Connaughton JF, Levine AS, Woodgate R. 1996. Identification of a DinB/UmuC homolog in the archeon *Sulfolobus solfataricus*. *Mutat Res* **357:** 245–253.

Kumari A, Minko IG, Harbut MB, Finkel SE, Goodman MF, Lloyd RS. 2008. Replication bypass of interstrand cross-link intermediates by *Escherichia coli* DNA polymerase IV. *J Biol Chem* **283:** 27433–27437.

Kuzminov A. 1999. Recombinational repair of DNA damage in *Escherichia coli* and bacteriophage λ. *Microbiol Mol Biol Rev* **63:** 751–813.

Lange SS, Takata K, Wood RD. 2011. DNA polymerases and cancer. *Nat Rev Cancer* **11:** 96–110.

Lawrence CW, Hinkle DC. 1996. DNA polymerase ζ and the control of DNA damage induced mutagenesis in eukaryotes. In *Cancer surveys: Genetic instability in cancer* (ed. Lindahl T), pp. 21–31. Cold Spring Harbor Laboratory Press, Cold Spring Harbor, NY.

Lemontt JF. 1971. Mutants of yeast defective in mutation induced by ultraviolet light. *Genetics* **68:** 21–33.

Lewis LK, Harlow GR, Gregg-Jolly LA, Mount DW. 1994. Identification of high affinity binding sites for LexA which define new DNA damage-inducible genes in *Escherichia coli*. *J Mol Biol* **241:** 507–523.

Lia G, Michel B, Allemand JF. 2012. Polymerase exchange during Okazaki fragment synthesis observed in living cells. *Science* **335:** 328–331.

Ling H, Boudsocq F, Woodgate R, Yang W. 2001. Crystal structure of a Y-family DNA polymerase in action: A mechanism for error-prone and lesion-bypass replication. *Cell* **107:** 91–102.

Ling H, Boudsocq F, Plosky BS, Woodgate R, Yang W. 2003. Replication of a *cis–syn* thymine dimer at atomic resolution. *Nature* **424:** 1083–1087.

Ling H, Boudsocq F, Woodgate R, Yang W. 2004a. Snapshots of replication through an abasic lesion; structural basis for base substitutions and frameshifts. *Mol Cell* **13:** 751–762.

Ling H, Sayer JM, Plosky BS, Yagi H, Boudsocq F, Woodgate R, Jerina DM, Yang W. 2004b. Crystal structure of a benzo[*a*]pyrene diol epoxide adduct in a ternary complex with a DNA polymerase. *Proc Natl Acad Sci* **101:** 2265–2269.

Little JW, Edmiston SH, Pacelli LZ, Mount DW. 1980. Cleavage of the *Escherichia coli* lexA protein by the recA protease. *Proc Natl Acad Sci* **77:** 3225–3229.

Maga G, Villani G, Crespan E, Wimmer U, Ferrari E, Bertocci B, Hubscher U. 2007. 8-Oxo-guanine bypass by human DNA polymerases in the presence of auxiliary proteins. *Nature* **447:** 606–608.

Makarova AV, Stodola JL, Burgers PM. 2012. A four-subunit DNA polymerase ζ complex containing Pol δ accessory subunits is essential for PCNA-mediated mutagenesis. *Nucleic Acids Res* doi: 10.1093/nar/gks948.

Masutani C, Araki M, Yamada A, Kusumoto R, Nogimori T, Maekawa T, Iwai S, Hanaoka F. 1999a. Xeroderma pigmentosum variant (XP-V) correcting protein from HeLa cells has a thymine dimer bypass DNA polymerase activity. *EMBO J* **18:** 3491–3501.

Masutani C, Kusumoto R, Yamada A, Dohmae N, Yokoi M, Yuasa M, Araki M, Iwai S, Takio K, Hanaoka F. 1999b. The *XPV* (xeroderma pigmentosum variant) gene encodes human DNA polymerase η. *Nature* **399:** 700–704.

Matsuda T, Bebenek K, Masutani C, Rogozin IB, Hanaoka F, Kunkel TA. 2001. Error rate and specificity of human and murine DNA polymerase η. *J Mol Biol* **312:** 335–346.

McCann JC, Springarn NE, Kobari J, Ames BN. 1975. Detection of carcinogens as mutagens: Bacterial tester strains with R factor plasmids. *Proc Natl Acad Sci* **72:** 979–983.

McDonald JP, Levine AS, Woodgate R. 1997. The *Saccharomyces cerevisiae RAD30* gene, a homologue of *Escherichia coli dinB* and *umuC*, is DNA damage inducible and functions in a novel error-free postreplication repair mechanism. *Genetics* **147:** 1557–1568.

McDonald JP, Rapic-Otrin V, Epstein JA, Broughton BC, Wang X, Lehmann AR, Wolgemuth DJ, Woodgate R. 1999. Novel human and mouse homologs of *Saccharomyces cerevisiae* DNA polymerase η. *Genomics* **60:** 20–30.

McDonald JP, Frank EG, Plosky BS, Rogozin IB, Masutani C, Hanaoka F, Woodgate R, Gearhart PJ. 2003. Identification of a nonsense mutation in DNA polymerase ι from 129-derived strains of mice and its effect on somatic hypermutation. *J Exp Med* **198:** 635–643.

McDonald JP, Hall A, Gasparutto D, Cadet J, Ballantyne J, Woodgate R. 2006. Novel thermostable Y-family polymerases: Applications for the PCR amplification of damaged or ancient DNAs. *Nucleic Acids Res* **34:** 1102–1111.

McHenry CS. 2011. DNA replicases from a bacterial perspective. *Annu Rev Biochem* **80:** 403–436.

McIlwraith MJ, Vaisman A, Liu Y, Fanning E, Woodgate R, West SC. 2005. Human DNA polymerase η promotes DNA synthesis from strand invasion intermediates (D-loops) of homologous recombination. *Mol Cell* **20:** 783–792.

Mead S, Vaisman A, Valjavec-Gratian M, Karata K, Vandewiele D, Woodgate R. 2007. Characterization of polV_{R391}: A Y-family polymerase encoded by *rumA′B* from the IncJ conjugative transposon, R391. *Mol Microbiol* **63:** 797–810.

Nair DT, Johnson RE, Prakash S, Prakash L, Aggarwal AK. 2004. Replication by human DNA polymerase ι occurs by Hoogsteen base-pairing. *Nature* **430:** 377–380.

Nair DT, Johnson RE, Prakash L, Prakash S, Aggarwal AK. 2005. Rev1 employs a novel mechanism of DNA synthesis using a protein template. *Science* **309:** 2219–2222.

Nair DT, Johnson RE, Prakash L, Prakash S, Aggarwal AK. 2006. An incoming nucleotide imposes an *anti* to *syn* conformational change on the templating purine in the human DNA polymerase-ι active site. *Structure* **14:** 749–755.

Neher SB, Sauer RT, Baker TA. 2003. Distinct peptide signals in the UmuD and UmuD′ subunits of UmuD/D′ mediate tethering and substrate processing by the ClpXP protease. *Proc Natl Acad Sci* **100:** 13219–13224.

Nelson JR, Lawrence CW, Hinkle DC. 1996a. Deoxycytidyl transferase activity of yeast REV1 protein. *Nature* **382:** 729–731.

Nelson JR, Lawrence CW, Hinkle DC. 1996b. Thymine–thymine dimer bypass by yeast DNA polymerase ζ. *Science* **272:** 1646–1649.

Nohmi T, Battista JR, Dodson LA, Walker GC. 1988. RecA-mediated cleavage activates UmuD for mutagenesis: Mechanistic relationship between transcriptional derepression and posttranslational activation. *Proc Natl Acad Sci* **85:** 1816–1820.

Ogi T, Kato TJ, Kato T, Ohmori H. 1999. Mutation enhancement by DINB1, a mammalian homologue of the *Escherichia coli* mutagenesis protein DinB. *Genes Cells* **4:** 607–618.

Ogi T, Shinkai Y, Tanaka K, Ohmori H. 2002. Polk protects mammalian cells against the lethal and mutagenic effects of polycyclic hydrocarbons. *Proc Natl Acad Sci* **99:** 15548–15553.

Ogi T, Limsirichaikul S, Overmeer RM, Volker M, Takenaka K, Cloney R, Nakazawa Y, Niimi A, Miki Y, Jaspers NG, et al. 2010. Three DNA polymerases, recruited by different mechanisms, carry out NER repair synthesis in human cells. *Mol Cell* **37:** 714–727.

Ohashi E, Ogi T, Kusumoto R, Iwai S, Masutani C, Hanaoka F, Ohmori H. 2000. Error-prone bypass of certain DNA lesions by the human DNA polymerase κ. *Genes Dev* **14:** 1589–1594.

Ohashi E, Murakumo Y, Kanjo N, Akagi J, Masutani C, Hanaoka F, Ohmori H. 2004. Interaction of hREV1 with three human Y-family DNA polymerases. *Genes Cells* **9:** 523–531.

Ohkumo T, Kondo Y, Yokoi M, Tsukamoto T, Yamada A, Sugimoto T, Kanao R, Higashi Y, Kondoh H, Tatematsu M, et al. 2006. UV-B radiation induces epithelial tumors in mice lacking DNA polymerase η and mesenchymal tumors in mice deficient for DNA polymerase ι. *Mol Cell Biol* **26:** 7696–7706.

Ohmori H, Hatada E, Qiao Y, Tsuji M, Fukuda R. 1995. *dinP*, a new gene in *Escherichia coli*, whose product shows similarities to UmuC and its homologues. *Mutat Res* **347:** 1–7.

Ohmori H, Friedberg EC, Fuchs RPP, Goodman MF, Hanaoka F, Hinkle D, Kunkel TA, Lawrence CW, Livneh Z, Nohmi T, et al. 2001. The Y-family of DNA polymerases. *Mol Cell* **8:** 7–8.

Pata JD. 2010. Structural diversity of the Y-family DNA polymerases. *Biochim Biophys Acta* **1804:** 1124–1135.

Patel M, Jiang Q, Woodgate R, Cox MM, Goodman MF. 2010. A new model for SOS-induced mutagenesis: How RecA protein activates DNA polymerase V. *Crit Rev Biochem Mol Biol* **45:** 171–184.

Peled JU, Kuang FL, Iglesias-Ussel MD, Roa S, Kalis SL, Goodman MF, Scharff MD. 2008. The biochemistry of somatic hypermutation. *Annu Rev Immunol* **26:** 481–511.

Perry KL, Walker GC. 1982. Identification of plasmid (pKM101) coded proteins involved in mutagenesis and UV resistance. *Nature* **300:** 278–281.

Petruska J, Goodman MF. 1995. Enthalpy–entropy compensation in DNA melting thermodynamics. *J Biol Chem* **270:** 746–750.

Petruska J, Sowers LC, Goodman MF. 1986. Comparison of nucleotide interactions in water, proteins, and vacuum: Model for DNA polymerase fidelity. *Proc Natl Acad Sci* **83:** 1559–1562.

Pham P, Bertram JG, O'Donnell M, Woodgate R, Goodman MF. 2001. A model for SOS-lesion targeted mutations in *E. coli* involving pol V, RecA, SSB and β sliding clamp. *Nature* **409:** 366–370.

Plosky BS, Vidal A, Fernández de Henestrosa AR, McLenigan MP, McDonald JP, Mead S, Woodgate R. 2006. Controlling the subcellular localization of DNA polymerases ι and η via interactions with ubiquitin. *EMBO J* **25:** 2847–2855.

Qiu Z, Goodman MF. 1997. The *Escherichia coli polB* locus is identical to *dinA*, the structural gene for DNA polymerase II. Characterization of Pol II purified from a *polB* mutant. *J Biol Chem* **272:** 8611–8617.

Radman M. 1974. Phenomenology of an inducible mutagenic DNA repair pathway in *Escherichia coli*: SOS repair hypothesis. In *Molecular and environmental aspects of mutagenesis* (ed. Prakash L, et al.), pp. 128–142. Charles C. Thomas, Springfield, IL.

Rajagopalan M, Lu C, Woodgate R, O'Donnell M, Goodman MF, Echols H. 1992. Activity of the purified mutagenesis proteins UmuC, UmuD′ and RecA in replicative bypass of an abasic DNA lesion by DNA polymerase III. *Proc Natl Acad Sci* **89:** 10777–10781.

Rangarajan S, Gudmundsson G, Qiu Z, Foster PL, Goodman MF. 1997. *Escherichia coli* DNA polymerase II catalyzes chromosomal and episomal DNA synthesis in vivo. *Proc Natl Acad Sci* **94:** 946–951.

Reuven NB, Arad G, Maor-Shoshani A, Livneh Z. 1999. The mutagenesis protein UmuC is a DNA polymerase activated by UmuD′, RecA, and SSB and is specialized for translesion replication. *J Biol Chem* **274:** 31763–31766.

Reyes-Lamothe R, Sherratt DJ, Leake MC. 2010. Stoichiometry and architecture of active DNA replication machinery in *Escherichia coli*. *Science* **328:** 498–501.

Rogozin IB, Pavlov YI, Bebenek K, Matsuda T, Kunkel TA. 2001. Somatic mutation hotspots correlate with DNA polymerase η error spectrum. *Nat Immunol* **2:** 530–536.

Roush AA, Suarez M, Friedberg EC, Radman M, Siede W. 1998. Deletion of the *Saccharomyces cerevisiae* gene *RAD30* encoding an *Escherichia coli* DinB homolog confers UV radiation sensitivity and altered mutability. *Mol Gen Genet* **257:** 686–692.

Sale JE, Lehmann AR, Woodgate R. 2012. Y-family DNA polymerases and their role in tolerance of cellular DNA damage. *Nat Rev Mol Cell Biol* **13:** 141–152.

Schenten D, Gerlach VL, Guo C, Velasco-Miguel S, Hladik CL, White CL, Friedberg EC, Rajewsky K, Esposito G. 2002. DNA polymerase κ deficiency does not affect somatic hypermutation in mice. *Eur J Immunol* **32:** 3152–3160.

Schlacher K, Goodman MF. 2007. Lessons from 50 years of SOS DNA-damage-induced mutagenesis. *Nat Rev Mol Cell Biol* **8:** 587–594.

Schlacher K, Cox MM, Woodgate R, Goodman MF. 2006. RecA acts in *trans* to allow replication of damaged DNA by DNA polymerase V. *Nature* **442:** 883–887.

Seki M, Wood RD. 2008. DNA polymerase θ (POLQ) can extend from mismatches and from bases opposite a (6–4) photoproduct. *DNA Repair* **7:** 119–127.

She Q, Singh RK, Confalonieri F, Zivanovic Y, Allard G, Awayez MJ, Chan-Weiher CC, Clausen IG, Curtis BA, De Moors A, et al. 2001. The complete genome of the crenarchaeon *Sulfolobus solfataricus* P2. *Proc Natl Acad Sci* **98:** 7835–7840.

Shinagawa H, Iwasaki H, Kato T, Nakata A. 1988. RecA protein-dependent cleavage of UmuD protein and SOS mutagenesis. *Proc Natl Acad Sci* **85:** 1806–1810.

Shtygasheva AA, Belousova EA, Rechkunova NI, Lebedeva NA, Lavrik OI. 2008. DNA polymerases β and λ as potential participants of TLS during genomic DNA replication on the lagging strand. *Biochemistry (Mosc)* **73:** 1207–1213.

Silvian LF, Toth EA, Pham P, Goodman MF, Ellenberger T. 2001. Crystal structure of a DinB family error-prone DNA polymerase from *Sulfolobus solfataricus*. *Nat Struct Biol* **8:** 984–989.

Skoneczna A, McIntyre J, Skoneczny M, Policinska Z, Sledziewska-Gojska E. 2007. Polymerase η is a short-lived, proteasomally degraded protein that is temporarily stabilized following UV irradiation in *Saccharomyces cerevisiae*. *J Mol Biol* **366:** 1074–1086.

Sommer S, Bailone A, Devoret R. 1993. The appearance of the UmuD′C protein complex in *Escherichia coli* switches repair from homologous recombination to SOS mutagenesis. *Mol Microbiol* **10:** 963–971.

Steinborn G. 1978. *Uvm* mutants of *Escherichia coli* K12 deficient in UV mutagenesis. I. Isolation of *uvm* mutants and their phenotypical characterization in DNA repair and mutagenesis. *Mol Gen Genet* **165:** 87–93.

Swan MK, Johnson RE, Prakash L, Prakash S, Aggarwal AK. 2009. Structure of the human Rev1–DNA–dNTP ternary complex. *J Mol Biol* **390:** 699–709.

Tang M, Bruck I, Eritja R, Turner J, Frank EG, Woodgate R, O'Donnell M, Goodman MF. 1998. Biochemical basis of SOS-induced mutagenesis in *Escherichia coli*: Reconstitution of in vitro lesion bypass dependent on the $UmuD'_2$-C mutagenic complex and RecA. *Proc Natl Acad Sci* **95:** 9755–9760.

Tang M, Shen X, Frank EG, O'Donnell M, Woodgate R, Goodman MF. 1999. $UmuD'_2C$ is an error-prone DNA polymerase, *Escherichia coli*, DNA pol V. *Proc Natl Acad Sci* **96:** 8919–8924.

Tissier A, McDonald JP, Frank EG, Woodgate R. 2000. polι, a remarkably error-prone human DNA polymerase. *Genes Dev* **14:** 1642–1650.

Tissier A, Kannouche P, Reck MP, Lehmann AR, Fuchs RP, Cordonnier A. 2004. Co-localization in replication foci and interaction of human Y-family members, DNA polymerase polη and REVl protein. *DNA Repair* **3:** 1503–1514.

Tompkins JD, Nelson JL, Hazel JC, Leugers SL, Stumpf JD, Foster PL. 2003. Error-prone polymerase, DNA polymerase IV, is responsible for transient hypermutation during adaptive mutation in *Escherichia coli*. *J Bacteriol* **185:** 3469–3472.

Trincao J, Johnson RE, Escalante CR, Prakash S, Prakash L, Aggarwal AK. 2001. Structure of the catalytic core of *S. cerevisiae* DNA polymerase η. Implications for translesion DNA synthesis. *Mol Cell* **8:** 417–426.

Vaisman A, McDonald JP, Woodgate R. 2012. Translesion DNA synthesis. In *EcoSal—Escherichia coli and Salmonella: Cellular and molecular biology* (ed. Böck A, Curtiss R III, Kaper JB, Karp PD, Neidhardt FC, Slauch JM, Squires CL). ASM Press, Washington, DC.

Villani G, Hubscher U, Gironis N, Parkkinen S, Pospiech H, Shevelev I, di Cicco G, Markkanen E, Syvaoja JE, Tanguy Le Gac N. 2011. In vitro gap-directed translesion DNA synthesis of an abasic site involving human DNA polymerases ε, λ, and β. *J Biol Chem* **286:** 32094–32104.

Wagner J, Gruz P, Kim SR, Yamada M, Matsui K, Fuchs RPP, Nohmi T. 1999. The *dinB* gene encodes a novel *Escherichia coli* DNA polymerase, DNA Pol IV, involved in mutagenesis. *Mol Cell* **4:** 281–286.

Wagner J, Fujii S, Gruz P, Nohmi T, Fuchs RP. 2001. The β clamp targets DNA polymerase IV to DNA and strongly increases its processivity. *EMBO Rep* **1:** 484–488.

Walker GC. 1984. Mutagenesis and inducible responses to deoxyribonucleic acid damage in *Escherichia coli*. *Microbiol Rev* **48:** 60–93.

Wang F, Yang W. 2009. Structural insight into translesion synthesis by DNA Pol II. *Cell* **139:** 1279–1289.

Warbrick E. 1998. PCNA binding through a conserved motif. *Bioessays* **20:** 195–199.

Warbrick E. 2000. The puzzle of PCNA's many partners. *Bioessays* **22:** 997–1006.

Waters LS, Minesinger BK, Wiltrout ME, D'Souza S, Woodruff RV, Walker GC. 2009. Eukaryotic translesion polymerases and their roles and regulation in DNA damage tolerance. *Microbiol Molec Biol Rev* **73:** 134–154.

Watson JD, Crick FH. 1953. Molecular structure of nucleic acids; A structure for deoxyribose nucleic acid. *Nature* **171:** 737–738.

Weigle JJ. 1953. Induction of mutation in a bacterial virus. *Proc Natl Acad Sci* **39:** 628–636.

Wilson RC, Pata JD. 2008. Structural insights into the generation of single-base deletions by the Y family DNA polymerase Dbh. *Mol Cell* **29:** 767–779.

Witkin EM. 1967. The radiation sensitivity of *Escherichia coli* B: A hypothesis relating filament formation and prophage induction. *Proc Natl Acad Sci* **57:** 1275–1279.

Witkin EM. 1976. Ultraviolet mutagenesis and inducible DNA repair in *Escherichia coli*. *Bacteriol Rev* **40:** 869–907.

Wittschieben J, Shivji MK, Lalani E, Jacobs MA, Marini F, Gearhart PJ, Rosewell I, Stamp G, Wood RD. 2000. Disruption of the developmentally regulated *REV3L* gene causes embryonic lethality. *Curr Biol* **10:** 1217–1220.

Wittschieben JP, Reshmi SC, Gollin SM, Wood RD. 2006. Loss of DNA polymerase ζ causes chromosomal instability in mammalian cells. *Cancer Res* **66:** 134–142.

Woodgate R, Rajagopalan M, Lu C, Echols H. 1989. UmuC mutagenesis protein of *Escherichia coli*: Purification and interaction with UmuD and UmuD′. *Proc Natl Acad Sci* **86:** 7301–7305.

Yamanaka K, Minko IG, Takata K, Kolbanovskiy A, Kozekov ID, Wood RD, Rizzo CJ, Lloyd RS. 2010. Novel enzymatic function of DNA polymerase ν in translesion DNA synthesis past major groove DNA–peptide and DNA–DNA cross-links. *Chem Res Toxicol* **23:** 689–695.

Yang W, Woodgate R. 2007. What a difference a decade makes: Insights into translesion DNA synthesis. *Proc Natl Acad Sci* **104:** 15591–15598.

Yeiser B, Pepper ED, Goodman MF, Finkel SE. 2002. SOS-induced DNA polymerases enhance long-term survival and evolutionary fitness. *Proc Natl Acad Sci* **99:** 8737–8741.

Zeng X, Winter DB, Kasmer C, Kraemer KH, Lehmann AR, Gearhart PJ. 2001. DNA polymerase η is an A-T mutator in somatic hypermutation of immunoglobulin variable genes. *Nat Immunol* **2:** 537–541.

Zhang Y, Yuan F, Wu X, Wang Z. 2000. Preferential incorporation of G opposite template T by the low-fidelity human DNA polymerase ι. *Mol Cell Biol* **20:** 7099–7108.

Zhang Y, Yuan F, Wu X, Taylor J-S, Wang Z. 2001. Response of human DNA polymerase ι to DNA lesions. *Nucleic Acids Res* **29:** 928–935.

Zhou B, Pata JD, Steitz TA. 2001. Crystal structure of a DinB lesion bypass DNA polymerase catalytic fragment reveals a classic polymerase catalytic domain. *Mol Cell* **8:** 427–437.

Rescuing Stalled or Damaged Replication Forks

Joseph T.P. Yeeles[1], Jérôme Poli[2], Kenneth J. Marians[1], and Philippe Pasero[2]

[1]Molecular Biology Program, Memorial Sloan-Kettering Cancer Center, New York, New York 10065

[2]Institute of Human Genetics, CNRS UPR 1142, 141 rue de la Cardonille, 34396 Montpellier Cedex 5, France

Correspondence: kmarians@sloankettering.edu; ppasero@igh.cnrs.fr

In recent years, an increasing number of studies have shown that prokaryotes and eukaryotes are armed with sophisticated mechanisms to restart stalled or collapsed replication forks. Although these processes are better understood in bacteria, major breakthroughs have also been made to explain how fork restart mechanisms operate in eukaryotic cells. In particular, repriming on the leading strand and fork regression are now established as critical for the maintenance and recovery of stalled forks in both systems. Despite the lack of conservation between the factors involved, these mechanisms are strikingly similar in eukaryotes and prokaryotes. However, they differ in that fork restart occurs in the context of chromatin in eukaryotes and is controlled by multiple regulatory pathways.

For DNA replication to be accurately completed, the replication fork must frequently overcome a multitude of structurally unrelated obstacles such as DNA lesions, transcribing RNA polymerases, and tightly bound protein–DNA complexes. As a consequence, numerous diverse mechanisms have evolved that either help minimize the frequency or impact of collisions or repair the damage that is left behind. This work will focus solely on the mechanisms that exist in prokaryotes and eukaryotes to facilitate replication on template DNA containing either leading- or lagging-strand polymerization-blocking lesions. Lesions of this type are generated frequently under normal growth conditions (Lindahl 1993), as well as being induced by exogenous genotoxic agents. While cells have mechanisms such as nucleotide excision repair (NER) and base excision repair that target and repair a vast array of DNA modifications (Freidberg 2005), it is inevitable that some damage will persist long enough to be encountered by the DNA replication machinery. To achieve the high fidelity required for genome duplication, the architecture and mechanism of replicative polymerases efficiently discriminate against the incorporation of mismatched bases (Kunkel 2004). As a consequence, even DNA lesions that do not significantly alter DNA structure often inhibit nascent chain elongation. Should the replisome encounter such damage, the template strand in which the damage is located impacts significantly on the mechanism by which it is overcome. It is generally accepted that lagging-strand template lesions present few obstacles to replication fork progression. The situation with leading-strand template damage is more complex, and as such, the events that occur following replisome collision remain the subject of considerable debate.

Cite this article as *Cold Spring Harb Perspect Biol* doi: 10.1101/cshperspect.a012815

LAGGING-STRAND TEMPLATE LESIONS

Multiple studies, both in vitro and in vivo, have shown that bacterial replisomes efficiently bypass lagging-strand template damage provided that progression of the replicative helicase—which translocates on the lagging-strand template—is not inhibited. Rolling circle replication assays on templates containing site-specific lagging-strand abasic sites, using both the *Escherichia coli* (McInerney and O'Donnell 2004) and bacteriophage T4 replisomes (Nelson and Benkovic 2010), showed that leading-strand replication was not affected by the lesion's presence. The ratio of leading- to lagging-strand replication products was also not altered significantly, indicating that coupled leading- and lagging-strand synthesis was maintained on the damage-containing templates. This is believed to be because the lagging strand is primed repeatedly for Okazaki fragment synthesis, providing an obvious mechanism by which lagging-strand reinitiation can occur. With the lagging-strand polymerase stalled at the site of damage, template unwinding and leading-strand synthesis continue. Lagging-strand synthesis is resumed downstream from the lesion once the stalled polymerase has dissociated and rebounded to a newly synthesized primer. Bypass of lesions in this manner leaves single-stranded (ss) DNA gaps in the lagging strand, which, using an *oriC*-based replication assay on a damage-containing template, have been estimated to be approximately 1–2 kb (Higuchi et al. 2003). Similarly, ssDNA gaps have been observed on the lagging strand in budding yeast and SV40 DNA after UV exposure (Mezzina et al. 1988; Lopes et al. 2006). Interestingly, these gaps are much smaller in eukaryotes (<400 nucleotides in budding yeast) than in *E. coli*, which may reflect differences in Okazaki fragment length. It is worth noting that unlike its bacterial counterpart, the eukaryotic replicative helicase complex translocates on the leading-strand template (Fu et al. 2011), which may facilitate the bypass of bulky DNA lesions on the lagging strand.

LEADING-STRAND TEMPLATE LESIONS

The debate surrounding leading-strand template lesions centers on whether the template damage presents an absolute block to replication that must be removed for replication to proceed or if it can be bypassed by reinitiating leading-strand synthesis downstream and then repaired postreplicatively. Following UV irradiation of NER-deficient (Rupp and Howard-Flanders 1968) or -proficient *E. coli* cells (Khidhir et al. 1985; Witkin et al. 1987; Courcelle et al. 2005; Belle et al. 2007; Rudolph et al. 2007), replication rates immediately postirradiation are reduced significantly (approximately 80%–90%) but do not appear to come to a complete halt. Replication then recovers in NER-proficient strains to the pre-UV rates over a period of time that correlates well with the time taken to remove the majority of pyrimidine dimers from the DNA (Courcelle et al. 1999; Rudolph et al. 2007). These data have been interpreted to mean that leading-strand lesions present a block to replication that must first be removed if replication is to continue. Consistent with these ideas, multiple accessory proteins that are involved in replisome remodeling and recombination are required for replication to recover following UV treatment (McGlynn and Lloyd 2002; Courcelle and Hanawalt 2003), some of which will be discussed later in this work.

Several of the above experiments were conducted using UV intensities sufficient to induce several hundred pyrimidine dimers per *E. coli* chromosome, equating to one dimer every 10–25 kbp (Rudolph et al. 2007). As the replisome travels at 500–1000 bp/sec (Chandler et al. 1975), and assuming that lagging-strand lesions are efficiently bypassed, replication forks would likely have encountered a leading-strand lesion within the first minute postirradiation, yet replication is seen to continue, albeit at a reduced rate, for considerably longer. Rudolph et al. argued that the majority of replication occurring after UV irradiation was in fact because of new rounds of replication initiation (Rudolph et al. 2007). While replication did take longer to recover in temperature-sensitive DnaA mutants unable to reinitiate replication at *oriC*, it did not stop completely following UV exposure. These observations therefore suggest that leading-strand template lesions may not form an absolute block to replication. Rupp and Howard-

Flanders analyzed the DNA synthesized in *E. coli* immediately following UV irradiation. Their studies revealed that the newly synthesized DNA contained single-stranded gaps that they estimated to be 1–2 kb (Iyer and Rupp 1971) and that all the nascent chains were considerably shorter than those in untreated controls (Rupp and Howard-Flanders 1968). To explain these observations, they hypothesized that replication could be reinitiated downstream from lesions in both the leading- and lagging-strand templates (Rupp and Howard-Flanders 1968; Rupp 1996).

The leading-strand reinitiation model initially received little support, as the consensus view was that leading-strand priming was restricted to the origin of replication. Furthermore, two studies of the *E. coli* replication machinery (Higuchi et al. 2003; Pages and Fuchs 2003) concluded that replication becomes uncoupled following an encounter with a leading-strand template lesion, with template unwinding but only lagging-strand synthesis continuing beyond the damage. Similar results were also observed with the bacteriophage T4 replisome (Nelson and Benkovic 2010). However, the studies conducted in *E. coli* were not designed to visualize leading-strand replication products generated downstream from the damage, and as such, they may have failed to observe leading-strand reinitiation (Higuchi et al. 2003; Pages and Fuchs 2003).

The discovery that primase could in fact prime the leading-strand template on model fork structures provided the first mechanistic evidence that leading-strand synthesis could be reinitiated outside of the origin of replication (Heller and Marians 2006a). This idea was developed in a recent study. Using an *oriC*-based replication system, the outcomes of collisions between the *E. coli* replisome and either a single, site-specific pyrimidine dimer or abasic site were investigated (Yeeles and Marians 2011). The data revealed that the replisome transiently stalls upon collision with the lesion but does not dissociate, remaining stably associated with the DNA template (Fig. 1A). Following a short lag, leading- and lagging-strand synthesis were reinitiated downstream from the damage via a de novo, DnaG-dependent priming event on the leading-strand template. The reaction proceeded independently of the replication restart proteins, demonstrating that the replisome has the inherent capacity to replicate beyond leading-strand template lesions by synthesizing the leading strand discontinuously. The precise details of this reaction are yet to be fully elucidated. It is presumed that template unwinding continues for some distance beyond the damage to expose a region of ssDNA on the leading-strand template where primer synthesis occurs. This region is estimated to be anywhere between tens of bases and several hundred, based on the distribution of leading-strand restart products that were observed (Yeeles and Marians 2011). Leading-strand synthesis would resume once the stalled polymerase had dissociated from the stall site and rebounded to the new leading-strand primer. Alternatively, as it has recently been shown that the *E. coli* replisome can contain three polymerases (McInerney et al. 2007; Reyes-Lamothe et al. 2010; Lia et al. 2012), the possibility exists that a third polymerase that is not bound to a template strand may bind to the new leading-strand primer to reinitiate leading-strand synthesis. The sequences with which these events take place and the rate-limiting step in the reaction are an interesting subject for future investigation.

Leading-strand reinitiation without replisome dissociation provides a potential explanation for the continued replication that is observed immediately post-UV irradiation in vivo. However, the observed leading-strand reinitiation was not 100% efficient, and in some instances template unwinding and only lagging-strand synthesis continued to the end of the template (4 kb). Thus it seems that although the replisome is able to bypass leading-strand template lesions, such lesions will also lead to the breakdown of the replication fork. Under conditions of replication stress—for example, following UV irradiation—multiple lesions in the chromosome will increase the chance of replisome dissociation, eventually leading to a replication arrest. In the absence of exogenous DNA-damaging agents, when template lesions are infrequently encountered, the ability to reinitiate replication downstream from lesions may be

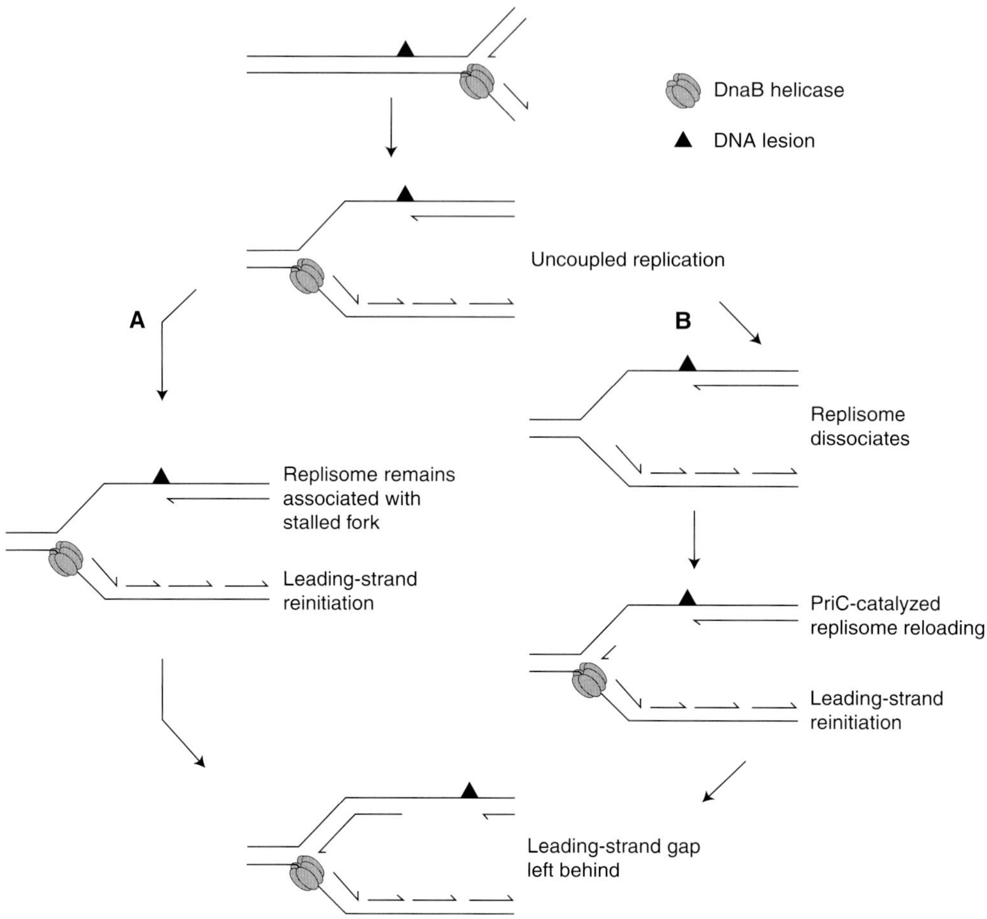

Figure 1. Bypass of leading-strand template damage by leading-strand repriming in *E. coli*. Following a collision with a leading-strand lesion, template unwinding and lagging-strand synthesis are believed to continue beyond the site of damage. (*A*) Repriming of the leading strand can occur downstream from the damage, which enables replication to continue without the replisome dissociating from the DNA (Yeeles and Marians 2011). (*B*) Should the replisome dissociate following the collision, the replication restart protein PriC can reload DnaB, which enables replication to be reinitiated by priming the leading strand downstream from the damage (Heller and Marians 2006b).

considered a general housekeeping function that prevents significant delays to replication fork movement. This will be particularly advantageous during periods of rapid bacterial growth when multiple replication forks are traversing the chromosome, while at the same time chromosome decatenation and cell division are also occurring. Furthermore, induction of the SOS response should be avoided when low levels of DNA damage are present, which in turn will prevent a cell division arrest and the induction of mutagenic translesion DNA polymerases (Goodman and Woodgate 2013).

In eukaryotes, it was long believed that leading-strand synthesis is continuous and is exclusively initiated at replication origins. Continuous synthesis through DNA lesions is ensured by specialized DNA polymerases called lesion-bypass DNA polymerases (Goodman and Woodgate 2013) through a process called translesion DNA synthesis (TLS). Recruitment of TLS polymerases to stalled forks depends on a conserved mechanism involving the Rad6- and Rad18-dependent ubiquitination of proliferating cell nuclear antigen (PCNA) (Fig. 2A). TLS polymerases are replaced with replicative polymerases once the damage is passed (Kannouche et al. 2004; Moldovan et al. 2007). Interestingly, discontinuities on both strands have been reported in budding yeast after UV exposure (Lopes et al. 2006), suggesting that repriming also occurs on the leading strand in eukaryotes, as may be the case in bacteria (Fig. 2B). These gaps are repaired postreplicatively by translesion synthesis (Fig. 2C) or by recombination-dependent template switch mechanisms (Fig. 2D) (Lehmann and Fuchs 2006; Branzei and Foiani 2010). Repriming and postreplicative gap filling is supported by recent studies showing that the *Saccharomyces cerevisiae* Rad6 pathway can be uncoupled from

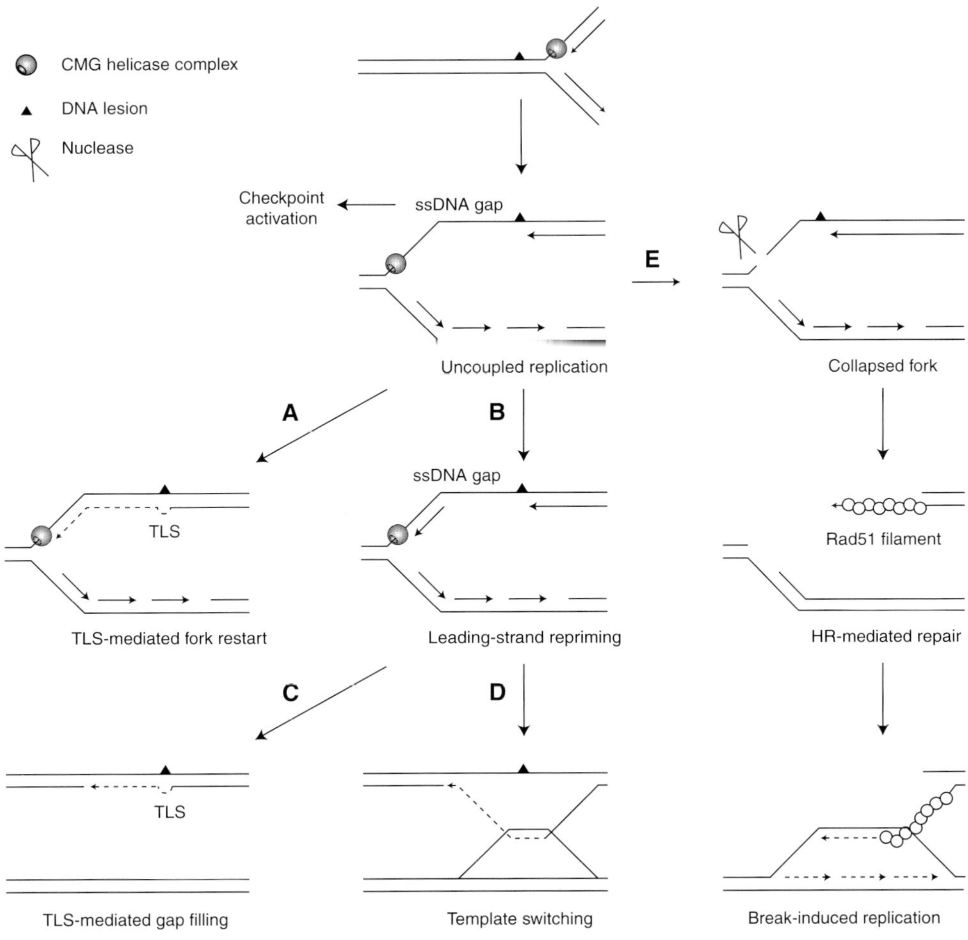

Figure 2. Bypass of leading-strand template damage in eukaryotes. As in bacteria, template unwinding and lagging-strand synthesis are believed to continue beyond the site of damage, resulting in the formation of an ssDNA gap. (*A*) Fork-associated lesion bypass allows the restart of leading-strand synthesis upon PCNA modification and the transient recruitment of a mutagenic TLS polymerase. (*B*) Repriming of the leading strand downstream from the lesion leaves an ssDNA gap. (*C*) This gap can be repaired postreplicatively by TLS polymerases. (*D*) Error-free lesion bypass can also be performed through a recombination-mediated mechanism called a template switch, which uses the newly synthesized sister chromatid as a template for primer elongation. Note that template switching can also occur after fork regression, as illustrated in Figure 4B. (*E*) Incomplete nucleotide excision repair of the DNA lesion or cleavage of the fork by endonucleases may also lead to the formation of a one-ended DSB, which can be repaired by an HR-related process called break-induced replication.

bulk DNA synthesis and can act in G_2/M to bypass UV lesions (Daigaku et al. 2010; Karras and Jentsch 2010; Ulrich 2011). In vertebrates, it is generally believed that replicative and postreplicative translesion syntheses coexist. Studies in chicken DT40 cells have shown that postreplicative TLS depends on PCNA ubiquitination, whereas TLS at stalled forks is independent of PCNA modifications but is promoted by a noncatalytic function of the Rev1 polymerase (Edmunds et al. 2008).

REPLICATION FORK BREAKDOWN

Failure of the bacterial replication machinery to reinitiate replication downstream from a leading-strand lesion results in arrest of replication fork progression and will likely lead to replisome dissociation, although it is worth noting that the fates of the original replisome components during such an event have yet to be described. It is also currently unclear how frequently this process occurs in the absence of exogenous DNA-damaging agents or precisely what structure the arrested replication fork adopts. Currently the consensus view is that lagging-strand replication continues beyond the damage for up to several thousand base pairs, generating an extensive region of ssDNA on the leading-strand template (Fig. 1B) (Higuchi et al. 2003; Pages and Fuchs 2003). However, as it has been recently shown that the replisome transiently stalls in response to leading-strand template damage (Yeeles and Marians 2011), an alternative possibility is that the replisome may dissociate close to the site of damage, before extensive template unwinding and uncoupled replication. In this scenario, little ssDNA would be generated on the leading-strand template, and the 3′ end of the blocked nascent leading strand would be situated in close proximity to the fork junction.

In eukaryotic cells, replication fork breakdown represents a major source of genomic instability and has been directly implicated in cancer development (Aguilera and Gomez-Gonzalez 2008; Hastings et al. 2009). To restrain this instability, eukaryotes have evolved complex mechanisms to maintain and signal stalled replisomes, which are absent in bacteria. The Mec1/ATR pathway plays a central role in this process and acts in various ways to protect stalled forks (Lopes et al. 2001; Tercero and Diffley 2001; see Makarova and Koonin 2013 for details). Although this crucial function has been extensively studied over the past decade, the underlying mechanisms remain poorly understood. Based on chromatin immunoprecipitation studies in budding yeast, it has been proposed that the replisome disassembles when checkpoint mutants of the Mec1 pathway are exposed to hydroxyurea (Katou et al. 2003; Cobb et al. 2005; Branzei and Foiani 2010). However, this view has recently been challenged by a biochemical study showing that the replisome remains associated with DNA in hydroxyurea-treated checkpoint mutants (De Piccoli et al. 2012). One way to reconcile these data is to assume that under these conditions, the replisome does not disassemble but moves away from sites of DNA synthesis. This would expose newly synthesized DNA to exonucleases (Cotta-Ramusino et al. 2005; Segurado and Diffley 2008) or to structure-specific endonucleases (Kai et al. 2005; Froget et al. 2008), leading to irreversible fork collapse. Further work is needed to clarify this important issue.

Besides checkpoint factors, homologous recombination (HR) proteins are also implicated in the protection of arrested forks, in a recombination-independent manner (Costanzo 2011). Studies in vertebrates have shown that the Rad51 recombinase is recruited to stalled forks through a mechanism that depends on Mre11 and Brca2 (Hashimoto et al. 2010; Schlacher et al. 2011; Sirbu et al. 2011). In current models the exonucleolytic activity of Mre11 enlarges ssDNA gaps left behind the replisome to facilitate postreplicative repair, while Brca2 loads Rad51 to limit the extension of ssDNA gaps and protect arrested forks (Costanzo 2011).

Multiple pathways for replication fork reactivation have been described. In bacteria the key step, common to all models, is the origin-independent assembly of the replisome, catalyzed by either the PriA- or PriC-dependent replisome loading systems that are defined by the different DNA structures that they recognize (Heller and Marians 2005). The reactivation pathways differ, however, when describing the sequence of events that occurs before replisome loading. Whereas some models envisage minimal fork processing, others posit extensive remodeling to enable the original lesion to be either repaired or bypassed before the resumption of replication. Recent evidence indicates that similar mechanisms exist in eukaryotes, despite the absence of PriA and PriC orthologs.

DIRECT REPLICATION RESTART

The first biochemical evidence for direct replication restart in bacteria was provided when primase was shown to catalyze the de novo synthesis of a primer on the leading-strand template (Fig. 1B) (Heller and Marians 2006a). Using a model forked substrate lacking nascent leading and lagging strands, and therefore mimicking a stalled fork generated by uncoupled replication downstream from a lesion, it was shown that PriC could direct the assembly of a replisome, which in turn led to the initiation of coupled replication via leading-strand priming. PriC facilitates the loading of DnaB to single-stranded DNA binding protein coated ssDNA on the lagging-strand template, with approximately 20 bp of ssDNA required for efficient loading. However, such a region of ssDNA may not always be available if initiation of the last Okazaki fragment occurs close to the fork junction. In this scenario the $3' \rightarrow 5'$ helicases Rep or PriA function to unwind the nascent lagging stand, exposing the region of ssDNA required for PriC-directed DnaB loading (Heller and Marians 2005a).

If the blocked nascent leading strand is located in close proximity to the fork junction, replication could be reinitiated via the PriA-dependent replication restart pathway (recently reviewed in Gabbai and Marians 2010), as PriA preferentially targets forked structures where the nascent leading strand is located at or close to the fork junction (Heller and Marians 2005b). Following PriA-directed replisome assembly, a leading-strand priming event would enable replication to proceed downstream from the original blockage in an analogous manner to the PriC-dependent pathway. A similar reaction has been fully reconstituted in vitro, in which replication was reinitiated downstream from a leading strand blocked with a chain-terminating dideoxynucleotide (Heller and Marians 2006a). Direct replication restart generates ssDNA gaps in the leading-strand template, in much the same way as those generated when the leading strand is reinitiated without replisome dissociation. To date these are the only models that are able to account for the discontinuous replication that has been observed following UV irradiation in vivo (Rupp and Howard-Flanders 1968).

Unlike bacterial genomes, eukaryotic chromosomes contain a large excess of licensed replication origins that can be used as backup origins to rescue terminally arrested forks (Ge et al. 2007; Ibarra et al. 2008). Consequently, the importance of replication restart pathways has long been disregarded in eukaryotes. With the emergence of DNA combing and other related DNA fiber assays (Tourrière et al. 2005; Petermann and Helleday 2010), it has become clearer that fork restart mechanisms operate in eukaryotic cells and are important for viability under replication stress conditions. For instance, it has been recently reported that human cells briefly exposed to hydroxyurea are able to restart paused forks through a Rad51-dependent mechanism that is distinct from classical double-strand break (DSB) repair (Petermann et al. 2010). Similarly, fork restart by recombination-dependent pathways has been reported in fission yeast, independently of DSB formation (Mizuno et al. 2009; Lambert et al. 2010). In budding yeast hydroxyurea induces a 10- to 20-fold reduction of fork rate (Sogo et al. 2002; Poli et al. 2012) and increases the amount of ssDNA gaps at paused forks by ~100 nucleotides (Sogo et al. 2002), which could be sufficient to allow repriming on the leading strand (Lambert et al. 2007). In higher eukaryotes much larger ssDNA gaps are generated as a consequence of extensive uncoupling between helicase and polymerase activities (Byun et al. 2005). In *Xenopus* egg extracts Pol α-primase is recruited to these ssDNA gaps and undergoes continued primer synthesis (Van et al. 2010). This recruitment depends on TopBP1, a replication initiation factor also involved in the activation of the ATR/Mec1 pathway (Yan and Michael 2009), indicating that the events signaling and reactivating arrested forks are functionally linked. Upon initiation, Pol α-primase interacts with other replisome components such as Mcm10 and Ctf4/And-1 (Ricke and Bielinsky 2004; Zhu et al. 2007; Gambus et al. 2009). These interactions are conserved from yeast to human and are important for both initiation and elongation (Tanaka et al. 2009; Kanke et al. 2012; van Deursen et al. 2012; Watase et al. 2012). Whether Pol α-primase binding to Ctf4 and Mcm10 is also required for repriming at stalled forks is currently unknown. However, recent evidence indicates that *S. cerevisiae* mutants that are unable to tether Pol α-primase to the replisome depend on the Mec1/ATR pathway for viability (Kilkenny et al. 2012).

REPLICATION FORK REMODELING

In contrast to models of direct replication restart, additional pathways for fork reactivation in *E. coli* have been proposed that require extensive remodeling to enable the original lesion to be either removed or bypassed before PriA-dependent replisome reloading (McGlynn and Lloyd 2002; Courcelle and Hanawalt 2003). Two of the major pathways describe the regression of the replication fork, either catalyzed by the helicase RecG (McGlynn and Lloyd 2000; McGlynn et al. 2001) or promoted by the strand exchange protein RecA (Fig. 3) (Robu et al. 2001; Lusetti and Cox 2002; Courcelle et al. 2003). Regression of a replication fork results in rewinding of the parental DNA and displacement of the nascent leading and lagging strands, which can subsequently base pair to form a four-way Holliday junction. Consequently, the original replication-blocking lesion is returned to a region of duplex DNA, enabling it to be removed by NER. Reverse branch migration, catalyzed by RecG (McGlynn and Lloyd 2000), or processing of the extruded nascent strands (Courcelle et al. 2003) enables a replication fork structure to be regenerated (Fig. 3A). Alternatively, the pairing of the extruded nascent leading and lagging strands can provide a template for extension of the leading strand across the site of damage in a reaction termed template switching (Fig. 3B) (Higgins et al. 1976). Resetting of the replication fork places the 3′ OH of the nascent leading strand beyond the site of damage so that replication can be resumed without a need for leading-strand repriming.

Whether fork regression also occurs in eukaryotic cells is a highly debated issue (Atkinson and McGlynn 2009). Because the unwinding of replication intermediates to form a four-way junction implies a disengagement of the replisome, it was initially believed that fork reversal is a pathological situation leading to irreversible fork breakdown, as observed in yeast checkpoint mutants exposed to hydroxyurea (Sogo et al. 2002; Bermejo et al. 2011). However, a recent study indicates that fork reversal is a physiological process that occurs at 15%–40% of the forks in different organisms on exposure to sublethal doses of Top1 inhibitors, such as camptothecin and topotecan. Interestingly, this process is mediated by poly(ADP-ribose) polymerase (PARP). PARP inactivation with chemical inhibitors or PARP1 genetic ablation impairs fork reversal and leads rather to the formation of DSBs, indicating that fork reversal prevents fork collapse at camptothecin-induced lesions (Ray Chaudhuri et al. 2012). An increasing number of eukaryotic factors able to drive fork reversal in vitro have been identified. These include the yeast Rad5 helicase (Blastyak et al. 2007; Minca and Kowalski 2010) and its human ortholog HLTF (Achar et al. 2011), as well as the Fanconi anemia helicase FANCM (Gari et al. 2008) and its fission yeast ortholog Fml1 (Sun et al. 2008). As illustrated in Figure 4, reversed forks can be restarted by exonucleolytic degradation, reverse branch migration catalyzed by the annealing helicase SMARCAL1 (Driscoll and Cimprich 2009; Betous et al. 2012), or HR-mediated mechanisms. How these mechanisms operate in vivo remains largely unexplored.

RECOMBINATION-MEDIATED FORK RESTART

When fork restart mechanisms fail or when the replisome encounters ssDNA nicks or gaps, replication forks can be converted into one-ended DSBs that are subsequently repaired by homology-mediated recombination mechanisms

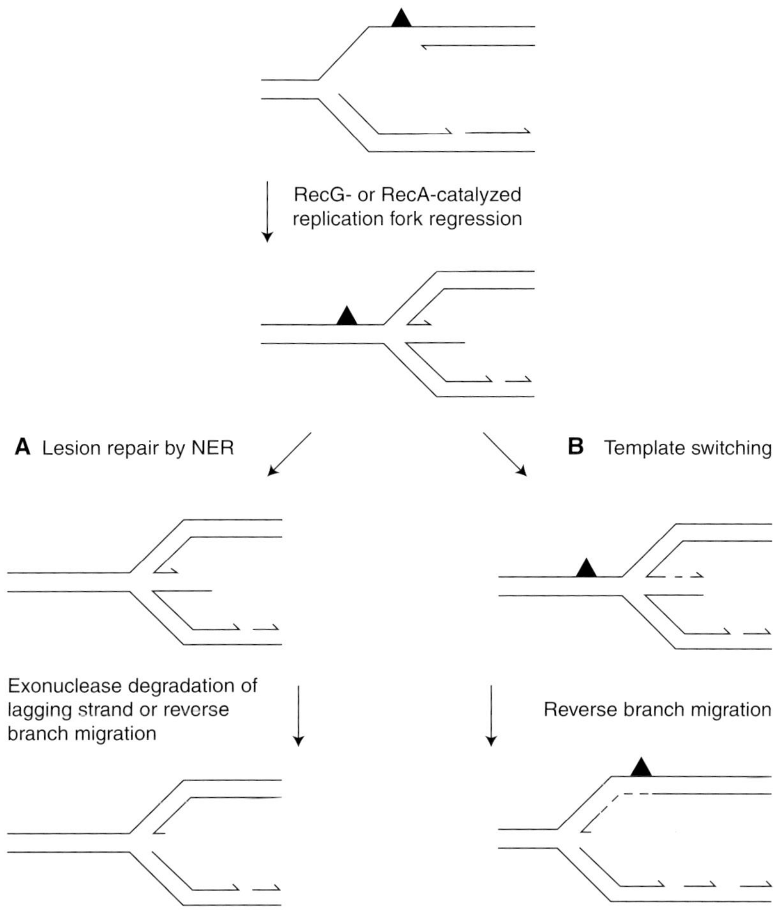

Figure 3. Models for replication fork regression at *E. coli* replication forks stalled by leading-strand template damage. The helicase RecG binds with high affinity to forks containing a leading-strand gap and unwinds the structure to generate a Holliday junction. A second pathway of fork regression involves RecA binding to the single-stranded region of the fork to drive fork regression and potentially form a Holliday junction. (*A*) Regression of the fork places the original lesion in a region of double-stranded DNA, enabling it to be repaired by NER. Exonucleolytic degradation of the lagging-strand extension resets the fork to enable replisome loading. (*B*) The nascent leading strand that is displaced by the regression reaction can be extended using the lagging strand as template. Rewinding of the Holliday junction, possibly by RecG, will place the nascent 3′ end of the leading strand beyond the site of damage to enable replication restart to occur without a need to reprime the leading strand.

related to break-induced replication (Fig. 2, pathway E; see de la Paz Sanchez et al. 2012 for details). In eukaryotes, this process depends on all the factors involved in processive DNA replication, with the exception of components of the prereplication complex such as ORC and Cdc6 (Lydeard et al. 2010). Evidence from *Xenopus* egg extracts indicates that when forks encounter ssDNA lesions, Pol ε and the helicase component GINS detach from the replisome and are reloaded in an Mre11- and Rad51-dependent manner. As for break-induced replication, fork restart involves PCNA modifications allowing Pol η–dependent strand extension (Hashimoto et al. 2012). In yeasts, fork restart depends on Rad52 and replication factor C, but not on factors involved in nonhomologous end joining. Interestingly, fork restart can also occur

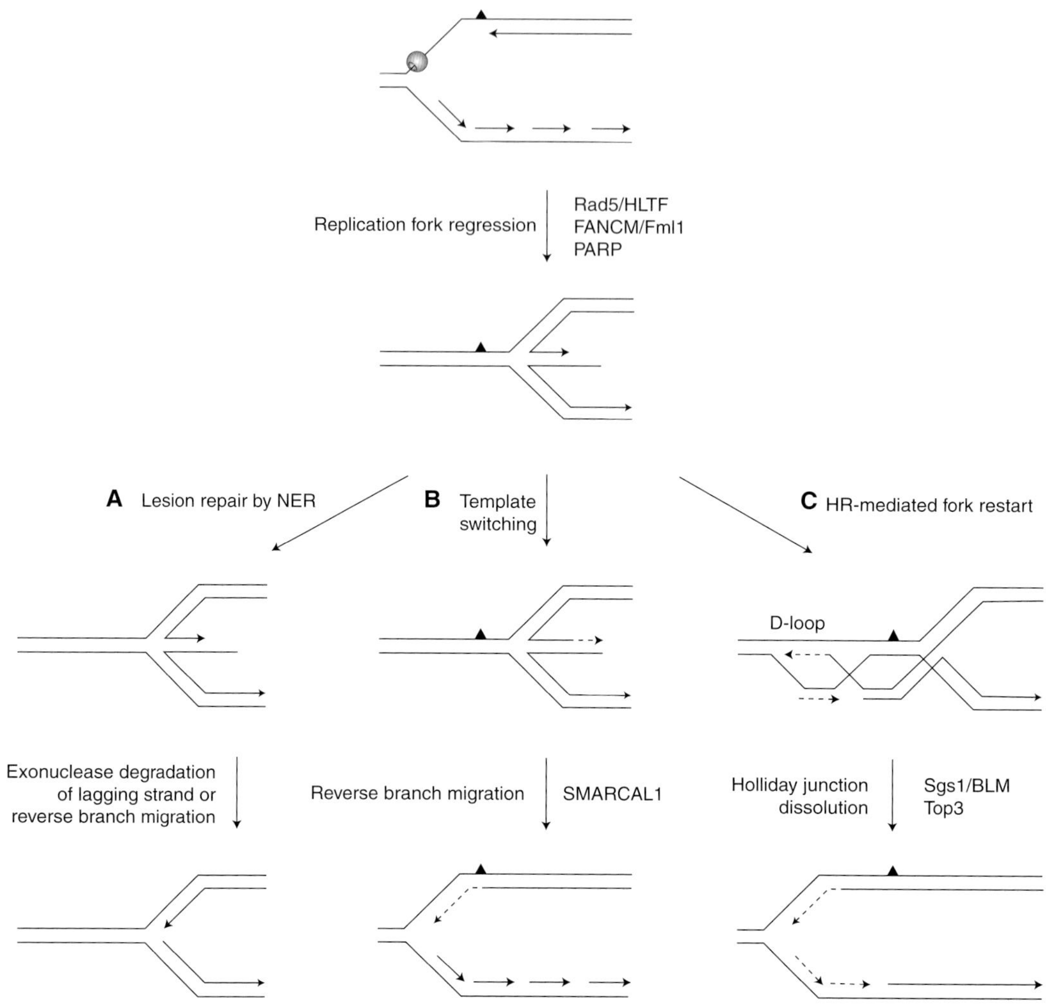

Figure 4. Models for replication fork regression at eukaryotic forks stalled by leading-strand template damage. Several DNA helicases have been shown to promote fork regression in vitro in yeast and vertebrates, including Rad5/HLTF and FANCM/Fml1. In vertebrates, fork regression in vivo also depends on poly(ADP-ribose) polymerase (PARP) (Ray Chaudhuri et al. 2012). Resumption of DNA replication after repair of the lesion (*A*) or template switching (*B*) is mediated by nucleolytic degradation of branched structures or reverse branch migration, as described for bacteria. (*C*) Fork restart can also occur after invasion of the duplex ahead of the lesion by the 3′ overhang of the forked structure. The resulting Holliday junctions are resolved by resolvases or dissolved by the combined action of the RecQ helicases BLM/Sgs1 and the type I topoisomerase Top3, whose function is conserved from yeasts to human.

in the absence of Rad51, through a mechanism that depends on the Pol δ subunit Pol32 (Moriel-Carretero and Aguilera 2010). Importantly, recombination-mediated fork restart also occurs in the absence of DSBs when the replisome encounters a replication fork barrier (Lambert et al. 2010).

Besides recombination factors, a wide variety of proteins have been implicated in the recombinational repair of stalled or broken replication forks. These include Slx4, a conserved scaffold protein that interacts with Mus81 and other structure-specific endonucleases (Flott et al. 2007; Andersen et al. 2009; Fekairi et al. 2009;

Muñoz et al. 2009; Svendsen et al. 2009). The mechanisms by which Slx4 and Mus81 promote fork recovery are currently unclear. An attractive possibility could be that Mus81 cleaves terminally arrested forks to convert them into structures that are more amenable to recombinational repair (Hanada et al. 2007). However, recent evidence indicates that terminally arrested forks are not efficiently restarted by Mus81 and are rescued primarily by the activation of dormant replication origins (Petermann et al. 2010).

In budding yeast, fork restart depends on Rtt107 (Roberts et al. 2008), a scaffold protein with BRCT domains that interacts with fork repair complexes such as Smc5/6 and Rtt101–Mms1–Mms22 (Luke et al. 2006; Ohouo et al. 2010), which are potentially conserved from yeasts to humans (De Piccoli et al. 2009; Piwko et al. 2011). In budding yeast, the Rtt101 complex promotes the recovery of MMS-arrested forks by sister-chromatid recombination (Duro et al. 2008). Interestingly, Rtt101 function also depends on the acetylation of histone H3 on lysine 56, which is found on newly incorporated histones and may help cells discriminate between parental and newly replicated DNA (Han et al. 2007; Wurtele et al. 2011). Deacetylation of H3K56 is prevented by the Mec1 kinase in response to replication stress in order to facilitate fork recovery (Masumoto et al. 2005). Remarkably, Mec1 also promotes the interaction between Dpb11/TopBP1 and the scaffold proteins Rtt107 and Slx4, which may target these factors to stalled forks (Ohouo et al. 2010). Finally, recent evidence indicates that fork restart also depends on cohesin (Tittel-Elmer et al. 2009) and on the chromatin remodeling complex INO80 (Papamichos-Chronakis and Peterson 2008; Shimada et al. 2008; Falbo et al. 2009). How these multiple factors cooperate to promote fork restart is currently unclear.

As is the case in eukaryotes, the bacterial homologous recombination machinery plays a critical role in salvaging broken replication forks—generated following collisions between the replisome and template nicks or via the cleavage of regressed forks (McGlynn and Lloyd 2002)—in a process termed recombination-dependent replication (Asai et al. 1994). The first step of the pathway involves the resection of the double-stranded DNA end by either an AddAB- or RecBCD-type helicase–nuclease to generate a 3′ ssDNA tail onto which the strand exchange protein RecA is loaded in a sequence-regulated manner (reviewed in Yeeles and Dillingham 2010). Subsequent strand invasion and homologous pairing with the intact chromosome arm generates a D-loop structure, which is then targeted by PriA for origin-independent replisome loading, leading to the establishment of a fully functional replisome (Xu and Marians 2003).

CONCLUDING REMARKS

Studies from many laboratories over the past 20 years have shown convincingly that stalling of replication fork progression can be a dangerous genome-destabilizing event, leading to unscheduled, toxic recombination; degradation of the nascent DNA; and the loss, rearrangement, and alteration of genetic information. The importance of stabilizing and restarting stalled replication forks is underscored by the many proteins involved in these processes that, when mutated, lead to DNA damage syndromes and cancer predisposition. Our understanding of these processes has improved considerably, but there is still much to do, particularly in modeling these events in vitro with the eukaryotic replisome.

ACKNOWLEDGMENTS

We thank Angelos Constantinou, Sarah Lambert, Massimo Lopes, and Benjamin Pardo for discussions and critical comments on the manuscript. Studies from K.J.M.'s laboratory were supported by National Institutes of Health grant GM34557. Studies from P.P.'s laboratory were supported by ANR and Fondation pour la Recherche Médicale (Equipe FRM).

REFERENCES

**Reference is also in this collection.*

Achar YJ, Balogh D, Haracska L. 2011. Coordinated protein and DNA remodeling by human HLTF on stalled replication fork. *Proc Natl Acad Sci* **108:** 14073–14078.

Aguilera A, Gomez-Gonzalez B. 2008. Genome instability: A mechanistic view of its causes and consequences. *Nat Rev Genet* 9: 204–217.

Andersen SL, Bergstralh DT, Kohl KP, LaRocque JR, Moore CB, Sekelsky J. 2009. *Drosophila* MUS312 and the vertebrate ortholog BTBD12 interact with DNA structure-specific endonucleases in DNA repair and recombination. *Mol Cell* **35:** 128–135.

Asai T, Bates DB, Kogoma T. 1994. DNA replication triggered by double-stranded breaks in *E. coli*: Dependence on homologous recombination functions. *Cell* **78:** 1051–1061.

Atkinson J, McGlynn P. 2009. Replication fork reversal and the maintenance of genome stability. *Nucleic Acids Res* **37:** 3475–3492.

Belle JJ, Casey A, Courcelle CT, Courcelle J. 2007. Inactivation of the DnaB helicase leads to the collapse and degradation of the replication fork: A comparison to UV-induced arrest. *J Bacteriol* **189:** 5452–5462.

Bermejo R, Capra T, Jossen R, Colosio A, Frattini C, Carotenuto W, Cocito A, Doksani Y, Klein H, Gomez-Gonzalez B, et al. 2011. The replication checkpoint protects fork stability by releasing transcribed genes from nuclear pores. *Cell* **146:** 233–246.

Betous R, Mason AC, Rambo RP, Bansbach CE, Badu-Nkansah A, Sirbu BM, Eichman BF, Cortez D. 2012. SMARCAL1 catalyzes fork regression and Holliday junction migration to maintain genome stability during DNA replication. *Genes Dev* **26:** 151–162.

Blastyak A, Pinter L, Unk I, Prakash L, Prakash S, Haracska L. 2007. Yeast Rad5 protein required for postreplication repair has a DNA helicase activity specific for replication fork regression. *Mol Cell* **28:** 167–175.

Branzei D, Foiani M. 2010. Maintaining genome stability at the replication fork. *Nat Rev Mol Cell Biol* **11:** 208–219.

Byun TS, Pacek M, Yee M-C, Walter JC, Cimprich KA. 2005. Functional uncoupling of MCM helicase and DNA polymerase activities activates the ATR-dependent checkpoint. *Genes Dev* **19:** 1040–1052.

Chandler M, Bird RE, Caro L. 1975. The replication time of the *Escherichia coli* K12 chromosome as a function of cell doubling time. *J Mol Biol* **94:** 127–132.

Cobb JA, Schleker T, Rojas V, Bjergbaek L, Tercero JA, Gasser SM. 2005. Replisome instability, fork collapse, and gross chromosomal rearrangements arise synergistically from Mec1 kinase and RecQ helicase mutations. *Genes Dev* **19:** 3055–3069.

Costanzo V. 2011. Brca2, Rad51 and Mre11: Performing balancing acts on replication forks. *DNA Repair (Amst)* **10:** 1060–1065.

Cotta-Ramusino C, Fachinetti D, Lucca C, Doksani Y, Lopes M, Sogo J, Foiani M. 2005. Exo1 processes stalled replication forks and counteracts fork reversal in checkpoint-defective cells. *Mol Cell* **17:** 153–159.

Courcelle J, Hanawalt PC. 2003. RecA-dependent recovery of arrested DNA replication forks. *Annu Rev Genet* **37:** 611–646.

Courcelle J, Crowley DJ, Hanawalt PC. 1999. Recovery of DNA replication in UV-irradiated *Escherichia coli* requires both excision repair and RecF protein function. *J Bacteriol* **181:** 916–922.

Courcelle J, Donaldson JR, Chow KH, Courcelle CT. 2003. DNA damage-induced replication fork regression and processing in *Escherichia coli*. *Science* **299:** 1064–1067.

Courcelle CT, Belle JJ, Courcelle J. 2005. Nucleotide excision repair or polymerase V-mediated lesion bypass can act to restore UV-arrested replication forks in *Escherichia coli*. *J Bacteriol* **187:** 6953–6961.

Daigaku Y, Davies AA, Ulrich HD. 2010. Ubiquitin-dependent DNA damage bypass is separable from genome replication. *Nature* **465:** 951–955.

* de la Paz Sanchez M, Costas C, Sequeira-Mendes J, Gutierrez C. 2012. Regulating DNA replication in plants. *Cold Spring Harb Perspect Biol* **4:** a010140.

De Piccoli G, Torres-Rosell J, Aragon L. 2009. The unnamed complex: What do we know about Smc5-Smc6? *Chromosome Res* **17:** 251–263.

De Piccoli G, Katou Y, Itoh T, Nakato R, Shirahige K, Labib K. 2012. Replisome stability at defective DNA replication forks is independent of S phase checkpoint kinases. *Mol Cell* **45:** 696–704.

Driscoll R, Cimprich KA. 2009. HARPing on about the DNA damage response during replication. *Genes Dev* **23:** 2359–2365.

Duro E, Vaisica JA, Brown GW, Rouse J. 2008. Budding yeast Mms22 and Mms1 regulate homologous recombination induced by replisome blockage. *DNA Repair (Amst)* **7:** 811–818.

Edmunds CE, Simpson LJ, Sale JE. 2008. PCNA ubiquitination and REV1 define temporally distinct mechanisms for controlling translesion synthesis in the avian cell line DT40. *Mol Cell* **30:** 519–529.

Falbo KB, Alabert C, Katou Y, Wu S, Han J, Wehr T, Xiao J, He X, Zhang Z, Shi Y, et al. 2009. Involvement of a chromatin remodeling complex in damage tolerance during DNA replication. *Nat Struct Mol Biol* **16:** 1167–1172.

Fekairi S, Scaglione S, Chahwan C, Taylor ER, Tissier A, Coulon S, Dong M-Q, Ruse C, Yates Iii JR, Russell P, et al. 2009. Human SLX4 is a Holliday junction resolvase subunit that binds multiple DNA repair/recombination endonucleases. *Cell* **138:** 78–89.

Flott S, Alabert C, Toh GW, Toth R, Sugawara N, Campbell DG, Haber JE, Pasero P, Rouse J. 2007. Phosphorylation of Slx4 by Mec1 and Tel1 regulates the single-strand annealing mode of DNA repair in budding yeast. *Mol Cell Biol* **27:** 6433–6445.

Freidberg EC, Walker GC, Siede W, Wood RD, Schultz RA, Ellenberger T. 2005. *DNA repair and mutagenesis*, 2nd ed. ASM Press, Washington.

Froget B, Blaisonneau J, Lambert S, Baldacci G. 2008. Cleavage of stalled forks by fission yeast Mus81/Eme1 in absence of DNA replication checkpoint. *Mol Biol Cell* **19:** 445–456.

Fu YV, Yardimci H, Long DT, Ho TV, Guainazzi A, Bermudez VP, Hurwitz J, van Oijen A, Scharer OD, Walter JC. 2011. Selective bypass of a lagging strand roadblock by the eukaryotic replicative DNA helicase. *Cell* **146:** 931–941.

Gabbai CB, Marians KJ. 2010. Recruitment to stalled replication forks of the PriA DNA helicase and replisome-loading activities is essential for survival. *DNA Repair (Amst)* **9:** 202–209.

Gambus A, van Deursen F, Polychronopoulos D, Foltman M, Jones RC, Edmondson RD, Calzada A, Labib K. 2009. A key role for Ctf4 in coupling the MCM2-7 helicase to DNA polymerase α within the eukaryotic replisome. *EMBO J* **28:** 2992–3004.

Gari K, Decaillet C, Stasiak AZ, Stasiak A, Constantinou A. 2008. The Fanconi anemia protein FANCM can promote branch migration of Holliday junctions and replication forks. *Mol Cell* **29:** 141–148.

Ge XQ, Jackson DA, Blow JJ. 2007. Dormant origins licensed by excess Mcm2–7 are required for human cells to survive replicative stress. *Genes Dev* **21:** 3331–3341.

* Goodman MF, Woodgate R. 2013. Translesion DNA polymerases. *Cold Spring Harb Perspect Biol* doi: 10.1101/cshperspect.a010363.

Han J, Zhou H, Horazdovsky B, Zhang K, Xu RM, Zhang Z. 2007. Rtt109 acetylates histone H3 lysine 56 and functions in DNA replication. *Science* **315:** 653–655.

Hanada K, Budzowska M, Davies SL, van Drunen E, Onizawa H, Beverloo HB, Maas A, Essers J, Hickson ID, Kanaar R. 2007. The structure-specific endonuclease Mus81 contributes to replication restart by generating double-strand DNA breaks. *Nat Struct Mol Biol* **14:** 1096–1104.

Hashimoto Y, Chaudhuri AR, Lopes M, Costanzo V. 2010. Rad51 protects nascent DNA from Mre11-dependent degradation and promotes continuous DNA synthesis. *Nat Struct Mol Biol* **17:** 1305–1311.

Hashimoto Y, Puddu F, Costanzo V. 2012. RAD51- and MRE11-dependent reassembly of uncoupled CMG helicase complex at collapsed replication forks. *Nat Struct Mol Biol* **19:** 17–24.

Hastings PJ, Ira G, Lupski JR. 2009. A microhomology-mediated break-induced replication model for the origin of human copy number variation. *PLoS Genet* **5:** e1000327.

Heller RC, Marians KJ. 2005a. Unwinding of the nascent lagging strand by Rep and PriA enables the direct restart of stalled replication forks. *J Biol Chem* **280:** 34143–34151.

Heller RC, Marians KJ. 2005b. The disposition of nascent strands at stalled replication forks dictates the pathway of replisome loading during restart. *Mol Cell* **17:** 733–743.

Heller RC, Marians KJ. 2006a. Replication fork reactivation downstream of a blocked nascent leading strand. *Nature* **439:** 557–562.

Heller RC, Marians KJ. 2006b. Replisome assembly and the direct restart of stalled replication forks. *Nat Rev Mol Cell Biol* **7:** 932–943.

Higgins NP, Kato K, Strauss B. 1976. A model for replication repair in mammalian cells. *J Mol Biol* **101:** 417–425.

Higuchi K, Katayama T, Iwai S, Hidaka M, Horiuchi T, Maki H. 2003. Fate of DNA replication fork encountering a single DNA lesion during *oriC* plasmid DNA replication *in vitro*. *Genes Cells* **8:** 437–449.

Ibarra A, Schwob E, Mendez J. 2008. Excess MCM proteins protect human cells from replicative stress by licensing backup origins of replication. *Proc Natl Acad Sci* **105:** 8956–8961.

Iyer VN, Rupp WD. 1971. Usefulness of benzoylated naphthoylated DEAE-cellulose to distinguish and fractionate double-stranded DNA bearing different extents of single-stranded regions. *Biochem Biophys Acta* **228:** 117–126.

Kai M, Boddy MN, Russell P, Wang TS. 2005. Replication checkpoint kinase Cds1 regulates Mus81 to preserve genome integrity during replication stress. *Genes Dev* **19:** 919–932.

Kanke M, Kodama Y, Takahashi TS, Nakagawa T, Masukata H. 2012. Mcm10 plays an essential role in origin DNA unwinding after loading of the CMG components. *EMBO J* **31:** 2182–2194.

Kannouche PL, Wing J, Lehmann AR. 2004. Interaction of human DNA polymerase η with monoubiquitinated PCNA: A possible mechanism for the polymerase switch in response to DNA damage. *Mol Cell* **14:** 491–500.

Karras GI, Jentsch S. 2010. The RAD6 DNA damage tolerance pathway operates uncoupled from the replication fork and is functional beyond S phase. *Cell* **141:** 255–267.

Katou Y, Kanoh Y, Bando M, Noguchi H, Tanaka H, Ashikari T, Sugimoto K, Shirahige K. 2003. S-phase checkpoint proteins Tof1 and Mrc1 form a stable replication pausing complex. *Nature* **424:** 1078–1083.

Khidhir MA, Casaregola S, Holland IB. 1985. Mechanism of transient inhibition of DNA synthesis in ultraviolet-irradiated *E. coli*: Inhibition is independent of *recA* whilst recovery requires RecA protein itself and an additional, inducible SOS function. *Mol Gen Genet* **199:** 133–140.

Kilkenny ML, De Piccoli G, Perera RL, Labib K, Pellegrini L. 2012. A conserved motif in the C-terminal tail of DNA polymerase α tethers primase to the eukaryotic replisome. *J Biol Chem* **287:** 23740–23747.

Kunkel TA. 2004. DNA replication fidelity. *J Biol Chem* **279:** 16895–16898.

Lambert S, Froget B, Carr AM. 2007. Arrested replication fork processing: Interplay between checkpoints and recombination. *DNA Repair (Amst)* **6:** 1042–1061.

Lambert S, Mizuno K, Blaisonneau J, Martineau S, Chanet R, Freon K, Murray JM, Carr AM, Baldacci G. 2010. Homologous recombination restarts blocked replication forks at the expense of genome rearrangements by template exchange. *Mol Cell* **39:** 346–359.

Lehmann AR, Fuchs RP. 2006. Gaps and forks in DNA replication: Rediscovering old models. *DNA Repair (Amst)* **5:** 1495–1498.

Lia G, Michel B, Allemand JF. 2012. Polymerase exchange during Okazaki fragment synthesis observed in living cells. *Science* **335:** 328–331.

Lindahl T. 1993. Instability and decay of the primary structure of DNA. *Nature* **362:** 709–715.

Lopes M, Foiani M, Sogo JM. 2006. Multiple mechanisms control chromosome integrity after replication fork uncoupling and restart at irreparable UV lesions. *Mol Cell* **21:** 15–27.

Luke B, Versini G, Jaquenoud M, Zaidi IW, Kurz T, Pintard L, Pasero P, Peter M. 2006. The cullin Rtt101p promotes replication fork progression through damaged DNA and natural pause sites. *Curr Biol* **16:** 786–792.

Lusetti SL, Cox MM. 2002. The bacterial RecA protein and the recombinational DNA repair of stalled replication forks. *Annu Rev Biochem* **71:** 71–100.

Lydeard JR, Lipkin-Moore Z, Sheu Y-J, Stillman B, Burgers PM, Haber JE. 2010. Break-induced replication requires all essential DNA replication factors except those specific for pre-RC assembly. *Genes Dev* **24:** 1133–1144.

* Makarova KS, Koonin EV. 2013. Archaeology of eukaryotic DNA replication. *Cold Spring Harb Perspect Biol* doi: 10.1101/cshperspect.a012963.

Masumoto H, Hawke D, Kobayashi R, Verreault A. 2005. A role for cell-cycle-regulated histone H3 lysine 56 acetylation in the DNA damage response. *Nature* **436:** 294–298.

McGlynn P, Lloyd RG. 2000. Modulation of RNA polymerase by (p)ppGpp reveals a RecG-dependent mechanism for replication fork progression. *Cell* **101:** 35–45.

McGlynn P, Lloyd RG. 2002. Recombinational repair and restart of damaged replication forks. *Nat Rev Mol Cell Biol* **3:** 859–870.

McGlynn P, Lloyd RG, Marians KJ. 2001. Formation of Holliday junctions by regression of nascent DNA in intermediates containing stalled replication forks: RecG stimulates regression even when the DNA is negatively supercoiled. *Proc Natl Acad Sci* **98:** 8235–8240.

McInerney P, O'Donnell M. 2004. Functional uncoupling of twin polymerases: Mechanism of polymerase dissociation from a lagging-strand block. *J Biol Chem* **279:** 21543–21551.

McInerney P, Johnson A, Katz F, O'Donnell M. 2007. Characterization of a triple DNA polymerase replisome. *Mol Cell* **27:** 527–538.

Mezzina M, Menck CF, Courtin P, Sarasin A. 1988. Replication of simian virus 40 DNA after UV irradiation: Evidence of growing fork blockage and single-stranded gaps in daughter strands. *J Virol* **62:** 4249–4258.

Minca EC, Kowalski D. 2010. Multiple Rad5 activities mediate sister chromatid recombination to bypass DNA damage at stalled replication forks. *Mol Cell* **38:** 649–661.

Mizuno K, Lambert S, Baldacci G, Murray JM, Carr AM. 2009. Nearby inverted repeats fuse to generate acentric and dicentric palindromic chromosomes by a replication template exchange mechanism. *Genes Dev* **23:** 2876–2886.

Moldovan G-L, Pfander B, Jentsch S. 2007. PCNA, the maestro of the replication fork. *Cell* **129:** 665–679.

Moriel-Carretero M, Aguilera A. 2010. A postincision-deficient TFIIH causes replication fork breakage and uncovers alternative Rad51- or Pol32-mediated restart mechanisms. *Mol Cell* **37:** 690–701.

Muñoz IM, Hain K, Déclais A-C, Gardiner M, Toh GW, Sanchez-Pulido L, Heuckmann JM, Toth R, Macartney T, Eppink B, et al. 2009. Coordination of structure-specific nucleases by human SLX4/BTBD12 is required for DNA repair. *Mol Cell* **35:** 116–127.

Nelson SW, Benkovic SJ. 2010. Response of the bacteriophage T4 replisome to noncoding lesions and regression of a stalled replication fork. *J Mol Biol* **401:** 743–756.

Ohouo PY, Bastos de Oliveira FM, Almeida BS, Smolka MB. 2010. DNA damage signaling recruits the Rtt107-Slx4 scaffolds via Dpb11 to mediate replication stress response. *Mol Cell* **39:** 300–306.

Pages V, Fuchs RP. 2003. Uncoupling of leading- and lagging-strand DNA replication during lesion bypass in vivo. *Science* **300:** 1300–1303.

Papamichos-Chronakis M, Peterson CL. 2008. The Ino80 chromatin-remodeling enzyme regulates replisome function and stability. *Nat Struct Mol Biol* **15:** 338–345.

Petermann E, Helleday T. 2010. Pathways of mammalian replication fork restart. *Nat Rev Mol Cell Biol* **11:** 683–687.

Petermann E, Orta ML, Issaeva N, Schultz N, Helleday T. 2010. Hydroxyurea-stalled replication forks become progressively inactivated and require two different RAD51-mediated pathways for restart and repair. *Mol Cell* **37:** 492–502.

Piwko W, Buser R, Peter M. 2011. Rescuing stalled replication forks: MMS22L-TONSL, a novel complex for DNA replication fork repair in human cells. *Cell Cycle* **10:** 1703–1705.

Poli J, Tsaponina O, Crabbe L, Keszthelyi A, Pantesco V, Chabes A, Lengronne A, Pasero P. 2012. dNTP pools determine fork progression and origin usage under replication stress. *EMBO J* **31:** 883–894.

Ray Chaudhuri A, Hashimoto Y, Herrador R, Neelsen KJ, Fachinetti D, Bermejo R, Cocito A, Costanzo V, Lopes M. 2012. Topoisomerase I poisoning results in PARP-mediated replication fork reversal. *Nat Struct Mol Biol* **19:** 417–423.

Reyes-Lamothe R, Sherratt DJ, Leake MC. 2010. Stoichiometry and architecture of active DNA replication machinery in *Escherichia coli*. *Science* **328:** 498–501.

Ricke RM, Bielinsky AK. 2004. Mcm10 regulates the stability and chromatin association of DNA polymerase-α. *Mol Cell* **16:** 173–185.

Roberts TM, Zaidi IW, Vaisica JA, Peter M, Brown GW. 2008. Regulation of Rtt107 recruitment to stalled DNA replication forks by the cullin Rtt101 and the Rtt109 acetyltransferase. *Mol Biol Cell* **19:** 171–180.

Robu ME, Inman RB, Cox MM. 2001. RecA protein promotes the regression of stalled replication forks *in vitro*. *Proc Natl Acad Sci* **98:** 8211–8218.

Rudolph CJ, Upton AL, Lloyd RG. 2007. Replication fork stalling and cell cycle arrest in UV-irradiated *Escherichia coli*. *Genes Dev* **21:** 668–681.

Rupp WD. 1996. DNA repair mechanisms. In Escherichia coli *and* Salmonella typhimurium: *Cellular and molecular biology* (ed. Neidhardt FC), pp. 2277–2294. American Society for Microbiology, Washington, DC.

Rupp WD, Howard-Flanders P. 1968. Discontinuities in the DNA synthesized in an excision-defective strain of *Escherichia coli* following ultraviolet irradiation. *J Mol Biol* **31:** 291–304.

Schlacher K, Christ N, Siaud N, Egashira A, Wu H, Jasin M. 2011. Double-strand break repair-independent role for BRCA2 in blocking stalled replication fork degradation by MRE11. *Cell* **145:** 529–542.

Segurado M, Diffley JFX. 2008. Separate roles for the DNA damage checkpoint protein kinases in stabilizing DNA replication forks. *Genes Dev* **22:** 1816–1827.

Shimada K, Oma Y, Schleker T, Kugou K, Ohta K, Harata M, Gasser SM. 2008. Ino80 chromatin remodeling complex promotes recovery of stalled replication forks. *Curr Biol* **18:** 566–575.

Sirbu BM, Couch FB, Feigerle JT, Bhaskara S, Hiebert SW, Cortez D. 2011. Analysis of protein dynamics at active, stalled, and collapsed replication forks. *Genes Dev* **25:** 1320–1327.

Sogo JM, Lopes M, Foiani M. 2002. Fork reversal and ssDNA accumulation at stalled replication forks owing to checkpoint defects. *Science* **297:** 599–602.

Sun W, Nandi S, Osman F, Ahn JS, Jakovleska J, Lorenz A, Whitby MC. 2008. The FANCM ortholog Fml1 promotes recombination at stalled replication forks and limits crossing over during DNA double-strand break repair. *Mol Cell* **32:** 118–128.

Svendsen JM, Smogorzewska A, Sowa ME, O'Connell BC, Gygi SP, Elledge SJ, Harper JW. 2009. Mammalian BTBD12/SLX4 assembles a Holliday junction resolvase and is required for DNA repair. *Cell* **138:** 63–77.

Tanaka H, Katou Y, Yagura M, Saitoh K, Itoh T, Araki H, Bando M, Shirahige K. 2009. Ctf4 coordinates the progression of helicase and DNA polymerase α. *Genes Cells* **14:** 807–820.

Tittel-Elmer M, Alabert C, Pasero P, Cobb JA. 2009. The MRX complex stabilizes the replisome independently of the S phase checkpoint during replication stress. *EMBO J* **28:** 1142–1156.

Tourrière H, Versini G, Cordón-Preciado V, Alabert C, Pasero P. 2005. Mrc1 and Tof1 promote replication fork progression and recovery independently of Rad53. *Mol Cell* **19:** 699–706.

Ulrich HD. 2011. Timing and spacing of ubiquitin-dependent DNA damage bypass. *FEBS Lett* **585:** 2861–2867.

Van C, Yan S, Michael WM, Waga S, Cimprich KA. 2010. Continued primer synthesis at stalled replication forks contributes to checkpoint activation. *J Cell Biol* **189:** 233–246.

van Deursen F, Sengupta S, De Piccoli G, Sanchez-Diaz A, Labib K. 2012. Mcm10 associates with the loaded DNA helicase at replication origins and defines a novel step in its activation. *EMBO J* **31:** 2195–2206.

Watase G, Takisawa H, Kanemaki MT. 2012. Mcm10 plays a role in functioning of the eukaryotic replicative DNA helicase, Cdc45-Mcm-GINS. *Curr Biol* **22:** 343–349.

Witkin EM, Roegner-Maniscalco V, Sweasy JB, McCall JO. 1987. Recovery from ultraviolet light-induced inhibition of DNA synthesis requires *umuDC* gene products in *recA718* mutant strains but not in *recA*$^+$ strains of *Escherichia coli*. *Proc Natl Acad Sci* **84:** 6805–6809.

Wurtele H, Bacal J, Schalck Kaiser G, St-Hilaire E, Lee EH, Dorn JF, Maddox P, Lisby M, Pasero P, Verreault A. 2011. Histone H3 lysine 56 acetylation and the response to DNA replication fork damage. *Mol Cell Biol* **32:** 154–172.

Xu L, Marians KJ. 2003. PriA mediates DNA replication pathway choice at recombination intermediates. *Mol Cell* **11:** 817–826.

Yan S, Michael WM. 2009. TopBP1 and DNA polymerase-α directly recruit the 9-1-1 complex to stalled DNA replication forks. *J Cell Biol* **184:** 793–804.

Yeeles JT, Dillingham MS. 2010. The processing of double-stranded DNA breaks for recombinational repair by helicase–nuclease complexes. *DNA Repair (Amst)* **9:** 276–285.

Yeeles JT, Marians KJ. 2011. The *Escherichia coli* replisome is inherently DNA damage tolerant. *Science* **334:** 235–238.

Zhu W, Ukomadu C, Jha S, Senga T, Dhar SK, Wohlschlegel JA, Nutt LK, Kornbluth S, Dutta A. 2007. Mcm10 and And-1/CTF4 recruit DNA polymerase α to chromatin for initiation of DNA replication. *Genes Dev* **21:** 2288–2299.

Replication of Telomeres and the Regulation of Telomerase

Verena Pfeiffer and Joachim Lingner

Swiss Institute for Experimental Cancer Research (ISREC), School of Life Sciences, Frontiers in Genetics National Center of Competence in Research, Ecole Polytechnique Fédérale de Lausanne (EPFL), 1015 Lausanne, Switzerland

Correspondence: joachim.lingner@epfl.ch

Telomeres are the physical ends of eukaryotic chromosomes. They protect chromosome ends from DNA degradation, recombination, and DNA end fusions, and they are important for nuclear architecture. Telomeres provide a mechanism for their replication by semiconservative DNA replication and length maintenance by telomerase. Through telomerase repression and induced telomere shortening, telomeres provide the means to regulate cellular life span. In this review, we introduce the current knowledge on telomere composition and structure. We then discuss in depth the current understanding of how telomere components mediate their function during semiconservative DNA replication and how telomerase is regulated at the end of the chromosome. We focus our discussion on the telomeres from mammals and the yeasts *Saccharomyces cerevisiae* and *Schizosaccharomyces pombe*.

Herman Muller and Barbara McClintock were probably the first scientists to recognize that the ends of eukaryotic chromosomes have special properties and crucial functions (reviewed by Blackburn 2006). Herman Muller studied *Drosophila* chromosomes and found that X rays could induce chromosome rearrangements. However, he never observed chromosomes carrying terminal deletions (Muller 1938). These results led him to conclude that "the terminal gene must have a special function, that of sealing the end of the chromosome, so to speak," and that "for some reason, a chromosome cannot persist indefinitely without having its ends thus sealed." Muller coined the term "telomere" for this terminal gene (from the Greek words *telos* = end and *meros* = part). Barbara McClintock analyzed maize strains in which dicentric chromosomes were produced with high frequency. Dicentric chromosomes break when the two centromeres are pulled toward opposite poles of the mitotic spindle during anaphase of the cell cycle. The broken chromosomal ends were unstable and fused with other broken ends with which they came in contact. However, when dicentric chromosomes were present in embryonic cells, the broken ends were somehow "healed" (McClintock 1941).

The molecular sequence of telomeric DNA was first determined by Elizabeth Blackburn and Joseph Gall in the ciliated protozoan *Tetrahymena thermophila* (Blackburn and Gall 1978). These experiments revealed that the DNA ends

Cite this article as *Cold Spring Harb Perspect Biol* doi: 10.1101/cshperspect.a010405

in this organism consist of 5′-T_2G_4-3′/5′-C_4A_2-3′ telomeric repeats with the G-rich strand running toward the 3′ end of the chromosome. Cloning and sequencing of telomeres from other eukaryotes revealed that the repetitive nature of telomeric repeats and the presence of a G-rich strand are common features of nearly all eukaryotes. Furthermore, it has been established that the 3′-end-containing strand protrudes at both ends of the chromosome. However, notable exceptions do exist. The roundworm *Caenorhabditis elegans* contains both 3′ G-overhangs as well as 5′ C-overhangs (Raices et al. 2008). Several genera of insects and plants do contain long DNA repeats (Martinez et al. 2001). *Drosophila* contains, instead of short telomeric repeats, retrotransposons at chromosome ends that have overtaken telomere functions (Pardue and Debaryshe 2011).

The formulation of the DNA end-replication problem by Watson and Olovnikov inferred that specialized mechanisms must exist to maintain telomere length (Olovnikov 1971; Watson 1972). Greider and Blackburn discovered telomerase activity in extracts from *Tetrahymena* (Greider and Blackburn 1985). They also identified the telomerase RNA moiety in this organism, which contains a sequence complementary to *Tetrahymena* telomeric repeats (Greider and Blackburn 1989). Mutation of this sequence proved that the telomerase RNA moiety provides the template for DNA repeat synthesis and that the telomerase functions as reverse transcriptase (Yu et al. 1990). Around the same time, Lundblad and Szostak were the first to discover an essential telomere maintenance protein gene in *Saccharomyces cerevisiae*. Dysfunction of the identified gene gave rise to an ever shorter telomeres (*est*) phenotype culminating in cellular senescence (Lundblad and Szostak 1989). The discovery of *EST1* was followed by the identification of *EST2*, *EST3*, and *EST4* in the Lundblad laboratory (Lendvay et al. 1996). Est2 turned out to be orthologous with the p123 telomerase reverse transcriptase (Tert) protein subunit (Lingner et al. 1997) that was identified in the ciliate *Euplotes aediculatus* upon biochemical purification of the telomerase enzyme (Lingner and Cech 1996). The identification of the first Tert subunits opened the door to identify the corresponding genes in humans and other organisms, allowing the study of telomerase during development and in cancer. Significantly, it was shown that ectopic expression of human Tert in various primary human cells is sufficient to reconstitute telomerase, rendering them immortal (Bodnar et al. 1998).

Ciliated protozoa were also instrumental in purifying and cloning the first eukaryotic telomere-binding proteins (Gottschling and Zakian 1986; Hicke et al. 1990; Gray et al. 1991). However, orthologs of these proteins were at first not recognized in other eukaryotes, and independent approaches were undertaken to identify the first telomeric proteins in various eukaryotes including yeasts and humans. Only much later was it realized that the ciliate telomere-binding proteins contain counterparts in a broad range of eukaryotes (Baumann and Cech 2001).

In this review, we introduce known basic components of telomeres and review the mechanisms of semiconservative DNA replication of telomeric DNA and the regulation of telomerase. For the discussion of telomere protection and the three-dimensional structures of telomere components, the reader is referred to excellent recent reviews (de Lange 2009; Jain and Cooper 2010; Lewis and Wuttke 2012).

TELOMERE COMPONENTS

In the following section, the stable constituents of telomeres that are implicated in regulating telomere replication are discussed. Abundant telomere-binding proteins as well as telomerase subunits from *S. cerevisiae*, *Schizosaccharomyces pombe*, and *Homo sapiens* described in detail in the text are summarized in Table 1.

S. cerevisiae

The telomeric DNA repeat consensus sequence for the 3′-end-containing strand from *S. cerevisiae* is 5′-$(TG)_{0-6}$TGGGTGTG(G)-3′ (Forstemann and Lingner 2001). The telomere length is around 300 bp. Native chromosome ends in *S. cerevisiae* have a number of subtelomeric

Table 1. Abundant telomere-binding proteins and telomerase proteins in *S. cerevisiae*, *S. pombe*, and *H. sapiens*

Organism	Subcomplex	Protein	Interaction partner	Ortholog	Function
S. cerevisiae		Rap1	Sir3, Sir4, or Rif1, Rif2	Rap1 (*S. p.*, *H. s.*)	Essential; major double-stranded telomere-binding protein; transcriptional regulation of telomeres and protein-coding genes; protection from NHEJ; negative regulator of telomere length
	Sirtuins	Sir2	Sir3, Sir4		NAD^{+}-dependent histone deacetylase; telomeric silencing
		Sir3	Sir2, Sir4, Rap1		Telomeric silencing
		Sir4	Sir2, Sir3, Rap1		Telomeric silencing
	Rifs	Rif1	Rif2, Rap1		Negative regulator of telomere length
		Rif2	Rif1, Rap1		Negative regulator of telomere length
	CST	Cdc13	Ten1, Stn1, Est1, Pol1	Ctc1 (*H. s.*; limited sequence similarity)	Essential G-strand single-stranded telomere-binding protein; protects from C-strand loss; recruits telomerase through interaction with Est1; essential for telomerase activity in vivo but not in vitro; interacts with Pol1, the catalytic subunit of DNA polymerase α-primase, promoting fill-in synthesis of telomerase-elongated telomeres
		Stn1	Ten1, Cdc13, Pol12	Stn1 (*S. p.*, *H. s.*)	Essential; protects from C-strand loss; negative regulator of telomere length; interacts with Pol12, the B subunit of DNA polymerase α-primase, promoting fill-in synthesis of telomerase-elongated telomeres
		Ten1	Stn1, Cdc13	Ten1 (*S. p.*, *H. s.*)	Essential; protects from C-strand loss; negative regulator of telomere length
	Telomerase	Est1	Tlc1, Est2, Cdc13	Est1 (*S. p.*, *H. s.* Est1A has limited sequence similarity)	*est* phenotype; binds Tlc1 and G-strand single-stranded telomeric DNA; recruits telomerase through interaction with Cdc13; essential for telomerase activity in vivo but not in vitro
		Est2	Tlc1, Est1	Tert (*H. s.*), Trt1 (*S. p.*)	*est* phenotype; telomerase reverse transcriptase; essential for telomerase activity in vivo and in vitro
		Est3	Est1, Est2		*est* phenotype; essential for telomerase activity in vivo but not in vitro

Continued

Table 1. *Continued*

Organism	Subcomplex	Protein	Interaction partner	Ortholog	Function
			Sm proteins	Smb1, Smd1, Smd2, Smd3, Sme1, Smx3, Smx2	Form heteroheptameric Sm ring that binds and stabilizes Tlc1
S. pombe		Taz1	Rif1, Rap1	Trf1, Trf2 (*H. s.*, *S. c.* Tbf1 has limited sequence similarity)	Major double-stranded telomere-binding protein; protection from NHEJ; negative regulator of telomere length; promotes telomere replication; telomeric silencing and TERRA repression; meiosis
		Rif1	Taz1		Negative regulator of telomere length; recruited by Taz1
		Rap1	Taz1, Poz1	Rap1 (*S. c.*, *H. s.*)	Negative regulator of telomere length; recruited by Taz1; telomeric silencing and TERRA repression
		Poz1	Rap1, Tpz1	Pot1 (*H. s.*)	Negative regulator of telomere length
		Tpz1	Poz1, Pot1, Ccq1	Tpp1 (*H. s.*)	Protects telomeric 3′ overhang; telomerase recruitment through interaction with Poz1 and Ccq1
		Pot1	Tpz1	Pot1 (*H. s.*)	Essential; single-stranded G-strand telomere-binding protein; protects telomeres from nucleolytic degradation
		Ccq1	Tpz1, Est1		Telomerase recruitment through interaction with Est1; meiosis
		Ten1	Stn1	Ten1 (*S. c.*, *H. s.*)	Telomere-capping protein; protects from telomere loss
		Stn1	Ten1	Stn1 (*S. c.*, *H. s.*)	Telomere-capping protein; needed for telomere maintenance
	Telomerase	Est1	Ter1, Ccq1, Trt1	Est1 (*S. c.*)	*est* phenotype; binds Ter1, which has an Sm protein-binding motif; recruits telomerase through interaction with Ccq1; essential for telomerase activity in vivo but not in vitro
		Trt1	Ter1, Est1		*est* phenotype; telomerase reverse transcriptase; essential for telomerase activity in vivo and in vitro
		Sm proteins	SmB, SmD1, SmD2, SmD3, SmE, SmF, SmG		Form heteroheptameric Sm ring that binds Ter1 precursor and promotes spliceosomal cleavage; promotes hypermethylation of the Ter1 5′ cap by Tgs1
		Lsm proteins	Lsm2–8		Bind mature Ter1 and promote association with Trt1; protect Ter1 from nucleolytic degradation

H. sapiens	Shelterin	Trf1	Tin2	Taz1 (*S. p.*)	Double-stranded homodimeric telomere-binding protein; negative regulator of telomere length; promotes telomere replication
		Trf2	Rap1, Tin2	Trf1 (*H. s.*), Taz1 (*S. p.*)	Double-stranded homodimeric telomere-binding protein; protection from NHEJ; repression of Atm
		Rap1	Trf2	Rap1 (*S. c.*, *S. p.*)	Recruited by Trf2; repression of telomere recombination and homology-directed repair
		Tin2	Trf1, Trf2, Tpp1		Stabilization of shelterin complex; recruits telomerase through interaction with Tpp1
		Tpp1	Pot1, Tin2, Tert	Tpz1 (*S. p.*)	Increased affinity of Pot1 for single-stranded telomeric DNA; stimulation of telomerase processivity in conjunction with Pot1; recruits telomerase through interaction with Tin2 and telomerase
		Pot1	Tpp1	Pot1 (*S. p.*)	Binds single-stranded G-strand telomeric DNA; inhibits telomerase in absence of Tpp1; Pot1–Tpp1 promotes telomerase processivity; repression of Atr
	CST	Ctc1	Stn1, Ten1	Cdc13 (*S. c.*; limited sequence similarity)	When assembled as CST, it binds single-stranded DNA with preference for telomeric G-strand sequence; inhibits telomerase activity through direct interaction with Pot1/Tpp1 and primer sequestration; increased telomere binding upon telomerase-mediated telomere elongation; competes with Pot1/Tpp1 for binding of telomeric DNA; with Stn1, Ctc1 stimulates affinity of DNA polymerase α-primase for template DNA; semiconservative DNA replication of specialized DNA sequences
		Stn1	Ten1, Ctc1	Stn1 (*S. p.*, *S. c.*)	See Ctc1
		Ten1	Ctc1, Stn1	Ten1 (*S. p.*, *S. c.*)	See Ctc1

Continued

Table 1. *Continued*

Organism	Subcomplex	Protein	Interaction partner	Ortholog	Function
	Telomerase	Tert	Terc, Tpp1		Telomerase reverse transcriptase; essential for telomerase activity in vivo and in vitro
		Tcab1			Telomerase localization to Cajal bodies and to telomeres in S phase; essential for telomerase activity in vivo but not in vitro
		Dyskerin	Nop10, Nhp2, Gar1		Binds H/ACA RNA motif–containing RNAs including Terc; Terc stability
		Nop10	Dyskerin, Nhp2		See Dyskerin
		Nhp2	Dyskerin, Nop10		See Dyskerin
		Gar1	Dyskerin		See Dyskerin
		Est1a	Terc, Tert	Est1 (*S. p.* and *S. c.* Est1 have limited sequence similarity; no evidence for conserved functional role in telomerase recruitment)	Negative regulator of TERRA at telomeres; depletion leads to stochastic telomere loss; overexpression leads to telomere uncapping and telomere fusions; involved in nonsense-mediated mRNA decay

NHEJ, nonhomologous end joining; TERRA, telomeric repeat–containing RNA; CST, Cdc13–Stn1–Ten1.

repeat elements termed X and Y′, adjacent to the telomeric repeats, which vary between ends and strains (Louis and Borts 1995). The 473-bp core X is found at all chromosome ends. It contains an ACS (ARS consensus sequence, the binding site for the yeast origin recognition complex) and an Abf1p-binding site at 31 out of 32 ends. There are also binding sites for the telomere-binding protein Tbf1 (Bilaud et al. 1996). So-called Y′ elements with a length of 6.7 kb or 5.2 kb are present at approximately half of *S. cerevesiae* telomeres.

The telomeric repeat sequence is bound by the double-stranded DNA-binding protein Rap1 (repressor activator protein 1) (Fig. 1) (Lustig et al. 1990). Rap1 also binds chromosome internal sites acting as positive transcriptional regulator of genes involved in growth control. Rap1 has two central Myb domains responsible for DNA binding and a Rap1 carboxy-terminal domain (RCT). RCT recruits the Rif proteins, which repress telomerase (see below), and it recruits the Sir3 and Sir4 proteins, which in turn interact with the Sir2 NAD^+-dependent histone deacetylase involved in transcriptional repression. Rap1 is also important for protection from end fusions (Marcand et al. 2008).

The telomeric 3′ overhang is bound by a trimeric complex consisting of Cdc13, Stn1, and Ten1 (CST) (Fig. 1) (Gao et al. 2007). Despite binding to the G-strand, the CST complex protects the C-strand from 5′-3′ resection. In addition, CST regulates telomere length through telomerase (see below). CST has been proposed

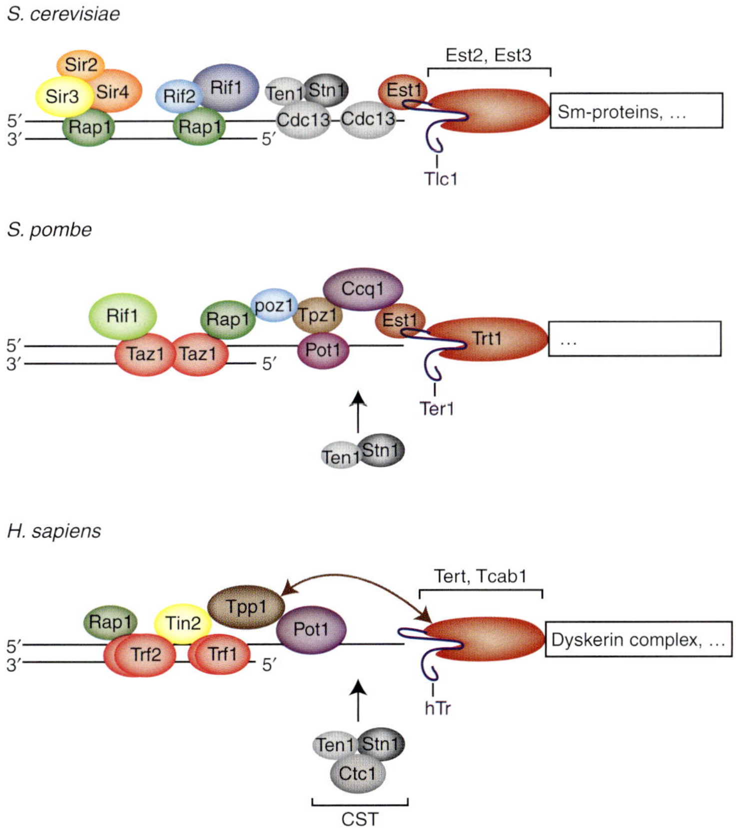

Figure 1. Telomeric proteins and telomerases at yeast and human telomeres. Only abundant telomere-binding proteins and telomerase components are indicated. Trf1 and Trf2 bind as homodimers to the double-stranded telomeric DNA repeats. Homologous proteins are shown in the same color.

to resemble in structure the replication protein A (RPA) complex. CST contains as RPA multiple oligosaccharide/oligonucleotide binding (OB)-folds. In particular, Stn1 and Ten1 proteins contain structural similarities with Rpa2 and Rpa3 (Gao et al. 2007).

Telomeres in *S. cerevisiae* are transcribed by RNA polymerase II into the long noncoding telomeric repeat-containing RNA (TERRA) (Luke et al. 2008). TERRA transcription starts in the subtelomeric DNA and proceeds approximately 100 nucleotides into the telomeric tract. *S. cerevisiae* TERRA carries a cap structure at the 5′ end and is polyadenylated. TERRA is degraded by the 5′-3′ exonuclease Rat1. TERRA transcription is negatively regulated by the Sir proteins at telomeres that contain only the X subtelomeric element at their chromosome ends. At Y′ elements containing telomeres, TERRA transcription is negatively regulated by the Rif proteins (Iglesias et al. 2011).

Telomeres in *S. cerevisiae* have been inferred to form fold-back or looped structures because Rap1 is associated with subtelomeric chromatin as well as with telomeric DNA (reviewed by Ottaviani et al. 2008). This fold-back structure has been proposed to be critical for the so-called telomere position effect, which refers to the repression of reporter genes inserted in the subtelomeric region (Gottschling et al. 1990). Silencing at telomeres is mediated by the silent information regulators Sir3, Sir4, and Sir2.

S. pombe

The telomeric DNA sequence from *S. pombe* is unusually heterogeneous with the consensus 5′-$C_{1-8}G_{0-1}T_{0-2}GTA_{1-3}$-3′ (Sugawara 1989). Telomere length, as in *S. cerevisiae*, is ~300 bp. The subtelomeric DNA in *S. pombe* contains specialized repeats (*dg* and *dh*), which are also found at other heterochromatic loci contributing to formation of heterochromatin by the RNA interference (RNAi)-induced transcriptional silencing machinery (Kanoh et al. 2005). The first identified telomere-binding protein, Taz1, was identified in a one-hybrid screen using the double-stranded telomeric DNA as bait (Cooper et al. 1997). Taz1 is orthologous with the mammalian telomere-binding proteins Trf1 and Trf2. Taz1 recruits Rap1 to telomeres (Fig. 1). Rap1 binds Poz1, Poz1 binds Tpz1, and Tpz1 interacts with Ccq1 and Pot1 (Miyoshi et al. 2008). Pot1 also binds the telomeric single-stranded 3′ overhang (Baumann and Cech 2001). Thus, multiple protein interactions bridge the double-stranded part of the telomeric DNA to the single-stranded 3′ overhang, which is reminiscent of the situation in mammals. *S. pombe* Pot1 also plays crucial roles in protecting chromosome ends from degradation, Ccq1 interacts with the telomerase subunit Est1, recruiting telomerase to telomeres (see below). In addition, Stn1 and Ten1 orthologs but not Cdc13 orthologs have been identified in fission yeast (Martin et al. 2007). They protect from rapid loss of telomeric DNA. TERRA and other telomeric transcripts have been recently characterized in fission yeast (Bah et al. 2012; Greenwood and Cooper 2012). Their expression is negatively regulated by Taz1 and Rap1.

Mammals

Vertebrate telomeric DNA consists of 5′-TTAGGG-3′ repeats. The telomere length in humans is considerably shorter (5–15 kb) than in *Mus musculus* (>40 kb), and the human subtelomeric sequences are not well annotated. They contain repetitive sequence variants of the 5′-TTAGGG-3′ repeat sequence and a patchwork of segmentally duplicated DNA tracts that are shared in some instances between different telomeres and are highly polymorphic (Riethman 2008). Biochemical purification identified the first human telomere-binding protein, Trf1 (Chong et al. 1995). Dimeric Trf1 binds to double-stranded telomeric DNA (Fig. 1). A second homologous protein termed Trf2 also binds directly to the double-stranded telomeric DNA as a dimer (Bilaud et al. 1997; Broccoli et al. 1997). Trf1 and Trf2 interact with Tin2, which in turn binds Tpp1. Tpp1 also binds the single telomere-binding protein Pot1 (de Lange 2005). Notably, whereas in humans there is only one Pot1 protein, the mouse contains two Pot1-like proteins termed Pot1a and Pot1b (Hockemeyer et al. 2006; Wu et al. 2006). Trf2 also recruits Rap1 to telomeres. Thus, as in fission yeast, Rap1 is tethered to telomeres via protein interaction, whereas in budding yeast it directly binds the double-stranded telomeric DNA. No mammalian protein that is orthologous with *S. pombe* Ccq1 has yet been identified. However, a human CST complex was identified that is present at telomeres and at other regions in the genome (Miyake et al. 2009; Surovtseva et al. 2009).

Human TERRA is transcribed from several or possibly all chromosome ends (Azzalin et al. 2007; Schoeftner and Blasco 2008). Transcription starts in the subtelomeric region and proceeds up to ~400 nucleotides into the pure TTAGGG tract (Porro et al. 2010). Poly(A) tails containing and lacking TERRA fractions can be distinguished. Approximately half of the poly(A)$^{-}$ TERRA fraction is chromatin-associated, colocalizing with telomeres. The less abundant poly(A)$^{+}$ TERRA is all nucleoplasmic (Porro et al. 2010).

Mammalian telomeric DNA has been found to fold into so-called T-loop structures in which the 3′ overhang of telomeres invades through strand displacement the double-stranded part of telomeric repeats (Griffith et al. 1999). T-loops have been proposed to provide a mechanism for telomere capping, but this awaits experimental testing.

TELOMERASE

The catalytic core of telomerase consists of Tert and the telomerase RNA. The Tert subunit reverse transcribes the template sequence of telomerase RNA in an iterative fashion at the ends of chromosomes. When the 3′ end of the telomere is extended to the 5′ end of the RNA template, telomerase may reposition the substrate to the other side of the template and add the next telomere repeat, or it may dissociate from the substrate. The propensity of some telomerases to add several repeats without dissociation is referred to as repeat addition processivity. The Tert subunit is well conserved throughout evolution. It contains canonical reverse transcriptase motifs also found in retroelements (Lingner et al. 1997; Nakamura and Cech 1998). In addition, Tert contains a telomerase essential amino-terminal domain (TEN), a telomerase RNA-binding domain (TRBD), and a carboxy-terminal extension (reviewed by Lewis and Wuttke 2012). Other telomerase subunits are less well conserved. These subunits are not involved per se in catalysis, but they are important for telomerase assembly or the recruitment to telomeres. The *S. cerevisiae* Est1, which recruits telomerase to telomeres (discussed below), has a clear counterpart in *S. pombe* (Beernink et al. 2003; Webb and Zakian 2012), but the orthologous relationship to human Est1-like proteins (referred to as Est1A/Smg6, Est1B/Smg5, and Est1C/Smg7) is uncertain (Reichenbach et al. 2003; Snow et al. 2003). Est1A interacts as the budding and fission yeast Est1 proteins with telomerase, but a role in telomerase recruitment has not been shown (Redon et al. 2007). Est1A depletion has a strong effect on TERRA abundance at telomeres (Azzalin et al. 2007). Furthermore, Est1A to -C are also involved in the nonsense-mediated mRNA decay in the cytoplasm (Isken and Maquat 2008).

S. cerevisiae Est3 is essential for telomerase action at chromosome ends, but its contribution is not clear. Est3 contains an OB-fold, which is similar to an OB-fold present in the mammalian Tpp1 protein (Lee et al. 2008). Sm proteins associate with *S. cerevisiae* telomerase RNA, contributing to telomerase maturation and stability (Seto et al. 1999). *S. pombe* telomerase RNA (Ter1) also contains an Sm consensus site (Leonardi et al. 2008; Webb and Zakian 2008). The canonical Sm ring and the related Lsm2–8 complex sequentially associate with *S. pombe* Ter1 (Tang et al. 2012). The Sm ring binds the Ter1 precursor, stimulating its maturation by spliceosomal cleavage. Subsequently, the Lsm2–8 complex associates with Ter1, promoting the association with catalytic moiety Trt1 (Tang et al. 2012). In mammals, the dyskerin protein complex may fulfill similar functions in telomerase assembly (Mitchell et al. 1999; Pogacic et al. 2000). Another crucial component of human telomerase is Tcab1, which is discussed further below (Venteicher et al. 2009).

TELOMERE REPLICATION

Semiconservative Replication of Telomeric DNA

The largest part of the telomeric DNA is replicated by semiconservative DNA replication. The G-strand serves as a template for lagging-strand synthesis and the C-strand for leading-strand synthesis (Fig. 2). Only in recent years did it become apparent that even the semiconservative replication of telomeric DNA is difficult, requiring specialized helicases and telomere-binding proteins. In *S. cerevisiae*, the Rrm3 5′-3′ helicase prevents replication fork stalling at telomeres (Ivessa et al. 2002; Makovets et al. 2004). In fission yeast, it was noted that efficient replication of telomeric DNA requires the telomere-binding protein Taz1 (Miller et al. 2006). In its absence, replication forks frequently stall and collapse, leading to rapid loss of entire telomeric tracts. Taz1-deleted cells can only survive because telomerase very efficiently reextends the transiently truncated telomeric sequences. Evidence for truncated and by telomerase reextended telomeres was also obtained in wild-type budding yeast cells, supporting the notion that the semiconservative replication of telomeric DNA is error-prone (Chang et al. 2007).

In mouse cells, it was shown that efficient replication requires the Taz1 ortholog Trf1 (Martinez et al. 2009; Sfeir et al. 2009). Upon Trf1 deletion in mouse cells, replication forks stall and telomeric DNA splits in multiple signals at metaphase chromosomes, indicative of telomeric DNA breakage or lack of condensation. The mechanism by which Trf1 promotes telomere replication involves recruitment of the Blm or Rtel1 helicases (Sfeir et al. 2009; Vannier et al. 2012). Both of these helicases belong to the highly conserved RecQ 3′-5′ helicase family,

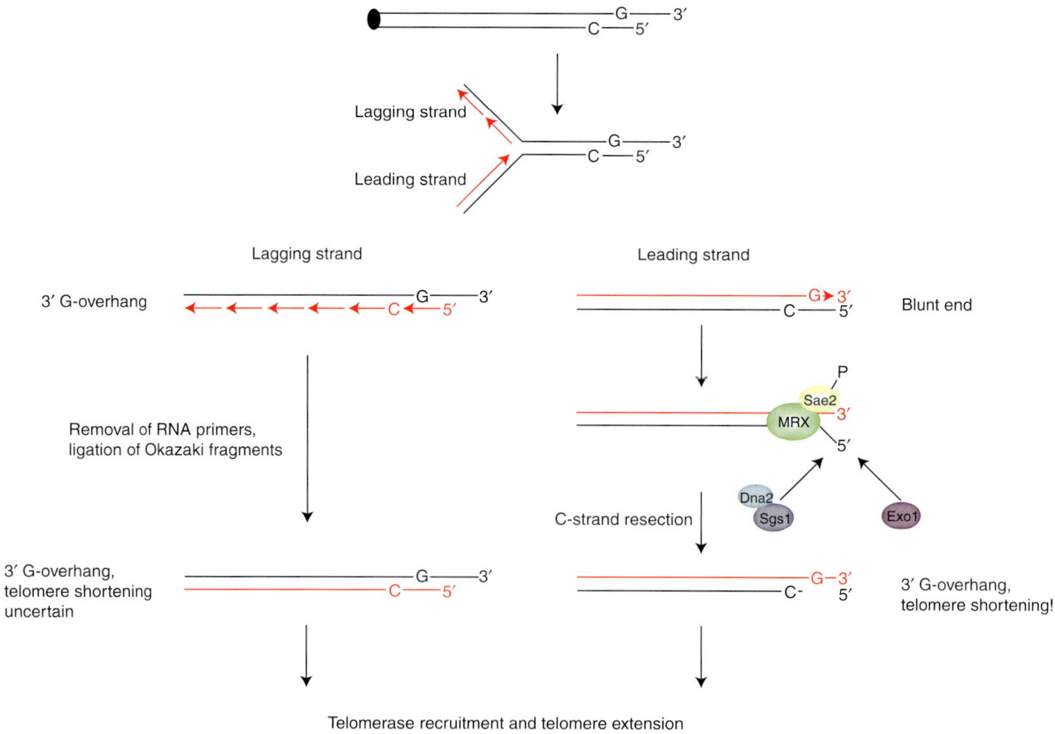

Figure 2. The end-replication problem and DNA end resection. The G-rich strand serves as a template for lagging-strand synthesis and the C-rich strand for leading-strand synthesis. The presumed blunt end intermediate at the leading-strand telomere is processed by 5′ end resection in order to recreate a 3′ overhang *(right)*. Therefore, the leading-strand telomere shortens.

which use ATP hydrolysis to drive the unwinding of RNA and DNA. Another member of this family, the Werner helicase (Wrn), is required for efficient telomere lagging-strand synthesis (Crabbe et al. 2004; Arnoult et al. 2009). Interestingly, the stochastic telomere loss in Wrn-deficient cells and mice can be healed by the expression of telomerase (Chang et al. 2004; Crabbe et al. 2004). Indeed, in mice, the Wrn phenotype, which is characterized by multiple signs of premature aging, only manifests itself upon concomitant deletion of telomerase (Chang et al. 2004). *S. cerevisiae* Pif1 is another DNA helicase important for replication of G-rich sequences being enriched at telomeres and other G-rich sequences throughout the genome that may form G-quadruplex structures (Paeschke et al. 2011).

The Upf1 DNA- or RNA-dependent ATPase and 5′-to-3′ helicase is best known for its roles in cytoplasmic RNA quality control. However, human Upf1 binds to telomeres in vivo (Azzalin et al. 2007), interacting with the shelterin component Tpp1 (Chawla et al. 2011). Upf1 depletion leads to frequent loss of the telomeres replicated by leading-strand synthesis (Chawla et al. 2011). Upf1 depletion also leads to an increase of the TERRA signal at telomeres, suggesting a link between this RNA and telomere loss.

The crucial functions of helicases during replication of telomeric DNA appear to be multiple. First, helicases may unfold G-quadruplex DNA structures (Vannier et al. 2012) that may form during the lagging-strand replication of telomeric DNA, during which single-stranded G-rich DNA forms transiently. The structures involve guanine tetrads that associate through Hoogsteen hydrogen bonding to form a square planar structure, and several tetrads stack on top of each other to form a G-quadruplex (Parkinson et al. 2002). Second, in the T-loop structures, the 3′ overhang of telomeres invades through strand displacement the double-stranded part of telomeric repeats (Griffith et al. 1999). It is conceivable that helicases are necessary to unwind T-loops during semiconservative DNA replication (Vannier et al. 2012). Third, TERRA is associated with telomeres. If TERRA and telomeric DNA form R-loops, which are characterized by RNA–DNA hybrids and the displacement of single-stranded DNA, they may need to be resolved by helicases in order to avoid formation of double-stranded DNA breaks (Huertas and Aguilera 2003; Chawla et al. 2011; Wahba et al. 2011).

Telomere End Resection

Telomere leading-strand synthesis is predicted to yield a double-stranded blunt end intermediate (Fig. 2, right panel). This presumed intermediate is not detectable, as it will be quickly processed to regenerate a 3′ overhang. Thus, the 5′-end-containing parental C-rich strand is resected (Fig. 2, right panel). The generated G-rich 3′ overhang is important in order to generate a substrate for telomerase in addition to providing a platform for the binding of telomeric proteins and the ability to invade the double-stranded part of the telomere in the T-loop configuration. Thus, the telomere will shorten at the leading-strand end because the parental C-rich strand is resected during each round of replication (Lingner et al. 1995). During telomere lagging-strand synthesis, short RNA primers of 8–14 nucleotides are synthesized by primase and extended (Fig. 2, left panel). It is unknown where on the G-rich strand the most telomere-proximal primer is laid down. Because the parental G-rich strand forms a 3′ overhang that is longer than the RNA primer, the end-replication problem may not manifest itself at this end of the chromosome. However, telomerase binds leading- and lagging-strand telomeres (see below). Telomere lagging-strand replication also generates a 3′ overhang. Whether this involves simply RNA primer removal or in addition nucleolytic processing is currently unknown.

The 5′ end resection of leading-strand telomeres involves redundant activities consisting of nucleases, DNA helicases, and associated proteins that also resect 5′ ends from DNA double-strand breaks, preparing them for homologous recombination (Mimitou and Symington 2009). The Mre11–Rad50–Xrs2 (MRX) complex plays a role in end processing, as in its absence the G-overhangs at *S. cerevisiae* telomeres that have a length of 12–14 bases are shorter

(Larrivee et al. 2004). Furthermore, the longer G-overhangs that accumulate in yeast in late S phase are also diminished. However, Mre11 has 3′-5′ exonuclease activity as well as endonuclease nicking activity but not the 5′-3′ exonuclease activity that should be required to generate a telomeric 3′ overhang. Furthermore, although MRX is important for wild-type telomere length maintenance, this function may at least partially relate to its activity as a Tel1 recruiter (see below). Indeed, a nuclease-deficient Mre11 mutant was proficient in telomere maintenance (Tsukamoto et al. 2001). This is in contrast to results obtained with an endonuclease-induced cut at telomere repeats at which resection and telomerase-mediated healing depends on MRX (Diede and Gottschling 2001; Bonetti et al. 2009). Sae2 and Sgs1 define parallel pathways required for processing of telomere ends (Bonetti et al. 2009). The Sae2 nuclease interacts with MRX and may start telomere end resection (Fig. 2). The Sgs1 helicase functions in conjunction with the Dna2 Flap endonuclease/helicase. Concomitant absence of Sae2 and Sgs1 triggered shorter telomeres but not cellular senescence, indicating that telomerase is still able to act (Bonetti et al. 2009). Indeed, exonuclease 1 (Exo1) can also contribute to telomere end resection, and telomere shortening in *sae2Δ/sgs1Δ* cells is partially suppressed by Exo1 overexpression (Bonetti et al. 2009). Cdk1 activity is required to promote 3′ overhang formation at *S. cevevisiae* telomeres (Frank et al. 2006; Vodenicharov and Wellinger 2006). This involves phosphorylation of Sae2 and possibly additional substrates. How phosphorylation activates Sae2 is not known.

In normal human fibroblasts, 3′ overhang length at leading-stand telomeres is around 60 nucleotides, versus 105 nucleotides at lagging-strand telomeres (Chai et al. 2006). Eighty percent of the C-rich strands end with the sequence 3′-CCAATC-5′, whereas the sequence at the 3′ terminus is less precise (Sfeir et al. 2005). The enzymatic activities that are involved in telomere and double-strand break end resection in *S. cerevisiae* are conserved throughout evolution. The MRN complex (Mre11–Rad50–Nbs1) interacts with Trf2 and binds to telomeres (Zhu et al. 2000). The role of MRN and the other processing activities discussed above in the processing of mammalian telomeres is not established. However, the Apollo 5′-3′ exonuclease is a member of the Snm1/Pso2 family of nucleases, binding to telomeres through its interaction with Trf2 (Lenain et al. 2006; van Overbeek and de Lange 2006). Apollo$^{-/-}$ cells have reduced 3′ overhangs at telomeres and they lose telomere protection at the leading-strand telomeres (Lam et al. 2010; Wu et al. 2010). Whether other nucleases in addition to Apollo contribute to telomere end processing at mammalian telomeres, as in yeast, is unclear. Notably also, Apollo has been proposed to facilitate semiconservative replication of telomeric DNA, relieving topological stress in conjunction with topoisomerase 2α (Ye et al. 2010).

TELOMERASE RECRUITMENT, TELOMERE EXTENSION, AND LENGTH HOMEOSTASIS

S. cerevisiae

S. cerevisiae telomerase extends telomeres in late S phase, presumably after semiconservative DNA replication and telomere end processing. Single-telomere extension analysis (STEX) revealed that telomerase extends only a small fraction (∼7%) of telomeres in a given cell cycle and that extension occurs most frequently at the shortest telomeres (Teixeira et al. 2004). Recognition of the shortest telomeres is mediated on one side by Tel1 (ortholog of mammalian Atm), which is recruited specifically to the shortest telomeres in an MRX complex-dependent manner (Fig. 3) (Bianchi and Shore 2007b; Hector et al. 2007; Sabourin et al. 2007). Tel1 function can be partially compensated for by Mec1 (ortholog of mammalian Atr) as *tel1Δ/mec1Δ* double-mutant cells but not the single mutants senesce (Ritchie et al. 1999). A second pathway that mediates recognition of the shortest telomeres requires Tbf1, a protein that carries sequence similarity with fission yeast Taz1 and mammalian Trf proteins (Arneric and Lingner 2007). Tbf1 binds to 5′-TTAGGG-3′ repeats that are present in the subtelomeric region of yeast telomeres (Koering et al. 2000).

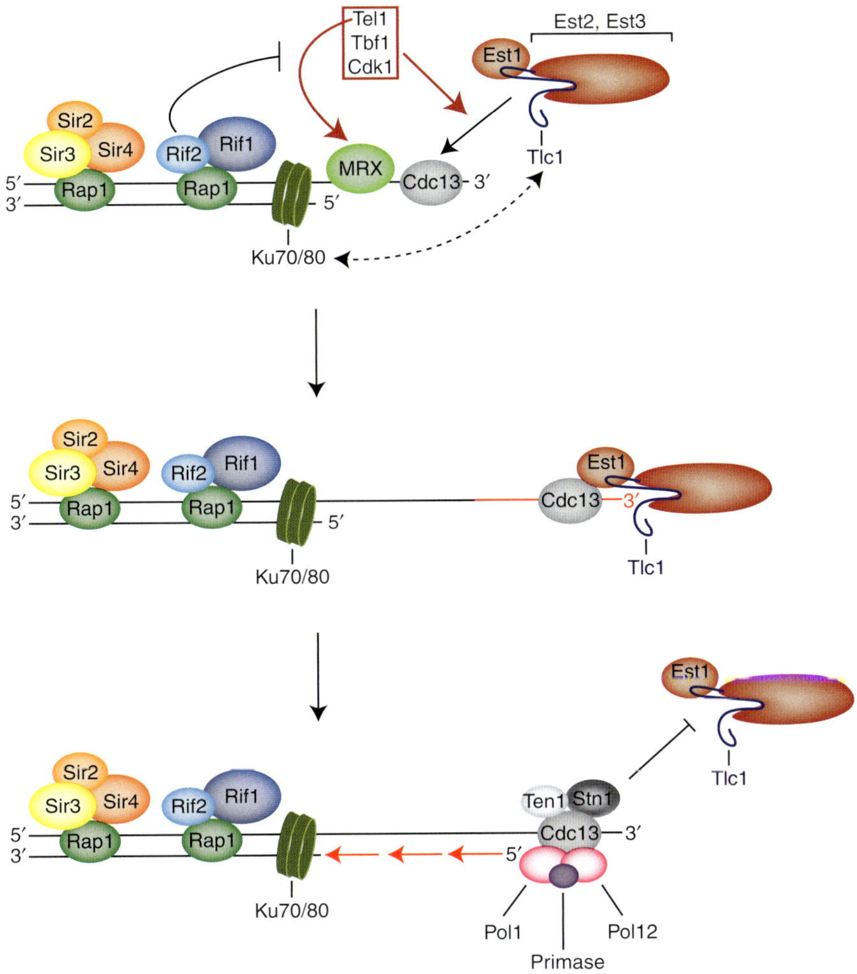

Figure 3. Telomerase recruitment and telomere extension in *S. cerevisiae*. MRX recruits Tel1 to the shortest telomeres, which promotes telomerase recruitment through the interaction of Cdc13 with Est1. Tbf1 can also promote preferential extension of the shortest telomeres in *tel1*Δ cells. Cdk1 contributes to the cell-cycle-dependent recruitment of telomerase through phosphorylation of Cdc13. Telomerase extends telomeres in a nonprocessive manner in *S. cerevisiae*. Human telomerase, on the other hand, adds ~60 nucleotides per telomere in a single binding and extension event (not shown; see text). Telomerase extension is terminated upon formation of the CST complex, which promotes recruitment of DNA Pol α-primase for fill-in synthesis.

Third, short telomeres replicate earlier in S phase. This therefore might provide more time for telomerase to mediate extension of the shortest telomeres (Bianchi and Shore 2007a).

The recruitment to telomeres in late S phase and telomere extension depend on a physical interaction between Cdc13 that binds to the telomeric 3′ overhang and the telomerase subunit Est1 (Fig. 3). The *cdc13-2* allele gives an *est* phenotype (Evans and Lundblad 1999). Cdc13-2 carries a single point mutation (E252L). Intriguingly, this mutation can be suppressed by a single point mutation in Est1 (L444E) (Pennock et al. 2001). Because the suppression involves the restoration of opposite charges in Cdc13 and Est1, this provides excellent genetic evidence that Cdc13 and Est1 interact directly and that the interaction involves these two amino acids. A second telomerase recruitment pathway in *S. cerevisiae* is mediated by the

Ku70/80 heterodimer, which binds telomeres and other DNA ends via a preformed channel (Fig. 3). In addition, Ku80 binds the telomerase RNA subunit Tlc1 (Peterson et al. 2001; Fisher et al. 2004). Ku enriches telomerase near telomeres throughout the cell cycle as assessed by chromatin immunoprecipitation and positively contributes to telomere length maintenance. Disruption of the Tlc1–Ku80 interaction reduces telomere length while not eliciting an *est1* phenotype. When assessing telomere association of telomerase by live-cell imaging, it seemed that the Tlc1–Ku80 interaction might only promote a transient interaction with telomeres, whereas the stable interaction in late S phase is mediated by Cdc13–Est1 (Gallardo et al. 2011). The Ku-mediated stimulation of telomerase may also involve its role in nuclear accumulation of Tlc1, which is reduced in Ku-deficient cells (Gallardo et al. 2008; Pfingsten et al. 2012). How Tel1 and Tbf1 promote the interaction of Cdc13 and Est1 and the preferential extension of the shortest telomeres is unclear. Cdc13 contains several consensus S/TQ sites for Tel1, but mutation of all these sites did not substantially impact on telomere length when the *cdc13* mutants were tested in their chromosomal context (Gao et al. 2010). However, opposite results were reported in a previous study by another group, in which the mutant *cdc13* alleles were expressed from a plasmid (Tseng et al. 2006). However, it seems that crucial Tel1 targets remain to be discovered. Cdk1 contributes to the cell-cycle-dependent recruitment of telomerase. The phosphorylation of Cdc13 at T308 by Cdk1 in late S phase and G_2 favors the interaction of Cdc13 with Est1 rather than the competing interaction of Cdc13 with Stn1–Ten1 (Li et al. 2009). Another modification, Cdc13 SUMOylation, peaks in early to mid S phase (Hang et al. 2011). Mutation of the SUMO site reduced the interaction with Stn1, promoting telomere elongation (Hang et al. 2011). Ku70/80 and Sir4 are other telomeric proteins modified by Siz2 E3 ligase-dependent SUMOylation, promoting perinuclear position of budding yeast telomeres (Ferreira et al. 2011). Because telomeres shift away from the nuclear envelope when elongating, this modification possibly links nuclear position with telomerase regulation.

STEX analysis of yeast cells that expressed two RNA subunits that differed in the telomeric sequence revealed that *S. cerevisiae* telomerase acts mostly in a nonprocessive manner in terms of telomere repeat addition (Chang et al. 2007). However, when very short telomeres, <125 nucleotides, are elongated, yeast telomerase gains in repeat addition processivity. The increased processivity requires Tel1. The telomere-shortening rate in the absence of telomerase is length-independent (Marcand et al. 1999). The observed increased frequency of telomere elongation by telomerase with telomere shortening is sufficient to explain telomere length homeostasis. The steady-state telomere length corresponds to the point at which the product of frequency of elongation and average extension length equals the shortening rate. The frequency of elongation is regulated by the number of Rap1/Rif1/Rif2 proteins that are bound to telomeres in dependence of their length (Marcand et al. 1997). Mechanistically, this may involve the competition of Rif1/2 with Tel1 for binding of the carboxyl terminus of Xrs2 (Hirano et al. 2009). Thus, at short telomeres that bind less Rap1/Rif/Rif2, Tel1 recruitment to the MRX complex may be favored, leading to preferential recruitment and activation of telomerase at short telomeres (Fig. 3).

Telomerase-mediated telomere extension must be followed by fill-in synthesis of the C-strand, presumably by classical replication enzymes involved in lagging-strand synthesis. Importantly, the CST complex may play also an important role here, as physical interactions have been detected between Cdc13 and Pol1, the catalytic subunit of DNA Pol α-primase (Qi and Zakian 2000); and between Stn1 and Pol12, the B subunit of DNA Pol α-primase (Fig. 3) (Grossi et al. 2004). Whether CST substitutes the related RPA stimulating the fill-in reaction is unknown.

S. pombe

Telomerase recruitment to telomeres in *S. pombe* occurs through the interaction of the telomerase subunit Est1 with Ccq1, which is tethered to telomeres through the interaction with Tpz1, which in turn interacts with Pot1 and Poz1 (Miyoshi et al. 2008; Tomita and Cooper 2008; Webb and Zakian 2012). Importantly, *S. pombe* Tel1 (ortholog of mammalian Atm) and Rad3 (ortholog of mammalian Atr) promote the direct interaction between Ccq1 and the Est1 subunit of telomerase (Moser et al. 2011; Yamazaki et al 2012). Ccq1 is phosphorylated on T93 by Tel1 and Rad3, and mutation of this site abolishes telomerase recruitment. A Cdc13 homolog has not been identified in *S. pombe*, and there is no known Ccq1 counterpart outside of fission yeast. Thus, telomerase recruitment seems to depend on different factors in fission and budding yeast. Telomere length regulation in *S. pombe* involves Taz1, Rap1, and Poz1, whose deletions lead to very substantial telomere elongation.

Mammals

Telomerase in humans is enriched in so-called Cajal bodies (CBs), dynamic subnuclear structures that are characterized by the coilin protein and that contain a variety of proteins and RNAs involved in the assembly and modification of small nuclear and nucleolar ribonucleoproteins (snRNPs and snoRNPs). CBs also harbor a class of CB-specific RNAs (scaRNAs) that are involved in methylation and pseudouridylation of small nuclear RNAs. In S phase of the cell cycle, a subset of CBs together with telomerase RNA is seen in association with telomeres in human cells (Jady et al. 2006; Tomlinson et al. 2006). As with other scaRNAs, human telomerase RNA contains a CAB-box motif that mediates association with CBs (Jady et al. 2004). Mutation of this motif does not prevent association of Tert with telomerase RNA (hTR), as catalytic telomerase activity is not impaired (Cristofari et al. 2007). However, CAB-box mutant telomerase does not efficiently associate with telomeres, and telomere extension is reduced (Cristofari et al. 2007). Tcab1 (Venteicher et al. 2009), also referred to as Wdr79 (Tycowski et al. 2009), was identified as a CAB-box binding protein and telomerase subunit. Tcab1 depletion from human cells leads to continuous telomere shortening (Venteicher et al. 2009). These results underline the importance of the CAB box and its interaction with Tcab1 for telomerase recruitment to telomeres. The CB association of telomerase has, in contrast to human cells, not been seen in mouse cells (Tomlinson et al. 2010). Thus, it will be interesting to determine if Tcab1 is also important for telomerase recruitment to telomeres in mice.

At the telomere, Tpp1 and Tin2 are important for telomerase recruitment or its retention, as depletion of either of these factors reduces (but does not abolish) telomerase association with telomeres (Abreu et al. 2010). Depletion and rescue studies further identified the amino-terminal OB-fold of Tpp1 as important for telomerase recruitment (Abreu et al. 2010), a domain that was found to also interact with telomerase in extracts (Xin et al. 2007). Whether telomerase recruitment is regulated in mammals by Atm or Atr, as in fission and budding yeast, is unknown, nor is it clear if in addition to Tpp1 other factors contribute to recruitment. The single-stranded telomere-binding protein Pot1 is an inhibitor of telomerase per se when binding to the 3′ end of the substrate (Kelleher et al. 2005; Lei et al. 2005). When associated with Tpp1, however, Pot1–Tpp1 decreases the dissociation of primer from telomerase and increases the efficiency of translocation (Wang et al. 2007; Latrick and Cech 2010). Therefore, the Pot1–Tpp1 complex increases repeat addition processivity of human telomerase severalfold. Thus, with the assistance of Pot1–Tpp1, telomerase can synthesize in vitro roughly with one or two binding and extension events the ~60 nucleotides that are typically added to a human telomere by telomerase in vivo (see below).

Using sophisticated molecular biological methods, the frequency of telomerase-mediated telomere elongation was measured to be 70%–100% at lagging-strand telomeres in HeLa and H1299 lung adenocarcinoma cells, and similar conclusions were drawn for the leading-strand telomeres (Zhao et al. 2009). This result was unexpected, as at lagging-strand telomeres the G-rich strand is parental and thus this strand is not predicted to shorten from incomplete end replication (Fig. 2, left panel). However, because this end is extended by telomerase, this finding implies that the parental G-rich strand must be shortened again subsequent to its elongation by telomerase, if not in every cell cycle at least in every couple of cell cycles. If not, the parental G-strand should suffer from continuous telomere elongation. The telomerase-mediated extension length was determined to be ~60 nucleotides, and experimental inhibition or depletion of telomerase reduced the frequency of extension events but not extension length (Zhao et al. 2011). This supports the notion that one single telomerase molecule acts at every telomere in every cell cycle.

The restriction of human telomerase to a single telomere binding and extension event may involve the human CST complex, which was identified recently (Miyake et al. 2009; Surovtseva et al. 2009). Human Stn1 and Ten1 carry clear sequence similarity with their budding yeast counterparts. Ctc1 (conserved telomere maintenance components 1), also referred to as AAF-132 (α-accessory factor) (Casteel et al. 2009), which assembles with Stn1 and Ten1 in the trimeric CST complex, bears little sequence similarity with *S. cerevisiae* Cdc13. Like budding yeast CST components, those of human CST contain putative OB-fold domains involved in binding single-stranded DNA and promoting protein interactions. Human CST inhibits telomerase activity through primer sequestration and physical interaction with the Pot1–Tpp1 telomerase processivity factor (Chen et al. 2012). CST binding at telomeres increases during late S/G_2 phase only upon telomerase action, coinciding with telomerase shutoff. CST depletion unleashes excessive telomerase activity, promoting telomere elongation. Increased binding of CST upon telomerase-mediated telomere elongation coincides with a switch from Pot1–Tpp1 binding with concomitant telomerase stimulation to CST binding and concomitant telomerase repression. Through binding of the telomerase-extended telomere, CST may limit telomerase action at individual telomeres to approximately one binding and extension event per cell cycle (Chen et al. 2012).

C-strand fill-in synthesis in human cells follows telomerase extension at the end of S phase (Zhao et al. 2009). Whether Pot1–Tpp1 or RPA stimulates this process is unclear. Possibly more relevant for this is the CST complex, which binds to the telomerase-extended telomeric 3′ end (Chen et al. 2012). Indeed, human Ctc1 and Stn1 stimulate DNA polymerase α-primase, increasing its affinity for template DNA (Casteel et al. 2009), and CST has been implicated in assisting lagging-strand synthesis at telomeres and other special DNA sequences throughout the genome (Miyake et al. 2009; Price et al. 2010; Gu et al. 2012). How the binding of single-stranded telomeric DNA by telomere-binding proteins, which include Pot1–Tpp1, CST, RPA, and possibly hnRNPA1, is regulated is unclear, nor is it known how T-loop formation may be influenced by these factors, possibly in a cell-cycle-dependent manner. It has been proposed that a switch from RPA binding in S phase to Pot1–Tpp1 after S phase is triggered by hnRNPA1 and TERRA (Flynn et al. 2011). In vitro, hnRNPA1 can compete with RPA for the binding of the single-stranded G-strand (Flynn et al. 2011). An increase of TERRA levels after S phase may sequester hnRNPA1 from telomeric DNA, as it has higher binding affinity for the 5′-UUAGGG-3′ repeats in TERRA than for the same 5′-TTAGGG-3′ telomeric DNA sequence. Binding of hnRNPA1 to TERRA may in turn allow binding of the telomeric 3′ overhang by Pot1–Tpp1. On the other hand, it has been argued that the higher local concentration of Pot1–Tpp1 near telomeres is sufficient to trigger Pot1–Tpp1 binding once semiconservative telomere replication is completed (Takai et al. 2011).

The above results on telomere extension in vivo cannot on their own explain telomere length homeostasis. Telomere length homeostasis requires that the frequency of telomere elongation, the extension length, or the shortening of telomeres occurs in a length-dependent manner. Telomerase preferentially elongated critically short telomeres in the offspring of mice to which telomerase was reintroduced after their ancestors grew in the absence of the telomerase RNA gene for several generations (Hemann et al. 2001). Thus, under nonhomeostatic conditions, telomerase is regulated in a telomere length-dependent manner, at least in mice. Telomere length-dependent regulation of telomerase in mammals is also supported by studies on the shelterin components, whose abundance at telomeres is length-dependent and whose experimental manipulation leads to telomere length changes (Smogorzewska and de Lange 2004; Hug and Lingner 2006). Generally, it has been observed that depletion of shelterin components by RNAi promotes telomerase-mediated telomere elongation, whereas their overexpression reduces telomere length. This may be at odds when considering the above-discussed positive roles of at least Tpp1 for telomerase recruitment and stimulation of processivity. However, it is important to consider that many telomere-binding proteins function like Swiss army knives. They may have multiple tasks in several processes at the chromosome end, such as telomere replication, regulation of telomerase, protection from nucleases, recombination and repair, and suppression of checkpoint signaling. Thus, depletion of a factor may highlight one function but may not allow detection of the multiple tools it may be equipped with to contribute to the correct functioning of chromosome ends.

CONCLUDING REMARKS

Elucidation of the complexity of telomere replication and the regulation of telomerase is a fascinating basic problem of biological research. It is clear that a better understanding of telomere replication and the regulation of telomerase in the future will come from studies in biochemistry, structural biology, genetics, and cell biology. Single-celled eukaryotes as well as animal model systems will continue to be of important value. For human telomere replication, the studies in induced pluripotent stem cells may become particularly revealing, as this system may allow learning about telomerase regulation and telomere replication in stem cells and during cellular differentiation. It is hoped that a much more thorough understanding of telomere maintenance may identify new targets that may become useful to inhibit telomerase action in cancer. After all these years (Kim et al. 1994), it has still not been possible to test if telomerase inhibition in human cancer is a valuable strategy to inhibit tumor growth. In addition, an increasing number of diseases have been linked to short telomeres (reviewed by Lansdorp 2009). Short-telomere diseases include dyskeratosis congenita (DC), idiopathic pulmonary fibrosis, the Werner syndrome, and the ICF (immunodeficiency, centromeric instability, facial anomalies) syndrome. DC patients die prematurely of bone marrow failure. ICF syndrome patients suffer from reduced immunoglobulin levels. Werner patients age prematurely and are predisposed to cancer. Although in some cases defective telomerase could be linked to short-telomere diseases (particularly in DC), in other cases the molecular mechanisms that underlie abnormal telomere shortening remain to be defined. In conclusion, it seems safe to assume that telomeres will not cease to unveil many hidden treasures.

ACKNOWLEDGMENTS

V.P. is supported by a long-term European Molecular Biology Organization postdoctoral fellowship. J.L.'s lab is supported by the Swiss National Science Foundation, a European Research Council advanced investigator grant (grant agreement number 232812), the Swiss Cancer League, and EPFL. We apologize to the researchers whose work could not be discussed and cited here because of tight space restrictions.

REFERENCES

Abreu E, Aritonovska E, Reichenbach P, Cristofari G, Culp B, Terns RM, Lingner J, Terns MP. 2010. TIN2-tethered TPP1 recruits human telomerase to telomeres in vivo. *Mol Cell Biol* **30:** 2971–2982.

Arneric M, Lingner J. 2007. Tel1 kinase and subtelomere-bound Tbf1 mediate preferential elongation of short telomeres by telomerase in yeast. *EMBO Rep* **8:** 1080–1085.

Arnoult N, Saintome C, Ourliac-Garnier I, Riou JF, Londono-Vallejo A. 2009. Human POT1 is required for efficient telomere C-rich strand replication in the absence of WRN. *Genes Dev* **23:** 2915–2924.

Azzalin CM, Reichenbach P, Khoriauli L, Giulotto E, Lingner J. 2007. Telomeric repeat containing RNA and RNA surveillance factors at mammalian chromosome ends. *Science* **318:** 798–801.

Bah A, Wischnewski H, Shchepachev V, Azzalin CM. 2012. The telomeric transcriptome of *Schizosaccharomyces pombe*. *Nucleic Acids Res* **40:** 2995–3005.

Baumann P, Cech T. 2001. Pot1, the putative telomere end-binding protein in fission yeast and humans. *Science* **292:** 1171–1175.

Beernink HT, Miller K, Deshpande A, Bucher P, Cooper JP. 2003. Telomere maintenance in fission yeast requires an Est1 ortholog. *Curr Biol* **13:** 575–580.

Bianchi A, Shore D. 2007a. Early replication of short telomeres in budding yeast. *Cell* **128:** 1051–1062.

Bianchi A, Shore D. 2007b. Increased association of telomerase with short telomeres in yeast. *Genes Dev* **21:** 1726–1730.

Bilaud T, Koering CE, Binet-Brasselet E, Ancelin K, Pollice A, Gasser SM, Gilson E. 1996. The telobox, a Myb-related telomeric DNA binding motif found in proteins from yeast, plants and human. *Nucleic Acids Res* **24:** 1294–1303.

Bilaud T, Brun C, Ancelin K, Koering CE, Laroche T, Gilson E. 1997. Telomeric localization of TRF2, a novel human telobox protein. *Nat Genet* **17:** 236–239.

Blackburn EH. 2006. A history of telomere biology. In *Telomeres*, 2nd ed. (ed. de Lange T, Lundblad V, Blackburn EH), pp. 1–19. Cold Spring Harbor Laboratory Press, Cold Spring Harbor, NY.

Blackburn E, Gall J. 1978. A tandemly repeated sequence at the termini of the extrachromosomal ribosomal RNA genes in *Tetrahymena*. *J Mol Biol* **120:** 33–53.

Bodnar AG, Ouellette M, Frolkis M, Holt SE, Chiu C-P, Morin GB, Harley CB, Shay JW, Lichtsteinter S, Wright WE. 1998. Extension of life-span by introduction of telomerase into normal human cells. *Science* **279:** 349–352.

Bonetti D, Martina M, Clerici M, Lucchini G, Longhese MP. 2009. Multiple pathways regulate 3′ overhang generation at *S. cerevisiae* telomeres. *Mol Cell* **35:** 70–81.

Broccoli D, Smogorzewska A, Chong L, de Lange T. 1997. Human telomeres contain two distinct Myb-related proteins, TRF1 and TRF2. *Nat Genet* **17:** 231–235.

Casteel DE, Zhuang S, Zeng Y, Perrino FW, Boss GR, Goulian M, Pilz RB. 2009. A DNA polymerase-α · primase cofactor with homology to replication protein A-32 regulates DNA replication in mammalian cells. *J Biol Chem* **284:** 5807–5818.

Chai W, Sfeir AJ, Hoshiyama H, Shay JW, Wright WE. 2006. The involvement of the Mre11/Rad50/Nbs1 complex in the generation of G-overhangs at human telomeres. *EMBO Rep* **7:** 225–230.

Chang S, Multani AS, Cabrera NG, Naylor ML, Laud P, Lombard D, Pathak S, Guarente L, DePinho RA. 2004. Essential role of limiting telomeres in the pathogenesis of Werner syndrome. *Nat Genet* **36:** 877–882.

Chang M, Arneric M, Lingner J. 2007. Telomerase repeat addition processivity is increased at critically short telomeres in a Tel1-dependent manner in *Saccharomyces cerevisiae*. *Genes Dev* **21:** 2485–2494.

Chawla R, Redon S, Raftopoulou C, Wischnewski H, Gagos S, Azzalin CM. 2011. Human UPF1 interacts with TPP1 and telomerase and sustains telomere leading-strand replication. *EMBO J* **30:** 4047–4058.

Chen LY, Redon S, Lingner J. 2012. The human CST complex is a terminator of telomerase activity. *Nature* **488:** 540–544.

Chong L, van Steensel B, Broccoli D, Erdjument-Bromage H, Hanish J, Tempst P, de Lange T. 1995. A human telomeric protein. *Science* **270:** 1663–1667.

Cooper JP, Nimmo ER, Allshire RC, Cech TR. 1997. Regulation of telomere length and function by a Myb-domain protein in fission yeast. *Nature* **385:** 744–747.

Crabbe L, Verdun RE, Haggblom CI, Karlseder J. 2004. Defective telomere lagging strand synthesis in cells lacking WRN helicase activity. *Science* **306:** 1951–1953.

Cristofari G, Adolf E, Reichenbach P, Sikora K, Terns RM, Terns MP, Lingner J. 2007. Human telomerase RNA accumulation in Cajal bodies facilitates telomerase recruitment to telomeres and telomere elongation. *Mol Cell* **27:** 882–889.

de Lange T. 2005. Shelterin: The protein complex that shapes and safeguards human telomeres. *Genes Dev* **19:** 2100–2110.

de Lange T. 2009. How telomeres solve the end-protection problem. *Science* **326:** 948–952.

Diede SJ, Gottschling DE. 2001. Exonuclease activity is required for sequence addition and Cdc13p loading at a de novo telomere. *Curr Biol* **11:** 1336–1340.

Evans SK, Lundblad V. 1999. Est1 and Cdc13 as comediators of telomerase access. *Science* **286:** 117–120.

Ferreira HC, Luke B, Schober H, Kalck V, Lingner J, Gasser SM. 2011. The PIAS homologue Siz2 regulates perinuclear telomere position and telomerase activity in budding yeast. *Nat Cell Biol* **13:** 867–874.

Fisher TS, Taggart AK, Zakian VA. 2004. Cell cycle-dependent regulation of yeast telomerase by Ku. *Nat Struct Mol Biol* **11:** 1198–1205.

Flynn RL, Centore RC, O'Sullivan RJ, Rai R, Tse A, Songyang Z, Chang S, Karlseder J, Zou L. 2011. TERRA and hnRNPA1 orchestrate an RPA-to-POT1 switch on telomeric single-stranded DNA. *Nature* **471:** 532–536.

Forstemann K, Lingner J. 2001. Molecular basis for telomere repeat divergence in budding yeast. *Mol Cell Biol* **21:** 7277–7286.

Frank CJ, Hyde M, Greider CW. 2006. Regulation of telomere elongation by the cyclin-dependent kinase CDK1. *Mol Cell* **24:** 423–432.

Gallardo F, Olivier C, Dandjinou AT, Wellinger RJ, Chartrand P. 2008. TLC1 RNA nucleo-cytoplasmic trafficking links telomerase biogenesis to its recruitment to telomeres. *EMBO J* **27:** 748–757.

Gallardo F, Laterreur N, Cusanelli E, Ouenzar F, Querido E, Wellinger RJ, Chartrand P. 2011. Live cell imaging of telomerase RNA dynamics reveals cell cycle-dependent clustering of telomerase at elongating telomeres. *Mol Cell* **44:** 819–827.

Gao H, Cervantes RB, Mandell EK, Otero JH, Lundblad V. 2007. RPA-like proteins mediate yeast telomere function. *Nat Struct Mol Biol* **14:** 208–214.

Gao H, Toro TB, Paschini M, Braunstein-Ballew B, Cervantes RB, Lundblad V. 2010. Telomerase recruitment in *Saccharomyces cerevisiae* is not dependent on Tel1-mediated phosphorylation of Cdc13. *Genetics* **186:** 1147–1159.

Gottschling D, Zakian V. 1986. Telomere proteins: Specific recognition and protection of the natural termini of *Oxytricha* macronuclear DNA. *Cell* **47:** 195–205.

Gottschling DE, Aparicio OM, Billington BL, Zakian VA. 1990. Position effect at *S. cerevisiae* telomeres: Reversible repression of Pol II transcription. *Cell* **63:** 751–762.

Gray JT, Celander DW, Price CM, Cech TR. 1991. Cloning and expression of genes for the *Oxytricha* telomere-binding protein: Specific subunit interactions in the telomeric complex. *Cell* **67:** 807–814.

Greenwood J, Cooper JP. 2012. Non-coding telomeric and subtelomeric transcripts are differentially regulated by telomeric and heterochromatin assembly factors in fission yeast. *Nucleic Acids Res* 40: 2956–2963.

Greider CW, Blackburn EH. 1985. Identification of a specific telomere terminal transferase activity in *Tetrahymena* extracts. *Cell* **43:** 405–413.

Greider CW, Blackburn EH. 1989. A telomeric sequence in the RNA of *Tetrahymena* telomerase required for telomere repeat synthesis. *Nature* **337:** 331–337.

Griffith JD, Comeau L, Rosenfield S, Stansel RM, Bianchi A, Moss H, de Lange T. 1999. Mammalian telomeres end in a large duplex loop. *Cell* **97:** 503–514.

Grossi S, Puglisi A, Dmitriev PV, Lopes M, Shore D. 2004. Pol12, the B subunit of DNA polymerase α, functions in both telomere capping and length regulation. *Genes Dev* **18:** 992–1006.

Gu P, Min JN, Wang Y, Huang C, Peng T, Chai W, Chang S. 2012. CTC1 deletion results in defective telomere replication, leading to catastrophic telomere loss and stem cell exhaustion. *EMBO J* **31:** 2309–2321.

Hang LE, Liu X, Cheung I, Yang Y, Zhao X. 2011. SUMOylation regulates telomere length homeostasis by targeting Cdc13. *Nat Struct Mol Biol* **18:** 920–926.

Hector RE, Shtofman RL, Ray A, Chen BR, Nyun T, Berkner KL, Runge KW. 2007. Tel1p preferentially associates with short telomeres to stimulate their elongation. *Mol Cell* **27:** 851–858.

Hemann MT, Strong MA, Hao LY, Greider CW. 2001. The shortest telomere, not average telomere length, is critical for cell viability and chromosome stability. *Cell* **107:** 67–77.

Hicke BJ, Celander DW, MacDonald GH, Price CM, Cech TR. 1990. Two versions of the gene encoding the 41-kilodalton subunit of the telomere binding protein of *Oxytricha nova*. *Proc Natl Acad Sci* **87:** 1481–1485.

Hirano Y, Fukunaga K, Sugimoto K. 2009. Rif1 and Rif2 inhibit localization of Tel1 to DNA ends. *Mol Cell* **33:** 312–322.

Hockemeyer D, Daniels JP, Takai H, de Lange T. 2006. Recent expansion of the telomeric complex in rodents: Two distinct POT1 proteins protect mouse telomeres. *Cell* **126:** 63–77.

Huertas P, Aguilera A. 2003. Cotranscriptionally formed DNA:RNA hybrids mediate transcription elongation impairment and transcription-associated recombination. *Mol Cell* **12:** 711–721.

Hug N, Lingner J. 2006. Telomere length homeostasis. *Chromosoma* **115:** 413–425.

Iglesias N, Redon S, Pfeiffer V, Dees M, Lingner J, Luke B. 2011. Subtelomeric repetitive elements determine TERRA regulation by Rap1/Rif and Rap1/Sir complexes in yeast. *EMBO Rep* **12:** 587–593.

Isken O, Maquat LE. 2008. The multiple lives of NMD factors: Balancing roles in gene and genome regulation. *Nat Rev Genet* **9:** 699–712.

Ivessa AS, Zhou JQ, Schulz VP, Monson EK, Zakian VA. 2002. *Saccharomyces* Rrm3p, a 5′ to 3′ DNA helicase that promotes replication fork progression through telomeric and subtelomeric DNA. *Genes Dev* **16:** 1383–1396.

Jady BE, Bertrand E, Kiss T. 2004. Human telomerase RNA and box H/ACA scaRNAs share a common Cajal body–specific localization signal. *J Cell Biol* **164:** 647–652.

Jady BE, Richard P, Bertrand E, Kiss T. 2006. Cell cycle-dependent recruitment of telomerase RNA and Cajal bodies to human telomeres. *Mol Biol Cell* **17:** 944–954.

Jain D, Cooper JP. 2010. Telomeric strategies: Means to an end. *Annu Rev Genet* **44:** 243–269.

Kanoh J, Sadaie M, Urano T, Ishikawa F. 2005. Telomere binding protein Taz1 establishes Swi6 heterochromatin independently of RNAi at telomeres. *Curr Biol* **15:** 1808–1819.

Kelleher C, Kurth I, Lingner J. 2005. Human protection of telomeres 1 (POT1) is a negative regulator of telomerase activity in vitro. *Mol Cell Biol* **25:** 808–818.

Kim NW, Piatyszek MA, Prowse KR, Harley CB, West MD, Ho PL, Coviello GM, Wright WE, Weinrich SL, Shay JW. 1994. Specific association of human telomerase activity with immortal cells and cancer. *Science* **266:** 2011–2015.

Koering CE, Fourel G, Binet-Brasselet E, Laroche T, Klein F, Gilson E. 2000. Identification of high affinity Tbf1p-binding sites within the budding yeast genome. *Nucleic Acids Res* **28:** 2519–2526.

Lam YC, Akhter S, Gu P, Ye J, Poulet A, Giraud-Panis MJ, Bailey SM, Gilson E, Legerski RJ, Chang S. 2010. SNMIB/Apollo protects leading-strand telomeres against NHEJ-mediated repair. *EMBO J* **29:** 2230–2241.

Lansdorp PM. 2009. Telomeres and disease. *EMBO J* **28:** 2532–2540.

Larrivee M, LeBel C, Wellinger RJ. 2004. The generation of proper constitutive G-tails on yeast telomeres is dependent on the MRX complex. *Genes Dev* **18:** 1391–1396.

Latrick CM, Cech TR. 2010. POT1–TPP1 enhances telomerase processivity by slowing primer dissociation and aiding translocation. *EMBO J* **29:** 924–933.

Lee J, Mandell EK, Tucey TM, Morris DK, Lundblad V. 2008. The Est3 protein associates with yeast telomerase through an OB-fold domain. *Nat Struct Mol Biol* **15:** 990–997.

Lei M, Zaug AJ, Podell ER, Cech TR. 2005. Switching human telomerase on and off with hPOT1 protein in vitro. *J Biol Chem* **280:** 20449–20456.

Lenain C, Bauwens S, Amiard S, Brunori M, Giraud-Panis MJ, Gilson E. 2006. The Apollo 5′ exonuclease functions together with TRF2 to protect telomeres from DNA repair. *Curr Biol* **16:** 1303–1310.

Lendvay TS, Morris DK, Sah J, Balasubramanian B, Lundblad V. 1996. Senescence mutants of *Saccharomyces cerevisiae* with a defect in telomere replication identify three additional *EST* genes. *Genetics* **144:** 1399–1412.

Leonardi J, Box JA, Bunch JT, Baumann P. 2008. TER1, the RNA subunit of fission yeast telomerase. *Nat Struct Mol Biol* **15:** 26–33.

Lewis KA, Wuttke DS. 2012. Telomerase and telomere-associated proteins: Structural insights into mechanism and evolution. *Structure* **20:** 28–39.

Li S, Makovets S, Matsuguchi T, Blethrow JD, Shokat KM, Blackburn EH. 2009. Cdk1-dependent phosphorylation of Cdc13 coordinates telomere elongation during cell-cycle progression. *Cell* **136:** 50–61.

Lingner J, Cech TR. 1996. Purification of telomerase from *Euplotes aediculatus*: Requirement of a primer 3′ overhang. *Proc Natl Acad Sci* **93:** 10712–10717.

Lingner J, Cooper JP, Cech TR. 1995. Telomerase and DNA end replication: No longer a lagging strand problem? *Science* **269:** 1533–1534.

Lingner J, Hughes TR, Shevchenko A, Mann M, Lundblad V, Cech TR. 1997. Reverse transcriptase motifs in the catalytic subunit of telomerase. *Science* **276:** 561–567.

Louis EJ, Borts RH. 1995. A complete set of marked telomeres in *Saccharomyces cerevisiae* for physical mapping and cloning. *Genetics* **139:** 125–136.

Luke B, Panza A, Redon S, Iglesias N, Li Z, Lingner J. 2008. The Rat1p 5′ to 3′ exonuclease degrades telomeric repeat-containing RNA and promotes telomere elongation in *Saccharomyces cerevisiae. Mol Cell* **32:** 465–477.

Lundblad V, Szostak JW. 1989. A mutant with a defect in telomere elongation leads to senescence in yeast. *Cell* **57:** 633–643.

Lustig AJ, Kurtz S, Shore D. 1990. Involvement of the silencer and UAS binding protein RAP1 in regulation of telomere length. *Science* **250:** 549–553.

Makovets S, Herskowitz I, Blackburn EH. 2004. Anatomy and dynamics of DNA replication fork movement in yeast telomeric regions. *Mol Cell Biol* **24:** 4019–4031.

Marcand S, Gilson E, Shore D. 1997. A protein-counting mechanism for telomere length regulation in yeast. *Science* **275:** 986–990.

Marcand S, Brevet V, Gilson E. 1999. Progressive *cis*-inhibition of telomerase upon telomere elongation. *EMBO J* **18:** 3509–3519.

Marcand S, Pardo B, Gratias A, Cahun S, Callebaut I. 2008. Multiple pathways inhibit NHEJ at telomeres. *Genes Dev* **22:** 1153–1158.

Martin V, Du LL, Rozenzhak S, Russell P. 2007. Protection of telomeres by a conserved Stn1–Ten1 complex. *Proc Natl Acad Sci* **104:** 14038–14043.

Martinez JL, Edstrom JE, Morcillo G, Diez JL. 2001. Telomeres in *Chironomus thummi* are characterized by different subfamilies of complex DNA repeats. *Chromosoma* **110:** 221–227.

Martinez P, Thanasoula M, Munoz P, Liao C, Tejera A, McNees C, Flores JM, Fernandez-Capetillo O, Tarsounas M, Blasco MA. 2009. Increased telomere fragility and fusions resulting from TRF1 deficiency lead to degenerative pathologies and increased cancer in mice. *Genes Dev* **23:** 2060–2075.

McClintock B. 1941. The stability of broken ends of chromosomes in *Zea mays. Genetics* **41:** 234–282.

Miller KM, Rog O, Cooper JP. 2006. Semi-conservative DNA replication through telomeres requires Taz1. *Nature* **440:** 824–828.

Mimitou EP, Symington LS. 2009. DNA end resection: Many nucleases make light work. *DNA Repair (Amst)* **8:** 983–995.

Mitchell JR, Cheng J, Collins K. 1999. A box H/ACA small nucleolar RNA-like domain at the human telomerase RNA 3′ end. *Mol Cell Biol* **19:** 567–576.

Miyake Y, Nakamura M, Nabetani A, Shimamura S, Tamura M, Yonehara S, Saito M, Ishikawa F. 2009. RPA-like mammalian Ctc1-Stn1-Ten1 complex binds to single-stranded DNA and protects telomeres independently of the Pot1 pathway. *Mol Cell* **36:** 193–206.

Miyoshi T, Kanoh J, Saito M, Ishikawa F. 2008. Fission yeast Pot1-Tpp1 protects telomeres and regulates telomere length. *Science* **320:** 1341–1344.

Moser BA, Chang YT, Kosti J, Nakamura TM. 2011. Tel1^{ATM} and Rad3^{ATR} kinases promote Ccq1-Est1 interaction to maintain telomeres in fission yeast. *Nat Struct Mol Biol* **18:** 1408–1413.

Muller HJ. 1938. The remaking of chromosomes. *Collecting Net (Woods Hole)* **13:** 181–198.

Nakamura TM, Cech TR. 1998. Reversing time—origin of telomerase. *Cell* **92:** 587–590.

Olovnikov AM. 1971. Principle of marginotomy in template synthesis of polynucleotides [in Russian]. *Dokl Akad Nauk SSSR* **201:** 1496–1499.

Ottaviani A, Gilson E, Magdinier F. 2008. Telomeric position effect: From the yeast paradigm to human pathologies? *Biochimie* **90:** 93–107.

Paeschke K, Capra JA, Zakian VA. 2011. DNA replication through G-quadruplex motifs is promoted by the *Saccharomyces cerevisiae* Pif1 DNA helicase. *Cell* **145:** 678–691.

Pardue ML, Debaryshe PG. 2011. Retrotransposons that maintain chromosome ends. *Proc Natl Acad Sci* **108:** 20317–20324.

Parkinson GN, Lee MP, Neidle S. 2002. Crystal structure of parallel quadruplexes from human telomeric DNA. *Nature* **417:** 876–880.

Pennock E, Buckley K, Lundblad V. 2001. Cdc13 delivers separate complexes to the telomere for end protection and replication. *Cell* **104:** 387–396.

Peterson S, Stellwagen A, Diede S, Singer M, Haimberger Z, Johnson C, Tzoneva M, Gottschling D. 2001. The function of a stem-loop in telomerase RNA is linked to the DNA repair protein Ku. *Nat Genet* **27:** 64–67.

Pfingsten JS, Goodrich KJ, Taabazuing C, Ouenzar F, Chartrand P, Cech TR. 2012. Mutually exclusive binding of telomerase RNA and DNA by Ku alters telomerase recruitment model. *Cell* **148:** 922–932.

Pogacic V, Dragon F, Filipowicz W. 2000. Human H/ACA small nucleolar RNPs and telomerase share evolutionarily conserved proteins NHP2 and NOP10. *Mol Cell Biol* **20:** 9028–9040.

Porro A, Feuerhahn S, Reichenbach P, Lingner J. 2010. Molecular dissection of TERRA biogenesis unveils the presence of distinct and multiple regulatory pathways. *Mol Cell Biol* **30:** 4808–4817.

Price CM, Boltz KA, Chaiken MF, Stewart JA, Beilstein MA, Shippen DE. 2010. Evolution of CST function in telomere maintenance. *Cell Cycle* **9:** 3157–3165.

Qi H, Zakian VA. 2000. The *Saccharomyces* telomere-binding protein Cdc13p interacts with both the catalytic subunit of DNA polymerase α and the telomerase-associated Est1 protein. *Genes Dev* **14:** 1777–1788.

Raices M, Verdun RE, Compton SA, Haggblom CI, Griffith JD, Dillin A, Karlseder J. 2008. *C. elegans* telomeres contain G-strand and C-strand overhangs that are bound by distinct proteins. *Cell* **132:** 745–757.

Redon S, Reichenbach P, Lingner J. 2007. Protein RNA and protein protein interactions mediate association of human EST1A/SMG6 with telomerase. *Nucleic Acids Res* **35:** 7011–7022.

Reichenbach P, Hoss M, Azzalin CM, Nabholz M, Bucher P, Lingner J. 2003. A human homolog of yeast Est1 associates with telomerase and uncaps chromosome ends when overexpressed. *Curr Biol* **13:** 568–574.

Riethman H. 2008. Human telomere structure and biology. *Annu Rev Genomics Hum Genet* **9:** 1–19.

Ritchie KB, Mallory JC, Petes TD. 1999. Interactions of TLC1 (which encodes the RNA subunit of telomerase), TEL1, and MEC1 in regulating telomere length in the yeast *Saccharomyces cerevisiae*. *Mol Cell Biol* **19:** 6065–6075.

Sabourin M, Tuzon CT, Zakian VA. 2007. Telomerase and Tel1p preferentially associate with short telomeres in *S. cerevisiae*. *Mol Cell* **27:** 550–561.

Schoeftner S, Blasco MA. 2008. Developmentally regulated transcription of mammalian telomeres by DNA-dependent RNA polymerase II. *Nat Cell Biol* **10:** 228–236.

Seto AG, Zaug AJ, Sobel SG, Wolin SL, Cech TR. 1999. *Saccharomyces cerevisiae* telomerase is an Sm small nuclear ribonucleoprotein particle. *Nature* **401:** 177–180.

Sfeir AJ, Chai W, Shay JW, Wright WE. 2005. Telomere-end processing the terminal nucleotides of human chromosomes. *Mol Cell* **18:** 131–138.

Sfeir A, Kosiyatrakul ST, Hockemeyer D, MacRae SL, Karlseder J, Schildkraut CL, de Lange T. 2009. Mammalian telomeres resemble fragile sites and require TRF1 for efficient replication. *Cell* **138:** 90–103.

Smogorzewska A, de Lange T. 2004. Regulation of telomerase by telomeric proteins. *Annu Rev Biochem* **73:** 177–208.

Snow BE, Erdmann N, Cruickshank J, Goldman H, Gill RM, Robinson MO, Harrington L. 2003. Functional conservation of the telomerase protein Est1p in humans. *Curr Biol* **13:** 698–704.

Sugawara N. 1989. DNA sequences at the telomeres of the fission yeast *S. pombe*. PhD thesis, Harvard University, Cambridge, MA.

Surovtseva YV, Churikov D, Boltz KA, Song X, Lamb JC, Warrington R, Leehy K, Heacock M, Price CM, Shippen DE. 2009. Conserved telomere maintenance component 1 interacts with STN1 and maintains chromosome ends in higher eukaryotes. *Mol Cell* **36:** 207–218.

Takai KK, Kibe T, Donigian JR, Frescas D, de Lange T. 2011. Telomere protection by TPP1/POT1 requires tethering to TIN2. *Mol Cell* **44:** 647–659.

Tang W, Kannan R, Blanchette M, Baumann P. 2012. Telomerase RNA biogenesis involves sequential binding by Sm and Lsm complexes. *Nature* **484:** 260–264.

Teixeira MT, Arneric M, Sperisen P, Lingner J. 2004. Telomere length homeostasis is achieved via a switch between telomerase-extendible and -nonextendible states. *Cell* **117:** 323–335.

Tomita K, Cooper JP. 2008. Fission yeast Ccq1 is telomerase recruiter and local checkpoint controller. *Genes Dev* **22:** 3461–3474.

Tomlinson RL, Ziegler TD, Supakorndej T, Terns RM, Terns MP. 2006. Cell cycle-regulated trafficking of human telomerase to telomeres. *Mol Biol Cell* **17:** 955–965.

Tomlinson RL, Li J, Culp BR, Terns RM, Terns MP. 2010. A Cajal body-independent pathway for telomerase trafficking in mice. *Exp Cell Res* **316:** 2797–2809.

Tseng SF, Lin JJ, Teng SC. 2006. The telomerase-recruitment domain of the telomere binding protein Cdc13 is regulated by Mec1p/Tel1p-dependent phosphorylation. *Nucleic Acids Res* **34:** 6327–6336.

Tsukamoto Y, Taggart AK, Zakian VA. 2001. The role of the Mre11-Rad50-Xrs2 complex in telomerase-mediated lengthening of *Saccharomyces cerevisiae* telomeres. *Curr Biol* **11:** 1328–1335.

Tycowski KT, Shu MD, Kukoyi A, Steitz JA. 2009. A conserved WD40 protein binds the Cajal body localization signal of scaRNP particles. *Mol Cell* **34:** 47–57.

van Overbeek M, de Lange T. 2006. Apollo, an Artemis-related nuclease, interacts with TRF2 and protects human telomeres in S phase. *Curr Biol* **16:** 1295–1302.

Vannier JB, Pavicic-Kaltenbrunner V, Petalcorin MI, Ding H, Boulton SJ. 2012. RTEL1 dismantles T loops and counteracts telomeric G4-DNA to maintain telomere integrity. *Cell* **149:** 795–806.

Venteicher AS, Abreu EB, Meng Z, McCann KE, Terns RM, Veenstra TD, Terns MP, Artandi SE. 2009. A human telomerase holoenzyme protein required for Cajal body localization and telomere synthesis. *Science* **323:** 644–648.

Vodenicharov MD, Wellinger RJ. 2006. DNA degradation at unprotected telomeres in yeast is regulated by the CDK1 (Cdc28/Clb) cell-cycle kinase. *Mol Cell* **24:** 127–137.

Wahba L, Amon JD, Koshland D, Vuica-Ross M. 2011. RNase H and multiple RNA biogenesis factors cooperate to prevent RNA:DNA hybrids from generating genome instability. *Mol Cell* **44:** 978–988.

Wang F, Podell ER, Zaug AJ, Yang Y, Baciu P, Cech TR, Lei M. 2007. The POT1-TPP1 telomere complex is a telomerase processivity factor. *Nature* **445:** 506–510.

Watson J. 1972. Origin of concatemeric T7 DNA. *Nat New Biol* **239:** 197–201.

Webb CJ, Zakian VA. 2008. Identification and characterization of the *Schizosaccharomyces pombe* TER1 telomerase RNA. *Nat Struct Mol Biol* **15:** 34–42.

Webb CJ, Zakian VA. 2012. *Schizosaccharomyces pombe* Ccq1 and TER1 bind the 14-3-3-like domain of Est1, which promotes and stabilizes telomerase-telomere association. *Genes Dev* **26:** 82–91.

Wu L, Multani AS, He H, Cosme-Blanco W, Deng Y, Deng JM, Bachilo O, Pathak S, Tahara H, Bailey SM, et al. 2006. Pot1 deficiency initiates DNA damage checkpoint activation and aberrant homologous recombination at telomeres. *Cell* **126:** 49–62.

Wu P, van Overbeek M, Rooney S, de Lange T. 2010. Apollo contributes to G overhang maintenance and protects leading-end telomeres. *Mol Cell* **39:** 606–617.

Xin H, Liu D, Wan M, Safari A, Kim H, Sun W, O'Connor MS, Songyang Z. 2007. TPP1 is a homologue of ciliate TEBP-β and interacts with POT1 to recruit telomerase. *Nature* **445:** 559–562.

Yamazaki H, Tarumoto Y, Ishikawa F. 2012. Tel1(ATM) and Rad3(ATR) phosphorylate the telomere protein Ccq1 to recruit telomerase and elongate telomeres in fission yeast. *Genes Dev* **26:** 241–246.

Ye J, Lenain C, Bauwens S, Rizzo A, Saint-Leger A, Poulet A, Benarroch D, Magdinier F, Morere J, Amiard S, et al. 2010. TRF2 and Apollo cooperate with topoisomerase 2α to protect human telomeres from replicative damage. *Cell* **142:** 230–242.

Yu GL, Bradley JD, Attardi LD, Blackburn EH. 1990. In vivo alteration of telomere sequences and senescence caused by mutated *Tetrahymena* telomerase RNAs. *Nature* **344:** 126–132.

Zhao Y, Sfeir AJ, Zou Y, Buseman CM, Chow TT, Shay JW, Wright WE. 2009. Telomere extension occurs at most chromosome ends and is uncoupled from fill-in in human cancer cells. *Cell* **138:** 463–475.

Zhao Y, Abreu E, Kim J, Stadler G, Eskiocak U, Terns MP, Terns RM, Shay JW, Wright WE. 2011. Processive and distributive extension of human telomeres by telomerase under homeostatic and nonequilibrium conditions. *Mol Cell* **42:** 297–307.

Zhu XD, Kuster B, Mann M, Petrini JH, Lange T. 2000. Cell--cycle-regulated association of RAD50/MRE11/NBS1 with TRF2 and human telomeres. *Nat Genet* **25:** 347–352.

Genomic Instability in Cancer

Tarek Abbas[1,2], Mignon A. Keaton[1,3], and Anindya Dutta[1]

[1]Department of Biochemistry and Molecular Genetics, University of Virginia, Charlottesville, Virginia 22908

[2]Department of Radiation Oncology, University of Virginia, Charlottesville, Virginia 22908

[3]Metabolon, Inc., Durham, North Carolina 27560

Correspondence: ad8q@cms.mail.virginia.edu

One of the fundamental challenges facing the cell is to accurately copy its genetic material to daughter cells. When this process goes awry, genomic instability ensues in which genetic alterations ranging from nucleotide changes to chromosomal translocations and aneuploidy occur. Organisms have developed multiple mechanisms that can be classified into two major classes to ensure the fidelity of DNA replication. The first class includes mechanisms that prevent premature initiation of DNA replication and ensure that the genome is fully replicated once and only once during each division cycle. These include cyclin-dependent kinase (CDK)-dependent mechanisms and CDK-independent mechanisms. Although CDK-dependent mechanisms are largely conserved in eukaryotes, higher eukaryotes have evolved additional mechanisms that seem to play a larger role in preventing aberrant DNA replication and genome instability. The second class ensures that cells are able to respond to various cues that continuously threaten the integrity of the genome by initiating DNA-damage-dependent "checkpoints" and coordinating DNA damage repair mechanisms. Defects in the ability to safeguard against aberrant DNA replication and to respond to DNA damage contribute to genomic instability and the development of human malignancy. In this article, we summarize our current knowledge of how genomic instability arises, with a particular emphasis on how the DNA replication process can give rise to such instability.

In eukaryotes, DNA replication initiates from hundreds of thousands of replication sites, termed origins of DNA replication (Leonard and Méchali 2013). Not only do cells need to initiate and terminate DNA replication at the right time during S phase, they must do so only once, at each of these replication origins, during each division cycle (Machida et al. 2005). Mechanisms that govern the initiation of DNA replication in eukaryotes are described in full detail in Bell and Kaguni (2013) and Tanaka and Araki (2013), whereas those that regulate DNA replication are described in Siddiqui et al. (2013) and Zielke et al. (2013). Rhind and Gilbert's (2013) work is dedicated to the mechanisms that control the timing of DNA replication. Here we focus on the mechanisms that prevent aberrant DNA replication.

Perturbations in DNA replication present cells with significant challenges. On one hand, incomplete genome duplication leads to cell inviability or, if cells survive, to aneuploidy. On

Cite this article as *Cold Spring Harb Perspect Biol* doi: 10.1101/cshperspect.a012914

the other hand, failure to restrict origin firing to once per replication origin per cell cycle, leads to overreplication. Reinitiation of DNA replication from the same origins of replication before the completion of S phase, commonly referred to as rereplication, is often associated with genome instability owing to the accumulation of replication intermediates, collapsed replication forks, and chromosomal breakages. In addition, defects in cytokinesis or mitotic regulation may lead to the complete reduplication of the genome. This latter process is reminiscent of endoreduplication, a physiological process that occurs in many metazoans during normal development and is characterized by multiple, discrete, and complete rounds of S phases without intervening mitosis. Finally, reinitiation of DNA replication from specific genomic loci is thought to be responsible for gene amplifications but the mechanisms underlying gene amplification are poorly understood.

How do these various mechanisms of genomic instability relate to cancer? An increase of copy number of chromosomes or genes allows cells to overexpress certain genes or mutate the extra copies to acquire growth, survival, or metastasis advantage. Equally important is the excessive DNA damage that is associated with these problems in DNA replication. When a cell attempts to segregate an underreplicated chromosome between two daughter cells, the result is often broken chromosomes and aneuploidy. Conversely, overreplication is marked by excessive DNA damage from collapsed replication forks. The repair processes are not perfect and so any increase in DNA damage leads to increased mutagenesis, and thus activation of oncogenes or inactivation of tumor suppressor genes, fueling malignant transformation and progression.

MECHANISMS PREVENTING REREPLICATION (ANTIREREPLICATION MECHANISMS)

The mechanisms that restrict DNA replication to one round per cell division were first elucidated in yeasts, both *Saccharomyces cerevisiae* and *Schizosaccharomyces pombe.* The simpler genomes, faster replication cycles, significant biochemical and genetic information, and facile experimental tools of the yeasts have significantly enriched our understanding of the network that maintains the fidelity and robustness of DNA replication. Subsequent work in more complex organisms has shown that many of the principles connecting replication regulation with genomic instability are largely conserved. Higher eukarotyes, however, have adapted additional mechanisms to safeguard against genomic instability.

Although the antirereplication mechanisms vary slightly from organism to organism, they are dependent on a common principle: the temporal and spatial uncoupling of the two steps involved in DNA replication, the licensing of origin of replication, and the initiation of DNA replication or the "firing" of replication origins (Machida et al. 2005; Zhu et al. 2005; Arias and Walter 2007; Hook et al. 2007). These mechanisms ensure that prereplication complexes (pre-RCs) are assembled immediately after the completion of mitosis until the G_1/S transition is reached and are prevented from assembling again until the completion of the next mitotic cycle (Fig. 1).

Inhibition of Pre-RC Assembly—A Universal Antirereplication Mechanism

During the G_1 phase of the cell cycle, transacting replication initiation factors recognize origins of replication and assemble prereplication complexes (pre-RCs) (Bell and Kaguni 2013; Tanaka and Araki 2013), which remain dormant until cells enter the S phase. Multiple mechanisms ensure that DNA is replicated once from each replication origin in a given cell cycle (Fig. 2).

Dbf4-Dependent Kinase—Master Initiator of DNA Replication

Two cell-cycle-regulated kinases, the Cdc7-Dbf4 kinase (also known as Dbf4-dependent kinase [DDK]) and the cyclin-dependent kinase (CDK), ensure that the assembly of pre-RC does not occur outside of G_1 (Labib 2010). DDK plays an important role in the initiation of

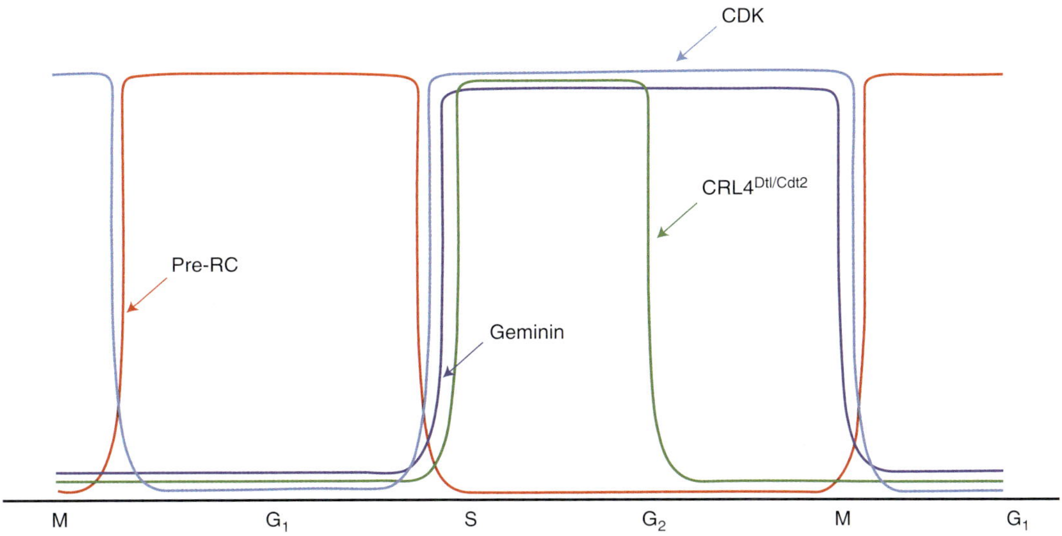

Figure 1. Cell-cycle-dependent regulation of prereplication complex (pre-RC). Pre-RCs are established immediately following exit from mitosis and are prevented from being assembled again from the G_1/S transition until exit from the next mitotic cycle. This regulation is dependent primarily on CDK, which is maintained at low levels early in G_1, but peaks to high levels at the G_1/S, during S, G_2, and M phases. Low CDK activity during G_1 results from the APC^{Cdc20}-dependent proteolysis of M-phase cyclins and the accumulation of CDK inhibitors. High CDK activity, mediated by S-, G_2-, or M-type cyclin, suppresses pre-RC through inactivation of multiple pre-RC components. In higher eukaryotes, pre-RC formation in S and G_2/M is further inhibited by geminin, a specific inhibitor of the replication-licensing factor Cdt1, as well as by an S-phase-specific ubiquitin ligase, $CRL4^{Dtl/Cdt2}$, which promotes the proteolysis of chromatin-bound Cdt1 and Set8.

DNA replication from individual replication origins. How DDK regulates origin firing and prevents new pre-RC assembly is not entirely clear. However, its ability to promote the phosphorylation of multiple components of the preinitiation complex may account for this activity. Early studies in yeast showed that DDK interacts with, phosphorylates, and activates the minichromosome maintenance subunit 2 protein (MCM2) of the MCM2-7 replicative helicase. The activation is significantly enhanced when MCM2 is prephosphorylated by CDK, suggesting that these two kinases work in concert to initiate DNA replication from individual origins (Lei et al. 1997; Masai et al. 2000). Subsequent studies, however, suggest that the *S. cerevisiae* MCM4 and MCM6 are the only targets of DDK essential for the recruitment to replication origins of GINS and Cdc45, two proteins required to activate the MCM2-7 helicase (Randell et al. 2010). It remains unclear, however, how the phosphorylation of MCM2-7 affects the recruitment of GINS and Cdc45, although it has been suggested that phosphorylation of MCM2-7 by DDK may alleviate some inhibitory activity in MCM4 (Sheu and Stillman 2010).

Additionally, in vitro and in vivo studies in *S. cerevisiae* revealed another cooperative activity of DDK and CDK, whereby DDK drives the initial recruitment of Sld3 and Cdc45, followed by CDK-dependent recruitment of GINS and the rest of the replisome to replication origins (Heller et al. 2011).

In response to DNA damage, such as that induced by hydroxyurea, the budding yeast Dbf4 (or its homolog in fission yeast, Dfp1/Him1) is hyperphosphorylated and inactivated through a Rad53-dependent mechanism triggering intra-S-phase checkpoint (Brown and Kelly 1999; Weinreich and Stillman 1999).

The DDK function in regulating origin firing and the DNA damage response is conserved

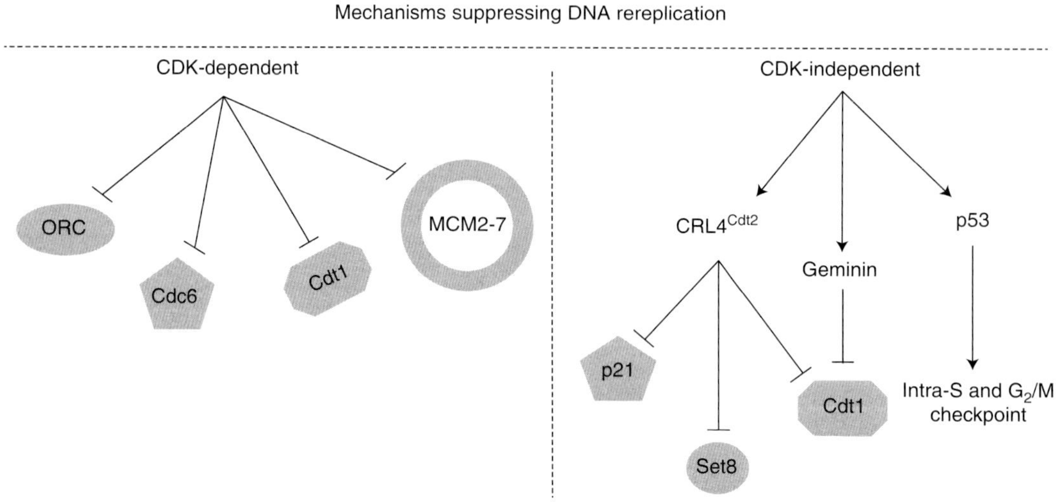

Figure 2. Multiple mechanisms suppress DNA rereplication. In all eukaryotic organisms, CDK activity, in S, G_2, and M phases of the cell cycle, inhibits rereplication by inactivating multiple components of the prereplication complex (pre-RC) to prevent reinitatiation of DNA replication. CDK directly phosphorylates origin recognition complex (ORC) proteins, Cdc6, Cdt1, as well as various subunits of the replicative helicase MCM2-7, resulting either in their targeted proteolysis or cytoplasmic sequestration and hence their inactivation (see text for details). Higher eukaryotes evolved additional CDK-independent mechanisms to suppress DNA rereplication in the more complex genomes. These include a specialized E3 ubiquitin ligase complex, $CRL4^{Dtl/Cdt2}$, which promotes the destruction of PCNA- and chromatin-bound Cdt1, Set8, and p21 during S phase of the cell cycle as well as in response to DNA damage. In addition, geminin, which is expressed in S and G_2 phases of the cell cycle, binds to and inhibits Cdt1 to prevent pre-RC assembly. The tumor suppressor p53 provides an additional barrier to DNA rereplication or to cell survival after rereplication, presumably owing to inhibition of S and G_2/M CDK kinase through the transcriptional up-regulation of p21, the induction of intra-S and G_2/M cell-cycle arrest, and the induction of apoptosis.

in higher eukaryotes. For example, both DDK and CDK are required for the formation of the Cdc45-Mcm2-7-GINS (CMG) complex and the activation of the replicative helicase at replication origins in *Drosophila melanogaster* (Ilves et al. 2010). The cooperative activation of DNA replication by DDK and CDK is also observed in *Xenopus laevis*, and here, DDK-mediated phosphorylation of MCM2 precedes that of CDK (Nougarede et al. 2000; Walter 2000). Additionally, and similar to yeast, the activation of the ATR pathway in *X. laevis* by single-stranded gaps generated in response to DNA damage impedes DDK association, thus inactivating the enzyme and resulting in the suppression of replication origins (Costanzo and Gautier 2003; Costanzo et al. 2003). In human cells, too, replication arrest with hydroxyurea is effected through checkpoint pathways that block the firing of origins of replication just before Cdc45 recruitment (Karnani and Dutta 2011). This is the very same step that is dependent on DDK and local CDK function. Thus, DDK seems to be a major target of inactivation in the cell's response to DNA damage to suppress origin firing. Whether DDK plays additional roles in restricting origin firing during normal S phase or in preventing refiring of replication origins in the same division cycle, remains to be determined.

CDK—Master Regulator of DNA Replication Initiation

The activity of CDK, the second enzyme regulating pre-RC assembly and the initiation of DNA replication, is under the regulation of the E3 ubiquitin ligase APC^{Cdc20}, which destroys the M-phase cyclin and thus maintains low CDK activity during G_1 (Fig. 1). CDK activity in various organisms is further inhibited in G_1 by a number of CDK inhibitors, such as Sic1 (an inhibitor of Clb-5/6-Cdk1/Cdc28 in *S. cerevisiae*) and Cdkn1a/p21/cip1 and Cdkn1b/p27/kip1 (inhibitors of cyclin E/A-Cdk2 in mammalian cells). At the G_1/S transition, in S, G_2, and M phases, cyclins reaccumulate and CDK inhibitors are degraded via ubiquitin-dependent proteolysis, resulting in high CDK activity (Zhu et al. 2005; Arias and Walter 2007; Hook et al. 2007).

In S, G_2, and M phases, CDK inhibits pre-RC by phosphorylating and inactivating multiple components of pre-RC (Fig. 2). In *S. cerevisiae*, high CDK activity in G_2/M suppresses pre-RC assembly (Dahmann et al. 1995; Tanaka et al. 1997; Tanaka and Araki 2010). Consequently, inactivation of G_2-CDK leads to rereplication, both in *S. cerevisiae*, and *S. pombe* (Correa-Bordes and Nurse 1995; Jallepalli and Kelly 1996; Noton and Diffley 2000). On the other hand, precocious activation of CDK in G_1, owing to overexpression of Clb2 or deletion of Sic1, inhibits pre-RC formation, results in fewer origin firing, prolongs S phase, and induces gross chromosomal rearrangement (Detweiler and Li 1998; Lengronne and Schwob 2002).

Origin Recognition Complex Proteins

The *S. cerevisiae* Cdc28 phosphorylates and inactivates Orc2 and Orc6 subunits of the origin recognition complex (ORC) (Nguyen et al. 2001). After DNA replication is initiated, the S-phase cyclin Clb5 binds Orc6 through a Cy motif and inhibits the assembly of new pre-RCs (Wilmes et al. 2004). Importantly, eliminating the Orc6 Cy motif results in rereplication, demonstrating the functional significance of cyclin-CDK-dependent inactivation of ORC proteins to prevent genomic instability (Wilmes et al. 2004).

In *S. pombe*, Cdk1/Cdc2, in complex with the mitotic cyclin Cdc13 (Cdc2-Cdc13), binds and phosphorylates Orp2 (the Orc2 homolog in fission yeast) at replication origins as cells enter S phase and prevents pre-RC assembly (Vas et al. 2001; Wuarin et al. 2002). Because Cdc13 does not function in S phase, the disruption of Cdc2-Cdc13 binding to Orp2 leads to endoreduplication rather than rereplication (Wuarin et al. 2002). This conclusion is consistent with the finding that mutations in Cdc13, the transient inhibition of Cdc13, or the overexpression of the Cdc2-Cdc13 inhibitor Rum1 causes endoreduplication (Correa-Bordes and Nurse 1995).

Inactivation of cyclin A-CDK in *Drosophila* Schneider D2 (SD2) cells, unlike the case in yeast, results in a complete duplication of the genome with 8N DNA content (Mihaylov et al. 2002). In *Xenopus* egg extract, cyclin A-CDK promotes the release of ORC proteins (along with Cdc6 and MCM) from chromatin and prevents pre-RC assembly (Findeisen et al. 1999). Orc1 is also selectively released from chromatin in mammalian cells in the S to M phase of the cell cycle where it is polyubiquitylated via the SCF^{Skp2} ubiquitin ligase before its proteolysis by the 26S proteasome (Fujita et al. 2002; Li and DePamphilis 2002; Mendez et al. 2002). Although mammalian Orc1 is also phosphorylated by cyclin A-CDK, it is unclear whether this phosphorylation targets the protein for proteolysis. However, CDK-mediated phosphorylation of mammalian Orc1 in mitosis prevents its association with chromatin and blocks pre-RC assembly (Li et al. 2004). Mammalian Orc2, on the other hand, is phosphorylated by the polo-like kinase 1 (Plk1) and this has been shown to promote DNA replication and is required to maintain functional pre-RC complexes under DNA replication stress (Song et al. 2011). A more recent study shows, however, that human CDK phosphorylates Orc2 on threonine 116 and threonine 226 during S and G_2 phases of the cell cycle leading to the dissociation of ORC2-5 from chromatin and replication origins, suggesting that CDK-mediated phosphorylation of Orc2 may inhibit ORC binding to newly replicated DNA (Lee et al. 2012).

Cdc6/Cdc18 and Cdc10-Dependent Transcript 1 (Cdt1)

The licensing activity of the replication initiation factor Cdc6 is also under the control of CDK. This is best illustrated in *S. cerevisiae*, in which Clb-Cdc28 controls Cdc6 activity through multiple mechanisms. First, Cdc28-dependent phosphorylation, and the consequent cytoplasmic sequestration of the transcriptional coactivator Swi5, inhibits Cdc6 transcription (Moll et al. 1991). Second, Clb-Cdc28 directly phosphorylates Cdc6 and promotes its proteolysis via the SCF^{Cdc4} ubiquitin ligase (Drury et al. 1997, 2000; Elsasser et al. 1999). Third, the stable association of Cdc6 with Clb2-Cdc28 upon phosphorylation of its amino-terminal CDK phosphorylation sites blocks its licensing activity (Mimura et al. 2004). In fission yeast, Cdc2 similarly phosphorylates the Cdc6 ortholog Cdc18 in S phase and promotes its destruction via the E3 ubiquitin ligase SCF^{Pop1} (Jallepalli et al. 1997; Kominami and Toda 1997). Significantly, ectopic expression of mutant Cdc18 lacking CDK phosphorylation sites results in significant rereplication. However, when Cdc18 mutant is expressed from its natural promoter, rereplication does not occur unless Cdt1 (Cdc10-dependent transcript 1) is coexpressed in these cells, suggesting that the simultaneous inactivation of Cdc18 and Cdt1 via CDK is required to suppress rereplication (Jallepalli et al. 1997; Gopalakrishnan et al. 2001). In higher eukaryotes, and as cells enter S phase, cyclin A-Cdk2 interacts with and phosphorylates Cdc6 leading to its cytoplasmic translocation and hence its inactivation (Saha et al. 1998; Jiang et al. 1999; Petersen et al. 1999; Delmolino et al. 2001).

CDK may also suppress rereplication by inactivating Cdt1. Cdt1 is phosphorylated by CDK in late G_1 and at the G_1/S transition and this results either in its exclusion from the nucleus, as is the case in budding yeast (Tanaka and Diffley 2002), or its destruction by the proteasome upon its ubiquitylation via the cullin-RING ubiquitin ligase 1 with the substrate-specific adaptor Skp2 ($CRL1^{Skp2}$; also known as SCF^{Skp2}) or via the cullin-RING ubiquitin ligases 4A/B with the substrate-specific adaptor Dtl/Cdt2 ($CRL4^{Dtl/Cdt2}$) as seen in fission yeast and metazoans (Nishitani et al. 2001; Li et al. 2003; Hu et al. 2004; Liu et al. 2004; Thomer et al. 2004; Arias and Walter 2006; Hu and Xiong 2006; Jin et al. 2006; Nishitani et al. 2006; Senga et al. 2006; Kim and Kipreos 2007; Guarino et al. 2011). Consistent with a significant role for CDK-mediated inactivation of Cdt1, mutations in the CDK phosphorylation sites in *Drosophila* Cdt1 (Dup) is sufficient to induce rereplication (Thomer et al. 2004).

Minichromosome Maintenance Proteins (MCM2-7)

The MCM2-7 replicative helicase is another target of inactivation by CDK. In *S. cereviseae*, Cdc28 phosphorylates various components of the MCM2-7 complex during G_2 and M and promotes its cytoplasmic translocation (Labib et al. 1999; Pasion and Forsburg 1999; Nguyen et al. 2000). However, in *S. pombe* and higher eukaryotes, the phosphorylation of MCM2-7 by CDK does not induce its cytoplasmic translocation but instead inhibits its chromatin binding (Kimura et al. 1994; Todorov et al. 1995; Coue et al. 1996; Pereverzeva et al. 2000; Lei and Tye 2001). An elegant study in yeast, however, showed that rereplication was not observed unless the inhibitory effect of CDK on ORC, CDC6, and MCM were simultaneously annulled (Nguyen et al. 2001). Annulling only one of the pathways resulted in weak rereplication, unlike that which was observed on the simultaneous inactivation of Cdc6 and Cdt1 described above. These findings show that inactivation of MCM by CDK is only one of several mechanisms that guard against rereplication.

Although inactivating MCM proteins by CDK is well established in yeast, there is little evidence in support for extensive direct regulation in mammalian cells. In vivo studies showed that cyclin B-Cdk1/Cdc2 phosphorylates MCM2 and MCM4 in human HeLa cells, and that inactivation of Cdc2 in the temperature-sensitive mutant murine FT210 cells resulted in sever impairment, both of G_2-associated MCM2/4 hyperphosphorylation and the release of the MCMs from chromatin (Fujita et al. 1998). On the other hand, in vitro studies showed that cyclin A-Cdk2 can phosphorylate multiple serine and threonine residues in the amino terminus of mouse MCM4 leading to inhibition of the Mcm4,6,7 subcomplex helicase activity (Ishimi et al. 2000; Ishimi and Komamura-Kohno 2001).

In addition to these pre-RC components, CDK also promotes DNA replication via phosphorylating additional proteins important for replication initiation. For example, Cdc28 phosphorylates Sld2 and Sld3 in S phase, an activity that is required for DNA replication in *S. cerevisеae* (Tanaka et al. 2007; Zegerman and Diffley 2007). Phosphorylation of Sld2 and Sld3 by Cdc28 is essential for the formation of the Sld2/Sld3/Dpb11 complex and the recruitment of the GINS complex to replication origins (Labib 2010). Furthermore, in vitro studies show that Cdc28, via phosphorylating Sld2, also stimulates the formation of an unstable DNA-independent complex between Sld2, Dpb11, DNA polymerase epsilon (Pol ε), and GINS (also known as the preloading complex [pre-LC]), which is required for replication initiation (Muramatsu et al. 2010). Whether CDK-mediated phosphorylation of these proteins prevents rereplication remains unclear.

$CRL4^{dtl/Cdt2}$, GEMININ, AND P53—CDK-INDEPENDENT SUPPRESSORS OF REREPLICATION AND GENOME INSTABILITY

Recent studies uncovered additional and CDK-independent mechanisms to prevent rereplication (Fig. 2). In mammalian cells, inactivation of the CDK-dependent proteolysis of Cdt1 via SCF^{Skp2} does not lead to rereplication, presumably because Cdt1 is still degraded via the $CRL4^{Dtl/Cdt2}$ pathway (Takeda et al. 2005). $CRL4^{Dtl/Cdt2}$ promotes the degradation of chromatin-bound Cdt1, both in S phase and after DNA damage, in a reaction that is dependent on its ability to bind PCNA through a specialized PCNA-interacting peptide (PIP-box) located in the amino terminus of Cdt1 (Arias and Walter 2006; Hu and Xiong 2006; Jin et al. 2006; Nishitani et al. 2006; Senga et al. 2006; Kim and Kipreos 2007; Guarino et al. 2011). Interestingly, inactivation of the $CRL4^{Dtl/Cdt2}$-dependent degradation of Cdt1 leads to rereplication in human cells and in zebrafish (Jin et al. 2006; Sansam et al. 2006). This important finding shows that (a) chromatin-bound Cdt1 is resistant to CDK-mediated phosphorylation and/or SCF^{Skp2}-mediated ubiquitylation and degradation, (b) Cdt1-PCNA interaction is not required for Cdt1 to promote DNA replication, and (c) chromatin- and PCNA-bound Cdt1 may be refractory to inhibition by geminin (see below).

$CRL4^{Dtl/Cdt2}$ may prevent rereplication via additional mechanisms. For example, $CRL4^{Dtl/Cdt2}$ promotes the ubiquitin-dependent degradation of the histone H4 lysine 20 (H4K20) monomethyl transferase SETD8/Pr-Set7/Set8 at the G_1/S and during S phase, and that inactivation of this pathway results in significant rereplication (Abbas et al. 2010; Tardat et al. 2010). $CRL4^{Dtl/Cdt2}$-insensitive but catalytically active mutant of Set8 is sufficient to nucleate DNA rereplication when tethered to specific genomic loci perhaps by preventing the assembly of pre-RC on chromatin (Tardat et al. 2010). Whether Set8 promotes rereplication through H4K20 methylation or through a yet-to-be-determined mechanism remains unclear. Also unclear is the relative contribution of Set8 and Cdt1 in promoting rereplication on $CRL4^{Dtl/Cdt2}$ inactivation.

Higher eukaryotes have evolved an additional mechanism to repress Cdt1 through its inactivation by geminin. Geminin was initially discovered in a screen for proteins degraded by the anaphase-promoting complex (APC) in mitosis and was found to inhibit DNA replication (McGarry and Kirschner 1998). Subsequent studies showed that geminin inhibits DNA replication through its ability to associate with and inhibit Cdt1 (Wohlschlegel et al. 2000; Tada et al. 2001; Lee et al. 2004; Saxena et al. 2004). Importantly, geminin depletion by small interfering RNA (siRNA) results in Cdt1-dependent rereplication, both in human and *Drosophila* cells (McGarry 2002; Mihaylov et al. 2002; Ballabeni et al. 2004; Melixetian et al. 2004; Zhu et al. 2004).

In human tumor-derived cells deficient of the tumor suppressor protein Tp53/p53, the mere overexpression of Cdt1 results in modest rereplication, which is significantly enhanced by the coexpression of Cdc6 (Vaziri et al. 2003).

This Cdt1+Cdc6-induced rereplication, however, was not observed in tumor cells with wild-type p53 or in primary untransformed cells. Inactivation of p53 renders cells susceptible to Cdt1+Cdc6-induced rereplication, demonstrating a critical function for p53 as a suppressor of rereplication (Vaziri et al. 2003). p53 was not actively suppressing rereplication but was modulating the cell's response to rereplication such that extensive rereplication is not seen.

Cdt1+Cdc6-induced rereplication, or rereplication induced by the depletion of geminin results in the activation of ATM and ATR and their downstream effectors and checkpoint kinases Chk2 and Chk1, respectively, which phosphorylate and activate p53 (Vaziri et al. 2003; Zhu et al. 2004; Lin and Dutta 2007). The activation of p53 leads to activation of p21 and a senescent response or leads to activation of proapoptotic genes and apoptosis (Vaziri et al. 2003; Zhu et al. 2004; Lin and Dutta 2007; Lin et al. 2010). Thus, the presence of p53 made the cells die in the presence of small amounts of rereplication so that it was difficult to detect cells with $>G_2$ DNA content.

Intriguingly, there are additional checkpoint pathways activated by rereplication upon $CRL4^{Dtl/Cdt2}$ inactivation or geminin depletion. In these instances, the ATR-Chk1-Cdc25C pathway is more active, leading to a G_2 block in which cells cannot enter mitosis but can still rereplicate their DNA. The difference between G_2 block with rereplication versus p53 activation and apoptosis may be dictated by the nature of rereplication or the accumulated replication intermediates induced under the various conditions, and consequently, the type of DNA damage generated (Vaziri et al. 2003; Zhu et al. 2004; Lin and Dutta 2007).

Geminin depletion allows rereplication in some human-derived cancer cells (e.g., HCT116 colon cancer cells), but not others (e.g., HeLa cervical cancer cells). The difference can be explained by the greater efficiency of the CDK-dependent pathway inhibiting rereplication in the HeLa cells. In such cells, the simultaneous inactivation of geminin and cyclin A-Cdk2, as, for example, by the depletion of Emi1 and premature activation of the APC, is sufficient to induce rereplication (Machida and Dutta 2007). Thus, although there are multiple pathways to inhibit rereplication, the relative importance of the different pathways appears to vary between cancer cell lines.

IS REREPLICATION A MAJOR DRIVER OF CANCER?

The increase in gene copy number that results from rereplication could sow the seeds of gene amplification, a major driving force in cancer. Indeed, careful perturbation of the pathways that prevent rereplication in *S. cerevisiae*, such that the rereplication is minimal and does not lead to cell death, can lead to gene amplification (Green et al. 2010). Therefore, theoretically it is possible that minor islands of rereplication lead to the gene amplification seen in human cancers. In practice, however, this has been difficult to show experimentally, mostly because all experimental approaches to induce rereplication in human cells inevitably produces extensive DNA damage and leads to cell death. In addition, mutations in the genes that prevent rereplication, aside from *TP53/p53*, have not yet been identified in cancers, although it is possible that transient epigenetic suppression of these genes is just sufficient to induce the local islands of rereplication that result in islands of stalled replication forks and DNA breaks, which lead to gene amplification, translocations, deletions, and other forms of cancer-promoting genome instability.

INDUCERS OF REREPLICATION: NOVEL CLASS OF ANTICANCER AGENTS

In contrast to the possibility that rereplication leads to cancer formation, the cytotoxicity that results from the induction of rereplication in human-derived cancer cells has surprisingly yielded a novel anticancer activity. Rereplication is invariably associated with spontaneous DNA damage and the activation of ATM/ATR-dependent intra-S-phase and G_2/M checkpoints. Bypassing rereplication-induced checkpoints leads to significant apoptosis or senescence (Melixetian et al. 2004; Zhu et al. 2004; Zhu and Dutta

2006; Machida and Dutta 2007). These findings suggest that inducing rereplication may be an effective strategy to inhibit the proliferation of cancer cells. This idea proved to be correct with the accidental discovery that a newly described drug, MLN4924, induces significant rereplication as well as apoptosis in a large number of human-derived cancer cell types and inhibited the growth of human tumor xenografts in mice (Soucy et al. 2009). MLN4924 was developed as a potent and specific inhibitor of the NEDD8-activating enzyme (NAE), an essential enzyme that controls the NEDD8 conjugation, which controls the activity of the cullin-RING subclass of E3 ubiquitin ligases including CRL4 and CRL1 ubiquitin ligases (Soucy et al. 2009). Rereplication and apoptosis induced in these cells correlated with the stabilization of Cdt1 and the induction of p21 owing to the inhibition of $CRL4^{Dtl/Cdt2}$ and $CRL1^{Skp2}$ ligases, and is accompanied by the induction of p21-dependent senescence (Lin et al. 2010; Jia et al. 2011). Interestingly, rereplication seems to selectively induce apoptosis in human cancer cells, but not in primary human-derived noncancer cells or cells immortalized by the SV40 large tumor antigen (Zhu and Depamphilis 2009). A recent study, using an image-based high-throughput screening for agents that induce rereplication, showed that two additional small bioactive agents are capable of inducing rereplication only in cancer cells, although the target of these molecules is yet to be determined (Zhu et al. 2011). These findings show that additional mechanisms exist in normal cells that safeguard against rereplication and that cancer cells may be selectively susceptible to the cytotoxic effects of rereplication-inducing agents.

DNA DAMAGE AND DNA REPLICATION

In addition to the mechanisms that operate during normal DNA replication, multiple mechanisms ensure that the genome is protected against various forms of DNA damage. Genomes are constantly experiencing DNA damage induced by normal metabolic by-products and by exogenous sources such as UV, ionizing radiation (IR), and chemical agents. DNA lesions act as roadblocks to DNA replication forks generating a precarious situation for the cell. If the fork is unable to bypass the lesion and restart replication, the cell must rely on an incoming replication fork from the opposite side to finish replication of the replicon. This may be particularly tricky in genomic regions with a paucity of replication origins that have a propensity for fork stalling or for forks that replicate at the end of S phase (Durkin and Glover 2007; Letessier et al. 2011). Such fork stalling lesions include bulky adducts from UV or chemical-induced damage, interstrand cross-links (ICLs), and intercalating drugs that distort the double helix. In addition, replication forks can stall owing to depletion of nucleotide pools, unusual DNA structures, and head-on collisions with transcription machinery (Branzei and Foiani 2010). Dissociation of the replication machinery at a stalled fork results in collapse and the generation of a one-ended double-strand break (DSB). Replication-induced DSBs can also result when a replication fork encounters a covalently linked topoisomerase (Topo I or II) that is generated by topoisomerase inhibitors that prevent religation of Topo-cleaved DNA (Liu et al. 2000; Deweese and Osheroff 2009).

Cellular Responses to DNA Damage

DSBs are highly detrimental to the cell and if left unrepaired, result in chromosomal rearrangements, deletions, and chromosomal loss. Thus, the cell has developed two parallel DNA damage response (DDR) pathways to respond to DSBs or stalled replication pathways mediated by the checkpoint kinases ATM or ATR, respectively (Fig. 3). Although ATM and ATR participate in parallel pathways that respond to different types of DNA damage, there is significant cross talk between the two pathways and the kinases converge on similar targets. For example, processing of DSBs produces single-stranded DNA that can activate ATR, whereas prolonged replication stress will result in DSB formation that subsequently activates ATM. Defects in the DDR result in unrepaired damage, which along with inaccurate DNA repair can lead to chromosomal

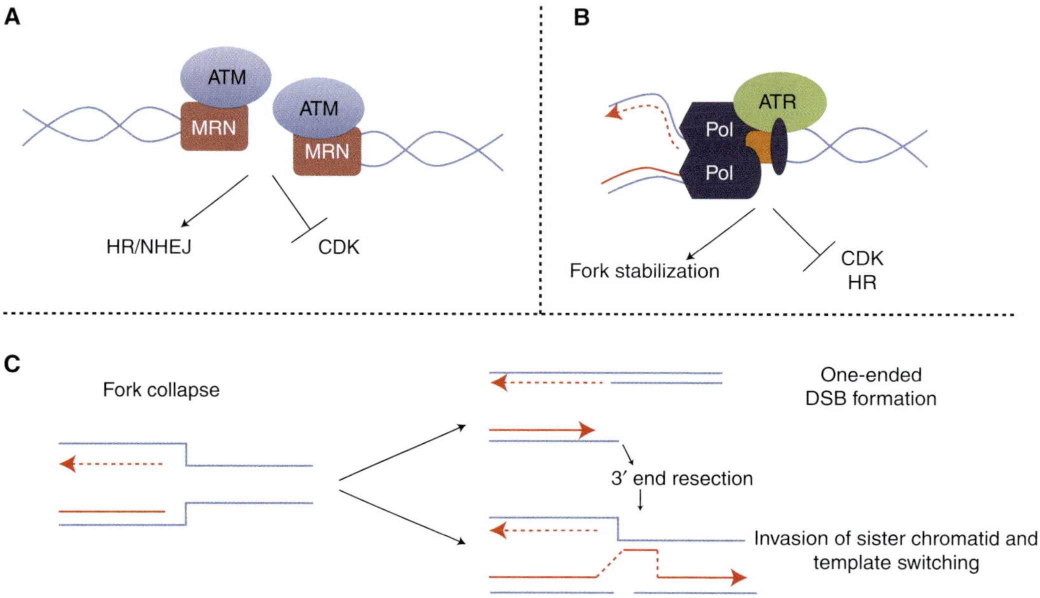

Figure 3. Model depicting cellular response to double-strand breaks (DSBs) and stalled replication forks. (*A*) Response to DSB: Mre11, Rad50, and Nbs1 complex (MRN) recognize DSBs and recruit ATM to the site where ATM signaling mediates repair by nonhomologous end joining (NHEJ) or homologous recombination (HR) while also inhibiting cyclin-dependent kinase (CDK) activity. (*B*) Response to stalled replication forks: Stalling of the replication machinery recruits ATR whose activity is required to prevent dissociation of the DNA polymerase complex. In addition, precocious HR is prevented and CDK activity at late firing replication origins is inhibited. (*C*) Dissociation of the replication machinery results in a collapsed fork. This may lead to a one-ended DSB, which can invade a sister chromatid to initiate replication restart.

alterations, a common cause of genomic instability and carcinogenesis.

Activation of the checkpoint in response to DSBs begins with binding of Mrell, Rad50, and Nbs1, the MRN complex, to the ends of the DSB where it recruits and subsequently activates the ATM kinase. ATM phosphorylates the histone variant H2AX initiating a signaling cascade involving phosphorylation and ubiquitylation to recruit a number of DDR and repair proteins to the site of the break. ATM also phosphorylates and activates the checkpoint kinase 2 (Chk2), which induces a G_2/M delay by sequestering Cdc25C in the cytoplasm, thus inhibiting M-phase CDK. ATM and Chk2 may also phosphorylate and activate p53 to induce prolonged cell-cycle arrest in G_1 or cell death by apoptosis in response to excessive DNA damage that overwhelms the repair pathways.

DNA damage during S phase results in replication fork stalling. This activates an ATR-dependent checkpoint to coordinate DNA repair and maintain DNA replication (Paulsen and Cimprich 2007; Nam and Cortez 2011). Uncoupling of the replicative helicase from DNA synthesis generates tracks of single-stranded DNA coated with RPA protein, which recruits ATR via its binding partner ATRIP. Rad17- and Rfc-mediated loading of the 911 complex (Rad9-Hus1-Rad1 trimeric clamp), together with TopBP1, activates ATR, which is required to recruit repair proteins and stabilize the replication fork to allow replication to restart once the block has been resolved. Maintaining the replication machinery at the fork prevents generation of DSBs. Additional clusters of replication origins in the vicinity of the stalled fork fire in order to complete replication. Late-replicating origins, however, are inhibited by ATR and its substrate and checkpoint kinase 1 (Chk1) (Debatisse et al. 2006; Ge and Blow 2010; Karnani and Dutta 2011). Chk1 (and Chk2) phosphorylates

Cdc25A on multiple serine residues resulting in its proteolytic degradation via the SCFbeta-TRCP pathway (Falck et al. 2001; Zhao et al. 2002; Busino et al. 2003; Hassepass et al. 2003). Chk1 additionally phosphorylates Cdc25A on serine 178 and threonine 507, thereby facilitating its binding to 14-3-3 and inhibiting its binding to cyclin B-Cdk1 (Chen et al. 2003). Cdc25A degradation/inhibition in response to DNA damage inactivates CDK2 and enforces intra-S-phase cell-cycle arrest critical for the repair of DNA damage before the resumption of DNA synthesis. Defective Chk1/Chk2-mediated phosphorylation and degradation of Cdc25A leads to radio-resistant DNA synthesis and genome instability.

Both *ATR* and *Chk1* are essential genes required during normal S phase, presumably because they are necessary to deal with endogenous replication stress such as troublesome secondary structures in the DNA or encounters with convergent transcription. This is evident from the ATR-mediated phosphorylation of H2AX at replication factories in unperturbed cells.

DSB Repair

Repair of DSBs is facilitated by nonhomologous end joining (NHEJ) and homologous recombination (HR). During NHEJ, Ku proteins bind to and protect the ends of the DSB and recruit the DNA-PK catalytic subunit to form the heterotrimeric DNA-PK holoenzyme. DNA-PK activity is required to process nonligatable ends before ligation by LigIV-Xrcc4 complex. In addition to the possibility of joining two unrelated DSB ends, this process involves the addition or removal of nucleotides at DSB ends and thus NHEJ is considered error prone. In contrast, HR uses sequence homology of a sister chromatid to regenerate the broken chromosome. During HR, the MRN complex recognizes the DSB leading to resection of the 5′ ends and generation of 3′ overhanging single-stranded DNA coated with RPA protein. RPA is exchanged for Rad51, which mediates strand invasion of the 3′ overhang into the undamaged sister chromatid searching for sequence homology. On establishment of sequence homology, the opposite strand of the sister chromatid is displaced and DNA polymerase extends the invading end using the sister chromatid as a template. A Holliday junction is formed and depending on how it is resolved, HR results either in crossover events (common during meiosis) in which chromosomal information is exchanged or noncrossover events (preferred during repair of DSBs). In the event the DSB occurs within a repetitive element, HR could occur between different chromosomes or within the same chromosome resulting in translocations, duplications, or deletions of genomic sequences.

Both NHEJ and HR compete with each other and it is still poorly understood what dictates which pathway is used. Cdk2 activity and S-phase-specific DNA repair pathways appear to promote HR. NHEJ operates primarily during G_1 phase (although it is active throughout the cell cycle) and defects in NHEJ result in sensitivity to DSB inducing agents. HR on the other hand is used during S and G_2 phases of the cell cycle when a sister chromatid is readily available and defective HR results in increased genomic instability and cancer predisposition (Kee and D'Andrea 2010).

Rescuing Stalled Replication Forks

Decoupling of the replicative helicase when polymerases stall generates tracks of single-stranded DNA that can promote undesired recombination events. Evidence suggests that ATR inhibits recombination at stabilized replication forks. However, once a fork has collapsed, HR is required to restart replication. Several models of replication restart for stalled or collapsed replication forks have been proposed, which include fork remodeling, template switching, and break-induced replication (Fig. 3). Fork remodeling requires nonreplicative helicases, such as Blm and Wrn, which have Holliday junction migration as well as DNA-annealing activities (Petermann and Helleday 2010). Template switching is invoked when the replication fork encounters slowed replication owing to difficult secondary structures or stalling induced by a DNA lesion. A Holliday junction can form in which one strand has invaded and replicated on the undamaged sister chromatid before switching back to the original template. On encountering a single-strand break, melting of the DNA strands during replication results in a DSB with only one end, which invokes break-induced replication. This may also arise by endonucleolytic cleavage of a lesion by Mus81-Eme1 at a collapsed fork. To repair the break, the 5′ end is resected, allowing invasion of the 3′ end into a sister chromatid, and restart of replication. Cycles of this break-induced replication ensue with dissociation of the replicating strand and reinvasion until a proper replication fork has been reestablished or the replicon has been completely replicated. Template switching and break-induced replication are relatively error-free as strand invasion is based on significant homology. However, repetitive sequences and microhomology between neighboring replisomes introduce opportunities for replication errors resulting in chromosomal duplications, insertions/deletions, and inversions.

ICLs pose a more complicated problem and invoke multiple repair mechanisms coordinated by the Fanconi anemia (FA) pathway (Kee and D'Andrea 2010). FA is a genetic disorder resulting in hypersensitivity to ICL-causing agents, genomic instability, and cancer. FA is caused by mutation in 13 different genes referred to as FANC genes. The FA pathway is dependent on the monoubiquitylation of FANCD2 and FANCI by the FA core complex. Recruitment and activation of the FA core complex at stalled forks requires ATR and the upstream FANCM protein, which recognizes damaged DNA during S phase. Monoubiquitylation of FANCD2-FANCI retains the heterodimer on chromatin where it coordinately regulates multiple repair processes via interactions with downstream proteins such as the endonuclease Fan1, Brca1, FAND1/Brca2, FANCJ/Bach1, and FANCN. Repair of ICLs is initiated by endonucleases (Xpf-Ercc1, Mus81-Eme1, or Fan1) "unhooking" the lesion from the Crick strand to generate a single-strand break on this strand that is soon converted to a DSB by replication. Translesion DNA synthesis (TLS) across the lesion on the Watson strand produces a double-stranded Watson chromatid on which nucleotide excision repair (NER) acts to repair the unhooked adduct. Finally, the DSB on the Crick chromatid can initiate HR by invading the repaired Watson chromatid to repair the DSB. In addition, FANCM recruits the Blm helicase and together they stabilize and remodel the replication fork to prevent deleterious recombination events. Although the FA pathway is most critical for repair of ICLs, the pathway is also activated in response to UV and replication-associated DSBs, as well as in response to rereplication (Zhu and Dutta 2006).

A number of genetic disorders characterized by genomic instability and cancer predisposition arise from mutations in DDR and HR proteins. These include FA, and ataxia telangiectasia, a genetic neurodegenerative disease that results from mutations in the *ATM* gene and is associated with extreme sensitivity to IR and predisposition to cancer. Mutation in Mre11 gives rise to similar disorders. Mutations in *BLM*, *WRN*, *RECQ4*, and RecQ family helicases involved in HR give rise to Bloom's syndrome, Werner's syndrome, and Rothmund-Thomson syndrome, respectively. Defects in Wrn and Recq4 result in decreased HR. Defects in Blm, in contrast, result in increased sister chromatid exchange and patients with Bloom's syndrome have an increased risk of loss of heterozygosity at tumor suppressor loci.

Point Mutations and Microsatellite Instability

The major replicative polymerases Pol δ and Pol ε that synthesize lagging and leading strands, respectively, have high fidelity rates (one error per 10^4–10^5 nucleotides) owing to their proofreading 3′ exonuclease activity (McCulloch and Kunkel 2008). Thus, with 6 billion base pairs in human diploid cells 100,000–1,000,000 mutations may be introduced during every division cycle owing to mistakes that escape proofreading. An additional increase in error rate occurs from mutations of DNA polymerases and alterations in dNTP levels. The mismatch repair (MMR) pathway is crucial for suppressing spontaneous mutations from errors in proofreading by recognizing mismatched nucleotides as well as insertions and deletions that arise from replication slippage. MMR proteins (MutS and MutL homologs) remove mismatched nucleotides via endonuclease and helicase activities allowing for reparative DNA synthesis. Defects in MMR result in a 100- to 1000-fold increase in replication errors including point mutations and expansion or contraction of simple repetitive sequences (microsatellites) that are prone to replication slippage (Arana and Kunkel 2010).

DNA damage owing to oxidation, alkylation, and hydrolysis of nucleotide bases, results in altered base-pairing leading to transversion mutations. An approximate 20,000 such insults to the mammalian genome occur per day (Preston et al. 2010). Most of these lesions are repaired by base excision repair (BER), a process in which DNA glycosylases recognize and remove the damaged base generating an apurinic or apyrimidinic (AP) site. AP endonuclease cleaves the AP site generating a single-strand break that is repaired by DNA synthesis (typically by Pol β) followed by DNA ligation. Typically, bulky DNA adducts, such as UV-induced cyclobutane pyrimidine dimers and 6–4 photoproducts, cause distortions to the double helix that are recognized and repaired by the NER pathway. Similar to BER, a NER pathway uses endonuclease and helicase activities to excise a small segment of DNA containing the lesion followed by DNA synthesis using the undamaged strand as a template. NER globally surveys the genome as well as being coupled to transcription of active genes.

In the event that unrepaired DNA damage, such as a UV-induced lesions or AP sites, is encountered by a replication fork, replication stalls and TLS is activated to allow replication across the lesion. This pathway involves recruitment of the ubiquitin ligase Rad18 to the stalled fork where it monoubiquitylates PCNA. This allows loading of specialized TLS polymerases, such as Pol η, Pol ξ, and Rev1/Rev3 that are able to bypass the lesion (Kannouche et al. 2004; Watanabe et al. 2004). Intriguingly, $CRL4^{Dtl/Cdt2}$ can also monoubiquitylate PCNA on the same residue that is modified by Rad18 (Terai et al. 2010), suggesting that there is some redundancy in the E3 ligase that initiates TLS. TLS polymerases lack proofreading ability and their fidelity rates differ greatly. Thus, TLS can be highly error prone depending on the type of lesion and the polymerase used (Hoffmann and Cazaux 2010). Deubiquitylation of PCNA by Usp1 or conversion of monoubiquitylation to polyubiquitylation prevents polymerase switching and repair of the lesion must rely on other mechanisms (Huang et al. 2006; Motegi et al. 2006; Unk et al. 2006).

Mutations that affect MMR, BER, and NER pathways also result in genomic instability and are associated with human cancer. For example, hereditary nonpolyposis colorectal cancer (also known as Lynch syndrome) arises from defects in MMR genes, such as *Msh2*, and is characterized by microsatellite instability. Similarly, biallelic inactivating germline mutations in the BER gene *MutYh*, an adenine DNA glycosylase, predispose to recessively inherited adenomatous polyposis (MUTYH-associated polyposis [MAP]) and colon cancer. Defects in repairing UV-induced lesions result in extreme sensitivity to sunlight and development of skin cancer. This occurs in xeroderma pigmentosum patients who have mutations in the NER pathway or the TLS polymerase Pol η. Mutations in these DNA repair pathways have also been documented in sporadic cancers (Negrini et al. 2010), as has the overexpression of error-prone polymerases (Hoffmann and Cazaux 2010).

CONCLUDING REMARKS

The last 20 years have witnessed significant advances in our understanding of the molecular events that are triggered by threats to the genome integrity. Key players have been identified and genetic and biochemical characterization of these genes and their encoded protein products now identify them as true "guardians of the genome." Dysregulation of these genes leads to unstable genomes. Excessive genomic instability usually leads to cell death, a factor in the utility of MLN4924 as an anticancer agent. However, because of the many redundant pathways for preventing overreplication or for repair, it is rare for mutations (or epigenetic silencing) affecting single pathways to create so much genomic instability as to cause cell death. Instead, inactivation of these single pathways simply increases the mutation burden without provoking cell death. The resulting heterogeneity in the genes of daughter cells allows the appearance of cells with growth or survival advantages, the driving force for cancer development. The same heterogeneity lies behind the resistance of cancers to many types of therapy. Thus, a major objective of our field should be to exploit known anomalies of repair and replication control in cancers to determine which redundant repair pathways should be therapeutically inhibited to produce "synthetic lethality" in the cancer cells. This will convert the mutation-generating machinery in the cancer into an Achilles heel by pushing the malignant cells selectively into extensive genomic instability and cell death.

ACKNOWLEDGMENTS

We apologize to authors whose primary work was not cited owing to space restrictions. Work in the authors' laboratory is supported by grants from the National Institutes of Health (NIH) (RO1CA60499 and GM084465). T.A. is supported by an NCI grant (RCA140774B). M.A.K. is supported by an NRSA from the NIH (5F32GM087843-03).

REFERENCES

*Reference is also in this collection.

Abbas T, Shibata E, Park J, Jha S, Karnani N, Dutta A. 2010. CRL4(Cdt2) regulates cell proliferation and histone gene expression by targeting PR-Set7/Set8 for degradation. *Mol Cell* **40:** 9–21.

Arana ME, Kunkel TA. 2010. Mutator phenotypes due to DNA replication infidelity. *Semin Cancer Biol* **20:** 304–311.

Arias EE, Walter JC. 2006. PCNA functions as a molecular platform to trigger Cdt1 destruction and prevent re-replication. *Nat Cell Biol* **8:** 84–90.

Arias EE, Walter JC. 2007. Strength in numbers: Preventing rereplication via multiple mechanisms in eukaryotic cells. *Genes Dev* **21:** 497–518.

Ballabeni A, Melixetian M, Zamponi R, Masiero L, Marinoni F, Helin K. 2004. Human geminin promotes pre-RC formation and DNA replication by stabilizing CDT1 in mitosis. *EMBO J* **23:** 3122–3132.

* Bell SP, Kaguni JM. 2013. Helicase loading at chromosomal origins of replication. *Cold Spring Harb Perspect Biol* doi: 10.1101/cshperspect.a010124.

Branzei D, Foiani M. 2010. Maintaining genome stability at the replication fork. *Nat Rev Mol Cell Biol* **11:** 208–219.

Brown GW, Kelly TJ. 1999. Cell cycle regulation of Dfp1, an activator of the Hsk1 protein kinase. *Proc Natl Acad Sci* **96:** 8443–8448.

Busino L, Donzelli M, Chiesa M, Guardavaccaro D, Ganoth D, Dorrello NV, Hershko A, Pagano M, Draetta GF. 2003. Degradation of Cdc25A by β-TrCP during S phase and in response to DNA damage. *Nature* **426:** 87–91.

Chen MS, Ryan CE, Piwnica-Worms H. 2003. Chk1 kinase negatively regulates mitotic function of Cdc25A phosphatase through 14-3-3 binding. *Mol Cell Biol* **23:** 7488–7497.

Correa-Bordes J, Nurse P. 1995. p25rum1 orders S phase and mitosis by acting as an inhibitor of the p34cdc2 mitotic kinase. *Cell* **83:** 1001–1009.

Costanzo V, Gautier J. 2003. Single-strand DNA gaps trigger an ATR- and Cdc7-dependent checkpoint. *Cell Cycle* **2:** 17.

Costanzo V, Shechter D, Lupardus PJ, Cimprich KA, Gottesman M, Gautier J. 2003. An ATR- and Cdc7-dependent DNA damage checkpoint that inhibits initiation of DNA replication. *Mol Cell* **11:** 203–213.

Coue M, Kearsey SE, Mechali M. 1996. Chromotin binding, nuclear localization and phosphorylation of *Xenopus* cdc21 are cell-cycle dependent and associated with the control of initiation of DNA replication. *EMBO J* **15:** 1085–1097.

Dahmann C, Diffley JF, Nasmyth KA. 1995. S-phase-promoting cyclin-dependent kinases prevent re-replication by inhibiting the transition of replication origins to a pre-replicative state. *Curr Biol* **5:** 1257–1269.

Debatisse M, El Achkar E, Dutrillaux B. 2006. Common fragile sites nested at the interfaces of early and late-replicating chromosome bands: *Cis* acting components of the G2/M checkpoint? *Cell Cycle* **5:** 578–581.

Delmolino LM, Saha P, Dutta A. 2001. Multiple mechanisms regulate subcellular localization of human CDC6. *J Biol Chem* **276:** 26947–26954.

Detweiler CS, Li JJ. 1998. Ectopic induction of Clb2 in early G1 phase is sufficient to block prereplicative complex formation in *Saccharomyces cerevisiae*. *Proc Natl Acad Sci* **95:** 2384–2389.

Deweese JE, Osheroff N. 2009. The DNA cleavage reaction of topoisomerase II: Wolf in sheep's clothing. *Nucleic Acids Res* **37:** 738–748.

Drury LS, Perkins G, Diffley JF. 1997. The Cdc4/34/53 pathway targets Cdc6p for proteolysis in budding yeast. *EMBO J* **16:** 5966–5976.

Drury LS, Perkins G, Diffley JF. 2000. The cyclin-dependent kinase Cdc28p regulates distinct modes of Cdc6p proteolysis during the budding yeast cell cycle. *Curr Biol* **10:** 231–240.

Durkin SG, Glover TW. 2007. Chromosome fragile sites. *Annu Rev Genet* **41:** 169–192.

Elsasser S, Chi Y, Yang P, Campbell JL. 1999. Phosphorylation controls timing of Cdc6p destruction: A biochemical analysis. *Mol Biol Cell* **10:** 3263–3277.

Falck J, Mailand N, Syljuasen RG, Bartek J, Lukas J. 2001. The ATM-Chk2-Cdc25A checkpoint pathway guards against radioresistant DNA synthesis. *Nature* **410:** 842–847.

Findeisen M, El-Denary M, Kapitza T, Graf R, Strausfeld U. 1999. Cyclin A-dependent kinase activity affects chromatin binding of ORC, Cdc6, and MCM in egg extracts of *Xenopus laevis*. *Eur J Biochem* **264:** 415–426.

Fujita M, Yamada C, Tsurumi T, Hanaoka F, Matsuzawa K, Inagaki M. 1998. Cell cycle- and chromatin binding state-dependent phosphorylation of human MCM heterohexameric complexes. A role for cdc2 kinase. *J Biol Chem* **273:** 17095–17101.

Fujita M, Ishimi Y, Nakamura H, Kiyono T, Tsurumi T. 2002. Nuclear organization of DNA replication initiation proteins in mammalian cells. *J Biol Chem* **277:** 10354–10361.

Ge XQ, Blow JJ. 2010. Chk1 inhibits replication factory activation but allows dormant origin firing in existing factories. *J Cell Biol* **191:** 1285–1297.

Gopalakrishnan V, Simancek P, Houchens C, Snaith HA, Frattini MG, Sazer S, Kelly TJ. 2001. Redundant control of rereplication in fission yeast. *Proc Natl Acad Sci* **98:** 13114–13119.

Green BM, Finn KJ, Li JJ. 2010. Loss of DNA replication control is a potent inducer of gene amplification. *Science* **329:** 943–946.

Guarino E, Shepherd ME, Salguero I, Hua H, Deegan RS, Kearsey SE. 2011. Cdt1 proteolysis is promoted by dual PIP degrons and is modulated by PCNA ubiquitylation. *Nucleic Acids Res* **39:** 5978–5990.

Hassepass I, Voit R, Hoffmann I. 2003. Phosphorylation at serine 75 is required for UV-mediated degradation of human Cdc25A phosphatase at the S-phase checkpoint. *J Biol Chem* **278:** 29824–29829.

Heller RC, Kang S, Lam WM, Chen S, Chan CS, Bell SP. 2011. Eukaryotic origin-dependent DNA replication in vitro reveals sequential action of DDK and S-CDK kinases. *Cell* **146:** 80–91.

Hoffmann JS, Cazaux C. 2010. Aberrant expression of alternative DNA polymerases: A source of mutator phenotype as well as replicative stress in cancer. *Semin Cancer Biol* **20:** 312–319.

Hook SS, Lin JJ, Dutta A. 2007. Mechanisms to control rereplication and implications for cancer. *Curr Opin Cell Biol* **19:** 663–671.

Hu J, Xiong Y. 2006. An evolutionarily conserved function of proliferating cell nuclear antigen for Cdt1 degradation by the Cul4-Ddb1 ubiquitin ligase in response to DNA damage. *J Biol Chem* **281:** 3753–3756.

Hu J, McCall CM, Ohta T, Xiong Y. 2004. Targeted ubiquitination of CDT1 by the DDB1-CUL4A-ROC1 ligase in response to DNA damage. *Nat Cell Biol* **6:** 1003–1009.

Huang TT, Nijman SM, Mirchandani KD, Galardy PJ, Cohn MA, Haas W, Gygi SP, Ploegh HL, Bernards R, D'Andrea AD. 2006. Regulation of monoubiquitinated PCNA by DUB autocleavage. *Nat Cell Biol* **8:** 339–347.

Ilves I, Petojevic T, Pesavento JJ, Botchan MR. 2010. Activation of the MCM2–7 helicase by association with Cdc45 and GINS proteins. *Mol Cell* **37:** 247–258.

Ishimi Y, Komamura-Kohno Y. 2001. Phosphorylation of Mcm4 at specific sites by cyclin-dependent kinase leads to loss of Mcm4,6,7 helicase activity. *J Biol Chem* **276:** 34428–34433.

Ishimi Y, Komamura-Kohno Y, You Z, Omori A, Kitagawa M. 2000. Inhibition of Mcm4,6,7 helicase activity by phosphorylation with cyclin A/Cdk2. *J Biol Chem* **275:** 16235–16241.

Jallepalli PV, Kelly TJ. 1996. Rum1 and Cdc18 link inhibition of cyclin-dependent kinase to the initiation of DNA replication in *Schizosaccharomyces pombe*. *Genes Dev* **10:** 541–552.

Jallepalli PV, Brown GW, Muzi-Falconi M, Tien D, Kelly TJ. 1997. Regulation of the replication initiator protein p65cdc18 by CDK phosphorylation. *Genes Dev* **11:** 2767–2779.

Jia L, Li H, Sun Y. 2011. Induction of p21-dependent senescence by an NAE inhibitor, MLN4924, as a mechanism of growth suppression. *Neoplasia* **13:** 561–569.

Jiang W, Wells NJ, Hunter T. 1999. Multistep regulation of DNA replication by Cdk phosphorylation of HsCdc6. *Proc Natl Acad Sci* **96:** 6193–6198.

Jin J, Arias EE, Chen J, Harper JW, Walter JC. 2006. A family of diverse Cul1 Ddb1 interacting proteins includes Cdt2, which is required for S phase destruction of the replication factor Cdt1. *Mol Cell* **23:** 709–721.

Kannouche PL, Wing J, Lehmann AR. 2004. Interaction of human DNA polymerase η with monoubiquitinated PCNA: A possible mechanism for the polymerase switch in response to DNA damage. *Mol Cell* **14:** 491–500.

Karnani N, Dutta A. 2011. The effect of the intra-S-phase checkpoint on origins of replication in human cells. *Genes Dev* **25:** 621–633.

Kee Y, D'Andrea AD. 2010. Expanded roles of the Fanconi anemia pathway in preserving genomic stability. *Genes Dev* **24:** 1680–1694.

Kim Y, Kipreos ET. 2007. The *Caenorhabditis elegans* replication licensing factor CDT-1 is targeted for degradation by the CUL-4/DDB-1 complex. *Mol Cell Biol* **27:** 1394–1406.

Kimura H, Nozaki N, Sugimoto K. 1994. DNA polymerase α associated protein P1, a murine homolog of yeast MCM3, changes its intranuclear distribution during the DNA synthetic period. *EMBO J* **13:** 4311–4320.

Kominami K, Toda T. 1997. Fission yeast WD-repeat protein pop1 regulates genome ploidy through ubiquitin-proteasome-mediated degradation of the CDK inhibitor Rum1 and the S-phase initiator Cdc18. *Genes Dev* **11:** 1548–1560.

Labib K. 2010. How do Cdc7 and cyclin-dependent kinases trigger the initiation of chromosome replication in eukaryotic cells? *Genes Dev* **24:** 1208–1219.

Labib K, Diffley JF, Kearsey SE. 1999. G1-phase and B-type cyclins exclude the DNA-replication factor Mcm4 from the nucleus. *Nat Cell Biol* **1:** 415–422.

Lee C, Hong B, Choi JM, Kim Y, Watanabe S, Ishimi Y, Enomoto T, Tada S, Cho Y. 2004. Structural basis for inhibition of the replication licensing factor Cdt1 by geminin. *Nature* **430:** 913–917.

Lee KY, Bang SW, Yoon SW, Lee SH, Yoon JB, Hwang DS. 2012. Phosphorylation of ORC2 protein dissociates origin recognition complex from chromatin and replication origins. *J Biol Chem* **287:** 11891–11898.

Lei M, Tye BK. 2001. Initiating DNA synthesis: From recruiting to activating the MCM complex. *J Cell Sci* **114:** 1447–1454.

Lei M, Kawasaki Y, Young MR, Kihara M, Sugino A, Tye BK. 1997. Mcm2 is a target of regulation by Cdc7-Dbf4 during the initiation of DNA synthesis. *Genes Dev* **11:** 3365–3374.

Lengronne A, Schwob E. 2002. The yeast CDK inhibitor Sic1 prevents genomic instability by promoting replication origin licensing in late G(1). *Mol Cell* **9:** 1067–1078.

* Leonard AC, Méchali M. 2013. DNA replication origins. *Cold Spring Harb Perspect Biol* doi: 10.1101/cshperspect.a010116.

Letessier A, Millot GA, Koundrioukoff S, Lachages AM, Vogt N, Hansen RS, Malfoy B, Brison O, Debatisse M. 2011. Cell-type-specific replication initiation programs set fragility of the FRA3B fragile site. *Nature* **470:** 120–123.

Li CJ, DePamphilis ML. 2002. Mammalian Orc1 protein is selectively released from chromatin and ubiquitinated during the S-to-M transition in the cell division cycle. *Mol Cell Biol* **22:** 105–116.

Li X, Zhao Q, Liao R, Sun P, Wu X. 2003. The SCF(Skp2) ubiquitin ligase complex interacts with the human replication licensing factor Cdt1 and regulates Cdt1 degradation. *J Biol Chem* **278:** 30854–30858.

Li CJ, Vassilev A, DePamphilis ML. 2004. Role for Cdk1 (Cdc2)/cyclin A in preventing the mammalian origin recognition complex's largest subunit (Orc1) from binding to chromatin during mitosis. *Mol Cell Biol* **24:** 5875–5886.

Lin JJ, Dutta A. 2007. ATR pathway is the primary pathway for activating G2/M checkpoint induction after re-replication. *J Biol Chem* **282:** 30357–30362.

Lin JJ, Milhollen MA, Smith PG, Narayanan U, Dutta A. 2010. NEDD8-targeting drug MLN4924 elicits DNA re-replication by stabilizing Cdt1 in S phase, triggering checkpoint activation, apoptosis, and senescence in cancer cells. *Cancer Res* **70:** 10310–10320.

Liu LF, Desai SD, Li TK, Mao Y, Sun M, Sim SP. 2000. Mechanism of action of camptothecin. *Ann NY Acad Sci* **922:** 1–10.

Liu E, Li X, Yan F, Zhao Q, Wu X. 2004. Cyclin-dependent kinases phosphorylate human Cdt1 and induce its degradation. *J Biol Chem* **279:** 17283–17288.

Machida YJ, Dutta A. 2007. The APC/C inhibitor, Emi1, is essential for prevention of rereplication. *Genes Dev* **21:** 184–194.

Machida YJ, Hamlin JL, Dutta A. 2005. Right place, right time, and only once: Replication initiation in metazoans. *Cell* **123:** 13–24.

Masai H, Matsui E, You Z, Ishimi Y, Tamai K, Arai K. 2000. Human Cdc7-related kinase complex. In vitro phosphorylation of MCM by concerted actions of Cdks and Cdc7 and that of a criticial threonine residue of Cdc7 bY Cdks. *J Biol Chem* **275:** 29042–29052.

McCulloch SD, Kunkel TA. 2008. The fidelity of DNA synthesis by eukaryotic replicative and translesion synthesis polymerases. *Cell Res* **18:** 148–161.

McGarry TJ. 2002. Geminin deficiency causes a Chk1-dependent G2 arrest in *Xenopus*. *Mol Biol Cell* **13:** 3662–3671.

McGarry TJ, Kirschner MW. 1998. Geminin, an inhibitor of DNA replication, is degraded during mitosis. *Cell* **93:** 1043–1053.

Melixetian M, Ballabeni A, Masiero L, Gasparini P, Zamponi R, Bartek J, Lukas J, Helin K. 2004. Loss of Geminin induces rereplication in the presence of functional p53. *J Cell Biol* **165:** 473–482.

Mendez J, Zou-Yang XH, Kim SY, Hidaka M, Tansey WP, Stillman B. 2002. Human origin recognition complex large subunit is degraded by ubiquitin-mediated proteolysis after initiation of DNA replication. *Mol Cell* **9:** 481–491.

Mihaylov IS, Kondo T, Jones L, Ryzhikov S, Tanaka J, Zheng J, Higa LA, Minamino N, Cooley L, Zhang H. 2002. Control of DNA replication and chromosome ploidy by geminin and cyclin A. *Mol Cell Biol* **22:** 1868–1880.

Mimura S, Seki T, Tanaka S, Diffley JF. 2004. Phosphorylation-dependent binding of mitotic cyclins to Cdc6 contributes to DNA replication control. *Nature* **431:** 1118–1123.

Moll T, Tebb G, Surana U, Robitsch H, Nasmyth K. 1991. The role of phosphorylation and the CDC28 protein kinase in cell cycle-regulated nuclear import of the *S. cerevisiae* transcription factor SWI5. *Cell* **66:** 743–758.

Motegi A, Sood R, Moinova H, Markowitz SD, Liu PP, Myung K. 2006. Human SHPRH suppresses genomic instability through proliferating cell nuclear antigen polyubiquitination. *J Cell Biol* **175:** 703–708.

Muramatsu S, Hirai K, Tak YS, Kamimura Y, Araki H. 2010. CDK-dependent complex formation between replication proteins Dpb11, Sld2, Pol ε, and GINS in budding yeast. *Genes Dev* **24:** 602–612.

Nam EA, Cortez D. 2011. ATR signalling: More than meeting at the fork. *Biochem J* **436:** 527–536.

Negrini S, Gorgoulis VG, Halazonetis TD. 2010. Genomic instability—An evolving hallmark of cancer. *Nat Rev Mol Cell Biol* **11:** 220–228.

Nguyen VQ, Co C, Irie K, Li JJ. 2000. Clb/Cdc28 kinases promote nuclear export of the replication initiator proteins Mcm2–7. *Curr Biol* **10:** 195–205.

Nguyen VQ, Co C, Li JJ. 2001. Cyclin-dependent kinases prevent DNA re-replication through multiple mechanisms. *Nature* **411:** 1068–1073.

Nishitani H, Taraviras S, Lygerou Z, Nishimoto T. 2001. The human licensing factor for DNA replication Cdt1 accumulates in G1 and is destabilized after initiation of S-phase. *J Biol Chem* **276:** 44905–44911.

Nishitani H, Sugimoto N, Roukos V, Nakanishi Y, Saijo M, Obuse C, Tsurimoto T, Nakayama KI, Nakayama K, Fujita M, et al. 2006. Two E3 ubiquitin ligases, SCF-Skp2 and DDB1-Cul4, target human Cdt1 for proteolysis. *EMBO J* **25:** 1126–1136.

Noton E, Diffley JF. 2000. CDK inactivation is the only essential function of the APC/C and the mitotic exit network proteins for origin resetting during mitosis. *Mol Cell* **5:** 85–95.

Nougarede R, Della Seta F, Zarzov P, Schwob E. 2000. Hierarchy of S-phase-promoting factors: Yeast Dbf4-Cdc7 kinase requires prior S-phase cyclin-dependent kinase activation. *Mol Cell Biol* **20:** 3795–3806.

Pasion SG, Forsburg SL. 1999. Nuclear localization of *Schizosaccharomyces pombe* Mcm2/Cdc19p requires MCM complex assembly. *Mol Biol Cell* **10:** 4043–4057.

Paulsen RD, Cimprich KA. 2007. The ATR pathway: Fine-tuning the fork. *DNA Repair (Amst)* **6:** 953–966.

Pereverzeva I, Whitmire E, Khan B, Coue M. 2000. Distinct phosphoisoforms of the *Xenopus* Mcm4 protein regulate the function of the Mcm complex. *Mol Cell Biol* **20:** 3667–3676.

Petermann E, Helleday T. 2010. Pathways of mammalian replication fork restart. *Nat Rev Mol Cell Biol* **11:** 683–687.

Petersen BO, Lukas J, Sorensen CS, Bartek J, Helin K. 1999. Phosphorylation of mammalian CDC6 by cyclin A/CDK2 regulates its subcellular localization. *EMBO J* **18:** 396–410.

Preston BD, Albertson TM, Herr AJ. 2010. DNA replication fidelity and cancer. *Semin Cancer Biol* **20:** 281–293.

Randell JC, Fan A, Chan C, Francis LI, Heller RC, Galani K, Bell SP. 2010. Mec1 is one of multiple kinases that prime the Mcm2–7 helicase for phosphorylation by Cdc7. *Mol Cell* **40:** 353–363.

* Rhind N, Gilbert DM. 2013. DNA replication timing. *Cold Spring Harb Perspect Biol* doi: 10.1101/cshperspect.a010132.

Saha P, Chen J, Thome KC, Lawlis SJ, Hou ZH, Hendricks M, Parvin JD, Dutta A. 1998. Human CDC6/Cdc18 associates with Orc1 and cyclin-cdk and is selectively eliminated from the nucleus at the onset of S phase. *Mol Cell Biol* **18:** 2758–2767.

Sansam CL, Shepard JL, Lai K, Ianari A, Danielian PS, Amsterdam A, Hopkins N, Lees JA. 2006. DTL/CDT2 is essential for both CDT1 regulation and the early G2/M checkpoint. *Genes Dev* **20:** 3117–3129.

Saxena S, Yuan P, Dhar SK, Senga T, Takeda D, Robinson H, Kornbluth S, Swaminathan K, Dutta A. 2004. A dimerized coiled-coil domain and an adjoining part of geminin interact with two sites on Cdt1 for replication inhibition. *Mol Cell* **15:** 245–258.

Senga T, Sivaprasad U, Zhu W, Park JH, Arias EE, Walter JC, Dutta A. 2006. PCNA is a cofactor for Cdt1 degradation by CUL4/DDB1-mediated N-terminal ubiquitination. *J Biol Chem* **281:** 6246–6252.

Sheu YJ, Stillman B. 2010. The Dbf4-Cdc7 kinase promotes S phase by alleviating an inhibitory activity in Mcm4. *Nature* **463:** 113–117.

* Siddiqui K, On KF, Diffley JFX. 2013. Regulating DNA replication in eukarya. *Cold Spring Harb Perspect Biol* doi: 10.1101/cshperspect.a012930.

Song B, Liu XS, Davis K, Liu X. 2011. Plk1 phosphorylation of Orc2 promotes DNA replication under conditions of stress. *Mol Cell Biol* **31:** 4844–4856.

Soucy TA, Smith PG, Milhollen MA, Berger AJ, Gavin JM, Adhikari S, Brownell JE, Burke KE, Cardin DP, Critchley S, et al. 2009. An inhibitor of NEDD8-activating enzyme as a new approach to treat cancer. *Nature* **458:** 732–736.

Tada S, Li A, Maiorano D, Mechali M, Blow JJ. 2001. Repression of origin assembly in metaphase depends on inhibition of RLF-B/Cdt1 by geminin. *Nat Cell Biol* **3:** 107–113.

Takeda DY, Parvin JD, Dutta A. 2005. Degradation of Cdt1 during S phase is Skp2-independent and is required for efficient progression of mammalian cells through S phase. *J Biol Chem* **280:** 23416–23423.

Tanaka S, Araki H. 2010. Regulation of the initiation step of DNA replication by cyclin-dependent kinases. *Chromosoma* **119:** 565–574.

* Tanaka S, Araki H. 2013. Helicase activation and establishment of replication forks at chromosomal origins of replication. *Cold Spring Harb Perspect Biol* doi: 10.1101/cshperspect.a010371.

Tanaka S, Diffley JF. 2002. Interdependent nuclear accumulation of budding yeast Cdt1 and Mcm2–7 during G1 phase. *Nat Cell Biol* **4:** 198–207.

Tanaka T, Knapp D, Nasmyth K. 1997. Loading of an Mcm protein onto DNA replication origins is regulated by Cdc6p and CDKs. *Cell* **90:** 649–660.

Tanaka S, Umemori T, Hirai K, Muramatsu S, Kamimura Y, Araki H. 2007. CDK-dependent phosphorylation of Sld2 and Sld3 initiates DNA replication in budding yeast. *Nature* **445:** 328–332.

Tardat M, Brustel J, Kirsh O, Lefevbre C, Callanan M, Sardet C, Julien E. 2010. The histone H4 Lys 20 methyltransferase PR-Set7 regulates replication origins in mammalian cells. *Nat Cell Biol* **12:** 1086–1093.

Terai K, Abbas T, Jazaeri AA, Dutta A. 2010. CRL4(Cdt2) E3 ubiquitin ligase monoubiquitinates PCNA to promote translesion DNA synthesis. *Mol Cell* **37:** 143–149.

Thomer M, May NR, Aggarwal BD, Kwok G, Calvi BR. 2004. *Drosophila* double-parked is sufficient to induce re-replication during development and is regulated by cyclin E/CDK2. *Development* **131:** 4807–4818.

Todorov IT, Attaran A, Kearsey SE. 1995. BM28, a human member of the MCM2-3-5 family, is displaced from chromatin during DNA replication. *J Cell Biol* **129:** 1433–1445.

Unk I, Hajdu I, Fatyol K, Szakal B, Blastyak A, Bermudez V, Hurwitz J, Prakash L, Prakash S, Haracska L. 2006. Human SHPRH is a ubiquitin ligase for Mms2-Ubc13-dependent polyubiquitylation of proliferating cell nuclear antigen. *Proc Natl Acad Sci* **103:** 18107–18112.

Vas A, Mok W, Leatherwood J. 2001. Control of DNA rereplication via Cdc2 phosphorylation sites in the origin recognition complex. *Mol Cell Biol* **21:** 5767–5777.

Vaziri C, Saxena S, Jeon Y, Lee C, Murata K, Machida Y, Wagle N, Hwang DS, Dutta A. 2003. A p53-dependent checkpoint pathway prevents rereplication. *Mol Cell* **11:** 997–1008.

Walter JC. 2000. Evidence for sequential action of cdc7 and cdk2 protein kinases during initiation of DNA replication in *Xenopus* egg extracts. *J Biol Chem* **275:** 39773–39778.

Watanabe K, Tateishi S, Kawasuji M, Tsurimoto T, Inoue H, Yamaizumi M. 2004. Rad18 guides polη to replication stalling sites through physical interaction and PCNA monoubiquitination. *EMBO J* **23:** 3886–3896.

Weinreich M, Stillman B. 1999. Cdc7p-Dbf4p kinase binds to chromatin during S phase and is regulated by both the APC and the RAD53 checkpoint pathway. *EMBO J* **18:** 5334–5346.

Wilmes GM, Archambault V, Austin RJ, Jacobson MD, Bell SP, Cross FR. 2004. Interaction of the S-phase cyclin Clb5 with an "RXL" docking sequence in the initiator protein Orc6 provides an origin-localized replication control switch. *Genes Dev* **18:** 981–991.

Wohlschlegel JA, Dwyer BT, Dhar SK, Cvetic C, Walter JC, Dutta A. 2000. Inhibition of eukaryotic DNA replication by geminin binding to Cdt1. *Science* **290:** 2309–2312.

Wuarin J, Buck V, Nurse P, Millar JB. 2002. Stable association of mitotic cyclin B/Cdc2 to replication origins prevents endoreduplication. *Cell* **111:** 419–431.

Zegerman P, Diffley JF. 2007. Phosphorylation of Sld2 and Sld3 by cyclin-dependent kinases promotes DNA replication in budding yeast. *Nature* **445:** 281–285.

Zhao H, Watkins JL, Piwnica-Worms H. 2002. Disruption of the checkpoint kinase 1/cell division cycle 25A pathway abrogates ionizing radiation-induced S and G2 checkpoints. *Proc Natl Acad Sci* **99:** 14795–14800.

Zhu W, Depamphilis ML. 2009. Selective killing of cancer cells by suppression of geminin activity. *Cancer Res* **69:** 4870–4877.

Zhu W, Dutta A. 2006. An ATR- and BRCA1-mediated Fanconi anemia pathway is required for activating the G2/M checkpoint and DNA damage repair upon rereplication. *Mol Cell Biol* **26:** 4601–4611.

Zhu W, Chen Y, Dutta A. 2004. Rereplication by depletion of geminin is seen regardless of p53 status and activates a G2/M checkpoint. *Mol Cell Biol* **24:** 7140–7150.

Zhu W, Abbas T, Dutta A. 2005. DNA replication and genomic instability. *Adv Exp Med Biol* **570:** 249–279.

Zhu W, Lee CY, Johnson RL, Wichterman J, Huang R, DePamphilis ML. 2011. An image-based, high-throughput screening assay for molecules that induce excess DNA replication in human cancer cells. *Mol Cancer Res* **9:** 294–310.

* Zielke N, Edgar BA, DePamphilis ML. 2013. Endoreplication. *Cold Spring Harb Perspect Biol* **5:** a012948.

Replication Proteins and Human Disease

Andrew P. Jackson[1], Ronald A. Laskey[2], and Nicholas Coleman[3]

[1]MRC Human Genetics Unit, Institute of Genetics and Molecular Medicine, University of Edinburgh, Edinburgh, United Kingdom

[2]Department of Zoology, University of Cambridge, Cambridge, United Kingdom

[3]Department of Pathology, University of Cambridge, Cambridge, United Kingdom

Correspondence: andrew.jackson@hgu.mrc.ac.uk; nc109@cam.ac.uk

In this article, we discuss the significance of DNA replication proteins in human disease. There is a broad range of mutations in genes encoding replication proteins, which result in several distinct clinical disorders that share common themes. One group of replication proteins, the MCMs, has emerged as effective biomarkers for early detection of a range of common cancers. They offer practical and theoretical advantages over other replication proteins and have been developed for widespread clinical use.

Semiconservative replication of DNA is essential for cellular proliferation. Therefore, mutation of genes encoding the replication machinery could be thought to be fundamentally harmful to an organism. However, inherited and acquired mutations in such genes do occur, resulting in a broad spectrum of disease. The first half of this article outlines the phenotypes associated with several classes of replication disorders, before focusing on the pre-replicative complex (pre-RC) and its involvement in both inherited and acquired human disease. In the second half, we discuss the utility of replication proteins as oncological disease markers, again with emphasis on pre-RC components. The function and mechanism of action of the replication proteins described here are covered in depth in other articles in this collection (Holt and Reyes 2012; Bell and Botchan 2013; Siddiqui et al. 2013).

INHERITED GENETIC DISORDERS OF REPLICATION

There are several clinically recognized genetic disorders resulting from mutations in replication genes. Although distinct from each other, there are common themes to their presentation. In particular, reduced growth and increased cancer risk are frequent features for genes involved in nuclear genome replication. In contrast, for mitochondrial replication genes, multisystem dysfunction of highly energy-dependent tissues is seen. Strikingly, particular developmental abnormalities are also observed in certain disorders. These are less easily attributable to core replication functions and include skeletal abnormalities involving the thumb and forearm (radial ray defects) and absence of the patella. In addition, skin rashes, premature aging, and immune and endocrine dysfunction can occur.

Cite this article as *Cold Spring Harb Perspect Biol* doi: 10.1101/cshperspect.a013060

recQ HELICASES: PREMATURE AGING, SKIN SIGNS, CANCER PREDISPOSITION

The recQ helicase family has important roles in replication through their functions in ATP-dependent unwinding of double-stranded nucleic acids. They have roles in repairing damaged replication forks by homologous recombination and in stabilizing stalled forks (Bachrati and Hickson 2008).

Autosomal-recessive mutations in three of the recQ helicases—WRN, BLM, and RECQ4L—cause distinct but overlapping syndromes (Monnat 2010), with reduced growth and cancer predisposition being common to all these conditions.

Bloom's Syndrome (BLM)

Alongside globally reduced growth, the defining clinical feature of Bloom's syndrome is a red telangiectatic facial rash that develops on the face in early childhood (Fig. 1). This occurs in a butterfly distribution involving the cheeks and nose (Bloom 1954). Growth is impaired from before birth and is accompanied by significant reduction in head size (microcephaly). Adult height is markedly reduced, reaching only 149 cm in men and 138 cm in women (~4 standard deviations [SD] below the population mean). There is a high risk of cancer, with both solid tumors and hematological malignancies being a major cause of early death (German 1997). Immunity is impaired with low immunoglobulin levels and an increased incidence of ear and chest infections in childhood, and with chronic chest disorders in adulthood. Diabetes and male infertility also occur. Cytogenetically, a high rate of sister-chromatid exchange (40–100 per metaphase) is pathognomonic for this condition. Mutations in the BLM gene are typically biallelic, truncating loss-of-function mutations (German et al. 2007).

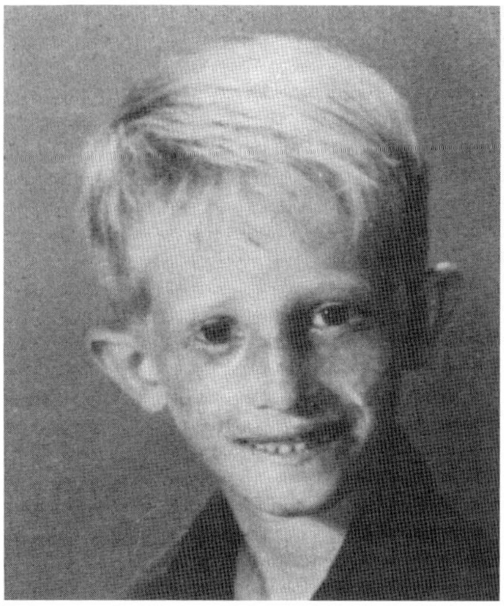

Figure 1. Bloom's syndrome with a red facial rash in a butterfly distribution, affecting cheeks and nose. (Reprinted from German 1969.)

Werner's Syndrome (WRN)

Werner's syndrome (Oto Werner 1904) (Muftuoglu et al. 2008; Oshima et al. 2012) is the archetypal "progeric" syndrome, with premature aging from early adulthood. In adults, there is early graying of hair, hair loss, and skin atrophy, leading to a tight, wizened facial appearance (Fig. 2). Accelerated aging also results in premature heart disease and strokes (atherosclerosis), as well as osteoporosis and diabetes. There are some similarities with the other RecQ disorders, with increased cancer predisposition and reduced growth in adolescence. Mutations in Werner syndrome are generally truncating loss of function, although occasional missense mutations destabilizing protein stability have been reported (Huang et al. 2006).

RECQ4L: Rothmund–Thompson, RAPADILLINO, and Ballier–Gerold Syndromes

Like Bloom's syndrome, dermatological signs are strong diagnostic features of Rothmund–Thompson syndrome (RTS) (Larizza et al. 2010). At a few months of age, reddening and blistering of the skin occur on the face (Fig. 3A), and then this spreads to the limbs and buttocks. With time, a reticulate pattern of over- and underpigmentation occurs, known as "poikiloderma" (Fig. 3B). Hair is sparse in this disorder and there are skeletal abnormalities, which can

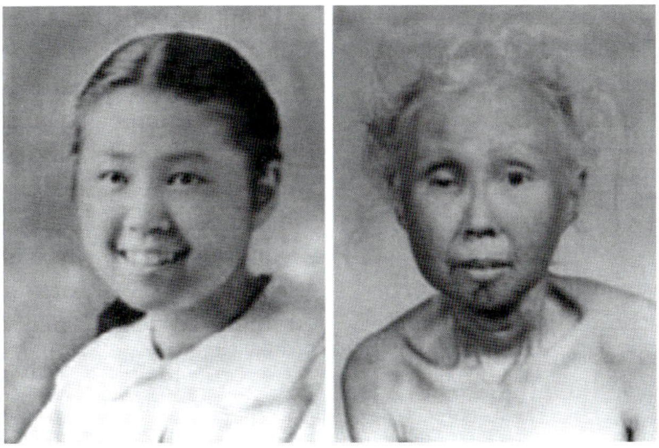

Figure 2. Premature aging in Werner's syndrome. (*Left*) Patient aged 15; (*right*) aged 48. (From Muftuoglu et al. 2008; reprinted, with kind permission, from Springer Science and Business Media.)

include radial ray defects (thumb abnormalities and/or absence of the radius in the forearm), as well as reduced size of knee caps (patella hypoplasia). Cataracts and chronic diarrhea are often also features. Growth is reduced prenatally and in childhood, but not as severely as in Bloom's syndrome (Vennos et al. 1992), and there is an increased rate of cancers, particularly osteosarcomas (Wang et al. 2001).

Two other conditions have also been found to have mutations in RecQ4L, and both have clinical features that substantially overlap with RTS (Sznajer et al. 2008). In RAPADILLINO syndrome (Jam et al. 1999), a cleft palate may be present and poikiloderma is said to be absent, whereas Ballier–Gerold syndrome is defined by premature fusion of skull sutures (coronal craniosynostosis), leading to abnormal skull shape and ocular proptosis (protrusion of the eyes).

Although most cases have biallelic truncating mutations, patient cells appear to have some residual protein activity (Siitonen et al. 2009).

MUTATIONS IN THE REPLICATIVE POLYMERASE γ AND TWINKLE HELICASE CAUSE MITOCHONDRIAL DNA DELETION AND DEPLETION SYNDROMES

Mitochondrial diseases are a heterogeneous group of disorders that all result from respiratory

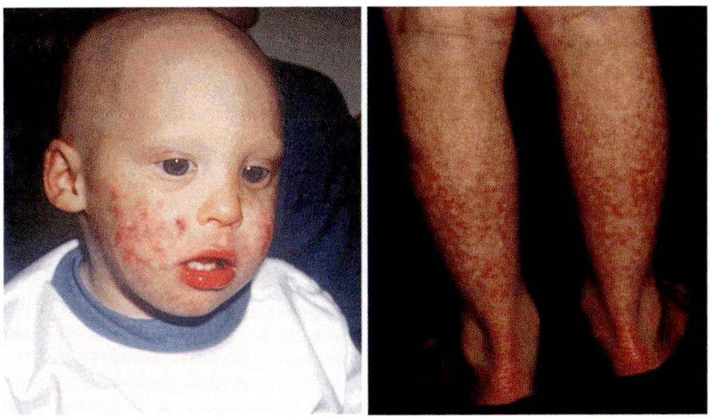

Figure 3. Rothmund–Thompson syndrome. (*A*) Facial rash. (*B*) Poikiloderma, a reticulate hypo/hyperpigmented rash with areas of skin atrophy and telangiectasia (small dilated blood vessels). (From Wang et al. 2001; reprinted, with permission.)

chain dysfunction. One or multiple organ systems may be involved, with the tissues involved being those with highest demands for oxidative metabolism. These include the brain, heart, muscle, retina, liver, and endocrine systems. In addition to mutations of genes encoded by the mitochondrial DNA (mtDNA), nuclear-encoded genes also cause mitochondrial diseases. The mitochondrial replicative polymerase POLG1 (Van Goethem et al. 2001; Hudson and Chinnery 2006; Copeland 2010), POLG2 genes (Longley et al. 2006; Young et al. 2011), and replicative helicase TWINKLE (Spelbrink et al. 2001) are three such genes. Mutations in these genes perturb replication and result in multiple large-scale deletions and point mutations in the mtDNA. In addition, they can also lead to an overall depletion of mtDNA in cells. Consequently the production of respiratory chain enzymes (encoded by the mtDNA) is impaired and leads to reduced mitochondrial oxidative function (Suomalainen and Isohanni 2010). As is often the case with mitochondrial disorders, different mutations within the same gene cause a spectrum of clinical disease that varies widely in severity, age of onset, and systems involved. Although distinct clinical syndromes exist, individual patients may not fit neatly into one category.

At the most severe, mitochondrial depletion syndromes (in which total cellular levels of mtDNA can be a few percent of normal) result in severe metabolic disturbance, with lactic acidosis and liver dysfunction. Alongside metabolic issues, muscle and neurological involvement occurs. Epilepsy, ataxia (incoordination), neuropathy, and deafness can all occur, and death generally ensues in early childhood (Suomalainen and Isohanni 2010). As well as autosomal-recessive mutations in POLG1 and TWINKLE, genes encoding many proteins regulating mitochondrial dNTP pools have also been implicated. These include thymidine kinase 2, deoxyguanosine kinase, and the P53-R2 subunit of ribonucleotide reductase (RRM2B) (Mandel et al. 2001; Saada et al. 2001; Bourdon et al. 2007).

At the other end of the spectrum are adult-onset disorders with neurological or muscular involvement. In such patients, Parkinsonism, skeletal muscle weakness, sensory neuropathy, cognitive decline, and/or ataxia occur. At its mildest, involvement may even simply be limited to the muscles surrounding the eye. In progressive external opthalmoplegia (PEO), there is impaired movement of the eyes, due to weak ocular muscles along with drooping eyelids (ptosis) (Fig. 4). Dominant or recessive mutations in

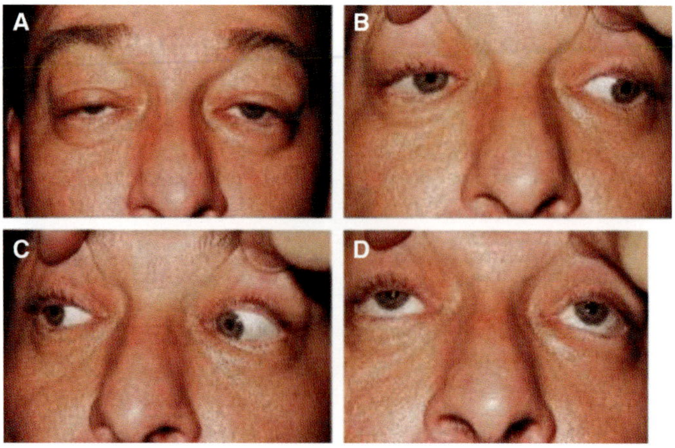

Figure 4. Chronic progressive external opthalmoplegia. (*A*) Drooping eyelids (ptosis) result from weakness in levator palpebrae muscles. (*B–D*) Weakness of external eye muscles leads to restricted eye movements. (Reprinted from Amato et al. 2006.)

POLG1, TWINKLE, and occasionally POLG2 cause PEO as the result of multiple mitochondrial deletions in muscle tissue.

THE PRE-REPLICATIVE COMPLEX, GROWTH, EARS, AND KNEE CAPS

Recently mutations have been identified in multiple components of the pre-replicative complex (pre-RC). The pre-RC forms at origins of replication. Firstly, the heterohexameric origin recognition complex (ORC1-6) becomes bound to DNA during early G_1 phase of the cell cycle. Then, in conjunction with the accessory proteins CDT1 and CDC6, it iteratively loads the MCM helicase to license the origin for replication. Recessive biallelic partial of loss-of-function mutations in ORC1, 4, 6, CDT1, and CDC6 have been identified in patients with Meier–Gorlin syndrome (Bicknell et al. 2011a,b; Guernsey et al. 2011). A mutation in MCM4 has also been reported in individuals with immune deficiency, genome instability, and adrenal failure (Gineau et al. 2012; Hughes et al. 2012).

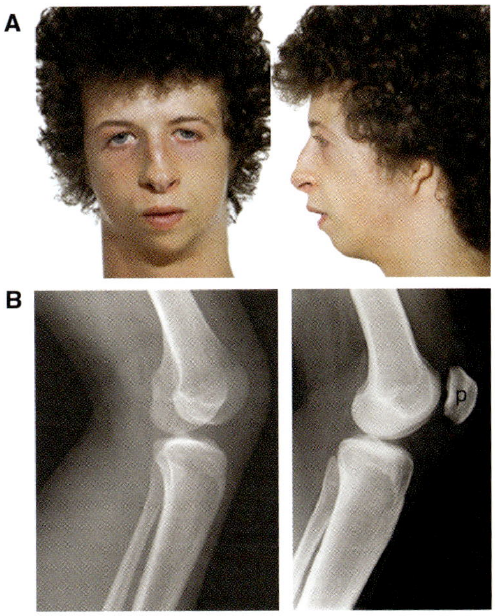

Figure 5. Meier–Gorlin syndrome. (*A*) Marked reduction in size of the external ear (pinna). (Adapted from Bicknell et al. 2011a; Images originally kindly provided by Dr. S. Aftimos.) (*B*) Absence of the patella. Lateral X ray of the knee of a patient with Meier–Gorlin syndrome (*left*), control subject (*right*). p, Patella. ([*Left*] From Guernsey et al. 2011; reprinted, with permission, © Nature Publishing Group. [*Right*] Courtesy of N. Morley.)

Meier–Gorlin Syndrome, ORC Complex, CDT1, and CDC6

Meier–Gorlin syndrome is an autosomal-recessive single-gene disorder defined by the triad of short stature, microtia (reduced external ear size), and hypoplasia/absence of the patella (Fig. 5) (Gorlin et al. 1975; Bongers et al. 2001). Growth in Meier–Gorlin syndrome is reduced before birth and postnatally, and accompanied by proportionate microcephaly. This growth pattern is similar to that seen in Bloom's syndrome. Height reduction of those with ORC1 mutations can be extreme ($\leq$6 SD), whereas mutations in the other genes can result in final height being at the lower end of the normal range (de Munnik et al. 2012). Intellect is generally normal.

Consistent with the pre-RC cell-essential function, mutations in the ORC1, 4, 6, CDC6, and CDT1 genes are thought to impair rather than abrogate pre-RC function (Bicknell et al. 2011a,b; Guernsey et al. 2011). Recurrent missense mutations are seen in evolutionarily conserved amino acids within domains important for pre-RC function. Additionally, localized (lobar) lung emphysema and retroflexion of the knees may be evident at birth, and in adult women, breast hypoplasia is evident. Although general failure in growth seems to be the result of impaired replication licensing, the reason for the microtia, absent patella, and additional clinical features is not readily evident. In this regard, it is interesting that Recq4L mutations can also specifically affect patella size.

MCM Helicase Mutations: Mice and Humans

The ENU-induced Chaos3 mutation in the *Mcm4* gene has been associated with genome instability and breast adenocarcinomas in mice (Shima et al. 2007). It is therefore surprising that there is no increased incidence of cancer in Meier–Gorlin syndrome. This and the lack of any reported mutations in the *MCM* genes in Meier–Gorlin syndrome suggest that these may result in a different phenotype. Recently, this has been shown to be the case, with two research groups reporting a new syndrome of adrenal insufficiency, natural killer (NK) cell deficiency, a chromosome instability that is associated with a homozygous splice-site mutation in the *MCM4* gene (Gineau et al. 2012; Hughes et al. 2012). This mutation has been found in multiple individuals in an extended consanguineous Irish Traveller family and results in a small amino-terminal truncation of the MCM4 protein. As with MGS, the human phenotype suggests that additional functions for this protein need to be established, because these features seem to beyond those expected from its core function as a component of the helicase that establishes the replication fork.

DNA REPLICATION AND CANCER DIAGNOSIS

Many of the commonest cancers can be treated more successfully when they are detected early, either as precursor lesions or early-stage malignancies. The choice of markers for early detection poses a dilemma. The aims are to distinguish normal cells from cancer cells, irrespective of the oncogenes or tumor-suppressor genes that the latter misexpress. Abundant evidence indicates that DNA replication proteins are exceptionally good markers for early detection of many of the common carcinomas and that they can also provide clinically useful prognostic information. This may seem surprising in view of the fact that these proteins are required for the replication of normal as well as cancer cells. However, if cells are recovered from body fluids after they have been exfoliated from the tissue surface, then replication proteins, such as the MCM proteins, can provide decisive clinical information. The reason for this is because MCM proteins are broken down before cells are shed from normal tissues into the body fluids. Therefore, cells shed from normal tissues should not contain MCM proteins, and this prediction is confirmed in practice. In contrast, malignant and premalignant (dysplastic) cells are not programmed to break down MCM proteins before they are lost from the tumor surface.

We have investigated the clinical value of antibodies against DNA replication proteins for detection of malignant and premalignant cells shed into body fluids. We have focused on those epithelia that are the sites of common cancers. MCM proteins are particularly valuable in this role, outperforming both PCNA and Ki67, both of which have also been used for this purpose. We discuss the properties of an ideal marker for early cancer screening and argue that MCM 2-7 proteins provide an exceptionally good match to those criteria.

IMPROVED CANCER SCREENING BY DETECTING MCMs

MCMs Are Sensitive Markers of Malignant and Dysplastic Cells

The essential role of MCMs 2–7 in proliferation led us to test the hypothesis that these proteins would be useful as biomarkers for early detection of cancer by screening. All six subunits are expressed through all phases of the cell cycle but are lost when cells exit the cycle into quiescence, senescence, or differentiation (Musahl et al. 1978; Madine et al. 2000; Stoeber et al. 2001). Although they bind to and detach from DNA according to cell cycle phase, they remain in the nucleus of higher eukaryotes throughout the cell cycle. Of greatest relevance to clinical applications, MCMs are rapidly lost from differentiating epithelial cells, for example, in the squamous epithelium of the cervix, glandular epithelium of the large bowel, and transitional epithelium of the urinary tract (Freeman et al. 1999; Gonzalez et al. 2005). The mechanisms involved are uncertain, although there is evidence of differentiation-associated posttranslational modification of MCM2, resulting in a cleaved fragment that may regulate pre-RC activity (Harada et al. 2008). In contrast, in malignant and premalignant lesions of differentiating epithelia, there is a substantial increase in the number of cells expressing MCMs (Hiraiwa et al. 1998; Todorov et al. 1998; Freeman

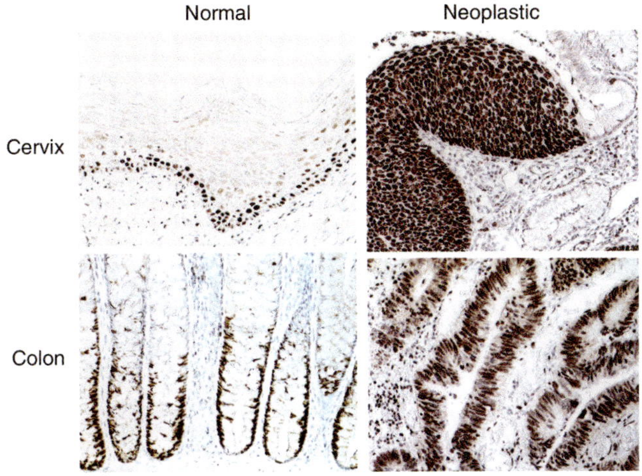

Figure 6. Expression of Mcm5 in normal and neoplastic tissue from cervix and colon. The protein is restricted to basal proliferative compartments in normal cervical squamous epithelium and normal large bowel crypts (brown immunoperoxidase staining) and is lost as cells differentiate as they move toward the epithelial surface. In contrast, neoplastic epithelia at the same sites (both malignant and premalignant) show full thickness expression of Mcm5. Numerous immunopositive cells are present at the epithelial surface, from which they can be sampled either actively or passively. Expression of MCMs 2–7 is essentially the same in all tissues examined. (From Freeman et al. 1999; reprinted, with permission, © AACR.)

et al. 1999; Gonzalez et al. 2005; Tachibana et al. 2005).

Approximately 80%–100% of cells are MCM immunopositive in malignant and high-grade premalignant lesions, for example, high-grade squamous intraepithelial lesions (SILs) of the cervix (Williams et al. 1998) and dysplastic adenomas of the large bowel (Fig. 6) (Freeman et al. 1999). We observed that there is essentially no difference between the expression patterns of each of the six pre-replication complex MCMs in a large number of tissue samples (Freeman et al. 1999). Based on these findings, we have argued that one of the characteristic features of dysplastic and malignant cells may be that they are continuously licensed for DNA replication.

An important consequence of the expansion of the proliferative compartment in neoplasia is that MCM-positive cells appear at the surface of epithelia, and will therefore be included in cytological samples. Such samples can be obtained actively, for example, by scraping, brushing, or washing, or passively, following spontaneous exfoliation into body fluids or solids, such as urine or stool (Fig. 7) (Stoeber et al. 1999, 2002; Davies et al. 2002; Gonzalez et al. 2005). MCMs are therefore very promising biomarkers for early detection of malignancy and premalignancy by population screening. Abnormal cells in cytological preparations generally stain with crisp nuclear signals, making them easy to identify, even at low magnification (Williams et al. 1998; Davies et al. 2002; Chatrath et al. 2003).

These properties make MCMs highly reliable for detecting abnormal cells in cytological samples. This is particularly important where abnormal cells are rarely represented, for example, following sampling of a small lesion or only a small part of a larger lesion. MCM-based tests consequently show high sensitivity[4] for detecting malignancy and premalignancy and can reduce the rate of false-negative results associated with conventional cytological screening.

[4]Sensitivity, the proportion of disease positives that are test positive; specificity, the proportion of disease negatives that are test negative; positive predictive value, the proportion of test positives that are disease positive; negative predictive value, the proportion of test negatives that are disease negative.

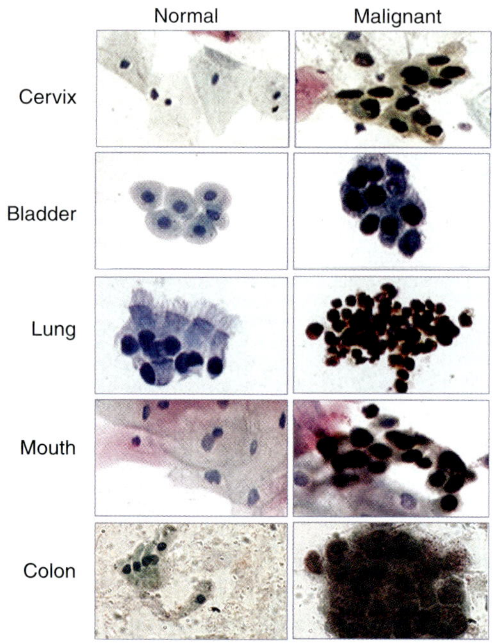

Figure 7. In cytological preparations, immunocytochemistry for Mcm5 enables discrimination between normal cells (*left*), which are immunonegative, and malignant (and premalignant) cells (*right*), which are immunopositive (brown nuclei). Equally good discrimination is obtained by staining for other pre-replication complex MCMs, particularly Mcm2. We have used antibodies against Mcm2 extensively in clinical studies, either singly or in combination with anti-Mcm5 antibodies. (From Laskey 2004; reprinted, with permission, © Cambridge University Press.)

MCMs Have Advantages over Other Markers

The sensitivity of MCM-based tests would be predicted to be superior to those using other currently used markers of cell cycle entry, such as Ki67 and PCNA. Ki67 was expressed by fewer cells than MCMs in most series of malignant and dysplastic lesions examined (Wharton et al. 2001; Chatrath et al. 2003; Scott et al. 2003; Davies et al. 2004; Dudderidge et al. 2005). This is in keeping with observations that Ki67 is not expressed by all cycling cells, including cells in S phase (Fig. 8) (Scholzen and Gerdes 2000). Indeed, the function of Ki67 in the cell cycle remains poorly understood. This is reflected in the threshold values used to define Ki67 overexpression in tissue samples, which are determined empirically and vary substantially between different centers (Yerushalmi et al. 2010).

PCNA shows wide variation in staining intensity in vivo, which is much greater than that seen for MCMs. This is consistent with fluctuations in expression levels of PCNA during the cell cycle, with greater expression in S than in G_1, G_2, and M (Celis and Celis 1985). Moreover, important roles for PCNA in DNA repair mean that it is still present in non-proliferating cells and therefore less useful as a specific marker of the deregulation of proliferation characteristic of malignancy and dysplasia (Toschi and Bravo 1988). Instead, semiarbitrary thresholds must be set for its use as a proliferation marker.

MCMs also offer theoretical advantages over other molecules suggested as being useful biomarkers for cancer detection. Many of the latter are involved in growth signaling pathways, which are frequently disrupted in neoplastic cells. However, there is an enormous degree of redundancy in such pathways, and detection of particular disruptions would be unlikely to provide adequate sensitivity for detection of neoplastic cells in clinical samples. Testing for MCMs represents a "reductionist" alternative, by detecting expression of proteins at the point where growth signaling pathways converge in the initiation of DNA replication (Fig. 9) (Gonzalez et al. 2005).

Far less literature is available regarding the potential translational value of other replication proteins. In our initial work (Williams et al. 1998), we attempted to use Cdc6 as a biomarker for early cancer detection. However, the relatively low nuclear abundance of Cdc6, compared with that of the MCMs, led to an inferior sensitivity/specificity balance that precluded further development. In contrast, antibodies against ORC proteins stained a wide range of normal cells, consistent with the chromatin-bound state of ORC proteins in quiescent mammalian nuclei (Madine et al. 2000). Cdt1 and geminin would not be expected to provide as much sensitivity for detecting abnormally proliferating cells as MCMs, because expression of each protein is restricted to particular cell cycle phases, rather than being abundant throughout the cell cycle (Gonzalez et al. 2004). On the other hand,

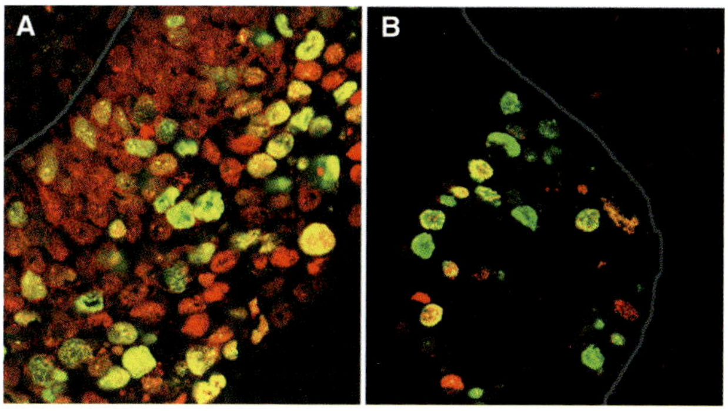

Figure 8. Dual immunofluorescence staining for Ki67 and other cell cycle markers in sections of premalignant cervical high-grade squamous intraepithelial lesions. In each image, the site of the basement membrane is indicated by a continuous line. (*A*) Staining for Ki67 (green) and Mcm5 (red). Although many nuclei express both proteins (appearing yellow), many others are red, indicating that Ki67 is not present in many cells expressing Mcm5. (*B*) Staining for Ki67 (green), together with our in situ DNA replication assay (Mills et al. 2000), in which S-phase nuclei are stained (here in red) as a result of their ability to incorporate labeled nucleotides. In situ replication is a sensitive and specific indicator of S-phase cells, not all of which are Ki67 positive. In the image shown, 13% of S-phase cells are Ki67 negative. (Image kindly provided by Tony Mills.)

these very features make geminin and Cdt1 potentially important markers for indicating tumor prognosis, based on immunohistochemical analysis of tissue sections (Gonzalez et al. 2004). Whether DNA replication initiation factors such as Mcm10, Cdc45, GINS complex members, and so on will offer any practical benefits over MCMs for cancer detection remains to be determined, although there is no strong theoretical basis for believing they would.

Detection of MCM-Positive Cells

A major advantage of using biomarkers in cancer screening is the potential for automated assessment of test results. The algorithm required to detect immunopositive cells by automated microscopy is straightforward (i.e., whether nuclei of a particular color are present or not) and much simpler than those used to assess morphological features (nuclear size, nuclear outline, staining intensity, etc.) in conventionally prepared slides. A simple algorithm should substantially increase throughput and reduce costs. The latter would potentially make automated screening affordable not only in industrialized countries but also in the developing world, where for many cancers the clinical need for a screening test is greatest.

An alternative approach to high-throughput test analysis is to lyse samples and detect the marker or markers of interest using a liquid phase assay such as DELFIA or ELISA. This approach has shown promise in samples in which neoplastic cells are not accompanied by unduly large numbers of normal cells, most notably screening for neoplasia in the urinary tract using urine sediments (Stoeber et al. 1999, 2002). In settings in which abnormal cells may be rare and/or accompanied by numerous normal cells (e.g., cervical smears, sputum, and fecal washings), the sensitivity of liquid phase assays will be substantially less than that of microscopic approaches in which signal is retained in abnormal cells rather than being diluted by lysis.

SPECIFIC APPLICATIONS OF MCM TESTING IN CANCER SCREENING

Cervical Cancer and Precancer

In a cervical smear test, cells are scraped from the cervical epithelial surface. An important premise underlying this approach is that changes at the surface of the cervical squamous epithelium (if detected correctly) can predict abnormalities in the entire epithelium (Baldwin et al. 2003). The current Papanicolaou (Pap)

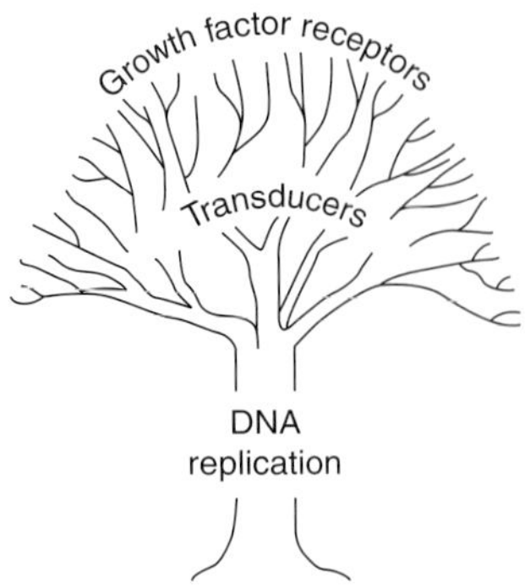

Figure 9. Cartoon illustrating the reductionist approach of using MCM proteins for early detection of malignancy and premalignancy. Other candidate tumor markers include molecules involved in signaling cell growth, many of which are mutated or otherwise disrupted in cancer. However, each growth signaling pathway is inherently redundant, and individual signaling molecules would not be expected to offer adequate sensitivity for detection of neoplastic cells. In contrast, DNA replication proteins, such as MCMs, represent the point of convergence of all signals leading to DNA replication, making them particularly sensitive markers for detecting neoplastic cells showing ectopic cell cycle entry at epithelial surfaces.

smear test involves subjective assessment of cell morphological features and requires highly trained observers. A liquid-based cytology cervical smear samples about 70,000 cells and in the United Kingdom is examined in 10–15 min, meaning that individual high-magnification microscopic fields are typically assessed for ~1–2 sec. Because the morphological abnormalities in neoplastic cells are often subtle, it is easy to miss cases of cervical cancer or precancer (SILs). Accordingly, despite expert cytoscreeners, the sensitivity of an individual cervical smear test for neoplastic disease is ~60% (Baldwin et al. 2003).

An important factor that contributes to this limited sensitivity is that most cervical smears are normal. In consequence, cases containing abnormal cells may be missed if the cytoscreener loses concentration and/or the abnormal cells are rare. This problem will worsen in years to come as the impact of human papillomavirus vaccines reduces the overall burden of cervical neoplastic disease. On the other hand, current vaccines are only predicted to prevent ~70% of cervical cancers (assuming 100% vaccine uptake) (Stanley 2010), so the requirement for regular cervical screening will remain for the foreseeable future. In view of these issues, there is an important clinical need for new biomarker-based approaches to cervical screening that permit objective disease detection and can be automated for high throughput.

Identifying the presence of MCM-containing cells in a smear is an exciting approach to cervical screening (Fig. 10). In our initial studies, MCM testing showed high sensitivity and specificity for cervical cancer/SILs in smears obtained in the United Kingdom and India (Williams et al. 1998; Baldwin et al. 2003; Mukherjee et al. 2007), providing greater accuracy and faster throughput than conventional Pap testing. The MCM detection technique was subsequently incorporated into diagnostic test kits for widespread clinical use. The first-generation product, ProExC (Becton Dickinson), combined antibodies versus MCM2 and topoisomerase-2A and has been shown in more than 30 studies in Europe and the United States to accurately identify neoplastic cells in cervical smears, with reported sensitivity and specificity values of up to 98% (Tambouret et al. 2008; Brown et al. 2012). A second product, SurePath Plus (Becton Dickinson), is based on combined antibodies against several MCM subunits and is currently undergoing large-scale clinical trials pending detailed evaluation by the U.S. Food and Drug Administration.

An important question is how to combine MCM-based tests with other clinical screening methods, for maximal clinical benefit. In Europe, MCM testing may best be used as a reflex investigation in HPV-positive smears, either in isolation or with other tests, such as additional biomarkers or Pap-based cytology. Indeed, an interesting recent study compared eight cervical screening strategies in more than 3000 smear samples and showed that the optimal screening approach was primary HPV testing, followed by ProExC triage of HPV-positive samples (Depuydt 2011). Elsewhere, including the United States, the order of using multiple biomarker-

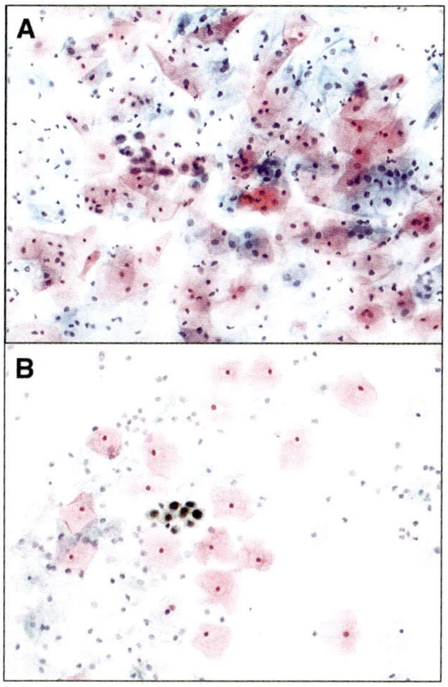

Figure 10. The high sensitivity of MCM immunocytochemistry is exemplified in the assessment of cervical smears. (*A*) A microscopic field from the cervical smear of a patient with a premalignant cervical high-grade squamous intraepithelial lesion, stained by the routine Papanicolaou (Pap) method. A field of this size has to be examined in routine cytological practice in a second or so. The premalignant cells near the center of the field are easy to miss because they do not stand out from the normal cells around them and can be identified only by highly trained cytoscreeeners and cytopathologists. (*B*) In contrast, cells from the same patient stained by immunocytochemistry for MCMs. By this approach, abnormal cells are stained a particular color (in this case, brown), enabling them to be recognized much more easily and quickly. Errors are considerably less likely to be made (particularly when the abnormal cells are rare), providing MCM testing with very high sensitivity for detecting disease in routine clinical samples. (From Baldwin et al. 2003; reprinted, with permission, © Nature Publishing Group.)

based tests may be different, with MCM and/or HPV tests being used as a reflex to triage smears with mild cytological abnormalities.

A further benefit of ProExC is in improving histological grading of cervical biopsies or resections. Histological examination of tissue sections stained with hematoxylin and eosin is complicated by other pathological or physiological processes that can affect cell morphology, such as inflammation or hormonal changes (including pregnancy). Accordingly, the interobserver and intraobserver reproducibilities of routine histological assessment are limited. Areas of particular difficulty include distinguishing some cases of non-neoplastic squamous metaplasia from squamous intraepithelial lesions (SILs) and some cases of low-grade SILs from high-grade SILs. It is important to resolve these difficulties, because clinical management differs substantially between the diagnostic categories. Immunohistochemical staining with ProExC gives characteristic staining patterns in cervical SILs, including increasing expression with increasing grade (Ozaki et al. 2011). The use of ProExC, alone or in combination, therefore provides a more objective and accurate method of distinguishing high-grade SILs from non-neoplastic mimics in tissue sections.

Colorectal Cancer

It is thought to take two to three years for an asymptomatic early colorectal cancer to develop into a symptomatic advanced lesion (Rhodes 2000). An effective strategy for early detection of colorectal cancer would produce very substantial benefits in overall survival. Current screening tests either detect the presence of occult blood in stool or identify gross abnormalities by endoscopy (Davies et al. 2005). All current tests are limited in their effectiveness and/or patient acceptability. There is a pressing need for new screening tests based on our increasing understanding of the biology and natural history of colorectal cancer. Stool testing is likely to be particularly valuable, because it represents a non-invasive method for screening all of the colon and rectum, without the need for bowel preparation. Methods for retrieving colonocytes from stool washings are now available (although stool samples should be processed within eight hours) (Davies et al. 2002), as are several assays for detecting neoplastic colonocytes. DNA testing is considered to be a promising approach, but clonal heterogeneity within tumors means that multitarget assays are required. These are very expensive and showed disappointing sensitivity for colorectal cancer in a recent large study of average-risk asymptomatic people (Imperiale et al. 2004). The "reductionist" approach offered by MCMs is an inexpensive alternative. In an initial clinical evaluation study, MCM-positive cells were retrieved from the stool of 37 of 40 patients with colorectal cancer, including all nine early-stage cancers, but from none of 25 control participants (Davies et al. 2002). Because colonocytes were often retrieved from stool in small numbers and obscured by fecal debris, we developed a new method of colonocyte retrieval, using stool-derived mucus retained by a 125-μm filter and subsequently processed in a fibrin clot (White et al. 2009). The method provided a >30-fold increase in yield of colonocytes, which are suitable for nucleic acid– and protein-based screening assays. Future work will aim to combine this cell retrieval technique with MCM staining in order to develop a test for high-throughput bowel cancer screening.

Other Cancers

MCM detection is also being developed for early diagnosis of other common cancers. The most advanced of these further applications is screening for bladder and prostate cancer using urinary sediments, based on immunocytochemistry of cytological preparations or liquid phase assays (Stoeber et al. 1999, 2002). Other promising cytological applications include screening for cancers and precancers of squamous epithelium at other sites (analogous to the cervical smear), particularly the oral cavity (Scott et al. 2006) and the anal canal (Scarpini et al. 2008).

IMPROVED PREDICTION OF CLINICAL OUTCOME BY DETECTING CELL CYCLE MARKERS

Immunohistochemistry for cell cycle proteins can address a further important challenge in diagnostic pathology, predicting the outcome of tumors using tissue sections of clinical samples. Markers that can be detected in routinely processed sections of formalin-fixed, paraffin-embedded tissue are particularly valuable. For many tumors, powerful prognostic information is provided by quantitative assessment of cell cycle entry and/or particular cell cycle phases.

As accurate indicators of cell cycle state, MCMs have been shown to predict survival in patients with a range of tumors, including malignancies of breast (Gonzalez et al. 2003), prostate (Meng et al. 2001), kidney (Rodins et al. 2002), bladder (Kruger et al. 2003), esophagus (Kato et al. 2003), mouth (Kodani et al. 2003), lung (Ramnath et al. 2001; Hashimoto et al. 2004), and brain (Wharton et al. 2001; Hunt et al. 2002; Scott et al. 2005). Because MCMs provide a strong nuclear signal in immunohistochemistry, assessment of MCM labeling is straightforward and in our hands highly reproducible. We showed by multivariate analysis that the MCM2 labeling index is superior to the Ki67 labeling index in predicting overall survival in breast cancer (Gonzalez et al. 2003). Moreover, for the data set analyzed, the MCM2 labeling index was also superior to clinico-pathological parameters that are currently widely used to predict outcome, namely, histological grade and lymph node stage.

In addition to information on cell cycle state in tumor cells provided by MCMs, markers of the advanced phases of the cell cycle (S, G_2, and M) indicate cell cycle progression and thereby provide a useful representation of cell cycle rate. A promising candidate marker in this regard is geminin (Wohlschlegel et al. 2000; Tada et al. 2001; Gonzalez et al. 2004), which is expressed in S, G_2, and M and down-regulated following cell cycle exit, making it a useful marker of cell cycle progression (McGarry and Kirschner 1998; Wohlschlegel et al. 2000). In breast cancer, geminin was a strong independent predictor of poor overall survival and the development of distant metastases. The geminin labeling index was superior to the Ki67 labeling index as to tumor grade, stage, and size (Gonzalez et al. 2004). The MCM labeling indices for the data set examined could not be included in multivariate analysis, because none of the patients with a low MCM index had died or developed metastatic disease at the time of their last follow-up. It was therefore not possible to compare the relative prognostic values of MCMs versus geminin, as indicators of cell cycle state and rate, respectively. It may prove to be the case that the greatest prognostic value is provided by the geminin/MCM fraction, which would represent the rate of proliferation of those tumor cells that are in cycle and therefore responsible for tumor growth.

CONCLUSIONS

Despite the replication machinery's central role in cellular proliferation, several inherited disorders of replication have been identified. Although these conditions often manifest phenotypes such as growth failure and cancer predisposition, distinct clinical disease entities are recognizable and associated with specific genes. It is therefore likely that in the future, additional replication proteins will be identified for further human genetic syndromes.

Initiation of DNA replication is a critical point of convergence of growth signaling pathways. MCMs are exceptionally promising candidate markers for early detection of malignancy and premalignancy at a wide range of anatomical sites, and clinical evaluation studies from numerous independent laboratories worldwide have provided strong evidence in support of this. The abundance of MCM proteins in the surface layers of dysplastic and malignant lesions and the strong nuclear signal produced in immunocytochemistry make MCMs suitable for high-throughput population screening or pre-screening for a range of common cancers. Automated microscopy is likely to be the most useful detection method across the broad range of potential clinical applications. Cell cycle markers that can be detected in routine formalin-fixed, paraffin-embedded tissue sections may refine existing clinico-pathological approaches for predicting outcome in the common cancers. MCMs and geminin are particularly promising as surrogate markers of cell cycle state and rate, respectively.

ACKNOWLEDGMENTS

We thank Dr. Tony Mills and Lesley Morris for some of the images used in the figures. We also thank Sally Hames for help with preparation of the manuscript. We thank Cancer Research UK for supporting the development of MCMs as cancer screening biomarkers and the Medical Research Council, Lister Institute for Preventative Medicine and European Research Council for funding.

Competing Interest Statement

Nicholas Coleman and Ronald A. Laskey are entitled to a share of royalties received by Cancer Research Technology Ltd. on sales of products related to the use of MCM detection in cancer diagnosis.

REFERENCES

*Reference is also in this collection.

Amato MM, Monheit B, Shore JW. 2006. Ptosis surgery. In *Duane's opthalmology* (ed. Tasman W, Jaeger FA), Chap. 78, Fig. 6. Lippincott Williams and Wilkins, London.

Bachrati CZ, Hickson ID. 2008. RecQ helicases: Guardian angels of the DNA replication fork. *Chromosoma* **117:** 219–233.

Baldwin P, Laskey R, Coleman N. 2003. Translational approaches to improving cervical screening. *Nat Rev Cancer* **3:** 217–226.

* Bell SD, Botchan MR. 2013. The minichromosome maintenance replicative helicase. *Cold Spring Perspect Biol* doi: 10.1101/cshperspect.a012807.

Bicknell LS, Bongers EM, Leitch A, Brown S, Schoots J, Harley ME, Aftimos S, Al-Aama JY, Bober M, Brown PA, et al. 2011a. Mutations in the pre-replication complex cause Meier–Gorlin syndrome. *Nat Genet* **43:** 356–359.

Bicknell LS, Walker S, Klingseisen A, Stiff T, Leitch A, Kerzendorfer C, Martin CA, Yeyati P, Al Sanna N, Bober M, et al. 2011b. Mutations in ORC1, encoding the largest subunit of the origin recognition complex, cause microcephalic primordial dwarfism resembling Meier–Gorlin syndrome. *Nat Genet* **43:** 350–355.

Bloom D. 1954. Congenital telangiectatic erythema resembling lupus erythematosus in dwarfs; probably a syndrome entity. *AMA Am J Dis Child* **88:** 754–758.

Bongers EM, Opitz JM, Fryer A, Sarda P, Hennekam RC, Hall BD, Superneau DW, Harbison M, Poss A, van Bokhoven H, et al. 2001. Meier–Gorlin syndrome: Report of eight additional cases and review. *Am J Med Genet* **102:** 115–124.

Bourdon A, Minai L, Serre V, Jais JP, Sarzi E, Aubert S, Chretien D, de Lonlay P, Paquis-Flucklinger V, Arakawa H, et al. 2007. Mutation of RRM2B, encoding p53-controlled ribonucleotide reductase (p53R2), causes severe mitochondrial DNA depletion. *Nat Genet* **39:** 776–780.

Brown CA, Bogers J, Sahebali S, Depuydt CE, De Prins F, Malinowski DP. 2012. Role of protein biomarkers in the detection of high-grade disease in cervical cancer screening programs. *J Oncol* **2012:** 289315.

Celis JE, Celis A. 1985. Cell cycle–dependent variations in the distribution of the nuclear protein cyclin proliferating cell nuclear antigen in cultured cells: Subdivision of S phase. *Proc Natl Acad Sci* **82:** 3262–3266.

Chatrath P, Scott IS, Morris LS, Davies RJ, Bird K, Vowler SL, Grant JW, Saeed IT, Howard D, Laskey RA, et al. 2003. Aberrant expression of minichromosome maintenance protein-2 and Ki67 in laryngeal squamous epithelial lesions. *Br J Cancer* **89:** 1048–1054.

Copeland WC. 2010. The mitochondrial DNA polymerase in health and disease. *Subcell Biochem* **50:** 211–222.

Davies RJ, Freeman A, Morris LS, Bingham S, Dilworth S, Scott I, Laskey RA, Miller R, Coleman N. 2002. Analysis of minichromosome maintenance proteins as a novel method for detection of colorectal cancer in stool. *Lancet* **359:** 1917–1919.

Davies RJ, Scott IS, Morris LS, Rushbrook SM, Bird K, Vowler SL, Arends MJ, Miller R, Coleman N. 2004. Increased expression of minichromosome maintenance protein 2 in active inflammatory bowel disease. *Colorectal Dis* **6:** 103–110.

Davies RJ, Miller R, Coleman N. 2005. Colorectal cancer screening: Prospects for molecular stool analysis. *Nat Rev Cancer* **5:** 199–209.

de Munnik SA, Bicknell LS, Aftimos S, Al-Aama JY, van Bever Y, Bober MB, Clayton-Smith J, Edrees AY, Feingold M, Fryer A, et al. 2012. Meier–Gorlin syndrome genotype–phenotype studies: 35 individuals with pre-replication complex gene mutations and 10 without molecular diagnosis. *Eur J Hum Genet* **20:** 598–606.

Depuydt CE, Makar AP, Ruymbeke MJ, Benoy IH, Vereecken AJ, Bogers JJ. 2011. BD-ProExC as adjunct molecular marker for improved detection of $CIN2^+$ after HPV primary screening. *Cancer Epidemiol Biomarkers Prev* **20:** 628–637.

Dudderidge TJ, Stoeber K, Loddo M, Atkinson G, Fanshawe T, Griffiths DF, Williams GH. 2005. Mcm2, Geminin, and KI67 define proliferative state and are prognostic markers in renal cell carcinoma. *Clin Cancer Res* **11:** 2510–2517.

Freeman A, Morris LS, Mills AD, Stoeber K, Laskey RA, Williams GH, Coleman N. 1999. Minichromosome maintenance proteins as biological markers of dysplasia and malignancy. *Clin Cancer Res* **5:** 2121–2132.

German J. 1969. Bloom's syndrome. I. Genetical and clinical observations in the first twenty-seven patients. *Am J Hum Genet* **21:** 196–227.

German J. 1997. Bloom's syndrome. XX. The first 100 cancers. *Cancer Genet Cytogenet* **93:** 100–106.

German J, Sanz MM, Ciocci S, Ye TZ, Ellis NA. 2007. Syndrome-causing mutations of the BLM gene in persons in the Bloom's Syndrome Registry. *Hum Mutat* **28:** 743–753.

Gineau L, Cognet C, Kara N, Lach FP, Dunne J, Veturi U, Picard C, Trouillet C, Eidenschenk C, Aoufouchi S, et al. 2012. Partial MCM4 deficiency in patients with growth retardation, adrenal insufficiency, and natural killer cell deficiency. *J Clin Invest* **122:** 821–832.

Gonzalez MA, Pinder SE, Callagy G, Vowler SL, Morris LS, Bird K, Bell JA, Laskey RA, Coleman N. 2003. Minichromosome maintenance protein 2 is a strong independent prognostic marker in breast cancer. *J Clin Oncol* **21:** 4306–4313.

Gonzalez MA, Tachibana KK, Chin S-F, Callagy G, Madine MA, Vowler SL, Pinder SE, Laskey RA, Coleman N. 2004. Geminin predicts adverse clinical outcome in breast cancer by reflecting cell-cycle progression. *J Pathol* **204:** 121–130.

Gonzalez MA, Tachibana KK, Laskey RA, Coleman N. 2005. Control of DNA replication and its potential clinical exploitation. *Nat Rev Cancer* **5:** 135–141.

Gorlin RJ, Cervenka J, Moller K, Horrobin M, Witkop CJ Jr. 1975. Malformation syndromes. A selected miscellany. *Birth Defects Orig Artic Ser* **11:** 39–50.

Guernsey DL, Matsuoka M, Jiang H, Evans S, Macgillivray C, Nightingale M, Perry S, Ferguson M, LeBlanc M, Paquette J, et al. 2011. Mutations in origin recognition complex gene ORC4 cause Meier–Gorlin syndrome. *Nat Genet* **43:** 360–364.

Harada H, Nakagawa H, Takaoka M, Lee J, Herlyn M, Diehl JA, Rustgi AK. 2008. Cleavage of MCM2 licensing protein fosters senescence in human keratinocytes. *Cell Cycle* **7:** 3534–3538.

Hashimoto K, Araki K, Osaki M, Nakamura H, Tomita K, Shimizu E, Ito H. 2004. MCM2 and Ki-67 expression in human lung adenocarcinoma: Prognostic implications. *Pathobiology* **71:** 193–200.

Hiraiwa A, Fujita M, Adachi A, Ono H, Nagasaka T, Matsumoto Y, Ohashi M, Tomita Y, Ishibashi M. 1998. Specific distribution patterns of hCDC47 expression in cutaneous diseases. *J Cutan Pathol* **25:** 285–290.

* Holt IJ, Reyes A. 2012. Human mitochondrial DNA replication. *Cold Spring Harb Perspect Biol* **4:** a012971.

Huang S, Lee L, Hanson NB, Lenaerts C, Hoehn H, Poot M, Rubin CD, Chen DF, Yang CC, Juch H, et al. 2006. The spectrum of WRN mutations in Werner syndrome patients. *Hum Mutat* **27:** 558–567.

Hudson G, Chinnery PF. 2006. Mitochondrial DNA polymerase-γ and human disease. *Hum Mol Genet* **15:** R244–R252.

Hughes CR, Guasti L, Meimaridou E, Chuang CH, Schimenti JC, King PJ, Costigan C, Clark AJ, Metherell LA. 2012. MCM4 mutation causes adrenal failure, short stature, and natural killer cell deficiency in humans. *J Clin Invest* **122:** 814–820.

Hunt DP, Freeman A, Morris LS, Burnet NG, Bird K, Davies TW, Laskey RA, Coleman N. 2002. Early recurrence of benign meningioma correlates with expression of mini-chromosome maintenance-2 protein. *Br J Neurosurg* **16:** 10–15.

Imperiale TF, Ransohoff DF, Itzkowitz SH, Turnbull BA, Ross ME. 2004. For the Colorectal Cancer Study Group. Fecal DNA versus fecal occult blood for colorectal-cancer screening in an average-risk population. *N Engl J Med* **351:** 2704–2714.

Jam K, Fox M, Crandall BF. 1999. RAPADILINO syndrome: A multiple malformation syndrome with radial and patellar aplasia. *Teratology* **60:** 37–38.

Kato H, Toki T, Shimizu M, Shiozawa T, Fujii S, Nikaido T, Konishi I. 2003. A new proliferation marker, minichromosome maintenance protein 2, is associated with tumor aggressiveness in esophageal squamous cell carcinoma. *J Surg Oncol* **84:** 24–30.

Kodani I, Osaki M, Shomori K, Araki K, Goto E, Ryoke K, Ito H. 2003. Minichromosome maintenance 2 expression is correlated with mode of invasion and prognosis in oral squamous cell carcinomas. *J Oral Pathol Med* **32:** 468–474.

Kruger S, Thorns C, Stocker W, Muller-Kunert E, Bohle A, Feller AC. 2003. Prognostic value of MCM2 immunoreactivity in stage T1 transitional cell carcinoma of the bladder. *Eur Urol* **43:** 138–145.

Larizza L, Roversi G, Volpi L. 2010. Rothmund–Thomson syndrome. *Orphanet J Rare Dis* **5:** 2.

Laskey RA. 2004. DNA and cancer. In *DNA: Changing science and society* (ed. Krude T), p. 88. Cambridge University Press, Cambridge.

Longley MJ, Clark S, Yu Wai Man C, Hudson G, Durham SE, Taylor RW, Nightingale S, Turnbull DM, Copeland WC, Chinnery PF. 2006. Mutant POLG2 disrupts DNA polymerase γ subunits and causes progressive external ophthalmoplegia. *Am J Hum Genet* **78:** 1026–1034.

Madine MA, Swietlik M, Pelizon C, Romanowski P, Mills AD, Laskey RA. 2000. The roles of the MCM, ORC, and Cdc6 proteins in determining the replication competence of chromatin in quiescent cells. *J Struct Biol* **129:** 198–210.

Mandel H, Szargel R, Labay V, Elpeleg O, Saada A, Shalata A, Anbinder Y, Berkowitz D, Hartman C, Barak M, et al. 2001. The deoxyguanosine kinase gene is mutated in individuals with depleted hepatocerebral mitochondrial DNA. *Nat Genet* **29:** 337–341.

McGarry TJ, Kirschner MW. 1998. Geminin, an inhibitor of DNA replication, is degraded during mitosis. *Cell* **93:** 1043–1053.

Meng MV, Grossfeld GD, Williams GH, Dilworth S, Stoeber K, Mulley TW, Weinberg V, Carroll PR, Tlsty TrD. 2001. Minichromosome maintenance protein 2 expression in prostate: Characterization and association with outcome after therapy for cancer. *Clin Cancer Res* **7:** 2712–2718.

Mills AD, Coleman N, Morris LS, Laskey RA, 2000. Detection of S phase cells in tissue sections by in situ DNA replication. *Nat Cell Biol* **2:** 244–245.

Monnat RJ Jr. 2010. Human RECQ helicases: Roles in DNA metabolism, mutagenesis and cancer biology. *Semin Cancer Biol* **20:** 329–339.

Muftuoglu M, Oshima J, von Kobbe C, Cheng WH, Leistritz DF, Bohr VA. 2008. The clinical characteristics of Werner syndrome: Molecular and biochemical diagnosis. *Hum Genet* **124:** 369–377.

Mukherjee G, Muralidhar B, Bafna UD, Laskey RA, Coleman N. 2007. MCM immunocytochemistry as a first line cervical screening test in developing countries: A prospective cohort study in a regional cancer centre in India. *Br J Cancer* **96:** 1107–1111.

Musahl C, Holthoff HP, Lesch R, Knippers R. 1978. Stability of the replicative Mcm3 protein in proliferating and differentiating human cells. *Exp Cell Res* **241:** 260–264.

Oshima J, Martin GM, Hisama FM. 2012. Werner syndrome. In *GeneReviews* (ed. Pagon RA, et al.). University of Washington, Seattle.

Ozaki S, Zen Y, Inoue M. 2011. Biomarker expression in cervical intraepithelial neoplasia: Potential progression predictive factors for low-grade lesions. *Hum Pathol* **42:** 1007–1012.

Ramnath N, Hernandez FJ, Tan DF, Huberman JA, Natarajan N, Beck AF, Hyland A, Todoro IT, Brooks JS, Bepler G. 2001. MCM2 is an independent predictor of survival in patients with non-small-cell lung cancer. *J Clin Oncol* **19:** 4259–4266.

Rhodes JM. 2000. Colorectal cancer screening in the UK: Joint position statement by the British Society of Gastroenterology, the Royal College of Physicians, and the Association of Coloproctology of Great Britain and Ireland. *Gut* **46:** 746–748.

Rodins K, Cheale M, Coleman N, Fox SB. 2002. Minichromosome maintenance protein 2 expression in normal kidney and renal cell carcinomas: Relationship to tumor dormancy and potential clinical utility. *Clin Cancer Res* **8:** 1075–1081.

Saada A, Shaag A, Mandel H, Nevo Y, Eriksson S, Elpeleg O. 2001. Mutant mitochondrial thymidine kinase in mitochondrial DNA depletion myopathy. *Nat Genet* **29:** 342–344.

Scarpini C, White V, Muralidhar B, Patterson A, Hickey N, Singh N, Mullerat J, Winslet M, Davies RJ, Phillips M-L, et al. 2008. Improved screening for anal neoplasia by immunocytochemical detection of minichromosome maintenance proteins. *Cancer Epidemiol Biomarkers Prev* **17:** 2855–2864.

Scholzen T, Gerdes J. 2000. The Ki-67 protein: From the known and the unknown. *J Cell Physiol* **182:** 311–322.

Scott IS, Morris LS, Bird K, Davies RJ, Vowler SL, Rushbrook M, Marshall AE, Laskey RA, Miller R, Arends MJ, et al. 2003. A novel immunohistochemical method to estimate cell-cycle phase distribution in archival tissue: Implications for the prediction of outcome in colorectal cancer. *J Pathol* **201:** 187–197.

Scott IS, Morris LS, Rushbrook SM, Bird K, Vowler SL, Burne NG, Coleman N. 2005. Immunohistochemical estimation of cell cycle entry and phase distribution in astrocytomas: Applications in diagnostic neuropathology. *Neuropathol Appl Neurobiol* **31:** 455–466.

Scott IS, Odell E, Chatrath P, Morris LS, Davies RJ, Vowler SL, Laskey RA, Coleman N. 2006. A minimally invasive immunocytochemical approach to early detection of oral squamous cell carcinoma and dysplasia. *Br J Cancer* **94:** 1170–1175.

Shima N, Alcaraz A, Liachko I, Buske TR, Andrews CA, Munroe RJ, Hartford SA, Tye BK, Schimenti JC. 2007. A viable allele of Mcm4 causes chromosome instability and mammary adenocarcinomas in mice. *Nat Genet* **39:** 93–98.

* Siddiqui K, On KF, Diffley JFX. 2013. Regulating DNA replication in eukarya. *Cold Spring Perspect Biol* doi: 10.1101/cshperspect.a012930.

Siitonen HA, Sotkasiira J, Biervliet M, Benmansour A, Capri Y, Cormier-Daire V, Crandall B, Hannula-Jouppi K, Hennekam R, Herzog D, et al. 2009. The mutation spectrum in RECQL4 diseases. *Eur J Hum Genet* **17:** 151–158.

Spelbrink JN, Li FY, Tiranti V, Nikali K, Yuan QP, Tariq M, Wanrooij S, Garrido N, Comi G, Morandi L, et al. 2001. Human mitochondrial DNA deletions associated with mutations in the gene encoding Twinkle, a phage T7 gene 4-like protein localized in mitochondria. *Nat Genet* **28:** 223–231.

Stanley M. 2010. HPV—Immune response to infection and vaccination. *Infect Agent Cancer* **5:** 19.

Stoeber K, Halsall I, Freeman A, Swinn R, Doble A, Morris L, Coleman N, Bullock N, Laskey RA, Hales CN, et al. 1999. Immunoassay for urothelial cancers that detects DNA replication protein Mcm5 in urine. *Lancet* **354:** 1524–1525.

Stoeber K, Tlsty RTD, Happerfield L, Thomas GA, Romanov S, Bobrow L, Williams ED, Williams GH. 2001. DNA replication licensing and human cell proliferation. *J Cell Sci* **114:** 2027–2041.

Stoeber K, Swinn R, Prevost AT, de Clive-Lowe P, Halsall I, Dilworth SM, Marr J, Turner WH, Bullock N, Doble A, et al. 2002. Diagnosis of genito-urinary tract cancer by detection of minichromosome maintenance 5 protein in urine sediments. *J Natl Cancer Inst* **94:** 1071–1079.

Suomalainen A, Isohanni P. 2010. Mitochondrial DNA depletion syndromes—Many genes, common mechanisms. *Neuromuscul Disord* **20:** 429–437.

Sznajer Y, Siitonen HA, Roversi G, Dangoisse C, Scaillon M, Ziereisen F, Tenoutasse S, Kestila M, Larizza L. 2008. Atypical Rothmund–Thomson syndrome in a patient with compound heterozygous mutations in RECQL4 gene and phenotypic features in RECQL4 syndromes. *Eur J Pediatr* **167:** 175–181.

Tachibana KK, Gonzalez MA, Coleman N. 2005. Cell cycle control of DNA replication and its relevance to cancer pathology. *J Pathol* **205:** 123–129.

Tada S, Li A, Maiorano D, Mechali M, Blow JJ. 2001. Repression of origin assembly in metaphase depends on inhibition of RLF-B/Cdt1 by geminin. *Nat Cell Biol* **3:** 107–113.

Tambouret RH, Misdraji J, Wilbur DC. 2008. Longitudinal clinical evaluation of a novel antibody cocktail for detection of high grade squamous intraepithelial lesions on cervical cytology specimens. *Arch Pathol Lab Med* **132:** 918–925.

Todorov IT, Werness BA, Wang HO, Buddharaju LN, Todorova PD, Slocum HK, Brooks JS, Huberman JA. 1998. HsMCM2/BM28: A novel proliferation marker for human tumors and normal tissues. *Lab Invest* **78:** 73–78.

Toschi L, Bravo R. 1988. Changes in cyclin/proliferating cell nuclear antigen distribution during DNA repair synthesis. *J Cell Biol* **107:** 1623–1628.

Van Goethem G, Dermaut B, Lofgren A, Martin JJ, Van Broeckhoven C. 2001. Mutation of POLG is associated with progressive external ophthalmoplegia characterized by mtDNA deletions. *Nat Genet* **28:** 211–212.

Vennos EM, Collins M, James WD. 1992. Rothmund–Thomson syndrome: Review of the world literature. *J Am Acad Dermatol* **27:** 750–762.

Wang LL, Levy ML, Lewis RA, Chintagumpala MM, Lev D, Rogers M, Plon SE. 2001. Clinical manifestations in a cohort of 41 Rothmund–Thomson syndrome patients. *Am J Med Genet* **102:** 11–17.

Wharton SB, Chan KK, Anderson JR, Stoeber K, Williams GH. 2001. Replicative Mcm2 protein as a novel proliferation marker in oligodendrogliomas and its relationship to Ki67 labelling index, histological grade and prognosis. *Neuropathol Appl Neurobiol* **27:** 305–313.

White V, Scarpini C, Barbosa-Morais NL, Ikelle E, Carter S, Laskey RA, Miller R, Coleman N. 2009. Isolation of stool-derived mucus provides a high yield of colonocytes suitable for early detection of colorectal carcinoma. *Cancer Epidemiol Biomarkers Prev* **18:** 2006–2013.

Williams GH, Romanowski P, Mills AD, Morris L, Stoeber K, Marr J, Laskey RA, Coleman NC. 1998. Improved cervical smear assessment using antibodies against proteins that regulate DNA replication. *Proc Natl Acad Sci* **95:** 14932–14937.

Wohlschlegel JA, Dwyer BT, Dhar SK, Cvetic C, Dutta A. 2000. Inhibition of eukaryotic DNA replication by geminin binding to Cdt1. *Science* **290:** 2309–2312.

Yerushalmi R, Woods R, Ravdin PM, Hayes MM, Gelmon KA. 2010. Ki67 in breast cancer: Prognostic and predictive potential. *Lancet Oncol* **11:** 174–183.

Young MJ, Longley MJ, Li FY, Kasiviswanathan R, Wong LJ, Copeland WC. 2011. Biochemical analysis of human POLG2 variants associated with mitochondrial disease. *Hum Mol Genet* **20:** 3052–3066.

Regulating DNA Replication in Bacteria

Kirsten Skarstad[1] and Tsutomu Katayama[2]

[1]Department of Cell Biology, Institute for Cancer Research, The Radium Hospital, Oslo University Hospital, 0310 Oslo, Norway

[2]Department of Molecular Biology, Graduate School of Pharmaceutical Sciences, Kyushu University, Fukuoka 812-8582, Japan

Correspondence: katayama@phar.kyushu-u.ac.jp

The replication origin and the initiator protein DnaA are the main targets for regulation of chromosome replication in bacteria. The origin bears multiple DnaA binding sites, while DnaA contains ATP/ADP-binding and DNA-binding domains. When enough ATP-DnaA has accumulated in the cell, an active initiation complex can be formed at the origin resulting in strand opening and recruitment of the replicative helicase. In *Escherichia coli, oriC* activity is directly regulated by DNA methylation and specific *oriC*-binding proteins. DnaA activity is regulated by proteins that stimulate ATP-DnaA hydrolysis, yielding inactive ADP-DnaA in a replication-coupled negative-feedback manner, and by DnaA-binding DNA elements that control the subcellular localization of DnaA or stimulate the ADP-to-ATP exchange of the DnaA-bound nucleotide. Regulation of *dnaA* gene expression is also important for initiation. The principle of replication-coupled negative regulation of DnaA found in *E. coli* is conserved in eukaryotes as well as in bacteria. Regulations by *oriC*-binding proteins and *dnaA* gene expression are also conserved in bacteria.

Bacteria typically contain few chromosomes, each carrying a defined origin of replication (Messer 2002). The model bacteria referred to in this work *(Escherichia coli, Bacillus subtilis, Caulobacter crescentus*, and *Helicobacter pylori*) all have a single circular chromosome that is replicated bidirectionally from the origin. Some bacteria, for instance *E. coli* and *B. subtilis*, grow with overlapping replication cycles in rich media (Fig. 1) (Kornberg and Baker 1992; Helmstetter 1996). This allows for cell doubling times that are shorter than the replication phase, and requires replication initiation to occur at 2, 4, or 8 origins, depending on the growth rate. Replication timing is maintained such that initiation occurs simultaneously at all origins once per generation (Skarstad et al. 1986).

The key protein responsible for DNA strand opening at the origin and for the recruitment of replisome components is the initiator protein, DnaA (Kornberg and Baker 1992; Messer 2002; Duderstadt and Berger 2008; Ozaki and Katayama 2009; Kaguni 2011; Leonard and Grimwade 2011). DnaA is an AAA^{+} type protein that binds ATP and ADP with high affinity. DnaA binds to high- and low-affinity sites in *oriC* and forms an oligomeric structure (Fig. 2) (Grimwade et al. 2000; Kawakami et al. 2005;

Cite this article as *Cold Spring Harb Perspect Biol* doi: 10.1101/cshperspect.a012922

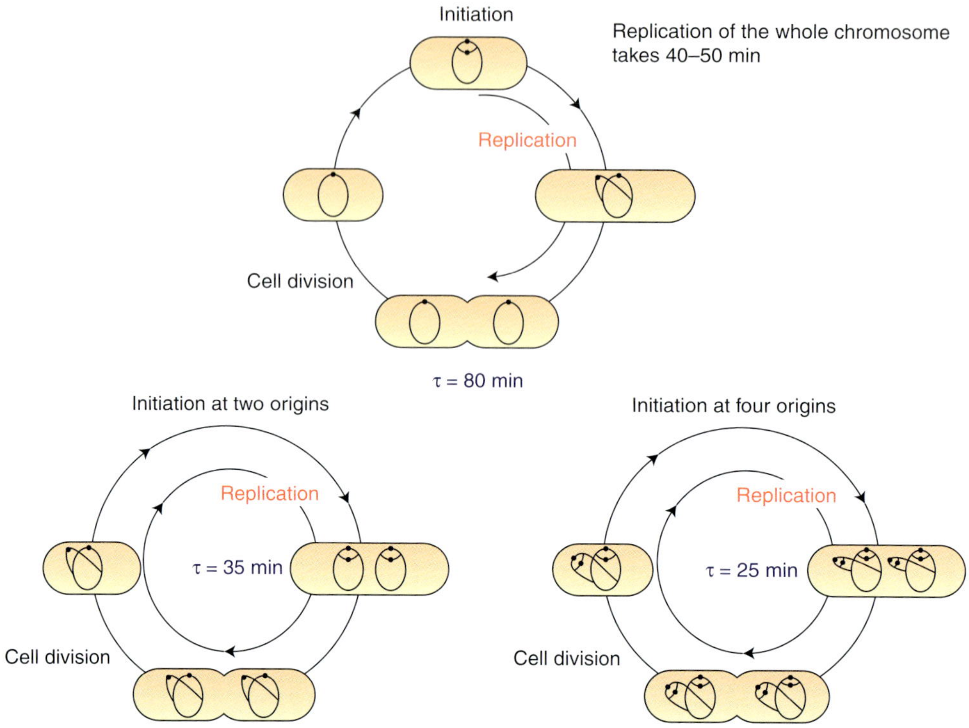

Figure 1. The replication cycle of the *E. coli* chromosome in slowly and rapidly growing cells. The chromosomal replication cycle and the cell division cycle in *E. coli* are shown. When cells are growing slowly (in this example, the doubling time τ is 80 min), cell division occurs after replication of the sister chromosomes is completed. When cells are growing rapidly (i.e., τ is 35 or 25 min), replication initiation simultaneously occurs at each *oriC* on the partially replicated chromosomes, and cell division occurs after the previous round of replication has been completed. Closed circles, *oriC*.

Erzberger et al. 2006; Kawakami and Katayama 2010) that involves two types of DnaA–DNA interactions, one with double-stranded and one with single-stranded DNA (Speck and Messer 2001; Fujikawa et al. 2003; Ozaki et al. 2008; Duderstadt et al. 2011; Ozaki and Katayama 2012). Only the ATP-bound form of DnaA is capable of binding to low-affinity sites (Fig. 2), which is supported by specific inter-DnaA interaction mediated through its AAA$^+$ domain, resulting in the nucleoprotein structures required for initiation activity (McGarry et al. 2004; Kawakami et al. 2005). Like ATP-DnaA, ADP-DnaA binds to high-affinity 9-mer DnaA boxes through its carboxy-terminal dsDNA binding domain, but forms multimers on *oriC* that are structurally distinct from those of ATP-DnaA. The binding of IHF, another positive regulator of initiation, to *oriC* induces a 180 degree bend in the DNA (Dillon and Dorman 2010) and plays an important role in forming an optimal initiation complex (Hwang and Kornberg 1992; Hiasa and Marians 1994; Cassler et al. 1995; Grimwade et al. 2000; Ryan et al. 2002; Leonard and Grimwade 2005; Ozaki and Katayama 2012). Other nucleoid-associated proteins (HU and Fis) also affect formation of the initiation complex (Fig. 3) (Gille et al. 1991; Hwang and Kornberg 1992; Hiasa and Marians 1994; Cassler et al. 1995; Wold et al. 1996; Margulies and Kaguni 1998; Ryan et al. 2002, 2004). Transcription by RNA polymerase is required for replication initiation (Kornberg and Baker 1992), though the mechanism by which this occurs remains unclear. Transcription in or around *oriC* is thought to facilitate opening of

 Cite this article as *Cold Spring Harb Perspect Biol* doi: 10.1101/cshperspect.a012922

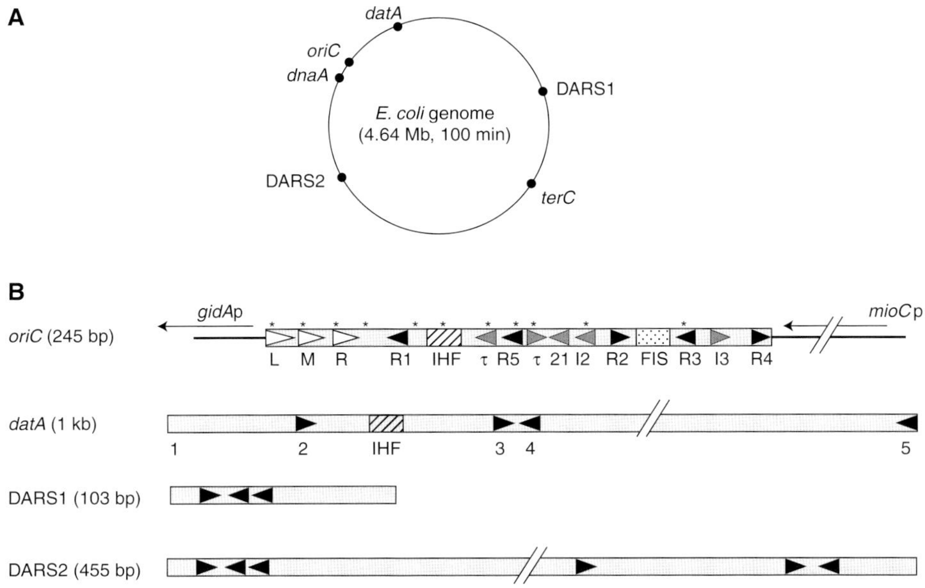

Figure 2. Structures of *oriC*, *datA*, and DARSs. (*A*) The *E. coli* chromosome is shown as a circle and the locations of *oriC* (at the genome map position of 84.6 min), *dnaA* (at 83.6 min), *datA* (at 94.6 min), DARS1 (at 17.5 min), DARS2 (at 64.0 min), and *terC* (around 36 min) are indicated. (*B*) Basic structures of *oriC*, *datA*, and DARSs are schematically shown. Closed triangle, DnaA box (9-nucleotide sequence). Gray triangle, I-sites, and τ-sites (both 6-nucleotide sequence). As for *datA*, DnaA boxes 2 and 3 are most crucial in repression of initiation (Ogawa et al. 2002). For details on low-affinity DnaA binding sites in *datA*, see Hansen et al. (2007). Open triangle, 13-mer AT-rich motif within DUE. IHF, IHF-binding site; FIS, FIS-binding site; *, GATC sequence within *oriC*.

the DNA strands, and may be a prerequisite for initiation in vivo under certain conditions, for instance, when nucleoid-associated proteins partially restrain negative superhelicity, thereby impeding strand opening (Baker and Kornberg 1988; Skarstad et al. 1990; Bates et al. 1997). In some circumstances, transcription through *oriC* may also be inhibitory (Su'etsugu et al. 2003) and for such transcription, the DnaA protein functions as a transcriptional regulator (Weigel and Messer 1997; Flåtten et al. 2009). Thus, DnaA has two roles at *oriC*, one as an initiator and the other as a transcription regulator.

After the DNA strands at *oriC* are separated, the DnaB helicase, bound to the helicase loader DnaC, is recruited to the initiation complex via interactions with *oriC*-bound DnaA (Sutton et al. 1998; Seitz et al. 2000; Messer 2002; Abe et al. 2007; Duderstadt and Berger 2008; Leonard and Grimwade 2010; Kaguni 2011). This stage may involve a reorganization of the DnaA-*oriC* complex and is influenced by the DiaA protein (Fig. 3) (Keyamura et al. 2009). DiaA binds to the amino terminus of DnaA and promotes initiation (Ishida et al. 2004; Keyamura et al. 2007). DiaA may also have a modulating effect on DnaB loading, because it binds to DnaA in the same place as DnaB, and inhibits loading of DnaB in vitro (Keyamura et al. 2009).

Regulation of initiation must fulfill two requirements. It must prevent extra initiation events, and it must ensure sufficient initiation so that one initiation event occurs per generation per origin. Several mechanisms ensure that extra initiation events do not occur (described below), but less is known about the timing of replication initiation, i.e., the rate-limiting steps, and whether the same factor(s) are required under all conditions. The frequency of replication must match the growth rate, otherwise the cellular DNA concentration will be altered. The cellular concentration of DnaA

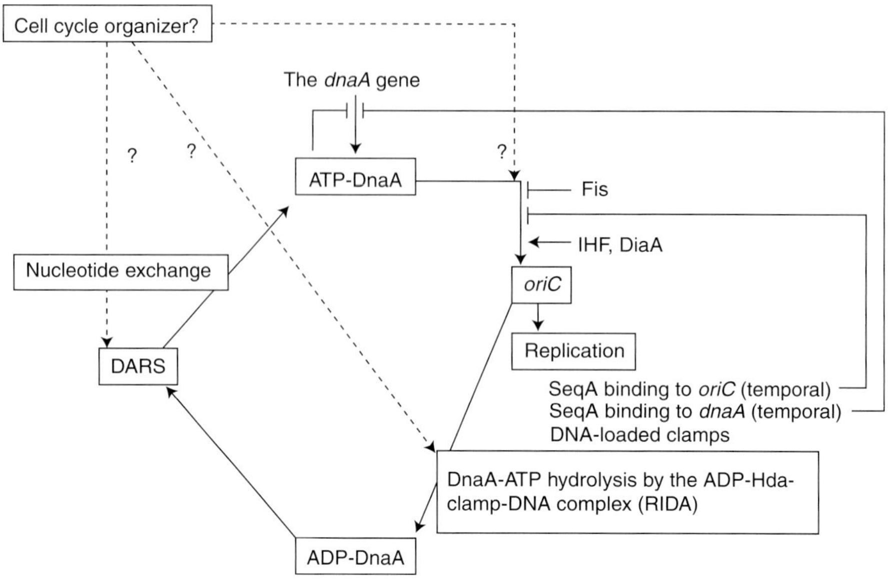

Figure 3. The DnaA cycle and regulation of *oriC* initiation. Transcription of the *dnaA* gene is autoregulated. ATP-DnaA is more active in inhibiting *dnaA* transcription than ADP-DnaA. Also, *dnaA* transcription increases before replication initiation and is repressed after it in a SeqA-dependent manner. Newly synthesized DnaA molecules adopt the ATP form. ATP-DnaA molecules are also generated by nucleotide exchange of ADP-DnaA in a DARS-dependent manner, though the cell cycle-dependent regulation of DARS remains unclear. With the assistance of DiaA and IHF, ATP-DnaA molecules form a multimeric complex on *oriC* in a cooperative manner. The resultant *oriC* complex unwinds DUE and is then engaged in replisome formation. The clamps remain on the synthesized DNA after Okazaki fragments are synthesized and form a complex with ADP-Hda, which interacts with and promotes the hydrolysis of ATP-DnaA (i.e., RIDA). Broken lines indicate expected but unrevealed pathways for cell cycle-coordinated regulation.

protein was found to be constant irrespective of the growth media (Hansen et al. 1991a) and a theoretical model has suggested that the production of DnaA couples growth rate to replication frequency (Hansen et al. 1991b). The ratio of ATP-DnaA to ADP-DnaA varies throughout the cell cycle and peaks right before replication initiation (Kurokawa et al. 1999). It has therefore been suggested that the frequency of initiation is determined by the accumulation of ATP-DnaA at *oriC* during steady-state growth (Donachie and Blakely 2003).

REGULATION AT THE ORIGIN SEQUENCE IN *E. coli*

Origin Sequestration

There are, in principle, two ways to prevent excess initiation of replication, one is to hinder origin usage, and the other is to inactivate the factors that perform initiation. In *E. coli*, origin usage is prevented by a process called origin sequestration that depends on the binding of the SeqA protein to newly replicated origins (Fig. 3; Table 1) (Lu et al. 1994; Slater et al. 1995; Brendler and Austin 1999; Waldminghaus and Skarstad 2009). Lack of SeqA causes untimely extra initiations and asynchrony in initiation of multiple origins (i.e., asynchrony phenotype).

SeqA discriminates between new and old origins by the methylation status of GATC sequences, which are present at high frequency in the *oriC* DNA sequence. The adenines of these sequences are recognized and methylated by Dam methylase. Chromosomal GATC sites remain hemimethylated for about a minute after the replication fork has passed (Campbell and Kleckner 1990). The high frequency of GATC sites in the origin (Fig. 2), however,

Table 1. Regulatory pathways for replication initiation

Regulation in *E. coli*	Target of regulation	Representative regulatory factor	Representative phenotype when regulation is disrupted	Possible analogs in other organisms
Sequestration	*oriC*	SeqA	Extra initiations, asynchrony phenotype	Regulation by Soj in *B. subtilis* Regulation by CtrA in *C. crescentus*
dnaA transcription	*dnaA* gene	DnaA, SeqA	Extra initiations, asynchrony phenotype (when SeqA function is inhibited or *dnaA* is overexpressed)	*dnaA* autoregulation in *B. subtilis* Cell cycle-coupled transcription in *C. crescentus*
Binding of DnaA to sites other than *oriC*	DnaA	*datA*	Extra initiations, changed timing in some growth conditions, asynchrony phenotype	Regulation by DnaA box clusters in *B. subtilis* and *Streptomyces coelicolor*
DnaA reactivation	ADP-DnaA	DARS	Inhibition of initiation, asynchrony phenotype	Unknown, but DARS-homologous sequences are conserved in certain bacterial genomes (Fujimitsu et al. 2009)
Stimulation of DnaA multimerization	ATP-DnaA	DiaA	Inhibition of initiation, asynchrony phenotype	Regulation by HobA in *H. pylori*
DnaA inactivation (RIDA)	ATP-DnaA	Hda	Severe amount of extra initiations, asynchrony phenotype Inhibition of cell division and cell proliferation	Regulation by YabA-clamp complex in *B. subtilis* Regulation by HdaA-clamp complex in *C. crescentus* Regulation by Cul4-DDB1-PCNA (clamp) complex in eukaryotes

results in origin DNA remaining hemimethylated for about a third of a generation after replication (Campbell and Kleckner 1990; Lu et al. 1994; Bach and Skarstad 2005). During this time, the origin is bound by SeqA, which, in addition to preventing GATC methylation by Dam methylase, inhibits replication initiation (Fig. 3; Table 1) (Lu et al. 1994; Boye et al. 1996; Guarne et al. 2002; Fujikawa et al. 2004). Also, DnaA itself contributes to origin sequestration, probably by hindering remethylation by Dam at some of the GATC sites in *oriC* (Lu et al. 1994; Bach et al. 2008). In vitro, SeqA forms fibers of head-to-head dimers (Guarne et al. 2005; Kang et al. 2005; Odsbu et al. 2005). Additionally, the binding of SeqA multimers to *oriC* can alter DNA superhelicity (Torheim et al. 1999) and inhibit ATP-DnaA binding to low-affinity sites (Nievera et al. 2006), both of which can inhibit the initiation reaction (Fig. 3; Table 1) (Lee et al. 2001; Waldminghaus and Skarstad 2009).

Microscopy of fluorescently labeled SeqA indicates that the majority of cellular SeqA forms relatively compact structures with the hemimethylated DNA at replication forks (Hiraga 1998, 2000; Brendler et al. 2000; Molina and Skarstad 2004; Yamazoe et al. 2005; Fossum et al. 2007). These structures are dynamic and trail the replication forks, and always bind to the most newly synthesized DNA (Yamazoe et al. 2005; Waldminghaus et al. 2012). Because extra SeqA foci representing sequestered origins have not been detected, it has been suggested that sequestered origins bound by SeqA are situated at, or near, the replication fork structures (Molina and Skarstad 2004; Bach and Skarstad 2005; Morigen et al. 2009).

Regulation by Transcription at and near the Origin

Transcription around and at *oriC* can affect replication initiation and may also contribute to regulation of initiation frequency. The *mioC* and *gidA* genes are located at the right and left side of *oriC*, respectively (Fig. 2). The *mioC* gene does not have a strong terminator and its transcripts read through *oriC* (Messer and Weigel 1997). The *gidA* gene is transcribed away from *oriC*. The *mioC* promoter region contains a DnaA box cluster and DnaA binding negatively regulates *mioC* gene transcription (Messer and Weigel 1997; Hansen et al. 2007). Transcription of *mioC* is repressed before replication initiation and is derepressed after it (Theisen et al. 1993; Ogawa and Okazaki 1994; Bogan and Helmstetter 1997). Constitutive transcription of *mioC* impedes initiation (Su'etsugu et al. 2003). The transcriptional pattern of *gidA* is opposite to that of *mioC* (Theisen et al. 1993; Ogawa and Okazaki 1994; Bogan and Helmstetter 1997). Deleting the promoters of these genes does not affect initiation regulation in wild type cells during steady-state growth, although the transcription of these genes can stimulate initiation in cells bearing a mutation impeding initiation (Bates et al. 1997).

REGULATION OF THE DnaA PROTEIN IN *E. coli*

Overview

The production and activity of *E. coli* DnaA protein is regulated in several ways: by transcription, intracellular localization, and conformation (Katayama et al. 2010; Leonard and Grimwade 2010; Kaguni 2011). DnaA protein is stable, but replication cycle-specific *dnaA* gene transcription is important for sustaining well-timed initiation of replication (Bogan and Helmstetter 1997; Riber and Løbner-Olesen 2005). This can be explained by the role of ATP-DnaA in activating initiation (Kurokawa et al. 1999; Nishida et al. 2002) and the idea that newly synthesized DnaA preferentially binds ATP, the cellular level of which is 10-fold higher than that of ADP (Fig. 3).

Also, timely initiation of replication during the cell cycle requires specific chromosomal regions termed DARS (DnaA-reactivating sequence) 1 and DARS2. These regions bind ADP-DnaA molecules and promote the regeneration of ATP-DnaA by nucleotide exchange (Figs. 2 and 3) (Fujimitsu et al. 2009).

It has been estimated that there are 500–2000 DnaA molecules per cell, depending on strain backgrounds and growth rates (Sekimizu et al. 1988; Chiaramello and Zyskind 1989; Hansen et al. 1991a). *oriC* can bind 10–20 DnaA molecules (Messer 2002). A considerable number of DnaA molecules can be titrated at a specific chromosomal locus termed *datA* that is required for repressing rifamcipin-resistant unregulated initiation events (Figs. 2 and 3) (Kitagawa et al. 1998; Morigen et al. 2005).

The ATP-DnaA level fluctuates during the replication cycle, peaking around the time of initiation (Kurokawa et al. 1999). ATP-DnaA hydrolysis is required to reduce ATP-DnaA levels after initiation (Fig. 3). This regulation, termed RIDA (regulatory inactivation of DnaA), is coupled with the action of the DNA polymerase III holoenzyme. RIDA is crucial for regulating initiation so that it occurs only once per generation (Katayama et al.1998; Kato and Katayama 2001; Su'etsugu et al. 2004; Camara et al. 2005).

Regulation of *DnaA* Gene Transcription

The cellular DnaA concentration was found to be constant irrespective of growth medium and the cell cycle (Hansen et al. 1991a). However, the transcription of the *dnaA* gene varies in a replication cycle-dependent manner (Bogan and Helmstetter 1997). The main reason for the fluctuation seems to be that the *dnaA* gene promoter is sequestered by SeqA for almost the same duration as the origin (Fig. 3; Table 1) (Campbell and Kleckner 1990; Lu et al. 1994; Riber and Løbner-Olesen 2005). During sequestration, the promoter is unavailable to the transcription machinery. Because the *dnaA* gene is situated

 Cite this article as *Cold Spring Harb Perspect Biol* doi: 10.1101/cshperspect.a012922

near *oriC*, this contributes to reducing the production of DnaA and thus the initiation potential (i.e., the ability to initiate replication) soon after new replication forks have been launched (Riber and Løbner-Olesen 2005). The *dnaA* promoter area also contains DnaA boxes and the promoter was found to be capable of autoregulation (Fig. 3; Table 1) (Messer and Weigel 1997; Hansen et al. 2007). Translation of *dnaA* mRNA has not been well characterized, but it is known that the start codon, GTG, functions inefficiently in *E. coli*.

Binding of DnaA to Sites Other than oriC

As mentioned, the DnaA protein functions not only as the initiator but also as a gene regulatory protein. There are about 300 high-affinity DnaA binding sites and a very large number of low-affinity sites around the chromosome (Kitagawa et al. 1996; Roth and Messer 1998; Hansen et al. 2007). As the chromosome is replicated, the DnaA binding sites are duplicated and contribute to titration of DnaA away from *oriC*. The main titration site, *datA*, is situated near *oriC* and is duplicated soon after initiation of replication (Kitagawa et al. 1996, 1998; Ogawa et al. 2002; Morigen et al. 2003, 2005). In vivo studies indicate that the *datA* site binds on average 60 DnaA molecules (Hansen et al. 2007). The *datA* site should therefore contribute to reducing the initiation potential at *oriC* when *oriC* is still in sequestration. The *datA* site is about 1 kb in size and contains five high-affinity DnaA binding sites and about 25 low-affinity sites (Fig. 2; Table 1) (Kitagawa et al. 1996, 1998; Hansen et al. 2007). High-affinity DnaA boxes 2 and 3 are crucial for efficient binding of DnaA to *datA*. It is possible that the DnaA bound to these sites function as a core for further cooperative DnaA binding (Ogawa et al. 2002). The production of DnaA protein as cells grow and the generation of binding sites as the chromosome is replicated, as well as sequestration and RIDA, has been simulated *in silico* (Hansen et al. 1991; Browning et al. 2004; Atlas et al. 2008) and may indicate that initiation occurs as soon as enough ATP-DnaA has accumulated at *oriC*. However, cells with similar numbers of origins and chromosomal DnaA binding sites but different growth conditions have been reported to contain different amounts of DnaA per origin at initiation (Torheim et al. 2000; Flåtten et al. 2009). So, although origin firing as soon as enough ATP-DnaA has accumulated is an attractive model of cell cycle regulation, a complete understanding of regulation encompassing different growth conditions is still lacking even in the well-characterized *E. coli* bacterium.

Regulation of the DnaA Nucleotide Form by DARS and Acidic Phospholipids

E. coli cells can convert ADP-DnaA to ATP-DnaA by nucleotide exchange (Fig. 3) (Kurokawa et al. 1999; Fujimitsu et al. 2009). The DARS1 and DARS2 sequences promote this reaction and are located halfway within the intergenic region between *oriC* and *terC*, to the right and left of *oriC*, respectively (Fig. 2A) (Fujimitsu et al. 2009). A common feature of DARS1 and DARS2 is the presence of a DnaA box cluster, in which three DnaA boxes are similarly oriented and are located at a similar distance (Fig. 2B). Multiple ADP-DnaA molecules can form complexes with DARS, which facilitates the release of ADP from DnaA. The resultant apo-DnaA molecules are likely released from DARS because of reduced complex formation activity, which allows the binding of ATP and DnaA reactivation. Increasing the cellular copy number of DARS1 or DARS2 increases the ATP-DnaA level, inducing extra initiation events, whereas deleting both DARS1 and DARS2 causes a delay in initiation (Fujimitsu et al. 2009). Newly translated DnaA protein binds ATP, providing cells with a basal level of ATP-DnaA. However, this alone is not enough to initiate replication in a timely manner. Thus, the function of DARSs is crucial for timely initiation (Fig. 3; Table 1). Functional regulation of DARS remains to be clarified, except for the fact that each DARS contains a regulatory region in addition to the common sequence bearing the DnaA box cluster (Fujimitsu et al. 2009; Leonard and Grimwade 2009). In addition to DARS, acidic phospholipids such as cardiolipin and phosphatidylglycerol play an important role in the nucleotide release of ADP-DnaA (Sekimizu and Kornberg 1988), and DnaA is reactivated by exchange of the bound nucleotide in vitro in the presence of *oriC* and ATP (Crooke et al. 1992). Reduction of the acidic phospholipids in vivo can inhibit initiation at *oriC* (Xia and Dowhan 1995), but exactly how acidic phospholipids affect DnaA activity during the cell cycle remains to be elucidated.

Regulation of DnaA Multimer Formation by DiaA

DiaA forms homotetramers and each protomer contains a specific site for binding to DnaA domain I (Fig. 3) (Ishida et al. 2004; Keyamura et al. 2007, 2009). These features allow a single DiaA tetramer to bind multiple molecules of DnaA, which can stimulate cooperative binding of DnaA to *oriC* and the unwinding reaction (Fig. 3; Table 1). The binding of ATP-DnaA to low-affinity binding sites within *oriC* is enhanced by DiaA (Fig. 1) (see Bell and Kaguni 2013; Leonard and Méchali 2013). In *diaA*-disrupted mutant cells, replication initiation is delayed and initiation at sister *oriC* copies occurs asynchronously in rapidly growing cells (Ishida et al. 2004; Keyamura et al. 2007, 2009). These data are consistent with the observation that replication is initiated asynchronously in mutants bearing DnaA box R4-deleted *oriC* (Bates et al. 1995), because the binding of DnaA to high-affinity DnaA box R4 enhances cooperative DnaA binding to low-affinity sites.

After DUE unwinding in *oriC* complexes, DiaA must be released from DnaA (Keyamura et al. 2009). The DiaA-binding site of DnaA domain I is also used to bind DnaB helicase. However, the mechanism of DiaA-DnaA dissociation has not yet been elucidated.

DiaA orthologs are evolutionarily conserved in bacterial species (Keyamura et al. 2007). In addition, the HobA protein of *Helicobacter pylori* (Hp), a member of ε-Proteobacteria, displays functional and structural similarity to DiaA, although there is no significant sequence similarity (Table 1) (Natrajan et al. 2007, 2009; Zakrzewsak-Czerwinska et al. 2007; Zawilak-Pawlik et al. 2007, 2011; Terradot et al. 2010).

Regulation of the DnaA Nucleotide Form by RIDA

After ATP-DnaA promotes replication initiation, it is hydrolyzed in a manner dependent on a complex consisting of ADP-Hda protein and the DNA-loaded clamp (Fig. 4; Table 1) (Katayama et al. 1998; Kato and Katayama 2001; Su'etsugu et al. 2008). The resultant ADP-DnaA is inactive in initiation. This system is termed RIDA (regulatory inactivation of DnaA). RIDA is crucial for DnaA inactivation and thereby effectively supports once-per-generation initiation (Kurokawa et al. 1999; Camara et al. 2005; Riber et al. 2009). The *hda* gene is required for promoting cell proliferation, decreasing cellular ATP-DnaA levels and repressing overinitiation (Kato and Katayama 2001; Fujimitsu et al. 2008; Charbon et al. 2011). Incubation of temperature-sensitive *hda* mutant cells at the restrictive temperature leads to overinitiation of replication and induces inhibition of cell division, producing filamentous cells (Fujimitsu et al. 2008). Inhibition of cell division is thought to be a consequence of checkpoint regulation, but the exact mechanism by which this occurs remains unknown. DnaA AAA$^+$ sensor II motif Arg-334 is specifically required for ATP-DnaA hydrolysis and expression of a DnaA R334A mutant protein causes overinitiation and inhibition of cell growth in an *oriC*-dependent manner (Table 1) (Nishida et al. 2002).

Hda protein consists of a short amino-terminal region containing the clamp-binding motif and an AAA$^+$ domain that is homologous to DnaA domain III (Fig. 4) (Dalrymple et al. 2001; Kato and Katayama 2001; Kurz et al. 2004; Su'etsugu et al. 2005; Xu et al. 2009). The Hda clamp-binding motif is commonly present in clamp-binding proteins such as DNA polymerase III core subunit α (Dalrymple et al. 2001). It binds to the hydrophobic pocket of the clamp, which is the same site to which DNA polymerase III subunit α binds (Dalrymple et al. 2001; Kurz et al. 2004; Su'etsugu et al. 2005). The Hda AAA$^+$ domain specifically binds ADP, but not ATP (Su'etsugu et al. 2008). ADP-Hda is monomeric and active in RIDA, whereas apo-Hda is multimeric and inactive in RIDA (Su'etsugu

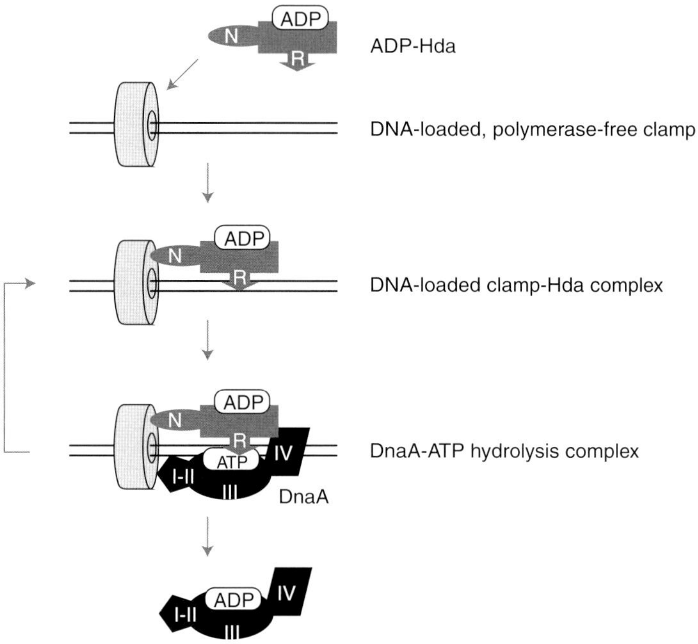

Figure 4. The basic mechanism of RIDA. When DNA polymerase III holoenzyme completes Okazaki fragment synthesis on the lagging strand, the clamp subunit is released from the DNA polymerase III core and remains on the synthesized DNA. ADP-Hda binds to the hydrophobic pocket of the DNA-loaded form of the clamp via the clamp-binding motifs in the amino terminus of Hda. The resultant ADP-Hda-clamp-DNA complex interacts with and promotes the DnaA-bound ATP hydrolysis, releasing ADP-DnaA back into the DnaA cycle. The interaction between the AAA$^+$ domains of Hda and DnaA is crucial, and is assisted by the interaction between Hda and DnaA domain IV (DNA-binding domain). N, Hda-amino terminus; I-II, DnaA domain I-II; III, DnaA domain III (AAA$^+$ domain); IV, DnaA domain IV (DNA-binding domain).

et al. 2008). The Hda AAA$^+$ domain bears a specific Arg residue (i.e., Arg finger) that is crucial for promoting DnaA-ATP hydrolysis (Su'etsugu et al. 2005). This residue may participate in forming the ATP hydrolysis catalytic center by direct DnaA-Hda interaction. This is a common characteristic of many AAA$^+$ proteins (Neuwald et al. 1999; Ogura et al. 2004; Indiani and O'Donnell 2006). In addition to the Arg finger, specific residues within the AAA$^+$ domains of DnaA and Hda are required for DnaA-Hda interaction and DnaA-ATP hydrolysis (Nakamura et al. 2010). DnaA domain IV (DNA binding domain) promotes this interaction by binding to Hda (Fig. 4) (Keyamura and Katayama 2011).

During DNA elongation, the clamps remain on the lagging strand after Okazaki fragments are synthesized and the DNA polymerase III core is released (Yuzhakov et al. 1996; Balakrishnan and Bambara 2013; Goodman and Woodgate 2013; Hedglin et al. 2013; MacAlpine and Almouzni 2013). DNA-loaded, DNA polymerase-free clamps bind ADP-Hda, resulting in the activation of RIDA (Su'etsugu et al. 2004, 2008). In addition, it is possible that because a clamp is a homodimer, Hda and DNA polymerase III subunit α bind to each protomer of the same clamp to allow Hda to hydrolyze DnaA-ATP as soon as replication forks are under way (Johnsen et al. 2011). DNA-free clamps are inactive in RIDA, although they can bind Hda, which ensures the timely and replication-coupled activation of RIDA. The dsDNA region flanking the clamp is required for RIDA, and may be recognized by DnaA (Fig. 4) (Su'etsugu et al. 2004).

The ADP-Hda-clamp-DNA complex is stable, whereas the affinity of this complex for DnaA is weak (Su'etsugu et al. 2008). This is consistent with the fact that the complex is reused for cyclically interacting with multiple ATP-DnaA molecules, and that there are only about 100 Hda molecules per cell (Katayama et al. 2010).

The main principle of RIDA, which is the use of DNA-loaded clamps for replication-coupled negative feedback to the initiator protein, is evolutionarily conserved from bacteria to eukaryotes, including yeast, *Xenopus*, and human cells (Table 1) (Katayama et al. 2010; Zielke et al. 2012).

Coordination of Regulation for *oriC* and DnaA

SeqA binds to hemimethylated *oriC* immediately after initiation and replication of the *oriC* region. In a culture with a doubling time of 30 min, the binding of SeqA to *oriC* lasts for about 10 min. The ratio of ATP-DnaA to ADP-DnaA peaks at initiation, and then undergoes a gradual decrease because of RIDA activity. It takes approximately 15 min to decrease this ratio to the lowest levels, and probably less than 10 min to decrease the ratio enough to prevent initiation from occurring when the origins are released after 10 min of sequestration. Thus, SeqA inactivates *oriC* early during the inter-initiation period, whereas RIDA represses the DnaA initiation activity for the remainder of this period. Coordination of these two steps is crucial for repressing untimely initiation events (Fig. 3) (Lu et al. 1994; Kurokawa et al. 1999; Skarstad and Løbner-Olesen 2003; Katayama et al. 2010). During multifork replication, in which replication goes on continuously and only the number of replication forks under way changes through the cycle (Fig. 1), it could be that RIDA needs to be temporarily inhibited to achieve a high enough ratio of ATP-DnaA to ADP-DnaA at initiation. Whether such a temporary inhibition occurs remains to be elucidated.

The balance between Dam methylase and SeqA activities governs the duration of the sequestration period (Bach et al. 2003). If Dam methylase is overexpressed, extra rounds of replication occur (Boye and Løbner-Olesen 1990), indicating that high levels of Dam methylase shorten the sequestration period, resulting in origin availability before the initiation potential has dropped (Skarstad and Løbner-Olesen 2003). Likewise, if DnaA is overexpressed or DnaA inactivation by RIDA has not occurred, extra rounds of DNA replication occur because the initiation window has stretched beyond the sequestration window (Atlung et al. 1987; Skarstad et al. 1989; Nishida et al. 2002; Camara et al. 2003).

Regulation of replication initiation in *E. coli* does not involve control over the concentration of origins. All origins present in the cell are initiated at the same time once per generation irrespective of how many there are. This can be seen when minichromosomes (plasmids dependent on an *oriC* sequence for replication initiation) are present in the cell; they initiate at the same time as the chromosomal *oriC*s (Helmstetter and Leonard 1987; Løbner-Olesen 1999). Thus, the mechanism is the following: all origins are initiated during a short time interval and all new origins are sequestered until initiation is no longer possible. Although replication fork speed can vary significantly (Morigen et al. 2009; Odsbu et al. 2009; Stokke et al. 2012), it is not known whether the speed of each individual replication fork is regulated. In *Bacillus subtilis,* it has been found that replication elongation can be regulated by ppGpp (Wang et al. 2007) and metabolic enzymes (Janniere et al. 2007).

Regulation of *oriC* and DnaA in *B. subtilis* and *C. crescentus*

Regulation by the Clamp-Binding Protein

Bacillus subtilis (*Bs*) is a representative of Gram-positive bacteria and is a member of Firmicutes, whereas *E. coli* is representative of Gram-negative bacteria and is a member of γ-Proteobacteria. This means that these species are evolutionarily distant. In *B. subtilis,* the *dnaA* gene is located between the clusters of the cognate DnaA boxes on the genome (Messer 2002) that form the replication origin (Moriya et al. 1992, 1994; Krause et al. 1997; and see Leonard and Méchali 2013) and *dnaA* gene transcription is autoregulated like in *E. coli* (Table 1) (Ogura et al. 2001).

Subcellular localization analysis indicates that a considerable number of BsDnaA molecules are titrated to the replication forks in a clamp- and YabA-dependent manner (Noirot-Gros et al. 2006; Soufo et al. 2008; Goranov et al. 2009). Although YabA has no sequence similarity to *E. coli* Hda, it binds the clamp and BsDnaA, like Hda (Table 1) (Noirot-Gros et al. 2006; Cho et al. 2008). The YabA-clamp system is required to repress overinitiation of replication (Noirot-Gros et al. 2006; Soufo et al. 2008; Goranov et al. 2009), which may be sustained by the sequestration of many DnaA molecules from *oriC*. Whether DNA-clamp-YabA complexes induce BsDnaA-ATP hydrolysis is not yet known.

Caulobacter crescentus (*Cc*) is an α-Proteobacterial species that is evolutionarily distant from both *E. coli* and *B. subtilis*. *C. crescentus* cells undergo asymmetrical cell division, yielding a stalked cell and a swarmer cell; only a stalked cell can replicate the chromosome whereas a swarmer cell must first differentiate into a stalked cell before replication can occur (Marczynski and Shapiro 2002). In *C. crescentus*, HdaA, the *E. coli* Hda ortholog, colocalizes with replisomes and is required to repress overinitiation (Table 1) (Collier and Shapiro 2009). *Cc*DnaA Arg-357 corresponds to the *E. coli* DnaA AAA$^+$ sensor II motif Arg-334, and, like *E. coli* DnaA R334A, *Cc*DnaA R357A can cause overinitiation of the cognate chromosomal replication (Fernandez-Fernandez et al. 2011). Taken together, the data indicate that replication-dependent DnaA-ATP hydrolysis is important for repressing *Cc*DnaA activity and extra initiation events.

Regulation by *oriC*-Binding and Various Factors

In *B. subtilis*, the chromosome partition proteins Soj and Spo0J play regulatory roles in replication initiation at *oriC* (Ogura et al. 2003; Lee and Grossman 2006; Murray and Errington 2008; Scholefield et al. 2011). Soj and Spo0J are members of the same protein family as ParA and ParB that are plasmid-partitioning proteins. The ParS region is a centromere-like region located near *oriC* and partitions the sister *oriC* regions by binding to both Soj and Spo0J. DNA-free Soj binds to DnaA assembled on *oriC*, and represses untimely initiation events (Table 1). ATP-bound Soj binds to ParS and stimulates initiation. Spo0J negatively regulates this Soj action by stimulating Soj-ATP hydrolysis and dissociation from the DNA (Murray and Errington 2008; Scholefield et al. 2011). In addition, the binding of BsDnaA to DnaA box clusters near *oriC* is also important for preventing premature initiation (Okumura et al. 2012). Similar role for DnaA box clusters is also found in *Streptomyces* (Table 1) (Smulczyk-Krawczyszyn et al. 2006).

When *B. subtilis* cells sporulate, the initiation of chromosome replication is inhibited by SirA (Wagner et al. 2009; Rahn-Lee et al. 2011). SirA is expressed specifically on sporulation and binds BsDnaA, which inhibits BsDnaA-*oriC* binding. *E. coli* does not contain homologs of Soj, Spo0J, and SirA, whereas *B. subtilis* does not contain homologs of SeqA and Dam (Table 1).

In *C. crescentus*, CtrA is a major cell cycle-dependent transcriptional regulator present in swarmer cells. It binds to the cognate *oriC*, thereby inhibiting replication initiation (Table 1) (Quan et al. 1998). ClpP degrades CtrA when swarmer cells differentiate into stalked cells (Gorbatyok and Marczynski 2005; McGrath et al. 2006). Homologs of SeqA and Dam are absent in this bacterium.

The Stringent response is activated on carbon starvation and primarily inhibits transcription (Strivatsan and Wang 2008). Carbon starvation activates the synthesis of ppGpp from GDP by RelA or SpoT. ppGpp directly binds to RNA polymerase and alters transcription modes, specifically repressing stable RNA (rRNA and tRNA) synthesis. In *E. coli*, the initiation of chromosomal replication is inhibited on carbon starvation, which may be caused by changes in RNA polymerase function (Zyskind and Smith 1992) but the entire mechanism is not revealed. In *B. subtilis*, elongation, but not initiation, is inhibited on carbon starvation, which is indicated to be caused by ppGpp-dependent inactivation of primase (Wang et al. 2007). Similar mechanisms may also exist in *E. coli* (Maciag et al. 2010). In *C. crescentus*, SpoT, the ppGpp synthetase, degrades DnaA on carbon starvation (Lesyley and Shapiro 2008).

CONCLUDING REMARKS

Multiple systems regulate *oriC* and DnaA in a concerted manner to ensure that replication initiation occurs only once per origin per generation. Some of these regulatory systems are coupled with specific events that are important for cell cycle regulation or chromosomal replication. In *E. coli*, *oriC* becomes hemimethylated temporarily just after initiation, which allows the binding of SeqA, an inhibitor of initiation. The DNA-loaded clamps bind Hda protein, and the resultant complexes stimulate DnaA-ATP hydrolysis, thereby inactivating DnaA. This clamp-dependent feedback regulation is also conserved in eukaryotes. Also, *dnaA* gene transcription is repressed in a SeqA-dependent and temporal manner. These three negative-feedback mechanisms regulate initiation in a replication-dependent manner in *E. coli*. In addition, excessive DnaA molecules are titrated to the *datA* locus on the chromosome, which adds another level of negative regulation. *dnaA* transcription is also autoregulated in both *E. coli* and *B. subtilis*. DARS reactivates DnaA by stimulating ADP-to-ATP exchange, thereby supporting timely initiation as well as the DnaA-DiaA interaction.

The *B. subtilis* clamp-YabA complex and the *C. crescentus* clamp-HdaA complex regulate initiation in a replication-dependent, negative-feedback manner. YabA and HdaA are considered to be the functional homologs of *E. coli* Hda, but only HdaA displays structural similarity with Hda. In *B. subtilis*, the ordered interaction of Soj, Spo0J, and the *parS* locus regulates the initiation activity and assembly of BsDnaA at *oriC* in a cell cycle-coordinated manner. Like *E. coli datA*, DnaA box clusters near *oriC* are also important for repressing initiation potential. SirA, which is expressed specifically before sporulation, inhibits the binding of BsDnaA to *oriC*. In *C. crescentus*, the cell cycle-specific binding of CtrA to the origin may regulate initiation. The stringent response, which is induced on carbon starvation and increases the level of ppGpp, an RNA polymerase inhibitor, stimulates DnaA degradation in *C. crescentus*, and inhibits elongation in *B. subitilis* through the ppGpp-primase interaction and initiation in *E. coli*.

REFERENCES

*Reference is also in this collection.

Abe Y, Jo T, Matsuda Y, Matsunaga C, Katayama T, Ueda T. 2007. Structure and function of DnaA N-terminal domains: Specific sites and mechanisms in inter-DnaA interaction and in DnaB helicase loading on *oriC*. *J Biol Chem* **282:** 17816–17827.

Atlas JC, Nikolaev EV, Browning ST, Shuler ML. 2008. Incorporating genome-wide DNA sequence information into a dynamic whole-cell model of *Escherichia coli*: Application to DNA replication. *IET Syst Biol* **2:** 369–382.

Atlung T, Lobner-Olesen A, Hansen FG. 1987. Overproduction of DnaA protein stimulates initiation of chromosome and minichromosome replication in *Escherichia coli*. *Mol Gen Genet* **206:** 51–59.

Bach T, Skarstad K. 2005. An *oriC*-like distribution of GATC sites mediates full sequestration of non-origin sequences in *Escherichia coli*. *J Mol Biol* **350:** 7–11.

Bach T, Krekling MA, Skarstad K. 2003. Excess SeqA prolongs sequestration of *oriC* and delays nucleoid segregation and cell division. *EMBO J* **22:** 315–323.

Bach T, Morigen, Skarstad K. 2008. The initiator protein DnaA contributes to keeping new origins inactivated by promoting the presence of hemimethylated DNA. *J Mol Biol* **384:** 1076–1085.

Baker TA, Kornberg A. 1988. Transcriptional activation of initiation of replication from the *E. coli* chromosome origin: An RNA-DNA hybrid near *oriC*. *Cell* **55:** 113–123.

* Balakrishnan L, Bambara RA. 2013. Okazaki fragment metabolism. *Cold Spring Harb Perspect Biol* doi: 10.1101/cshperspect.a010173.

Bates DB, Asai T, Cao Y, Chambers MW, Cadwell GW, Boye E, Kogoma T. 1995. The DnaA box R4 in the minimal *oriC* is dispensable for initiation of *Escherichia coli* chromosome replication. *Nucleic Acids Res* **23:** 3119–3125.

Bates DB, Boye E, Asai T, Kogoma T. 1997. The absence of effect of *gid* or *mioC* transcription on the initiation of chromosomal replication in *Escherichia coli*. *Proc Natl Acad Sci* **94:** 12497–12502.

* Bell SP, Kaguni JM. 2013. Helicase loading at chromosomal origins of replication. *Cold Spring Harb Perspect Biol* doi: 10.1101/cshperspect.a010124.

Bogan JA, Helmstetter CE. 1997. DNA sequestration and transcription in the *oriC* region of *Escherichia coli*. *Mol Microbiol* **26:** 889–896.

Boye E, Lobner-Olesen A. 1990. The role of *dam* methyltransferase in the control of DNA replication in *E. coli*. *Cell* **62:** 981–989.

Boye E, Stokke T, Kleckner N, Skarstad K. 1996. Coordinating DNA replication initiation with cell growth: Differential roles for DnaA and SeqA proteins. *Proc Natl Acad Sci* **93:** 12206–12211.

Brendler T, Austin S. 1999. Binding of SeqA protein to DNA requires interaction between two or more complexes bound to separate hemimethylated GATC sequences. *EMBO J* **18:** 2304–2310.

Brendler T, Sawitzke J, Sergueev K, Austin S. 2000. A case for sliding SeqA tracts at anchored replication forks during

Escherichia coli chromosome replication and segregation. *EMBO J* **19:** 6249–6258.

Browning ST, Castellanos M, Shuler ML. 2004. Robust control of initiation of prokaryotic chromosome replication: Essential considerations for a minimal cell. *Biotechnol Bioeng* **88:** 575–584.

Camara JE, Breier AM, Brendler T, Austin S, Cozzarelli NR, Crooke E. 2005. Hda inactivation of DnaA is the predominant mechanism preventing hyperinitiation of *Escherichia coli* DNA replication. *EMBO Rep* **6:** 736–741.

Campbell JL, Kleckner N. 1990. *E coli oriC* and the *dnaA* gene promoter are sequestered from *dam* methyltransferase following the passage of the chromosomal replication fork. *Cell* **62:** 967–979.

Cassler MR, Grimwade JE, Leonard AC. 1995. Cell cycle-specific changes in nucleoprotein complexes at a chromosomal replication origin. *EMBO J* **14:** 5833–5841.

Charbon G, Riber L, Cohen M, Skovgaard O, Fujimitsu K, Katayama T, Lobner-Olesen A. 2011. Suppressors of DnaA(ATP) imposed overinitiation in *Escherichia coli*. *Mol Microbiol* **79:** 914–928.

Chiaramello A, Zyskind JW. 1989. Expression of *Escherichia coli dnaA* and *mioC* genes as a function of growth rate. *J Bacteriol* **171:** 4272–4280.

Cho E, Ogasawara N, Ishikawa S. 2008. The functional analysis of YabA, which interacts with DnaA and regulates initiation of chromosome replication in *Bacillus subtilis*. *Genes Genet Syst* **83:** 111–125.

Collier J, Shapiro L. 2009. Feedback control of DnaA-mediated replication initiation by replisome-associated HdaA protein in *Caulobacter*. *J Bacteriol* **191:** 5706–5716.

Crooke E, Castuma CE, Kornberg A. 1992. The chromosome origin of *Escherichia coli* stabilizes DnaA protein during rejuvenation by phospholipids. *J Biol Chem* **267:** 16779–16782.

Dalrymple BP, Kongsuwan K, Wijffels G, Dixon NE, Jennings PA. 2001. A universal protein–protein interaction motif in the eubacterial DNA replication and repair system. *Proc Natl Acad Sci* **98:** 11627–11632.

Dillon SC, Dorman CJ. 2010. Bacterial nucleoid-associated proteins, nucleoid structure and gene expression. *Nat Rev Microbiol* **8:** 185–195.

Donachie WD, Blakely GW. 2003. Coupling the initiation of chromosome replication to cell size in *Escherichia coli*. *Curr Opin Microbiol* **6:** 146–150.

Duderstadt KE, Berger JM. 2008. AAA^+ ATPases in the initiation of DNA replication. *Crit Rev Biochem Mol Biol* **43:** 163–187.

Duderstadt KE, Chuang K, Berger JM. 2011. DNA stretching by bacterial initiators promotes replication origin opening. *Nature* **478:** 209–213.

Erzberger JP, Mott ML, Berger JM. 2006. Structural basis for ATP-dependent DnaA assembly and replication-origin remodeling. *Nat Struct Mol Biol* **13:** 676–683.

Fernandez-Fernandez C, Gonzalez D, Collier J. 2011. Regulation of the activity of the dual-function DnaA protein in *Caulobacter crescentus*. *PLoS ONE* **6:** e26028.

Flatten I, Morigen, Skarstad K. 2009. DnaA protein interacts with RNA polymerase and partially protects it from the effect of rifampicin. *Mol Microbiol* **71:** 1018–1030.

Fossum S, Crooke E, Skarstad K. 2007. Organization of sister origins and replisomes during multifork DNA replication in *Escherichia coli*. *EMBO J* **26:** 4514–4522.

Fujikawa N, Kurumizaka H, Nureki O, Terada T, Shirouzu M, Katayama T, Yokoyama S. 2003. Structural basis of replication origin recognition by the DnaA protein. *Nucleic Acids Res* **31:** 2077–2086.

Fujikawa N, Kurumizaka H, Nureki O, Tanaka Y, Yamazoe M, Hiraga S, Yokoyama S. 2004. Structural and biochemical analyses of hemimethylated DNA binding by the SeqA protein. *Nucleic Acids Res* **32:** 82–92.

Fujimitsu K, Su'etsugu M, Yamaguchi Y, Mazda K, Fu N, Kawakami H, Katayama T. 2008. Modes of overinitiation, *dnaA* gene expression, and inhibition of cell division in a novel cold-sensitive *hda* mutant of *Escherichia coli*. *J Bacteriol* **190:** 5368–5381.

Fujimitsu K, Senriuchi T, Katayama T. 2009. Specific genomic sequences of *E. coli* promote replicational initiation by directly reactivating ADP-DnaA. *Genes Dev* **23:** 1221–1233.

Gille H, Egan JB, Roth A, Messer W. 1991. The FIS protein binds and bends the origin of chromosomal DNA replication, *oriC*, of *Escherichia coli*. *Nucleic Acids Res* **19:** 4167–4172.

* Goodman MF, Woodgate R. 2013. Translesion DNA polymerases. *Cold Spring Harb Perspect Biol* doi: 10.1101/cshperspect.a010363.

Goranov AI, Breier AM, Merrikh H, Grossman AD. 2009. YabA of *Bacillus subtilis* controls DnaA-mediated replication initiation but not the transcriptional response to replication stress. *Mol Microbiol* **74:** 454–466.

Gorbatyuk B, Marczynski GT. 2005. Regulated degradation of chromosome replication proteins DnaA and CtrA in *Caulobacter crescentus*. *Mol Microbiol* **55:** 1233–1245.

Grimwade JE, Ryan VT, Leonard AC. 2000. IHF redistributes bound initiator protein, DnaA, on supercoiled *oriC* of *Escherichia coli*. *Mol Microbiol* **35:** 835–844.

Guarne A, Zhao Q, Ghirlando R, Yang W. 2002. Insights into negative modulation of *E. coli* replication initiation from the structure of SeqA-hemimethylated DNA complex. *Nat Struct Biol* **9:** 839–843.

Guarne A, Brendler T, Zhao Q, Ghirlando R, Austin S, Yang W. 2005. Crystal structure of a SeqA-N filament: Implications for DNA replication and chromosome organization. *EMBO J* **24:** 1502–1511.

Hansen FG, Atlung T, Braun RE, Wright A, Hughes P, Kohiyama M. 1991a. Initiator (DnaA) protein concentration as a function of growth rate in *Escherichia coli* and *Salmonella typhimurium*. *J Bacteriol* **173:** 5194–5199.

Hansen FG, Christensen BB, Atlung T. 1991b. The initiator titration model: Computer simulation of chromosome and minichromosome control. *Res Microbiol* **142:** 161–167.

Hansen FG, Christensen BB, Atlung T. 2007. Sequence characteristics required for cooperative binding and efficient *in vivo* titration of the replication initiator protein DnaA in *E. coli*. *J Mol Biol* **367:** 942–952.

* Hedglin M, Kumar R, Benkovic SJ. 2013. Replication clamps and clamp loaders. *Cold Spring Harb Perspect Biol* doi: 10.1101/cshperspect.a010165.

Helmstetter C. 1996. Timing of synthetic activities in the cell cycle. In *Escherichia coli and Salmonella: Cellular and molecular biology*, 2nd ed. (ed. Neidhardt FC, Curtiss R III, Ingraham JL, Lin ECC, Low KB, Magasanik B, Reznikoff WS, Riley M, Schaechter M, Umbarger HE), pp. 1627–1639. ASM Press, Washington, DC.

Helmstetter CE, Leonard AC. 1987. Coordinate initiation of chromosome and minichromosome replication in *Escherichia coli*. *J Bacteriol* **169:** 3489–3494.

Hiasa H, Marians KJ. 1994. Fis cannot support *oriC* DNA replication *in vitro*. *J Biol Chem* **269:** 24999–25003.

Hiraga S. 2000. Dynamic localization of bacterial and plasmid chromosomes. *Annu Rev Genet* **34:** 21–59.

Hiraga S, Ichinose C, Niki H, Yamazoe M. 1998. Cell cycle-dependent duplication and bidirectional migration of SeqA-associated DNA-protein complexes in *E. coli*. *Mol Cell* **1:** 381–387.

Hwang DS, Kornberg A. 1992. Opening of the replication origin of *Escherichia coli* by DnaA protein with protein HU or IHF. *J Biol Chem* **267:** 23083–23086.

Indiani C, O'Donnell M. 2006. The replication clamp-loading machine at work in the three domains of life. *Nat Rev Mol Cell Biol* **7:** 751–761.

Ishida T, Akimitsu N, Kashioka T, Hatano M, Kubota T, Ogata Y, Sekimizu K, Katayama T. 2004. DiaA, a novel DnaA-binding protein, ensures the timely initiation of *Escherichia coli* chromosome replication. *J Biol Chem* **279:** 45546–45555.

Janniere L, Canceill D, Suski C, Kanga S, Dalmais B, Lestini R, Monnier AF, Chapuis J, Bolotin A, Titok M, et al. 2007. Genetic evidence for a link between glycolysis and DNA replication. *PLoS ONE* **2:** e447.

Johnsen L, Flatten I, Morigen, Dalhus B, Bjoras M, Waldminghaus T, Skarstad K. 2011. The G157C mutation in the *Escherichia coli* sliding clamp specifically affects initiation of replication. *Mol Microbiol* **79:** 433–446.

Kaguni JM. 2011. Replication initiation at the *Escherichia coli* chromosomal origin. *Curr Opin Chem Biol* **15:** 606–613.

Kang S, Han JS, Kim KP, Yang HY, Lee KY, Hong CB, Hwang DS. 2005. Dimeric configuration of SeqA protein bound to a pair of hemi-methylated GATC sequences. *Nucleic Acids Res* **33:** 1524–1531.

Katayama T, Kubota T, Kurokawa K, Crooke E, Sekimizu K. 1998. The initiator function of DnaA protein is negatively regulated by the sliding clamp of the *E. coli* chromosomal replicase. *Cell* **94:** 61–71.

Katayama T, Ozaki S, Keyamura K, Fujimitsu K. 2010. Regulation of the replication cycle: conserved and diverse regulatory systems for DnaA and *oriC*. *Nat Rev Microbiol* **8:** 163–170.

Kato J, Katayama T. 2001. Hda, a novel DnaA-related protein, regulates the replication cycle in *Escherichia coli*. *EMBO J* **20:** 4253–4262.

Kawakami H, Katayama T. 2010. DnaA, ORC, and Cdc6: Similarity beyond the domains of life and diversity. *Biochem Cell Biol* **88:** 49–62.

Kawakami H, Keyamura K, Katayama T. 2005. Formation of an ATP-DnaA-specific initiation complex requires DnaA Arginine 285, a conserved motif in the AAA$^+$ protein family. *J Biol Chem* **280:** 27420–27430.

Keyamura K, Katayama T. 2011. DnaA protein DNA-binding domain binds to Hda protein to promote inter-AAA$^+$ domain interaction involved in regulatory inactivation of DnaA. *J Biol Chem* **286:** 29336–29346.

Keyamura K, Fujikawa N, Ishida T, Ozaki S, Su'etsugu M, Fujimitsu K, Kagawa W, Yokoyama S, Kurumizaka H, Katayama T. 2007. The interaction of DiaA and DnaA regulates the replication cycle in *E. coli* by directly promoting ATP DnaA-specific initiation complexes. *Genes Dev* **21:** 2083–2099.

Keyamura K, Abe Y, Higashi M, Ueda T, Katayama T. 2009. DiaA dynamics are coupled with changes in initial origin complexes leading to helicase loading. *J Biol Chem* **284:** 25038–25050.

Kitagawa R, Mitsuki H, Okazaki T, Ogawa T. 1996. A novel DnaA protein-binding site at 94.7 min on the *Escherichia coli* chromosome. *Mol Microbiol* **19:** 1137–1147.

Kitagawa R, Ozaki T, Moriya S, Ogawa T. 1998. Negative control of replication initiation by a novel chromosomal locus exhibiting exceptional affinity for *Escherichia coli* DnaA protein. *Genes Dev* **12:** 3032–3043.

Kornberg A, Baker TA. 1992. *DNA replication*, 2nd ed. W. H. Freeman, New York.

Krause M, Ruckert B, Lurz R, Messer W. 1997. Complexes at the replication origin of *Bacillus subtilis* with homologous and heterologous DnaA protein. *J Mol Biol* **274:** 365–380.

Kurokawa K, Nishida S, Emoto A, Sekimizu K, Katayama T. 1999. Replication cycle-coordinated change of the adenine nucleotide-bound forms of DnaA protein in *Escherichia coli*. *EMBO J* **18:** 6642–6652.

Kurz M, Dalrymple B, Wijffels G, Kongsuwan K. 2004. Interaction of the sliding clamp β-subunit and Hda, a DnaA-related protein. *J Bacteriol* **186:** 3508–3515.

Lee PS, Grossman AD. 2006. The chromosome partitioning proteins Soj (ParA) and Spo0J (ParB) contribute to accurate chromosome partitioning, separation of replicated sister origins, and regulation of replication initiation in *Bacillus subtilis*. *Mol Microbiol* **60:** 853–869.

Lee H, Kang S, Bae SH, Choi BS, Hwang DS. 2001. SeqA protein aggregation is necessary for SeqA function. *J Biol Chem* **276:** 34600–34606.

Leonard AC, Grimwade JE. 2005. Building a bacterial orisome: Emergence of new regulatory features for replication origin unwinding. *Mol Microbiol* **55:** 978–985.

Leonard AC, Grimwade JE. 2009. Initiating chromosome replication in *E. coli*: It makes sense to recycle. *Genes Dev* **23:** 1145–1150.

Leonard AC, Grimwade JE. 2011. Regulation of DnaA assembly and activity: Taking directions from the genome. *Annu Rev Microbiol* **65:** 19–35.

* Leonard AC, Méchali M. 2013. DNA replication origins. *Cold Spring Harb Perspect Biol* doi: 10.1101/cshperspect.a010116.

Lesley JA, Shapiro L. 2008. SpoT regulates DnaA stability and initiation of DNA replication in carbon-starved *Caulobacter crescentus*. *J Bacteriol* **190:** 6867–6880.

Lobner-Olesen A. 1999. Distribution of minichromosomes in individual *Escherichia coli* cells: Implications for replication control. *EMBO J* **18:** 1712–1721.

Lu M, Campbell JL, Boye E, Kleckner N. 1994. SeqA: A negative modulator of replication initiation in *E. coli*. *Cell* **77:** 413–426.

* MacAlpine DM, Almouzni G. 2013. Chromatin and DNA replication. *Cold Spring Harb Perspect Biol* doi: 10.1101/cshperspect.a010207.

Maciag M, Kochanowska M, Lyzen R, Wegrzyn G, Szalewska-Palasz A. 2010. ppGpp inhibits the activity of *Escherichia coli* DnaG primase. *Plasmid* **63:** 61–67.

Marczynski GT, Shapiro L. 2002. Control of chromosome replication in *Caulobacter crescentus*. *Annu Rev Microbiol* **56:** 625–656.

Margulies C, Kaguni JM. 1998. The FIS protein fails to block the binding of DnaA protein to *oriC*, the *Escherichia coli* chromosomal origin. *Nucleic Acids Res* **26:** 5170–5175.

McGarry KC, Ryan VT, Grimwade JE, Leonard AC. 2004. Two discriminatory binding sites in the *Escherichia coli* replication origin are required for DNA strand opening by initiator DnaA-ATP. *Proc Natl Acad Sci* **101:** 2811–2816.

McGrath PT, Iniesta AA, Ryan KR, Shapiro L, McAdams HH. 2006. A dynamically localized protease complex and a polar specificity factor control a cell cycle master regulator. *Cell* **124:** 535–547.

Messer W. 2002. The bacterial replication initiator DnaA. DnaA and *oriC*, the bacterial mode to initiate DNA replication. *FEMS Microbiol Rev* **26:** 355–374.

Messer W, Weigel C. 1997. DnaA initiator—also a transcription factor. *Mol Microbiol* **24:** 1–6.

Molina F, Skarstad K. 2004. Replication fork and SeqA focus distributions in *Escherichia coli* suggest a replication hyperstructure dependent on nucleotide metabolism. *Mol Microbiol* **52:** 1597–1612.

Morigen, Lobner-Olesen A, Skarstad K. 2003. Titration of the *Escherichia coli* DnaA protein to excess *datA* sites causes destabilization of replication forks, delayed replication initiation and delayed cell division. *Mol Microbiol* **50:** 349–362.

Morigen, Molina F, Skarstad K. 2005. Deletion of the *datA* site does not affect once-per-cell-cycle timing but induces rifampin-resistant replication. *J Bacteriol* **187:** 3913–3920.

Morigen, Odsbu I, Skarstad K. 2009. Growth rate dependent numbers of SeqA structures organize the multiple replication forks in rapidly growing *Escherichia coli*. *Genes Cells* **14:** 643–657.

Moriya S, Atlung T, Hansen FG, Yoshikawa H, Ogasawara N. 1992. Cloning of an autonomously replicating sequence (ars) from the *Bacillus subtilis* chromosome. *Mol Microbiol* **6:** 309–315.

Moriya S, Firshein W, Yoshikawa H, Ogasawara N. 1994. Replication of a *Bacillus subtilis oriC* plasmid *in vitro*. *Mol Microbiol* **12:** 469–478.

Murray H, Errington J. 2008. Dynamic control of the DNA replication initiation protein DnaA by Soj/ParA. *Cell* **135:** 74–84.

Nakamura K, Katayama T. 2010. Novel essential residues of Hda for interaction with DnaA in the regulatory inactivation of DnaA: Unique roles for Hda AAA Box VI and VII motifs. *Mol Microbiol* **76:** 302–317.

Natrajan G, Hall DR, Thompson AC, Gutsche I, Terradot L. 2007. Structural similarity between the DnaA-binding proteins HobA (HP1230) from *Helicobacter pylori* and DiaA from Escherichia coli. *Mol Microbiol* **65:** 995–1005.

Natrajan G, Noirot-Gros MF, Zawilak-Pawlik A, Kapp U, Terradot L. 2009. The structure of a DnaA/HobA complex from Helicobacter pylori provides insight into regulation of DNA replication in bacteria. *Proc Natl Acad Sci* **106:** 21115–21120.

Neuwald AF, Aravid L, Spouge JL, Koonin EV. 1999. AAA$^+$: A class of chaperon-like ATPases associated with the assembly, operation, and disassembly of protein complexes. *Genome Res* **9:** 27–43.

Nievera C, Torgue JJ, Grimwade JE, Leonard AC. 2006. SeqA blocking of DnaA-*oriC* interactions ensures staged assembly of the *E. coli* pre-RC. *Mol Cell* **24:** 581–592.

Nishida S, Fujimitsu K, Sekimizu K, Ohmura T, Ueda T, Katayama T. 2002. A nucleotide switch in the *Escherichia coli* DnaA protein initiates chromosomal replication: Evidence from a mutant DnaA protein defective in regulatory ATP hydrolysis in vitro and in vivo. *J Biol Chem* **277:** 14986–14995.

Noirot-Gros MF, Velten M, Yoshimura M, McGovern S, Morimoto T, Ehrlich SD, Ogasawara N, Polard P, Noirot P. 2006. Functional dissection of YabA, a negative regulator of DNA replication initiation in *Bacillus subtilis*. *Proc Natl Acad Sci* **103:** 2368–2373.

Odsbu I, Klungsoyr HK, Fossum S, Skarstad K. 2005. Specific N-terminal interactions of the *Escherichia coli* SeqA protein are required to form multimers that restrain negative supercoils and form foci. *Genes Cells* **10:** 1039–1049.

Odsbu I, Morigen, Skarstad K. 2009. A reduction in ribonucleotide reductase activity slows down the chromosome replication fork but does not change its localization. *PLoS ONE* **4:** e7617.

Ogawa T, Okazaki T. 1994. Cell cycle-dependent transcription from the *gid* and *mioC* promoters of *Escherichia coli*. *J Bacteriol* **176:** 1609–1615.

Ogawa T, Yamada Y, Kuroda T, Kishi T, Moriya S. 2002. The *datA* locus predominantly contributes to the initiator titration mechanism in the control of replication initiation in *Escherichia coli*. *Mol Microbiol* **44:** 1367–1375.

Ogura Y, Imai Y, Ogasawara N, Moriya S. 2001. Autoregulation of the *dnaA-dnaN* operon and effects of DnaA protein levels on replication initiation in *Bacillus subtilis*. *J Bacteriol* **183:** 3833–3841.

Ogura Y, Ogasawara N, Harry EJ, Moriya S. 2003. Increasing the ratio of Soj to Spo0J promotes replication initiation in *Bacillus subtilis*. *J Bacteriol* **185:** 6316–6324.

Ogura T, Whiteheart SW, Wilkinson AJ. 2004. Conserved arginine residues implicated in ATP hydrolysis, nucleotide-sensing, and inter-subunit interactions in AAA and AAA$^+$ ATPases. *J Struct Biol* **146:** 106–112.

Okumura H, Yoshimura M, Ueki M, Oshima T, Ogasawara N, Ishikawa S. 2012. Regulation of chromosomal replication initiation by *oriC*-proximal DnaA-box clusters in *Bacillus subtilis*. *Nucleic Acids Res* **40:** 220–234.

Ozaki S, Katayama T. 2009. DnaA structure, function, and dynamics in the initiation at the chromosomal origin. *Plasmid* **62:** 71–82.

Ozaki S, Katayama T. 2012. Highly organized DnaA-*oriC* complexes recruit the single-stranded DNA for replication initiation. *Nucleic Acids Res* **40:** 1648–1665.

Ozaki S, Kawakami H, Nakamura K, Fujikawa N, Kagawa W, Park SY, Yokoyama S, Kurumizaka H, Katayama T. 2008. A common mechanism for the ATP-DnaA-dependent formation of open complexes at the replication origin. *J Biol Chem* **283:** 8351–8362.

Quon KC, Yang B, Domian IJ, Shapiro L, Marczynski GT. 1998. Negative control of bacterial DNA replication by a cell cycle regulatory protein that binds at the chromosome origin. *Proc Natl Acad Sci* **95:** 120–125.

Rahn-Lee L, Merrikh H, Grossman AD, Losick R. 2011. The sporulation protein SirA inhibits the binding of DnaA to the origin of replication by contacting a patch of clustered amino acids. *J Bacteriol* **193:** 1302–1307.

Riber L, Lobner-Olesen A. 2005. Coordinated replication and sequestration of *oriC* and *dnaA* are required for maintaining controlled once-per-cell-cycle initiation in *Escherichia coli. J Bacteriol* **187:** 5605–5613.

Riber L, Fujimitsu K, Katayama T, Lobner-Olesen A. 2009. Loss of Hda activity stimulates replication initiation from I-box, but not R4 mutant origins in *Escherichia coli. Mol Microbiol* **71:** 107–122.

Roth A, Messer W. 1998. High-affinity binding sites for the initiator protein DnaA on the chromosome of *Escherichia coli. Mol Microbiol* **28:** 395–401.

Ryan VT, Grimwade JE, Nievera CJ, Leonard AC. 2002. IHF and HU stimulate assembly of pre-replication complexes at *Escherichia coli oriC* by two different mechanisms. *Mol Microbiol* **46:** 113–124.

Ryan VT, Grimwade JE, Camara JE, Crooke E, Leonard AC. 2004. *Escherichia coli* prereplication complex assembly is regulated by dynamic interplay among Fis, IHF and DnaA. *Mol Microbiol* **51:** 1347–1359.

Scholefield G, Whiting R, Errington J, Murray H. 2011. Spo0J regulates the oligomeric state of Soj to trigger its switch from an activator to an inhibitor of DNA replication initiation. *Mol Microbiol* **79:** 1089–1100.

Seitz H, Weigel C, Messer W. 2000. The interaction domains of the DnaA and DnaB replication proteins of *Escherichia coli. Mol Microbiol* **37:** 1270–1279.

Sekimizu K, Kornberg A. 1988. Cardiolipin activation of *dnaA* protein, the initiation protein of replication in *Escherichia coli. J Biol Chem* **263:** 7131–7135.

Sekimizu K, Yung BY-M, Kornberg A. 1988. The *dnaA* protein of *Escherichia coli*: Abundance, improved purification, and membrane binding. *J Biol Chem* **263:** 7136–7140.

Skarstad K, Lobner-Olesen A. 2003. Stable co-existence of separate replicons in *Escherichia coli* is dependent on once-per-cell-cycle initiation. *EMBO J* **22:** 140–150.

Skarstad K, Boye E, Steen HB. 1986. Timing of initiation of chromosome replication in individual *Escherichia coli* cells. *EMBO J* **5:** 1711–1717.

Skarstad K, Lobner-Olesen A, Atlung T, von Meyenburg K, Boye E. 1989. Initiation of DNA replication in *Escherichia coli* after overproduction of the DnaA protein. *Mol Gen Genet* **218:** 50–56.

Skarstad K, Baker TA, Kornberg A. 1990. Strand separation required for initiation of replication at the chromosomal origin of *E.coli* is facilitated by a distant RNA–DNA hybrid. *EMBO J* **9:** 2341–2348.

Slater S, Wold S, Lu M, Boye E, Skarstad K, Kleckner N. 1995. *E coli* SeqA protein binds *oriC* in two different methyl-modulated reactions appropriate to its roles in DNA replication initiation and origin sequestration. *Cell* **82:** 927–936.

Smulczyk-Krawczyszyn A, Jakimowicz D, Ruban-Osmialowska B, Zawilak-Pawlik A, Majka J, Chater K, Zakrzewska-Czerwinska J. 2006. Cluster of DnaA boxes involved in regulation of *Streptomyces* chromosome replication: From *in silico* to *in vivo* studies. *J Bacteriol* **188:** 6184–6194.

Soufo CD, Soufo HJ, Noirot-Gros MF, Steindorf A, Noirot P, Graumann PL. 2008. Cell-cycle-dependent spatial sequestration of the DnaA replication initiator protein in *Bacillus subtilis. Dev Cell* **15:** 935–941.

Speck C, Messer W. 2001. Mechanism of origin unwinding: Sequential binding of DnaA to double- and single-stranded DNA. *EMBO J* **20:** 1469–1476.

Srivatsan A, Wang JD. 2008. Control of bacterial transcription, translation and replication by (p)ppGpp. *Curr Opin Microbiol* **11:** 100–105.

Stokke C, Flåtten I, Skarstad K. 2012. An easy-to-use simulation program demonstrates variations in bacterial cell cycle parameters depending on medium and temperature. *PloS ONE* **7:** e30981.

Su'etsugu M, Emoto A, Fujimitsu K, Keyamura K, Katayama T. 2003. Transcriptional control for initiation of chromosomal replication in *Escherichia coli*: Fluctuation of the level of origin transcription ensures timely initiation. *Genes Cells* **8:** 731–745.

Su'etsugu M, Takata M, Kubota T, Matsuda Y, Katayama T. 2004. Molecular mechanism of DNA replication-coupled inactivation of the initiator protein in *Escherichia coli*: Interaction of DnaA with the sliding clamp-loaded DNA and the sliding clamp-Hda complex. *Genes Cells* **9:** 509–522.

Su'etsugu M, Shimuta TR, Ishida T, Kawakami H, Katayama T. 2005. Protein associations in DnaA-ATP hydrolysis mediated by the Hda-replicase clamp complex. *J Biol Chem* **280:** 6528–6536.

Su'etsugu M, Nakamura K, Keyamura K, Kudo Y, Katayama T. 2008. Hda monomerization by ADP binding promotes replicase clamp-mediated DnaA-ATP hydrolysis. *J Biol Chem* **283:** 36118–36131.

Sutton MD, Carr KM, Vicente M, Kaguni JM. 1998. *Escherichia coli* DnaA protein. The N-terminal domain and loading of DnaB helicase at the *E. coli* chromosomal origin. *J Biol Chem* **273:** 34255–34262.

Terradot L, Zawilak-Pawlik A. 2010. Structural insight into *Helicobacter pylori* DNA replication initiation. *Gut Microbes* **1:** 330–334.

Theisen PW, Grimwade JE, Leonard AC, Bogan JA, Helmstetter CE. 1993. Correlation of gene transcription with the time of initiation of chromosome replication in *Escherichia coli. Mol Microbiol* **10:** 575–584.

Torheim NK, Skarstad K. 1999. *Escherichia coli* SeqA protein affects DNA topology and inhibits open complex formatiom at *oriC. EMBO J* **18:** 4882–4888.

Torheim NK, Boye E, Lobner-Olesen A, Stokke T, Skarstad K. 2000. The *Escherichia coli* SeqA protein destabilizes mutant DnaA204 protein. *Mol Microbiol* **37:** 629–638.

Wagner JK, Marquis KA, Rudner DZ. 2009. SirA enforces diploidy by inhibiting the replication initiator DnaA during spore formation in *Bacillus subtilis. Mol Microbiol* **73:** 963–974.

Waldminghaus T, Skarstad K. 2009. The *Escherichia coli* SeqA protein. *Plasmid* **61:** 141–150.

Waldminghaus T, Weigel C, Skarstad K. 2012. Replication fork movement and methylation govern SeqA binding to the *Escherichia coli* chromosome. *Nucleic Acids Res* **40:** 5465–5476.

Wang JD, Sanders GM, Grossman AD. 2007. Nutritional control of elongation of DNA replication by (p)ppGpp. *Cell* **128:** 865–875.

Wold S, Crooke E, Skarstad K. 1996. The *Escherichia coli* Fis protein prevents initiation of DNA replication from *oriC in vitro. Nucleic Acids Res* **24:** 3527–3532.

Xia W, Dowhan W. 1995. *In vivo* evidence for the involvement of anionic phospholipid in initiation of DNA replication in *Escherichia coli. Proc Natl Acad Sci* **92:** 783–787.

Xu Q, McMullan D, Abdubek P, Astakhova T, Carlton D, Chen C, Chiu HJ, Clayton T, Das D, Deller MC, et al. 2009. A structural basis for the regulatory inactivation of DnaA. *J Mol Biol* **385:** 368–380.

Yamazoe M, Adachi S, Kanaya S, Ohsumi K, Hiraga S. 2005. Sequential binding of SeqA protein to nascent DNA segments at replication forks in synchronized cultures of *Escherichia coli. Mol Microbiol* **55:** 289–298.

Yuzhakov A, Turner J, O'Donnell M. 1996. Replisome assembly reveals the basis for asymmetric function in leading and lagging strand replication. *Cell* **86:** 877–886.

Zakrzewska-Czerwinska J, Jakimowicz D, Zawilak-Pawlik A, Messer W. 2007. Regulation of the initiation of chromosomal replication in bacteria. *FEMS Microbiol Rev* **31:** 378–387.

Zawilak-Pawlik A, Kois A, Stingl K, Boneca IG, Skrobuk P, Piotr J, Lurz R, Zakrzewska-Czerwinska J, Labigne A. 2007. HobA—a novel protein involved in initiation of chromosomal replication in *Helicobacter pylori. Mol Microbiol* **65:** 979–994.

Zawilak-Pawlik A, Donczew R, Szafranski S, Mackiewicz P, Terradot L, Zakrzewska-Czerwinska J. 2011. DiaA/HobA and DnaA: A pair of proteins co-evolved to cooperate during bacterial orisome assembly. *J Mol Biol* **408:** 238–251.

* Zielke N, Edgar BA, DePamphilis ML. 2013. Endoreplication. *Cold Spring Harb Perspect Biol* **5:** a012948.

Zyskind JW, Smith DW. 1992. DNA replication, the bacterial cell cycle, and cell growth. *Cell* **69:** 5–8.

Regulating DNA Replication in Eukarya

Khalid Siddiqui, Kin Fan On, and John F.X. Diffley

Cancer Research UK, London Research Institute, Clare Hall Laboratories, South Mimms, Herts EN6 3LD, United Kingdom

Correspondence: John.Diffley@cancer.org.uk

DNA replication is tightly controlled in eukaryotic cells to ensure that an exact copy of the genetic material is inherited by both daughter cells. Oscillating waves of cyclin-dependent kinase (CDK) and anaphase-promoting complex/cyclosome (APC/C) activities provide a binary switch that permits the replication of each chromosome exactly once per cell cycle. Work from several organisms has revealed a conserved strategy whereby inactive replication complexes are assembled onto DNA during periods of low CDK and high APC activity but are competent to execute genome duplication only when these activities are reversed. Periods of high CDK and low APC/C serve an essential function by blocking reassembly of replication complexes, thereby preventing rereplication. Higher eukaryotes have evolved additional CDK-independent mechanisms for preventing rereplication.

The *Eukarya* include a wide spectrum of organisms, with genome sizes ranging from $\sim 10^7$ bp in yeasts to $\sim 10^{12}$ bp in protozoa. Rapid duplication of large genomes is achieved by distribution of the genetic material across several chromosomes. Each of these chromosomes initiates replication from sites called replication origins, which must fire no more than once per cell cycle to ensure a single error-free copy of the genome. Generating replication forks from an origin more than once leads to rereplication, an event that creates multiple copies of a single genomic region within a single cell. This leads to gene amplification and promotes genome instability (Green et al. 2010), a phenomenon observed in many human cancers (Lengauer et al. 1998). The process of genome duplication is therefore under stringent control to ensure that few, if any, defects are transmitted from one generation to the next.

GENERAL STRATEGY FOR INITIATION OF EUKARYOTIC DNA REPLICATION

Origin Licensing: Loading of the Replicative Helicase

Eukaryotic cells initiate DNA replication in two discrete steps. First, an inactive form of the replicative helicase is assembled onto double-stranded DNA (dsDNA) in a process called origin licensing. This occurs during late mitosis and G_1 phase of the cell cycle. The six-subunit origin-recognition complex (ORC) binds to DNA sequences called origins of replication and recruits the Cdc6 and Cdt1 proteins. Together these three **licensing factors** direct the loading of the helicase, the minichromosome maintenance (MCM) complex, around dsDNA. The MCM complex thus loaded is topologically linked to DNA and forms a double hexamer (Donovan et al. 1997; Rowles et al. 1999; Seki

Cite this article as *Cold Spring Harb Perspect Biol* doi: 10.1101/cshperspect.a012930

and Diffley 2000; Evrin et al. 2009; Remus et al. 2009; Gambus et al. 2011). This form of the inactive helicase is also referred to as the pre-replicative complex (pre-RC).

Origin Firing: Activation of the Replicative Helicase

During S phase, the inactive pre-RC is converted into an active helicase that unwinds dsDNA, thus allowing DNA polymerases to access and copy the two template strands. This second step of origin firing involves the formation of the CMG complex, named after its components: Cdc45, the MCM proteins, and the GINS complex (Moyer et al. 2006; Aparicio et al. 2009). The active CMG helicase is then coupled to a DNA polymerase, either Pol ε for the leading strand or Pol δ for the lagging strand (Kunkel and Burgers 2008). This process requires the activity of the Sld2, Sld3, Sld7, and Dpb11 proteins as well as the two protein kinases cyclin-dependent kinase (CDK) and Dbf4-dependent kinase (DDK) (Bousset and Diffley 1998; Kamimura et al. 1998; Zou and Stillman 1998, 2000; Kamimura et al. 2001; Masumoto et al. 2002; Tanaka et al. 2007; Zegerman and Diffley 2007; Tanaka et al. 2011b). These six **firing factors** are essential for initiating DNA synthesis from licensed origins.

CELL-CYCLE CONTROL OF THE INITIATION OF DNA REPLICATION

Control of Initiation of Replication during an Unperturbed Cell Cycle

The two steps of initiation described above are isolated from each other in different stages of the cell cycle. No origin firing can be allowed in G_1 while pre-RC complexes are assembled, lest there be regions of the genome that have not been properly licensed. Conversely, no origin licensing can be permitted during S phase while origin activation is triggered (Fig. 1). This ensures that multiple replication forks do not initiate from the same origin, thus preventing re-replication. Insulation of these two steps is achieved by the concerted action of two enzyme complexes: the CDK and the anaphase-promoting complex/cyclosome (APC/C).

Eukaryotes express different cyclin proteins during different stages of the cell cycle, leading to cell cycle stage-specific cyclin–CDK complexes

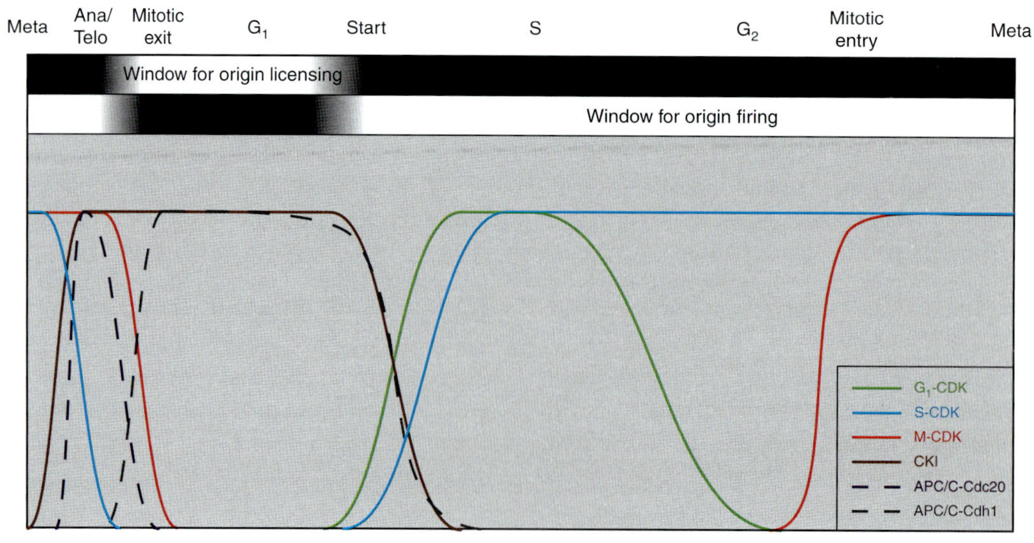

Figure 1. Regulation of DNA replication in the cell cycle. A generic diagram summarizing the oscillation of CDK activity in the cell cycle in response to the fluctuation of APC/C activity and the presence of CDK inhibitor (CKI). The details of regulation in different organisms are described in the corresponding sections of the article.

 Cite this article as *Cold Spring Harb Perspect Biol* doi: 10.1101/cshperspect.a012930

(Table 1). The cyclin subunit of the CDK contributes to determining substrate specificity, thus resulting in cell cycle stage-dependent phosphorylation of different target proteins. G_1-phase cyclin–CDKs (G_1-CDKs) phosphorylate proteins to promote S-phase entry, S-phase cyclin–CDKs (S-CDKs) are required to activate DNA replication, and mitotic cyclin–CDKs (M-CDKs) regulate accurate chromosome segregation through mitosis. Although different cyclins can confer some substrate specificity, experiments in fission yeast have shown that a single cyclin–CDK fusion can support a near-normal cell cycle (Coudreuse and Nurse 2010).

The APC/C is a multisubunit E3 ubiquitin ligase that polyubiquitinates different proteins targeted to it by a substrate adaptor (e.g., Cdc20 or Cdh1) (Peters 2006). The resulting protein–ubiquitin conjugates are then degraded by the proteasome. The S-phase and G_2/M-phase cyclins are targeted for APC/C-mediated degradation. G_1-phase cyclins are resistant to such regulation. Instead, they are processed for degradation during S phase by a different E3 ligase, the Skp1–Cul1–F-box protein (SCF) complex (Ang and Wade Harper 2005).

M-CDK-dependent phosphorylation of the APC/C results in activation of this E3 ligase (Rudner and Murray 2000; Kraft et al. 2003). This results in ubiquitination of S-phase and G_2/M-phase cyclins, leading to their degradation and a subsequent loss of associated S-CDK and M-CDK activity. Mitosis ensues and is followed by a period of high APC/C-Cdh1 activity that defines the G_1 phase in cycling cells and the G_0 phase in differentiated/quiescent cells. In actively proliferating cells, growth factor stimulation induces transcription of G_1-phase cyclins, resulting in an accumulation of G_1-CDK activity. Phosphorylation of the APC/C adaptor Cdh1 by G_1-CDK prevents it from binding to and activating the APC/C, effectively inhibiting its function and allowing S-CDK activity to accumulate (Zachariae et al. 1998; Jaspersen et al. 1999). S-CDK then inhibits and switches off the APC/C during S and G_2 phases. Hence, CDK and APC/C enzymes regulate each other, and their peak activity times are mutually exclusive in the cell cycle, setting up biphasic oscillations (Fig. 1). Vertebrate cells have additionally evolved proteins that inhibit APC/C function during S phase, such as Emi1 and Emi2 (Reimann et al. 2001; Tung et al. 2005). During prometaphase, Cdc2 and Plk1 cooperate to phosphorylate these inhibitors, presenting them for $SCF^{\beta\text{-TrCP}}$-mediated degradation, thus relieving the inhibition of the APC/C and promoting mitosis (Nakayama and Nakayama 2006).

Origin licensing occurs exclusively during late mitosis and G_1, when APC/C activity is high and S-CDK activity is low. S-CDK phosphorylation inhibits pre-RC assembly during S, G_2, and M phases, and thus all origins must be licensed before cells can enter S phase, because high CDK activity in the rest of the cell cycle would prevent any further licensing. If large regions of a chromosome are left unlicensed, it is possible that adjacent replication forks are unable to travel far enough to fuse into each other, leaving unreplicated stretches of DNA in between (Blow et al. 2011). Hence, several origins are licensed but only subsets are used to generate replication forks in every cell cycle, leaving clusters of dormant origins as a backup system. This ensures that every last base of the parental DNA gets copied. Pre-RCs that do not fire normally get displaced by passing replication forks, thus marking chromosome regions that have already been duplicated. Because many more origins need to be licensed than are used per cell divi-

Table 1. Regulators of cyclin-dependent kinase (CDK) in different organisms

	Saccharomyces cerevisiae	*Schizosaccharomyces pombe*	Metazoans
G_1-CDK	Cln1, Cln2, Cln3-Cdc28	Cig1, Puc1-Cdc2	Cyclin D–CDK4, cyclin E–CDK2
S-CDK	Clb5, Clb6-Cdc28	Cig2-Cdc2	Cyclin A–CDK2
M-CDK	Clb1, Clb2, Clb3, Clb4-Cdc28	Cig2, Cdc13-Cdc2	Cyclin A–CDK1, cyclin B–CDK1
CKI	Sic1	Rum1	p21, p27

CKI, CDK inhibitor.

sion, it is proposed that cells are prevented from entering S phase until complete assembly of replication complexes (Shreeram et al. 2002) and that conditions that decrease the total number of active replication complexes may result in genome instability (Kawabata et al. 2011).

Control of DNA Replication on Exposure to Genotoxic Stress

Cells have evolved checkpoint pathways to respond to external genotoxic insults by arresting the cell cycle, so that the cell can pause replication and correct mutations and lesions rather than segregating chromosomes with errors to daughter cells (Harper and Elledge 2007). The S-phase checkpoint acts to prevent licensed origins from firing and also stabilizes ongoing replication forks so that polymerases may resume activity once the damage has been corrected; these details are discussed in a separate article.

REGULATION OF REPLICATION INITIATION IN *Saccharomyces cerevisiae*

Cell-Cycle Regulation of Pre-RC Assembly

On exit from mitosis, CDK activity is lowered in two ways: by ubiquitin-mediated degradation of mitotic cyclin Clb2 by the 26S proteasome (Schwab et al. 1997; Visintin et al. 1997) and by Sic1 inhibition of G_1-CDK activity (Nugroho and Mendenhall 1994; Schwob et al. 1994). During this period, the Cdc14 phosphatase also promotes pre-RC assembly. First, it dephosphorylates Cdh1, thereby promoting its association with APC/C (Visintin et al. 1998; Zachariae et al. 1998; Jaspersen et al. 1999). Cdc14 also dephosphorylates the transcription factor Swi5, promoting its nuclear localization to activate transcription of Sic1 and Cdc6 (Knapp et al. 1996; Visintin et al. 1998). Finally, Cdc14 dephosphorylates Sic1, stabilizing it from SCF^{Cdc4}-mediated degradation (Visintin et al. 1998). Cdc6 also acts as an M-CDK inhibitor by direct binding to Clb2 (Elsasser et al. 1996; Calzada et al. 2001).

During G_1 phase a cell begins the next complete round of cell cycle on reaching a sufficiently large cell size (Skotheim et al. 2008). Passage through the point of no return, or "Start," is initiated by the G_1-CDK Cln-Cdc28 (Tyers et al. 1993; Dirick et al. 1995; Stuart and Wittenberg 1995). Two transcription factor complexes, SBF (Swi4–Swi6) and MBF (Mbp1–Swi6), are activated by phosphorylation of their allosteric inhibitor Whi5 (Costanzo et al. 2004; de Bruin et al. 2004). Whereas SBF activates transcription of Cln cyclins (Nasmyth and Dirick 1991; Spellman et al. 1998; Eser et al. 2011), MBF promotes transcription of Clb5 along with other replication genes (Lowndes et al. 1992; Koch et al. 1993; Spellman et al. 1998; Eser et al. 2011).

A key barrier to origin firing during G_1 phase is the CDK inhibitor Sic1, which must be degraded before cells can initiate DNA synthesis (Donovan et al. 1994; Schwob et al. 1994; Schneider et al. 1996). Simultaneous phosphorylation on multiple CDK consensus sites by Cln-Cdc28 and Clb-Cdc28 targets Sic1 for SCF-mediated polyubiquitination and proteolysis (Feldman et al. 1997; Verma et al. 1997a,b; Koivomagi et al. 2011a).

Another barrier to origin firing in G_1 phase is APC/C-Cdh1 activity, which actively degrades Clb cyclins. During late G_1 phase Cln2-Cdc28 and Clb5-Cdc28 phosphorylate Cdh1 and prevent its association with APC/C (Zachariae et al. 1998; Jaspersen et al. 1999). This allows accumulation of Clb5-Cdc28 activity essential for origin firing.

CDK Control of Origin Licensing

CDK phosphorylates several initiation proteins to inhibit pre-RC assembly. Clb-Cdc28 is recruited to ORC in an RxL-dependent manner (Weinreich et al. 2001; Wilmes et al. 2004) and phosphorylates Orc2 and Orc6 (Nguyen et al. 2001). These mechanisms inhibit interaction between ORC and Cdt1 (Chen et al. 2007), impeding loading of MCM complexes onto DNA (Chen and Bell 2011).

Cln-Cdc28 and Clb-Cdc28 phosphorylate Cdc6 to promote its subsequent degradation by the SCF^{Cdc4} complex (Drury et al. 1997, 2000; Elsasser et al. 1999; Perkins et al. 2001). During mitosis, binding of Clb2 to phosphorylated Cdc6 not only protects itself from SCF-mediated degradation but also prevents Cdc6 from interacting with ORC, rendering Cdc6 inactive for pre-RC assembly (Mimura et al. 2004). On mitotic exit, degradation of Clb2 by the APC/C releases Cdc6 to promote pre-RC assembly in the subsequent G_1 phase.

As a result of a stable interaction with the MCM complex, Cdt1 and MCM are regulated as a single unit in budding yeasts (Tanaka and Diffley 2002). Phosphorylation of CDK consensus sites in Mcm3 results in soluble Cdt1/Mcm2–7 being transported out of the nucleus (Labib et al. 1999; Nguyen et al. 2000; Liku et al. 2005), where they are retained during G_2. Interestingly, every cell cycle requires a new round of MCM gene transcription, and "old" MCM proteins are degraded by ubiquitin-mediated proteolysis at the end of mitosis (Cheng et al. 2002; Braun and Breeden 2007).

CDK phosphorylation of pre-RC components is the primary barrier to rereplication in budding yeasts. The inhibition of ORC, Cdc6, and Cdt1 creates redundant pathways that must be simultaneously deregulated to result in significant overreplication of DNA in a single cell (Nguyen et al. 2001). Yeast strains that are mutated in either one or two of these modules show undetectable or mild rereplication phenotypes (Green et al. 2006), suggesting that CDK-mediated prevention mechanisms work together to ensure that rereplication becomes an extremely rare event (Diffley 2011).

CDK Control of Origin Firing

Phosphorylation of Sld2 and Sld3 by S-CDK promotes interaction of these proteins with the amino-terminal and carboxy-terminal BRCT domains of Dpb11, respectively (Tanaka et al. 2007; Zegerman and Diffley 2007; Muramatsu et al. 2010). Cln-Cdc28 cannot phosphorylate Sld2 and Sld3 during G_1 phase, possibly because of substrate specificity conferred by cyclin (Koivomagi et al. 2011b).

Yeast strains expressing an Sld3–Dpb11 fusion protein in combination with a phosphomimetic mutant of Sld2 can promote CDK-independent DNA synthesis in G_1-arrested cells, thereby bypassing the requirement for S-CDKs (Zegerman and Diffley 2007). This indicates that Sld2 and Sld3 are the minimal set of CDK substrates required for DNA replication. Interestingly, the *Jet1* allele of Cdc45 bypasses the Sld3 phosphorylation requirement for cell survival, thus implicating Cdc45 in the interaction between Sld3 and Dpb11 (Tanaka et al. 2007).

DDK Control of Origin Firing

The first evidence that DDK promotes initiation via Mcm2–7 came from the isolation of a mutant allele of Mcm5 (*mcm5-bob1*), which bypassed the requirement for DDK (Hardy et al. 1997). Purified DDK phosphorylates Mcm2, Mcm4, and Mcm6 subunits within double hexamers bound to DNA and has weak or no activity toward subunits within soluble hexamers (Francis et al. 2009). DDK phosphorylation of an amino-terminal region in Mcm4 facilitates Mcm–Cdc45 complex formation during S phase (Sheu and Stillman 2006). Mutational analysis of Mcm4 reveals that the unstructured amino terminus of this protein contains an inhibitory activity that is alleviated on DDK phosphorylation. Accordingly, deletion of this region results in an Mcm4 protein that promotes DDK-independent DNA synthesis in cells arrested at G_1 phase and rescues the viability of strains lacking functional DDK (Sheu and Stillman 2010). These data suggest that Mcm2, -4, and -6 are essential substrates for DDK in vivo. The exact mechanism by which DDK promotes origin firing is currently unclear and may involve the recruitment of firing factors such as Sld3, Sld7, and Cdc45 to origins (Tanaka et al. 2011a) perhaps via a DDK-phosphorylated MCM complex.

Activity of DDK is restricted to S phase as a result of APC/C-mediated degradation of the regulatory Dbf4 subunit (Oshiro et al. 1999; Weinreich and Stillman 1999; Ferreira et al. 2000). This has been proposed to prevent premature firing of origins during G_1 phase. Although expression of a nondegradable Dbf4 mutant does not induce significant rereplication (Ferreira et al. 2000), overproduction of Dbf4 in the CDK bypass yeast strain (see above) is lethal (Zegerman and Diffley 2007). This observation highlights the importance of regulated origin firing to prevent premature initiation events during G_1 phase.

REGULATION OF INITIATION IN *Schizosaccharomyces pombe*

Origin Licensing in Fission Yeasts

In *Schizosaccharomyces pombe* origin licensing requires the combined actions of ORC (Orp1 to -6), Cdc18, Cdt1, and the MCM complex (Hofmann and Beach 1994; Nishitani and Nurse 1995; Grallert and Nurse 1996; Maiorano et al. 1996; Muzi Falconi et al. 1996; Ogawa et al. 1999; Nishitani et al. 2000). No consensus autonomously replicating sequence has been defined for this species, and ORC binds to AT-rich sequences in the genome by virtue of nine AT-hook motifs on the amino terminus of Orp4 (Chuang and Kelly 1999; Kong and DePamphilis 2001; Lee et al. 2001; Hayashi et al. 2007). Accordingly, the binding to AT-rich sequences by Orp4 is ATP-independent (Chuang and Kelly 1999), whereas ORC in budding yeasts binds to origin DNA in the ATP-bound state (Bell and Stillman 1992; Klemm and Bell 2001). However, the evolutionary conservation of the ATP-binding sites in ORC suggests an essential role of the ATPase activity in licensing, most likely during the loading of the MCM helicase.

Origin Firing in Fission Yeasts

In fission yeasts Hsk1 activity is required for Sld3 recruitment to origins during G_1 phase (Nakajima and Masukata 2002), but Sld3 associates with origins independently of Sna41 (Yamada et al. 2004). This is followed by the sequential recruitment of Cut5, GINS, and Sna41 (Dolan et al. 2004; Yabuuchi et al. 2006). Cdc23 is also required for Sna41 association within the pre-initiation complex (pre-IC) (Gregan et al. 2003). However, the details of how the replicative helicase is activated currently remain unclear.

CDK Control of Origin Licensing

Orp2 is phosphorylated by Cdc2 in vivo, and a fission yeast strain expressing a nonphosphorylatable Orp2 protein is sensitized to rereplication (Vas et al. 2001). Overexpression of Cdc18 in such a strain results in more rereplication than that observed on overexpression in a wild-type strain. This suggests that CDK phosphorylation of Orp2 is redundant with Cdc18 regulation in rereplication control in fission yeasts. The mitotic cyclin Cdc13 localizes to replication origins in an ORC-dependent manner during G_2 phase and early mitosis. A yeast strain expressing a tagged Orp2 that reduces Cdc13 origin association shows hypersensitivity to endoreduplication, suggesting a role for Orp2–Cdc13 association in rereplication control (Wuarin et al. 2002). The mechanism behind this regulation is currently unknown, but it is possible that direct binding of Cdc13 to Orp2 somehow reduces the accessibility of other pre-RC assembly factors to ORC. This scenario is similar to the Clb5-Orc6 and Clb2-Cdc6 interactions in budding yeasts, where direct binding may play a role in inhibiting factors involved in pre-RC assembly (see above).

Cdc18 is an unstable protein, and its levels are regulated throughout the cell cycle. Cdc18 accumulation begins in late mitosis and decreases during S phase. Cdc10 controls the transcription of Cdc18, which accumulates during G_1 phase (Kelly et al. 1993; Nishitani and Nurse 1995; Muzi Falconi et al. 1996; Baum et al. 1998). On S-phase entry, Cig2-Cdc2 phosphorylates Cdc18 and targets it for polyubiquitination by the SCF complex and degradation by the proteasome (Jallepalli et al. 1997, 1998; Kominami and Toda 1997; Kominami et al. 1998; Lopez-Girona et al. 1998; Wolf et al. 1999). Overproduction of wild-type Cdc18 alone induces rereplication (Nishitani and Nurse 1995; Muzi Falconi et al. 1996), and Cdc18 mutants lacking CDK consensus sites promote rereplication even more efficiently than the wild-type proteins (Jallepalli et al. 1997; Lopez-Girona et al. 1998). Expression of either of these mutants at a low level, however, is not sufficient to induce rereplication.

Cdt1 levels peak during late M phase as a consequence of Cdc10-mediated transcriptional control, and protein levels decline during S phase as a result of proteolysis (Hofmann and

Beach 1994; Nishitani et al. 2000). Cdt1 degradation is mediated by CRL4-Cdt2 in a proliferating cell nuclear antigen (PCNA)-dependent manner (Hu and Xiong 2006; Ralph et al. 2006; Guarino et al. 2011), and chronic, low-level expression of both Cdt1 and Cdc18 is required to induce rereplication, suggesting that these two proteins are redundantly regulated (Gopalakrishnan et al. 2001).

Although fission yeast MCMs are constitutively nuclear (Yanow et al. 2001), their association with chromatin is cell-cycle-regulated. MCM complex assembly may also be regulated by the action of Mcb1 (MCM-binding protein) during the cell cycle (Ding and Forsburg 2011).

CDK Control of Origin Firing

Fission yeast Drc1 and Sld3 are phosphorylated by S-phase CDK to promote formation of a ternary complex with Cut5 (Nakajima and Masukata 2002; Noguchi et al. 2002; Fukuura et al. 2011). The interaction recruits the complex to chromatin and may generate a platform for the formation of CMG complex. An Sld7 homolog in fission yeast has not been identified, and it is not clear if pre-IC factors are limiting for replication initiation in fission yeast or if they are targeted during checkpoint-dependent inhibition of late origin firing.

REGULATION OF REPLICATION INITIATION IN METAZOANS

Current knowledge on DNA replication in metazoans is based on studies performed primarily using three model systems: *Xenopus* egg extracts, *Drosophila* embryos and cell lines, and immortalized or cancerous mammalian cell lines. Advances in RNA interference (RNAi) and transgenesis have enabled genetic studies in cell culture or whole animals to elucidate regulation of these pathways. The details of origin recognition, origin licensing, and origin firing in these systems are summarized below. Plant DNA replication is discussed in detail elsewhere and is not included here.

Regulation of Origin Recognition in Metazoans

The replicon model proposed that initiation of DNA replication is determined by the binding of initiator proteins to a specific sequence of DNA at the origin of replication, termed the replicator (Jacob and Brenner 1963). Although this model holds true for prokaryotes and certain animal viruses, replicators in eukaryotes do not share a consensus sequence. Budding yeast ORC is the only known eukaryotic initiator that displays sequence specificity and binds to an 11-bp autonomously replicating sequence element in vivo. However, in vitro, ORC from all species has intrinsic nonspecific DNA-binding activity and is capable of assembling pre-RCs onto diverse DNA sequences. Recruitment of a Gal4–ORC fusion protein to a plasmid in cultured human cells converts the sequence adjacent to the Gal4-binding sites into an origin of replication and confers on it the property of once-per-cell-cycle replication (Takeda et al. 2005b). Thus, ORC binding to DNA is a primary requirement for any sequence to function as an origin. In vivo, ORC can be recruited by different sequence-specific binding proteins to chromosomal loci, for example, to telomeres by interaction with telomeric repeat-binding factor 2 (TRF2) (Tatsumi et al. 2008) or to the chorion gene cluster in *Drosophila* by Myb (Beall et al. 2002). Such a mechanism may also be used by viral genomes, for example, Epstein–Barr virus, that can recruit ORC to *oriP* in an Epstein–Barr nuclear antigen 1 (EBNA1)-dependent process (Dhar et al. 2001). There is increasing evidence that CpG islands and G-rich elements that can form G-quadruplexes influence origin recognition in vivo, and these are discussed in greater detail in a separate article (see article by Leonard and Méchali 2013, and references therein).

ORC binds to nucleosome-free regions of DNA in vivo (Sequeira-Mendes et al. 2009; Karnani et al. 2010; MacAlpine et al. 2010), and the nature of the chromatin around chromosomal ORC-binding sites influences origin licensing. Recruitment of ORC to different regions of the genome is necessary but not sufficient for pre-RC formation in vivo. For example, ORC binding to HP1 protein has no apparent role in origin determination; rather, it has been implicated in gene silencing and heterochromatin formation in many species (Pak et al. 1997; Shareef et al. 2001; Prasanth et al. 2010).

The retinoblastoma protein interacts with ORC in human and fly cells and negatively regulates DNA replication (Bosco et al. 2001; Mendoza-Maldonado et al. 2010), likely by recruiting histone deacetylase (HDAC) activities that generate repressive chromatin marks (Brehm et al. 1998). In support of this idea, stage 10 mosaic embryos derived from flies with mutations in Rpd3/Hdac1 are capable of replication across the entire genome in follicle cells, whereas this is restricted to the chorion genes in wild type (Aggarwal and Calvi 2004). Hdac11 has also been implicated as an inhibitor of origin licensing in mammalian cells (Wong et al. 2010).

To counter these deacetylating activities, ORC and Cdt1 recruit Hbo1, a histone acetyltransferase that binds to origins and acetylates surrounding chromatin during G_1 to promote MCM loading (Miotto and Struhl 2008, 2010). Reduction of Hbo1 levels by small interfering RNA (siRNA) treatment results in defective licensing and cell-cycle arrest. Increased acetylation and open chromatin is thus a common feature of most origins in both flies and human cells, as revealed by genome-wide mapping of origins in fly and human cell lines (MacAlpine et al. 2010; Mesner et al. 2011).

The Set8 histone methyltransferase also regulates origin licensing in human cells (Tardat et al. 2010). Targeting of Gal4–Set8 fusion protein to Gal4-binding sites on plasmid DNA in cultured cells is sufficient to promote MCM loading on adjacent sequences. Set8 is a substrate of the CRL4-Cdt2-dependent degradation pathway in S phase (see below), and expression of a nondegradable Set8 protein results in rereplication in cultured human cells.

In rapidly dividing *Xenopus* embryos, DNA synthesis initiates at intervals of ~10 kb in specific clusters (Blow et al. 2001). This allows a large amount of DNA to be replicated in a relatively short phase of 20 min. Although exogenous AT-rich asymmetric sequences can outcompete replication complexes (Stanojcic et al. 2008), the nature of the DNA sequences at the initiation sites and the composition of histone modifications around the chromosomal initiation sites are unknown.

Regulation of Origin Licensing in Metazoans

ORC, Cdc6, and Cdt1 are essential for origin licensing in *Xenopus* egg extracts, *Drosophila* embryos, and cultured mammalian cells (Whittaker et al. 2000; Gillespie et al. 2001; Rialland et al. 2002; Mailand and Diffley 2005; Svitin and Chesnokov 2010; Gambus et al. 2011). Whereas fly ORC contains six subunits, similar to yeast, ORC in both vertebrates exists as a stable Orc1–Orc5 assembly, with little or no Orc6 protein associated. This five-subunit complex is functional for MCM loading and sequence-independent replication of DNA substrates in a reconstituted system (Gillespie et al. 2001; Vashee et al. 2003). The absence of stoichiometric amounts of Orc6 subunit suggests that the mechanism of MCM loading in higher eukaryotes may be different from that in budding yeasts, where Orc6–Cdt1 interactions are critical for licensing (Chen et al. 2007; Chen and Bell 2011). Despite significant homology between metazoan Orc6 proteins (Dhar and Dutta 2000), human Orc6 interacts weakly with ORC, compared with its fly counterpart (Chesnokov et al. 1999; Vashee et al. 2003; Siddiqui and Stillman 2007). However, it is still required for DNA synthesis and may have evolved to perform other roles in origin recognition (Prasanth et al. 2002; Balasov et al. 2007; Thomae et al. 2008, 2011).

The function of ORC is restricted to G_1 phase by regulated ORC–DNA interactions across the cell cycle. *Xenopus* ORC is released from chromatin on licensing (Sun et al. 2002) and is only weakly associated with chromatin later in G_2/M (Rowles et al. 1999). Mammalian ORC undergoes complex disassembly during S phase as a consequence of ubiquitination of the Orc1 subunit. Ubiquitination results in Orc1 degradation in human cells (Mendez et al. 2002; Tatsumi et al. 2003) but, interestingly, not in hamster nuclei (Natale et al. 2000; Li and DePamphilis 2002). Human Orc1 ubiquitination is mediated by the SCF^{Skp2} ubiquitin ligase, which is known to act on phosphorylated substrates. Hence, it has been suggested that Orc1 degradation in human cells may be promoted by CDK phosphorylation, representing a CDK-dependent mechanism to prevent relicensing. *Drosophila* cells uniquely regulate ORC and the APC/C ubiquitinates Orc1 for degradation after mitosis, in a situation similar to human Cdc6 (see below).

In metazoan cells ORC exhibits cell-cycle-dependent complex assembly. Human Orc4 protein requires an intact ATP-binding site for complex assembly in vitro, and this subunit does not associate stably with ORC across the cell cycle in vivo (Ranjan and Gossen 2006; Siddiqui and Stillman 2007). Immunofluorescence studies show that as cells proceed from G_1 to G_2/M, the chromatin-bound fraction of Orc2 and Orc3 decreases significantly (Prasanth et al. 2004; Siddiqui and Stillman 2007), and chromatin immunoprecipitation reveals a loss of ORC subunits at origins across the cell cycle (Gerhardt et al. 2006). This may also be promoted via CDK phosphorylation of Orc2, because a nonphosphorylatable Orc2 protein is found associated with origins during G_2/M (Lee et al. 2012). It is therefore possible that assembly of ORC itself may be one mechanism to prevent unscheduled licensing.

Both *Xenopus* and human Cdc6 are phosphorylated by S-CDK, and ectopically expressed protein, on phosphorylation, is transported out of the nucleus into the cytoplasm (Petersen et al. 1999; Coverley et al. 2000; Pelizon et al. 2000; Delmolino et al. 2001). This is postulated to be a CDK-dependent control to prevent relicensing and may be redundant with other mechanisms, as expression of a nonphosphorylatable mutant of Cdc6 alone does not result in significant rereplication. It has been reported, however, that a significant fraction of native phosphorylated Cdc6 is retained on chromatin across S and G_2/M in human cells (Coverley et al. 2000; Mendez and Stillman 2000; Alexandrow and Hamlin 2004) and may regulate entry into mitosis (Clay-Farrace et al. 2003; Lau et al. 2006). At the end of mitosis, human Cdc6 is targeted for APC/C-Cdh1-mediated degradation (Petersen et al. 2000) and degraded in early G_1 phase. The consequence of having undegraded Cdc6 in cycling human cells is, at present, unknown. However, APC/C-mediated degradation may be a way of preventing unscheduled licensing in quiescent G_0-phase cells. On cell-cycle reentry in cultured human cells, cyclin E–CDK2-mediated phosphorylation of Cdc6 blocks Cdh1 binding and stabilizes Cdc6 earlier than geminin, generating a window of opportunity to license origins (Duursma and Agami 2005; Mailand and Diffley 2005). We note that the APC/C-dependent degradation sequence in *Drosophila* Orc1 is bounded by a consensus CDK site, similar to the situation in human Cdc6. It is possible that the cyclin E-dependent association of *Drosophila* MCMs with chromatin may be via Orc1 stabilization (Su and O'Farrell 1997, 1998).

Metazoan Cdt1 is regulated by multiple pathways. Chromatin-bound Cdt1 is ubiquitinated during S phase by CRL4-Cdt2 ubiquitin ligase and targeted for degradation (Arias and Walter 2005; Li and Blow 2005). This pathway is dependent on its interaction with PCNA and is essential to prevent rereplication during S phase (Arias and Walter 2006; Senga et al. 2006).

Additionally, a second SCF^{Skp2}-dependent pathway also promotes Cdt1 degradation throughout the cell cycle (Li et al. 2003; Nishitani et al. 2006). Although Cdt1 binds cyclin A in an RxL-dependent manner and is phosphorylated at consensus CDK sites (Li et al. 2003; Sugimoto et al. 2004), mutations of these motifs do not result in significant rereplication (Takeda et al. 2005a; Nishitani et al. 2006), implying that the SCF^{Skp2}-dependent proteolysis is a minor pathway. The CRL4-Cdt2 pathway is essential for Cdt1 degradation from fission yeasts to metazoans (Jin et al. 2006; Guarino et al. 2011). However, a role for the SCF^{Skp2} in Cdt1 degradation has only been shown in human cells, and this pathway may have arisen recently in evolution (Kim and Kipreos 2007).

Cdt1 activity is also regulated by its interaction with geminin (Wohlschlegel et al. 2000; Tada et al. 2001), a protein discovered as an inhibitor of DNA replication in *Xenopus* (McGarry and Kirschner 1998). Geminin is targeted for degradation by APC/C-Cdh1 and is hence absent from late mitosis until the end of G_1, when licensing occurs (Nishitani et al. 2001). In *Drosophila* and mammalian cells, a major CDK-independent block to rereplication appears to require geminin, and reduction of protein levels by RNAi is sufficient to promote rereplication in many different cell lines (Quinn et al. 2001; Mihaylov et al. 2002; Melixetian et al. 2004; Zhu et al. 2004). Reduction of geminin levels by immunodepletion from *Xenopus* egg extracts or by injecting antisense oligonucleotides in embryos does not induce rereplication (McGarry and Kirschner 1998; McGarry 2002). However, expression of nondegradable Cdt1 promotes more rereplication in a geminin-depleted extract, suggesting that both these pathways may be important for rereplication control in this species (Kerns et al. 2007).

Similar to yeast and *Xenopus*, mammalian MCMs are loaded onto the chromatin at the end of mitosis and are removed from chromatin as the cells pass through S phase (Mendez and Stillman 2000). The MCM-BP protein is imported into the nucleus late in S phase and interacts strongly with Mcm7 and may promote disassembly of the CMG complex (Nishiyama et al. 2011). Treatment of HeLa cells with siRNAs against MCM-BP results in G_2/M-arrested cells with MCMs persisting on chromatin for longer periods of time. The interaction between MCM-BP and MCMs appears to be replication-dependent, as nuclear MCM-BP does not stimulate disassembly of MCMs in aphidicolin-arrested cells.

Regulation of Origin Firing in Metazoans

Homologs of essential firing factors are known in metazoans, and CDK-dependent activation of replication origins has been verified, with roles for cyclin E in initiation of S phase and cyclin A during the elongation phase (Strausfeld et al. 1996; Mahbubani et al. 1997; Coverley et al. 2002). Cyclin E is essential only in quiescent mouse fibroblasts that are reentering the cell cycle on growth factor stimulation (Geng et al. 2003) and may be redundant with cyclin A in most other cases (Kalaszczynska et al. 2009). DDK activity is similarly required for G_1/S transition and DNA synthesis in mammalian cells and in *Xenopus* (Strausfeld et al. 1994; Jackson et al. 1995; Jiang et al. 1999; Walter 2000; Jares et al. 2004)

Despite highly divergent sequences to the budding yeast counterparts, the vertebrate homologs of Sld2 (RecQL4) and Sld3 (Treslin/Ticrr) have been identified and are essential for replication initiation (Sangrithi et al. 2005; Kumagai et al. 2010; Sanchez-Pulido et al. 2010; Sansam et al. 2010; Boos et al. 2011). Analysis of TopBP1 (Dpb11 homolog) has revealed that it contains nine BRCT domains, referred to as BRCT0 to BRCTVIII (Makiniemi et al. 2001; Huo et al. 2010). Treslin binds to BRCTI/II domains of TopB1 in a CDK-dependent manner (Boos et al. 2011; Kumagai et al. 2011), similar to budding yeast Sld3–Dpb11 interaction. Based on homology between BRCTIII/IV of Dpb11 and BRCTIV/V of TopBP1, RecQL4 is expected to interact with TopBP1 via this domain. In contrast to the yeast proteins, the RecQL4–TopBP1 interaction is CDK-independent (Matsuno et al. 2006). Also, an amino-terminal fragment of TopBP1 containing BRCTI-III repeats is necessary and sufficient for Treslin function in *Xenopus* extracts (Kumagai et al. 2010). It is presently unclear if RecQL4 can interact with this amino-terminal fragment, perhaps via BRCTIII or other proteins that may be essential for initiation. Recently, GEMC1 and DUE-B have been identified as proteins that are phosphorylated in vivo that interact with Cdc45 and TopBP1 and have essential roles in vertebrate DNA synthesis (Balestrini et al. 2010; Chowdhury et al. 2010). It is unknown if these interactions are regulated across the cell cycle and what their specific roles are in activating the replicative helicase. RecQL4 may be a part of the active helicase in mammals, owing to its stable interaction with the CMG complex during S phase (Xu et al. 2009). These observations suggest significant differences in the regulation of yeast and metazoan DNA replication, and further work is required to elucidate these mechanisms.

The Mcm10 protein is essential during early steps of DNA synthesis. Recombinant Mcm10 interacts with single-stranded DNA (ssDNA) via an evolutionarily conserved oligonucleotide/oligosaccharide binding (OB)-fold domain. Mcm10 also recruits polymerase α–primase complex (Zhu et al. 2007), which is required to initiate de novo replication. This recruitment is dependent on the And-1/Ctf4 protein. Based on its known interaction partners, it is possible that Mcm10 either mediates the initial melting reaction or stabilizes ssDNA generated during the initial unwinding reaction. If this is the case, then the activity of Mcm10 must be regulated because it is reported to bind chromatin independently of CDK or Cdc7 activities. (Van Hatten et al. 2002; Wohlschlegel et al. 2002). Mcm10 appears to be important for CMG complex assembly (Im et al. 2009) and may couple helicase to polymerase in replisomes in a Ctf4-dependent process (Zhu et al. 2007). Interestingly, Mcm10 stabilizes interactions between CMG and RecQL4 (Xu et al. 2009), and CDK phosphorylation of Mcm10 may be required to release RecQL4 during origin firing. These data raise the possibility that Mcm10 may be partly responsible for determining the temporal order of replication timing in metazoan cells, and the mechanisms underlying regulation of these interactions will be an important focus of future studies.

CDK-Dependent Control of Replication Licensing

CDK phosphorylation of Cdc6 plays a positive role in promoting licensing during $G_0 \rightarrow G_1$ transition in mammalian cells (Mailand and Diffley 2005). Unlike in yeasts, it is unclear if CDK phosphorylation of the pre-RC proteins prevents rereplication during S phase in metazoans. Mammalian ORC, Cdc6, and Cdt1 all bind cyclin–CDKs directly and are substrates of CDKs in vitro (Saha et al. 1998; Petersen et al. 1999; Mendez et al. 2002; Sugimoto et al. 2004; Hemerly et al. 2009). However, studies using phosphorylation mutants of these proteins in vivo have not conclusively shown a role for such modifications in preventing rereplication during the cell cycle. Deletion of the amino-terminal region of Orc1 abolishes CDK phosphorylation in vitro but has no effect on ubiquitination in vivo (Mendez et al. 2002). Overexpression of Cdt1 mutants that are unable to bind cyclin or are not phosphorylatable shows more rereplication than wild-type protein (Takeda et al. 2005a), but because such mutants are degraded normally during S phase, it is likely that they promote relicensing in other cell-cycle stages (Nishitani et al. 2006). The essential function of CDKs in preventing relicensing, therefore, appears to be during the G_2/M phase of the cell cycle. Chemical inhibition of CDK activity during this period results in relicensing of chromatin even in the presence of geminin (Ballabeni et al. 2004; Vassilev et al. 2006). The essential targets of CDK for this G_2/M-specific inhibition are unknown; however, both Cdt1 and geminin are phosphorylated by CDKs in nocodazole-treated extracts and are potential candidates for this CDK-mediated regulation.

CDK-Independent Control of Replication Licensing

Metazoans have evolved CDK-independent pathways to prevent rereplication, and these are largely devoted to Cdt1 regulation. In contrast to yeast, deregulation of metazoan Cdt1 alone is sufficient to induce significant rereplication during S phase (Melixetian et al. 2004; Zhu et al. 2004; Jin et al. 2006; Lovejoy et al. 2006; Sansam et al. 2006). Consequently, Cdt1 protein levels and activity are tightly regulated. This is achieved by a combination of Cdt2-Ddb1-dependent degradation of chromatin-bound Cdt1 during S phase and interaction of Cdt1 with geminin during S and G_2/M phases. Degrading geminin during S phase by premature activation of the APC/C leads to significant rereplication, and this can be suppressed by expression of a nondegradable mutant of geminin (Di Fiore and Pines 2007; Machida and Dutta 2007), highlighting the important role of this protein in maintaining genome ploidy.

Geminin inhibits licensing by blocking the Cdt1–MCM interaction (Wohlschlegel et al. 2000; Yanagi et al. 2002). Crystal structures have suggested a cell-cycle-dependent transition between two possible conformations of Cdt1-geminin (Lee et al. 2004; De Marco et al. 2009).

A heterotrimeric (Cdt1-geminin-geminin) state exposes the surface on Cdt1 that interacts with the MCMs and thus permits licensing. This alternates with a heterohexameric [Cdt1-(geminin)$_2$]$_2$ state, in which the Cdt1–MCM interaction surface is hidden and is thus inhibitory to origin licensing. These observations are explained by a feedback model, whereby geminin functions cooperatively to inhibit licensing at multiple origins, therefore creating a critical threshold resulting in all-or-none inhibition of replication licensing (Ode et al. 2011). It is proposed that geminin and Cdt1 interact at individual origins and the ability of these two proteins to oligomerize promotes interactions between proteins at adjacent origins, thereby propagating the inhibition of licensing and clustering origins into subnuclear foci. This clustering does not affect MCM complexes that are already loaded onto origins during G_1. Based on the expression profiles of these proteins, geminin serves its essential function in preventing rereplication primarily during G_2/M, when it not only inhibits Cdt1 degradation but also releases it for licensing in the subsequent G_1 (Ballabeni et al. 2004), much like Clb2-Cdc6 in budding yeasts.

CONCLUDING REMARKS

Although CDK-dependent pathways are primarily responsible for maintaining genome stability in budding yeasts, it is clear that CDK-independent mechanisms play a critical role in maintaining genome stability in multicellular organisms. Several cyclin subunits and CDK2 can be knocked out in mice (Sherr and Roberts 2004; Malumbres and Barbacid 2009), suggesting significant redundancy among the functions of cyclin–CDK complexes. Redundant mechanisms targeting ORC, Cdc6, and Cdt1-MCM inhibit rereplication in yeasts. In contrast, it appears that some mammalian cancer cell lines are particularly sensitive to Cdt1 deregulation alone, because RNAi-mediated silencing of geminin or Cdt2 is sufficient to induce significant rereplication in some cell lines but not others (Melixetian et al. 2004; Zhu et al. 2004; Jin et al. 2006; Lovejoy et al. 2006; Sansam et al. 2006). Although deregulating either ORC or Cdc6 alone does not have a similar outcome, overexpression of these proteins can enhance the rereplication seen on Cdt1 overexpression (Vaziri et al. 2003; Sugimoto et al. 2009).

Deregulated licensing has emerged as a sensitive and early indicator of tumor development in human cancers (Freeman et al. 1999; Davies et al. 2002). Recent work has proposed that cancer cells may respond differently to licensing inhibition than primary cells (Shreeram et al. 2002; Zhu and DePamphilis 2009), and this could be exploited in designing therapies that selectively target cancer cells. The next few years will witness the identification of new replication factors (e.g., vertebrate Sld7) and lead to a better understanding as to how the divergence of proteins such as Orc6, Treslin, and RecQL4 confers unique properties to replication control in metazoan cells.

ACKNOWLEDGMENTS

We thank Anne Early for helpful comments on the manuscript. We apologize to authors whose work was not included or cited because of space constraints.

REFERENCES

*Reference is also in this collection.

Aggarwal BD, Calvi BR. 2004. Chromatin regulates origin activity in *Drosophila* follicle cells. *Nature* **430:** 372–376.

Alexandrow MG, Hamlin JL. 2004. Cdc6 chromatin affinity is unaffected by serine-54 phosphorylation, S-phase progression, and overexpression of cyclin A. *Mol Cell Biol* **24:** 1614–1627.

Ang XL, Wade Harper J. 2005. SCF-mediated protein degradation and cell cycle control. *Oncogene* **24:** 2860–2870.

Aparicio T, Guillou E, Coloma J, Montoya G, Mendez J. 2009. The human GINS complex associates with Cdc45 and MCM and is essential for DNA replication. *Nucleic Acids Res* **37:** 2087–2095.

Arias EE, Walter JC. 2005. Replication-dependent destruction of Cdt1 limits DNA replication to a single round per cell cycle in *Xenopus* egg extracts. *Genes Dev* **19:** 114–126.

Arias EE, Walter JC. 2006. PCNA functions as a molecular platform to trigger Cdt1 destruction and prevent re-replication. *Nat Cell Biol* **8:** 84–90.

Balasov M, Huijbregts RP, Chesnokov I. 2007. Role of the Orc6 protein in origin recognition complex-dependent DNA binding and replication in *Drosophila melanogaster*. *Mol Cell Biol* **27:** 3143–3153.

Balestrini A, Cosentino C, Errico A, Garner E, Costanzo V. 2010. GEMC1 is a TopBP1-interacting protein required for chromosomal DNA replication. *Nat Cell Biol* **12:** 484–491.

Ballabeni A, Melixetian M, Zamponi R, Masiero L, Marinoni F, Helin K. 2004. Human Geminin promotes pre-RC formation and DNA replication by stabilizing CDT1 in mitosis. *EMBO J* **23:** 3122–3132.

Baum B, Nishitani H, Yanow S, Nurse P. 1998. Cdc18 transcription and proteolysis couple S phase to passage through mitosis. *EMBO J* **17:** 5689–5698.

Beall EL, Manak JR, Zhou S, Bell M, Lipsick JS, Botchan MR. 2002. Role for a *Drosophila* Myb-containing protein complex in site-specific DNA replication. *Nature* **420:** 833–837.

Bell SP, Stillman B. 1992. ATP-dependent recognition of eukaryotic origins of DNA replication by a multiprotein complex. *Nature* **357:** 128–134.

Blow JJ, Gillespie PJ, Francis D, Jackson DA. 2001. Replication origins in *Xenopus* egg extract are 5–15 kilobases apart and are activated in clusters that fire at different times. *J Cell Biol* **152:** 15–25.

Blow JJ, Ge XQ, Jackson DA. 2011. How dormant origins promote complete genome replication. *Trends Biochem Sci* **36:** 405–414.

Boos D, Sanchez-Pulido L, Rappas M, Pearl LH, Oliver AW, Ponting CP, Diffley JF. 2011. Regulation of DNA replication through Sld3-Dpb11 interaction is conserved from yeast to humans. *Curr Biol* **21:** 1152–1157.

Bosco G, Du W, Orr-Weaver TL. 2001. DNA replication control through interaction of E2F-RB and the origin recognition complex. *Nat Cell Biol* **3:** 289–295.

Bousset K, Diffley JF. 1998. The Cdc7 protein kinase is required for origin firing during S phase. *Genes Dev* **12:** 480–490.

Braun KA, Breeden LL. 2007. Nascent transcription of *MCM2-7* is important for nuclear localization of the minichromosome maintenance complex in G_1. *Mol Biol Cell* **18:** 1447–1456.

Brehm A, Miska EA, McCance DJ, Reid JL, Bannister AJ, Kouzarides T. 1998. Retinoblastoma protein recruits histone deacetylase to repress transcription. *Nature* **391:** 597–601.

Calzada A, Sacristan M, Sanchez E, Bueno A. 2001. Cdc6 cooperates with Sic1 and Hct1 to inactivate mitotic cyclin-dependent kinases. *Nature* **412:** 355–358.

Chen S, Bell SP. 2011. CDK prevents Mcm2–7 helicase loading by inhibiting Cdt1 interaction with Orc6. *Genes Dev* **25:** 363–372.

Chen S, de Vries MA, Bell SP. 2007. Orc6 is required for dynamic recruitment of Cdt1 during repeated Mcm2–7 loading. *Genes Dev* **21:** 2897–2907.

Cheng IH, Roberts LA, Tye BK. 2002. Mcm3 is polyubiquitinated during mitosis before establishment of the pre-replication complex. *J Biol Chem* **277:** 41706–41714.

Chesnokov I, Gossen M, Remus D, Botchan M. 1999. Assembly of functionally active *Drosophila* origin recognition complex from recombinant proteins. *Genes Dev* **13:** 1289–1296.

Chowdhury A, Liu G, Kemp M, Chen X, Katrangi N, Myers S, Ghosh M, Yao J, Gao Y, Bubulya P, et al. 2010. The DNA unwinding element binding protein DUE-B interacts with Cdc45 in preinitiation complex formation. *Mol Cell Biol* **30:** 1495–1507.

Chuang RY, Kelly TJ. 1999. The fission yeast homologue of Orc4p binds to replication origin DNA via multiple AT-hooks. *Proc Natl Acad Sci* **96:** 2656–2661.

Clay-Farrace L, Pelizon C, Santamaria D, Pines J, Laskey RA. 2003. Human replication protein Cdc6 prevents mitosis through a checkpoint mechanism that implicates Chk1. *EMBO J* **22:** 704–712.

Costanzo M, Nishikawa JL, Tang X, Millman JS, Schub O, Breitkreuz K, Dewar D, Rupes I, Andrews B, Tyers M. 2004. CDK activity antagonizes Whi5, an inhibitor of G1/S transcription in yeast. *Cell* **117:** 899–913.

Coudreuse D, Nurse P. 2010. Driving the cell cycle with a minimal CDK control network. *Nature* **468:** 1074–1079.

Coverley D, Pelizon C, Trewick S, Laskey RA. 2000. Chromatin-bound Cdc6 persists in S and G_2 phases in human cells, while soluble Cdc6 is destroyed in a cyclin A-cdk2 dependent process. *J Cell Sci* **113:** 1929–1938.

Coverley D, Laman H, Laskey RA. 2002. Distinct roles for cyclins E and A during DNA replication complex assembly and activation. *Nat Cell Biol* **4:** 523–528.

Davies RJ, Freeman A, Morris LS, Bingham S, Dilworth S, Scott I, Laskey RA, Miller R, Coleman N. 2002. Analysis of minichromosome maintenance proteins as a novel method for detection of colorectal cancer in stool. *Lancet* **359:** 1917–1919.

de Bruin RA, McDonald WH, Kalashnikova TI, Yates J III, Wittenberg C. 2004. Cln3 activates G1-specific transcription via phosphorylation of the SBF bound repressor Whi5. *Cell* **117:** 887–898.

Delmolino LM, Saha P, Dutta A. 2001. Multiple mechanisms regulate subcellular localization of human CDC6. *J Biol Chem* **276:** 26947–26954.

De Marco V, Gillespie PJ, Li A, Karantzelis N, Christodoulou E, Klompmaker R, van Gerwen S, Fish A, Petoukhov MV, Iliou MS, et al. 2009. Quaternary structure of the human Cdt1-Geminin complex regulates DNA replication licensing. *Proc Natl Acad Sci* **106:** 19807–19812.

Dhar SK, Dutta A. 2000. Identification and characterization of the human ORC6 homolog. *J Biol Chem* **275:** 34983–34988.

Dhar SK, Yoshida K, Machida Y, Khaira P, Chaudhuri B, Wohlschlegel JA, Leffak M, Yates J, Dutta A. 2001. Replication from oriP of Epstein-Barr virus requires human ORC and is inhibited by geminin. *Cell* **106:** 287–296.

Diffley JF. 2011. Quality control in the initiation of eukaryotic DNA replication. *Philos Trans R Soc Lond B Biol Sci* **366:** 3545–3553.

Di Fiore B, Pines J. 2007. Emi1 is needed to couple DNA replication with mitosis but does not regulate activation of the mitotic APC/C. *J Cell Biol* **177:** 425–437.

Ding L, Forsburg SL. 2011. *Schizosaccharomyces pombe* minichromosome maintenance-binding protein (MCM-BP) antagonizes MCM helicase. *J Biol Chem* **286:** 32918–32930.

Dirick L, Bohm T, Nasmyth K. 1995. Roles and regulation of Cln–Cdc28 kinases at the start of the cell cycle of *Saccharomyces cerevisiae*. *EMBO J* **14:** 4803–4813.

Dolan WP, Sherman DA, Forsburg SL. 2004. *Schizosaccharomyces pombe* replication protein Cdc45/Sna41 requires Hsk1/Cdc7 and Rad4/Cut5 for chromatin binding. *Chromosoma* **113:** 145–156.

Donovan JD, Toyn JH, Johnson AL, Johnston LH. 1994. $P40^{SDB25}$, a putative CDK inhibitor, has a role in the M/G_1 transition in *Saccharomyces cerevisiae*. *Genes Dev* **8:** 1640–1653.

Donovan S, Harwood J, Drury LS, Diffley JF. 1997. Cdc6p-dependent loading of Mcm proteins onto pre-replicative chromatin in budding yeast. *Proc Natl Acad Sci* **94:** 5611–5616.

Drury LS, Perkins G, Diffley JF. 1997. The Cdc4/34/53 pathway targets Cdc6p for proteolysis in budding yeast. *EMBO J* **16:** 5966–5976.

Drury LS, Perkins G, Diffley JF. 2000. The cyclin-dependent kinase Cdc28p regulates distinct modes of Cdc6p proteolysis during the budding yeast cell cycle. *Curr Biol* **10:** 231–240.

Duursma A, Agami R. 2005. p53-dependent regulation of Cdc6 protein stability controls cellular proliferation. *Mol Cell Biol* **25:** 6937–6947.

Elsasser S, Lou F, Wang B, Campbell JL, Jong A. 1996. Interaction between yeast Cdc6 protein and B-type cyclin/Cdc28 kinases. *Mol Biol Cell* **7:** 1723–1735.

Elsasser S, Chi Y, Yang P, Campbell JL. 1999. Phosphorylation controls timing of Cdc6p destruction: A biochemical analysis. *Mol Biol Cell* **10:** 3263–3277.

Eser U, Falleur-Fettig M, Johnson A, Skotheim JM. 2011. Commitment to a cellular transition precedes genome-wide transcriptional change. *Mol Cell* **43:** 515–527.

Evrin C, Clarke P, Zech J, Lurz R, Sun J, Uhle S, Li H, Stillman B, Speck C. 2009. A double-hexameric MCM2-7 complex is loaded onto origin DNA during licensing of eukaryotic DNA replication. *Proc Natl Acad Sci* **106:** 20240–20245.

Feldman RM, Correll CC, Kaplan KB, Deshaies RJ. 1997. A complex of Cdc4p, Skp1p, and Cdc53p/cullin catalyzes ubiquitination of the phosphorylated CDK inhibitor Sic1p. *Cell* **91:** 221–230.

Ferreira MF, Santocanale C, Drury LS, Diffley JF. 2000. Dbf4p, an essential S phase-promoting factor, is targeted for degradation by the anaphase-promoting complex. *Mol Cell Biol* **20:** 242–248.

Francis LI, Randell JC, Takara TJ, Uchima L, Bell SP. 2009. Incorporation into the prereplicative complex activates the Mcm2–7 helicase for Cdc7–Dbf4 phosphorylation. *Genes Dev* **23:** 643–654.

Freeman A, Morris LS, Mills AD, Stoeber K, Laskey RA, Williams GH, Coleman N. 1999. Minichromosome maintenance proteins as biological markers of dysplasia and malignancy. *Clin Cancer Res* **5:** 2121–2132.

Fukuura M, Nagao K, Obuse C, Takahashi TS, Nakagawa T, Masukata H. 2011. CDK promotes interactions of Sld3 and Drc1 with Cut5 for initiation of DNA replication in fission yeast. *Mol Biol Cell* **22:** 2620–2633.

Gambus A, Khoudoli GA, Jones RC, Blow JJ. 2011. MCM2-7 form double hexamers at licensed origins in *Xenopus* egg extract. *J Biol Chem* **286:** 11855–11864.

Geng Y, Yu Q, Sicinska E, Das M, Schneider JE, Bhattacharya S, Rideout WM, Bronson RT, Gardner H, Sicinski P. 2003. Cyclin E ablation in the mouse. *Cell* **114:** 431–443.

Gerhardt J, Jafar S, Spindler MP, Ott E, Schepers A. 2006. Identification of new human origins of DNA replication by an origin-trapping assay. *Mol Cell Biol* **26:** 7731–7746.

Gillespie PJ, Li A, Blow JJ. 2001. Reconstitution of licensed replication origins on *Xenopus* sperm nuclei using purified proteins. *BMC Biochem* **2:** 15.

Gopalakrishnan V, Simancek P, Houchens C, Snaith HA, Frattini MG, Sazer S, Kelly TJ. 2001. Redundant control of rereplication in fission yeast. *Proc Natl Acad Sci* **98:** 13114–13119.

Grallert B, Nurse P. 1996. The ORC1 homolog orp1 in fission yeast plays a key role in regulating onset of S phase. *Genes Dev* **10:** 2644–2654.

Green BM, Morreale RJ, Ozaydin B, Derisi JL, Li JJ. 2006. Genome-wide mapping of DNA synthesis in *Saccharomyces cerevisiae* reveals that mechanisms preventing re-initiation of DNA replication are not redundant. *Mol Biol Cell* **17:** 2401–2414.

Green BM, Finn KJ, Li JJ. 2010. Loss of DNA replication control is a potent inducer of gene amplification. *Science* **329:** 943–946.

Gregan J, Lindner K, Brimage L, Franklin R, Namdar M, Hart EA, Aves SJ, Kearsey SE. 2003. Fission yeast Cdc23/Mcm10 functions after pre-replicative complex formation to promote Cdc45 chromatin binding. *Mol Biol Cell* **14:** 3876–3887.

Guarino E, Shepherd ME, Salguero I, Hua H, Deegan RS, Kearsey SE. 2011. Cdt1 proteolysis is promoted by dual PIP degrons and is modulated by PCNA ubiquitylation. *Nucleic Acids Res* **39:** 5978–5990.

Hardy CF, Dryga O, Seematter S, Pahl PM, Sclafani RA. 1997. *mcm5/cdc46-bob1* bypasses the requirement for the S phase activator Cdc7p. *Proc Natl Acad Sci* **94:** 3151–3155.

Harper JW, Elledge SJ. 2007. The DNA damage response: Ten years after. *Mol Cell* **28:** 739–745.

Hayashi M, Katou Y, Itoh T, Tazumi A, Yamada Y, Takahashi T, Nakagawa T, Shirahige K, Masukata H. 2007. Genome-wide localization of pre-RC sites and identification of replication origins in fission yeast. *EMBO J* **26:** 1327–1339.

Hemerly AS, Prasanth SG, Siddiqui K, Stillman B. 2009. Orc1 controls centriole and centrosome copy number in human cells. *Science* **323:** 789–793.

Hofmann JF, Beach D. 1994. *cdt1* is an essential target of the Cdc10/Sct1 transcription factor: Requirement for DNA replication and inhibition of mitosis. *EMBO J* **13:** 425–434.

Hu J, Xiong Y. 2006. An evolutionarily conserved function of proliferating cell nuclear antigen for Cdt1 degradation by the Cul4-Ddb1 ubiquitin ligase in response to DNA damage. *J Biol Chem* **281:** 3753–3756.

Huo YG, Bai L, Xu M, Jiang T. 2010. Crystal structure of the N-terminal region of human topoisomerase IIβ binding protein 1. *Biochem Biophys Res Commun* **401:** 401–405.

Im JS, Ki SH, Farina A, Jung DS, Hurwitz J, Lee JK. 2009. Assembly of the Cdc45-Mcm2–7-GINS complex in human cells requires the Ctf4/And-1, RecQL4, and Mcm10 proteins. *Proc Natl Acad Sci* **106:** 15628–15632.

Jackson PK, Chevalier S, Philippe M, Kirschner MW. 1995. Early events in DNA replication require cyclin E and are blocked by p21[CIP1]. *J Cell Biol* **130:** 755–769.

Jacob F, Brenner S. 1963. [On the regulation of DNA synthesis in bacteria: The hypothesis of the replicon]. *C R Hebd Seances Acad Sci* **256:** 298–300.

Jallepalli PV, Brown GW, Muzi-Falconi M, Tien D, Kelly TJ. 1997. Regulation of the replication initiator protein p65*[cdc18]* by CDK phosphorylation. *Genes Dev* **11:** 2767–2779.

Jallepalli PV, Tien D, Kelly TJ. 1998. *sud1*[+] targets cyclin-dependent kinase-phosphorylated Cdc18 and Rum1 proteins for degradation and stops unwanted diploidization in fission yeast. *Proc Natl Acad Sci* **95:** 8159–8164.

Jares P, Luciani MG, Blow JJ. 2004. A *Xenopus* Dbf4 homolog is required for Cdc7 chromatin binding and DNA replication. *BMC Mol Biol* **5:** 5.

Jaspersen SL, Charles JF, Morgan DO. 1999. Inhibitory phosphorylation of the APC regulator Hct1 is controlled by the kinase Cdc28 and the phosphatase Cdc14. *Curr Biol* **9:** 227–236.

Jiang W, McDonald D, Hope TJ, Hunter T. 1999. Mammalian Cdc7–Dbf4 protein kinase complex is essential for initiation of DNA replication. *EMBO J* **18:** 5703–5713.

Jin J, Arias EE, Chen J, Harper JW, Walter JC. 2006. A family of diverse Cul4-Ddb1-interacting proteins includes Cdt2, which is required for S phase destruction of the replication factor Cdt1. *Mol Cell* **23:** 709–721.

Kalaszczynska I, Geng Y, Iino T, Mizuno S, Choi Y, Kondratiuk I, Silver DP, Wolgemuth DJ, Akashi K, Sicinski P. 2009. Cyclin A is redundant in fibroblasts but essential in hematopoietic and embryonic stem cells. *Cell* **138:** 352–365.

Kamimura Y, Masumoto H, Sugino A, Araki H. 1998. Sld2, which interacts with Dpb11 in *Saccharomyces cerevisiae*, is required for chromosomal DNA replication. *Mol Cell Biol* **18:** 6102–6109.

Kamimura Y, Tak YS, Sugino A, Araki H. 2001. Sld3, which interacts with Cdc45 (Sld4), functions for chromosomal DNA replication in *Saccharomyces cerevisiae*. *EMBO J* **20:** 2097–2107.

Karnani N, Taylor CM, Malhotra A, Dutta A. 2010. Genomic study of replication initiation in human chromosomes reveals the influence of transcription regulation and chromatin structure on origin selection. *Mol Biol Cell* **21:** 393–404.

Kawabata T, Luebben SW, Yamaguchi S, Ilves I, Matise I, Buske T, Botchan MR, Shima N. 2011. Stalled fork rescue via dormant replication origins in unchallenged S phase promotes proper chromosome segregation and tumor suppression. *Mol Cell* **41:** 543–553.

Kelly TJ, Martin GS, Forsburg SL, Stephen RJ, Russo A, Nurse P. 1993. The fission yeast *cdc18*[+] gene product couples S phase to START and mitosis. *Cell* **74:** 371–382.

Kerns SL, Torke SJ, Benjamin JM, McGarry TJ. 2007. Geminin prevents rereplication during *Xenopus* development. *J Biol Chem* **282:** 5514–5521.

Kim Y, Kipreos ET. 2007. Cdt1 degradation to prevent DNA re-replication: Conserved and non-conserved pathways. *Cell Div* **2:** 18.

Klemm RD, Bell SP. 2001. ATP bound to the origin recognition complex is important for preRC formation. *Proc Natl Acad Sci* **98:** 8361–8367.

Knapp D, Bhoite L, Stillman DJ, Nasmyth K. 1996. The transcription factor Swi5 regulates expression of the cyclin kinase inhibitor p40*[SIC1]*. *Mol Cell Biol* **16:** 5701–5707.

Koch C, Moll T, Neuberg M, Ahorn H, Nasmyth K. 1993. A role for the transcription factors Mbp1 and Swi4 in progression from G1 to S phase. *Science* **261:** 1551–1557.

Koivomagi M, Valk E, Venta R, Iofik A, Lepiku M, Balog ER, Rubin SM, Morgan DO, Loog M. 2011a. Cascades of multisite phosphorylation control Sic1 destruction at the onset of S phase. *Nature* **480:** 128–131.

Koivomagi M, Valk E, Venta R, Iofik A, Lepiku M, Morgan DO, Loog M. 2011b. Dynamics of Cdk1 substrate specificity during the cell cycle. *Mol Cell* **42:** 610–623.

Kominami K, Toda T. 1997. Fission yeast WD-repeat protein Pop1 regulates genome ploidy through ubiquitin–proteasome-mediated degradation of the CDK inhibitor Rum1 and the S-phase initiator Cdc18. *Genes Dev* **11:** 1548–1560.

Kominami K, Ochotorena I, Toda T. 1998. Two F-box/WD-repeat proteins Pop1 and Pop2 form hetero- and homo-complexes together with cullin-1 in the fission yeast SCF (Skp1-Cullin-1-F-box) ubiquitin ligase. *Genes Cells* **3:** 721–735.

Kong D, DePamphilis ML. 2001. Site-specific DNA binding of the *Schizosaccharomyces pombe* origin recognition complex is determined by the Orc4 subunit. *Mol Cell Biol* **21:** 8095–8103.

Kraft C, Herzog F, Gieffers C, Mechtler K, Hagting A, Pines J, Peters JM. 2003. Mitotic regulation of the human anaphase-promoting complex by phosphorylation. *EMBO J* **22:** 6598–6609.

Kumagai A, Shevchenko A, Shevchenko A, Dunphy WG. 2010. Treslin collaborates with TopBP1 in triggering the initiation of DNA replication. *Cell* **140:** 349–359.

Kumagai A, Shevchenko A, Shevchenko A, Dunphy WG. 2011. Direct regulation of Treslin by cyclin-dependent kinase is essential for the onset of DNA replication. *J Cell Biol* **193:** 995–1007.

Kunkel TA, Burgers PM. 2008. Dividing the workload at a eukaryotic replication fork. *Trends Cell Biol* **18:** 521–527.

Labib K, Diffley JF, Kearsey SE. 1999. G1-phase and B-type cyclins exclude the DNA-replication factor Mcm4 from the nucleus. *Nat Cell Biol* **1:** 415–422.

Lau E, Zhu C, Abraham RT, Jiang W. 2006. The functional role of Cdc6 in S–G2/M in mammalian cells. *EMBO Rep* **7:** 425–430.

Lee JK, Moon KY, Jiang Y, Hurwitz J. 2001. The *Schizosaccharomyces pombe* origin recognition complex interacts with multiple AT-rich regions of the replication origin DNA by means of the AT-hook domains of the spOrc4 protein. *Proc Natl Acad Sci* **98:** 13589–13594.

Lee C, Hong B, Choi JM, Kim Y, Watanabe S, Ishimi Y, Enomoto T, Tada S, Kim Y, Cho Y. 2004. Structural basis for inhibition of the replication licensing factor Cdt1 by geminin. *Nature* **430:** 913–917.

Lee KY, Bang SW, Yoon SW, Lee SH, Yoon JB, Hwang DS. 2012. Phosphorylation of ORC2 protein dissociates origin recognition complex from chromatin and replication origins. *J Biol Chem* **287:** 11891–11898.

Lengauer C, Kinzler KW, Vogelstein B. 1998. Genetic instabilities in human cancers. *Nature* **396:** 643–649.

* Leonard AC, Méchali M. 2013. DNA replication origins. *Cold Spring Harb Perspect Biol* doi: 10.1101/cshperspect.a010116.

Li A, Blow JJ. 2005. Cdt1 downregulation by proteolysis and geminin inhibition prevents DNA re-replication in *Xenopus*. *EMBO J* **24:** 395–404.

Li CJ, DePamphilis ML. 2002. Mammalian Orc1 protein is selectively released from chromatin and ubiquitinated during the S-to-M transition in the cell division cycle. *Mol Cell Biol* **22:** 105–116.

Li X, Zhao Q, Liao R, Sun P, Wu X. 2003. The SCF^{Skp2} ubiquitin ligase complex interacts with the human replication licensing factor Cdt1 and regulates Cdt1 degradation. *J Biol Chem* **278:** 30854–30858.

Liku ME, Nguyen VQ, Rosales AW, Irie K, Li JJ. 2005. CDK phosphorylation of a novel NLS-NES module distributed between two subunits of the Mcm2–7 complex prevents chromosomal rereplication. *Mol Biol Cell* **16:** 5026–5039.

Lopez-Girona A, Mondesert O, Leatherwood J, Russell P. 1998. Negative regulation of Cdc18 DNA replication protein by Cdc2. *Mol Biol Cell* **9:** 63–73.

Lovejoy CA, Lock K, Yenamandra A, Cortez D. 2006. DDB1 maintains genome integrity through regulation of Cdt1. *Mol Cell Biol* **26:** 7977–7990.

Lowndes NF, Johnson AL, Breeden L, Johnston LH. 1992. SWI6 protein is required for transcription of the periodically expressed DNA synthesis genes in budding yeast. *Nature* **357:** 505–508.

MacAlpine HK, Gordan R, Powell SK, Hartemink AJ, MacAlpine DM. 2010. *Drosophila* ORC localizes to open chromatin and marks sites of cohesin complex loading. *Genome Res* **20:** 201–211.

Machida YJ, Dutta A. 2007. The APC/C inhibitor, Emi1, is essential for prevention of rereplication. *Genes Dev* **21:** 184–194.

Mahbubani HM, Chong JP, Chevalier S, Thommes P, Blow JJ. 1997. Cell cycle regulation of the replication licensing system: Involvement of a Cdk-dependent inhibitor. *J Cell Biol* **136:** 125–135.

Mailand N, Diffley JF. 2005. CDKs promote DNA replication origin licensing in human cells by protecting Cdc6 from APC/C-dependent proteolysis. *Cell* **122:** 915–926.

Maiorano D, Van Assendelft GB, Kearsey SE. 1996. Fission yeast cdc21, a member of the MCM protein family, is required for onset of S phase and is located in the nucleus throughout the cell cycle. *EMBO J* **15:** 861–872.

Makiniemi M, Hillukkala T, Tuusa J, Reini K, Vaara M, Huang D, Pospiech H, Majuri I, Westerling T, Makela TP, et al. 2001. BRCT domain-containing protein TopBP1 functions in DNA replication and damage response. *J Biol Chem* **276:** 30399–30406.

Malumbres M, Barbacid M. 2009. Cell cycle, CDKs and cancer: A changing paradigm. *Nat Rev Cancer* **9:** 153–166.

Masumoto H, Muramatsu S, Kamimura Y, Araki H. 2002. S-Cdk-dependent phosphorylation of Sld2 essential for chromosomal DNA replication in budding yeast. *Nature* **415:** 651–655.

Matsuno K, Kumano M, Kubota Y, Hashimoto Y, Takisawa H. 2006. The N-terminal noncatalytic region of *Xenopus* RecQ4 is required for chromatin binding of DNA polymerase α in the initiation of DNA replication. *Mol Cell Biol* **26:** 4843–4852.

McGarry TJ. 2002. Geminin deficiency causes a Chk1-dependent G2 arrest in *Xenopus*. *Mol Biol Cell* **13:** 3662–3671.

McGarry TJ, Kirschner MW. 1998. Geminin, an inhibitor of DNA replication, is degraded during mitosis. *Cell* **93:** 1043–1053.

Melixetian M, Ballabeni A, Masiero L, Gasparini P, Zamponi R, Bartek J, Lukas J, Helin K. 2004. Loss of Geminin induces rereplication in the presence of functional p53. *J Cell Biol* **165:** 473–482.

Mendez J, Stillman B. 2000. Chromatin association of human origin recognition complex, Cdc6, and minichromosome maintenance proteins during the cell cycle: Assembly of prereplication complexes in late mitosis. *Mol Cell Biol* **20:** 8602–8612.

Mendez J, Zou-Yang XH, Kim SY, Hidaka M, Tansey WP, Stillman B. 2002. Human origin recognition complex large subunit is degraded by ubiquitin-mediated proteolysis after initiation of DNA replication. *Mol Cell* **9:** 481–491.

Mendoza-Maldonado R, Paolinelli R, Galbiati L, Giadrossi S, Giacca M. 2010. Interaction of the retinoblastoma protein with Orc1 and its recruitment to human origins of DNA replication. *PloS ONE* **5:** e13720.

Mesner LD, Valsakumar V, Karnani N, Dutta A, Hamlin JL, Bekiranov S. 2011. Bubble-chip analysis of human origin distributions demonstrates on a genomic scale significant clustering into zones and significant association with transcription. *Genome Res* **21:** 377–389.

Mihaylov IS, Kondo T, Jones L, Ryzhikov S, Tanaka J, Zheng J, Higa LA, Minamino N, Cooley L, Zhang H. 2002. Control of DNA replication and chromosome ploidy by geminin and cyclin A. *Mol Cell Biol* **22:** 1868–1880.

Mimura S, Seki T, Tanaka S, Diffley JFX. 2004. Phosphorylation-dependent binding of mitotic cyclins to Cdc6 contributes to DNA replication control. *Nature* **431:** 1118–1123.

Miotto B, Struhl K. 2008. HBO1 histone acetylase is a coactivator of the replication licensing factor Cdt1. *Genes Dev* **22:** 2633–2638.

Miotto B, Struhl K. 2010. HBO1 histone acetylase activity is essential for DNA replication licensing and inhibited by Geminin. *Mol Cell* **37:** 57–66.

Moyer SE, Lewis PW, Botchan MR. 2006. Isolation of the Cdc45/Mcm2–7/GINS (CMG) complex, a candidate for the eukaryotic DNA replication fork helicase. *Proc Natl Acad Sci* **103:** 10236–10241.

Muramatsu S, Hirai K, Tak YS, Kamimura Y, Araki H. 2010. CDK-dependent complex formation between replication proteins Dpb11, Sld2, Pol ε, and GINS in budding yeast. *Genes Dev* **24:** 602–612.

Muzi Falconi M, Brown GW, Kelly TJ. 1996. *cdc18*$^{+}$ regulates initiation of DNA replication in *Schizosaccharomyces pombe*. *Proc Natl Acad Sci* **93:** 1566–1570.

Nakajima R, Masukata H. 2002. SpSld3 is required for loading and maintenance of SpCdc45 on chromatin in DNA replication in fission yeast. *Mol Biol Cell* **13:** 1462–1472.

Nakayama KI, Nakayama K. 2006. Ubiquitin ligases: Cell-cycle control and cancer. *Nat Rev Cancer* **6:** 369–381.

Nasmyth K, Dirick L. 1991. The role of *SWI4* and *SWI6* in the activity of G1 cyclins in yeast. *Cell* **66:** 995–1013.

Natale DA, Li CJ, Sun WH, DePamphilis ML. 2000. Selective instability of Orc1 protein accounts for the absence of functional origin recognition complexes during the M–G_1 transition in mammals. *EMBO J* **19:** 2728–2738.

Nguyen VQ, Co C, Irie K, Li JJ. 2000. Clb/Cdc28 kinases promote nuclear export of the replication initiator proteins Mcm2–7. *Curr Biol* **10:** 195–205.

Nguyen VQ, Co C, Li JJ. 2001. Cyclin-dependent kinases prevent DNA re-replication through multiple mechanisms. *Nature* **411:** 1068–1073.

Nishitani H, Nurse P. 1995. p65^{cdc18} plays a major role controlling the initiation of DNA replication in fission yeast. *Cell* **83:** 397–405.

Nishitani H, Lygerou Z, Nishimoto T, Nurse P. 2000. The Cdt1 protein is required to license DNA for replication in fission yeast. *Nature* **404:** 625–628.

Nishitani H, Taraviras S, Lygerou Z, Nishimoto T. 2001. The human licensing factor for DNA replication Cdt1 accumulates in G_1 and is destabilized after initiation of S-phase. *J Biol Chem* **276:** 44905–44911.

Nishitani H, Sugimoto N, Roukos V, Nakanishi Y, Saijo M, Obuse C, Tsurimoto T, Nakayama KI, Nakayama K, Fujita M, et al. 2006. Two E3 ubiquitin ligases, SCF-Skp2 and DDB1-Cul4, target human Cdt1 for proteolysis. *EMBO J* **25:** 1126–1136.

Nishiyama A, Frappier L, Mechali M. 2011. MCM-BP regulates unloading of the MCM2–7 helicase in late S phase. *Genes Dev* **25:** 165–175.

Noguchi E, Shanahan P, Noguchi C, Russell P. 2002. CDK phosphorylation of Drc1 regulates DNA replication in fission yeast. *Curr Biol* **12:** 599–605.

Nugroho TT, Mendenhall MD. 1994. An inhibitor of yeast cyclin-dependent protein kinase plays an important role in ensuring the genomic integrity of daughter cells. *Mol Cell Biol* **14:** 3320–3328.

Ode KL, Fujimoto K, Kubota Y, Takisawa H. 2011. Inter-origin cooperativity of geminin action establishes an all-or-none switch for replication origin licensing. *Genes Cells* **16:** 380–396.

Ogawa Y, Takahashi T, Masukata H. 1999. Association of fission yeast Orp1 and Mcm6 proteins with chromosomal replication origins. *Mol Cell Biol* **19:** 7228–7236.

Oshiro G, Owens JC, Shellman Y, Sclafani RA, Li JJ. 1999. Cell cycle control of Cdc7p kinase activity through regulation of Dbf4p stability. *Mol Cell Biol* **19:** 4888–4896.

Pak DT, Pflumm M, Chesnokov I, Huang DW, Kellum R, Marr J, Romanowski P, Botchan MR. 1997. Association of the origin recognition complex with heterochromatin and HP1 in higher eukaryotes. *Cell* **91:** 311–323.

Pelizon C, Madine MA, Romanowski P, Laskey RA. 2000. Unphosphorylatable mutants of Cdc6 disrupt its nuclear export but still support DNA replication once per cell cycle. *Genes Dev* **14:** 2526–2533.

Perkins G, Drury LS, Diffley JF. 2001. Separate SCF^{CDC4} recognition elements target Cdc6 for proteolysis in S phase and mitosis. *EMBO J* **20:** 4836–4845.

Peters JM. 2006. The anaphase promoting complex/cyclosome: A machine designed to destroy. *Nat Rev Mol Cell Biol* **7:** 644–656.

Petersen BO, Lukas J, Sorensen CS, Bartek J, Helin K. 1999. Phosphorylation of mammalian CDC6 by cyclin A/CDK2 regulates its subcellular localization. *EMBO J* **18:** 396–410.

Petersen BO, Wagener C, Marinoni F, Kramer ER, Melixetian M, Lazzerini Denchi E, Gieffers C, Matteucci C, Peters JM, Helin K. 2000. Cell cycle- and cell growth-regulated proteolysis of mammalian CDC6 is dependent on APC–CDH1. *Genes Dev* **14:** 2330–2343.

Prasanth SG, Prasanth KV, Stillman B. 2002. Orc6 involved in DNA replication, chromosome segregation, and cytokinesis. *Science* **297:** 1026–1031.

Prasanth SG, Prasanth KV, Siddiqui K, Spector DL, Stillman B. 2004. Human Orc2 localizes to centrosomes, centromeres and heterochromatin during chromosome inheritance. *EMBO J* **23:** 2651–2663.

Prasanth SG, Shen Z, Prasanth KV, Stillman B. 2010. Human origin recognition complex is essential for HP1 binding to chromatin and heterochromatin organization. *Proc Natl Acad Sci* **107:** 15093–15098.

Quinn LM, Herr A, McGarry TJ, Richardson H. 2001. The *Drosophila* Geminin homolog: Roles for Geminin in limiting DNA replication, in anaphase and in neurogenesis. *Genes Dev* **15:** 2741–2754.

Ralph E, Boye E, Kearsey SE. 2006. DNA damage induces Cdt1 proteolysis in fission yeast through a pathway dependent on Cdt2 and Ddb1. *EMBO Rep* **7:** 1134–1139.

Ranjan A, Gossen M. 2006. A structural role for ATP in the formation and stability of the human origin recognition complex. *Proc Natl Acad Sci* **103:** 4864–4869.

Reimann JD, Freed E, Hsu JY, Kramer ER, Peters JM, Jackson PK. 2001. Emi1 is a mitotic regulator that interacts with Cdc20 and inhibits the anaphase promoting complex. *Cell* **105:** 645–655.

Remus D, Beuron F, Tolun G, Griffith JD, Morris EP, Diffley JF. 2009. Concerted loading of Mcm2–7 double hexamers around DNA during DNA replication origin licensing. *Cell* **139:** 719–730.

Rialland M, Sola F, Santocanale C. 2002. Essential role of human CDT1 in DNA replication and chromatin licensing. *J Cell Sci* **115:** 1435–1440.

Rowles A, Tada S, Blow JJ. 1999. Changes in association of the *Xenopus* origin recognition complex with chromatin on licensing of replication origins. *J Cell Sci* **112:** 2011–2018.

Rudner AD, Murray AW. 2000. Phosphorylation by Cdc28 activates the Cdc20-dependent activity of the anaphase-promoting complex. *J Cell Biol* **149:** 1377–1390.

Saha P, Chen J, Thome KC, Lawlis SJ, Hou ZH, Hendricks M, Parvin JD, Dutta A. 1998. Human CDC6/Cdc18 associates with Orc1 and cyclin-cdk and is selectively eliminated from the nucleus at the onset of S phase. *Mol Cell Biol* **18:** 2758–2767.

Sanchez-Pulido L, Diffley JF, Ponting CP. 2010. Homology explains the functional similarities of Treslin/Ticrr and Sld3. *Curr Biol* **20:** R509–R510.

Sangrithi MN, Bernal JA, Madine M, Philpott A, Lee J, Dunphy WG, Venkitaraman AR. 2005. Initiation of DNA replication requires the RECQL4 protein mutated in Rothmund-Thomson syndrome. *Cell* **121:** 887–898.

Sansam CL, Shepard JL, Lai K, Ianari A, Danielian PS, Amsterdam A, Hopkins N, Lees JA. 2006. DTL/CDT2 is essential for both CDT1 regulation and the early G2/M checkpoint. *Genes Dev* **20:** 3117–3129.

Sansam CL, Cruz NM, Danielian PS, Amsterdam A, Lau ML, Hopkins N, Lees JA. 2010. A vertebrate gene, *ticrr*, is an essential checkpoint and replication regulator. *Genes Dev* **24:** 183–194.

Schneider BL, Yang QH, Futcher AB. 1996. Linkage of replication to start by the Cdk inhibitor Sic1. *Science* **272:** 560–562.

Schwab M, Lutum AS, Seufert W. 1997. Yeast Hct1 is a regulator of Clb2 cyclin proteolysis. *Cell* **90:** 683–693.

Schwob E, Bohm T, Mendenhall MD, Nasmyth K. 1994. The B-type cyclin kinase inhibitor $p40^{SIC1}$ controls the G1 to S transition in *S. cerevisiae*. *Cell* **79:** 233–244.

Seki T, Diffley JF. 2000. Stepwise assembly of initiation proteins at budding yeast replication origins in vitro. *Proc Natl Acad Sci* **97:** 14115–14120.

Senga T, Sivaprasad U, Zhu W, Park JH, Arias EE, Walter JC, Dutta A. 2006. PCNA is a cofactor for Cdt1 degradation by CUL4/DDB1-mediated N-terminal ubiquitination. *J Biol Chem* **281:** 6246–6252.

Sequeira-Mendes J, Diaz-Uriarte R, Apedaile A, Huntley D, Brockdorff N, Gomez M. 2009. Transcription initiation activity sets replication origin efficiency in mammalian cells. *PLoS Genet* **5:** e1000446.

Shareef MM, King C, Damaj M, Badagu R, Huang DW, Kellum R. 2001. *Drosophila* heterochromatin protein 1 (HP1)/origin recognition complex (ORC) protein is associated with HP1 and ORC and functions in heterochromatin-induced silencing. *Mol Biol Cell* **12:** 1671–1685.

Sherr CJ, Roberts JM. 2004. Living with or without cyclins and cyclin-dependent kinases. *Genes Dev* **18:** 2699–2711.

Sheu YJ, Stillman B. 2006. Cdc7-Dbf4 phosphorylates MCM proteins via a docking site-mediated mechanism to promote S phase progression. *Mol Cell* **24:** 101–113.

Sheu YJ, Stillman B. 2010. The Dbf4–Cdc7 kinase promotes S phase by alleviating an inhibitory activity in Mcm4. *Nature* **463:** 113–117.

Shreeram S, Sparks A, Lane DP, Blow JJ. 2002. Cell type-specific responses of human cells to inhibition of replication licensing. *Oncogene* **21:** 6624–6632.

Siddiqui K, Stillman B. 2007. ATP-dependent assembly of the human origin recognition complex. *J Biol Chem* **282:** 32370–32383.

Skotheim JM, Di Talia S, Siggia ED, Cross FR. 2008. Positive feedback of G1 cyclins ensures coherent cell cycle entry. *Nature* **454:** 291–296.

Spellman PT, Sherlock G, Zhang MQ, Iyer VR, Anders K, Eisen MB, Brown PO, Botstein D, Futcher B. 1998. Comprehensive identification of cell cycle-regulated genes of the yeast *Saccharomyces cerevisiae* by microarray hybridization. *Mol Biol Cell* **9:** 3273–3297.

Stanojcic S, Lemaitre JM, Brodolin K, Danis E, Mechali M. 2008. In *Xenopus* egg extracts, DNA replication initiates preferentially at or near asymmetric AT sequences. *Mol Cell Biol* **28:** 5265–5274.

Strausfeld UP, Howell M, Rempel R, Maller JL, Hunt T, Blow JJ. 1994. Cip1 blocks the initiation of DNA replication in *Xenopus* extracts by inhibition of cyclin-dependent kinases. *Curr Biol* **4:** 876–883.

Strausfeld UP, Howell M, Descombes P, Chevalier S, Rempel RE, Adamczewski J, Maller JL, Hunt T, Blow JJ. 1996. Both cyclin A and cyclin E have S-phase promoting (SPF) activity in *Xenopus* egg extracts. *J Cell Sci* **109:** 1555–1563.

Stuart D, Wittenberg C. 1995. CLN3, not positive feedback, determines the timing of CLN2 transcription in cycling cells. *Genes Dev* **9:** 2780–2794.

Su TT, O'Farrell PH. 1997. Chromosome association of minichromosome maintenance proteins in *Drosophila* mitotic cycles. *J Cell Biol* **139:** 13–21.

Su TT, O'Farrell PH. 1998. Chromosome association of minichromosome maintenance proteins in *Drosophila* endoreplication cycles. *J Cell Biol* **140:** 451–460.

Sugimoto N, Tatsumi Y, Tsurumi T, Matsukage A, Kiyono T, Nishitani H, Fujita M. 2004. Cdt1 phosphorylation by cyclin A-dependent kinases negatively regulates its function without affecting geminin binding. *J Biol Chem* **279:** 19691–19697.

Sugimoto N, Yoshida K, Tatsumi Y, Yugawa T, Narisawa-Saito M, Waga S, Kiyono T, Fujita M. 2009. Redundant and differential regulation of multiple licensing factors ensures prevention of re-replication in normal human cells. *J Cell Sci* **122:** 1184–1191.

Sun WH, Coleman TR, DePamphilis ML. 2002. Cell cycle-dependent regulation of the association between origin recognition proteins and somatic cell chromatin. *EMBO J* **21:** 1437–1446.

Svitin A, Chesnokov I. 2010. Study of DNA replication in *Drosophila* using cell free in vitro system. *Cell Cycle* **9:** 815–819.

Tada S, Li A, Maiorano D, Mechali M, Blow JJ. 2001. Repression of origin assembly in metaphase depends on inhibition of RLF-B/Cdt1 by geminin. *Nat Cell Biol* **3:** 107–113.

Takeda DY, Parvin JD, Dutta A. 2005a. Degradation of Cdt1 during S phase is Skp2-independent and is required for efficient progression of mammalian cells through S phase. *J Biol Chem* **280:** 23416–23423.

Takeda DY, Shibata Y, Parvin JD, Dutta A. 2005b. Recruitment of ORC or CDC6 to DNA is sufficient to create an artificial origin of replication in mammalian cells. *Genes Dev* **19:** 2827–2836.

Tanaka S, Diffley JFX. 2002. Interdependent nuclear accumulation of budding yeast Cdt1 and Mcm2–7 during G1 phase. *Nat Cell Biol* **4:** 198–207.

Tanaka S, Umemori T, Hirai K, Muramatsu S, Kamimura Y, Araki H. 2007. CDK-dependent phosphorylation of Sld2 and Sld3 initiates DNA replication in budding yeast. *Nature* **445:** 328–332.

Tanaka S, Nakato R, Katou Y, Shirahige K, Araki H. 2011a. Origin association of Sld3, Sld7, and Cdc45 proteins is a key step for determination of origin-firing timing. *Curr Biol* **21:** 2055–2063.

Tanaka T, Umemori T, Endo S, Muramatsu S, Kanemaki M, Kamimura Y, Obuse C, Araki H. 2011b. Sld7, an Sld3-associated protein required for efficient chromosomal DNA replication in budding yeast. *EMBO J* **30:** 2019–2030.

Tardat M, Brustel J, Kirsh O, Lefevbre C, Callanan M, Sardet C, Julien E. 2010. The histone H4 Lys 20 methyltransferase PR-Set7 regulates replication origins in mammalian cells. *Nat Cell Biol* **12:** 1086–1093.

Tatsumi Y, Ohta S, Kimura H, Tsurimoto T, Obuse C. 2003. The ORC1 cycle in human cells: I. Cell cycle-regulated oscillation of human ORC1. *J Biol Chem* **278:** 41528–41534.

Tatsumi Y, Ezura K, Yoshida K, Yugawa T, Narisawa-Saito M, Kiyono T, Ohta S, Obuse C, Fujita M. 2008. Involvement of human ORC and TRF2 in pre-replication complex assembly at telomeres. *Genes Cells* **13:** 1045–1059.

Thomae AW, Pich D, Brocher J, Spindler MP, Berens C, Hock R, Hammerschmidt W, Schepers A. 2008. Interaction between HMGA1a and the origin recognition complex creates site-specific replication origins. *Proc Natl Acad Sci* **105:** 1692–1697.

Thomae AW, Baltin J, Pich D, Deutsch MJ, Ravasz M, Zeller K, Gossen M, Hammerschmidt W, Schepers A. 2011. Different roles of the human Orc6 protein in the replication initiation process. *Cell Mol Life Sci* **68:** 3741–3756.

Tung JJ, Hansen DV, Ban KH, Loktev AV, Summers MK, Adler JR III, Jackson PK. 2005. A role for the anaphase-promoting complex inhibitor Emi2/XErp1, a homolog of early mitotic inhibitor 1, in cytostatic factor arrest of *Xenopus* eggs. *Proc Natl Acad Sci* **102:** 4318–4323.

Tyers M, Tokiwa G, Futcher B. 1993. Comparison of the *Saccharomyces cerevisiae* G_1 cyclins: Cln3 may be an upstream activator of Cln1, Cln2 and other cyclins. *EMBO J* **12:** 1955–1968.

Van Hatten RA, Tutter AV, Holway AH, Khederian AM, Walter JC, Michael WM. 2002. The *Xenopus* Xmus101 protein is required for the recruitment of Cdc45 to origins of DNA replication. *J Cell Biol* **159:** 541–547.

Vas A, Mok W, Leatherwood J. 2001. Control of DNA rereplication via Cdc2 phosphorylation sites in the origin recognition complex. *Mol Cell Biol* **21:** 5767–5777.

Vashee S, Cvetic C, Lu W, Simancek P, Kelly TJ, Walter JC. 2003. Sequence-independent DNA binding and replication initiation by the human origin recognition complex. *Genes Dev* **17:** 1894–1908.

Vassilev LT, Tovar C, Chen S, Knezevic D, Zhao X, Sun H, Heimbrook DC, Chen L. 2006. Selective small-molecule inhibitor reveals critical mitotic functions of human CDK1. *Proc Natl Acad Sci* **103:** 10660–10665.

Vaziri C, Saxena S, Jeon Y, Lee C, Murata K, Machida Y, Wagle N, Hwang DS, Dutta A. 2003. A p53-dependent checkpoint pathway prevents rereplication. *Mol Cell* **11:** 997–1008.

Verma R, Annan RS, Huddleston MJ, Carr SA, Reynard G, Deshaies RJ. 1997a. Phosphorylation of Sic1p by G_1 Cdk required for its degradation and entry into S phase. *Science* **278:** 455–460.

Verma R, Feldman RM, Deshaies RJ. 1997b. SIC1 is ubiquitinated in vitro by a pathway that requires CDC4, CDC34, and cyclin/CDK activities. *Mol Biol Cell* **8:** 1427–1437.

Visintin R, Prinz S, Amon A. 1997. CDC20 and CDH1: A family of substrate-specific activators of APC-dependent proteolysis. *Science* **278:** 460–463.

Visintin R, Craig K, Hwang ES, Prinz S, Tyers M, Amon A. 1998. The phosphatase Cdc14 triggers mitotic exit by reversal of Cdk-dependent phosphorylation. *Mol Cell* **2:** 709–718.

Walter JC. 2000. Evidence for sequential action of cdc7 and cdk2 protein kinases during initiation of DNA replication in *Xenopus* egg extracts. *J Biol Chem* **275:** 39773–39778.

Weinreich M, Stillman B. 1999. Cdc7p–Dbf4p kinase binds to chromatin during S phase and is regulated by both the APC and the *RAD53* checkpoint pathway. *EMBO J* **18:** 5334–5346.

Weinreich M, Liang C, Chen HH, Stillman B. 2001. Binding of cyclin-dependent kinases to ORC and Cdc6p regulates the chromosome replication cycle. *Proc Natl Acad Sci* **98:** 11211–11217.

Whittaker AJ, Royzman I, Orr-Weaver TL. 2000. *Drosophila* double parked: A conserved, essential replication protein that colocalizes with the origin recognition complex and links DNA replication with mitosis and the down-regulation of S phase transcripts. *Genes Dev* **14:** 1765–1776.

Wilmes GM, Archambault V, Austin RJ, Jacobson MD, Bell SP, Cross FR. 2004. Interaction of the S-phase cyclin Clb5 with an "RXL" docking sequence in the initiator protein Orc6 provides an origin-localized replication control switch. *Genes Dev* **18:** 981–991.

Wohlschlegel JA, Dwyer BT, Dhar SK, Cvetic C, Walter JC, Dutta A. 2000. Inhibition of eukaryotic DNA replication by geminin binding to Cdt1. *Science* **290:** 2309–2312.

Wohlschlegel JA, Dhar SK, Prokhorova TA, Dutta A, Walter JC. 2002. *Xenopus* Mcm10 binds to origins of DNA replication after Mcm2–7 and stimulates origin binding of Cdc45. *Mol Cell* **9:** 233–240.

Wolf DA, McKeon F, Jackson PK. 1999. Budding yeast Cdc6p induces re-replication in fission yeast by inhibition of SCF^{Pop}-mediated proteolysis. *Mol Gen Genet* **262:** 473–480.

Wong PG, Glozak MA, Cao TV, Vaziri C, Seto E, Alexandrow M. 2010. Chromatin unfolding by Cdt1 regulates MCM loading via opposing functions of HBO1 and HDAC11-geminin. *Cell Cycle* **9:** 4351–4363.

Wuarin J, Buck V, Nurse P, Millar JB. 2002. Stable association of mitotic cyclin B/Cdc2 to replication origins prevents endoreduplication. *Cell* **111:** 419–431.

Xu X, Rochette PJ, Feyissa EA, Su TV, Liu Y. 2009. MCM10 mediates RECQ4 association with MCM2-7 helicase complex during DNA replication. *EMBO J* **28:** 3005–3014.

Yabuuchi H, Yamada Y, Uchida T, Sunathvanichkul T, Nakagawa T, Masukata H. 2006. Ordered assembly of Sld3, GINS and Cdc45 is distinctly regulated by DDK and CDK for activation of replication origins. *EMBO J* **25:** 4663–4674.

Yamada Y, Nakagawa T, Masukata H. 2004. A novel intermediate in initiation complex assembly for fission yeast DNA replication. *Mol Biol Cell* **15:** 3740–3750.

Yanagi K, Mizuno T, You Z, Hanaoka F. 2002. Mouse geminin inhibits not only Cdt1-MCM6 interactions but also a novel intrinsic Cdt1 DNA binding activity. *J Biol Chem* **277:** 40871–40880.

Yanow SK, Lygerou Z, Nurse P. 2001. Expression of Cdc18/Cdc6 and Cdt1 during G_2 phase induces initiation of DNA replication. *EMBO J* **20:** 4648–4656.

Zachariae W, Schwab M, Nasmyth K, Seufert W. 1998. Control of cyclin ubiquitination by CDK-regulated binding of Hct1 to the anaphase promoting complex. *Science* **282:** 1721–1724.

Zegerman P, Diffley JF. 2007. Phosphorylation of Sld2 and Sld3 by cyclin-dependent kinases promotes DNA replication in budding yeast. *Nature* **445:** 281–285.

Zhu W, DePamphilis ML. 2009. Selective killing of cancer cells by suppression of geminin activity. *Cancer Res* **69:** 4870–4877.

Zhu W, Chen Y, Dutta A. 2004. Rereplication by depletion of geminin is seen regardless of p53 status and activates a G_2/M checkpoint. *Mol Cell Biol* **24:** 7140–7150.

Zhu W, Ukomadu C, Jha S, Senga T, Dhar SK, Wohlschlegel JA, Nutt LK, Kornbluth S, Dutta A. 2007. Mcm10 and And-1/CTF4 recruit DNA polymerase α to chromatin for initiation of DNA replication. *Genes Dev* **21:** 2288–2299.

Zou L, Stillman B. 1998. Formation of a preinitiation complex by S-phase cyclin CDK-dependent loading of Cdc45p onto chromatin. *Science* **280:** 593–596.

Zou L, Stillman B. 2000. Assembly of a complex containing Cdc45p, replication protein A, and Mcm2p at replication origins controlled by S-phase cyclin-dependent kinases and Cdc7p-Dbf4p kinase. *Mol Cell Biol* **20:** 3086–3096.

Regulating DNA Replication in Plants

Maria de la Paz Sanchez[1,2,3], Celina Costas[1,2], Joana Sequeira-Mendes[1,2], and Crisanto Gutierrez[1]

[1]Centro de Biologia Molecular "Severo Ochoa," CSIC-UAM, Nicolas Cabrera 1, Cantoblanco, 28049 Madrid, Spain

Correspondence: cgutierrez@cbm.uam.es

Chromosomal DNA replication in plants has requirements and constraints similar to those in other eukaryotes. However, some aspects are plant-specific. Studies of DNA replication control in plants, which have unique developmental strategies, can offer unparalleled opportunities of comparing regulatory processes with yeast and, particularly, metazoa to identify common trends and basic rules. In addition to the comparative molecular and biochemical studies, genomic studies in plants that started with *Arabidopsis thaliana* in the year 2000 have now expanded to several dozens of species. This, together with the applicability of genomic approaches and the availability of a large collection of mutants, underscores the enormous potential to study DNA replication control in a whole developing organism. Recent advances in this field with particular focus on the DNA replication proteins, the nature of replication origins and their epigenetic landscape, and the control of endoreplication will be reviewed.

Faithful genome duplication during the S phase of the cell cycle uses strategies largely conserved in all eukaryotes (DePamphilis and Bell 2011) and is pivotal to preserve genome integrity. Genome duplication in dividing plant cells has the same requirements and constraints than in animal cells, including the strict rule of occurring once and only once every cell cycle. The initial discoveries of several basic biological processes were performed in studies with plant cells, e.g., transposons, telomeres, RNA interference, to cite a few. DNA replication in eukaryotes is not an exception. Pioneering work in the mid-1950s showed the semiconservative nature of chromosomal DNA replication in the common bean *Vicia faba* (Taylor et al. 1957). Since then, plant DNA replication studies have focused primarily on defining temporal patterns of DNA replication at the chromosomal level along the S phase. DNA fiber autoradiography was first used in plants to determine replicon size at the single molecule level (Nitta and Nagata 1976; Van't Hof 1976) and genome organization in early and late replicon families (Van't Hof et al. 1978; Van't Hof and Bjerknes 1981; reviewed in Bryant 2010; Costas et al. 2011a).

In this work, we will focus on: (1) plant DNA replication proteins (see also Appendix, Table 3);

[2]These authors share the first authorship.

[3]Present address: Instituto de Ecología, Universidad Nacional Autónoma de México, 3er Circuito Exterior, Cd. Universitaria, México DF 04510, Mexico.

Cite this article as *Cold Spring Harb Perspect Biol* doi: 10.1101/cshperspect.a010140

(2) plant DNA replication origins and their epigenetic landscape; (3) novel licensing mechanisms; (4) the relevance of DNA replication proteins in the control of the endoreplication cycle during plant development; and, finally, (5) a brief overview of duplication of plant DNA viruses.

DNA REPLICATION PROTEINS

The Prereplication Complex

Homologs of the major pre-RC components (ORC, CDC6, CDT1, and MCM2-7) (see Bell and Kaguni 2013) have been identified in several plant species (Gutierrez 2006; Shultz et al. 2007). Some of them, notably ORC1, CDC6, and CDT1, are encoded by duplicated genes in *Arabidopsis*. Although the coding regions of duplicated pre-RC proteins are largely conserved, their promoter regions differ considerably, suggesting that they are subjected to distinct transcriptional regulation.

ORC transcripts are abundant in cultured cells and in proliferating cells within the adult plant body. In addition, *Arabidopsis ORC5* and *ORC6* (Diaz-Trivino et al. 2005) and maize *ORC3* (Witmer et al. 2003) are also expressed in nondividing cells in which they may play specific roles, still unknown, as it has been described for some metazoan ORC genes (Sasaki and Gilbert 2007). The interaction map of plant ORC subunits is also quite conserved (Witmer et al. 2003; Diaz-Trivino et al. 2005). A distinctive feature of plant ORC1 is its function as a transcriptional coactivator (Sanchez and Gutierrez 2009), contrary to the silencing role of ORC1 in yeast (Hickman and Rusche 2010). Plant ORC1 contains in its amino-terminal half a plant homeodomain (PHD) that seems to mediate interaction with trimethylated histone H3 at lysine 4 (H3K4me3) at the promoter of target genes (Sanchez and Gutierrez 2009). Interestingly, some of these ORC1-binding sites coincide with replication origins (Costas et al. 2011b), pointing to coordination between replication and transcription. It is worth noting that the Orc1 interaction with methylated histones also has been described in animals. Among the ORC subunits in eukarya, only Orc1 has a BAH domain that mediates protein–protein interactions. The BAH domain in human Orc1 facilitates the ability of Orc1 to activate replication origins in vivo (Noguchi et al. 2006), and links H4K20me2 to DNA replication licensing and the Meier-Gorlin syndrome, primordial dwarfism that is related directly to mutations in prereplication proteins (Bicknell et al. 2011; Kuo et al. 2012).

Plant CDC6 and CDT1 conserve the major domains of these proteins, although each of the two homologs of CDC6 and CDT1 proteins show significant divergence, in particular CDT1 (Castellano et al. 2001, 2004). Another major difference with the metazoan proteins is the absence in plants of a homolog of metazoan geminin (Gmnn), the CDT1-interacting protein that inhibits its activity (Caro and Gutierrez 2007; Miotto and Struhl 2010). Remarkably, plants contain a completely unrelated CDT1-interacting protein called GEM (GL2-EXPRESSION MODULATOR [Caro et al. 2007]). It is not known yet whether GEM has a role in DNA replication. Expression of both *CDT1a* and *CDT1b* depends on a complex regulation. Whereas both *CDT1a* and *CDT1b* are E2F targets (Castellano et al. 2004), only *CDT1a* is activated by ORC1 (Sanchez and Gutierrez 2009). Both *CDT1a* and *CDT1b* genes are also repressed by ABAP1 (ARMADILLO BTB ARABIDOPSIS PROTEIN1), a novel factor that works in association with the TCP24 transcription factor (Masuda et al. 2008).

Plant MCM2-7 proteins contain the typical conserved motifs (Sabelli et al. 1996, 1999; Stevens et al. 2002; Masuda et al. 2004; Dresselhaus et al. 2006; Shultz et al. 2009). An evolutionary conserved protein called ETG1 (for E2F TARGET1) that interacts with the MCM2-7 complex has been identified in *Arabidopsis* (Takahashi et al. 2008). ETG1 is necessary for efficient DNA replication and its absence leads to G_2 checkpoint activation and loss of chromatid cohesion, a phenotype that is enhanced in the absence of CTF18, another factor necessary for chromatid cohesion that has been confirmed in mammalian cells (Takahashi et al. 2010).

 Cite this article as *Cold Spring Harb Perspect Biol* doi: 10.1101/cshperspect.a010140

Replication Fork Assembly

MCM8, MCM9, and MCM10

MCM8 is highly conserved among eukaryotes, including *Arabidopsis*, rice, and other plants (Dresselhaus et al. 2006; Shultz et al. 2007). The major protein motifs are conserved, suggesting similar biochemical functions such as stabilization of RPA and DNA polymerase α. However, a ~60 amino acid stretch at the amino terminus and a ~20 amino acid motif, present in animal MCM8, are missing in *Arabidopsis* MCM8 (Shultz et al. 2007). Their physiological relevance, if any, remains to be determined.

Genes encoding MCM9 have been identified in several plant species (Dresselhaus et al. 2006; Shultz et al. 2007) and, frequently, they are transcribed at very low levels and in different spliced forms (Schmid et al. 2005; Shultz et al. 2007). Interestingly, plant MCM9 lacks the carboxy-terminal domain present in animal MCM9, although the functional relevance of this protein organization is not known. It must be pointed out that Mcm9 is not essential for DNA replication in mammals, but the Mcm8–Mcm9 protein complex is involved in DNA recombination (Hartford et al. 2011; Lutzmann et al. 2012; Nishimura et al. 2012), an aspect that has not be addressed in plants.

A homolog of MCM10, a protein essential in yeast for unwinding of origin DNA and loading of DNA polymerase α (Kanke et al. 2012; van Deursen et al. 2012; Watase et al. 2012), has been identified in various plant species (Shultz et al. 2007). The degree of similarity is small compared with other replication proteins, although, remarkably, plant MCM10 conserves the CCCH-type $Zn2^{+}$ finger that mediates its homomultimerization.

CDC45

The function of plant CDC45 is probably to serve as a switch from the pre-RC complex to the initiation complex, because *Arabidopsis* CDC45 shows a >80% similarity with *S. pombe* and human CDC45. Like in yeast and mammalian cells, *Arabidopsis* CDC45 gene expression is cell cycle regulated, showing high levels at the G_1/S transition (Stevens et al. 2004). In spite of these similarities, its role in plant DNA replication has not been shown. *Arabidopsis* CDC45 is unable to complement a yeast *cdc45* temperature-sensitive mutant and *Arabidopsis* RNAi plant lines with reduced *CDC45* transcripts did not show defects in DNA synthesis (Stevens et al. 2004).

GINS Complex

The genes encoding GINS subunits have been identified in the genomic sequence of *Arabidopsis*, rice, maize, and soybean, including a duplicated *PSF3* gene (Gutierrez 2006; Shultz et al. 2007). Interestingly, PSF2 contains a motif conserved in vertebrate but not in yeast, reinforcing the idea of a closer evolutionary relationship between *Arabidopsis* and human proteins. The plant GINS complex has been poorly characterized biochemically, but probably in association with MCM2-7, CDC45, and DNA polymerase ε, participates in initiation and fork progression, like in other eukaryotes (Gambus et al. 2006; Moyer et al. 2006).

Replication Fork Progression

DNA Polymerases

Eukaryotic DNA polymerases have been grouped in five different families, of which family B contains the replicative polymerases (Johansson and Dixon 2013). Like in animals, plant replicative DNA polymerases have been classified in three-major types: DNA polymerase α (POLA), DNA polymerase δ (POLD), and DNA polymerase ε (POLE).

POLA

The presence of DNA polymerase α (POLA) activity in plants has been shown biochemically in maize, wheat, pea, and cauliflower (Bryant 1980; Castroviejo et al. 1990; Laquel et al. 1990; Balmukhanov et al. 1992; Bryant et al. 1992; Coello et al. 1992; Coello and Vazquez-Ramos 1995b; Garcia et al. 1997; Luque et al. 1998). More recently, the genome sequence analysis of *Arabidopsis* and rice allowed the identification of the four putative subunits of POLA complex (POLA1 to 4). The largest subunit (POLA1) contains the DNA polymerase activity whereas the two smallest subunits have the DNA primase activity (POLA3 and POLA4, also known as PRI1 and PRI2, respectively).

Interestingly, the *POLA1* gene, also known as *INCURVATA2* (*ICU2*), was originally identified in a genetic screen for abnormal leaf morphology and has a role in transcriptional silencing (Barrero et al. 2007; Liu et al. 2010a). The sequence features of POLA1 are conserved with the exception of a YYRRLFP motif of unknown function present in yeast and animal POLA4 subunit but not in plants (Shultz et al. 2007). Conservation of the remaining essential motifs suggests a similar function of POLA subunits in plants. In agreement with this, the plant DNA POLA complex contains a DNA polymerase activity with low processivity and a primase activity. POLA activity is regulated by phosphorylation because it increases when all subunits are phosphorylated (Coello and Vazquez-Ramos 1995a,b; Garcia et al. 1997).

POLD and POLE

Plant POLD has been described in rice, wheat, maize, and *Arabidopsis* (Richard et al. 1991; Garcia et al. 1997). The *Arabidopsis* POLD complex contains four subunits (POLD1 to D4) whereas in rice it contains five subunits (Shultz et al. 2007). Sequence analysis of *Arabidopsis*, soybean, rice, and maize indicated that the plant's largest subunits are highly conserved. The essential motif that mediates the interaction of POLD2 and POLD3 subunits with PCNA is conserved as well. The expression pattern of *POLD1* correlates with proliferative tissues and with PCNA expression (Uchiyama et al. 2002; Garcia et al. 2006). The biochemical activity of plant DNA POLD has been assessed in wheat, and also contains an associated $3'-5'$ exonuclease activity, which confers a proofreading ability, and is highly stimulated by PCNA (Richard et al. 1991; Laquel et al. 1993; Luque et al. 2000). Rice POLD1 and POLD2 genes are regulated by sucrose and UV treatment (Uchiyama et al. 2002), indicating that during evolution some replication genes have become direct targets of the environmental response.

Plant POLE has been studied in *Arabidopsis* and rice, including genetic analyses that have not been performed in other multicellular organisms (Jenik et al. 2005; Ronceret et al. 2005; Yin et al. 2009; del Olmo et al. 2010). *Arabidopsis* POLE holoenzyme possesses two catalytic subunits (POLE1A and POLE1B, also called AtPOL2a and AtPOL2b), and the regulatory subunits POLE2 (or DPB2), DPB4, and DBP3a. Genetic analysis suggests that POLE1A and/or POLE2 play a crucial role in DNA replication, whereas POLE1B is not required when POLE1A is present. In fact, the catalytic activity of POLE1A is necessary during embryogenesis, floral transition, and in meristems, whereas POLE1B is expressed at detectable levels preferentially under certain stress conditions (Jenik et al. 2005; Ronceret et al. 2005). Genes encoding the POLE3 and POLE4 subunits belong to a multigenic family and are difficult to identify unambiguously in the *Arabidopsis* genome. Hypomorphic mutants of POLE1 (*esd7/abo4/til1*) show an early flowering phenotype because of changes in the expression of key flowering genes, a process that occurs through a genetic interaction of *POLE1* with *LHP1* (*LIKE-HETEROCHROMATIN PROTEIN1*) (Yin et al. 2009; del Olmo et al. 2010).

PCNA

PCNA is an essential protein, highly conserved in all eukaryotes, including plants (Suzuka et al. 1989; Kodama et al. 1991; Hata et al. 1992; Matsumoto et al. 1994; Lopez et al. 1995; Shimizu and Mori 1998; Strzalka et al. 2010). It is frequent that plant genomes contain two PCNA genes, in contrast to PCNA pseudogenes normally found in mammalian cells. Plant PCNA forms a homotrimeric complex (Sanchez et al. 2002; Strzalka et al. 2009) that binds to various cyclin–CDK complexes (Herrera et al. 2000; Sanchez et al. 2005). In agreement with the hypothesis of serving as a sliding platform to other proteins, in vitro assays showed that PCNA-associated cyclin/CDK activity phosphorylates

DNA polymerase α. Because this phosphorylation does not affect DNA polymerase activity, PCNA could recruit regulatory proteins (Gomez Roig and Vazquez-Ramos 2003). Unfortunately, a direct function on the processivity and activity of the plant DNA polymerases δ or ε is so far unknown. However, recombinant rice and bean PCNA stimulate human DNA polymerase δ (Matsumoto et al. 1994; Strzalka et al. 2009). In addition, two wheat DNA polymerases δ have been identified and both increase their in vitro activity and processivity when stimulated with calf thymus PCNA (Laquel et al. 1993).

The participation of plant PCNA in DNA damage processing has been shown in *Arabidopsis*. Thus, POLH (AtPol η) and POLL (AtPol λ), members of the Y and X families, respectively, interact with PCNA2 (Anderson et al. 2008; Amoroso et al. 2011) to catalyze translesion synthesis and restore replication fork progression when DNA is damaged.

RFC

All five replication factor C (RFC1–5) subunits have been identified in plants and their sequence analysis suggests that they play roles similar to other eukaryotes (Luque et al. 2002; Furukawa et al. 2003; Shultz et al. 2007; Hedglin et al. 2013). Like other DNA replication proteins, plant *RFC* expression occurs preferentially in proliferative tissues and is cell cycle regulated (Luque et al. 2002; Furukawa et al. 2003). RFC1 is required for replication of the ssDNA-containing wheat dwarf virus (WDV) genome to stimulate binding of the initiator Rep protein to 3′-OH containing DNA substrates (Luque et al. 2002). Besides to its role as a clamp loader, RFC1 also contributes to transcriptional gene silencing of certain genes and transposons (Liu et al. 2010b).

RPA

The eukaryotic single-stranded DNA binding (SSB) protein (RPA) is required for pre-RC activation and for ordered loading of essential initiator factors at origins (Sakaguchi et al. 2009). Each of the three RPA subunits is encoded by several genes in plant genomes. This allows the possibility to form different species of RPA complexes (Ishibashi et al. 2006; Shultz et al. 2007). The function of these complexes remains to be determined, but, based on their localization, one complex may act in the chloroplast, whereas other complexes could participate in nuclear processes (Ishibashi et al. 2006). Moreover, studies with mutant plants revealed that *Arabidopsis* RPA1A and RPA1C (nuclear proteins) are essential for DNA replication, whereas RPA1B (chloroplast and nuclear) is not essential and is required only during DNA repair (Sakaguchi et al. 2009). Apart from its function in DNA replication, RPA2 is also required to maintain transcriptional silencing of a subset of transposon elements (Elmayan et al. 2005; Kapoor et al. 2005; Xia et al. 2006).

FEN1

The specific flap endonuclease 1 (FEN1) is conserved in plants and has been characterized so far in cauliflower, wheat, rice, and *Arabidopsis* (Kimura et al. 1997, 2000, 2003; Przykorska et al. 2004; Shultz et al. 2007). Interestingly, *Arabidopsis* has only one *FEN1* gene whereas rice possesses two. These are expressed in proliferating cells as well as in shoot and root apical meristems, but *FEN1B* is also expressed in mature organs (Kimura et al. 2000, 2003). FEN1A complements the yeast mutant strain *Δrad27* (Kimura et al. 2003) and associates with PCNA, whereas FEN1B lacks a PCNA binding motif (Kimura et al. 2001), a difference that might be important for its function. The biochemical characterization of plant FEN1 indicates that this enzyme has 5′→3′-double-strand-specific exonuclease activity and cleaves endonucleolytically at the 3′ end of the unpaired region in stem-loop structures (Kimura et al. 1997, 2001).

LIG1

Plant DNA ligase 1 (LIG1) catalyzes the formation of a phosphodiester bond at single and double strand at various stages of DNA metabolism (Howell and Stern 1971; Kessler 1971).

The *Arabidopsis* LIG1 protein can complement the *cdc9* temperature sensitive mutation in yeast and is expressed constitutively in all tissues analyzed (Taylor et al. 1998).

DNA REPLICATION ORIGINS AND THEIR EPIGENETIC LANDSCAPE

In eukaryotic organisms, there are thousands of sites scattered across the genome in which DNA replication begins (Méchali 2010; Leonard and Méchali 2013). However, the usage of all replication origins must be controlled by a strict regulatory mechanism to avoid genome instability. Origin specification, marked by the assembly of pre-RC in G_1, is the first level of regulation. All the potential origins that could be used during a particular S phase are licensed at this point. The second key step is origin firing, which occurs at the G_1/S transition and during S phase, and whereby only a small percentage of all the licensed origins become active. The strict regulation of the firing mechanism ensures that the replication of the DNA takes place once and only once per cell cycle (DePamphilis and Bell 2011).

The molecular mechanisms responsible for origin specification are only starting to be elucidated. In plants this issue is even more elusive and only during the last years significant advances have been made (Lee et al. 2010; Costas et al. 2011a; Liu and Gong 2011). An earlier approach consisted of scanning the ribosomal DNA (rDNA) region of cultured root meristem pea (*Pisum sativum*) cells to identify a replication origin within 1.5 kb at the nontranscribed spacer (NTS) of the rDNA (Van't Hof et al. 1987; Hernandez et al. 1993), a feature shared with other eukaryotes (DePamphilis and Bell 2011). These data were further confirmed by 2D-gel electrophoresis (Van't Hof and Lamm 1992). Recently, the development of genomic tools, including massive sequencing of newly synthesized DNA, yielded a genome-wide map of replication origins in *Arabidopsis* cultured cells (Costas et al. 2011b).

In multicellular organisms the DNA sequence is not sufficient for regulating the usage and function of the replication origins and other factors such as DNA methylation and epigenetic modifications of histones may play important roles (Fig. 1). The regulation and availability of the effectors of these chromatin changes (the epigenetic modifiers such as chromatin remodeling complexes, histone chaperones and histone modification enzymes) are crucial for proper DNA replication and maintenance of genome integrity. The analysis of *Arabidopsis* origins reveals that most of them are preferentially located at the 5′ ends of gene bodies and associated with highly expressed genes, as it has been described for human origins (Valenzuela et al. 2011). Additionally, they show a relatively low CG methylation whereas they are enriched in the histone variant H2A.Z (Costas et al. 2011b) and histone H3.3 (Stroud et al. 2012b). This is consistent with the depletion in CG methylation found in early and mid-S-phase replicons of *Arabidopsis* chromosome 4 (Lee et al. 2010). Moreover, the higher than average C + G content found in *Arabidopsis* origin regions (midpoint $\pm$ 100bp), associated with a low CG methylation might favor a more open chromatin that facilitates pre-RC assembly and origin activation, in agreement with results obtained in *Drosophila* (Macalpine et al. 2010) and mammalian cells (Karnani et al. 2010; Cayrou et al. 2011; Valenzuela et al. 2011).

The genome-wide maps of several histone modifications analyzed show that *Arabidopsis* origins are associated with H3K4me2, H3K4-me3, and H4K5ac, and depleted of H3K4me1 and H3K9me2 (Costas et al. 2011b), a typical mark of heterochromatin in plants (Fuchs et al. 2006). Histone acetylation can affect origin specification and replication timing in other eukaryotes (Schwaiger et al. 2009; Karnani et al. 2010; Casas-Delucchi et al. 2011). In *Arabidopsis*, it is possible that acetylation in other histone residues, besides H4K5ac, is relevant for origin activation as suggested by the increase in histone H3K18ac and H4K16ac observed as S phase progresses (Jasencakova et al. 2003) and by the presence of H3K56ac in early replicons (Lee et al. 2010; Roudier et al. 2011).

The observation that the *Arabidopsis* genome was organized in two large replicon families (Van't Hof et al. 1978), strongly suggested a temporal control of DNA replication (Jackson

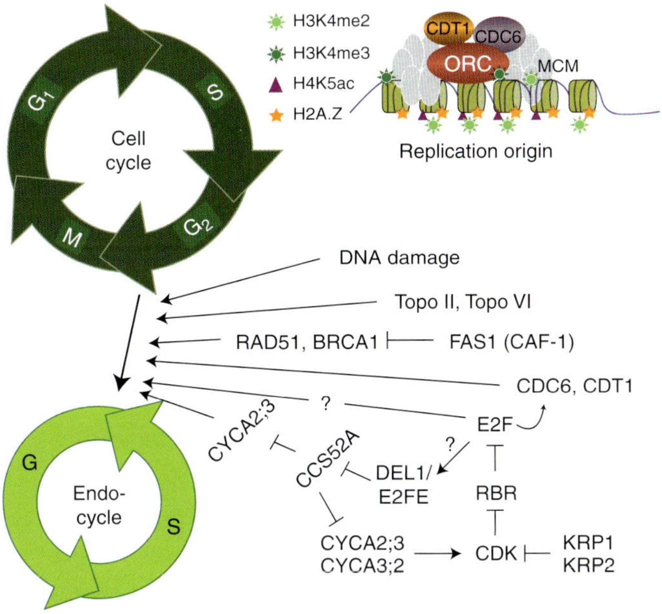

Figure 1. The transition from the cell cycle to the endoreplication cycle in plants. Replication origins during the S phase of the cell cycle tend to be enriched in histone H2A.Z and some epigenetic marks, such as H3K4me2, H3K4me3, and H4K5ac. Nucleosome occupancy across the replication origin is not considered here. Different pathways, including the direct action of replication proteins, affect the transition from the cell cycle to the endoreplication cycle. See the text for further explanation. Topo, topoisomerase.

et al. 2012). More recently, the combination of cell sorting of early, mid-, and late S-phase nuclei of culture cells with tiling microarray hybridization, revealed a biphasic mode of replication, with the majority of the euchromatin replicating in early/mid-S phase and the heterochromatin and remaining euchromatin replicating late (Lee et al. 2010). The early/mid-replicating domains colocalize with regions of open chromatin enriched in actively transcribed genes and H3K56ac. Conversely, late replicating domains are enriched in transposon elements and display repressive epigenetic marks (H3K9me2 and 5mC). These observations are in strong agreement with the existing reports for metazoan systems (MacAlpine et al. 2004; Hiratani et al. 2009). Interestingly, H3K56ac has a fundamental role in nucleosome reassembly following DNA replication and repair (Chen and Tyler 2008) as well as in replication-independent histone turnover at promoters (Rufiange et al. 2007). However, no direct link of H3K56ac with origin activation has been provided to date.

DNA REPLICATION LICENSING

Misregulation of origin licensing has consequences on genome stability as a result of uncontrolled rereplication. Interfering with the activity of pre-RC bound to origins that have fired in a given S phase is one strategy to prevent rereplication (Arias and Walter 2007). Several mechanisms, discussed elsewhere in this volume, have evolved to prevent rereplication, such as: (1) selective proteolysis of pre-RC components; (2) changes in their subcellular localization; and (3) binding of pre-RC inhibitors, such as geminin (Gmnn). Plants contain homologs of all pre-RC proteins and other factors that participate in licensing, except Gmnn (Caro and Gutierrez 2007; Shultz et al. 2007). Interestingly, they contain GEM, a CDT1-interacting protein that is structurally unrelated to metazoan Gmnn (Caro et al. 2007), but its role in licensing remains to be experimentally addressed.

Mechanistic studies of plant pre-RC dynamics are scarce. However, it is known that the

MCM complex remains bound to chromatin throughout the cell cycle, except during mitosis (Shultz et al. 2009). Also, increased levels of CDC6 and/or CDT1 trigger extra endoreplication cycles, which implies full-genome duplication, instead of limited rereplication (Castellano et al. 2001, 2004). These differences with other model organisms support the idea that plant-specific mechanisms of rereplication control may have evolved (Costas et al. 2011a).

In addition to pre-RC availability, particular chromatin states are also involved in replication licensing control. The level of H3K27me1 is crucial to prevent rereplication in *Arabidopsis* heterochromatin. Thus, ATXR5 and ATXR6, encoded by E2F target genes, are monomethyl H3K27 transferases that interact with PCNA (Raynaud et al. 2006; Jacob et al. 2009). Massive sequencing of sorted 2–16C nuclei of an *atxr5 atxr6* double mutant revealed, surprisingly, a significant increase (>30%) in the amount of pericentromeric heterochromatic sequences (Jacob et al. 2010). In fact, the PCNA-interacting protein (PIP) motif of ATXR5 and ATXR6 is required to prevent the rereplication phenotype (Jacob et al. 2010). Moreover, the rereplicated sequences are preferentially located at specific loci, suggesting a heterochromatin-specific loss of licensing control, although dispersed heterochromatic regions across chromosome arms also show rereplicated material. Furthermore, the rereplication phenotype of *atxr5 atxr6* mutants requires factors involved in DNA methylation, as revealed by the reduced rereplication observed in *ddm1*, *met1* and, to a lesser extent, *cmt3* DNA methylation mutant backgrounds (Stroud et al. 2012a), although H3K27me1 is not affected by the DNA methylation level (Mathieu et al. 2005). It is worth noting that a heterochromatin-specific rereplication control depends on geminin in *Drosophila* (Ding and MacAlpine 2010). In animal cells, the level of H4K20me1 is crucial to restrain rereplication (Tardat et al. 2010). The existence of H4K20 methylation in plants has not been shown yet by mass spectrometry (Zhang et al. 2007). However, immunofluorescence and ChIP experiments show important differences with animal cells, because H4K20me1 is typical of pericentromeric heterochromatin whereas H4K20me3 associates with active euchromatic promoters (Fischer et al. 2006; Fuchs et al. 2006). In fact, H4K20me3 colocalizes with a subset of ORC1-binding sites, some of which act as replication origins (Sanchez and Gutierrez 2009).

DNA REPLICATION AND ENDOREPLICATION DURING DEVELOPMENT

One characteristic feature of plant development is that organogenesis is an indeterminate and postembryonic process that occurs continuously during the life of the adult plant (Gutierrez 2005; Inze and De Veylder 2006). This growth pattern implies a high coordination to maintain the balance between pools of proliferating and arrested cells that eventually initiate differentiation and organ formation. In response to certain developmental cues, as well as to environmental factors, proliferating cells exit the cell cycle to enter the endocycle, whereby full genome duplication occurs several times in the absence of an intervening mitosis (Caro et al. 2008; Breuer et al. 2010; De Veylder et al. 2011). Endoreplication is a prerequisite for certain differentiation pathways and is regulated at the level of initiation, maintenance and number of replication rounds to be developed. It must be emphasized that endoreplication is not equivalent to a loss of the licensing control (Zielke et al. 2013), because it involves duplication of the entire genome during each endoreplication cycle instead of uncontrolled reinitiation events within a genomic locus.

As in the case of metazoa, endocycling in plant cells faces the challenge of maintaining CDK activity sufficiently below a threshold to prevent triggering of mitosis. This is achieved by controlling the levels of CDK activity through regulating the amount of CDK (protein and activity), the availability of cyclins, and the levels of CDK inhibitors (De Veylder et al. 2011). Thus, several mechanisms have evolved, in some cases independently of those acting in metazoa. It must be kept in mind that plants contain several non-PSTAIRE CDK involved in cell cycle progression and a large set of cyclins of the A-,

B-, and D types, but they lack a typical E-type cyclin (Gutierrez 2009).

The APC/C degrades several cyclins including CYCB1;1, CYCB1;2, CYCA2;3, and CYCA3;1 (De Veylder et al. 2011). In *Arabidopsis*, homologs of both Cdc20/Fzy and Cdh1/Fzr APC/C activators exist and participate in controlling the onset of endocycles by targeting various cyclins (Gutierrez 2009). The *Arabidopsis* homolog of Cdh1/Fzr is known as CCS52A (CELL CYCLE SWITCH 52A), and its loss of function produces a reduction in ploidy (Cebolla et al. 1999; Larson-Rabin et al. 2009). Interestingly, CCS52A is a target of E2FE/DEL1, one of the three atypical E2F encoded in the *Arabidopsis* genome (Ramirez-Parra et al. 2007; Lammens et al. 2008) and homolog of the human E2F7-9. Thus, the availability of APC/C activators is crucial for the endocycle switch. This is controlled in *Arabidopsis* by the UVI4 and GIG1/OSD1 (GIGAS CELLS 1/OMISSION OF SECOND DIVISION 1) proteins that interfere with the activity of the CDC20/FZY and CCS52A activators, respectively (Iwata et al. 2011). Remarkably, altering GIG1/OSD1 function increases ploidy through endomitosis whereas in the case of UVI4 is typically though endoreplication. CYCA2;3 levels are independently regulated by various proteins, such as ILP1 (Yoshizumi et al. 2006), a homolog of the mammalian GC-binding protein, and TCP15, a transcription factor that induces extra endocycles when it is overexpressed in cells genetically programmed to endoreplicate (Li et al. 2012).

Reduced activity of CDKB1;1, a non-PSTAIRE plant-specific CDK partner of CYCA2;3, increases endoreplication (Boudolf et al. 2009). CDKB1;1/CYCA2;3 activity is also required to phosphorylate several MYB transcription factors that activate the expression of G_2/M-specific genes (Araki et al. 2004; Kato et al. 2009). Plants contain two major classes of CDK inhibitors. One includes proteins with a domain conserved in animal p27 and is known as the KRP family, which in *Arabidopsis* contains seven members (Gutierrez 2009). The other is a plant-specific class constituted by SIAMESE (SIM) and SIM-RELATED (SMR) inhibitors (Walker et al. 2000; Churchman et al. 2006). Overexpression of KRP1/ICK1 and KRP2/ICK2, which are preferentially expressed in endoreplicating cells, affects the switch to endocycles in *Arabidopsis* (Wang et al. 2000; De Veylder et al. 2001; Verkest et al. 2005; Weinl et al. 2005), maize (Coelho et al. 2005), and tomato (Bisbis et al. 2006), whereas endoreplication is repressed in the *sim* mutants (Churchman et al. 2006). Interestingly, SIM interacts with the CDKA;1, the canonical PSTAIRE-type CDK, and with cyclin D, implicating this cyclin not only in the G_1/S transition but also in control of endocycles. Consistent with this, CYCD3 overexpression induces hyperplasia and reduction of the number of endoreplicated cells in various *Arabidopsis* organs (Dewitte et al. 2003) whereas a triple *cycd3;1-2-3* mutant contains reduced cell number and increased endoreplication levels (Dewitte et al. 2007). CDKA/CYCD3 are responsible for the phosphorylation of RBR (RETINOBLASTOMA-RELATED), the homolog of the mammalian RB protein. In fact, the RBR pathway regulates the number of endocycles once cells have exited the cell cycle (Park et al. 2005; Desvoyes et al. 2006; Borghi et al. 2010). Whether it is a direct effect or mediated by downstream RBR/E2F targets is not clear. For example, maintaining high levels of CDC6 and/or CDT1, which are regulated by RBR/E2F, during leaf development increases cellular ploidy (Castellano et al. 2001, 2004). Together, these data show a direct role of RBR/E2F, some of their targets, and CDK/cyclin in controlling endoreplication during development.

Several topoisomerases are also crucial for the endoreplication cycles (Fig. 1). Thus, mutations in the A and B subunits of topoisomerase VI, encoded by the *ROOT HAIRLESS2* (*RHL2*) and *HYPOCOTYL6* (*HYP6*) genes, prevent endocycle progression beyond the 8C DNA level (Hartung et al. 2002; Sugimoto-Shirasu et al. 2002). Likewise, cells of the *root hairless1/hypocotyl7* mutant that lack topoisomerase II show abnormal endocycle progression (Sugimoto-Shirasu et al. 2005). Therefore, functional topoisomerases are not needed for initiation of the endoreplication program, but they are at later stages, likely to handle topological constraints produced when the ploidy level increases. A systematic evaluation of the possible role of other DNA replication factors in endoreplication control would be enlightening.

A correct DNA replication-dependent deposition of histones H3 and H4 is required in *Arabidopsis* to prevent activation of the G_2 checkpoint (Fig. 1). Mutants in the *FASCIATA1* (*FAS1*) gene, encoding the large subunit of the H3-H4 chaperone chromatin assembly factor 1 (CAF-1), show a constitutive activation of the G_2 checkpoint, including transcriptional activation of *RAD51* and *BRCA1* genes, among others (Ramirez-Parra and Gutierrez 2007a). Mutant cells, instead of becoming permanently arrested or forced to enter a death pathway, trigger the endoreplication program ectopically and at developmental times in which it should not occur. It has been speculated that plants have evolved the ability to switch to the endocycle as a way to escape permanent cell cycle arrest during development (Ramirez-Parra and Gutierrez 2007b).

Although the endoreplication field is advancing quickly several key questions still remain unanswered. These include: (1) how licensing occurs in the absence of mitosis; (2) what is the replication timing and origin usage in relation to the S phase of the cell cycle; (3) what are the origins active in each endoreplication round and whether they are the same as those used during the cell cycle. Because the experimental approaches are now available, it is likely that studies in whole plants will contribute to significant advances in the coming years.

GEMINIVIRUS DNA REPLICATION

Geminiviruses are a family of plant viruses that have a significant impact in agricultural production worldwide. Their genome consists of 1 or 2 molecules of single-stranded DNA (ssDNA), ~2.4–3.1 kb in length, depending on the virus genera (Mastre-, Curto-, Topocu-, and Begomovirus). A comprehensive review of geminivirus biology has appeared recently (Jeske 2009). Others have covered aspects more related to DNA replication (Hanley-Bowdoin et al. 2004; Gutierrez 2006; Ascencio-Ibanez et al. 2008).

The replication cycle includes a first stage in which ssDNA is converted into dsDNA, a process performed entirely by host proteins. The dsDNA is used as a substrate in which 13–16 superhelical turns are introduced by association with histone octamers (Pilartz and Jeske 1992), much in the same way as it occurs with SV40 DNA. In the second stage, new viral ssDNA molecules are produced by a rolling-circle mechanism (Gutierrez 1999; Saunders 2008). The viral initiator protein (Rep) is a multifunctional protein with origin-specific DNA binding, homo- and heteroligomerization, and helicase and endonuclease activities (reviewed in Jeske 2009). Interestingly, Rep-binding sites at the replication origin are located in close proximity or partially overlapping the two divergent promoters responsible for early and late viral gene expression (reviewed in Gutierrez 2002). In addition to the rolling-circle replication pathway, a recombination-dependent replication mechanism that does not rely on a sequence-specific replication origin is also used (Jeske 2009).

A rich variety of interactions of viral early proteins with host factors have been identified as necessary for efficient viral DNA replication. These include: (1) inactivation of the retinoblastoma-related pathway (Xie et al. 1995; Kong et al. 2000), the G_2/M transition (Kong and Hanley-Bowdoin 2002; Nagar et al. 2002), and the host antiviral defense and silencing mechanisms (Vanitharani et al. 2005); (2) interaction with cell differentiation transcription factors (Xie et al. 1999; Selth et al. 2005), pathogen response (Ascencio-Ibanez et al. 2008), the ubiquitination (Lai et al. 2009; Zhang et al. 2011), and the sumoylation system (Castillo et al. 2004; Sanchez-Duran et al. 2011); and (3) recruitment of DNA replication factors such as RFC1 (Luque et al. 2002), PCNA (Castillo et al. 2003), RPA2 (Singh et al. 2007), and POLA2 and POLD2 (Lozano-Duran et al. 2011).

ACKNOWLEDGMENTS

We are indebted to the C.G. laboratory members for insightful discussions. C.C. and J.S.-M. are recipients of JAE-Doc (CSIC) and Juan de la Cierva (MICINN) contracts, respectively. Research is supported by grants BFU2009-9783 and CSD2007-00057-B (Ministry of Science and Education), and by an institutional grant from Fundación Ramón Areces.

REFERENCES

*Reference is also in this collection.

Amoroso A, Concia L, Maggio C, Raynaud C, Bergounioux C, Crespan E, Cella R, Maga G. 2011. Oxidative DNA damage bypass in *Arabidopsis thaliana* requires DNA polymerase lambda and proliferating cell nuclear antigen 2. *Plant Cell* **23:** 806–822.

Anderson HJ, Vonarx EJ, Pastushok L, Nakagawa M, Katafuchi A, Gruz P, Di Rubbo A, Grice DM, Osmond MJ, Sakamoto AN, et al. 2008. *Arabidopsis thaliana* Y-family DNA polymerase eta catalyses translesion synthesis and interacts functionally with PCNA2. *Plant J* **55:** 895–908.

Araki S, Ito M, Soyano T, Nishihama R, Machida Y. 2004. Mitotic cyclins stimulate the activity of c-Myb-like factors for transactivation of G2/M phase-specific genes in tobacco. *J Biol Chem* **279:** 32979–32988.

Arias EE, Walter JC. 2007. Strength in numbers: Preventing rereplication via multiple mechanisms in eukaryotic cells. *Genes Dev* **21:** 497–518.

Ascencio-Ibanez JT, Sozzani R, Lee TJ, Chu TM, Wolfinger RD, Cella R, Hanley-Bowdoin L. 2008. Global analysis of *Arabidopsis* gene expression uncovers a complex array of changes impacting pathogen response and cell cycle during geminivirus infection. *Plant Physiol* **148:** 436–454.

Balmukhanov TS, Erekenov AM, Ajtkhozhina NA. 1992. Isolation of DNA polymerase α from germinated wheat embryos. *Mol Biol Rep* **16:** 11–16.

Barrero JM, Gonzalez-Bayon R, del Pozo JC, Ponce MR, Micol JL. 2007. INCURVATA2 encodes the catalytic subunit of DNA Polymerase α and interacts with genes involved in chromatin-mediated cellular memory in *Arabidopsis thaliana*. *Plant Cell* **19:** 2822–2838.

* Bell SP, Kaguni JM. 2013. Helicase loading at chromosomal origins of replication. *Cold Spring Harb Perspect Biol* doi: 10.1101/cshperspect.a010124.

Bicknell LS, Walker S, Klingseisen A, Stiff T, Leitch A, Kerzendorfer C, Martin CA, Yeyati P, Al Sanna N, Bober M, et al. 2011. Mutations in ORC1, encoding the largest subunit of the origin recognition complex, cause microcephalic primordial dwarfism resembling Meier-Gorlin syndrome. *Nature Genet* **43:** 350–355.

Bisbis B, Delmas F, Joubes J, Sicard A, Hernould M, Inze D, Mouras A, Chevalier C. 2006. Cyclin-dependent kinase (CDK) inhibitors regulate the CDK-cyclin complex activities in endoreduplicating cells of developing tomato fruit. *J Biol Chem* **281:** 7374–7383.

Borghi L, Gutzat R, Futterer J, Laizet Y, Hennig L, Gruissem W. 2010. *Arabidopsis* RETINOBLASTOMA-RELATED is required for stem cell maintenance, cell differentiation, and lateral organ production. *Plant Cell* **22:** 1792–1811.

Boudolf V, Lammens T, Boruc J, Van Leene J, Van Den Daele H, Maes S, Van Isterdael G, Russinova E, Kondorosi E, Witters E, et al. 2009. CDKB1;1 forms a functional complex with CYCA2;3 to suppress endocycle onset. *Plant Physiol* **150:** 1482–1493.

Breuer C, Ishida T, Sugimoto K. 2010. Developmental control of endocycles and cell growth in plants. *Curr Opin Plant Biol* **13:** 654–660.

Bryant JA. 1980. Biochemical aspects of DNA replication with particular reference to plants. *Biol Rev Camb Philos Soc* **55:** 237–284.

Bryant JA. 2010. Replication of nuclear DNA. *Prog Bot* **71:** 25–60.

Bryant J, Fitchett P, Hughes S, Sibson D. 1992. DNA polymerase a in pea is part of a large multiprotein complex. *J Exp Bot* **43:** 31–40.

Caro E, Gutierrez C. 2007. A green GEM: Intriguing analogies with animal geminin. *Trends Cell Biol* **17:** 580–585.

Caro E, Castellano MM, Gutierrez C. 2007. A chromatin link that couples cell division to root epidermis patterning in *Arabidopsis*. *Nature* **447:** 213–217.

Caro E, Desvoyes B, Ramirez-Parra E, Sanchez Mde L, Gutierrez C. 2008. Endoreduplication control during plant development. *SEB Exp Biol Ser* **59:** 167–187.

Casas-Delucchi CS, van Bemmel JG, Haase S, Herce HD, Nowak D, Meilinger D, Stear JH, Leonhardt H, Cardoso MC. 2011. Histone hypoacetylation is required to maintain late replication timing of constitutive heterochromatin. *Nucleic Acids Res* doi: 10.1093/nar/gkr723.

Castellano MM, del Pozo JC, Ramirez-Parra E, Brown S, Gutierrez C. 2001. Expression and stability of *Arabidopsis* CDC6 are associated with endoreplication. *Plant Cell* **13:** 2671–2686.

Castellano MM, Boniotti MB, Caro E, Schnittger A, Gutierrez C. 2004. DNA replication licensing affects cell proliferation or endoreplication in a cell type-specific manner. *Plant Cell* **16:** 2380–2393.

Castillo AG, Collinet D, Deret S, Kashoggi A, Bejarano ER. 2003. Dual interaction of plant PCNA with geminivirus replication accessory protein (Ren) and viral replication protein (Rep). *Virology* **312:** 381–394.

Castillo AG, Kong LJ, Hanley-Bowdoin L, Bejarano ER. 2004. Interaction between a geminivirus replication protein and the plant sumoylation system. *J Virol* **78:** 2758–2769.

Castroviejo M, Gatius MT, Litvak S. 1990. A low molecular weight DNA polymerase from wheat embryos. *Plant Mol Biol* **15:** 383–397.

Cayrou C, Coulombe P, Vigneron A, Stanojcic S, Ganier O, Peiffer I, Rivals E, Puy A, Laurent-Chabalier S, Desprat R, et al. 2011. Genome-scale analysis of metazoan replication origins reveals their organization in specific but flexible sites defined by conserved features. *Genome Res* **21:** 1438–1449.

Cebolla A, Vinardell JM, Kiss E, Olah B, Roudier F, Kondorosi A, Kondorosi E. 1999. The mitotic inhibitor ccs52 is required for endoreduplication and ploidy-dependent cell enlargement in plants. *EMBO J* **18:** 4476–4484.

Chen CC, Tyler J. 2008. Chromatin reassembly signals the end of DNA repair. *Cell Cycle* **7:** 3792–3797.

Churchman ML, Brown ML, Kato N, Kirik V, Hulskamp M, Inze D, De Veylder L, Walker JD, Zheng Z, Oppenheimer DG, et al. 2006. SIAMESE, a plant-specific cell cycle regulator, controls endoreplication onset in *Arabidopsis thaliana. Plant Cell* **18:** 3145–3157.

Coello P, Vazquez-Ramos JM. 1995a. Maize DNA polymerase 2 is a phosphoprotein with increasing activity during germination. *Eur J Biochem* **231:** 99–103.

Coello P, Vazquez-Ramos JM. 1995b. Studies on the processivity of maize DNA polymerase 2, an [α]-type enzyme. *Plant Physiol* **109:** 645–650.

Coello P, Rodriquez R, Garcia E, Vazquez-Ramos JM. 1992. A DNA polymerase from maize axes: its purification and possible role. *Plant Mol Biol* **20:** 1159–1168.

Coelho CM, Dante RA, Sabelli PA, Sun Y, Dilkes BP, Gordon-Kamm WJ, Larkins BA. 2005. Cyclin-dependent kinase inhibitors in maize endosperm and their potential role in endoreduplication. *Plant Physiol* **138:** 2323–2336.

Costas C, Sanchez MP, Sequeira-Mendes J, Gutierrez C. 2011a. Progress in understanding DNA replication control. *Plant Scie (Shannon, Ireland)* **181:** 203–209.

Costas C, Sanchez MP, Stroud H, Yu Y, Oliveros JC, Feng S, Benguria A, Lopez-Vidriero I, Zhang X, Solano R, et al. 2011b. Genome-wide mapping of *Arabidopsis* origins of DNA replication and their associated epigenetic marks. *Nat Struct Mol Biol* **18:** 395–400.

De Veylder L, Beemster GT, Beeckman T, Inze D. 2001. CKS1At overexpression in *Arabidopsis thaliana* inhibits growth by reducing meristem size and inhibiting cell-cycle progression. *Plant J* **25:** 617–626.

De Veylder L, Larkin JC, Schnittger A. 2011. Molecular control and function of endoreplication in development and physiology. *Trends Plant Sci* **16:** 624–634.

del Olmo I, Lopez-Gonzalez L, Martin-Trillo MM, Martinez-Zapater JM, Pineiro M, Jarillo JA. 2010. EARLY IN SHORT DAYS 7 (ESD7) encodes the catalytic subunit of DNA polymerase epsilon and is required for flowering repression through a mechanism involving epigenetic gene silencing. *Plant J* **61:** 623–636.

DePamphilis M, Bell S. 2011. *Genome duplication. Concepts, mechanisms, evolution, and disease.* Garland Science, Taylor & Francis Group London.

Desvoyes B, Ramirez-Parra E, Xie Q, Chua NH, Gutierrez C. 2006. Cell type-specific role of the retinoblastoma/E2F pathway during *Arabidopsis* leaf development. *Plant Physiol* **140:** 67–80.

Dewitte W, Riou-Khamlichi C, Scofield S, Healy JM, Jacqmard A, Kilby NJ, Murray JA. 2003. Altered cell cycle distribution, hyperplasia, and inhibited differentiation in *Arabidopsis* caused by the D-type cyclin CYCD3. *Plant Cell* **15:** 79–92.

Dewitte W, Scofield S, Alcasabas AA, Maughan SC, Menges M, Braun N, Collins C, Nieuwland J, Prinsen E, Sundaresan V, et al. 2007. *Arabidopsis* CYCD3 D-type cyclins link cell proliferation and endocycles and are rate-limiting for cytokinin responses. *Proc Natl Acad Sci* **104:** 14537–14542.

Diaz-Trivino S, Castellano MM, Sanchez MP, Ramirez-Parra E, Desvoyes B, Gutierrez C. 2005. The genes encoding *Arabidopsis* ORC subunits are E2F targets and the two ORC1 genes are differently expressed in proliferating and endoreplicating cells. *Nucleic Acids Res* **33:** 5404–5414.

Ding Q, MacAlpine DM. 2010. Preferential re-replication of *Drosophila* heterochromatin in the absence of geminin. *PLoS Genet* **6:** e1001112.

Dresselhaus T, Srilunchang KO, Leljak-Levanic D, Schreiber DN, Garg P. 2006. The fertilization-induced DNA replication factor MCM6 of maize shuttles between cytoplasm and nucleus, and is essential for plant growth and development. *Plant Physiol* **140:** 512–527.

Elmayan T, Proux F, Vaucheret H. 2005. *Arabidopsis* RPA2: A genetic link among transcriptional gene silencing, DNA repair, and DNA replication. *Curr Biol* **15:** 1919–1925.

Fischer A, Hofmann I, Naumann K, Reuter G. 2006. Heterochromatin proteins and the control of heterochromatic gene silencing in *Arabidopsis. J Plant Physiol* **163:** 358–368.

Fuchs J, Demidov D, Houben A, Schubert I. 2006. Chromosomal histone modification patterns–from conservation to diversity. *Trends Plant Sci* **11:** 199–208.

Furukawa T, Ishibashi T, Kimura S, Tanaka H, Hashimoto J, Sakaguchi K. 2003. Characterization of all the subunits of replication factor C from a higher plant, rice (*Oryza sativa* L.), and their relation to development. *Plant Mol Biol* **53:** 15–25.

Gambus A, Jones RC, Sanchez-Diaz A, Kanemaki M, van Deursen F, Edmondson RD, Labib K. 2006. GINS maintains association of Cdc45 with MCM in replisome progression complexes at eukaryotic DNA replication forks. *Nat Cell Biol* **8:** 358–366.

Garcia E, Orjuela D, Camacho Y, Zuniga JJ, Plasencia J, Vazquez-Ramos JM. 1997. Comparison among DNA polymerases 1, 2 and 3 from maize embryo axes. A DNA primase activity copurifies with DNA polymerase 2. *Plant Mol Biol* **33:** 445–455.

Garcia E, Quiroz F, Uchiyama Y, Sakaguchi K, Vazquez-Ramos JM. 2006. Expression of a maize δ-type DNA polymerase during seed germination. *Physiol Plant* **127:** 268–276.

Gomez Roig E, Vazquez-Ramos JM. 2003. Maize DNA polymerase α is phosphorylated by a PCNA-associated cyclin/Cdk complex: Effect of benzyladenine. *J Plant Physiol* **160:** 983–990.

Gutierrez C. 1999. Geminivirus DNA replication. *Cell Mol Life Sci* **56:** 313–329.

Gutierrez C. 2002. Strategies for geminivirus DNA replication and cell cycle interference. *Physiol Plant Mol Biol* **69:** 219–230.

Gutierrez C. 2005. Coupling cell proliferation and development in plants. *Nat Cell Biol* **7:** 535–541.

Gutierrez C. 2006. Plant cell and viruses. In *DNA replication and human disease* (ed. DePamphilis ML), pp. 257–271. Cold Spring Harbor Laboratory Press, Cold Spring Harbor, NY.

Gutierrez C. 2009. The *Arabidopsis* cell cycle. In *The arabidopsis book.* ASPB, http://www.aspb.org/publications/arabidopsis/.

Hanley-Bowdoin L, Settlage SB, Robertson D. 2004. Reprogramming plant gene expression: A prerequisite to geminivirus DNA replication. *Mol Plant Pathol* **5:** 149–156.

Hartford SA, Luo Y, Southard TL, Min IM, Lis JT, Schimenti JC. 2011. Minichromosome maintenance helicase paralog MCM9 is dispensable for DNA replication but functions in germ-line stem cells and tumor suppression. *Proc Natl Acad Sci* **108:** 17702–17707.

Hartung F, Angelis KJ, Meister A, Schubert I, Melzer M, Puchta H. 2002. An archaebacterial topoisomerase homolog not present in other eukaryotes is indispensable for cell proliferation of plants. *Curr Biol* **12:** 1787–1791.

Hata S, Tsukamoto T, Osumi T, Hashimoto J, Suzuka I. 1992. Analysis of carrot genes for proliferating cell nuclear antigen homologs with the aid of the polymerase chain reaction. *Biochem Biophys Res Commun* **184:** 576–581.

* Hedglin M, Kumar R, Benkovic SJ. 2013. Replication clamps and clamp loaders. *Cold Spring Harb Perspect Biol* doi: 10.1101/cshperspect.a010165.

Hernandez P, Martin-Parras L, Martinez-Robles ML, Schvartzman JB. 1993. Conserved features in the mode of replication of eukaryotic ribosomal RNA genes. *EMBO J* **12:** 1475–1485.

Herrera I, Sánchez MP, Molina J, Plasencia J, Vázquez-Ramos J. 2000. Proliferating cell nuclear antigen expression in maize seed development and germination: Regulation by phytohormones and its association with putative cell cycle proteins. *Physiol Plant* **110:** 127–134.

Hickman MA, Rusche LN. 2010. Transcriptional silencing functions of the yeast protein Orc1/Sir3 subfunctionalized after gene duplication. *Proc Natl Acad Sci* **107:** 19384–19389.

Hiratani I, Takebayashi S, Lu J, Gilbert DM. 2009. Replication timing and transcriptional control: Beyond cause and effect—part II. *Curr Opin Genet Dev* **19:** 142–149.

Howell SH, Stern H. 1971. The appearance of DNA breakage and repair activities in the synchronous meiotic cycle of Lilium. *J Mol Biol* **55:** 357–378.

Inze D, De Veylder L. 2006. Cell cycle regulation in plant development. *Annu Rev Genet* **40:** 77–105.

Ishibashi T, Kimura S, Sakaguchi K. 2006. A higher plant has three different types of RPA heterotrimeric complex. *J Biochem* **139:** 99–104.

Iwata E, Ikeda S, Matsunaga S, Kurata M, Yoshioka Y, Criqui MC, Genschik P, Ito M. 2011. GIGAS CELL1, a novel negative regulator of the anaphase-promoting complex/cyclosome, is required for proper mitotic progression and cell fate determination in *Arabidopsis*. *Plant Cell* **23:** 4382–4393.

* Jackson D, Wang X, Rudner DZ. 2012. Spatio-temporal organization of replication in bacteria and eukaryotes (nucleoids and nuclei). *Cold Spring Harb Perspect Biol* **4:** a010389.

Jacob Y, Feng S, LeBlanc CA, Bernatavichute YV, Stroud H, Cokus S, Johnson LM, Pellegrini M, Jacobsen SE, Michaels SD. 2009. ATXR5 and ATXR6 are H3K27 monomethyltransferases required for chromatin structure and gene silencing. *Nat Struct Mol Biol* **16:** 763–768.

Jacob Y, Stroud H, Leblanc C, Feng S, Zhuo L, Caro E, Hassel C, Gutierrez C, Michaels SD, Jacobsen SE. 2010. Regulation of heterochromatic DNA replication by histone H3 lysine 27 methyltransferases. *Nature* **466:** 987–991.

Jasencakova Z, Soppe WJ, Meister A, Gernand D, Turner BM, Schubert I. 2003. Histone modifications in *Arabidopsis*—high methylation of H3 lysine 9 is dispensable for constitutive heterochromatin. *Plant J* **33:** 471–480.

Jenik PD, Jurkuta RE, Barton MK. 2005. Interactions between the cell cycle and embryonic patterning in *Arabidopsis* uncovered by a mutation in DNA polymerase epsilon. *Plant Cell* **17:** 3362–3377.

Jeske H. 2009. Geminiviruses. *Curr Top Microbiol Immunol* **331:** 185–226.

* Johansson E, Dixon N. 2013. Replicative DNA polymerases. *Cold Spring Harb Perspect Biol* doi: 10.1101/cshperspect.a012799.

Kanke M, Kodama Y, Takahashi TS, Nakagawa T, Masukata H. 2012. Mcm10 plays an essential role in origin DNA unwinding after loading of the CMG components. *EMBO J* **31:** 2182–2194.

Kapoor A, Agarwal M, Andreucci A, Zheng X, Gong Z, Hasegawa PM, Bressan RA, Zhu JK. 2005. Mutations in a conserved replication protein suppress transcriptional gene silencing in a DNA-methylation-independent manner in *Arabidopsis*. *Curr Biol* **15:** 1912–1918.

Karnani N, Taylor CM, Malhotra A, Dutta A. 2010. Genomic study of replication initiation in human chromosomes reveals the influence of transcription regulation and chromatin structure on origin selection. *Mol Biol Cell* **21:** 393–404.

Kato K, Galis I, Suzuki S, Araki S, Demura T, Criqui MC, Potuschak T, Genschik P, Fukuda H, Matsuoka K, et al. 2009. Preferential up-regulation of G2/M phase-specific genes by overexpression of the hyperactive form of NtmybA2 lacking its negative regulation domain in tobacco BY-2 cells. *Plant Physiol* **149:** 1945–1957.

Kessler B. 1971. Isolation, characterization and distribution of a DNA ligase from higher plants. *Biochim Biophys Acta* **240:** 496–505.

Kimura S, Kai M, Kobayashi H, Suzuki A, Morioka H, Otsuka E, Sakaguchi K. 1997. A structure-specific endonuclease from cauliflower (Brassica oleracea var. botrytis) inflorescence. *Nucleic Acids Res* **25:** 4970–4976.

Kimura S, Ueda T, Hatanaka M, Takenouchi M, Hashimoto J, Sakaguchi K. 2000. Plant homologue of flap endonuclease-1: Molecular cloning, characterization, and evidence of expression in meristematic tissues. *Plant Mol Biol* **42:** 415–427.

Kimura S, Suzuki T, Yanagawa Y, Yamamoto T, Nakagawa H, Tanaka I, Hashimoto J, Sakaguchi K. 2001. Characterization of plant proliferating cell nuclear antigen (PCNA) and flap endonuclease-1 (FEN-1), and their distribution in mitotic and meiotic cell cycles. *Plant J* **28:** 643–653.

Kimura S, Furukawa T, Kasai N, Mori Y, Kitamoto HK, Sugawara F, Hashimoto J, Sakaguchi K. 2003. Functional characterization of two flap endonuclease-1 homologues in rice. *Gene* **314:** 63–71.

Kodama H, Ito M, Ohnishi N, Suzuka I, Komamine A. 1991. Molecular cloning of the gene for plant proliferating-cell nuclear antigen and expression of this gene during the cell cycle in synchronized cultures of Catharanthus roseus cells. *Eur J Biochem* **197:** 495–503.

Kong LJ, Hanley-Bowdoin L. 2002. A geminivirus replication protein interacts with a protein kinase and a motor protein that display different expression patterns during plant development and infection. *Plant Cell* **14:** 1817–1832.

Kong LJ, Orozco BM, Roe JL, Nagar S, Ou S, Feiler HS, Durfee T, Miller AB, Gruissem W, Robertson D, et al.

2000. A geminivirus replication protein interacts with the retinoblastoma protein through a novel domain to determine symptoms and tissue specificity of infection in plants. *EMBO J* **19:** 3485–3495.

Kuo AJ, Song J, Cheung P, Ishibe-Murakami S, Yamazoe S, Chen JK, Patel DJ, Gozani O. 2012. The BAH domain of ORC1 links H4K20me2 to DNA replication licensing and Meier-Gorlin syndrome. *Nature* **484:** 115–119.

Lai J, Chen H, Teng K, Zhao Q, Zhang Z, Li Y, Liang L, Xia R, Wu Y, Guo H, et al. 2009. RKP, a RING finger E3 ligase induced by BSCTV C4 protein, affects geminivirus infection by regulation of the plant cell cycle. *Plant J* **57:** 905–917.

Lammens T, Boudolf V, Kheibarshekan L, Zalmas LP, Gaamouche T, Maes S, Vanstraelen M, Kondorosi E, La Thangue NB, Govaerts W, et al. 2008. Atypical E2F activity restrains APC/CCCS52A2 function obligatory for endocycle onset. *Proc Natl Acad Sci* **105:** 14721–14726.

Laquel P, Castroviejo M, Litvak S. 1990. Further biochemical characterization of wheat DNA primase: Possible functional implication of copurification with DNA polymerase A. *Nucleic Acids Res* **18:** 4867–4876.

Laquel P, Litvak S, Castroviejo M. 1993. Mammalian proliferating cell nuclear antigen stimulates the processivity of two wheat embryo DNA polymerases. *Plant Physiol* **102:** 107–114.

Larson-Rabin Z, Li Z, Masson PH, Day CD. 2009. FZR2/CCS52A1 expression is a determinant of endoreduplication and cell expansion in *Arabidopsis*. *Plant Physiol* **149:** 874–884.

Lee TJ, Pascuzzi PE, Settlage SB, Shultz RW, Tanurdzic M, Rabinowicz PD, Menges M, Zheng P, Main D, Murray JA, et al. 2010. *Arabidopsis thaliana* chromosome 4 replicates in two phases that correlate with chromatin state. *PLoS Genet* **6:** e1000982.

* Leonard AC, Méchali M. 2013. DNA replication origins. *Cold Spring Harb Perspect Biol* doi: 10.1101/cshperspect.a010116.

Li ZY, Li B, Dong AW. 2012. The *Arabidopsis* transcription factor AtTCP15 regulates endoreduplication by modulating expression of key cell-cycle genes. *Mol Plant* **5:** 270–280.

Liu Q, Gong Z. 2011. The coupling of epigenome replication with DNA replication. *Curr Opin Plant Biol* **14:** 187–194.

Liu J, Ren X, Yin H, Wang Y, Xia R, Wang Y, Gong Z. 2010a. Mutation in the catalytic subunit of DNA polymerase α influences transcriptional gene silencing and homologous recombination in *Arabidopsis*. *Plant J* **61:** 36–45.

Liu Q, Wang J, Miki D, Xia R, Yu W, He J, Zheng Z, Zhu JK, Gong Z. 2010b. DNA replication factor C1 mediates genomic stability and transcriptional gene silencing in *Arabidopsis*. *Plant Cell* **22:** 2336–2352.

Lopez I, Khan S, Vazquez-Ramos J, Hussey PJ. 1995. Molecular cloning of a maize cDNA clone encoding a putative proliferating cell nuclear antigen. *Biochim Biophys Acta* **1260:** 119–121.

Lozano-Duran R, Rosas-Diaz T, Luna AP, Bejarano ER. 2011. Identification of host genes involved in geminivirus infection using a reverse genetics approach. *PLoS ONE* **6:** e22383.

Luque AE, Benedetto JP, Castroviejo M. 1998. Wheat DNA polymerase CI: A homologue of rat DNA polymerase β. *Plant Mol Biol* **38:** 647–654.

Luque AE, Toueille M, Benedetto JP, Castroviejo M. 2000. Subunit composition of DNA polymerases A and B from wheat cell. *Plant Physiol Biochem* **38:** 551–558.

Luque A, Sanz-Burgos AP, Ramirez-Parra E, Castellano MM, Gutierrez C. 2002. Interaction of geminivirus Rep protein with replication factor C and its potential role during geminivirus DNA replication. *Virology* **302:** 83–94.

Lutzmann M, Grey C, Traver S, Ganier O, Maya-Mendoza A, Ranisavljevic N, Bernex F, Nishiyama A, Montel N, Gavois E, et al. 2012. MCM8- and MCM9-deficient mice reveal gametogenesis defects and genome instability due to impaired homologous recombination. *Mol Cell* **47:** 523–534.

MacAlpine DM, Rodriguez HK, Bell SP. 2004. Coordination of replication and transcription along a *Drosophila* chromosome. *Genes Dev* **18:** 3094–3105.

Macalpine HK, Gordan R, Powell SK, Hartemink AJ, Macalpine DM. 2010. *Drosophila* ORC localizes to open chromatin and marks sites of cohesin complex loading. *Genome Res* **20:** 201–211.

Masuda HP, Ramos GB, de Almeida-Engler J, Cabral LM, Coqueiro VM, Macrini CM, Ferreira PC, Hemerly AS. 2004. Genome based identification and analysis of the pre-replicative complex of *Arabidopsis thaliana*. *FEBS Lett* **574:** 192–202.

Masuda HP, Cabral LM, De Veylder L, Tanurdzic M, de Almeida Engler J, Geelen D, Inze D, Martienssen RA, Ferreira PC, Hemerly AS. 2008. ABAP1 is a novel plant Armadillo BTB protein involved in DNA replication and transcription. *EMBO J* **27:** 2746–2756.

Mathieu O, Probst AV, Paszkowski J. 2005. Distinct regulation of histone H3 methylation at lysines 27 and 9 by CpG methylation in *Arabidopsis*. *EMBO J* **24:** 2783–2791.

Matsumoto T, Hata S, Suzuka I, Hashimoto J. 1994. Expression of functional proliferating-cell nuclear antigen from rice (Oryza sativa) in Escherichia coli. Activity in association with human DNA polymerase δ. *Eur J Biochem* **223:** 179–187.

Mechali M. 2010. Eukaryotic DNA replication origins: Many choices for appropriate answers. *Nat Rev* **11:** 728–738.

Miotto B, Struhl K. 2010. HBO1 histone acetylase activity is essential for DNA replication licensing and inhibited by Geminin. *Mol Cell* **37:** 57–66.

Moyer SE, Lewis PW, Botchan MR. 2006. Isolation of the Cdc45/Mcm2–7/GINS (CMG) complex, a candidate for the eukaryotic DNA replication fork helicase. *Proc Natl Acad Sci* **103:** 10236–10241.

Nagar S, Hanley-Bowdoin L, Robertson D. 2002. Host DNA replication is induced by geminivirus infection of differentiated plant cells. *Plant Cell* **14:** 2995–3007.

Nishimura K, Ishiai M, Horikawa K, Fukagawa T, Takata M, Takisawa H, Kanemaki MT. 2012. Mcm8 and Mcm9 form a complex that functions in homologous recombination repair induced by DNA interstrand crosslinks. *Mol Cell* **47:** 511–522.

Nitta T, Nagata T. 1976. DNA fiber autoradiography of tobacco mesophyll protoplasts. *Chromosoma* **59:** 83–88.

Noguchi K, Vassilev A, Ghosh S, Yates JL, DePamphilis ML. 2006. The BAH domain facilitates the ability of human Orc1 protein to activate replication origins in vivo. *EMBO J* **25:** 5372–5382.

Park JA, Ahn JW, Kim YK, Kim SJ, Kim JK, Kim WT, Pai HS. 2005. Retinoblastoma protein regulates cell proliferation, differentiation, and endoreduplication in plants. *Plant J* **42:** 153–163.

Pilartz M, Jeske H. 1992. Abutilon mosaic geminivirus double-stranded DNA is packed into minichromosomes. *Virology* **189:** 800–802.

Przykorska A, Solecka K, Olszak K, Keith G, Nawrot B, Kuligowska E. 2004. Wheat (Triticum vulgare) chloroplast nuclease ChSI exhibits 5′ flap structure-specific endonuclease activity. *Biochemistry* **43:** 11283–11294.

Ramirez-Parra E, Gutierrez C. 2007a. E2F regulates FASCIATA1, a chromatin assembly gene whose loss switches on the endocycle and activates gene expression by changing the epigenetic status. *Plant Physiol* **144:** 105–120.

Ramirez-Parra E, Gutierrez C. 2007b. The many faces of chromatin assembly factor 1. *Trends Plant Sci* **12:** 570–576.

Ramirez-Parra E, del Pozo JC, Desvoyes B, Sanchez MP, Gutierrez C. 2007. E2F-DP transcription factors. In *Cell cycle control and plant development* (ed. Inzé D), pp. 138–162. Blackwell, Oxford.

Raynaud C, Sozzani R, Glab N, Domenichini S, Perennes C, Cella R, Kondorosi E, Bergounioux C. 2006. Two cell-cycle regulated SET-domain proteins interact with proliferating cell nuclear antigen (PCNA) in *Arabidopsis*. *Plant J* **47:** 395–407.

Richard MC, Litvak S, Castroviejo M. 1991. DNA polymerase B from wheat embryos: A plant δ-like DNA polymerase. *Arch Biochem Biophys* **287:** 141–150.

Ronceret A, Guilleminot J, Lincker F, Gadea-Vacas J, Delorme V, Bechtold N, Pelletier G, Delseny M, Chaboute ME, Devic M. 2005. Genetic analysis of two *Arabidopsis* DNA polymerase epsilon subunits during early embryogenesis. *Plant J* **44:** 223–236.

Roudier F, Ahmed I, Berard C, Sarazin A, Mary-Huard T, Cortijo S, Bouyer D, Caillieux E, Duvernois-Berthet E, Al-Shikhley L, et al. 2011. Integrative epigenomic mapping defines four main chromatin states in *Arabidopsis*. *EMBO J* **30:** 1928–1938.

Rufiange A, Jacques PE, Bhat W, Robert F, Nourani A. 2007. Genome-wide replication-independent histone H3 exchange occurs predominantly at promoters and implicates H3 K56 acetylation and Asf1. *Mol Cell* **27:** 393–405.

Sabelli PA, Burgess SR, Kush AK, Young MR, Shewry PR. 1996. cDNA cloning and characterisation of a maize homologue of the MCM proteins required for the initiation of DNA replication. *Mol Gen Genet* **252:** 125–136.

Sabelli PA, Burgess SR, Valasek L, Shewry PR. 1999. Molecular cloning and characterisation of a maize cDNA for a homologue of the large subunit of the eukaryotic initiation factor 3 (eIF3). *Mol Gen Genet* **261:** 820–830.

Sakaguchi K, Ishibashi T, Uchiyama Y, Iwabata K. 2009. The multi-replication protein A (RPA) system—a new perspective. *FEBS J* **276:** 943–963.

Sanchez MP, Gutierrez C. 2009. *Arabidopsis* ORC1 is a PHD-containing H3K4me3 effector that regulates transcription. *Proc Natl Acad Sci* **106:** 2065–2070.

Sanchez MP, Torres A, Boniotti MB, Gutierrez C, Vazquez-Ramo JM. 2002. PCNA protein associates to Cdk-A type protein kinases in germinating maize. *Plant Mol Biol* **50:** 167–175.

Sanchez MP, Gurusinghe SH, Bradford KJ, Vazquez-Ramos JM. 2005. Differential response of PCNA and Cdk-A proteins and associated kinase activities to benzyladenine and abscisic acid during maize seed germination. *J Exp Bot* **56:** 515–523.

Sanchez-Duran MA, Dallas MB, Ascencio-Ibanez JT, Reyes MI, Arroyo-Mateos M, Ruiz-Albert J, Hanley-Bowdoin L, Bejarano ER. 2011. Interaction between geminivirus replication protein and the SUMO-conjugating enzyme is required for viral infection. *J Virol* **85:** 9789–9800.

Sasaki T, Gilbert DM. 2007. The many faces of the origin recognition complex. *Curr Opin Cell Biol* **19:** 337–343.

Saunders K. 2008. Analysis of geminivirus DNA replication by 2-D gel. *Methods Mol Biol* **51:** 135–143.

Schmid M, Davison TS, Henz SR, Pape UJ, Demar M, Vingron M, Scholkopf B, Weigel D, Lohmann JU. 2005. A gene expression map of *Arabidopsis thaliana* development. *Nat Genet* **37:** 501–506.

Schwaiger M, Stadler MB, Bell O, Kohler H, Oakeley EJ, Schubeler D. 2009. Chromatin state marks cell-type- and gender-specific replication of the *Drosophila* genome. *Genes Dev* **23:** 589–601.

Selth LA, Dogra SC, Rasheed MS, Healy H, Randles JW, Rezaian MA. 2005. A NAC domain protein interacts with tomato leaf curl virus replication accessory protein and enhances viral replication. *Plant Cell* **17:** 311–325.

Shimizu S, Mori H. 1998. Analysis of cycles of dormancy and growth in pea axillary buds based on mRNA accumulation patterns of cell cycle-related genes. *Plant Cell Physiol* **39:** 255–262.

Shultz RW, Tatineni VM, Hanley-Bowdoin L, Thompson WF. 2007. Genome-wide analysis of the core DNA replication machinery in the higher plants *Arabidopsis* and rice. *Plant Physiol* **144:** 1697–1714.

Shultz RW, Lee TJ, Allen GC, Thompson WF, Hanley-Bowdoin L. 2009. Dynamic localization of the DNA replication proteins MCM5 and MCM7 in plants. *Plant Physiol* **150:** 658–669.

Singh DK, Islam MN, Choudhury NR, Karjee S, Mukherjee SK. 2007. The 32 kDa subunit of replication protein A (RPA) participates in the DNA replication of Mung bean yellow mosaic India virus (MYMIV) by interacting with the viral Rep protein. *Nucleic Acids Res* **35:** 755–770.

Stevens R, Mariconti L, Rossignol P, Perennes C, Cella R, Bergounioux C. 2002. Two E2F sites in the *Arabidopsis* MCM3 promoter have different roles in cell cycle activation and meristematic expression. *J Biol Chem* **277:** 32978–32984.

Stevens R, Grelon M, Vezon D, Oh J, Meyer P, Perennes C, Domenichini S, Bergounioux C. 2004. A CDC45 homolog in *Arabidopsis* is essential for meiosis, as shown by RNA interference-induced gene silencing. *Plant Cell* **16:** 99–113.

Stroud H, Hale CJ, Feng S, Caro E, Jacob Y, Michaels SD, Jacobsen SE. 2012a. DNA methyltransferases are required to induce heterochromatic re-replication in *Arabidopsis*. *PLoS Genet* **8:** e1002808.

Stroud H, Otero S, Desvoyes B, Ramírez-Parra E, Jacobsen SE, Gutierrez C. 2012b. Genome-wide analysis of histone H3.1 and H3.3 variants in *Arabidopsis thaliana*. *Proc Natl Acad Sci* **109:** 5379–5375.

Strzalka W, Oyama T, Tori K, Morikawa K. 2009. Crystal structures of the *Arabidopsis thaliana* proliferating cell nuclear antigen 1 and 2 proteins complexed with the human p21 C-terminal segment. *Protein Sci* **18:** 1072–1080.

Strzalka W, Kaczmarek A, Naganowska B, Ziemienowicz A. 2010. Identification and functional analysis of PCNA1 and PCNA-like1 genes of Phaseolus coccineus. *J Exp Bot* **61:** 873–888.

Sugimoto-Shirasu K, Stacey NJ, Corsar J, Roberts K, McCann MC. 2002. DNA topoisomerase VI is essential for endoreduplication in *Arabidopsis*. *Curr Biol* **12:** 1782–1786.

Sugimoto-Shirasu K, Roberts GR, Stacey NJ, McCann MC, Maxwell A, Roberts K. 2005. RHL1 is an essential component of the plant DNA topoisomerase VI complex and is required for ploidy-dependent cell growth. *Proc Natl Acad Sci* **102:** 18736–18741.

Suzuka I, Daidoji H, Matsuoka M, Kadowaki K, Takasaki Y, Nakane PK, Moriuchi T. 1989. Gene for proliferating-cell nuclear antigen (DNA polymerase δ auxiliary protein) is present in both mammalian and higher plant genomes. *Proc Natl Acad Sci* **86:** 3189–3193.

Takahashi N, Lammens T, Boudolf V, Maes S, Yoshizumi T, De Jaeger G, Witters E, Inze D, De Veylder L. 2008. The DNA replication checkpoint aids survival of plants deficient in the novel replisome factor ETG1. *EMBO J* **27:** 1840–1851.

Takahashi N, Quimbaya M, Schubert V, Lammens T, Vandepoele K, Schubert I, Matsui M, Inze D, Berx G, De Veylder L. 2010. The MCM-binding protein ETG1 aids sister chromatid cohesion required for postreplicative homologous recombination repair. *PLoS Genet* **6:** e1000817.

Tardat M, Brustel J, Kirsh O, Lefevbre C, Callanan M, Sardet C, Julien E. 2010. The histone H4 Lys 20 methyltransferase PR-Set7 regulates replication origins in mammalian cells. *Nat Cell Biol* **12:** 1086–1093.

Taylor JH, Woods PS, Hughes WL. 1957. The organization and duplication of chromosomes as revealed by autoradiographic studies using tritium-labeled Thymidinee. *Proc Natl Acad Sci* **43:** 122–128.

Taylor RM, Hamer MJ, Rosamond J, Bray CM. 1998. Molecular cloning and functional analysis of the *Arabidopsis thaliana* DNA ligase I homologue. *Plant J* **14:** 75–81.

Uchiyama Y, Hatanaka M, Kimura S, Ishibashi T, Ueda T, Sakakibara Y, Matsumoto T, Furukawa T, Hashimoto J, Sakaguchi K. 2002. Characterization of DNA polymerase δ from a higher plant, rice (Oryza sativa L.). *Gene* **295:** 19–26.

Valenzuela MS, Chen Y, Davis S, Yang F, Walker RL, Bilke S, Lueders J, Martin MM, Aladjem MI, Massion PP, et al. 2011. Preferential localization of human origins of DNA replication at the 5′-ends of expressed genes and at evolutionarily conserved DNA sequences. *PLoS ONE* **6:** e17308.

van Deursen F, Sengupta S, De Piccoli G, Sanchez-Diaz A, Labib K. 2012. Mcm10 associates with the loaded DNA helicase at replication origins and defines a novel step in its activation. *EMBO J* **31:** 2195–2206.

Vanitharani R, Chellappan P, Fauquet CM. 2005. Geminiviruses and RNA silencing. *Trends Plant Sci* **10:** 144–151.

Van't Hof J. 1976. Replicon size and rate of fork movement in early S of higher plant cells (*Pisum sativum*). *Exp Cell Res* **103:** 395–403.

Van't Hof J, Bjerknes CA. 1981. Similar replicon properties of higher plant cells with different S periods and genome sizes. *Exp Cell Res* **136:** 461–465.

Van't Hof J, Lamm SS. 1992. Site of initiation of replication of the ribosomal genes of pea (Pisum sativum) detected by two-dimensional gel electrophoresis. *Plant Mol Biol* **20:** 377–382.

Van't Hof J, Kuniyuki A, Bjerknes CA. 1978. The size and number of replicon families of chromosomal DNA of *Arabidopsis thaliana*. *Chromosoma* **68:** 269–285.

Van't Hof J, Hernandez P, Bjerknes CA, Lamm SS. 1987. Location of the replication origin in the 9 kb repeat size class of rDNA in pea (Pisum sativum). *Plant Mol Biol* **9:** 87–96.

Verkest A, Manes CL, Vercruysse S, Maes S, Van Der Schueren E, Beeckman T, Genschik P, Kuiper M, Inze D, De Veylder L. 2005. The cyclin-dependent kinase inhibitor KRP2 controls the onset of the endoreduplication cycle during *Arabidopsis* leaf development through inhibition of mitotic CDKA;1 kinase complexes. *Plant Cell* **17:** 1723–1736.

Walker JD, Oppenheimer DG, Concienne J, Larkin JC. 2000. SIAMESE, a gene controlling the endoreduplication cell cycle in *Arabidopsis thaliana* trichomes. *Development* **127:** 3931–3940.

Wang H, Zhou Y, Gilmer S, Whitwill S, Fowke LC. 2000. Expression of the plant cyclin-dependent kinase inhibitor ICK1 affects cell division, plant growth and morphology. *Plant J* **24:** 613–623.

Watase G, Takisawa H, Kanemaki MT. 2012. Mcm10 plays a role in functioning of the eukaryotic replicative DNA helicase, Cdc45-Mcm-GINS. *Curr Biol* **22:** 343–349.

Weinl C, Marquardt S, Kuijt SJ, Nowack MK, Jakoby MJ, Hulskamp M, Schnittger A. 2005. Novel functions of plant cyclin-dependent kinase inhibitors, ICK1/KRP1, can act non-cell-autonomously and inhibit entry into mitosis. *Plant Cell* **17:** 1704–1722.

Witmer X, Alvarez-Venegas R, San-Miguel P, Danilevskaya O, Avramova Z. 2003. Putative subunits of the maize origin of replication recognition complex ZmORC1-ZmORC5. *Nucleic Acids Res* **31:** 619–628.

Xia R, Wang J, Liu C, Wang Y, Wang Y, Zhai J, Liu J, Hong X, Cao X, Zhu JK, et al. 2006. ROR1/RPA2A, a putative replication protein A2, functions in epigenetic gene silencing and in regulation of meristem development in *Arabidopsis*. *Plant Cell* **18:** 85–103.

Xie Q, Suarez-Lopez P, Gutierrez C. 1995. Identification and analysis of a retinoblastoma binding motif in the replication protein of a plant DNA virus: Requirement for efficient viral DNA replication. *EMBO J* **14:** 4073–4082.

Xie Q, Sanz-Burgos AP, Guo H, Garcia JA, Gutierrez C. 1999. GRAB proteins, novel members of the NAC domain family, isolated by their interaction with a geminivirus protein. *Plant Mol Biol* **39:** 647–656.

Yin H, Zhang X, Liu J, Wang Y, He J, Yang T, Hong X, Yang Q, Gong Z. 2009. Epigenetic regulation, somatic homologous recombination, and abscisic acid signaling are influenced by DNA polymerase epsilon mutation in *Arabidopsis*. *Plant Cell* **21:** 386–402.

Yoshizumi T, Tsumoto Y, Takiguchi T, Nagata N, Yamamoto YY, Kawashima M, Ichikawa T, Nakazawa M, Yamamoto N, et al. 2006. Increased level of polyploidy1, a conserved repressor of CYCLINA2 transcription, controls endoreduplication in *Arabidopsis*. *Plant Cell* **18:** 2452–2468.

Zhang K, Sridhar VV, Zhu J, Kapoor A, Zhu JK. 2007. Distinctive core histone post-translational modification patterns in *Arabidopsis thaliana*. *PLoS ONE* **2:** e1210.

Zhang Z, Chen H, Huang X, Xia R, Zhao Q, Lai J, Teng K, Li Y, Liang L, Du Q, et al. 2011. BSCTV C2 attenuates the degradation of SAMDC1 to suppress DNA methylation-mediated gene silencing in *Arabidopsis*. *Plant Cell* **23:** 273–288.

* Zielke N, Edgar BA, DePamphilis ML. 2013. Endoreplication. *Cold Spring Harb Perspect Biol* **5:** a012948.

Endoreplication

Norman Zielke[1], Bruce A. Edgar[1], and Melvin L. DePamphilis[2]

[1]Deutsches Krebsforschungszentrum (DKFZ)-Zentrum für Molekulare Biologie der Universität Heidelberg (ZMBH) Allianz, Im Neuenheimer Feld 282, 69120 Heidelberg, Germany

[2]National Institute of Child Health and Human Development, National Institutes of Health, Bethesda, Maryland 20892-2753

Correspondence: b.edgar@Dkfz-Heidelberg.de; n.zielke@DKFZ-heidelberg.de

Developmentally programmed polyploidy occurs by at least four different mechanisms, two of which (endoreduplication and endomitosis) involve switching from mitotic cell cycles to endocycles by the selective loss of mitotic cyclin-dependent kinase (CDK) activity and bypassing many of the processes of mitosis. Here we review the mechanisms of endoreplication, focusing on recent results from *Drosophila* and mice.

Eukaryotic cells proliferate by undergoing a sequence of events termed the "mitotic cell cycle" in which the genome is duplicated once and only once between cell divisions. The result is a population of cells with two copies of each chromosome (diploid, or 2C). Agents that interfere with the mechanisms that govern genome duplication frequently induce reinitiation of nuclear DNA replication during S phase. This phenomenon, termed "DNA rereplication," is an aberrant event that produces a population of cells with a heterogeneous DNA content that reflects incomplete chromosome duplication, stalled replication forks, and DNA damage. In most cells, these events can lead to inducing the cell's DNA damage response and can lead to apoptosis (Lee et al. 2010).

Remarkably, some cells are developmentally programmed to exit their mitotic cell cycle in response either to environmental signals or to injury or stress, and then differentiate into nonproliferating, viable, polyploid cells. This phenomenon, termed "developmentally programmed polyploidy," is a normal part of animal and plant development that occurs frequently in ferns, flowering plants, mollusks, arthropods, amphibians, and fish, although rarely in mammals. In contrast to DNA rereplication, developmentally programmed polyploidy produces cells with a DNA content of >4C, but in integral multiples of 4C (e.g., 8C, 16C, 32C, etc.), consistent with multiple S phases in the absence of cytokinesis. These cells typically stop proliferating but remain viable in a terminally differentiated state that may serve to regulate tissue size or organization, to trigger cell differentiation or morphogenesis, to increase the number of genes dedicated to tissue-specific functions without increasing the number of cells, or to adapt to environmental conditions. Mitotic divisions of polyploid cells are common for plant species, but they are rarely found in animals. Although known for decades, polyploid mitosis in insects remained mostly unstudied until it was recently

Cite this article as *Cold Spring Harb Perspect Biol* doi: 10.1101/cshperspect.a012948

shown that the cells of the rectal papilla in *Drosophila* undergo mitosis after executing two or more endocycles (Fox et al. 2010). Thus, polyploidy is not an irreversible process, although the benefit of this cell cycle variant remains to be elucidated.

Developmentally programmed polyploidy occurs by at least four different mechanisms (Ullah et al. 2009). Proliferating cells in the syncytial blastoderm of *Drosophila* embryos and some hepatocytes in the postnatal liver of mammals become multinucleated and therefore polyploid by failing to undergo cytokinesis after mitosis ("acytokinetic mitosis"). Differentiation of skeletal muscle myoblasts into myotubes, monocytes into osteoclasts, and formation of placental syncytiotrophoblasts involves "cell fusion" to produce multinucleated, terminally differentiated cells that are similarly polyploid. Alternatively, cells may exit their mitotic cell cycle by arresting mitosis during anaphase and failing to undergo cytokinesis. This phenomenon, termed "endomitosis," produces cells with a single giant nucleus that may subsequently fragment into a multinuclear appearance. Endomitosis occurs in mammals when megakaryoblasts differentiate into megakaryocytes (Bluteau et al. 2009), and in some plant cells (Weingartner et al. 2004). However, the primary mechanism for developmentally programmed polyploidy in arthropods (Smith and Orr-Weaver 1991; Edgar and Orr-Weaver 2001), plants (de la Paz Sanchez et al. 2012), and possibly mammals (Ullah et al. 2009) is "endoreplication" (also referred to as "endoreduplication"). Endoreplication occurs when a cell exits the mitotic cell cycle in G_2 phase and undergoes multiple S phases without entering mitosis and undergoing cytokinesis. The result is a giant cell with a single, enlarged, polyploid nucleus.

ENDOCYCLING CELL TYPES

"Endocycles" refer to multiple rounds of nuclear genome duplication in cells undergoing either endoreplication or endomitosis. Endocycles typically include Gap (G) phases between each S phase and use the same molecular machinery as mitotic cell cycles to regulate successive rounds of DNA replication. In *Drosophila*, larval growth is achieved primarily via endoreplication (Edgar and Nijhout 2004). Consequently, most larval tissues are composed mainly of polyploid cells. These tissues include the salivary glands, fat body, epidermis, gut, trachea, and Malpighian or renal tubules (Lee et al. 2009a). The best studied of these are the giant salivary gland cells that undergo about 10 endocycles, resulting in a final ploidy of ~1350C (Hammond and Laird 1985b). In the female germline, nurse cells, which provide the developing oocyte with mRNA and proteins, achieve a final ploidy of ~1500C (Hammond and Laird 1985a), whereas the somatic follicle cells that surround the egg undergo only three endocycles. Loss of the ability to endoreplicate in nurse cells results in sterility, indicating that endocycling is essential for oogenesis (Lilly and Spradling 1996). During pupal development, shaft and socket cells that form parts of the mechanosory bristle undergo two or three endocycles to produce cells with 8C or 16C DNA (Audibert et al. 2005). Finally, the subperineural glia increases in size by endoreplication, allowing it to accommodate growing neurons, while simultaneously maintaining the blood–brain barrier (Unhavaithaya and Orr-Weaver 2012).

Mouse development includes many examples of terminal cell differentiation (Ullah et al. 2009), but in only four of these cases does polyploidy occur through endoreplication. Cardiac myoblasts, basal epithelial cells, and primitive podocytes endoreplicate when placed under stress by either injury or infection. Trophoblast stem (TS) cells are the paradigm for endoreplication in mammals (Hu and Cross 2010). They are derived from the trophectoderm of blastocysts, the outer layer of epithelial cells that give rise exclusively to the various cells composing the placenta. TS cells proliferate in vitro when cultured in the presence of fibroblast growth factor 4 (FGF4) and conditioned medium, but when cultured without these factors, they differentiate into the trophoblast giant (TG) cells that are essential for implantation and placentation. When reestablished in a blastocyst, TS cells contribute exclusively to the placenta, giving rise to all of the cell lineages derived from the trophectoderm (Oda et al. 2010). However, the DNA content of TG cells generated in vitro seldom exceeds 64C, whereas TG cells in vivo have been reported to reach >1000C.

SUSTAINING ENDOCYCLES

Oscillation of Cyclin E·Cdk2 Activity Drives Endocycles

Cyclin E (CycE) and its kinase partner cyclin-dependent kinase 2 (Cdk2, encoded by *cdc2c*) are the major regulators of S-phase entry in *Drosophila*. In *cycE*$^{-/-}$ embryos, DNA synthesis is abolished in cells undergoing mitotic cell cycles or endocycles (Knoblich et al. 1994). Similarly, endocycles in salivary glands are eliminated by tissue-specific removal of Cdk2/*cdc2c* (Zielke et al. 2011). Moreover, overexpression of CycE is sufficient to induce DNA replication in endoreplicative cells arrested by starvation (Britton and Edgar 1998). Likewise, ablation of both cyclin E (CcnE) alleles in mice prevents endocycles in TG cells and megakaryocytes (Geng et al. 2003; Parisi et al. 2003; Eliades et al. 2010), and ablation of Cdk2 prevents endoreplication in TS cells (Ullah et al. 2008). Together, these results showed that CycE/CcnE·Cdk2 is essential in flies and mammals for the execution of S phases during endocycling. Studies in yeast and frog eggs (Bell and Kaguni 2013; Tanaka and Araki 2013) revealed that CycE·Cdk2 triggers S phase by phosphorylating Sld2 and Sld3, which then recruit Dpb11 to replication origins, thereby activating the Mcm2-7 DNA helicase. Mammalian orthologs of Sld2, Sld3, and Dpb11, and *Drosophila* orthologs of Sld2 and Dpb11 have been identified (see Appendix, Table 3), suggesting that the same mechanism applies to these organisms as well.

Continuous overexpression of CycE blocks endocycles in *Drosophila* salivary glands (Follette et al. 1998; Weiss et al. 1998), implying that the oscillation of CycE·Cdk2 activity is essential for endocycle progression. CycE is an unstable protein (Lilly and Spradling 1996; Weng et al. 2003) that is targeted for ubiquitin-dependent degradation by the E3 ubiquitin ligase CRL1·Ago (Moberg et al. 2001). The product of the *Drosophila minus* (*mi*) gene—not conserved in vertebrates—physically interacts with CRL1·Ago and confers its substrate specificity toward CycE (Szuplewski et al. 2009). Salivary gland and follicle cells that are either *ago*$^{-/-}$ or *mi*$^{-/-}$ fail to undergo DNA replication, implying that the degradation of CycE is essential for endocycle progression (Shcherbata et al. 2004; Szuplewski et al. 2009; Zielke et al. 2011).

In mammals, cyclin E is phosphorylated by cyclin A (CcnA)·Cdk2 soon after S phase begins during a mitotic cell cycle. This phosphorylated form of CcnE is then ubiquitinated by CRL1·Fbw7 and degraded by the 26S proteasome (Koepp et al. 2001; Strohmaier et al. 2001). Moreover, *Fbw7*$^{-/-}$ mouse embryos and placentas have increased levels of cyclin E protein, which persist throughout G and S phases (Tetzlaff et al. 2004), suggesting that this function is conserved between flies and mammals. However, there is no evidence that CRL1·Fbw7 activity oscillates during either mitotic cell cycles or endocycles. Therefore, its function may be to render CycE constitutively unstable, rather than to function as part of the endocycle core oscillator.

A "Core Oscillator" for Endocycles

Why CycE·Cdk2 activity must oscillate in the endocycle and how this oscillation is orchestrated have been critical questions driving recent research on *Drosophila* endoreplication. In metazoa, S phase depends on many genes that rely on the E2F transcription factors for their expression (van den Heuvel and Dyson 2008). Cyclin E is one such gene. Mammals contain nine E2F proteins, three of which activate transcription by forming heterodimeric complexes with DP proteins that allow site-specific binding to DNA. The smaller fly genome contains only a single activator termed E2F1 and a single repressor, E2F2, both of which dimerize with the same DP protein. Activator E2F·DP complexes are required for cyclin E expression in both mammals and flies.

E2F1 is essential for both the expression and oscillation of CycE protein in *Drosophila* (Zielke et al. 2011, and references therein).

Consequently, the deletion of E2F1 causes rapid endocycle arrest in *Drosophila* cells. Whereas E2F1 transcripts are expressed ubiquitously in actively cycling cells, E2F1 protein accumulates during G phases and is destroyed during S phase. This oscillation of E2F1 protein is mediated by the CRL4·Cdt2 E3 ubiquitin ligase, whose activity requires ongoing DNA replication (Arias and Walter 2006; Shibutani et al. 2007, 2008). S-phase-specific degradation of E2F1 relies on the fact that activation of CRL4·Cdt2 requires chromatin-bound PCNA, which exists only at active replication forks (Havens and Walter 2009). Substrates of CRL4·Cdt2 contain a conserved PCNA-interacting protein (PIP) motif that mediates the interaction between PCNA and its substrate (Arias and Walter 2006). Consequently, mutation of the amino-terminal PIP box in E2F1 inhibits its turnover (Shibutani et al. 2008). Tellingly, either depletion of CRL4·Cdt2 or stabilization of E2F1 by PIP box mutations prevents endocycling in salivary glands (Zielke et al. 2011). Therefore, E2F1 and CRL4·Cdt2 are essential components of an autoregulatory feedback loop that mediates the oscillations of CycE activity, which drive endocycling in *Drosophila* salivary glands (Fig. 1).

Although both E2F and CRL4·Cdt2 are also essential for endoreplication in *Arabidopsis* (Roodbarkelari et al. 2010), the periodic destruction of E2F1 by CRL4·Cdt2 has not yet been shown in either mammals or plants, and therefore the E2F1–CRL4·Cdt2 oscillator proposed by Zielke and coworkers may be unique to flies or perhaps arthropods in general. As previously noted, an E2F-independent S-phase oscillator is theoretically possible. This might comprise CRL4·Cdt2 and another of its well-characterized substrates, such as the Cdt1 ortholog *double-parked* (Dup), an essential component of the origin licensing mechanism (Zielke et al. 2011).

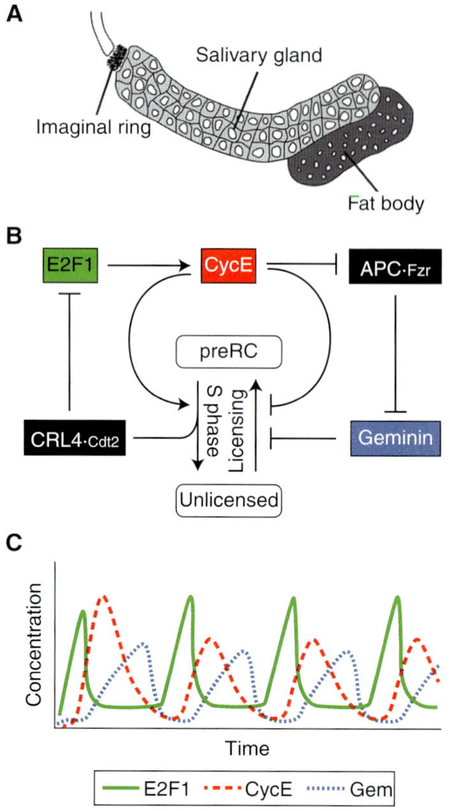

Figure 1. Sustaining endocycles in *Drosophila* salivary glands. (*A*) Schematic of a larval salivary gland with attached fat body. Both salivary glands and fat bodies undergo endocycles resulting in cells with DNA contents of ~1350C and ~256C, respectively, whereas the diploid cells of the imaginal ring proliferate mitotically. (*B*) Diagram of the regulatory network proposed to control salivary gland endocycles. The counterplay of E2F1 and its antagonist CRL4·Cdt2 ensures that CycE activity peaks in late G phase. This initiates DNA replication and inhibits the APC, so that Geminin (Gem) can accumulate and prevent relicensing of replication origins during S phase. (*C*) Endocycle oscillations in wild-type salivary glands predicted by computational modeling. Peaks of E2F1 protein were always followed by increases in CycE concentration that subsequently led to Geminin accumulation.

Given that endoreplicating cells generally grow to giant proportions (Britton and Edgar 1998; Pierce et al. 2004), they are presumably dependent on extracellular growth signals. Consistent with this hypothesis, E2F1 protein accumulates in salivary glands in which the nutrient- and insulin-signaling-dependent kinase mTOR is active (Zielke et al. 2011). Moreover, E2F1 mRNA is selectively recruited to polyribosomes in the presence of growth signals. Thus, growth factor-dependent translation of E2F1 might couple cell growth to endoreplication.

The Rb–E2F pathway is also critical to placentation. The loss of Rb in TS cells results in up-regulation of the activator E2F3, which causes overproduction of trophoblast cells, disruption of placental architecture, and ultimately fetal death (Wenzel et al. 2007; Chong et al. 2009). The same phenotype results from ablation of the repressors E2F7 and E2F8 in trophoblast cell lineages (Ouseph et al. 2012). Notably, cells lacking both E2F3 activator and E2F7/E2F8 repressors undergo fairly normal endocycles (Ouseph et al. 2012), a phenomenon previously reported in flies (Frolov et al. 2001). Altogether, these data indicate that the Rb–E2F pathway is also essential for execution of endocycles. One specific target of this regulation may be the transcription factor p73β. E2F1 up-regulates p73β, which, in turn, activates transcription of the CDK-specific inhibitor p57 (Blint et al. 2002; Ma and Cress 2007) that is essential for initiating endoreplication in TS cells (discussed below).

Suppressing DNA Rereplication during Endocycles

DNA rereplication during mitotic cell cycles causes aneuploidy and genomic instability that leads to cancer (Storchova and Pellman 2004). However, because polyploid cells do not proliferate, the consequences of DNA rereplication in endoreplicating tissues are less severe. Nevertheless, with the exception of a few underreplicated regions, the fly genome is duplicated precisely during each endocycle (Nordman et al. 2011; Sher et al. 2012). Endocycling cells use the same mechanisms as proliferating cells to avoid premature initiation events, but several changes are required to compensate for the absence of mitosis.

In mitotic cell cycles, prereplication complexes (preRCs) are assembled at DNA replication origins distributed throughout the genome (Bell and Kaguni 2013). Restricting genome duplication to once per cell division requires that cells do not reinitiate nuclear DNA replication within regions that have already replicated until cell division is completed. Hence, preRC assembly occurs only when CDK activity is suppressed during the anaphase to G_1-phase transition, and preRC activation occurs only when CDK and DDK activities are up-regulated during the G_1 to S-phase transition (Bell and Kaguni 2013). To prevent DNA rereplication, both flies and mammals use CDKs and ubiquitin ligases to inactivate one or more preRC proteins.

The anaphase promoting complex (APC) is a ubiquitin ligase that regulates the transition from mitosis to G_1 phase in eukarya (Pesin and Orr-Weaver 2008). It contains one of two activator proteins, either Fzr/Cdc20 or Frz/Cdh1, both of which target cyclin A, cyclin B, and Geminin in multicellular eukarya. Geminin prevents DNA rereplication, a role that is shared by CyclinA·Cdk2 (Zhu and Depamphilis 2009). $APC^{Fzy/Cdc20}$, which operates when CDK activity is high, triggers exit from metaphase into anaphase. $APC^{Fzr/Cdh1}$, which operates when CDK activity is low, maintains G_1 phase by suppressing CDK activity and the expression of Geminin.

Fzr/Cdh1 is essential for endocycles in both *Drosophila* and mice (Sigrist and Lehner 1997; Schaeffer et al. 2004; García-Higuera et al. 2008), although not in *Arabidopsis* (Roodbarkelari et al. 2010). APC^{Fzr} activity oscillates during *Drosophila* endocycles (Narbonne-Reveau et al. 2008; Zielke et al. 2008) because CycE·Cdk2 phosphorylates and inactivates Fzr (Sigrist and Lehner 1997; Reber et al. 2006). The same appears to be true in mammals (Keck et al. 2007). Thus, low CycE·Cdk2 activity allows high $APC^{Fzr/Cdh1}$ activity, which, in turn, degrades mitotic cyclins and Geminin, thereby allowing assembly of preRCs. Conversely, high CycE·Cdk2 activity inactivates $APC^{Fzr/Cdh1}$, thereby allowing the onset of S phase in the presence of Geminin (Fig. 2). Thus, Geminin protein levels oscillate in endoreplicating salivary glands as they do in proliferating cells. Moreover, this oscillation appears critical for endocycle progression, because constitutive expression of Geminin protein in salivary glands inhibits endocycles (Zielke et al. 2008). Nevertheless, Geminin is not essential for the development of salivary glands (Zielke et al. 2011), although it is essential for cell proliferation (Quinn et al.

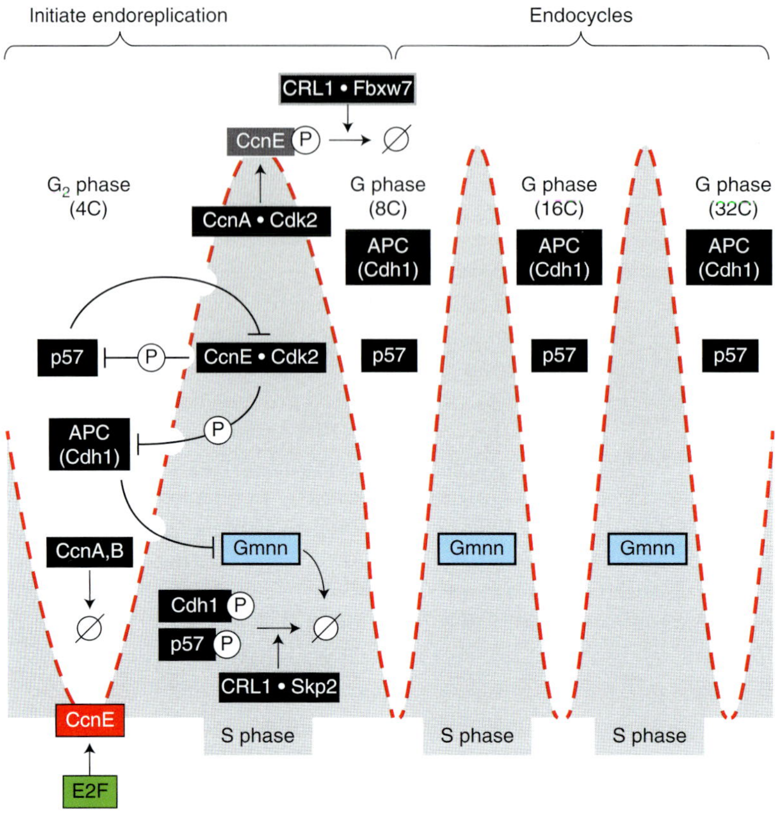

Figure 2. Sustaining endocycles in mouse TG cells. Oscillation of APC activity and the levels of CDK-specific inhibitors and Geminin (Gmnn) are inversely related to oscillation of cyclin E (broken red line). APC activity and CDK inhibitor levels are high in G phases but low in S phases, whereas cyclin E is low in G phases but high in S phases. CcnE·Cdk2 activity is required to begin S phase. Endocycles result from a sequence of feedback loops, resulting from phosphorylation events by CcnE·Cdk2 and ubiquitination events by CRL1 and the APC. Ubiquitination by CRL1 requires CDK-dependent phosphorylation of its substrate. These events inhibit the activity of their protein targets and cause them to be degraded by the 26S proteasome (→). Protein symbols are those for mammals.

2001), suggesting that other mechanisms can prevent DNA rereplication in endocycling cells.

Suppressing CDK Activity during G Phase

During mitotic cell cycles, preRCs are assembled in the absence of CDK activity and activated in the presence of CDK activity (Bell and Kaguni 2013). This transition is facilitated by oscillations in the CIP/KIP family of cyclin-dependent kinase inhibitors (CKIs), which antagonize CycE·Cdk2 activity, as well as by the oscillation of APC activity. The same appears to be true for some endocycling cells as well.

Drosophila has only one CKI, "Dacapo" (Dap). Dap protein oscillations closely follow those of CycE in endoreplicating nurse cells of the ovary, and CycE promotes the accumulation of Dap RNA and protein (de Nooij et al. 2000). In the absence of Dap, nurse cells have reduced levels of the chromatin-bound MCM2–7 complex, S phase is extended and DNA damage accumulates, consistent with reduced preRC assembly during G phase (Hong et al. 2007). Dap, however, is not essential for endoreplication in flies. Mutation of *dap* in the germline allowed fairly normal nurse cell endoreplication (Hong et al. 2003), and endoreplication defects were not detected in *dap* mutant salivary glands (Zielke et al. 2011). Moreover, Dap protein was not detected in endoreplicating follicle cells, salivary glands, or socket and shaft cells of the bristle lineage (Shcherbata et al. 2004; Audibert et al. 2005; Zielke et al. 2011). Hence, Dap appears to promote replication licensing during *Drosophila* endocycles by reinforcing low CDK activity during G phase in some cell types, but not in others. Where it is used, its activity appears to extend the window for DNA licensing, thereby increasing the number of active replication origins. This hypothesis would account for the fact that late-replicating heterochromatic sequences in *Drosophila* are underrepresented in some polyploid tissues (Nordman et al. 2011; Sher et al. 2012), but in *dap*$^{-/-}$ nurse cells where CycE·Cdk2 activity is extended, these sequences are nearly fully replicated (Hong et al. 2007).

Mammals contain three CKIs that target Cdk1 and Cdk2 during cell proliferation and differentiation (Besson et al. 2008). CKIs p27^{Kip1} and p57^{Kip2} are structurally related proteins that share a conserved CDK-binding domain that targets them for CDK-dependent phosphorylation and ubiquitin-dependent degradation during S phase. However, their phenotypes are distinctly different. The mammalian ortholog of Dacapo, p27^{Kip1}, prevents premature entrance into S phase during mitotic cell cycles. Thus, p27$^{-/-}$ mice are viable but show hyperproliferative characteristics expected for an inhibitor of cell proliferation. However, p57$^{-/-}$ mice suffer developmental defects in multiple tissues leading to neonatal death (Zhang et al. 1997, and references therein). Remarkably, most of the developmental defects characteristic of p57$^{-/-}$ mice can be corrected by expressing p27 protein from the p57 promoter, suggesting that p27 can substitute for p57 when necessary (Susaki et al. 2009). The p21^{Cip1} protein participates in DNA-damage-induced arrest of cell proliferation in G_1 and G_2 phases. Mice lacking p21 are viable and fertile, but sensitive to radiation.

Both p57 and p21 are up-regulated in TG cells, but only p57 levels oscillate during endocycles. Whereas p57 mRNA is expressed constitutively in TG cells, the p57 protein accumulates during the G phase of endocycles and is degraded during S phase (Fig. 2) (Hattori et al. 2000; Ullah et al. 2008). Degradation is triggered by CcnA·Cdk2-dependent phosphorylation of p57, which targets it for ubiquitination by CRL1·Skp2 followed by degradation by the 26S proteasome (Hattori et al. 2000; Kamura et al. 2003). Ectopic expression of a stable variant of p57 blocks DNA synthesis in TG cells (Hattori et al. 2000), indicating that the oscillation of p57 is crucial for endocycle progression. The fact that endocycles occur in *p57*$^{-/-}$ TG cells when Cdk1 activity is selectively inhibited by RO3306 (Ullah et al. 2008) suggests that p27 is sufficient to sustain endocycles in the absence of p57. TS cells that lack p57 continue to proliferate in the absence of FGF4 until they become multinucleated TG cells (Ullah et al. 2008), which would account for the observation that trophoblasts in p57$^{-/-}$ mice proliferate more extensively and still produce polyploid TG cells (Zhang et al. 1998).

The Role of ORC

The origin recognition complex (ORC) is a heterohexamer that is essential for origin licensing in eukaryotic cells (Bell and Kaguni 2013) and that is highly regulated during mitotic cell cycles in both flies and mammals (DePamphilis 2005). For example, DmOrc1 is degraded in an APC^{Fzr}-dependent manner and then resynthesized in response to E2F activation during late G_1 phase (Araki et al. 2005). Therefore, it is surprising that neither Orc1 nor Orc2 appeared to be essential for endoreplication in salivary glands (Park and Asano 2008). However, dramatic changes in the pattern of DNA replication in Orc2 mutants suggest that ORC is crucial for normal endoreplication and that the mild phenotype found in ORC mutants might be due to a small pool of ORCs that persists on the chromatin (Sher et al. 2012).

Cyclin A Plays a Role in Some *Drosophila* Endocycles

In contrast to mammals, where S phase is regulated by both cyclin A- and cyclin E-dependent Cdk2 activity, *Drosophila* CycA forms complexes exclusively with Cdk1 (Knoblich et al. 1994) and is believed to function specifically in mitosis. Nevertheless, ectopic CycA can induce S phases in several *Drosophila* cell types, suggesting a possible S-phase function (Dong et al. 1997; Sprenger et al. 1997). Although the *cycA* and *cdk1* genes are not expressed in most *Drosophila* endocycles, CycA protein oscillates in endoreplicating cells of the mechanosensory bristle lineage, such that it is undetectable in G and early S phase, but accumulates during late S phase (Salle et al. 2012). Either increasing or decreasing the level of CycA reduces ploidy, delays S phase, and dramatically affects association of Orc2 with heterochromatic regions. This suggests that CycA·Cdk1 can regulate Orc2 localization and thereby modulate replication timing in endoreplicating bristle cells. Whether or not these effects of CycA during late S phase are limited to the bristle cell lineage remains to be determined. Expression of the mitotic regulators Cdk1, Stg/Cdc25, and CycA are decreased >1000-fold in endoreplicating salivary glands relative to mitotic cells, and these genes are therefore unlikely to regulate these endocycles (Zielke et al. 2008; Maqbool et al. 2010).

INITIATING ENDOREPLICATION

Flies

In the ovarian follicle cells and other endoreplicating tissues of *Drosophila*, the down-regulation of Cdk1/cdc2 is essential for the switch from mitotic cycling to endoreplication (Fig. 3). Indeed, this seems to be a universal requirement, because even fission yeast cells can be forced to undergo endocycles by mutating the p34/cdc2·p56/cdc13 complex (Hayles et al. 1994). Similarly, loss of *cdc2* (Cdk1) function in mitotically proliferating *Drosophila* cells can dramatically increase their DNA content (Hayashi 1996; Weigmann et al. 1997), although it was never determined whether these enlarged cells result from endocycles with distinct G phases or are produced via a more aberrant form of continuous rereplication.

A key upstream regulator of the mitotic-to-endocycle transition in ovarian follicle cells is the Notch signaling pathway (Deng et al. 2001; Lopez-Schier and St Johnston 2001). Oocytes express the ligand Delta, which activates the Notch receptor in follicle cells. This results in Dacapo and String/Cdc25 being repressed, while the APC activator Fzr/Cdh1 is up-regulated (Schaeffer et al. 2004; Shcherbata et al. 2004). String/Cdc25 is a protein phosphatase that removes inhibitory phosphates from Cdk1, allowing it to associate with CycA and CycB and initiate mitosis. Thus, repression of String prevents mitosis. Up-regulating Fzr allows APC^{Fzr} to degrade CycA, CycB, thereby reinforcing the block to mitosis. Concomitant suppression of the CDK inhibitor Dacapo releases CycE/Cdk2, which is required to initiate S phase. Thus, in response to Notch signaling, follicle cells arrest in G_2 phase, assemble preRCs and then activate the preRCs when CycE·Cdk2 activity reaches the level needed to initiate endoreplication.

Notch signaling in the follicle cells activates the transcription factor *Hindsight* (*Hnt*), which represses String/Cdc25 and the transcription factor *Cut* (Sun and Deng 2007). Down-regulation of Cut allows Fzr/Cdh1 to accumulate. Thus, the switch to endoreplication occurs in two steps: First, Hnt-mediated down-regulation of String/Cdc25 arrests the cell in G_2 phase, and then down-regulation of Cut and subsequent derepression of Fzr/Cdh1 allow the cell to bypass mitosis and enter a G_1-like state that allows PreRC formation (Sun and Deng 2007). Interestingly, Delta/Notch signaling also promotes the switch from mitotic to endocycles during differentiation of enterocytes in the adult *Drosophila* intestine (J Xiang and BA Edgar, unpubl.), and thus this mechanism of developmental control may be widely used.

Some interesting tissue-specific differences exist. For example, the mitotic cyclins in endoreplicating follicle cells are regulated only by ubiquitin-dependent degradation, whereas in salivary glands, transcription of mitotic cyclins is terminated concomitantly with up-regulation of Fzr (Zielke et al. 2008; Maqbool et al. 2010). Furthermore, suppression of mitotic regulators in endoreplicating salivary glands is mediated, at least in part, by the transcriptional repressor E2F2 (Zielke et al. 2011).

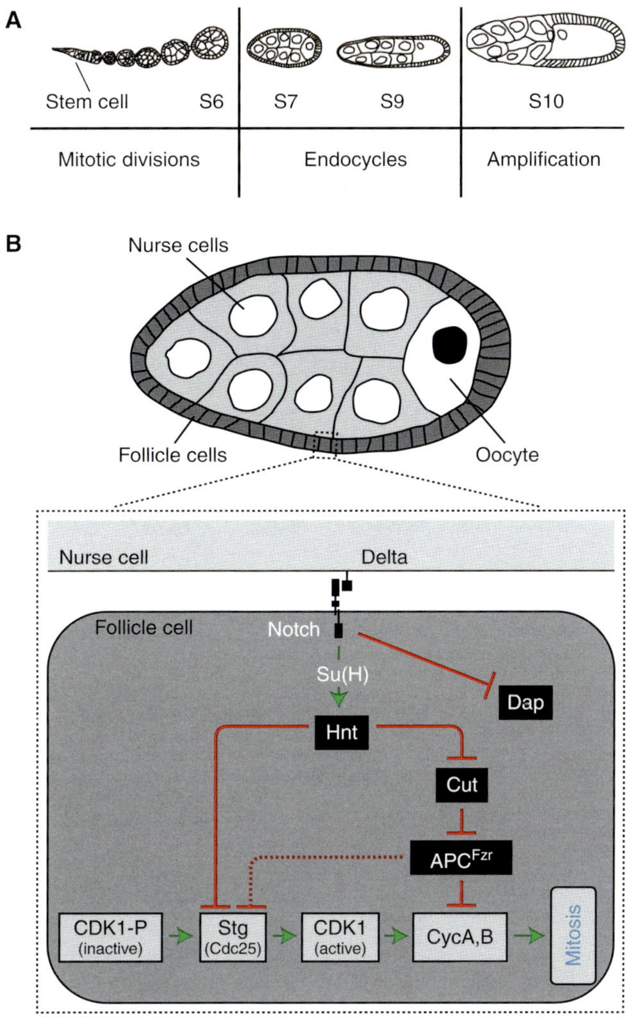

Figure 3. Initiation of endoreplication in *Drosophila* follicle cells. (*A*) The follicle cells that encapsulate the ovary originate from stem cells and undergo three different cell cycle variants during their differentiation. Up to stage 6, follicle cells undergo normal mitotic divisions resulting in a final number of approximately 650 follicle cells. During stages 7–10A, follicle cells execute three endocycles. At stage 10B, follicle cells switch from endoreplication to gene amplification, in which specific loci become amplified by repeated origin activation. (*B*) At stages 6–7, up-regulation of the Notch ligand Delta (Dl) in germline cells activates the Notch pathway in follicle cells, which results in inhibition of Stg/Cdc25 and activation of the transcription factor Hnt. Activation of Hnt prevents the transcriptional repressor Cut from inhibiting APC^{Fzr}, thereby promoting proteasomal degradation of mitotic cyclins and Stg/Cdc25. Down-regulation of Stg/Cdc25 arrests the cell in G_2 phase, whereas the subsequent degradation of mitotic cyclins suppresses the initiation of mitosis.

Initiation of endoreplication not only involves changes in the expression of cell cycle genes, but it also affects the response to DNA damage. Promoters of pro-apoptotic genes are silenced in endoreplicating cells, rendering them insensitive to genotoxic stress caused, for example, by overexpression of Dup/Cdt1 (Mehrotra et al. 2008). This modification might be attributed to the fact that *Drosophila* polyploid cells include underreplicated regions containing stalled replication forks that would otherwise trigger a DNA damage response.

Mammals

Endoreplication in TS cells is triggered by depletion of FGF4 (Fig. 4). Deprivation of this mitogen rapidly induces expression of the p57^{Kip2} and p21^{Cip1} proteins in TS cells with a concomitant loss of Cdk1 activity (Ullah et al. 2008). The decrease in Cdk1 activity could not be accounted for by corresponding changes in expression of either Cdk1 protein or cyclins A, B or E, suggesting that either p57 or p21 or both selectively inhibits Cdk1, but not Cdk2. If both Cdk2 and

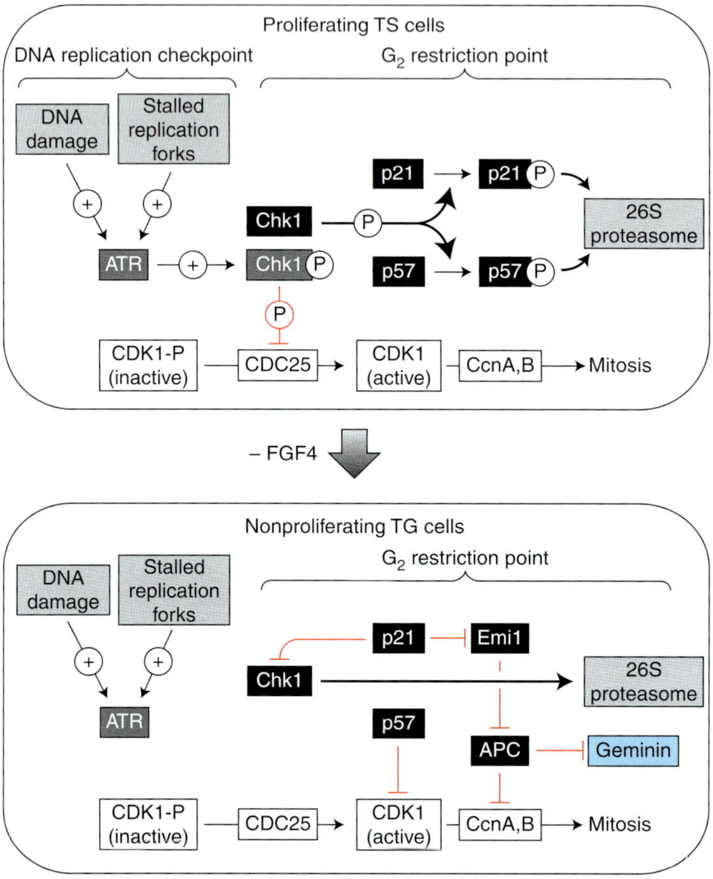

Figure 4. Initiation of endoreplication in the mouse trophectoderm lineage. Chk1 is the effector kinase for two checkpoints. The "DNA replication checkpoint" prevents premature mitosis in response to DNA damage and stalled replication forks. These events activate (+) the ATR kinase that activates the Chk1 kinase that inhibits (⊥) the CDC25 phosphatase through site-specific phosphorylation. Because CDC25 is required to activate Cdk1, inhibition of CDC25 prevents entrance into mitosis. The "G_2 restriction point" prevents expression of p57 and p21 in response to mitogen stimulation. In the case of TS cells, the mitogen is FGF4. In the presence of FGF4, TS cells express the p21 and p57 genes, but Chk1 phosphorylates the p21 and p57 proteins, thereby targeting them for degradation (Ø) by the 26S proteasome. This prevents TS cells from exiting their mitotic cell cycle. In the absence of FGF4, Chk1 protein is degraded, thereby up-regulating p57 and p21. P21 may constitute a feedback loop that sustains suppression of Chk1 expression in TG cells (Gottifredi et al. 2001). Inhibition of Cdk1 by p57 triggers endoreduplication and differentiation into TG cells. P21 could facilitate this event by down-regulating Emi1, a specific inhibitor of the APC (Lee et al. 2009b). The APC targets mitotic cyclins and Geminin for degradation, thereby promoting origin licensing. The absence of Chk1 in TG cells allows p57 protein levels to oscillate. As CcnE levels increase, Cdk2·CcnE will eventually phosphorylate p57 at a CDK-specific site, thereby targeting it for degradation.

Cdk1 were inhibited, then DNA replication could not occur, because Cdk2 and Cdk1 are the only CDKs that trigger S phase during either mitotic cell cycles or endocycles (Ullah et al. 2008). In fact, endoreplication can be initiated in proliferating TS cells, as in other metazoan cells, by selectively inhibiting Cdk1 activity with a chemical analog (Hochegger et al. 2007; Ullah et al. 2008; Ma et al. 2009). Selective inhibition of Cdk1 reversibly arrests cells in G2 phase without inducing endoreplication, whereas cells arrested in prometaphase endoreplicate, because loss of Cdk1 activity then activates the APC, which degrades targets Geminin, cyclin A, and cyclin B for degradation. TS cells are unique, however, in that selective inhibition of Cdk1 by this method induces terminal differentiation in TG cells, whereas the same treatment induces apoptosis in embryonic stem cells. Thus, cells that are developmentally programmed for terminal differentiation avoid the apoptosis associated with DNA replication stress. In TG cells, this function, as well as others (Fig. 4), appears to be played by p21 (Ullah et al. 2008).

Only p57 is essential for switching from mitotic cell cycles to endocycles in TS cells. Not only are $p21^{-/-}$ mice viable and fertile, but FGF4 deprivation of TS($p21^{-/-}$) cells triggers endoreplication. In contrast, p57-deficient mice suffer multiple developmental abnormalities, and FGF4 deprivation of TS($p57^{-/-}$) does not trigger endoreplication. Instead, TS($p57^{-/-}$) cells undergo several rounds of cell division, express TG-cell-specific genes, and eventually form multinucleated TG cells, the consequence of multiple mitosis in the absence of cytokinesis (Ullah et al. 2008). Therefore, p57 is not essential either for expression of TG-specific genes or for inhibition of cytokinesis, and endoreplication can be replaced by acytokinetic mitoses. Nevertheless, TS($p57^{-/-}$) cells are capable of endoreplication, because selective inhibition of Cdk1 in these cells induces endocycles. Suppressing either expression of the endogenous p57 gene in wild-type TG cells or expression of a recombinant Cdk1 protein restores mitosis and results in multinucleated cells (Ullah et al. 2011). The ability of TS($p57^{-/-}$) cells to overproduce multinucleated TG cells could account for the placentomegaly and preeclampsia observed in p57-deficient mice and humans (Knox and Baker 2007).

What links FGF4 deprivation to expression of p57 and p21? Remarkably, this involves Checkpoint kinase-1 (Chk1), a protein well known for its role in arresting cell proliferation in response to DNA damage or stalled replication forks. Although p21, p27, and p57 are transcribed and translated in TS cells, Chk1 phosphorylates p21 and p57 at specific sites, thereby selectively targeting them for degradation by the 26S proteasome (Ullah et al. 2011). When TS cells are deprived of FGF4, expression of Chk1 is suppressed. This allows unphosphorylated p57 and p21 proteins to accumulate and trigger endoreplication. The absence of Chk1 in TG cells also allows p57 levels to oscillate in response to changes in CcnE·Cdk2 activity. Ectopic expression of Chk1 in endocycling TG cells suppresses expression of p57 during G phases, thereby restoring mitosis, but not cytokinesis. Thus, in addition to its role in the DNA damage checkpoint, Chk1 serves as a mitogen-dependent kinase that prevents proliferating TS cells from exiting the mitotic cell cycle by premature expression of p57. The mechanism by which Chk1 expression is suppressed in response to FGF4 deprivation of TS cells remains to be elucidated, although it likely involves one of the ubiquitin ligases that target Chk1 for proteasomal degradation (Leung-Pineda et al. 2009; Zhang et al. 2009).

The function of Chk1 in TS cells is the equivalent of a "G_2 restriction point" (Fig. 4), in that Chk1 responds to changes in mitogenic stimuli by regulating cell proliferation. This dual role for Chk1 in sensing both DNA replication stress and mitogenic stress would account for the fact that Chk1 is essential for mammalian cell proliferation (Tang et al. 2006, and references therein). Because the ability of Chk1 to prevent p57 expression is not restricted to TS cells (Ullah et al. 2011), the "G_2 restriction point" presumably exists in all cells that are developmentally programmed for terminal differentiation by upregulating either p57 or p21.

Suppression of Geminin activity also has been suggested as a mechanism for inducing differentiation of pluripotent blastomeres into polyploid trophoblast cells, because *Gmnn*$^{-/-}$ embryos overreplicate their DNA at the 8-cell stage and express trophoblast-specific genes (Gonzalez et al. 2006). Moreover, Geminin knockdown in mouse embryonal carcinoma and embryonic stem cells results in loss of stem cell identity and expression of trophoblast genes (Yang et al. 2011). However, others report that *Gmnn*$^{-/-}$ embryos undergo DNA rereplication and apoptosis at the 8-cell stage (Hara et al. 2006) and that Geminin knockdown does not affect the ability of embryonic stem (ES) cells to maintain or exit pluripotency, although it does impair their ability to acquire a neural fate (Yellajoshyula et al. 2011). Given that the overall level of Geminin remains constant during differentiation of TS cells into TG cells (Ullah et al. 2008) and that Geminin oscillates during endocycles in *Drosophila* cells as it does during mitotic cell cycles (Zielke et al. 2008), the simplest explanation is that Geminin prevents DNA rereplication during endocycles, as it does during mitotic cell cycles. In addition, Geminin may affect gene expression during preimplantation development as it does during postimplantation development (Yellajoshyula et al. 2011).

CONCLUSION

The control mechanisms that govern endocycle progression in mammals, insects, and even plants share much in common, but given that all of these endocycles use the same machinery used to control S phase in mitotic cell cycles, this should not come as a surprise. However, distinct differences in regulatory modes—for instance, the use of CKIs—are also apparent between cell types and organisms. Although further research is required to determine how fundamental these differences really are, it seems likely that endocycles have arisen multiple times during evolution and therefore will show substantial diversity.

ACKNOWLEDGMENTS

We thank Terry Orr-Weaver and Michel Gho for helpful comments. Furthermore, we apologize to the authors whose work could not be included due to space restrictions. The research in the Edgar Laboratory is supported by the DKFZ and the National Institutes of Health (grant GM51186). M. DePamphilis thanks the NICHD/NIH intramural program for funding.

REFERENCES

*Reference is also in this collection.

Araki M, Yu H, Asano M. 2005. A novel motif governs APC-dependent degradation of *Drosophila* ORC1 in vivo. *Genes Dev* **19:** 2458–2465.

Arias EE, Walter JC. 2006. PCNA functions as a molecular platform to trigger Cdt1 destruction and prevent re-replication. *Nat Cell Biol* **8:** 84–90.

Audibert A, Simon F, Gho M. 2005. Cell cycle diversity involves differential regulation of Cyclin E activity in the *Drosophila* bristle cell lineage. *Development* **132:** 2287–2297.

* Bell SP, Kaguni JM. 2013. Helicase loading at chromosomal origins of replication. *Cold Spring Harb Perspect Biol* doi: 10.1101/cshperspect.a010124.

Besson A, Dowdy SF, Roberts JM. 2008. CDK inhibitors: Cell cycle regulators and beyond. *Dev Cell* **14:** 159–169.

Blint E, Phillips AC, Kozlov S, Stewart CL, Vousden KH. 2002. Induction of $p57^{KIP2}$ expression by p73β. *Proc Natl Acad Sci* **99:** 3529–3534.

Bluteau D, Lordier L, Di Stefano A, Chang Y, Raslova H, Debili N, Vainchenker W. 2009. Regulation of megakaryocyte maturation and platelet formation. *J Thromb Haemost* **7:** 227–234.

Britton JS, Edgar BA. 1998. Environmental control of the cell cycle in *Drosophila*: Nutrition activates mitotic and endoreplicative cells by distinct mechanisms. *Development* **125:** 2149–2158.

Chong JL, Tsai SY, Sharma N, Opavsky R, Price R, Wu L, Fernandez SA, Leone G. 2009. E2f3a and E2f3b contribute to the control of cell proliferation and mouse development. *Mol Cell Biol* **29:** 414–424.

* de la Paz Sanchez M, Costas C, Sequeira-Mendes J, Gutierrez C. 2012. Regulating DNA replication in plants. *Cold Spring Harb Perspect Biol* **4:** a010140.

Deng WM, Althauser C, Ruohola-Baker H. 2001. Notch–Delta signaling induces a transition from mitotic cell cycle to endocycle in *Drosophila* follicle cells. *Development* **128:** 4737–4746.

de Nooij JC, Graber KH, Hariharan IK. 2000. Expression of the cyclin-dependent kinase inhibitor Dacapo is regulated by cyclin E. *Mech Dev* **97:** 73–83.

DePamphilis ML. 2005. Cell cycle dependent regulation of the origin recognition complex. *Cell Cycle* **4:** 70–79.

Dong X, Zavitz KH, Thomas BJ, Lin M, Campbell S, Zipursky SL. 1997. Control of G_1 in the developing *Drosophila* eye: Rca1 regulates Cyclin A. *Genes Dev* **11:** 94–105.

Edgar BA, Nijhout HF. 2004. Growth and cell cycle control in *Drosophila*. In *Cell growth—Control of cell size* (ed.

Hall MN, et al.), pp. 23–83. Cold Spring Harbor Laboratory Press, Cold Spring Harbor, NY.

Edgar BA, Orr-Weaver TL. 2001. Endoreplication cell cycles: More for less. *Cell* **105:** 297–306.

Eliades A, Papadantonakis N, Ravid K. 2010. New roles for cyclin E in megakaryocytic polyploidization. *J Biol Chem* **285:** 18909–18917.

Follette PJ, Duronio RJ, O'Farrell PH. 1998. Fluctuations in cyclin E levels are required for multiple rounds of endocycle S phase in *Drosophila*. *Curr Biol* **8:** 235–238.

Fox DT, Gall JG, Spradling AC. 2010. Error-prone polyploid mitosis during normal *Drosophila* development. *Genes Dev* **24:** 2294–2302.

Frolov MV, Huen DS, Stevaux O, Dimova D, Balczarek-Strang K, Elsdon M, Dyson NJ. 2001. Functional antagonism between E2F family members. *Genes Dev* **15:** 2146–2160.

García-Higuera I, Manchado E, Dubus P, Cañamero M, Méndez J, Moreno S, Malumbres M. 2008. Genomic stability and tumour suppression by the APC/C cofactor Cdh1. *Nat Cell Biol* **10:** 802–811.

Geng Y, Yu Q, Sicinska E, Das M, Schneider JE, Bhattacharya S, Rideout WM, Bronson RT, Gardner H, Sicinski P. 2003. Cyclin E ablation in the mouse. *Cell* **114:** 431–443.

Gonzalez MA, Tachibana KE, Adams DJ, van der Weyden L, Hemberger M, Coleman N, Bradley A, Laskey RA. 2006. Geminin is essential to prevent endoreduplication and to form pluripotent cells during mammalian development. *Genes Dev* **20:** 1880–1884.

Gottifredi V, Karni-Schmidt O, Shieh SS, Prives C. 2001. p53 down-regulates CHK1 through p21 and the retinoblastoma protein. *Mol Cell Biol* **21:** 1066–1076.

Hammond MP, Laird CD. 1985a. Chromosome structure and DNA replication in nurse and follicle cells of *Drosophila melanogaster*. *Chromosoma* **91:** 267–278.

Hammond MP, Laird CD. 1985b. Control of DNA replication and spatial distribution of defined DNA sequences in salivary gland cells of *Drosophila melanogaster*. *Chromosoma* **91:** 279–286.

Hara K, Nakayama KI, Nakayama K. 2006. Geminin is essential for the development of preimplantation mouse embryos. *Genes Cells* **11:** 1281–1293.

Hattori N, Davies TC, Anson-Cartwright L, Cross JC. 2000. Periodic expression of the cyclin-dependent kinase inhibitor $p57^{Kip2}$ in trophoblast giant cells defines a G_2-like gap phase of the endocycle. *Mol Biol Cell* **11:** 1037–1045.

Havens CG, Walter JC. 2009. Docking of a specialized PIP Box onto chromatin-bound PCNA creates a degron for the ubiquitin ligase CRL4Cdt2. *Mol Cell* **35:** 93–104.

Hayashi S. 1996. A Cdc2 dependent checkpoint maintains diploidy in *Drosophila*. *Development* **122:** 1051–1058.

Hayles J, Fisher D, Woollard A, Nurse P. 1994. Temporal order of S phase and mitosis in fission yeast is determined by the state of the p34cdc2–mitotic B cyclin complex. *Cell* **78:** 813–822.

Hochegger H, Dejsuphong D, Sonoda E, Saberi A, Rajendra E, Kirk J, Hunt T, Takeda S. 2007. An essential role for Cdk1 in S phase control is revealed via chemical genetics in vertebrate cells. *J Cell Biol* **178:** 257–268.

Hong A, Lee-Kong S, Iida T, Sugimura I, Lilly MA. 2003. The $p27^{cip/kip}$ ortholog *dacapo* maintains the *Drosophila* oocyte in prophase of meiosis I. *Development* **130:** 1235–1242.

Hong A, Narbonne-Reveau K, Riesgo-Escovar J, Fu H, Aladjem MI, Lilly MA. 2007. The cyclin-dependent kinase inhibitor Dacapo promotes replication licensing during *Drosophila* endocycles. *EMBO J* **26:** 2071–2082.

Hu D, Cross JC. 2010. Development and function of trophoblast giant cells in the rodent placenta. *Int J Dev Biol* **54:** 341–354.

Kamura T, Hara T, Kotoshiba S, Yada M, Ishida N, Imaki H, Hatakeyama S, Nakayama K, Nakayama KI. 2003. Degradation of $p57^{Kip2}$ mediated by SCFSkp2-dependent ubiquitylation. *Proc Natl Acad Sci* **100:** 10231–10236.

Keck JM, Summers MK, Tedesco D, Ekholm-Reed S, Chuang LC, Jackson PK, Reed SI. 2007. Cyclin E overexpression impairs progression through mitosis by inhibiting APC(Cdh1). *J Cell Biol* **178:** 371–385.

Knoblich JA, Sauer K, Jones L, Richardson H, Saint R, Lehner CF. 1994. Cyclin E controls S phase progression and its down-regulation during *Drosophila* embryogenesis is required for the arrest of cell proliferation. *Cell* **77:** 107–120.

Knox KS, Baker JC. 2007. Genome-wide expression profiling of placentas in the $p57^{Kip2}$ model of pre-eclampsia. *Mol Hum Reprod* **13:** 251–263.

Koepp DM, Schaefer LK, Ye X, Keyomarsi K, Chu C, Harper JW, Elledge SJ. 2001. Phosphorylation-dependent ubiquitination of cyclin E by the SCFFbw7 ubiquitin ligase. *Science* **294:** 173–177.

Lee HO, Davidson JM, Duronio RJ. 2009a. Endoreplication: Polyploidy with purpose. *Genes Dev* **23:** 2461–2477.

Lee J, Kim JA, Barbier V, Fotedar A, Fotedar R. 2009b. DNA damage triggers p21/WAF1-dependent Emi1 down-regulation that maintains G_2 arrest. *Mol Biol Cell* **20:** 1891–1902.

Lee J, Kim JA, Margolis RL, Fotedar R. 2010. Substrate degradation by the anaphase promoting complex occurs during mitotic slippage. *Cell Cycle* **9:** 1792–1801.

Leung-Pineda V, Huh J, Piwnica-Worms H. 2009. DDB1 targets Chk1 to the Cul4 E3 ligase complex in normal cycling cells and in cells experiencing replication stress. *Cancer Res* **69:** 2630–2637.

Lilly MA, Spradling AC. 1996. The *Drosophila* endocycle is controlled by Cyclin E and lacks a checkpoint ensuring S-phase completion. *Genes Dev* **10:** 2514–2526.

Lopez-Schier H, St Johnston D. 2001. Delta signaling from the germ line controls the proliferation and differentiation of the somatic follicle cells during *Drosophila* oogenesis. *Genes Dev* **15:** 1393–1405.

Ma Y, Cress WD. 2007. Transcriptional upregulation of $p57^{Kip2}$ by the cyclin-dependent kinase inhibitor BMS-387032 is E2F dependent and serves as a negative feedback loop limiting cytotoxicity. *Oncogene* **26:** 3532–3540.

Ma HT, Tsang YH, Marxer M, Poon RY. 2009. Cyclin A2-cyclin-dependent kinase 2 cooperates with the PLK1-SCFβ-TrCP1-EMI1-anaphase-promoting complex/cyclosome axis to promote genome reduplication in the absence of mitosis. *Mol Cell Biol* **29:** 6500–6514.

Maqbool SB, Mehrotra S, Kolpakas A, Durden C, Zhang B, Zhong H, Calvi BR. 2010. Dampened activity of E2F1-DP and Myb-MuvB transcription factors in *Drosophila* endocycling cells. *J Cell Sci* **123:** 4095–4106.

Mehrotra S, Maqbool SB, Kolpakas A, Murnen K, Calvi BR. 2008. Endocycling cells do not apoptose in response to DNA rereplication genotoxic stress. *Genes Dev* **22:** 3158–3171.

Moberg KH, Bell DW, Wahrer DC, Haber DA, Hariharan IK. 2001. Archipelago regulates Cyclin E levels in *Drosophila* and is mutated in human cancer cell lines. *Nature* **413:** 311–316.

Narbonne-Reveau K, Senger S, Pal M, Herr A, Richardson HE, Asano M, Deak P, Lilly MA. 2008. APC/CFzr/Cdh1 promotes cell cycle progression during the *Drosophila* endocycle. *Development* **135:** 1451–1461.

Nordman J, Li S, Eng T, Macalpine D, Orr-Weaver TL. 2011. Developmental control of the DNA replication and transcription programs. *Genome Res* **21:** 175–181.

Oda M, Shiota K, Tanaka S. 2010. Trophoblast cell lineage in cloned mouse embryos. *Dev Growth Differ* **52:** 285–291.

Ouseph MM, Li J, Chen HZ, Pecot T, Wenzel P, Thompson JC, Comstock G, Chokshi V, Byrne M, Forde B, et al. 2012. Atypical E2F repressors and activators coordinate placental development. *Dev Cell* **22:** 849–862.

Parisi T, Beck AR, Rougier N, McNeil T, Lucian L, Werb Z, Amati B. 2003. Cyclins E1 and E2 are required for endoreplication in placental trophoblast giant cells. *EMBO J* **22:** 4794–4803.

Park SY, Asano M. 2008. The origin recognition complex is dispensable for endoreplication in *Drosophila*. *Proc Natl Acad Sci* **105:** 12343–12348.

Pesin JA, Orr-Weaver TL. 2008. Regulation of APC/C activators in mitosis and meiosis. *Annu Rev Cell Dev Biol* **24:** 475–499.

Pierce SB, Yost C, Britton JS, Loo LW, Flynn EM, Edgar BA, Eisenman RN. 2004. dMyc is required for larval growth and endoreplication in *Drosophila*. *Development* **131:** 2317–2327.

Quinn LM, Herr A, McGarry TJ, Richardson H. 2001. The *Drosophila* Geminin homolog: Roles for Geminin in limiting DNA replication, in anaphase and in neurogenesis. *Genes Dev* **15:** 2741–2754.

Reber A, Lehner CF, Jacobs HW. 2006. Terminal mitoses require negative regulation of Fzr/Cdh1 by Cyclin A, preventing premature degradation of mitotic cyclins and String/Cdc25. *Development* **133:** 3201–3211.

Roodbarkelari F, Bramsiepe J, Weinl C, Marquardt S, Novak B, Jakoby MJ, Lechner E, Genschik P, Schnittger A. 2010. Cullin 4-ring finger-ligase plays a key role in the control of endoreplication cycles in *Arabidopsis* trichomes. *Proc Natl Acad Sci* **107:** 15275–15280.

Salle J, Campbell SD, Gho M, Audibert A. 2012. CycA is involved in the control of endoreplication dynamics in the *Drosophila* bristle lineage. *Development* **139:** 547–557.

Schaeffer V, Althauser C, Shcherbata HR, Deng WM, Ruohola-Baker H. 2004. Notch-dependent Fizzy-related/Hec1/Cdh1 expression is required for the mitotic-to-endocycle transition in *Drosophila* follicle cells. *Curr Biol* **14:** 630–636.

Shcherbata HR, Althauser C, Findley SD, Ruohola-Baker H. 2004. The mitotic-to-endocycle switch in *Drosophila* follicle cells is executed by Notch-dependent regulation of G_1/S, G_2/M and M/G_1 cell-cycle transitions. *Development* **131:** 3169–3181.

Sher N, Bell GW, Li S, Nordman J, Eng T, Eaton ML, Macalpine DM, Orr-Weaver TL. 2012. Developmental control of gene copy number by repression of replication initiation and fork progression. *Genome Res* **22:** 64–75.

Shibutani S, Swanhart LM, Duronio RJ. 2007. Rbf1-independent termination of E2f1-target gene expression during early *Drosophila* embryogenesis. *Development* **134:** 467–478.

Shibutani ST, de la Cruz AF, Tran V, Turbyfill WJ 3rd, Reis T, Edgar BA, Duronio RJ. 2008. Intrinsic negative cell cycle regulation provided by PIP box- and Cul4Cdt2-mediated destruction of E2f1 during S phase. *Dev Cell* **15:** 890–900.

Sigrist SJ, Lehner CF. 1997. Drosophila *fizzy-related* down-regulates mitotic cyclins and is required for cell proliferation arrest and entry into endocycles. *Cell* **90:** 671–681.

Smith AV, Orr-Weaver TL. 1991. The regulation of the cell cycle during *Drosophila* embryogenesis: The transition to polyteny. *Development* **112:** 997–1008.

Sprenger F, Yakubovich N, O'Farrell PH. 1997. S-phase function of *Drosophila* cyclin A and its downregulation in G_1 phase. *Curr Biol* **7:** 488–499.

Storchova Z, Pellman D. 2004. From polyploidy to aneuploidy, genome instability and cancer. *Nat Rev Mol Cell Biol* **5:** 45–54.

Strohmaier H, Spruck CH, Kaiser P, Won KA, Sangfelt O, Reed SI. 2001. Human F-box protein hCdc4 targets cyclin E for proteolysis and is mutated in a breast cancer cell line. *Nature* **413:** 316–322.

Sun J, Deng WM. 2007. Hindsight mediates the role of notch in suppressing Hedgehog signaling and cell proliferation. *Dev Cell* **12:** 431–442.

Susaki E, Nakayama K, Yamasaki L, Nakayama KI. 2009. Common and specific roles of the related CDK inhibitors p27 and p57 revealed by a knock-in mouse model. *Proc Natl Acad Sci* **106:** 5192–5197.

Szuplewski S, Sandmann T, Hietakangas V, Cohen SM. 2009. *Drosophila* Minus is required for cell proliferation and influences Cyclin E turnover. *Genes Dev* **23:** 1998–2003.

* Tanaka S, Araki H. 2013. Helicase activation and establishment of replication forks at chromosomal origins of replication. *Cold Spring Harb Perspect Biol* doi: 10.1101/cshperspect.a010371.

Tang J, Erikson RL, Liu X. 2006. Checkpoint kinase 1 (Chk1) is required for mitotic progression through negative regulation of Polo-like kinase 1 (Plk1). *Proc Natl Acad Sci* **103:** 11964–11969.

Tetzlaff MT, Yu W, Li M, Zhang P, Finegold M, Mahon K, Harper JW, Schwartz RJ, Elledge SJ. 2004. Defective cardiovascular development and elevated cyclin E and Notch proteins in mice lacking the Fbw7 F-box protein. *Proc Natl Acad Sci* **101:** 3338–3345.

Ullah Z, Kohn MJ, Yagi R, Vassilev LT, DePamphilis ML. 2008. Differentiation of trophoblast stem cells into giant cells is triggered by p57/Kip2 inhibition of CDK1 activity. *Genes Dev* **22:** 3024–3036.

Ullah Z, Lee CY, Depamphilis ML. 2009. Cip/Kip cyclin-dependent protein kinase inhibitors and the road to polyploidy. *Cell Div* **4:** 10.

Ullah Z, de Renty C, DePamphilis ML. 2011. Checkpoint kinase 1 prevents cell cycle exit linked to terminal cell differentiation. *Mol Cell Biol* **31:** 4129–4143.

Unhavaithaya Y, Orr-Weaver TL. 2012. Polyploidization of glia in neural development links tissue growth to blood–brain barrier integrity. *Genes Dev* **26:** 31–36.

van den Heuvel S, Dyson NJ. 2008. Conserved functions of the pRB and E2F families. *Nat Rev Mol Cell Biol* **9:** 713–724.

Weigmann K, Cohen SM, Lehner CF. 1997. Cell cycle progression, growth and patterning in imaginal discs despite inhibition of cell division after inactivation of *Drosophila* Cdc2 kinase. *Development* **124:** 3555–3563.

Weingartner M, Criqui MC, Meszaros T, Binarova P, Schmit AC, Helfer A, Derevier A, Erhardt M, Bogre L, Genschik P. 2004. Expression of a nondegradable cyclin B1 affects plant development and leads to endomitosis by inhibiting the formation of a phragmoplast. *Plant Cell* **16:** 643–657.

Weiss A, Herzig A, Jacobs H, Lehner CF. 1998. Continuous Cyclin E expression inhibits progression through endoreduplication cycles in *Drosophila*. *Curr Biol* **8:** 239–242.

Weng L, Zhu C, Xu J, Du W. 2003. Critical role of active repression by E2F and Rb proteins in endoreplication during *Drosophila* development. *EMBO J* **22:** 3865–3875.

Wenzel PL, Wu L, de Bruin A, Chong JL, Chen WY, Dureska G, Sites E, Pan T, Sharma A, Huang K, et al. 2007. Rb is critical in a mammalian tissue stem cell population. *Genes Dev* **21:** 85–97.

Yang VS, Carter SA, Hyland SJ, Tachibana-Konwalski K, Laskey RA, Gonzalez MA. 2011. Geminin escapes degradation in G_1 of mouse pluripotent cells and mediates the expression of Oct4, Sox2, and Nanog. *Curr Biol* **21:** 692–699.

Yellajoshyula D, Patterson ES, Elitt MS, Kroll KL. 2011. Geminin promotes neural fate acquisition of embryonic stem cells by maintaining chromatin in an accessible and hyperacetylated state. *Proc Natl Acad Sci* **108:** 3294–3299.

Zhang P, Liegeois NJ, Wong C, Finegold M, Hou H, Thompson JC, Silverman A, Harper JW, DePinho RA, Elledge SJ. 1997. Altered cell differentiation and proliferation in mice lacking $p57^{KIP2}$ indicates a role in Beckwith-Wiedemann syndrome. *Nature* **387:** 151–158.

Zhang P, Wong C, DePinho RA, Harper JW, Elledge SJ. 1998. Cooperation between the Cdk inhibitors $p27^{KIP1}$ and $p57^{KIP2}$ in the control of tissue growth and development. *Genes Dev* **12:** 3162–3167.

Zhang YW, Brognard J, Coughlin C, You Z, Dolled-Filhart M, Aslanian A, Manning G, Abraham RT, Hunter T. 2009. The F box protein Fbx6 regulates Chk1 stability and cellular sensitivity to replication stress. *Mol Cell* **35:** 442–453.

Zhu W, Depamphilis ML. 2009. Selective killing of cancer cells by suppression of geminin activity. *Cancer Res* **69:** 4870–4877.

Zielke N, Querings S, Rottig C, Lehner C, Sprenger F. 2008. The anaphase-promoting complex/cyclosome (APC/C) is required for rereplication control in endoreplication cycles. *Genes Dev* **22:** 1690–1703.

Zielke N, Kim KJ, Tran V, Shibutani ST, Bravo MJ, Nagarajan S, van Straaten M, Woods B, von Dassow G, Rottig C, et al. 2011. Control of *Drosophila* endocycles by E2F and CRL4(CDT2). *Nature* **480:** 123–127.

Archaeology of Eukaryotic DNA Replication

Kira S. Makarova and Eugene V. Koonin

National Center for Biotechnology Information, National Library of Medicine, National Institutes of Health, Bethesda, Maryland 20894

Correspondence: koonin@ncbi.nlm.nih.gov

Recent advances in the characterization of the archaeal DNA replication system together with comparative genomic analysis have led to the identification of several previously uncharacterized archaeal proteins involved in replication and currently reveal a nearly complete correspondence between the components of the archaeal and eukaryotic replication machineries. It can be inferred that the archaeal ancestor of eukaryotes and even the last common ancestor of all extant archaea possessed replication machineries that were comparable in complexity to the eukaryotic replication system. The eukaryotic replication system encompasses multiple paralogs of ancestral components such that heteromeric complexes in eukaryotes replace archaeal homomeric complexes, apparently along with subfunctionalization of the eukaryotic complex subunits. In the archaea, parallel, lineage-specific duplications of many genes encoding replication machinery components are detectable as well; most of these archaeal paralogs remain to be functionally characterized. The archaeal replication system shows remarkable plasticity whereby even some essential components such as DNA polymerase and single-stranded DNA-binding protein are displaced by unrelated proteins with analogous activities in some lineages.

Double-stranded DNA is the molecule that carries genetic information in all cellular life-forms; thus, replication of this genetic material is a fundamental physiological process that requires high accuracy and efficiency (Kornberg and Baker 2005). The general mechanism and principles of DNA replication are common in all three domains of life—archaea, bacteria, and eukaryotes—and include recognition of defined origins, melting DNA with the aid of dedicated helicases, RNA priming by the dedicated primase, recruitment of DNA polymerases and processivity factors, replication fork formation, and simultaneous replication of leading and lagging strands, the latter via Okazaki fragments (Kornberg and Baker 2005; Barry and Bell 2006; Hamdan and Richardson 2009; Hamdan and van Oijen 2010). Thus, it was a major surprise when it became clear that the protein machineries responsible for this complex process are drastically different, especially in bacteria compared with archaea and eukarya. The core components of the bacterial replication systems, such as DNA polymerase, primase, and replication helicase, are unrelated or only distantly related to their counterparts in the archaeal/eukaryotic replication apparatus (Edgell 1997; Leipe et al. 1999).

The existence of two distinct molecular machines for genome replication has raised

Cite this article as *Cold Spring Harb Perspect Biol* doi: 10.1101/cshperspect.a012963

obvious questions on the nature of the replication system in the last universal common ancestor (LUCA) of all extant cellular life-forms, and three groups of hypotheses have been proposed (Leipe et al. 1999; Forterre 2002; Koonin 2005, 2006, 2009; Glansdorff 2008; McGeoch and Bell 2008): (1) The replication systems in Bacteria and in the archaeo–eukaryotic lineage originated independently from an RNA-genome LUCA or from a noncellular ancestral state that encompassed a mix of genetic elements with diverse replication strategies and molecular machineries. (2) The LUCA was a typical cellular life-form that possessed either the archaeal or the bacterial replication apparatus in which several key components have been replaced in the other major cellular lineage. (3) The LUCA was a complex cellular life-form that possessed both replication systems, so that the differentiation of the bacterial and the archaeo–eukaryotic replication machineries occurred as a result of genome streamlining in both lines of descent that was accompanied by differential loss of components. With regard to the possible substitution of replication systems, a plausible mechanism could be replicon takeover (Forterre 2006; McGeoch and Bell 2008). Under the replicon takeover hypothesis, mobile elements introduce into cells a new replication system or its components, which can displace the original replication system through one or several instances of integration of the given element into the host genome accompanied by inactivation of the host replication genes and/or origins of replication. This scenario is compatible with the experimental results showing that DNA replication DNA in *Escherichia coli* with an inactivated *DnaA*gene or origin of replication can be rescued by the replication apparatus of R1 or F1 plasmids integrated into the bacterial chromosome (Bernander et al. 1991; Koppes 1992). Furthermore, genome analysis suggests frequent replicon fusion in archaea and bacteria (McGeoch and Bell 2008); in particular, such events are implied by the observation that in archaeal genomes, genes encoding multiple paralogs of the replication helicase MCM and origins of replication are associated with mobile elements (Robinson and Bell 2007; Krupovic et al. 2010). Replicon fusion also is a plausible path from a single origin of replication that is typical of bacteria to multiple origins present in archaea and eukaryotes. However, all the evidence in support of frequent replicon fusion and the plausibility of replicon takeover notwithstanding, there is no evidence of displacement of the bacterial replication apparatus with the archaeal version introduced by mobile elements, or vice versa, displacement of the archaeal machinery with the bacterial version, despite the rapid accumulation of diverse bacterial and archaeal genome sequences. Thus, the displacement scenarios of DNA replication machinery evolution are so far not supported by comparative genomic data.

Regardless of the nature of the DNA replication system (if any) in the LUCA and the underlying causes of the archaeo–bacterial dichotomy of replication machineries, the similarity between the archaeal and eukaryotic replication systems is striking (Table 1). Generally, the archaeal replication protein core appears to be the same as the eukaryotic core, but eukaryotes typically possess multiple paralogous subunits within complexes that are homomeric in archaea, many additional components interacting with the core ones and more complex regulation (Leipe et al. 1999; Bell and Dutta 2002; Bohlke et al. 2002; Kelman and White 2005; Barry and Bell 2006). Thus, the archaeal replication system appears to be an ancestral version of the eukaryotic system and hence a good model for functional and structural studies aimed at gaining mechanistic insights into eukaryotic replication.

In the last few years, there has been substantial progress in the study of the archaeal replication systems that has led to an apparently complete delineation of all proteins that are essential for replication (Berquist et al. 2007; Beattie and Bell 2011a; MacNeill 2011). The combination of experimental, structural, and bioinformatics studies has led to the discovery of archaeal homologs (orthologs) for several components of the replication system that have been previously deemed specific for eukaryotes (Barry and Bell 2006; MacNeill 2010, 2011; Makarova et al. 2012). Furthermore, complex evolutionary events that involve multiple lineage-specific

Table 1. The relationship between archaeal and eukaryotic replication systems

Archaea (projection for LACA)	Eukaryotes (projection for LECA)	Comments
ORC complex		
arORC1	Orc1, Cdc6	In LACA the ORC/Cdc6 complex probably consisted of two distinct subunits, and in LECA of six distinct.
arORC2	Orc2, Orc3, Orc4, Orc5	
TFIIB or homolog[a]	Orc6	
WhiP or other wHTH protein[a]	Cdt1	Both complexes might possess additional Orc6 and Cdt1 components.
CMG complex		
Archaeal Cdc45/RecJ	Cdc45	In many archaea and eukaryotes, CDC45/RecJ apparently contain inactive DHH phosphoesterase domains.
Mcm	Mcm2, Mcm3, Mcm4, Mcm5, Mcm6, Mcm7	
Gins23	Gins2, Gins3	The RecJ family is triplicated in euryarchaea, and some of the paralogs could be involved in repair.
Gins15	Gins1, Gins5	
Inactivated MCM homolog[a]	Mcm10	MCM is independently duplicated in several lineages of euryarchaea.
CMG activation factors		
—	RecQ/Sld2	There is no evidence that kinases and phosphatases in archaea are directly involved in replication, although they probably regulate cell division.
—	Treslin/Sld3	
—	TopBP1/Dpb11	
STK	CDK, DDK	
PP2C	PP2C	
Primases		
Prim1/p48	PriS	In eukaryotes, Pol α is involved in priming by adding short DNA fragments to RNA primers.
Prim2a/p58	PriL	
DnaG	—	In archaea, DnaG might be involved in priming specifically on the lagging strand.
Polymerases		
PolB3	Pol α, Pol δ, Pol ζ	No eukaryotic homologs of DP2 are known, but Zn fingers of Pol ε are apparently derived from DP2.
PolB1	Pol ε	
DP1	B subunits of Pol α, Pol δ, Pol ζ, Pol ε	
DP2	—	
DNA polymerase sliding clamp and clamp loader		
RFCL	RFC1	Eukaryotes have additional duplications of both RFCs and PCNA involved in checkpoint complexes (Rad27 and Rad1, Rad9, Hus1, respectively).
RFCS	RFC2, RFC3, RFC4, RFC4	
PCNA	PCNA	
Primer removal and gap closure		
RNase H2	RNase II	There is a triplication of ligases (LigI, LigIII, LigIV) in eukaryotes, but only LigI is directly involved in replication.
Fen1	Fen1/EXO1, Rad2, Rad27	
Lig1	Lig1	In a few *Halobacteria*, ATP-dependent ligase is replaced by NAD-dependent ligase.
SSB		
arRPA1_long	Rpa1	In Thermoproteales, RPA is displaced by the non-homologous ThermoSSB; two short RPA forms in many euryarchaea; expansion of short RPA forms in *Halobacteria*.
arRPA1_short and RPA2	Rpa2	
arCOG05741[a]	Rpa3	

For eukaryotic genes in *Homo sapiens* and *Saccharomyces cerevisiae*, gene names are indicated. Archaeal genes are denoted as in Barry and Bell (2006) or as introduced here.

[a]Not confidently traced to LACA.

duplications, domain rearrangements, and gene loss, and in part seem to parallel the evolution of the evolution of the replication system in eukaryotes, have been delineated for a variety of replication proteins in several archaeal lineages (Tahirov et al. 2009; Chia et al. 2010; Krupovic et al. 2010). Here we summarize these findings and present several additional case studies that show the complexity of evolutionary scenarios for the components of the archaeal replication machinery and new aspects of their relationship with the eukaryotic replication system.

PREREPLICATION COMPLEX

DNA replication begins at specific sites in the genome known as the origins of replication. Some archaea possess a single and others possess several replication origins, whereas in all eukaryotes replication starts from numerous, independent origins (Robinson and Bell 2005; Coker et al. 2009). Typically, replication origins encompass a distinct sequence signature, the AT-rich box, that can be used to predict origins in silico (Zhang and Zhang 2005).

The origin recognition complex (ORC) is a hexameric protein complex that binds the origins of DNA replication and recruits additional replication factors resulting in the formation of the prereplicative complex (Fig. 1). Eukaryotic Orc1-5 and the closely related CDC6 are paralogous proteins that belong to the DnaA/ORC clade of AAA^+ ATPases (Iyer et al. 2004) and often also contain a carboxy-terminal helix–turn–helix (HTH) DNA-binding domain. All the six families of eukaryotic ORC subunits have been traced back to the last eukaryotic common ancestor (LECA) (Makarova et al. 2005). The Orc2 and Orc3 families are highly diverged and contain inactivated ATPase domains.

Phylogenetic analysis of archaeal ORC homologs reveals a complex evolutionary scenario of gene duplications, losses, and extreme divergence (Fig. 2). All this complexity notwithstanding, the phylogenetic tree clearly shows the existence of two major groups of ORC proteins, namely, the slow-evolving arORC1 and the fast-evolving arORC2. The vast majority of archaea encode representatives of each of these groups.

Thermoproteales, Nanoarchaeum equitans, and several Thaumarchaeota possess only one ORC homolog that, in the latter two cases, belongs to the arORC1 group. Haloarchaea show evidence of ancestral bursts of duplication events in both arORC1 and arORC2 clades (Fig. 2). The distinction between the two ancestral ORC subfamilies persists, but most of the lineage-specific paralogs show accelerated evolutionary rates, and some have inactivated ATPase domains. Further duplications are detectable in several *Halobacterial* lineages giving rise to up to 20 ORC paralogs in *Haloterrigena turkmenica*. It appears that paralogization of ORC genes is driven by the appearance of additional origins of replication including acquisition and integration of extra-chromosomal elements (McGeoch and Bell 2008). In *Sulfolobus*, it has been shown that the two origins are recognized by distinct monomers or dimers of Orc1/Cdc6 paralogs homologs and that the three Orc1/Cdc6 genes are differentially expressed in different growth phases and cell cycle stages, suggesting a role for these proteins in modulating the activity of the origins (Robinson et al. 2004). However, the functional differences and the causes behind the distinct evolutionary constraints in the two major groups of archaeal ORCs remain poorly understood.

The Euryarchaeota of the order Methanococcales are currently thought to lack ORC proteins, and Thermococcales are assumed to possess only one ORC that belongs to the arORC1 subfamily (Barry and Bell 2006). We identified a subfamily of archaeal AAA^+ ATPases (arCOG00472) that are fused to the carboxy-terminal wHTH domain and clusters with the arORC2 branch in the phylogenetic tree. The two major lineages represented in this subfamily, Methanococcales and Thermococcales, are exactly those that were missing from the arORC2 clade. Moreover, the genetic context of these genes in Methanococcales suggests their possible role in replication (Fig. 3). The only archaeon in which we were unable to identify an ORC candidate is *Methanopyrus kandleri* AV19. Thus, it is most likely that the last common ancestor

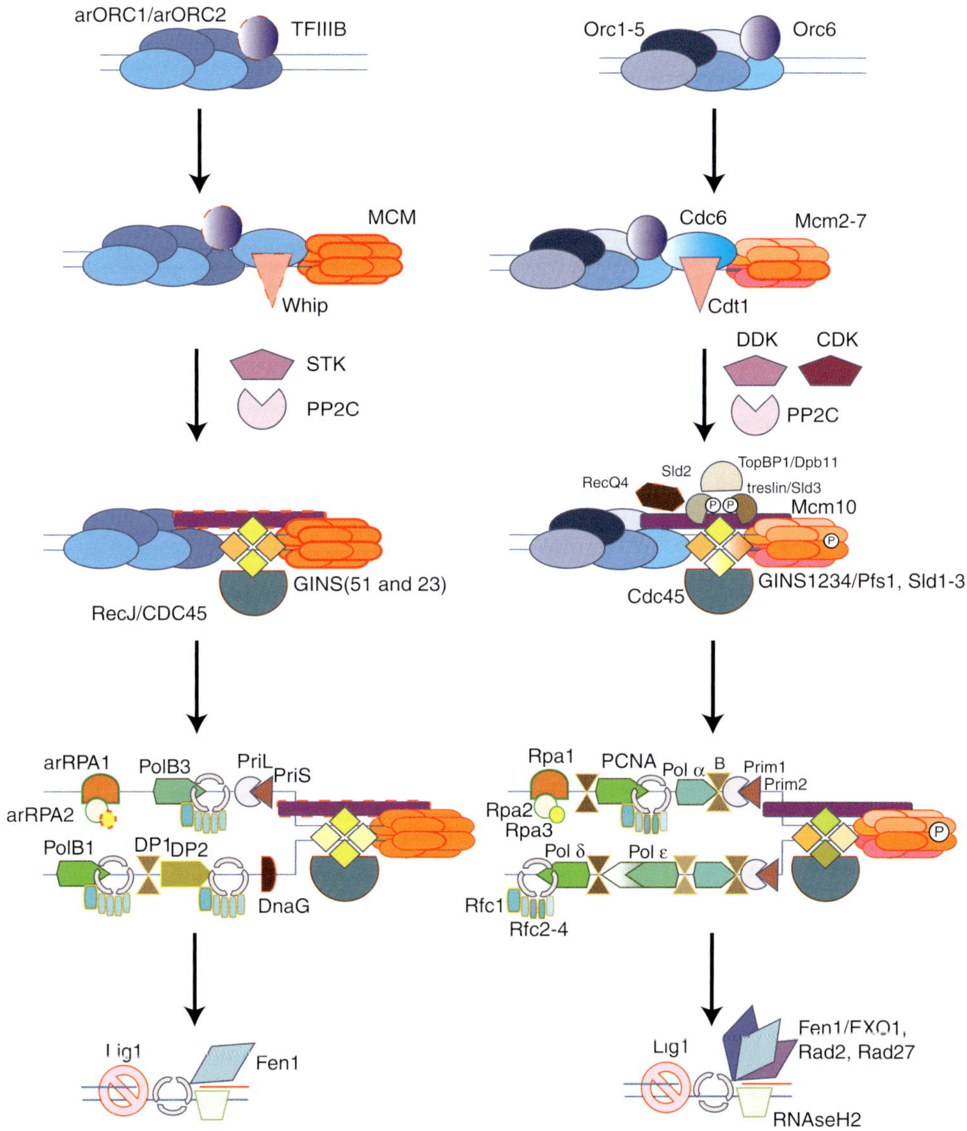

Figure 1. Comparison of the archaeal (reconstructed LACA) and eukaryotic replication systems (reconstructed LECA). Orthologs are shown by shapes of the same color, and paralogs are denoted by similar colors. The gene product names are indicated as follows: for most eukaryotic genes, the *Homo sapiens* nomenclature is used, but in several cases, both *H. sapiens* and *Saccharomyces cerevisiae* gene names are given for clarity. Archaeal genes are named as in Barry and Bell (2006) or as discussed in the text. The dotted outline in some shapes indicates components that could not be confidently traced to LACA. Small circles with "P" inside indicate phosphorylation.

of the extant archaea and the last archaeo–eukaryotic ancestor (whatever the exact nature of this entity) already encoded at least two paralogous ORC proteins. The nature of the functional distinctions between the slow-evolving and fast-evolving ORC paralogs remains unclear. In *S. solfataricus*, both the slow-evolving arORC1 and two fast-evolving paralogs from the arORC2 branch all recognize origins of replication albeit with different specificities (Robinson

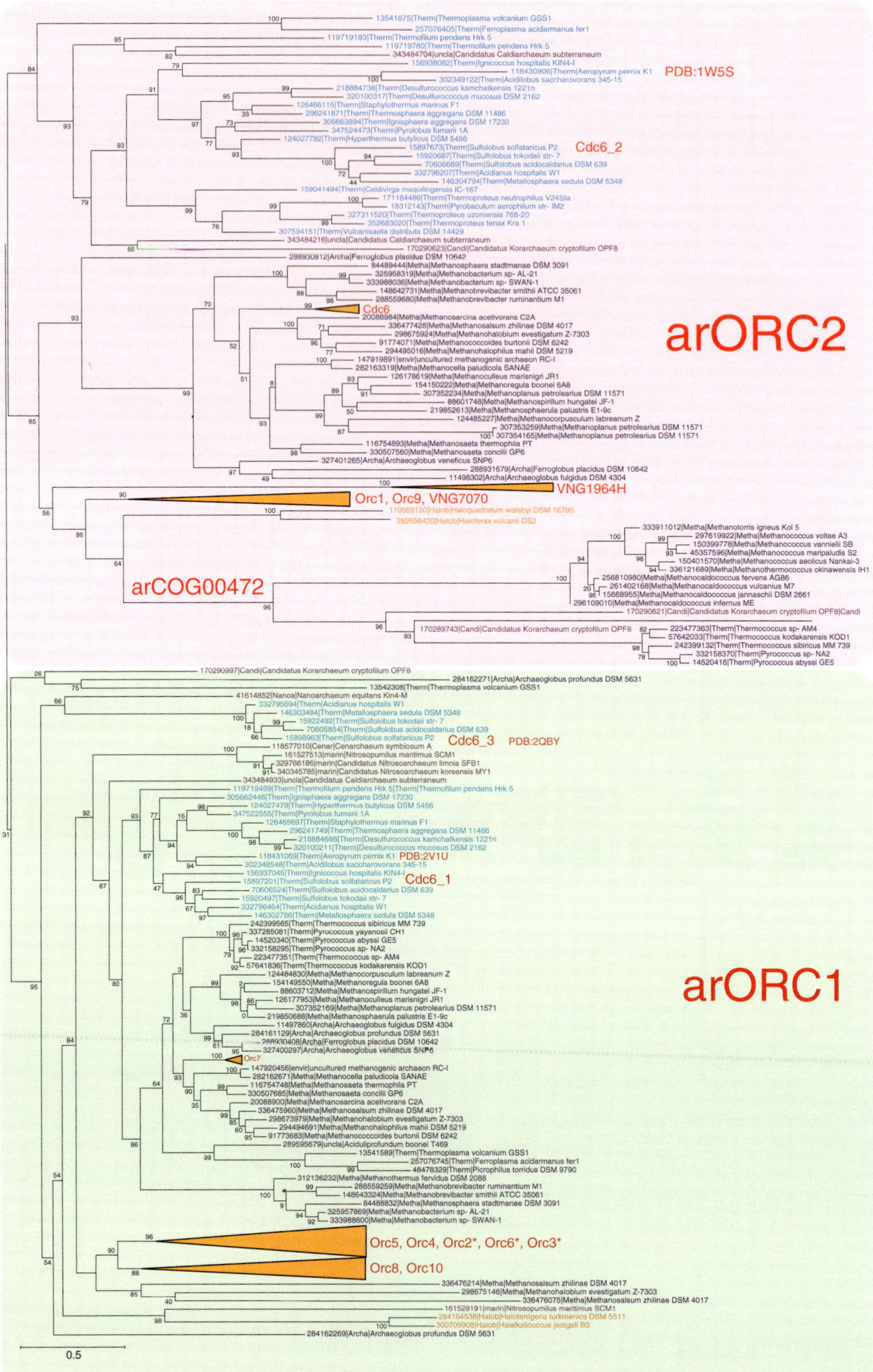

Figure 2. Phylogenetic analysis of the ORC family in archaea and eukaryotes. (Gray) Eukaryotes; (dark blue) Euryarchaeota, with the exception of (orange) *Halobacteria*; (light blue) Crenarchaeota; (purple) deeply branched archaeal lineages (Thaumarchaeota, Korarchaeota, Nanoarchaeota). (*Legend continues on following page.*)

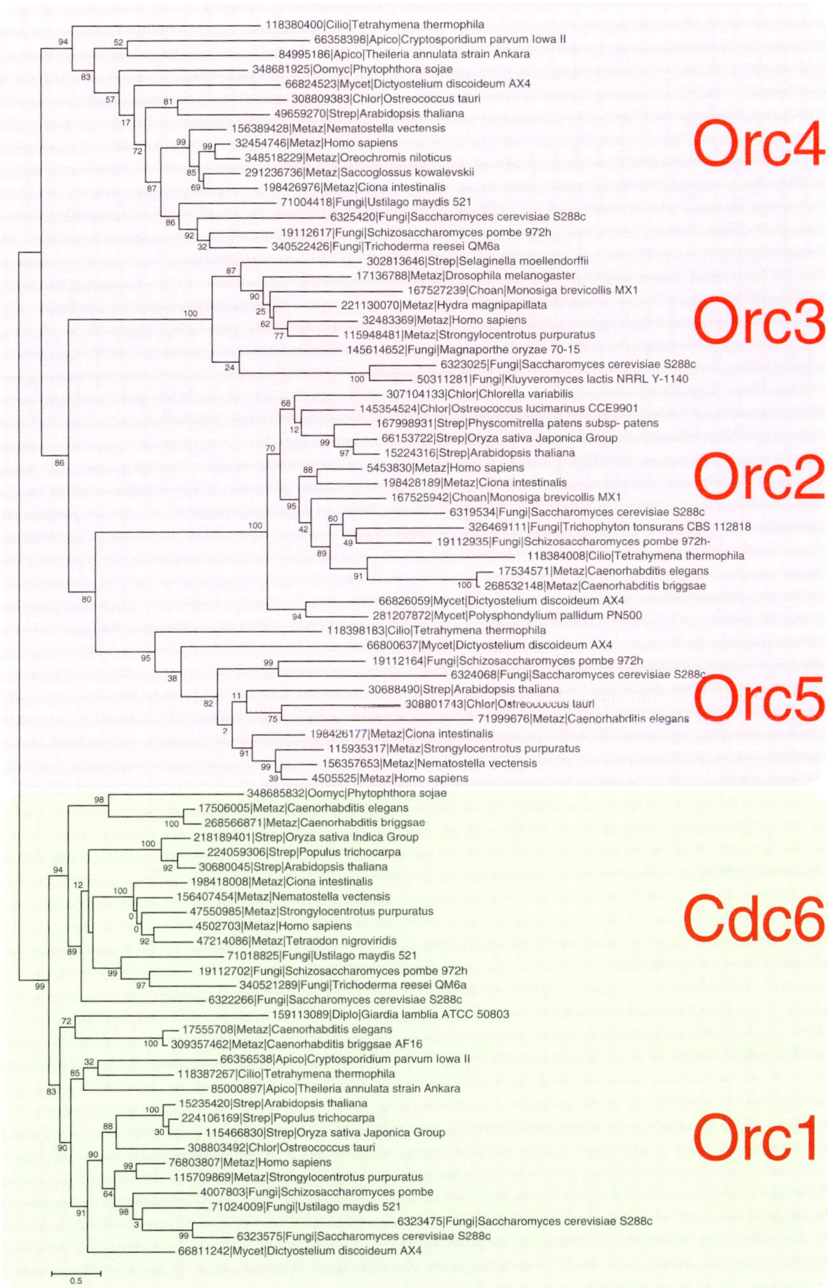

Figure 2. (*Continued*) The MUSCLE program (Edgar 2004) was used for construction of sequence alignments. The tree was reconstructed using the FastTree program (Price et al. 2010) (293 sequences and 186 aligned positions for the archaeal tree and 84 sequences and 380 aligned positions for the eukaryotic tree). For *Halobacteria*, the branches are collapsed. For characterized genes of *Halobacteria*, *Sulfolobus*, and yeast, the conventional gene name or protein identifies are indicated in red on the right of the corresponding sequence or the corresponding branch. For two archaeal sequences for which the structure is solved, its PDB code is indicated. (*) The corresponding genes are missing in some *Halobacteria*. (Red) Fast-evolving branches; (green) slow-evolving branches.

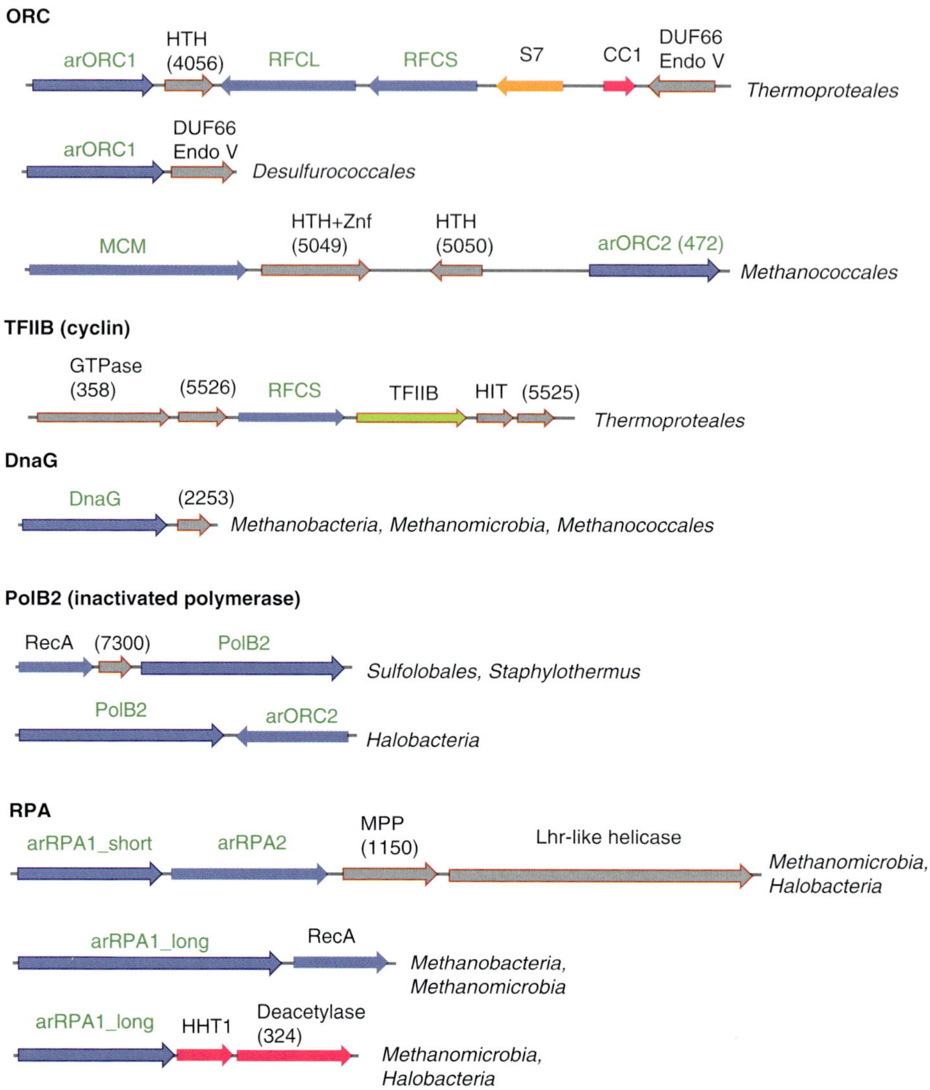

Figure 3. Genomic context of selected genes for proteins involved in replication in archaea. Homologous genes are shown by arrows of the same color; genes are shown approximately to scale. (Blue) Replication genes; (magenta) genes coding for chromatin-binding protein and the respective modification enzymes; (orange) genes for translation system components; (gray) uncharacterized genes. Genes for which the neighborhood is discussed in the text are marked by an outline. The arCOG numbers to which uncharacterized proteins are assigned are shown in parentheses. HTH, helix–turn–helix domain; S7, ribosomal protein S7; CC1, DNA-binding protein CC1; Endo V, homolog of endonuclease V; Znf, Zn finger; TFIIB, transcription initiation factor TFIIB; HIT, HIT family hydrolase; MPP, metallophosphatase superfamily protein; HHT1, histone.

et al. 2004). Furthermore, the loss of ORC1 and the colocalization of the *ORC2* gene with replication machinery components in Methanococcales imply that on some occasions in the evolution of archaea, ORC2 could take over the essential function of ORC1 in replication. In addition, the absence of recognizable ORC homologs in *M. kandleri* suggests that alternative, uncharacterized mechanisms of archaeal replication initiation might exist.

The phylogenetic tree of the eukaryotic ORC/CDC6 family also divides into two branches: the slow-evolving ORC1/CDC6 and fast-evolving ORC2-3-4-5 (Fig. 2). Owing to the extreme sequence divergence, it is difficult to precisely decipher the relationships between archaeal and eukaryotic ORC families, but it seems likely that arORC1 is the ancestor of Orc1/CDC6, whereas arORC2 is the ancestor of the ORC2-3-4-5 branch. However, in eukaryotes, the functional division between the slow-evolving and fast-evolving ORC subunits is different from that in archaea because all of these proteins are components of the heteromeric complex involved in replication.

The sixth component of the ORC complex, Orc6, is shown to be required for DNA replication in eukaryotes, and it is critical for ORC function (Duncker et al. 2009). Orc6 does not display any sequence similarity with ATPases. Structural modeling of the amino-terminal domain of metazoan Orc6 that is essential for replication (Balasov et al. 2009) and the recently resolved structure of the middle portion of human Orc6 both show that Orc6 contains two cyclin-like domains similar to those in the transcription factor TFIIB that belongs to a conserved achaeo–eukaryotic protein family (Liu et al. 2011). Unlike Orc6, the TFIIB family proteins, in addition to cyclin domains, also contain an amino-terminal Zn finger. However, in several archaea, the Zn-binding ligands are lacking or the finger module is lost completely. Interestingly, in several genomes of Thermoproteales, the gene for TFIIB is encoded in a predicted operon with the small subunit of clamp loader, replication factor C (RFC) (Fig. 3). Furthermore, several archaeal lineages encode additional, functionally uncharacterized TFIIB paralogs that typically contain a single cyclin-like domain. Therefore, it cannot be ruled out that TFIIB and/or its paralog perform dual functions in both transcription and replication (Fig. 1).

In eukaryotes, the hexameric ORC recruits two additional components, namely, Cdc6, which is an apparent product of ancestral duplication of Orc1 gene in eukaryotes (Fig. 2), and Cdt1 (Duncker et al. 2009). Both the middle and the carboxy-terminal, MCM-interacting, domains of Cdt1 adopt the winged HTH fold (Lee et al. 2004; De Marco et al. 2009). In the Crenarchaeota *Aeropyrum pernix* and several *Sulfolobus* species, a gene adjacent to the origin of replication region (ORI) encodes a winged-helix initiator protein (WhiP) (Robinson and Bell 2007). Given that WhiP contains two winged HTH domains, it has been proposed that this protein is the ortholog and functional analog of Cdt1 (Robinson and Bell 2007). Typical WhiP proteins with two HTH domains so far have been detected only in Desulfurococcales, although many other archaea encode closely related homologs of the amino-terminal HTH domain of WhiP, and some of these proteins might interact with ORI regions. Moreover, in several archaeal genomes, HTH-containing proteins are encoded adjacent to ORC subunits (Fig. 3), suggesting that these proteins also could perform specific roles in replication.

The resulting complete ORC–Cdc6–Cdt1 complex is responsible for the ATP-dependent loading of the replicative helicase, the minichromosome maintenance (MCM) complex, in eukaryotes and probably in some if not most archaea and by inference in LACA (Fig. 1).

THE CMG COMPLEX

The eukaryotic MCM complex consists of six paralogous proteins (Mcm2–7), all of which are essential for cell viability and are required for the initiation of DNA replication and replication fork progression (Bochman and Schwacha 2009). Additional eukaryotic MCM protein families, Mcm8 and Mcm9, might be ancestral but are not present in all species, and their function in replication is unclear (Bochman and Schwacha 2009). Genetic, biochemical, and structural studies have shown that the hexameric MCM complex is the replicative helicase that separates DNA strands during chromosomal replication (Bochman and Schwacha 2009). The conserved core structure of all MCM proteins consists of an α-helical amino-terminal subdomain, an OB fold domain with an inserted Zn finger, an AAA^+ ATPase domain, and a carboxy-terminal HTH domain. The

HTH domain can be confidently detected only in Mcm6. The structure of this domain has been solved, and it has been shown to bind Cdt1 (Wei et al. 2010). The same core structure is conserved in MCM proteins from archaea including the carboxy-terminal HTH domain. It appears that all eukaryotic MCM paralogs are products of ancestral duplications arising from a single ancestral MCM protein (Chia et al. 2010; Krupovic et al. 2010). All archaea encode at least one MCM protein homolog, and serial duplication of the MCM genes occurred in multiple lineages. At least two paralogs could be traced back to common ancestor of Halobacteriales and independently to the common ancestor of Methanococcales (Chia et al. 2010; Krupovic et al. 2010). Other duplications and losses followed the ancestral duplication in Methanococcales giving rise to up to eight MCM paralogs in *Methanococcus maripaludis* C6 (Chia et al. 2010; Walters and Chong 2010). Duplications occurred also within a few other narrow archaeal lineages. It has been shown that all four MCM paralogs of *M. maripaludis* S2 could form a heterohexameric complex (Walters and Chong 2010). In many archaea, some of the paralogous MCM genes are associated with mobile elements or viruses and show accelerated evolution (Chia et al. 2010). To date, a single genetic study has been published showing that only one of the three MCM paralogs in the euryarchaeon *T. kodakarensis* is essential for cell viability (Pan et al. 2011b). Interestingly, an intein inserted into one of the MCM genes, apparently early in the evolution of euryarchaea, followed by inactivation of the ATPase domain in several intein-containing MCM homologs; all archaea that possess this inactivated MCM protein also encode an intact paralog.

In eukaryotes, in vitro and in vivo experiments have shown that the MCM complex, on its own, is not the active helicase but requires the association with two accessory factors, the tetrameric GINS complex (Sld5, Psf1-3) and the Cdc45 protein (Moyer et al. 2006; Pacek et al. 2006; Labib and Gambus 2007). This complex is referred to as the CMG (Cdc45, MCM, GINS) complex and is thought to be the active replicative helicase unit in vivo (Moyer et al. 2006; Pacek et al. 2006; Labib and Gambus 2007). In addition to binding MCM proteins, the GINS complex also associates with Pol α-primase, the protein complex that synthesizes the primers on the lagging strand, and with the leading and lagging strands polymerases, Pol ε and Pol δ, respectively (MacNeill 2010). In contrast to the eukaryotic Mcm2–7 complex, which does not show helicase activity without the associated GINS and Cd45 proteins, in vitro experiments with the archaeal MCM homohexamers from several organisms have revealed robust helicase activity that did not require additional subunits (Sakakibara et al. 2009).

Highly diverged archaeal GINS homologs have been originally identified in silico (Makarova et al. 2005) and have been subsequently shown to interact with MCM (Marinsek et al. 2006; Bell 2011). Archaea encode two distinct forms of the GINS proteins, one of which appears to have been derived from the other by circular permutation of a small domain (Marinsek et al. 2006). These two forms are also present in eukaryotes, which have two ancestral genes with one domain configuration (Psf2 and Psf3) and two genes with the permuted domain arrangement (Psf1 and Sld5). Although many archaea encode only one GINS protein that belongs to the Psf1/Sld5 subfamily, Crenarchaeota, Thaumarchaeota, Korarchaeaon, and *Thermococci* encode proteins of both subfamilies, suggesting that eukaryotes inherited both forms from their archaeal ancestor (Marinsek et al. 2006). Recently, the structure of two GINS protein from *T. kodakaraensis* has been solved, revealing that the backbone structure of each subunit and the tetrameric assembly closely resemble those of the human GINS complex (Oyama et al. 2011).

Until very recently, it was believed that Cdc45, which is present in all eukaryotes in one copy, is specific to eukaryotes (MacNeill 2010). However, an in-depth computational analysis has shown that the amino-terminal region of Cdc45 contains a DHH phosphoestarase domain, suggesting that Cdc45 is the eukaryotic ortholog of archaeal and bacterial RecJ nucleases (Sanchez-Pulido and Ponting 2011). Furthermore, the GINS complex of

S. solfataricus, in addition to the association with MCM, binds a protein denoted RecJdbd (RecJ-like DNA-binding domain), which is homologous to the carboxy-terminal domain of bacterial RecJ but lacks the nuclease domain (Marinsek et al. 2006).

Recently, the GINS complex of the euryarchaeon *T. kodakarensis* has been shown to interact with primase, MCM, DNA polymerase D, PCNA, and the GINS-associated nuclease (GAN) (Li et al. 2010, 2011). Unlike the RecJdbd of *S. solfataricus*, GAN is a bona fide ortholog of bacterial RecJ containing a DHH phosphoesterase domain with all the essential catalytic residues. Recent phylogenomic analysis of RecJ homologs in archaea led to the identification of previously unsuspected homologs in Thermoproteales and revealed a complex scenario of RecJ family evolution in archaea and the origin of Cdc45 (Fig. 4A) (Makarova et al. 2012).

Under this scenario, the last common ancestor of all extant archaea possessed a single RecJ ortholog that was encoded in the conserved neighborhood including also the S15, S3, and Pcc1 genes (Fig. 4B). This ancestral protein was an active DHH nuclease and an essential component of the replication machinery. However, the experiments with the *S. solfataricus* replication system indicate that the nuclease domain is not required for replication; thus, the ancestral RecJ protein might have performed additional functions, for example, in repair, that required the nuclease activity. In Crenarchaeota, the DHH domain partially deteriorated, losing the nuclease activity, and the RecJ homolog apparently became a dedicated replication system component; the subsequent routes of evolution were notably different between the three major crenarchaeal branches—Sulfolobales, Desulfurococcales, and Themrorpoteales—resulting in extreme sequence divergence. The archaeal ancestor of eukaryotes, the exact nature of which remains elusive, also retained the RecJ ortholog (Cdc45) in which some but not all catalytic residues of the DHH domain are conserved; so far, to our knowledge, there are no experimental data showing a nuclease activity of Cdc45. In Euryarchaeota, the *RecJ* gene seems to have undergone triplication. One clade evolved very fast and developed some specialized function (arCOG00429, COG1107). Two other clades (within arCOG00427) retained significant levels of sequence similarity. Most of these proteins contain an active DHH nuclease domain, suggesting that they are active nucleases. Only one of these paralogs (arCOG00427_II) is often encoded in a conserved neighborhood with the ribosomal proteins S15 and S3, and the Pcc1 subunit of the KEOPS complex (Fig. 4B). Thermococci might have lost one paralog (arCOG00427_I). Methanococcales encode an additional paralog (arCOG00433) that has lost the DHH domain. These genes are located in genomic neighborhoods that encode no other proteins involved in replication; thus, it is unclear whether they are components of replication systems. In *Halobacteria*, the RecJ orthologs (arCOG00428), which are encoded in the same conserved neighborhood, contain an inactivated DHH domain. Thus, inactivation of the RecJ-like nuclease that apparently became a dedicated replication protein seems to have occurred at least twice independently in different archaeal lineages (Fig. 4C).

In eukaryotes, MCM10 is an essential replication protein that interacts with the assembled preRC and is required for the recruitment of Cdc45 (Wohlschlegel et al. 2002). It stays in the active replisome complex throughout the S phase and appears to be important for both replisome assembly and fork progression. Mcm10 physically interacts with both PCNA and Pol α (Chattopadhyay and Bielinsky 2007). The structure of the conserved middle domain of Mcm10 revealed a unique arrangement of a CCCH-type Zn-binding module and an OB-fold that jointly form a DNA-binding interface (Warren et al. 2008). The carboxy-terminal domain of MCM10 is composed of a second, CCCC-type Zn-binding module that is not involved in DNA binding and a HTH domain. The CCCC-type Zn-binding module of MCM10 is structurally similar to a corresponding module in the amino-terminal oligomerization domain of eukaryotic and archaeal MCM helicases, some of which also possess a carboxy-terminal HTH domain (Robertson et al. 2010). In several fungi, Mcm10 additionally contains an amino-

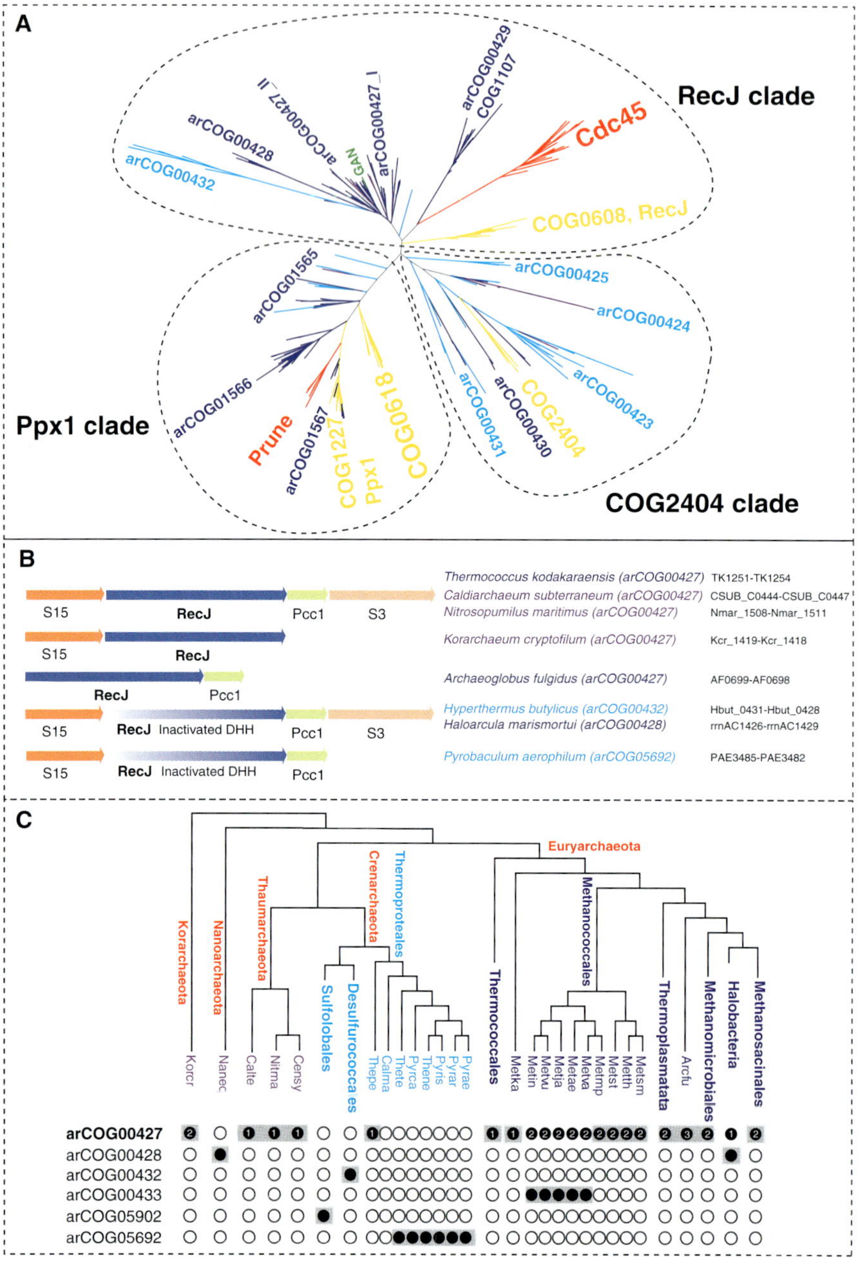

Figure 4. RecJ homologs in archaea. (Red) Eukaryotes; (yellow) bacteria; the color code for archaea is as in Figure 2. (*A*) Phylogeny of the DHH superfamily of phosphoesterases. The tree was reconstructed using aligned blocks corresponding to the DHH domain (Makarova et al. 2012). The MUSCLE program (Edgar 2004) was used for the construction of sequence alignments (229 sequences in total, 83 aligned positions). The maximum likelihood (ML) phylogenetic tree was constructed by using the MOLPHY program (Adachi and Hasegawa 1992) with the JTT substitution matrix to perform local rearrangement of an original Fitch tree (Fitch and Margoliash 1967). The MOLPHY program was also used to compute RELL bootstrap values. arCOG or COG numbers or family names (for eukaryotes) are indicated for the corresponding branches. (Green) The location of the GAN protein. (*B*) Genomic context of the RecJ homologs in selected archaea. The designations are as in Figure 3. The arCOGs to which RecJ-like proteins are assigned are indicated in parentheses. (*Legend continues on following page.*)

terminal P-loop ATPase domain (Fien et al. 2004; Fien and Hurwitz 2006). The human MCM10 protein forms a hexameric structure similar to the structure formed by other MCM proteins (Okorokov et al. 2007). Taken together, these observations suggest that Mcm10 might be derived from the MCM protein family. Archaea also encode several families of modified MCM homologs. For example, inactivated MCM paralogs originated independently in *M. kandleri* and in *Thermococci* (arCOG05761). In both cases, Zn-binding amino acids are also substituted, but the carboxy-terminal HTH domain is present. The function of these MCM homologs in archaea is unknown. Several Methanosarcinales and Halobacteriales possess a protein family with a CCCC-type Zn finger clearly related to that in MCM and an amino-terminal HTH domain (arCOG02259 and arCOG02260, respectively), resembling the configuration of the carboxy-terminal portion of eukaryotic MCM10. Some of these proteins also contain an additional carboxy-terminal RepH/I/J family domain that is involved in replication of plasmid pNRC100 in *Haloferax volcanii* (Ng and DasSarma 1993). Thus, a homolog of Mcm10 might have been present already in the archaeal ancestor of eukaryotes (Fig. 1).

In eukaryotes, activation of the replicative helicase and loading of the replisome depend on two serine–threonine kinases, CDK (cyclin-dependent kinase) and DDK (Dbf4-dependent kinase) (Tanaka et al. 2007; Araki 2010; Sheu and Stillman 2010). The DDK phosphorylates one of the Mcm2/4/6 proteins (Sheu and Stillman 2010), whereas CDK phosphorylates Mcm5, Sld2, Sld3, and Mcm10 (Tanaka et al. 2007). This phosphorylation results in binding of Cdc45 and Dpb11 proteins and activation of the Mcm2–7 helicase (Sclafani and Holzen 2007). The Sld2, Sld3, and Dpb11 proteins are conserved in most eukaryotes and are essential regulators of replication initiation (Pospiech et al. 2010).

Dpb11 (known as TopBP1 in human) is a member of the BRCT domain superfamily that contains several BRCT repeats (Bork et al. 1997; Garcia et al. 2005). In prokaryotes, the BRCT domain is present only in NAD-dependent DNA ligase (Bork et al. 1997). The NAD-dependent ligase is an essential enzyme of DNA replication and repair that is present in almost all bacteria (except for some intracellular symbionts with the smallest genomes) but only in a few archaea, which probably acquired this gene via horizontal transfer from bacteria. The eukaryotic BRCT domains most likely evolved from the bacterial ones.

The Sld3-like proteins show poor sequence conservation even among eukaryotes. Recently, however, it has been shown that an α-helical conserved domain of Sld3 is present in the majority of eukaryotes (in animals and plants, this protein is called treslin); however, homologs in prokaryotes so far have not been detected (Sanchez-Pulido et al. 2010).

Figure 4. (*Continued*) The protein IDs for this region in the corresponding genomes are indicated. (*C*) Phyletic patterns of RecJ-related arCOGs. The phyletic patterns for indicated arCOGs ([filled circles] presence; [empty circles] absence) are superimposed over the phylogenetic tree of archaea. The circles for proteins implicated in replication are shaded. The number of paralogs is indicated inside the circles for arCOG00427 (all other subfamilies have one representative in each genome). The tree is a modified version of the consensus phylogeny of archaea (Makarova et al. 2010) with *Caldiarchaeum subterraneum* included in the Thaumarchaeota branch and several branches with the same distribution of RecJ-related subfamilies collapsed. Arcfu, *Archaeoglobus fulgidus*; Metsm, *Methanobrevibacter smithii* ATCC 35061; Metth, *Methanothermobacter thermautotrophicus*; Metst, *Methanosphaera stadtmanae*; Metmp, *Methanococcus maripaludis* S2; Metva, *Methanococcus vannielii* SB; Metae, *Methanococcus aeolicus* Nankai-3; Metja, *Methanocaldococcus jannaschii*; Metvu, *Methanocaldococcus vulcanius* M7; Metin, *Methanocaldococcus infernus* ME; Metka, *Methanopyrus kandleri*; Pyrae, *Pyrobaculum aerophilum*; Pyrar, *Pyrobaculum arsenaticum* DSM 13514; Pyris, *Pyrobaculum islandicum* DSM 4184; Thene, *Thermoproteus neutrophilus* V24Sta; Pyrca, *Pyrobaculum calidifontis* JCM 11548; Thete, *Thermoproteus tenax*; Calma, *Caldivirga maquilingensis* IC-167; Thepe, *Thermofilum pendens* Hrk 5; Censy, *Cenarchaeum symbiosum*; Nitma, *Nitrosopumilus maritimus* SCM1; Naneq, *Nanoarchaeum equitans*; Korcr, *Candidatus Korarchaeum cryptofilum* OPF8; Calte, *Candidatus Caldiarchaeum subterraneum*.

The Sld2 sequence is poorly conserved as well, and this protein so far has been identified only in Fungi. However, the animal RecQ4 helicase contains an amino-terminal Sld2 domain (Capp et al. 2010). We found that the amino-terminal domain of plant RecQ (e.g., GI: 308804427) shows weak sequence similarity and compatible secondary structure with Sld2 (KS Makarova, unpubl.). This observation suggests that the Sld2 domain is involved in the regulation of prereplication complex formation in all eukaryotes. Moreover, the LECA probably contained the Sld2 domain fused with the RecQ-like helicase that might be directly involved in replication (Capp et al. 2010; Pospiech et al. 2010). Most bacteria encode a RecQ helicase, whereas only a few archaea do. In *Methanomicrobia*, the *recQ* genes are clearly transferred from bacteria. In contrast, *Acidilobus saccharovorans*, *Aeropyrum pernix*, and *Korarchaeum cryptofilum* possess a small subfamily of RecQ homologs that are highly diverged and contain a long amino-terminal region that in *K. cryptofilum* encompasses a DEDDh-like $3'-5'$ exonuclease domain and in the other two species probably is an inactivated derivative of this nuclease. This domain architecture resembles the animal Werner syndrome RecQ-like helicase (Mushegian et al. 1997; Bernstein et al. 2010). It seems likely that this particular archaeal subfamily of RecQ helicases is ancestral to the eukaryotic RecQ. An interesting question that probably can be addressed only by structure comparison is whether the extended amino-terminal region of these archaeal RecQ-like proteins also contains a domain homologous to Sld2 and if they are involved in replication.

Homologs of CDK and DDK (COG0515) serine–threonine kinases are present in most major archaeal lineages (with the exception of Thaumarchaeota and Nanoarchaeota) and thus can be confidently projected to the archaeal ancestor of eukaryotes as well. Most likely, serine/threonine protein phosphatase(s) might be present in this ancestral form as well. A functional link between kinase and phosphatase and cell division and membrane-remodeling systems in archaea is strongly suggested by conserved genomic contexts (Makarova and Koonin 2010; Makarova et al. 2010), but at this point, there is no evidence of the involvement of these kinases in regulation of archaeal DNA replication.

THE ACTIVE REPLISOME

Once the CMG complex is assembled and activated, it begins separating the two DNA strands and forming a replication "bubble" from which the two replication forks, each containing a leading and a lagging strand, move away (Moyer et al. 2006; Pacek et al. 2006; Labib and Gambus 2007; MacNeill 2010). Once replication is in progress, other proteins required for the bulk DNA synthesis and protection of single-stranded DNA can be recruited to form the full replisome complex at each replication fork (Fig. 1).

Primases

Similarly to the eukaryotic replication system, archaeal DNA replication involves a two-subunit primase (PriS, the catalytic subunit, and PriL, the polymerase-interacting subunit) that synthesizes an 8- to 12-nt RNA primer (Kuchta and Stengel 2010). In eukaryotes, the primer is then elongated by Pol α complexed with the cognate B (small) subunit to form a covalent DNA–RNA hybrid that is required for the replicative polymerase to start elongation (Kuchta and Stengel 2010). In archaea, the RNA primer seems to be sufficient for the activation of the replicative polymerase (Barry and Bell 2006). The evolution of these proteins in both eukaryotes and archaea appears to have involved primarily simple vertical descent because the majority of archaeal and eukaryotic genomes encompass a single gene for each subunit. The only deviation from this simple pattern is found in *N. equitans*, which encodes a hybrid protein containing regions approximately corresponding to the amino-terminal domain of the small subunit and the carboxy-terminal domain of the large subunit (NEQ395).

In addition to the large and small subunits of the archaeo–eukaryotic primase (PriSL), all archaea possess at least one gene for the bacterial-type DnaG primase, although in contrast to the bacterial homologs, archaeal DnaG lacks the amino-terminal Zn-finger domain and the carboxy-terminal domain that binds the bacterial replicative helicase DnaB (Makarova et al. 1999). The recently characterized DnaG protein of *S. solfataricus* has been shown to possess similar properties to the *Escherichia coli* primase and shows a fourfold faster rate of DNA priming than SsoPriSL (Zuo et al. 2010). It has been proposed that DnaG might be involved in the initiation of Okazaki fragment synthesis on the lagging strand (Zuo et al. 2010). The *dnaG* gene is often colocalized with the gene for the major form of archaeal superfamily B DNA polymerase, although these genes might not be co-regulated given that they are oriented convergently (Fig. 5). In many methanogens, *dnaG* is encoded in a predicted operon with uncharacterized α-helical, potentially metal-binding protein of arCOG02254 that by inference might have a role in replication (Fig. 3). Most eukaryotes (but not animals in which this gene has been inactivated) encode a DnaG protein that is typically fused to a DnaB-like helicase domain and is not directly related to archaeal DnaG. Instead, this gene appears to have been derived from the mitochondria and has a dedicated function in mitochondrial genome replication (Shutt and Gray 2006).

DNA Polymerase Sliding Clamp and Clamp Loader

The PCNA (proliferating cell nuclear antigen) functions as the sliding clamp, a trimer that encircles double-stranded DNA and freely slides along the DNA molecule (Beattie and Bell 2011b). The sliding clamp is required for the processivity of the DNA polymerase and coordinates the function of multiple binding partners that are also required for replication and repair processes. The sliding clamp is one of the few universally conserved proteins involved in replication (the bacterial ortholog is known as the DNA polymerase β subunit). In eukaryotes and many archaea, there is a single PCNA gene the product of which forms a homotrimer; this gene can be projected back to the last common ancestor of archaea and the archaeal ancestor of eukaryotes. However, in eukaryotes and independently in Crenarchaea, the PCNA gene is duplicated. Apparently there had been an ancestral duplication in Crenarchaea

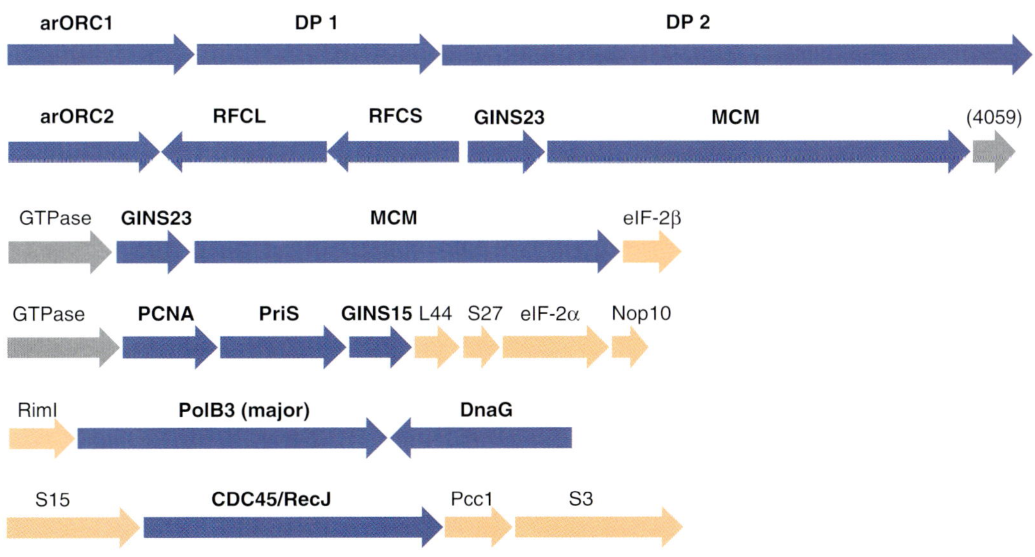

Figure 5. Conserved genomic context of selected genes encoding replication proteins. The designations are as in Figure 3. S3, S15, S27, L44, The respective ribosomal proteins; RimI, GNAT family acetyltransferase; eIF-2α and eIF-2β, the respective translation initiation factors; Nop10, an RNA-binding protein; Pcc1, subunit of the KEOPS complex.

followed by an additional duplication in the Desulfurococcales lineage (Chia et al. 2010). Some species of Crenarchaea encode multiple copies of PCNA that can form either homotrimers or heterotrimers (Barry and Bell 2006; Pan et al. 2011a). In eukaryotes, the ancestral duplication yielded at least three additional PCNA-like families, all of which are subunits of the checkpoint 9-1-1 complex (components HUS1, RAD1, and RAD9), which is involved in DNA damage checkpoint control (Aravind et al. 1999; Majka and Burgers 2004).

Replication factor C, the clamp loader, is another universally conserved protein that is required for PCNA assembly around a DNA molecule at template–primer junctions (Bloom 2009). In both archaea and eukaryotes, RFC is a pentamer that consists of one large subunit (RFCL) encoded by a single gene and four small subunits (RFCS) that are identical in most archaea but are represented by four distinct paralogs encoded by ancestral genes in eukaryotes (Makarova et al. 2005; Chia et al. 2010). Both large and small subunits are paralogs and belong to the AAA^{+} superfamily of ATPases (Iyer et al. 2004). In addition, early in eukaryotic evolution, further duplications gave rise to other families, such as Rad24, which were recruited for distinct roles in checkpoint complexes in parallel with the duplication of the corresponding PCNAs and Chl12, a specific chromatid cohesion clamp loader (Majka and Burgers 2004). In archaea, independent duplications of RFCS occurred in *Methanomicrobia/Halobacteria*, Thermoproteales, and a few smaller lineages independently, often followed by acceleration of the evolutionary rate of the "extra" copies (Chia et al. 2010).

Notably, in archaeal genomes, the "original" paralogs of both PCNA and RFC that retain the ancestral roles in replication are encoded in conserved gene contexts adjacent to other proteins involved in replication, although some additional paralogs are encoded in suggestive contexts as well (Figs. 3 and 5).

DNA Polymerases

In archaea and eukaryotes, the main replicative polymerases belong to the B family of Palm-domain polymerases (Burgers et al. 2001). In addition to the polymerase core domain, all these proteins contain an amino-terminal 3′–5′ exonuclease domain. All eukaryotes possess four paralogous B-family polymerases denoted Pol α, Pol δ, Pol ε, and Pol ζ that are involved in both DNA replication and repair (Makarova et al. 2005; Kunkel and Burgers 2008). Archaea encode at least two paralogous B-family polymerases: the "major" one (present in all archaea), and PolB3 and the minor one (several lineages of methanogens lack this gene), which are both projected to LACA PolB1 (Edgell et al. 1998; Rogozin et al. 2008; Tahirov et al. 2009). A small subset of archaea possesses another, divergent B-family polymerase (arCOG04926), which seems to contain active exonuclease and Palm domains. This family is represented by three paralogs in *Methanococcoides burtonii* and appears to be prone to duplication and HGT. In addition, several archaea encode a derivative B-family polymerase, PolB2, in which both the exonuclease and the Palm domain appear to be inactivated (Edgell et al. 1998; Rogozin et al. 2008). In Crenarachaeota, this gene often colocalizes with genes encoding a small α-helical protein from arCOG07300 and RadA protein, whereas in Euryarchaeota, it colocalizes with the arORC2 gene, suggestive of involvement in replication (Fig. 3).

In addition to the B-family polymerases, many archaea encode the unique D-family polymerase (Cann et al. 1998), which is absent in Crenarachaea but present in all deeply branching lineages (Thaumarchaea, Nanoarchaeon, Korarchaeon), suggesting that this polymerase was present in LACA. The D-family polymerases consist of two subunits. The large subunit DP2 is a large multidomain protein that forms a homodimer that is responsible for the polymerase activity (Shen et al. 2001; Matsui et al. 2011). The DP2 protein does not display any sequence similarity with other protein families (except for two Zn fingers), but examination of conserved motifs suggests that it might be a highly diverged Palm-domain polymerase (Cann et al. 1998). The small subunit DP1 contains at least two domains, an ssDNA-binding OB-fold and the 3′–5′ exonuclease domain of the metallophosphatase (MPP) family. The DP1 protein is the ancestor of the small B subunit of eukaryotic replicative polymerases of the B family that, however, have lost the catalytic amino acid residues of the 3′–5′ exonuclease (Aravind and Koonin 1998; Klinge et al. 2009). Evidence has been presented that the D-family polymerase specializes in the replication of the lagging strand, whereas the B-family polymerase is involved in the leading-strand replication (Henneke et al. 2005). Interestingly, all Crenarachaea that have no family-D polymerase possess at least one additional active polymerase of the B family, suggesting that the two distinct B-family polymerases specialize on the leading- and lagging-strand replication, respectively, as is the case in eukaryotes.

Phylogenetic analysis of archaeal, eukaryotic, and bacterial B-family DNA polymerases combined with an analysis of domain architectures indicates that eukaryotic B-family polymerases, most likely, originated from two distinct archaeal ancestors, the major archaeal form that gave rise to the catalytically active amino-terminal domain of Pol ε, and the minor form that became the common ancestor of Pol α, Pol δ, and Pol ζ. All eukaryotic B-family polymerases contain two carboxy-terminal Zn-finger modules. Interestingly, the carboxy-terminal module appears to be derived from the Zn finger in the DP2 subunit of archaeal D-family DNA polymerases that are unrelated to the B family, at least as judged by sequence comparison (Tahirov et al. 2009). The Zn finger of Pol ε shows greater similarity to the counterpart in archaeal DP2 than the Zn fingers of other eukaryotic B-family polymerases. The carboxy-terminal portion of eukaryotic Pol ε consists of two additional polymerase and exonuclease domains, both inactivated; there are indications that this module could be of bacterial or bacteriophage origin (Tahirov et al. 2009). The presence of an inactivated exonuclease–polymerase module in Pol ε parallels a similar inactivation of both enzymatic domains in a distinct subfamily of inactivated archaeal B-family polymerases (Rogozin et al. 2008).

The apparent derivation of the large subunits of eukaryotic B-family polymerases from the major (PolB3) and minor (PolB1) forms of archaeal polymerases, and the origin of the small subunit from DP1 and the origin of the carboxy-terminal Zn finger from DP2 emphasize the joint contributions of the B-family and D-family archaeal polymerases to the evolution of the eukaryotic replication machinery. Furthermore, these findings suggest that the archaeal ancestor of eukaryotes possessed a highly complex replication apparatus, possibly more complex than any known extant archaeon.

Primer Removal and Gap Closure

In both eukaryotes and archaea, RNase HII and FEN1 flap endonuclease are responsible for removal of RNA primers during replication (Fig. 1) (Barry and Bell 2006). Both enzymes are present in all archaea with only a few duplications observed. Eukaryotes have a single gene for RNase II, but ancestral duplications have led to at least three FEN1-like families (Rad2, Rad27, EXO1), which, as shown in yeast, possess essentially the same activities and can complement each other in both replication and repair processes (Sun et al. 2003).

Another essential replication enzyme is DNA ligase, which is responsible for joining of the Okazaki fragments during lagging-strand replication. The main ligase involved in DNA replication in archaea is an ATP-dependent ligase (arCOG01347), which is encoded in the vast majority of archaeal genomes and is an apparent ortholog of the three eukaryotic ligases (I, III, and IV) that evolved through ancestral duplications (Ellenberger and Tomkinson 2008; Yutin and Koonin 2009). These enzymes share a common DNA-binding domain, a catalytic nucleotidyltransferase domain, and an OB-fold domain (Martin and MacNeill 2002; Ellenberger and Tomkinson 2008). Only Ligase I in eukaryotes is directly involved in replication, whereas Ligase III and Ligase IV have adopted different roles in DNA repair and homologous recombination. Ligase III and Ligase IV contain carboxy-terminal BRCT domains that are involved in protein–protein interactions via phosphoserine recognition (Martin and MacNeill 2002). As pointed out above, a BRCT domain is also present in bacterial NAD-dependent DNA ligase, which is also found in several archaea. So far only two archaea, *Halalkalicoccus jeotgali* and *Halorhabdus utahensis*, lack the ATP-dependent ligase but encode the NAD-dependent ligase, which in these organisms can be predicted to function in replication.

Single-Stranded DNA-Binding Proteins

Single-stranded DNA-binding proteins (SSBs) are essential components of the replication machinery that prevent reannealing of the growing DNA chain with the template and protect ssDNA from degradation. In all three domains of life, the major SSB proteins possess an OB-fold (Murzin 1993). In eukaryotes, the SSB, known as RPA, is a heterotrimer composed of three subunits of 70, 32, and 14 kDa (denoted RPA1, RPA2, and RPA3 in yeast) that contain four, one, and one OB-fold units, respectively. All three subunits are essential for the formation of a stable, functional RPA complex (Bochkarev and Bochkareva 2004). Archaea show multiple variations of RPA domain organization, number of homologs, and modes of interaction. Among the experimentally characterized RPA complexes in archaea, there is a homotetramer containing a single OB-fold in *S. solfataricus* (Wadsworth and White 2001); single-subunit RPAs containing multiple OB-folds in methanogens *Methanosarcina acetivorans*, *Methanocaldococcus jannaschii*, and *Methanothermobacter thermautotrophicus* (Robbins et al. 2005, and references therein); and a heterotrimeric RPA in *Pyrococcus furiosus* that consists of three distantly related subunits: RPA41 (COG1599), RPA32 (COG3390), and RPA14 (arCOG05741), each containing an OB-fold; in addition, RPA41 contains a Zn-finger-like motif (Komori and Ishino 2001).

To characterize general trends in the evolution of RPA in archaea, we reconstructed phylogenetic trees for COG1599 and COG3390 (Fig. 6A). The COG1599 subtree (which we here denote arRPA1) divides into two clades: arRPA1_long with several OB-fold domains and often a Zn finger homologous to that in eukaryotic RPA1; and arRPA1_short, typically with a single OB-fold domain.

The diverged COG3390 (arRPA2) consists of short proteins most of which contain a single OB-fold domain and a carboxy-terminal wHTH domain, the same domain arrangement as in the eukaryotic RPA2 protein (Mer et al. 2000). So far, RPA2 has been detected only in Euryarchaeota (Fig. 6B). It seems likely that arRPA2 evolved by duplication of the short form of arRPA1 followed by accelerated evolution. Similar events have been detected in eukaryotes. For example, fungal CDC13, Stn1, and Ten1, the subunits of a heterotrimeric complex essential for telomere maintenance, are paralogs of the Rpa1–Rpa2–Rpa3 complex subunits but show only remote sequence similarity to RPA1, RPA2, and RPA3, respectively (Sun et al. 2009, 2011).

Thus, most archaea possess at least two RPA genes, one long (arRPA1_long) and one or several short ones (arRPA1_short or arRPA2), which are apparent ancestors of eukaryotic RPA1 and RPA2, respectively, and can be confidently projected to LACA (Fig. 6B). Whether or not LACA encoded a third short RPA (arCOG05741) remains uncertain. but, given the presence of this gene in Korarchaeaon and several recently sequenced genomes of deeply branching archaea (Fig. 6B) (Narasingarao et al. 2012), it seems likely that this RPA is ancestral in archaea as well (Fig. 1). The major exceptions from this pattern of multiple RPA forms are Sulfolobales/Desulfurococcales, which possess only RPA1_short, and Thermoproteales, which are the only known organisms without detectable OB-fold-containing SSBs.

The apparent absence of a canonical RPA/SSB in Thermoproteales stimulated the search for an alternative, which resulted in the identification of a distinct ssDNA-binding protein in *T. tenax* that is unrelated to RPA and has been denoted ThermoSSB (Paytubi et al. 2012). ThermoSSB (arCOG05578) contains an ssDNA-binding domain with a novel fold and a leucine-zipper domain that mediates dimerization of this protein (Paytubi et al. 2012). ThermoSSB is present in all Thermoproteales with the exception of *T. pendens*, which encodes two RPA-like proteins. Thus, ThermoSSB perfectly complements

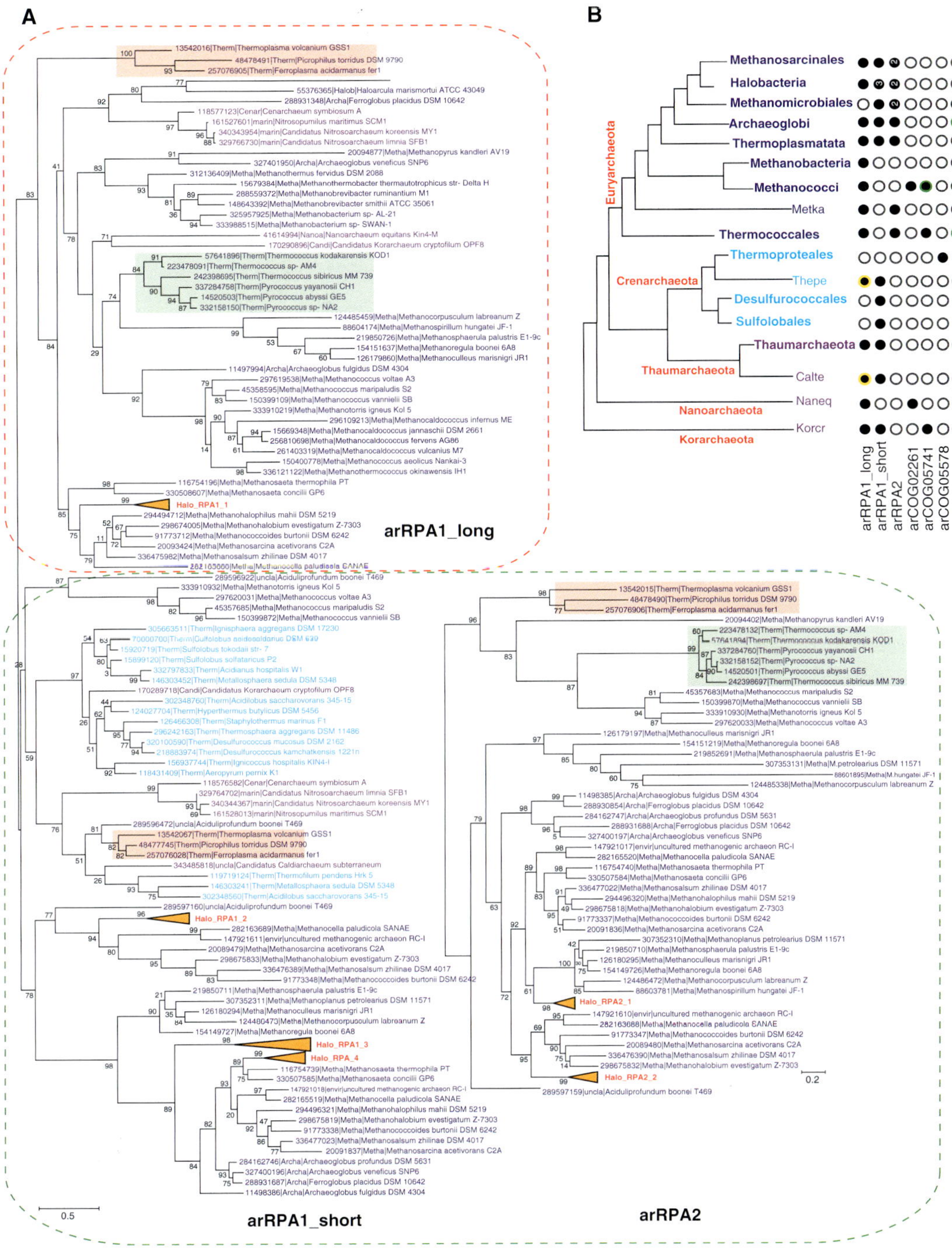

Figure 6. Phylogenetic analysis of the RPA family in archaea and eukarya. The designations and the method of tree reconstruction are as in Figure 2. (Green) The Thermococcales branches; (pink) the Thermoplasmata branches. Halobacterial branches are collapsed and numbered as follows: for arRPA1 from Halo_RPA1_1 to Halo_RPA1_4 and for arRPA2 from Halo_RPA2_1 to Halo_RPA2_2. "Long" RPAs are outlined by the red dotted line and "short" RPAs by the green dotted line. (*A*) Phylogenetic trees of the RPA1 and RPA2 families. RPA1 corresponds to COG1599 (167 sequences in total, 89 aligned positions), and RPA2 corresponds to COG03390 (76 sequences in total, 149 aligned positions). (*B*) The phyletic patterns of SSB/RPA proteins and their homologs in archaea. The designations are as in Figure 4C. (In panel *B*, circles with a yellow outline denote) RPA proteins from several organisms that could not be confidently aligned and thus are not present in the corresponding tree but included into the phyletic pattern. (Green outline) Those that do not include all representatives in the corresponding lineage.

the phyletic pattern of RPA such that SSB proteins now have been identified in all archaea.

Some potential archaeal SSBs remain unassigned. In particular, an uncharacterized paralog of ThermoSSB (arCOG03772) shows a broader phyletic distribution being present, in addition to Thermoproteales, in Sulfolobales/Desulfurococcales, Thermococcales, and Archaeoglobi (Paytubi et al. 2012). In addition, there is at least one more protein family containing OB-fold domains and distantly similar to archaeal RPA2 (arCOG02261) that is present in most Methanococcales and *Nanoarchaeum equitans*.

GENOMIC CONTEXT AND PREDICTION OF NEW COMPONENTS OF THE REPLICATION MACHINERY

The results of comparative genomics outlined here show that despite the fundamental importance of DNA replication, the protein machinery responsible for this process shows remarkable evolutionary plasticity that involves multiple events of gene loss, non-orthologous displacement, and lineage-specific duplication, even among the genes encoding core replication proteins. Examples of such events leading to complex evolutionary scenarios are the substitution a B-family polymerase for the D-family polymerase in Crenarchaeota and substitution of ThermoSSB for RPA in Thermoproteales. Moreover, some components of replication systems evolve rapidly and often lose sequence similarity, which obscures their origin. However, in many such cases, the rapidly evolving gene remains in a conserved and functionally coherent genomic neighborhood (Figs. 3 and 5). One such example is the extremely divergent RecJ homolog in Thermoproteales and another is the numerous GINS proteins that are still annotated as "hypothetical" in sequenced archaeal genomes. These highly divergent proteins are encoded in the same neighborhoods as their better conserved homologs, which facilitates functional prediction. Moreover, using the "guilt by association" principle (Aravind 2000; Galperin and Koonin 2000), conservation of gene neighborhoods can be used for prediction of new genes associated with replication. However, this approach requires caution because some housekeeping genes encoded in the same locus might not be functionally related but rather are highly expressed genes that "hitchhike" with genes that are not directly functional but are expressed similarly (Rogozin et al. 2002). In particular, genes involved in replication are often associated with genes coding for components of the translation system (Fig. 5) (Berthon et al. 2008). Thus, uncharacterized genes in the respective operons might not be involved in replication directly but are nevertheless interesting targets for experimental study (Figs. 3 and 5; Table 1). Recently, for example, analysis of gene neighborhoods led to the prediction that such genes as PACE12, a member of the GPN-loop GTPase family; and NudF, a NUDIX pyrophosphatase family member are involved in DNA replication and/or repair in archaea (Berthon et al. 2008). Here we also identified several uncharacterized genes that could be involved in these processes (Figs. 3 and 5; Table 1). In addition, because replication is a complex, metabolically costly process, many other proteins and cellular systems could be directly or indirectly involved in replication. Recent analysis of proteins interacting with core replication proteins in *T. kodakarensis* suggests that there could be dozens of such partners, and for many of these, the specific function is not known (Li et al. 2010). Thus, all recent advances notwithstanding, the current understanding of the archaeal replication system is still far from being complete.

GENERAL TRENDS IN THE EVOLUTION OF ARCHAEAL AND EUKARYOTIC REPLICATION SYSTEMS

The study of archaeal replication is currently a dynamic research field that continues to identify previously uncharacterized protein components of the replication machinery, most of which show new connections with the eukaryotic replication machinery. Comparative genomic analysis reveals a (nearly) precise correspondence between the components of the archaeal and eukaryotic replication systems, with a few notable exceptions such as the CMG-activating proteins for which archaeal counterparts have not been detected (Fig. 1; Table 1). Thus, it appears most likely that the archaeal ancestor of eukaryotes possessed a DNA replication apparatus that was as complex in its main features as the eukaryotic replication apparatus. Given this essentially precise correspondence between the archaeal and eukaryotic replication systems, it can be expected that additional replication proteins shared between archaea and eukaryotes eventually will be discovered (Fig. 1; Table 1). A major innovation in eukaryotes could be the regulation of replication by phosphorylation of replisome subunits catalyzed by dedicated kinases. Whether a counterpart to this regulatory circuit exists in archaea remains to be determined.

The close correspondence between the components of the archaeal and eukaryotic replication systems notwithstanding, eukaryotic replication does show increased complexity compared with archaeal replication. The principal modality of this greater complexity is duplication of the replication machinery components followed by subfunctionalization (Lynch and Force 2000; Ward and Durrett 2004) whereby heteromeric complexes in eukaryotes replace homomeric complexes in archaea. The obvious cases in point are the ORC, MCM, GINS, and RFC complexes. This evolutionary trend in the replication machinery is part of a general relationship between eukaryotes and their prokaryotic ancestors as exemplified by the proteasome, the exosome, and many other macromolecular complexes (Makarova et al. 2005).

Although, on the whole, the eukaryotic replication system is more complex than the archaeal counterparts, complicated histories of lineage-specific expansion of gene families accompanied by functional diversification abound in archaea. Unlike eukaryotes, where the main route to increased complexity is serial intragenomic gene duplication, in archaea, replicon fusion and more generally horizontal gene transfer are major factors of evolution, leading in particular to pseudoparalogy (Makarova et al. 2005). Strikingly, in some families of proteins involved in replication, archaea attain far greater diversity than eukaryotes, the RecJ family being the prime example. In other gene families, multiple, independent duplications are traceable in eukaryotes and in different lineages of archaea. The degree of paralogization of the archaeal replication system remains underappreciated, and many divergent paralogs still await functional characterization.

In addition to the diversification via paralogization, archaeal replication systems show less uniformity and a greater plasticity than the eukaryotic counterparts. The striking examples include non-orthologous displacement of DNA polymerases, SSB, and ligases. The actual scope of this plasticity is still unexplored because even among the relatively few archaeal genomes sequenced to date, some gaps within the core of the replication machinery are apparent such as the lack of ORC subunits in *M. kandleri*, suggestive of additional displacements of essential components.

In summary, the recent experimental and phylogenomic advances in the study of eukaryotic and archaeal replication systems clearly complement each other and jointly are yielding an increasingly complete picture of the organization and evolution of the core replication machinery. This obvious progress notwithstanding, the identity and roles of more peripheral components of the replication apparatus, the functions of paralogs of replicative genes, and especially, the ultimate origins of the complex replication machinery that is inferred to have existed in LACA, remain poorly understood and are targets for future investigation.

ACKNOWLEDGMENTS

The authors' research is supported by the intramural funds of the U.S. Department of Health and Human Services (to the National Library of Medicine).

REFERENCES

Adachi J, Hasegawa M. 1992. MOLPHY: Programs for molecular phylogenetics. In *Computer science monographs 27*. Institute of Statistical Mathematics, Tokyo.

Araki H. 2010. Cyclin-dependent kinase-dependent initiation of chromosomal DNA replication. *Curr Opin Cell Biol* **22:** 766–771.

Aravind L. 2000. Guilt by association: Contextual information in genome analysis. *Genome Res* **10:** 1074–1077.

Aravind L, Koonin EV. 1998. Phosphoesterase domains associated with DNA polymerases of diverse origins. *Nucleic Acids Res* **26:** 3746–3752.

Aravind L, Walker DR, Koonin EV. 1999. Conserved domains in DNA repair proteins and evolution of repair systems. *Nucleic Acids Res* **27:** 1223–1242.

Balasov M, Huijbregts RP, Chesnokov I. 2009. Functional analysis of an Orc6 mutant in *Drosophila*. *Proc Natl Acad Sci* **106:** 10672–10677.

Barry ER, Bell SD. 2006. DNA replication in the Archaea. *Microbiol Mol Biol Rev* **70: 15:** 614–619.

Beattie TR, Bell SD. 2011a. Molecular machines in archaeal DNA replication. *Curr Opin Chem Biol* **15:** 614–619.

Beattie TR, Bell SD. 2011b. The role of the DNA sliding clamp in Okazaki fragment maturation in archaea and eukaryotes. *Biochem Soc Trans* **39:** 70–76.

Bell SD. 2011. DNA replication: Archaeal oriGINS. *BMC Biol* **9:** 36.

Bell SP, Dutta A. 2002. DNA replication in eukaryotic cells. *Annu Rev Biochem* **71:** 333–374.

Bernander R, Dasgupta S, Nordstrom K. 1991. The *E. coli* cell cycle and the plasmid R1 replication cycle in the absence of the DnaA protein. *Cell* **64:** 1145–1153.

Bernstein KA, Gangloff S, Rothstein R. 2010. The RecQ DNA helicases in DNA repair. *Annu Rev Genet* **44:** 393–417.

Berquist BR, DasSarma P, DasSarma S. 2007. Essential and non-essential DNA replication genes in the model halophilic Archaeon, *Halobacterium* sp. NRC-1. *BMC Genet* **8:** 31.

Berthon J, Cortez D, Forterre P. 2008. Genomic context analysis in Archaea suggests previously unrecognized links between DNA replication and translation. *Genome Biol* **9:** R71.

Bloom LB. 2009. Loading clamps for DNA replication and repair. *DNA Repair (Amst)* **8:** 570–578.

Bochkarev A, Bochkareva E. 2004. From RPA to BRCA2: Lessons from single-stranded DNA binding by the OB-fold. *Curr Opin Struct Biol* **14:** 36–42.

Bochman ML, Schwacha A. 2009. The Mcm complex: Unwinding the mechanism of a replicative helicase. *Microbiol Mol Biol Rev* **73:** 652–683.

Bohlke K, Pisani FM, Rossi M, Antranikian G. 2002. Archaeal DNA replication: Spotlight on a rapidly moving field. *Extremophiles* **6:** 1–14.

Bork P, Hofmann K, Bucher P, Neuwald AF, Altschul SF, Koonin EV. 1997. A superfamily of conserved domains in DNA damage-responsive cell cycle checkpoint proteins. *FASEB J* **11:** 68–76.

Burgers PM, Koonin EV, Bruford E, Blanco L, Burtis KC, Christman MF, Copeland WC, Friedberg EC, Hanaoka F, Hinkle DC, et al. 2001. Eukaryotic DNA polymerases: Proposal for a revised nomenclature. *J Biol Chem* **276:** 43487–43490.

Cann IK, Komori K, Toh H, Kanai S, Ishino Y. 1998. A heterodimeric DNA polymerase: Evidence that members of Euryarchaeota possess a distinct DNA polymerase. *Proc Natl Acad Sci* **95:** 14250–14255.

Capp C, Wu J, Hsieh TS. 2010. RecQ4: The second replicative helicase? *Crit Rev Biochem Mol Biol* **45:** 233–242.

Chattopadhyay S, Bielinsky AK. 2007. Human Mcm10 regulates the catalytic subunit of DNA polymerase-α and prevents DNA damage during replication. *Mol Biol Cell* **18:** 4085–4095.

Chia N, Cann I, Olsen GJ. 2010. Evolution of DNA replication protein complexes in eukaryotes and Archaea. *PLoS ONE* **5:** e10866.

Coker JA, DasSarma P, Capes M, Wallace T, McGarrity K, Gessler R, Liu J, Xiang H, Tatusov R, Berquist BR, et al. 2009. Multiple replication origins of *Halobacterium* sp. strain NRC-1: Properties of the conserved orc7-dependent oriC1. *J Bacteriol* **191:** 5253–5261.

De Marco V, Gillespie PJ, Li A, Karantzelis N, Christodoulou E, Klompmaker R, van Gerwen S, Fish A, Petoukhov MV, Iliou MS, et al. 2009. Quaternary structure of the human Cdt1–Geminin complex regulates DNA replication licensing. *Proc Natl Acad Sci* **106:** 19807–19812.

Duncker BP, Chesnokov IN, McConkey BJ. 2009. The origin recognition complex protein family. *Genome Biol* **10:** 214.

Edgar RC. 2004. MUSCLE: Multiple sequence alignment with high accuracy and high throughput. *Nucleic Acids Res* **32:** 1792–1797.

Edgell DR, Doolittle WF. 1997. Archaea and the origin(s) of DNA replication proteins. *Cell* **89:** 995–998.

Edgell DR, Malik SB, Doolittle WF. 1998. Evidence of independent gene duplications during the evolution of archaeal and eukaryotic family B DNA polymerases. *Mol Biol Evol* **15:** 1207–1217.

Ellenberger T, Tomkinson AE. 2008. Eukaryotic DNA ligases: Structural and functional insights. *Annu Rev Biochem* **77:** 313–338.

Fien K, Hurwitz J. 2006. Fission yeast Mcm10p contains primase activity. *J Biol Chem* **281:** 22248–22260.

Fien K, Cho YS, Lee JK, Raychaudhuri S, Tappin I, Hurwitz J. 2004. Primer utilization by DNA polymerase α primase is influenced by its interaction with Mcm10p. *J Biol Chem* **279:** 16144–16153.

Fitch WM, Margoliash E. 1967. Construction of phylogenetic trees. *Science* **155:** 279–284.

Forterre P. 2002. The orgin of DNA genomes and DNA replication proteins. *Curr Opin Microbiol* **5:** 525–532.

Forterre P. 2006. Three RNA cells for ribosomal lineages and three DNA viruses to replicate their genomes: A hypothesis for the origin of cellular domain. *Proc Natl Acad Sci* **103:** 3669–3674.

Galperin MY, Koonin EV. 2000. Who's your neighbor? New computational approaches for functional genomics. *Nat Biotechnol* **18:** 609–613.

Garcia V, Furuya K, Carr AM. 2005. Identification and functional analysis of TopBP1 and its homologs. *DNA Repair (Amst)* **4:** 1227–1239.

Glansdorff N, Xu Y, Labedan B. 2008. The last universal common ancestor: Emergence, constitution and genetic legacy of an elusive forerunner. *Biol Direct* **3:** 29.

Hamdan SM, Richardson CC. 2009. Motors, switches, and contacts in the replisome. *Annu Rev Biochem* **78:** 205–243.

Hamdan SM, van Oijen AM. 2010. Timing, coordination, and rhythm: Acrobatics at the DNA replication fork. *J Biol Chem* **285:** 18979–18983.

Henneke G, Flament D, Hubscher U, Querellou J, Raffin JP. 2005. The hyperthermophilic euryarchaeota *Pyrococcus abyssi* likely requires the two DNA polymerases D and B for DNA replication. *J Mol Biol* **350:** 53–64.

Iyer LM, Leipe DD, Koonin EV, Aravind L. 2004. Evolutionary history and higher order classification of AAA$^+$ ATPases. *J Struct Biol* **146:** 11–31.

Kelman Z, White MF. 2005. Archaeal DNA replication and repair. *Curr Opin Microbiol* **8:** 669–676.

Klinge S, Nunez-Ramirez R, Llorca O, Pellegrini L. 2009. 3D architecture of DNA Pol α reveals the functional core of multi-subunit replicative polymerases. *EMBO J* **28:** 1978–1987.

Komori K, Ishino Y. 2001. Replication protein A in *Pyrococcus furiosus* is involved in homologous DNA recombination. *J Biol Chem* **276:** 25654–25660.

Koonin EV. 2006. Temporal order of evolution of DNA replication systems inferred by comparision of cellular and viral DNA polymerases. *Biol Direct* **1:** 39.

Koonin EV. 2009. On the origin of cells and viruses: primordial virus world scenario. *Ann NY Acad Sci* **1178:** 47–64.

Koonin EV, Martin W. 2005. On the origin of genomes and cells within inorganic compartments. *Trends Genet* **21:** 647–654.

Koppes LJ. 1992. Nonrandom F-plasmid replication in *Escherichia coli* K-12. *J Bacteriol* **174:** 2121–2123.

Kornberg A, Baker TA. 2005. *DNA replication*, 2nd ed. University Science Books, Sausalito, CA.

Krupovic M, Gribaldo S, Bamford DH, Forterre P. 2010. The evolutionary history of archaeal MCM helicases: A case study of vertical evolution combined with hitchhiking of mobile genetic elements. *Mol Biol Evol* **27:** 2716–2732.

Kuchta RD, Stengel G. 2010. Mechanism and evolution of DNA primases. *Biochim Biophys Acta* **1804:** 1180–1189.

Kunkel TA, Burgers PM. 2008. Dividing the workload at a eukaryotic replication fork. *Trends Cell Biol* **18:** 521–527.

Labib K, Gambus A. 2007. A key role for the GINS complex at DNA replication forks. *Trends Cell Biol* **17:** 271–278.

Lee C, Hong B, Choi JM, Kim Y, Watanabe S, Ishimi Y, Enomoto T, Tada S, Cho Y. 2004. Structural basis for inhibition of the replication licensing factor Cdt1 by geminin. *Nature* **430:** 913–917.

Leipe DD, Aravind L, Koonin EV. 1999. Did DNA replication evolve twice independently? *Nucleic Acids Res* **27:** 3389–3401.

Li Z, Santangelo TJ, Cubonova L, Reeve JN, Kelman Z. 2010. Affinity purification of an archaeal DNA replication protein network. *MBio* **1:** e00221-10.

Li Z, Pan M, Santangelo TJ, Chemnitz W, Yuan W, Edwards JL, Hurwitz J, Reeve JN, Kelman Z. 2011. A novel DNA nuclease is stimulated by association with the GINS complex. *Nucleic Acids Res* **39:** 6114–6123.

Liu S, Balasov M, Wang H, Wu L, Chesnokov IN, Liu Y. 2011. Structural analysis of human Orc6 protein reveals a homology with transcription factor TFIIB. *Proc Natl Acad Sci* **108:** 7373–7378.

Lynch M, Force A. 2000. The probability of duplicate gene preservation by subfunctionalization. *Genetics* **154:** 459–473.

MacNeill SA. 2010. Structure and function of the GINS complex, a key component of the eukaryotic replisome. *Biochem J* **425:** 489–500.

MacNeill SA. 2011. Protein–protein interactions in the archaeal core replisome. *Biochem Soc Trans* **39:** 163–168.

Majka J, Burgers PM. 2004. The PCNA-RFC families of DNA clamps and clamp loaders. *Prog Nucleic Acid Res Mol Biol* **78:** 227–260.

Makarova KS, Koonin EV. 2010. Two new families of the FtsZ-tubulin protein superfamily implicated in membrane remodeling in diverse bacteria and archaea. *Biol Direct* **5:** 33.

Makarova KS, Aravind L, Galperin MY, Grishin NV, Tatusov RL, Wolf YI, Koonin EV. 1999. Comparative genomics of the Archaea (Euryarchaeota): Evolution of conserved protein families, the stable core, and the variable shell. *Genome Res* **9:** 608–628.

Makarova KS, Wolf YI, Mekhedov SL, Mirkin BG, Koonin EV. 2005. Ancestral paralogs and pseudoparalogs and their role in the emergence of the eukaryotic cell. *Nucleic Acids Res* **33:** 4626–4638.

Makarova KS, Yutin N, Bell SD, Koonin EV. 2010. Evolution of diverse cell division and vesicle formation systems in archaea. *Nat Rev Microbiol* **8:** 731–741.

Makarova KS, Koonin EV, Kelman Z. 2012. The archaeal CMG (CDC45/RecJ, MCM, GINS) complex is a conserved component of the DNA replication system in all archaea and eukaryotes. *Biol Direct* **7:** 7.

Marinsek N, Barry ER, Makarova KS, Dionne I, Koonin EV, Bell SD. 2006. GINS, a central nexus in the archaeal DNA replication fork. *EMBO Rep* **7:** 539–545.

Martin IV, MacNeill SA. 2002. ATP-dependent DNA ligases. *Genome Biol* **3:** REVIEWS3005.

Matsui I, Urushibata Y, Shen Y, Matsui E, Yokoyama H. 2011. Novel structure of an N-terminal domain that is crucial for the dimeric assembly and DNA-binding of an archaeal DNA polymerase D large subunit from *Pyrococcus horikoshii*. *FEBS Lett* **585:** 452–458.

McGeoch AT, Bell SD. 2008. Extra-chromosomal elements and the evolution of cellular DNA replication machineries. *Nat Rev Mol Cell Biol* **9:** 569–574.

Mer G, Bochkarev A, Gupta R, Bochkareva E, Frappier L, Ingles CJ, Edwards AM, Chazin WJ. 2000. Structural basis for the recognition of DNA repair proteins UNG2, XPA, and RAD52 by replication factor RPA. *Cell* **103:** 449–456.

Moyer SE, Lewis PW, Botchan MR. 2006. Isolation of the Cdc45/Mcm2–7/GINS (CMG) complex, a candidate for the eukaryotic DNA replication fork helicase. *Proc Natl Acad Sci* **103:** 10236–10241.

Murzin AG. 1993. OB(oligonucleotide/oligosaccharide binding)-fold: Common structural and functional solution for non-homologous sequences. *EMBO J* **12:** 861–867.

Mushegian AR, Bassett DE Jr, Boguski MS, Bork P, Koonin EV. 1997. Positionally cloned human disease genes: Patterns of evolutionary conservation and functional motifs. *Proc Natl Acad Sci* **94:** 5831–5836.

Narasingarao P, Podell S, Ugalde JA, Brochier-Armanet C, Emerson JB, Brocks JJ, Heidelberg KB, Banfield JF, Allen EE. 2012. De novo metagenomic assembly reveals abundant novel major lineage of Archaea in hypersaline microbial communities. *ISME J* **6:** 81–93.

Ng WL, DasSarma S. 1993. Minimal replication origin of the 200-kilobase *Halobacterium* plasmid pNRC100. *J Bacteriol* **175:** 4584–4596.

Okorokov AL, Waugh A, Hodgkinson J, Murthy A, Hong HK, Leo E, Sherman MB, Stoeber K, Orlova EV, Williams GH. 2007. Hexameric ring structure of human MCM10 DNA replication factor. *EMBO Rep* **8:** 925–930.

Oyama T, Ishino S, Fujino S, Ogino H, Shirai T, Mayanagi K, Saito M, Nagasawa N, Ishino Y, Morikawa K. 2011. Architectures of archaeal GINS complexes, essential DNA replication initiation factors. *BMC Biol* **9:** 28.

Pacek M, Tutter AV, Kubota Y, Takisawa H, Walter JC. 2006. Localization of MCM2–7, Cdc45, and GINS to the site of DNA unwinding during eukaryotic DNA replication. *Mol Cell* **21:** 581–587.

Pan M, Kelman LM, Kelman Z. 2011a. The archaeal PCNA proteins. *Biochem Soc Trans* **39:** 20–24.

Pan M, Santangelo TJ, Li Z, Reeve JN, Kelman Z. 2011b. *Thermococcus kodakarensis* encodes three MCM homologs but only one is essential. *Nucleic Acids Res* **39:** 9671–9680.

Paytubi S, McMahon SA, Graham S, Liu H, Botting CH, Makarova KS, Koonin EV, Naismith JH, White MF. 2012. Displacement of the canonical single-stranded DNA-binding protein in the Thermoproteales. *Proc Natl Acad Sci* **109:** E398–E405.

Pospiech H, Grosse F, Pisani FM. 2010. The initiation step of eukaryotic DNA replication. *Subcell Biochem* **50:** 79–104.

Price MN, Dehal PS, Arkin AP. 2010. FastTree 2—Approximately maximum-likelihood trees for large alignments. *PLoS ONE* **5:** e9490.

Robbins JB, McKinney MC, Guzman CE, Sriratana B, Fitz-Gibbon S, Ha T, Cann IK. 2005. The Euryarchaeota, nature's medium for engineering of single-stranded DNA-binding proteins. *J Biol Chem* **280:** 15325–15339.

Robertson PD, Chagot B, Chazin WJ, Eichman BF. 2010. Solution NMR structure of the C-terminal DNA binding domain of Mcm10 reveals a conserved MCM motif. *J Biol Chem* **285:** 22942–22949.

Robinson NP, Bell SD. 2005. Origins of DNA replication in the three domains of life. *FEBS J* **272:** 3757–3766.

Robinson NP, Bell SD. 2007. Extrachromosomal element capture and the evolution of multiple replication origins in archaeal chromosomes. *Proc Natl Acad Sci* **104:** 5806–5811.

Robinson NP, Dionne I, Lundgren M, Marsh VL, Bernander R, Bell SD. 2004. Identification of two origins of replication in the single chromosome of the archaeon *Sulfolobus solfataricus*. *Cell* **116:** 25–38.

Rogozin IB, Makarova KS, Murvai J, Czabarka E, Wolf YI, Tatusov RL, Szekely LA, Koonin EV. 2002. Connected gene neighborhoods in prokaryotic genomes. *Nucleic Acids Res* **30:** 2212–2223.

Rogozin IB, Makarova KS, Pavlov YI, Koonin EV. 2008. A highly conserved family of inactivated archaeal B family DNA polymerases. *Biol Direct* **3:** 32.

Sakakibara N, Kelman LM, Kelman Z. 2009. Unwinding the structure and function of the archaeal MCM helicase. *Mol Microbiol* **72:** 286–296.

Sanchez-Pulido L, Ponting CP. 2011. Cdc45: The missing RecJ ortholog in eukaryotes? *Bioinformatics* **27:** 1885–1888.

Sanchez-Pulido L, Diffley JF, Ponting CP. 2010. Homology explains the functional similarities of Treslin/Ticrr and Sld3. *Curr Biol* **20:** R509–R510.

Sclafani RA, Holzen TM. 2007. Cell cycle regulation of DNA replication. *Annu Rev Genet* **41:** 237–280.

Shen Y, Musti K, Hiramoto M, Kikuchi H, Kawarabayashi Y, Matsui I. 2001. Invariant Asp-1122 and Asp-1124 are essential residues for polymerization catalysis of family D DNA polymerase from *Pyrococcus horikoshii*. *J Biol Chem* **276:** 27376–27383.

Sheu YJ, Stillman B. 2010. The Dbf4–Cdc7 kinase promotes S phase by alleviating an inhibitory activity in Mcm4. *Nature* **463:** 113–117.

Shutt TE, Gray MW. 2006. Twinkle, the mitochondrial replicative DNA helicase, is widespread in the eukaryotic radiation and may also be the mitochondrial DNA primase in most eukaryotes. *J Mol Evol* **62:** 588–599.

Sun X, Thrower D, Qiu J, Wu P, Zheng L, Zhou M, Bachant J, Wilson DM 3rd, Shen B. 2003. Complementary functions of the *Saccharomyces cerevisiae* Rad2 family nucleases in Okazaki fragment maturation, mutation avoidance, and chromosome stability. *DNA Repair (Amst)* **2:** 925–940.

Sun J, Yu EY, Yang Y, Confer LA, Sun SH, Wan K, Lue NF, Lei M. 2009. Stn1–Ten1 is an Rpa2–Rpa3–like complex at telomeres. *Genes Dev* **23:** 2900–2914.

Sun J, Yang Y, Wan K, Mao N, Yu TY, Lin YC, DeZwaan DC, Freeman BC, Lin JJ, Lue NF, et al. 2011. Structural bases of dimerization of yeast telomere protein Cdc13 and its interaction with the catalytic subunit of DNA polymerase alpha. *Cell Res* **21:** 258–274.

Tahirov TH, Makarova KS, Rogozin IB, Pavlov YI, Koonin EV. 2009. Evolution of DNA polymerases: An inactivated polymerase–exonuclease module in Pol ε and a chimeric origin of eukaryotic polymerases from two classes of archaeal ancestors. *Biol Direct* **4:** 11.

Tanaka S, Umemori T, Hirai K, Muramatsu S, Kamimura Y, Araki H. 2007. CDK-dependent phosphorylation of Sld2 and Sld3 initiates DNA replication in budding yeast. *Nature* **445:** 328–332.

Wadsworth RI, White MF. 2001. Identification and properties of the Crenarchaeal single-stranded DNA binding protein from *Sulfolobus solfataricus*. *Nucleic Acids Res* **29:** 914–920.

Walters AD, Chong JP. 2010. An archaeal order with multiple minichromosome maintenance genes. *Microbiology* **156:** 1405–1414.

Ward R, Durrett R. 2004. Subfunctionalization: How often does it occur? How long does it take? *Theor Popul Biol* **66:** 93–100.

Warren EM, Vaithiyalingam S, Haworth J, Greer B, Bielinsky AK, Chazin WJ, Eichman BF. 2008. Structural basis for DNA binding by replication initiator Mcm10. *Structure* **16:** 1892–1901.

Wei Z, Liu C, Wu X, Xu N, Zhou B, Liang C, Zhu G. 2010. Characterization and structure determination of the Cdt1 binding domain of human minichromosome maintenance (Mcm) 6. *J Biol Chem* **285:** 12469–12473.

Wohlschlegel JA, Dhar SK, Prokhorova TA, Dutta A, Walter JC. 2002. *Xenopus* Mcm10 binds to origins of DNA replication after Mcm2–7 and stimulates origin binding of Cdc45. *Mol Cell* **9:** 233–240.

Yutin N, Koonin EV. 2009. Evolution of DNA ligases of nucleo-cytoplasmic large DNA viruses of eukaryotes: A case of hidden complexity. *Biol Direct* **4:** 51.

Zhang R, Zhang CT. 2005. Identification of replication origins in archaeal genomes based on the Z-curve method. *Archaea* **1:** 335–346.

Zuo Z, Rodgers CJ, Mikheikin AL, Trakselis MA. 2010. Characterization of a functional DnaG-type primase in Archaea: Implications for a dual-primase system. *J Mol Biol* **397:** 664–676.

Human Mitochondrial DNA Replication

Ian J. Holt and Aurelio Reyes

MRC Mitochondrial Biology Unit, Cambridge, United Kingdom

Correspondence: ih@mrc-mbu.cam.ac.uk

Elucidation of the process of DNA replication in mitochondria is in its infancy. For many years, maintenance of the mitochondrial genome was regarded as greatly simplified compared to the nucleus. Mammalian mitochondria were reported to lack all DNA repair systems, to eschew DNA recombination, and to possess but a single DNA polymerase, polymerase γ. Pol γ was said to replicate mitochondrial DNA exclusively via one mechanism, involving only two priming events and a handful of proteins. In this "strand-displacement model," leading strand DNA synthesis begins at a specific site and advances approximately two-thirds of the way around the molecule before DNA synthesis is initiated on the "lagging" strand. Although the displaced strand was long-held to be coated with protein, RNA has more recently been proposed in its place. Furthermore, mitochondrial DNA molecules with all the features of products of conventional bidirectional replication have been documented, suggesting that the process and regulation of replication in mitochondria is complex, as befits a genome that is a core factor in human health and longevity.

Human mitochondrial DNA (mtDNA) typically forms 5-μm circles, or 16.5 kilobases, 1 genome in length. Outside the animal kingdom, mtDNA varies prodigiously in size and form, yet it invariably exists to provide the cell with essential components of the respiratory chain (and ATP synthase). Hence, mutations of mtDNA can cause respiratory chain dysfunction, which in turn causes rare diseases in plants (Levings and Pring 1976; Newton and Coe 1986) and humans (Holt et al. 1988). Budding yeast has the option of growing anaerobically and so can manage without mtDNA (Goldring et al. 1970), and remarkably vertebrate cells cultivated in the laboratory can achieve the same feat, on a diet of glucose far less concentrated than a typical soda (Desjardins et al. 1985; King and Attardi 1989). Nevertheless, not even a simple free-living multicellular organism, such as the nematode worm, can survive without mtDNA (Bratic et al. 2009).

The mtDNA of all vertebrates is very similar to that of humans, almost all the noncoding DNA is concentrated in one region, of little more than a kilobase, known as the major noncoding region, or NCR (Fig. 1). Although most of the mtDNA sequence variation among humans is found in the NCR, it is not evenly distributed but concentrated in three hypervariable sections (Ingman et al. 2000). In the early 1970s, two groups reported that many mtDNA molecules contain a triple-stranded region, which forms a displacement or D-loop (Arnberg et al. 1971; Kasamatsu et al. 1971). The D-loop

Cite this article as *Cold Spring Harb Perspect Biol* doi: 10.1101/cshperspect.a012971

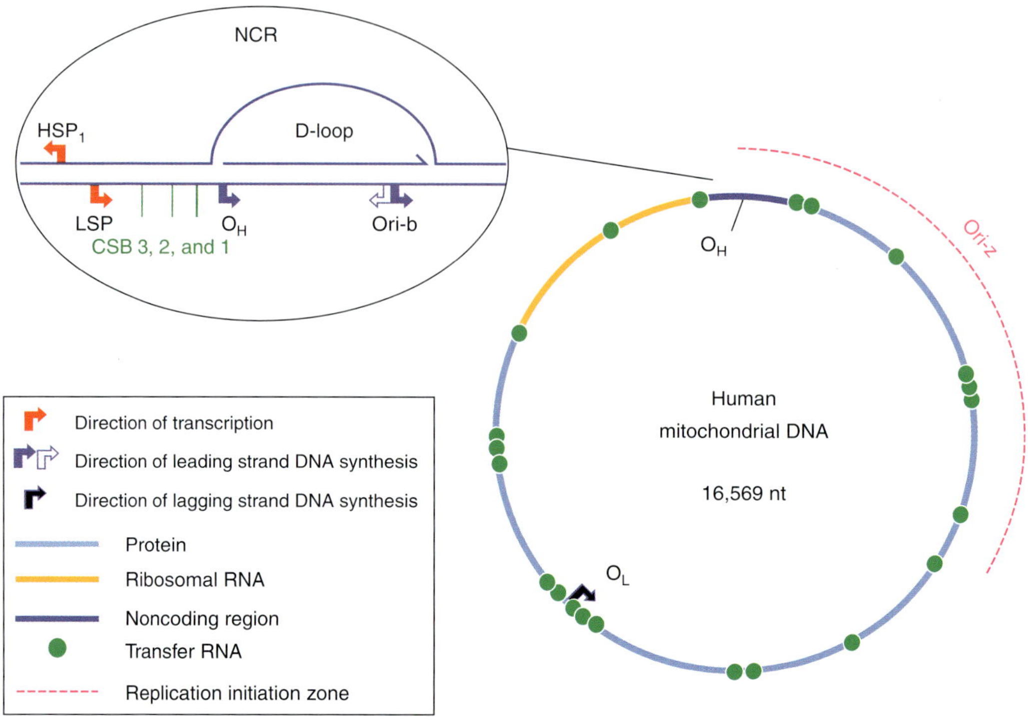

Figure 1. Schematic diagram of the human mitochondrial genome. Human mtDNA encodes 13 proteins that are essential for oxidative phosphorylation, and the ribosomal and transfer RNAs necessary for their translation. RITOLS replication initiates in the major noncoding region, at O_H and Ori-b, and O_L is a major initiation site of lagging strand DNA synthesis in mammals. Conversely, conventional bidirectional replication of mtDNA can initiate almost anywhere in the mitochondrial genome, with most such events occurring in a region of several kilobases adjacent to the major noncoding region (NCR), Ori-z (see main text for details). LSP, light strand promoter; HSP, heavy strand promoter; CSBs, conserved sequence blocks.

occupies much but not all, of the NCR, and so the terms are not interchangeable. The NCR is also known as the control region as it contains important *cis*-elements for transcription and two putative origins of replication. The critical importance of the NCR to genome maintenance is attested to by the fact that all, of the many, partially deleted mtDNAs that have been characterized in humans retain the NCR (Samuels et al. 2004), although at least a quarter of the D-loop region appears to be dispensable (Behar et al. 2008).

Many researchers date the discovery of DNA in mitochondria to the 1960s, when Nass published electron micrograph images of structures with the staining properties of DNA in chicken liver mitochondria, and Schatz isolated DNA from yeast mitochondria (Nass and Nass 1963; Schatz 1963); yet the existence of mitochondrial DNA (mtDNA), was inferred much earlier, when some heritable traits were shown to be transmitted via the cytoplasm, rather than the nucleus (Ephrussi 1953; Caspari 1955).

MECHANISMS OF MITOCHONDRIAL DNA REPLICATION

Strand-Displacement Replication

A mechanism of replication for mammalian mtDNA was elaborated in the early 1970s from EM images of DNA isolated from rodent mitochondria by cesium chloride density centrifugation (Kasamatsu and Vinograd 1973). The mechanism is quite different from that of bacterial chromosomal replication and DNA replication in the nucleus, as almost all the "theta"-like molecules had several kilobases of single-stranded DNA on one branch. Thus, the EM images of mouse mtDNA suggested a model whereby leading strand synthesis started at a specific site and advanced approximately two-thirds of the way around the molecule before second-strand DNA synthesis initiated (Kasamatsu and Vinograd 1973). The mechanism has enjoyed a number of names: asynchronous, asymmetric, and strand-displacement replication, but none is entirely satisfactory: uncoupled (leading and lagging strand DNA synthesis) replication is perhaps a more informative name, but we will continue to employ the acronym SDM (strand-displacement mechanism), for reasons of continuity. In apparent support of SDM, free 5′ ends of DNA, were mapped ~11 kilobases apart and on opposite strands (Clayton 1982). These are candidate initiation sites for leading (O_H)[1] and lagging (O_L)[1] strand DNA synthesis (see Fig. 1), although as discussed below free 5′ ends of DNA are not necessarily synonymous with replication initiation sites. Although starkly different from nuclear DNA replication, SDM was not unprecedented, as several plasmids and viruses have a protracted delay between the initiation of leading and lagging strand DNA synthesis (Khan 1997).

David Wolstenholme and colleagues produced electron micrographs of replicating molecules of rat liver mtDNA that were fully duplex, as long ago as the late 1960s (Kirschner et al. 1968), although the same group later reported that other preparations of mammalian mtDNA contained many partially single-stranded replicating molecules, leading them to propose a step-wise mode of replication, in addition to SDM. In the former case, repeated pausing of leading strand DNA synthesis at approximately 2.5 kb intervals acts as the prelude to the initiation of second-strand DNA synthesis (Koike and Wolstenholme 1974; Wolstenholme et al. 1974). Today we would recognize this as a potential strand-switching mode of replication (Michel et al. 2004). The fact that other articles from Wolstenholme's own group emphasized the novelty of the partially single-stranded intermediates of replicating mtDNA (e.g., Wolstenholme et al. 1974), doubtless contributed to the fully duplex species being largely ignored for three decades. Moreover, SDM was elaborated in some detail over the course of the next decade (Clayton and Larsson 2006, and references therein), whereas no new data emerged to support the idea of multiple sites of initiation on the lagging strand. Thus, it wasn't until restriction digested human and rodent mtDNA were shown to form arcs of replication intermediates characteristic of fully duplex DNA, using neutral two-dimensional agarose gel electrophoresis (2D-AGE), that the dominance of the strand-displacement model was seriously challenged (Holt et al. 2000). 2D-AGE resolves DNA molecules on the basis of a combination of shape and mass; aficionados of DNA replication love the method, because it captures the entire process and provides detailed mechanistic insight (Brewer and Fangman 1987, 1991; Friedman and Brewer 1995). However, it is generally loathed by the wider world because it takes considerable time and effort to learn to interpret the complex arcs and spots to which the method gives rise: it says it all that the children of one of the investigators referred to the autoradiographs of replicating mtDNA scattered about the house as the "squashed-fly pictures." Actually, the arcs comprise a complete series of intermediates from all stages of the replication cycle. Fortunately, for nonspecialists and specialists alike, some nucleic acid modifying enzymes dramatically alter the patterns of replication intermediates (RIs), greatly aiding interpretation. In the first such study of mammalian mtDNA replication (Holt et al. 2000), the use of single-stranded nuclease was especially helpful as it degraded all partially single-stranded species (Fig. 2); thus, the surviving mitochondrial RIs did not simply have the shape and mobility of fully duplex intermediates, they proved resistant to enzymes that degrade single-stranded DNA. In mitochondria

[1]The two strands of mitochondrial DNA have different buoyant densities on denaturing caesium chloride gradients because of differences in their nucleotide composition and so they are denoted H (heavy) and L (light) strands; the putative initiation sites were named the heavy and light strand origins of replication (O_H and O_L, respectively), accordingly.

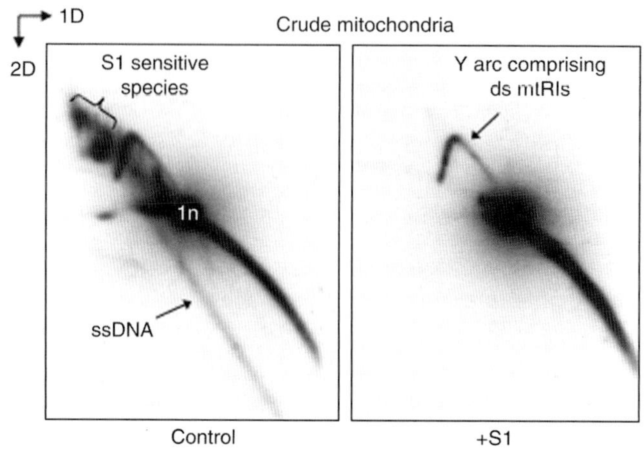

Figure 2. Some mitochondrial replication intermediates form standard replication fork arcs and are resistant to single-strand nuclease. In the example shown mouse liver mtDNA was digested with *Dra*I and half the digested was additionally treated with S1 nuclease prior to 2D-AGE. After blotting the *Dra*I filter was hybridized with probe that detected a 4.5 kb fragment of mouse mtDNA, np 5 276–9 817. ssDNA, single-stranded DNA; ds mtRIs, double-stranded mitochondrial replication intermediates.

of human cells and placenta, and mouse liver, the signal along the length of many of the replication arcs was sufficiently even to exclude not only SDM, but also the regimented step-wise replication model of Wolstenholme (Koike and Wolstenholme 1974). Instead, the single-strand nuclease resistant mtRIs were proposed to be the products of coupled leading and lagging strand DNA synthesis (Holt et al. 2000), which is discussed in more detail in the next section but one.

RNA INCORPORATION DURING mtDNA REPLICATION: RITOLS

The presumed products of strand-coupled DNA synthesis appeared to be in the minority, relative to the partially single-stranded species, and the latter were at that time attributed to SDM (Holt et al. 2000). However, later more highly purified mitochondria yielded unusually large DNA molecules forming slow-moving arcs of replication intermediates, in place of the partially single-stranded species (Fig. 3A) (Yang et al. 2002). The components of the slow-moving arcs corresponded to molecules that the restriction enzymes were unable to cut at one or more sites on one branch.[2] At first inspection, the phenomenon of the slow-moving arcs appeared compatible with SDM, because most restriction enzymes cannot cut single-stranded DNA. However, several other features and properties of the slow-moving arcs could not be reconciled with SDM. Based on their masses, the species of the slow-moving arcs were fully duplex (Fig. 3A) and (Yasukawa et al. 2006). A more clear-cut distinction was provided by the few restriction endonucleases that can cut single-stranded, as well as duplex, DNA. Crucially, such enzymes would not produce slow-moving replication fork arcs in the case of SDM, instead they would create arcs that begin at the usual point immediately adjacent to nonreplicating fragments of DNA (termed 1n species) but end prematurely, at 1.5n rather than the usual 2n. In practice, digesting gradient-purified mtDNA with enzymes that cut both duplex and single-stranded DNA gave rise to the same slow-moving Yarcs, as

[2]Enzymes that cut a replicating molecule on both branches produce so-called replication fork, or Y arcs, enzymes cutting neither branch will yield bubble structures; in the case of mtDNA, enzymes were frequently cutting only one branch of the bubble (Yasukawa et al. 2006 and Figure 3).

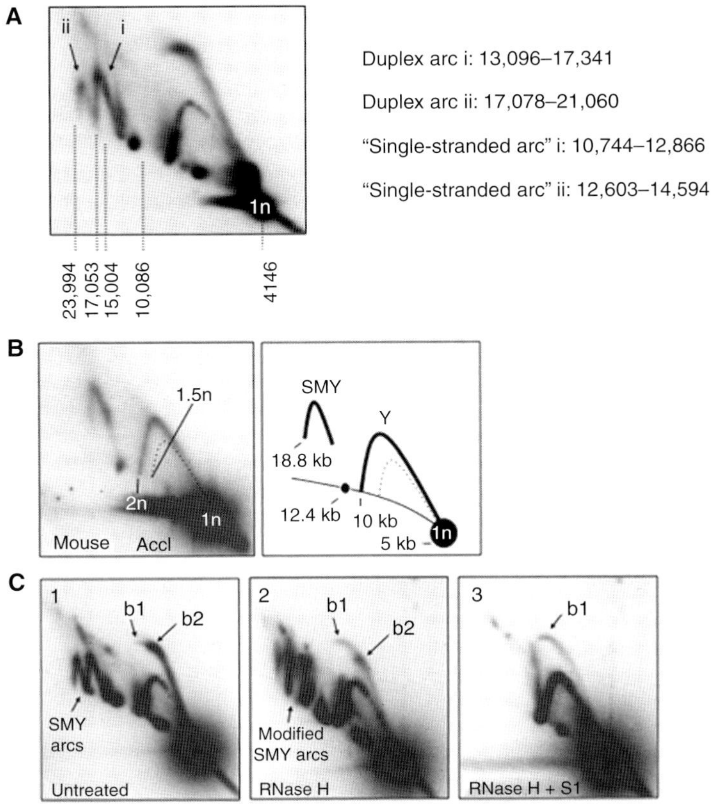

Figure 3. Replicating molecules of mtDNA produce duplex slow-moving arcs, which are sensitive to RNase H. (*A*) 2D-AGE of *Bcl*I digested mouse liver mtDNA reveals a pair of slow-moving replication fork (SMY) arcs, after hybridization to a probe detecting the 4 kb fragment nucleotides 12,027–16,174. SMY arcs result from blocked restriction sites and are essentially duplex, based on their mass, arc i has a theoretical mass of 13,096 bp initially, increasing to 17,341 bp if fully duplex, whereas it would begin at 10,744 and end at 12,866 bp if it contained a single-stranded DNA branch. Arc ii spans 17,078–21,060 if duplex; 12,603–14,594 if it follows SDM. The spans of the arcs predicted by SDM are indicated in gray font, as no such arcs were detected when gradient-purified mtDNA was analyzed. (*B*) 2D-AGE of *Acc*I digested mouse liver mtDNA reveals a slow-moving replication fork (SMY) arc, after hybridization to a probe detecting the 5-kb fragment, nucleotides 9602–14,654. N.B., *Acc*I cleaves both single-stranded and double-stranded DNA. Dashed line corresponds to the expected arc in the case of SDM replication. (*C*) The same digest of purified mouse liver mtDNA as shown in panel *A* was incubated with and without RNase H and single-stranded (S1) nuclease, and analyzed by 2D-AGE. SMY arcs are inferred to have ribonucleotides at one or more restriction sites, in part because the SMY arcs are grossly modified by RNase H. Similarly, bubble arc b2 is also sensitive to RNase H and therefore contains significant amounts of RNA, as opposed to bubble arc b1; see also main text and Yasukawa et al. (2006) for details.

enzymes that cannot cut single-stranded DNA (Fig. 3B) (Yasukawa et al. 2006). If the novel arcs were not partially single-stranded, what then could be responsible for these unusual species? Based on their mass and shape they were indistinguishable from duplex DNA; nevertheless, they were not cut by REs on one branch. Because the samples had been treated thoroughly with protease and phenol-chloroform extracted, sometimes repeatedly, protein was unlikely to account for the slow-moving arcs. Lipid and carbohydrate seemed even more improbable, and so that left RNA.

Nature had thoughtfully provided the precise tool to test this hypothesis. RNase H degrades RNA only if it is hybridized to DNA, and it proved highly effective at modifying mitochondrial RIs (Fig. 3C) (Yang et al. 2002). The presence of RNA complementary to the lagging strand template was confirmed by extracting mtRIs from 2D gels, subjecting them to denaturation, 1D-AGE and hybridization to reveal L-strand RNAs of 200–600 nucleotides in length (Yasukawa et al. 2006). Therefore, we proposed that RNA was incorporated throughout the lagging strand during the replication of mtDNA: the RITOLS model of mtDNA replication (Fig. 4). The source of the RNA has not yet been established, but two possible models were mooted: the RNA might be the product of an exuberant primase, synthesizing long RNA "primers," or the RNA could be derived from preformed transcripts which were threaded onto the lagging strand template, the so-called bootlace model (Yasukawa et al. 2006).

The RITOLS model of replication has much in common with the SDM. Both models predict that the two strands of DNA are not synthesized concurrently and both mark "O_L" as a major initiation site of second-strand DNA synthesis in mammals. Strip away the RNA from RITOLS intermediates and one is left with molecules that conform, to a first approximation, to intermediates of SDM. Or put another way SDM + lagging strand RNA = RITOLS replication. The RITOLS mechanism also retains the major noncoding region (NCR) as the origin-containing region, although there appear to be not one, but two origins in this region (Yasukawa et al. 2006), roughly speaking at either end of the D-loop.

One potential advantage of RITOLS over SDM is avoiding exposing direct repeats as single-stranded DNA, which in bacterial plasmids increases the frequency of deletion formation (Bron et al. 1991). Many pathological partial deletions of mtDNA have a residual direct repeat at the deletion junction (Mita et al. 1990), and these might be much more frequent without RNA to protect the lagging strand template. It has been suggested that the truth may lie

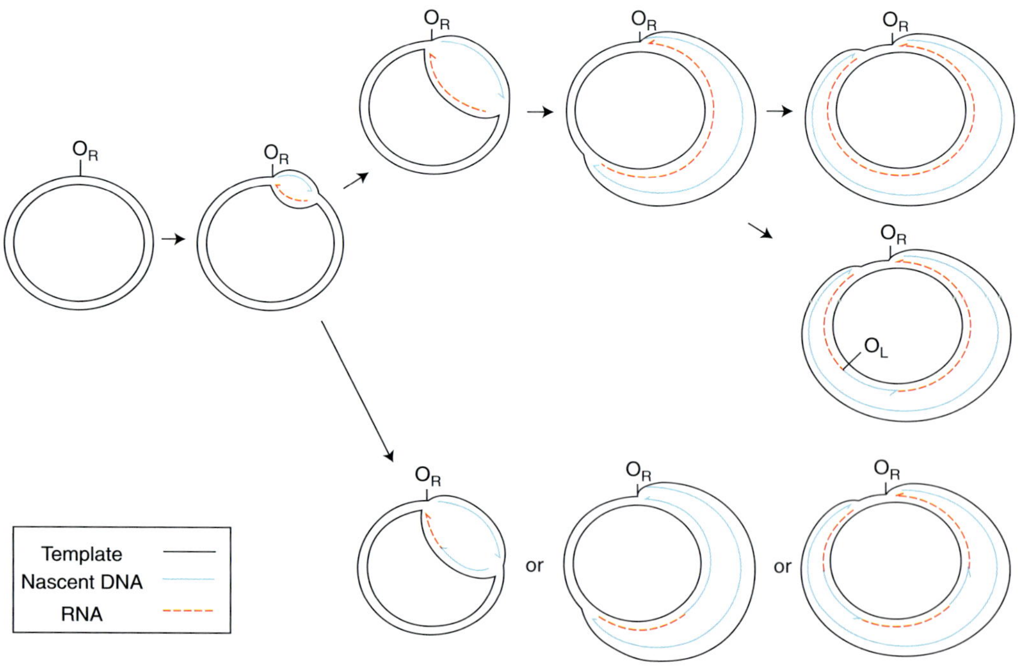

Figure 4. The RITOLS model of mitochondrial DNA replication. Replication initiates at one of two sites (O_H or Ori-b), here conflated to O_R, for origin of replication. Leading strand DNA synthesis progresses with concurrent incorporation of RNA on the lagging strand. At some point (frequently O_L), lagging strand DNA synthesis initiates and the lagging strand RNA is replaced by, or converted to, DNA.

somewhere between the SDM and the RITOLS models; i.e., some sections of the lagging strand template might be hybridized to RNA, whereas other sections might be coated with the mitochondrial single-stranded binding protein (mtSSB) (Wanrooij and Falkenberg 2010). Of course it would come as no surprise if there were some mtSSB at the replication fork, as per conventional strand-coupled DNA replication, yet the data firmly predict that RNA coats much, if not essentially all, of the lagging strand. First, the slow-moving arcs accounted for the same proportion of mitochondrial RIs irrespective of an enzyme's ability to cut ssDNA; second, the masses of the mitochondrial RIs corresponded to fully duplex DNA, within the limits of the technique (Fig. 3A) (Yang et al. 2002; Yasukawa et al. 2006). Thus, we estimate RNA coverage of the lagging strand to be at least 80% based on 2D-AGE analysis. More recent EM images revealed no gaps whatsoever in the vast majority of replicating molecules of mtDNA (Pohjoismaki et al. 2010). This result suggests that single-stranded regions account for no more than 100–200 nucleotides of mtDNA. There is a good reason to expect the coverage to be all but total; RNA has a major theoretical advantage over protein. Although the fidelity of RNA synthesis is lower than DNA synthesis, RNA and DNA contain the same genetic information, and so RNA could be used to repair any damage that occurs to the lagging strand template during mtDNA replication; no protein could perform this role. If this is the reason that RNA is preferred to mtSSB, the entire displaced lagging strand template will be hybridized to RNA. Ironically, given the predictions of extensive single-strandedness that are inherent to SDM, RNA incorporation affords the possibility of having even less single-strandedness than conventional replication mechanisms, if it follows the aforementioned bootlace model. This is because the use of preformed transcripts would dispense with the need to loop out the lagging strand for Okazaki fragment synthesis. If true, then this might explain why RITOLS evolved, as it could reduce the risk of DNA damage imposed by the formation of single-stranded DNA at the replication fork. This may be critically important in the crucible of the mitochondrion, in which reactive oxygen species are all too frequent by-products of respiration. Alternatively or additionally, RITOLS may be designed to prevent the deleterious consequences of collisions between transcription and replication complexes (Pomerantz and O'Donnell 2010), either by inhibiting transcription altogether on replicating molecules (Bentin et al. 2005) or to reduce the speed of replication, thereby, quite literally, reducing the impact of any such collisions.

Why then hasn't RITOLS been more widely adopted? It may well be that it is simply too slow—mtDNA replication being notoriously laggardly, at least in cell culture (Bogenhagen and Clayton 1977). Moreover, if RITOLS uses preformed transcripts (the bootlace model) then it could only be adopted for DNA sequences that are constitutively transcribed. This would exclude many bacterial genomes in which different genes are expressed under different conditions, which would create gaps were RITOLS replication to be used, and the nuclear genome, which comprises mainly noncoding sequences. Nevertheless, it might be worthwhile to inspect closely the replication of nuclear ribosomal DNA genes. The long lengths of noncoding DNA in plant mtDNAs also suggest that it will eschew RITOLS replication, for the most part.

COUPLED LEADING AND LAGGING STRAND DNA SYNTHESIS

The first 2D-AGE study of mammalian mtDNA replication concluded that a proportion of mitochondrial RIs comprised fully duplex DNA and so they were proposed to be the products of coupled leading and lagging strand DNA replication (Holt et al. 2000). This was a tentative conclusion, although it did suggest as a minimum that second-strand DNA synthesis could initiate at many positions on the circle. A later study (Bowmaker et al. 2003) provided further insight, yet was handicapped by the fact that the RITOLS mechanism (Yasukawa et al. 2006) had yet to be elaborated. For the benefit of readers, we will attempt to reconstruct events with the benefit of hindsight. The 2003 study mapped the majority of replication initiation sites to a zone of several kilobases outside the NCR, suggesting that initiation of replication rarely starts at the widely recognized unidirectional origin, O_H. In apparent contradiction of these results, the 2006 study mapped the start of replication to two tightly defined areas both within the *NCR*, one of which was O_H. The explanation for this apparent discrepancy is that the two reports delineated two different mechanisms of replication, the later one earmarking the NCR as the exclusive start region for RITOLS replication (Fig. 4). In contrast, the 2003 report (Bowmaker et al. 2003) characterized fully duplex DNA intermediates, the presumed products of coupled leading and lagging strand DNA synthesis, although nowhere is this explicitly stated in the manuscript for the aforementioned reasons. Another contrasting feature between the intermediates of RITOLS replication and strand-coupled DNA synthesis is directionality: RITOLS is substantially unidirectional (2006), whereas the replication initiation zone gave rise to bidirectional replication forks (Bowmaker et al. 2003). Further evidence of bidirectional replication of mtDNA came from the study of mitochondrial RIs of birds (Reyes et al. 2005). Unfortunately, this conclusion relied on an understanding of fork-direction gels (Friedman and Brewer 1995), whose interpretation is even more challenging than the standard 2D gel system. In essence, an in-gel digestion step trims the replication forks at one or other end, which generates one of two modified arcs, dependent on whether the fork entered the fragment at one end or the other—predicting the effects of different in-gel digests makes a great parlor game that all the family can enjoy! Based on a series of such digests, we were able to establish that replication forks enter many fragments of mtDNA from either end, again indicating that replication initiation occurs across a broad zone of several kilobases. Thus, both birds and mammals operate bidirectional strand-coupled DNA replication, initiating across a broad initiation zone (Reyes et al. 2005).

Currently the replication initiation zone is the only firm evidence we have for conventional coupled leading and lagging strand DNA synthesis in mitochondria, because mitochondrial RIs with duplex DNA on all branches can be derivatives of RITOLS replication. This was highlighted by the use of that most popular of drugs among mtDNA researchers, ethidium bromide. Without going into detail, ethidium bromide accumulates in mitochondria, and so low doses inhibit mtDNA replication, with little effect on nuclear DNA. Hence, three days exposure of growing cells to 125-μM ethidium bromide results in a tenfold decrease in mtDNA copy number, with no change in the generation time. Removal of the drug is followed by a burst of mtDNA replication to restore copy number to normal. Harvesting mtDNA during the recovery phase revealed fully duplex mitochondrial RIs, which were suggestive of coupled leading and lagging strand DNA replication originating at a bidirectional origin within the NCR (Yasukawa et al. 2005). At the time, we suggested this could be the result of either a contraction of the previously defined initiation zone, or rapid conversion of the "RITOLS" RNA to DNA (Fig. 5A) (Yasukawa et al. 2005). The rapid conversion of RNA to DNA, or early maturation hypothesis was strengthened when initiation of RITOLS replication was mapped to two sites in the NCR (Yasukawa et al. 2006), one of which corresponded to the novel origin identified in the ethidium bromide study (see below for more on origins of replication). It is probably the case that residual ethidium bromide causes stalling of leading strand DNA synthesis and that stalling promotes initiation of second-strand DNA synthesis. If this were to occur via strand-switching, it would breath fresh life into the early Wolstenholme model of mtDNA replication (Koike and Wolstenholme 1974). It is also noteworthy that mutant forms of the mitochondrial DNA helicase, Twinkle, produce substantially the same effect on the pattern of mtRIs as recovery from transient depletion, although the stalling is so severe that it causes mtDNA depletion (Wanrooij et al. 2007; Goffart et al. 2009).

LET US START AT THE VERY BEGINNING

The initiation of replication must necessarily create a piece of nucleic acid with a free 5′ end; typically this is the first base of a short RNA primer, which is conjugated to DNA. The RNA

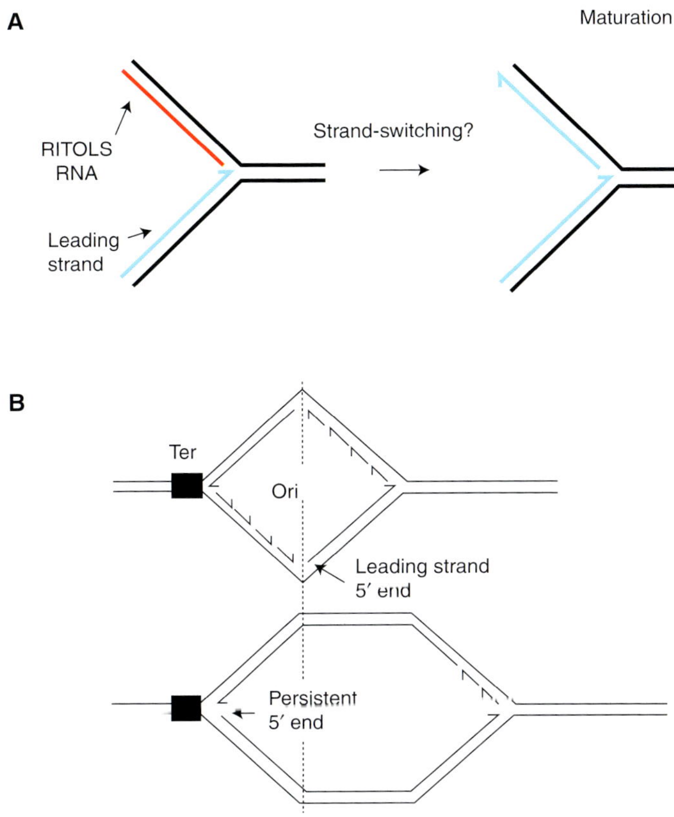

Figure 5. Maturation of RITOLS RNA, and the formation of free 5′ ends of DNA during DNA replication. Replication stalling is associated with an increase in fully duplex DNA intermediates, which could be the result of RNA replacement with DNA (*A*). This could be the result of strand-switching by the replicative DNA polymerase. (*B*) One of a pair of replication forks derived from a bidirectional origin of replication has just reached the terminus (Ter) causing arrest of leading strand DNA synthesis; the Okazaki fragments are joined to make the lagging strand continuous, leaving a single free 5′ end remaining at the terminus, whereas the origin is no longer detectable.

primer is chemically more labile than DNA and so it may degrade during nucleic acid extraction and processing. Even if the RNA primer is preserved, many DNA polymerases will not read through RNA, therefore primer extension can be used to map free 5′ ends of DNA, and this procedure was for many years the method of choice for defining origins of replication. In this way, the origin of replication of mtDNA was mapped to specific nucleotides in the NCR, especially nucleotide 191 of human mtDNA, aka O_H (Crews et al. 1979). Later RNA species that spanned the region from the light strand promoter (LSP) to, approximately, O_H were identified, leading to the suggestion that transcripts initiated at LSP could be processed to generate primers for the initiation of DNA synthesis, and so RNA processing of L-strand transcripts would be the first critical step in the initiation of mtDNA replication (Chang et al. 1985). It has been suggested that a catalytic (7–2/MRP) RNA is responsible for this RNA processing step (Topper et al. 1992), although there appears to be little if any 7–2/MRP in some mitochondrial preparations (Kiss and Filipowicz 1992).

Many features of this model may prove to be correct, but in light of recent history in the field, it is worthwhile considering several caveats. First, there are abundant 7S DNA molecules of mitochondrial D-loops, the 5′ ends of these molecules map to a cluster of nucleotides in the NCR including nucleotide 191. Ordinarily, they are far more abundant than RIs, and so (labeled) primers should be placed outside the NCR to be confident that they are not contributing to the results. Second, 5′ ends of DNA need not necessarily mark an initiation site: persistent nicks will give the same output. Third, origins dispersed across a zone of several kilobases will produce variable 5′ ends of DNA, which will almost certainly be below the limits of detection because, fourth, initiation sites of bidirectional replication are ephemeral creatures (Fig. 5B). Fifth, a free 5′ end of DNA may mark a replication pause or termination site, both of which will generate far more prominent 5′ ends of DNA than even a discrete origin of replication (Fig. 5B) (Santamaria et al. 1998; Hyvarinen et al. 2007), unless, that is, it is a unidirectional origin of replication.

Currently there are several candidate initiation sites for leading strand mtDNA synthesis. Early reports agreed that the most abundant free 5′ ends of DNA lay at or close to nucleotide 191 of human mtDNA (e.g., Crews et al. 1979). Much later, Giuseppe Attardi revisited the question, and concluded that the major start site was a hundred or so nucleotides downstream of nucleotide 191, at nucleotide 57 (Fish et al. 2004). In contrast, Falkenberg and colleagues RNase H treated their samples and found prominent primer extension products mapping upstream of nucleotide 191, at the second of three conserved sequence blocks in the NCR, CSB2 (i.e., ~nucleotide 300) (Pham et al. 2006). Because the sites identified in these studies were all in the vicinity of one another they were all compatible with the early EM data (Kasamatsu and Vinograd 1973). Our mapping of free 5′ ends of DNA relied on ligation-mediated PCR, which we readily acknowledge is far from perfect, as the amplification process is at best semi-quantitative. However, the fine mapping was coupled to 2D-AGE analysis. The bubble arcs on 2D gels, which are a signature of replication origins, mapped a "RITOLS" origin (or terminus) approximately to O_H, and LM-PCR revealed free 5′ ends of DNA at nucleotides 54 and 191 (as well as a few points in between); although we did not detect ends at CSB2, our samples were not treated with RNase H. No final ruling can be made as to the precise location of O_H, but there is little doubt an origin (and terminus) is located in the vicinity of nucleotide 191. The second origin in the NCR revealed by the ethidium bromide study (Yasukawa et al. 2005) maps to the other end of the D-loop, around nucleotide 16,197 (aka Ori-b). This discovery raised the intriguing possibility that Ori-b may be the major or primary origin, and O_H is the replication terminus. In this model, bidirectional replication from Ori-b produces a nascent leading strand of DNA that encounters the terminus after a mere half a kilobase. Concurrent Okazaki fragment synthesis fills in on the lagging strand template back to Ori-b (Fig. 5), and so the newly synthesized H-strand DNA from O_H to Ori-b is technically *lagging strand,* not leading strand, DNA. However, when RNA is incorporated on the L-strand template, the same piece of nascent H-strand DNA (O_H-Ori-b) cannot be said to be lagging strand, it then becomes a second initiation event on the leading strand; notwithstanding, this remains a subsidiary event to initiation at Ori-b.

THE ENDGAME

Almost nothing is known about the termination of mtDNA replication beyond its location. Terminal replication intermediates contain two forks, one or both of which is advancing toward the other, creating so-called double Y arcs on 2D gels. When the two forks meet, the molecule mimics a four-way junction (or X structure), before separating into the two daughter molecules. In most cells and tissues, double Y arcs of mtDNA are the exclusive preserve of fragments that contain the NCR, and the termination site maps to O_H. In bacteria, termination of replication and resolution of replication intermediates is a well-orchestrated process, which requires a site-specific recombination apparatus (Grainge et al. 2011). To date, we merely have the suggestion that three proteins of the mTERF family, mTERF, mTERFD1, and mTERFD3, in some way contribute to the termination of DNA replication. Overexpression of mTERF induces an increase in 5′ ends at O_H without altering mtDNA copy number (Hyvarinen et al. 2007), which is suggestive of prolonged termination (Pohjoismaki and Goffart 2011). And gene-silencing of mTERFD1 and mTERFD3 leads to the accumulation of X species, centered on O_H (Hyvarinen et al. 2011).

THE MACHINERY OF MITOCHONDRIAL DNA REPLICATION: WHERE'S THE BEEF?

Remarkably few enzymes are known to be critical for mtDNA replication in mammals, those known or inferred to be essential are mitochondrial DNA polymerase γ (Hance et al. 2005), Twinkle DNA helicase (Tyynismaa et al. 2004), mitochondrial RNA polymerase, single-stranded DNA binding protein, RNase H1 (Cerritelli et al. 2003), DNA ligase III (Simsek et al. 2011), and topoisomerases. We confidently predict that this is very far from a full list, despite the fact that the first two, with mtSSB and TFAM, can synthesize pieces of DNA up to 16 kilobases in length, in vitro (Korhonen et al. 2004).

If conventional strand-coupled DNA replication occurs in mitochondria then there must be the apparatus for Okazaki fragment maturation, the sealing of the discontinuous segments of DNA that comprise the nascent lagging strand of DNA. One firm possibility is that the enzymes involved are shared with the nucleus (Holt 2009). PIF1 (Futami et al. 2007), FEN1 (Liu et al. 2008), and DNA2 (Duxin et al. 2009) have each been reported to be present in mitochondria, as well as in the nucleus, but the last two might be used exclusively in DNA repair, equally PIF1 could have functions unrelated to DNA replication. Nevertheless, it is noteworthy that all the apparatus of Okazaki fragment maturation appears to be present in mitochondria — it would be a pity to ignore it.

If, as the data suggest, mitochondria operate two, or more, mechanisms of replication then there is no reason a priori why they should restrict themselves to a single DNA polymerase. We forgot at our peril that the field of bacterial DNA replication blithely believed Pol I was the replicative DNA polymerase for 15 years. Moreover, mitochondria of Trypanosomes have four DNA polymerases (Klingbeil et al. 2002) and while it is true they have a bizarre arrangement of mtDNA, we doubt that more sophisticated mammals ration themselves to a single enzyme, given the importance of the mitochondrial genome.

REPLICATION AND SELECTION OF MUTANT MITOCHONDRIAL DNA

Interest in mitochondrial DNA replication stems in part from the human diseases caused by mutant mtDNA. Pathological mutations in human mtDNA were first described early in 1988; these were partial deletions that resulted in the loss of multiple genes (Holt et al. 1988). A key distinction from nuclear DNA mutations is that mutant mtDNA does not follow Mendelian genetics, both because mtDNA is maternally inherited and because there are typically thousands of copies of mtDNA per cell. Thus, a variant of mtDNA may account for anywhere between 0.1% and 99.9% of the mitochondrial genotype. In most individuals, the vast majority of mtDNA molecules share the same sequence (homoplasmy), whereas deleterious mutants often coexist with wild-type molecules (heteroplasmy). Clearly this suggests competition between different mitochondrial genotypes. Mutants, such as partial deletions, that result in complete loss of function can never achieve total dominance (homoplasmy) in complex organisms, because aerobic energy production is essential for life, yet there is no question that mutant mtDNA can accumulate to levels that seriously impair respiration and produce devastating diseases, as was clear from the very first report (Holt et al. 1988). Although we are still grappling with the forces that drive selection of mutant mtDNA, the underlying problem of replicative advantage is becoming clear, at least in some cases. The danger for the cell is that any mutation that confers on mtDNA a replicative advantage will result in the selection of the mutant genotype irrespective of its effect on mitochondrial function.

What do we mean by the rather nebulous concept of replicative advantage? As has been widely recognized, partial deletions of mtDNA will complete the replication cycle ahead of their full-length, wild-type counterparts, and so they will be the first daughter molecules available for subsequent rounds of replication (Diaz et al. 2002). Partial duplications have more rather than less mtDNA to copy and so should have a marked disadvantage; however, this is more than compensated for if, as is invariably the case in patients, the extra DNA contains an additional origin(s) of replication (Holt et al. 1997). The point mutation most frequently associated with disease in humans resides at nucleotide position 3243 (Goto et al. 1990), far from any predicted origin of replication, yet it too is thought to confer a selective advantage to its owner. Nucleotide 3243 is located in the binding site for a protein known as the mitochondrial transcription termination factor, mTERF. Overexpression of mTERF enhances replication pausing around nucleotide 3243 (Hyvarinen et al. 2007). Crucially, the m.3243G mutation reduces the affinity of binding of mTERF (Hess et al. 1991), and so it is predicted to reduce replication pausing; hence, m.3243G can confer a replicative advantage. Although it is not clear if every one of the hundreds of pathological mtDNA mutations confers a direct replicative advantage, there is currently no other hypothesis to explain the preferential accumulation of deleterious mitochondrial variants. Although impaired mitochondrial clearance (mitophagy) could permit mitochondria containing mutant DNA to thrive, such a defect should result in any and every mutant mtDNA variant persisting, rather than the fixation of specific mutants, as is seen in mitochondrial diseases.

NUCLEAR DNA MUTATIONS AFFECTING MITOCHONDRIAL DNA

Mitochondrial DNA of humans and most other animals is entirely reliant on the nucleus for its existence. This is amply shown by the many disease-causing mutations in genes required for mtDNA maintenance: POLG1, Twinkle (Spelbrink et al. 2001; Van Goethem et al. 2001). However, this is not a reliable method of assessing a protein's credentials for a role in replication, or mtDNA maintenance. TFAM, which is absolutely essential for mtDNA maintenance (Larsson et al. 1998), has thus far never been linked to a human pathology. Moreover, if proteins are shared between the mitochondria and the nucleus many mutations will prove lethal because of their effects on nuclear DNA, or if they permit survival they may produce unusual phenotypes; that said, given the huge diversity of phenotypes associated with mitochondrial dysfunction it is hard to picture an atypical set of signs and symptoms.

CONCLUDING REMARKS

As the opening sentence of the abstract indicated, there is much work to be performed, and the full text should have left the reader in no doubt as to the many gaps in our knowledge, which are unlikely all to be filled by mtSSB. The single most pressing issue is to confirm or refute the presence of RNA in mitochondrial replication intermediates, as this lies at the heart of the competing RITOLS and SD mechanisms. If confirmed, what is the source of the RITOLS RNA, and does it function to slow replication or inhibit transcription, as here proposed? Can mtDNA replication proceed via SDM in the absence of the RITOLS RNA, and if not, why not? Mitochondrial RNA polymerase appears to have a multiple roles in mtDNA replication, for example (Fuste et al. 2010), and further study of it, and the relationship between transcription and replication, are also likely to yield further insight into the mechanisms of mtDNA replication. We anticipate the identification of new enzymes involved in mtDNA replication will help enormously to unravel the mechanics of the process. As if two (or three?) mechanisms weren't enough, there is additionally evidence of recombination-dependent replication operating in some solid tissues of adults (Pohjoismaki et al. 2009), which awaits detailed investigation. Major topics such as replication restart and fork-rescue have barely been touched on in mitochondria, and even if the mechanics of initiation of leading strand DNA synthesis are as simple as some have suggested, we still know almost nothing of how it is regulated.

Although the prospects for therapies that rely on intervening directly in the process of mtDNA replication seem bleak given the paucity of knowledge of the basic mechanisms, segregation bias may provide an indirect means of solving the problem. Also the burgeoning evidence of DNA transacting proteins shared between the nucleus and mitochondria should act as a stimulus for further research, and as previously noted, elucidation of mtDNA replication may even aid understanding of the same process in its giant neighbor, the nucleus (Holt 2009).

ACKNOWLEDGMENTS

We gratefully acknowledge the many colleagues and collaborators who have worked with us on mitochondrial DNA replication over the course of the past decade and a half, and the Medical Research Council and the European Union for funding. In addition, we thank Jaakko Pohjoismaki for a critical reading of the manuscript.

REFERENCES

Arnberg A, van Bruggen EF, Borst P. 1971. The presence of DNA molecules with a displacement loop in standard mitochondrial DNA preparations. *Biochim Biophys Acta* **246:** 353–357.

Behar DM, Blue-Smith J, Soria-Hernanz DF, Tzur S, Hadid Y, Bormans C, Moen A, Tyler-Smith C, Quintana-Murci L, Wells RS. 2008. A novel 154-bp deletion in the human mitochondrial DNA control region in healthy individuals. *Hum Mutat* **29:** 1387–1391.

Bentin T, Cherny D, Larsen HJ, Nielsen PE. 2005. Transcription arrest caused by long nascent RNA chains. *Biochim Biophys Acta* **1727:** 97–105.

Bogenhagen D, Clayton DA. 1977. Mouse L cell mitochondrial DNA molecules are selected randomly for replication throughout the cell cycle. *Cell* **11:** 719–727.

Bowmaker M, Yang MY, Yasukawa T, Reyes A, Jacobs HT, Huberman JA, Holt IJ. 2003. Mammalian mitochondrial DNA replicates bidirectionally from an initiation zone. *J Biol Chem* **278:** 50961–50969.

Bratic I, Hench J, Henriksson J, Antebi A, Burglin TR, Trifunovic A. 2009. Mitochondrial DNA level, but not active replicase, is essential for *Caenorhabditis elegans* development. *Nucleic Acids Res* **37:** 1817–1828.

Brewer BJ, Fangman WL. 1987. The localization of replication origins on ARS plasmids in *S. cerevisiae*. *Cell* **51:** 463–471.

Brewer BJ, Fangman WL. 1991. Mapping replication origins in yeast chromosomes. *Bioessays* **13:** 317–322.

Bron S, Holsappel S, Venema G, Peeters BP. 1991. Plasmid deletion formation between short direct repeats in *Bacillus subtilis* is stimulated by single-stranded rolling-circle replication intermediates. *Mol Gen Genet* **226:** 88–96.

Caspari E. 1955. The role of genes and cytoplasmic particles in differentiation. *Ann NY Acad Sci* **60:** 1026–1037.

Cerritelli SM, Frolova EG, Feng C, Grinberg A, Love PE, Crouch RJ. 2003. Failure to produce mitochondrial DNA results in embryonic lethality in Rnaseh1 null mice. *Mol Cell* **11:** 807–815.

Chang DD, Hauswirth WW, Clayton DA. 1985. Replication priming and transcription initiate from precisely the same site in mouse mitochondrial DNA. *EMBO J* **4:** 1559–1567.

Clayton DA. 1982. Replication of animal mitochondrial DNA. *Cell* **28:** 693–705.

Clayton DA, Larsson NG. 2006. Mitochondrial DNA replication and human disease. In *DNA replication and human disease* (ed. Depamphilis ML), pp. 547–560. Cold Spring Harbor Laboratory Press, Cold Spring Harbor, NY.

Crews S, Ojala D, Posakony J, Nishiguchi J, Attardi G. 1979. Nucleotide sequence of a region of human mitochondrial DNA containing the precisely identified origin of replication. *Nature* **277:** 192–198.

Desjardins P, Frost E, Morais R. 1985. Ethidium bromide-induced loss of mitochondrial DNA from primary chicken embryo fibroblasts. *Mol Cell Biol* **5:** 1163–1169.

Diaz F, Bayona-Bafaluy MP, Rana M, Mora M, Hao H, Moraes CT. 2002. Human mitochondrial DNA with large deletions repopulates organelles faster than full-length genomes under relaxed copy number control. *Nucleic Acids Res* **30:** 4626–4633.

Duxin JP, Dao B, Martinsson P, Rajala N, Guittat L, Campbell JL, Spelbrink JN, Stewart SA. 2009. Human Dna2 is a nuclear and mitochondrial DNA maintenance protein. *Mol Cell Biol* **29:** 4274–4282.

Ephrussi B. 1953. *Nucleo-cytoplasmic relations in microorganisms*. Clarendon, New York.

Fish J, Raule N, Attardi G. 2004. Discovery of a major D-loop replication origin reveals two modes of human mtDNA synthesis. *Science* **306:** 2098–2101.

Friedman KL, Brewer BJ. 1995. Analysis of replication intermediates by two-dimensional agarose gel electrophoresis. *Methods Enzymol* **262:** 613–627.

Fuste JM, Wanrooij S, Jemt E, Granycome CE, Cluett TJ, Shi Y, Atanassova N, Holt IJ, Gustafsson CM, Falkenberg M. 2010. Mitochondrial RNA polymerase is needed for activation of the origin of light-strand DNA replication. *Mol Cell* **37:** 67–78.

Futami K, Shimamoto A, Furuichi Y. 2007. Mitochondrial and nuclear localization of human Pif1 helicase. *Biol Pharm Bull* **30:** 1685–1692.

Goffart S, Cooper HM, Tyynismaa H, Wanrooij S, Suomalainen A, Spelbrink JN. 2009. Twinkle mutations associated with autosomal dominant progressive external ophthalmoplegia lead to impaired helicase function and in vivo mtDNA replication stalling. *Hum Mol Genet* **18:** 328–340.

Goldring ES, Grossman LI, Krupnick D, Cryer DR, Marmur J. 1970. The petite mutation in yeast. Loss of mitochondrial deoxyribonucleic acid during induction of petites with ethidium bromide. *J Mol Biol* **52:** 323–335.

Goto Y, Nonaka I, Horai S. 1990. A mutation in the tRNA (Leu)(UUR) gene associated with the MELAS subgroup of mitochondrial encephalomyopathies. *Nature* **348:** 651–653.

Grainge I, Lesterlin C, Sherratt DJ. 2011. Activation of XerCD-dif recombination by the FtsK DNA translocase. *Nucleic Acids Res* **39:** 5140–5148.

Hance N, Ekstrand MI, Trifunovic A. 2005. Mitochondrial DNA polymerase γ is essential for mammalian embryogenesis. *Hum Mol Genet* **14:** 1775–1783.

Hess JF, Parisi MA, Bennett JL, Clayton DA. 1991. Impairment of mitochondrial transcription termination by a point mutation associated with the MELAS subgroup of mitochondrial encephalomyopathies. *Nature* **351:** 236–239.

Holt IJ. 2009. Mitochondrial DNA replication and repair: All a flap. *Trends Biochem Sci* **34:** 358–365.

Holt IJ, Harding AE, Morgan-Hughes JA. 1988. Deletions of muscle mitochondrial DNA in patients with mitochondrial myopathies. *Nature* **331:** 717–719.

Holt IJ, Dunbar DR, Jacobs HT. 1997. Behaviour of a population of partially duplicated mitochondrial DNA molecules in cell culture: Segregation, maintenance and recombination dependent upon nuclear background. *Hum Mol Genet* **6:** 1251–1260.

Holt IJ, Lorimer HE, Jacobs HT. 2000. Coupled leading- and lagging-strand synthesis of mammalian mitochondrial DNA. *Cell* **100:** 515–524.

Hyvarinen AK, Pohjoismaki JL, Reyes A, Wanrooij S, Yasukawa T, Karhunen PJ, Spelbrink JN, Holt IJ, Jacobs HT. 2007. The mitochondrial transcription termination factor mTERF modulates replication pausing in human mitochondrial DNA. *Nucleic Acids Res* **35:** 6458–6474.

Hyvarinen AK, Pohjoismaki JL, Holt IJ, Jacobs HT. 2011. Overexpression of MTERFD1 or MTERFD3 impairs the completion of mitochondrial DNA replication. *Mol Biol Rep* **38:** 1321–1328.

Ingman M, Kaessmann H, Paabo S, Gyllensten U. 2000. Mitochondrial genome variation and the origin of modern humans. *Nature* **408:** 708–713.

Kasamatsu H, Vinograd J. 1973. Unidirectionality of replication in mouse mitochondrial DNA. *Nat New Biol* **241:** 103–105.

Kasamatsu H, Robberson DL, Vinograd J. 1971. A novel closed-circular mitochondrial DNA with properties of a replicating intermediate. *Proc Natl Acad Sci* **68:** 2252–2257.

Khan SA. 1997. Rolling-circle replication of bacterial plasmids. *Microbiol Mol Biol Rev* **61:** 442–455.

King MP, Attardi G. 1989. Human cells lacking mtDNA: Repopulation with exogenous mitochondria by complementation. *Science* **246:** 500–503.

Kirschner RH, Wolstenholme DR, Gross NJ. 1968. Replicating molecules of circular mitochondrial DNA. *Proc Natl Acad Sci* **60:** 1466–1472.

Kiss T, Filipowicz W. 1992. Evidence against a mitochondrial location of the 7–2/MRP RNA in mammalian cells. *Cell* **70:** 11–16.

Klingbeil MM, Motyka SA, Englund PT. 2002. Multiple mitochondrial DNA polymerases in *Trypanosoma brucei*. *Mol Cell* **10:** 175–186.

Koike K, Wolstenholme DR. 1974. Evidence for discontinuous replication of circular mitochondrial DNA molecules from Novikoff rat ascites hepatoma cells. *J Cell Biol* **61:** 14–25.

Korhonen JA, Pham XH, Pellegrini M, Falkenberg M. 2004. Reconstitution of a minimal mtDNA replisome in vitro. *EMBO J* **23:** 2423–2429.

Larsson NG, Wang J, Wilhelmsson H, Oldfors A, Rustin P, Lewandoski M, Barsh GS, Clayton DA. 1998. Mitochondrial transcription factor A is necessary for mtDNA maintenance and embryogenesis in mice. *Nat Genet* **18:** 231–236.

Levings CS 3rd, Pring DR. 1976. Restriction endonuclease analysis of mitochondrial DNA from normal and Texas cytoplasmic male-sterile maize. *Science* **193:** 158–160.

Liu P, Qian L, Sung JS, de Souza-Pinto NC, Zheng L, Bogenhagen DF, Bohr VA, Wilson DM III, Shen B, Demple B. 2008. Removal of oxidative DNA damage via FEN1-dependent long-patch base excision repair in human cell mitochondria. *Mol Cell Biol* **28:** 4975–4987.

Michel B, Grompone G, Flores MJ, Bidnenko V. 2004. Multiple pathways process stalled replication forks. *Proc Natl Acad Sci* **101:** 12783–12788.

Mita S, Rizzuto R, Moraes CT, Shanske S, Arnaudo E, Fabrizi GM, Koga Y, DiMauro S, Schon EA. 1990. Recombination via flanking direct repeats is a major cause of large-scale deletions of human mitochondrial DNA. *Nucleic Acids Res* **18:** 561–567.

Nass MM, Nass S. 1963. Intramitochondrial fibers with DNA characteristics. I. Fixation and electron staining reactions. *J Cell Biol* **19:** 593–611.

Newton KJ, Coe EH. 1986. Mitochondrial DNA changes in abnormal growth (nonchromosomal stripe) mutants of maize. *Proc Natl Acad Sci* **83:** 7363–7366.

Pham XH, Farge G, Shi Y, Gaspari M, Gustafsson CM, Falkenberg M. 2006. Conserved sequence box II directs transcription termination and primer formation in mitochondria. *J Biol Chem* **281:** 24647–24652.

Pohjoismaki JL, Goffart S. 2011. Of circles, forks and humanity: Topological organisation and replication of mammalian mitochondrial DNA. *Bioessays* **33:** 290–299.

Pohjoismaki JL, Goffart S, Tyynismaa H, Willcox S, Ide T, Kang D, Suomalainen A, Karhunen PJ, Griffith JD, Holt IJ, et al. 2009. Human heart mitochondrial DNA is organized in complex catenated networks containing abundant four-way junctions and replication forks. *J Biol Chem* **284:** 21446–21457.

Pohjoismaki JL, Holmes JB, Wood SR, Yang MY, Yasukawa T, Reyes A, Bailey LJ, Cluett TJ, Goffart S, Willcox S, et al. 2010. Mammalian mitochondrial DNA replication intermediates are essentially duplex but contain extensive tracts of RNA/DNA hybrid. *J Mol Biol* **397:** 1144–1155.

Pomerantz RT, O'Donnell M. 2010. What happens when replication and transcription complexes collide? *Cell Cycle* **9:** 2537–2543.

Reyes A, Yang MY, Bowmaker M, Holt IJ. 2005. Bidirectional replication initiates at sites throughout the mitochondrial genome of birds. *J Biol Chem* **280:** 3242–3250.

Samuels DC, Schon EA, Chinnery PF. 2004. Two direct repeats cause most human mtDNA deletions. *Trends Genet* **20:** 393–398.

Santamaria D, de la Cueva G, Martinez-Robles ML, Krimer DB, Hernandez P, Schvartzman JB. 1998. DnaB helicase is unable to dissociate RNA-DNA hybrids. Its implication in the polar pausing of replication forks at ColE1 origins. *J Biol Chem* **273:** 33386–33396.

Schatz G. 1963. The isolation of possible mitochondrial precursor structures from aerobically grown baker's yeast. *Biochem Biophys Res Commun* **12:** 448–451.

Simsek D, Furda A, Gao Y, Artus J, Brunet E, Hadjantonakis AK, Van Houten B, Shuman S, McKinnon PJ, Jasin M. 2011. Crucial role for DNA ligase III in mitochondria but not in Xrcc1-dependent repair. *Nature* **471:** 245–248.

Spelbrink JN, Li FY, Tiranti V, Nikali K, Yuan QP, Tariq M, Wanrooij S, Garrido N, Comi G, Morandi L, et al. 2001. Human mitochondrial DNA deletions associated with mutations in the gene encoding Twinkle, a phage T7 gene 4-like protein localized in mitochondria. *Nat Genet* **28:** 223–231.

Topper JN, Bennett JL, Clayton DA. 1992. A role for RNAase MRP in mitochondrial RNA processing. *Cell* **70:** 16–20.

Tyynismaa H, Sembongi H, Bokori-Brown M, Granycome C, Ashley N, Poulton J, Jalanko A, Spelbrink JN, Holt IJ, Suomalainen A. 2004. Twinkle helicase is essential for mtDNA maintenance and regulates mtDNA copy number. *Hum Mol Genet* **13:** 3219–3227.

Van Goethem G, Dermaut B, Lofgren A, Martin JJ, Van Broeckhoven C. 2001. Mutation of POLG is associated with progressive external ophthalmoplegia characterized by mtDNA deletions. *Nat Genet* **28:** 211–212.

Wanrooij S, Falkenberg M. 2010. The human mitochondrial replication fork in health and disease. *Biochim Biophys Acta* **1797:** 1378–1388.

Wanrooij S, Goffart S, Pohjoismaki JL, Yasukawa T, Spelbrink JN. 2007. Expression of catalytic mutants of the mtDNA helicase Twinkle and polymerase POLG causes distinct replication stalling phenotypes. *Nucleic Acids Res* **35:** 3238–3251.

Wolstenholme DR, Koike K, Cochran-Fouts P. 1974. Replication of mitochondrial DNA: replicative forms of molecules from rat tissues and evidence for discontinuous replication. *Cold Spring Harb Symp Quant Biol* **38:** 267–280.

Yang MY, Bowmaker M, Reyes A, Vergani L, Angeli P, Gringeri E, Jacobs HT, Holt IJ. 2002. Biased incorporation of ribonucleotides on the mitochondrial L-strand accounts for apparent strand-asymmetric DNA replication. *Cell* **111:** 495–505.

Yasukawa T, Yang MY, Jacobs HT, Holt IJ. 2005. A bidirectional origin of replication maps to the major noncoding region of human mitochondrial DNA. *Mol Cell* **18:** 651–662.

Yasukawa T, Reyes A, Cluett TJ, Yang MY, Bowmaker M, Jacobs HT, Holt IJ. 2006. Replication of vertebrate mitochondrial DNA entails transient ribonucleotide incorporation throughout the lagging strand. *EMBO J* **25:** 5358–5371.

Parvovirus Diversity and DNA Damage Responses

Susan F. Cotmore[1] and Peter Tattersall[12]

[1]Department of Laboratory Medicine, Yale University School of Medicine, New Haven, Connecticut 06510

[2]Department of Genetics, Yale University School of Medicine, New Haven, Connecticut 06510

Correspondence: peter.tattersall@yale.edu

Parvoviruses have a linear single-stranded DNA genome, around 5 kb in length, with short imperfect terminal palindromes that fold back on themselves to form duplex hairpin telomeres. These contain most of the *cis*-acting information required for viral "rolling hairpin" DNA replication, an evolutionary adaptation of rolling-circle synthesis in which the hairpins create duplex replication origins, prime complementary strand synthesis, and act as hinges to reverse the direction of the unidirectional cellular fork. Genomes are packaged vectorially into small, rugged protein capsids ~260 Å in diameter, which mediate their delivery directly into the cell nucleus, where they await their host cell's entry into S phase under its own cell cycle control. Here we focus on genus-specific variations in genome structure and replication, and review host cell responses that modulate the nuclear environment.

Viruses from the family *Parvoviridae* are unique in having a linear single-stranded DNA genome, ~5 kb in length, which terminates in short (120–420 base) imperfect palindromes that fold into self-priming hairpin telomeres. These viruses replicate via a "rolling hairpin" mechanism, with strong evolutionary and mechanistic links to "rolling-circle" replication, as reviewed in detail in previous editions of this work (Cotmore and Tattersall 1996, 2006). Rolling hairpin synthesis relies on the ability of each hairpin to give rise to a duplex origin sequence, which can be nicked by a viral initiator nuclease to create a base-paired DNA primer, and to function as a hinge, allowing quasi-circular synthesis by alternately folding and unfolding to shuttle a continuous unidirectional replication fork back and forth along linear DNA. Together with a few adjacent nucleotides, these palindromes provide all of the *cis*-acting information required for viral DNA replication and packaging. However, the size, sequence, and predicted structures of the hairpins can vary substantially between genera, or even between the two ends of a single genome, and they appear to have adapted to fulfill multiple additional roles in the life cycle of specific viruses. Parvoviral DNA amplification proceeds via a unidirectional single-strand displacement mechanism through a series of monomeric and concatemeric duplex replicative-form (RF) intermediates, and while the viral initiator protein, variously called NS1 or Rep, serves both as a site- and strand-specific nickase and as a 3′-to-5′ helicase, all other replicative functions are co-opted from the host cell. This mechanism benefits from

Cite this article as *Cold Spring Harb Perspect Biol* doi: 10.1101/cshperspect.a012989

suppression of host DNA synthesis, and generates long stretches of single-stranded DNA with alien terminal structures that invoke host damage responses, which impact both positively and negatively on viral replication. Although details of the mechanisms that mediate parvoviral replication have received relatively little attention since our previous review (Cotmore and Tattersall 2006), recognition that infection is invariably associated with host DNA damage responses (DDRs), some of which are specifically required for efficient viral DNA amplification, has led to significant reappraisal of the nuclear environment and replicative machinery available to these viruses. In parallel, major advances have occurred in our knowledge of genome diversity and cell specificity in this ever-expanding family, which provide novel insight into mechanisms of replication control.

The International Committee on Taxonomy of Viruses (ICTV, Tijssen et al. 2011), classifies viruses in the family *Parvoviridae* that infect vertebrates as the subfamily *Parvovirinae*, which currently contains just five genera: the *Parvoviruses*, *Dependoviruses*, *Amdoviruses*, *Erythroviruses*, and *Bocaviruses*, although there are at least two additional genera, tentatively called *Partetraviruses* and *Avetalviruses*, that await ICTV recognition. This reflects a major jump in known virus diversity, with many new species and genera first identified in clinical or veterinary samples using PCR-based virus discovery methods (Allander et al. 2005; Jones et al. 2005; Day and Zsak 2010). Potential human pathogens that are still pending recognition include genetic variants of Human Bocavirus (HBoV 1–4), which are particularly common in the respiratory and gastrointestinal tracts of children (Kapoor et al. 2010; Kantola et al. 2011), and two broadly distributed genotypes of a "PARV4"-based genus (the aforementioned *Partetraviruses*), parenterally transmitted among injecting drug users, hemophiliacs, and polytransfused individuals (Sharp et al. 2009; Lahtinen et al. 2011). Most recently, viruses from another potential genus, with sequences resembling both *Parvoviruses* and *Amdoviruses*, were detected in fecal samples from children in Burkina Faso, and tentatively named *Bufaviruses* (Phan et al. 2012). In vitro culture systems or high titer clinical samples are often not available for new members, which can thus only be studied by PCR amplification from patient tissue.

Although the vast majority of parvoviruses replicate without the aid of a helper virus, the adeno-associated viruses (AAVs) from the genus *Dependovirus* establish latent infections that only become productive when cells are coinfected with a more complex virus, typically an *adeno-* or *herpesvirus*. To date the replication mechanisms of adeno-associated virus 2 (AAV2) and minute virus of mice (MVM), from the genus *Parvovirus*, have been studied in detail, as documented in previous editions of this work. Here we adopt a broader perspective, discussing inter-genera variations that shed light on replication control, and reviewing host cell responses to viral infection that modulate the nuclear environment.

GENOMIC DIVERSITY IN THE PARVOVIRINAE

Parvoviruses package a single copy of their nonpermuted, single-stranded, linear genome into robust $T = 1$ icosahedral protein capsids 18–26 nm in diameter, which lack histones or accessory proteins. Viruses from the *Dependovirus* and *Erythrovirus* genera are homotelomeric, meaning that their terminal hairpins form part of inverted terminal repeats (ITRs), positioning the same DNA origin sequence at both ends of the genome. These origins are then resolved by equivalent "terminal resolution" mechanisms, ultimately leading to the displacement and encapsidation of DNA strands of both polarities. During terminal resolution the sequence of the hairpin is inverted with each round of synthesis, generating equal numbers of termini with two inverted complementary sequence orientations, termed "flip" and "flop," as detailed previously (Cotmore and Tattersall 1996, 2006). In contrast, members of the *Parvo-, Amdo-, and Bocavirus* genera are "heterotelomeric," with very different palindromes at the two ends of their genome, which are each processed by different mechanisms and at different rates. In such viruses, the hairpin at the 5′ end of the negative-sense strand, by convention called the "right end," is typically resolved by terminal resolution, yielding progeny telomeres in flip and flop orientations. Termini at genomic "left ends," however, are generated from palindromic dimer RF molecules by "asymmetric junction resolution," through a heterocruciform intermediate, a mechanism that conserves a single sequence orientation, denoted flip, at the left end of negative-strand progeny genomes (Cotmore and Tattersall 2003). Although the unique flip sequence could theoretically serve additional critical roles in the viral life cycle, recent studies with viruses carrying axial hairpin mutations indicate that both sequence and orientation changes in the hairpin ears can be tolerated, suggesting that maintaining the flip orientation of these structures is a consequence of, but not the reason for, asymmetric left-end processing. However, the same studies indicated that asymmetric dimer junction resolution is critical for effective MVM replication (Li et al. 2012). In heterotelomeric viruses, imbalance of origin firing typically leads to the preferential displacement of negative-sense progeny single strands from the more potent right-end DNA replication origin, which are then packaged without further selection (Cotmore and Tattersall 2005). Thus, the relative frequency of plus and minus strands and flip and flop termini in packaged DNA provides a useful indicator of the type of resolution mechanism(s) used for their excision.

Genome structures of the type species for recognized genera, and their encoded proteins, are diagrammed in Figure 1. Genera are differentiated primarily by DNA sequence phylogeny, but they are also distinguishable by telomere structure and by the nature and role of the ancillary proteins. Viruses use from one to three transcriptional promoters and one to two polyadenylation signals to coordinate the expression of two major gene complexes, a nonstructural (NS or *rep*) gene that encodes the viral initiator protein (NS1 or Rep68/78) and a structural (VP or *cap*) gene. Each genus also encodes a few small ancillary proteins, with disparate functions, variably disposed throughout the genome. In viral origins, the initiator proteins bind site-specifically to two to five variably spaced reiterations of a duplex cognate tetranucleotide motif via oligomeric interactions detailed in Hickman et al. (2004). For viruses in the genus *Parvovirus*, these clusters of 5′-TGGT-3′ repeats are also reiterated in complex patterns at ~100 base pair intervals throughout the genome, so that, in vivo, NS1 is bound throughout MVM duplex RF DNA (Cotmore et al. 2007). A somewhat similar pattern of extensive 5′-TGGT-3′ reiteration is also apparent in the newly identified *Bufaviruses*, suggesting that they share this cognate binding site and chromatin-evasion strategy, but Rep binding site clustering is not seen throughout AAV genomes. In other genera selection experiments to identify preferred oligonucleotide motifs have yet to be reported.

Despite their limited genetic complexity, most parvoviruses replicate productively without the aid of a helper virus, although the absence of viral gene products that induce resting cells to enter S phase restricts their replication to actively dividing cell populations. To support DNA replication, all telomeres must encode multipartite replication origins and engage in hinge-like hairpin rearrangements, which require local melting induced by site-specific binding of the initiator proteins (Willwand et al. 2002; Lou et al. 2004). In addition, axial asymmetries in the hairpins, most commonly manifest as T- or Y-shaped structures, likely facilitating hairpin unfolding and refolding by lowering the free energy of the duplex. Nevertheless, the complexity and diversity of these hairpins suggests that they fulfill additional roles in the life cycle of specific viruses. For example, in MVM, and potentially in other heterotelomeric viruses, the left-end hairpin contains both symmetric and asymmetric binding sites for transcription factors that modulate expression from the adjacent P4 promoter (Faisst et al. 1994; Gu et al. 1995; Deleu et al. 1999). More recently, the hairpins of AAV have been shown to bind MRE11/Rad50/NBS1 (MRN) complexes (Schwartz et al. 2007) and Ku70/80 heterodimers (Choi et al. 2010), which are involved in the sensing and repair of DNA damage, raising the possibility that they are tailored to trigger and usurp specific damage responses.

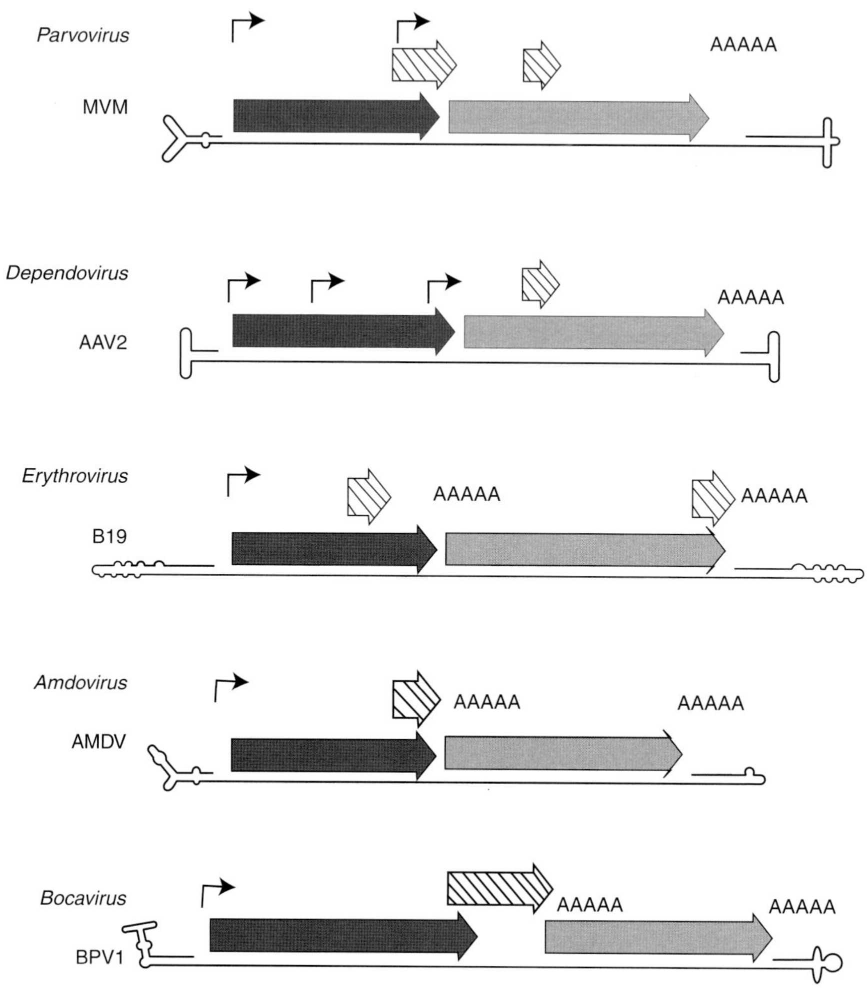

Figure 1. Genetic strategies of representative viruses from genera in the subfamily *Parvovirinae*. Genomes from the type species of each genus are denoted as a single line terminating in hairpin structures. The hairpins, drawn to represent their predicted structures, are scaled about 20× with respect to the rest of the genome. Open reading frames are represented by arrowed boxes, shaded dark gray for the major SF3 domain-containing replication initiator protein, light gray for the structural proteins of the capsid, and cross-hatched for sequences unique to the ancillary nonstructural proteins. Ancillary proteins for MVM are NS2 (*left*) and SAT (*right*) (Zádori et al. 2005; Ruiz et al. 2011), AAP for AAV2 (Sonntag et al. 2011), 7.5 kDa (*left*) and 11 kDa (*right*) for B19 (Zhi et al. 2006), NS2 for AMDV (Oleksiewicz et al. 1996), and NP1 for BPV (Chen et al. 1986). Transcriptional promoters are indicated by solid arrows and polyadenylation sites by the AAAAA sequence block.

The *Erythrovirus* and *Bocavirus* genera are of particular interest because of their relevance for human disease. As seen in Figure 1, the human erythrovirus B19 encodes three nonstructural proteins: the initiator protein NS1, an early 7.5 kDa species of uncertain function, and an essential 11 kDa late protein (Zhi et al. 2006). It has a homotelomeric genome, with ITRs of 381 nucleotides that terminate in novel, imperfect, 361 nucleotide palindromes. These can potentially fold into long linear duplexes with a few unpaired nucleotides creating a series of small, but highly conserved, mismatched bubbles. B19 shows fastidious tissue specificity, with productive infection restricted to human erythroid progenitor cells (EPCs) between BFU-E (burst-forming unit-erythroid) and CFU-E (colony-forming unit-erythroid) stages (Ozawa et al. 1986). Although high titer viremias often occur in vivo, culture systems capable of supporting similar levels of replication remain elusive. Some replication can be induced in nonpermissive 293 cells by cotransfection with adenovirus E2a, E4ORF6, and VA RNA genes (Guan et al. 2009), which are the helper genes typically required for AAV replication. However, these same adenovirus genes enhance replication of many parvovirus-based vectors in 293T cells (El-Andaloussi et al. 2011; L Li and P Tattersall, unpubl.), suggesting that they likely elicit common helper mechanisms. More recently, in vivo derived EPCs, expanded in culture to generate $CD36^{+}$ cells, have been shown to support substantial B19 replication when reexposed to erythropoietin (Epo). In these cells, stimulation of the Epo/Epo receptor/Janus kinase (Jak2) pathway, or constitutive expression of activated Epo receptors alone, proved sufficient to activate replication of internalized virus (Chen et al. 2010), although progeny virions that could expand infection were still not generated at useful levels. However, culture under hypoxic conditions designed to mimic the normal bone marrow environment enhances both viral DNA replication and progeny production via a HIFα-independent, STAT5A-mediated pathway (Chen et al. 2011). Exactly how this signaling cascade licenses DNA synthesis remains uncertain, but promises to be of substantial interest.

In contrast to the erythroviruses, the heterotelomeric genomes of bocaviruses in many ways resemble those of the parvoviruses, although NS2 homologs are generally absent and instead bocaviruses encode an abundant ~22.5 kDa nuclear phosphoprotein, called NP1, which is essential for replication. The founding members of this genus, bovine parvovirus (BPV) and minute virus of canines (MVC), can both be propagated efficiently in culture using established cell lines. Following transfection, replication of MVC genomes with a null mutation in NP1 are nonviable, but these can be rescued in *trans* by NP1 genes from MVC, BPV, or HBoV1, indicating tightly conserved roles for this protein in different host species (Sun et al. 2009). Bocaviruses appear to encapsidate predominantly (90%) negative-sense DNA with left-end termini that are chiefly (90%) in the flip orientation (Chen et al. 1988), suggesting a mixture of resolution strategies in which junction resolution predominates. MVC and BPV genomes have unique hairpins of around 183 and 198 nucleotides at their left and right ends, respectively. When folded to minimize free energy, the MVC left-end hairpin assumes a structure that is somewhat reminiscent of the MVM left end, with a duplex stem in which unpaired nucleotides create a nine residue bubble, a single asymmetric thymidine nucleotide, and axial T-shaped ears, whereas those of BPV contain an additional bubble in place of the asymmetric T. Also akin to MVM, the right-end telomeres of both viruses can be folded into simple duplex structures, which contain internal palindromes that could potentially rearrange to create small cruciform branches near their major dyad axis, as illustrated in Figure 1.

Productive culture systems for human HBoV variants initially appeared elusive, although an important clue for HBoV-1 came from the observation that low levels of viral DNA, detectable by quantitative PCR, were released from pseudostratified human airway epithelium cells following inoculation with HBoV1 PCR-positive nasopharyngeal washes (Dijkman et al. 2009). Remarkably, the entire HBoV-1 genome, complete with hairpins resembling those of MVC and BPV, has now been cloned into bacterial plasmids, and shown to support viral replication and the production of single-round virus stocks following transfection into producer 293 cells (Huang et al. 2012). Such stocks are able to productively infect polarized primary human airway epithelia (HAE) from the apical surface, giving rise to infected HAE that show multiple hallmarks of lung airway-tract injury. Significantly, the 293 cell-derived stocks could also productively infect an immortalized airway epithelial cell line, CuFi-8, but only after these cells had been experimentally induced to polarize. This intriguing and robust reverse genetics system should thus ensure rapid dissection of productive viral replication routes. Given the similarities between the bocaviruses and parvoviruses, it will be intriguing to see to what extent their very different NS2 and NP1 ancillary proteins overlap in function.

ESTABLISHING PRODUCTIVE INFECTION

Parvoviral virions are exceptionally rugged, and while they undergo sequential conformational rearrangements during trafficking through the host cell, most evidence suggests they are ultimately delivered intact directly into the host nucleus (Sonntag et al. 2006). Virions are dense, with single-stranded DNA packed tightly into preformed capsids under control of the viral 3′-to-5′ helicase (King et al. 2001). Although it has been suggested that nuclear factors mediate disassembly of AAV virions (Thomas et al. 2004), data for MVM suggests that the genome is simply ejected from the otherwise intact capsid, again in a 3′-to-5′ direction, and then remains tightly associated with it via the genomic 5′ terminus (Cotmore et al. 2010), potentially allowing the capsid to play additional roles in the viral life cycle. For MVM, analysis of capsid mutants suggests that both genome packaging and uncoating are mediated by narrow pore-bearing cylindrical structures that surround the icosahedral fivefold axes of the capsid (Plevka et al. 2011; Cotmore and Tattersall 2012), suggesting a highly organized "two portal" model of genome packaging and release, outlined in Figure 2.

Incoming parvovirus virions lack a duplex transcription template or accessory proteins capable of manipulating the cell cycle, and therefore remain quiescent within host cell nuclei until complementary strand DNA synthesis creates the first viral transcription template. For MVM, this duplex is synthesized when the cell enters S phase under its own cell cycle control, and is primed from the left-end hairpin of the negative-sense genome, which is recognized and extended by cellular machinery. The first viral transcripts, driven from the S-phase-responsive P4 promoter in MVM, then allow synthesis of the initiator protein, which ultimately mediates DNA amplification but also acts alongside genus-specific accessory proteins,

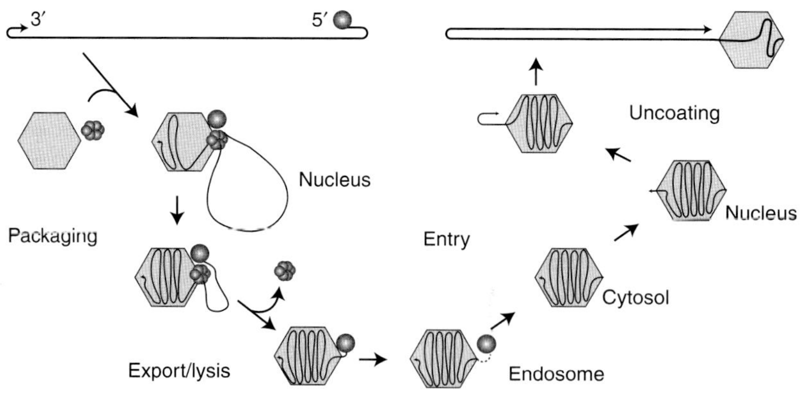

Figure 2. "Two portal" model for genome packaging and uncoating. (*Top left*) A newly displaced single-stranded progeny genome is depicted 3′-to-5′, left-to-right, with a sphere denoting a covalently bound NS1 molecule. Beneath this, the DNA is shown being introduced into a preassembled capsid in a 3′-to-5′ direction, likely via the helicase activity of an NS1 peptide oligomer (depicted as a hexameric structure, by analogy with AAV) (Yoon-Robarts et al. 2004), with the 3′ end of the entering genome lodging near the capsid shell, proximal to its destined exit pore. The rest of the strand is then pumped into the particle under increasing pressure, adopting a topology that will allow it to be subsequently unraveled in a 3′-to-5′ direction. NS1 is removed from the 5′ end of the DNA during cell entry. Ca^{2+}/Mg^{2+} ions in the virion are required to stabilize the pressurized structure, but when these are depleted in vitro, perhaps mimicking a specific host factor trigger in vivo, the 3′ end of the DNA, followed by the coding sequences, is ejected. This uncoated DNA can support complementary strand synthesis, while the 5′ end of the negative strand genome remains threaded with the capsid.

such as NS2, to modify the host environment in ways that allow viral replication to proceed.

Where, and in what form, these viruses persist during G_0/G_1 remains enigmatic, because exposed incoming genomes of this type would be expected to invoke cellular DDRs, leading to delays in the G_1-S transition and restructuring of the viral DNA. One interesting possibility is that infecting genomes are retained within the virion until specifically induced to uncoat at the start of S phase, thus potentially allowing the capsid to separate the genome from nuclear surveillance mechanisms. Although parvoviruses cannot induce cells to enter S phase, when A9 cells were experimentally synchronized at G_1/S using a procedure that allows virus to accumulate in the nucleus before the start of S, cellular DNA synthesis rates in MVM-infected cells initially rose, paralleling those in uninfected cells for the first 2 h, but then declined precipitously (Cotmore and Tattersall 1987). Meanwhile, DNA synthesis in uninfected cells peaked between 3 and 5 h after release, and followed a sharp biphasic curve, beginning to exit S phase after ~8 h. In contrast, in infected cells duplex viral RF first became detectable at ~6 h after the start of S phase, reached maximal levels by 10 h, and maintained these levels for at least 20 h. This suggests that the presence of virus was detected within 2 h of entry into S phase, leading to the cessation of cellular replication. Following a lag period in which the viral nonstructural proteins accumulated, viral DNA amplification then reached peak levels, which were maintained long after uninfected cells entered G_2, creating a "pseudo S phase" (Davy and Doorbar 2007) in which viral DNA synthesis continued although host replication had terminated. When DNA content is measured by flow cytometry, this pseudo S phase reads out as sustained supra-diploid DNA levels typically scored as "S phase" or "G_2/M" cell cycle arrest. Accordingly, S and G_2/M phase arrest has been reported following infection with many parvovirus genera, but because a similar flow phenotype can be induced following expression of certain viral nonstructural proteins alone, without ongoing viral DNA replication, interpretation of such data can be complex, as discussed below.

INDUCED DNA DAMAGE RESPONSES

During infection, replicating viral DNA is typically first observed in a series of discrete foci scattered throughout the nucleus, which are highly enriched for the viral initiator protein NS1/Rep. In synchronized MVM-infected cells NS1 morphology progresses through a series of distinct patterns, as illustrated in Figure 3. Replicating viral DNA first becomes apparent at ~6–8 h into S phase, in foci with type II NS1 morphology (Fig. 3C), creating nuclear subdomains called autonomous parvovirus-associated replication (APAR) bodies. These foci recruit cyclin A and various cellular replication proteins known to be involved in viral DNA amplification such as PCNA, RPA, and DNA polymerase δ, but they also sequester DNA polymerase α–primase complexes that are not required for MVM replication, so that how specific proteins are recruited to, or excluded from, these sites remains unclear (Cziepluch et al. 2000; Bashir et al. 2001). During this early phase, activated components of various cellular DDR pathways also accumulate within the APAR compartment, as illustrated in Figure 3G–I, which shows the recruitment of DNA-damage foci containing MDC1 (mediator of DNA-damage checkpoint 1) to the vicinity of APAR bodies in MVM-infected mouse embryonic fibroblasts. Such damage foci also contain γH2AX, which rapidly becomes concentrated within the APAR compartment although MDC1 does not. Similar discrete, but relatively transient, viral replication domains sequestering a spectrum of activated DDR proteins have been observed for many parvoviruses, but exactly which proteins are recruited appears to vary between virus species, and even between cell types. Such foci ultimately expand and coalesce, until accumulating NS1 and viral DNA occupies most of the enlarged nucleus (type IV morphology, Fig. 3E), leaving cellular chromatin compacted and marginated (Ihalainen et al. 2007, 2009).

In general, parvovirus-induced DDRs are characterized by the complex phosphorylation of RPA32, by induction of γH2AX, and by retention of these modified proteins within

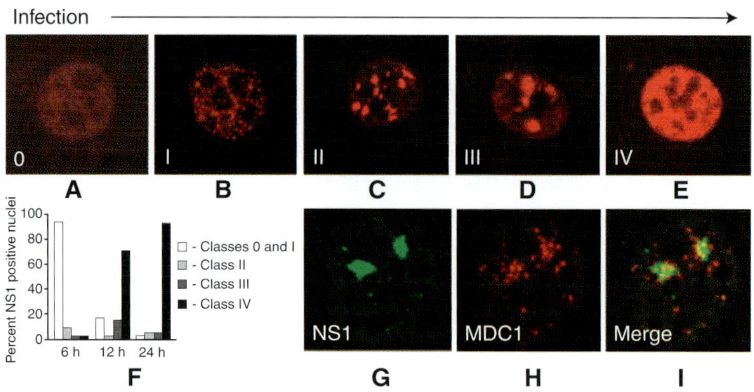

Figure 3. Progressive development of viral replication centers. (*A–F*) A classification scheme, with classes 0–IV, shows the progressive changes in NS1 distribution that emerge with time after synchronized A9 cells infected with wild-type MVMp are released into S phase. Classes 0 and I were first observed 3 h into S phase, class II at 6 h, and classes III and IV at 12 h, although the number of cells showing the latter class morphologies progressively increased with time (quantified in *F*, where the distribution of NS1 morphology classes is plotted against time in seconds). (*G–I*) Accumulation of MDC1 DDR foci around APAR bodies (stained for NS1) in MVM-infected wild-type mouse embryonic fibroblasts. (From Ruiz et al. 2011; reproduced, with permission, from Elsevier © 2011.)

viral replication centers. Each of the three major phosphatidylinositol 3-kinase-like-kinases (PIKKs), ATM, ATR, and DNA-PK, have been implicated in such responses, although details vary. Where tested, pronounced activation of at least one pathway appears required for efficient viral replication, although it is not clear how much this augmentation reflects the direct action of specific DDR components on viral replication mechanisms, and how much simply results from competitive advantages gained by suppression of cellular replication. Thus, for example, infection of murine fibroblasts with MVM potently activates the ATM pathway (Adeyemi et al. 2010; Ruiz et al. 2011), leading to phosphorylation of a number of its downstream substrates and their sequestration within APAR foci. In embryonic mouse fibroblasts, activated ATR and its checkpoint kinase, Chk1, also colocalize with viral replication centers, whereas in A9 fibroblasts, ATR signaling is suppressed in response to infection (Ruiz et al. 2011). Moreover, in human cells, the DNA-PK catalytic subunit and its accessory proteins, Ku70 and Ku86, accumulate in APAR foci. Despite this battery of potential responses, studies with pharmacological inhibitors show that inhibition of the major ATM response, but not DNA-PK, results in impaired viral DNA replication in MVM-infected A9 cells (Adeyemi et al. 2010).

Similarly, bocavirus MVC infection of canine cells results in activation of both ATM and ATR cascades, but pharmacological inhibition and siRNA knockdown studies show that only ATM activation is required for efficient viral replication and progeny production (Luo et al. 2011a). This could suggest that ATM simply induces a critically intense checkpoint response in these cells, but may also indicate that particular component(s) of the ATM pathway are specifically required. During such infections, MRN complexes were shown to colocalize with replicating viral DNA, and silencing MRE11 inhibited viral replication, which was surprising because MRN complexes have been reported to impair AAV replication (Schwartz et al. 2007). However, MRN fulfills multiple roles in DDR cascades, so that timing of the response may be critical. Accordingly, in MVM infections, MRE11 appeared to colocalize with NS1 at early times, but to be down-regulated via a proteasome-dependent mechanism as infection progressed (Adeyemi et al. 2010), perhaps indicating that MRN plays a positive role in activating signaling, but exerts a negative effect during subsequent viral DNA amplification.

By contrast, erythrovirus B19 infection of CD36$^+$ EPCs resulted in activation of all three major PIKK cascades, leading to recruitment of these kinases and their downstream transducers, Chk2, Chk1, and Ku70/80, to the viral replication compartment. However, in this instance, viral DNA synthesis was impaired by pharmacological inhibitors and shRNAs directed against ATR and DNA-PK, and not by inhibition of ATM (Luo et al. 2011b). Similarly, direct inhibition of Chk1, the ATR-induced checkpoint kinase, sufficed to impair B19 DNA amplification, although whether a dominant negative form of Chk1 could, by itself, promote viral replication during ATR suppression, remains to be explored.

Because viruses rapidly evolve to exploit whatever environments they encounter, it seems likely that specific components of evoked host responses may eventually be co-opted to serve novel roles in viral replication. A possible example of this is seen for AAV. Damage responses invoked by AAVs have been studied in some detail, and found to vary according to the presence or absence of helper virus genes, cell type, and whether the infecting particle contains a viral genome or vector sequence, or has been UV-inactivated (Jurvansuu et al. 2005; Cervelli et al. 2008; Fragkos et al. 2008; Collaco et al. 2009; Schwartz et al. 2009; Choi et al. 2010). However, when replicating productively in the presence of adenovirus helper proteins, they typically induce a DNA-PK-dominated response, suppression of which results in decreased AAV replication. Attempts to identify the cellular factors directly involved in viral replication in vitro, by fractionating crude extracts from adenovirus-infected 293 cells, had suggested that components of the minichromosome maintenance (MCM) complex were absolutely required, perhaps functioning as a replicative helicase (Nash et al. 2008). However, subsequent studies by this group showed that AAV Rep interacts with the Ku70/80 complex (Nash et al. 2009), which serves as the downstream transducer for DNA-PK, and that depletion or inactivation of Ku complexes in crude replication extracts impaired AAV DNA synthesis. These complexes, which can also bind directly to the AAV telomeres, were then shown to support strand-displacement synthesis in vitro, partially substituting for the candidate replicative helicase, MCM. Such observations clearly suggest that current ideas about viral fork structure need to be carefully reexamined in the context of their altered host environment.

CONCLUDING REMARKS

Previous research efforts, predominantly on MVM and AAV2, focused on viral structures, proteins, and the molecular mechanisms they use to support their unique linear "rolling hairpin" form of unidirectional single-strand displacement synthesis, and on the repertoire of constitutive host factors that underpin this process. More recently, a growing understanding of how cells respond to alien viral DNA structures has shifted this emphasis, refocusing attention on how these inert single-stranded genomes elude recognition while allowing the cell to progress into S phase, and then evoke, evade, and exploit cellular DNA replication and damage responses to promote their own preferential DNA synthesis. Future efforts will likely address how the incoming viral genome establishes transcription and replication complexes, while resisting cellular chromatinization and silencing, and how they usurp control of the S phase replicative and damage-activated pathways to suppress cellular replication while reorganizing the nucleus for optimal virus production. In particular, it should become possible to differentiate between direct effects on the cells metabolic machinery induced by the small number of nonstructural proteins encoded by these viruses, and those resulting from activation of cellular damage pathways evoked by ongoing viral DNA replication.

Modern advances in virus detection methods have revealed the existence of many previously unknown parvoviruses, including entire unsuspected genera and several important human pathogens. Research emphasis will thus likely shift to our expanded knowledge of the viral genomic repertoire that these represent.

Although some may prove a challenge to propagate in cell culture, the rich diversity of their structures, life cycles, and host cell specificity can be expected to provide opportunities for many new insights into the interactions between this family of simple viral parasites and their host cells.

ACKNOWLEDGMENTS

This work is supported by Public Health Service grant AI026109 from the National Institutes of Health.

REFERENCES

Adeyemi RO, Landry S, Davis ME, Weitzman MD, Pintel DJ. 2010. Parvovirus minute virus of mice induces a DNA damage response that facilitates viral replication. *PLoS Pathog* **6:** e1001141.

Allander T, Tammi MT, Eriksson M, Bjerkner A, Tiveljung-Lindell A, Andersson B. 2005. Cloning of a human parvovirus by molecular screening of respiratory tract samples. *Proc Natl Acad Sci* **102:** 12891–12896.

Bashir T, Rommelaere J, Cziepluch C. 2001. In vivo accumulation of cyclin A and cellular replication factors in autonomous parvovirus minute virus of mice-associated replication bodies. *J Virol* **75:** 4394–4398.

Cervelli T, Palacios JA, Zentilin L, Mano M, Schwartz RA, Weitzman MD, Giacca M. 2008. Processing of recombinant AAV genomes occurs in specific nuclear structures that overlap with foci of DNA-damage-response proteins. *J Cell Science* **121:** 349–357.

Chen KC, Shull BC, Moses EA, Lederman M, Stout ER, Bates RC. 1986. Complete nucleotide sequence and genome organization of bovine parvovirus. *J Virol* **60:** 1085–1097.

Chen KC, Shull BC, Lederman M, Stout ER, Bates RC. 1988. Analysis of the termini of the DNA of bovine parvovirus: Demonstration of sequence inversion at the left terminus and its implication for the replication model. *J Virol* **62:** 3807–3813.

Chen AY, Guan W, Lou S, Liu Z, Kleiboeker S, Qiu J. 2010. Role of erythropoietin receptor signaling in parvovirus B19 replication in human erythroid progenitor cells. *J Virol* **84:** 12385–12396.

Chen AY, Kleiboeker S, Qiu J. 2011. Productive parvovirus B19 infection of primary human erythroid progenitor cells at hypoxia is regulated by STAT5A and MEK signaling but not HIFα. *PLoS Pathog* **7:** e1002088.

Choi Y-K, Nash K, Byrne BJ, Muzyczka N, Song S. 2010. The effect of DNA-dependent protein kinase on adeno-associated virus replication. *PLoS ONE* **5:** e15073.

Collaco RF, Bevington JM, Bhrigu V, Kalman-Maltese V, Trempe JP. 2009. Adeno-associated virus and adenovirus coinfection induces a cellular DNA damage and repair response via redundant phosphatidylinositol 3-like kinase pathways. *Virology* **392:** 24–33.

Cotmore SF, Tattersall P. 1987. The autonomously replicating parvoviruses of vertebrates. *Adv Virus Res* **33:** 91–174.

Cotmore SF, Tattersall P. 1996. Parvovirus DNA replication. In *DNA replication in eukaryotic cells* (ed. DePamphilis M), pp. 799–813. Cold Spring Harbor Laboratory Press, Cold Spring Harbor, NY.

Cotmore SF, Tattersall P. 2003. Resolution of parvovirus dimer junctions proceeds through a novel heterocruciform intermediate. *J Virol* **77:** 6245–6254.

Cotmore SF, Tattersall P. 2005. Genome packaging sense is controlled by the efficiency of the nick site in the right-end replication origin of parvoviruses minute virus of mice and LuIII. *J Virol* **79:** 2287–2300.

Cotmore SF, Tattersall P. 2006. Parvoviruses. In *DNA replication and human disease* (ed. DePamphilis M), pp. 593–608. Cold Spring Harbor Laboratory Press, Cold Spring Harbor, NY.

Cotmore SF, Tattersall P. 2012. Mutations at the base of the icosahedral five-fold cylinders of minute virus of mice induce 3′-to-5′ genome uncoating and critically impair entry functions. *J Virol* **86:** 69–80.

Cotmore SF, Gottlieb RL, Tattersall P. 2007. Replication initiator protein NS1 of the parvovirus minute virus of mice binds to modular divergent sites distributed throughout duplex viral DNA. *J Virol* **81:** 13015–13027.

Cotmore SF, Hafenstein S, Tattersall P. 2010. Depletion of virion-associated divalent cations induces parvovirus minute virus of mice to eject its genome in a 3′-to-5′ direction from an otherwise intact viral particle. *J Virol* **84:** 1945–1956.

Cziepluch C, Lampel S, Grewenig A, Grund C, Lichter P, Rommelaere J. 2000. H-1 parvovirus-associated replication bodies: A distinct virus-induced nuclear structure. *J Virol* **74:** 4807–4815.

Davy C, Doorbar J. 2007. G2/M cell cycle arrest in the life cycle of viruses. *Virology* **368:** 219–226.

Day JM, Zsak L. 2010. Determination and analysis of the full-length chicken parvovirus genome. *Virology* **399:** 59–64.

Deleu L, Pujol A, Faisst SR, Rommelaere J. 1999. Activation of promoter P4 of the autonomous parvovirus minute virus of mice at early S phase is required for productive infection. *J Virol* **73:** 3877–3885.

Dijkman R, Koekkoek SM, Molenkamp R, Schildgen O, van der Hoek L. 2009. Human bocavirus can be cultured in differentiated human airway epithelial cells. *J Virol* **83:** 7739–7748.

El-Andaloussi N, Endele M, Leuchs B, Bonifati S, Kleinschmidt J, Rommelaere J, Marchini A. 2011. Novel adenovirus-based helper system to support production of recombinant parvovirus. *Cancer Gene Ther* **18:** 240–249.

Faisst SR, Perros M, Deleu L, Spruyt N, Rommelaere J. 1994. Mapping of upstream regulatory elements in the P4 promoter of parvovirus minute virus of mice. *Virology* **202:** 466–470.

Fragkos M, Breuleux M, Clément N, Beard P. 2008. Recombinant adeno-associated viral vectors are deficient in provoking a DNA damage response. *J Virol* **82:** 7379–7387.

Gu Z, Plaza S, Perros M, Cziepluch C, Rommelaere J, Cornelis JJ. 1995. NF-Y controls transcription of the minute virus of mice P4 promoter through interaction with an unusual binding site. *J Virol* **69:** 239–246.

Guan W, Wong S, Zhi N, Qiu J. 2009. The genome of human parvovirus b19 can replicate in nonpermissive cells with the help of adenovirus genes and produces infectious virus. *J Virol* **83:** 9541–9553.

Hickman AB, Ronning DR, Perez ZN, Kotin RM, Dyda F. 2004. The nuclease domain of adeno-associated virus rep coordinates replication initiation using two distinct DNA recognition interfaces. *Mol Cell* **13:** 403–414.

Huang Q, Deng X, Yan Z, Cheng F, Luo Y, Shen W, Lei-Butters DCM, Chen AY, Li Y, Tang L, et al. 2012. Establishment of a reverse genetic system for studying human Bocavirus in human airway epithelia. *PLoS Pathog* **8:** e1002899.

Ihalainen TO, Niskanen EA, Jylhävä J, Turpeinen T, Rinne J, Timonen J, Vihinen-Ranta M. 2007. Dynamics and interactions of parvoviral NS1 protein in the nucleus. *Cell Microbiol* **9:** 1946–1959.

Ihalainen TO, Niskanen EA, Jylhävä J, Paloheimo O, Dross N, Smolander H, Langowski J, Timonen J, Vihinen-Ranta M. 2009. Parvovirus induced alterations in nuclear architecture and dynamics. *PLoS ONE* **4:** e5948.

Jones MS, Kapoor A, Lukashov VV, Simmonds P, Hecht F, Delwart E. 2005. New DNA viruses identified in patients with acute viral infection syndrome. *J Virol* **79:** 8230–8236.

Jurvansuu J, Raj K, Stasiak A, Beard P. 2005. Viral transport of DNA damage that mimics a stalled replication fork. *J Virol* **79:** 569–580.

Kantola K, Hedman L, Arthur J, Alibeto A, Delwart E, Jartti T, Ruuskanen O, Hedman K, Söderlund-Venermo M. 2011. Seroepidemiology of human bocaviruses 1–4. *J Infect Dis* **204:** 1403–1412.

Kapoor A, Simmonds P, Slikas E, Li L, Bodhidatta L, Sethabutr O, Triki H, Bahri O, Oderinde BS, Baba MM, et al. 2010. Human bocaviruses are highly diverse, dispersed, recombination prone, and prevalent in enteric infections. *J Infect Dis* **201:** 1633–1643.

King JA, Dubielzig R, Grimm D, Kleinschmidt JA. 2001. DNA helicase-mediated packaging of adeno-associated virus type 2 genomes into preformed capsids. *EMBO J* **20:** 3282–3291.

Lahtinen A, Kivelä P, Hedman L, Kumar A, Kantele A, Lappalainen M, Liitsola K, Ristola M, Delwart E, Sharp C, et al. 2011. Serodiagnosis of primary infections with human parvovirus 4, Finland. *Emerg Infect Dis* **17:** 79–82.

Li L, Cotmore SF, Tattersall P. 2012. Maintenance of the flip sequence orientation of the ears in the parvoviral left-end hairpin ears is a non-essential consequence of the critical asymmetry in the hairpin stem. *J Virol* doi:10.1128/JVI.01450-12.

Lou HJ, Brister JR, Li JJ, Chen W, Muzyczka N, Tan W. 2004. Adeno-associated virus Rep78/Rep68 promotes localized melting of the rep binding element in the absence of adenosine triphosphate. *Chembiochem* **5:** 324–332.

Luo Y, Chen AY, Qiu J. 2011a. Bocavirus infection induces a DNA damage response that facilitates viral DNA replication and mediates cell death. *J Virol* **85:** 133–145.

Luo Y, Lou S, Deng X, Liu Z, Li Y, Kleiboeker S, Qiu J. 2011b. Parvovirus B19 infection of human primary erythroid progenitor cells triggers ATR-Chk1 signaling, which promotes B19 virus replication. *J Virol* **85:** 8046–8055.

Nash K, Chen W, Muzyczka N. 2008. Complete in vitro reconstitution of adeno-associated virus DNA replication requires the minichromosome maintenance complex proteins. *J Virol* **82:** 1458–1464.

Nash K, Chen W, Salganik M, Muzyczka N. 2009. Identification of cellular proteins that interact with the adeno-associated virus rep protein. *J Virol* **83:** 454–469.

Oleksiewicz MB, Costello F, Huhtanen M, Wolfinbarger JB, Alexandersen S, Bloom ME. 1996. Subcellular localization of Aleutian mink disease parvovirus proteins and DNA during permissive infection of Crandell feline kidney cells. *J Virol* **70:** 3242–3247.

Ozawa K, Kurtzman G, Young N. 1986. Replication of the B19 parvovirus in human bone marrow cell cultures. *Science* **233:** 883–886.

Phan TG, Nguyen PVO, Bonkoungou IJO, Kapoor A, Barro N, O'Ryan M, Kapusinszky B, Wang C, Delwart E. 2012. Acute diarrhea in West-African children: Diverse enteric viruses and a novel parvovirus genus. *J Virol* doi: 10.1128/JVI.01427-12.

Plevka P, Hafenstein S, Li L, D'Abramo A, Cotmore SF, Rossmann MG, Tattersall P. 2011. Structure of a packaging-defective mutant of minute virus of mice indicates that the genome is packaged via a pore at a 5-fold axis. *J Virol* **85:** 4822–4827.

Ruiz Z, Mihaylov IS, Cotmore SF, Tattersall P. 2011. Recruitment of DNA replication and damage response proteins to viral replication centers during infection with NS2 mutants of minute virus of mice (MVM). *Virology* **410:** 375–384.

Schwartz RA, Palacios JA, Cassell GD, Adam S, Giacca M, Weitzman MD. 2007. The Mre11/Rad50/Nbs1 complex limits adeno-associated virus transduction and replication. *J Virol* **81:** 12936–12945.

Schwartz RA, Carson CT, Schuberth C, Weitzman MD. 2009. Adeno-associated virus replication induces a DNA damage response coordinated by DNA-dependent protein kinase. *J Virol* **83:** 6269–6278.

Sharp CP, Lail A, Donfield S, Simmons R, Leen C, Klenerman P, Delwart E, Gomperts ED, Simmonds P. 2009. High frequencies of exposure to the novel human parvovirus PARV4 in hemophiliacs and injection drug users, as detected by a serological assay for PARV4 antibodies. *J Infect Dis* **200:** 1119–1125.

Sonntag F, Bleker S, Leuchs B, Fischer R, Kleinschmidt JA. 2006. Adeno-associated virus type 2 capsids with externalized VP1/VP2 trafficking domains are generated prior to passage through the cytoplasm and are maintained until uncoating occurs in the nucleus. *J Virol* **80:** 11040–11054.

Sonntag F, Köther K, Schmidt K, Weghofer M, Raupp C, Nieto K, Kuck A, Gerlach B, Böttcher B, Müller OJ, et al. 2011. The assembly-activating protein promotes capsid assembly of different adeno-associated virus serotypes. *J Virol* **85:** 12686–12697.

Sun Y, Chen AY, Cheng F, Guan W, Johnson FB, Qiu J. 2009. Molecular characterization of infectious clones of the minute virus of canines reveals unique features of bocaviruses. *J Virol* **83:** 3956–3967.

Thomas CE, Storm TA, Huang Z, Kay MA. 2004. Rapid uncoating of vector genomes is the key to efficient liver transduction with pseudotyped adeno-associated virus vectors. *J Virol* **78:** 3110–3122.

Tijssen P, Agbandje-McKenna M, Almendral JM, Bergoin M, Flegel TW, Hedman K, Kleinschmidt JA, Li Y, Pintel DJ, Tattersall P. 2011. Parvoviridae. In *Virus taxonomy: Classification and nomenclature of viruses: Ninth report of the international committee on taxonomy of viruses* (ed. King AMQ, Adams MJ, Carstens E, Lefkowitz EJ), pp. 375–395. Elsevier, San Diego.

Willwand K, Moroianu A, Horlein R, Stremmel W, Rommelaere J. 2002. Specific interaction of the nonstructural protein NS1 of minute virus of mice (MVM) with [ACCA](2) motifs in the centre of the right-end MVM DNA palindrome induces hairpin-primed viral DNA replication. *J Gen Virol* **83:** 1659–1664.

Yoon-Robarts M, Blouin AG, Bleker S, Kleinschmidt JA, Aggarwal AK, Escalante CR, Linden RM. 2004. Residues within the B′ motif are critical for DNA binding by the superfamily 3 helicase Rep40 of adeno-associated virus type 2. *J Biol Chem* **279:** 50472–50481.

Zádori Z, Szelei J, Tijssen P. 2005. SAT: A late NS protein of porcine parvovirus. *J Virol* **79:** 13129–13138.

Zhi N, Mills IP, Lu J, Wong S, Filippone C, Brown KE. 2006. Molecular and functional analyses of a human parvovirus B19 infectious clone demonstrates essential roles for NS1, VP1, and the 11-kilodalton protein in virus replication and infectivity. *J Virol* **80:** 5941–5950.

Human Papillomavirus Infections: Warts or Cancer?

Louise T. Chow and Thomas R. Broker

University of Alabama at Birmingham, Department of Biochemistry and Molecular Genetics, Birmingham, Alabama 35294-0005

Correspondence: ltchow@uab.edu

Human papillomaviruses (HPVs) are prevalent pathogens of mucosal and cutaneous epithelia. Productive infections of squamous epithelia lead to benign hyperproliferative warts, condylomata, or papillomas. Persistent infections of the anogenital mucosa by high-risk HPV genotypes 16 and 18 and closely related types can infrequently progress to high-grade intraepithelial neoplasias, carcinomas-in-situ, and invasive cancers in women and men. HPV-16 is also associated with a fraction of head and neck cancers. We discuss the interactions of the mucosotropic HPVs with the host regulatory proteins and pathways that lead to benign coexistence and enable HPV DNA amplification or, alternatively, to cancers that no longer support viral production.

Human papillomaviruses (HPVs) are ubiquitous small DNA viruses that comprise a family of more than 150 genotypes. Closely related types show a predilection for particular epithelial tissues and have similar pathogenic properties (de Villiers et al. 2004; Bravo et al. 2010). The mucosotropic HPVs can be sexually transmitted. Infections are frequently asymptomatic and are cleared by the immune systems. Active cases are manifested as hyperproliferative lesions but often regress into subclinical persistency within a year. Latent infections can reactivate following immunosuppression or upon undergoing repeated cycles of wounding and healing (Fig. 1). A small fraction of persistent infections by the high-risk (HR) HPV genotypes leads to cancers (reviewed by zur Hausen 2009). Worldwide, each year there are about 500,000 new cases of cervical cancer and 275,000 deaths. More than 99% of cervical cancers are caused by high-risk (HR) HPVs, and types 16 and 18 are responsible for about 70% of the cases (Walboomers et al. 1999). The HR HPVs are also responsible for a high percentage of cancers of other anogenital sites in men and women. Moreover, HPV-16 is associated with a fraction of the head and neck neoplasias, in particular tonsillar and oro-pharyngeal cancers (reviewed by Syrjänen 2010). In contrast, infections by the low-risk (LR) types 6 and 11 cause 90% of genital warts and essentially all laryngeal papillomas (recurrent respiratory papillomatosis or RRP) (reviewed by Derkay and Wiatrak 2008) but they rarely result in carcinomas. Juvenile-onset RRP results from transmission in utero or during passage through the birth canal of mothers with active condylomata, yet infections may become symptomatic years later. Recent studies

Cite this article as *Cold Spring Harb Perspect Biol* doi: 10.1101/cshperspect.a012997

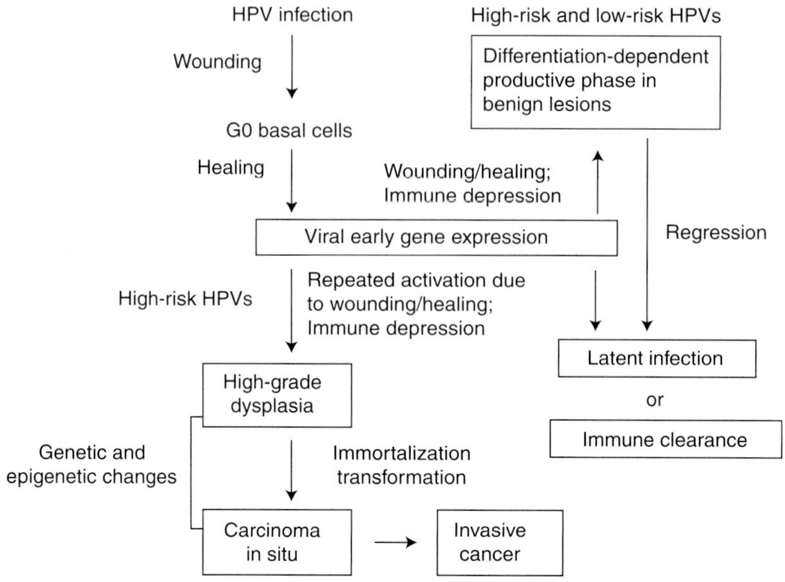

Figure 1. Natural history of HPV infections. Most infections are subclinical and are quickly cleared by host immune surveillance. Some infections become active but regress to a latent state or are cleared within a year. Immune suppression or cycles of wounding and healing, conditions that induce viral early gene expression in proliferating undifferentiated cells, may reactivate the infections. In tissues persistently infected by high-risk HPVs, repeated wounding and healing lead to recurring and extended inactivation of tumor suppressor proteins by the viral oncoproteins and hence excessive basal cell proliferation and progression to high-grade lesions. These cells may accumulate mutations and epigenetic changes, resulting over time in cell immortalization and transformation, integration of the viral DNA into host chromosomes (in most cases), and carcinomas.

indicate that HPVs can also replicate in trophoblasts and produce infectious particles and that HPVs can be detected in spontaneous aborted tissues (Gomez et al. 2008; You et al. 2008). Together, these observations add considerable medical significance to HPV infections during pregnancy.

There are no consistently effective drugs for treating HPV infections. Surgical removal or local ablation of lesions are the standards of care. The implementation of Pap smear screening in the developed countries over the past 60 years has enabled early detection of cytological changes caused by HPV infections of the cervix and has dramatically reduced morbidity and mortality. Nucleic acids-based HPV detection and genotyping have greatly improved screening sensitivity and specificity. However, in resource-poor countries where early detection and intervention are not generally available, cervical cancers remain a major malignancy among women.

How would a normally benign papillomavirus infection become cancerous? The answer resides in the virus–host interactions necessary to support the viral replicative cycle. Regardless of their oncogenic potential, the production of progeny virions occurs only in postmitotic, differentiated cells of squamous epithelia (reviewed by Chow et al. 2010). Thus, the viruses encode E7 and E6 proteins to reestablish a milieu conducive for viral DNA amplification. E7 does so by destabilizing p130, a member of the retinoblastoma (pRB) tumor suppressor protein family that controls cell cycle entry and progression (Litovchick et al. 2011, and references therein), whereas E6 inactivates the transcription-transactivation function of the tumor suppressor TP53 (Thomas and Chiang 2005), which functions to maintain genome stability by controlling critical cellular processes such as proliferation, DNA repair, senescence, and apoptosis. Unique to the HR HPVs, their E7 and E6 proteins also destabilize pRB and TP53, respectively (for reviews, see McLaughlin-Drubin and Münger 2009; Moody and Laimins 2010; Korzeniewski et al. 2011). This ability of the HR HPV oncoproteins to override the tumor suppressors then becomes a double-edged sword should they become up-regulated in undifferentiated cells as a result of multiple episodes of wounding and healing. Over time, critical host regulatory genes can incur mutations or become dysregulated because of epigenetic changes, resulting eventually in progression to oncogenic transformation. It is not understood why the HR HPVs acquired these unique properties, in view of the fact that progeny virions are no longer produced in cancers. In this chapter, we will discuss virus–host interactions, leading to divergent outcomes.

THE PERMISSIVE HOST TISSUES

Squamous epithelia turn over and are renewed every few weeks (for a review, Fuchs 1990). The basal reserve cells rarely divide except when replacing the cells immediately above, called parabasal transit amplifying (TA) cells. The TA cells have committed to differentiation but divide daily for up to 80 cell divisions whereupon they undergo terminal differentiation. With each asymmetrical cell division, one daughter cell remains in the TA layer while the other is pushed upward, exits the cell cycle, and differentiates into spinous cells that comprise the thickest strata of the epithelium. In the cutaneous skin, the upper spinous cells further differentiate into granulocytes. The superficial cells undergo programmed death and slough off as cornified envelopes.

Papillomavirions gain entry into basal cells through a wound or microlesion. The virus depends on the cells reentering into the cell cycle during wound healing to establish infection (Pyeon et al. 2009). Active HPV infections stimulate cell cycling and increase the thickness of the parabasal and spinous strata, resulting in warty growth. The viral genome is maintained as low copy number nuclear plasmids in lower strata in which viral DNA and RNA are present at or below the sensitivity of detection. Up-regulation of viral RNA transcription and DNA amplification occurs in a subset of mid- to upper-spinous cells. Progeny virus particles are produced in a small fraction of superficial cells that are shed during desquamation (Chow et al. 2010). This differentiation-dependent viral gene expression is controlled by transcription factors that bind to the viral enhancer and promoter sequences (reviewed by Bernard 2002) and by chromatin remodeling (Zhao et al. 1999; Kim et al. 2003; Wu et al. 2006; Wooldrige and Laimins 2008; Jha et al. 2010).

Three-dimensional organotypic culture systems were developed to achieve squamous differentiation, starting with native epithelial tissue fragments, with isolated primary human keratinocytes (PHKs) (reviewed by Fuchs 1990; Chow and Broker 1997), or with certain immortalized epithelial cell lines (Lambert et al. 2005). In these systems, epithelial cells are supported on a dermal equivalent consisting of rat tail type 1 collagen and fibroblasts and are held for 10 to 21 d at the air–medium interface, where the drying effects of the air help promote squamous differentiation. Using such "raft" cultures, the complete HPV reproductive program has been recapitulated.

VIRAL GENOME ORGANIZATION

The double-stranded circular DNA genome is approximately 7900 bp long. All genes are encoded along the same template strand (Fig. 2). Transcription regulatory elements and the origin of replication (*ori*) are situated in the upstream regulatory region (URR) (also called the long control region or LCR). Promoters are located in the URR and within the early regions. The E and L gene blocks are each followed by a polyadenylation site. The usage of different promoters and polyA sites as well as alternative mRNA splicing provides access to each of the ORFs and also enables the fusions of portions of different ORFs. Some messages are bicistronic (Wang et al. 2011; reviewed by Chow and Broker 2007). This organization is conserved among human and animal papillomaviruses except that many cutaneous HPVs do not have an obvious E5 ORF. The functions of the URR and

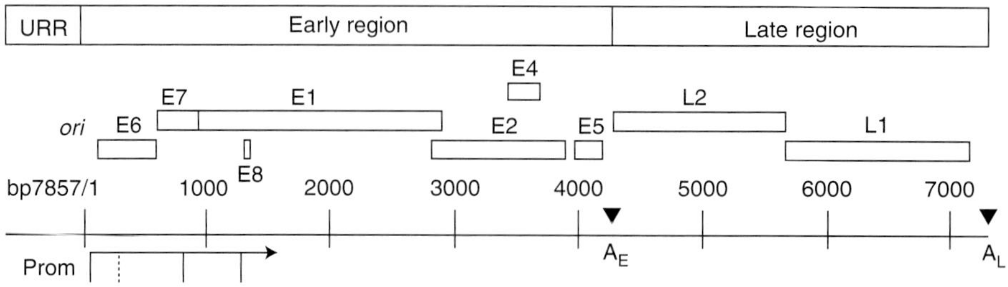

Figure 2. The genome organization of HPV-18. The upstream regulatory region (URR) contains transcription regulatory elements as well as the origin of DNA replication *(ori)*, which overlaps the major promoter preceding *E6*. Major promoters (Prom) are indicated (solid vertical lines connected to an arrow, which denotes the direction of transcription). Open reading frames (ORFs) are depicted by open boxes. The early (E) and late (L) gene blocks are each followed by a polyadenylation site (A_E and A_L). This genome organization is largely conserved among HPVs. Some mucosotropic genotypes have one or two E5 ORFs whereas many cutaneous HPVs have no recognizable E5 ORF. The HR HPVs use the E6 promoter to express both *E6* and *E7* and possibly *E1* and *E2*. Alternative *E6* intragenic mRNA splicing allows efficient expression of the E7 protein. The LR HPV types use a separate promoter located in *E6* (dotted vertical line) to express *E7*. Downstream ORFs are accessed by alternative RNA splicing. Transcripts from the promoter located within *E7* encode E1, E2, E1M^E2C (a fusion of the amino-terminal portion of E1 to the middle of E2), E1^E4 (a fusion of the amino terminus of E1 to E4) and E5, as well as L2 and L1. The minor promoter embedded within *E1* generates a short E8 exon, which is spliced to the middle of the E2 ORF, creating E8^E2C. The expression of late genes from the promoter in *E7* occurs only in the superficial squamous cells after viral DNA amplification and requires elongation of transcripts through A_E to A_L and stabilization of the late 3′ UTR. The functions of encoded proteins and regulatory elements are detailed in Table 1.

proteins are summarized in Table 1. Briefly, the E proteins function directly or indirectly to support viral DNA maintenance, amplification (E1, E2, E8^E2C, E1M^E2C, E5, E6, and E7) and to modulate the host immune responses (E5, E6, and E7). The E1^E4 fusion protein binds the cytokeratin network and may weaken the cellular envelopes and facilitate egress of virions from the shed squames (Bryan and Brown 2000; Khan et al. 2011). L1 and L2 are the major and minor capsid proteins necessary for encapsidating the viral genome. L2 also facilitates viral genome trafficking into the nucleus (Bienkowska-Haba et al. 2012). At high concentrations, L1 self-assembles into empty virus-like particles (VLPs). VLPs are the immunogens in type-restricted prophylactic vaccines that effectively prevent new infections (reviewed by Schiller and Lowy 2006).

THE FUNCTIONS OF E1 AND E2 PROTEINS IN VIRAL DNA REPLICATION

Replication requires the viral *ori*, the *ori* binding protein E2, the E1 replicative DNA helicase, and host replication proteins (reviewed by Chow and Broker 2006). The *ori* consists of a cluster of three E2 binding sites (BSs) flanking a series of overlapping E1 BSs. Three E2 dimers bind to the *ori* to form a toroidal ring, presumably partially denaturing the AT-rich E1 BSs in the negatively supercoiled DNA (Sim et al. 2008). E2 interacts with E1 and recruits it to *ori*. E1 assembles into a pair of hexameric rings (Liu et al. 1998; reviewed by Schuck and Stenlund 2005) while E2 is being released (Abbate et al. 2004). In the presence of topoisomerase I, RPA, and ATP, the E1 dihexamer is potent bidirectional helicase on a supercoiled plasmid (Lin et al. 2002). It is required throughout replication initiation and elongation (Liu et al. 1995), as it also recruits the host DNA polymerase α and the single-stranded DNA binding protein RPA to initiate and sustain replication.

Being a highly active DNA helicase with low sequence specificity, the E1 protein must be tightly controlled to avoid unintended DNA strand separation or damage (Fradet-Turcotte et al. 2011; Sakakibara et al. 2011), and this is achieved by a cell-cycle regulated E1 nuclear

Table 1. Summary of HPV protein functions and regulatory elements

URR	Upstream regulatory region containing the replication *ori*, transcription promoters and regulatory protein binding sites.
E6	Inactivation of TP53. HR HPV E6 also up-regulates telomerase and destabilizes TP53 and many PDZ domain-containing host proteins in a complex with E6AP and other ubiquitin ligases.
E7	Destabilization of p130, leading to S phase-entry in differentiated and quiescent cells. HR HPV E7 also destabilizes pRB.
E1	DNA helicase for viral DNA replication. Recruitment of cellular replication proteins to the origin.
E2	Origin binding protein and recruitment of E1 to the *ori*. Interaction with mitotic apparatus or chromosomes to enable viral genome partitioning during mitosis. Negative regulation of the URR promoter.
E4	Association with cytokeratins. It might weaken the cellular envelope, facilitating virion release.
E5	Membrane protein, enhancing signal transduction of receptor tyrosine kinases.
L2	Minor capsid protein, essential for DNA packaging.
L1	Major capsid protein. It self-assembles into virus-like particles, antigens for prophylactic vaccines.
E5, E6, E7	Down-regulation of host immune surveillance.
HR E6, E7	Immortalization of primary human keratinocytes in vitro. Essential to maintain transformed cell phenotypes.

localization. Near the E1 protein amino terminus is a localization regulatory region (LRR), which controls reversible nucleo-cytoplasmic shuttling. Efficient nuclear entry of HPV-11 E1 requires a bipartite, positively charged nuclear localization sequence (NLS) as well as phosphorylation by mitogen-activated protein kinases (MAPKs), primarily ERK1/2, on serine residue(s) within the LRR (Yu et al. 2007). MAP kinase docking motifs are located in the carboxyl-terminal helicase domain. The default position of HPV-11 E1 in the cytoplasm is attributable to a dominant nuclear export sequence (NES) and the nuclear exportin Crm1. The NES is inactivated by cyclin-dependent kinases (cdks) that phosphorylate a serine residue within NES (Deng et al. 2004). Indeed, cyclin E/cdk2 is required for efficient amplification of HPV-11 *ori* plasmids (Ma et al. 1999; Lin et al. 2000). Similarly, cdk2 also regulates the nucleo-cytoplasmic location of HPV-31 E1 and is required for genome maintenance (Fradet-Turcotte et al. 2010). LRR sequences and the MAPK docking motifs are conserved among human and animal PVs. In light of the crucial role of MAP kinases, the ability of the HPV E5 protein to enhance signal transduction from receptor tyrosine kinases (Pedroza-Saavedra et al. 2010; Suprynowicz et al. 2010; Belleudi et al. 2011) that function upstream of MEK1/2 and ERK1/2 would predict that it plays an important role in viral DNA amplification, and indeed it does in organotypic cultures (J-H Yu, TR Broker, and LT Chow, unpubl.).

The E2 protein is additionally responsible for maintenance of the plasmid in dividing cells to ensure equitable partitioning by associating with the mitotic apparatus or with mitotic chromosomes (Van Tine et al. 2004a; Dao et al. 2006; Parish et al. 2006; reviewed by Kadaja et al. 2009). Two fusion proteins E1M^E2C (Chiang et al. 1991, 1992) and E8^E2C (Zobel et al. 2003; Lace et al. 2008), each with an alternative amino terminus from the full-length E2 while retaining the DNA binding and protein dimerization domain, negatively control viral DNA amplification. The E2BS are immediately adjacent to the URR promoter, which controls *E6* and *E7* expression, and binding of the E2 or the E2-related proteins suppresses the URR promoter activity, constituting a feedback regulation.

THE INDUCTION OF S-PHASE REENTRY IN THE DIFFERENTIATED KERATINOCYTES BY E7

The tumor suppressor protein that confers susceptibility to retinoblastomas (pRB) is the gatekeeper of the G_1 to S-phase transition during the cell cycle. It is inactivated via sequential phosphorylation by cyclin D/cdk4 or cyclin D/cdk6 and by cyclin E/cdk2 to initiate transcription activation of a large suite of E2F-dependent genes, thus promoting cell cycle entry and progression (Sherr and Roberts 2004). The HR but not the LR HPV E7 proteins effectively destabilize the under-phosphorylated form of pRB (McLaughlin-Drubin and Münger 2009; Moody and Laimins 2010), bypassing the need for cyclin D/cdk4 or cyclin D/cdk6.

However, in normal squamous epithelia, p130, rather than pRB, is principally responsible for maintaining quiescence of basal cells and the homeostasis of the postmitotic differentiated cells. In contrast, pRB is primarily detected in actively cycling basal and TA cells (Genovese et al. 2008). In organotypic cultures of PHKs acutely transduced with a retrovirus expressing *E7* of HPV-6, -11, -16, -18, or the plantar wart virus HPV-1, suprabasal S-phase reentry is induced stochastically, and HR HPV E7 proteins are more effective than the nononcogenic HPV E7 (Banerjee et al. 2006; Genovese et al. 2008, 2011). This is because E7 proteins are able to destabilize p130 to different avidity (Zhang et al. 2006; Genovese et al. 2008, 2011).

The E7 protein binds to the pRB family of pocket proteins via the LxCxE motif in the conserved region 2 (CR2), which is also present in adenovirus E1A protein and SV40/PY large T-antigen. Phosphorylation of E7 by casein kinase II (CKII) on one or two serine residues that closely follow the LxCxE motif increases its affinity for the pocket proteins and is critical for efficient p130 destabilization and for S-phase reentry (Chien et al. 2000; Genovese et al. 2008). Remarkably, mutations of just three amino acid residues that are within or downstream of CR2 in the LR HPV-11 E7 can increase its affinity for the pocket proteins and its ability to induce S-phase reentry to the levels achieved by the HR HPV E7 proteins (Genovese et al. 2011).

THE PROPERTIES OF THE HR HPV E6 ONCOPROTEIN

The best-known property of the HR HPV E6 protein is TP53 degradation via the proteasome pathway. E6 and a host protein E6AP function as a ubiquitin ligase of p53. In complexes with E6AP or other ubiquitin ligases, the HR HPV E6 also destabilizes many PDZ domain-containing host proteins that regulate signal transduction, proliferation, cell–cell communication, cell polarity, and differentiation (reviewed by Thomas et al. 2008 and by Howie et al. 2009). Moreover, E6 activates transcription of the human telomerase catalytic subunit (hTERT) (Klingelhutz et al. 1996). In contrast, the LR HPV E6 proteins do not have these activities, although both the HR and LR HPV E6 proteins can inactivate the transcription activation activity of TP53 (Thomas and Chiang 2005). These major distinctions in E6 and E7 protein properties between the HR and LR HPV genotypes largely account for their differential oncogenic potentials in vivo. In vitro, only the HR HPV *E6* and *E7* can immortalize primary human keratinocytes (Halbert et al. 1992, and references therein).

Unique to the HR HPVs, the great majority of the *E6* transcripts undergo intragenic or intergenic alternative splicing. The spliced mRNAs encode truncated E6* peptides that lack carboxy-terminal residues can still destabilize certain PDZ proteins but not TP53 (Pim et al. 2009). The intragenic mRNA splice is thought to facilitate translation of the downstream E7, E1, and E2 ORFs (Hubert and Laimins 2002; Tang et al. 2006). The reduction of HR HPV E6 protein by RNA splicing would also ensure the presence of a certain level of TP53 and other targeted host proteins needed to maintain host genome stability or for some unknown aspect of the viral life cycle. For the LR HPVs, a promoter located within the *E6* gene is responsible for *E7* transcription (DiLorenzo and Steinberg 1995) (Fig. 2), negating the need for intragenic RNA splicing.

RECAPITULATION OF THE VIRAL PRODUCTIVE PROGRAM IN ORGANOTYPIC CULTURES OF EPITHELIAL CELL LINES

The productive phase of HPV infections was initially recapitulated in organotypic raft cultures of explanted HPV-11 infected human epithelial tissue (Dollard et al. 1992) and in TPA-treated raft cultures of an HPV-31 plasmid-containing epithelial cell line derived from a dysplasia (Meyers et al. 1992). For genetic analyses in the context of an autonomously replicating viral genome in the raft cultures, HR HPV DNA is excised from recombinant plasmids and used to transfect and immortalize PHKs or, alternatively, to transfect an already immortalized epithelial cell line (reviewed by Lambert et al. 2005; Wilson and Laimins 2005). However, raft cultures of immortalized cells usually do not support a highly productive viral program, thus compromising the analyses. Moreover, many functionally significant HR HPV E6 mutants cannot be studied by using this approach because they cannot immortalize PHKs and are not stably maintained as plasmids in transfected cells (Park and Androphy 2002; Lee et al. 2007, and references therein).

PSEUDOVIRIONS

As an alternative to raft cultures, high titers of pseudovirions can be produced in human 293 TT cells (Buck et al. 2004; Pyeon et al. 2005). The cells are cotransfected with an L1 and L2 expression vector and a reporter plasmid or HPV DNA excised from a recombinant plasmid. Packaging of plasmid DNA of 8 kbp or shorter is not sequence-specific. However, pseudovirions have low infectivity in PHKs and have not been used to study the virus life cycle. In contrast, pseudovirions containing a reporter gene are used extensively to elucidate the initial virus–host cell interactions in the infection processes in cell lines and in a mouse genital tract model (Roberts et al. 2007; reviewed by Sapp and Bienkowska-Haba 2009; Schiller et al. 2010). They also have been highly instrumental in developing and validating prophylactic HPV vaccines and candidates (Jagu et al. 2009; Lin et al. 2010).

HPV PRODUCTION IN ORGANOTYPIC CULTURES OF PRIMARY HUMAN KERATINOCYTES

Robust production of infectious HPV-18 virus has recently been achieved in raft cultures of PHKs from neonatal foreskin (Chow et al. 2009; Wang et al. 2009a). In this system, HPV-18 genomic plasmids are efficiently generated in transfected PHKs via excision from supercoiled recombinant plasmids using Cre-LoxP site-specific recombination. 20%–30% of the cells survive the 4-day selection for the drug-resistance gene expressed from the residual vector, and there is no need for selection based on the immortalization functions of the HR HPVs. Within a week of transfection, the cells are used to develop raft cultures. The highly productive viral life cycle resembles the infected tissue xenografts in athymic mice (Stoler et al. 1990), as neither has the adaptive immune system to curb the viral activities. The abundant yield of virus particles recovered elicited productive infection of naïve PHKs in raft cultures.

INITIATION OF VIRAL DNA AMPLIFICATION IN G_2 PHASE

The high HPV-18 productivity in PHK raft cultures affords an opportunity to analyze the viral life cycle in detail (Wang et al. 2009a). Unexpectedly, host DNA replication and viral DNA amplification are temporally separate events. In situ assays showed that host chromosomal DNA replicates first in the spinous cells, as denoted by BrdU incorporation and the nuclear S-phase cyclin A. In contrast, viral DNA amplification initiates during a protracted G_2 phase, as revealed by elevated cytoplasmic G_2 cyclin B1 (Fig. 3). Pulse-chase experiments have verified this time course. As the viral DNA amplifies, the E7 activity is reduced and eventually extinguished such that p130 reappears and cells exit the cell cycle. The infectious program then switches to the late phase in which capsid proteins are synthesized and progeny virions are assembled in the superficial keratinocytes and finally mature as these cells die.

Interestingly, E7 alone is responsible for the prolonged G_2 phase by inducing or activating host proteins typically associated with the DNA damage response (DDR) (Banerjee et al. 2011). Briefly, cdc2 is elevated but inactivated following hyperphosphorylation by Wee1 and Myt1 kinases and is sequestered in the cytoplasm. Moreover, the cdc25 phosphatase, which activates cdc2, is itself phosphorylated and inactivated

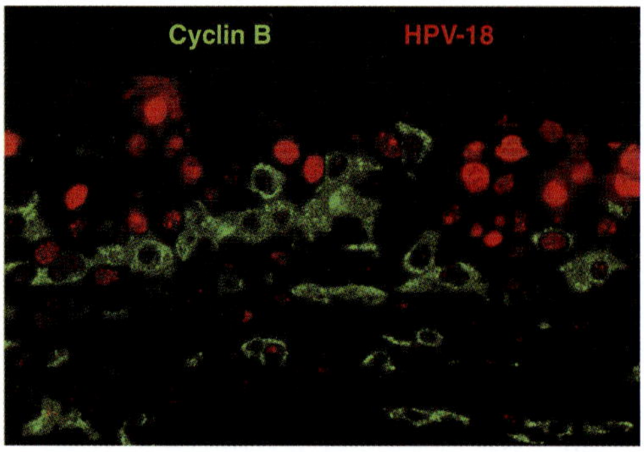

Figure 3. The initiation of viral DNA amplification in G_2 phase cells. A formalin-fixed, paraffin-embedded section of an HPV-18 containing PHK raft culture held at the air–medium interface for 12 days was simultaneously probed for viral DNA (red) by in situ hybridization and cyclin B1 protein (green) with an antibody. Low viral DNA signal was detected in cells with cytoplasmic G_2 cyclin B1. Cells with high viral DNA signal were negative for cyclin B1 (and cyclin A, not shown) as these cells have already exit the cell cycle. (Adapted from Wang et al. 2009a.)

by activated ATM, Chk1, Chk2, or JNK, also accumulates in the cytoplasm (Fig. 4). The ATM DDR was also reported in cells harboring HPV-31 plasmids (Moody and Laimins 2009).

HPV-16 *E7* also activates the Fanconi anemia (FA) pathway involved in DNA repair (Spardy et al. 2007). FA patients with a mutated component gene are prone to developing squamous cell carcinomas at multiple body sites. In transgenic mice in which HPV-16 *E7* expression was targeted to the basal cells, *FancD2* knockout increased the frequency of head and neck cancers relative to the control mice with wild-type *FancD2* (Park et al. 2010). In organotypic cultures of HPV-16 *E6/E7* immortalized keratinocytes, knockdown of a FA component leads to hyperplasia, consistent with a role of the FA pathway in HPV pathogenesis (Hoskins et al. 2009).

THE ROLE OF E6 IN THE PRODUCTIVE LIFE CYCLE

In PHK raft cultures, the expression of the HR HPV *E6* alone does not promote S-phase reentry in differentiated cells, whereas expression of *E7* alone induces S-phase reentry and TP53 stabilization, via phosphorylation by kinases activated by E7 just described. When *E6* and *E7* are coexpressed, TP53 is abolished (Jian et al. 1998), consistent with known biochemical properties of the HR HPV E6 protein. The Cre-loxP excisional recombination system to generate HPV-18 whole genomic plasmids enabled the examination of an HPV-18 E6*I mutant not previously possible. In this mutant, the *E6* gene is deleted of the E6*I intron coding sequence and hence only expresses the truncated E6*I peptide because of premature translation determination after the splice. In PHK raft cultures, the mutant DNA amplification is greatly reduced and L1 is not expressed. Notably, elevated TP53 accumulates for prolonged periods of time in numerous basal and suprabasal cells without inducing apoptosis (Wang et al. 2009a). The *E6* mutant could be partially complemented in *trans* by a retrovirus delivering

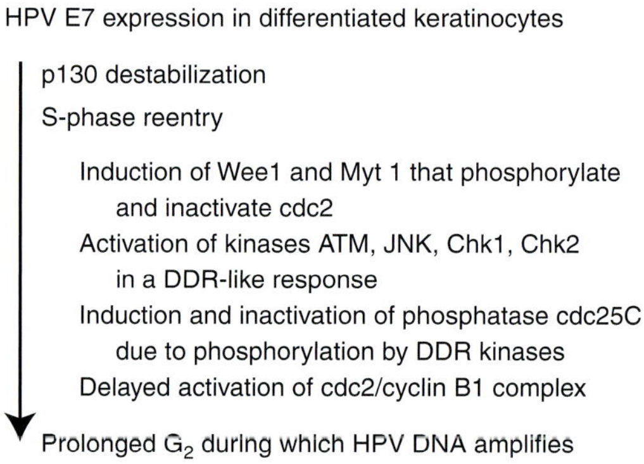

Figure 4. The activities of E7 in differentiated keratinocytes. The expression of E7 in the differentiated keratinocytes destabilizes p130, leading to S-phase reentry. Wee1 and Myt 1 kinases are also activated, hyperphosphorylating and inactivating cdc2. Moreover, E7 further induces a DNA damage repair (DDR)-like response in that ATM, Chk1, Chk2, and JNK are each activated, resulting in the inactivation of the cdc25C phosphatase, disabling it from removing the inhibitory phosphates from cdc2. The inactive cdc2 and cyclin B1 accumulate in the cytoplasm in a prolonged G_2 phase during which viral DNA amplifies (see Fig. 3). However, an antibody to γH2AX detected only sporadic signals (LT Chow and TR Broker, unpubl.), inconsistent with widespread DNA damage.

HPV-18 URR-*E6*, increasing viral DNA amplification and L1 expression, whereas TP53 protein was no longer detected (EY Kho, HK Wang, NS Banerjee, TR Broker, and LT Chow, unpubl.). These observations support the hypothesis that TP53 is an E6 target during the viral life cycle. Ectopic TP53 inhibits transient viral ori-dependent replication in transfected cells, mediated by TP53 interacting with the E2 protein (Lepik et al. 1998; Massimi et al. 1999). Detailed examination of missense *E6* mutants is in progress to verify this conclusion and to identify additional E6 targets critical for the productive cycle.

MODULATION OF THE PRODUCTIVE PROGRAM BY p21CIP1 AND p27KIP1

Although the HPV E7 protein reestablishes an S-phase potential in practically all the differentiated cells, as suggested by the general loss of p130 and the widespread induction of the proliferating cell nuclear antigen (PCNA), only a fraction of these cells reenters S phase. This outcome appears to be determined by the levels of p27kip1, an inhibitor of cdks (CKI) (Fig. 2) (Noya et al. 2001; Chien et al. 2002). Briefly, in suprabasal cells of normal PHK raft cultures, p27kip1 is variably but stably expressed, whereas the related p21cip1 is constitutively expressed, but is quickly degraded by proteasomes. When *E7* was expressed, suprabasal cells that reentered S phase were negative for p27kip1, p21cip1, and cyclin E, whereas cells that did not reenter S phase were positive for all three. We suggest that, if the differentiated cells contain low levels of p27kip1, *E7* expression induces S-phase genes, including *cyclin E* and *cdk2,* and p27kip1 is phosphorylated and degraded, allowing the cells to enter into S phase. The normal feedback regulation then prevents further expression of *cyclin E.* If, however, the p27kip1 protein is already high, it would bind to and inhibit cyclin E/cdk2, preventing S-phase entry. Cyclin E continues to be expressed and is then costabilized in an inactive complex with p21cip1/cdk2, unable to reenter S phase. This reciprocal relationship between p21cip1 stabilization and HPV-11 DNA amplification was observed in RRP patient specimens (Schmidt-Grimminger et al. 1998; Jian et al. 1999). Thus, the pair of CKIs appears to serve as a defense against viral DNA amplification in a fraction of the infected cells.

Nonetheless, in a study of benign genital lesions, this relationship was not consistently observed in cells with high HR or LR HPV DNA (Zehbe et al. 1999). Similarly, in the highly productive HPV-18 PHK raft cultures, some cells with amplified viral DNA were positive for p21cip1 (AA Duffy, HK Wang, TR Broker, and LT Chow, unpubl.). One interpretation would have CKI accumulation occurring following viral DNA amplification, as it did following host DNA replication in *E7*-transduced raft cultures (Chien et al. 2002). Alternatively, HR but not the LR HPV E7 can overcome the inhibitory effects of CKI (Jones et al. 1997; Helt et al. 2002; Shin et al. 2009). This issue of host defense deserves further investigation.

ATTRIBUTES OF HR HPV IN CANCERS AND MODEL SYSTEMS

The trade-off for restricting the HPV productive phase to postmitotic differentiated cells is the occasional development of cancers that no longer support progeny virus production. Cancers caused by HR HPVs have several attributes. First, they harbor wild-type *TP53* and *pRB* but are phenotypically negative for the encoded proteins. *TP53* mutations are rare in cervical cancers. Thus, cancers that are frequently associated with mutated *TP53* (and therefore with more stable and elevated TP53) are not likely to be caused by HR HPVs. Second, HPV DNA is often, but not always, integrated into host chromosomes. Third, the HR HPV *E6* and *E7* mRNAs are consistently expressed at elevated levels. Experimentally, reenabling the pRB or TP53 pathways in cancer cell lines by repressing the viral *E7* or *E6* expression leads to senescence or apoptosis, underscoring the addiction to the viral oncoproteins for sustaining the transformed phenotypes (DeFilippis et al. 2003; Wells et al. 2003).

The initiation of viral carcinogenesis is attributable to repeated dysregulation of the viral oncogenes in the basal reserve or stem cells when the tissue is subjected to cycles of wounding and healing, events known to activate viral early genes. Diminished immune capability is also a contributing factor. The loss of pRB and TP53 promotes excessive cell cycling while increasing the probability of the cell acquiring mutations and incurring chromosome instability (Fig. 1). Detailed mechanisms have been expertly reviewed recently (Howie et al. 2009; Moody and Laimins 2010; Korzeniewski et al. 2011) and will not be repeated here.

HPV-associated cancer development usually takes years or decades until the reserve or stem cells have accumulated combinations of genetic and epigenetic changes to become transformed. Gain or loss of host chromosomal loci has been reported (Wilting et al. 2009, and references therein). Epigenetic changes in certain host genes increase with lesion severity (Wilting et al. 2010; Hesselink et al. 2011; van der Meide et al. 2011). The reprogramming of *p16INK4a* and *Hox* has been attributed to viral oncoproteins (McLaughlin-Drubin et al. 2011). In fact, the accumulation of p16INK4a is one of the validated biomarkers for HPV-induced high-grade dysplasias and cancers (Roelens et al. 2012) along with S-phase biomarkers such as Ki-67, MCM7, and PCNA in the suprabasal cells of a squamous epithelium. Furthermore, HPV oncoproteins also affect the expression of oncogenic or tumor suppressive miRNAs (Martinez et al. 2008; Wang et al. 2008, 2009b; Wald et al. 2011). These viral activities very likely contribute to the productive life cycle or to viral carcinogenesis. It remains to be seen whether some of these events might be potential biomarkers predictive for risk of progression to cancers.

HR HPV *E6*/*E7* immortalized cells are excellent models to characterize molecular and phenotypic changes associated with progression of lesions in vivo. In organotypic cultures, the immortalized cells resemble various grades of dysplasias. Upon extended cell passage, the raft cultures show increasingly dysplastic histology, consistent with additional mutations in host genes (Hurlin et al. 1991; Merrick et al. 1992; Steenbergen et al. 1998). For example, immortalized cells that acquire anchorage-independent growth have up-regulated PIK3CA, which is upstream in the AKT pathway. Indeed, up-regulated PIK3CA is consistently observed in cervical cancers (Henken et al. 2011, and references therein).

Transgenic mice models for cervical cancer, head and neck cancer, and anal cancers have been established by targeting the expression of HPV-16 *E7*, *E6*, or *E5* alone or in combination to basal cells of the epithelia. Long-term exposure to estrogen is required. These studies have verified the importance of destabilization of pRB, TP53, and PDZ domain proteins in cancer development (Chung et al. 2010; Jabbar et al. 2010; Maufort et al. 2010; Stelzer et al. 2010, and references therein).

INTEGRATION OF HPV DNA IN CERVICAL CANCERS AND CLONAL SELECTION OF CANCER CELLS

In cervical cancers and cell lines derived therefrom, as well as in many HPV DNA-immortalized cell lines, viral DNA integration and clonal expansion appear to be important early events (Klaes et al. 1999; Park et al. 2003; Hopman et al. 2004; Melsheimer et al. 2004; Pett et al. 2004; Vinokurova et al. 2008). The HPV DNA is integrated as single copies or as tandem repetitions, and it along with flanking chromosomal DNA can be translocated to multiple sites (Macville et al. 1999). Integration appears to favor the open chromatin of expressed host genes (Kraus et al. 2008). The viral DNA breakage and integration site invariably occurs within the *E1* or *E2* genes. There is no sequence specificity. One predominant *E6*/*E7*-host chimeric transcript is derived from the viral copy located at the downstream integration junction, whereas all upstream copies are silenced by DNA methylation, resulting in the loss of expression from all other viral genes (Wentzensen et al. 2002; Van Tine et al. 2004b). Consequently, viral DNA methylation is high in carcinomas and cell lines thereof and low in benign lesions (Kalantari et al. 2004; Brandsma et al. 2009; Ding et al. 2009; Vinokurova and von Knebel Doeberitz 2011). The host 3′ untranslated region (UTR) and polyA site in the chimeric *E6*/*E7* transcript may confer increased stability (Fig. 5). Moreover, the loss

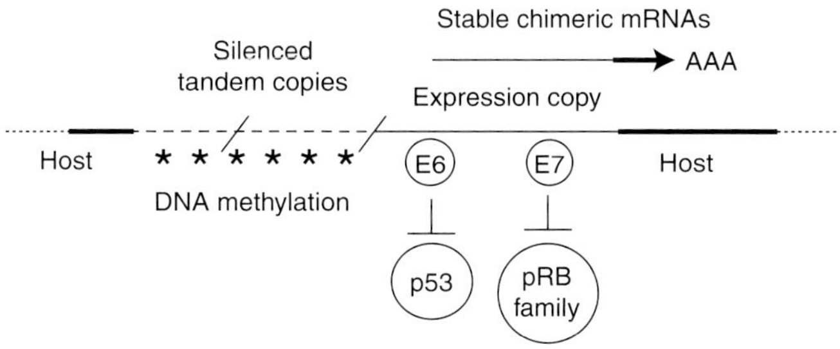

Figure 5. Epigenetic regulation of HPV transcription in a clonally selected cancer cell with tandemly integrated viral DNA. Thick dotted lines represent host sequences. Thin dashed lines represent HPV genomes integrated in tandem arrays. The predominant viral transcripts are chimeric. They originate from the URR promoter of the downstream integration junction copy, which is disrupted in *E1* or *E2*. The E6/E7 transcripts have 3′ UTR and polyA site sequence derived from the host. All upstream viral DNA copies are silenced by DNA methylation (Wentzensen et al. 2002; Van Tine et al. 2004b) such that other viral genes are no longer expressed.

of *E2* expression eliminates the negative feedback regulation of the URR promoter, elevating the expression of *E6* and *E7* (Broker et al. 1989). Collectively, these mechanisms would provide the cells with a growth advantage (Jeon and Lambert 1995; Jeon et al. 1995).

THE POTENTIAL ROLES OF HPV DNA REPLICATION, INTEGRATION, AND E1 AND E2 PROTEINS IN ONCOGENESIS

What leads to the integration of the viral DNA is a matter of conjecture and ongoing research. Nonhomologous recombination between a broken viral DNA and the host chromosome would result in integration. As discussed, E2 proteins of mucosotropic HPVs associate with mitotic apparatus or mitotic chromosomes. The tight association of E2 to *ori* in the integrated HPV genomes might then act as a viro-centromere (Van Tine et al. 2004a), which, in combination with the host chromosome centromere, could establish a multicentric chromosome, thereafter initiating rounds of the breakage-fusion-bridge cycle (McClintock 1951). In cells with integrated viral genomes, continued synthesis of E1 and E2 proteins would also lead to repeated reinitiation of replication from the viral origin to form "onion-skin" replication bubbles, making the host chromosome highly vulnerable to breakage. Newly replicated viral DNA fragments could integrate into additional chromosomal sites, increasing host genome instability (reviewed by Kadaja et al. 2009). To avoid mitotic catastrophe, the expression of *E2* protein from intact copies of the gene must cease via epigenetic silencing in the tumor cell that eventually emerges. The absence of E2 and E2-related proteins would deregulate the viral oncogene expression, further driving the process of oncogenesis, as already discussed.

DOWN-REGULATION OF HOST IMMUNE SURVEILLANCE BY HPV ONCOPROTEINS

Virus production is usually sporadic and low, unless the individual is compromised in their immune systems such as during pregnancy, in patients who are taking immunosuppressive drugs, in cancer patients undergoing chemotherapy and radiation, and in patients with HIV-AIDS. Minimizing gene expression in cycling cells and delaying elevated protein expression to the mid to upper differentiated strata are viral strategies to avoid detection by the host immune systems. In addition, HPVs have developed elaborate schemes to undermine host immune surveillance (reviewed by Bhat et al. 2011). The E6 and E7 proteins affect both innate and adaptive immunity whereas the E5 protein down-regulates class I and II HLA molecules. Collectively, these strategies allow HPVs and host cells to establish a benign co-existence. In the long run, the virus spreads and casts a wider net than otherwise possible. In cancers, the HR HPV *E6* and *E7* genes are expressed at elevated levels. Thus, the ability to impair the host immune responses exacerbates viral carcinogenesis and additionally presents a challenge to the development of therapeutic vaccines that target the viral oncoproteins (reviewed by Su et al. 2010).

CONCLUDING REMARKS

The inability to propagate HPVs in conventional cell cultures has been a blessing in disguise. It forced investigators to develop model systems that recapitulate the spectrum of patient lesions. The viral oncoproteins have been invaluable tools to probe the functions of the two major host tumor suppressor proteins, pRB and TP53. The HPV system continues to provide fertile opportunities for investigations of chromosome dynamics, cellular and molecular responses to DNA damage, signal transduction, host immune response, and cancer biology.

Within 30 years since the cloning of the HR HPV genomes from cervical cancers, intensive basic research made tremendous progress in prevention and diagnosis. The development and licensing of highly effective prophylactic vaccines against the oncogenic HPV-16 and HPV-18 (Gardasil, Cervarix), as well as against nononcogenic HPV-6 and HPV-11 (Gardasil), is the biggest recent advancement that impacts on the public health (Schiller and Lowy 2006). Less expensive and cross-reactive vaccines are being developed (Jagu et al. 2009; Lin et al. 2010). Probe technologies for HPV genotypes and RNA transcripts are leading the transition from cytology to molecular-based screening. Identification of protein biomarkers for staging of HPV lesions is ongoing and will be important for guiding optimal treatments. Looking ahead, a thorough investigation of the functions of the viral proteins during productive infections and elucidation of the mechanisms of their regulation remain as high priorities. These studies will lead to the identification of therapeutic targets for pharmacologic treatment of active papillomavirus infections that, despite concerted efforts at prevention, will continue to occur globally for decades to come.

ACKNOWLEDGMENTS

The authors are supported by USPHS/NIH/National Cancer Institute grants CA36200, CA83679, and CA107338 and gratefully acknowledge the invaluable contributions of our students, fellows, and staff members as well as many collaborators. We thank Carolyn S. Ashworth, MD and the nurses in the Newborn Nursery at UAB Hospital for collecting foreskins from elective circumcision that are critical for our research. We regret a less than comprehensive description and citation of the numerous relevant publications by colleagues because of page limits.

REFERENCES

Abbate EA, Berger JM, Botchan MR. 2004. The X-ray structure of the papillomavirus helicase in complex with its molecular matchmaker E2. *Genes Dev* **18:** 1981–1986.

Banerjee NS, Genovese N, Noya F, Chien W-M, Broker TR, Chow LT. 2006. Inducible E7 proteins of high-risk and low-risk HPVs promote S phase re-entry in post-mitotic differentiated keratinocytes in organotypic cultures. *J Virol* **80:** 6517–6524.

Banerjee NS, Wang H-K, Broker TR, Chow LT. 2011. HPV E7 protein induces prolonged G2 following S-phase reentry in differentiated human keratinocytes. *J Biol Chem* **286:** 11473–11482.

Belleudi F, Leone L, Purpura V, Cannella F, Scrofani C, Torrisi MR. 2011. HPV16 E5 affects the KGFR/FGFR2b-mediated epithelial growth through alteration of the receptor expression, signaling and endocytic traffic. *Oncogene* **30:** 4963–4876.

Bernard H-U. 2002. Gene expression of genital human papillomaviruses and considerations on potential antiviral approaches. *Antiviral Ther* **7:** 219–237.

Bhat P, Mattarollo SR, Gosmann C, Frazer IH, Leggatt GR. 2011. Regulation of immune responses to HPV infection and during HPV-directed immunotherapy. *Immunol Rev* **239:** 85–98.

Bienkowska-Haba M, Williams C, Kim SM, Garcea RL, Sapp M. 2012. Cyclophilins facilitate dissociation of the human papillomavirus type 16 capsid protein L1 from the L2/DNA complex following virus entry. *J Virol* **86:** 9875–9887.

Brandsma JL, Sun Y, Lizardi PM, Tuck DP, Zelterman D, Haines GK 3rd, Martel M, Harigopal M, Schofield K, Neapolitano M. 2009. Distinct human papillomavirus type 16 methylomes in cervical cells at different stages of premalignancy. *Virology* **389:** 100–107.

Bravo IG, de Sanjose S, Gottschling M. 2010. The clinical importance of understanding the evolution of papillomaviruses. *Trends Microbiol* **18:** 432–438.

Broker TR, Chow LT, Chin MT, Rhodes CR, Wolinsky SM, Whitbeck A, Stoler MH. 1989. A molecular portrait of human papillomavirus carcinogenesis. *Cancer Cells* **7:** 197–208.

Bryan JT, Brown DR. 2000. Association of the human papillomavirus type 11 E1^E4 protein with cornified cell envelopes derived from infected genital epithelium. *Virology* **277:** 262–269.

Buck CB, Pastrana DV, Lowy DR, Schiller JT. 2004. Efficient intracellular assembly of papillomaviral vectors. *J Virol* **78:** 751–757.

Chiang C-M, Broker TR, Chow LT. 1991. An E1M^E2C fusion protein encoded by human papillomavirus type 11 is a sequence-specific transcription repressor. *J Virol* **65:** 3317–3329.

Chiang C-M, Dong G, Broker TR, Chow LT. 1992. Control of human papillomavirus type 11 origin replication by the E2 family of transcriptional regulatory proteins. *J Virol* **66:** 5224–5231.

Chien W-M, Parker JN, Schmidt-Grimminger D-C, Broker TR, Chow LT. 2000. Casein kinase II phosphorylation of the human papillomavirus-18 E7 protein is critical for promoting S phase entry. *Cell Growth Diff* **11:** 425–435.

Chien W-M, Noya F, Benedict-Hamilton HM, Broker TR, Chow LT. 2002. Alternative fates of keratinocytes transduced by human papillomavirus type 18 E7 during squamous differentiation. *J Virol* **76:** 2964–2972.

Chow LT, Broker TR. 1997. In vitro experimental systems for HPV: Epithelial raft cultures for viral reproduction and pathogenesis and for genetic analyses of viral proteins and regulatory sequences. *Clin Dermatol* **15:** 217–227.

Chow LT, Broker TR. 2006. Mechanisms and regulation of papillomavirus DNA replication. In *Recent advances in papillomavirus research* (ed. Campo MS), pp. 53–71. Caister Academic, Norfolk, UK.

Chow LT, Broker TR. 2007. Human papillomavirus RNA transcription. In *Papillomaviruses* (ed. Garcea RL, DiMaio D), pp. 109–144. Springer, New York.

Chow LT, Duffy AA, Wang H-K, Broker TR. 2009. A highly efficient system to produce infectious human papillomavirus. Elucidation of natural virus-host interactions. *Cell Cycle* **8:** 1319–1323.

Chow LT, Broker TR, Steinberg BM. 2010. The natural history of human papillomavirus infections of the mucosal epithelia. *APMIS* **118:** 422–449.

Chung SH, Franceschi S, Lambert PF. 2010. Estrogen and ERα: Culprits in cervical cancer? *Trends Endocrinol Metab* **21:** 504–511.

Dao LD, Duffy AA, Van Tine BA, Wu S-Y, Chiang C-M, Broker TR, Chow LT. 2006. Dynamic localization of the HPV-11 origin binding protein E2 during mitosis while in association with the spindle apparatus. *J Virol* **80:** 4792–4800.

DeFilippis RA, Goodwin EC, Wu L, DiMaio D. 2003. Endogenous human papillomavirus E6 and E7 proteins differentially regulate proliferation, senescence, and apoptosis in HeLa cervical carcinoma cells. *J Virol* **77:** 1551–1563.

Deng W, Lin BY, Jin G, Wheeler C, Ma T, Harper JW, Broker TR, Chow LT. 2004. Cyclin/CDK regulates the nucleo-cytoplasmic localization of the human papillomavirus E1 DNA helicase. *J Virol* **78:** 13954–13965.

Derkay CS, Wiatrak B. 2008. Recurrent respiratory papillomatosis: A review. *Laryngoscope* **118:** 1236–1247.

de Villiers E-M, Fauquet C, Broker TR, Bernard H-U, zur Hausen H. 2004. Classification of papillomaviruses. *Virology* **324:** 17–27.

DiLorenzo TP, Steinberg BM. 1995. Differential regulation of human papillomavirus type 6 and 11 early promoters in cultured cells derived from laryngeal papillomas. *J Virol* **69:** 6865–6872.

Ding DC, Chiang MH, Lai HC, Hsiung CA, Hsieh CY, Chu TY. 2009. Methylation of the long control region of HPV16 is related to the severity of cervical neoplasia. *Eur J Obstet Gynecol Reprod Biol* **147:** 215–220.

Dollard SC, Wilson JL, Demeter LM, Bonnez W, Reichman RC, Broker TR, Chow LT. 1992. Production of human papillomavirus and modulation of the infectious program in epithelial raft cultures. *Genes Dev.* **6:** 1131–1142.

Fradet-Turcotte A, Moody C, Laimins LA, Archambault J. 2010. Nuclear export of human papillomavirus type 31 E1 is regulated by Cdk2 phosphorylation and required for viral genome maintenance. *J Virol* **84:** 11747–11760.

Fradet-Turcotte A, Bergeron-Labrecque F, Moody CA, Lehoux M, Laimins LA, Archambault J. 2011. Nuclear accumulation of the papillomavirus E1 helicase blocks S-phase progression and triggers an ATM-dependent DNA damage response. *J Virol* **85:** 8996–9012.

Fuchs E. 1990. Epidermal differentiation: The bare essentials. *J Cell Biol* **111:** 2807–2814.

Genovese NJ, Banerjee NS, Broker TR, Chow LT. 2008. Casein kinase II motif-dependent phosphorylation of HPV E7 protein promotes p130 degradation and S-phase induction in differentiated human keratinocytes. *J Virol* **82:** 4862–4873.

Genovese NJ, Broker TR, Chow LT. 2011. Non-conserved lysine residues attenuate the biological function of the low-risk human papillomavirus E7 protein. *J. Virol.* **85:** 5546–5554.

Gomez LM, Ma Y, Ho C, McGrath CM, Nelson DB, Parry S. 2008. Placental infection with human papillomavirus is associated with spontaneous preterm delivery. *Hum Reprod* **23:** 709–715.

Halbert CL, Demers GW, Galloway DA. 1992. The E6 and E7 genes of human papillomavirus type 6 have weak immortalizing activity in human epithelial cells. *J Virol* **66:** 2125–2134.

Helt AM, Funk JO, Galloway DA. 2002. Inactivation of both the retinoblastoma tumor suppressor and p21 by the human papillomavirus type 16 E7 oncoprotein is necessary to inhibit cell cycle arrest in human epithelial cells. *J Virol* **76:** 10559–10568.

Henken FE, Banerjee NS, Snijders PJF, Meijer CJLM, De-Castro Arce J, Rösl F, Broker TR, Chow LT, Steenbergen RDM. 2011. PIK3CA-mediated PI3-kinase signalling is essential for HPV-induced transformation in vitro. *Mol Cancer* **10:** doi: 10.1186/1476-4598-10-71.

Hesselink AT, Heideman DAM, Steenbergen RDM, Coupé VM, Overmeer RM, Rijkaart D, Berkhof J, Meijer CJLM, Snijders PJF. 2011. Combined promoter methylation analysis of CADM1 and MAL: An objective triage tool for high-risk human papillomavirus DNA-positive women. *Clin Cancer Res* **17:** 2459–2465.

Hopman AHN, Smedts F, Dignef W, Ummelen M, Sonke G, Mravunac M, Vooijs GP, Speel EJ, Ramaekers FC. 2004. Transition of high-grade cervical intraepithelial neoplasia to micro-invasive carcinoma is characterized by integration of HPV 16/18 and numerical chromosome abnormalities. *J Pathol* **202:** 23–33.

Hoskins EE, Morris TA, Higginbotham JM, Spardy N, Cha E, Kelly P, Williams DA, Wikenheiser-Brokamp KA, Duensing S, Wells SI. 2009. Fanconi anemia deficiency stimulates HPV-associated hyperplastic growth in organotypic epithelial raft culture. *Oncogene* **28:** 674–685.

Howie HL, Katzenellenbogen RA, Galloway DA. 2009. Papillomavirus E6 proteins. *Virology* **384:** 324–334.

Hubert WG, Laimins LA. 2002. Human papillomavirus type 31 replication modes during the early phases of the viral life cycle depend on transcriptional and posttranscriptional regulation of E1 and E2 expression. *J Virol* **76:** 2263–2273.

Hurlin PJ, Kaur P, Smith PP, Perez-Reyes N, Blanton RA, McDougall JK. 1991. Progression of human papillomavirus type 18-immortalized human keratinocytes to a malignant phenotype. *Proc Natl Acad Sci* **88:** 570–574.

Jabbar S, Strati K, Shin MK, Pitot HC, Lambert PF. 2010. Human papillomavirus type 16 E6 and E7 oncoproteins act synergistically to cause head and neck cancer in mice. *Virology* **407:** 60–67.

Jagu S, Karanam B, Gambhira R, Chivukula SV, Chaganti RJ, Lowy DR, Schiller JT, Roden RB. 2009. Concatenated multitype L2 fusion proteins as candidate prophylactic pan-human papillomavirus vaccines. *J Natl Cancer Inst* **101:** 782–792.

Jeon S, Lambert PF. 1995. Integration of human papillomavirus type 16 DNA into the human genome leads to increased stability of E6 and E7 mRNAs: Implications for cervical carcinogenesis. *Proc. Natl Acad Sci USA* **92:** 1654–1658.

Jeon S, Allen-Hoffmann BL, Lambert PF. 1995. Integration of human papillomavirus type 16 into the human genome correlates with a selective growth advantage of cells. *J Virol* **69:** 2989–2997.

Jha S, Vande Pol S, Banerjee NS, Dutta AB, Chow LT, Dutta A. 2010. Destabilization of TIP60 by human papillomavirus E6 results in attenuation of TIP60 dependent transcriptional regulation and apoptotic pathway. *Mol Cell* **38:** 700–711.

Jian Y, Schmidt-Grimminger D-C, Chien W-M, Wu X, Broker TR, Chow LT. 1998. Post-transcriptional induction of p21cip1 protein by HPV E7 in differentiated epithelial cells inhibits reactivated unscheduled DNA synthesis. *Oncogene* **17:** 2027–2038.

Jian Y, Van Tine BA, Chien W-M, Shaw GM, Broker TR, Chow LT. 1999. Concordant induction of cyclin E and p21cip1 in differentiated keratinocytes by the HPV E7 protein inhibits cellular and viral DNA synthesis. *Cell Growth Diff* **10:** 101–111.

Jones DL, Alani RM, Münger K. 1997. The human papillomavirus E7 oncoprotein can uncouple cellular differentiation and proliferation in human keratinocytes by abrogating p21Cip1–mediated inhibition of cdk2. *Genes Dev* **11:** 2101–2111.

Kadaja M, Silla T, Ustav E, Ustav M. 2009. Papillomavirus DNA replication—from initiation to genomic instability. *Virology* **384:** 360–368.

Kalantari M, Calleja-Macias IE, Tewari D, Hagmar B, Lie K, Barrera-Saldana HA, Wiley DJ, Bernard H-U. 2004. Conserved methylation patterns of human papillomavirus type 16 DNA in asymptomatic infection and cervical neoplasia. *J Virol* **78:** 12762–12772.

Khan J, Davy CE, McIntosh PB, Jackson DJ, Hinz S, Wang Q, Doorbar J. 2011. Role of calpain in the formation of human papillomavirus type 16 E1^E4 amyloid fibers and reorganization of the keratin network. *J Virol* **85:** 9984–9997.

Kim K, Garner-Hamrick PA, Fisher C, Lee D, Lambert PF. 2003. Methylation patterns of papillomavirus DNA, its influence on E2 function, and implications in viral infection. *J Virol* **77:** 12450–12459.

Klaes R, Woerner SM, Ridder R, Wentzensen N, Dürst M, Schneider A, Lotz B, Melsheimer P, von Knebel Doeberitz M. 1999. Detection of high-risk cervical intraepithelial neoplasia and cervical cancer by amplification of transcripts derived from integrated papillomavirus oncogenes. *Cancer Res* **59:** 6132–6136.

Klingelhut AJ, Foster SA, McDougall JK. 1996. Telomerase activation by the E6 gene product of human papillomavirus type 16. *Nature* **380:** 79–82.

Korzeniewski N, Spardy N, Duensing A, Duensing S. 2011. Genomic instability and cancer: Lessons learned from human papillomaviruses. *Cancer Lett* **305:** 113–122.

Kraus I, Driesch C, Vinokurova S, Hovig E, Schneider A, von Knebel Doeberitz M, Dürst M. 2008. The majority of viral-cellular fusion transcripts in cervical carcinomas cotranscribe cellular sequences of known or predicted genes. *Cancer Res* **68:** 2514–2522.

Lace MJ, Anson JR, Thomas GS, Turek LP, Haugen TH. 2008. The E8^E2 gene product of human papillomavirus type 16 represses early transcription and replication but is dispensable for viral plasmid persistence in keratinocytes. *J Virol* **82:** 10841–10853.

Lambert PF, Ozbun MA, Collins A, Holmgren S, Lee D, Nakahara T. 2005. Using an immortalized cell line to study the HPV life cycle in organotypic "raft" cultures. *Methods Mol Med* **119:** 141–155.

Lee C, Wooldridge TR, Laimins LA. 2007. Analysis of the roles of E6 binding to E6TP1 and nuclear localization in the human papillomavirus type 31 life cycle. *Virology* **358:** 201–210.

Lepik D, Ilves I, Kristjuhan A, Maimets T, Ustav M. 1998. p53 protein is a suppressor of papillomavirus DNA amplificational replication. *J Virol* **72:** 6822–6831.

Lin BY, Ma T, Liu J-S, Kuo S-R, Jin G, Broker TR, Harper JW, Chow LT. 2000. HeLa cells are phenotypically limiting in cyclin E/CDK2 for efficient human papillomavirus DNA replication. *J Biol Chem* **275:** 6167–6174.

Lin BY, Makhov AM, Griffith JD, Broker TR, Chow LT. 2002. Human chaperone proteins abrogate the human papillomavirus E2 protein inhibition of DNA unwinding by the replicative helicase. *Mol Cell Biol* **22:** 6591–6603.

Lin K, Doolan K, Hung CF, Wu T.-C. 2010. Perspectives for preventive and therapeutic HPV vaccines. *J Formos Med Assoc* **109:** 4–24.

Litovchick L, Florens LA, Swanson SK, Washburn MP, DeCaprio JA. 2011. DYRK1A protein kinase promotes quiescence and senescence through DREAM complex assembly. *Genes Dev* **25:** 801–813.

Liu J-S, Kuo S-R, Broker TR, Chow LT. 1995. The functions of human papillomavirus type 11 E1, E2, and E2C proteins in cell-free DNA replication. *J Biol Chem* **270:** 27283–27291.

Liu J-S, Kuo S-R, Makhov AM, Cyr DM, Griffith JD, Broker TR, Chow LT. 1998. Human Hsp70 and Hsp40 chaperone proteins facilitate HPV-11 E1 protein binding to the origin and stimulate cell-free replication. *J Biol Chem* **273:** 30704–30712.

Ma T, Zou N, Lin BY, Chow LT, Harper JW. 1999. Interaction between cyclin-dependent kinases and human papillomavirus replication initiation protein E1 is required for efficient viral replication. *Proc Natl Acad Sci* **96:** 382–387.

Macville M, Schröck E, Padilla-Nash H, Keck C, Ghadimi BM, Zimonjic D, Popescu N, Ried T. 1999. Comprehensive and definitive molecular cytogenetic characterization of HeLa cells by spectral karyotyping. *Cancer Res* **59:** 141–150.

Martinez I, Gardiner AS, Board KF, Monzon FA, Edwards RP, Khan SA. 2008. Human papillomavirus type 16 reduces the expression of microRNA-218 in cervical carcinoma cells. *Oncogene* **27:** 2575–2582.

Massimi P, Pim D, Bertoli C, Bouvard V, Banks L. 1999. Interaction between the HPV 16 E2 transcriptional activator and p53. *Oncogene* **18:** 7748–7754.

Maufort JP, Shai A, Pitot HC, Lambert PF. 2010. A role for HPV16 E5 in cervical carcinogenesis. *Cancer Res* **70:** 2924–2931.

McClintock B. 1951. Chromosome organization and genic expression. *Cold Spring Harb Symp Quant Biol* **16:** 13–47.

McLaughlin-Drubin ME, Münger K. 2009. Oncogenic activities of human papillomaviruses. *Virus Res* **143:** 195–208.

McLaughlin-Drubin ME, Crum CP, Münger K. 2011. Human papillomavirus E7 oncoprotein induces KDM6A and KDM6B histone demethylase expression and causes epigenetic reprogramming. *Proc Natl Acad Sci* **108:** 2130–2135.

Melsheimer P, Vinokurova S, Wentzensen N, Bastert G, von Knebel Doeberitz M. 2004. DNA aneuploidy and integration of human papillomavirus type 16 E6/E7 oncogenes in intraepithelial neoplasia and invasive squamous cell carcinoma of the cervix uteri. *Clin Cancer Res* **10:** 3059–3063.

Merrick DT, Blanton RA, Gown AM, McDougall JK. 1992. Altered expression of proliferation and differentiation markers in human papillomavirus 16 and 18 immortalized epithelial cells grown in organotypic culture. *Am J Pathol* **140:** 167–177.

Meyers C, Frattini MG, Hudson JB, Laimins LA. 1992. Biosynthesis of human papillomavirus from a continuous cell line upon epithelial differentiation. *Science* **257:** 971–973.

Moody CA, Laimins LA. 2009. Human papillomaviruses activate the ATM DNA damage pathway for viral genome amplification upon differentiation. *PLoS Pathog* **5:** e1000605.

Moody CA, Laimins LA. 2010. Human papillomavirus oncoproteins: Pathways to transformation. *Nat Rev Cancer* **10:** 550–560.

Noya F, Chien W-M, Broker TR, Chow LT. 2001. p21cip1 degradation in differentiated keratinocytes is abrogated by co-stabilization with cyclin E induced by HPV E7. *J Virol* **75:** 6121–6134.

Parish JL, Bean AM, Park RB, Androphy EJ. 2006. ChlR1 is required for loading papillomavirus E2 onto mitotic chromosomes and viral genome maintenance. *Mol Cell* **24:** 867–876.

Park RB, Androphy EJ. 2002. Genetic analysis of high-risk E6 in episomal maintenance of human papillomavirus genomes in primary human keratinocytes. *J Virol* **76:** 11359–11364.

Park TW, Riethdorf S, Schulz G, Riethdorf L, Wright T, Loning T. 2003. Clonal expansion and HPV-induced immortalization are early molecular alterations in cervical carcinogenesis. *Anticancer Res* **23:** 155–160.

Park JW, Pitot HC, Strati K, Spardy N, Duensing S, Grompe M, Lambert PF. 2010. Deficiencies in the Fanconi anemia DNA damage response pathway increase sensitivity to HPV-associated head and neck cancer. *Cancer Res* **70:** 9959–9968.

Pedroza-Saavedra A, Lam EW, Esquivel-Guadarrama F, Gutierrez-Xicotencatl L. 2010. The human papillomavirus type 16 E5 oncoprotein synergizes with EGF receptor signaling to enhance cell cycle progression and the down-regulation of p27(Kip1). *Virology* **400:** 44–52.

Pett MR, Alazawi WO, Roberts I, Dowen S, Smith DI, Stanley MA, Coleman N. 2004. Acquisition of high-level chromosomal instability is associated with integration of human papillomavirus type 16 in cervical keratinocytes. *Cancer Res* **64:** 1359–1368.

Pim D, Tomaic V, Banks L. 2009. The human papillomavirus (HPV) E6* proteins from high-risk, mucosal HPVs can direct degradation of cellular proteins in the absence of full-length E6 protein. *J Virol* **83:** 9863–9874.

Pyeon D, Lambert PF, Ahlquist P. 2005. Production of infectious human papillomavirus independently of viral replication and epithelial cell differentiation. *Proc Natl Acad Sci* **102:** 9311–9316.

Pyeon D, Pearce SM, Lank SM, Ahlquist P, Lambert PF. 2009. Establishment of human papillomavirus infection requires cell cycle progression. *PLoS Pathog* **5:** e1000318.

Roberts JN, Buck CB, Thompson CD, Kines R, Bernardo M, Choyke PL, Lowy DR, Schiller JT. 2007. Genital transmission of HPV in a mouse model is potentiated by nonoxynol-9 and inhibited by carrageenan. *Nat Med* **13:** 857–861.

Roelens J, Reuschenbach M, von Knebel Doeberitz M, Wentzensen N, Bergeron C, Arbyn M. 2012. p16 (INK4a) immunocytochemistry versus human papillomavirus testing for triage of women with minor cytologic abnormalities: A systematic review and meta-analysis. *Cancer Cytopathol* doi: 10.1002/cncy.21205.

Sakakibara N, Mitra R, McBride AA. 2011. The papillomavirus E1 helicase activates a cellular DNA damage response in viral replication foci. *J Virol* **85:** 8981–8995.

Sapp M, Bienkowska-Haba M. 2009. Viral entry mechanisms: Human papillomavirus and a long journey from extracellular matrix to the nucleus. *FEBS J* **276:** 7206–7216.

Schiller JT, Lowy DR. 2006. Prospects for cervical cancer prevention by human papillomavirus vaccination. *Cancer Res* **66:** 10229–10232.

Schiller JT, Day PM, Kines RC. 2010. Current understanding of the mechanism of HPV infection. *Gynecol Oncol* **118:** S12–S17.

Schmidt-Grimminger D-C, Wu X, Jian Y, Broker TR, Chow LT. 1998. Post-transcriptional induction of p21cip1 in condylomata and dysplasias is inversely related to human papillomavirus activities. *Am J Pathol* **152:** 1015–1024.

Schuck S, Stenlund A. 2005. Assembly of a double hexameric helicase. *Mol Cell* **20:** 377–389.

Sherr CJ, Roberts JM. 2004. Living with or without cyclins and cyclin-dependent kinases. *Genes Dev* **18:** 2699–2711.

Shin MK, Balsitis S, Brake T, Lambert PF. 2009. Human papillomavirus E7 oncoprotein overrides the tumor suppressor activity of p21Cip1 in cervical carcinogenesis. *Cancer Res* **69:** 5656–5663.

Sim J, Ozgur S, Lin BY, Yu JH, Broker TR, Chow LT, Griffith JD. 2008. Remodeling of the human papillomavirus type 11 replication origin into discrete nucleoprotein particles and looped structures by the E2 protein. *J Mol Biol* **375:** 1165–1177.

Spardy N, Duensing A, Charles D, Haines N, Nakahara T, Lambert PF, Duensing S. 2007. The human papillomavirus type 16 E7 oncoprotein activates the Fanconi anemia (FA) pathway and causes accelerated chromosomal instability in FA cells. *J Virol* **81:** 13265–13270.

Steenbergen RDM, Parker JN, Isern S, Walboomers JMM, Meijer CJLM, Snijders PJF, Broker TR, Chow LT. 1998. Viral E6/E7 transcription in the basal layers of organotypic cultures without apparent p21*cip1* expression precedes immortalization of keratinocytes by transfected HPV16 and HPV18 DNAs. *J Virol* **72:** 749–757.

Stelzer MK, Pitot HC, Liem A, Schweizer J, Mahoney C, Lambert PF. 2010. A mouse model for human anal cancer. *Cancer Prev Res (Phila)* **3:** 1534–1541.

Stoler MH, Whitbeck A, Wolinsky SM, Broker TR, Chow LT, Howett MK, Kreider J. 1990. Infectious cycle of human papillomavirus type 11 in human foreskin xenografts in nude mice. *J Virol* **64:** 3310–3318.

Su JH, Wu A, Scotney E, Ma B, Monie A, Hung CF, Wu TC. 2010. Immunotherapy for cervical cancer: Research status and clinical potential. *BioDrugs* **4:** 109–129.

Suprynowicz FA, Krawczyk E, Hebert JD, Sudarshan SR, Simic V, Kamonjoh CM, Schlegel R. 2010. The human papillomavirus type 16 E5 oncoprotein inhibits epidermal growth factor trafficking independently of endosome acidification. *J Virol* **84:** 10619–10629.

Syrjänen S. 2010. The role of human papillomavirus infection in head and neck cancers. *Ann Oncol* **21:** vii243–245.

Tang S, Tao M, McCoy JP Jr, Zheng ZM. 2006. The E7 oncoprotein is translated from spliced E6*I transcripts in high-risk human papillomavirus type 16- or type 18-positive cervical cancer cell lines via translation reinitiation. *J Virol* **80:** 4249–4263.

Thomas MC, Chiang C-M. 2005. E6 oncoprotein represses p53–dependent gene activation via inhibition of protein acetylation independently of inducing p53 degradation. *Mol Cell* **17:** 251–264.

Thomas M, Narayan N, Pim D, Tomaić V, Massimi P, Nagasaka K, Kranjec C, Gammoh N, Banks L. 2008. Human papillomaviruses, cervical cancer and cell polarity. *Oncogene* **27:** 7018–7030.

Ueda Y, Enomoto T, Miyatake T, Ozaki K, Yoshizaki T, Kanao H, Ueno Y, Nakashima R, Shroyer KR, Murata Y. 2004. Monoclonal expansion with integration of high-risk type human papillomaviruses is an initial step for cervical carcinogenesis: Association of clonal status and human papillomavirus infection with clinical outcome in cervical intraepithelial neoplasia. *Lab Invest* **83:** 1517–1527.

van der Meide WF, Snellenberg S, Meijer CJLM, Baalbergen A, Helmerhorst TJM, van der Sluis WB, Snijders PJF, Steenbergen RDM. 2011. Promoter methylation analysis of WNT/β-catenin signaling pathway regulators to detect adenocarcinoma or its precursor lesion of the cervix. *Gynecol Oncol* **123:** 116–122.

Van Tine BA, Dao LD, Wu S-Y, Sonbuchner TM, Lin BY, Zou N, Chiang C-M, Broker TR, Chow LT. 2004a. HPV origin binding protein associates with mitotic spindles to enable viral DNA partitioning. *Proc Natl Acad Sci* **101:** 4030–4035.

Van Tine BA, Kappes JC, Banerjee NS, Knops J, Lai L, Steenbergen RDM, Meijer CLJM, Snijders PJF, Chatis P, Broker TR, et al. 2004b. Clonal selection for transcriptionally active viral oncogenes during progression to cancer by DNA methylation-mediated silencing. *J Virol* **78:** 11172–11186.

Vinokurova S, von Knebel Doeberitz M. 2011. Differential methylation of the HPV 16 upstream regulatory region during epithelial differentiation and neoplastic transformation. *PLoS ONE* **6:** e24451.

Vinokurova S, Wentzensen N, Kraus I, Klaes R, Driesch C, Melsheimer P, Kisseljov F, Dürst M, Schneider A, von Knebel Doeberitz M. 2008. Type-dependent integration frequency of human papillomavirus genomes in cervical lesions. *Cancer Res* **68:** 307–313.

Walboomers JMM, Jacobs MV, Manos MM, Bosch FX, Kummer JA, Shah KV, Snijders PJF, Peto J, Meijer CJLM, Muñoz N. 1999. Human papillomavirus is a necessary cause of invasive cervical cancer worldwide. *Pathology* **189:** 12–19.

Wald AI, Hoskins EE, Wells SI, Ferris RL, Khan SA. 2011. Alteration of microRNA profiles in squamous cell carcinoma of the head and neck cell lines by human papillomavirus. *Head Neck* **33:** 504–512.

Wang X, Tang S, Le SY, Lu R, Rader JS, Meyers C, Zheng ZM. 2008. Aberrant expression of oncogenic and tumor-suppressive microRNAs in cervical cancer is required for cancer cell growth. *PLoS ONE* **3:** e2557.

Wang H-K, Duffy AA, Broker TR, Chow LT. 2009a. Robust production and passaging of infectious HPV in squamous epithelium of primary human keratinocytes. *Genes Dev* **23:** 181–194.

Wang X, Wang H-K, McCoy JP, Banerjee NS, Rader JS, Broker TR, Meyers C, Chow LT, Zheng ZM. 2009b. Oncogenic HPV infection interrupts the expression of tumor-suppressive miR-34a through viral oncoprotein E6. *RNA* **15:** 637–647.

Wang X, Meyers C, Wang H-K, Chow LT, Zheng ZM. 2011. Construction of a full transcription map of human papillomavirus type 18 during productive viral infection. *J Virol* **85:** 8080–8092.

Wells SI, Aronow BJ, Wise TM, Williams SS, Couget JA, Howley PM. 2003. Transcriptome signature of irreversible senescence in human papillomavirus-positive cervical cancer cells. *Proc. Natl Acad Sci* **100:** 7093–7098.

Wentzensen N, Ridder R, Klaes R, Vinokurova S, Schaefer U, von Knebel Doeberitz M. 2002. Characterization of viral-cellular fusion transcripts in a large series of HPV16 and 18 positive anogenital lesions. *Oncogene* **21:** 419–426.

Wilson R, Laimins LA. 2005. Differentiation of HPV-containing cells using organotypic "raft" culture or methylcellulose. *Methods Mol Med* **119:** 157–169.

Wilting SM, Smeets SJ, Snijders PJF, van Wieringen WN, van de Wiel MA, Meijer GA, Ylstra B, Leemans CR, Meijer CJLM, Brakenhoff RH, et al. 2009. Genomic profiling identifies common HPV-associated chromosomal alterations in squamous cell carcinomas of cervix and head and neck. *BMC Med Genomics* **2**: doi: 10.1186/1755-8794-2-32.

Wilting SM, van Boerdonk RAA, Henken FE, Meijer CJLM, Diosdado B, Meijer GA, le Sage C, Agami R, Snijders PJF, Steenbergen RDM. 2010. Methylation-mediated silencing and tumour suppressive function of hsa-miR-124 in cervical cancer. *Mol Cancer* **9**: doi: 10.1186/1476-4598-9-167.

Wooldridge TR, Laimins LA. 2008. Regulation of human papillomavirus type 31 gene expression during the differentiation-dependent life cycle through histone modifications and transcription factor binding. *Virology* **374:** 371–380.

Wu SY, Lee AY, Hou SY, Kemper JK, Erdjument-Bromage H, Tempst P, Chiang C-M. 2006. Brd4 links chromatin targeting to HPV transcriptional silencing. *Genes Dev* **20:** 2383–2396.

You H, Liu Y, Agrawal N, Prasad CK, Edwards JL, Osborne AF, Korourian S, Lowery CL, Hermonat PL. 2008. Multiple human papillomavirus types replicate in 3A trophoblasts. *Placenta* **29:** 30–38.

Yu J-H, Lin BY, Deng W, Broker TR, Chow LT. 2007. Mitogen-activated protein kinases activate the nuclear localization sequence of the human papillomavirus type 11 E1 DNA helicase to promote efficient nuclear import. *J Virol* **81:** 5066–5078.

Zehbe I, Ratsch A, Alunni-Fabbroni M, Burzlaff A, Bakos E, Dürst M, Wilander E, Tommasino M. 1999. Overriding of cyclin-dependent kinase inhibitors by high and low risk human papillomavirus types: Evidence for an in vivo role in cervical lesions. *Oncogene* **18:** 2201–2211.

Zhang B, Chen W, Roman A. 2006. The E7 proteins of low- and high-risk human papillomaviruses share the ability to target the pRB family member p130 for degradation. *Proc Natl Acad Sci* **103:** 437–442.

Zhao W, Noya F, Chen WY, Townes TM, Chow LT, Broker TR. 1999. Trichostatin A up-regulates HPV-11 URR-E6 promoter activity in undifferentiated primary human keratinocytes. *J Virol* **73:** 5026–5033.

Zobel T, Iftner T, Stubenrauch F. 2003. The papillomavirus E8-E2C protein represses DNA replication from extrachromosomal origins. *Mol Cell Biol* **23:** 8352–8362.

zur Hausen H. 2009. Papillomaviruses in the causation of human cancers—A brief historical account. *Virology* **384:** 260–265.

Adenovirus DNA Replication

Rob C. Hoeben[1] **and Taco G. Uil**[2]

[1]Department of Molecular Cell Biology, Leiden University Medical Centre, 2333 ZC Leiden, The Netherlands
[2]Crucell, 2333 CN Leiden, The Netherlands

Correspondence: r.c.hoeben@lumc.nl

Adenoviruses have attracted much attention as probes to study biological processes such as DNA replication, transcription, splicing, and cellular transformation. More recently these viruses have been used as gene-transfer vectors and oncolytic agents. On the other hand, adenoviruses are notorious pathogens in people with compromised immune functions. This article will briefly summarize the basic replication strategy of adenoviruses and the key proteins involved and will deal with the new developments since 2006. In addition, we will cover the development of antivirals that interfere with human adenovirus (HAdV) replication and the impact of HAdV on human disease.

Members of the family *Adenoviridae* are nonenveloped icosahedral viruses that replicate in the nucleus. In recent years, new DNA amplification techniques and sequencing efforts have led to the identification of many new human and nonhuman adenoviruses (AdV). The human adenoviruses (HAdV) are grouped as seven species (formerly called "subgroups"), HAdV-A–G, in the Mastadenovirus genus. Within the species, distinct "types" (formerly called "serotypes") are recognized based on the absence of serological cross-neutralization. With the advent of high-throughput sequencing of new virus isolates, a debate sparked around the question how sequence data should be used to define types and species (Aoki et al. 2011; Seto et al. 2011).

The successful propagation in tissue culture of small DNA tumor viruses like adenoviruses, polyoma virus, and SV40 opened up the field of studies on the replication and transcription of the genomes of these viruses in the 1960s and 1970s (Tooze 1980; Yaniv 2009). These studies yielded many of the basic principles of mammalian DNA replication. AdV DNA replication was reviewed in detail in Hay (1966), Van der Vliet (1996), and Van der Vliet and Hoeben (2006). The mechanisms of AdV DNA replication have now been firmly established, and many of the roles of the key actors in the process have been elucidated in detail. During the last decade, the advancement in acquiring insight in the molecular processes involved in AdV DNA replication slowed down as the interest of many groups shifted from the study of viral DNA replication to the study of cellular processes.

A DETAILED MODEL OF ADENOVIRUS REPLICATION

HAdV contain a linear double-stranded genome of about 36 kb with inverted terminal

Cite this article as *Cold Spring Harb Perspect Biol* doi: 10.1101/cshperspect.a013003

repetitions of ~100 bp. Both 5′ termini are covalently attached to a 55-kDa terminal protein (TP). AdV DNA replication is a very efficient process. Within 40 h, an infected cell produces approximately one million copies of viral DNA. Much information on the replication mechanism came from an in vitro system that replicates HAdV-5 and -2 DNA with purified proteins. Two identical origins of replication are located within the inverted terminal repeats, covering ~1–50 bp. The terminal 18 bp form the minimal origin and the remainder act as the auxiliary origin (Fig. 1A). Protein-primed DNA synthesis starts by covalent addition of a dCMP

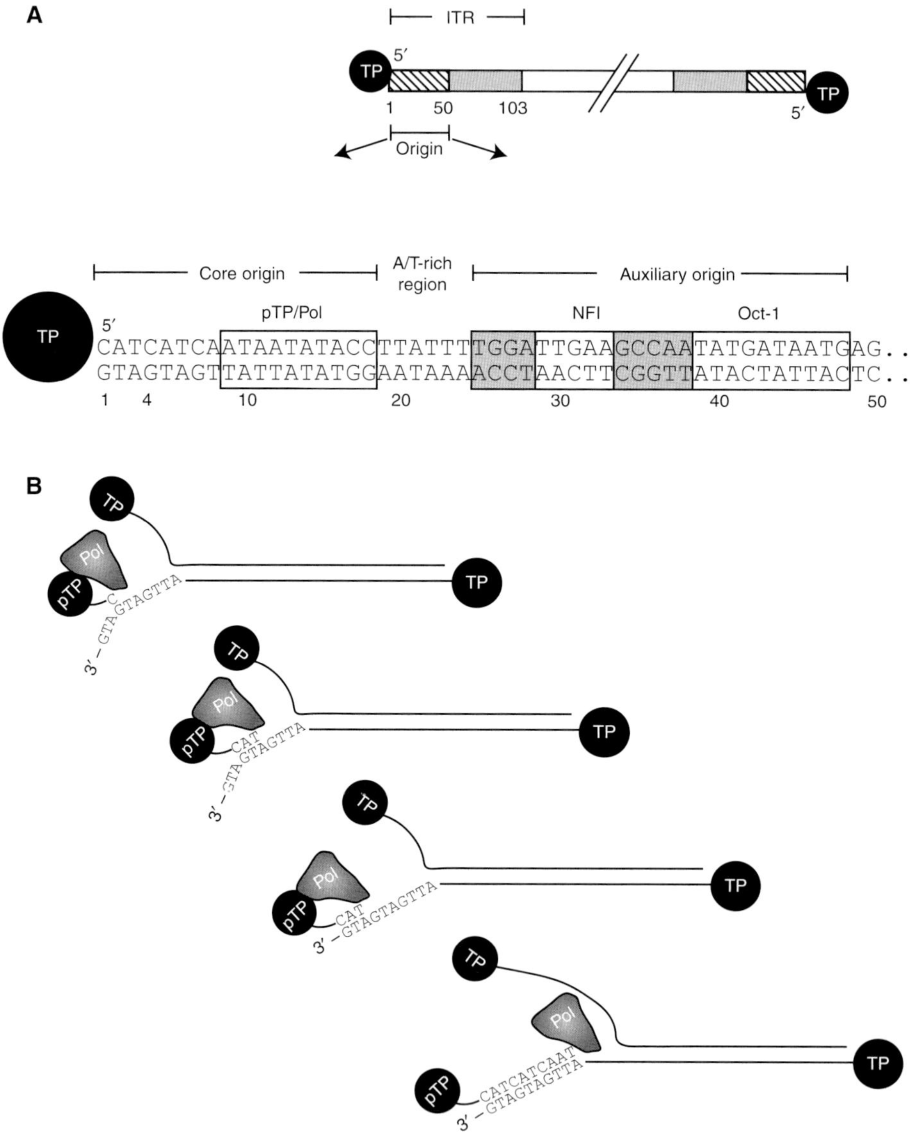

Figure 1. (*A*) Structure of the adenovirus type 5 origin and regions required for interaction with viral and cellular proteins. (*B*) Schematic representation of the initiation of DNA replication illustrating how the nucleotides 4–6 are used as a template for the formation of the pTP-CAT complex. The complex subsequently displaces to pair with nucleotides 1–3 of the template stand, after which the elongation continues.

residue to an 80-kDa precursor of the TP (pTP). Replication requires three viral proteins encoded by E2 genes: pTP, AdV DNA polymerase (AdV Pol), and the DNA-binding protein (DBP). Two cellular transcription factors, NFI and Oct-1, bind the auxiliary origin and thereby enhance initiation several hundredfold. In the final stages of replication, pTP is cleaved by a viral protease to TP, resulting in progeny DNA, which is subsequently packaged in virions.

Multiple interactions between the three viral and two cellular proteins and the origin orchestrate the formation of a preinitiation complex. This step is accompanied by subtle changes in the origin structure by DBP and by extensive origin bending, mainly induced by NFI and Oct-1. Initiation occurs at position 4 of the template (sequence 3′-G_1TAG_4TA) followed by formation of a pTP-trinucleotide intermediate, pTP-CAT (Fig. 1B). This pTP-CAT intermediate jumps back three bases and becomes paired with template residues 1–3 (King and Van der Vliet 1994). This jumping-back mechanism explains the presence of a short 3-bp (or sometimes 2- or 4-bp) repeat sequence in the first 10 bp of all AdV origins (De Jong et al. 2003). For replication to begin at an internal site, rather than at the genome termini, appears to be a common feature among protein-primed replication systems (see below). A specific amino acid in pTP, Ser580 in HAdV-5, is used as a primer for covalent binding of dCMP by AdV Pol, generating pTP-C (Desiderio and Kelly 1981; Smart and Stillman 1982). During, or shortly after the three-nucleotide jump, AdV Pol dissociates from pTP (King and van der Vliet 1994). This leads to an increased rate of polymerization and proofreading. Further disassembly of the preinitiation complex takes place when NFI dissociates early in initiation, whereas Oct-1 dissociates only when the recognition site has become single stranded on passing of the replication fork (van Leeuwen et al. 1997). The pTP-CAT intermediate is efficiently elongated by the combined action of AdV Pol and DBP, leading to the formation of a new duplex genome and the displacement of the nontemplate strand. The ITRs of displaced nontemplate strands may anneal together (either intra- or intermolecularly) to restore functional origins that could serve as substrates for new replication rounds. Newly generated Ad genome duplexes (formed either by replication or the annealing of complementary displaced single strands) may also enter subsequent replication rounds or be processed and encapsidated into virions. A schematic representation of the protein–protein interactions and the protein–template interactions can be found in an earlier version of this article (van der Vliet and Hoeben 2006).

Although the mechanism of adenovirus DNA replication is unique among the eukaryotic viruses, it bears many similarities with the protein-primed DNA replication of several bacteriophages such as the *Bacillus subtilis* bacteriophages ϕ29 and GA-1, the *Escherichia coli* bacteriophage PRD1, and the *Streptococcus pneumoniae* phage CP-1. Similar to AdV, these phages contain linear double-stranded DNA genomes, which are replicated by protein primed DNA replication mechanisms. The replication leaves a TP covalently coupled to the viral DNA (Blanco et al. 1989; Salas 1991; Illana et al 1996). Also in the replication of these bacteriophages, sliding-back or jumping-back mechanisms are used in which the nucleotides other than the 3′ terminal nucleotides are used as a template for the 5′ terminal nucleotide that remains covalently coupled to the phage's TP (Salas 1991; Méndez et al. 1992; Caldentey et al. 1993; Illana et al. 1996; Martín et al. 1996).

VIRAL REPLICATION PROTEINS

The Precursor TP

The precursor TP (pTP) is an 80-kDa protein that functions as the primer for initiation. pTP binds both single-strand DNA (ssDNA) and double-strand DNA (dsDNA), and the affinity of pTP for ssDNA might function in stabilizing AdV Pol on partially unwound origin DNA (de Jong et al. 2003). During initiation, AdV Pol covalently couples the first dCMP residue to Ser580 of pTP. The initiation of replication by pTP-binding dCMP is well conserved in AdV.

Hence almost all adenoviruses characterized to date have a C residue at the 5′ end of their genome, suggesting an inherent preference of pTP to bind dCMP. The single exception to this rule has been found in the fowl adenovirus-A (FAdV-A). In the FAdV-A isolates Phelps and KUR, initiation starts with incorporation of a dGMP. Interestingly, the highly similar FAdV-A strain OTE is conventional and starts initiation by binding of dCMP (Rademaker et al. 2006). Studies with minireplicons showed that the pTP and Pol of CELO-OTE and CELO-Phelps do not have different specificities, but rather show relaxed template specificity. This allows these viruses to use 3′-GTAGTAG as well as 3′-CTACTAC as templates for replication initiation (Rademaker et al. 2006). This implies the efficient formation of pTP-GMP complexes. This is in contrast to HAdV-5, in which mutation of G4 in the template strand fully blocked replication initiation in vitro. For HAdV-5, even in the absence of any template DNA, pTP-C-Pol complexes are formed. The preference of HAdV-5 pTP to bind dCMP was reduced in the presence of manganese ions in the replication-initiation assay, suggesting that structural factors govern the template specificity (King and van der Vliet 1994).

The AdV DNA Polymerase

The AdV DNA polymerase (AdV Pol) belongs to the distinct group of family B (or α-like) DNA polymerases that uses a protein primer for the initiation of replication. The crystal structure of AdV Pol has not been determined, but sequence comparisons with polymerases of known structure as well as biochemical and mutagenesis studies have revealed AdV Pol's general domain organization (Fig. 2) (Joung et al. 1991; Joung and Engler 1992; Knopf 1998; Liu et al. 2000; Brenkman et al. 2001, 2002; Liu et al. 2003). Besides consisting of the common polymerase subdomains palm, fingers, and thumb, which are associated with 5′–3′ polymerase activity, AdV Pol carries a 3′–5′ exonuclease domain that is necessary for its intrinsic proofreading ability (King et al. 1997; Brenkman et al. 2002; Liu et al. 2003). Furthermore, sequence comparisons have suggested that AdV Pol contains certain insertions, termed TPR1 and TPR2, which are specific for the protein-primed family B polymerases (Blasco et al. 1990; Dufour et al. 2000). Although no information is available on the roles of TPR1 and TPR2 in AdV Pol, studies on the bacteriophage ϕ29 provided some insights into their function. The crystal structure

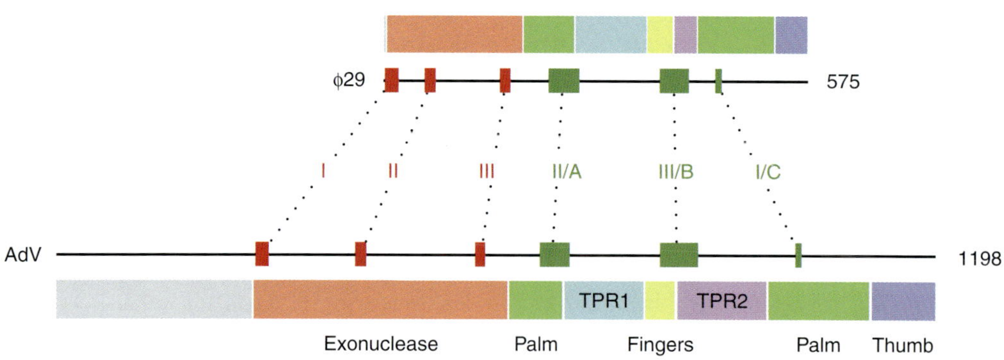

Figure 2. Representations of AdV and ϕ29 polymerases showing the locations of conserved sequence motifs and structural domains. The domain organization shown for ϕ29 Pol was described previously by Kamtekar et al. (2004) and is based on structural data. The domain organization depicted for AdV Pol is inferred from that of ϕ29 Pol and is based on the relative location of motifs shared among (many of) the DNA polymerases. Indicated for both polymerases are the most highly conserved of these sequence motifs within the exonuclease domain ("Exo" motifs I, II, and III) and the polymerase region ("Pol" motifs I, II, and III, also known as, respectively, motifs C, A, and B) (Wong et al. 1988; Bernad et al. 1989; Delarue et al. 1990). All representations are to scale.

of the ϕ29 polymerase has been determined (Kamtekar et al. 2004, 2006; Berman et al. 2007). In the phage ϕ29 polymerase, TPR1 and TPR2 form discrete subdomains. Based on these structural data as well as biochemical studies on ϕ29 polymerase and its protein primer (Rodríguez et al. 2005), the TPR structures are known to be responsible for specific binding to the protein primer (TPR1) and/or suspected to contribute to ϕ29 Pol's intrinsic strand-displacement capacity (TPR2) and/or high processivity (TPR1 and TPR2). They would contribute to the latter two functionalities by making part of the tunnels encircling, respectively, the downstream template strand and the upstream duplex product. These tunnels would topologically and functionally resemble, respectively, the helicases and sliding clamp proteins of replisomes. Interestingly, sequence comparisons with the other protein-primed family B polymerases indicate that the putative TPR2 domain of AdV Pol is rather large (Fig. 1) (Uil et al. 2011), which might suggest that it has an additional or different function.

Previously, AdV Pol function has mostly been studied by biochemical assays using purified proteins or cell extracts (Liu et al. 2003). Recent work, however, gave insight into the nucleotide selection and proofreading functions of AdV Pol in the context of full-fledged viral replication. In a study that aimed to make mutator AdV polymerases that could serve in viral directed evolution procedures, AdV Pol mutants were made that carried single-amino-acid substitutions in either the nucleotide binding pocket or the exonuclease active site (Uil et al. 2011). Using a polymerase complementation approach (in which a polymerase-defective virus is replicated on cells stably expressing the AdV Pol), these mutants were tested for their ability to support productive adenovirus replication. Furthermore, they were tested for mutator activity by performing deep sequencing on pools of replicated viruses.

Mutation of highly conserved residues within the nucleotide binding pocket generally had drastic effects on the ability to support Ad replication. For most of the mutants this would have likely been owing to a decrease in viral genome replication efficiency (rather than an excessive mutation rate). Also most of the exonuclease domain mutants tested were unable to support AdV replication. Interestingly, all substitutions of the most highly conserved exonuclease residues (those that contribute to the metal ion binding framework of the exonuclease active site) were found to be lethal for virus replication. Mutations applied to the homologs of these residues in other viral polymerases are known to effectively knock out 3′–5′ exonuclease activity and therewith proofreading activity (Bernad et al. 1989; Esteban et al. 1994; Abdus Sattar et al. 1996; Kuhn and Knopf 1996), but are not necessarily lethal in the context of productive virus replication, as has been found for RB69, HSV, and CMV (Hwang et al. 1997; Bebenek et al. 2001; Chou and Marousek 2008). Therefore, barring nonspecific structural effects, the lethality of these mutations for adenovirus might point to an additional role of these exonucleolytic residues. Perhaps related to this issue is that homologous mutations in ϕ29 polymerase result in a reduced velocity of DNA polymerization on dsDNA templates (Soengas et al. 1992; Esteban et al. 1994).

Of the viable Adv Pol mutants, several caused excess mutations during viral passaging, confirming the importance of the respective mutated motifs for incoming nucleotide selection and proofreading. For example, mutation of a nonhighly conserved residue located in a loop between two α helices of the fingers domain (D827A) led to a degree of mutator activity. As the concerning residue is not thought to directly contact the incoming nucleotide, it may be that its substitution indirectly affects nucleotide selection, for example, by changing the relative positions of the two α helices. The strongest mutator AdV Pols identified in the study were those of two mutants of the exonuclease motifs I and II (Fig. 1). These mutants, T286I and F421Y, carry substitutions of residues whose homologs in ϕ29 are known to play a role in primer terminus stabilization at the exonuclease active site during proofreading (de Vega et al. 1996, 1998). The mutator activities found for these mutants would support a similar role of these highly conserved residues in AdV Pol.

The AdV DBP

On infection the major viral replication protein produced is DBP. DBP binds dsDNA and ssDNA, the latter in a cooperative fashion, and it changes the DNA structure drastically (Kanellopoulos et al. 1995). DBP is involved in several steps during DNA replication. Besides protecting ssDNA against nucleases, DBP acts during initiation by stimulating binding of AdV Pol and NFI, presumably by slight changes in the configuration of origin DNA (Hay 1966; van Breukelen et al. 2003; van der Vliet and Hoeben 2006). Moreover, DBP lowers the Km for coupling of the first nucleotide to pTP. During elongation, DBP enhances the rate and processivity of AdV DNA polymerase and changes its sensitivity to inhibitors. DBP, which has a helix destabilizing activity, has an essential role in DNA unwinding, which in vitro is an ATP-independent process not requiring a helicase. DBP drives template unwinding by formation of DBP multimers.

Other Viral Proteins

In addition to the viral proteins directly involved in DNA replication, HAdV encodes several proteins that affect cellular processes that are involved in the down-regulation of cellular antiviral responses, stimulate cell-cycle progression, or inhibit the DNA-damage responses. These factors facilitate and promote efficient viral DNA replication in the cells. Many such viral factors and processes have been reviewed recently (Giberson et al. 2012).

Cellular Factors Optimize Initiation

The pTP-Pol complex binds weakly to the origin and can catalyze only minimal initiation. In vitro replication assays facilitated the identification of nuclear factors (NF) that enhance AdV replication considerably, up to the levels observed in vivo. NFI has been identified to be identical to CTF-1, a cellular transcription factor, NFII is a topoisomerase type 1 required for the synthesis of genome-length molecules, and NFIII is identical to the transcription factor Oct-1. NFI and Oct-1 form part of the preinitiation complex and both interact with AdV DNA and with one of the viral replication proteins (van der Vliet 1996; de Jong and van der Vliet 1999; van der Vliet and Hoeben 2006).

NFI encompasses a family of cellular transcription factors that also enhances initiation of AdV DNA replication up to 60-fold. NFI binds to position 25–38 in the origin of HAdV-5. Mutation of this sequence reduces replication of the virus in infected cells. The efficiency of initiation is highly dependent on the AdV Pol concentration, suggesting that AdV Pol is recruited to the origin by an interaction with NFI. A second way in which NFI stimulates initiation is by changing the origin structure. NFI bends the origin 60° as shown by scanning force microscopy (Mysiak et al. 2004a). The bending is dependent on the conserved AT-rich sequence directly preceding the core recognition site of NFI.

Oct-1 belongs to the family of octamer-binding transcription factors containing a POU DNA-binding domain. The POU domain has a bipartite structure with two subdomains, POU specific (POUs) and POU homeodomain (POUhd), that both contribute to DNA binding and to the stimulation of initiation. Oct-1 binds next to the NFI recognition sequence at origin position 39–47.

Whereas NFI interacts with AdV Pol, Oct-1 interacts with pTP and tethers it to the origin as shown by a reduced dissociation rate of the pTP-Pol-origin interaction (van Leeuwen et al. 1997). Like in the case of NFI, stimulation of initiation depends on the structural changes in the origin induced by Oct-1. Scanning force microscopy showed a 42°-bend angle in origin DNA on binding the Oct-1 POU domain, confirming earlier biochemical data. Simultaneous binding of NFI and Oct-1 induces an 82° bend indicating that both bends are oriented toward each other (Mysiak et al. 2004b). Because both NFI and Oct-1 act synergistically to enhance initiation, this strengthens the notion that origin bending is essential for initiation, presumably by facilitating the optimal assembly of a preinitiation complex.

In conclusion, the two cellular transcription factors stimulate initiation in two different ways: recruitment of the viral replication proteins as well as bending the origin. Combined with the role of DBP in modifying the origin structure and facilitating the kinetics of initiation, a picture arises in which all proteins act in a concerted fashion, contributing to optimal assembly of a preinitiation complex as well as to the kinetics of the actual initiation and elongation reactions.

Human Adenoviruses and Disease

Although in immune-competent persons most HAdV infections occur subclinically, AdV infections are related to a wide range of malignancies (reviewed by Horwitz 2001; Lenaerts et al. 2008). The gastrointestinal disease caused by enteric AdV (i.e., HAdV-F) is considered a major contributing factor in childhood diarrhea in underdeveloped and high-population density areas. In addition, the HAdV are the most common infectious cause of ocular disease worldwide. AdV are isolated frequently from patients with a compromised immune system (Kojaoghlanian et al. 2003; Walls et al. 2003; Lenaerts et al. 2008). They are a frequent cause of morbidity and mortality after allogeneic stem cell transplantation (Symeonidis et al. 2007). Remarkably, the incidence of HAdV infections in pediatric stem cell recipients is rising sharply (Bruno et al. 2003; van Tol et al. 2005). Another group that is at risk is HIV-infected individuals.

Despite these disease associations, HAdV-derived vectors have been very popular as gene-transfer vehicles in experimental gene therapy. Initially, replication-deficient vectors were used. In these vectors, a deletion of the E1 genes blocks expression of the viral genes and replication, and provides the space needed to insert a gene of interest. Although the gene transfer is efficient, immune-competent hosts mount an immune response against viral and transgene antigens (reviewed by Bangari and Mittal 2006). This leads to the elimination of virally transduced cells in immune-competent hosts. In the newer generations of AdV vectors the residual AdV genes were removed (reviewed by Seiler et al. 2007). The propensity of the E1-deleted HAdV vectors to induce potent vector and transgene product-specific immune responses led to their use as live-virus vaccine carriers (Liu et al. 2009). In recent years, HAdV attracted a renewed interest as oncolytic agents in cancer therapy. Although the use of replication-competent HAdV in cancer therapy has been explored already more than 50 years ago (Kelly and Russell 2007), the current approaches use modifications in the vector genomes that limit viral replication to tumor cells. In this respect, the vectors HAdV-5 mutant "ONYX-015" and HAdV-5 Δ24 can serve as archetypes. The AdV ONYX-015 virus has a deletion in the region coding for the E1B 55 kDa protein. As a result, the virus is incapable of down-regulating the p53 activity. The HAdV Δ24 virus carries a deletion in the viral E1A gene that encompasses the region responsible for binding the RB protein. As a result, the Δ24 E1A protein cannot bind RB to relieve RB repression of E2F transcription factors. As a result, replication of HAdV-5 Δ24 viruses can only occur in tumor cells in which RB is inactivated (Berk 2005). RB and p53 are two critical transcription regulatory proteins and the pathways in which they operate are disturbed in nearly all human cancers (Polager and Ginsberg 2009). This should lead to preferential replication in tumor cells with alterations in the p53 pathway. These viruses have been evaluated in several clinical studies for a wide variety of tumor types with promising results (reviewed by Jiang et al. 2009). In 2006, the AdV "H101," which is very similar to AdV ONYX-015, was registered in China for clinical use.

AdV Antivirals

There is a considerable morbidity and mortality associated with HAdV infections. This would clearly be limited by the availability of effective antiviral drugs. In addition, the increased clinical use of HAdV-derived replication-competent vectors makes the availability of effective antivirals against HAdV infections desirable. To date, there is no FDA-approved antiviral for the treatment of HAdV infections available. Several phosphonyl acyclic nucleotide analogs have activity against HAdV. Cidofovir, (S)-9-(3-hydroxy-2-phosphonomethoxy propyl)cytosine dehydrate, is among the most potent of antivirals. It has been registered by the FDA for treatment of human cytomegalovirus-induced retinitis in AIDS patients. Mechanistically, cidofovir mimics the monophosphate form of deoxycytosine and targets the AdV Pol directly. This has been elegantly confirmed by selecting cidofovir-resistant HAdV-5 mutants. All these mutations are located in the AdV Pol-coding region (Kinchington et al. 2002). Cidofovir inhibits all HAdV types tested (Kinchington et al. 2005). This is in contrast to, for instance, Ribavirin, which inhibits replication in all HAdV-C isolates, but had variable activity in isolates of HAdV-A, B, and D (Morfin et al. 2009).

So far the clinical efficacy of cidofovir has been promising but not unequivocal (reviewed by Ljungman 2004; Lenaerts et al. 2008). Current efforts are concentrated on optimizing the formulation of cidofovir and the development of new derivatives (Toth et al. 2008; Paolino et al. 2011; Ruiz et al. 2011). In addition, high-throughput screening for inhibitors of viral replication can identify new compounds that function as broad-spectrum antiviral agents. Already a compound was identified that is a potent inhibitor of dihydroorotate dehydrogenase, which is a key enzyme of the de novo pyrimidine synthesis pathway. This compound inhibits the replication of several viruses, including AdV (Hoffmann et al. 2011).

In parallel, immunotherapy protocols are developed involving infusion of AdV-specific T lymphocytes. The feasibility of this approach has been established in patients undergoing hematopoietic stem cell transplantation (HSCT) (Feuchtinger et al. 2006). This approach may be facilitated by the cross-reactivity of Ad-specific T cells (Heemskerk et al. 2003; Leen et al. 2006; Veltrop-Duits et al. 2006). Although promising, the immunotherapy approach is very labor intensive and expensive, restricting its application. Therefore it will remain a major challenge to develop effective antivirals for treatment for AdV infections that have activity against the wide diversity of HAdV serotypes. The detailed insight in key aspects of AdV DNA replication may facilitate rational development of new AdV-specific antivirals.

REFERENCES

Abdus Sattar AK, Lin TC, Jones C, Konigsberg WH. 1996. Functional consequences and exonuclease kinetic parameters of point mutations in bacteriophage T4 DNA polymerase. *Biochemistry* **35:** 16621–16629.

Aoki K, Benko M, Davison AJ, Echavarria M, Erdman DD, Harrach B, Kajon AE, Schnurr D, Wadell G. 2011. Toward an integrated human adenovirus designation system that utilizes molecular and serological data and serves both clinical and fundamental virology. *J Virol* **85:** 5703–5704.

Bangari DS, Mittal SK. 2006. Current strategies and future directions for eluding adenoviral vector immunity. *Curr Gene Ther* **6:** 215–226.

Bebenek A, Dressman HK, Carver GT, Ng S, Petrov V, Yang G, Konigsberg WH, Karam JD, Drake JW. 2001. Interacting fidelity defects in the replicative DNA polymerase of bacteriophage RB69. *J Biol Chem* **276:** 10387–10397.

Berk AJ. 2005. Recent lessons in gene expression, cell cycle control, and cell biology from adenovirus. *Oncogene* **24:** 7673–7685.

Berman AJ, Kamtekar S, Goodman JL, Lázaro JM, de Vega M, Blanco L, Salas M, Steitz TA. 2007. Structures of ϕ29 DNA polymerase complexed with substrate: The mechanism of translocation in B-family polymerases. *EMBO J* **26:** 3494–3505.

Bernad A, Blanco L, Lázaro JM, Martín G, Salas M. 1989. A conserved $3' \rightarrow 5'$ exonuclease active site in prokaryotic and eukaryotic DNA polymerases. *Cell* **59:** 219–228.

Blanco L, Bernad A, Lázaro JM, Martin G, Garmendia C, Salas M. 1989. Highly efficient DNA synthesis by the phage φ29 DNA polymerase. Symmetrical mode of DNA replication. *J Biol Chem* **264:** 8935–8940.

Blasco MA, Blanco L, Parés E, Salas M, Bernad A. 1990. Structural and functional analysis of temperature-sensitive mutants of the phage ϕ29 DNA polymerase. *Nucleic Acids Res* **18:** 4763–4770.

Brenkman AB, Heideman MR, Truniger V, Salas M, Van der Vliet PC. 2001. The (I/Y)XGG motif of adenovirus DNA polymerase affects template DNA binding and the transition from initiation to elongation. *J Biol Chem* **276:** 29846–29853.

Brenkman AB, Breure EC, Van der Vliet PC. 2002. Molecular architecture of adenovirus DNA polymerase and location of the protein primer. *J Virol* **76:** 8200–8207.

Bruno B, Gooley T, Hackman RC, Davis C, Corey L, Boeckh M. 2003. Adenovirus infection in hematopoietic stem cell transplantation: Effect of ganciclovir and impact on survival. *Biol Blood Marrow Transplant* **9:** 341–352.

Caldentey J, Blanco L, Bamford DH, Salas M. 1993. In vitro replication of bacteriophage PRD1 DNA. Characterization of the protein-primed initiation site. *Nucleic Acids Res* **21:** 3725–3730.

Chou S, Marousek GI. 2008. Accelerated evolution of maribavir resistance in a cytomegalovirus exonuclease domain II mutant. *J Virol* **82:** 246–253.

de Jong RN, Van der Vliet PC. 1999. Mechanism of DNA replication in eukaryotic cells: Cellular host factors stimulating adenovirus DNA replication. *Gene* **236:** 1–12.

de Jong RN, Van der Vliet PC, Brenkman AB. 2003. Adenovirus DNA replication: Protein priming, jumping back and the role of the DNA binding protein DBP. *Curr Top Microbiol Immunol* **272:** 187–211.

de Vega M, Lazaro JM, Salas M, Blanco L. 1996. Primer-terminus stabilization at the 3′–5′ exonuclease active site of ϕ29 DNA polymerase. Involvement of two amino acid residues highly conserved in proofreading DNA polymerases. *EMBO J* **15:** 1182–1192.

de Vega M, Lázaro JM, Salas M, Blanco L. 1998. Mutational analysis of ϕ29 DNA polymerase residues acting as ssDNA ligands for 3′–5′ exonucleolysis. *J Mol Biol* **279:** 807–822.

Delarue M, Poch O, Tordo N, Moras D, Argos P. 1990. An attempt to unify the structure of polymerases. *Protein Eng* **3:** 461–467.

Desiderio SV, Kelly TJ Jr. 1981. Structure of the linkage between adenovirus DNA and the 55,000 molecular weight terminal protein. *J Mol Biol* **145:** 319–337.

Dufour E, Méndez J, Lázaro JM, de Vega M, Blanco L, Salas M. 2000. An aspartic acid residue in TPR-1, a specific region of protein-priming DNA polymerases, is required for the functional interaction with primer terminal protein. *J Mol Biol* **304:** 289–300.

Esteban JA, Soengas MS, Salas M, Blanco L. 1994. 3′→5′ Exonuclease active site of ϕ29 DNA polymerase. Evidence favoring a metal ion-assisted reaction mechanism. *J Biol Chem* **269:** 31946–31954.

Feuchtinger T, Matthes-Martin S, Richard C, Lion T, Fuhrer M, Hamprecht K, Handgretinger R, Peters C, Schuster FR, Beck R, et al. 2006. Safe adoptive transfer of virus-specific T-cell immunity for the treatment of systemic adenovirus infection after allogeneic stem cell transplantation. *Br J Haematol* **134:** 64–76.

Giberson AN, Davidson AR, Parks RJ. 2012. Chromatin structure of adenovirus DNA throughout infection. *Nucleic Acids Res* **40:** 2369–2376.

Hay RT. 1966. Adenovirus DNA replication. In *DNA replication in eukaryotic cells* (ed. DePamphilis ML), pp. 699–721. Cold Spring Harbor Laboratory Press, Cold Spring Harbor, NY.

Heemskerk B, Veltrop-Duits LA, Van Vreeswijk T, Ten Dam MM, Heidt S, Toes RE, Van Tol MJ, Schilham MW. 2003. Extensive cross-reactivity of CD4$^+$ adenovirus-specific T cells: Implications for immunotherapy and gene therapy. *J Virol* **77:** 6562–6566.

Hoffmann HH, Kunz A, Simon VA, Palese P, Shaw ML. 2011. Broad-spectrum antiviral that interferes with de novo pyrimidine biosynthesis. *Proc Natl Acad Sci* **108:** 5777–5782.

Horwitz MS. 2001. Adenoviruses, Philadelphia. In *Field's virology* (ed. Knipe DM, Howley PM, Griffin DE, Lamb RA, Martin MA, Roizman B, Strauss SE), pp. 2301–2325. Lippincott, Williams & Wilkins, Philadelphia, PA.

Hwang YT, Liu BY, Coen DM, Hwang CB. 1997. Effects of mutations in the Exo III motif of the herpes simplex virus DNA polymerase gene on enzyme activities, viral replication, and replication fidelity. *J Virol* **71:** 7791–7798.

Illana B, Blanco L, Salas M. 1996. Functional characterization of the genes coding for the terminal protein and DNA polymerase from bacteriophage GA-1. Evidence for a sliding-back mechanism during protein-primed GA-1 DNA replication. *J Mol Biol* **264:** 453–464.

Jiang H, Gomez-Manzano C, Lang FF, Alemany R, Fueyo J. 2009. Oncolytic adenovirus: Preclinical and clinical studies in patients with human malignant gliomas. *Curr Gene Ther* **9:** 422–427.

Joung I, Engler JA. 1992. Mutations in two cysteine-histidine-rich clusters in adenovirus type 2 DNA polymerase affect DNA binding. *J Virol* **66:** 5788–5796.

Joung I, Horwitz MS, Engler JA. 1991. Mutagenesis of conserved region I in the DNA polymerase from human adenovirus serotype 2. *Virology* **184:** 235–241.

Kamtekar S, Berman AJ, Wang J, Lázaro JM, de Vega M, Blanco L, Salas M, Steitz TA. 2004. Insights into strand displacement and processivity from the crystal structure of the protein-primed DNA polymerase of bacteriophage ϕ29. *Mol Cell* **16:** 609–618.

Kamtekar S, Berman AJ, Wang J, Lázaro JM, de Vega M, Blanco L, Salas M, Steitz TA. 2006. The ϕ29 DNA polymerase:protein-primer structure suggests a model for the initiation to elongation transition. *EMBO J* **25:** 1335–1343.

Kanellopoulos PN, Van der Zandt H, Tsernoglou D, Van der Vliet PC, Tucker PA. 1995. Crystallization and preliminary x-ray crystallographic studies on the adenovirus ssDNA binding protein in complex with ssDNA. *J Struct Biol* **115:** 113–116.

Kelly E, Russell SJ. 2007. History of oncolytic viruses: Genesis to genetic engineering. *Mol Ther* **15:** 651–659.

Kinchington PR, Araullo-Cruz T, Vergnes JP, Yates K, Gordon YJ. 2002. Sequence changes in the human adenovirus type 5 DNA polymerase associated with resistance to the broad spectrum antiviral cidofovir. *Antiviral Res* **56:** 73–84.

Kinchington PR, Romanowski EG, Jerold GY. 2005. Prospects for adenovirus antivirals. *J Antimicrob Chemother* **55:** 424–429.

King AJ, Van der Vliet PC. 1994. A precursor terminal protein-trinucleotide intermediate during initiation of adenovirus DNA replication: Regeneration of molecular ends in vitro by a jumping back mechanism. *EMBO J* **13:** 5786–5792.

King AJ, Teertstra WR, Van der Vliet PC. 1997. Dissociation of the protein primer and DNA polymerase after initiation of adenovirus DNA replication. *J Biol Chem* **272:** 24617–24623.

Knopf CW. 1998. Evolution of viral DNA-dependent DNA polymerases. *Virus Genes* **16:** 47–58.

Kojaoghlanian T, Flomenberg P, Horwitz MS. 2003. The impact of adenovirus infection on the immunocompromised host. *Rev Med Virol* **13:** 155–171.

Kühn FJ, Knopf CW. 1996. Herpes simplex virus type 1 DNA polymerase. Mutational analysis of the 3′-5′-exonuclease domain. *J Biol Chem* **271:** 29245–29254.

Leen AM, Bollard CM, Myers GD, Rooney CM. 2006. Adenoviral infections in hematopoietic stem cell transplantation. *Biol Blood Marrow Transplant* **12:** 243–251.

Lenaerts L, De Clercq E, Naesens L. 2008. Clinical features and treatment of adenovirus infections. *Rev Med Virol* **18:** 357–374.

Liu H, Naismith JH, Hay RT. 2000. Identification of conserved residues contributing to the activities of adenovirus DNA polymerase. *J Virol* **74:** 11681–11689.

Liu H, Naismith JH, Hay RT. 2003. Adenovirus DNA replication. *Curr Top Microbiol Immunol* **272:** 131–164.

Liu J, O'Brien KL, Lynch DM, Simmons NL, La PA, Riggs AM, Abbink P, Coffey RT, Grandpre LE, Seaman MS, et al. 2009. Immune control of an SIV challenge by a T-cell-based vaccine in rhesus monkeys. *Nature* **457:** 87–91.

Ljungman P. 2004. Treatment of adenovirus infections in the immunocompromised host. *Eur J Clin Microbiol Infect Dis* **23:** 583–588.

Martín AC, Blanco L, García P, Salas M, Méndez J. 1996. In vitro initiation of pneumococcal phage Cp-1 DNA replication occurs at the third 3′ nucleotide of the linear template: A stepwise sliding-back mechanism. *J Mol Biol* **260:** 369–377.

Méndez J, Blanco L, Esteban JA, Bernad A, Salas M. 1992. Initiation of φ29 DNA replication occurs at the second 3′ nucleotide of the linear template: A sliding-back mechanism for protein-primed DNA replication. *Proc Natl Acad Sci* **89:** 9579–9583.

Morfin F, Dupuis-Girod S, Frobert E, Mundweiler S, Carrington D, Sedlacek P, Bierings M, Cetkovsky P, Kroes AC, Van Tol MJ, et al. 2009. Differential susceptibility of adenovirus clinical isolates to cidofovir and ribavirin is not related to species alone. *Antivir Ther* **14:** 55–61.

Mysiak ME, Bleijenberg MH, Wyman C, Holthuizen PE, Van der Vliet PC. 2004a. Bending of adenovirus origin DNA by nuclear factor I as shown by scanning force microscopy is required for optimal DNA replication. *J Virol* **78:** 1928–1935.

Mysiak ME, Wyman C, Holthuizen PE, Van der Vliet PC. 2004b. NFI and Oct-1 bend the Ad5 origin in the same direction leading to optimal DNA replication. *Nucleic Acids Res* **32:** 6218–6225.

Paolino K, Sande J, Perez E, Loechelt B, Jantausch B, Painter W, Anderson M, Tippin T, Lanier ER, Fry T, et al. 2011. Eradication of disseminated adenovirus infection in a pediatric hematopoietic stem cell transplantation recipient using the novel antiviral agent CMX001. *J Clin Virol* **50:** 167–170.

Polager S, Ginsberg D. 2009. p53 and E2f: Partners in life and death. *Nat Rev Cancer* **9:** 738–748.

Rademaker HJ, Fallaux FJ, Van den Wollenberg DJ, De Jong RN, Van der Vliet PC, Hoeben RC. 2006. Relaxed template specificity in fowl adenovirus 1 DNA replication initiation. *J Gen Virol* **87:** 553–562.

Rodríguez I, Lázaro JM, Blanco L, Kamtekar S, Berman AJ, Wang J, Steitz TA, Salas M, de Vega M. 2005. A specific subdomain in φ29 DNA polymerase confers both processivity and strand-displacement capacity. *Proc Natl Acad Sci* **102:** 6407–6412.

Ruiz J, Beadle JR, Buller RM, Schreiwer J, Prichard MN, Keith KA, Lewis KC, Hostetler KY. 2011. Synthesis, metabolic stability and antiviral evaluation of various alkoxyalkyl esters of cidofovir and 9-(S)-[3-hydroxy-2-(phosphonomethoxy)propyl]adenine. *Bioorg Med Chem* **19:** 2950–2958.

Salas M. 1991. Protein-priming of DNA replication. *Annu Rev Biochem* **60:** 39–71.

Seiler MP, Cerullo V, Lee B. 2007. Immune response to helper dependent adenoviral mediated liver gene therapy: Challenges and prospects. *Curr Gene Ther* **7:** 297–305.

Seto D, Chodosh J, Brister JR, Jones MS. 2011. Using the whole-genome sequence to characterize and name human adenoviruses. *J Virol* **85:** 5701–5702.

Smart JE, Stillman BW. 1982. Adenovirus terminal protein precursor. Partial amino acid sequence and the site of covalent linkage to virus DNA. *J Biol Chem* **257:** 13499–13506.

Soengas MS, Esteban JA, Lázaro JM, Bernad A, Blasco MA, Salas M, Blanco L. 1992. Site-directed mutagenesis at the Exo III motif of ϕ29 DNA polymerase; overlapping structural domains for the 3′–5′ exonuclease and strand-displacement activities. *EMBO J* **11:** 4227–4237.

Symeonidis N, Jakubowski A, Pierre-Louis S, Jaffe D, Pamer E, Sepkowitz K, O'Reilly RJ, Papanicolaou GA. 2007. Invasive adenoviral infections in T-cell-depleted allogeneic hematopoietic stem cell transplantation: High mortality in the era of cidofovir. *Transpl Infect Dis* **9:** 108–113.

Tooze J. 1980. *DNA tumor viruses.* Cold Spring Harbor Laboratory Press, Cold Spring Harbor, NY.

Toth K, Spencer JF, Dhar D, Sagartz JE, Buller RM, Painter GR, Wold WS. 2008. Hexadecyloxypropyl-cidofovir, CMX001, prevents adenovirus-induced mortality in a permissive, immunosuppressed animal model. *Proc Natl Acad Sci* **105:** 7293–7297.

Uil TG, Vellinga J, de Vrij J, Van den Hengel SK, Rabelink MJ, Cramer SJ, Eekels JJ, Ariyurek Y, Van Galen M, Hoeben RC. 2011. Directed adenovirus evolution using engineered mutator viral polymerases. *Nucleic Acids Res* **39:** e30.

Van Breukelen B, Brenkman AB, Holthuizen PE, Van der Vliet PC. 2003. Adenovirus type 5 DNA binding protein stimulates binding of DNA polymerase to the replication origin. *J Virol* **77:** 915–922.

van der Vliet PC. 1996. Roles of transcription factors in DNA replication. In *DNA replication in eukaryotic cells.* (ed. DePamphilis ML), pp. 87–119. Cold Spring Harbor Laboratory Press, Cold Spring Harbor, NY.

van der Vliet PC, Hoeben RC. 2006. Adenovirus. In *DNA replication and human disease* (ed. DePamphilis ML), pp. 645–661. Cold Spring Harbor Laboratory Press, Cold Spring Harbor, NY.

van Leeuwen HC, Rensen M, Van der Vliet PC. 1997. The Oct-1 POU homeodomain stabilizes the adenovirus preinitiation complex via a direct interaction with the priming protein and is displaced when the replication fork passes. *J Biol Chem* **272:** 3398–3405.

van Tol MJ, Claas EC, Heemskerk B, Veltrop-Duits LA, De Brouwer CS, van Vreeswijk T, Sombroek CC, Kroes AC, Beersma MF, de Klerk EP, et al. 2005. Adenovirus infection in children after allogeneic stem cell transplantation: Diagnosis, treatment and immunity. *Bone Marrow Transplant* **35** (Suppl 1)**:** S73–S76.

Veltrop-Duits LA, Heemskerk B, Sombroek CC, van Vreeswijk T, Gubbels S, Toes RE, Melief CJ, Franken KL, Havenga M, van Tol MJ, et al. 2006. Human CD4$^+$ T cells stimulated by conserved adenovirus 5 hexon peptides recognize cells infected with different species of human adenovirus. *Eur J Immunol* **36:** 2410–2423.

Walls T, Shankar AG, Shingadia D. 2003. Adenovirus: An increasingly important pathogen in paediatric bone marrow transplant patients. *Lancet Infect Dis* **3:** 79–86.

Wong SW, Wahl AF, Yuan PM, Arai N, Pearson BE, Arai K, Korn D, Hunkapiller MW, Wang TS. 1988. Human DNA polymerase α gene expression is cell proliferation dependent and its primary structure is similar to both prokaryotic and eukaryotic replicative DNA polymerases. *EMBO J* **7:** 37–47.

Yaniv M. 2009. Small DNA tumour viruses and their contributions to our understanding of transcription control. *Virology* **384:** 369–374.

Herpes Simplex Virus DNA Replication

Sandra K. Weller[1] and Donald M. Coen[2]

[1]Department of Molecular, Microbial and Structural Biology, University of Connecticut Health Center, Farmington, Connecticut 06030-3205

[2]Department of Biological Chemistry and Molecular Pharmacology, Harvard Medical School, Boston, Massachusetts 02115

Correspondence: weller@nso2.uchc.edu

Herpes simplex virus (HSV) encodes seven proteins necessary for viral DNA synthesis—UL9 (origin-binding protein), ICP8 (single-strand DNA [ssDNA]-binding protein), UL30/UL42 (polymerase), and UL5/UL8/UL52 (helicase/primase). It is our intention to provide an up-to-date analysis of our understanding of the structures of these replication proteins and how they function during HSV replication. The potential roles of host repair and recombination proteins will also be discussed.

The Herpesviridae are a large family of double-stranded DNA viruses responsible for many human and veterinary diseases. Although members of this family differ in tissue tropism and many aspects of their interactions with their hosts, the mechanisms by which they replicate their DNA during productive ("lytic") infection are largely conserved. The molecular mechanisms involved in herpesvirus DNA replication and its regulation are of interest as they provide important models for the study of eukaryotic DNA replication. Many of the replication proteins encoded by herpesviruses represent functional analogs of the eukaryotic DNA replication machinery, with informative similarities and differences. In addition, viral enzymes involved in DNA replication have provided a rich store of useful targets for antiviral therapy. This work will focus primarily on DNA replication of herpes simplex virus 1 and 2 (HSV-1 and HSV-2), but will refer, on occasion, to findings from other herpesviruses. Because this work is intended to update the work on HSV DNA replication written for the previous edition of *DNA Replication and Human Disease* (Weller and Coen 2006), it draws primarily from work published during the last 6 years. Topics covered in detail in the previous work by the authors (Weller and Coen 2006) and reviewed elsewhere will be summarized only briefly (see Coen 2009; Weller 2010; Livingston and Kyratsous 2011; Ward and Weller 2011; Weitzman and Weller 2011).

HSV-1 GENOME AND ORIGINS OF DNA REPLICATION

The HSV genome is large (152 kbp) and structurally complex, consisting of two unique regions (UL and US) flanked by inverted repeat

Cite this article as *Cold Spring Harb Perspect Biol* doi: 10.1101/cshperspect.a013011

sequences arranged as ab-UL-b′a′c′-US-ca (reviewed in Ward and Weller 2011, and references therein). The HSV genome contains three origins of replication: OriS is present twice in the viral genome in the repeated c region and OriL is present in UL. Both OriL and OriS are located in the promoter-regulatory regions of divergently transcribed genes: OriL between genes encoding replication proteins, ICP8 (UL29) and the catalytic subunit of the polymerase (Pol) and OriS between genes encoding immediate early proteins ICP4 and either ICP22 or ICP47. OriL and OriS contain an A/T-rich region surrounded by recognition sites for the origin-binding protein (OBP), UL9. OriS contains a 45-bp imperfect palindrome in which the A/T-rich region is flanked by two recognition sites for the OBP (box I and box II) (see Fig. 1 from Ward and Weller 2011). A third weaker binding site (box III) is located adjacent to box I. Box I and box III are located on opposite strands and have been reported to adopt an unusual hairpin structure on single-strand (ssDNA) (OriS*) (Aslani et al. 2002). HSV OriL contains a 144-bp perfect palindrome that includes four recognition sites for UL9 such that hairpins could theoretically form on both sides of the AT-rich region in OriL. A recent report suggests that the varicella zoster virus (VZV) genome also contains two copies of OriS and that, as in HSV-1 and HSV-2, each OriS contains three origin-binding sites (Khalil et al. 2011). The configuration, however, differs in that the OBP binding sites are all on the same strand and oriented in the same direction, suggesting that the hairpin structure that can form in HSV OriS cannot form in the VZV version. Although HSV-1 mutants lacking one or two origins are competent for lytic replication in cell culture (Polvino-Bodnar et al. 1987; Igarashi et al. 1993), the in vivo behavior of mutant viruses containing point mutations in site I of OriL or OriS may indicate that OriL plays a role in in vivo replication or pathogenesis (Balliet and Schaffer 2006). Interestingly it appears that both OriL and OriS encode micro RNAs (miRNAs) (Jurak et al. 2010), so it cannot be excluded that the mutations in OriL that alter pathogenicity exert their influence through these miRNAs.

TRANS-ACTING PROTEINS: ESSENTIAL REPLICATION PROTEINS

HSV encodes seven essential replication proteins (see Table 1). Six of these play "core" replication roles at the replication fork and are conserved in all known herpesviruses: a single-strand DNA-binding protein (known as ICP8 or UL29), a two-subunit DNA polymerase (catalytic subunit Pol and processivity subunit UL42), and a three-subunit helicase/primase complex (H/P:UL5, UL8, and UL52) (reviewed in Ward and Weller 2011). In addition to the core replication proteins, HSV-1 and HSV-2 also encode an origin-binding protein (UL9). The origins of replication in β and γ herpesviruses are more complex, and no clear UL9 homologs have been identified. Thus, although the overall strategy of lytic viral DNA replication appears to be conserved in all herpesviruses, the mechanism and regulation of initiation of viral DNA synthesis may differ among the three herpesvirus subfamilies.

ICP8, THE MAJOR ssDNA-BINDING PROTEIN (UL29)

ICP8, encoded by the UL29 gene, originally identified as the major HSV single-strand DNA-binding protein (SSB) is believed to play a role in viral DNA synthesis, control of viral gene expression, the formation of prereplicative sites and replication compartments, and in the high level of recombination shown during HSV infection (as reviewed in Weller and Coen 2006; Ward and Weller 2011). This multifunctional 130-kDa zinc metalloprotein preferentially binds ssDNA in a nonsequence-specific and cooperative manner and shows helix-destabilizing activities that are consistent with a role in DNA synthesis. ICP8 has been reported to contact many viral and cellular proteins to carry out its multiple functions. Viral interaction partners include UL9, HSV polymerase, helicase/primase (H/P), UL12, ICP4, and ICP27 (reviewed in Chattopadhyay et al. 2006; Weller and Coen 2006; Ward and Weller 2011). Many of these interactions result in stimulation of activities reviewed below. Putative cellular interaction partners are discussed below.

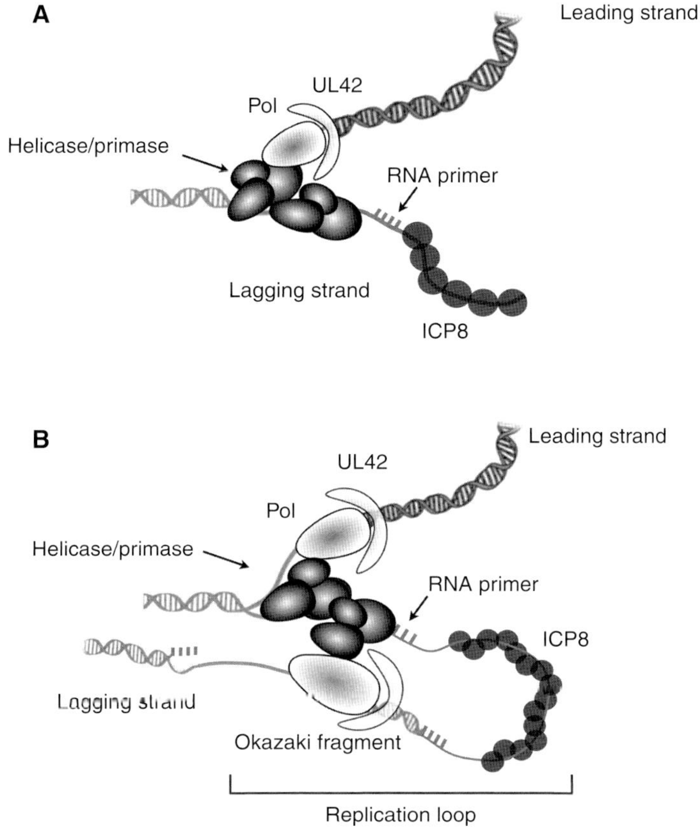

Figure 1. Organization of the HSV-1 replication fork. (*A*) Herpes simplex virus (HSV) replication fork. Two heterotrimeric complexes of the helicase-primase (H/P) (dark gray spheres with white centers) are shown at the replication fork. This model is based on a recent report that to achieve efficient primase activity at a forked substrate, two (or more) heterotrimers of the helicase-primase complex are required (Chen et al. 2011). In this diagram, a duplex DNA molecule is unwound by one molecule of H/P. The HSV polymerase (white shapes with gray centers in which UL30 is shown as an oval and UL42 as a crescent) performs leading-strand DNA synthesis on the top strand. On the bottom strand the second molecule of H/P is shown laying down an RNA primer depicted as hatched bars. The single strand DNA (ssDNA) extruded behind the helicase is coated by the ssDNA-binding protein ICP8 (uniformly dark gray circles). (*B*) Coupled leading- and lagging-strand synthesis. Coupled leading- and lagging-strand DNA synthesis was recently reconstituted in vitro and requires HSV polymerase, helicase-primase, and ICP8 (Stengel and Kuchta 2011). In this model, coupled leading- and lagging-strand DNA synthesis requires two heterotrimers of helicase/primase and two molecules of the DNA polymerase. The leading- and lagging-strand polymerase molecules are proposed to communicate with each other mediated by interactions between UL30 and the UL5 and UL8 subunits of the H/P (reviewed in the text). A replication loop is formed in the lagging strand to align it with the leading strand. The lagging-strand DNA polymerase initiates Okazaki fragment synthesis using RNA primers (hatched bars).

The structure of ICP8 lacking its carboxy-terminal 60 residues was published in 2005 (Mapelli et al. 2005) and details were reviewed in Weller and Coen (2006) and Ward and Weller (2011). Purified ICP8 forms dimers and higher-order filaments in vitro and when bound to ssDNA holds the DNA in an extended conformation (Mapelli et al. 2000, 2005; Makhov et al. 2009). Recent three-dimensional (3D) reconstruction of an ICP8-ssDNA complex sheds light on the role of ICP8 as a single-strand annealing protein (SSAP) (G Tolun, AM Makhov,

SJ Ludtke, et al., in prep.). Fitting the crystal structure into this 3D reconstruction allows predictions concerning the mechanism of ICP8-driven annealing.

Pol/UL42, the DNA Polymerase

The crystal structure of HSV Pol reveals that this enzyme contains structural domains typical of most polymerases including thumb, palm, and fingers, along with a 3′–5′ exonuclease domain similar to those of other α-like DNA polymerases. However, the structure revealed two other domains: a pre-NH_2-terminal domain (1–140) and an NH_2-terminal domain defined by Liu et al. (2006) as residues 141–362 and 594–639. The pre-NH_2-terminal domain is conserved among herpesvirus, but not other α-like DNA polymerases. A conserved motif in this domain is not important for DNA polymerase activity, but is important for viral DNA synthesis in infected cells (S Terrell and DM Coen, submitted). The NH_2-terminal domain contains some structures and motifs found in herpesvirus and certain other α-like DNA polymerases. The more amino-terminal segment of this domain resembles certain RNA-binding proteins, leading to the revival of the idea that HSV Pol contains a 5′–3′ RNase H activity important for removal of primers (Liu et al. 2006). On the other hand, although HSV Pol has RNase H activity, it is not clear if this activity is separate from the 3′–5′ exonuclease activity (reviewed in Coen 1996). Additionally, the cellular protein, flap endonuclease-1 (Fen-1) can cooperate with HSV Pol to remove primers from model lagging-strand intermediates in vitro, whereas these primers were not removed by Pol alone (Zhu et al. 2010). An alternative role for the amino-terminal domain—that it has a role in displacement of RNA primers for degradation by Fen-1—was proposed in a paper describing the structure of yeast DNA polymerase δ (Swan et al. 2009).

Unexpectedly, HSV Pol contains apurinic/apyrmidinic and 5′-deoxyribose phosphate lyase activities, normally associated with repair polymerases (Bogani and Boehmer 2008). The DNA lyase activity active site appears to reside somewhere in the thumb, palm, or fingers domains (Bogani and Boehmer 2008). HSV Pol also binds to the HSV uracil-DNA-glycosylase, UL2 (Bogani et al. 2010). UL2 binding does not require the extreme carboxyl terminus of Pol (Bogani et al. 2010). A similar result was obtained with the human cytomegalovirus (HCMV) homologs of these proteins (Strang et al. 2010a). UL2 is not required for viral replication (Mullaney et al. 1989; Pyles and Thompson 1994) raising the possibility that this protein–protein interaction is important for coordinating DNA synthesis and repair.

HSV Pol has also been reported to interact with cellular proteins. Pol appears to be a “client” of heat shock protein (Hsp) 90 (Burch and Weller 2005). Pol can interact with the transcriptional coactivator HCF-1 in yeast two-hybrid and transfection-coimmunoprecipitation assays. HCF-1 can interact with the histone chaperone Asf1b in similar assays, and Asf1b is important for viral DNA replication (Peng et al. 2010). Whether these interactions occur during HSV infection remains under investigation.

As previously reviewed (Weller and Coen 2006), UL42 closely resembles a monomer of the “sliding clamp” processivity factor, PCNA. Interestingly, the related proteins of β and γ herpesviruses form head-to-head dimers that partially encircle DNA (Appleton et al. 2004, 2006; Komazin-Meredith et al. 2008b; Baltz et al. 2009; Murayama et al. 2009). One surface of UL42 binds to Pol, whereas the other surface mediates DNA binding (Randell et al. 2005; Komazin-Meredith et al. 2008c) and is important for viral replication, DNA synthesis, and replication fidelity (Jiang et al. 2007a,b). Too tight binding to DNA reduces viral replication and replication fidelity (Jiang et al. 2009). UL42 can diffuse rapidly on DNA, with diffusion coefficients approaching 2×10^5 bp^2/sec (Komazin-Meredith et al. 2008a). Remarkably, unlike PCNA (Kochaniak et al. 2009), UL42 does not “slide” on DNA. Rather, it hops; i.e., it microscopically dissociates from DNA and then reassociates with either the same stretch of DNA or an adjacent stretch (Komazin-Meredith et al. 2008a). Thus, UL42 promotes processivity by tethering UL30 to DNA through direct DNA binding, and diffuses on DNA behind UL30 as it synthesizes long DNA chains.

Like PCNA (Maga and Hubscher 2003; Moldovan et al. 2007), herpesvirus processivity subunits interact with multiple proteins. Aside from previously described interactions (reviewed in Weller and Coen 2006), UL42 coimmunoprecipitates with Fen-1 from HSV-infected cells (Zhu et al. 2010). The HCMV homolog of UL42 (UL44) coimmunoprecipitates with a variety of viral and cellular proteins, including, surprisingly, the nucleolar protein nucleolin, which was found to be important for maintenance of replication compartment architecture and viral DNA synthesis (Gao et al. 2008; Kim and Ahn 2010; Strang and Coen 2010; Strang et al. 2010a,b, 2012).

HELICASE/PRIMASE (UL5/UL8/UL52)

The HSV-1 helicase/primase (H/P) is a heterotrimer consisting of the products of the UL5, UL8, and UL52 genes (reviewed in Chattopadhyay et al. 2006; Ward and Weller 2011). A subcomplex of UL5 and UL52 shows DNA-dependent ATPase, primase, and helicase activities, and UL8 interacts with other components of the replication machinery possibly functioning to coordinate replication fork progress. The relationship between UL5 and UL52 reflects a complex interdependence between these two subunits. For instance, mutations in the putative zinc finger at the carboxyl terminus of UL52 abrogate not only primase but also ATPase, helicase, and DNA-binding activities of the UL5/UL52 subcomplex (Chen et al. 2005; Cavanaugh and Kuchta 2009; Cavanaugh et al. 2009). Unlike all other replicative helicases that form hexameric rings around ssDNA, the HSV H/P is unlikely to form a hexamer. UL5 is a member of the SF1 helicase superfamily generally thought to exist as monomers in solution. The HSV H/P behaves as a single heterotrimer by gel filtration (Dodson and Lehman 1991); however, as described below, two or more H/P heterotrimers are needed on partially double-stranded DNA substrates to observe efficient primase activity (Fig. 1A) (Chen et al. 2011). This apparent DNA-dependent multimerization is reminiscent of the behavior of other SF1 family helicases, which show unwinding of double-stranded DNA (dsDNA) only if at least two helicase polypeptides are present (Maluf et al. 2003; Niedziela-Majka et al. 2007; Yang et al. 2008). Direct protein–protein interactions between the helicase protomers appear unnecessary to achieve enhanced unwinding. By analogy with other SF1 family members, it is possible that two HSV heterotrimeric complexes bind at the replication fork even if there is no direct contact between them and that the presence of two or more complexes enhances the processivity of the H/P complex.

UL9, OBP

UL9, the HSV OBP, forms a dimer in solution and shows several biochemical activities including nucleoside triphosphatase, DNA helicase on partially double-stranded substrates, nonspecific ssDNA binding, and cooperative binding at an origin of replication reviewed in Ward and Weller (2011). Seven conserved helicase motifs characteristic of the SF2 family of helicases reside in the amino-terminal domain of UL9 (between residues 1–534) and are essential both in vivo and for the in vitro ATPase and helicase activities reviewed in Ward and Weller (2011). The domain responsible for specific origin binding (at repeated GTTCGCAC sites) lies within the carboxyl terminus of UL9 (residues 564–832). More specifically, a conserved sequence element in UL9, RVKNL (amino acids 756–760), was recently shown to be required for origin-specific DNA binding (Olsson et al. 2009). Interestingly, residues in other regions of the UL9 protein are capable of modulating origin-specific DNA-binding activity of UL9 (Chattopadhyay and Weller 2006, 2007; Olsson et al. 2009). Although the carboxyl terminus of UL9 can bind origin DNA with a similar footprint to that of full-length UL9, it cannot induce the A/T-rich distorted structure seen with full-length UL9 suggesting that sequences important for cooperativity reside in the amino terminus (reviewed in Ward and Weller 2011). Based on these observations, it had been assumed that the amino-terminal 534 residues contain a region necessary for efficient dimerization and that UL9 dimers adopt a head-to-head configuration. However, UL9 dimers may be oriented in a head-to-tail arrangement in which the amino terminus is in contact with the carboxyl terminus (Chattopadhyay and Weller 2007). Electron microscopy (EM) studies of UL9 bound to oriS suggest the existence of a pair of dimers assembled on oriS (Makhov et al. 1996). It has been proposed that one subunit of the dimer contacts a pentanucleotide repeat of the UL9 recognition site in the major groove of one strand, whereas the other subunit contacts the partially overlapping pentanucleotide site on the other strand (Aslani et al. 2002; Macao et al. 2004).

In addition to binding and unwinding duplex DNA at the origins of replication, UL9 has been reported to interact with other viral and cellular proteins. For instance, UL9 has been reported to interact with ICP8, UL8, and UL42 (reviewed in Ward and Weller 2011). ICP8 has been reported to stimulate the helicase and ATPase activities of UL9, and a 13-amino acid sequence (WPXXXGAXXFXXL) in the extreme carboxyl terminus of UL9 has been shown to be responsible for this interaction (Olsson et al. 2009). UL42 has also been reported to enhance the 3′-to-5′ helicase activity of UL9; however, the functional significance of the UL9–UL8 interaction has not been explored. In addition, although several cellular proteins have been reported to interact with UL9 including the large subunit of DNA polymerase α, human neural F-box 42-kDa protein (NFB42 or FBX2), and some heat shock proteins (reviewed in Ward and Weller 2011), it is not clear whether these interactions are required in the context of HSV infection.

Auxiliary Factors Not Essential for Viral DNA Replication

The HSV-1 genome encodes at least six proteins that play a nonessential auxiliary role in viral DNA synthesis (Table 1). These include several enzymes involved in nucleotide biosynthesis and DNA metabolism: thymidine kinase, ribonucleotide reductase, deoxyuridine triphosphatase, and uracil-DNA-glycolsylase (Table 1). Thymidine kinase is well conserved in the α and γ herpesviruses and functions to phosphorylate thymidine and other nucleosides including antiviral compounds such as acyclovir. Viral thymidine kinase (TK) is not required for viral growth in cell culture or in certain tissues in

Table 1. HSV DNA replication genes

Gene	Alternate abbreviation	Major function	Essential for DNA synthesis in cultured cells
UL9	OBP	Origin-binding protein, helicase	Yes
UL30	Pol	Catalytic subunit of DNA polymerase	Yes
UL42		Processivity subunit of DNA polymerase	Yes
UL5		Subunit of helicase/primase; contains helicase motifs	Yes
UL8		Subunit of helicase/primase; interacts with other proteins	Yes
UL52		Subunit of helicase/primase; contains primase motifs	Yes
UL29	ICP8	Single-strand DNA-binding protein Single-strand annealing protein	Yes
UL23	TK	Thymidine kinase	No
UL39	RR1, ICP6	Large subunit of ribonucleotide reductase	No
UL40	RR2	Small subunit of ribonucleotide reductase	No
UL2	UNG	Uracil-DNA glycosylase	No
UL50	dUTPase	Deoxyuridine triphosphatase	No
UL12	5′–3′ Exo	Alkaline nuclease; interacts with ICP8 Viral recombinase	No

vivo. Ribonucleotide reductase functions to generate deoxyribonucleotides important for DNA synthesis and consists of two subunits, RR1 and RR2, both of which are needed for enzymatic activity. In addition to the established role for RR1 as the catalytic subunit of ribonucleotide reductase responsible for generating deoxyribonucleotides, the RR1 subunit (also known as ICP6) has been shown to show other activities. For instance the RR1 subunit from both HSV-1 and HSV-2 has been shown to protect cells against apoptosis by interacting with caspase 8 receptor-interacting protein 1 (RIP1) thus interfering with apoptotic host defense mechanisms (Chabaud et al. 2007; Dufour et al. 2010). Moreover, RR1 appears to show chaperone activity similar to that of small heat shock proteins (Chabaud et al. 2003). Lastly, HSV-1 RR1 promotes the assembly of the translational initiation machinery suggesting that ICP6 can function as an eIF4F-assembly chaperone (Walsh and Mohr 2006). These observations underscore the complex evolution of viral proteins such that they are capable of multiple roles during infection.

Other auxiliary functions encoded by HSV-1 include the deoxyuridine triphosphatase (dUTPase) and uracil-DNA-glycosylase (UL2), which are not required for viral growth in cell culture but appear to show more severe defects in nondividing cells and in vivo (reviewed in Ward and Weller 2011). The recently reported interaction between the HSV-1 UL2 and Pol described above may indicate a possible role for base excision repair during DNA replication (Bogani et al. 2010).

The HSV alkaline nuclease (UL12) is unusual in that it is not required for viral DNA synthesis per se but is essential for the production of infectious virus. Thus, although newly synthesized viral DNA can be detected in cells infected with HSV-1 mutants defective in UL12, its structure is aberrant leading to defects in packaging. UL12 is well conserved in all herpesvirus genomes and evolutionarily related to a larger family of $5'-3'$ exonucleases that play important roles in DNA viruses of bacterial, fungi, mammals, and insects. UL12 interacts with HSV ICP8 as well as several components of the host DNA repair machinery (Balasubramanian et al. 2010). Possible roles of UL12 in the virus life cycle will be discussed below.

OVERALL MODEL FOR HSV DNA REPLICATION

Initiation of Viral DNA Synthesis

HSV DNA synthesis is believed to initiate at one of the three viral origins of replication, OriL or one of the two copies of OriS with UL9 and ICP8 acting to distort the AT-rich origin spacer region. Complexes containing ICP8 and UL9 can be visualized on negatively supercoiled plasmids containing OriS (Makhov et al. 2003) suggesting that negative supercoiling may facilitate unwinding at the origin. In another in vitro experiment, UL9 and ICP8 were found to unwind an 80-bp double-stranded oligonucleotide containing minimal OriS leading to a model in which UL9 and ICP8 go through a series of conformational changes leading to distortion of an origin. In the first step, two UL9 dimers appear to bind cooperatively to boxes I and II; in a second step, UL9, in conjunction with ICP8 in the presence of ATP, can induce the formation of the OriS* hairpin (Aslani et al. 2002; Macao et al. 2004; Olsson et al. 2009). This step is accompanied by conformational changes in UL9 that shift the binding specificity from duplex origin binding to nonspecific ssDNA binding (Macao et al. 2004; Olsson et al. 2009). Because origin binding is specified by the carboxyl terminus and the ssDNA-binding domain requires motif Ia within the amino terminus (Marintcheva and Weller 2003b), the conformational changes in UL9 would be major. The exposure of single-stranded DNA may also cause a conformational change in ICP8 that releases it from UL9 and positions it onto single-stranded DNA where it may act to prevent reannealing of complementary strands. Aspects of this model have been supported by a biophysical analysis of the carboxy-terminal domain of UL9 and a truncated version of ICP8 in complex with a 15-mer double-stranded DNA-containing box I (Manolaridis et al. 2009). This study suggests that a conformational switch of the

UL9-binding domain occurs on binding to box I followed by the recruitment of a UL9–ICP8 complex. Several questions remain about how well these in vitro experiments mimic the conditions that arise in an infected cell. Other viral or cellular factors may be required for origin activation on the viral genome in the context of viral infection.

Experiments in infected cells also support the notion that conformational changes in UL9 play critical roles during infection. For instance, UL9 is required at the earliest times postinfection, but appears not to be required at late times. In fact, at late times, UL9 is inhibitory (reviewed in Ward and Weller 2011). The inhibitory properties of UL9 correlate at least in part with its ability to bind to the origins of replication, because inhibition can be relieved by mutations that abrogate DNA binding (Marintcheva and Weller 2003a; Chattopadhyay and Weller 2006). Binding of UL9 to the origin of replication is presumably desirable to initiate DNA synthesis but less desirable at later times. Conformational changes in UL9 may be caused by interaction with viral proteins such as ICP8 as described above. Interestingly, other posttranslational modifications of UL9 have also been reported such as phosphorylation and cleavage by cathepsins (Isler and Schaffer 2001; Link et al. 2007). It will be of interest to analyze structural and conformational properties of full-length and truncated versions of UL9 in infected cells to determine the mechanistic basis for the regulation of UL9 during infection.

Elongation

Once ICP8 and UL9 have initiated the distortion or destabilization at a viral origin as described above, it is thought that the H/P complex is recruited to unwind the duplex DNA and synthesize short RNA primers to initiate DNA replication (Fig. 1A). Aspects of this reaction can be reconstituted in vitro. Recombinant H/P complex can unwind duplex DNA but only if a single-stranded region of at least six nucleotides is provided (Chen et al. 2011). Primase activity can be detected in assays that use a single-strand oligonucleotide as a substrate; however, primase activity is inefficient in assays using forked substrates (KL Graves-Woodward and SK Weller, unpubl.; KA Ramirez-Aguilar and RD Kuchta, unpubl.). A possible reason for this discrepancy has recently been revealed. Although it has previously been reported that in the absence of DNA, the H/P complex exists as a monomer in solution, electrophoretic shift and surface plasmon resonance analysis have now indicated that in the presence of forked DNA, higher-order complexes can form between the H/P and a forked substrate (Chen et al. 2011). Electrophoretic mobility shift assays reveal two discrete complexes with different mobilities only when H/P is bound to DNA containing a single-stranded region, and surface plasmon resonance analysis confirms larger amounts of the complex bound to forked substrates than to single-overhang substrates. In addition, primase activity shows a cooperative dependence on protein concentration, whereas ATPase and helicase activities do not (Chen et al. 2011). Taken together, these data suggest that the primase activity of the H/P requires formation of a dimer or higher-order structure, whereas ATPase activity does not. As depicted in Figure 1A, the functional form of the H/P complex likely contains at least two copies of the H/P complex.

The last of the herpes replication proteins recruited to the fork appears to be the two-subunit polymerase, Pol and UL42. In infected cells, recruitment of Pol requires the presence of an active primase, indicating that conformational changes that occur in an active H/P complex and/or the RNA primer itself are important to bring the polymerase to viral DNA (Carrington-Lawrence and Weller 2003). Recruitment of polymerase to the fork may involve direct interactions between Pol and the H/P. The UL8 subunit of H/P has been reported to interact with Pol (Marsden et al. 1996). In addition, the UL5 subunit has recently been found to interact with Pol based on coimmunoprecipitation and yeast two-hybrid analysis (P Bai, G Liu, J Liu, et al., in prep.). Once the polymerase complex is recruited to the replication fork, it is believed to catalyze leading- and lagging-strand DNA synthesis; however, initial attempts to reconstitute HSV DNA synthesis on primed substrates showed efficient leading-strand synthesis but much less efficient lagging-strand synthesis (Graves-Woodward et al. 1997; Falkenberg et al. 2000). Coupled leading- and lagging-strand DNA synthesis has now for the first time been reconstituted on synthetic minicircular DNA templates with the H/P, ICP8, and the viral polymerase (Pol/UL42) (Stengel and Kuchta 2011). To achieve efficient leading- and lagging-strand synthesis, it was necessary to provide high H/P concentrations and a lagging-strand template whose sequence resembled that of the viral DNA (Stengel and Kuchta 2011). The presence of two H/P complexes at the replication fork may provide a simple mechanism for recruiting two polymerases to the replication fork (Fig. 1B). It is anticipated that this system will facilitate future experiments to examine how other viral and cellular proteins will affect DNA synthesis. It is hoped that eventually it will be possible to develop a system to reconstitute origin-dependent replication.

Formation of Concatemeric DNA

Production of HSV concatemeric DNA is an essential step for the generation of progeny virus as the packaging machinery must recognize longer-than-unit-length concatemers during encapsidation. Although it has been proposed that the viral genome circularizes and rolling circle replication leads to the formation of concatemers, several lines of evidence suggest that HSV DNA replication is more complex and may involve recombination-dependent replication.

By analogy with λ phage, HSV may use viral and/or cellular recombination proteins during DNA replication. Herpesviruses have evolved a complex relationship with host DNA damage response pathways (reviewed in Weitzman and Weller 2011). Several cellular factors involved in double-strand-break (DSB) repair including Mre11, Rad50, Nbs1, and Rad51 are recruited to viral prereplicative sites and replication compartments. Several of these are important for efficient virus production leading to the suggestion that one or more of the DSB repair pathways may be used during HSV infection. Using chromosomally integrated reporter assays designed to distinguish between these pathways, it was found that single-strand annealing (SSA) was increased in HSV-infected cells, whereas homologous recombination (HR), nonhomologous end joining (NHEJ), and alternative nonhomologous end joining (A-NHEJ) were decreased (Schumacher et al. 2012). The increase in SSA was abolished when cells were infected with a viral mutant lacking UL12. Moreover, expression of UL12 alone caused an increase in SSA. UL12 and ICP8 are reminiscent of the λ phage recombination system that has been used as a tool for stimulating recombination-mediated genetic engineering in bacteria (Szczepanska 2009, and references therein). In addition, this system plays an important role in the production of viral DNA concatemers necessary for encapsidation and the production of infectious progeny (Lo Piano et al. 2011). The similarities between λ and HSV DNA replication raise the possibility that concatemer formation during HSV infection involves recombination-dependent replication.

In addition to the UL12/ICP8 viral recombination system and the potential role of the SSA pathway during DNA replication, several other repair pathways have been implicated as required for efficient replication (Muylaert and Elias 2007, 2010; Muylaert et al. 2011). DNA ligase IV/XRCC4 have been reported to be important for efficient virus replication (Muylaert and Elias 2007) suggesting a beneficial role for NHEJ; however, in cells deficient for DNA-PK or Ku70, HSV-1 grows better than in their presence suggesting this conclusion may be oversimplified (Parkinson et al. 1999; K Mohni and S Weller, unpubl.). As described earlier, the interaction between the HSV-1 polymerase and HSV uracil-DNA-glycosylase (UL2) may indicate a role for base excision repair (Bogani and Boehmer 2008; Bogani et al. 2009, 2010). Recent reports that MSH2 and MLH1 are required for efficient replication of HSV-1 in normal human cells and are localized to viral replication compartments may indicate a role for mismatch repair in HSV replication (Mohni et al. 2011). Further investigation of the complex relationship between HSV and host repair/recombination pathways will be required to elucidate the role of various repair pathways in HSV DNA replication.

FORMATION OF REPLICATION COMPARTMENTS, SITES OF VIRAL DNA SYNTHESIS

HSV infection results in the dramatic reorganization of the infected cell nucleus involving relocalization of cellular proteins and the ordered assembly of replication compartments, large globular domains within the nucleus of infected cells in which gene expression, DNA replication, and cleavage and packaging are thought to occur (reviewed in Weller 2010; Ward and Weller 2011). Following entry of viral genomes into the nucleus, nucleoprotein complexes containing viral immediate early proteins ICP4 and ICP27 can be detected followed by the recruitment of ND10 proteins such as PML to sites adjacent to the ICP4/ICP27 nucleoprotein complexes (Everett et al. 2004; Everett and Murray 2005; Livingston et al. 2008). The recruitment of ND10 proteins to incoming viral genomes is thought to be repressive to viral gene expression, and these newly formed ND10-like foci are disrupted during lytic infection by a mechanism mediated by the HSV immediate early protein ICP0. The earliest stage prereplicative sites (ICP8 microfoci) contain ICP8, H/P, and UL9 and their formation requires the oligomerization of the immediate early protein ICP4 on viral genomes, the disruption of ND10-like structures, and ICP8 itself (Livingston et al. 2008, and other references). Recruitment of HSV Pol (UL30/42) requires the presence of an active primase (Carrington-Lawrence and Weller 2003), and once this occurs, small replication compartments form, at which time ICP8 foci appear to merge with ICP4/27 foci (Livingston et al. 2008). Recent studies that monitor the location of incoming genomes have indicated that each RC forms from a single genome and that only a limited number of incoming genomes are actively expressed and go on to be replicated (Kobiler et al. 2011). Interestingly, the separation between different RCs is maintained even when the nucleus is filled with viral DNA. Small RCs are known to grow in size, move, and coalesce. Chang et al. (2011) have recently reported that the movement of small RCs occurs by directed motion requiring actin and myosin. The late RCs appear to represent transcriptionally active RCs that have merged with nuclear speckles that promote export of late viral messenger RNAs (mRNAs).

Efforts to Combat Viral Infections by Targeting DNA Replication

Most of the antiviral agents developed to combat herpesvirus infections have targeted the viral polymerase. The mechanisms of action and resistance of these drugs, the most successful of which is acyclovir, have been reviewed (Weller and Coen 2006; Coen 2009). A new insight into the mechanism of the antiherpesvirus drug, foscarnet, has come from a crystallographic study suggesting that foscarnet stabilizes the untranslocated state of the polymerase, thereby stalling the enzyme (Zahn et al. 2011).

Despite its tremendous success, acyclovir has several drawbacks, including limited potency and efficacy, particularly against herpesviruses other than HSV and VZV, and the frequency at which immunocompromised patients develop resistant infections. This has led to the development of several new compounds, some of which have entered clinical trials. These new compounds include the nucleoside analogs valomaciclovir, a prodrug of a compound with a mechanism of action similar to acyclovir; FV-100, which is active against VZV, but may not be a polymerase inhibitor; CMX001, an orally available version of the approved anti-HCMV drug cidofovir; and cyclopropavir, which is active against HCMV and other β herpesviruses (reviewed in Price and Prichard 2011). Development has also continued on compounds that target the UL5/UL8/UL52 H/P, which were reviewed previously (Weller and Coen 2006). In particular, AIC316 has completed a phase II trial. An interesting antiviral agent, maribavir, inhibits the HCMV protein kinase, UL97, which is important for viral DNA synthesis, particularly in nondividing cells, because it phosphorylates retinoblastoma protein family members (Wolf et al. 2001; Biron et al. 2002; Hume et al. 2008; Prichard et al. 2008; Kamil et al. 2009). This drug showed promise in phase II studies, but failed in phase III trials against HCMV infections in transplant patients (Marty et al. 2011). Thus, there is reason for optimism but also for caution regarding the prospects of these newer DNA replication inhibitors.

CONCLUDING REMARKS

In summary, HSV DNA replication is a complex process involving at least seven viral DNA replication proteins and possibly several host proteins as well. A major impediment to deeper understanding remains the lack of an in vitro system that reconstitutes origin-dependent replication. More incremental goals for the future are to determine the role of UL9 and other viral and cellular proteins in the initiation of replication, understand the mechanism of replication and the role of repair and recombination in the generation of replication products, and continue probing the structure, function, and functional interactions of the viral replication proteins. It is hoped that the results of these efforts will help us develop more effective strategies for antiviral therapy to prevent and treat all human herpesviruses.

ACKNOWLEDGMENTS

We thank members of our laboratories for helpful suggestions, especially Samantha Marques for help with Figure 1. This work is supported by National Institutes of Health (NIH) grants AI21747 and AI069136 to S.K.W., Connecticut stem cell research grant 09-SCA16 UCHC-34 to April Schumacher in the Weller laboratory, and NIH grants AI19838, AI26077, and AI26126 to D.M.C.

REFERENCES

Appleton BA, Loregian A, Filman DJ, Coen DM, Hogle JM. 2004. The cytomegalovirus DNA polymerase subunit UL44 forms a C clamp-shaped dimer. *Mol Cell* **15:** 233–244.

Appleton BA, Brooks J, Loregian A, Filman DJ, Coen DM, Hogle JM. 2006. The crystal structure of the cytomegalovirus DNA polymerase subunit with the C-terminus from the catalytic subunit: Differences in structure and function relative to unliganded UL44. *J Biol Chem* **281:** 5224–5232.

Aslani A, Olsson M, Elias P. 2002. ATP-dependent unwinding of a minimal origin of DNA replication by the origin-binding protein and the single-strand DNA-binding protein ICP8 from herpes simplex virus type I. *J Biol Chem* **277:** 41204–41212.

Balasubramanian N, Bai P, Buchek G, Korza G, Weller SK. 2010. Physical interaction between the herpes simplex virus type 1 exonuclease, UL12, and the DNA double-strand break-sensing MRN complex. *J Virol* **84:** 12504–12514.

Balliet JW, Schaffer PA. 2006. Point mutations in herpes simplex virus type 1 oriL, but not in oriS, reduce pathogenesis during acute infection of mice and impair reactivation from latency. *J Virol* **80:** 440–450.

Baltz JL, Filman DJ, Ciustea M, Silverman JEY, Lautenschlager CL, Coen DM, Ricciardi RP, Hogle JM. 2009. The crystal structure of PF-8, the DNA polymerase accessory subunit from Kaposi's sarcoma-associated herpesvirus. *J Virol* **83:** 12215–12228.

Biron KK, Harvey RJ, Chamberlain SC, Good SS, Smith AA, Davis MG, Talarico CL, Ferris R, Dornsife RE, Stanat SC, et al. 2002. Potent and selective inhibition of human cytomegalovirus replication by 1263W94, a benzimidazole L-riboside with a unique mode of action. *Antimicrob Agents Chemother* **46:** 2365–2372.

Bogani F, Boehmer PE. 2008. The replicative DNA polymerase of herpes simplex virus 1 exhibits apurinic/apyrimidinic and 5′-deoxyribose phosphate lyase activities. *Proc Natl Acad Sci* **105:** 11709–11714.

Bogani F, Chua CN, Boehmer PE. 2009. Reconstitution of uracil DNA glycosylase-initiated base excision repair in herpes simplex virus-1. *J Biol Chem* **284:** 16784–16790.

Bogani F, Corredeira I, Fernandez V, Sattler U, Rutvisuttinunt W, Defais M, Boehmer PE. 2010. Association between the herpes simplex virus-1 DNA polymerase and uracil DNA glycosylase. *J Biol Chem* **285:** 27664–27672.

Burch AD, Weller SK. 2005. Herpes simplex virus type 1 DNA polymerase requires the mammalian chaperone hsp90 for proper localization to the nucleus. *J Virol* **79:** 10740–10749.

Carrington-Lawrence SD, Weller SK. 2003. Recruitment of polymerase to herpes simplex virus type 1 replication foci in cells expressing mutant primase (UL52) proteins. *J Virol* **77:** 4237–4247.

Cavanaugh NA, Kuchta RD. 2009. Initiation of new DNA strands by the herpes simplex virus-1 primase-helicase complex and either herpes DNA polymerase or human DNA polymerase α. *J Biol Chem* **284:** 1523–1532.

Cavanaugh NA, Ramirez-Aguilar KA, Urban M, Kuchta RD. 2009. Herpes simplex virus-1 helicase-primase: Roles of each subunit in DNA binding and phosphodiester bond formation. *Biochemistry* **48:** 10199–10207.

Chabaud S, Lambert H, Sasseville AM-J, Lavoie H, Guilbault C, Massie B, Landry J, Langelier Y. 2003. The R1 subunit of herpes simplex virus ribonucleotide reductase has chaperone-like activity similar to Hsp27. *FEBS Lett* **545:** 213–218.

Chabaud S, Sasseville AM-J, Elahi SM, Caron A, Dufour F, Massie B, Langelier Y. 2007. The ribonucleotide reductase domain of the R1 subunit of herpes simplex virus type 2 ribonucleotide reductase is essential for R1 antiapoptotic function. *J Gen Virol* **88:** 384–394.

Chang L, Godinez WJ, Kim IH, Tektonidis M, de Lanerolle P, Eils R, Rohr K, Knipe DM. 2011. Herpesviral replication compartments move and coalesce at nuclear speckles to enhance export of viral late mRNA. *Proc Natl Acad Sci* **108:** E136–E144.

Chattopadhyay S, Weller SK. 2006. DNA binding activity of the herpes simplex virus type 1 origin binding protein, UL9, can be modulated by sequences in the N terminus: Correlation between transdominance and DNA binding. *J Virol* **80:** 4491–4500.

Chattopadhyay S, Weller SK. 2007. Direct interaction between the N- and C-terminal portions of the herpes simplex virus type 1 origin binding protein UL9 implies the formation of a head-to-tail dimer. *J Virol* **81:** 13659–13667.

Chattopadhyay S, Chen Y, Weller SK. 2006. The two helicases of herpes simplex virus type 1 (HSV-1). *Front Biosci* **11:** 2213–2223.

Chen Y, Carrington-Lawrence SD, Bai P, Weller SK. 2005. Mutations in the putative zinc-binding motif of UL52 demonstrate a complex interdependence between the UL5 and UL52 subunits of the human herpes simplex virus type 1 helicase/primase complex. *J Virol* **79:** 9088–9096.

Chen Y, Bai P, Mackay S, Korza G, Carson JH, Kuchta RD, Weller SK. 2011. Herpes simplex virus type 1 helicase-primase: DNA binding and consequent protein oligomerization and primase activation. *J Virol* **85:** 968–978.

Coen DM. 1996. Viral DNA polymerases. In *DNA replication in eukaryotic cells* (ed. DePamphilis ML), pp. 495–523. Cold Spring Harbor Laboratory Press, Cold Spring Harbor, NY.

Coen DM. 2009. Antiherpesviral DNA polymerase inhibitors. In *Antiviral research: Strategies in antiviral drug discovery* (ed. LaFemina R), pp. 1–18. ASM Press, Washington, DC.

Dodson MS, Lehman IR. 1991. Association of DNA helicase and primase activities with a subassembly of the herpes simplex virus 1 helicase-primase composed of the UL5 and UL52 gene products. *Proc Natl Acad Sci* **88:** 1105–1109.

Dufour F, Sasseville A, Chabaud S. 2010. The ribonucleotide reductase R1 subunits of herpes simplex virus types 1 and 2 protect cells against TNFα-and FasL-induced apoptosis by interacting with caspase-8. *Apoptosis* **16:** 256–271.

Everett RD, Murray J. 2005. ND10 components relocate to sites associated with herpes simplex virus type 1 nucleoprotein complexes during virus infection. *J Virol* **79:** 5078–5089.

Everett RD, Sourvinos G, Leiper C, Clements JB, Orr A. 2004. Formation of nuclear foci of the Herpes Simplex Virus Type 1 regulatory protein ICP4 at early times of infection: Localization, dynamics, recruitment of ICP27, and evidence for the de novo induction of ND10-like complexes. *J Virol* **78:** 1903–1917.

Falkenberg M, Lehman IR, Elias P. 2000. Leading and lagging strand DNA synthesis in vitro by a reconstituted herpes simplex virus type 1 replisome. *Proc Natl Acad Sci* **97:** 3896–3900.

Gao Y, Colletti K, Pari GS. 2008. Identification of human cytomegalovirus UL84 virus- and cell-encoded binding partners by using proteomics analysis. *J Virol* **82:** 96–104.

Graves-Woodward KL, Gottlieb J, Challberg MD, Weller SK. 1997. Biochemical analyses of mutations in the HSV-1 helicase-primase that alter ATP hydrolysis, DNA unwinding, and coupling between hydrolysis and unwinding. *J Biol Chem* **272:** 4623–4630.

Hume AJ, Finkel JS, Kamil JP, Coen DM, Culbertson MR, Kalejta RF. 2008. Phosphorylation of retinoblastoma protein by viral protein with cyclin-dependent kinase function. *Science* **320:** 797–799.

Igarashi K, Fawl R, Roller RJ, Roizman B. 1993. Construction and properties of a recombinant herpes simplex virus 1 lacking both S-component origins of DNA synthesis. *J Virol* **67:** 2123–2132.

Isler JA, Schaffer PA. 2001. Phosphorylation of the herpes sVirus type 1 origin binding protein. *J Virol* **75:** 628–637.

Jiang C, Hwang YT, Randell JCW, Coen DM, Hwang CBC. 2007a. Mutations that decrease DNA binding of the processivity factor of the herpes simplex virus DNA polymerase reduce viral yield, alter the kinetics of viral DNA replication, and decrease fidelity of DNA replication. *J Virol* **81:** 3495–3502.

Jiang C, Hwang YT, Wang G, Randell JCW, Coen DM, Hwang CBC. 2007b. Herpes simplex virus mutants with multiple substitutions affecting DNA binding of UL42 are impaired for viral replication and DNA synthesis. *J Virol* **81:** 12077–12079.

Jiang C, Komazin-Meredith G, Tian W, Coen DM, Hwang CB-C. 2009. Mutations that increase DNA binding by the processivity factor of herpes simplex virus affect virus production and DNA replication fidelity. *J Virol* **83:** 7573–7580.

Jurak I, Kramer MF, Mellor JC, van Lint AL, Roth FP, Knipe DM, Coen DM. 2010. Numerous conserved and divergent microRNAs expressed by herpes simplex viruses 1 and 2. *J Virol* **84:** 4659–4672.

Kamil JP, Hume AJ, Jurak I, Münger K, Kalejta RF, Coen DM. 2009. Human paillomavirus 16 E7 inactivator of retinoblastoma family proteins complements human cytomegalovirus lacking UL97 protein kinase. *Proc Natl Acad Sci* **106:** 16823–16828.

Khalil MI, Arvin A, Jones J, Ruyechan WT. 2011. A sequence within the varicella-zoster virus (VZV) OriS is a negative regulator of DNA replication and is bound by a protein complex containing the VZV ORF29 protein. *J Virol* **85:** 12188–12200.

Kim YE, Ahn JH. 2010. Role of the specific interaction of UL112-113 p84 with UL44 DNA polymerase processivity factor in promoting DNA replication of human cytomegalovirus. *J Virol* **84:** 8409–8421.

Kobiler O, Brodersen P, Taylor MP, Ludmir EB, Enquist LW. 2011. Herpesvirus replication compartments originate with single incoming viral genomes. *mBio* **2:** e00278-11.

Kochaniak AB, Habuchi S, Loparo JJ, Chang DJ, Cimprich KA, Walter JC, van Oijen AM. 2009. Proliferating cell nuclear antigen uses two distinct modes to move along DNA. *J Biol Chem* **284:** 17700–17710.

Komazin-Meredith G, Mirchev R, Golan DE, van Oijen AM, Coen DM. 2008a. Hopping of a processivity factor on DNA revealed by single-molecule assays of diffusion. *Proc Natl Acad Sci* **105:** 10721–10726.

Komazin-Meredith G, Petrella RJ, Santos WL, Filman DJ, Hogle JM, Verdine GL, Karplus M, Coen DM. 2008b. The human cytomegalovirus UL44 C clamp wraps around DNA. *Structure* **16:** 1214–1225.

Komazin-Meredith G, Santos WL, Filman DJ, Hogle JM, Verdine GL, Coen DM. 2008c. The positively charged surface of herpes simplex virus UL42 mediates DNA binding. *J Biol Chem* **283:** 6154–6161.

Link MA, Silva LA, Schaffer PA. 2007. Cathepsin B mediates cleavage of herpes simplex virus type 1 origin binding protein (OBP) to yield OBPC-1, and cleavage is dependent upon viral DNA replication. *J Virol* **81:** 9175–9182.

Liu S, Knafels JD, Chang JS, Waszak GA, Baldwin ET, Deibel MR Jr, Thomsen DR, Homa FL, Wells PA, Tory MC, et al. 2006. Crystal structure of the herpes simplex virus 1 DNA polymerase. *J Biol Chem* **281:** 18193–18200.

Livingston C, Kyratsous C. 2011. Molecular chaperones and alphaherpesvirus infection. In *Alphaherpesviruses: Molecular virology* (ed. Weller SK), Chap. 14, pp. 237–256. Caister Academic, Norfolk, UK.

Livingston CM, DeLuca NA, Wilkinson DE, Weller SK. 2008. Oligomerization of ICP4 and rearrangement of heat shock proteins may be important for herpes simplex virus type 1 prereplicative site formation. *J Virol* **82:** 6324–6336.

Lo Piano A, Martinez-Jimenez MI, Zecchi L, Ayora S. 2011. Recombination-dependent concatemeric viral DNA replication. *Virus Res* **160:** 1–14.

Macao B, Olsson M, Elias P. 2004. Functional properties of the herpes simplex virus type I origin-binding protein are controlled by precise interactions with the activated form of the origin of DNA replication. *J Biochem* **279:** 29211–29217.

Maga G, Hubscher U. 2003. Proliferating cell nuclear antigen (PCNA): A dancer with many partners. *J Cell Sci* **116:** 3051–3060.

Makhov AM, Boehmer PE, Lehman IR, Griffith JD. 1996. The herpes simplex virus type 1 origin-binding protein carries out origin specific DNA unwinding and forms stem-loop structures. *EMBO J* **15:** 1742–1750.

Makhov AM, Lee SS, Lehman IR, Griffith JD. 2003. Origin-specific unwinding of herpes simplex virus 1 DNA by the viral UL9 and ICP8 proteins: Visualization of a specific preunwinding complex. *Proc Natl Acad Sci* **100:** 898–903.

Makhov AM, Sen A, Yu X, Simon MN, Griffith JD, Egelman EH. 2009. The bipolar filaments formed by herpes simplex virus type 1 SSB/recombination protein (ICP8) suggest a mechanism for DNA annealing. *J Mol Biol* **386:** 273–279.

Maluf NK, Fischer CJ, Lohman TM. 2003. A dimer of *Escherichia coli* UvrD is the active form of the helicase in vitro. *J Mol Biol* **325:** 913–935.

Manolaridis I, Mumtsidu E, Konarev P, Makhov AM, Fullerton SW, Sinz A, Kalkhof S, McGeehan JE, Cary PD, Griffith JD, et al. 2009. Structural and biophysical characterization of the proteins interacting with the herpes simplex virus 1 origin of replication. *J Biochem* **284:** 16343–16353.

Mapelli M, Muhleisen M, Persico G, van Der Zandt H, Tucker PA. 2000. The 60-residue C-terminal region of the single-stranded DNA binding protein of herpes simplex virus type 1 is required for cooperative DNA binding. *J Virol* **74:** 8812–8822.

Mapelli M, Panjikar S, Tucker PA. 2005. The crystal structure of the herpes simplex virus 1 ssDNA-binding protein suggests the structural basis for flexible, cooperative single-stranded DNA binding. *J Biol Chem* **280:** 2990–2997.

Marintcheva B, Weller SK. 2003a. Existence of transdominant and potentiating mutants of UL9, the herpes simplex virus type 1 origin-binding protein, suggests that levels of UL9 protein may be regulated during infection. *J Virol* **77:** 9639–9651.

Marintcheva B, Weller SK. 2003b. Helicase motif Ia is involved in single-strand DNA-binding and helicase activities of the herpes simplex virus type 1 origin-binding protein, UL9. *J Virol* **77:** 2477–2488.

Marsden HS, Cross AM, Francis GJ, Patel AH, MacEachran K, Murphy M, McVey G, Haydon D, Abbotts A, Stow ND. 1996. The herpes simplex virus type 1 UL8 protein influences the intracellular localization of the UL52 but not the ICP8 or POL replication proteins in virus-infected cells. *J Gen Virol* **77:** 2241–2249.

Marty FM, Ljungman P, Papanicolaou GA, Winston DJ, Chemaly RF, Strasfeld L, Young JA, Rodriguez T, Maertens J, Schmitt M, et al. 2011. Maribavir prophylaxis for prevention of cytomegalovirus disease in recipients of allogeneic stem-cell transplants: A phase 3, double-blind, placebo-controlled, randomised trial. *Lancet Infect Dis* **11:** 284–292.

Mohni KN, Mastrocola AS, Bai P, Weller SK, Heinen CD. 2011. DNA mismatch repair proteins are required for efficient herpes simplex virus 1 replication. *J Virol* **85:** 12241–12253.

Moldovan GL, Pfander B, Jentsch S. 2007. PCNA, the maestro of the replication fork. *Cell* **129:** 665–679.

Mullaney J, Moss HW, McGeoch DJ. 1989. Gene UL2 of herpes simplex virus type 1 encodes a uracil-DNA glycosylase. *J Gen Virol* **70:** 449–454.

Murayama K, Nakayama S, Kato-Murayama M, Akasaka R, Ohbayshi N, Kamewari-Hayami Y, Terada T, Shirouzu M, Tsurumi T, Yokoyama S. 2009. Crystal structure of Epstein-Barr virus DNA polymerase processivity factor BMRF1. *J Biol Chem* **284:** 35896–35905.

Muylaert I, Elias P. 2007. Knockdown of DNA ligase IV/XRCC4 by RNA interference inhibits herpes simplex virus type I DNA replication. *J Biochem* **282:** 10865–10872.

Muylaert I, Elias P. 2010. Contributions of nucleotide excision repair, DNA polymerase η, and homologous recombination to replication of UV-irradiated herpes simplex virus type 1. *J Biol Chem* **285:** 13761–13768.

Muylaert I, Tang K-W, Elias P. 2011. Replication and recombination of herpes simplex virus DNA. *J Biol Chem* **286:** 15619–15624.

Niedziela-Majka A, Chesnik MA, Tomko EJ, Lohman TM. 2007. *Bacillus stearothermophilus* PcrA monomer is a single-stranded DNA translocase but not a processive helicase in vitro. *J Biol Chem* **282:** 27076–27085.

Olsson M, Tang K-W, Persson C, Wilhelmsson LM, Billeter M, Elias P. 2009. Stepwise evolution of the herpes simplex virus origin binding protein and origin of replication. *J Biochem* **284:** 16246–16255.

Parkinson J, Lees-Miller SP, Everett RD. 1999. Herpes simplex virus type 1 immediate-early protein vmw110 induces the proteasome-dependent degradation of the catalytic subunit of DNA-dependent protein kinase. *J Virol* **73:** 650–657.

Peng H, Nogueira ML, Vogel JL, Kristie TM. 2010. Transcriptional coactivator HCF-1 couples the histone chaperone Asf1b to HSV-1 DNA replication compartments. *Proc Natl Acad Sci* **107:** 2461–2466.

Polvino-Bodnar M, Orberg PK, Schaffer PA. 1987. Herpes simplex virus type 1 oriL is not required for virus replication or for the establishment and reactivation of latent infection in mice. *J Virol* **61:** 3528–3535.

Price NB, Prichard MN. 2011. Progress in the development of new therapies for herpesvirus infections. *Curr Opin Virol* **1:** 548–554.

Prichard MN, Sztul E, Daily SL, Perry AL, Frederick SL, Gill RB, Hartline CB, Streblow DN, Varnum SM, Smith RD, et al. 2008. Human cytomegalovirus UL97 kinase is required for the hyperphosphorylation of retinoblastoma protein and inhibits the formation of nuclear aggresomes. *J Virol* **82:** 5054–5067.

Pyles RB, Thompson RL. 1994. Evidence that the herpes simplex virus type 1 uracil DNA glycosylase is required for efficient viral replication and latency in the murine nervous system. *J Virol* **68:** 4963–4972.

Randell JCW, Komazin G, Jiang C, Hwang CBC, Coen DM. 2005. Effects of substitutions of arginine residues on the basic surface of herpes simplex virus UL42 support a role for DNA binding in processive DNA synthesis. *J Virol* **79:** 12025–12034.

Schumacher AJ, Mohni KN, Kan Y, Hendrickson E, Stark JM, Weller SK. 2012. The HSV-1 exonuclease, UL12, stimulates recombination by a single strand annealing mechanism. *PLoS Pathog* (under revision).

Stengel G, Kuchta RD. 2011. Coordinated leading and lagging strand DNA synthesis by using the herpes simplex virus 1 replication complex and minicircle DNA templates. *J Virol* **85:** 957–967.

Strang BL, Coen DM. 2010. Interaction of the human cytomegalovirus uracil DNA glycosylase UL114 with the viral DNA polymerase catalytic subunit UL54. *J Gen Virol* **91:** 2029–2033.

Strang BL, Boulant S, Coen DM. 2010a. Nucleolin associates with the human cytomegalovirus DNA polymerase accessory subunit UL44 and is necessary for efficient viral replication. *J Virol* **84:** 1771–1784.

Strang BL, Geballe AP, Coen DM. 2010b. Association of human cytomegalovirus proteins IRS1 and TRS1 with the viral DNA polymerase accessory subunit UL44. *J Gen Virol* **91:** 2167–2175.

Strang BL, Boulant S, Kirchhausen T, Coen DM. 2012. Host cell nucleolin is required to maintain the architecture of human cytomegalovirus replication compartments. *mBio* **3:** e00301–e00311.

Swan MK, Johnson RE, Prakash L, Prakash S, Aggarwal AK. 2009. Structural basis of high-fidelity DNA synthesis by yeast DNA polymerase delta. *Nat Struct Mol Biol* **16:** 979–986.

Szczepanska AK. 2009. Bacteriophage-encoded functions engaged in initiation of homologous recombination events. *Crit Rev Microbiol* **35:** 197–220.

Walsh D, Mohr I. 2006. Assembly of an active translation initiation factor complex by a viral protein. *Genes Dev* **20:** 461–472.

Ward S, Weller S. 2011. HSV-1 DNA replication. In *Alphaherpesviruses: Molecular virology* (ed. Weller SK), Chap. 6, pp. 89–112. Caister Academic, Norfolk, UK.

Weitzman M, Weller S. 2011. Interactions between HSV-1 and the DNA damage response. In *Alphaherpesviruses: Molecular virology* (ed. Weller SK), Chap. 15, pp. 257–268. Caister Academic, Norfolk, UK.

Weller SK. 2010. Herpes simplex virus reorganizes the cellular DNA repair and protein quality control machinery. *PLoS Pathog* **6:** e1001105.

Weller SK, Coen DM. 2006. Herpes simplex virus. In *DNA replication and human disease* (ed. DePamphilis ML), pp. 663–686. Cold Spring Harbor Laboratory Press, Cold Spring Harbor, NY.

Wolf DG, Courcelle CT, Prichard MN, Mocarski ES. 2001. Distinct and separate roles for herpesvirus-conserved UL97 kinase in cytomegalovirus DNA synthesis and encapsidation. *Proc Natl Acad Sci* **98:** 1895–1900.

Yang Y, Dou SX, Ren H, Wang PY, Zhang XD, Qian M, Pan BY, Xi XG. 2008. Evidence for a functional dimeric form of the PcrA helicase in DNA unwinding. *Nucl Acids Res* **36:** 1976–1989.

Zahn KE, Tchesnokov EP, Gotte M, Doublie S. 2011. Phosphonoformic acid inhibits viral replication by trapping the closed form of the DNA polymerase. *J Biol Chem* **286:** 25246–25255.

Zhu Y, Wu X, Cardoso MC, Parris DS. 2010. Processing of lagging-strand intermediates *in vitro* by herpes simplex virus type 1 DNA polymerase. *J Virol* **84:** 7459–7472.

Epstein–Barr Virus DNA Replication

Wolfgang Hammerschmidt[1] and Bill Sugden[2]

[1]Department of Gene Vectors, Helmholtz Zentrum München, German Research Center for Environmental Health, Marchioninistr. 25, D-81377 Munich, Germany

[2]Department of Oncology, McArdle Laboratory for Cancer Research, University of Wisconsin-Madison, Madison, Wisconsin 53706

Correspondence: sugden@oncology.wisc.edu

Epstein–Barr virus (EBV) is a paradigm for human tumor viruses: It is the first virus recognized to cause cancer in people; it causes both lymphomas and carcinomas; yet these tumors arise infrequently given that most people in the world are infected with the virus. EBV is maintained extrachromosomally in infected normal and tumor cells. Eighty-four percent of these viral plasmids replicate each S phase, are licensed, require a single viral protein for their synthesis, and can use two functionally distinct origins of DNA replication, *oriP,* and Raji ori. Eighty-eight percent of newly synthesized plasmids are segregated faithfully to the daughter cells. Infectious viral particles are not synthesized under these conditions of latent infection. This plasmid replication is consistent with survival of EBV's host cells. Rare cells in an infected population either spontaneously or following exogenous induction support EBV's lytic cycle, which is lethal for the cell. In this case, the viral DNA replicates 100-fold or more, uses a third kind of viral origin of DNA replication, *oriLyt,* and many viral proteins. Here we shall describe the three modes of EBV's replication as a function of the viral origins used and the viral and cellular proteins that mediate the DNA synthesis from these origins focusing, where practical, on recent advances in our understanding.

oriP

A single contiguous fragment of EBV DNA with a length of 1.7 kbps supports autonomous extrachromosomal replication and maintenance of recombinant plasmids in human cells (Yates 1996) when the Epstein–Barr nuclear protein 1 (EBNA1) is provided in *trans* (Yates 1996) (Fig. 1). This subgenomic EBV DNA is termed *oriP* for "origin of plasmid replication." Similar to the parental EBV genome (Yates 1996), initiation of DNA replication within *oriP* is synchronized with the timed and strictly regulated replication of chromosomal DNA (Yates 1996; Shirakata et al. 1999), and mediated by the cellular replication machinery acting at *oriP.* The viral protein EBNA1 binds to *oriP* site-specifically and recruits the cellular DNA replication machinery (Chaudhuri et al. 2001; Dhar et al. 2001; Schepers et al. 2001; Ritzi et al. 2003). Thus, EBNA1 is essential but not sufficient for *oriP*'s function. In resting and proliferating cells, *oriP* confers stable extrachromosomal maintenance of recombinant plasmids, which also relies on EBNA1 in *trans* independent of its role in DNA replication. An EBV genome

Cite this article as *Cold Spring Harb Perspect Biol* doi: 10.1101/cshperspect.a013029

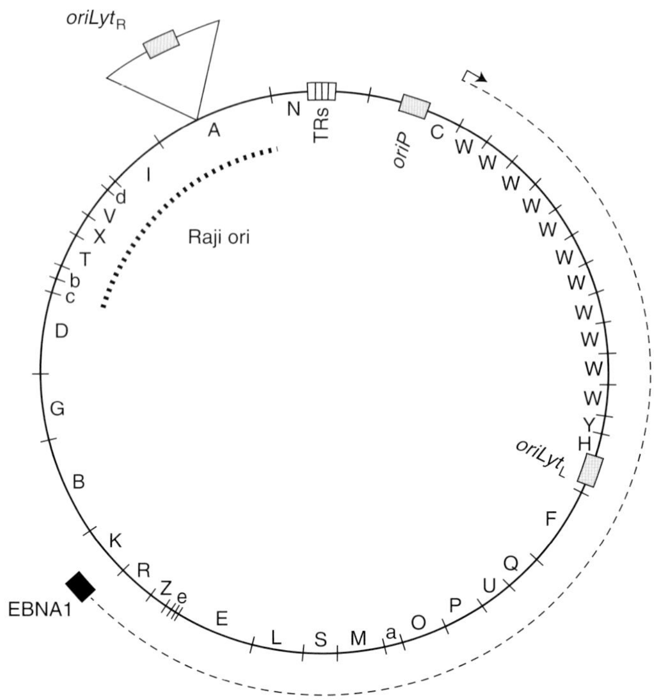

Figure 1. A physical map of the B95-8 laboratory strain of EBV DNA of 165 kbp is shown. The letters on the inner surface of the circle denote the fragments generated by digestion of this DNA with the *Bam*H1 endonuclease and used to mark its transcripts (Baer et al. 1984). The primary transcript for EBNA1 is denoted by the dashed line. The white box marked TRs indicates the terminal repeats of linear virion DNA that mediate circularization of the viral DNA on infection of cells. The sites for *oriP* and *oriLyt* (L, left) are shown with the approximate location of DNA deleted from the B95-8 strain, present in all other analyzed strains, that contains a second copy of *oriLyt* (R, right). Raji ori, a region that is defined only approximately maps to sequences spanning the deletion found in the B95-8 strain.

with a deletion in EBNA1 has been found only to be integrated into the cellular chromosome in virus-infected B cells, consistent with its being unable to be maintained extrachromosomally (Humme et al. 2003).

oriP's replicator is coincident with or located near the origin of DNA replication (Yates 1996) and is operationally termed the "dyad symmetry" (DS) element (Fig. 2A). About 1 kbp upstream of DS, an array of tandem repeats 650 bps in total size, termed the "family of repeats" (FR), provides 20 sequence motifs to which EBNA1 molecules bind with high affinity. The FR element does not contribute to DNA synthesis but is mandatory for the stable maintenance and nuclear retention of *oriP* plasmids and genomic EBV DNA.

FR, the Plasmid Maintenance Element

The family of repeats, termed FR, is an array of 21 imperfectly conserved, 30 bp, direct repeats (Fig. 2A). EBNA1 binds as homotypic dimers to 20 sequence motifs within FR although only seven EBNA1-binding sites are minimally needed to function efficiently (Yates 1996). One essential role of FR is to prevent plasmid loss from proliferating cells. This plasmid maintenance/nuclear retention function appears to act by tethering the FR element via EBNA1 to condensed mitotic chromosomes as they segregate during mitosis leading to comigration of FR-carrying DNA molecules (Yates 1996; Marechal et al. 1999; Hung et al. 2001). This tethering by EBNA1 is accomplished through its carboxyl

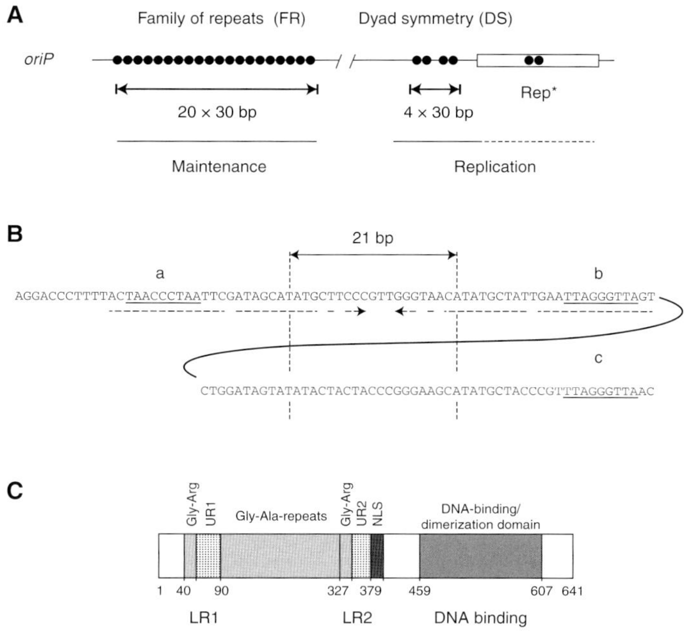

Figure 2. Shown are depictions of *oriP* and EBNA1. (*A*) *oriP* consists of multiple sets of binding sites for EBNA1. The family of repeats in the B95-8 strain of EBV has 20 binding sites for EBNA1, which together mediate maintenance of *oriP* plasmids. DS has two pairs of binding sites and is the site or close to the site at which DNA synthesis initiates within *oriP*. Rep* can substitute for DS functionally but inefficiently. It too contains a pair of EBNA1-binding sites with the same spacing as each pair in DS. When multimerized eightfold, Rep* functions as efficiently as does DS to support initiation of DNA synthesis. (*B*) The sequence and spacing of EBNA1-binding sites in DS of *oriP* are shown. The two sites of each pair are separated by 21 bps allowing similar contacts on one face of helical DNA. The nonomer repeats that can bind telomere-associated proteins are underlined and denoted a, b, and c. (*C*) The domains of EBNA1 from the B95-8 strain of EBV are represented and are composed of 641 residues. LR1 and LR2 are linking regions rich in Gly-, Arg residues, which when fused to protein DNA-binding domains, can link the DNAs presumably by the AT-hook activities intrinsic to LR1 and LR2. LR1 and LR2 each contain unique sequences, UR1 and UR2. UR1 is involved in transcription regulated by EBNA1; UR2 does not have a known function. The Gly, Ala repeats have various activities ascribed to them but little or no effect on EBNA1's functions in cell culture. EBNA1 has one identified nuclear localization sequence (NLS) (Yates 1996). EBNA1's carboxy-terminal one-third comprises a dimerization and DNA-binding domain.

terminus binding site, specifically to FR and its linking regions in its amino terminus binding AT-rich sequences in host chromosomes (see below). Live-cell imaging has shown that newly synthesized daughter molecules remain colocalized as pairs until anaphase when 88% segregate one-to-one to daughter cells and the remaining 12% segregate randomly (Nanbo et al. 2007). The mechanism of this nonrandom segregation, which must be established in S phase, is not understood. FR in conjunction with EBNA1 also can act as an EBNA1-dependent, transcriptional enhancer (Kennedy and Sugden 2003).

DS, the Origin of DNA Replication

The DS element is an EBNA1-dependent replicator for the EBV genome. An approximately 120 bp region of *oriP* containing the minimal replicator has been termed DS because it contains an element of dyad symmetry (DS) 65 bp in length (Fig. 2B). Bidirectional DNA synthesis initiates at or close to DS (Yates 1996) and DS does support EBNA1-dependent DNA replication in the absence of the FR element (Yates et al. 2000). The DS element consists of two pairs of EBNA1-binding sites (Fig. 2B). The EBNA1-

binding sites within one pair are separated by 21 bp such that they are in the same helical phase, which is essential for their acting as a replicator (Bashaw and Yates 2001). The binding sites within FR for EBNA1 do not require the same spatially strict arrangement for FR's functions.

It has been shown that one pair of correctly spaced EBNA1-binding sites is sufficient to recruit the origin recognition complex (see below) to the replicator (Julien et al. 2004) indicating that the correct binding of EBNA1 is a critical parameter for origin activation. This minimal replicator shows only a fraction of the activity of the intact DS element. Auxiliary elements are likely to contribute to the efficiency of initiation of DNA synthesis, but their genetic dissection is still incomplete. These auxiliary elements include 14 bp repeats in which are embedded nonamer motifs (Fig. 2B) that bind some telomere-associated proteins (see below). The binding of these proteins increases the efficiency of DS as a replicator (Deng et al. 2002, 2003; Lindner et al. 2008).

An element termed Rep* of about 300 bps, which is located downstream from DS was identified to be an alternative EBNA1-dependent replicator with reduced activity (Fig. 2A) (Kirchmaier and Sugden 1998). Although Rep* supports only short-term plasmid replication, it may contribute to the overall replication efficiency of *oriP* because multimers of Rep* support long-term plasmid replication as efficiently as does *oriP* (Wang et al. 2006). Initiation of DNA replication within an *oriP* replicon is likely to be influenced by its chromatin structure. Nucleosomes surrounding DS appear to be phased and modified in a cell-cycle-dependent manner (Wensing et al. 2001; Avolio-Hunter and Frappier 2003; Zhou et al. 2005; Tempera and Lieberman 2010). The extent to which the chromatin structure controls the replicator or is controlled by the machinery at the replicator is not yet clear.

Proteins that Support the Functions of *oriP*

oriP consists largely of two stretches of binding sites for EBV's EBNA1 protein that differ in their numbers, spacing, and affinities. EBNA1 is an essential participant in *oriP*'s functions. EBNA1 can now be viewed as an assembly of at least four functional elements that support EBV as an extrachromosomal replicon (Fig. 2C).

The carboxy-terminal quarter of EBNA1 encodes two functions essential for all *trans*-acting phenotypes ascribed to EBNA1. This region provides a structure that forms a dimer that binds DNA site-specifically. This structure closely resembles that of the dimerization and DNA-binding domain of the E2 protein encoded by papillomaviruses (Hegde et al. 1992; Bochkarev et al. 1996). The dimerization and DNA-binding domain of EBNA1 alone bends DNAs it binds site-specifically as does that of the E2 proteins (Hegde et al. 1992; Kim et al. 2000; Bashaw and Yates 2001; Wang et al. 2006). It also acts as a dominant negative derivative for EBNA1's known functions such that its expression in EBV-positive tumors forces the loss of the virus from cells to reveal the phenotypes provided them by EBV (Vereide and Sugden 2011).

A stretch of glycine, alanine repeats that spans more than 200 residues in EBNA1 of the B95-8 laboratory strain of EBV has been found to inhibit antigen processing (Levitskaya et al. 1997). This same repeated element has been found to limit translation in *cis*, which might also underlie EBNA1's being poorly recognized by cytotoxic T-cells (Yin et al. 2003). These repeats may contribute to EBNA1's being recognized by $CD4^+$ T cells following its autophagic degradation and the display of its resulting peptides on MHC class II molecules (Paludan et al. 2005). EBNA-1 is the only viral protein consistently expressed in proliferating, infected cells. Its atypical immune recognition may be essential for EBV's long-term success as a human parasite.

A third functional element of EBNA1 consists of a stretch of 25 amino acids termed UR1 (unique region 1) positioned between a stretch of glycine, arginine and the glycine, alanine repeats. It is essential for EBV's transformation of human B cells (Altmann et al. 2006). UR1 supports EBNA1's activation of transcription but not replication (Wu et al. 2002; Kennedy and Sugden 2003). This element contains two cysteines, which allow two monomers of EBNA1 to bind Zinc coordinately. This zinc coordination is necessary for EBNA1's transcriptional transactivation (Aras et al. 2009) and is reminiscent of the amino terminus of the E2 protein of papillomaviruses, which can form intradimers within one pair of monomers or interdimers between two pairs of monomers (Antson et al. 2000).

Another functional moiety of EBNA1 consists of two redundant elements termed LR1 and LR2 (linking region 1 and 2). LR1 and LR2 are rich in arginine and glycine residues and were identified by their ability to loop regions of DNA together to which they were bound in *cis* (Yates 1996) or to link such elements in *trans* (Yates 1996). Genetic analyses of LR1 and LR2 indicate that both contribute to EBNA1's support of transcription and of replication (Mackey and Sugden 1999). One mechanism by which these linking regions contribute to EBNA1's support of replication of *oriP* has emerged. Both LR1 and LR2 have AT-hook activities derived from their arginine, glycine repeats, which promote the association of EBNA1 to AT-rich stretches of DNA (Sears et al. 2004). Importantly, fusions of the cellular protein HMGA1a, which has three AT-hooks to the dimerization and DNA-binding domain of EBNA1 support the replication of *oriP* plasmids in human cells stably (Hung et al. 2001; Sears et al. 2003). Fusions of EBNA1's dimerization and DNA-binding domain to HMG1 or EBP2, both of which appear to lack AT-hooks, fail to support the replication of *oriP* stably in human cells (Sears et al. 2003, 2004). The AT-hooks of EBNA1's LR1 and LR2 thus are important for its support of the maintenance of *oriP* in proliferating cells. They likely provide EBV another, associated function. The tethering of EBV plasmid DNAs to AT-rich chromosomal sites may insure that the viral replicators home to special sites in the nucleus where they can function when their attached chromosomal replicators initiate synthesis. A third proposed contribution of the linking regions to replication is their binding G-rich RNA and through this binding apparently recruiting ORC to DS bound by EBNA1 (Norseen et al. 2008). This role is supported by GST-fusions to LR1 and LR2 binding ORC from cell extracts in an RNAse-susceptible manner (Norseen et al. 2008). This described role is perplexing though, because it is not easily reconciled with the requirement for the specific spacing of EBNA1's DNA-binding sites needed to recruit ORC (Bashaw and Yates 2001; Wang et al. 2006). In addition, Cdc6 binds EBNA1 directly in vitro in an RNAse-resistant manner and contributes to EBNA1's recruitment of ORC to DS (Moriyama et al. 2012).

Proteins other than EBNA1 and ORC can bind to or associate with *oriP* to facilitate its functions. The MCM complex associates with DS in a cell cycle-dependent manner; MCM2, 3, and 7 have been detected at DS by chromatin immunoprecipitation and to be enriched there during G_1 and early S phases (Chaudhuri et al. 2001; Ritzi et al. 2003). Additional cellular proteins have been found to associate with DS dependent on EBNA1 in cell extracts and in vivo. The three nonamer repeats (TTAGGGTTA) abutting EBNA1-binding sites in DS are similar to telomeric repeats (TTAGGG) and bind proteins in vivo as measured by genomic footprinting (Fig. 2B) (Niller et al. 1995). DNA-affinity chromatography with DS coupled to a matrix enriches for several telomere-associated proteins only from EBNA1-positive cells (Deng et al. 2002, 2003). These enriched proteins include TRF1 and TRF2 (telomeric repeat binding factors 1 and 2) and hRap1 (homolog of repressor activator protein 1), which bind the nonamers directly (TRF1 and TRF2) or indirectly by binding TRF2 (Deng et al. 2002, 2003). The roles these proteins have in *oriP*-mediated replication are not clear, though. Derivatives of *oriP* with mutated nonamers that fail to bind these telomere-associated proteins are maintained stably in cells at copy numbers one-half of that of wild-type *oriP* (Deng et al. 2003). One simple model for TRF2, is based on its binding ORC1 directly (Atanasiu et al. 2006). Binding of TRF2 to DS should aid in recruiting ORC to DS. This simple model is consistent with analyses of Rep*. The two binding sites for EBNA1 in Rep* do not have neighboring nonamer sequences and support DNA replication poorly (Wang et al. 2006). When these EBNA1-binding sites are embedded in neutral lambda DNA and octamerized, they support DNA synthesis as efficiently as does wild-type DS (Wang et al. 2006). This result is consistent with the finding that the overall avidity of EBNA1 for derivatives of DS correlates with the efficiency of replication of those derivatives (Lindner et al. 2008).

Chromatin immunoprecipitation experiments have identified additional proteins present at *oriP* in cells that likely affect its replication. Two proteins implicated in yeast in stabilizing replication forks, Timeless (Tim) and Timeless interacting protein, have been detected there (Dheekollu and Lieberman 2011). Inhibiting Tim leads to the accumulation of linear EBV DNA in cells, consistent with these proteins stabilizing replication forks at *oriP* (Dheekollu and Lieberman 2011). The deubiquitylating enzyme, USP7, binds EBNA1 in vitro and localizes to *oriP* in vivo (Sarkari et al. 2009). This association affects the levels of monoubiquitylated histone H2B in the vicinity of *oriP* and is thought to affect transactivation of transcription by EBNA1 bound to FR (Sarkari et al. 2009).

Raji ori, AN ALTERNATE LICENSED ORIGIN OF EBV

Although DS of *oriP* was the only origin of plasmid DNA synthesis identified by screening cloned fragments of EBV DNA for "ARS" activity in EBV-positive cells (Yates 1996), work of Carl Schildkraut and his colleagues has identified a new class of origins that function in extrachromosomal EBV (Yates 1996; Norio and Schildkraut 2004). They studied replicative intermediates of EBV DNA in Raji cells with two dimensional gel analyses and showed that DNA synthesis in Raji EBV often originated 25 kbp away from its *oriP* (Fig. 1) (Yates 1996). Multiple observations indicate that this origin, which we shall refer to as "Raji ori," is akin to chromosomal origins consisting of "zones of replication" such as that in the DHFR locus, but Raji ori functions in an extrachromosomal replicon.

Raji ori can support DNA synthesis efficiently and is not unique to the EBV strain in Raji cells. EBV DNA synthesis has been characterized in Raji cells and in the Mutu1 cell line using "single molecule analysis of replicated DNA" (Norio and Schildkraut 2004). This approach has revealed that DNA synthesis can originate throughout much or all of EBV DNA in both Raji cells and Mutu1 cells, that initiation is most common at Raji ori in Raji cells, and is less frequent but does occur in the Raji ori region of the EBV DNA in Mutu1 cells. The frequency of initiation within Raji ori in Raji cells does not reflect a genetic defect in the DS element of its *oriP*. This DS has been isolated, sequenced, and found not to have significant differences from the wild type sequence (Koons et al. 2001). DS has been deleted from a third strain of EBV, the recombinant virus introduced into an EBV-negative, B-cell line, and replicative intermediates analyzed by two dimensional gel analysis. The Raji ori of this recombinant strain was found to support the initiation of DNA synthesis efficiently, although other regions showed the presence of bubble arcs too (Norio et al. 2000).

Raji ori shares at least two properties with some cellular origins. First, it is licensed. The EBV DNA in Raji cells was studied in density shift experiments and shown to incorporate BrdU at the same rate as did host chromosomal DNA showing that EBV DNA replicates semiconservatively, once per cell cycle in these cells (Yates 1996), consistent with its being licensed. Second, Raji ori also appears to have multiple sites at which DNA synthesis can initiate (Yates 1996) as has been found, for example, with the cellular DHFR locus (Kobayashi et al. 1998).

Both EBV and the host cell contribute *trans*-acting factors necessary for the function of Raji ori. Multiple observations indicate that EBNA1 in *trans* in conjunction with FR in *cis* is essential for the maintenance of replicons using Raji ori in proliferating cells (Yates 1996). EBNA1 does not contribute directly to origin function of Raji ori, though. EBNA1 does not bind detectably to Raji ori as measured by gel shift assays using 40 overlapping fragments of 600 bp that span this zone of initiation (Wang and Sugden 2008). Given that Raji ori is a zone with multiple sites for the initiation of licensed DNA replication, ORC likely binds those sites but the means by which it does so is unknown.

One question, "Why does EBV have *oriP* given it has Raji ori?" has been answered. The DNA replication mediated by DS at *oriP* is more efficient than that of Raji ori with FR in *cis* and EBNA1 in *trans* when initially introduced into Raji cells. This more efficient replication allows plasmids with DS plus FR to become "established" (Wang and Sugden 2008). Only 1%–10% of EBV's plasmid replicons that are introduced into cells and initially replicate progress to be maintained stably after 15 generations or so, a process termed "establishment" (Leight and Sugden 2001). EBV plasmids are present in all examined clones of cells in a wide distribution of numbers of plasmids per cell, an equilibrium that reflects both defects in their synthesis and partitioning (Nanbo et al. 2007). The efficient initial replication of replicons with DS plus FR allows them to achieve this distribution needed to be established. Replicons with Raji ori and FR fail to be established but can be maintained in Raji cells once established by virtue of having had DS in *cis* (Wang and Sugden 2008). Thus, the DS replicator is peculiarly efficient on being introduced into cells, a function required during virus infection to allow EBV to be established. Once established chromosomal-like replicators can take over and sustain replication of this viral extrachromosomal replicon.

oriLyt

DNA replication during the lytic phase of EBV's life cycle is uncoupled from and independent of that during its latent phase. Both the *cis*-acting elements and *trans*-acting factors involved in latent and lytic viral DNA replication differ, indicative of their different mechanisms. Although DNA replication during the latent phase ensures the faithful duplication of 84% of the viral genomes in each cell cycle (Nanbo et al. 2007), viral genomes replicate independently of such constraints during EBV's lytic cycle and are amplified several hundred-fold within one to two days (Hammerschmidt and Sugden 1988). The products of lytic DNA replication are long concatemers (Bloss and Sugden 1994). They become the substrate for further processing eventually yielding cleaved, packaged, linear genome units bound in their preformed capsids with polyamines (Gibson and Roizman 1971). These "naked" viral DNAs are unmethylated, circularize in the recipient cell following infection, and eventually reside in the nucleus in which they are organized into chromatin and become methylated at cytosines in CpG dinucleotides (Shaw et al. 1979; Fernandez et al. 2009; Kalla et al. 2010).

The *cis*-acting element, which acts as the lytic origin of replication of EBV, is termed *oriLyt* (Yates 1996). Two copies, which are about 100 kbps apart and therefore located opposite to each other in the circular genome, are present in all EBV strains examined with the exception of the B95-8 laboratory strain (Fig. 1). EBV's *oriLyt* shares limited sequence homology with lytic origins of members of the β- and γ-herpesvirus family but not with those of α-herpesviruses such as herpes simplex virus type 1 (HSV-1). *oriLyt* is characterized by a duplicated 1055-bp long core element, which is virtually identical in both copies of *oriLyt* (Fig. 3). Within this core element there are two essential components separated by about 530 bps. Loosely defined auxiliary components that flank the two essential upstream and downstream components also contribute to *oriLyt*'s activity (Hammerschmidt and Sugden 1988; Schepers et al. 1993b). These nonessential but auxiliary components greatly enhance the activity of the core components (Hammerschmidt and Sugden 1988; Schepers et al. 1993b).

The products of lytic DNA replication of all herpesviruses are DNA concatemers in which the single genome units are arranged head to tail. This arrangement has also been shown for EBV (Hammerschmidt and Sugden 1988; Schepers et al. 1993b) and is consistent with circularized monomeric DNA molecules serving as templates for DNA replication via a rolling-circle mechanism. However, a rolling-circle mode alone poorly accounts for the rapid accumulation of progeny DNA. For example, in HSV-1-infected cells, about 30 min is needed to complete the replication of one viral molecule, but this herpesviral DNA is replicated several hundredfold during a short period of time to yield up to 1000 genomic copies per cell (Jacob and Roizman 1977). A pure rolling-circle mechanism synthesizes progeny DNA only linearly,

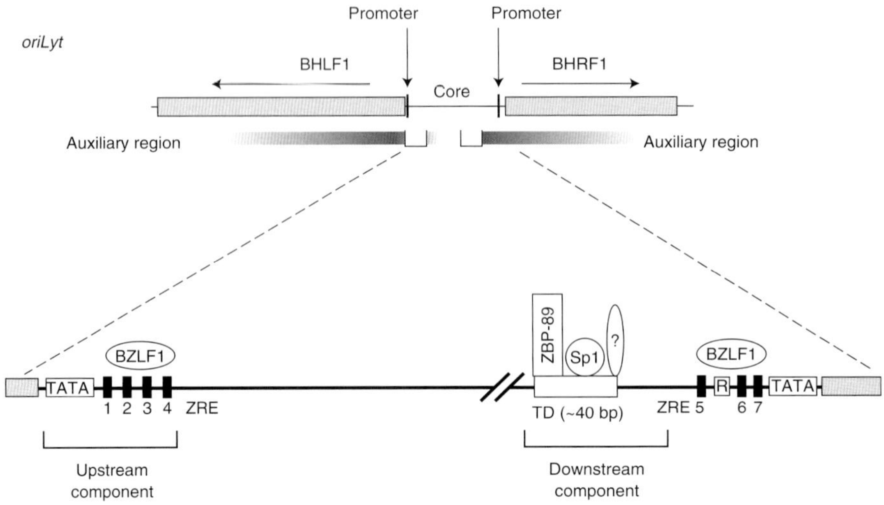

Figure 3. *oriLyt* and its expanded core domain are shown. *oriLyt* spans about 7700 bp of the B95-8 strain of EBV including two genes BHLF1 and BHRF1 and their promoters. Sequences in these genes contribute to but are not essential for the function of *oriLyt*. The essential or core domain of *oriLyt* includes the promoters for and intergenic region of these genes. Two essential components (upstream and downstream) of *oriLyt* bind viral (BZLF1) at BZLF1-responsive elements (ZRE) and cellular factors (ZBP-89, Sp1, and likely additional unknown proteins) are required for *oriLyt*'s function. The BZLF1 protein also binds to promoter elements of the BHRF1 gene to enhance *oriLyt*'s activity. This domain (ZRE sites 5, 6, and 7) can be replaced by a heterologous enhancer to support *oriLyt*'s function. The boxed element marked R binds the viral BRLF1 transactivator, which does not contribute directly to EBV's lytic DNA replication.

not exponentially with time, and appears inadequate to explain the rapid amplification of herpesviral DNA. In fact, early during EBV's lytic cycle viral replicative DNA intermediates were found to replicate semi-conservatively, to be amplified exponentially, and to be covalently closed, circular DNAs of parental length. These products of the first phase of EBV's lytic DNA synthesis likely provide the many circular templates for DNA synthesis via a rolling-circle mechanism needed to yield efficient DNA amplification during the second phase (Pfuller and Hammerschmidt 1996). This biphasic mode of herpesviral DNA replication is consistent with the notion that the origin-binding proteins of herpesviruses are needed only to initiate but not to support continued lytic DNA synthesis (Schildgen et al. 2005).

Proteins that Support the Functions of *oriLyt*

All herpesviruses encode replication proteins that specifically interact with their origins of lytic DNA replication to provide key functions, including polymerases, helicases, primases, DNA-binding proteins, and associated factors, as well as enzymatic activities involved in nucleotide synthesis and phosphorylation (Challberg 1986; Challberg and Kelly 1989). Six EBV genes, which encode essential lytic functions, have been identified (Challberg 1986; Tsurumi 2001). With the exception of the virus-specific DNA-binding proteins essential for the activation of lytic origins of herpesviral DNA replication, these lytic gene products share considerable primary amino acid sequence and function among herpesviruses (Table 1) (Tsurumi 2001). These genes of EBV are under the control of two viral transactivators, BZLF1 and BRLF1, which thus orchestrate transcriptionally the expression of EBV's replication machinery (Feederle et al. 2000).

BZLF1, a key transcriptional regulator, is EBV's lytic origin-binding protein (Schepers et al. 1993a, 1996). BZLF1 has been found to be required for wild-type levels of lytic DNA

Table 1. Viral proteins acting at EBV's characterized origins and their functions

Origin	Protein	Function
oriP	EBNA1	1. Origin-binding protein at the DS element; recruits ORC for licensed DNA replication 2. Binds FR element to mediate maintenance of the plasmid replicon
oriLyt	BZLF1	1. Origin-binding protein of *oriLyt* that can bind multiple viral replication proteins 2. Viral transcription factor required to initiate the lytic phase of EBV and support expression of viral factors of lytic DNA amplification
	BALF5	DNA polymerase catalytic subunit; interacts with the helicase–primase complex (Fujii et al. 2000)
	BALF2	Single-stranded DNA-binding protein
	BMRF1	DNA polymerase accessory subunit with a PCNA-like structure (Murayama et al. 2009; Nakayama et al. 2010); interacts with BALF5 (Tsurumi et al. 1993); can act as a transcriptional coactivator of BZLF1 (Nakayama et al. 2009)
	BBLF4	DNA helicase
	BSLF1	Primase
	BBLF2/3	Primase-associated factor. BBLF4, BSLF1, and BBLF2/3 form the helicase–primase complex

replication when it is bound to four sites within the essential upstream component of *oriLyt* (Schepers et al. 1993a, 1996), although it binds additional sites within the *oriLyt* sequence (Fig. 3). These four sites are an intrinsic part of the BHLF1 promoter; deletion experiments indicated that promoter elements in addition to the BZFL1-binding motifs contribute to lytic DNA replication (Schepers et al. 1993b; Rennekamp and Lieberman 2011). One of the putative factors could be the BHLF1 transcript itself that is expressed from this promoter (Fig. 3) (Rennekamp et al. 2010; Rennekamp and Lieberman 2011), but efficient transcription from this promoter is not sufficient to support replication (Schepers et al. 1993a, 1996) because a viral mutant lacking all BZLF1-binding-sites within *oriLyt* supported a low level of lytic replication (Feederle and Delecluse 2004).

BZLF1 may act independently of EBV's six lytic gene products to support the initiation of DNA replication at *oriLyt* (Yates 1996). This hypothesis suggests that cellular proteins contribute to the formation of early replicative DNA intermediates, which subsequently become the substrate for lytic DNA amplification. In striking contrast to the origin-binding protein of HSV-1, UL9, which is a sequence specific helicase, BZLF1 does not contribute an intrinsic enzymatic function to DNA synthesis.

BZLF1 contributes to the replication complex at *oriLyt* presumably in part by associating with some members of the complex. BZLF1 has been reported to interact with the viral helicase–primase complex (Table 1) (Liao et al. 2001, 2005; El-Guindy et al. 2010) and the viral polymerase accessory factor BMRF1 (Takagi et al. 1991; Daikoku et al. 2005; Nakayama et al. 2009). BMRF1 bears structural similarities with cellular PCNA (Murayama et al. 2009; Nakayama et al. 2010) and could potentially provide an additional tethering function for the replication complex (Zhang et al. 1997; Baumann et al. 1999). In addition, the primase-associated factor might serve a similar tethering function in conjunction with cellular, *oriLyt*-binding proteins (see below and Liao et al. 2005).

The essential downstream component of *oriLyt* is highly sensitive to sequence alterations within a stretch of about 40 bps, termed the "TD" element (Gruffat et al. 1995). TD was found to be the binding site for several cellular proteins (Gruffat et al. 1995). The transcription factors Sp1, Sp3, and ZBP-89, have been identified and shown to make essential and direct contributions to *oriLyt*'s function(s) (Gruffat et al. 1995; Baumann et al. 1999). They interact with EBV's DNA polymerase and its processivity factor and likely tether viral replication proteins to *oriLyt* via direct protein–protein interactions at

TD. These findings have a parallel in ZBRK1 and KAP-1, a zinc-finger DNA binding protein and its corepressor, which bind to a site located about 200 bps downstream of TD (Liao et al. 2005) within the previously identified *oriLyt* enhancer region (Yates 1996) and colocalize EBV's helicase–primase complex to operationally defined replication compartments in lytically induced cells. Replication compartments are sites to which components of the cellular homologous recombination pathway are recruited (Wilkinson and Weller 2004, and references therein) together with various *cis*-acting elements of lytic replicons of herpesviruses. Lytic DNA replication involves both replication and homologous recombination of DNA, which are two interdependent processes active during the lytic phase of EBV's life cycle (Pfuller and Hammerschmidt 1996). Several cellular recombination and DNA repair factors have been recently found recruited to EBV's replication compartments (Daikoku et al. 2006; Kudoh et al. 2009; Sugimoto et al. 2011) and induction of EBV's lytic phase induces a genuine DNA-damage response signal (Kudoh et al. 2005; Sato et al. 2010). It now seems that *oriLyt* is complex not only in mediating mechanistically distinct biphasic DNA replication but also in using a complex repertoire of viral and cellular proteins to carry out these two modes of replication.

IN SUMMARY

EBV is an enormously successful human parasite, having infected more than 6.5 billion people in the world today. Its success is intimately tied to its having evolved as an extrachromosomal replicon, inducing the infected cell to cycle, and successfully usurping the cell's machinery to carry out its licensed DNA synthesis. It has built on a common herpesviral property of amplifying viral DNA during productive infections through dedicated viral origins of DNA synthesis and an array of viral proteins that mediate this unlicensed synthesis. It carries out these different modes of DNA replication in human B cells by affecting the differentiated state of it B-cell host.

REFERENCES

Altmann M, Pich D, Ruiss R, Wang J, Sugden B, Hammerschmidt W. 2006. Transcriptional activation by EBV nuclear antigen 1 is essential for the expression of EBV's transforming genes. *Proc Natl Acad Sci* **103:** 14188–14193.

Antson AA, Burns JE, Moroz OV, Scott DJ, Sanders CM, Bronstein IB, Dodson GG, Wilson KS, Maitland NJ. 2000. Structure of the intact transactivation domain of the human papillomavirus E2 protein. *Nature* **403:** 805–809.

Aras S, Singh G, Johnston K, Foster T, Aiyar A. 2009. Zinc coordination is required for and regulates transcription activation by Epstein–Barr nuclear antigen 1. *PLoS Pathog* **5:** e1000469.

Atanasiu C, Deng Z, Wiedmer A, Norseen J, Lieberman PM. 2006. ORC binding to TRF2 stimulates OriP replication. *EMBO Rep* **7:** 716–721.

Avolio-Hunter TM, Frappier L. 2003. EBNA1 efficiently assembles on chromatin containing the Epstein–Barr virus latent origin of replication. *Virology* **315:** 398–408.

Baer R, Bankier AT, Biggin MD, Deininger PL, Farrell PJ, Gibson TJ, Hatfull G, Hudson GS, Satchwell SC, Seguin C, et al. 1984. DNA sequence and expression of the B95-8 Epstein–Barr virus genome. *Nature* **310:** 207–211.

Bashaw JM, Yates JL. 2001. Replication from oriP of Epstein–Barr virus requires exact spacing of two bound dimers of EBNA1 which bend DNA. *J Virol* **75:** 10603–10611.

Baumann M, Feederle R, Kremmer E, Hammerschmidt W. 1999. Cellular transcription factors recruit viral replication proteins to activate the Epstein–Barr virus origin of lytic DNA replication, oriLyt. *EMBO J* **18:** 6095–6105.

Bloss TA, Sugden B. 1994. Optimal lengths for DNAs encapsidated by Epstein–Barr virus. *J Virol* **68:** 8217–8222.

Bochkarev A, Barwell JA, Pfuetzner RA, Bochkareva E, Frappier L, Edwards AM. 1996. Crystal structure of the DNA-binding domain of the Epstein–Barr virus origin-binding protein, EBNA1, bound to DNA. *Cell* **84:** 791–800.

Challberg MD. 1986. A method for identifying the viral genes required for herpesvirus DNA replication. *Proc Natl Acad Sci* **83:** 9094–9098.

Challberg MD, Kelly TJ. 1989. Animal virus DNA replication. *Annu Rev Biochem* **58:** 671–717.

Chaudhuri B, Xu H, Todorov I, Dutta A, Yates JL. 2001. Human DNA replication initiation factors, ORC and MCM, associate with oriP of Epstein–Barr virus. *Proc Natl Acad Sci* **98:** 10085–10089.

Daikoku T, Kudoh A, Fujita M, Sugaya Y, Isomura H, Shirata N, Tsurumi T. 2005. Architecture of replication compartments formed during Epstein–Barr virus lytic replication. *J Virol* **79:** 3409–3418.

Daikoku T, Kudoh A, Sugaya Y, Iwahori S, Shirata N, Isomura H, Tsurumi T. 2006. Postreplicative mismatch repair factors are recruited to Epstein–Barr virus replication compartments. *J Biol Chem* **281:** 11422–11430.

Deng Z, Lezina L, Chen CJ, Shtivelband S, So W, Lieberman PM. 2002. Telomeric proteins regulate episomal maintenance of Epstein–Barr virus origin of plasmid replication. *Mol Cell* **9:** 493–503.

Deng Z, Atanasiu C, Burg JS, Broccoli D, Lieberman PM. 2003. Telomere repeat binding factors TRF1, TRF2, and hRAP1 modulate replication of Epstein–Barr virus OriP. *J Virol* **77:** 11992–12001.

Dhar SK, Yoshida K, Machida Y, Khaira P, Chaudhuri B, Wohlschlegel JA, Leffak M, Yates J, Dutta A. 2001. Replication from oriP of Epstein–Barr virus requires human ORC and is inhibited by geminin. *Cell* **106:** 287–296.

Dheekollu J, Lieberman PM. 2011. The replisome pausing factor Timeless is required for episomal maintenance of latent Epstein–Barr virus. *J Virol* **85:** 5853–5863.

El-Guindy A, Heston L, Miller G. 2010. A subset of replication proteins enhances origin recognition and lytic replication by the Epstein–Barr virus ZEBRA protein. *PLoS Pathog* **6:** e1001054.

Feederle R, Delecluse HJ. 2004. Low level of lytic replication in a recombinant Epstein–Barr virus carrying an origin of replication devoid of BZLF1-binding sites. *J Virol* **78:** 12082–12084.

Feederle R, Kost M, Baumann M, Janz A, Drouet E, Hammerschmidt W, Delecluse HJ. 2000. The Epstein–Barr virus lytic program is controlled by the co-operative functions of two transactivators. *EMBO J* **19:** 3080–3089.

Fernandez AF, Rosales C, Lopez-Nieva P, Grana O, Ballestar E, Ropero S, Espada J, Melo SA, Lujambio A, Fraga MF, et al. 2009. The dynamic DNA methylomes of double-stranded DNA viruses associated with human cancer. *Genome Res* **19:** 438–451.

Fujii K, Yokoyama N, Kiyono T, Kuzushima K, Homma M, Nishiyama Y, Fujita M, Tsurumi T. 2000. The Epstein–Barr virus pol catalytic subunit physically interacts with the BBLF4-BSLF1-BBLF2/3 complex. *J Virol* **74:** 2550–2557.

Gibson W, Roizman B. 1971. Compartmentalization of spermine and spermidine in the herpes simplex virion. *Proc Natl Acad Sci* **68:** 2818–2821.

Gruffat H, Renner O, Pich D, Hammerschmidt W. 1995. Cellular proteins bind to the downstream component of the lytic origin of DNA replication of Epstein–Barr virus. *J Virol* **69:** 1878–1886.

Hammerschmidt W, Sugden B. 1988. Identification and characterization of oriLyt, a lytic origin of DNA replication of Epstein–Barr virus. *Cell* **55:** 427–433.

Hegde RS, Grossman SR, Laimins LA, Sigler PB. 1992. Crystal structure at 1.7 A of the bovine papillomavirus-1 E2 DNA-binding domain bound to its DNA target. *Nature* **359:** 505–512.

Humme S, Reisbach G, Feederle R, Delecluse HJ, Bousset K, Hammerschmidt W, Schepers A. 2003. The EBV nuclear antigen 1 (EBNA1) enhances B cell immortalization several thousandfold. *Proc Natl Acad Sci* **100:** 10989–10994.

Hung SC, Kang MS, Kieff E. 2001. Maintenance of Epstein–Barr virus (EBV) oriP-based episomes requires EBV-encoded nuclear antigen-1 chromosome-binding domains, which can be replaced by high-mobility group-I or histone H1. *Proc Natl Acad Sci* **98:** 1865–1870.

Jacob RJ, Roizman B. 1977. Anatomy of herpes simplex virus DNA VIII. Properties of the replicating DNA. *J Virol* **23:** 394–411.

Julien MD, Polonskaya Z, Hearing J. 2004. Protein and sequence requirements for the recruitment of the human origin recognition complex to the latent cycle origin of DNA replication of Epstein–Barr virus oriP. *Virology* **326:** 317–328.

Kalla M, Schmeinck A, Bergbauer M, Pich D, Hammerschmidt W. 2010. AP-1 homolog BZLF1 of Epstein–Barr virus has two essential functions dependent on the epigenetic state of the viral genome. *Proc Natl Acad Sci* **107:** 850–855.

Kennedy G, Sugden B. 2003. EBNA-1, a bifunctional transcriptional activator. *Mol Cell Biol* **23:** 6901–6908.

Kim SS, Tam JK, Wang AF, Hegde RS. 2000. The structural basis of DNA target discrimination by papillomavirus E2 proteins. *J Biol Chem* **275:** 31245–31254.

Kirchmaier AL, Sugden B. 1998. Rep*: A viral element that can partially replace the origin of plasmid DNA synthesis of Epstein–Barr virus. *J Virol* **72:** 4657–4666.

Kobayashi T, Rein T, DePamphilis ML. 1998. Identification of primary initiation sites for DNA replication in the hamster dihydrofolate reductase gene initiation zone. *Mol Cell Biol* **18:** 3266–3277.

Koons MD, Van Scoy S, Hearing J. 2001. The replicator of the Epstein–Barr virus latent cycle origin of DNA replication, oriP, is composed of multiple functional elements. *J Virol* **75:** 10582–10592.

Kudoh A, Fujita M, Zhang L, Shirata N, Daikoku T, Sugaya Y, Isomura H, Nishiyama Y, Tsurumi T. 2005. Epstein–Barr virus lytic replication elicits ATM checkpoint signal transduction while providing an S-phase-like cellular environment. *J Biol Chem* **280:** 8156–8163.

Kudoh A, Iwahori S, Sato Y, Nakayama S, Isomura H, Murata T, Tsurumi T. 2009. Homologous recombinational repair factors are recruited and loaded onto the viral DNA genome in Epstein–Barr virus replication compartments. *J Virol* **83:** 6641–6651.

Leight ER, Sugden B. 2001. Establishment of an oriP replicon is dependent upon an infrequent, epigenetic event. *Mol Cell Biol* **21:** 4149–4161.

Levitskaya J, Sharipo A, Leonchiks A, Ciechanover A, Masucci MG. 1997. Inhibition of ubiquitin/proteasome-dependent protein degradation by the Gly-Ala repeat domain of the Epstein–Barr virus nuclear antigen 1. *Proc Natl Acad Sci* **94:** 12616–12621.

Liao G, Wu FY, Hayward SD. 2001. Interaction with the Epstein–Barr virus helicase targets Zta to DNA replication compartments. *J Virol* **75:** 8792–8802.

Liao G, Huang J, Fixman ED, Hayward SD. 2005. The Epstein–Barr virus replication protein BBLF2/3 provides an origin-tethering function through interaction with the zinc finger DNA binding protein ZBRK1 and the KAP-1 corepressor. *J Virol* **79:** 245–256.

Lindner SE, Zeller K, Schepers A, Sugden B. 2008. The affinity of EBNA1 for its origin of DNA synthesis is a determinant of the origin's replicative efficiency. *J Virol* **82:** 5693–6702.

Mackey D, Sugden B. 1999. The linking regions of EBNA1 are essential for its support of replication and transcription. *Mol Cell Biol* **19:** 3349–3359.

Marechal V, Dehee A, Chikhi-Brachet R, Piolot T, Coppey-Moisan M, Nicolas JC. 1999. Mapping EBNA-1 domains involved in binding to metaphase chromosomes. *J Virol* **73:** 4385–4392.

Moriyama K, Yoshizawa-Sugata N, Obuse C, Tsurimoto T, Masai H. 2012. Epstein–Barr nuclear antigen 1 (EBNA-1)- dependent recruitment of origin recognition complex (Orc) on oriP of Epstein–Barr virus with purified proteins: Stimulation by Cdc6 through its direct interaction with EBNA1. *J Biol Chem* **287:** 23977–23994.

Murayama K, Nakayama S, Kato-Murayama M, Akasaka R, Ohbayashi N, Kamewari-Hayami Y, Terada T, Shirouzu M, Tsurumi T, Yokoyama S. 2009. Crystal structure of Epstein–Barr virus DNA polymerase processivity factor BMRF1. *J Biol Chem* **284:** 35896–35905.

Nakayama S, Murata T, Murayama K, Yasui Y, Sato Y, Kudoh A, Iwahori S, Isomura H, Kanda T, Tsurumi T. 2009. Epstein–Barr virus polymerase processivity factor enhances BALF2 promoter transcription as a coactivator for the BZLF1 immediate-early protein. *J Biol Chem* **284:** 21557–21568.

Nakayama S, Murata T, Yasui Y, Murayama K, Isomura H, Kanda T, Tsurumi T. 2010. Tetrameric ring formation of Epstein–Barr virus polymerase processivity factor is crucial for viral replication. *J Virol* **84:** 12589–12598.

Nanbo A, Sugden A, Sugden B. 2007. The coupling of synthesis and partitioning of EBV's plasmid replicon is revealed in live cells. *EMBO J* **26:** 4252–4262.

Niller HH, Glaser G, Knuchel R, Wolf H. 1995. Nucleoprotein complexes and DNA 5′-ends at oriP of Epstein–Barr virus. *J Biol Chem* **270:** 12864–12868.

Norio P, Schildkraut CL. 2004. Plasticity of DNA replication initiation in Epstein–Barr virus episomes. *PLoS Biol* **2:** e152.

Norio P, Schildkraut CL, Yates JL. 2000. Initiation of DNA replication within oriP is dispensable for stable replication of the latent Epstein–Barr virus chromosome after infection of established cell lines. *J Virol* **74:** 8563–8574.

Norseen J, Thomae A, Sridharan V, Aiyar A, Schepers A, Lieberman PM. 2008. RNA-dependent recruitment of the origin recognition complex. *EMBO J* **27:** 3024–3035.

Paludan C, Schmid D, Landthaler M, Vockerodt M, Kube D, Tuschl T, Münz C. 2005. Endogenous MHC class II processing of a viral nuclear antigen after autophagy. *Science* **307:** 593–596.

Pfuller R, Hammerschmidt W. 1996. Plasmid-like replicative intermediates of the Epstein–Barr virus lytic origin of DNA replication. *J Virol* **70:** 3423–3431.

Rennekamp AJ, Lieberman PM. 2011. Initiation of Epstein–Barr virus lytic replication requires transcription and the formation of a stable RNA–DNA hybrid molecule at OriLyt. *J Virol* **85:** 2837–2850.

Rennekamp AJ, Wang P, Lieberman PM. 2010. Evidence for DNA hairpin recognition by Zta at the Epstein–Barr virus origin of lytic replication. *J Virol* **84:** 7073–7082.

Ritzi M, Tillack K, Gerhardt J, Ott E, Humme S, Kremmer E, Hammerschmidt W, Schepers A. 2003. Complex protein-DNA dynamics at the latent origin of DNA replication of Epstein–Barr virus. *J Cell Sci* **116:** 3971–3984.

Sarkari F, Sanchez-Alcaraz T, Wang S, Holowaty MN, Sheng Y, Frappier L. 2009. EBNA1-mediated recruitment of a histone H2B deubiquitylating complex to the Epstein–Barr virus latent origin of DNA replication. *PLoS Pathog* **5:** e1000624.

Sato Y, Shirata N, Murata T, Nakasu S, Kudoh A, Iwahori S, Nakayama S, Chiba S, Isomura H, Kanda T, et al. 2010. Transient increases in p53-responsible gene expression at early stages of Epstein–Barr virus productive replication. *Cell Cycle* **9:** 807–814.

Schepers A, Pich D, Hammerschmidt W. 1993a. A transcription factor with homology to the AP-1 family links RNA transcription and DNA replication in the lytic cycle of Epstein–Barr virus. *EMBO J* **12:** 3921–3929.

Schepers A, Pich D, Mankertz J, Hammerschmidt W. 1993b. *cis*-acting elements in the lytic origin of DNA replication of Epstein–Barr virus. *J Virol* **67:** 4237–4245.

Schepers A, Pich D, Hammerschmidt W. 1996. Activation of oriLyt, the lytic origin of DNA replication of Epstein–Barr virus, by BZLF1. *Virology* **220:** 367–376.

Schepers A, Ritzi M, Bousset K, Kremmer E, Yates JL, Harwood J, Diffley JF, Hammerschmidt W. 2001. Human origin recognition complex binds to the region of the latent origin of DNA replication of Epstein–Barr virus. *EMBO J* **20:** 4588–4602.

Schildgen O, Graper S, Blumel J, Matz B. 2005. Genome replication and progeny virion production of herpes simplex virus type 1 mutants with temperature-sensitive lesions in the origin-binding protein. *J Virol* **79:** 7273–7278.

Sears J, Kolman J, Wahl GM, Aiyar A. 2003. Metaphase chromosome tethering is necessary for the DNA synthesis and maintenance of oriP plasmids but is insufficient for transcription activation by Epstein–Barr nuclear antigen 1. *J Virol* **77:** 11767–11780. Erratum in *J Virol* **78:** 5531, 2004.

Sears J, Ujihara M, Wong S, Ott C, Middeldorp J, Aiyar A. 2004. The amino terminus of Epstein–Barr virus (EBV) nuclear antigen 1 contains AT hooks that facilitate the replication and partitioning of latent EBV genomes by tethering them to cellular chromosomes. *J Virol* **78:** 11487–11505.

Shaw JE, Levinger LF, Carter CW Jr. 1979. Nucleosomal structure of Epstein–Barr virus DNA in transformed cell lines. *J Virol* **29:** 657–665.

Shirakata M, Imadome KI, Hirai K. 1999. Requirement of replication licensing for the dyad symmetry element-dependent replication of the Epstein–Barr virus oriP minichromosome. *Virology* **263:** 42–54.

Sugimoto A, Kanda T, Yamashita Y, Murata T, Saito S, Kawashima D, Isomura H, Nishiyama Y, Tsurumi T. 2011. Spatiotemporally different DNA repair systems participate in Epstein–Barr virus genome maturation. *J Virol* **85:** 6127–6135.

Takagi S, Takada K, Sairenji T. 1991. Formation of intranuclear replication compartments of Epstein–Barr virus with redistribution of BZLF1 and BMRF1 gene products. *Virology* **185:** 309–315.

Tempera I, Lieberman PM. 2010. Chromatin organization of gammaherpesvirus latent genomes. *Biochim Biophys Acta* **1799:** 236–245.

Tsurumi T. 2001. EBV replication enzymes. *Curr Top Microbiol Immunol* **258:** 65–87.

Tsurumi T, Kobayashi A, Tamai K, Daikoku T, Kurachi R, Nishiyama Y. 1993. Functional expression and characterization of the Epstein–Barr virus DNA polymerase catalytic subunit. *J Virol* **67:** 4651–4658.

Vereide DT, Sugden B. 2011. Lymphomas differ in their dependence on Epstein–Barr virus. *Blood* **117:** 1977–1985.

Wang CY, Sugden B. 2008. Identifying a property of origins of DNA synthesis required to support plasmids stably in human cells. *Proc Natl Acad Sci* **105:** 9639–9644.

Wang J, Lindner SE, Leight ER, Sugden B. 2006. Essential elements of a licensed, mammalian plasmid origin of DNA synthesis. *Mol Cell Biol* **26:** 1124–1134.

Wensing B, Stuhler A, Jenkins P, Hollyoake M, Karstegl CE, Farrell PJ. 2001. Variant chromatin structure of the oriP region of Epstein–Barr virus and regulation of EBER1 expression by upstream sequences and oriP. *J Virol* **75:** 6235–6241.

Wilkinson DE, Weller SK. 2004. Recruitment of cellular recombination and repair proteins to sites of herpes simplex virus type 1 DNA replication is dependent on the composition of viral proteins within prereplicative sites and correlates with the induction of the DNA damage response. *J Virol* **78:** 4783–4796.

Wu H, Kapoor P, Frappier L. 2002. Separation of the DNA replication, segregation, and transcriptional activation functions of Epstein–Barr nuclear antigen 1. *J Virol* **76:** 2480–2490.

Yates JL. 1996. Epstein–Barr virus DNA replication. In *DNA replication in eukaryotic cells* (ed. DePamphilis ML), pp. 751–774. Cold Spring Harbor Laboratory Press, Cold Spring Harbor, NY.

Yates JL, Camiolo SM, Bashaw JM. 2000. The minimal replicator of Epstein–Barr virus oriP. *J Virol* **74:** 4512–4522.

Yin Y, Manoury B, Fahraeus R. 2003. Self-inhibition of synthesis and antigen presentation by Epstein–Barr virus-encoded EBNA1. *Science* **301:** 1371–1374.

Zhang Q, Holley-Guthrie E, Ge JQ, Dorsky D, Kenney S. 1997. The Epstein–Barr virus (EBV) DNA polymerase accessory protein, BMRF1, activates the essential downstream component of the EBV oriLyt. *Virology* **230:** 22–34.

Zhou J, Chau CM, Deng Z, Shiekhattar R, Spindler MP, Schepers A, Lieberman PM. 2005. Cell cycle regulation of chromatin at an origin of DNA replication. *EMBO J* **24:** 1406–1417.

Poxvirus DNA Replication

Bernard Moss

Laboratory of Viral Diseases, National Institute of Allergy and Infectious Diseases, National Institutes of Health, Bethesda, Maryland 20892

Correspondence: bmoss@nih.gov

Poxviruses are large, enveloped viruses that replicate in the cytoplasm and encode proteins for DNA replication and gene expression. Hairpin ends link the two strands of the linear, double-stranded DNA genome. Viral proteins involved in DNA synthesis include a 117-kDa polymerase, a helicase–primase, a uracil DNA glycosylase, a processivity factor, a single-stranded DNA-binding protein, a protein kinase, and a DNA ligase. A viral FEN1 family protein participates in double-strand break repair. The DNA is replicated as long concatemers that are resolved by a viral Holliday junction endonuclease.

Poxviruses are large, enveloped, DNA viruses that infect vertebrate and invertebrate species and replicate entirely in the cytoplasm (Moss 2007). Two poxviruses are human-specific: variola virus and molluscum contagiosum virus. The former causes smallpox, a severe disease with high mortality that was eradicated more than two decades ago; the latter is distributed worldwide and produces discrete benign skin lesions in infants and extensive disease in immunocompromised individuals. Other poxviruses that infect nonhuman species can be transmitted to humans; monkeypox virus, for one, causes lethal disease in Africa and has the potential to spread geographically.

Vaccinia virus (VACV), the vaccine used to prevent smallpox, is the prototype of the family, and cited experiments relate to this virus unless stated otherwise. The VACV genome, together with viral enzymes and factors required for transcription of the early subset of genes, is packaged in the core of infectious virus particles (Moss 2007). The DNA replication proteins, in contrast to those involved in early transcription, are not packaged in virions but are translated from viral early mRNAs. DNA replication occurs following release of the genome from the core, and progeny DNA serves as the template for transcription of intermediate- and late-stage genes (Yang et al. 2011).

TIMING AND LOCATION OF VIRAL DNA REPLICATION

Poxvirus DNA synthesis can usually be detected within 2 h after infection and occurs in the cytoplasm within discrete juxtanuclear sites called factories that can easily be visualized by staining with a fluorescent dye (Fig. 1). A virus factory can form from a single virion, and the number of factories is proportional to the multiplicity of infection (Cairns 1960; Katsafanas and Moss 2007). However, coalescence of individual factories frequently occurs with time

Cite this article as *Cold Spring Harb Perspect Biol* doi: 10.1101/cshperspect.a010199

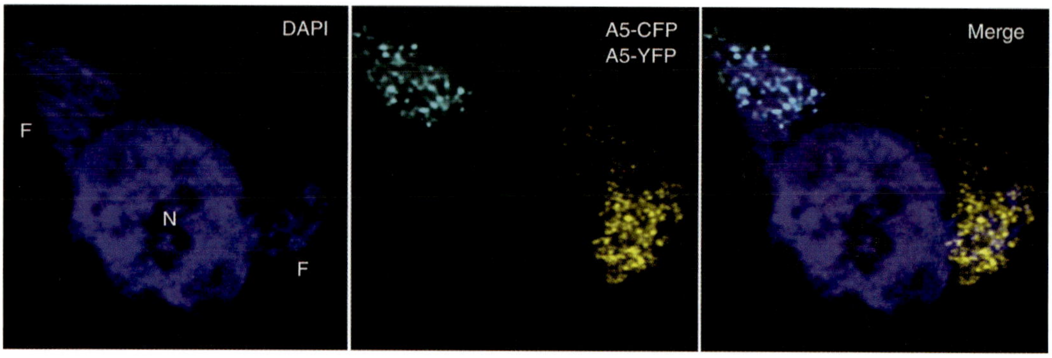

Figure 1. Poxvirus cytoplasmic factories. HeLa cells were simultaneously infected with two recombinant VACVs. One expresses the A5 core protein fused to cyan fluorescent protein (A5-CFP) and the other expresses the A5 core protein fused to yellow fluorescent protein (A5-YFP). The nuclei (N) and virus factories (F) were stained blue with 4′,6-diamidino-2-phenylindole (DAPI). A confocal microscopy image of a single cell in which individual factories arose from separate viruses is shown. (Reprinted, with permission, from Katsafanas and Moss 2007.)

(Katsafanas and Moss 2007; Lin and Evans 2010). Factories are initially compact and surrounded by endoplasmic reticulum membranes (Tolonen et al. 2001); a role for membranes in DNA replication has been suggested (Schramm and Krijnse Locker 2005). The factory is also the site of transcription and translation of viral mRNAs in addition to virion assembly (Katsafanas and Moss 2007).

POXVIRUS GENOME

Poxviruses have linear, double-stranded DNA genomes that vary from 130 to 230 kbp. As depicted in Fig. 2A, the two DNA strands are connected at their termini and form a continuous polynucleotide chain (Geshelin and Berns 1974; Baroudy et al. 1982a). The hairpin termini exist in inverted and complementary forms that are incompletely base-paired and AT-rich (Fig. 2B). In the case of VACV, the hairpins are 104 nucleotides in length and contain a 4-nucleotide loop and 10 extrahelical bases on one strand and 2 on the other. The maintenance of the extrahelical bases posits the existence of a double-stranded replication intermediate that is subsequently resolved to form the hairpin. The hairpin is at the end of a long inverted terminal repetition containing sets of short, tandemly repeated sequences (Fig. 2A) (Wittek and Moss 1980; Baroudy and Moss 1982). A region of <100 bp, between the end of the hairpin and start of the short repeats, contains a conserved sequence that is necessary for concatemer resolution (see below).

DNA PRECURSOR METABOLISM

Many poxviruses encode one or more enzymes involved in the synthesis of deoxyribonucleotides, presumably to enhance genome replication in resting cells with suboptimal precursor pools. The list (Table 1) includes thymidine kinase, thymidylate kinase, thymidylate synthetase, ribonucleotide reductase, dUTPase, and deoxycytidine kinase (Moss 2007). The corresponding deletion mutants are viable, although they are generally attenuated in vivo.

INITIATION OF DNA REPLICATION

Experiments designed to locate putative origins of replication instead showed that any circular DNA molecule transfected into cells infected with VACV or Shope fibroma virus replicates; furthermore, replication is not enhanced by insertion of any viral DNA fragment (DeLange and McFadden 1986; Merchlinsky and Moss 1988). The replicated DNA is in the form of unbranched head-to-tail concatemers, which

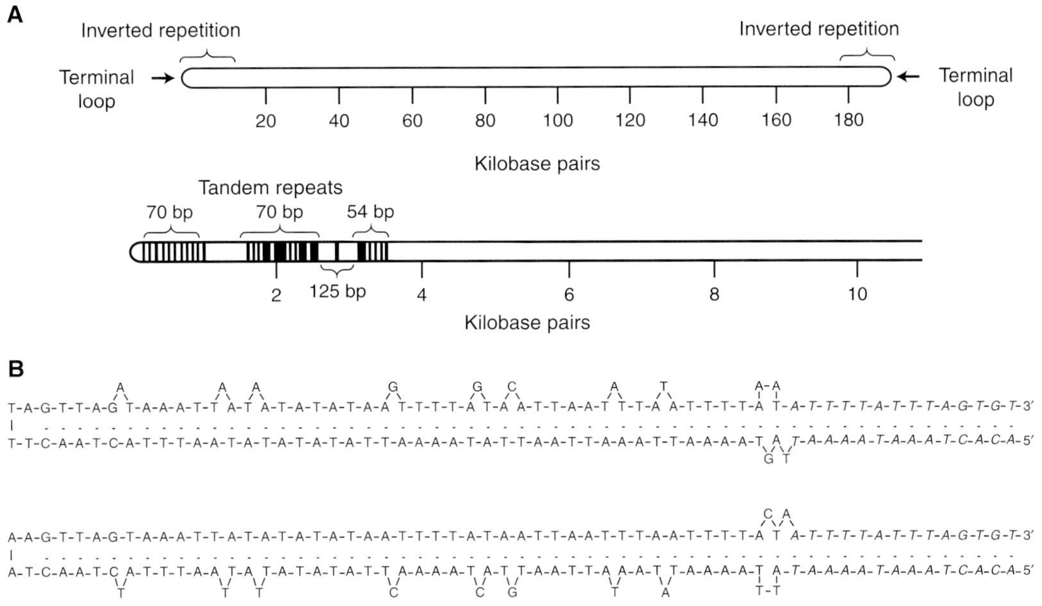

Figure 2. Structure of VACV genomic DNA. (*A*) Depiction of entire genome and an expansion of the 10-kbp inverted terminal repetition showing tandem direct repeats. (*B*) Nucleotide sequences of inverted and complementary forms of the terminal loops. (*A*, Reprinted, with permission, from Moss et al. 1983; *B*, reprinted, with permission, from Baroudy et al. 1982a.)

could form by a rolling circle mechanism. "Origin-independent" plasmid replication occurs within viral DNA factories and depends on each of the known viral proteins needed for genome replication (De Silva and Moss 2005).

Analysis of $[^3H]$thymidine incorporation following a shift from nonpermissive to permissive temperature in cells infected with a VACV DNA-negative mutant suggested that synthesis begins near the genome termini (Pogo et al. 1984). Experiments performed by transfecting linear DNA minichromosomes with hairpin ends indicated the presence of *cis*-acting replication-enhancing sequences within the terminal 200 bp (Du and Traktman 1996). The latter result suggested a specific replication initiation site within the conserved sequence between the hairpin loop and the direct repeats. However, the latter region contains a well-defined sequence required for concatemer resolution (described below), raising the possibility that cleavage of concatemers enhances replication or recovery of linear minichromosomes. It may be that a random nick can serve as a replication origin but that nicking preferentially occurs near the termini of intact genomes for reasons of specific structure or sequence.

DNA REPLICATION PROTEINS

A characteristic of the replication proteins, as well as the nucleotide metabolism enzymes, is their early synthesis. This feature distinguishes proteins involved in DNA synthesis from those made at intermediate or late times that have roles in DNA processing and packaging. We refer to VACV genes by a letter and a number corresponding to the open reading frame (ORF), followed by R or L indicating the direction of transcription; the direction is omitted when referring to the corresponding protein.

DNA Pol

The VACV *E9L* gene encodes a 117-kDa DNA polymerase (Pol) (Jones and Moss 1984; Trakt-

Table 1. Poxvirus DNA replication, processing, and packaging proteins

Protein	Mass[a] (kDa)	Conservation[b] (VACV ORF)[c]	Expression	Essential?[d]
Precursor metabolism				
Thymidine kinase	20	S (*J2R*)	Early	No
Thymidylate kinase	23	S (*A48R*)	Early	No
Ribonucleotide reductase	37, 88	S (*F4L, I4L*)	Early	No
dUTPase	16	S (*F2L*)	Early	No
Replication				
DNA polymerase	117	A (*E9L*)	Early	Yes
Helicase–primase	90	A (*D5R*)	Early	Yes
Uracil DNA glycosylase	25	A (*D4R*)	Early	Yes
Processivity factor	49	A (*A20R*)	Early	Yes
Protein kinase	35	S (*B1R*)	Early	Host-dependent
Single-stranded DNA-binding protein	30	C (*I3L*)	Early	Yes
DNA ligase	63	S (*A50R*)	Early	Host-dependent
FEN1-like nuclease	50	A (*G5R*)	Early	Impaired
DNA processing				
Holliday junction resolvase	21	A (*A22R*)	Intermediate	Yes
Topoisomerase	37	A (*H6R*)	Late	Impaired
DNA packaging				
ATPase	34	A (*A32L*)	Intermediate	Yes
Telomere-binding protein 1	36	C (*I1L*)	Intermediate	Yes

[a]Calculated mass of protein encoded by vaccinia virus (VACV) genome.
[b]A, all poxviruses; C, all chordopoxviruses; S, some poxviruses.
[c]Open reading frame (ORF) names according to Copenhagen strain of VACV (http://www.poxvirus.org).
[d]In most cases, essential nature is determined by severe defect of conditional mutant in tissue culture cells, and in some cases by inability to isolate a deletion mutant; severity of defect is host- and temperature-dependent for protein kinase mutant. Topoisomerase and FEN1-like nuclease deletion mutants replicate very poorly. Viral ligase can be complemented by cellular ligase I.

man et al. 1984; Earl et al. 1986), orthologs of which are found in all sequenced poxviruses. Poxvirus DNA Pol is homologous with the catalytic subunits of the DNA Pol of other viruses as well as eukaryotic α and δ DNA Pol (Wang et al. 1989). Purified VACV DNA Pol catalyzes primer- and template-dependent DNA synthesis and possesses 3′–5′ proofreading exonuclease activity (Challberg and Englund 1979). The activity of purified DNA Pol is inherently distributive, adding <10 nucleotides per primer–template binding event, although a more processive form exists in the cytoplasm (McDonald et al. 1997). The DNA Pol can also catalyze single-strand DNA annealing, which could generate branched molecules and thereby link DNA synthesis and recombination (Willer et al. 1999, 2000; Hamilton and Evans 2005).

Helicase–Primase

Temperature-sensitive (*ts*) mutants that map to the *D5R* gene of VACV show a fast-stop DNA replication phenotype (Roseman and Hruby 1987; Evans and Traktman 1992; Boyle et al. 2007). The *D5R* ORF encodes a 90-kDa protein that has nucleic-acid-independent ability to catalyze the hydrolysis of ribonucleotide triphosphates (rNTPs) and deoxyribonucleotide triphosphates (dNTPs) (Evans et al. 1995). D5 contains an ATP/GTP-binding motif in the carboxy-terminal segment of the protein, which is conserved in poxviruses and distantly related large DNA viruses (Gorbalenya and Koonin 1989; Iyer et al. 2001). Targeted mutations within the Walker A or B domains, the superfamily III helicase motif C, or the AAA^+ motif prevented in vivo complementation of VACV infectivity, and the purified proteins were defective in ATP hydrolysis. Multimerization of D5 appeared to be a prerequisite for enzymatic activity (Boyle et al. 2007).

An additional motif, present in the archaeoeukaryotic primase superfamily, was discovered in the amino-terminal domain of VACV D5 and its orthologs in other poxviruses (Iyer et al. 2005). Individual conserved aspartic acid residues in the predicted primase active site were required for in vivo complementation of infectious virus formation as well as genome and plasmid replication (De Silva et al. 2007). Furthermore, purified recombinant D5 protein was shown to catalyze the synthesis of oligoribonucleotides in vitro without stringent template specificity. Mutagenesis studies showed that the primase and NTPase activities of the recombinant D5 protein could be independently inactivated (De Silva et al. 2007). The absence of stringent template specificity suggests a role for this enzyme in discontinuous lagging-strand DNA synthesis (De Silva et al. 2009).

Uracil DNA Glycosylase

In eukaryotic and prokaryotic cells, uracil DNA glycosylase (UDG) participates in the removal of uracil from DNA, which can arise through misincorporation of deoxyuridine monophosphate (dUMP) or deamination of cytosine. UDG initiates base-excision repair by hydrolyzing the glycosylic bond linking uracil to a deoxyribose sugar. Shope fibroma virus (Upton et al. 1993) and VACV (Stuart et al. 1993) encode enzymatically active UDGs of 25 kDa, and orthologs of the protein are conserved in all sequenced members of the poxvirus family. Because cellular UDGs function in repair and are typically not essential for viability, it was surprising to find that VACV *D4R ts* mutants are unable to replicate DNA at the nonpermissive temperature (Stuart et al. 1993; Millns et al. 1994) and that a *D4R* deletion mutant can only replicate in a complementing cell line expressing the viral protein (Holzer and Falkner 1997). VACV mutants with enzymatically inactive UDG show no defect in DNA replication, however, demonstrating that the essential role of the protein is unrelated to the repair function (De Silva and Moss 2003). Nevertheless, mutants with enzymatically inactive UDG are attenuated in mice, indicating that the repair function is beneficial. D4 interacts with another essential replication protein, A20, suggesting a role as part of a multisubunit replication–repair complex (McCraith et al. 2000; Stanitsa et al. 2006; Boyle et al. 2011).

Processivity Factor

The VACV *A20R* gene encodes a 49-kDa protein that is conserved in all sequenced poxviruses but has no cellular homolog. A VACV genome-wide yeast two-hybrid analysis showed that the product of the *A20R* ORF interacts with proteins encoded by the *D4R*, *D5R*, and *H5R* ORFs, two of which were known to be involved in DNA replication (McCraith et al. 2000). Additional coimmunoprecipitation studies showed that nonoverlapping regions of A20 bind D4, D5, and H5, suggesting that these proteins can interact simultaneously to form a multicomponent complex (Ishii and Moss 2002). The early expression of the protein and the phenotype of conditional lethal A20 *ts* mutants supported a role in DNA replication (Ishii and Moss 2001; Punjabi et al. 2001). Cytoplasmic extracts of cells infected with a *ts* A20 mutant show a defect in processive DNA polymerase activity, and the A20 protein copurifies with a processive form of DNA Pol (Klemperer et al. 2001; Punjabi et al. 2001). Further studies suggest that A20 collaborates with D4 to enable processivity of DNA Pol (Stanitsa et al. 2006)

Protein Kinase

The VACV B1R ORF encodes a 35-kDa serine/threonine protein kinase that is packaged in virions (Banham and Smith 1992; Lin et al. 1992; Rempel and Traktman 1992). *ts* mutants that show a DNA replication-negative phenotype were mapped to the B1R gene (Rempel et al. 1990). Comparative genomic analyses identified two sequence-related human putative serine/threonine protein kinases, VRK1 and VRK2

(Nezu et al. 1997; Nichols and Traktman 2004). Human and mouse VRK1 genes can complement the replication defect of a VACV B1 mutant (Boyle and Traktman 2004). A cellular cytoplasmic protein called barrier to autointegration factor (BAF) acts as a potent inhibitor of poxvirus DNA replication unless its DNA-binding activity is blocked by B1-mediated phosphorylation (Nichols et al. 2006; Wiebe and Traktman 2007). The DNA-binding and dimerization capabilities of BAF are essential for its antipoxviral function (Ibrahim et al. 2011). B1 may have an additional role in DNA replication related to its interaction and phosphorylation of the VACV H5 protein (Beaud et al. 1995; McCraith et al. 2000), which interacts with the processivity factor A20 (McCraith et al. 2000; Ishii and Moss 2002).

Single-Stranded DNA-Binding Protein

An early 30-kDa phosphoprotein encoded by the VACV *I3L* gene preferentially binds single-stranded DNA (Davis and Mathews 1993; Rochester and Traktman 1998). I3 forms octameric complexes on DNA, similar to the *Escherichia coli* single-stranded DNA-binding protein (Tseng et al. 1999). Participation of I3 in DNA replication is suggested by the presence of the protein in punctate cytoplasmic inclusions containing parental viral genomes (Domi and Beaud 2000; Welsch et al. 2003). The inability to isolate viable *I3L* deletion mutants indicates that the corresponding protein is essential (Rochester and Traktman 1998), although direct evidence for a role in DNA replication and recombination has not yet been reported.

DNA Ligase

An ATP-dependent DNA ligase encoded by the VACV *A50R* ORF is conserved in many but not all chordopoxviruses (Kerr and Smith 1989). The ligase repairs nicked duplex DNA substrates consisting of a 5′-phosphate-terminated strand and a 3′-hydroxyl-terminated strand annealed to a bridging template strand (Shuman 1995). Adenylyltransferase and DNA ligation activities are in the amino-terminal one-third of the VACV ligase (Sekiguchi and Shuman 1997). Poxvirus mutants with deleted ligase genes are able to replicate (Colinas et al. 1990; Kerr and Smith 1991) but show host-range defects (Parks et al. 1998) and reduced pathogenicity (Kerr et al. 1991). The ability of VACV to replicate in the absence of the viral DNA ligase was recently shown by small interfering RNA knockdown experiments to depend on cellular DNA ligase I but not other cellular ligases (Paran et al. 2009). Ligase I is recruited to the viral DNA factories in the cytoplasm. Furthermore, replication of ligase-deficient VACV was greatly reduced and delayed in quiescent cells, which have low levels of ligase I. However, synthesis of ligase I was induced within several hours after infection (Paran et al. 2009). By encoding its own ligase, VACV can accelerate DNA synthesis in resting cells and enhance replication.

CONCATEMER RESOLUTION

Head-to-head and tail-to-tail telomere junctions, detected by restriction enzyme analysis of viral DNA from infected cells, indicated the presence of concatemers (Moyer and Graves 1981; Baroudy et al. 1982b). The junction contains a precise duplex copy of the hairpin loop at the ends of the mature viral genome (Merchlinsky et al. 1988). When transfected into cells infected with VACV or Shope fibroma virus, circular plasmids containing concatemer junctions are resolved into linear minichromosomes with hairpin termini (Delange et al. 1986; Merchlinsky and Moss 1986). The sequence 5′-T_6-N_{7-9}-T/C-A_3-T/A-3′ in an inverted repeat orientation on either side of a concatemer junction was found to be required for resolution (DeLange and McFadden 1987; Merchlinsky 1990). In supercoiled plasmids, the junction can form a cruciform structure resembling a four-way Holliday junction (HJ) recombination intermediate (Dickie et al. 1987; Merchlinsky et al. 1988). Bioinformatic analyses led to the discovery of motifs and structural elements that are critical for activity of *E. coli* RuvC HJ resolvase in ORFs that are conserved in all poxvirus genomes (Aravind et al. 2000; Garcia et al. 2000). The RuvC homolog encoded by the

VACV *A22R* gene was found to cleave four- and three-stranded junctions and also have branch nuclease activity (Garcia et al. 2000, 2006; Culyba et al. 2006, 2007). Mutation of either of two conserved acidic amino acids abrogated the catalytic activity of the viral protein without affecting HJ DNA binding (Garcia et al. 2000). Like RuvC, the A22 protein is a dimer in solution and when bound to HJ structures (Garcia et al. 2006). A VACV-inducible A22 null mutant was defective in processing concatemers into unit-length genomes with hairpin ends under nonpermissive conditions, indicating that the enzyme is required for resolution, although additional proteins may also be involved in determining site specificity (Garcia and Moss 2001).

The VACV-encoded type 1 topoisomerase (Shaffer and Traktman 1987; Shuman and Moss 1987) cleaves and ligates a variety of DNA structures in vitro, including an HJ (Sekiguchi et al. 1996; Palaniyar et al. 1999). Studies with a VACV topoisomerase deletion mutant suggest a role for the enzyme in enhancing early gene expression within the confines of the virus core (Da Fonseca and Moss 2003).

DNA REPLICATION MODELS

Self-priming, primer-dependent, and recombination models of poxvirus DNA replication have been considered. DNA hairpins at the ends of the poxvirus genome suggest a self-priming model of DNA replication (Fig. 3), resembling the rolling hairpin strand-displacement mechanism for parvoviruses (Muzyczka and Berns 2001). In the poxvirus variation of this model, a nick on one strand proximal to the hairpin by an unidentified nuclease generates a 3′ OH end to which deoxynucleotides can be added (Moyer and Graves, 1981; Baroudy et al. 1982a,b). Because of self-complementarity, the strands fold back and the replication complex continues deoxynucleotide addition to the distal hairpin and around it. The result is the formation of a concatemer, and reiteration of the process could lead to higher-order concatemers. Finally, resolution of the concatemers by the HJ resolvase results in unit genomes.

Reports of VACV DNA covalently linked to RNA and the chasing of short DNA into larger molecules raises the possibilities of RNA priming and semidiscontinuous DNA replication (Olgiati et al. 1976; Esteban and Holowczak 1977). This model has acquired new interest with the recent findings of a poxvirus primase (De Silva et al. 2007) and the requirement for either a viral or cellular DNA ligase (Paran et al. 2009).

DNA RECOMBINATION

There is evidence that the DNA polymerase participates directly in recombination and that the 3′ exonuclease and DNA-joining activities are involved (Hamilton and Evans 2005; Hamilton et al. 2007; Gammon and Evans 2009). In addition, the protein encoded by the *G5R* gene, which is conserved in all poxviruses and expressed early in infection, belongs to the FEN1 family of exo-/endonucleases (Da Silva et al. 2006; Iyer et al. 2006). Although the amount of viral DNA produced in the absence of G5 is similar to that made by wild-type virus, the mean size is approximately one-fourth of the genome length and the defective virions package little or no DNA (Senkevich et al. 2009). Experiments with transfected plasmids show that G5 is required for double-strand break repair by homologous recombination, suggesting a similar role during VACV genome replication (Senkevich et al. 2009). Double-strand breaks commonly occur at replication forks, consistent with the semidiscontinuous model of poxvirus DNA replication.

The chordopoxviruses, except for avipoxviruses, encode homologs of serine recombinases (Afonso et al. 2006). However, only the crocodile poxvirus contains all amino acids that comprise the catalytic active site, and deletion of the VACV F16 homolog had no effect on replication in cultured cells (Senkevich et al. 2011).

DNA PACKAGING

Electron microscopic images suggest that DNA nucleoids enter the spherical immature virion before their closure (Morgan 1976; Griffiths

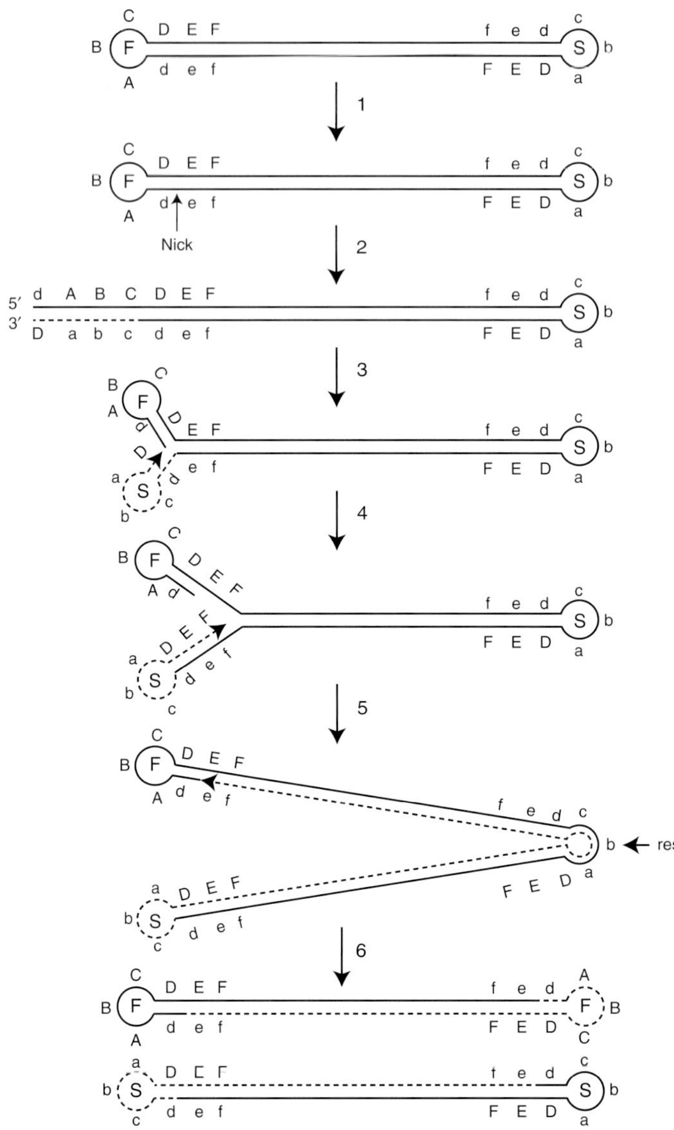

Figure 3. Self-priming model of poxvirus DNA replication. F and S within the hairpin loops refer to the fast and slow electrophoretic mobilities of DNA fragments containing inverted and complementary hairpin sequences. Dashed lines indicate newly synthesized DNA with arrowheads at the 3′ OH ends. Complementary sequences are depicted by upper- and lowercase letters. The resolution site (res) within the concatemer junction is indicated. (Modified from Moss et al. 1983.)

et al. 2001). The protein encoded by the *A32L* gene has sequence similarity to the products of gene *I* of filamentous single-stranded DNA bacteriophages and to the *IVa2* gene of adenovirus, both of which are ATPases involved in DNA packaging (Koonin et al. 1993). Repression of A32 synthesis prevents genome packaging (Cassetti et al. 1998). DNA packaging also fails to occur under nonpermissive conditions in cells infected with a *ts* mutant mapped to the I6 telomere-binding protein of VACV (DeMasi et al. 2001; Grubisha and Traktman 2003). Although concatemer resolution is required for DNA packaging (Garcia and Moss 2001), the two processes are not coupled (Merchlinsky and Moss 1989) and neither the A32 nor I1 mutant has a defect in concatemer resolution. It seems likely that binding of I6 occurs after resolution and precedes interaction of the genome with A32 or other packaging proteins.

DNA REPLICATION TARGETS FOR ANTIVIRAL THERAPY

Poxvirus-encoded DNA replication proteins are prime targets for antivirals. Cidofovir and oral derivatives are acyclic nucleoside analogs that are incorporated into the growing DNA strand and inhibit the 5′-to-3′ chain extension and 3′-to-5′ exonuclease activities of poxvirus DNA polymerase (Andrei and Snoeck 2010; Rice et al. 2011).

FUTURE DIRECTIONS

The following are but a few of the many outstanding questions. (1) Does DNA replication initiate by a self-priming mechanism at a nick, via a primer, or both? If nicking occurs, is there sequence or structural specificity to the site of the nick and what enzyme catalyzes nick formation? (2) Is the entire genome replicated by strand displacement, or does discontinuous synthesis occur, as suggested by the findings of a role for primase and DNA ligase? (3) Does DNA replication become recombination-dependent at late stages? (4) Is the HJ endonuclease sufficient for concatemer resolution? (5) How is DNA packaged?

ACKNOWLEDGMENTS

I thank George Katsafanas for help with the figures. The review was written with the support of the Division of Intramural Research, National Institute of Allergy and Infectious Diseases, National Institutes of Health.

REFERENCES

Afonso CL, Tulman ER, Delhon G, Lu Z, Viljoen GJ, Wallace DB, Kutish GF, Rock DL. 2006. Genome of crocodilepox virus. *J Virol* **80:** 4978–4991.

Andrei G, Snoeck R. 2010. Cidofovir activity against poxvirus infections. *Viruses* **2:** 2803–2830.

Aravind L, Makarova KS, Koonin EV. 2000. Holliday junction resolvases and related nucleases: Identification of new families, phyletic distribution and evolutionary trajectories. *Nucleic Acids Res* **28:** 3417–3432.

Banham A, Smith GL. 1992. Vaccinia virus gene B1R encodes a 34-kDa serine/threonine protein kinase that localizes in cytoplasmic factories and is packaged into virions. *Virology* **191:** 803–812.

Baroudy BM, Moss B. 1982. Sequence homologies of diverse length tandem repetitions near ends of vaccinia virus genome suggest unequal crossing over. *Nucleic Acids Res* **10:** 5673–5679.

Baroudy BM, Venkatesan S, Moss B. 1982a. Incompletely base-paired flip-flop terminal loops link the two DNA strands of the vaccinia virus genome into one uninterrupted polynucleotide chain. *Cell* **28:** 315–324.

Baroudy BM, Venkatesan S, Moss B. 1982b. Structure and replication of vaccinia virus telomeres. *Cold Spring Harb Symp Quant Biol* **47:** 723–729.

Beaud G, Beaud R, Leader DP. 1995. Vaccinia virus gene H5R encodes a protein that is phosphorylated by the multisubstrate vaccinia virus B1R protein kinase. *J Virol* **69:** 1819–1826.

Boyle KA, Traktman P. 2004. Members of a novel family of mammalian protein kinases complement the DNA-negative phenotype of a vaccinia virus *ts* mutant defective in the B1 kinase. *J Virol* **78:** 1992–2005.

Boyle KA, Arps L, Traktman P. 2007. Biochemical and genetic analysis of the vaccinia virus D5 protein: Multimerization-dependent ATPase activity is required to support viral DNA replication. *J Virol* **81:** 844–859.

Boyle KA, Stanitsa ES, Greseth MD, Lindgren JK, Traktman P. 2011. Evaluation of the role of the vaccinia virus uracil DNA glycosylase and A20 proteins as intrinsic components of the DNA polymerase holoenzyme. *J Biol Chem* **286:** 24702–24713.

Cairns J. 1960. The initiation of vaccinia infection. *Virology* **11:** 603–623.

Cassetti MC, Merchlinsky M, Wolffe EJ, Weisberg AS, Moss B. 1998. DNA packaging mutant: Repression of the vaccinia virus A32 gene results in noninfectious, DNA-deficient, spherical, enveloped particles. *J Virol* **72:** 5769–5780.

Challberg MD, Englund PT. 1979. Purification and properties of the deoxyribonucleic acid polymerase induced by vaccinia virus. *J Biol Chem* **254:** 7812–7819.

Colinas RJ, Goebel SJ, Davis SW, Johnson GP, Norton EK, Paoletti E. 1990. A DNA ligase gene in the Copenhagen strain of vaccinia virus is nonessential for viral replication and recombination. *Virology* **179:** 267–275.

Culyba MJ, Harrison JE, Hwang Y, Bushman FD. 2006. DNA cleavage by the A22R resolvase of vaccinia virus. *Virology* **352:** 466–476.

Culyba MJ, Minkah N, Hwang Y, Benhamou OM, Bushman FD. 2007. DNA branch nuclease activity of vaccinia A22 resolvase. *J Biol Chem* **282:** 34644–34652.

Da Fonseca F, Moss B. 2003. Poxvirus DNA topoisomerase knockout mutant exhibits decreased infectivity associated with reduced early transcription. *Proc Nat Acad Sci* **100:** 11291–11296.

Da Silva M, Shen L, Tcherepanov V, Watson C, Upton C. 2006. Predicted function of the vaccinia virus G5R protein. *Bioinformatics* **22:** 2846–2850.

Davis RE, Mathews CK. 1993. Acidic C terminus of vaccinia virus DNA-binding protein interacts with ribonucleotide reductase. *Proc Natl Acad Sci* **90:** 745–749.

DeLange AM, McFadden G. 1986. Sequence-nonspecific replication of transfected plasmid DNA in poxvirus-infected cells. *Proc Natl Acad Sci* **83:** 614–618.

DeLange AM, McFadden G. 1987. Efficient resolution of replicated poxvirus telomeres to native hairpin structures requires two inverted symmetrical copies of a core target DNA sequence. *J Virol* **61:** 1957–1963.

Delange AM, Reddy M, Scraba D, Upton C, McFadden G. 1986. Replication and resolution of cloned poxvirus telomeres in vivo generates linear minichromosomes with intact viral hairpin termini. *J Virol* **59:** 249–259.

DeMasi J, Du S, Lennon D, Traktman P. 2001. Vaccinia virus telomeres: Interaction with the viral I1, I6, and K4 proteins. *J Virol* **75:** 10090–10105.

De Silva FS, Moss B. 2003. Vaccinia virus uracil DNA glycosylase has an essential role in DNA synthesis that is independent of its glycosylase activity: Catalytic site mutations reduce virulence but not virus replication in cultured cells. *J Virol* **77:** 159–166.

De Silva FS, Moss B. 2005. Origin-independent plasmid replication occurs in vaccinia virus cytoplasmic factories and requires all five known poxvirus replication factors. *Virol J* **2:** 23.

De Silva FS, Lewis W, Berglund P, Koonin EV, Moss B. 2007. Poxvirus DNA primase. *Proc Natl Acad Sci* **104:** 18724–18729.

De Silva FS, Paran N, Moss B. 2009. Products and substrate/template usage of vaccinia virus DNA primase. *Virology* **383:** 136–141.

Dickie P, Morgan AR, McFadden G. 1987. Cruciform extrusion in plasmids bearing the replicative intermediate configuration of a poxvirus telomere. *J Mol Biol* **196:** 541–558.

Domi A, Beaud G. 2000. The punctate sites of accumulation of vaccinia virus early proteins are precursors of sites of viral DNA synthesis. *J Gen Virol* **81:** 1231–1235.

Du S, Traktman P. 1996. Vaccinia virus DNA replication: Two hundred base pairs of telomeric sequence confer optimal replication efficiency on minichromosome templates. *Proc Natl Acad Sci* **93:** 9693–9698.

Earl PL, Jones EV, Moss B. 1986. Homology between DNA polymerases of poxviruses, herpesviruses, and adenoviruses: Nucleotide sequence of the vaccinia virus DNA polymerase gene. *Proc Natl Acad Sci* **83:** 3659–3663.

Esteban M, Holowczak JA. 1977. Replication of vaccinia DNA in mouse L cells. I. In vivo DNA synthesis. *Virology* **78:** 57–75.

Evans E, Traktman P. 1992. Characterization of vaccinia virus DNA replication mutants with lesions in the D5 gene. *Chromosoma* **102:** S72–S82.

Evans E, Klemperer N, Ghosh R, Traktman P. 1995. The vaccinia virus D5 protein, which is required for DNA replication, is a nucleic acid-independent nucleotide triphosphatase. *J Virol* **69:** 5353–5361.

Gammon DB, Evans DH. 2009. The 3′-to-5′ exonuclease activity of vaccinia virus DNA polymerase is essential and plays a role in promoting virus genetic recombination. *J Virol* **83:** 4236–4250.

Garcia AD, Moss B. 2001. Repression of vaccinia virus Holliday junction resolvase inhibits processing of viral DNA into unit-length genomes. *J Virol* **75:** 6460–6471.

Garcia AD, Aravind L, Koonin EV, Moss B. 2000. Bacterial-type DNA Holliday junction resolvases in eukaryotic viruses. *Proc Natl Acad Sci* **97:** 8926–8931.

Garcia AD, Otero J, Lebowitz J, Schuck P, Moss B. 2006. Quartenary structure and cleavage specificity of a poxvirus Holliday junction resolvase. *J Biol Chem* **281:** 11618–11626.

Geshelin P, Berns KI. 1974. Characterization and localization of the naturally occurring cross-links in vaccinia virus DNA. *J Mol Biol* **88:** 785–796.

Gorbalenya AE, Koonin EV. 1989. Viral proteins containing the purine NTP-binding sequence pattern. *Nucleic Acids Res* **17:** 8413–8440.

Griffiths G, Roos N, Schleich S, Krijnse Locker J. 2001. Structure and assembly of intracellular mature vaccinia virus: Thin-section analyses. *J Virol* **75:** 11056–11070.

Grubisha O, Traktman P. 2003. Genetic analysis of the vaccinia virus I6 telomere-binding protein uncovers a key role in genome encapsidation. *J Virol* **77:** 10929–10942.

Hamilton MD, Evans DH. 2005. Enzymatic processing of replication and recombination intermediates by the vaccinia virus DNA polymerase. *Nucleic Acids Res* **33:** 2259–2268.

Hamilton MD, Nuara AA, Gammon DB, Buller RM, Evans DH. 2007. Duplex strand joining reactions catalyzed by vaccinia virus DNA polymerase. *Nucleic Acids Res* **35:** 143–151.

Holzer G, Falkner FG. 1997. Construction of a vaccinia virus deficient in the essential DNA repair enzyme uracil DNA glycosylase by a complementing cell line. *J Virol* **71:** 4997–5002.

Ibrahim N, Wicklund A, Wiebe MS. 2011. Molecular characterization of the host defense activity of the barrier to autointegration factor against vaccinia virus. *J Virol* **85:** 11588–11600.

Ishii K, Moss B. 2001. Role of vaccinia virus A20R protein in DNA replication: Construction and characterization of temperature-sensitive mutants. *J Virol* **75:** 1656–1663.

Ishii K, Moss B. 2002. Mapping interaction sites of the A20R protein component of the vaccinia virus DNA replication complex. *Virology* **303:** 232–239.

Iyer LM, Aravind L, Koonin EV. 2001. Common origin of four diverse families of large eukaryotic DNA viruses. *J Virol* **75:** 11720–11734.

Iyer LM, Koonin EV, Leipe DD, Aravind L. 2005. Origin and evolution of the archaeo-eukaryotic primase superfamily and related palm-domain proteins: Structural insights and new members. *Nucleic Acids Res* **33:** 3875–3896.

Iyer LM, Balaji S, Koonin EV, Aravind L. 2006. Evolutionary genomics of nucleo-cytoplasmic large DNA viruses. *Virus Res* **117:** 156–184.

Jones EV, Moss B. 1984. Mapping of the vaccinia virus DNA polymerase gene by marker rescue and cell-free translation of selected mRNA. *J Virol* **49:** 72–77.

Katsafanas GC, Moss B. 2007. Colocalization of transcription and translation within cytoplasmic poxvirus factories coordinates viral expression and subjugates host functions. *Cell Host Microbe* **2:** 221–228.

Kerr SM, Smith GL. 1989. Vaccinia virus encodes a polypeptide with DNA ligase activity. *Nucleic Acids Res* **17:** 9039–9050.

Kerr SM, Smith GL. 1991. Vaccinia virus DNA ligase is nonessential for virus replication: Recovery of plasmids from virus-infected cells. *Virology* **180:** 625–632.

Kerr SM, Johnston LH, Odell M, Duncan SA, Law KM, Smith GL. 1991. Vaccinia DNA ligase complements *Saccharomyces cerevisiae* Cdc9, localizes in cytoplasmic factories and affects virulence and virus sensitivity to DNA damaging agents. *EMBO J* **10:** 4343–4350.

Klemperer N, McDonald W, Boyle K, Unger B, Traktman P. 2001. The A20R protein is a stoichiometric component of the processive form of vaccinia virus DNA polymerase. *J Virol* **75:** 12298–12307.

Koonin EV, Senkevich TG, Chernos VI. 1993. Gene A32 protein product of vaccinia virus may be an ATPase involved in viral DNA packaging as indicated by sequence comparisons with other putative viral ATPases. *Virus Genes* **7:** 89–94.

Lin YCJ, Evans DH. 2010. Vaccinia virus particles mix inefficiently, and in a way that would restrict viral recombination, in coinfected cells. *J Virol* **84:** 2432–2443.

Lin S, Chen W, Broyles SS. 1992. The vaccinia virus B1R gene product is a serine/threonine protein kinase. *J Virol* **66:** 2717–2723.

McCraith S, Holtzman T, Moss B, Fields S. 2000. Genome-wide analysis of vaccinia virus protein–protein interactions. *Proc Natl Acad Sci* **97:** 4879–4884.

McDonald WF, Klemperer N, Traktman P. 1997. Characterization of a processive form of the vaccinia virus DNA polymerase. *Virology* **234:** 168–175.

Merchlinsky M. 1990. Mutational analysis of the resolution sequence of vaccinia virus DNA: Essential sequence consists of two separate AT-rich regions highly conserved among poxviruses. *J Virol* **64:** 5029–5035.

Merchlinsky M, Moss B. 1986. Resolution of linear minichromosomes with hairpin ends from circular plasmids containing vaccinia virus concatemer junctions. *Cell* **45:** 879–884.

Merchlinsky M, Moss B. 1988. Sequence-independent replication and sequence-specific resolution of plasmids containing the vaccinia virus concatemer junction: Requirements for early and late *trans*-acting factors. In *Cancer cells 6: Eukaryotic DNA replication* (ed. Kelly T, Stillman B), pp. 87–93. Cold Spring Harbor Laboratory Press, Cold Spring Harbor, NY.

Merchlinsky M, Moss B. 1989. Resolution of vaccinia virus DNA concatemer junctions requires late gene expression. *J Virol* **63:** 1595–1603.

Merchlinsky M, Garon C, Moss B. 1988. Molecular cloning and sequence of the concatemer junction from vaccinia virus replicative DNA: Viral nuclease cleavage sites in cruciform structures. *J Mol Biol* **199:** 399–413.

Millns AK, Carpenter MS, DeLange AM. 1994. The vaccinia virus-encoded uracil DNA glycosylase has an essential role in viral DNA replication. *Virology* **198:** 504–513.

Morgan C. 1976. The insertion of DNA into vaccinia virus. *Science* **193:** 591–592.

Moss B. 2007. Poxviridae: The viruses and their replication. In *Fields virology* (ed. Knipe DM, Howley PM), pp. 2905–2946. Lippincott Williams & Wilkins, Philadelphia.

Moss B, Winters E, Jones EV. 1983. Replication of vaccinia virus. In *Mechanics of DNA replication and recombination* (ed. Cozzarelli N), pp. 449–461. A. Liss, New York.

Moyer RW, Graves RL. 1981. The mechanism of cytoplasmic orthopoxvirus DNA replication. *Cell* **27:** 391–401.

Muzyczka N, Berns KI. 2001. Parvoviridae: The viruses and their replication. In *Fields virology* (ed. Knipe DM, Howley PM), pp. 2327–2359. Lippincott, Williams & Wilkins, Philadelphia.

Nezu J-I, Oku A, Jones MH, Shimane M. 1997. Identification of two novel human putative serine/threonine kinases, VRK1 and VRK2, with structural similarity to vaccinia virus B1R kinase. *Genomics* **45:** 327–331.

Nichols RJ, Traktman P. 2004. Characterization of three paralogous members of the mammalian vaccinia related kinase family. *J Biol Chem* **279:** 7934–7946.

Nichols RJ, Wiebe MS, Traktman P. 2006. The vaccinia-related kinases phosphorylate the N′ terminus of BAF, regulating its interaction with DNA and its retention in the nucleus. *Mol Biol Cell* **17:** 2451–2464.

Olgiati DD, Pogo BG, Dales S. 1976. Evidence for RNA linked to nascent DNA in HeLa cells. *J Cell Biol* **68:** 557–566.

Palaniyar N, Gerasimopoulos E, Evans DH. 1999. Shope fibroma virus DNA topoisomerase catalyses Holliday junction resolution and hairpin formation in vitro. *J Mol Biol* **287:** 9–20.

Paran N, De Silva FS, Senkevich TG, Moss B. 2009. Cellular DNA ligase I is recruited to cytoplasmic vaccinia virus factories and masks the role of the vaccinia ligase in viral DNA replication. *Cell Host Microbe* **6:** 563–569.

Parks RJ, Winchcombe-Forhan C, DeLange AM, Xing X, Evans DH. 1998. DNA ligase gene disruptions can depress viral growth and replication in poxvirus-infected cells. *Virus Res* **56:** 135–147.

Pogo BG, Berkowitz EM, Dales S. 1984. Investigation of vaccinia virus DNA replication employing a conditional lethal mutant defective in DNA. *Virology* **132:** 436–444.

Punjabi A, Boyle K, DeMasi J, Grubisha O, Unger B, Khanna M, Traktman P. 2001. Clustered charge-to-alanine mutagenesis of the vaccinia virus A20 gene: Temperature-sensitive mutants have a DNA-minus phenotype and are defective in the production of processive DNA polymerase activity. *J Virol* **75:** 12308–12318.

Rempel RE, Traktman P. 1992. Vaccinia virus B1 kinase: Phenotypic analysis of temperature-sensitive mutants and enzymatic characterization of recombinant proteins. *J Virol* **66:** 4413–4426.

Rempel RE, Anderson MK, Evans E, Traktman P. 1990. Temperature-sensitive vaccinia virus mutants identify a gene with an essential role in viral replication. *J Virol* **64:** 574–583.

Rice AD, Adams MM, Lampert B, Foster S, Lanier R, Robertson A, Painter G, Moyer RW. 2011. Efficacy of CMX001 as a prophylactic and presymptomatic antiviral agent in New Zealand white rabbits infected with rabbit-pox virus, a model for orthopoxvirus infections of humans. *Viruses* **3:** 63–82.

Rochester SC, Traktman P. 1998. Characterization of the single-stranded DNA binding protein encoded by the vaccinia virus I3 gene. *J Virol* **72:** 2917–2926.

Roseman NA, Hruby DE. 1987. Nucleotide sequence and transcript organization of a region of the vaccinia virus genome which encodes a constitutively expressed gene required for DNA replication. *J Virol* **61:** 1398–1406.

Schramm B, Krijnse Locker J. 2005. Cytoplasmic organization of poxvirus DNA replication. *Traffic* **6:** 839–846.

Sekiguchi J, Shuman S. 1997. Domain structure of vaccinia DNA ligase. *Nucleic Acids Res* **25:** 727–734.

Sekiguchi J, Seeman NC, Shuman S. 1996. Resolution of Holliday junctions by eukaryotic DNA topoisomerase I. *Proc Natl Acad Sci* **93:** 785–789.

Senkevich TG, Koonin EV, Moss B. 2009. Predicted poxvirus FEN1-like nuclease required for homologous recombination, double-strand break repair and full-size genome formation. *Proc Natl Acad Sci* **106:** 17921–17926.

Senkevich TG, Koonin EV, Moss B. 2011. Vaccinia virus F16 protein, a predicted catalytically inactive member of the prokaryotic serine recombinase superfamily, is targeted to nucleoli. *Virology* **417:** 334–342.

Shaffer R, Traktman P. 1987. Vaccinia virus encapsidates a novel topoisomerase with the properties of a eucaryotic type I enzyme. *J Biol Chem* **262:** 9309–9315.

Shuman S. 1995. Vaccinia virus DNA ligase: Specificity, fidelity, and inhibition. *Biochemistry* **34:** 16138–16147.

Shuman S, Moss B. 1987. Identification of a vaccinia virus gene encoding a type I DNA topoisomerase. *Proc Natl Acad Sci* **84:** 7478–7482.

Stanitsa ES, Arps L, Traktman P. 2006. Vaccinia virus uracil DNA glycosylase interacts with the A20 protein to form a heterodimeric processivity factor for the viral DNA polymerase. *J Biol Chem* **281:** 3439–3451.

Stuart DT, Upton C, Higman MA, Niles EG, McFadden G. 1993. A poxvirus-encoded uracil DNA glycosylase is essential for virus viability. *J Virol* **67:** 2503–2512.

Tolonen N, Doglio L, Schleich S, Krijnse Locker J. 2001. Vaccinia virus DNA replication occurs in endoplasmic reticulum-enclosed cytoplasmic mini-nuclei. *Mol Biol Cell* **12:** 2031–2046.

Traktman P, Sridhar RC, Roberts BE. 1984. Transcriptional mapping of the DNA polymerase gene of vaccinia virus. *J Virol* **49:** 125–131.

Tseng M, Palaniyar N, Zhang WD, Evans DH. 1999. DNA binding and aggregation properties of the vaccinia virus I3L gene product. *J Biol Chem* **274:** 21637–21644.

Upton C, Stuart DT, McFadden G. 1993. Identification of a poxvirus gene encoding a uracil DNA glycosylase. *Proc Natl Acad Sci* **90:** 4518–4522.

Wang TS, Wong SW, Korn D. 1989. Human DNA polymerase α: Predicted functional domains and relationships with viral DNA polymerases. *FASEB J* **3:** 14–21.

Welsch S, Doglio L, Schleich S, Krijnse Locker J. 2003. The vaccinia virus I3L gene product is localized to a complex endoplasmic reticulum-associated structure that contains the viral parental DNA. *J Virol* **77:** 6014–6028.

Wiebe MS, Traktman P. 2007. Poxviral B1 kinase overcomes barrier to autointegration factor, a host defense against virus replication. *Cell Host Microbe* **1:** 187–197.

Willer DO, Mann MJ, Zhang WD, Evans DH. 1999. Vaccinia virus DNA polymerase promotes DNA pairing and strand-transfer reactions. *Virology* **257:** 511–523.

Willer DO, Yao XD, Mann MJ, Evans DH. 2000. In vitro concatemer formation catalyzed by vaccinia virus DNA polymerase. *Virology* **278:** 562–569.

Wittek R, Moss B. 1980. Tandem repeats within the inverted terminal repetition of vaccinia virus DNA. *Cell* **21:** 277–284.

Yang Z, Reynolds SE, Martens CA, Bruno DP, Porcella SF, Moss B. 2011. Expression profiling of the intermediate and late stages of poxvirus replication. *J Virol* **85:** 9899–9908.

Appendix

Table 1. Databases for identification of genes in different organisms

Organism	Source
All organisms	National Center for Biotechnology Information, http://www.ncbi.nlm.nih.gov/ EMBL European Bioinformatics Institute
Homo sapiens	Human Gene Nomenclature Committee, http://www.genenames.org/ NCBI Genomic Biology (*Homo sapiens*), http://www.ncbi.nlm.nih.gov/projects/genome/guide/human/
Mus musculus	Mouse Genome Informatics, http://www.informatics.jax.org/
Saccharomyces cerevisiae	*Saccharomyces* Genome Database, http://www.yeastgenome.org
Schizosaccharomyces pombe	PomBase, http://www.pombase.org/
Drosophila melanogaster	FlyBase, http://flybase.bio.indiana.edu The Interactive Fly, http://www.sdbonline.org/fly/aimain/5zygotic.htm
Caenorhabditis elegans	WormBase, http://www.wormbase.org/
Xenopus laevis	Xenbase, http://www.xenbase.org/common/
Arabidopsis thaliana	The *Arabidopsis* Information Resource, http://www.arabidopsis.org/
Sulfolobus solfataricus	*Sulfolobus solfataricus* P2 Complete Genome Sequencing Project, http://www-archbac.u-psud.fr/projects/sulfolobus/
Escherichia coli	PortEco site, http://www.ecolihub.org/
Bacillus subtilis	*Bacillus subtilis* Genome Database, http://bacillus.genome.ad.jp/ GenoList, http://genodb.pasteur.fr/cgi-bin/WebObjects/GenoList.woa/14/wa/goToTaxoRank?level=Bacillussubtilis168

Cite this article as *Cold Spring Harb Perspect Biol* doi: 10.1101/cshperspect.a015727

Table 2. Style conventions for gene and protein nomenclature

Eukaryotic organisms	Convention
H. sapiens, X. laevis genes	*CDC6* = wild type *CDC6-V441I* = mutant allele
M. musculus genes	*Cdc6* = wild type $Cdc6^{23}$ = mutant allele
A. thaliana genes	*CDC6* = wild type *cdc6-1* = mutant allele *cdc6S234A* = specific point mutation
S. cerevisiae genes	*CDC6* = wild type *CDC6-23* = dominant mutant allele *cdc6-23* = recessive mutant allele
S. pombe genes	*cdc18* = wild type *cdc18-K46* = mutant allele *cdc18-S123A* = specific point mutation
D. melanogaster genes	*Cdc6* = wild type when gene named for dominant allele *dap* = wild type when gene named for recessive allele dap^{Ml} = dominant mutant allele dap^{ml} = recessive mutant allele
H. sapiens, S. cerevisiae, S. pombe, D. melanogaster and *X. laevis* proteins	Use "Title Case" (Orc1)
M. musculus, A. thaliana proteins	Use "UPPER CASE" (ORC1)
Protein complexes	Use "UPPER CASE" [origin recognition complex (ORC)]
Different names for the same gene or protein	Separate with "/" (*S. pombe* Dfp1/Him1/Rad35)
Different genes or proteins	Separate with "," (Orc1, Orc2, Orc3)
Eukaryotic genes and proteins with the same name, but from different species can be distinguished using a two-letter prefix: Human Mcm3 protein is HsMcm3. Human Mcm3 gene is *HsMcm3*.	Hs = *Homo sapiens* Mm = *Mus musculus* Xl = *Xenopus lavis* Sc = *Saccharomyces cerevisiae* Sp = *Schizosaccharomyces pombe* At = *Arabidopsis thaliana* Dm = *Drosophila melanogaster* Ce = *Caenorhabditis elegans*
Prokaryotic organisms	**Convention**
Prokaryotic gene names	*Italicized* in lower case *(dnaA)*
Prokaryotic protein names	Normal type in Title Case (DnaA)
Prokaryotic genes and proteins with the same name, but from different species can be distinguished using a three-letter prefix: *E. coli* DnaA protein is EcoDnaA. *E. coli dnaA* gene is *EcodnaA*.	Eco = *Escherischia coli* Bsu = *Bacillus subtilis* Sau = *Staphylococcus aureus* Sso = *Sulfolobus solfataricus*
Prokaryotic redundancies	Define three-letter prefix on first use to avoid redundancy. For example, *Acinetobacter baylyi* and *A. baumannii* can be distinguished as *Aby* and *Abm*.

Gene names are always *italicized*. Protein names are always in normal type.

Table 3. Nomenclature for proteins and protein complexes in different organisms

Mammals	Budding yeast	Fission yeast	Flies	Plants	Archaea	Bacteria
Prereplication complex assembly						
H. sapiens	*S. cerevisiae*	*S. pombe*	*D. melanogaster*	*A. thaliana*	*S. solfataricus*	*E. coli*
Hs	Sc	Sp	Dm	At	Sso	Eco
ORC	ORC	ORC	ORC	ORC	[Orc1/Cdc6]-1, 2, 3	DnaA
Orc1/p97	Orc1/p104	Orc1/Orp1/p81	Orc1/p103	Orc1a, Orc1b		
Orc2/p82	Orc2/p71	Orc2/Orp2/p61	Orc2/p69	Orc2		
Orc3/p66	Orc3/p72	Orc3/Orp3/p80	Orc3/Lat/p82	Orc3		
Orc4/p50	Orc4/p61	Orc4/Orp4/p109	Orc4/p52	Orc4		
Orc5L/p50	Orc5/p55	Orc5/Orp5/p52	Orc5/p52	Orc5		
Orc6/p28	Orc6/p50	Orc6/Orp6/p31	Orc6/p29	Orc6		
Cdc6	Cdc6	Cdc18	Cdc6	Cdc6a, Cdc6b	[Orc1/Cdc6]-1, 2, 3	DnaC
Cdt1/Rlf-B	Tah11/Sid2/Cdt1	Cdt1	Dup/Cdt1	Cdt1a, Cdt1b	Whip	
MCM helicase	MCM helicase	MCM helicase	MCM helicase	MCM helicase	Mcm	DnaB
Mcm2	Mcm2	Mcm2/Nda1/Cdc19	Mcm2	Mcm2		
Mcm3	Mcm3	Mcm3	Mcm3	Mcm3		
Mcm4	Mcm4/Cdc54	Mcm4/Cdc21	Mcm4/Dpa	Mcm4		
Mcm5	Mcm5/Cdc46/Bob1	Mcm5/Nda4	Mcm5	Mcm5		
Mcm6	Mcm6	Mcm6/Mis5	Mcm6	Mcm6		
Mcm7	Mcm7/Cdc47	Mcm7	Mcm7	Mcm7/Prolifera		
Gmnn/Geminin			Geminin			
Mcm9				Mcm9		
Hbo1			Chm/Hat1	Ham1		
				Ham2		
						DiaA
						Ihfα
						Ihfβ
						Fis
						SeqA
Replication fork assembly						
Hs	Sc	Sp	Dm	At	Sso	Eco
Mcm8			Rec/Mcm8	Mcm8		
Mcm10	Mcm10/Dna43	Mcm10/Cdc23	Mcm10	Mcm10		
DDK complex	DDK complex	DDK complex	DDK complex			
Cdc7	Cdc7	Hsk1	l(1)G0148	Hsk1-like		
Dbf4/Ask	Dbf4	Dfp1/Him1/Rad35	Chif/chiffon			
Drf1						

Continued

Replication fork assembly *(Continued)*

Hs	Sc	Sp	Dm	At	Sso	Eco
CDK complex	CDK complex	CDK complex	CDK complex	CDK complex		
Cdk1	Cdc28/Cdk1	Cdc2/Cdk1	Cdc2	CdkA		
Cdk2			Cdc2c			
CcnA1, A2			CycA	CycA1, A2, A3		
CcnB1, B2, B3	Clb1 to Clb6	Cig2/Cyc17, Cig1, Cdc13	CycB, B3	CycB1, B2, B3		
CcnE1, E2			CycE			
Cdc45	Cdc45	Cdc45/Sna41/Goa1	Cdc45	Cdc45		
PP2A	PP2A	PP2A	PP2A	PP2A		
RecQL4	Sld2/Drc1	Drc1	RecQL4			
Treslin	Sld3	Sld3		Sld3		
TopBP1	Dpb11	Rad4/Cut5	Mus101	Topbp1		
GINS complex	GINS complex	GINS complex	GINS complex	GINS complex		
Gins4	Sld5/Cdc105	Sld5	Sld5	Sld5	Gins 23	
Gins1	Psf1/Cdc101	Psf1	Psf1	Psf1	Gins 15	
Gins2	Psf2/Cdc102	Psf2	Psf2	Psf2		
Gins3	Psf3/Cdc103	Psf3	Psf3	Psf3a, Psf3b		

DNA synthesis at replication forks

Hs	Sc	Sp	Dm	At	Sso	Eco
DNA Pol α	DNA Pol α	DNA Pol α	DNA Pol α	DNA Pol α	DNA Pol B	DNA Pol III core enzyme
PolA/p180	Pol1/Cdc17/Crt5/Lrs9/Hpr3	Pol1	DNA Pol α180	Pola1		
PolA2/p68	Pol12	Spb70/Pol12	DNA Pol α60	Pola2		
DNA primase	DNA primase	DNA primase	DNA primase	DNA primase	DNA primase	DnaG
Prim1/p48	Pri1	Spp1	DNA Pol α50	Pola3/Pri1	PriS/p38	
Prim2a/p58	Pri2	Spp2	DNA Pol α73	Pola4/Pri2	PriL/p36	
DNA Pol δ	DNA Pol δ	DNA Pol δ	DNA Pol δ	DNA Pol δ	DNA Pol D[b]	DNA Pol III holoenzyme
Pold1/p125	Cdc2/Pol3/Hpr6/Tex1	Cdc6/Pol3/Mis10	DNA Pol δ138	Pold1		
Pold2/p50	Hys2/Pol31/Hus2/Sdp5	Cdc1/Mis1	DNA Pol δ47	Pold2		
Pold3/p66	Pol32	Cdc27		Pold3		
Pold4/p12		Cdm1		Pold4		
DNA Pol ε	DNA Pol ε	DNA Pol ε	DNA Pol ε	DNA Pol ε	DNA Pol B	DNA Pol III holoenzyme
Pole/p261	Pol2/Dun2	Cdc20/Pol2	DNA Pol ε	Pole1a, Pole1b		
Pole2/p59	Dpb2	Dpb2		Pole2		
Pole3/p17	Dpb3	Dpb3		Pole3		
Pole4/p12	Dpb4	Dpb4		Pole4		

Hs	Sc	Sp	Dm	At	Sso	Eco
DNA Pol γ Polg/p140 Polg2/p55	DNA Pol γ Mip1	DNA Pol γ Mip1	DNA Pol γ Tam/Tamas/ γ125 DNA Pol γ35			
RPA complex Rpa1/p70 Rpa2/p32 Rpa3/p14	RPA complex Rfa1/Buf2/Fun3/Rpa1 Rfa2/Buf1/Rpa2 Rfa3	RPA complex Ssb1/Rad11/Rpa1 Ssb2 Ssb3/Rpa3	RPA complex Rpa70 Rpa30 Rpa8	RPA complex Rpa1a, Rpa1b Rpa1c, Rpa1d Rpa2a, Rpa2b Rpa3a, Rpa3b	SSB	SSB
PCNA	Pol30	Pcn1	Mus209	PCNA1 PCNA2	PCNA1 PCNA2 PCNA3	β-clamp
RFC clamp loader	RFC clamp loader	RFC clamp loader	RFC clamp loader	RFC clamp loader	RFC clamp loader	Fork clamp loader
RNaseH2 complex Rnaseh2a Rnaseh2b Rnaseh2c	RNaseH2 complex Rnh201/Rnh35 Rnh202 Rnh203	Rnh201	Rnh 1	RnaseH2	RnaseH2	RNaseH
Dna2	Dna2/Web2	Dna2	CG2990	Dna2		DNA Pol I
Fen1	Rad27/Erc11/Rth1	Rad2/Fen1	Fen1	Fen1	Fen1	DNA Pol I
Lig1	Cdc9/Mms8	Cdc17	DNA-ligI	Lig1a, Lig1b	Lig1	Lig
Top1/Topo I	Top1/Mak1/Mak17	Top1	Top1	Top1a, Top1b	Topo I,[c] Topo III	DNA gyrase
Top2/Topo II	Top2/Tor3/Trf3	Top2/Ptr11	Top2	Top2	DNA gyrase[c] Topo VI	DNA gyrase Topo IV
CDK inhibitors Cdkn1a/p21/Cip1 Cdkn1b/p27/Kip1 Cdkn1c/p57/Kip2	Sic1/p40/Sdb25[a]	Rum1/p25[a]	Dap/Dacapo/p26	Krp1 to Krp7		

Replication fork progression and stability

Hs	Sc	Sp	Dm	At	Sso	Eco
Clspn/Claspin	Mrc1	Mrc1/Huc1				
Timeless	Tof1	Swi1	Tim/Timeless	Tim1		
Tipin	Csm3	Swi3				
Elg1-RFC clamp loader	Elg1-RFC clamp loader	Elg1-RFC clamp loader				
Brca2/FANCd1				Brca2a, Brca2b		

Continued

Replication fork progression and stability *(Continued)*

Hs	Sc	Sp	Dm	At	Sso	Eco
RecQ-like helicases				RecQ-like		RecQ
Blm	Sgs1	Rqh1	Mus309/DmBlm			
Wrn						
RecQL4				RecQ4a		
Top3α	Top3	Top3	Top3			
Rmi1	Rmi1/Nec4	Rmi1		Rmi1		
Rmi2						

Replication-dependent chromatin assembly

Hs	Sc	Sp	Dm	At	Sso	Eco
CHAF1/CAF1 complex	CAF1 complex	CAF1 complex	CAF1 complex	CAF1 complex		
p150	Rlf2/Cac1	Pcf1	p180	Fas1		
p60	Cac2	Pcf2	p105	Fas2		
p48	Msi1/Cac3	Pcf3	p55	Msi1 to Msi4		
FACT complex	FACT complex	FACT complex	FACT complex	FACT complex		
Supt16h/Spt16	Spt16/Cdc68	Spt16	dre4/spt16	Spt16		
Ssrp1	Pob3	Pob3	Ssrp	Ssrp1		
	Nhp6	Nhp6				
Asf1a/Cia	Asf1	Cia1	Asf1	Asf1a/Sga2		
Asf1b/Cia2				Asf1b/Sga1		
Nap1L1 to Nap1L5	Nap1	Nap1	Nap1	Nap1 to Nap4		
				Nrp1		
				Nrp2		
Npm1/nucleoplasmin			Nlp/nucleoplasmin			
Npm2/nucleoplasmin						
Npm3/nucleoplasmin						
Nasp	Hif1	Sim3				
			Df31			
ACF complex	ISW2 complex		ACF complex			
Baz1a/Acf1	Itc1		Acf1	Acf1		
Smarca5/Snf2h	Isw2		Isw1			
RSF complex						
Hbxap/Rsf1						
Smarca5/hSnf2h						

Chd1	Chd1	Hrp1/Chd1	Chd			
CHRAC complex Baz1a/Acf1 Smarca5/Snf2h Chrac1/Chrac15 Pole3/Chrac17			CHRAC complex Acf1 Iswi Top2 Chrac-14 Chrac-16			
						Ihfα, Ihfβ
H2a (H2afb1, b2, b3, j, v, y, z)[e]	Hta1, Hta2	Hta1, Hta2	His2A	H2a (Hta1, 2, 6, 7, 10, 12, 13)	Alba	Huα, Huβ
H2afx/H2ax				H2ax (Hta3, Hta5)		
H2afz/H2az	Htz1/H2az	Htb1	His2Av/H2AZ	H2a.z (Hta4, 8, 9, 11)		
H2b (H2bfm, s, wt)	Htb1, Htb2		His2B	H2b (Htb1-11)		
H3 (H3f3a, H3f3c)	Hht1, Hht2	Hht1, Hht2	His3	H3.1 (Htr1-3, 9, 11, 13)	Cren7	
H3.3b (H3f3b)		Hht3		H3.3 (Htr4-5, 8)		
H4 (Hist1h4a – Hist1H4l)	Hhf1, Hhf2	Hhf1, Hhf2, Hhf3	His4	H4 (Hfo2 – 8)	Sul7d	
H1 (H1fo, nt, oo, x)	Hho1		His1	H1 (Hon1, 2, 3)		Fis

Sister chromatid cohesion

Hs	Sc	Sp	Dm	At	Sso	Eco
Wdhd1/And-1	Ctf4/Pob1/Chl15	Mcl1	Ctf4			
Esco1	Eco1/Ctf7	Eso1	Eco1			
Ctf18-RFC clamp loader	Ctf18-RFC clamp loader	Ctf18-RFC clamp loader				
Cohesin complex	Cohesin complex	Cohesin complex	Cohesin complex			
Smc1	Smc1	Psm1	Smc1	Smc1/TTN8		
Smc3	Smc3	Psm3	Smc3	Smc3/TTN7		
Rad21	Scc1/Mcd1	Rad21	Rad21	Rad21.1, Rad21.2, Rad21.3		
Stag1, Stag2	Irr1/Scc3	Psc3	Sa			
CdcA5/Sororin			Dmt/Dalmatian			
Wfdc5/Wap1						
Pds5	Pds5	Pds5	Pds5			

DNA damage response

Hs	Sc	Sp	Dm	At	Sso	Eco
ATM	Tel1	Tel1	Tefu/telomere fusion/ATM	ATM		RecA
MRN complex	MRX complex	MRN complex	MRN complex	MRN complex	MRN complex	RecBCD complex

Continued

DNA damage response (*Continued*)

Hs	Sc	Sp	Dm	At	Sso	Eco
ATR	Mec1/Esr1/Sad3	Rad3	Mei-41/Mus-103	ATR		RecA
Atrip	Lcd1/Ddc2/Pie1	Rad26	Mus304	Atrip		
Chek1/Chk1	Chk1	Chk1/Rad27	Grp/Grapes/Chk1			
Chek2/Chk2	Rad53/Lsd1/Mec2/Spk1	Cds1	Lok/Loki/Chk2			
Cdc25	Mih1	Cdc25/Sal2	Stg/String/Cdc25	Cdc25-like		
Checkpoint clamp Rad9 Rad1 Hus1	Checkpoint clamp Ddc1 Rad17 Mec3	Checkpoint clamp Rad9 Rad1 Hus1	Checkpoint clamp Rad9 Rad1 Hus1	Checkpoint clamp Rad9 Rad1 Hus1		
Rad17-RFC clamp loader	Rad24-RFC clamp loader	Rad17-RFC clamp loader	Rad17-RFC clamp loader	Rad17-RFC clamp loader		
Ctf18-RFC clamp loader	Ctf18-RFC clamp loader	Ctf18-RFC clamp loader				
Tp53/p53			p53			
Tp53bp1						
Mdm2				Mdm2-like		
Msh2	Msh2/Pms5	Msh2	Spel1	Msh2		
Msh3	Msh3	Swi4		Msh3		
Msh6	Msh6/Pms3	Msh6	Msh6	Msh6		
Mlh1	Mlh1/Pms2	Mlh1	Dmlh1	Mlh1		
Mlh3	Mlh3					
Pms1	Pms1	Pms1		Pms1		
Pms2			Pms2			
Dclre1c/Artemis						
Brca1	Rad9	Rhp9/Crb2		Brca1		

Homologous recombination at stalled replication forks

Hs	Sc	Sp	Dm	At	Sso	Eco
MRN complex Mre11 Rad50 Nbs1/Nibrin	MRX complex Mre11 Rad50 Xrs2	MRN complex Rad32 Rad50 Nbs1	MRN complex Mre11 Rad50 Nbs1	MRN complex Mre11 Rad50 Nbs1	MRN complex Mre11 Rad50 HerA, NurA	RecBCD complex RecB RecC RecD
Uba2	Sae2	Fub2	Uba2	Sae2		

Rad51	Rad51	Rhp51	Spn-a	Rad51a, b, c, d	RadA	RecA
Rad52	Rad52	Rhp52/Rad22, Rti1				RecFOR complex
						REcF
						RecO
						RecR
Rad54b	Rad54	Rhp54	Okr/Okra	Rad54		
Rad51C	Rad55	Rhp55		Rad51c		
Xrcc3	Rad57	Rhp57	Spn-b	Xrcc3		
	Rad59					
	Rdh54					
Exo1	Exo1	Exo1	Tos/Tosca			ExoII
FANCi						
FANCd2						

Clamp loaders

Hs	Sc	Sp	Dm	At	Sso	Eco
RFC clamp loader	RFC clamp loader	RFC clamp loader	RFC clamp loader	RFC clamp loader	RFC clamp loader	Fork clamp loader
Rfc1/p140	Rfc1/Cdc44	Rfc1	Rfc140/Gnf1	Rfc1	Rf-CL	τ1
Rfc2/p40	Rfc2	Rfc2	Rfc40/Rfc2	Rfc2	$(Rf\text{-}CS)_4$	τ2
Rfc3/p38	Rfc3	Rfc3	Rfc38	Rfc3		τ3
Rfc4/p37	Rfc4	Rfc4	Rfc4	Rfc4		δ
Rfc5/p36	Rfc5	Rfc5		Rfc5		δ′
						χ
						ψ
Ctf18-RFC clamp loader	Ctf18-RFC clamp loader	Ctf18-RFC clamp loader	Ctf18-RFC clamp loader	Ctf18-RFC clamp loader		γ-complex
						γ1
Chtf18	Ctf18	Ctf18	Cutlet	Ctf18		γ2
Dcc1	Dcc1	Dcc1	CG11788			γ3
Derpc	Ctf8	Ctf8				δ
Rfc2, 3, 4, 5	Rfc2, 3, 4, 5	Rfc2, 3, 4, 5	Rfc3, 4, 38			δ′
Rad17-RFC clamp loader	Rad24-RFC clamp loader	Rad17-RFC clamp loader	Rad17-RFC clamp loader	Rad17-RFC clamp loader		
			Rad17			
Rad17	Rad24	Rad17	Rfc2, 3, 4, 5	Rad17		
Rfc2, 3, 4, 5	Rfc2, 3, 4, 5	Rfc2, 3, 4, 5		Rfc2, 3, 4, 5		
Elg1-RFC clamp loader	Elg1-RFC clamp loader	Elg1-RFC clamp loader				
Atad5	Elg1	Elg1				
Rfc2, 3, 4, 5	Rfc2, 3, 4, 5	Rfc2, 3, 4, 5				

Continued

Telomere maintenance

Hs	Sc	Sp	Dm	At	Sso	Eco
Tert	Est2	Trt1	Het-A Tart Tahre	Tert		
Est1A, B, C	Est1	Est1		Est1		
	Cdc13					
	Est3					
	Stn1			Stn1		
	Ten1					
Terf1/Trf1		Taz1		Trfl6, Trfl3, Trb1, Trb2, Trfl10, Trfl5, Trfl4, Trfl8, Trb3[d]		
Terf2/Trf2				Trfl3, Trb1, Trb2, Trb3, Trfl5, Trfl6, Trfl8, Trfl7, Trfl4, Trfl10[d]		
Pot1		Pot1		Pot1		
Tinf2/Tin2						
TPP1/Tint1/Ptop/Pip1						
Rap1	Rap1	Rap1				
Tnk1/Tank1						
Tnk2/Tank2						
Xrcc6/Ku70	Yku70	Pku70	Irbp	Ku70, Ku70-like		
Xrcc5/Ku80	Yku80	Pku80	Ku80	Ku80		
Rif1	Rif1					
	Rif2					
SMC5/6 complex	Smc5/6 complex	SMC5/6 complex				
Smc5	Smc5	Smc5	Smc5	Smc5		
Smc6	Smc6	Smc6		Smc6a, Smc6b		
Nse1	Nse1	Nse1				
Nse2	Mms21/Nse2	Nse2		Mms21		
Nse3	Nse3	Nse3				
Nse4	Nse4	Nse4/Rad62				
Nse5	Nse5	Nse5				
Nse6	Kre29/Nse6	Nse6				

Cite this article as *Cold Spring Harb Perspect Biol* doi: 10.1101/cshperspect.a015727

Linking DNA replication to cell division

Hs	Sc	Sp	Dm	At	Sso	Eco
CDK complex	CDK complex	CDK complex	CDK complex			
Cdk4	Cdc28/Cdk1	Cdc2/Cdk1	Cdk4			
Cdk6						
CcnD1,D2,D3	Cln3	Puc1	CycD	CycD		
Cdc25A, B, C	Mih1	Cdc25	Stg/String			
			Twe/Twine			
APC	APC	APC	APC	APC		
Anapc1	Apc1	Cut4/Apc1		Apc1		
Anapc2	Apc2	Apc2		Apc2		
Cdc27/Apc3	Cdc27/Apc3	Nuc2/Apc3	Cdc27/Apc3	Apc3a/Cdc27a		
				Apc3b/Cdc27b		
Anapc4	Apc4	Cut20/Lid1/Apc4	Apc4	Apc4		
Anapc5	Apc5	Apc5	Ida/Apc5	Apc5		
Cdc16/Apc6	Cdc16/Apc6	Cut9/Dre1/Apc6	Cdc16/Apc6	Apc6/Cdc16		
Anapc7			Apc7	Apc7		
Cdc23/Apc8	Cdc23/Apc8	Cut23/Apc8	Cdc23/Apc8	Apc8/Cdc23		
	Mnd2					
	Apc9					
Anapc10	Doc1/Apc10	Apc10	Apc10	Apc10		
Anapc11	Apc11	Apc11	Lmg/Apc11	Apc11		
Cdc26/Apc12	Cdc26	Hcn1				
Anapc13	Swm1/Apc13	Apc13	CG33981/Apc13			
		Apc14				
		Apc15				
Anapc16						
Cdc20	Cdc20	Slp1	Fzy	Cdc20		
Cdh1	Cdh1	Ste9	Fzr	Fzr1/Ccs52a2		
				Ccs52a1		
Emi1			Rca1			
Emi2						
CRL1/SCF2 Ub ligase	CRL1/SCF Ub ligase	CRL1/SCF Ub ligase	CRL1/SCF Ub ligase	CRL1/SCF Ub ligase		
				Cul1		
Cul1	Cdc53	Pcu1	Lin19	Rbx1		
Rbx1	Hrt1/Rbx1/Roc1/	Rbx1		Skp1		
Skp1	Skp1	Skp1	SkpA	Rub1		
Nedd8	Rub1	Ned8	Nedd8	Skp2a, Skp2b		
Skp2	Skp2		CG9772			
Fbxw7	Cdc4	Cdc4	ago/Archipelago			

Continued

Linking DNA replication to cell division *(Continued)*

Hs	Sc	Sp	Dm	At	Sso	Eco
CRL4 Ub ligase	CRL4 Ub ligase	CRL4 Ub ligase	CRL4 Ub ligase	CRL4 Ub ligase		
Cul4	Rtt101/Cul4	Pcu4	Cul4	Cul4		
Rbx1	Hrt1/Rbx1/Roc1/	Rbx1		Rbx1		
Ddb1	Mms1/Ddb1	SPBC17D11.08	Pic	Ddb1a, Ddb1b		
Nedd8	Rub1	Ned8	Nedd8	Rub1		
Dtl/Cdt2		Cdt2	L2dtl			

DNA replication proteins in different organisms that are homologs, orthologs, paralogs, or analogs are aligned. **Homologs** are proteins related by descent from a common ancestral DNA sequence. **Orthologs** are proteins that evolved from a common ancestral gene and retain the same function in different species. **Paralogs** are proteins related by duplication within a genome, but that have evolved a new function. **Analogs** are proteins that have a similar function but unrelated sequences.

Style conventions for gene and protein nomenclature vary extensively among organisms (Table 2). Therefore, to facilitate recognition of related proteins among different organisms, the names of proteins are in Title Case and the names of protein complexes are in CAPITALS. However, as some proteins are so familiar in CAPITALS that they are not immediately recognized in Title Case, exceptions were made: PCNA (proliferating cell nuclear antigen), ATM (ataxia telangiectasia mutated), ATR (ataxia telangiectasia and Rad3 related), FANC (Fanconi anemia proteins), and SSB (single-strand DNA binding protein).

Multiple names for the same protein are separated by "/" with the first name the one recommended by the standard database (Table 1).

A comma separates names of different proteins of the same family.

Molecular weights based on amino acid composition are indicated as p(number) [e.g., p27, protein 27 kDa in size].

ACF, ATP-utilizing chromatin assembly and remodeling factor; APC, anaphase promoting complex; CDK, cyclin-dependent protein kinase; CHAF1/CAF1, chromosome assembly factor 1; CHRAC, chromatin accessibility; CRL, Cullin-RING ubiquitin ligase; DDK, Dbf4-dependent protein kinase; FACT, facilitates chromatin transcription; GINS, Go, Ichi, Ni, and San; MCM, minichromosome maintenance; MRN, Mre11, Rad50, and Nbs1; ORC, origin recognition complex; RFC, replication factor C; RPA, replication protein A; RSF, remodeling and spacing factor.

[a]Yeast Sic1 and Rum1 share a structurally conserved inhibitory domain with mammalian p27(Kip1), although they do not inhibit Cdk·G1 cyclins in yeast.

[b]Although not found in the Crenarchaeon Sulfolobus, Euryarchaeota also contain Pol D DNA polymerases. Pol D contains 3′-5′ exonuclease activity, uses RNA primers more efficiently, and has a higher intrinsic processivity than Pol B DNA polymerases.

[c]Topo I and Topo II homologs have been detected in some, but not all species of Archaea, for example, members of the Sulfolobales do not possess gyrase homologs.

[d]The Arabidopsis homologs to human Terf1 and to Terf2 are listed in decreasing order of similarity.

[e]The 31 human histones are divided into eight groups: H2A, H2B, H3, and H4 form the nucleosome core. H2AX is phosphorylated in response to double-strand DNA breaks. H2AZ and H3.3 are enriched at transcriptionally active sites. H1 links core particles together. Subgroups are in parentheses.

Cite this article as *Cold Spring Harb Perspect Biol* doi: 10.1101/cshperspect.a015727

Index

E

F

G

P

R

S